Preparation for Calculus

Preparation for Calculus:
Functions and How They Change

Bruce Crauder
Oklahoma State University

Benny Evans
Oklahoma State University

Alan Noell
Oklahoma State University

macmillan learning

Austin • Boston • New York • Plymouth

Senior Vice President, STEM: Daryl Fox
Program Director, Math, Statistics, and Physical Sciences: Andrew Dunaway
Program Manager: Nikki Miller Dworsky
Marketing Manager: Leah Christians
Executive Content Development Manager, STEM: Debbie Hardin
Senior Development Editors: Debbie Hardin, Susan Teahan
Executive Project Manager, Content, STEM: Katrina Mangold
Editorial Project Manager: Karen Misler
Director of Content, Math and Statistics: Daniel Lauve
Executive Media Editor: Catriona Kaplan
Senior Media Editor: Holly Floyd
Editorial Assistant: Justin Jones
Marketing Assistant: Morgan Psiuk
Senior Director, Content Management Enhancement: Tracey Kuehn
Senior Managing Editor: Lisa Kinne
Senior Content Project Manager: Kerry O'Shaughnessy
Senior Workflow Project Manager: Paul Rohloff
Director of Design, Content Management: Diana Blume
Design Services Manager: Natasha Wolfe
Cover Design Manager: John Callahan
Cover Designer: Joseph DePinho
Text Designer: Tamara Newnam
Illustration Coordinator: Janice Donnola
Illustrations: Network Graphics
Director of Digital Production: Keri deManigold
Media Project Manager: Hanna Squire
Executive Permissions Editor: Cecilia Varas
Rights and Billing Editor: Alexis Gargin
Composition: Lumina Datamatics, Inc.
Printing and Binding: LSC Communications
Title Page Image: Photograph by Anthony Acosta, graph created using the Desmos graphing calculator

Library of Congress Control Number: 2020946009

Student Edition Paperback:
ISBN-13: 978-1-4641-1581-3
ISBN-10: 1-4641-1581-8

International Edition Paperback:
ISBN-13: 978-1-319-46636-7
ISBN-10: 1-319-46636-2

Student Edition Loose-leaf:
ISBN-13: 978-1-4641-8669-1
ISBN-10: 1-4641-8669-3

In 1946, William Freeman founded W. H. Freeman and Company and published Linus Pauling's *General Chemistry*, which revolutionized the chemistry curriculum and established the prototype for a Freeman text. W. H. Freeman quickly became a publishing house where leading researchers can make significant contributions to mathematics and science. In 1996, W. H. Freeman joined Macmillan and we have since proudly continued the legacy of providing revolutionary, quality educational tools for teaching and learning in STEM.

For Doug, Robbie, and Lindsay,
and especially Laurie
—Bruce

For all my grandchildren
—Benny

To the memory of my parents
—Alan

ABOUT THE AUTHORS

Oklahoma State University

Bruce Crauder received his Ph.D. from Columbia University in 1981. After postdoctoral positions at the Institute for Advanced Study and elsewhere, he has been at Oklahoma State University since 1986, where he is a professor of mathematics. Bruce's research interests are in algebraic geometry and the challenge of the beginning college math curriculum, which combines scholarship with his passion for teaching.

Bruce has two sons and a daughter-in-law. He and his wife, Laurie, enjoy bird watching, traveling, and their chickens.

Courtesy Benny Evans

Benny Evans received his Ph.D. in mathematics from the University of Michigan. He is Professor Emeritus of Mathematics at Oklahoma State University, where he has served as undergraduate director, learning center director, associate head, and department head. He has held visiting appointments at the Institute for Advanced Study, Rice University, University of Nevada Reno, and Texas A&M. His research interests are topology and mathematics education.

Elizabeth Noell

Alan Noell received his Ph.D. in mathematics from Princeton University in 1983. After a postdoctoral position at Caltech, he joined the faculty of Oklahoma State University in 1985. He is currently a professor of mathematics. His scholarly activities include research in complex analysis and curriculum development.

Alan and his wife, Liz, have four children, two daughters-in-law, two sons-in-law, and two grandchildren. They are very involved with their church and also enjoy reading and cats.

CONTENTS

Preface   xiii

Acknowledgments   xviii

ON0   **Algebra Review***

ON0.1   Numbers, Products, and Factors
    The Real Number Line
    The Laws of Algebra
    Integer Exponents
    Products of Algebraic Expressions
    Factoring Algebraic Expressions

ON0.2   Absolute Value, Roots, and Radicals
    Absolute Value
    Absolute Value as Distance
    Roots and Radicals
    Rational Exponents
    Models and Applications: Terminal Velocity

ON0.3   Algebraic Fractions
    Factoring and Cancellation
    Adding and Subtracting Fractions
    Complex Fractions and Radicals
    Models and Applications: Earthquakes

ON0.4   Solving Equations
    Linear Equations
    Quadratic Equations: Factoring
    Quadratic Equations: Completing the Square and the Quadratic Formula
    Equations Involving Radicals
    Models and Applications: Temperature Conversion

    CHAPTER STUDY GUIDE
    CHAPTER QUIZ
    CHAPTER REVIEW EXERCISES

Basics of Functions

P   **The Coordinate Plane and Inequalities**   1

P.1   The Coordinate Plane: Lines and Circles   2
    Coordinates in the Plane   2
    The Distance and Midpoint Formulas   3
    Lines and Their Slopes   6
    Common Forms for Equations of Lines   8
    Circles   10
    Models and Applications: Locating Through Triangulation   12

P.2   Linear Inequalities   18
    Inequalities and Intervals   18
    Solving Inequalities   20
    Absolute Value Inequalities   22
    Models and Applications: Supply and Demand   25

P.3   Nonlinear Inequalities   28
    Quadratic Inequalities   29
    Inequalities Involving More Than Two Factors   33
    Inequalities Involving Fractions   35
    Models and Applications: Profit   37

    CHAPTER ROADMAP AND STUDY GUIDE   43
    CHAPTER QUIZ   44
    CHAPTER REVIEW EXERCISES   45

1   **Functions and How They Change**   48

1.1   The Basics of Functions   49
    Definition of a Function   50
    Evaluating Functions   50
    Domain and Range   52
    Equations and Functions   55
    Graphs and Functions   56
    Models and Applications: Representing Costs of Everyday Purchases   58

1.2   Average Rates of Change   65
    Secant Lines   65
    Models and Applications: Ants, Ebola, and Temperature   68

1.3   Graphs and Rates of Change   77
    An Intuitive Description of Rates of Change   78
    Increasing and Decreasing Functions   79
    Maximum and Minimum Values   81
    Concavity and Rates of Change   84
    Models and Applications: Sales Charts   87

1.4   Limits and End Behavior of Graphs   97
    An Intuitive Discussion of Limits at Infinity   97
    Examples of Limits at Infinity   98
    Elementary Limit Calculations   101
    Breaking Limit Calculations into Simpler Steps   102
    Models and Applications: Long-Term Behavior   105

    CHAPTER ROADMAP AND STUDY GUIDE   114
    CHAPTER QUIZ   116
    CHAPTER REVIEW EXERCISES   117

2   **Operations on Functions and Graphs**   122

2.1   New Functions from Old   123
    Arithmetic Combinations of Functions   124
    Composition of Functions   127
    Function Decomposition   129
    Models and Applications: Decomposing Functions in Applications   130

2.2   Inverse Functions   135
    Defining Inverse Functions   136
    Finding Formulas and Graphs for Inverse Functions   138
    The Horizontal Line Test for Inverses   140
    One-to-One Functions   142
    Restricting the Domain to Find Inverse Functions   142
    Models and Applications: A School Trip   144

2.3   Transformations of Graphs   150
    Shifting of Graphs   150
    Vertical Stretching, Compressing, and Reflecting of Graphs   155
    Horizontal Stretching, Compressing, and Reflecting of Graphs   157
    Even and Odd Functions   163
    Models and Applications: Starting Times   164

    CHAPTER ROADMAP AND STUDY GUIDE   173
    CHAPTER QUIZ   174
    CHAPTER REVIEW EXERCISES   176

*ON refers to material that can be found online with registration in Achieve: achieve.macmillanlearning.com.

Four Major Classes of Functions

 3 Linear and Exponential Functions 181

3.1 Linear Functions: Constant Rate of Change 182
Constant Rate of Change 182
Graphs of Linear Functions 184
Models and Applications: Temperature Conversion and More 187
Models and Applications: Regression Lines 190

3.2 Exponential Functions: Constant Proportional Change 198
Constant Proportional Change 198
Graphs of Exponential Functions 201
The Special Number e 204
Models and Applications: Exponential Growth and Decay 205
Rapidity of Exponential Growth 207

3.3 Modeling with Exponential Functions 214
Constant Percentage Change 215
Compound Interest 219
Installment Loans 222
Logistic Models 223
CHAPTER ROADMAP AND STUDY GUIDE 234
CHAPTER QUIZ 235
CHAPTER REVIEW EXERCISES 237

 4 Logarithms 240

4.1 Logarithms: Definition and Fundamental Properties 241
The Logarithm as an Inverse Function 241
Graphs of Logarithms 243
Common and Natural Logarithms 244
Long-Term Behavior of the Logarithm 248
Change-of-Base Formula 249
Models and Applications: Spell Checkers 251

4.2 Laws of Logarithms 259
Deriving the Laws of Logarithms 259
Using Laws of Logarithms 261
Models and Applications: Earthquakes 265

4.3 Solving Exponential and Logarithmic Equations 273
Solving Exponential Equations 274
Solving Logarithmic Equations 277
Mixed Equations 279
Models and Applications: Radiocarbon Dating 281
CHAPTER ROADMAP AND STUDY GUIDE 287
CHAPTER QUIZ 288
CHAPTER REVIEW EXERCISES 289

5 Polynomials and Rational Functions 293

5.1 Quadratic Functions 294
Zeros of Quadratic Functions and the Discriminant 295
Complex Zeros of Quadratic Functions 298
The Factor Theorem for Quadratic Polynomials 299
Graphs of Quadratic Functions 301
Models and Applications: Maximum Height of Objects Propelled Upward 304

5.2 Long Division and the Factor Theorem 313
Long Division 314
The Remainder and Factor Theorems 316
Synthetic Division 318

5.3 Zeros of Higher-Degree Polynomials 324
Factoring: Methods and Formulas 324
Using the Factor Theorem to Solve Equations 328
Rational Zeros 329
Models and Applications: Poiseuille's Law 333

5.4 Graphs of Polynomials 338
Fundamental Theorem of Algebra and Complete Factorization 339
Polynomials with Real Coefficients 340
Long-Term Behavior 342
Local Behavior 344
Local Maxima and Minima 349
Models and Applications: Changing Value of Investments 350

5.5 Rational Functions 358
Domain and Poles 359
Local Behavior: Vertical Asymptotes 360
Long-Term Behavior of Rational Functions 363
End Behavior of Graphs: Horizontal Asymptotes 364
Models and Applications: Traffic Flow 368
CHAPTER ROADMAP AND STUDY GUIDE 377
CHAPTER QUIZ 379
CHAPTER REVIEW EXERCISES 381

Trigonometry

6C Introduction to Trigonometry: A Unit Circle Approach† 387

6C.1 Angles 388
Conventions for Angles 388
Degree Measure 389
Radian Measure 390
Converting Angle Measures 392
Arc Length and Area 394
Similar Triangles 396
Models and Applications: Shadows 399

6C.2 The Unit Circle 405
Finding Points on the Unit Circle 406
Compass Points on the Unit Circle 407
Special Angles 410
Quadrants and Reference Numbers 412
Using Reference Numbers to Find Coordinates of Trigonometric Points 414

6C.3 The Trigonometric Functions 422
Defining the Trigonometric Functions 423
Trigonometric Functions of Special Angles 423
Trigonometric Functions of Compass Points 426
Signs of Trigonometric Functions in the Four Quadrants 429
Using Reference Numbers to Evaluate Trigonometric Functions 430
The Pythagorean Identity 432

6C.4 Right Triangle Trigonometry 438
Trigonometric Functions of Acute Angles 438
Using Right Triangles to Evaluate Trigonometric Functions 439
Representing Angles 441

†Either Chapter 6C or Chapter 6T can be used; the intention is for instructors to assign only one of these two chapters.

Using Representative Triangles to Evaluate
   Trigonometric Functions   442
Solving Right Triangles   445
A New Formula for the Area of a Triangle   448
Models and Applications: Transits and Shadows   449

CHAPTER ROADMAP AND STUDY GUIDE   458
CHAPTER QUIZ   460
CHAPTER REVIEW EXERCISES   462

OR

6T   Introduction to Trigonometry: A Right Triangle Approach†   469

6T.1   Angles   470
Conventions for Angles   470
Degree Measure   471
Radian Measure   472
Converting Angle Measures   474
Arc Length and Area   476
Similar Triangles   478
Models and Applications: Shadows   481

6T.2   Definition of the Trigonometric Functions Using Right Triangles   487
Definitions of Trigonometric Functions of Acute
   Angles   488
Evaluating Trigonometric Functions Using Right
   Triangles   489
Consistency of the Definitions   490
Special Angles   492
Representing Angles   494
Using Representative Triangles to Evaluate
   Trigonometric Functions   495

6T.3   Analysis of Right Triangles   502
Solving Right Triangles   502
A New Formula for the Area of a Triangle   505
Models and Applications: Transits and
   Shadows   507

6T.4   Extending the Trigonometric Functions: The Unit Circle   514
The Unit Circle   515
Extending the Trigonometric Functions:
   Trigonometric Points   516
Trigonometric Functions of Compass Points   518
Quadrants   521
Reference Numbers   522
Using Reference Numbers to Evaluate
   Trigonometric Functions   524
The Pythagorean Identity   529

CHAPTER ROADMAP AND STUDY GUIDE   535
CHAPTER QUIZ   537
CHAPTER REVIEW EXERCISES   539

7   Graphs and Periodicity of Trigonometric Functions   546

7.1   Graphs of the Sine and Cosine   547
Periodicity of the Sine and Cosine   547
The Graph of the Sine Function   549
The Graph of the Cosine Function   550
Amplitude   552
The Period   554
Phase Shift   555
Even and Odd Functions   559

7.2   Graphs of Other Trigonometric Functions   567
Periods of the Other Trigonometric Functions   568
Graphs of the Tangent and Cotangent   570
Graphs of the Secant and Cosecant   572
Transformations of Other Trigonometric
   Functions   574

CHAPTER ROADMAP AND STUDY GUIDE   585
CHAPTER QUIZ   586
CHAPTER REVIEW EXERCISES   587

ON7.3   Modeling with Periodic Functions*
Adjusting Sine Waves
Simple Harmonic Motion
Sound Waves
Spring–Mass Systems

8   The Algebra of Trigonometric Functions   591

8.1   Trigonometric Identities   592
Basic Identities   593
Cofunction Identities   595
Even and Odd Identities   597
Algebraic Verification of Trigonometric
   Identities   598

8.2   Sum and Difference Formulas   606
Sum and Difference Formulas for Sine and
   Cosine   607
Sum and Difference Formulas for the Tangent and
   Cotangent   611
Product-to-Sum and Sum-to-Product
   Formulas   613

8.3   Double-Angle and Half-Angle Formulas   622
Double-Angle Formulas   623
Half-Angle Formulas   625
Combining Sum and Multiple-Angle
   Formulas   629
Power Reduction Formulas   630

8.4   Inverse Trigonometric Functions   638
Arcsine and Arccosine   639
Graphs of the Arcsine and Arccosine   644
The Arctangent Function   646
Models and Applications: Flying and
   Basketball   650
The Other Inverse Trigonometric
   Functions: Arccosecant, Arcsecant, and
   Arccotangent   651

8.5   Solving Trigonometric Equations   662
Basic Trigonometric Equations   663
Solving by Factoring   671
Analogy with Quadratic Equations   673

CHAPTER ROADMAP AND STUDY GUIDE   679
CHAPTER QUIZ   681
CHAPTER REVIEW EXERCISES   682

Geometry

9   Topics in Geometry   690
9.1   Law of Cosines   691
Recalling Congruence Criteria   692
Derivation of the Law of Cosines   692

†Either Chapter 6C or Chapter 6T can be used; the intention is for instructors to assign only one of these two chapters.
*ON refers to material that can be found online with registration in Achieve: achieve.macmillanlearning.com.

Using the Law of Cosines to Solve Triangles   693
Heron's Formula for the Area of a Triangle   698
Models and Applications: A Soccer Player   699

9.2   **Law of Sines**   709
Derivation of the Law of Sines   710
Solving Triangles Using the Law of Sines   711
The Ambiguous Case: Side-Side-Angle   714
Choosing a Law to Solve a Triangle   718
Models and Applications: A Lost Airplane   720

9.3   **Vectors**   728
Overview of Vectors   729
Scalar Multiplication   729
Vector Addition   731
Components of Vectors   734
Models and Applications: An Airplane   736

9.4   **Vectors in the Plane and in Three Dimensions**   745
Standard Position in the Plane   746
Dot Product   748
Three-Dimensional Vectors   751
Models and Applications: Rotating an Antenna   754

CHAPTER ROADMAP AND STUDY GUIDE   761
CHAPTER QUIZ   763
CHAPTER REVIEW EXERCISES   765

ON9.5   **Complex Numbers and de Moivre's Theorem***
Graphical Representation of Complex Numbers
Trigonometric Form
De Moivre's Theorem: Products, Powers, and Quotients in Trigonometric Form
Roots of Complex Numbers

ON9.6   **Polar Coordinates***
Defining Polar Coordinates
Converting Between Polar and Rectangular Coordinates
Graphs of Polar Equations
Converting Between Polar and Rectangular Graphs
Models and Applications: Firing Projectiles
Exotic Graphs

ON9.7   **Parametric Equations***
Parametric Curves: Vector and Parametric Forms
Graphs of Parametric Equations
Linear Parametrizations
Relationships Between Parametric and Rectangular Forms
Models and Applications: A Light on a Wheel

Rates of Change

 10 **A Qualitative Exploration of Rates of Change**   770

10.1   **An Introduction to the Rate of Change as a Function**   771
Units of the Rate of Change   772
Sketching the Graph of the Rate of Change   773
Concavity Revisited   777
Using Rates of Change to Analyze Functions   780

10.2   **Change Equations: Linear and Exponential Functions**   789
Constant Rate of Change   789
The Exponential Case   791

10.3   **Graphical Solutions of Change Equations**   800
Change Equations: Equilibrium Solutions   801
Sketching Graphs of Solutions   803

CHAPTER ROADMAP AND STUDY GUIDE   816
CHAPTER QUIZ   817
CHAPTER REVIEW EXERCISES   822

 Systems of Equations and Matrices*

ON11.1   Systems of Linear Equations
Solutions of Systems of Linear Equations
Gaussian Elimination
The Case of No Solution
The Case of Infinitely Many Solutions
Models and Applications: Competition Between Populations and Mixing Ingredients

ON11.2   Augmented Matrices
Augmented Matrices
Row-Echelon Form for Augmented Matrices
Using Row-Echelon Form to Solve Systems of Equations
Gauss-Jordan Elimination
Models and Applications: Investment Strategies

ON11.3   Matrix Algebra
Matrix Addition and Scalar Multiplication
Matrix Multiplication
Properties of Matrix Algebra
Representing Systems of Equations Using Matrix Multiplication
Models and Applications: Cost and a Disease

ON11.4   Determinants
Two by Two Determinants
Three by Three Determinants
Cramer's Rule
Models and Applications: Areas and Volumes

ON11.5   Systems of Nonlinear Equations
Solving by Substitution
Elimination Method
Areas
Models and Applications: Building an Enclosure

ON11.6   Systems of Inequalities
Systems of Linear Inequalities
Nonlinear Systems of Inequalities
Models and Applications: A Car Dealer

ON11.7   The Method of Partial Fractions
Preparing for Partial Fractions
Partial Fractions, Part 1: Distinct Linear Factors
Partial Fractions, Part 2: Irreducible Quadratic Factors
Partial Fractions, Part 3: Repeated Factors

CHAPTER ROADMAP AND STUDY GUIDE
CHAPTER QUIZ
CHAPTER REVIEW EXERCISES

Analytic Geometry of Conic Sections*

ON12.1   Circles and Ellipses
Review of Circles
Ellipses Centered at the Origin
Eccentricity of an Ellipse
Moving the Center of Ellipses
Models and Applications: Area of an Ellipse

ON12.2 Parabolas and Hyperbolas
Parabolas
Hyperbolas Centered at the Origin
Shifting Parabolas and Hyperbolas
Completing the Square to Obtain Standard
    Forms

ON12.3 Polar and Parametric Equations of
Conic Sections
Unified Definition of Conic Sections
Polar Equations of Conic Sections
Parametric Equations for Conic Sections
Models and Applications: Orbits of Planets

ON12.4 Rotations of Conic Sections
Rotating Conic Sections: Polar Equations
Rotating Conic Sections: Rectangular Graphs
The Discriminant
Models and Applications: A Planetary Collision?

CHAPTER ROADMAP AND STUDY GUIDE
CHAPTER QUIZ
CHAPTER REVIEW EXERCISES

 ON13 Sequences, Sums, and the Binomial Theorem*

ON13.1 Sequences
The Basics of Sequences
Arithmetic and Geometric Sequences
Sequences Defined by a Recurrence Relation
Models and Applications: Discrete Dynamical
    Systems

ON13.2 Sums
The Sigma Notation
Arithmetic Sums
Geometric Sums
Geometric Series
Riemann Sums
Models and Applications: The Great Pyramid
    of Giza

ON13.3 Counting and the Binomial Theorem
Permutations: When Order Matters
Combinations: When Order Does Not Matter
The Binomial Theorem
Pascal's Triangle

CHAPTER ROADMAP AND STUDY GUIDE
CHAPTER QUIZ
CHAPTER REVIEW EXERCISES

Appendices APP-1

Appendix 1 Using Technology to Solve
    Equations   APP-1

Appendix 2 Complex Numbers   APP-3

Appendix 3 The Complex Exponential Function   APP-6

Appendix 4 Derivation of the Monthly Payment
    Formula   APP-9

Appendix 5 Circles and the Law of Sines   APP-11

Appendix 6 Matrix Inverses   APP-13

Appendix 7 Induction   APP-17

Appendix 8 Probability   APP-22

Appendix 9 Vector Equations of Lines   APP-26

Appendix 10 Cofactor Definition of
    Determinants   APP-28

Answers ANS-1

Glossary G-1

Index I-1

*ON refers to material that can be found online with registration in Achieve: achieve.macmillanlearning.com.

PREFACE

Our Vision for This Textbook

Calculus serves as the mathematical entry to science, engineering, and often business, yet for many it stands as an impediment rather than a gateway. The resulting challenge may lead to second- or third-choice majors. Our simple goal with this textbook is to make calculus accessible to every student who wishes to discover its beauty and power. Keys to successful preparation for calculus are the development of calculation skills and growth in mathematical sophistication. We present these as an integrated whole because neither alone is sufficient. Fostering success in mathematics remains our passion, and we hope you see this passion in every page of the textbook.

Course Challenges for Precalculus

Challenge: Students entering calculus have inadequate calculation skills.

> **Response:** We hone the calculation skills of students through a wide variety of examples and exercises. Each example includes a thorough solution and the *Try It Yourself* feature, which allows students to practice calculations themselves rather than just view the authors' work. This active participation in the learning process is crucial to the proper development of calculation skills.

Challenge: Students entering calculus are not properly prepared for the intellectual challenges they face. The fundamental ideas of calculus are far more sophisticated than most students have encountered in earlier courses.

> **Response:** Our textbook includes an informal and intuitive presentation of limits and rates of change, which we interpret in terms of the graph. These ideas are not an add-on at the end of the book. Rather, they are introduced early and are incorporated wherever appropriate throughout the text. Students are presented from the beginning with the dynamic view of functions and their graphs that is essential in any calculus course.

Challenge: Precalculus students face a bewildering variety of topics that may seem unrelated.

> **Response:** Rates of change, limits, and a dynamic view of functions tie together the topics in our textbook and drive their organization. Further, we confine our presentation to the topics essential for calculus and in so doing present students with a more succinct, less intimidating text. Finally, Chapter 10 serves as a capstone that summarizes the course, showing the unifying concepts and serving as a springboard into calculus.

Challenge: Many precalculus students do not see the utility of the course.

> **Response:** Our book has an extensive collection of applications exercises. Many of these are taken from the scientific literature and include appropriate references, showing actual applications of the concepts developed in the text. In addition, most sections of the text include a subsection presenting models and applications.

Engage Every Student with Achieve for *Preparation for Calculus: Functions and How They Change*

Achieve includes a full learning path of content for pre-class preparation, in-class active learning, and post-class engagement and assessment. Built with learning science research and an intuitive interface, Achieve is the perfect solution for engaging all students, no matter how you teach.

What's Inside Achieve? 🎓 Achieve

- **Insights and Reports** provide powerful analytics on class and student performance based on assignments, units, and learning objectives.
- **Dynamic Figures**, powered by Desmos, are interactive versions of select figures from the text. They are designed to help students visualize key concepts by providing hands-on experience while exploring graphical relationships.
- **Instructor Resources** are available with Achieve and *Preparation for Calculus*, such as lecture and image slides, clicker questions, the test bank, and a solutions manual.
- The **e-book** is fully accessible and boasts handy features like note taking, highlighting, offline reading, and screen reader functionality.

Pre-Class Assessment

- **Guided Learn and Practice** assignments serve as a tutorial for students for major concepts throughout the course. Embedded video and interactive resources, combined with assessment, can be used for class preparation or pre-exam study.
- **LearningCurve adaptive quizzing** encourages students to read the text with a question in mind and test their own comprehension.

In-Class Activities

- **Active learning resources**, including clicker questions and easy-to-implement Instructor Activity Guides, help foster student engagement in any type of class. Student access to the iClicker app is included for free with Achieve.

Post-Class Assessment

- **Homework exercises** in Achieve feature built-in coaching tools—detailed and error-specific feedback, hints, and fully worked solutions—meant to guide students toward the correct answers and a deeper conceptual understanding. Our proprietary grading algorithm combines our homegrown parser and the computer algebra system, SymPy. It is programmed to accept every valid equivalent answer and to trigger warnings for answers entered in an incorrect format.
- **Graded Graph assessment modules**, powered by Desmos, allow students to graph functions as part of their homework assignments and receive instant feedback.

A Letter to the Student

One of the most important events in the history of mathematics was the invention of calculus by Isaac Newton (1642–1727) and Wilhelm Leibniz (1646–1716). Calculus, which can be defined as the science of change, opened the way for the development of physics and modern science and engineering. We wrote this textbook as a roadmap that will serve as your gateway to this most important of all mathematical innovations.

Your earlier courses in mathematics likely had a static focus: Variables like x and y were unchanging throughout a given problem, and you were expected to manipulate them according to certain rules. In this textbook, by contrast, you will experience a dynamic view of mathematics. By this we mean that we will ask how the variable y changes when the variable x changes, and so a function and its graph will be studied to illustrate how the variables change. An integral step in preparing for calculus is understanding the fundamental ideas embodied in a dynamic view of functions. In Chapter 1 of this textbook, we introduce the concepts of limit and rates of change. Although these ideas are sophisticated, we present them in a way that we believe will be intuitive and understandable, and we revisit them throughout the chapters. For example, straight lines—which will be familiar to students—are presented here as graphs of functions having a constant rate of change. Once you master the concepts of limit and rates of change, you will have access to new and more effective ways to understand functions and their graphs and you will be ready for the challenges of calculus.

This textbook, like any other, can only help you if you *use* it. We have written this book for you, not your instructor, and you can be successful only if you actively participate in the learning process. The roadmap shows how the chapter topic fits into the big picture of the book and explains *why* a topic is covered and *what* is covered. This visual feature will help you to read carefully and thoughtfully.

Each section begins with a list of learning objectives and is organized to facilitate effective study. The narrative is punctuated with three types of highlighted boxes:

- *Concepts to Remember* summarize key ideas for easy reference. You will find these useful when you attack the exercises at the end of the section.
- *Laws of Mathematics* display important mathematical theorems. These often codify the results of mathematical derivations in the narrative.
- *Step-by-Step Strategy* provides a clear path for solving specific types of problems.

You will find many worked examples for each section that illustrate the ideas presented there and that provide real-world applications. These worked examples are followed by *Try It Yourself* exercises, which give you an opportunity to work through a similar problem on your own. Answers are provided at the end of the section, so you can check your success as you work through the material. The *Try It Yourself* exercises should be considered as stop signs that you should not pass until you have successfully completed the exercises.

As you proceed through the section, you will encounter *Extend Your Reach* questions. These are not intended as questions you should be able to answer easily on your own. Rather, they are designed to promote active learning and lively classroom discussion.

Each section ends with a generous collection of exercises of several types. Your instructor will select from these as a key part of your homework:

- **Check Your Understanding** segments present short-answer questions that are designed to check your readiness to proceed.
- **Skill Building** exercises provide straightforward practice with the ideas presented in the section.
- **Problems** focus on mastery of the section content.
- **Models and Applications** are real-world applications that use the ideas in the section. Many of these are taken from scientific literature.
- **Challenge Exercises for Individuals or Groups** are intended for you to solve with your friends and classmates.
- **Review and Refresh** is a collection of exercises from previous sections, designed to keep earlier material fresh in your mind.

A religiously followed study schedule devoted to the exercises assigned by your instructor is the key to success in this, and any other, mathematics course. Beyond regular class attendance, such a schedule should include specific time after *each* class meeting set aside for study and work on exercises. Many will find it profitable to work with a group of friends.

We hope you enjoy your experience with this text, and we are always interested in hearing about ways that it can be improved.

—Bruce Crauder, Benny Evans, Alan Noell

A Letter to the Instructor

Students entering calculus find their calculation skills challenged at the same time they are faced with new, sophisticated mathematical ideas. Although every precalculus textbook, including this one, addresses calculation issues, this book also prepares students for the new ideas they will encounter in calculus. For example, limits are a complicated mathematical tool, and a rigorous development of rates of change depends on an understanding of the limit. To make this easier for students first encountering the concepts, in this book we present rates of change and limits at infinity in an intuitive way that will be accessible to precalculus students. These bedrock concepts, then, pave the way to success in understanding their more sophisticated cousins in calculus.

There are no difficult limit calculations in this textbook, nor does the rigorous definition appear. Rather, in Chapter 1 limits at infinity are introduced as a descriptive tool. For example, $\frac{1}{x} \to 0$ as $x \to \infty$ means that the right tail end of the graph of $y = \frac{1}{x}$ approaches the x-axis. The concept of a limit is used throughout the textbook. For example, horizontal asymptotes of rational functions are naturally associated with limits, and the long-term behavior of the natural logarithm is described using $\ln x \to \infty$ as $x \to \infty$.

Derivatives are presented in an intuitive way, not calculated. Instead, rates of change are presented as slopes of intuitively understood tangent lines. With this approach, students can easily grasp the idea that a positive rate of change indicates an increasing function, a negative rate of change indicates a decreasing function, and extrema may occur where the rate of change is zero. To enforce the primacy of the concept, we incorporate ideas from rates of change wherever appropriate in Chapters P–9. The book concludes with a unique, capstone Chapter 10, which serves as a launch point into calculus. Chapter 10 highlights the fact that properties of functions can be deduced from information about their rate of change.

Furthermore, rates of change drive the organization of this textbook. In particular, linear functions are presented in Chapter 3 as functions that have a constant rate of change. This organization allows for a natural progression that groups linear functions with exponential functions that have a constant proportional rate of change. In simplistic terms, linear functions change by constant sums, whereas exponential functions change by constant multiples; we use these features to show these functions as models of natural phenomena.

Another unique feature of this textbook is the flexibility of its presentation of trigonometry. Some textbooks use right triangles to introduce trigonometric functions and then extend the functions via the unit circle. Other textbooks use the reverse order. We present both versions of the chapter in this textbook, allowing instructors to choose the path they find appropriate. Instructors who prefer to begin with the unit circle will use Chapter 6C, and those who prefer to begin with right triangles will use Chapter 6T instead. (No one will use both.)

As much as is practical, throughout the textbook we provide applications that are taken from real situations, and we often include references to scientific literature. Most sections include at least one application example, and the exercises make up an extensive collection.

Our aim was to avoid an overwhelming, encyclopedic approach to precalculus and instead focus on only what a student truly needs to prepare for calculus. In the print book, we include lively, concise coverage common to all precalculus courses. A wealth of additional material is available online, including a review of algebra (Chapter 0), systems of equations and matrices (Chapter 11), conic sections (Chapter 12), and sequences, sums, and the binomial theorem (Chapter 13).

We have worked tirelessly on this project for more than five years, and we are very proud of the results. We hope you will agree that this presents a unique approach to what can be a difficult course to teach. Please let us know what you think. We would love to hear from you about ways the book can be improved.

—Bruce Crauder, Benny Evans, Alan Noell

ACKNOWLEDGMENTS

We wish to thank Nikki Miller Dworsky and Debbie Hardin, our intrepid editors, who embraced our vision of how to truly prepare students for calculus and who suffered through the many revisions and edits needed to make the vision a reality.

This project has undergone rigorous review and accuracy checking, thanks to myriad instructors who graciously offered their time and expertise. Any remaining errors (few though we hope they are) remain ours.

Reviewers

Marwan Abu-Sawwa, *Florida State College at Jacksonville*

Nirmal Aggarwal, *Embry Riddle Aero. University*

Wendy Ahrendsen, *South Dakota State University*

Allison W. Arnold, *University of Georgia*

Dionne T. Bailey, *Angelo State University*

Henry C. Bailey, III, *Missouri University of Science and Technology*

Paul Bailey, *California State University, Chico*

Mario Barrientos, *Angelo State University*

Man Basnet, *Iowa State University*

Sam Bazzi, *Henry Ford College*

Phil Bergonio, *University of Georgia*

Kelly Black, *University of Georgia*

Julie Blinder, *Lake Michigan College—Benton Harbor*

Latrice N. Bowman, *University of Alaska Fairbanks*

Troy Brachey, *Tennessee Technological University*

Dave Bregenzer, *Utah State University*

Lee Ann Brown, *Oklahoma State University*

Jonathan Brucks, *University of Texas at San Antonio*

Tim Chappell, *Metropolitan Community College-Penn Valley*

Hongwei Chen, *Christopher Newport University*

Carey Childers, *Clarion University*

Caroline Cochran, *Acadia University*

Charles Conrad, *Volunteer State Community College*

Michael Cotton, *Miami University*

Ana-Maria Croicu, *Kennesaw State University*

Vinh Dang, *Lone Star College—North Harris*

Gregory T. Daubenmire, *Las Positas College*

Judy Fethe, *Pellissippi State Community College*

Michael J. Fisher, *West Chester University*

Tim Flood, *Pittsburg State University*

Mary Jule Gabiou, *North Idaho College*

William Girton, *Florida Institute of Technology*

Lourdes Gonzalez, *Miami-Dade Community College*

Suzie Goss, *Lone Star College Kingwood*

Vladimir Grupchev, *University of South Florida*

Shuvra Gupta, *Iowa State University*

Boyko Georgiev Gyurov, *Georgia Gwinnett College*

Victoria Hamlin, *Middle Tennessee State University*

Yu-ing Hargett, *Jefferson State Community College*

Sheyleah Harris-Plant, *South Plains College*

H. M. Hassani, *Columbus State University*

Caleb D. Holloway, *University of Arkansas*

Miles Hubbard, *St. Cloud State University*

Jeff Igo, *University of Michigan—Dearborn*

Mohamed Jamaloodeen, *Georgia Gwinnett College*

Yvette Janecek, *Blinn College*

Jacqueline Jensen-Vallin, *Lamar University*

M. Kazemi, *University of North Carolina—Charlotte*

Susan Keith, *Georgia State University, Perimeter*

Kathryn Kozak, *Coconino Community College*

Elena Kravchuk, *University of Alabama at Birmingham*

Thurai Kugan, *John Jay College of Criminal Justice, CUNY*

Debra Lackey, *Odessa College*

Alexander Lavrentiev, *University of Wisconsin Fox Valley*

Joseph Liddle, *University of Alaska Southeast*

Richard M. Low, *San Jose State University*

Jeffrey W. Lyons, *University of Hawaii*

Phil McCartney, *Northern Kentucky University*

Timothy H. McNicholl, *Iowa State University*

Edward T. Migliore, *University of California Santa Cruz*

Michelle Moravec, *Baylor University*

Larry Musolino, *Penn State University*

Charlie Nazemian, *University of Nevada, Reno*

Yves Nievergelt, *Eastern Washington University*

Solomon A. Osifodunrin, *Houston Community College*

Stan Perrine, *Georgia Gwinnett College*

Shawn Peterson, *Texas State University*

Maria Consuelo Pickle, *College of Southern Nevada*

Daniel Pinzon, *Georgia Gwinnett College*

Katherine Pinzon, *Georgia Gwinnett College*

Charlotte Pisors, *Baylor University*

L. Potaniec, *DePaul University*

Michael Price, *University of Oregon*

Karin Pringle, *The University of Tennessee, Knoxville*

Rajendra Ra, *University of Texas Rio Grande Valley*

Traci M. Reed, *St. Johns River State College*

Candice L. Ridlon, *University of Maryland Eastern Shore*

Steven Riley, *Chattahoochee Technical College*

Maryan Rogers, *Kennesaw State University*

David Royster, *University of Kentucky*

Lynne Ryan, *Blue Ridge Community College*

Tori Ryburn, *University of Arkansas*

Haazim Sabree, *Perimeter College at Georgia State University*

Sutandra Sarkar, *Georgia State University*

Victoria Savatorova, *University of Nevada, Las Vegas*

Michael William Schroeder, *Marshall University*

Alicia Serfaty de Markus, *Miami Dade College*

Tim Sheng, *Baylor University*

Brenda Shepard, *Lake Michigan College*

Sharon Sledge, *San Jacinto College*

Chris Southworth, *University of Nevada, Las Vegas*

Alin Stancu, *Columbus State University*

Catherine Stevens, *Lone Star College—North Harris*

Bob Strozak, *Old Dominion University*

Frances Tishkevich, *Massachusetts Maritime Academy*

Abby S. Train, *Texas State*

Laura Urbanski, *Northern Kentucky University*

Diane Valade, *Piedmont Virginia Community College*

Philip Veer, *Johnson County Community College*

Mina Vora, *East Georgia State College*

Mike Weimerskirch, *University of Minnesota*

Scott Wilde, *Baylor University*

Marissa Wolfe, *Tulsa Community College*

Alan Yang, *Columbus State Community College*

Justin Young, *The Ohio State University at Newark*

Ahmed I. Zyed, *DePaul University*

Victor Zinger, *University of Alaska Fairbanks*

THE COORDINATE PLANE AND INEQUALITIES

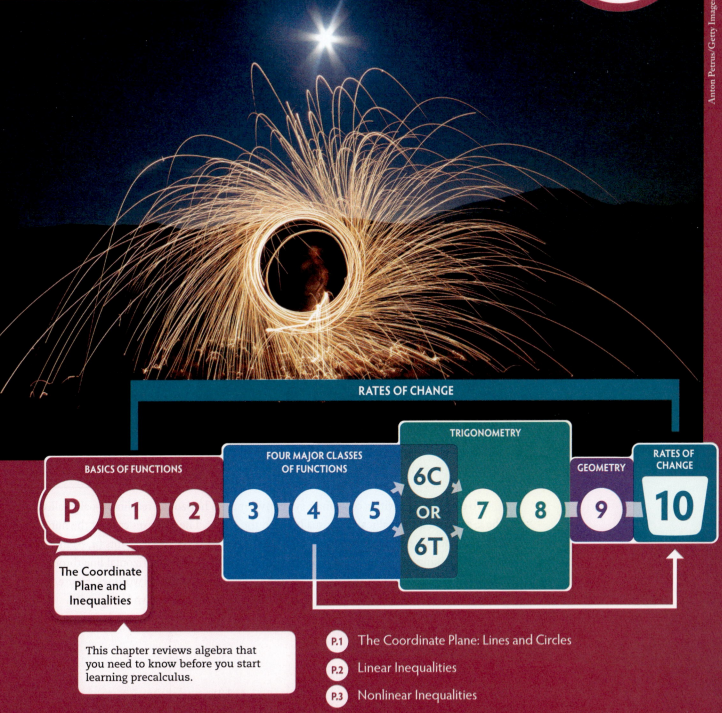

The Coordinate Plane and Inequalities

This chapter reviews algebra that you need to know before you start learning precalculus.

P.1   The Coordinate Plane: Lines and Circles

P.2   Linear Inequalities

P.3   Nonlinear Inequalities

When students have difficulty in calculus, it's often because of their algebra skills. This is why review is so important. This preliminary chapter is a refresher on the elementary concepts of coordinate geometry in the plane and inequalities. Even well-prepared students may need to review the graphical properties of lines and circles as well as methods for solving inequalities. The topics in this chapter are revisited in more detail throughout the book.

P.1   The Coordinate Plane: Lines and Circles

P.2   Linear Inequalities

P.3   Nonlinear Inequalities

P.1   The Coordinate Plane: Lines and Circles

> Using the coordinate plane allows us to visualize mathematical concepts.

In this section, you will learn to:

1. Locate points and sketch the graph of equations in the coordinate plane.
2. Apply the distance and midpoint formulas in calculations.
3. Explain the slope and y-intercept of a line in terms of the graph.
4. Calculate the slope of a line.
5. Express the equation of a line in the slope-intercept, point-slope, and two-point forms.
6. Express the equation of a circle in terms of its center and radius.
7. Determine the center and radius of a circle from its equation, and sketch the graph.
8. Solve applied problems involving circles.

Equipping a plane with a coordinate system allows geometric objects such as lines and circles to be described by algebraic formulas. Conversely, this connection between algebra and geometry allows us to visualize equations.

Coordinates in the Plane

> Graphs are pictures of equations.

To equip a plane with a coordinate system, we make a horizontal real line, usually called the x-axis, and a vertical real line, usually called the y-axis. The two axes cross at the number 0 on each line. This crossing point is called the origin. We locate points (a, b) in the plane by making a vertical line through $x = a$ on the x-axis and a horizontal line through $y = b$ on the y-axis. The point (a, b) is the intersection of these two lines. **Figure P.1** shows the origin, which is the point $(0, 0)$. Additional points are shown in **Figure P.2**.

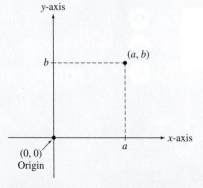

Figure P.1 Locating points in the plane

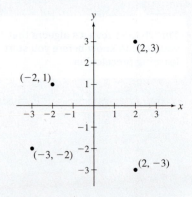

Figure P.2 Various points in the plane

The **graph** of an equation is the result of plotting many points. Let's sketch the graph of the equation $y = x^2 - 1$. We begin with a brief table of values. Next, we plot the corresponding points in the coordinate plane, as shown in **Figure P.3**. Connecting these points with a smooth curve gives the graph in **Figure P.4**.

The **graph** of an equation in x and y consists of all points (x, y) in the coordinate plane that satisfy the equation.

x	$y = x^2 - 1$
-2	3
-1	0
0	-1
1	0
2	3

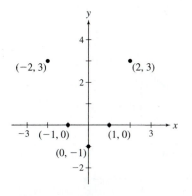

Figure P.3 Plotting points determined by $y = x^2 - 1$

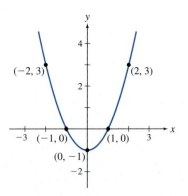

Figure P.4 The completed graph

We note that graphing calculators and computers make graphs using this method, except that they can plot many more points with much greater accuracy. But in many settings there are better methods for constructing graphs. This is certainly the case for lines and circles.

The Distance and Midpoint Formulas

> Distances in the plane can be calculated using the Pythagorean theorem.

The basic tool we need to find the distance between two points in the plane is the Pythagorean theorem.

LAWS OF MATHEMATICS: Pythagorean Theorem

Suppose that the legs of a right triangle have lengths A and B and the hypotenuse has length C. Then

$$A^2 + B^2 = C^2$$

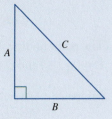

For example, we can use the Pythagorean theorem to find the length C of the hypotenuse in **Figure P.5**:

$$C^2 = A^2 + B^2$$
$$C^2 = 3^2 + 4^2$$
$$C^2 = 25$$
$$C = 5$$

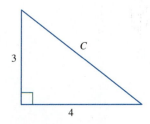

Figure P.5 An unknown hypotenuse

To find the distance between two given points $P = (x_1, y_1)$ and $Q = (x_2, y_2)$, we first plot them in the plane, as shown in **Figure P.6**. For simplicity we assume that $x_1 > x_2$ and $y_1 > y_2$. From these points we get a right triangle with sides of length $A = x_1 - x_2$ and $B = y_1 - y_2$. The distance between the two points is the length C of the hypotenuse, and using the Pythagorean theorem to calculate C gives the distance formula:

$$C^2 = A^2 + B^2 \qquad \blacktriangleleft \text{ Use the Pythagorean theorem.}$$

$$C^2 = (x_1 - x_2)^2 + (y_1 - y_2)^2 \qquad \blacktriangleleft \text{ Substitute values for } A \text{ and } B.$$

$$C = \sqrt{(x_1 - x_2)^2 + (y_1 - y_2)^2} \qquad \blacktriangleleft \text{ Take square roots.}$$

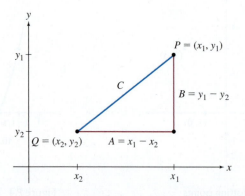

Figure P.6 Deriving the distance formula

LAWS OF MATHEMATICS: Distance Formula

The distance between the points $P = (x_1, y_1)$ and $Q = (x_2, y_2)$ is

$$\text{Distance} = \sqrt{(x_1 - x_2)^2 + (y_1 - y_2)^2}$$

We use the term *midpoint* for the point midway between two given points on the line segment joining them. We can locate the midpoint using the midpoint formula.

LAWS OF MATHEMATICS: Midpoint Formula

The point midway between the points $P = (x_1, y_1)$ and $Q = (x_2, y_2)$ is

Midpoint = (Average of x-coordinates, Average of y-coordinates)

$$= \left(\frac{x_1 + x_2}{2}, \frac{y_1 + y_2}{2} \right)$$

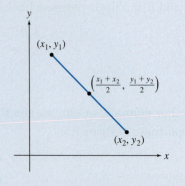

EXAMPLE P.1 Using the Distance and Midpoint Formulas

a. Locate the points $P = (2, 1)$ and $Q = (3, -1)$ on a coordinate plane. Find their midpoint, and locate it on the coordinate plane. Next, find the distance between P and Q.

b. Let t be any real number. Calculate the distance between the points $P = (t, t + 3)$ and $Q = (1, 2)$. Simplify your answer.

SOLUTION

a. We have located the points in **Figure P.7.**

The midpoint is

$$\text{Midpoint} = \left(\frac{2+3}{2}, \frac{1+(-1)}{2} \right)$$
$$= \left(\frac{5}{2}, 0 \right)$$

In **Figure P.8** we have added this point to the coordinate plane.

We calculate the distance between the points using the distance formula:

$$\text{Distance} = \sqrt{(2-3)^2 + (1-(-1))^2} \quad \blacktriangleleft \text{ Apply the distance formula.}$$
$$= \sqrt{(-1)^2 + 2^2} \quad \blacktriangleleft \text{ Simplify.}$$
$$= \sqrt{5} \quad \blacktriangleleft \text{ Simplify.}$$

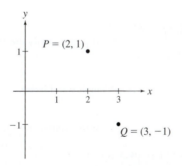

Figure P.7 Using the midpoint formula: the points $(2, 1)$ and $(3, -1)$

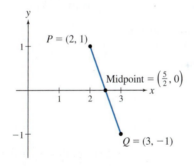

Figure P.8 Using the midpoint formula: the midpoint added

b. We use the distance formula:

$$\text{Distance} = \sqrt{(t-1)^2 + ((t+3)-2)^2} \quad \blacktriangleleft \text{ Apply the distance formula.}$$
$$= \sqrt{(t-1)^2 + (t+1)^2} \quad \blacktriangleleft \text{ Simplify.}$$
$$= \sqrt{t^2 - 2t + 1 + t^2 + 2t + 1} \quad \blacktriangleleft \text{ Expand squares.}$$
$$= \sqrt{2t^2 + 2} \quad \blacktriangleleft \text{ Simplify.}$$

TRY IT YOURSELF P.1 Brief answers provided at the end of the section.

Let t be a real number. Find the distance between $P = (t + 3, t + 6)$ and $Q = (t, t + 2)$, and calculate their midpoint.

Lines and Their Slopes

> Linear equations are represented by straight lines in the plane.

A **linear equation** in x and y is an equation of the form $y = mx + b$.

The **slope** of the line $y = mx + b$ is the constant m.

The **y-intercept** of the line $y = mx + b$ is the constant b.

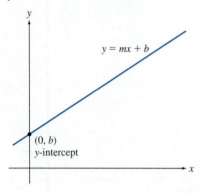

Figure P.9 y-intercept of a line

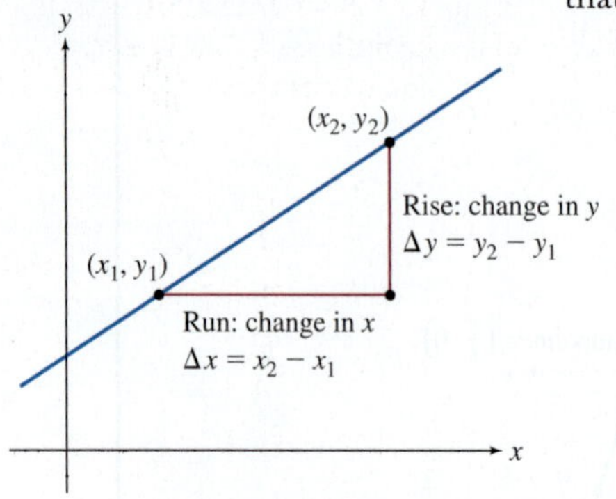

Perhaps the most familiar graph, and certainly the simplest, is the straight line. The equation of a straight line that is not vertical is the **linear equation** $y = mx + b$. The constant m is the **slope** of the line, and the constant b is the **y-intercept** of the line because the graph of the line crosses the y-axis at the point $(0, b)$ (**Figure P.9**). The y-intercept is also called the vertical intercept.

The equation of a vertical line has the form $x = a$. In this section we will focus on lines that are not vertical.

If (x_1, y_1) and (x_2, y_2) are two points on the graph of the linear equation $y = mx + b$, we refer to the difference in x-values, namely $x_2 - x_1$, as the change in x. Similar terminology applies to the y-coordinate, and these changes are identified by various names.

- The change in x, also called the run, is denoted by Δx. It is calculated by $\Delta x = x_2 - x_1$.

- The change in y, also called the rise, is denoted by Δy. It is calculated by $\Delta y = y_2 - y_1$.

These terms are illustrated in **Figure P.10**. The rise and run can be used to calculate the slope m of the line $y = mx + b$ using any two distinct points (x_1, y_1) and (x_2, y_2) that lie on the line. We use the fact that $y_1 = mx_1 + b$ and $y_2 = mx_2 + b$:

$$\frac{\text{Rise}}{\text{Run}} = \frac{\Delta y}{\Delta x} = \frac{y_2 - y_1}{x_2 - x_1}$$

$$= \frac{(mx_2 + b) - (mx_1 + b)}{x_2 - x_1} \qquad \blacktriangleleft \text{ The points satisfy the equation } y = mx + b.$$

$$= \frac{m(x_2 - x_1)}{x_2 - x_1} \qquad \blacktriangleleft \text{ Simplify and factor.}$$

$$= m \qquad \blacktriangleleft \text{ Cancel common terms.}$$

Figure P.10 Rise and run

The result is that the slope is the rise over the run.

EXAMPLE P.2 Calculating the Slope of a Line

Calculate the slope of the line through the points $(2, 5)$ and $(4, -1)$.

SOLUTION

The rise is the change in the y-value:

$$\text{Rise} = \Delta y = -1 - 5 = -6$$

The run is the change in the x-value:

$$\text{Run} = \Delta x = 4 - 2 = 2$$

This gives

$$\text{Slope} = \frac{\text{Rise}}{\text{Run}} = \frac{\Delta y}{\Delta x} = \frac{-6}{2} = -3$$

TRY IT YOURSELF P.2 Brief answers provided at the end of the section.

Find the slope of the line through $(1, 4)$ and $(-2, 7)$.

We note that in calculating the slope of the line through two points (x_1, y_1) and (x_2, y_2) we can calculate the difference in coordinates in either order we please *as long as we are consistent*:

$$\text{Slope} = \frac{y_2 - y_1}{x_2 - x_1} = \frac{y_1 - y_2}{x_1 - x_2}$$

But we can't use one order in the numerator and another in the denominator—that would give the negative of the correct answer.

The slope of a line measures how steep it is. Let's look, for example, at the graph of $y = 2x + 3$, which has a slope of 2. Start at the point $(4, 11)$ shown in **Figure P.11**. If we move 1 unit to the right so that the x-value is now 5, the y-value is $y = 2 \times 5 + 3 = 13$. Thus, 1 unit of run results in 2 units of rise. You can easily verify that for this line, 1 unit of run always results in 2 (the slope) units of rise.

In general, if we increase the x-value by 1 unit, what happens to the y-value? Our linear formula gives the answer:

$$\text{New } y\text{-value} = m(x + 1) + b = mx + m + b = (mx + b) + m = \text{Old } y\text{-value} + m$$

So the graph of $y = mx + b$ changes by m vertical units (the slope) for each horizontal unit of increase.

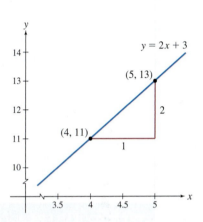

Figure P.11 Rise versus run: One unit of run for the line $y = 2x + 3$ results in 2 (the slope) units of rise.

CONCEPTS TO REMEMBER: Slope of a Line

For a linear equation $y = mx + b$:

- The graph is a straight line with y-intercept b and slope m.
- Each unit of run (or Δx) corresponds to m units of rise (or Δy).
- The slope is given by

$$\text{Slope} = \frac{\text{Change in } y}{\text{Change in } x} = \frac{\text{Rise}}{\text{Run}} = \frac{\Delta y}{\Delta x}$$

Explicitly, if (x_1, y_1) and (x_2, y_2) are any two distinct points on the line, the slope is $\frac{y_2 - y_1}{x_2 - x_1}$.

If the slope is positive, the line is inclined upward as we move from left to right; the larger the slope, the steeper the line. If the slope is negative, the line is inclined downward as we move from left to right; the greater the magnitude of the slope, the steeper the line. A line with zero slope is horizontal. Lines with various slopes are shown in **Figure P.12**.

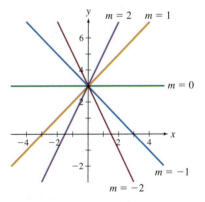

Figure P.12 Lines with various slopes

EXAMPLE P.3 Sketching a Line and Finding Its Equation

Sketch the line with y-intercept $b = 2$ and slope $m = 1$, and write its linear equation.

SOLUTION

A line with y-intercept 2 must pass through the point $(0, 2)$, as shown in **Figure P.13**. A slope of 1 means that for a run of 1 unit the rise is 1 unit, which gives a second point $(1, 3)$. This point is also illustrated in Figure P.13. The completed graph is shown in **Figure P.14**. Because this line has slope $m = 1$ and y-intercept 2, the corresponding linear equation $y = mx + b$ is $y = x + 2$.

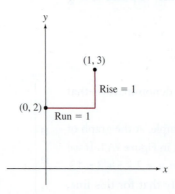

Figure P.13 Preparing to sketch a line: y-intercept 2 and slope 1

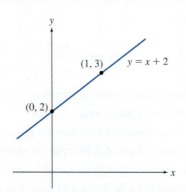

Figure P.14 Completing the graph: the line with y-intercept 2 and slope 1

TRY IT YOURSELF P.3   Brief answers provided at the end of the section.

Write the equation of a line with y-intercept -1 and slope 2, and sketch its graph on a coordinate plane.

EXTEND YOUR REACH

a. Sketch the lines $y = 10x$, $y = 50x$, and $y = 100x$ on the same coordinate axes.

b. How would you describe these lines? What is their relationship?

c. Earlier we defined the slope only for lines that are not vertical. What do these three lines suggest about defining the slope of a vertical line?

d. Recall that the equation of a vertical line has the form $x = a$. Determine the equation of the vertical line that passes through the point $(1, 0)$.

Common Forms for Equations of Lines

> We can find the equation of a line given the slope and the y-intercept, the slope and a point, or two points.

Often lines are given in terms of one point and the slope or in terms of two points. There's a shortcut for finding the equation of a line in each case.

Suppose we know that the slope of a line is m and that it passes through the point (x_1, y_1). If (x, y) is an arbitrary point on the line, then we can use these two points to calculate the slope:

$$m = \frac{y - y_1}{x - x_1}$$

Multiplying by $x - x_1$ yields the point-slope form of the equation of a line:

$$y - y_1 = m(x - x_1)$$

Suppose that we are given distinct points (x_1, y_1) and (x_2, y_2) that lie on a line. Again, we let (x, y) denote an arbitrary point on the line, and we equate two calculations of the slope:

$$\frac{y - y_1}{x - x_1} = \frac{y_2 - y_1}{x_2 - x_1}$$

Multiplying by $x - x_1$ yields the two-point form of the equation of a line:

$$y - y_1 = \frac{y_2 - y_1}{x_2 - x_1}(x - x_1)$$

CONCEPTS TO REMEMBER: Formulas for Lines

The slope-intercept form of the equation of the line with slope m and y-intercept b is $y = mx + b$.

The point-slope form of the equation of the line with slope m passing through the point (x_1, y_1) is $y - y_1 = m(x - x_1)$.

The two-point form of the equation of the line passing through distinct points (x_1, y_1) and (x_2, y_2) is $y - y_1 = \dfrac{y_2 - y_1}{x_2 - x_1}(x - x_1)$.

EXAMPLE P.4 Equations of Lines

a. Determine the equation of the line that passes through $(2, -1)$ and has slope 3.

b. Determine the equation of the line that passes through $(-1, 3)$ and $(2, 5)$.

SOLUTION

a. We use the point-slope form with $m = 3$ and $(x_1, y_1) = (2, -1)$:

$$y - y_1 = m(x - x_1)$$

$$y - (-1) = 3(x - 2)$$

This simplifies to $y + 1 = 3(x - 2)$ or $y = 3x - 7$.

b. We use the two-point form with $(x_1, y_1) = (-1, 3)$ and $(x_2, y_2) = (2, 5)$:

$$y - y_1 = \frac{y_2 - y_1}{x_2 - x_1}(x - x_1)$$

$$y - 3 = \frac{5 - 3}{2 - (-1)}(x - (-1))$$

This simplifies to $y - 3 = \dfrac{2}{3}(x + 1)$ or $y = \dfrac{2}{3}x + \dfrac{11}{3}$.

TRY IT YOURSELF P.4 Brief answers provided at the end of the section.

Determine the equation of the line through $(1, -4)$ and $(-2, 5)$.

EXTEND YOUR REACH

a. Determine the equation of the line that passes through $(1, 2)$ and has slope 2.

b. Determine the equation of the line that passes through $(1, 3)$ and also has slope 2.

c. Sketch the lines from parts a and b on the same coordinate axes.

d. How do you think lines with the same slope compare? Can you justify your answer by thinking of the slope as rise over run?

EXAMPLE P.5 Parallel Lines

Distinct lines are parallel when they have the same slope. Determine the equation of the line through $(1, 5)$ that is parallel to $y = 5x + 4$.

SOLUTION

We know that the point $(1, 5)$ is on the line. The slope of the line is the same as the slope of $y = 5x + 4$, which is 5. We use the point-slope form with $m = 5$ and $(x_1, y_1) = (1, 5)$:

$$y - y_1 = m(x - x_1)$$
$$y - 5 = 5(x - 1)$$

This simplifies to $y = 5x$.

TRY IT YOURSELF P.5   Brief answers provided at the end of the section.

Determine the equation of the line through $(-1, -5)$ that is parallel to $y = -5x + 4$.

Circles

> The equation of a circle identifies its center and radius.

A circle is the collection of points in the plane that lie at a fixed distance (called the radius) from a given point (called the center). The distance formula allows us to find the equation of a circle with center (x_0, y_0) and radius r. A point (x, y) lies on the circle provided the distance from (x, y) to the center (x_0, y_0) equals the radius r of the circle:

$$\text{Distance from } (x, y) \text{ to } (x_0, y_0) = r$$

$$\sqrt{(x - x_0)^2 + (y - y_0)^2} = r$$

If we square both sides of this equation, we obtain the following standard form for the equation of a circle.

LAWS OF MATHEMATICS: Equation of a Circle

The equation of a circle with center (x_0, y_0) and radius r is

$$(x - x_0)^2 + (y - y_0)^2 = r^2$$

EXAMPLE P.6 Equations and Graphs of Circles

a. Determine the equation of the circle with center $(2, -1)$ and radius 2.

b. Identify the center and radius of the circle with equation $(x+1)^2 + (y-1)^2 = 9$. Then plot its graph.

SOLUTION

a. In the equation of a circle, we use $(x_0, y_0) = (2, -1)$ and $r = 2$:

$$(x - 2)^2 + (y - (-1))^2 = 2^2$$

$$(x - 2)^2 + (y + 1)^2 = 4$$

b. The idea is to put the equation into the standard form $(x - x_0)^2 + (y - y_0)^2 = r^2$:

$$(x + 1)^2 + (y - 1)^2 = 9 \text{ becomes}$$

$$(x - (-1))^2 + (y - 1)^2 = 3^2$$

We see that the center of the circle is $(-1, 1)$ and the radius is 3. The graph is shown in **Figure P.15**.

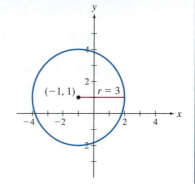

Figure P.15 A circle with center $(-1, 1)$ and radius 3

TRY IT YOURSELF P.6 Brief answers provided at the end of the section.

Identify the center and radius of the circle with equation $(x + 2)^2 + y^2 = 1$, and plot its graph.

EXTEND YOUR REACH

a. There are three possibilities for the way in which two distinct circles in the plane intersect. What are these three possibilities?

b. Find equations of pairs of circles in the plane that illustrate each of the three possibilities from part a. You will exhibit three pairs of circle equations.

Sometimes equations for circles need to be rewritten before we can recognize them as equations of circles. An important technique for accomplishing this is the process of completing the square, which we recall from algebra.

Consider the equation $x^2 + 4x + y^2 = 1 + 2y$. It doesn't look like the equation of a circle, but we can rewrite the equation to make it so. We begin by moving all terms except the constant term to the left:

$$x^2 + 4x + y^2 - 2y = 1$$

$$(x^2 + 4x + 4) + y^2 - 2y = 1 + 4 \qquad \blacktriangleleft \text{ Add } \frac{4^2}{4} = 4 \text{ to complete the square for } x\text{–terms.}$$

$$(x^2 + 4x + 4) + (y^2 - 2y + 1) = 1 + 4 + 1 \qquad \blacktriangleleft \text{ Add } \frac{(-2)^2}{4} = 1 \text{ to complete the square for } y\text{–terms.}$$

$$(x + 2)^2 + (y - 1)^2 = 6 \qquad \blacktriangleleft \text{ Factor as squares.}$$

$$(x - (-2))^2 + (y - 1)^2 = \sqrt{6}^2 \qquad \blacktriangleleft \text{ Write in standard form.}$$

Thus, this is the equation of a circle with center $(-2, 1)$ and radius $\sqrt{6}$.

EXAMPLE P.7 Completing the Square to Identify a Circle

Find the center and radius of the circle whose equation is $x^2 + 6x - 3 = 4y - y^2$. Then plot its graph.

SOLUTION

We proceed as before:

$$x^2 + 6x - 3 = 4y - y^2$$

$$x^2 + 6x + y^2 - 4y = 3 \qquad \blacktriangleleft \text{ Move variable terms left and constants right.}$$

$$(x^2 + 6x + 9) + (y^2 - 4y + 4) = 3 + 9 + 4 \qquad \blacktriangleleft \text{ Complete the squares.}$$

$$(x + 3)^2 + (y - 2)^2 = 16 \qquad \blacktriangleleft \text{ Simplify.}$$

$$(x - (-3))^2 + (y - 2)^2 = 4^2 \qquad \blacktriangleleft \text{ Put in standard form.}$$

This is the equation of a circle with center $(-3, 2)$ and radius 4. The graph is shown in **Figure P.16**.

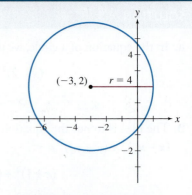

Figure P.16 A circle with center $(-3, 2)$ and radius 4

TRY IT YOURSELF P.7 Brief answers provided at the end of the section.

Find the center and radius of the circle whose equation is $2x + y^2 = 3 - x^2$.

MODELS AND APPLICATIONS   Locating Through Triangulation

EXAMPLE P.8 Locating an Airplane

An airplane is lost in a heavy fog. Radar station 1 finds the airplane to be 10,000 feet away, east of the station and flying due east. Radar station 2, which is 21,000 feet due east of station 1, finds the distance from the airplane to be 17,000 feet. This situation is illustrated in **Figure P.17**, which also gives coordinate axes to use. (Distances are measured in thousands of feet, and the vertical axis indicates altitude.) The procedure presented here is known as triangulation.

a. The airplane lies on a circle centered at radar station 1. Using this station as the origin, find the equation of this circle.

b. Radar station 2 is located at the point $(21, 0)$. The airplane lies on a circle centered at this radar station. Determine the equation of this circle.

c. How far east of radar station 1 is the airplane?

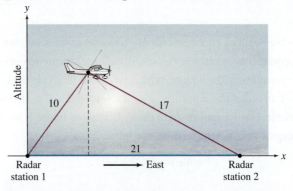

Figure P.17 Distances from an airplane to radar stations

SOLUTION

a. The airplane is at a distance of 10,000 feet from the origin, so it lies on the circle centered at the origin with radius 10:

$$x^2 + y^2 = 10^2$$

b. The circle is centered at $(21, 0)$ and has radius 17. The required equation is

$$(x - 21)^2 + y^2 = 17^2$$

c. The airplane lies on the intersection of the circles from parts a and b. Thus, it satisfies both equations. If we subtract the equation in part a from the equation in part b, the variable y is eliminated:

$$(x - 21)^2 - x^2 = 17^2 - 10^2 \quad \blacktriangleleft \text{ Subtract equations.}$$

$$x^2 - 42x + 21^2 - x^2 = 189 \quad \blacktriangleleft \text{ Expand the square.}$$

$$-42x + 441 = 189 \quad \blacktriangleleft \text{ Cancel } x^2.$$

$$-42x = -252 \quad \blacktriangleleft \text{ Move constants right.}$$

$$6 = x \quad \blacktriangleleft \text{ Divide by } -42.$$

We find that the airplane is 6000 feet east of radar station 1.

TRY IT YOURSELF P.8 Brief answers provided at the end of the section.

What is the altitude of the airplane? (Assume that both radar stations have an altitude of zero.)

TRY IT YOURSELF ANSWERS

P.1 The distance is 5. The midpoint is $\left(t + \dfrac{3}{2}, t + 4\right)$.

P.2 -1

P.3 $y = 2x - 1$

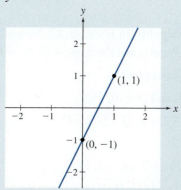

P.4 $y = -3x - 1$

P.5 $y = -5x - 10$

P.6 The center is $(-2, 0)$, and the radius is 1.

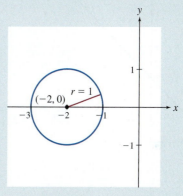

P.7 The center is $(-1, 0)$, and the radius is 2.

P.8 8000 feet

EXERCISE SET P.1

CHECK YOUR UNDERSTANDING

1. The equation $(x - a)^2 + (y - b)^2 = r^2$ is the equation of _____.

2. If $r > 0$, then the equation $(x + a)^2 + (y + b)^2 = r$ is the equation of _____.

3. Describe the graph of the equation $(x - 1)^2 + (y - 2)^2 = 0$.

4. Describe the graph of the equation $x^2 + y^2 = -4$.

5. What geometric quantity is represented by $\sqrt{(5 - 2)^2 + (8 - 3)^2}$?

6. Does a vertical line have a slope? If so, what is it?

7. What information is given by the slope of a line?

8. What are the possibilities for the intersection of a line with a circle?

9. What are the possibilities for the intersection of two distinct circles?

10. What can you conclude about the circles $x^2 + y^2 = r_1^2$ and $x^2 + y^2 = r_2^2$ if r_1 is positive but smaller than r_2?

PROBLEMS

Matching. In Exercises 11 through 16, match the given graph with one of these equations:

A. $x^2 + y^2 = 1$ D. $(x+1)^2 + y^2 = 1$

B. $x + y = 2$ E. $(x-1)^2 + y^2 = 1$

C. $-x + y = 2$ F. $x^2 + (y-1)^2 = 1$

11.

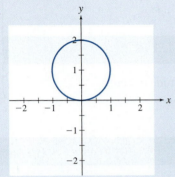

12.

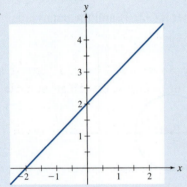

13.

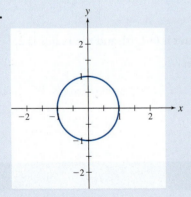

14.

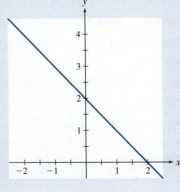

15.

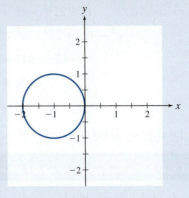

16.

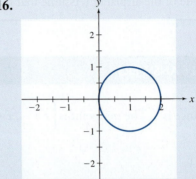

Distance and midpoint. In Exercises 17 through 25, find the midpoint of P and Q and the distance between P and Q.

17. $P = (1, 3)$ and $Q = (-1, 3)$

18. $P = (6, 0)$ and $Q = (0, 5)$

19. $P = (-1, -2)$ and $Q = (-3, -4)$

20. $P = (-1, 2)$ and $Q = (-2, 1)$

21. $P = (t, t)$ and $Q = (t-1, 1)$

22. $P = (0, t)$ and $Q = (t, 1)$

23. $P = (x, y)$ and $Q = (x+y, x-y)$

24. $P = (x+y, y)$ and $Q = (x-y, y)$

25. $P = (x, y)$ and $Q = (y, x)$

26. **Distance.** Use a calculation to show that the distance from (x, y) to (z, w) is the same as the distance from $(-x, -y)$ to $(-z, -w)$.

Slope of a line. In Exercises 27 through 33, use the given information to find the slope of the line.

27. The line passes through $(1, 1)$ and $(2, -1)$.

28. The line passes through $(-1, 1)$ and $(2, 1)$.

29. The line passes through $(-7, -4)$ and $(-3, 5)$.

30. The line passes through (x, y) and $(x+1, y+1)$.

31. The line passes through $(x, 1)$ and $(2x, 1)$, $x \neq 0$.

32. The line passes through $(1, x)$ and $(2, 2x)$.

33. The line passes through $(1, -4)$ and (x, y), $x \neq 1$.

Equations of lines. In Exercises 34 through 46, use the given information to find the equation of the line.

34. The y-intercept is $b = 3$, and the slope is $m = -1$.

35. The y-intercept is $b = -5$, and the slope is $m = 2$.

36. The y-intercept is $b = s$, and the slope is $m = t$.

37. The line passes through $(1, -4)$ and has slope 7.

38. The line passes through $(4, 1)$ and has slope -2.

39. The line passes through $(0, 0)$ and has slope 5.

40. The line passes through $(1, t)$ and has slope 2.

41. The line passes through $(t, 2t)$ and has slope $3t$.

42. The line passes through $(1, -4)$ and $(2, -3)$.

43. The line passes through $(2, 1)$ and $(3, 3)$.

44. The line passes through $(3, 0)$ and $(0, 3)$.

45. The line passes through (p, q) and (q, p), $p \neq q$.

46. The line passes through $(p, 1)$ and $(q, 2)$, $p \neq q$.

47. Making lines with various slopes. In parts a through c, write the linear equation corresponding to each line.

    a. The line with y-intercept $b = 2$ and slope $m = 0$

    b. The line with y-intercept $b = 2$ and slope $m = -1$

    c. The line with y-intercept $b = 3$ and slope $m = 1$

    d. Sketch the lines in parts a through c on the same coordinate plane along with the line in Example P.3. Comment on any patterns you see.

Equations of circles. In Exercises 48 through 52, use the given information to find the equation of the circle.

48. Center $(1, 2)$ and radius 2

49. Center $(3, 0)$ and radius 1

50. Center $(2, -1)$ and radius 4

51. Center (p, p) and radius p. Here $p \neq 0$.

52. Center $\left(-1, \frac{1}{p}\right)$ and radius $\sqrt{p}$. Here $p > 0$.

Finding center and radius. In Exercises 53 through 62, find the center and radius of the circle with the given equation.

53. $(x + 2)^2 + y^2 = 5$

54. $x^2 + y^2 = 1$

55. $x^2 + y^2 = 3t^2$, $t \neq 0$

56. $x^2 - 2x + y^2 - 3 = 0$

57. $x^2 + y^2 - y = 0$

58. $x^2 - x + y^2 = 4y$

59. $x^2 + y^2 = 4 + 2x + 4y$

60. $x^2 + 4x + y^2 + 6y + 12 = 0$

61. $x^2 + y^2 = 2px$, $p \neq 0$

62. $x^2 + y^2 + 2y + 2 = 2tx + 2x + 2ty$, $t > 0$

Graphing circles. In Exercises 63 through 66, identify the center and radius, then plot the graph of the circle.

63. $x^2 + (y - 1)^2 = 4$

64. $(x - 3)^2 + (y + 1)^2 = 9$

65. $x^2 + y^2 + 4 = 2x + 4y$

66. $x^2 + 2x + y^2 + 2y = 2$

67. Equation of a circle. Determine the equation of the circle for which $(-1, -2)$ and $(3, 2)$ are the endpoints of a diameter. *Suggestion:* Use the midpoint formula to find the center, then use the distance formula to find the radius.

Intersections of circles and lines. In Exercises 68 through 71, find the points where the given circle and line intersect.

68. The circle $(x - 1)^2 + (y - 2)^2 = 1$ and the line $y = -8x + 3$.

69. The circle $x^2 + y^2 = 5$ and the line $x + 2y = 5$.

70. The circle with center $(2, 0)$ and radius 2, and the line $y = 2x$.

71. The circle with center $(1, -1)$ and radius 2, and the line $y = 2x + 1$.

Locating points. In Exercises 72 through 76, find the indicated points.

72. Find all points on the line $y = 2x + 3$ that are at a distance 3 from $(0, 0)$.

73. Find all points on the line $y = -3x + 4$ that are at a distance 9 from $(3, 1)$.

74. Find all points on the line $y = 2x$ that are at a distance $\sqrt{2}\,p$ from (p, p).

75. Find all points on the line $y = 2x + 3$ that are at a distance 5 from $(3, 1)$.

76. Find all points on the line $y = 2x + 3$ that are a distance 2 from $(1, 12)$. *Suggestion:* Keep in mind that some quadratic equations have no real solutions.

Working with parallel lines. Recall from Example P.5 that distinct lines are parallel when they have the same slope. Use this fact to solve Exercises 77 through 80.

77. Determine the equation of the line through $(2, -1)$ that is parallel to $y = -3x + 1$.

78. Determine the equation of the line through $(1, 5)$ that is parallel to the line through $(1, 4)$ and $(2, 5)$.

79. Determine the equation of the line through $(7, 2)$ that is parallel to the line through $(2, -3)$ and $(6, -2)$.

80. Determine the equation of the line through $(2, 7)$ that is parallel to the x-axis.

Working with perpendicular lines. A later exercise shows that a line with slope m_1 is perpendicular to a line with slope m_2 if and only if $m_1 = -\dfrac{1}{m_2}$. Use this fact to solve Exercises 81 through 86.

81. Determine the equation of the line through $(3, 5)$ and perpendicular to the line $y = 3x + 5$.

82. Determine the equation of the line through $(-2, 4)$ and perpendicular to the line $y = -3x + 7$.

83. Determine the equation of the line through $(0, 7)$ and perpendicular to the line $y = 9x - 4$.

84. Determine the equation of the line through $(2, -1)$ and perpendicular to the line through $(1, 0)$ and $(2, 5)$.

85. Determine the equation of the line through $(-1, -1)$ and perpendicular to the line through $(-1, 4)$ and $(3, -2)$.

86. Determine the equation of the line through $(0, 0)$ and perpendicular to the line through $(2, 5)$ and $(7, 0)$.

MODELS AND APPLICATIONS

87. **Roadway grade.** The concept of slope is used to measure the steepness of highways and railroad beds. Often the term *gradient* or just *grade* is used to describe roadbeds. For example, a railroad track with a 3% grade rises (or falls) 3 feet for each 100 horizontal feet, as shown in **Figure P.18**. What is the slope of a track with 3% grade?

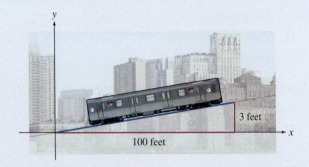

Figure P.18 Rise and run for a roadway. Tutti Frutti/Shutterstock

88. **Coast Guard.** A ship is lost in a dense fog. The ship is 13 miles away from Coast Guard station 1. Coast Guard station 2 is 14 miles due north of station 1 across a strait. The ship is 15 miles from Coast Guard station 2. Place the origin at station 1, let north correspond to the positive y-axis, and let east correspond to the positive x-axis (**Figure P.19**).

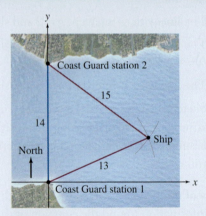

Figure P.19 Distances from a ship to Coast Guard stations

a. The ship lies on a circle centered at Coast Guard station 1. Determine the equation of this circle.

b. The ship lies on a circle centered at Coast Guard station 2. Determine the equation of this circle.

c. There are two possible locations of the ship. Give both in terms of distance north and distance east or west of Coast Guard station 1.

89. **The Mississippi River.** The Mississippi River has its headwaters at Lake Itasca, Minnesota, which has an elevation of 1475 feet. Let's model the river as a straight line from Lake Itasca to the Gulf of Mexico 2340 miles to the south. Think of south as

pointing to the right along the positive x-axis and y as representing elevation (above sea level).

a. What is the slope of the line representing the Mississippi River? *Suggestion:* What is the elevation of the river when it reaches the Gulf of Mexico?

b. St. Louis is 784 miles south of Lake Itasca. What is the elevation of the river as it passes St. Louis?

90. **A hiking trail.** A hiking trail is modeled by a straight line. You pass a sign that says *Elevation 4300 feet.* One horizontal mile farther along the trail, a sign reads *Elevation 4500 feet.* The trail ends at an elevation of 5000 feet.

This situation is illustrated in **Figure P.20.** What is the horizontal distance from the first sign to the end of the trail?

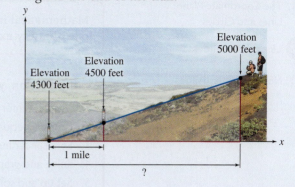

Figure P.20 A hiking trail

CHALLENGE EXERCISES FOR INDIVIDUALS OR GROUPS

91. **Parallel lines 1.** Show that if two distinct lines have the same slope, then they are parallel. *Suggestion:* By definition, two lines $y = m_1 x + b_1$ and $y = m_2 x + b_2$ are parallel if they have no points in common. Thus, you are being asked to show that if $m_1 = m_2$ and the lines are distinct (that is, not the same line), so $b_1 \neq b_2$, then there is no solution to the system of equations

$$y = m_1 x + b_1$$
$$y = m_1 x + b_2$$

92. **Parallel lines 2.** Show that if two lines are parallel, then they have the same slope. *Suggestion:* Show that if two lines have different slopes, they are not parallel. By definition, two lines $y = m_1 x + b_1$ and $y = m_2 x + b_2$ are not parallel if they have a point in common. Thus, you are being asked to show that if $m_1 \neq m_2$, then there is a solution to the system of equations

$$y = m_1 x + b_1$$
$$y = m_2 x + b_2$$

93. **Perpendicular lines.** The goal of this exercise is to show that $y = m_1 x + b_1$ is perpendicular to $y = m_2 x + b_2$ precisely when $m_1 = -\dfrac{1}{m_2}$. (We assume that m_1 and m_2 are nonzero.) We will need a stronger form of the Pythagorean theorem: If a triangle has sides of lengths A, B, and C, then the angle between the sides of lengths A and B is a right angle if and only if $A^2 + B^2 = C^2$. Using this fact, we reach the goal in several steps.

a. Explain why the two given lines are perpendicular precisely when the lines $y = m_1 x$ and $y = m_2 x$ are perpendicular. *Suggestion:* Use the results of the two preceding exercises.

b. In the coordinate plane, make the triangle with vertices

   i. The origin

   ii. The point on the line $y = m_1 x$ at $x = 1$

   iii. The point on the line $y = m_2 x$ at $x = 1$ (see **Figure P.21**)

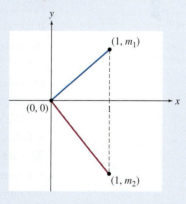

Figure P.21 A triangle for Exercise 93

c. Calculate the lengths of each of the three sides of the triangle you made in part b.

d. Use your calculations in part c to show that the lines are perpendicular precisely when $m_1 = -\dfrac{1}{m_2}$.

94. **Walking along a line.** Suppose we are at a point on a line with slope m. Suppose we move a distance d along the line (going to the right). By how much does our y-coordinate change?

P.1 The Coordinate Plane:
     Lines and Circles

P.2 Linear Inequalities

P.3 Nonlinear Inequalities

P.2 Linear Inequalities

> Mathematical inequalities allow us to determine over what range of values one quantity exceeds another.

In this section, you will learn to:

1. Describe intervals of real numbers using interval notation.
2. Express inequalities using unions and complements of intervals.
3. Solve linear inequalities using the rules for manipulating inequalities.
4. Solve and graph inequalities involving absolute values.
5. Set up and solve applied problems involving linear inequalities.

In this section we examine the simplest kinds of inequalities encountered in calculus.

Inequalities and Intervals

> Intervals are represented by segments or rays on the number line.

The two mathematical statements $x < y$ and $y > x$ are inequalities. They both mean that x is less than y, or (to say it another way) that $y - x$ is a positive number. The expressions $x \le y$ and $y \ge x$ both mean that x is less than or equal to y. Note that in all four cases the inequality symbol points to the smaller number.

When $a < b$, we use (a, b) to indicate the set of all real numbers between a and b, including neither a nor b.[1] We use the term *interval* for such sets, so (a, b) is the interval from a to b. Other intervals may include one or more endpoints. If we wish to include an endpoint in an interval, we use a square bracket, as in $[a, b)$. For example, $(-1, 3)$ denotes the interval of all numbers greater than -1 and less than 3, and $(2, 7]$ denotes the interval of all numbers greater than 2 and less than or equal to 7. We write the set consisting of a single number a as $\{a\}$.

We will be using the symbol ∞. It is read as "infinity," and $-\infty$ is read as "negative infinity." If we want to indicate the set of all numbers less than a, we use $(-\infty, a)$. The set of all numbers greater than or equal to a is denoted by $[a, \infty)$. The set of all real numbers, $\mathbb{R}$, may also be denoted by $(-\infty, \infty)$. Note that the symbols $-\infty$ and ∞ are *not* numbers and have no meaning if used out of context. They are just handy symbols indicating that the interval extends forever in the negative direction (for $-\infty$) or in the positive direction (for ∞).

---

[1]Note that (a, b) represents the numbers between a and b, but this notation is also used to represent a point in the plane with coordinates a and b. Multiple usage is common in English and mathematics. The meaning is determined by the context.

Sometimes it is convenient to describe intervals and other sets using set-builder notation. For example, the interval (a, b) can be written as $\{x : a < x < b\}$. We read this as "The set of all x such that $a < x < b$."

It is common to string inequalities together when they all point in the same direction, as in the expression $a < x < b$. Do not string inequalities together like this if some point one way and some the other. For example, $0 < x > 1$ makes no sense.

CONCEPTS TO REMEMBER: Interval Notation

Intervals of real numbers are represented as follows:

- (a, b) is the set of all numbers x such that $a < x$ and $x < b$. These two inequalities can be combined as $a < x < b$.
- $[a, b)$ is the set of all numbers x such that $a \leq x < b$.
- $(a, b]$ is the set of all numbers x such that $a < x \leq b$.
- $[a, b]$ is the set of all numbers x such that $a \leq x \leq b$.
- (a, ∞) is the set of all numbers x such that $a < x$.
- $[a, \infty)$ is the set of all numbers x such that $a \leq x$.
- $(-\infty, b)$ is the set of all numbers x such that $x < b$.
- $(-\infty, b]$ is the set of all numbers x such that $x \leq b$.
- $(-\infty, \infty)$ is the set of all real numbers.

Sometimes we want to refer to the set of all numbers *not* belonging to a certain interval. This is called the **complement** of the interval.

For example, the interval $(-\infty, 2]$ consists of all numbers less than or equal to 2, so its complement consists of all numbers greater than 2. This is the interval $(2, \infty)$.

Sometimes the complement of an interval consists of two intervals. We call such a set the **union** of the intervals.

We use the notation $\cup$ to express the union. Thus, the union of the intervals $(0, 1)$ and $(3, 5]$ is denoted $(0, 1) \cup (3, 5]$.

As an example, consider the complement of the interval $[4, 5)$. The interval $[4, 5)$ consists of all numbers between 4 and 5, including 4 but not 5. Thus, the complement contains all numbers less than 4 and all numbers greater than 5, but we need to think a bit more about the endpoints. Because 4 is included in the interval, 4 is not included in the complement. Also, 5 is not in the interval, so 5 *is* included in the complement. Therefore, the complement consists of the two intervals $(-\infty, 4)$ and $[5, \infty)$, and we write the complement of $[4, 5)$ as $(-\infty, 4) \cup [5, \infty)$. Graphs of the interval and its complement are shown in **Figure P.22** and **Figure P.23**. A closed dot indicates that the endpoint is included in the interval, and an open dot indicates that the endpoint is not included in the interval.

The **complement** of an interval consists of all real numbers that are *not* in the interval.

The **union** of two sets consists of all elements belonging to one or both of the sets.

Figure P.22 The graph of the interval $[4, 5)$

Figure P.23 The graph of $(-\infty, 4) \cup [5, \infty)$, the complement of $[4, 5)$

EXAMPLE P.9 Relating Inequalities and Intervals

a. Use interval notation to express the set of numbers x that satisfy the inequality $3 < x \le 5$. Then graph the interval.

b. Graph the interval $(-2, \infty)$, and express the interval using inequalities.

c. Use interval notation to express the complement of $[0, 1)$. Graph both the interval and its complement.

SOLUTION

a. The inequality $3 < x \le 5$ does not include 3 but does include 5. Hence, the set of numbers x that satisfy it is the interval $(3, 5]$. The graph is shown in **Figure P.24**.

Figure P.24 The graph of the interval $(3, 5]$

b. The interval $(-2, \infty)$ is the set of all numbers x greater than -2. This is represented by the set of all numbers x satisfying the inequality $-2 < x$. The graph of the interval is shown in **Figure P.25**.

Figure P.25 The graph of the interval $(-2, \infty)$

c. The interval $[0, 1)$ consists of all numbers between 0 and 1, including 0 but not 1. The complement of $[0, 1)$ consists of all numbers not in the interval, so it contains all numbers less than 0 and those greater than 1, but we still need to consider the endpoints. The number 0 is included in the interval, so 0 is not included in the complement. Also, 1 is not in the interval, so 1 is included in the complement. Thus, the complement consists of the union of the two intervals $(-\infty, 0)$ and $[1, \infty)$. We write the complement as $(-\infty, 0) \cup [1, \infty)$. The graphs of the intervals are shown in **Figure P.26** and **Figure P.27**.

Figure P.26 The graph of the interval $[0, 1)$

Figure P.27 The complement of the interval $[0, 1)$

TRY IT YOURSELF P.9 Brief answers provided at the end of the section.

Use interval notation to express the complement of $(1, \infty)$. Graph the interval and its complement.

Solving Inequalities

> Linear inequalities are solved using the strategy for solving linear equations.

To solve an equation means to find all values of the variable that make the equation true. Similarly, to solve an inequality means to find all values of the variable that make the inequality true. We can solve many inequalities in much the same way as we solve equations. For example, we can add to, and subtract from, both sides of the inequality (which allows us to move terms to the other side if we change the sign). But when it comes to multiplication or division we must exercise great care. Multiplication or division by a negative number reverses the direction of an inequality. These rules follow directly from the fact that $x > y$ if and only if $x - y > 0$.

LAWS OF MATHEMATICS: Rules for Manipulating Inequalities

1. Adding or subtracting the same quantity on both sides preserves the inequality. As a consequence of this rule, we can move terms to the other side of the inequality if we change their sign.
2. Multiplying or dividing by the same *positive* number preserves the inequality.
3. Multiplying or dividing by the same *negative* number *reverses* the inequality.

A **linear inequality** is the same as a linear equation where the equal sign is replaced by an inequality sign ($<, \leq, >,$ or $\geq$). These are the easiest inequalities to solve.

A **linear inequality** is one that can be put in the form $ax + b > 0$ or $ax + b \geq 0$ using the preceding rules. Here $a \neq 0$.

With the preceding rules in mind, let's solve the linear inequality $2 - 3x > 14$:

$2 - 3x > 14$

$-3x > 14 - 2$ ◀ Subtract 2 from both sides.

$-3x > 12$ ◀ Simplify.

$x < -4$ ◀ Divide by −3 (because −3 is negative, we reverse the inequality).

In interval notation, the solution of this inequality is $(-\infty, -4)$.

EXAMPLE P.10 Solving Linear Inequalities

Solve the following inequalities.

a. $4(x - 5) \geq 2x + 6$

b. $2 - x < x - 2$

SOLUTION

a. First we expand both sides:

$4(x - 5) \geq 2x + 6$

$4x - 20 \geq 2x + 6$ ◀ Apply the distributive law.

$4x - 2x \geq 20 + 6$ ◀ Move variables left and constants right.

$2x \geq 26$ ◀ Simplify.

$x \geq 13$ ◀ Divide by 2 (inequality does not reverse).

In interval notation, the solution is $[13, \infty)$.

b. This time our approach will require that we reverse the inequality:

$$2 - x < x - 2$$

$$-x - x < -2 - 2 \qquad \blacktriangleleft \textbf{Move varibles left and constants right.}$$

$$-2x < -4 \qquad \blacktriangleleft \textbf{Simplify.}$$

$$x > \frac{-4}{-2} \qquad \blacktriangleleft \textbf{Divide by } -2 \textbf{, and reverse inequality.}$$

$$x > 2 \qquad \blacktriangleleft \textbf{Simplify.}$$

In interval notation, the solution is $(2, \infty)$.

TRY IT YOURSELF P.10   Brief answers provided at the end of the section.

Solve the inequality $x - 4 > 3(x + 8)$. Write your answer both as an inequality and as an interval.

EXTEND YOUR REACH

a. Solve each of the following inequalities.

- $3x + 2 < 2x + 1$
- $3x + 2 < 3x + 1$
- $3x + 2 < 3x + 5$

b. Under what conditions does the inequality $ax + b < cx + d$ have a solution? Justify your answer, and give the solution in the case that one exists.

STEP-BY-STEP STRATEGY: Solving Linear Inequalities

Step 1 Move all terms that involve the variable to one side of the equation (changing signs as necessary).

Step 2 Move all terms that do not involve the variable to the other side of the equation (changing signs as necessary).

Step 3 Divide by the coefficient of the variable, being careful to reverse the inequality if the coefficient is negative.

Absolute Value Inequalities

> **Absolute values are often used to represent distances.**

Recall that $|x|$ can be thought of as the distance from x to 0. More generally, $|x - y|$ represents the distance between the two points x and y on the number line. For example:

- $|5 - 2| = 3$, which is the distance between 5 and 2 on the number line.
- $|-3 - 7| = 10$, which is the distance between -3 and 7 on the number line.
- $|4 + 5| = |4 - (-5)| = 9$, which is the distance between 4 and -5 on the number line.

The inequality $|x| < 2$ says that the distance between x and 0 is less than 2. In general, the inequality $|x| < a$ tells us that the distance between x and 0 is less than a, so $-a < x < a$. Thus, the set of such values of x is exactly the interval $(-a, a)$. This observation yields the following important fact regarding absolute value inequalities.

CONCEPTS TO REMEMBER: Absolute Value Inequalities

For a positive real number a,

$$|x| < a \text{ if and only if } -a < x < a$$

Similarly,

$$|x| \le a \text{ if and only if } -a \le x \le a$$

The solution of the inequality $|x| > a$ consists of all numbers that are not solutions of the complementary inequality $|x| \le a$. Thus,

$$|x| > a \text{ if and only if } x < -a \text{ or } x > a$$

$$|x| \ge a \text{ if and only if } x \le -a \text{ or } x \ge a$$

Figure P.28 The solution of $|x| < a$

The solutions of two absolute value inequalities are illustrated in **Figure P.28** and **Figure P.29**.

The following example shows how to use this kind of geometric reasoning to solve certain absolute value inequalities.

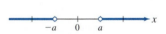

Figure P.29 The solution of $|x| > a$

EXAMPLE P.11 Absolute Value and Distance

a. Interpret the meaning of $|x - 5| < 2$ in terms of distance.

b. Solve the inequality $|x - 5| < 2$.

SOLUTION

a. The inequality says that the distance from x to 5 is less than 2.

b. We rewrite the inequality as an interval:

$$|x - 5| < 2$$

$$-2 < x - 5 < 2 \qquad \blacktriangleleft \text{Use: } |x| < a \text{ is equivalent to } -a < x < a.$$

$$-2 + 5 < x < 2 + 5 \qquad \blacktriangleleft \text{Add 5 to each term.}$$

$$3 < x < 7 \qquad \blacktriangleleft \text{Simplify.}$$

In interval notation, the answer is $(3, 7)$.

Here's another way to think about this inequality: The solution of the inequality is the set of all numbers within a distance of 2 from 5. The number 7 is 2 units to the right of 5, and the number 3 is 2 units to the left of 5. Therefore, the solution consists of the numbers in the interval $(3, 7)$. A visual representation is shown in **Figure P.30**.

Figure P.30 Numbers at a distance less than 2 from 5

TRY IT YOURSELF P.11   Brief answers provided at the end of the section.

Solve the inequality $|x + 2| < 3$. Report your answer using interval notation.

EXAMPLE P.12 Solving Absolute Value Inequalities

Solve the inequality $|3x - 2| \le 5$.

SOLUTION

The given absolute value inequality is equivalent to

$$-5 \le 3x - 2 \le 5$$

We transform this using the rules for solving linear inequalities:

$$-5 \le 3x - 2 \le 5$$
$$-3 \le 3x \le 7 \quad \text{◀ Add 2 to each term.}$$
$$-1 \le x \le \frac{7}{3} \quad \text{◀ Divide by 3.}$$

In interval notation, the solution is $\left[-1, \dfrac{7}{3}\right]$.

TRY IT YOURSELF P.12   Brief answers provided at the end of the section.

Solve the inequality $|5 - 2x| \le 7$.

EXTEND YOUR REACH

a. The inequality $|x - 3| < |x - 7|$ is difficult to solve algebraically, but it is easier to solve if you think in terms of distance. *Suggestion*: First fill in the blank in the statement, "The inequality says that x is closer to 3 than it is to _____."

b. Solve the inequality $|x - 3| < |x + 7|$. *Suggestion*: If we rewrite $|x + 7|$ as $|x - (-7)|$, then we can think in terms of distance.

EXAMPLE P.13 Solving Using Complements

Solve the inequality $|4 - 3x| \ge 7$.

SOLUTION

Sometimes when we solve an inequality, it is easier to find the complement of the solution set. That is, we find all values of x for which the given inequality is *not* true. In this case we want to solve $|4 - 3x| < 7$.

This complementary inequality is equivalent to $-7 < 4 - 3x < 7$, which we transform using the rules for solving inequalities:

$$-7 < 4 - 3x < 7$$

$$-11 < -3x < 3 \qquad \blacktriangleleft \text{ Subtract 4 from each term.}$$

$$\frac{11}{3} > x > -1 \qquad \blacktriangleleft \text{ Divide by } -3 \text{, which reverses the inequalities.}$$

The solution of the complementary inequality is then $\left(-1, \dfrac{11}{3}\right)$. This is the interval where our original inequality is *not* true. So $|4 - 3x| \geq 7$ *is* true on the complement of this interval. Thus, the solution of the original inequality is $(-\infty, -1] \cup \left[\dfrac{11}{3}, \infty\right)$.

TRY IT YOURSELF P.13 Brief answers provided at the end of the section.

Solve the inequality $6 \leq |4 - 2t|$. Report your answer using interval notation.

STEP-BY-STEP STRATEGY: Solving Inequalities Involving Absolute Values

No single strategy will solve all absolute value inequalities, but the following steps are useful.

Step 1 If possible, rewrite the inequality in the form $|A| < b$. Then convert to $-b < A < b$.

Step 2 If possible, rewrite the inequality in the form $|A| > b$. Then solve the complementary inequality $|A| \leq b$, and take the complement of the answer.

Step 3 Use the fact that some absolute value inequalities can be solved if they are interpreted in terms of distance.

MODELS AND APPLICATIONS   Supply and Demand

There are practical applications of solving linear inequalities.

EXAMPLE P.14 Supply of Trucks

The quantity S of light trucks, in millions, that manufacturers are willing to produce and offer for sale at a price P, in thousands of dollars, is determined by $P = 5S + 22$. The price of light trucks is expected to stay above \$32,000 and below \$35,000. What can we expect to be true about the quantity of light trucks offered for sale?

SOLUTION

If the price of light trucks is strictly between \$32,000 and \$35,000, we have $32 < P < 35$. Putting in the relationship between P and S gives

$$32 < 5S + 22 < 35$$

We solve this using the preceding rules:

$$32 < 5S + 22 < 35$$

$$10 < \quad 5S \quad < 13 \quad \blacktriangleleft \text{ Subtract 22 from each term and simplify.}$$

$$2 < \quad S \quad < 2.6 \quad \blacktriangleleft \text{ Divide by 5 (inequality does not reverse).}$$

We can expect that the quantity of light trucks offered for sale will be greater than 2 million and fewer than 2.6 million.

TRY IT YOURSELF P.14 Brief answers provided at the end of the section.

The quantity D of light trucks, in millions, that consumers are willing to purchase at a price P, in thousands of dollars, is determined by $P = 37 - 2.5D$. The price of light trucks is expected to stay above \$32,000. What can we expect to be true about the quantity of light trucks purchased?

TRY IT YOURSELF ANSWERS

P.9 The complement is $(-\infty, 1]$.

Graph of the interval:

Graph of the complement:

P.10 $x < -14$, or in interval notation $(-\infty, -14)$

P.11 $(-5, 1)$

P.12 $[-1, 6]$

P.13 $(-\infty, -1] \cup [5, \infty)$

P.14 We can expect that the quantity of light trucks purchased will be fewer than 2 million.

EXERCISE SET P.2

CHECK YOUR UNDERSTANDING

1. How does multiplying both sides of an inequality by a nonzero number affect the inequality?

2. How does adding a number to both sides of an inequality affect the inequality?

3. How does dividing both sides of an inequality by a nonzero number affect the inequality?

4. How does subtracting a number from both sides of an inequality affect the inequality?

5. Is an inequality preserved if we apply the absolute value to each side?

6. If $x^2 < y^2$, under what conditions is it true that $x < y$?

7. If a is positive, then the inequality $|x| < a$ is equivalent to which of the following?

    a. $x < a$

    b. $x > -a$

    c. $-a < x < a$

    d. $a < x < -a$

8. Solve the inequality $|x| \leq -3$.

9. If $|x| < 3$, what can you conclude about $|x - 1|$? *Suggestion*: Think in terms of distance.

10. If y is a negative number, rewrite the inequality $|x| \leq |y|$ so that no absolute values occur.

11. For which values of x is $x \leq x^2$?

12. Imagine that your friend says that to solve $|x - 5| < 4$, you just add 5 to get $|x| < 9$, so the answer is $(-9, 9)$. Where did your friend go wrong?

13. If $0 < x < y < 1$, how is $\dfrac{1}{x}$ related to $\dfrac{1}{y}$?

PROBLEMS

Interval notation. In Exercises 14 through 21, use interval notation to express the set of x that satisfy the given inequality.

14. $-15 < x < 4$

15. $11 \le x < 13$

16. $-1 < x \le 1$

17. $-2 \le x \le 1$

18. $x < 4$

19. $x \ge 15$

20. $x \le 3$

21. $x > 7$

Complement. In Exercises 22 through 26, use interval notation to express the complement of the given interval.

22. $(-\infty, 7]$

23. $(-5, \infty)$

24. $(1, 3]$

25. $(-1, 4)$

26. $[-7, -2]$

Graphical representations. In Exercises 27 through 36, make a graphical representation of the given interval or inequality.

27. $(0, 4)$

28. $(-2, 3]$

29. $[1, 5]$

30. $[-3, 1)$

31. $4 < x \le 5$

32. $2 \le x < 5$

33. $[-3, \infty)$

34. $(-\infty, 1)$

35. $2 > x$

36. $1 \le x$

Solving linear inequalities. In Exercises 37 through 48, solve the given inequality. Give your answer in terms of intervals.

37. $3x + 1 < 4$

38. $x \ge 7x - 3$

39. $3x - 5 \le 2 - x$

40. $1 - 2x > 5 - x$

41. $5 - x < 3 - x$

42. $3 + x < 4 + x$

43. $2x \le 3x$

44. $-6 < 4 - 5x < 7$

45. $x - 1 \le 5 + 4x \le 4 + x$

46. $4 - x < 2 + 2x < 6 - x$

47. $x(x - 1) \le (x + 2)(x - 2)$. *Suggestion*: Expand each side, and cancel terms.

48. $(x + 1)(x + 2) \le (x + 3)(x + 4)$

True or false? In Exercises 49 through 51, determine whether the given statement is always true. If it is sometimes false, give an example.

49. If $x \le y$, then $x^2 \le y^2$.

50. $|3 + x| < |4 + x|$

51. $|x + y| \le |x| + |y|$

Distance. In Exercises 52 through 58, rephrase the given statement in terms of distance.

52. $|x - 3| = 5$

53. $|x - 4| < 7$

54. $|x - 1| \ge 4$

55. $|x - y| < |x - z|$

56. If $|x - 1| > 2$ then $|y - 4| > 3$.

57. If $|x - 2| < 3$ then $|y - 4| < 5$.

58. If $|x - a| < \delta$ then $|y - b| < \varepsilon$.

Solving absolute value inequalities. Solve the inequalities in Exercises 59 through 68. Give your answer in terms of intervals. You may find some of the inequalities easier to solve if you interpret them in terms of distance.

59. $|x - 5| < 4$

60. $|7 - x| < 3$

61. $|x + 4| \le 9$

62. $|x - 4| > 9$

63. $|x + 1| \ge 1$

64. $|2x - 5| < 11$

65. $|5 + 3x| \le 9$

66. $|4x + 1| > 6$

67. $|3 - 2x| \ge 5$

68. $6 < |5x + 9|$

Graphs and inequalities. The inequalities in Exercises 69 through 72 are difficult to solve by hand. Use technology to make an appropriate graph, and use the graph to find the solution.

69. $|x + 3| + |x + 2| - |x| < 0$

70. $|x + 1| - |x + 2| + x + 1 \ge 0$

71. $|x| - |2 - |x|| < 0$

72. $|2x - 3| < x$. *Suggestion*: Plot $y = |2x - 3| - x$.

MODELS AND APPLICATIONS

73. Concentration of a drug. The concentration of a certain drug in the bloodstream declines after injection. The concentration C at the time t hours after injection is given by

$$C = 100 - 15t \text{ milligrams per deciliter}$$

The drug is effective as long as there are at least 50 milligrams per deciliter in the bloodstream. During what period is the drug effective?

74. **Buying gold.** For t years after a certain date, the price of gold was given by

$$P = 1150 + 80t \text{ dollars per ounce}$$

During what time period was the price between $1250 and $1350 per ounce?

75. **Making a profit.** A company produces and sells belts for $30 each. The expense of producing n such belts is given by

$$\text{Expense} = 500 + 20n \text{ dollars}$$

The company makes a profit when income exceeds expenses. How many belts must the company produce and sell to be profitable?

76. **A rescue pet.** You are looking at a variety of puppies with the intention of adopting. You want an average-sized dog, not too big or too small. You figure that a dog that weighs between 40 and 60 pounds when full-grown would be ideal. For medium-sized breeds, the following formula predicts the adult weight W of a puppy that weighs w pounds after t weeks:

$$W = 52\,\frac{w}{t}$$

The puppies you are looking at are 12 weeks old. What is the weight range for the puppy you will choose?

CHALLENGE EXERCISES FOR INDIVIDUALS OR GROUPS

77. **The triangle inequality.** The following steps will lead to a proof of the triangle inequality: If x and y are real numbers, then $|x + y| \le |x| + |y|$. The proof makes use of the facts that $A \le |A|$ and that $|A| \le B$ if and only if $-B \le A \le B$.

   Step 1: Show that $-|x| \le x \le |x|$ and similarly that $-|y| \le y \le |y|$.

   Step 2: Add the two inequalities from step 1 to show that $-(|x| + |y|) \le x + y \le |x| + |y|$.

   Step 3: Use step 2 to conclude that $|x + y| \le |x| + |y|$.

78. **A consequence of the triangle inequality.** Use the triangle inequality established in the preceding exercise to show that $|A - B| \ge |A| - |B|$.

Suggestion: Apply the triangle inequality with $x = A - B$ and $y = B$.

79. **Another consequence of the triangle inequality.** Show that $|x + y + z| \le |x| + |y| + |z|$.

Challenging inequalities. Solve the inequalities in Exercises 80 through 83.

80. $|x - 3| < |x - 4|$ *Suggestion*: Think in terms of distance, and make a picture on the number line.

81. $|x - 6| < x$

82. $|x + 5| \le |x + 1| + 4$. *Suggestion*: Write $|x + 5| = |(x + 1) + 4|$. Then see what the triangle inequality $|A + B| \le |A| + |B|$ tells you.

83. $|2x + 9| > |x + 5| + |x + 4|$. *Suggestion*: See the preceding exercise.

P.1 The Coordinate Plane: Lines and Circles

P.2 Linear Inequalities

P.3 Nonlinear Inequalities

P.3   Nonlinear Inequalities

Nonlinear inequalities arise in applications and require alternative methods for arriving at a solution.

In this section, you will learn to:

1. Solve quadratic inequalities using sign tables and sign diagrams.
2. Interpret inequalities using a graph.
3. Solve inequalities involving several factors or fractions.
4. Solve applied problems involving nonlinear inequalities.

In this section, we will apply the factoring techniques that helped us solve quadratic equations to help us solve nonlinear inequalities. Solving these inequalities is useful in calculus for graphing polynomial and rational functions.

Quadratic Inequalities

> **Factoring is the key to solving quadratic inequalities.**

The first type of inequality we solve is **quadratic**.

When a quadratic inequality appears in factored form, we can solve it by considering the signs of the factors. Let's illustrate the method with this quadratic inequality:

$$x^2 - 5x + 4 < 0$$

First, factor the quadratic expression to get

$$(x - 1)(x - 4) < 0$$

Our goal is to solve the inequality by deciding when the product is positive or negative based on the signs of the factors. We next discuss two common ways in which the work to accomplish this is normally presented.

An inequality is **quadratic** if the rules for manipulating inequalities can be used to put it in the form $ax^2 + bx + c < 0$ or $ax^2 + bx + c \leq 0$ with $a \neq 0$.

Method 1: The Sign Table

We show how to make the sign table in several steps. With just a bit of practice, you will find the table can be completed quickly. We use the points where the factors are zero, in this case at $x = 1$ and $x = 4$, to make a table of signs as follows:

Region	$x < 1$	$1 < x < 4$	$4 < x$
Sign of $x - 1$			
Sign of $x - 4$			
Sign of $(x - 1)(x - 4)$			

Next, fill in the rows corresponding to $x - 1$ and $x - 4$ by indicating their signs in each of the regions.

In the first row, corresponding to the factor $x - 1$, we observe that $x - 1$ is negative when x is less than 1 and positive when x is greater than 1. We fill in the $x - 1$ row in the table accordingly:

Region	$x < 1$	$1 < x < 4$	$4 < x$
Sign of $x - 1$	$-$	$+$	$+$
Sign of $x - 4$			
Sign of $(x - 1)(x - 4)$			

In the second row, corresponding to the factor $x - 4$, we observe that $x - 4$ is negative when x is less than 4 and positive when x is greater than 4. We fill in the $x - 4$ row in the table accordingly:

Region	$x < 1$	$1 < x < 4$	$4 < x$
Sign of $x - 1$	$-$	$+$	$+$
Sign of $x - 4$	$-$	$-$	$+$
Sign of $(x - 1)(x - 4)$			

When the signs of the factors $(x - 1)$ and $(x - 4)$ are the same, their product is positive, and when their signs are different, their product is negative. We fill in the last row of the table using this idea.

Region	$x < 1$	$1 < x < 4$	$4 < x$
Sign of $x - 1$	–	+	+
Sign of $x - 4$	–	–	+
Sign of $(x - 1)(x - 4)$	$(-)(-) = +$	$(+)(-) = -$	$(+)(+) = +$

The final row of the table allows us to solve the original inequality. The inequality $(x - 1)(x - 4) < 0$ requires that $(x - 1)(x - 4)$ be negative. The last row of the sign table shows that this is true on the interval from 1 to 4. (The endpoints, 1 and 4, of the interval both make the product zero. Hence, they are not part of the solution.) The final answer, in interval notation, is $(1, 4)$.

Now, suppose that we had been asked to solve the inequality $(x - 1)(x - 4) \le 0$. We would use the same table as before and just observe that the endpoints of the interval $(1, 4)$ should be included in the solution, which would be $[1, 4]$.

Finally, suppose that we had been asked to solve the inequality: $(x - 1)(x - 4) > 0$. We would again use the same table as before and just observe that in the last row the plus signs indicate the solution, which would be $(-\infty, 1) \cup (4, \infty)$.

If you are unsure of the sign of one of the linear factors in a given region, you can check using a test number in the region. This is because the sign of a factor $x - a$ can change only at $x = a$.

Method 2: The Sign Diagram

Another device for solving the inequality $(x - 1)(x - 4) < 0$ is to draw the number line, mark 1 and 4 on it, and put plus and minus signs for the factors $(x - 1)$ and $(x - 4)$ on each of the intervals $(-\infty, 1), (1, 4)$, and $(4, \infty)$. This is a sign diagram.

$$(x - 1)(x - 4)$$

$$(-)(-) = + \quad | \quad (+)(-) = - \quad | \quad (+)(+) = +$$

$$\overline{\qquad\qquad\qquad\qquad\qquad\qquad\qquad\qquad}$$

$$1 \qquad\qquad\qquad 4$$

We arrive at the same solution as with the sign table, namely $1 < x < 4$, but this is a more compact, visual representation. Use whichever one you prefer; we will include both in the examples that follow.

Finally, a graphical display can help us visualize the solution of inequalities of this sort. **Figure P.31** shows the graph of $y = x^2 - 5x + 4$ and highlights in red where the graph is below the x-axis. These values correspond to solutions of the inequality.

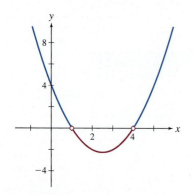

Figure P.31 Using a graphical display: The inequality $x^2 - 5x + 4 < 0$ is true when the graph is below the x-axis.

STEP-BY-STEP STRATEGY: Solving Quadratic Inequalities

Step 1 Move all terms to one side.

Step 2 Factor.

Step 3 Use a sign table or sign diagram to determine the intervals where the product is positive and where it is negative.

Step 4 Determine which endpoints satisfy the inequality, and incorporate the information from step 3 to find the final solution.

EXAMPLE P.15 Solving Quadratic Inequalities

Solve the quadratic inequality $x < x^2$.

SOLUTION

We move all terms to the left side of the equation and then factor:

$$x < x^2$$

$$x - x^2 < 0 \quad \blacktriangleleft \text{ Move all terms left.}$$

$$x(1 - x) < 0 \quad \blacktriangleleft \text{ Factor.}$$

The factors are zero at $x = 0$ and $x = 1$, so we use these points to make the sign table. Be careful with the factor $1 - x$. It is *positive* when x is less than 1 and *negative* when x is greater than 1.

Region	$x < 0$	$0 < x < 1$	$1 < x$
Sign of x	$-$	$+$	$+$
Sign of $1 - x$	$+$	$+$	$-$
Sign of $x(1 - x)$	$(-)(+) = -$	$(+)(+) = +$	$(+)(-) = -$

The corresponding sign diagram is:

$$x(1 - x)$$

$$(-)(+) = - \quad | \quad (+)(+) = + \quad | \quad (+)(-) = -$$

$$\overline{\hspace{4cm}}$$
$$\qquad 0 \qquad\qquad 1$$

We see that the inequality is satisfied when $x < 0$ or $x > 1$. We do not include the endpoints because they make the product $x(1 - x)$ zero. In interval notation, the solution is $(-\infty, 0) \cup (1, \infty)$. The region of the graph where the inequality is true is shown in red in **Figure P.32**.

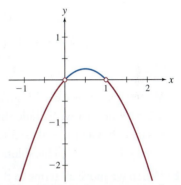

Figure P.32 The inequality $x < x^2$: This inequality holds where the graph of $y = x - x^2$ is below the x-axis.

TRY IT YOURSELF P.15 Brief answers provided at the end of the section.

Solve the inequality $x^2 + 6 \leq 5x$.

Let's look again at Example P.15, where we solve the inequality $x < x^2$. A common error is to proceed as follows:

$$x < x^2$$

$$\frac{x}{x} < \frac{x^2}{x} \quad \blacktriangleleft \text{ Divide by } x. \text{ (ERROR)}$$

$$1 < x \quad \blacktriangleleft \text{ Simplify.}$$

Remember that when we multiply or divide an inequality by a number x, the inequality must be reversed if x is negative but kept the same if x is positive. Because we don't know the sign of x, we don't know whether or not to reverse the inequality symbol after dividing by x. This error must be avoided. The correct method is to factor and use a sign table or sign diagram.

EXAMPLE P.16 More Quadratic Inequalities

Solve the following quadratic inequalities.

a. $x^2 \geq 5 + 4x$ **b.** $x^2 + 1 > 2x$

SOLUTION

a. We proceed as in the previous example:

$$x^2 \geq 5 + 4x$$

$$x^2 - 4x - 5 \geq 0 \qquad \blacktriangleleft \text{Move all terms left.}$$

$$(x+1)(x-5) \geq 0 \qquad \blacktriangleleft \text{Factor.}$$

The factors are zero at $x = -1$ and $x = 5$. We use these numbers to make the sign table.

Region	$x < -1$	$-1 < x < 5$	$5 < x$
Sign of $x + 1$	–	+	+
Sign of $x - 5$	–	–	+
Sign of $(x+1)(x-5)$	$(-)(-) = +$	$(+)(-) = -$	$(+)(+) = +$

The corresponding sign diagram is

$$(x+1)(x-5)$$

$$(-)(-) = + \quad | \quad (+)(-) = - \quad | \quad (+)(+) = +$$

$$\overline{}$$

$$-1 \qquad\qquad 5$$

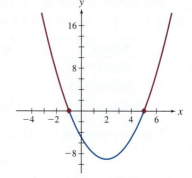

Figure P.33 The inequality $x^2 - 4x - 5 \geq 0$: The solution corresponds to points where the graph of $y = x^2 - 4x - 5$ is on or above the x-axis.

We see that $x^2 - 4x - 5$ is greater than or equal to zero when $x \leq -1$ or $x \geq 5$. Note that we include the endpoints here because the inequality is satisfied when $x^2 - 4x - 5$ is zero. In interval notation, the solution is $(-\infty, -1] \cup [5, \infty)$. This solution is illustrated in **Figure P.33**.

b. When we move all terms to the left and factor, we get

$$(x-1)^2 > 0$$

Because a square is never negative, the only time the inequality fails is when $(x-1)^2$ is zero. That occurs when $x = 1$. Thus, the inequality is true for all real numbers except 1. That gives the solution $(-\infty, 1) \cup (1, \infty)$.

TRY IT YOURSELF P.16 Brief answers provided at the end of the section.

Solve the inequality $x^2 \leq -2x$.

EXTEND YOUR REACH

a. The following inequalities illustrate that sometimes the solution of a quadratic inequality is neither an interval nor a union of intervals of the kind we have seen so far. Solve both inequalities.

- $x^2 + 1 > 0$
- $x^2 \leq 0$

b. The preceding inequalities show two possibilities for the solution of a quadratic inequality that is not of the kind we have seen so far. There is a third possibility. Identify it, and write an illustrative example.

c. Is it possible that the solution of the inequality $x^2 + ax + b > 0$ is an interval of the form (p, ∞) for some number p? *Suggestion:* To answer this question, recall that the graph of a quadratic expression is a parabola.

Inequalities Involving More Than Two Factors

> Sign tables and sign diagrams are used to solve many inequalities.

The step-by-step strategy for solving quadratic inequalities applies to products of more than two factors. The sign tables get longer, but the idea remains the same: If an odd number of factors are negative, their product is negative. If an even number of factors are negative, their product is positive. For example, let's look at the inequality

$$x(1-x)(x-3) < 0$$

There are three factors here, and they are zero at $x = 0$, $x = 1$ and $x = 3$. We use these three numbers to make the sign table.

Region	$x < 0$	$0 < x < 1$	$1 < x < 3$	$3 < x$
Sign of x	$-$	$+$	$+$	$+$
Sign of $1-x$	$+$	$+$	$-$	$-$
Sign of $x-3$	$-$	$-$	$-$	$+$
Sign of $x(1-x)(x-3)$	$(-)(+)(-) = +$	$(+)(+)(-) = -$	$(+)(-)(-) = +$	$(+)(-)(+) = -$

Here is the corresponding sign diagram:

$$x(1-x)(x-3)$$

$(-)(+)(-) = +$ | $(+)(+)(-) = -$ | $(+)(-)(-) = +$ | $(+)(-)(+) = -$

$0 \qquad 1 \qquad 3$

The inequality $x(1-x)(x-3) < 0$ is satisfied when the product is negative. Hence, the solution is $(0, 1) \cup (3, \infty)$ (**Figure P.34**).

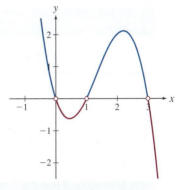

Figure P.34 The inequality $x(1-x)(x-3) < 0$: This inequality holds where the graph of $y = x(1-x)(x-3)$ is below the x-axis.

EXAMPLE P.17 Solving Inequalities Involving Several Factors

Solve the inequality $x^3(x+1)^2(x-3) \geq 0$.

SOLUTION

The three factors are zero at $x = -1$, $x = 0$, and $x = 3$. We make the sign table using these numbers. Pay special attention to the factors x^3 and $(x+1)^2$. Because 3 is odd, x^3 has the same sign as x. The factor $(x+1)^2$ is zero when $x = -1$, but it is a square and so is never negative. This means that the row in the table for $(x+1)^2$ has all plus signs.

Region	$x < -1$	$-1 < x < 0$	$0 < x < 3$	$3 < x$
Sign of x^3	$-$	$-$	$+$	$+$
Sign of $(x+1)^2$	$+$	$+$	$+$	$+$
Sign of $x-3$	$-$	$-$	$-$	$+$
Sign of $x^3(x+1)^2(x-3)$	$(-)(+)(-) = +$	$(-)(+)(-) = +$	$(+)(+)(-) = -$	$(+)(+)(+) = +$

Here is the corresponding sign diagram:

$$x^3(x+1)^2(x-3)$$

$(-)(+)(-) = +$ | $(-)(+)(-) = +$ | $(+)(+)(-) = -$ | $(+)(+)(+) = +$

$-1 \qquad 0 \qquad 3$

The inequality is satisfied when the product is positive or equal to zero. Thus, $x = -1$, $x = 0$, and $x = 3$ are all included in the solution. The final answer is $(-\infty, 0] \cup [3, \infty)$ (**Figure P.35**).

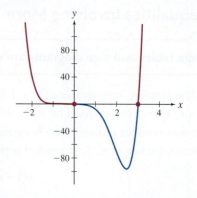

Figure P.35 The inequality $x^3(x+1)^2(x-3) \geq 0$: This inequality holds where the graph of $y = x^3(x+1)^2(x-3)$ is on or above the x-axis.

TRY IT YOURSELF P.17 Brief answers provided at the end of the section.

Solve the inequality $(x-1)(x-3)(x-5) \geq 0$.

Let's change the problem from the previous example slightly to $x^3(x+1)^2(x-3) > 0$. The sign diagram for this inequality is the same as that in the preceding solution. The difference is that we must exclude points that make the product zero. Thus, we must exclude $x = -1$, $x = 0$, and $x = 3$. The solution for this inequality is $(-\infty, -1) \cup (-1, 0) \cup (3, \infty)$.

EXAMPLE P.18 Factoring and Solving Inequalities

Solve the inequality $x^3 < x$.

SOLUTION

We can solve this inequality in a fashion similar to that used in the preceding example if we first move all terms to the left and factor:

$$x^3 < x$$

$$x^3 - x < 0 \quad \blacktriangleleft \text{ Move all terms left.}$$

$$x(x^2 - 1) < 0 \quad \blacktriangleleft \text{ Begin factoring.}$$

$$x(x-1)(x+1) < 0 \quad \blacktriangleleft \text{ Complete factoring.}$$

The factors are zero at $x = -1$, $x = 0$, and $x = 1$. We use these to make the sign table.

Region	$x < -1$	$-1 < x < 0$	$0 < x < 1$	$1 < x$
Sign of x	$-$	$-$	$+$	$+$
Sign of $x - 1$	$-$	$-$	$-$	$+$
Sign of $x + 1$	$-$	$+$	$+$	$+$
Sign of $x(x-1)(x+1)$	$(-)(-)(-) = -$	$(-)(-)(+) = +$	$(+)(-)(+) = -$	$(+)(+)(+) = +$

Here is the corresponding sign diagram:

$$x(x-1)(x+1)$$

$$(-)(-)(-) = - \quad | \quad (-)(-)(+) = + \quad | \quad (+)(-)(+) = - \quad | \quad (+)(+)(+) = +$$

$$\underline{\hspace{3cm}}$$
$$-1 \qquad\qquad 0 \qquad\qquad 1$$

The final answer is $(-\infty, -1) \cup (0, 1)$, because the inequality is satisfied when the product is negative (**Figure P.36**).

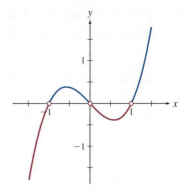

Figure P.36 The inequality $x(x-1)(x+1) < 0$: The solution corresponds to points where the graph of $y = x(x-1)(x+1)$ is below the x-axis.

TRY IT YOURSELF P.18 Brief answers provided at the end of the section.

Solve the inequality $x^3 + 6x > 5x^2$.

Inequalities Involving Fractions

> When fractions are involved, numbers that result in division by zero must be excluded from the solution.

We can use the preceding strategy to solve some inequalities involving fractions. The only additional concern is that we must avoid division by zero.

EXAMPLE P.19 Inequalities Involving Fractions

Solve the inequality $\dfrac{x}{(x-1)(x+1)} \geq 0$.

SOLUTION

We make a sign table using the factors just as in the earlier examples, but this time we include both the numerator and the denominator. The factors are zero at $x = 0$, $x = 1$, and $x = -1$.

Region	$x < -1$	$-1 < x < 0$	$0 < x < 1$	$1 < x$
Sign of x	$-$	$-$	$+$	$+$
Sign of $x - 1$	$-$	$-$	$-$	$+$
Sign of $x + 1$	$-$	$+$	$+$	$+$
Sign of $\dfrac{x}{(x-1)(x+1)}$	$\dfrac{(-)}{(-)(-)} = -$	$\dfrac{(-)}{(-)(+)} = +$	$\dfrac{(+)}{(-)(+)} = -$	$\dfrac{(+)}{(+)(+)} = +$

Here is the corresponding sign diagram:

$$\frac{x}{(x-1)(x+1)}$$

$$\frac{(-)}{(-)(-)} = - \quad | \quad \frac{(-)}{(-)(+)} = + \quad | \quad \frac{(+)}{(-)(+)} = - \quad | \quad \frac{(+)}{(+)(+)} = +$$

$$\boxed{-1} \qquad 0 \qquad \boxed{1}$$

We must never include −1 or 1 in the solution because these numbers cause division by zero. In the sign diagram they are in boxes to alert us to this fact.

The inequality is satisfied when the fraction is zero or positive. We include $x = 0$ in the solution because it makes the fraction zero. The solution is $(-1, 0] \cup (1, \infty)$ (**Figure P.37**).

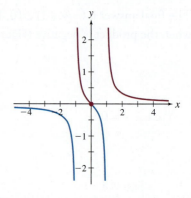

Figure P.37 The inequality $\dfrac{x}{(x-1)(x+1)} \geq 0$: The solution corresponds to points where the graph of $y = \dfrac{x}{(x-1)(x+1)}$ is above or on the x-axis.

TRY IT YOURSELF P.19   Brief answers provided at the end of the section.

Solve the inequality $\dfrac{x+1}{x-1} < 0$.

STEP-BY-STEP STRATEGY: Solving Inequalities Involving Algebraic Fractions

Step 1 Move all terms to one side.

Step 2 Combine algebraic fractions into a single fraction.

Step 3 Simplify and factor the combined fraction.

Step 4 Use a sign table or sign diagram to determine the intervals where the combined fraction is positive and where it is negative.

Step 5 Determine which endpoints satisfy the inequality, and incorporate the information from step 4 to find the final solution. Be sure to exclude endpoints that would lead to division by zero.

EXAMPLE P.20 More Inequalities with Fractions

Solve the inequality $\dfrac{1}{x} < \dfrac{1}{x-1}$.

SOLUTION

We can solve this inequality using the preceding step-by-step strategy:

$$\frac{1}{x} < \frac{1}{x-1}$$

$$\frac{1}{x} - \frac{1}{x-1} < 0 \qquad \blacktriangleleft \text{ Move all terms left.}$$

$$\frac{(x-1) - x}{x(x-1)} < 0 \qquad \blacktriangleleft \text{ Combine into single fraction.}$$

$$\frac{-1}{x(x-1)} < 0 \quad \blacktriangleleft \text{ Simplify.}$$

$$\frac{1}{x(x-1)} > 0 \quad \blacktriangleleft \text{ Multiply by } -1, \text{ and reverse inequality sign.}$$

We use $x = 0$ and $x = 1$ to make the sign table.

Region	$x < 0$	$0 < x < 1$	$1 < x$
Sign of x	$-$	$+$	$+$
Sign of $x - 1$	$-$	$-$	$+$
Sign of $\dfrac{1}{x(x-1)}$	$\dfrac{(+)}{(-)(-)} = +$	$\dfrac{(+)}{(+)(-)} = -$	$\dfrac{(+)}{(+)(+)} = +$

The corresponding sign diagram is

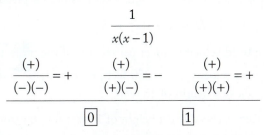

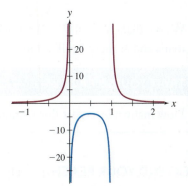

Figure P.38 The inequality $\dfrac{1}{x(x-1)} > 0$: The solution corresponds to points where the graph of $y = \dfrac{1}{x(x-1)}$ is above the x-axis.

The inequality is true when the fraction is positive. Hence, the solution is $(-\infty, 0) \cup (1, \infty)$. This solution corresponds to the places where the graph in **Figure P.38** is above the x-axis.

TRY IT YOURSELF P.20   Brief answers provided at the end of the section.

Solve the inequality $\dfrac{x}{x-1} + \dfrac{x}{x+1} \geq 0$.

MODELS AND APPLICATIONS   Profit

EXAMPLE P.21   Profit

The profit P, in millions of dollars, that a manufacturer makes depends on the number N, in millions, of items produced in a year. The relationship is

$$P = -N^2 + 11N - 8$$

How many items must the manufacturer produce in a year so that the profit will be \$20 million or more?

SOLUTION

If the profit is \$20 million or more, then $20 \leq P$. Putting in the relationship between P and N gives

$$20 \leq -N^2 + 11N - 8$$

To solve this quadratic inequality, proceed as before:

$$20 \leq -N^2 + 11N - 8$$

$$N^2 - 11N + 28 \leq 0 \qquad \blacktriangleleft \text{ Move all terms left and simplify.}$$

$$(N - 4)(N - 7) \leq 0 \qquad \blacktriangleleft \text{ Factor.}$$

The factors are zero at $N = 4$ and $N = 7$. We use these numbers to make the sign diagram:

$$(N - 4)(N - 7)$$

$$\frac{(-)(-) = + \quad | \quad (+)(-) = - \quad | \quad (+)(+) = +}{4 \qquad\qquad\qquad 7}$$

We see that $(N - 4)(N - 7) \leq 0$ when $4 \leq N \leq 7$. Thus, the manufacturer must produce at least 4 million items and at most 7 million items in a year so that the profit will be \$20 million or more.

TRY IT YOURSELF P.21   Brief answers provided at the end of the section.

How many items must this manufacturer produce in a year so that the profit will be more than \$16 million?

EXTEND YOUR REACH   The height H of a projectile, in feet, t seconds after it is fired is given by

$$H = vt - 16t^2 + 15$$

Here v is the initial upward velocity, in feet per second, of the projectile.

a. If the initial upward velocity is $v = 250$ feet per second, for how long will the projectile be above a height of 15 feet?

b. The projectile needs to be above a height of 15 feet for 20 seconds. What initial upward velocity must be used?

TRY IT YOURSELF ANSWERS

P.15 $[2, 3]$

P.16 $[-2, 0]$

P.17 $[1, 3] \cup [5, \infty)$

P.18 $(0, 2) \cup (3, \infty)$

P.19 $(-1, 1)$

P.20 $(-\infty, -1) \cup (1, \infty)$

P.21 The manufacturer must produce more than 3 million items and fewer than 8 million items in a year.

EXERCISE SET P.3

CHECK YOUR UNDERSTANDING

1. Critique the following: To solve $x^2 < 3x$, we divide by x to obtain $x < 3$.

2. **True or false:** It is a good practice to multiply both sides of an inequality by a variable expression in order to clear denominators.

3. To make a sign diagram, we must first do the following:

   a. Move all terms to one side and factor.

   b. Factor both sides of the inequality.

   c. Find the least common denominator.

   d. None of the above.

4. **True or false:** A sign diagram can be made for any factored expression.

5. **True or false:** A sign table yields more information than a sign diagram.

6. If $0 < x < y < 1$, how does $\dfrac{1}{x}$ compare with $\dfrac{1}{y}$?

7. **True or false:** If the denominator of a fraction is always positive, then it need not be considered when making a sign diagram.

8. Make up an inequality that has no solution.

9. Make up an inequality that is true for all real numbers.

10. How many intervals occur in the solution of $(x - 1)(x - 2)(x - 3)\cdots(x - 10) < 0$?

PROBLEMS

Making sign tables. In Exercises 11 through 16, make a sign table for the given expression.

11. $(x-4)(x+1)$

12. $(4-x)(x+1)$

13. $\dfrac{x-4}{x+1}$

14. $\dfrac{x-2}{(x-1)(x-3)}$

15. $\dfrac{x-2}{(x-1)^2(x-3)}$

16. $\dfrac{x}{x^2-5x+6}$

Using sign tables. In Exercises 17 through 20, solve the inequality using the following sign table.

Region	$x<-1$	$-1<x<4$	$4<x$
Sign of $x-4$	–	–	+
Sign of $x+1$	–	+	+
Sign of $(x-4)(x+1)$	+	–	+

17. $(x-4)(x+1)<0$

18. $(x-4)(x+1)\le 0$

19. $(x-4)(x+1)>0$

20. $(x-4)(x+1)\ge 0$

Using sign tables. In Exercises 21 through 24, solve the inequality using the following sign table.

Region	$x<-1$	$-1<x<4$	$4<x$
Sign of $4-x$	+	+	–
Sign of $x+1$	–	+	+
Sign of $(4-x)(x+1)$	–	+	–

21. $(4-x)(x+1)>0$

22. $(4-x)(x+1)\ge 0$

23. $(4-x)(x+1)<0$

24. $(4-x)(x+1)\le 0$

Using sign tables. In Exercises 25 through 28, solve the inequality using the following sign table.

Region	$x<-1$	$-1<x<4$	$4<x$
Sign of $x+1$	–	+	+
Sign of $x-4$	–	–	+
Sign of $\dfrac{x-4}{x+1}$	+	–	+

25. $\dfrac{x-4}{x+1}<0$

26. $\dfrac{x-4}{x+1}\le 0$

27. $\dfrac{x-4}{x+1}>0$

28. $\dfrac{x-4}{x+1}\ge 0$

Using sign tables. In Exercises 29 through 32, solve the inequality using the following sign table.

Region	$x<1$	$1<x<2$	$2<x<3$	$x>3$
Sign of $x-1$	–	+	+	+
Sign of $x-2$	–	–	+	+
Sign of $x-3$	–	–	–	+
Sign of $\dfrac{x-2}{(x-1)(x-3)}$	–	+	–	+

29. $\dfrac{x-2}{(x-1)(x-3)}>0$

30. $\dfrac{x-2}{(x-1)(x-3)}\ge 0$

31. $\dfrac{x-2}{(x-1)(x-3)}<0$

32. $\dfrac{x-2}{(x-1)(x-3)}\le 0$

Using sign tables. In Exercises 33 through 36, solve the inequality using the following sign table.

Region	$x<1$	$1<x<2$	$2<x<3$	$x>3$
Sign of $(x-1)^2$	+	+	+	+
Sign of $x-2$	–	–	+	+
Sign of $x-3$	–	–	–	+
Sign of $\dfrac{x-2}{(x-1)^2(x-3)}$	+	+	–	+

33. $\dfrac{x-2}{(x-1)^2(x-3)}\ge 0$

34. $\dfrac{x-2}{(x-1)^2(x-3)}>0$

35. $\dfrac{x-2}{(x-1)^2(x-3)}<0$

36. $\dfrac{x-2}{(x-1)^2(x-3)}\le 0$

Using sign tables. In Exercises 37 through 40, solve the inequality using the following sign table.

Region	$x<0$	$0<x<2$	$2<x<3$	$3<x$
Sign of x	–	+	+	+
Sign of $x-2$	–	–	+	+
Sign of $x-3$	–	–	–	+
Sign of $\dfrac{x}{(x-2)(x-3)}$	–	+	–	+

37. $\dfrac{x}{x^2-5x+6}\ge 0$

38. $\dfrac{x}{x^2-5x+6}>0$

39. $\dfrac{x}{x^2-5x+6}\le 0$

40. $\dfrac{x}{x^2-5x+6}<0$

Making sign diagrams. In Exercises 41 through 45, construct the sign diagram for the given expression.

41. $(x-2)(x-5)$

42. $(x-2)(5-x)$

43. $\dfrac{x-2}{x-5}$

45. $\dfrac{(x-1)^2(x-2)^3}{x}$

44. $(x-2)^2(5-x)$

Using sign diagrams. In Exercises 46 through 49, solve the inequality using the following sign diagram.

$$(x-2)(x-5)$$

$$\underline{\quad (-)(-) = + \quad | \quad (+)(-) = - \quad | \quad (+)(+) = + \quad}$$
$$\qquad\qquad 2 \qquad\qquad\qquad 5$$

46. $(x-2)(x-5) > 0$ **48.** $(x-2)(x-5) < 0$

47. $(x-2)(x-5) \geq 0$ **49.** $(x-2)(x-5) \leq 0$

Using sign diagrams. In Exercises 50 through 53, solve the inequality using the following sign diagram.

$$(x-2)(5-x)$$

$$\underline{\quad (-)(+) = - \quad | \quad (+)(+) = + \quad | \quad (+)(-) = - \quad}$$
$$\qquad\qquad 2 \qquad\qquad\qquad 5$$

50. $(x-2)(5-x) > 0$ **52.** $(x-2)(5-x) < 0$

51. $(x-2)(5-x) \geq 0$ **53.** $(x-2)(5-x) \leq 0$

Using sign diagrams. In Exercises 54 through 57, solve the inequality using the following sign diagram.

$$\dfrac{x-2}{x-5}$$

$$\underline{\quad \dfrac{(-)}{(-)} = + \quad | \quad \dfrac{(+)}{(-)} = - \quad | \quad \dfrac{(+)}{(+)} = + \quad}$$
$$\qquad\qquad 2 \qquad\qquad\quad \boxed{5}$$

54. $\dfrac{x-2}{x-5} > 0$ **56.** $\dfrac{x-2}{x-5} < 0$

55. $\dfrac{x-2}{x-5} \geq 0$ **57.** $\dfrac{x-2}{x-5} \leq 0$

Using sign diagrams. In Exercises 58 through 61, solve the inequality using the following sign diagram.

$$(x-2)^2(5-x)$$

$$\underline{\quad (+)(+) = + \quad | \quad (+)(+) = + \quad | \quad (+)(-) = - \quad}$$
$$\qquad\qquad 2 \qquad\qquad\qquad 5$$

58. $(x-2)^2(5-x) > 0$ **60.** $(x-2)^2(5-x) < 0$

59. $(x-2)^2(5-x) \geq 0$ **61.** $(x-2)^2(5-x) \leq 0$

Using sign diagrams. In Exercises 62 through 65, solve the inequality using the following sign diagram.

$$\dfrac{(x-1)(x-2)}{x}$$

$$\underline{\quad \dfrac{(-)(-)}{(-)} = - \quad | \quad \dfrac{(-)(-)}{(+)} = + \quad | \quad \dfrac{(+)(-)}{(+)} = - \quad | \quad \dfrac{(+)(+)}{(+)} = + \quad}$$
$$\qquad \boxed{0} \qquad\qquad 1 \qquad\qquad 2$$

62. $\dfrac{(x-1)(x-2)}{x} > 0$ **64.** $\dfrac{(x-1)(x-2)}{x} < 0$

63. $\dfrac{(x-1)(x-2)}{x} \geq 0$ **65.** $\dfrac{(x-1)(x-2)}{x} \leq 0$

Using sign diagrams. In Exercises 66 through 69, solve the inequality using the following sign diagram.

$$\dfrac{(x-1)^2(x-2)^3}{x}$$

$$\underline{\quad \dfrac{(+)(-)}{(-)} = + \quad | \quad \dfrac{(+)(-)}{(+)} = - \quad | \quad \dfrac{(+)(-)}{(+)} = - \quad | \quad \dfrac{(+)(+)}{(+)} = + \quad}$$
$$\qquad \boxed{0} \qquad\qquad 1 \qquad\qquad 2$$

66. $\dfrac{(x-1)^2(x-2)^3}{x} > 0$ **68.** $\dfrac{(x-1)^2(x-2)^3}{x} < 0$

67. $\dfrac{(x-1)^2(x-2)^3}{x} \geq 0$ **69.** $\dfrac{(x-1)^2(x-2)^3}{x} \leq 0$

Quadratic inequalities. Solve the inequalities in Exercises 70 through 80.

70. $x^2 - 3x + 2 > 0$ **76.** $6x^2 + x < 2$

71. $x^2 - 5x + 6 \leq 0$ **77.** $3x^2 \geq 2$

72. $x^2 + 11x + 30 < 0$ **78.** $6x^2 < 7x$

73. $x^2 - 4x + 4 \geq 0$ **79.** $20x^2 \leq 9x - 1$

74. $8x^2 + 23x + 15 \leq 0$ **80.** $4x^2 + 20x \leq -25$

75. $12x^2 > 1 + x$

More complicated inequalities. Solve the inequalities in Exercises 81 through 91.

81. $x(x-1)(x-2) > 0$

82. $x^3 < x$

83. $x^3 \geq x^2$

84. $x^3 - 5x^2 + 6x > 0$

85. $x^4 - 5x^3 + 6x^2 \leq 0$

86. $x^4(x-2)^3(x-3) > 0$

87. $x^2(x-2)^4(x-3)^6 > 0$

88. $x^4(x-2)^6(x-3)^2 \geq 0$

89. $\dfrac{2}{x} > \dfrac{1}{x-1}$

90. $\dfrac{1}{x+4} + \dfrac{1}{x+2} \geq 0$

91. $\dfrac{5}{x-5} < \dfrac{1}{x-1}$

Some common errors. In Exercises 92 through 96, you are asked to identify some common errors.

92. A student, Mateo, claimed to another student, Suchita, that $(1, \infty]$ was the answer to a problem involving inequalities. Without even seeing the problem, Suchita told Mateo that his answer was wrong. What was Mateo's obvious error?

93. A student, Lisa, claimed to another student, Minh, that $(2, 1)$ was the answer to a problem involving inequalities. Without even seeing the problem, Minh told Lisa that her answer was wrong. What was Lisa's obvious error?

94. To solve $\dfrac{2}{x} > 1$, a student simplifies to get $2 > x$ and says that the answer is $(-\infty, 2)$. What did the student do wrong, and what is the correct answer?

95. A student wrote: "The set of solutions of $(x-1)(x-2) > 0$ consists of all numbers x that are less than 1 $(x < 1)$ and greater than 2 $(x > 2)$." What is wrong with this statement?

96. A student solved $(x-2)(x-5) > 0$ and expressed his answer by saying it was the set of all numbers x such that $2 > x > 5$. What is wrong with this statement?

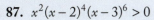 **Using a graph to solve inequalities.** Exercises 97 through 100 are inequalities that are difficult to solve by hand. Use a graphing utility to make an appropriate graph, and then give the solution correct to two decimal places.

97. $\dfrac{1}{x} + \sqrt{x} - 3 < 0$

98. $x^3 - |x+4| > 0$

99. $\dfrac{1}{x^2+x+1} > x^2$. *Suggestion:* Plot $\dfrac{1}{x^2+x+1} - x^2$.

100. $\left(\dfrac{1}{x-3} - x\right)^2 < x+1$. *Suggestion:* Plot $\left(\dfrac{1}{x-3} - x\right)^2 - (x+1)$.

MODELS AND APPLICATIONS

101. **Wave height.** The following formula gives the approximate wave height h in feet corresponding to a wind speed w in miles per hour:

$$h = 0.02w^2$$

Waters are safe for a small boat as long as the wave height does not exceed 5 feet. What wind speeds indicate safe sailing for this craft?

102. **A tossed ball.** The height H in feet of a ball t seconds after it is tossed upward is given by

$$H = 28t - 16t^2$$

During what time period is the ball 10 feet or higher above the ground?

103. **Logistic growth.** For a seal population, the population growth rate G, measured in hundreds per year, depends on the population N, measured in thousands. The relationship is

$$G = N\left(1 - \dfrac{N}{3}\right)$$

The graph of this relationship is shown in **Figure P.39**. This is an example of logistic growth. What population levels give a growth rate of $\dfrac{2}{3}$ hundred per year or higher?

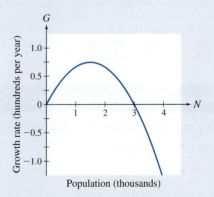

Figure P.39 Growth rate versus population size

104. Logistic growth with a threshold. For a whale population, the population growth rate G, measured in hundreds per year, depends on the population N, measured in thousands. The relationship is

$$G = -0.05N\left(1 - \frac{N}{10}\right)\left(1 - \frac{N}{70}\right)$$

A graph of this relationship is shown in **Figure P.40**.

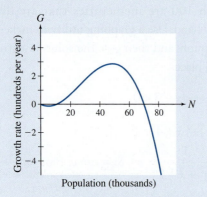

Population (thousands)

Figure P.40 Growth rate in the presence of a population threshold

a. The population is growing when the growth rate is positive. For which population levels is the population growing?

b. Explain what is happening to the population when the population is less than 10,000 or greater than 70,000.

105. Reading level. The Flesch-Kincaid Grade Level Formula calculates the grade level K of a passage of text.[2] In the formula, w is the number of words

_____

[2]*Derivation of New Readability Formulas (Automated Readability Index, Fog Count and Flesch Reading Ease Formula) for Navy Enlisted Personnel,* J. P. Kincaid, R. P. Fisburne, Jr., R. L. Rogers, and B. S. Chissom, Research Branch Report 8-75, Memphis-Millington, TN: Naval Technical Training Command, U.S. Naval Air Station, 1975.

in the passage, s is the number of sentences, and n is the number of syllables. The formula is

$$K = 0.39\frac{w}{s} + 11.8\frac{n}{w} - 15.59$$

Consider passages with 20 sentences and 450 syllables.

a. Use a graphing utility to make the graph of K versus w.

b. What number of words gives a reading level of 9 to 12? Use whole numbers in reporting your answer.

106. Kinetic energy. A mass of m kilograms with a velocity of v meters per second has a kinetic energy of

$$E = \frac{1}{2}mv^2 \text{ joules}$$

For a mass of 10 kilograms, what range of velocity results in a kinetic energy of between 125 and 320 joules?

107. Saving for college. You want to begin making investments of $600 each month for your child's college education 16 years in the future. If the investment is earning a monthly interest rate of r as a decimal, then the value of your investment after 16 years will be

$$V = \frac{600}{r}((1+r)^{192} - 1)$$

a. Plot the graph of V versus r using a range of $r = 0$ to $r = 0.01$.

b. You figure you will need between $200,000 and $240,000 to cover all expenses. What monthly interest rate range will accommodate that goal? Round your answer to four decimal places.

CHALLENGE EXERCISES FOR INDIVIDUALS OR GROUPS

Challenging inequalities. For Exercises 108 through 111, solve the inequalities.

108. $|3x + 4| > x^2$

109. $x > \dfrac{1}{3 - x}$

110. $\dfrac{x^2}{x^2 + 1} + (x - 1)^2 \geq 0$. Note that squares are never negative.

111. $\dfrac{x^2 - 3x + 2}{1 + |x|} \leq 0$. *Suggestion*: Use a sign diagram.

112. Proving inequalities. Suppose $|x - 1| < \dfrac{1}{2}$.

a. Show that $|x + 1| < \dfrac{5}{2}$.

b. Show that $|x^2 - 1| < \dfrac{5}{4}$.

CHAPTER ROADMAP AND STUDY GUIDE

P.1   The Coordinate Plane: Lines and Circles

Using the coordinate plane allows us to visualize mathematical concepts.

Coordinates in the Plane
Graphs are pictures of equations.

The Distance and Midpoint Formulas
Distances in the plane can be calculated using the Pythagorean theorem.

Lines and Their Slopes
Linear equations are represented by straight lines in the plane.

Common Forms for Equations of Lines
We can find the equation of a line given the slope and the y-intercept, the slope and a point, or two points.

Circles
The equation of a circle identifies its center and radius.

MODELS AND APPLICATIONS
Locating Through Triangulation

P.2   Linear Inequalities

Mathematical inequalities allow us to determine over what range of values one quantity exceeds another.

Inequalities and Intervals
Intervals are represented by segments or rays on the number line.

Solving Inequalities
Linear inequalities are solved using the strategy for solving linear equations.

Absolute Value Inequalities
Absolute values are often used to represent distances.

MODELS AND APPLICATIONS
Supply and Demand

P.3   Nonlinear Inequalities

Nonlinear inequalities arise in applications and require alternative methods for arriving at a solution.

Quadratic Inequalities
Factoring is the key to solving quadratic inequalities.

Inequalities Involving More Than Two Factors
Sign tables and sign diagrams are used to solve many inequalities.

Inequalities Involving Fractions
When fractions are involved, numbers that result in division by zero must be excluded from the solution.

MODELS AND APPLICATIONS
Profit

CHAPTER QUIZ

1. Calculate the distance and midpoint between the points $(2, 5)$ and $(6, 1)$.

   Answer: Distance $4\sqrt{2}$. Midpoint $(4, 3)$.

   XR Example P.1

2. Sketch the line with slope -1 and y-intercept 3.

   Answer:

   XR Example P.3

   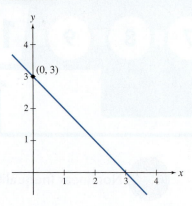

3. **a.** Find the equation of the line with slope 4 that passes through $(1, 2)$.

   b. Find the equation of the line through the points $(2, 1)$ and $(1, 0)$.

   XR Example P.4

   Answer:

   a. $y = 4x - 2$

   b. $y = x - 1$

4. Find the center and radius of the circle whose equation is $x^2 + 2x + y^2 = 4y + 4$. Then plot its graph.

   XR Example P.7

   Answer: Center $(-1, 2)$. Radius 3.

   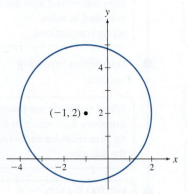

5. Solve the inequality $3(x - 2) \leq 12 + x$.

   XR Example P.10

   Answer: $x \leq 9$ or $(-\infty, 9]$

6. Solve the following inequalities.

   a. $|2x + 1| < 5$

   b. $|x + 4| \geq 5$

   XR Example P.12

Answer:

a. $(-3, 2)$

b. $(-\infty, -9] \cup [1, \infty)$

7. Solve the inequality $2x - x^2 \le 0$. Include either a sign table or a sign diagram with your solution.

XR **Example P.15**

Answer:

Region	$x < 0$	$0 < x < 2$	$2 < x$
Sign of x	$-$	$+$	$+$
Sign of $2 - x$	$+$	$+$	$-$
Sign of $x(2 - x)$	$-$	$+$	$-$

$$x(2 - x)$$

$$(-)(+) = - \quad | \quad (+)(+) = + \quad | \quad (+)(-) = -$$
$$\overline{\qquad\qquad\quad 0 \qquad\qquad\quad 2 \qquad\qquad\quad}$$

The solution is $(-\infty, 0] \cup [2, \infty)$.

8. Solve the inequality $x^4 < x^2$.

XR **Example P.18**

Answer: $(-1, 0) \cup (0, 1)$

9. Solve the inequality $\dfrac{1}{x} \ge x$.

XR **Example P.19**

Answer: $(-\infty, -1] \cup (0, 1]$

CHAPTER REVIEW EXERCISES

Section P.1

Distance and midpoint. In Exercises 1 through 4, find the distance and midpoint between P and Q.

1. $P = (1, 3)$ $Q = (3, 7)$

Answer: Distance $2\sqrt{5}$.

Midpoint $= (2, 5)$.

2. $P = (4, 5),\ Q = (7, 0)$

Answer: Distance $\sqrt{34}$.

Midpoint $\left(\dfrac{11}{2}, \dfrac{5}{2}\right)$.

3. $P = (-2, 3),\ Q = (1, 7)$

Answer: Distance 5.

Midpoint $\left(-\dfrac{1}{2}, 5\right)$.

4. $P = (-4, -1),\ Q = (-2, -3)$

Answer: Distance $2\sqrt{2}$.

Midpoint $(-3, -2)$.

Equations of lines and circles. In Exercises 5 through 11, find the required equation.

5. The line through $(1, 3)$ and $(2, 4)$

Answer: $y = x + 2$

6. The line through $(1, 5)$ and $(3, 9)$

Answer: $y = 2x + 3$

7. The line with slope -2 through $(7, 8)$

Answer: $y = -2x + 22$

8. The line with slope -3 and y-intercept 5

Answer: $y = -3x + 5$

9. The circle with center at the origin and radius 3

Answer: $x^2 + y^2 = 9$

10. The circle with center $(4, 1)$ and radius 2

Answer: $(x - 4)^2 + (y - 1)^2 = 4$

11. The circle with center $(-2, 1)$ and radius $\sqrt{7}$

Answer: $(x + 2)^2 + (y - 1)^2 = 7$

Graphs of lines and circles. Plot the graphs of the equations in Exercises 12 through 15. In the case of circles, identify the center and radius.

12. $y = 3x - 2$

Answer:

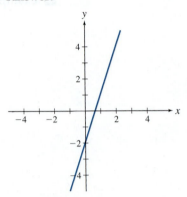

13. $y = -2x + 3$

Answer:

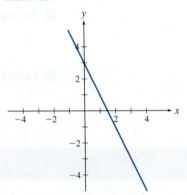

14. $x^2 + y^2 + 1 = 2x + 4y$

Answer: Circle with center at $(1, 2)$ and radius 2.

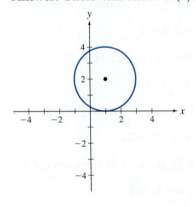

15. $x^2 + 2x + y^2 = 8$

Answer: Circle with center at $(-1, 0)$ and radius 3.

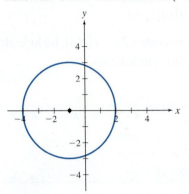

Section P.2

Solving equalities. In Exercises 16 through 27, solve the given inequality.

16. $3x + 1 < 13$

Answer: $(-\infty, 4)$

17. $4x - 3 \geq 21$

Answer: $[6, \infty)$

18. $1 - 2x > 7 + x$

Answer: $(-\infty, -2)$

19. $2(3 - x) \leq 3 + x$

Answer: $[1, \infty)$

20. $x(x - 1) > (x + 2)(x - 2)$

Answer: $(-\infty, 4)$

21. $(x - 1)(x - 2) > (x - 5)(x - 6)$

Answer: $\left(\frac{7}{2}, \infty\right)$

22. $|x - 5| < 7$

Answer: $(-2, 12)$

23. $|x + 8| \leq 12$

Answer: $[-20, 4]$

24. $|x - 4| > 9$

Answer: $(-\infty, -5) \cup (13, \infty)$

25. $|x + 7| \geq 1$

Answer: $(-\infty, -8] \cup [-6, \infty)$

26. $|3x - 5| < 4$

Answer: $\left(\dfrac{1}{3}, 3 \right)$

27. $|2 - 5x| > 3$

Answer: $\left(-\infty, -\dfrac{1}{5} \right) \cup (1, \infty)$

Section P.3

Solving nonlinear inequalities. Solve the inequalities in Exercises 28 through 37.

28. $(x - 1)(x - 2) < 0$

Answer: $(1, 2)$

29. $x(x + 4) \geq 0$

Answer: $(-\infty, -4] \cup [0, \infty)$

30. $x^2 + 15 < 8x$

Answer: $(3, 5)$

31. $x^2 \leq 3x + 4$

Answer: $[-1, 4]$

32. $\dfrac{x - 1}{x - 2} \leq 0$

Answer: $[1, 2)$

33. $x(x - 1)(x - 2) > 0$

Answer: $(0, 1) \cup (2, \infty)$

34. $\dfrac{x}{x^2 - 4} \geq 0$

Answer: $(-2, 0] \cup (2, \infty)$

35. $\dfrac{x^2 - 3x - 10}{x^2 - 4x + 3} \leq 0$

Answer: $[-2, 1) \cup (3, 5]$

36. $x(x - 1)^2(x^2 + 1) \geq 0$

Answer: $[0, \infty)$

37. $\dfrac{x^2 - x}{(x - 2)(x - 3)(x - 4)} < 0$

Answer: $(-\infty, 0) \cup (1, 2) \cup (3, 4)$

FUNCTIONS AND HOW THEY CHANGE

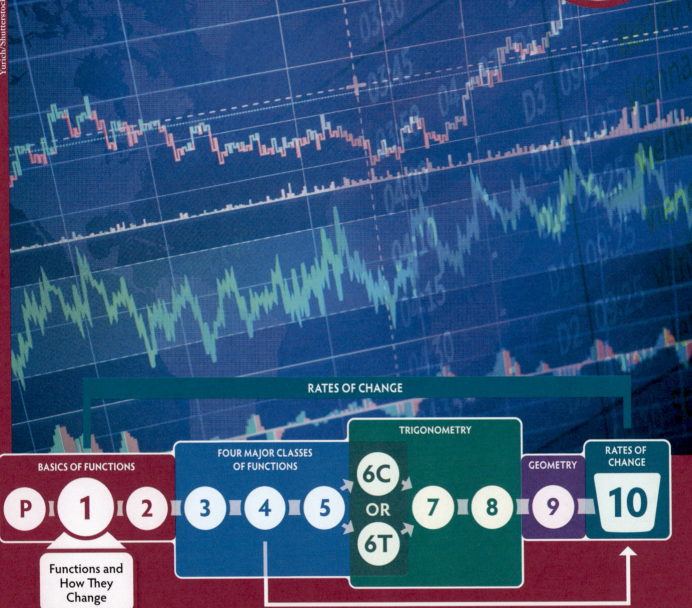

RATES OF CHANGE

BASICS OF FUNCTIONS

P 1 2

Functions and How They Change

FOUR MAJOR CLASSES OF FUNCTIONS

3 4 5

TRIGONOMETRY

6C OR 6T

7 8

GEOMETRY

9

RATES OF CHANGE

10

This chapter presents the basics of functions and how they change, foreshadowing the in-depth exploration of the major classes of functions that will come in later chapters. You will learn about the rate of change of a function, the shape of a graph, and the limiting behavior of functions.

1.1 The Basics of Functions

1.2 Average Rates of Change

1.3 Graphs and Rates of Change

1.4 Limits and End Behavior of Graphs

Yurich/Shutterstock

On a very basic level, calculus is simply the study of functions and how they change. A function is a description of how one quantity depends on another—for example, how our weekly pay depends on the number of hours we work. We can use graphs to visualize these relationships. A graph that may receive more scrutiny than any other is associated with the Dow Jones Industrial Average (**Figure 1.1**), which illustrates how the value of stocks depends on time.

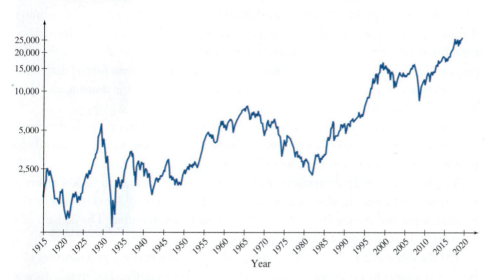

Figure 1.1 Dow-Jones averages

This graph shows a change in relationship (in other words, whether stocks are increasing or decreasing in value). Understanding how the market changes is the key to profitable, short-term investments. But investment counselors advise a focus on the long-term behavior of the market rather than on its daily fluctuations. When the graph is viewed over the long term, it is possible to see that the Dow Jones Industrial Average (an average value of 30 leading U.S. stocks) shows a steady increase over many years. Because of this long-term view, the stock market has been a cornerstone of many long-term investment plans such as retirement accounts. The importance of a dynamic view (contrasted with a static view) of the stock market carries over to the study of virtually all functions and their corresponding graphs. Functions, graphs, rates of change, and long-term behavior are central themes of calculus, of this chapter, and of this book.

1.1 The Basics of Functions

Functions are fundamental tools for modeling real-world phenomena.

1.1 The Basics of Functions

1.2 Average Rates of Change

1.3 Graphs and Rates of Change

1.4 Limits and End Behavior of Graphs

In this section, you will learn to:

1. Evaluate functions given by formulas, words, or tables of values.
2. Interpret functional notation.
3. Identify the domain and range of a given function.
4. Determine whether a given equation defines a function.
5. Determine whether a given graph defines a function.
6. Solve applied problems using the concept of a function.

Definition of a Function

> A function gives a rule for assigning some objects to others.

A **function** is a rule that assigns to each object in one collection *exactly one* object in another collection.

The **domain** of a function is the collection of objects for which this rule of assignment is defined.

The term *function* is common in English usage. For example, we may say that weight is a function of calorie intake to indicate that one's weight depends on one's intake of calories. Similarly, we may say that profit is a function of items sold, meaning that our profit is determined by the number of items sold. Earlier we defined functions informally by saying they describe how one quantity depends on another. To prepare for the use of functions in calculus, we need to give a more formal definition of a **function** as a rule assigning to each object in one collection, the **domain**, an object in another collection.

The rules of assignment we will encounter are usually given by a formula whose domain consists of numbers. However, a function might also assign passwords to people, or prices to TV sets, or desks to dorm rooms.

The formula $y = 2x$ determines y as a function of x because to each number x we assign exactly one number y, namely $y = 2x$. In this setting, we refer to x as the function input and the resulting value $y = 2x$ as the function output. The domain of a function is the collection of all allowed inputs.

On the other hand, the formula $y = \pm\sqrt{x}$ does not determine y as a function of x because it assigns to a number x both $\sqrt{x}$ and $-\sqrt{x}$. (For example, the formula assigns to $x = 4$ both $y = 2$ and $y = -2$. Thus, the input $x = 4$ gives two outputs, $y = 2$ and $y = -2$.) A function cannot assign two outputs to any single input.

A classic representation of a function as the process that produces the output from the input is shown in **Figure 1.2**.

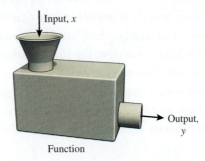

Input, x

Output, y

Function

Figure 1.2 An illustration of a function: A function is the process that produces outputs from given inputs.

Evaluating Functions

> Functions are evaluated by replacing the independent variable by the desired quantity.

A typical example of a function given by a formula is $y = x^2 + 1$, and we say "y is a function of x." If we are given a value for x, the equation tells us how to find the resulting value of y. For example, if $x = 4$, then putting 4 in place of x in the equation gives $y = 4^2 + 1 = 17$. Because the value of y depends on what value is assigned to x, we call y the dependent variable, and we call x the independent variable.

To emphasize the relationship, the notation $f(x) = x^2 + 1$ is commonly used in place of $y = x^2 + 1$. The expression $f(x)$ is referred to as functional notation and is read "f of x." The name of the function is f, and $f(x)$ represents the value of the function given the independent variable x. In the preceding example where $x = 4$ resulted in $y = 17$, we would say that $f(4) = 17$.

This idea holds even for more complicated expressions, as **Table 1.1** shows.

Table 1.1 Finding the Value of a Function

		For $f(x) = x^2 + 1$		
Functional notation	Meaning	Method of calculation	Result	
$f(5)$	f evaluated at 5	Replace x by 5	$f(5) = 5^2 + 1 = 26$	
$f(y)$	f evaluated at y	Replace x by y	$f(y) = y^2 + 1$	
$f(x + 2)$	f evaluated at $x + 2$	Replace x by $x + 2$	$f(x + 2) = (x + 2)^2 + 1 = x^2 + 4x + 5$	
$f(x^2)$	f evaluated at x^2	Replace x by x^2	$f(x^2) = (x^2)^2 + 1 = x^4 + 1$	
$f(\text{Apple})$	f evaluated at Apple	Replace x by Apple	$f(\text{Apple}) = \text{Apple}^2 + 1$	

EXAMPLE 1.1 Finding Function Values

a. If $f(x) = x^2 - x - 1$, find the value of $f(4)$.

b. If $h(s) = \dfrac{s}{s + 1}$, find and simplify $h(s + 1)$.

SOLUTION

a. To find the value of $f(4)$, we replace each occurrence of x by 4 in the formula $f(x) = x^2 - x - 1$:

$$f(x) = x^2 - x - 1 \quad \blacktriangleleft \text{ Use the function definition.}$$

$$f(4) = 4^2 - 4 - 1 \quad \blacktriangleleft \text{ Replace } x \text{ by 4.}$$

$$f(4) = 11 \quad \blacktriangleleft \text{ Simplify.}$$

b. To find $h(s + 1)$, we replace each occurrence of s by $s + 1$ in the formula:

$$h(s) = \frac{s}{s + 1} \quad \blacktriangleleft \text{ Use the function definition.}$$

$$h(s + 1) = \frac{(s + 1)}{(s + 1) + 1} \quad \blacktriangleleft \text{ Replace each occurrence of } s \text{ by } s + 1.$$

$$= \frac{s + 1}{s + 2} \quad \blacktriangleleft \text{ Simplify.}$$

TRY IT YOURSELF 1.1 Brief answers provided at the end of the section.

a. If $g(x) = 4x + \sqrt{x}$, find the value of $g(9)$.

b. If $h(t) = t^2 - 1$, find and simplify $h(t - 1)$.

Functions are often defined by formulas. But graphs, tables of values, and verbal descriptions are also common presentations of functions.

EXAMPLE 1.2 Alternative Function Presentations

a. A technician measures the voltage $V(t)$, in volts, of a battery that is being charged. Here t is the time, in minutes, since the process of charging began. The results are in the table. Find the value of $V(20)$, and explain what it means.

b. If $G(n)$ is the grade you receive on math exam number n, explain the meaning of $G(2)$ in terms of your math exams.

t	V(t)
0	3.4
20	3.9
40	3.9
60	4.0

SOLUTION

a. To find the value of $V(20)$, we locate the row of the table corresponding to $t = 20$ and look in the second column. We see that $V(20) = 3.9$ volts. This says that the voltage measured after 20 minutes is 3.9 volts.

b. The expression $G(2)$ means the grade you receive on the second math exam.

TRY IT YOURSELF 1.2 Brief answers provided at the end of the section.

Use the table in the example to find the value of $V(60)$, and explain what it means.

EXAMPLE 1.3 Piecewise-Defined Functions

Sometimes functions are defined by different formulas for different x-values. These are called piecewise-defined functions. For example, consider the function f defined by

$$f(x) = \begin{cases} 2x & \text{if } x < 0 \\ -5x & \text{if } x \geq 0 \end{cases}$$

Calculate the values of $f(-3)$ and $f(3)$.

SOLUTION

Because $-3 < 0$, to calculate $f(-3)$ we use the part of the function definition that says $f(x) = 2x$. This gives $f(-3) = 2 \times (-3) = -6$. Because $3 \geq 0$, to calculate $f(3)$ we use the part of the definition that says $f(x) = -5x$. Thus, $f(3) = -5 \times 3 = -15$.

TRY IT YOURSELF 1.3 Brief answers provided at the end of the section.

Find the value of $f(-1)$.

Domain and Range

> The domain comprises the allowable inputs for a function, and the range comprises the outputs.

Sometimes the domain of a function, the allowable inputs, is specifically stated. When a function is given by a formula but the domain is not stated, the domain is assumed to be the set of all inputs for which the formula makes sense. That is, the domain is the set of all inputs for which the formula gives a real number as the output.

For example, 3 is in the domain of $f(x) = \dfrac{2+x}{x}$ because if we put 3 into the formula we get a perfectly reasonable answer:

$$f(3) = \frac{2+3}{3} = \frac{5}{3}$$

On the other hand, 0 is not in the domain of f because insertion of 0 into the formula results in division by 0, which is never allowed. In fact, we get a meaningful output as long as we do not use the input $x = 0$. Hence the domain of f is

$$(-\infty, 0) \cup (0, \infty)$$

STEP-BY-STEP STRATEGY: Finding the Domain of a Function

When a function is given by a formula and no domain is specified, the domain is understood to be all real numbers for which the formula makes sense. That is, the domain is the set of all inputs for which the formula gives a real number as the output.

Step 1 Exclude all real numbers for which the formula does not make sense (for example, numbers that would result in division by 0, square roots of negative numbers, or 0 to the 0 power).

Step 2 The domain consists of all real numbers not excluded in the previous step.

EXAMPLE 1.4 Finding the Domain

In each part, find the domain of the given function.

a. $g(t) = 3t^2 + 2$

b. $f(x) = \dfrac{\sqrt{x-3}}{x-7}$

c. $k(t) = \dfrac{t^2}{9t}$

SOLUTION

a. The formula $g(t) = 3t^2 + 2$ makes sense for all values of t. Thus, the domain of g is the set $\mathbb{R}$ of all real numbers.

b. We look for values of x for which the formula does not make sense and exclude them. We note first the expression $x - 7$ in the denominator. The denominator of a fraction cannot be 0, so we must exclude $x = 7$ from the domain. Now look at the numerator. We cannot take the square root of a negative number, so we must have $x - 3 \geq 0$ or $x \geq 3$. We conclude that the formula makes sense for all real numbers greater than or equal to 3 but not equal to 7. Using interval notation, we write the domain as

$$[3, 7) \cup (7, \infty)$$

c. The formula does not make sense when the denominator is 0, so we exclude $t = 0$ from the domain. There are no other restrictions, so the domain is

$$(-\infty, 0) \cup (0, \infty)$$

TRY IT YOURSELF 1.4 Brief answers provided at the end of the section.

Find the domain of the function $h(x) = \sqrt{x-1}$.

EXTEND YOUR REACH

a. Here is a "solution" of part c of the example that contains a significant error. Cancel the common t from both numerator and denominator to obtain $\dfrac{t^2}{9t} = \dfrac{t}{9}$. Because this last expression makes sense for all real numbers, we conclude that the domain of $k(t)$ is the set of all real numbers. Explain what is wrong with this "solution."

b. Bearing in mind that the expression $\dfrac{t^2}{9t}$ is not defined when $t = 0$, sketch the graph of the equation $y = \dfrac{t^2}{9t}$.

c. Does the graph you made in part b suggest a reasonable way to extend the definition of $k(t)$ so that its domain is the set of all real numbers?

The **range** of a function is the set of all function outputs.

Thus far we have focused on the inputs for a function. Now we focus on the function outputs, which constitute the **range**. These are the values taken by the function as the input varies through the domain.

For example, 8 is in the range of $g(x) = x^2 + 7$ because $g(1) = 8$. On the other hand, 5 is not in the range of g: There is no x so that $x^2 + 7 = 5$ (because $x^2 \geq 0$ for all real x).[1]

EXAMPLE 1.5 Finding the Domain and Range

In each part, find the domain and range of the given function.

a. $g(s) = s^2 + 5$

b. The constant function $h(x) = 3$

SOLUTION

a. The formula makes sense for all choices of s, so the domain is $\mathbb{R}$. The range is the set of all function outputs. Because $s^2 \geq 0$, the formula $g(s) = s^2 + 5$ gives real numbers that are 5 or larger. In fact, every real number that is 5 or larger is a value of the function. We conclude that the range of g is $[5, \infty)$.

b. Note that h is a function whose value $h(x)$ is the number 3, regardless of the value of x. So $h(0) = 3$, $h(3) = 3$, and $h(x^2 + 1) = 3$ for every x. Because the formula makes sense for every value of x, the domain is $\mathbb{R}$. The only output is the single number 3, so the range is $\{3\}$.

TRY IT YOURSELF 1.5 Brief answers provided at the end of the section.

Find the range of the function $h(x) = \sqrt{x-1}$.

---

[1] The equation does have a solution if we allow complex numbers, but in general, we will restrict our attention to real numbers unless we state otherwise.

Equations and Functions

> An equation may determine a function.

Some equations define functions, and some do not. For example, the equation $x + y^2 - 2y = 0$ does not determine y as a function of x. To understand this, consider $x = 0$. Observe that both $y = 0$ and $y = 2$ satisfy the equation when $x = 0$. So the formula assigns to $x = 0$ more than one y-value. On the other hand, the equation $x + y = 7$ does define y as a function of x, namely the function $y = 7 - x$.

STEP-BY-STEP STRATEGY: Deciding Whether an Equation Determines a Function

To determine whether an equation in x and y determines y as a function of x, proceed as follows:

Step 1 Solve the equation for y.

Step 2 The equation determines y as a function of x provided the equation has at most one solution for each value of x.

EXAMPLE 1.6 Deciding Whether an Equation Defines a Function

Decide whether the following equations define y as a function of x. Justify your answer.

a. $x^2 + y^2 = 1$

b. $y + x^2 = 3 - y$

SOLUTION

a. We solve the equation for y:

$$x^2 + y^2 = 1$$

$$y^2 = 1 - x^2 \qquad \text{Move } x^2 \text{ right.}$$

$$y = \pm\sqrt{1 - x^2} \qquad \text{Take square roots.}$$

We can see here that for some values of x there are two solutions; hence, some inputs have more than one output.

An alternative approach is to let $x = 0$ in the equation $x^2 + y^2 = 1$. We get $y^2 = 1$, which has two solutions, $y = 1$ and $y = -1$. Because this x-value does not result in a unique y-value, the equation does not define y as a function of x.

b. We can solve the equation for y:

$$y + x^2 = 3 - y$$

$$2y = 3 - x^2 \qquad \text{Add } y \text{ and subtract } x^2.$$

$$y = \frac{3 - x^2}{2} \qquad \text{Divide by 2.}$$

This equation *does* define y as a function of x because it assigns to each value of x exactly one value of y, namely $y = \dfrac{3 - x^2}{2}$.

TRY IT YOURSELF 1.6   Brief answers provided at the end of the section.

Does the equation $(3y - x)(2y - x) = 0$ define y as a function of x? Justify your answer.

EXTEND YOUR REACH   Does the equation $x + |y| = 1$ define y as a function of x? Does it define x as a function of y?

Graphs and Functions

> Graphs are visual representations of functions.

The **graph of a function** f is the graph of the equation $y = f(x)$.

The **graph of a function** is the most common and important example of a graph of an equation. For example, the graph of $f(x) = x^3$ is just the graph of the equation $y = x^3$ shown in **Figure 1.3**.

Some graphs of equations define functions, but some do not. For example, the graph in **Figure 1.4** defines y as a function of x. Note the dashed vertical line in the figure. A graph defines a function precisely when each vertical line hits the graph no more than once. This is because the number of times the vertical line through x crosses the graph determines the number of y-values assigned to x. This is a useful visual test for deciding whether a graph represents a function.

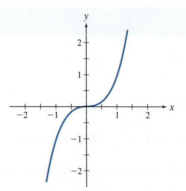

Figure 1.3 The graph of the function $f(x) = x^3$

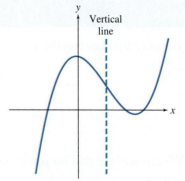

Figure 1.4 The vertical line test: This graph defines a function because each vertical line intersects the graph exactly once (if at all).

LAWS OF MATHEMATICS: The Vertical Line Test

A graph represents a function provided that each vertical line intersects the graph no more than once. If there is *any* vertical line that intersects the graph more than once, the graph does not represent a function.

EXAMPLE 1.7 Graphs and Functions

Determine whether the graph in **Figure 1.5** defines y as a function of x.

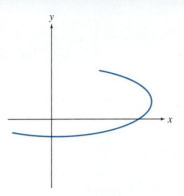

Figure 1.5 Graph for Example 1.7

SOLUTION

The dashed vertical line shown in **Figure 1.6** intersects the graph twice, so the graph fails the vertical line test. It does not define y as a function of x. It is worth noting that some vertical lines do intersect the graph in Figure 1.5 exactly once, but this is not sufficient for a function to be defined. For a graph to pass the vertical line test, *every* vertical line must intersect the graph at most once.

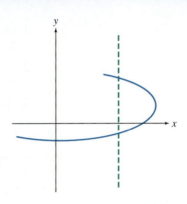

Figure 1.6 The vertical line test: This graph fails the vertical line test.

TRY IT YOURSELF 1.7 Brief answers provided at the end of the section.

Does the graph in **Figure 1.7** define y as a function of x?

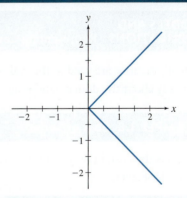

DF Figure 1.7 Graph for Try It Yourself 1.7

EXTEND YOUR REACH

a. Does the graph in **Figure 1.8** define y as a function of x? Note that some vertical lines will not meet the graph at all.

b. Start with the graph of $y = x$. For every rational number x, replace the point (x, x) on the graph by $(x, -x)$. Does this graph pass the vertical line test? Can you think of a way to draw this graph?

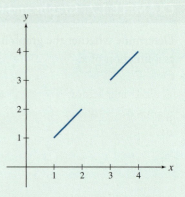

Figure 1.8 Applying the vertical line test

CONCEPTS TO REMEMBER: The Basics of Functions

1. A function consists of a domain and a rule that assigns to each element of the domain *exactly* one element of some collection.
2. A function $y = f(x)$ may be defined by a formula, words, a graph, or a table of values.
3. For a function $f(x)$ given by a formula, the domain is assumed to be the set of all x-values for which the formula makes sense, unless specifically stated otherwise.
4. The range of a function is the set of all values taken by the function. For a function given by a formula $y = f(x)$, the range is the set of all values of c for which the equation $f(x) = c$ has a solution x in the domain of f.
5. A graph in terms of x and y determines y as a function of x precisely when each vertical line does not intersect the graph more than once.

MODELS AND APPLICATIONS   Representing Costs of Everyday Purchases

When functions are used in the real world, it is important to use functional notation properly and think carefully about the domain and range.

EXAMPLE 1.8 Cost of Sodas

If $C(s)$ is the cost, in dollars, of purchasing s sodas, use functional notation to indicate the cost of purchasing a six-pack of sodas.

SOLUTION

In functional notation, the cost, in dollars, of purchasing six sodas is $C(6)$. It is important to emphasize that the expression $C(6)$ has a perfectly valid meaning, even if we cannot calculate its value.

TRY IT YOURSELF 1.8   Brief answers provided at the end of the section.

If $P(f)$ is the number of pints of paint needed to paint f square feet, explain the meaning of $P(21)$ in terms of pints of paint.

When a function is given by a formula, it is understood that the domain and range are determined strictly by the formula. But in applications, additional constraints may apply.

EXAMPLE 1.9 Domain and Range in an Application

A child sells lemonade in a neighborhood of 200 people. Her total income I depends on the number n of glasses of lemonade sold. The relationship is $I = 0.50n$ dollars. Assuming that the maximum possible number of glasses sold is the population of the neighborhood, determine the domain and range of I.

SOLUTION

She can sell at most 200 glasses of lemonade. Thus, the domain is $[0, 200]$. Her total income varies from 0 dollars to $0.50 \times 200 = 100$ dollars. That gives a range, in dollars, of $[0, 100]$.

TRY IT YOURSELF 1.9 Brief answers provided at the end of the section.

What are the domain and range if she is guaranteed the sale of a glass to each of her parents and both of her brothers?

TRY IT YOURSELF ANSWERS

1.1 **a.** $g(9) = 39$
 b. $h(t - 1) = (t - 1)^2 - 1 = t^2 - 2t$

1.2 $V(60) = 4.0$ volts. The voltage measured after 60 minutes is 4.0 volts.

1.3 $f(-1) = -2$

1.4 The domain is $[1, \infty)$.

1.5 The range is $[0, \infty)$.

1.6 The equation does not define y as a function of x because it assigns to each value of x both $y = \dfrac{x}{2}$ and $y = \dfrac{x}{3}$.

1.7 The graph does not define y as a function of x because it fails the vertical line test: Some vertical lines intersect the graph more than once.

1.8 The expression $P(21)$ is the number of pints of paint needed to paint 21 square feet.

1.9 Domain $[4, 200]$. Range $[2, 100]$.

EXERCISE SET 1.1

CHECK YOUR UNDERSTANDING

1. How many outputs can a function assign to a given input that is in the domain?

2. **True or false:** Functional notation such as $f(7)$ makes sense only if we can calculate its value.

3. What test is used to determine whether a graph is the graph of a function?

4. **True or false:** All functions are defined by formulas.

5. To say that t is in the domain of a function f means _____.

6. The graph of a function f is the graph of an equation. Which equation?

7. **True or false:** If a graph passes the vertical line test, then it is the graph of a function.

8. A certain equation has the same solution for y if $x = 4$ or $x = 9$. Does this mean the equation does not define y as a function of x?

SKILL BUILDING

Finding the domain. In Exercises 9 through 18, state the domain of the given function.

9. $f(x) = \dfrac{x}{x-1}$

10. $f(x) = \dfrac{1}{\sqrt{x}}$

11. $f(x) = x^2 + \dfrac{x}{3}$

12. $f(x) = \dfrac{1}{x} + \dfrac{1}{x+1}$

13. $h(x) = \dfrac{1}{x} + \dfrac{1}{x+1}, x \geq -\dfrac{1}{2}$

14. $f(x) = \sqrt{x-7}$

15. $f(x) = 2^{1-x} + 3^{x-1}$

16. $y = \dfrac{x}{x^2 - 3x + 2}$

17. $y = \dfrac{x^2}{x}$

18. $y = \dfrac{1}{x^3 - 3x^2 + 2x}$

Calculating function values. In Exercises 19 through 32, calculate the indicated function value. Simplify your answer.

19. $f(x) = \dfrac{x}{x+1}$. Calculate $f(2)$.

20. $f(x) = x^2 + x^{-2}$. Calculate $f(-2)$.

21. $g(t) = \sqrt{t^2 - 5}$. Find $g(3)$.

22. $h(x) = (x-1)^2$. Find $h(x+1)$.

23. $k(\omega) = \sqrt{\omega}$. Find $k(\omega^2)$.

24. $x(t) = \dfrac{t}{t+1}$. Find $x\left(\dfrac{1}{t-1}\right)$, $t \neq 0$.

25. $h(y) = \dfrac{1}{\sqrt{y}}$. Find $h\left(\dfrac{1}{z^2}\right)$, $z \neq 0$.

26. $\sigma(s) = (s-2)(s-4) + 1$. Find $\sigma(s+3)$.

27. $f(x) = x - \dfrac{1}{x}$. Calculate $f\left(\dfrac{1}{x}\right)$, $x \neq 0$.

28. $f(x) = x^2$. Calculate $f(x^2)$.

29. $f(x) = \sqrt{x+1}$. Calculate $f(x^2 + 2x)$, assuming that x is positive.

30. $f(x) = x^2 - h^2$. Calculate $f(x+h)$.

31. $f(x) = \dfrac{x}{x+1}$. Calculate $f(f(x))$, $x \neq -1$.

32. $f(x) = \dfrac{x+1}{x-2}$. Calculate $f\left(\dfrac{2x+1}{x-1}\right)$, $x \neq 1$.

Difference quotients. Exercises 33 through 39 involve difference quotients, which are fractions involving function values. They arise in the calculation of derivatives in calculus. Calculate the quotient and simplify.

33. $f(x) = 2x + 1$. Simplify $\dfrac{f(x) - f(a)}{x - a}$, $x \neq a$.

34. $f(x) = 2x + 1$. Simplify $\dfrac{f(x+h) - f(x)}{h}$, $h \neq 0$.

35. $f(x) = x^2$. Simplify $\dfrac{f(x+h) - f(x)}{h}$, $h \neq 0$.

36. $f(x) = x^2$. Simplify $\dfrac{f(x) - f(a)}{x - a}$, $x \neq a$.

37. $f(x) = \sqrt{x}$. Show that $\dfrac{f(x) - f(a)}{x - a} = \dfrac{1}{f(x) + f(a)}$ if $x \neq a$ and both x and a are positive.

38. $f(x) = \sqrt{x}$. Show that $\dfrac{f(x+h) - f(x)}{h} = \dfrac{1}{\sqrt{x+h} + \sqrt{h}}$ if both x and h are positive.

39. $f(x) = \dfrac{1}{x}$. Simplify $\dfrac{f(x) - f(y)}{x - y}$ if $x \neq 0$, $y \neq 0$, and $x \neq y$.

Graphs and functions. In Exercises 40 through 45, determine whether the given graph is the graph of a function.

40.

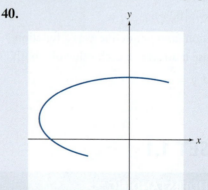

41.

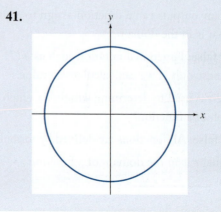

42.

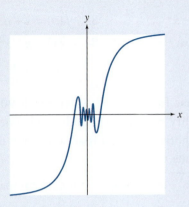

43.

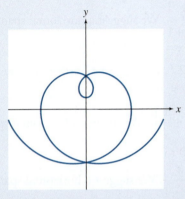

44.

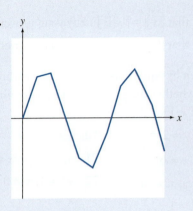

45.

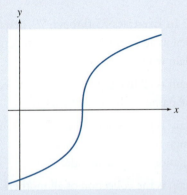

PROBLEMS

Range of a function. In Exercises 46 through 54, find the range of the given function.

46. $f(x) = 7 + 3x$

47. $f(x) = \sqrt{x}$

48. $f(x) = \sqrt{x+1}$

49. $f(x) = 1 + \dfrac{1}{x}$

50. $f(x) = x^3 + 1$

51. $f(x) = \dfrac{x^2}{x}$

52. $f(x) = \dfrac{x+1}{x}$

53. $f(x) = \sqrt{x^2 + 1}$

54. $f(x) = (x+1)^2$

Piecewise-defined functions.

55. Define $g(x) = \begin{cases} x^3 + 1 & \text{if } x \le 2 \\ x^2 & \text{if } x > 2. \end{cases}$

Calculate $g(7)$, $g(1)$, and $g(2)$.

56. Bill says that g defined by

$$g(x) = \begin{cases} x+3 & \text{if } x \le 0 \\ x+1 & \text{if } x \ge 0 \end{cases}$$

is a piecewise-defined function, but Alice says it is not. Who is right, and why?

57. Define $f(t) = \begin{cases} \dfrac{t^2 - 1}{t - 1} & \text{if } t \ne 1 \\ 2 & \text{if } t = 1. \end{cases}$

Find $f(1)$ and $f(2)$.

58. Define $h(x) = \begin{cases} x - 1 & \text{if } x < 2 \\ 9 & \text{if } x = 2 \\ x + 1 & \text{if } x > 2. \end{cases}$

Find $h(1)$, $h(2)$, and $h(3)$.

59. Define $f(x) = \begin{cases} x+1 & \text{if } x \le 5 \\ x+6 & \text{if } x > 5. \end{cases}$

Calculate $f(x+1)$.

The greatest integer function. The greatest integer function $[x]$ is the integer n such that $n \le x < n+1$. It gives the largest integer less than or equal to x. For example, $[3.1] = 3$ and $[-3.1] = -4$.

60. Calculate $[5.8]$.

61. Calculate $[\sqrt{72}]$.

62. Calculate $[-\sqrt{44}]$.

63. Calculate $[29]$.

64. Is it true that $\lfloor |x| \rfloor = |\lfloor x \rfloor|$? Justify your answer.

65. Is it true that $\lfloor \lfloor x \rfloor \rfloor = \lfloor x \rfloor$? Justify your answer.

Functions given by equations. In Exercises 66 through 72, determine whether the given equation determines y as a function of x. Explain your answer.

66. $2x^2 y + 5y = 2x^2$

67. $\sqrt{xy} = 1,\ x \neq 0$

68. $x - y = xy,\ x \neq -1$

69. $y^2 - 2x^2 = 1$

70. $x + y^2 = 4$

71. $xy + 1 = x,\ x \neq 0$

72. $x^y = 1,\ x \neq 0$ and $x \neq 1$

Is it a function? In Exercises 73 through 79, determine whether the given assignment determines a function. If it does not, explain why.

73. The domain is all living people. Assign to each person the names of his or her uncles.

74. The domain is the decades from 1790 through 2020. Assign to each decade the average U.S. population over the decade.

75. The domain is the collection of all mothers. Assign to each mother her children.

76. The domain is the collection of all living children. Assign to each child its birth mother.

77. The domain is the set of all real numbers. The range is the set of all real numbers. Assign to each number all numbers no farther away than a distance of 1 from the given number.

78. The domain consists of all allowable weights, in ounces, of first-class domestic letters. The range is a collection of prices. The assignment $p(x)$ is the current postage required to mail a first-class domestic letter that weighs x ounces.

79. The domain is the collection of all U.S. presidents. The range is all years from 1790 to the present. Assign to each president the years served.

Graphing functions. In Exercises 80 through 84, graph the given function. You may want to consult a table of values.

80. $f(x) = 1 - x^2$

81. $f(x) = \sqrt{x}$

82. $f(x) = \dfrac{1}{x}$

83. $f(x) = x + \sqrt{x}$

84. $f(x) = x - x^2$

Making graphs. In Exercises 85 through 90, use a graphing utility to graph the indicated function. The suggested horizontal span is an indication of the x-values to include.

85. $g(x) = \sqrt{x^3 - x}$. We suggest a horizontal span from $x = -1$ to $x = 2$.

86. $h(x) = \dfrac{x}{x^2 + 1}$. We suggest a horizontal span from $x = -3$ to $x = 3$.

87. $f(t) = t^3 2^{-t}$. We suggest a horizontal span from $t = -1$ to $t = 10$.

88. $f(x) = \left(1 + \dfrac{1}{x}\right)^x$. We suggest a horizontal span from $x = 0$ to $x = 10$.

89. $f(x) = \dfrac{x}{x^2 - 1}$. We suggest a horizontal span from $x = -5$ to $x = 5$.

90. The π function. The function $\pi(x)$ gives the number of primes less than or equal to x. Plot the graph of $\pi(x)$ on a span of 0 to 100. (Note that you will need a computer algebra system such as Maple, Mathematica, or MATLAB to make this graph.) When would we expect $\pi(x)$ to be different from $\pi(x + 1)$?

Using a graph to find range. In Exercises 91 through 94, use a graphing utility to plot the graph of the given function. Use the graph to find the range of the function. Pay attention to the given restrictions on the domain.

91. $f(x) = \dfrac{2x}{x^2 + 1},\quad -5 \leq x \leq 5$

92. $f(x) = 3^x - 2^x,\quad 0 \leq x \leq 2$

93. $f(x) = x(1 + 2^{-x}),\quad 0 \leq x \leq 2$

94. $\dfrac{7 + 2x^2}{11 + x^2},\quad -2 \leq x \leq 2$

MODELS AND APPLICATIONS

Determining domain and range from practical constraints.

95. A rock falling. If we drop a rock from 256 feet above the surface of Earth and ignore air resistance, after t seconds the rock will fall $D(t) = 16t^2$ feet.

This formula is valid until the rock strikes the ground.

a. The rock strikes the ground when $D(t) = 256$ feet. What is the value of t when the rock strikes the ground?

b. What is the domain of this function?

c. What is the range of this function?

96. **Projectile path.** A projectile is fired from the origin and follows the path of the parabola $y = \dfrac{x(400 - x)}{400}$ until it strikes the ground. (Both x and y are measured in feet.) The graph of $y = \dfrac{x(400 - x)}{400}$ is shown in **Figure 1.9**. Let $H(x)$ denote the height, in feet, of the projectile x feet downrange. Find the domain and range of H.

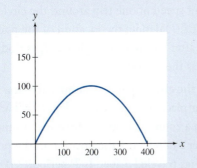

Figure 1.9 Graph of $y = x(400 - x)/400$

97. **A weekly newspaper.** A weekly newspaper is sold locally for $1.25 per copy in a town with a population of 13,000. Let $R(n)$ be the amount, in dollars, received from the sale of n newspapers this week. Practically speaking, what is the domain of R? (Strictly speaking, n takes values in the integers, but you may assume that n can take other values.) What is the range of R in the best possible case for the paper?

98. **An investment.** A man invests $5000 in an account that draws interest. He leaves the money in the account for 25 years and then closes the account. The balance B of the account after t years is given by

$$B(t) = 5000 \times 1.06^t \text{ dollars}$$

Find the domain and range of B.

99. **Another investment.** A woman invests in an account that draws interest. Let $B(t)$ denote the balance, in dollars, of the account after t years. Explain in terms of the balance of the account what $B(5)$ means.

100. **Cooking a yam.** A yam is placed in the oven to bake. Let $Y(t)$ denote the temperature, in degrees Fahrenheit, of the yam after t minutes in the oven. Explain what $Y(3)$ means in terms of the temperature of the yam.

101. **A roof line.** The automobile shop shown in **Figure 1.10** is 50 feet wide. If we take $t = 0$ to be the location of the west wall, then the height of the roof at a distance of t horizontal feet from the west wall is given by $H = 11 + 0.5t$ feet. This formula is valid until we reach the peak, 25 feet from the west wall. If instead we think of $t = 0$ as the location of the peak, then the roof from the peak to east wall follows $H = 23.5 - 0.5t$ feet, where t is the horizontal distance in feet from the peak. This is valid up to the east wall 25 feet from the peak. Using $T = 0$ as the horizontal distance in feet from the west wall, find a piecewise-defined function that gives the height of the roof in terms of T.

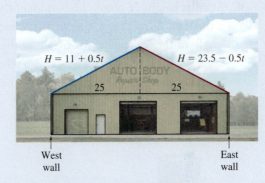

Figure 1.10 An automobile shop

102. **A rising river.** The measurement of snowfall in the mountains allows city planners to model the projected depth of a river running through the city over the next year. The model is

$$D = \frac{10(t - 175)^2 + 75{,}000}{2500 + (t - 175)^2}$$

where D is the depth in feet and t is the time in days since January 1.

a. Plot the graph of river depth over a 1-year period.

b. Flood danger occurs when the river reaches a depth of 33 feet. Does the city need to prepare for a flood this year?

CHALLENGE EXERCISES FOR INDIVIDUALS OR GROUPS

103. A leaky can. Suppose that a cylindrical can is 10 inches tall and full of tea. It springs a leak in the side, and the tea begins to stream out. The depth H, in inches, of the tea remaining in the can is a function of the distance D in inches (measured from the base of the can) at which the stream of tea strikes the ground. This situation is shown in **Figure 1.11.** The relationship is given by

$$H(D) = \frac{D^2}{4} + 1.$$ This formula is valid only until

the height of the tea has decreased to the height of the leak.

   a. Do larger values of D correspond to larger or smaller values of H?

   b. What is the value of D when $H = 10$?

   c. What is the domain of H? *Suggestion:* Your answer from part b should be helpful.

   d. At what height H above the bottom of the can is the leak? *Suggestion:* Observe that, when the height of the tea reaches the height of the leak, the tea will stop flowing. Thus, $D = 0$ when the height of the tea reaches the height of the leak.

   e. What is range of H?

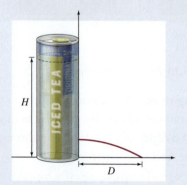

Figure 1.11 A leaking can of tea

Periodic functions. Exercises 104 through 111 are concerned with periodic functions. A function f is said to be periodic if there is a number p such that $f(x + p) = f(x)$ for all x in the domain of f. The period of such a function is the smallest positive value of p that makes the preceding formula valid (assuming such a value exists). Intuitively, a periodic function is one that repeats in a predictable fashion. The period is the smallest time that it takes to repeat. For example, the display on a digital clock representing minutes starts

at 0, increases to 59, and then starts over at 0. If $M(t)$ is the minute part of the time shown by a digital clock after t minutes starting at 0, then $M(t + 60) = M(t)$, and M is periodic with period 60 minutes.

104. How long until my birthday? For this problem, ignore leap years and assume that each year is exactly 365 days long. For a given day of the year, let B be the number of days left until your next birthday. Is B periodic? If so, what is its period?

105. Around a circle. A point moves around a circle at a constant speed of one trip around the circle each hour. Let $L(t)$ denote the location of the point t minutes after the motion begins. Is L periodic? If so, what is the period of L?

106. Around the circle again. A point starts at a particular place on a circle of radius 1 foot and moves in a counterclockwise direction around the circle. Let $P(d)$ denote the position of the point when it has traveled d feet. Is P periodic? If so, what is its period?

107. Around the circle yet again. A point begins at a particular location on a circle. It moves counterclockwise around the circle. Its speed increases by 1 unit each time it completes a revolution of the circle. Let $L(t)$ be the location of the point t minutes after motion began. Explain why L is not periodic.

108. More repetition. Assume that f is a periodic function with period p. How does $f(x)$ compare with $f(x + 2p)$? How about $f(x + np)$ for a positive integer n?

109. Periodic or not. Suppose f is periodic with period p. Which of the following functions are periodic? If they are periodic, what is the period?

   a. $g(x) = f(x) + 1$ **d.** $g(x) = f(2x)$

   b. $g(x) = -f(x)$ **e.** $g(x) = f(x - 1)$

   c. $g(x) = 2f(x)$

110. Division. For a positive integer n, let $R(n)$ denote the remainder when n is divided by 17. Is R periodic? If so, what is its period?

111. A variant of the greatest integer function. The greatest integer function $[x]$ is the integer n such that $n \le x < n + 1$. Is the function $f(x) = x - [x]$ periodic? If so, what is its period?

112. From Section P.1: Find the equation of the line that passes through $(3, -2)$ and has slope 3.

Answer: $y = 3x - 11$

113. From Section P.1: Find the center and radius of the circle with the equation $x^2 + y^2 = 6 + 4x + 2y$.

Answer: Center $(2, 1)$ and radius $\sqrt{11}$

114. From Section P.2: Solve the inequality $3x - 1 > x + 5$.

Answer: $(3, \infty)$

115. From Section P.2: Solve the inequality $|x - 8| < 5$.

Answer: $(3, 13)$

116. From Section P.3: Solve the inequality $x^2 + 4x - 12 > 0$.

Answer: $(-\infty, -6) \cup (2, \infty)$

117. From Section P.3: Solve the inequality $x^3 < x^2$.

Answer: $(-\infty, 0) \cup (0, 1)$

1.2 Average Rates of Change

The average rate of change is one way to measure change over an interval.

1.1 The Basics of Functions

1.2 Average Rates of Change

1.3 Graphs and Rates of Change

1.4 Limits and End Behavior of Graphs

In this section, you will learn to:

1. Calculate the average rate of change of a function on an interval.
2. Interpret the average rate of change in applied settings.
3. Use the average rate of change to estimate function values in applied settings.

If a child grew 10 inches over a period of 10 years, we would immediately conclude that the child had grown by an average of 1 inch per year. Of course, that does not mean that growth was exactly 1 inch each year—that just isn't how people grow. But it is a reasonable way of describing the growth of the child based on limited information. A growth of 1 inch per year is an average rate of change, and it is an entirely intuitive notion.

It is also useful. How much growth would be expected next year? Well, 1 inch is a reasonable guess, and the best guess you can make, given the information you have. You don't expect your guess to be exactly right, and indeed, you know that the child will stop growing someday.

The average rate of change for any function works just like the average rate of change for the growing child. Calculating rates of change for functions is fundamental in both mathematics and its applications to real-world problems.

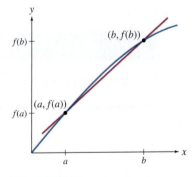

Figure 1.12 The secant line

Secant Lines

A secant line can be used to measure the rate of change of a function.

We can associate a **secant line** to every pair of points on the graph of a function. **Figure 1.12** shows the secant line in relation to the graph.

The **average rate of change** of a function on an interval is the slope of the associated secant line.

The **secant line** for a function f for the interval $[a, b]$ is the line through the points $(a, f(a))$ and $(b, f(b))$.

The **average rate of change** of a function $y = f(x)$ on the interval $[a, b]$ is the slope of the secant line for $[a, b]$.

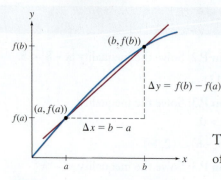

DF **Figure 1.13** The average rate of change

The formula for the average rate of change is

$$\text{Average rate of change} = \frac{\Delta y}{\Delta x}$$

$$= \frac{f(b) - f(a)}{b - a}$$

This is also called the average rate of change from $x = a$ to $x = b$. The average rate of change is the change in y over the change in x. **Figure 1.13** illustrates the numerator and denominator in this definition. The expression $\dfrac{f(b) - f(a)}{b - a}$ is called a difference quotient.

EXAMPLE 1.10 Calculating the Average Rate of Change

Calculate the average rate of change of $f(x) = x^2$ on the interval $[2, 4]$.

SOLUTION

Because $f(2) = 2^2 = 4$ and $f(4) = 4^2 = 16$, we have

$$\text{Average rate of change} = \frac{f(b) - f(a)}{b - a}$$

$$= \frac{f(4) - f(2)}{4 - 2}$$

$$= \frac{16 - 4}{4 - 2}$$

$$= 6$$

This average rate of change is the slope of the secant line for the interval $[2, 4]$. The graph of f along with the secant line is shown in **Figure 1.14**.

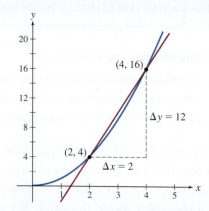

Figure 1.14 A secant line: The slope of the secant line is the average rate of change.

TRY IT YOURSELF 1.10 Brief answers provided at the end of the section.

Calculate the average rate of change of $f(x) = \dfrac{1}{x}$ from $x = 1$ to $x = 3$.

The next example shows how the average rate of change depends on the length of the interval. This is a key step in calculating the derivative in calculus.

EXAMPLE 1.11 More on Calculating Average Rates of Change

Let $f(x) = \sqrt{x}$. Show that if both a and h are positive, then the average rate of change of f on the interval $[a, a + h]$ is $\dfrac{1}{\sqrt{a + h} + \sqrt{a}}$.

SOLUTION

We calculate the average rate of change as follows:

$$\text{Average rate of change} = \frac{f(a+h) - f(a)}{(a+h) - a}$$

$$= \frac{\sqrt{a+h} - \sqrt{a}}{(a+h) - a}$$

$$= \frac{\sqrt{a+h} - \sqrt{a}}{h}$$

To change this expression to the required form, we rationalize the numerator of our fraction by multiplying the numerator and denominator by $\sqrt{a+h} + \sqrt{a}$, the conjugate of the numerator. Then we use the fact (based on factoring the difference of squares) that

$$(\sqrt{A} - \sqrt{B})(\sqrt{A} + \sqrt{B}) = A - B$$

We find

$$\frac{\sqrt{a+h} - \sqrt{a}}{h} = \frac{(\sqrt{a+h} - \sqrt{a})(\sqrt{a+h} + \sqrt{a})}{h(\sqrt{a+h} + \sqrt{a})}$$ ◀ Multiply top and bottom by conjugate.

$$= \frac{(a+h) - a}{h(\sqrt{a+h} + \sqrt{a})}$$ ◀ Use $(\sqrt{A} - \sqrt{B})(\sqrt{A} + \sqrt{B}) = A - B$.

$$= \frac{\cancel{h}}{\cancel{h}(\sqrt{a+h} + \sqrt{a})}$$ ◀ Cancel common factors.

$$= \frac{1}{\sqrt{a+h} + \sqrt{a}}$$ ◀ Simplify.

TRY IT YOURSELF 1.11   Brief answers provided at the end of the section.

Show that if both a and b are nonzero, the average rate of change of $g(x) = \dfrac{1}{x^2}$ on the interval $[a, b]$ is $-\dfrac{a+b}{a^2 b^2}$.

EXTEND YOUR REACH

a.   For the function $y = x^2$, calculate the average rate of change on each of the following intervals. In each case, give your answer correct to three decimal places.

- From $x = 1$ to $x = 2$
- From $x = 1$ to $x = 1.5$
- From $x = 1$ to $x = 1.25$
- From $x = 1$ to $x = 1.1$

b.   As the interval gets shorter, does it appear that the average rate of change is approaching a specific number? If so, identify that number.

MODELS AND APPLICATIONS   Ants, Ebola, and Temperature

The idea of an average rate of change is common in daily life. For example, if you drive 30 miles in half an hour, your average speed is

$$\frac{\text{Change in distance}}{\text{Change in time}} = \frac{30}{\frac{1}{2}} = 60 \text{ miles per hour}$$

In this context, the average rate of change is just your average speed. There are always units associated with the average rate of change: the units of the function divided by the units of the independent variable. In this case, the units are miles per hour. Units are always important, and getting the units right can often facilitate the calculation and use of the average rate of change.

EXAMPLE 1.12   Units and Interpreting Average Rates of Change: An Ant Colony

An ant colony is monitored in a laboratory. It is found that the population N of the colony after t weeks is given by

$$N(t) = 200 \times 1.2^t \text{ ants}$$

a. Calculate the average rate of change of the population from week 3 to week 5. Round your answer to the nearest whole number, and be sure to include proper units.

b. Explain the meaning of the calculation you made in part a in terms of the ant colony.

SOLUTION

a. We note that formally the units for the average rate of change are the units of the function divided by the units of the independent variable, so in this case, the units are ants per week. But the formality is really not needed because the average rate of change is the change in population divided by the number of weeks — that is, ants per week.

We want to find the average rate of change of N on the interval $[3, 5]$:

$$\text{Average rate of change} = \frac{N(5) - N(3)}{5 - 3}$$

$$= \frac{(200 \times 1.2^5) - (200 \times 1.2^3)}{2}$$

$$\approx 76 \text{ ants per week}$$

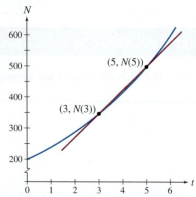

Figure 1.15 Graph of ant population and secant line

b. The average rate of change of 76 ants per week is the average increase in the ant population per week over the 2-week period. In other words, during this period, the population grew by an average of 76 ants each week. The graph of the ant population along with the secant line we used to calculate the average rate of change is shown in **Figure 1.15**.

TRY IT YOURSELF 1.12   Brief answers provided at the end of the section.

Find the average rate of change per week of the ant population from week 3 to week 7. Round your answer to the nearest whole number, and be sure to include proper units.

EXAMPLE 1.13   Using the Average Rate of Change from a Table: Ebola

The accompanying table shows the cumulative number of cases in West Africa of Ebola, a deadly virus, reported by the World Health Organization for the given dates in 2015.

Day	October 1	October 16	October 22	October 26
Cumulative cases	28,408	28,468	28,504	28,528

Let $N(d)$ denote the cumulative number of Ebola cases reported d days after October 1. The plot of N is shown in **Figure 1.16**. Note that the vertical axis begins at 28,000.

a. Calculate the average rate of change in the cumulative number of cases from October 1 to October 16. Be sure to give proper units.

b. Explain the meaning of the number you calculated in part a in terms of the number of new cases of Ebola.

c. Use the average rate of change to estimate the cumulative number of cases reported by October 9.

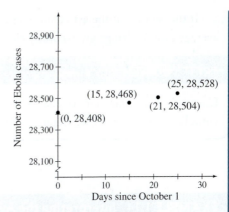

Figure 1.16 Plot of Ebola data

SOLUTION

a. October 1 corresponds to $d = 0$, and October 16 corresponds to $d = 15$. The change in the cumulative number of cases N over this time period is

$$28,468 - 28,408 = 60 \text{ new cases reported}$$

This change occurred over a 15-day period. Therefore,

$$\text{Average rate of change} = \frac{\text{Change in } N}{\text{Change in } d}$$
$$= \frac{60}{15}$$
$$= 4 \text{ new cases per day}$$

b. During the period from October 1 to October 16, there were, on average, four new cases of Ebola reported each day.

c. During the period from October 1 to October 16, there were about four new cases per day. Over the 8-day period from October 1 to October 9, that gives an expected increase of 8×4 cases. Using this result and the fact that there were 28,408 cases by October 1, we can estimate the total for October 9 as

$$\text{Estimated cumulative cases by October 9} = \text{Number of cases by October 1} + \text{Number of new cases}$$
$$N(8) \approx N(0) + 8 \times \text{Average rate of change}$$
$$= 28,408 + 8 \times 4$$
$$= 28,440$$

This estimate is illustrated in **Figure 1.17**, where the point (8, 28,440) lies on the line joining the nearest two data points.

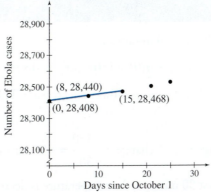

Figure 1.17 Ebola data: The estimate for October 9 lies on the line joining data from October 1 and October 16.

It turns out that the actual number reported by October 9, 2015 was 28,429. Our estimate using the average rate of change got us pretty close to the right number.

TRY IT YOURSELF 1.13 Brief answers provided at the end of the section.

Calculate the average rate of change from October 22 to October 26. Use your answer to estimate the cumulative number of cases reported by October 25.

EXAMPLE 1.14   Interpreting the Average Rate of Change: Newton's Law of Cooling

Newton's law of cooling describes how the temperature of an object changes over time. Suppose an aluminum pan is removed from an oven and left to cool in a room where the air temperature is constant. The table gives the temperature $A(t)$ in degrees Fahrenheit of the aluminum as a function of the time t in minutes since it was removed from the oven. The plot of the data shown in **Figure 1.18** illustrates the rapid initial decrease in temperature.

t = Time in minutes	A = Temperature in degrees Fahrenheit
0	300
30	150
60	100
90	79
120	73
150	72
180	71
210	71

a. Calculate the average rate of change from $t = 0$ to $t = 30$. Explain the meaning of the number you calculated in terms of the temperature of the aluminum.

b. Calculate the average rate of change from $t = 90$ to $t = 120$.

c. Use the results in parts a and b to answer the following question: Does a hot object cool more rapidly when the difference between its temperature and that of the air is larger or when that difference is smaller?

d. Use the average rate of change to estimate the temperature of the aluminum 100 minutes after it is removed from the oven.

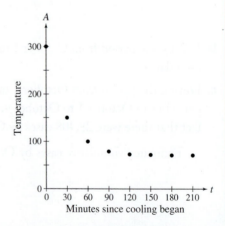

Figure 1.18 Plot of temperature data

SOLUTION

a. The change in temperature over the first 30 minutes is

$$A(30) - A(0) = 150 - 300 = -150°$$

The negative sign reflects the fact that the temperature is decreasing. This drop occurred over a 30-minute period, so

$$\text{Average rate of change} = \frac{-150}{30} = -5° \text{ per minute}$$

This result means that over the first 30 minutes, the temperature is decreasing by an average of 5° each minute.

b. We calculate the same way as in part a:

$$\text{Average rate of change} = \frac{A(120) - A(90)}{120 - 90}$$

$$= \frac{73 - 79}{120 - 90}$$

$$= \frac{-6}{30}$$

$$= -0.2° \text{ per minute}$$

c. The temperature drops at an average rate of 5° per minute during the time the aluminum is hotter, and it drops at an average rate of only 0.2° per minute when the temperature of the aluminum is nearer that of the surrounding air. As we might expect, a larger temperature difference means more rapid cooling.

d. Between $t = 90$ and $t = 120$, the aluminum cools at an average rate of 0.2° per minute. At 90 minutes the temperature is 79°. We estimate its temperature 10 minutes later using

$$\text{Temperature at 100 min.} \approx \text{Temperature at 90 min.} + 10 \times \text{Average rate of change}$$

$$A(100) \approx A(90) + 10 \times \text{Average rate of change}$$

$$= 79 + 10 \times (-0.2)$$

$$= 77°$$

This gives an estimated temperature of 77° after 100 minutes of cooling. **Figure 1.19** shows that the temperature estimate for 100 minutes is on the line joining the data points for 90 and 120 minutes.

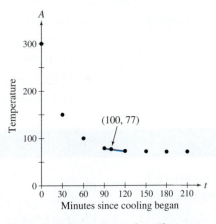

Figure 1.19 Temperature data: The point (100, 77) lies on the line joining the nearest two data points.

TRY IT YOURSELF 1.14 Brief answers provided at the end of the section.

Use the average rate of change in temperature from $t = 30$ to $t = 60$ to estimate the temperature at $t = 35$. Round your answer to the nearest degree.

EXTEND YOUR REACH

In the example, what can you say about the average rate of change in temperature over a single minute after the aluminum has cooled for a very long time?

TRY IT YOURSELF ANSWERS

1.10 $-\dfrac{1}{3}$

1.11 Simplify $\dfrac{\dfrac{1}{b^2}-\dfrac{1}{a^2}}{b-a}$.

1.12 93 ants per week

1.13 The average rate of change is six new cases per day. That gives an estimate of 28,522 cases reported by October 25.

1.14 About $142°$

EXERCISE SET 1.2

CHECK YOUR UNDERSTANDING

1. **True or false:** The value of the average rate of change depends both on the function and on the interval over which it is calculated.

2. For a function f, the expression $\dfrac{f(7)-f(4)}{3}$ gives _____.

3. The line that is related to the average rate of change is called _____.

4. **True or false:** The average rate of change is always positive.

5. If $f(a) < f(b)$, then the average rate of change of f over the interval $[a, b]$:

   a. is positive

   b. is negative

   c. is zero

   d. cannot be determined without further information

6. If the average rate of change of f on the interval $[a, b]$ is 0, then which is true?

   a. $f(a) > f(b)$

   b. $f(a) < f(b)$

   c. $f(a) = f(b)$

   d. The relationship between $f(a)$ and $f(b)$ cannot be determined without further information.

7. The average rate of change can be calculated for an interval of any positive length. Can it be calculated for an interval of length 0?

8. What is the average rate of change on any interval of a constant function?

SKILL BUILDING

Calculating average rates of change. In Exercises 9 through 19, calculate the average rate of change indicated.

9. $f(x) = 2x + 1$ from $x = 1$ to $x = 4$

10. $f(x) = x^2$ from $x = 2$ to $x = 5$

11. $f(x) = \dfrac{1}{x}$ from $x = 1$ to $x = 2$

12. $f(x) = x^3 + x + 1$ on $[2, 3]$

13. $f(x) = \dfrac{x+1}{x-1}$ on $[3, 5]$

14. $f(x) = \sqrt{x}$ on $[0, 10]$

15. $f(x) = \dfrac{1}{\sqrt{x}}$ on $[1, 4]$

16. $f(x) = x - x^2$ on $[0, 1]$

17. $f(x) = x^4 + x^2$ on $[-2, 2]$

18. $f(x) = \dfrac{1}{2x+1}$ on $[3, 4]$

19. $f(x) = |x|$ on $[-3, -2]$

Units associated with average rate of change and their meaning. In Exercises 20 through 30, state the units associated with the average rate of change of the given function, and explain what the average rate of change means in terms of the variables of the exercise.

20. $F(d)$ is the cumulative number of flu cases reported as of day d.

21. $B(t)$ is the balance in dollars of a savings account after t months.

22. $A(d)$ is the cumulative number of traffic accidents at a certain intersection as of day d.

23. $M(h)$ is the total miles driven after h hours.

24. $W(y)$ is the weight in pounds of a growing child who is y years old.

25. $V(t)$ is the speed in miles per hour of an airplane t minutes after takeoff.

26. $T(I)$ is your income tax liability in dollars if your income is I dollars.

27. $P(n)$ is the profit in dollars your company earns if you produce n items.

28. $H(s)$ is the heart rate in beats per minute of an athlete who is running at s feet per second.

29. $A(d)$ is the area in square feet of a circle of diameter d measured in feet.

30. $E(m)$ is the elevation in feet along a hiking trail m miles from its beginning.

PROBLEMS

Average rates of change over general intervals. In Exercises 31 through 42, calculate the average rate of change of the given function on the interval $[a, a + h]$ (assuming $h > 0$). Simplify your answers as much as possible.

31. $f(x) = c$, where c is a constant.

32. $f(x) = x$

33. $f(x) = 2 - 5x$

34. $f(x) = mx + b$, where m and b are constants.

35. $f(x) = x^2$

36. $f(x) = 3x^2 + 4x$

37. $f(x) = 2 - x^2$

38. $f(x) = \sqrt{x + 1}$ (Assume $a \geq -1$.)

39. $f(x) = 2\sqrt{x}$ (Assume $a \geq 0$.)

40. $f(x) = x^3$ *Suggestion*: Recall that $(A + B)^3 = A^3 + 3A^2B + 3AB^2 + B^3$.

41. $f(x) = \dfrac{1}{x}$ (Assume $a > 0$.)

42. $f(x) = \dfrac{1}{\sqrt{x}}$ (Assume $a > 0$.) *Suggestion*: If you have difficulty, look at Example 1.11.

43. **Constants and the average rate of change.** How does the average rate of change of $f(x) + c$ compare with the average rate of change of f?

Estimating unknown function values. In Exercises 44 through 50, use the following information. If $a + h$ lies in the interval $(a, b]$ and $f(a)$ and $f(b)$ are known, then we can use the average rate of change to approximate $f(a + h)$:

$$f(a + h) \approx f(a) + h \frac{f(b) - f(a)}{b - a}$$

You will be led to a derivation of this formula in a later exercise showing that the formula is using a secant line to approximate the graph.

44. For $f(x) = \sqrt{x}$, use the values of $f(100)$ and $f(121)$ to estimate the value of $\sqrt{107}$.

45. For $f(x) = \sqrt[3]{x}$, use the values of $f(8)$ and $f(27)$ to estimate the value of $\sqrt[3]{20}$.

46. For $f(x) = \sqrt{x}$, use the values of $f(25)$ and $f(36)$ to estimate the value of $\sqrt{30}$.

47. For $f(x) = 2^x$, use the values of $f(2)$ and $f(3)$ to estimate the value of $2^{2.4}$.

48. Suppose that $f(x)$ is a function with $f(4) = 3$ and $f(7) = 4$. Estimate $f(5)$.

49. Suppose that $f(x)$ is a function with $f(3) = 9$ and $f(7) = 5$. Estimate $f(5)$.

50. **Using different intervals.**

　a. Use an average rate of change over the interval $[4, 25]$ to estimate the value of $\sqrt{7}$.

　b. Use an average rate of change over the interval $[4, 9]$ to estimate the value of $\sqrt{7}$.

　c. Which of the two preceding estimates is nearer the actual value of $\sqrt{7}$?

　d. Based on the result of this exercise, do you think better results are obtained when using shorter or longer intervals?

51. **Average rate of change for the absolute value function.** Let $f(x) = |x|$. Show that if $0 < a < b$, then the average rate of change of f on $[a, b]$ is 1. Show that if $a < b < 0$, then the average rate of change of f on $[a, b]$ is -1. Show that if $a > 0$, then the average rate of change of f on $[-a, a]$ is 0.

MODELS AND APPLICATIONS

52. Carbon-14. Carbon-14 is a radioactive substance that decays over time. It is used to date fossils and other objects containing organic matter. In the following table, $C(t)$ is the amount in grams of carbon-14 remaining at time t, measured in thousands of years.

t = time in thousands of years	0	10	15	20
C = amount in grams remaining	5	1.5	0.8	0.4

a. Use the average rate of change to estimate the amount of carbon-14 remaining after 12,000 years. Be careful about the units.

b. We can reverse the roles of the variables and think of t as depending on C. Call this function g. For example, $g(0.8) = 15$ because $t = 15$ when $C = 0.8$.

    i. Use the table to calculate the average rate of change for g from 0.8 gram to 1.5 grams.

    ii. Use your answer to the preceding part to estimate the time when there are 1.3 grams remaining.

53. Prescription drugs. The following table shows the number N (in billions) of prescriptions filled in the United States in the given year.[2]

Year	2007	2013	2014	2015
Number of prescriptions	3.50	3.99	4.08	4.17

a. Make a table showing, for each of the periods in the table, the average rate of change per year in the number of prescriptions.

b. During which of these periods was the number of prescriptions growing at its smallest rate?

c. Use the average rate of change from 2014 to 2015 to estimate the number of prescriptions in 2020.

54. Arterial blood flow. A relatively small increase in the radius of an artery corresponds to a relatively large increase in blood flow. For example, if one artery has a radius 5% larger than another, the blood flow rate is 1.22 times as large. The accompanying table gives more information.

Increase in radius	5%	10%	15%	20%
Times greater blood flow rate	1.22	1.46	1.75	2.07

Use the average rate of change to estimate how many times greater the blood flow rate is in an artery that has a radius 7% larger than another.

55. Belk revenues. The retailer Belk's annual reports show the following revenues in billions of dollars.

Year	2009	2012	2015	2018
Revenue	3.50	3.70	4.11	3.60

a. During which 3-year period was the average rate of change in revenue negative?

b. Calculate the average rate of change in revenue from 2015 to 2018. Be sure to give proper units.

c. Use your answer from part b to give an estimated revenue for 2020.

56. Scooter rides. The number N of scooter rides in a city t weeks after the launch of a scooter ride business is given by

$$N = \frac{12,500}{1 + 20 \times 0.87^{t}}$$

Round values of N to the nearest integer.

a. Calculate the average rate of change per week in rides for

• Week 50 to week 55

• Week 55 to week 60

• Week 60 to week 65

b. As of week 65, you are considering investing in the scooter business. The profitability of the investment depends on the growth in scooter rides. Do the calculations from part a indicate that this is a wise investment?

57. Choosing a bat. A 26-inch baseball bat is recommended for a child who is 38 inches tall. As the child grows, up to 62 inches, the average rate of change in recommended bat length is 0.25 inch of bat length per inch of growth. Bat lengths are in whole numbers of inches. How many bats will you buy from the time your child is 38 inches tall through the time your child is 62 inches tall?

[2]See http://www.statista.com/statistics/261303/total-number-of-retail-prescriptions-filled-annually-in-the-us.

58. Yellowfin tuna. The following table shows data from a study of yellowfin tuna.[3]

L = length (centimeters)	100	110	120	130	140	160
W = weight (pounds)	42.5	56.8	74.1	94.7	119.0	179.0

a. What is the average rate of change, in pounds per centimeter, for lengths from 100 to 110 centimeters? From 140 to 160 centimeters?

b. Two fishermen tell tales slightly exaggerating the lengths of a yellowfin tuna they caught, with one claiming to have caught a 100-centimeter fish and the other a 150-centimeter fish. Which fish tale indicates the larger exaggeration in weight?

c. Estimate the weight of a 108-centimeter yellowfin tuna.

d. Estimate the length of a yellowfin tuna weighing 155 pounds.

59. Puppy weight. A woman wants to estimate the adult weight of her puppy based on the age at which it reaches a critical weight. She knows that puppies reaching that critical weight at age 5 weeks will weigh 52 pounds as adults, and puppies reaching that critical weight at age 12 weeks will weigh 22 pounds as adults. If her puppy reaches that critical weight at age 7 weeks, estimate the adult weight of her puppy. Round your answer to the nearest pound.

60. Home equity. When a bank loans money for a home, it effectively owns the home. Each payment you make is partly interest and partly purchase price for a small share in home ownership. The ownership share that you have purchased is your equity. If you secure a 30-year home mortgage for $350,000 at an annual percentage rate, or APR, of 6%, then after k monthly payments your equity is

$$E = \frac{350,000(1.005^k - 1)}{1.005^{360} - 1} \text{ dollars}$$

a. Do you own half the home after half of the payments (that is, 180 payments) have been made?

b. Calculate the average rate of change per month for the following periods.

- From $k = 0$ to $k = 12$ (the first year of payments)
- From $k = 108$ to $k = 120$ (the 10th year)
- From $k = 228$ to $k = 240$ (the 20th year)
- From $k = 348$ to $k = 360$ (the 30th year)

c. Does your equity grow faster early or late in the term of the loan?

61. Spending on video games. The following table shows the amount, in billions of dollars, spent on video game content in the United States in the given year.[4]

Year	2013	2014	2015	2016	2017
Spent (billions of dollars)	20.2	21.4	23.2	24.5	29.1

a. Make a table showing the average rate of change *per month* for each period shown in the table. Round your answers to two decimal places.

b. Over which period is the average rate of change the largest?

62. A falling rock. A rock dropped near the surface of Earth falls $D(t) = 16t^2$ feet in t seconds. **Figure 1.20** shows how the distance fallen varies with time. The aim of this exercise is to determine the speed of the rock at $t = 2$ seconds—a determination that is more difficult than it appears at first sight. We calculate the speed over a time interval using the average rate of change for the function D:

$$\text{Speed} = \frac{\text{Distance traveled}}{\text{Elapsed time}}$$

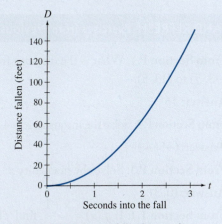

Figure 1.20 Distance fallen by a rock

[3]*Mathematics for the Biosciences,* Michael R. Cullen, TechBooks, Fairfax, VA, 1983.

[4]See http://www.theesa.com/about-esa/essential-facts-computer-video-game-industry.

We ask: How is it possible to find the speed when the elapsed time is 0?

a. Calculate the speed over the following time intervals, and report your answers correct to four decimal places.

   i. From $t = 2$ to $t = 3$ **ii.** From $t = 2$ to $t = 2.5$

iii. From $t = 2$ to $t = 2.1$ **v.** From $t = 2$ to $t = 2.001$

iv. From $t = 2$ to $t = 2.01$ **vi.** From $t = 2$ to $t = 2.0001$

b. Without calculating further, and based on your calculations in part a, what do you think the speed at $t = 2$ is?

CHALLENGE EXERCISES FOR INDIVIDUALS OR GROUPS

63. **A formula for estimating unknown function values.** In this exercise, you are led to derive a formula for approximating unknown function values in terms of the average rate of change. In the process, you will see that the formula is using a secant line to approximate the graph. Suppose we know the values of $f(a)$ and $f(b)$, but for some $a + h$ in the interval $[a, b]$ the function value $f(a + h)$ is unknown.

a. Use the fact that the slopes of the lines in **Figure 1.21** are nearly the same to obtain the approximate equality

$$\frac{f(a + h) - f(a)}{h} \approx \frac{f(b) - f(a)}{b - a}$$

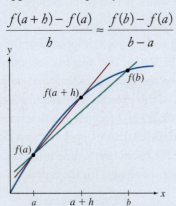

Figure 1.21 Two lines with nearly the same slopes

b. Use the results from part a to obtain the approximate equality

$$f(a + h) \approx f(a) + h\frac{f(b) - f(a)}{b - a}$$

This approximation is illustrated in **Figure 1.22**. Note that we are approximating the graph of f by the secant line.

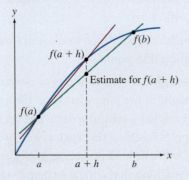

Figure 1.22 Approximate value for $f(a + h)$

REVIEW AND REFRESH: Exercises from Previous Sections

64. **From Section P.1:** What is the distance from $(-3, 5)$ to $(13, 5)$?

Answer: 16

65. **From Section P.2:** Solve the inequality $x - 4 < 3x + 6$.

Answer: $(-5, \infty)$

66. **From Section P.3:** Solve the inequality $x^2 < 3x$.

Answer: $(0, 3)$

67. **From Section 1.1:** If $f(x) = \dfrac{x + 1}{x}$, find $f(x + 1)$.

Answer: $\dfrac{x + 2}{x + 1}$

68. **From Section 1.1:** If $f(x) = x^2 - 1$, find $f(\sqrt{x})$.

Answer: $x - 1$

69. **From Section 1.1:** Find the domain of $f(x) = \dfrac{x + 1}{\sqrt{x - 1}}$.

Answer: $(1, \infty)$

70. **From Section 1.1:** Find the domain of $f(x) = \dfrac{x}{x^{1/4}}$.

Answer: $(0, \infty)$

71. **From Section 1.1:** Find the range of $g(t) = t^4 + 7$.

Answer: $[7, \infty)$

1.3 Graphs and Rates of Change

> Rates of change are the key to a dynamic view of functions.

1.1 The Basics of Functions

1.2 Average Rates of Change

1.3 Graphs and Rates of Change

1.4 Limits and End Behavior of Graphs

In this section, you will learn to:

1. Calculate the rate of change of a function at a point given information about the tangent line.
2. Locate the intervals on which a function is increasing or decreasing.
3. Relate the intervals on which a function is increasing or decreasing to the sign of the rate of change.
4. Identify local and absolute maximum and minimum values of a function.
5. Locate the intervals on which a graph is concave up or concave down.
6. Interpret concavity in terms of rates of change.
7. Identify the inflection points of a graph.
8. Sketch graphs having specified properties.
9. Interpret the shape of a graph in applications.

A familiar formula for calculating velocity is

$$\text{Velocity} = \frac{\text{Distance traveled}}{\text{Elapsed time}}$$

This formula gives the average velocity, the average rate of change in distance with respect to time. But the speedometer in your car (pictured in **Figure 1.23**) shows a different type of velocity. If you accelerate on a narrow road to pass a slow-moving truck, the speedometer needle might move smoothly from 40 to 65 miles per hour and just as smoothly slow to 60 miles per hour. As the needle moves, it shows your velocity at each instant of time: your instantaneous velocity.

Figure 1.23 Car speedometer: The speedometer in your car shows instantaneous velocity. Jamesbowyer/Getty Images

The same idea applies to any function. We can calculate an average rate of change, but there is also the instantaneous rate of change or just the rate of change. Just as the speedometer in your car provides important information regarding your driving, the instantaneous rate of change provides invaluable information regarding the function.

The rate of change of a function is a central idea in calculus and its applications, and for many good reasons. Among them is the fact that the rate of change of a function describing a natural phenomenon is often easier to understand than the phenomenon itself. For example, if a rock is dropped, its velocity as a function of time is difficult to measure. But the rate of change in the velocity of the rock, which is the acceleration due to gravity, is easier to measure because it is constant. Calculus then shows us how to use acceleration to find velocity.

In this section, we focus on graphs and their relation to the rate of change. The graph can reveal how functions change, show where functions increase or decrease, and identify maximum and minimum values of a function. Dynamic properties such as increase, decrease, and concavity provide a preview of concepts that will be important in calculus.

An Intuitive Description of Rates of Change

> The rate of change is the slope of a line tangent to the graph.

In **Figure 1.24**, we have plotted secant lines for the graph of a function $y = f(x)$ through the point (x, y) and over ever shorter intervals. Note how these secant lines approach a line that appears to be tangent to the curve (similar to the way a line may be tangent to a circle). This line is aptly named the tangent line to the graph at (x, y). Just as the secant lines approach the tangent line, the slopes of the secant lines, the average rates of change, approach the slope of the tangent line. The slope of the tangent line is the instantaneous rate of change or simply **rate of change**[5] at x of the function f. (In calculus the rate of change is called the derivative.)

In **Figure 1.25**, we have plotted a number of tangent lines to the graph of a function $y = g(x)$. The slopes of these tangent lines give the rate of change of the function at the corresponding x-value. Note how the tangent lines track the path of the graph. That is, tangent lines with a positive slope indicate a rising graph, and tangent lines with a negative slope indicate a falling graph. Note also that steeper tangent lines correspond to steeper sections of the graph. Thus, we can associate the rate of change at a point x with the "direction" of the graph of g at the corresponding point.

> The **rate of change** at x of the function f is the slope of the tangent line to the graph $y = f(x)$ at the point $(x, f(x))$.

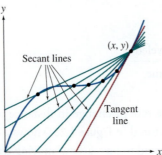

DF **Figure 1.24** Secant lines and tangent lines: Over ever shorter intervals, the secant lines approach the tangent line.

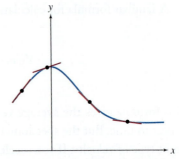

DF **Figure 1.25** Tangent lines: These superimposed lines track the path of the graph, indicating its direction.

CONCEPTS TO REMEMBER: Properties of the Rate of Change of a Function

- Secant lines for the graph through a given point on the graph and over ever shorter intervals approach the tangent line, as is illustrated in Figure 1.24.

- The slope of the tangent line is the instantaneous rate of change or just the rate of change of the function at the given value of the independent variable.

- The average rate of change gives an approximation to the (instantaneous) rate of change. Average rates of change over shorter and shorter intervals give better and better approximations of the rate of change.

- Positive rates of change indicate regions where the graph is rising.

- Negative rates of change indicate regions where the graph is falling.

- Steeper tangent lines correspond to steeper portions of the graph.

[5]Not all functions have a well-defined rate of change at every point. But for the classes of functions we will consider later (linear, exponential, etc.), the rate of change is defined at each point in the domain. The discussion in this section is intuitive, and we assume that the rate of change exists.

EXAMPLE 1.15 Finding a Rate of Change

The graph of $f(x) = x^2$ and the tangent line to the graph for $x = 1$ are shown in **Figure 1.26**. The tangent line also passes through the point $(2, 3)$. Find the rate of change of f at $x = 1$.

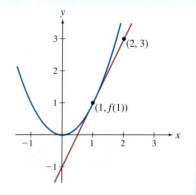

Figure 1.26 The graph of $f(x) = x^2$ and its tangent line for $x = 1$

SOLUTION

The rate of change of f at $x = 1$ is the slope of the tangent line. The tangent line passes through the points $(1, f(1)) = (1, 1)$ and $(2, 3)$. Recall how we use these two points to calculate the slope:

$$\text{Rate of change} = \text{Slope of tangent line}$$

$$= \frac{\Delta y}{\Delta x}$$

$$= \frac{3 - 1}{2 - 1}$$

$$= 2$$

Thus, the rate of change of f at $x = 1$ is 2.

TRY IT YOURSELF 1.15 Brief answers provided at the end of the section.

The tangent line to the graph of $f(x) = x^2$ at $x = 2$ passes through the points $(2, f(2))$ and $(3, 8)$. Find the rate of change of f at $x = 2$.

Increasing and Decreasing Functions

Rates of change can show where functions are increasing and where they are decreasing.

Where the graph of a function is rising, it is natural to think of the function as an **increasing function** there. Similarly, where the graph falls, we think of the function as a **decreasing function** there.

In words, a function $f(x)$ is increasing if larger values of x yield larger values of f. Similarly, it is decreasing if larger values of x yield smaller values of f.

Regions of increase or decrease for a function may not be apparent from a formula, but they can be seen from a graph. For example, it is by no means apparent where the function $f(x) = \dfrac{x^3 - 48x}{128}$ is increasing or where it is decreasing. We show the graph in

An **increasing function** on an interval I is a function f such that $f(x) < f(y)$ whenever $x < y$ and both x and y are in the interval I.

A **decreasing function** on an interval I is a function f such that $f(x) > f(y)$ whenever $x < y$ and both x and y are in the interval I.

Figure 1.27. In **Figure 1.28**, we have marked the regions where the graph rises and where it falls. We conclude that f is increasing on $(-\infty, -4]$ and $[4, \infty)$ and is decreasing on $[-4, 4]$.

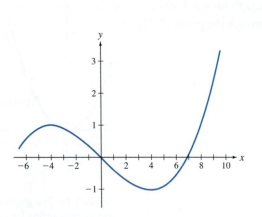

Figure 1.27 The graph of $f(x) = \dfrac{x^3 - 48x}{128}$.

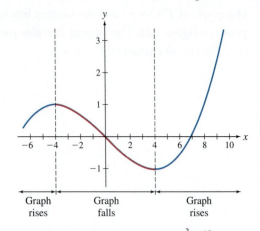

DF **Figure 1.28** The graph of $f(x) = \dfrac{x^3 - 48x}{128}$: Regions where the graph rises and where it falls are marked.

In **Figure 1.29**, we have added a few tangent lines. We see that their slopes, the rates of change, give additional information. The figure shows that positive rates of change indicate regions of increase, and negative rates of change indicate regions of decrease.

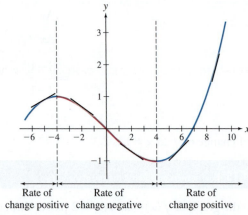

Figure 1.29 The graph of $f(x) = \dfrac{x^3 - 48x}{128}$: Positive rates of change indicate regions of increase, and negative rates of change indicate regions of decrease.

CONCEPTS TO REMEMBER: How Rates of Change Determine Increase or Decrease

- A positive rate of change indicates a rising graph and thus a region of increase for a function. Conversely, if a function is increasing on an interval, then the rate of change is nonnegative there.

- A negative rate of change indicates a falling graph and thus a region of decrease for a function. Conversely, if a function is decreasing on an interval, then the rate of change is nonpositive there.

- Rates of change that are larger in magnitude indicate a steeper graph.

EXAMPLE 1.16 Sketching a Graph

Sketch the graph of a function that has a positive rate of change on $(-\infty, 0)$ and a negative rate of change on $(0, \infty)$.

SOLUTION

Many graphs satisfy the given information, so there are many correct answers. We provide one here. The rate of change is used to determine where the graph rises and where it falls.

- The function has a positive rate of change on $(-\infty, 0)$. Consequently, the function is increasing, and the graph rises on this interval. This fact shows us how to begin the graph, as we have done in **Figure 1.30**.

- The function has a negative rate of change on $(0, \infty)$. Consequently, the function is decreasing, and the graph falls on this interval. This fact allows us to complete the graph as shown in **Figure 1.31**.

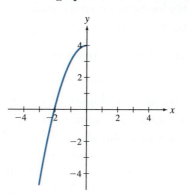

Figure 1.30 Positive rate of change: The graph rises on $(-\infty, 0)$.

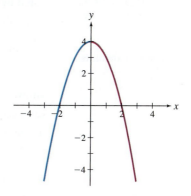

Figure 1.31 Negative rate of change: The fall on $(0, \infty)$ completes the graph.

TRY IT YOURSELF 1.16 Brief answers provided at the end of the section.

Sketch the graph of a function with a negative rate of change on $(-\infty, 0)$ and a positive rate of change on $(0, \infty)$.

Maximum and Minimum Values

> Functions may reach maximum or minimum values where the rate of change is zero.

Let's examine the graph of $f(x) = \dfrac{x^3 - 48x}{128}$ further. Other interesting features of the graph include the peak at $x = -4$ and the valley at $x = 4$, which are shown in **Figure 1.32**.

The peak at $x = -4$ corresponds to a **local maximum** value of $y = 1$ for f, and the valley at $x = 4$ corresponds to a **local minimum** value of $y = -1$.

We use the term *local* here because the peak at $x = -4$ is not the highest point on the entire graph over $\mathbb{R}$ — it is only the highest among nearby points. Likewise, the valley at $x = 4$ is not the lowest point on the entire graph — it is only the lowest among nearby points. The terms *relative maximum* and *relative minimum* are also used.

In **Figure 1.33**, we have added tangent lines at both the local maximum and the local minimum. Both of these lines are horizontal and so have slope 0. This illustrates

A function f reaches a **local maximum** value of $f(x_0)$ at $x = x_0$ if there is an interval (a, b) containing x_0 so that $f(x_0) \geq f(x)$ for every x in the interval (a, b).

A function f reaches a **local minimum** value of $f(x_0)$ at $x = x_0$ if there is an interval (a, b) containing x_0 so that $f(x_0) \leq f(x)$ for every x in the interval (a, b).

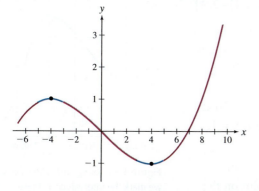

Figure 1.32 Peaks and valleys: The peak at $x = -4$ is the highest among nearby points (blue) on the graph. The valley at $x = 4$ is the lowest among nearby points (blue) on the graph.

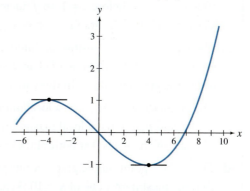

Figure 1.33 Tangent lines added: The rate of change is typically 0 at local maxima and minima.

the fact that at local maxima (plural of maximum) or minima, the rate of change is typically 0.

If there is a highest point on the entire graph, we say that point represents an **absolute maximum** value. Similarly, the **absolute minimum** value, if it exists, is indicated by the lowest point on the entire graph. Such points may occur at local maxima or minima, or at endpoints of graphs.

For example, the function $f(x) = \dfrac{x^3 - 48x}{128}$ has neither an absolute maximum nor an absolute minimum. But if we restrict the domain of f to the interval $[-6, 10]$ shown by the graph in Figure 1.32, then it has both. The function with a restricted domain has an absolute minimum value of $y = -1$, which occurs at $x = 4$. It has an absolute maximum value of $y = 4.0625$, which occurs at the endpoint $x = 10$.

Note the difference between a local maximum value and the place where it occurs. In the preceding example $f(x) = \dfrac{x^3 - 48x}{128}$, the local maximum value is $f(-4) = 1$, and it occurs at $x = -4$. Note also that the definitions are structured so that neither a local maximum nor a local minimum can occur at an endpoint of a graph.

> A function f reaches an **absolute maximum** value of $f(x_0)$ at $x = x_0$ if $f(x_0) \geq f(x)$ for every x in the domain of f.
>
> A function f reaches an **absolute minimum** value of $f(x_0)$ at $x = x_0$ if $f(x_0) \leq f(x)$ for every x in the domain of f.

EXAMPLE 1.17 Identifying Features of a Graph

Figure 1.34 shows the graph of a function $y = f(x)$ whose domain is $[-3, 3]$.

a. Identify the intervals of increase and decrease.

b. Identify the regions where the rate of change is positive and the regions where it is negative. (Do not include the endpoints of the domain.)

c. Determine any local maxima or minima. What is the rate of change at these points?

d. Find any absolute maxima or minima.

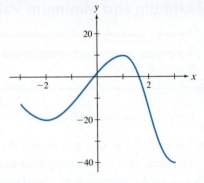

Figure 1.34 Graph for Example 1.17: The domain is $[-3, 3]$.

SOLUTION

a. As we move from left to right, the graph falls until $x = -2$ and after $x = 1$. Hence, f decreases on $[-3, -2]$ and on $[1, 3]$. The graph rises after $x = -2$ and before $x = 1$, so f increases on the interval $[-2, 1]$. These features of the graph are shown in **Figure 1.35**.

b. The rate of change is positive on intervals where the graph is rising, in this case $(-2, 1)$. The rate of change is negative on intervals where the graph is falling, in this case $(-3, -2)$ and $(1, 3)$.

c. The graph shows that f has a local minimum value of $y = -20$ at $x = -2$ and that f has a local maximum value of $y = 10$ at $x = 1$. The rate of change is 0 at both of these values of x.

d. The highest point on the graph occurs at $x = 1$, and f has an absolute maximum value of $y = 10$ there. The lowest point on the graph is at the right endpoint. Thus, the absolute minimum of f occurs at $x = 3$, and the value is $y = -40$.

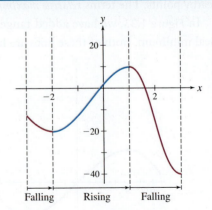

Figure 1.35 Rising and falling: For part a, we mark the intervals of increase and decrease.

TRY IT YOURSELF 1.17 Brief answers provided at the end of the section.

Figure 1.36 shows the graph of a function g whose domain is $[-3, 3]$. Identify the intervals of increase and decrease, and determine any local maxima or minima.

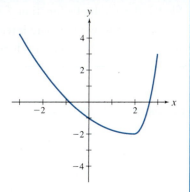

Figure 1.36 Graph for Try It Yourself 1.17: The domain is $[-3, 3]$.

EXAMPLE 1.18 Showing Maxima and Minima on a Graph

Sketch the graph of a function $y = f(x)$ with the following properties:

a. The domain of f is $[-3, 5]$.

b. The rate of change of f is negative on the intervals $(-3, -2)$ and $(2, 5)$, and f is increasing elsewhere.

c. f has an absolute maximum at $x = 2$.

d. f has a local minimum at $x = -2$, but this is not the location of an absolute minimum.

SOLUTION

We show how each of the preceding properties contributes to the construction of the graph.

a. We use the domain $[-3, 5]$ to determine the horizontal extent of the graph.

b. We know that f is decreasing on the intervals $(-3, -2)$ and $(2, 5)$ where the rate of change is negative, and it is increasing on $(-2, 2)$. The intervals of increase and decrease tell us how to start the graph. It falls from $x = -3$ to $x = -2$, rises from $x = -2$ to $x = 2$, and starts to fall again after that. These features are indicated in **Figure 1.37**.

c. We want to be sure that the highest point on the graph occurs at $x = 2$ because that is the location of the absolute maximum.

d. There will be a local minimum at $x = -2$, but because this is not supposed to be the location of an absolute minimum, we must be sure that it is not the lowest point on the entire graph. We can arrange that by making the right endpoint the absolute minimum.

The completed graph is shown in **Figure 1.38**, where we have also marked both local and absolute maxima and minima.

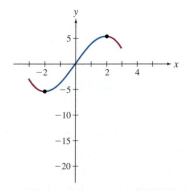

Figure 1.37 Beginning the graph: Regions of increase and decrease are indicated.

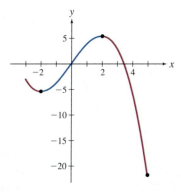

Figure 1.38 Completing the graph: Ensure an absolute maximum at $x = 2$ and an absolute minimum at $x = 5$.

TRY IT YOURSELF 1.18 Brief answers provided at the end of the section.

Sketch the graph of a function f with the following properties. The domain of f is $[-4, 4]$, and f decreases from $x = -4$ to $x = 0$, where it reaches both a local minimum and an absolute minimum. There is an absolute maximum at each endpoint. There is no local maximum.

EXTEND YOUR REACH

a. Is it possible to make the graph of a function with the following properties? Its domain is the set of all real numbers, it is increasing on $(-\infty, 0)$ and has a negative rate of change on $(0, \infty)$, but it has no local maximum. *Suggestion:* The graph you make may have breaks.

b. Is it possible to make the graph of a function that has a zero rate of change at $x = 0$ but reaches neither a maximum nor a minimum at $x = 0$? *Suggestion:* Sketch the tangent line to the graph of $y = x^3$ at $x = 0$.

Concavity and Rates of Change

> Concavity describes the shape of a graph and can be characterized in terms of rates of change.

A graph of a function is **concave up** if it looks like a piece of wire whose ends are bent upward.

A graph of a function is **concave down** if it looks like a piece of wire whose ends are bent downward.

An **inflection point** on the graph of a function is a point where the direction of bending changes.

Let's focus once more on the graph of $f(x) = \dfrac{x^3 - 48x}{128}$, as shown in **Figure 1.39**. The part of the graph to the left of $x = 0$ has the shape of a piece of wire whose ends are bent downward. To the right of $x = 0$, the graph has the shape of a wire whose ends are bent upward. At $x = 0$, the direction of bending changes. This concept of direction of bending is important enough to have its own terminology: **concave up**, **concave down**, and **inflection point**.

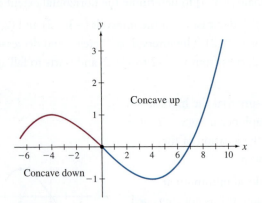

DF **Figure 1.39** Direction of bending: The graph is bent downward before $x = 0$ and bent upward after.

These definitions are only intuitive. Precise definitions can be given using basic geometry or ideas from calculus. A straight-line graph is not bent, so it is neither concave up nor concave down.

For example, the graph in Figure 1.39 is concave down on $(-\infty, 0)$ and concave up on $(0, \infty)$. The direction of bending changes at the origin, so $(0, 0)$ is a point of inflection.

Rates of change can help us understand concavity. In **Figure 1.40**, we show the graph of a function that is increasing and concave up. Note that the upward concavity

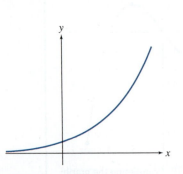

Figure 1.40 Graph of a function: This graph represents a function that is increasing and concave up.

causes the graph to get steeper as we move from left to right. Remember that steeper graphs indicate larger rates of change. We conclude that the function is increasing at an increasing rate. This type of observation is useful for any function and is of particular importance in applications.

EXAMPLE 1.19 Concavity and Rates of Change

Graphs are shown in **Figure 1.41** and **Figure 1.42**. For each graph, determine which of the following statements applies.

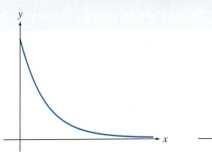

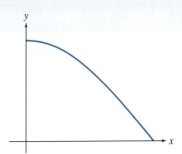

Figure 1.41 Graph of a function: This graph represents a function that is decreasing and concave up.

Figure 1.42 Graph of a function: This graph represents a function that is decreasing and concave down.

a. The corresponding function is decreasing at an increasing rate (that is, the rate of decrease is increasing).

b. The corresponding function is decreasing at a decreasing rate (that is, the rate of decrease is decreasing).

c. The corresponding function is increasing at an increasing rate.

d. The corresponding function is increasing at a decreasing rate.

SOLUTION

Figure 1.41: The graph flattens out as we move to the right. Thus, the function decreases rapidly at first but more slowly as we move to the right. The function is decreasing at a decreasing rate. Thus, statement b applies to this graph.

Figure 1.42: The graph gets steeper as we move to the right. Thus, the function decreases more rapidly as we move to the right. The function is decreasing at an increasing rate. We conclude that statement a applies to this graph.

TRY IT YOURSELF 1.19 Brief answers provided at the end of the section.

A graph that is increasing and concave down is shown in **Figure 1.43**. Which of the preceding statements a through d applies to this graph?

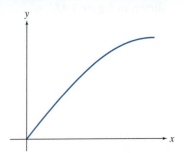

Figure 1.43 Graph of a function: This graph represents a function that is increasing and concave down.

EXTEND YOUR REACH

   a. Can you make the graph of a function that is concave down on $(-\infty, \infty)$ yet has no maximum and no minimum?

   b. Explain in terms of rates of change why a graph that is a straight line is neither concave up nor concave down.

EXAMPLE 1.20 Sketching Graphs with Given Concavity

Draw the graph of a function $y = f(x)$ that satisfies all of the following conditions, and identify the inflection points:

a. The domain is $[0, 10]$.

b. The function is increasing and concave up on $[0, 3)$.

c. The function is increasing and concave down on $(3, 5)$.

d. The function is decreasing and concave down on $(5, 7)$.

e. The function is decreasing and concave up on $(7, 10]$.

SOLUTION

The given conditions do not completely determine the graph, so there is no single correct answer. But the conditions do determine the general shape of the graph, and we show how each feature contributes to the construction of the graph.

a. The domain is $[0, 10]$, and that determines the horizontal extent of the graph.

b. We begin at $x = 0$ with an increasing function. Because it is concave up, the graph should get steeper as we move toward $x = 3$. This piece of the graph is shown in **Figure 1.44**.

c. Between $x = 3$ and $x = 5$, the graph is increasing and concave down. That makes the graph flatten out. In **Figure 1.45**, we have added this feature to the graph.

d. The function starts to decrease after $x = 5$. The downward concavity makes the graph steeper until we get to $x = 7$. This fact is shown in **Figure 1.46**.

e. From $x = 7$ on, the function continues to decrease, but because the graph is concave up, it begins to level off. The completed graph is shown in **Figure 1.47**.

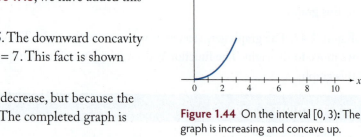

Figure 1.44 On the interval $[0, 3)$: The graph is increasing and concave up.

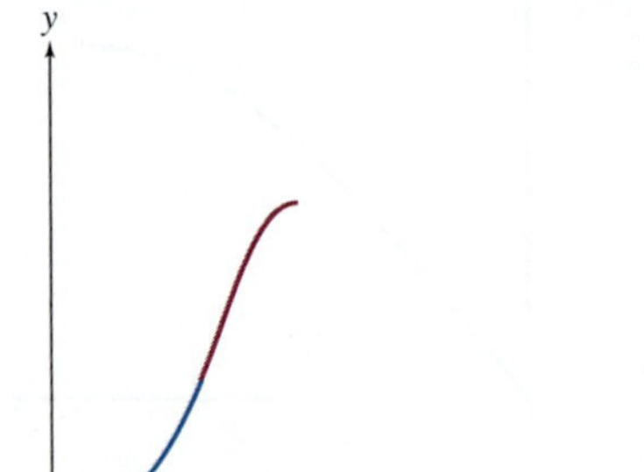

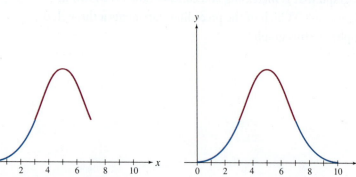

Figure 1.45 On the interval $(3, 5)$: The graph is increasing and concave down.

Figure 1.46 On the interval $(5, 7)$: The graph is decreasing and concave down.

Figure 1.47 On the interval $(7, 10]$: The graph is decreasing and concave up.

The inflection points occur where the concavity changes. There are two such points, $(3, f(3))$ and $(7, f(7))$.

TRY IT YOURSELF 1.20 Brief answers provided at the end of the section.

Sketch the graph of a function that has a positive rate of change for $x < 0$, has a negative rate of change for $x > 0$, and is always concave down.

MODELS AND APPLICATIONS Sales Charts

Business models often include charts or graphs. The following example shows how mathematics applies to interpreting such figures.

EXAMPLE 1.21 Interpreting the Shape of Graphs: Sales

The graph in **Figure 1.48** shows the daily sales S in dollars for a certain company over a 1-year period. The horizontal axis shows the time t in months since the beginning of the year.

a. When did sales reach a (local) minimum?

b. Over what period were sales increasing?

c. Compare the growth in sales over the 4-month period from $t = 4$ to $t = 8$ with the growth in sales over the 4-month period from $t = 8$ to $t = 12$. Interpret your answer in terms of the shape of the graph.

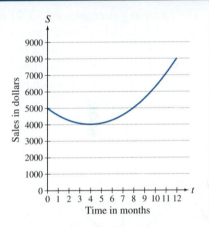

Figure 1.48 Sales as a function of time

SOLUTION

a. Sales reach a minimum where the graph is at a valley. That occurred at $t = 4$ (the beginning of May).

b. Sales are increasing when the graph is rising. That is from $t = 4$ to $t = 12$.

c. From $t = 4$ to $t = 8$, sales grew from $4000 to $5000, which is a growth of $1000. From $t = 8$ to $t = 12$, sales grew from $5000 to $8000, which is a growth of $3000. Sales grew faster over the later 4-month period. We expect this result because the graph gets steeper as we move to the right from $t = 4$: the graph is increasing and concave up.

TRY IT YOURSELF 1.21 Brief answers provided at the end of the section.

The graph in **Figure 1.49** shows the population of an endangered species over time. Over what period was the rate of decline in population slowing?

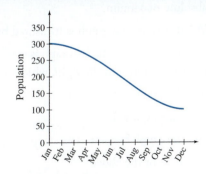

Figure 1.49 Population of an endangered species

TRY IT YOURSELF ANSWERS

1.15 The rate of change is 4.

1.16

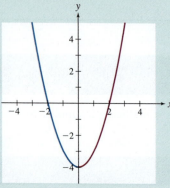

1.17 The function g decreases on $[-3, 2]$ and increases on $[2, 3]$. It has a local minimum value of $y = -2$ at $x = 2$. This is also an absolute minimum. There is no local maximum. The function has an absolute maximum value of $y = 4$, which occurs at $x = -3$.

1.18

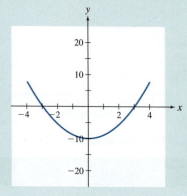

1.19 Statement d applies: the function is increasing at a decreasing rate.

1.20

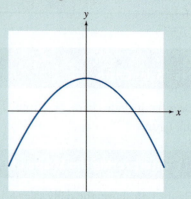

1.21 The rate of decline is slowing where the function is decreasing and concave up. That is from July through December.

EXERCISE SET 1.3

CHECK YOUR UNDERSTANDING

1. **True or false:** A local maximum may also be an absolute maximum.

2. What feature of a graph is measured by concavity?

3. If f is increasing and its graph is concave down, then as we move to the right, which is true?

   a. The graph gets steeper.

   b. The graph falls.

   c. The graph becomes less steep.

   d. None of the above.

4. **True or false:** The absolute minimum of a graph may occur at an endpoint.

5. A graph may have:

   a. more than one local maximum

   b. no local maximum

   c. more than one local minimum

   d. any of the above

6. What can be said about the rate of change of an increasing function on an interval?

7. A graph that has the shape of a frown is:

   a. concave up   **c.** increasing

   b. concave down   **d.** decreasing

8. Let f be a decreasing function with domain $[0, 2]$. Identify the location of the absolute minimum and the absolute maximum of f.

9. **True or false:** A decreasing function cannot have a local minimum.

10. **True or false:** If the graph of an increasing function is concave down, then the function is increasing at a decreasing rate.

11. The rate of change at a local maximum or local minimum is typically _____.

SKILL BUILDING

12. **Maximum?** A friend says that $x = -1$ is a local maximum value of the function $f(x)$ whose graph is shown in **Figure 1.50**. Is she correct?

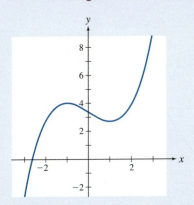

Figure 1.50 Possible local maximum value

Coloring graphs. In Exercises 13 through 23, a graph of a function f is given. Color the graphs as directed. You may trace the graph to color it.

13. Color green the portion of the graph in **Figure 1.51** where f is decreasing.

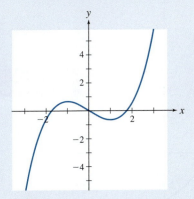

Figure 1.51 Graph for Exercises 13 through 15

14. Color green the portion of the graph in Figure 1.51 where f is concave up.

15. Color green the portion of the graph in Figure 1.51 that is between a local maximum and a local minimum.

16. Color green the portion of the graph in **Figure 1.52** where the rate of change of f is negative.

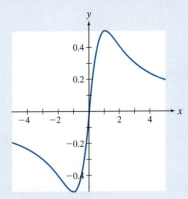

Figure 1.52 Graph for Exercises 16 through 18

17. Color green the portion of the graph in Figure 1.52 where the rate of change of f is positive.

18. Color green the portion of the graph in Figure 1.52 where the graph is concave up.

19. Color green the portion of the graph in **Figure 1.53** where f is increasing.

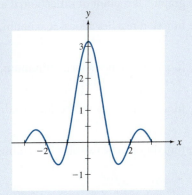

Figure 1.53 Graph for Exercises 19 through 21

20. Color green the region of the graph in Figure 1.53 containing the absolute maximum where the graph is concave down.

21. Color green the portion of the graph in Figure 1.53 where the graph is concave up.

22. Color green the portion of the graph in **Figure 1.54** where the function is increasing at a decreasing rate.

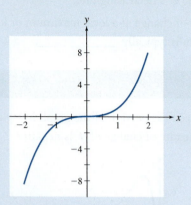

Figure 1.54 Graph for Exercises 22 and 23

23. Color green the portion of the graph in Figure 1.54 where the function is increasing at an increasing rate.

Features of graphs. The graph of a function $y = f(x)$ is shown with each of Exercises 24 through 32. You may assume that the domain of the function is indicated by the extent of the horizontal axis. Determine, using the given graph, each of the following:

a. Where the function is increasing, if anywhere

b. Where the function is decreasing, if anywhere

c. Where the graph is concave up, if anywhere

d. Where the graph is concave down, if anywhere

e. Where the graph has a point of inflection, if any

f. Where the function reaches a local maximum, if any

g. Where the function reaches a local minimum, if any

h. Where the function reaches an absolute maximum, if any

i. Where the function reaches an absolute minimum, if any

Remember that absolute maxima and minima may occur at endpoints of a graph.

24.

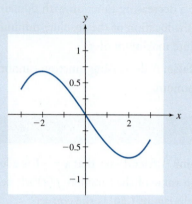

25.

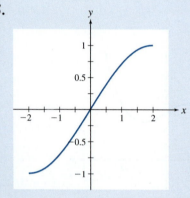

26.

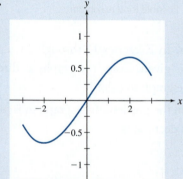

27.

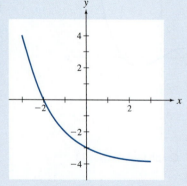

28.

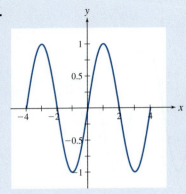

For this graph, you will need to *estimate* the intervals of concavity and the location of the inflection points.

32.

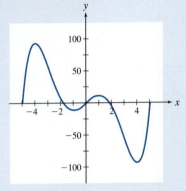

29.

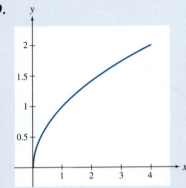

Identifying rates of change. Exercises 33 through 38 refer to the graph of a function $y = f(x)$. For each function, estimate where the rate of change is:

a. positive

b. negative

c. 0

In each case, you may assume the domain of the function is indicated by the extent of the x-axis. (Do not include the endpoints of the domain in your answers.)

30.

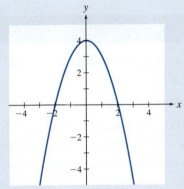

33.

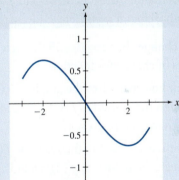

31.

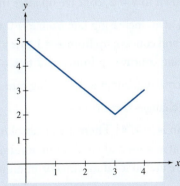

34.

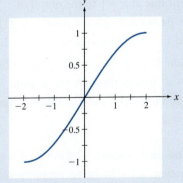

35.

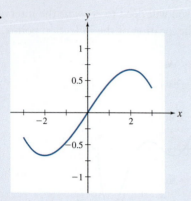

36.

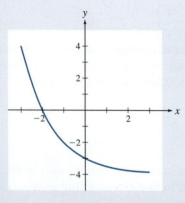

37.

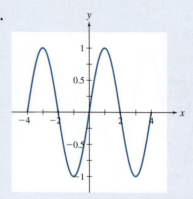

38.

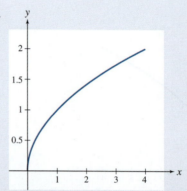

PROBLEMS

Finding rates of change. In Exercises 39 through 42, you will calculate rates of change for functions.

39. The tangent line at $x = 1$ to the graph of $f(x) = x^3$ passes through the points $(1, f(1))$ and $(2, 4)$. Find the rate of change of f at $x = 1$.

40. The tangent line at $x = 2$ to the graph of $f(x) = x^3$ passes through the points $(2, f(2))$ and $(3, 20)$. Find the rate of change of f at $x = 2$.

41. The tangent line at $x = 1$ to the graph of $f(x) = \sqrt{x}$ passes through the points $(1, f(1))$ and $(5, 3)$. Find the rate of change of f at $x = 1$.

42. The tangent line at $x = 2$ to the graph of $f(x) = x - x^2$ passes through the points $(2, f(2))$ and $(4, -8)$. Find the rate of change of f at $x = 2$.

Sketching graphs. In Exercises 43 through 53, sketch a graph of a function with the given properties. There should be one graph for each exercise. For each exercise, there is more than one correct answer. Answers provided may serve as examples.

43. Increasing from $x = 0$ to $x = 2$, and decreasing after that

44. Local minimum at $x = 1$ and at $x = 3$, local maximum at $x = 2$

45. Point of inflection at $(2, 1)$

46. The rate of change is negative from $x = 0$ to $x = 2$ and from $x = 3$ to $x = 4$. The rate of change is positive elsewhere (except at $x = 0, x = 2, x = 3,$ and $x = 4$).

47. Concave down to $x = 2$, concave up after that

48. Decreasing at an increasing rate from $x = 0$ to $x = 1$, decreasing and concave up from $x = 1$ to $x = 2$, and increasing and concave up from $x = 2$ to $x = 3$

49. Always decreasing, but at a decreasing rate

50. Concavity changes three times

51. The domain is $[-4, 4]$. There are an absolute maximum at $x = -4$ and an absolute minimum at $x = 4$. There are no local maxima or minima.

52. There are many (perhaps infinitely many) local maxima and local minima.

53. The domain of f is $[-2, 2]$ excluding the point $x = 0$. Also, f decreases on $[-2, 0)$ and on $(0, 2]$, but $f(-1) < f(1)$. *Suggestion*: Think in terms of a piecewise-defined function.

Familiar graphs. Exercises 54 through 63 involve functions whose graphs can be constructed easily by hand. The domain is assumed to be the set of all real numbers for which the given formula is defined.

54. Find the local and absolute maxima and minima of $f(x) = x^2$.

55. For $f(x) = x^2$, determine the regions where the rate of change is positive and where the rate of change is negative.

56. For $f(x) = x^2$, determine where the rate of change is zero.

57. Does the graph of $y = x^2$ have any points of inflection?

58. Let $y = 3x + 1$. Where is this function increasing, and where is it decreasing? Find all maxima and minima.

59. Find all points where the rate of change of $f(x) = x - 2$ is positive.

60. Let $y = 1 - x^2$. Find all local and absolute maxima and minima. Where is the graph concave up, and where is it concave down?

61. Let $y = x^3$. Locate all points of inflection.

62. For $f(x) = x^3$, are there any points where the rate of change is negative?

63. Let $y = \dfrac{1}{x}$. Determine the concavity.

64. **Is there a maximum?** Restrict the domain of $f(x) = \dfrac{1}{x}$ to the interval $(0, 1)$. Does f have an absolute maximum? Does it have an absolute minimum?

Finding maxima and minima. In Exercises 65 through 71, use a graphing utility to make the graph. Then give (approximately) any local maxima and minima as well as absolute maxima and minima. The answers that are provided are accurate to two decimal places. The accuracy of your answers will depend on the specific technology used and the preference of your instructor.

65. $f(x) = \dfrac{2^x}{1 + x^2}$. The domain is $[-2, 4]$.

66. $f(x) = 2x^3 + 3x^2 - 12x + 1$. The domain is $[-3, 3]$.

67. $f(x) = \dfrac{1}{\sqrt{x^2 + x + 3}}$. The domain is $[-3, 3]$.

68. $f(x) = \dfrac{x^2}{1 + x^4}$. The domain is $[-3, 3]$.

69. $f(x) = 6\sqrt{x} - x$. The domain is $[0, 25]$.

70. $f(x) = 4\sqrt{x} - x$. The domain is $[0, 25]$.

71. $f(x) = x^3 + x + 1$. The domain is $[-3, 3]$.

MODELS AND APPLICATIONS

72. **A sunflower.** Figure 1.55 shows the height in centimeters of a sunflower as a function of the time in days since germination. Over what (approximate) period is the graph concave up? What is happening to the growth rate of the flower over this period?

73. **Photosynthesis.** Figure 1.56 shows the net carbon dioxide exchange rates for 5-week-old rice plants as a function of the amount of light (in thousands of foot-candles) at $40°, 60°$, and $80°$. Values above the horizontal axis indicate absorption by the plant, and values below the axis indicate expulsion of carbon dioxide.

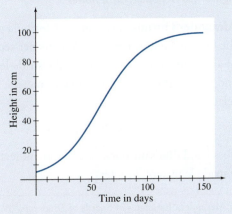

Figure 1.55 Growth of a sunflower

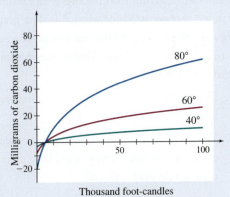

Figure 1.56 Carbon dioxide exchange

a. At which temperature do the rice plants show the least sensitivity to light?

b. At 80°, under which light condition would a change in light have a greater effect on the carbon dioxide exchange: dimmer light or brighter light?

74. Going for ice cream. The graph in **Figure 1.57** shows your distance west of home, in yards, as a function of the time in minutes since you left home. You walk at a steady pace to the Dairy-berry store, where you enjoy an ice cream cone. You then start back home but stop to visit a friend.

a. How far from home is the Dairy-berry store?

b. How long did you spend eating ice cream?

c. How far is your friend's house from the ice cream store?

d. How is the "steady" pace of your walk to the store reflected in the shape of the graph over that period?

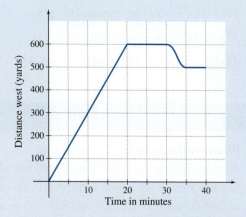

Figure 1.57 A trip for ice cream

75. Windchill. Wind speed affects perceived temperature, giving the effective temperature or temperature adjusted for windchill. The graph in **Figure 1.58** shows the effective temperature as a function of the wind speed when the thermometer reads 20° Fahrenheit.

a. Does a higher wind speed cause a higher or a lower effective temperature?

b. Does a change in wind speed have a greater effect on effective temperature on a calm day or on a windy day?

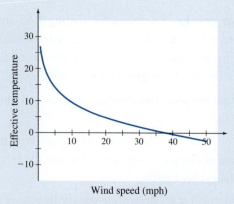

Figure 1.58 Windchill when the thermometer reads 20°

76. A population of wolves. The graph in **Figure 1.59** shows the population N of wolves in a protected area t years after observation of the population began.

a. What (approximately) was the largest wolf population during the observation period, and when did that occur?

b. An extended drought severely limited the food supply available for wolves. Identify (approximately) the time period of the drought.

c. Was the wolf population growing faster early in the observation period or later?

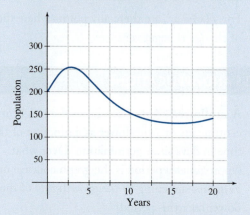

Figure 1.59 A population of wolves

77. Alexander's formula. Alexander's formula gives the (approximate) speed v, in meters per second, of a running animal in terms of the stride length s and hip height h, both measured in meters.[6] The relationship is

$$v = 0.78s^{1.67}h^{-1.17}$$

Fossilized dinosaur tracks are found, indicating a stride length of 3 meters.

[6]R. McNeill Alexander, "Estimates of Speeds of Dinosaurs," *Nature* 261: 129–130 (1976).

a. Plot the graph of speed versus hip height for hip heights up to 4 meters.

b. Suppose we estimate the hip height h and calculate the running speed. Does an error in hip height estimate lead to a greater error in the running speed calculation for dinosaurs of longer or shorter hip height?

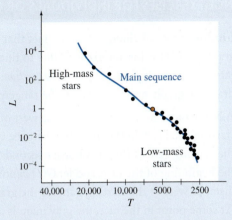

Figure 1.60 H-R diagram for main-sequence stars

78. Red shift. Many stellar objects are moving away from us at sufficient radial velocity to produce a significant red shift in the spectrum. The radial velocity v can be calculated from the red shift z using

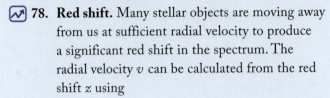

$$v = c\left(\frac{(z+1)^2 - 1}{(z+1)^2 + 1}\right)$$

where c is the speed of light.

a. The red shift of the quasar 3C 273 is $z = 0.158$. (This means that the wavelengths of its spectral lines are stretched by 15.8%.) How fast is this quasar moving away from us? Report your answer as a percentage of the speed of light.

b. Use a graphing utility to make a graph of

$$y = \frac{(z+1)^2 - 1}{(z+1)^2 + 1}$$

c. Does the graph show an increasing function or a decreasing function? What is the concavity of the graph?

d. What does the information from part c reveal about the relationship between the red shift and the radial velocity?

79. Main-sequence stars. Main-sequence stars are common stars like our own sun. The Hertzsprung–Russell diagram, or H-R diagram, in **Figure 1.60** shows for main-sequence stars the relation between the surface temperature T, in kelvins, and the relative luminosity L, which is the ratio of a star's luminosity to that of the sun. Thus, the sun has relative luminosity 1, and a star that is four times as bright as the sun has a relative luminosity of 4. Be careful in using this graph: the horizonal axis is in reverse order, as is standard for the H-R diagram.

a. What is the surface temperature of the sun?

b. Is the relative luminosity for a main sequence star an increasing function of surface temperature or a decreasing function of surface temperature?

80. Supply and demand curves. A supply curve is a graph that shows the quantity of a product that is made available by suppliers as a function of the price. Similarly, a demand curve is a graph that shows the quantity of a product consumers are willing to purchase as a function of the price. For example, the red graph in **Figure 1.61** is a supply curve, and the blue graph is a demand curve.

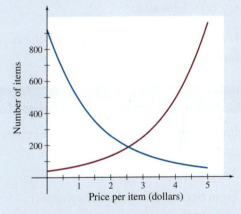

Figure 1.61 Supply (red) and demand (blue) curves

a. Is the supply curve in Figure 1.61 increasing or decreasing? Explain your answer in terms of the availability of items as the price increases.

b. Is the demand curve in Figure 1.61 increasing or decreasing? Explain your answer in terms of the quantity of items in demand as the price increases.

c. The equilibrium price is the price where demand matches supply. What is the equilibrium price in this case?

CHALLENGE EXERCISES FOR INDIVIDUALS OR GROUPS

81. Finding a rate of change. Use the fact that the graph of $f(x) = 1$ is a horizontal line to find the rate of change of f at any point. *Suggestion*: First draw the graph, and then sketch the tangent line.

82. A horse race. A racecourse runs due west of the starting gate. Let $R(t)$ denote the distance west of the starting gate of the roan horse t seconds after the beginning of the race, and let $G(t)$ denote the distance west of the starting gate for the gray mare. At some point in the race, the gray mare is ahead of the roan. From that time on, the rate of change of $G(t)$ is at least as large as the rate of change of $R(t)$. Bearing in mind that the rate of change of distance west is the velocity, which horse wins the race? Explain your reasoning.

83. A possible graph? Is it possible to have a function f with all of the following properties?

- The domain of f is $[0, \infty)$.

- f is increasing.

- $f(x) < 1$ for all x.

If not, explain why. If so, produce the required function and graph.

84. Another possible graph? Find a function g with the same properties as in Exercise 83 except that the domain is $(-\infty, \infty)$. *Suggestion*: Think in terms of a piecewise-defined graph.

85. Calculating a rate of change. Recall that the rate of change of a function f at x is defined in terms of the average rate of change of f on shorter and shorter intervals with x as an endpoint. In this exercise, we use this definition to calculate the rate of change of $f(x) = x^2$ at the point $x = 1$.

  a. Show that the average rate of change of $f(x) = x^2$ on the interval $[1, 1 + h]$ equals $2 + h$. (Assume that $h > 0$.)

  b. Describe what happens to the average rate of change from part a as the length of the interval gets very small—that is, when h is near 0. What does your answer give for the rate of change of $f(x) = x^2$ at the point $x = 1$?

  c. Use your answer to part b to find the tangent line to the graph $y = x^2$ corresponding to the point $x = 1$. Then make a careful sketch of the graph and the tangent line.

REVIEW AND REFRESH: Exercises from Previous Sections

86. From Section P.1: Find the distance between the points $(1, 4)$ and $(2, 5)$.

  Answer: $\sqrt{2}$

87. From Section P.2: Solve the inequality $|x - 4| < 3$.

  Answer: $(1, 7)$

88. From Section P.3: Solve the inequality $x(x + 1)(x - 3) > 0$.

  Answer: $(-1, 0) \cup (3, \infty)$

89. From Section 1.1: If $f(x) = x^2$, simplify $\dfrac{f(x + h) - f(x)}{h}$ for $h \neq 0$.

  Answer: $2x + h$

90. From Section 1.1: Find the domain of $\dfrac{\sqrt{x}}{x^2 - x - 2}$.

  Answer: $[0, 2) \cup (2, \infty)$

91. From Section 1.2: If $f(x) = \dfrac{1}{x^2}$, find the average rate of change of $f(x)$ from $x = 1$ to $x = 2$.

  Answer: $-\dfrac{3}{4}$

92. From Section 1.2: If $f(x) = x^2 - 3$, calculate the average rate of change of $f(x)$ on the interval $[x, x + h]$. Assume that $h > 0$, and simplify your answer.

  Answer: $2x + h$

1.4 Limits and End Behavior of Graphs

> The limit at infinity describes the long-term behavior of functions.

1.1 The Basics of Functions

1.2 Average Rates of Change

1.3 Graphs and Rates of Change

1.4 Limits and End Behavior of Graphs

In this section, you will learn to:

1. Explain the concept of the long-term behavior, or limit, of a function.
2. Use the appropriate notation to represent the limit of a function.
3. Calculate elementary limits.
4. Estimate the limit of a function using a graph or a table.
5. Interpret the limit of a function in applications.

When a skydiver (**Figure 1.62**) jumps from an airplane, gravity causes her downward velocity to increase rapidly. But air resistance works against the downward pull of gravity, and the faster she goes, the greater the opposing force due to air resistance. Eventually, air resistance almost matches the pull of gravity, and the skydiver's downward velocity levels out at a terminal velocity of about 120 miles per hour. This is an example of how the limiting behavior of processes is important in life.

Limiting values are also important in medicine. When your doctor gives you an injection of an antibiotic, the drug level in your bloodstream increases rapidly. But after a time, it reaches a maximum, and then it slowly decays. Eventually, the drug level approaches the normal, limiting value of zero.

Consider also the study of population growth. With plentiful resources, an animal or human population may grow rapidly. But as the population increases, there is greater and greater competition for resources that are finite. The result is that population growth eventually slows, and population levels stabilize.

Figure 1.62 A skydiver Brian Erler/Getty Images

An Intuitive Discussion of Limits at Infinity

> The limit at infinity is understood in terms of the end behavior of the graph.

Let's look more closely at the skydiver. Her downward velocity t seconds after she has jumped from the airplane, and before the parachute opens, is given by

$$V(t) = 120(1 - 0.75^t) \text{ miles per hour}$$

The accompanying table of values and the graph in **Figure 1.63** help us understand what the formula means to the skydiver.

Time t	Velocity $V(t)$
0	0
5	91.52
10	113.24
15	118.40
20	119.62
25	119.91

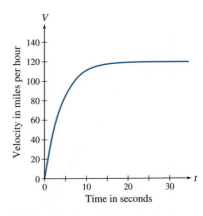

Figure 1.63 The velocity of a skydiver

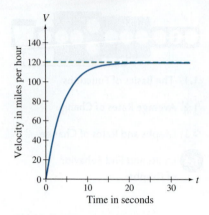

Figure 1.64 Limiting value: The skydiver's velocity approaches 120 miles per hour.

Both the table of values and the graph indicate that, as time passes, the skydiver's velocity stabilizes as it approaches 120 miles per hour. To further illustrate the point, in **Figure 1.64** we have added the horizontal line $y = 120$ to the graph of $V(t)$. This line clearly shows the long-term behavior or end behavior of the function $V(t)$ and its graph.

We use the notation $V(t) \to 120$ as $t \to \infty$ to indicate that, for large values of t, $V(t)$ approaches the value 120. Many calculus texts use the notation $\lim\limits_{t \to \infty} V(t) = 120$. Both indicate the limiting value of $V(t)$ or the limit as t goes to infinity of $V(t)$.

Examples of Limits at Infinity

> Some common patterns emerge in the end behavior of graphs.

A number of different things may happen at the tail ends of graphs. A few examples of long-term behavior are illustrated by the graphs of $y = x^2$, $y = -x^2$, and $y = \dfrac{1}{x^2}$.

Long-term behavior of $y = x^2$: When x gets larger and larger, x^2 gets larger too. This fact is illustrated by both the accompanying table of values and the graph in **Figure 1.65**, which rises without any bound as we move to the right.

x	$y = x^2$
10	100
100	10,000
1000	1,000,000
100,000	10,000,000,000

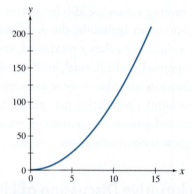

DF Figure 1.65 $x^2 \to \infty$ as $x \to \infty$: The function $y = x^2$ increases without bound as x increases.

We indicate this fact by writing

$$x^2 \to \infty \text{ as } x \to \infty$$

This notation is read "x^2 approaches infinity as x approaches infinity." Many calculus texts use the notation

$$\lim\limits_{x \to \infty} f(x) = \infty$$

Long-term behavior of $y = -x^2$: When x gets larger and larger, $-x^2$ gets larger in magnitude too, but with a minus sign. Thus, the function decreases without bound. This fact is illustrated by the accompanying table of values and the graph in **Figure 1.66**, which falls indefinitely as we move to the right.

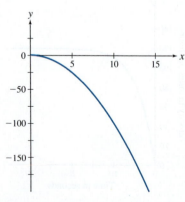

Figure 1.66 $-x^2 \to -\infty$ as $x \to \infty$: The function $y = -x^2$ decreases without bound as x increases.

x	$y = -x^2$
10	−100
100	−10,000
1000	−1,000,000
100,000	−10,000,000,000

We indicate this fact by writing

$$-x^2 \to -\infty \text{ as } x \to \infty$$

This notation is read "$-x^2$ approaches negative infinity as x approaches infinity."

Long-term behavior of $y = \dfrac{1}{x^2}$: A large denominator (in relation to the size of the numerator) means a small fraction. For example, 10,000 is a very large number, so $1/10{,}000 = 0.0001$ is a very small number. This example suggests that when x is large, $y = \dfrac{1}{x^2}$ is very small. That is, as x gets larger and larger, $\dfrac{1}{x^2}$ gets closer and closer to zero. In symbols, $\dfrac{1}{x^2} \to 0$ as $x \to \infty$. This notation is read "$\dfrac{1}{x^2}$ approaches zero as x approaches infinity." In calculus texts, this fact is often written as $\displaystyle\lim_{x\to\infty} \dfrac{1}{x^2} = 0$.

The fact that the limit is 0 means the graph shown in **Figure 1.67** approaches the x-axis at its right tail end. The behavior of $\dfrac{1}{x^2}$ for increasing x can also be seen in the accompanying table of values.

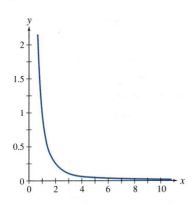

x	$\dfrac{1}{x^2}$
10	0.01
100	0.0001
1000	0.000001
100,000	0.0000000001

Figure 1.67 $\dfrac{1}{x^2} \to 0$ as $x \to \infty$: The graph of $\dfrac{1}{x^2}$ gets close to the x-axis as x increases.

In our discussion, we have considered the right tail end of the graph of f in connection with the long-term behavior of $f(x)$. In a similar way, we can study the left tail end of the graph of f. In this situation, we ask about the behavior of f when x is large in size but negative, and we write $x \to -\infty$.

CONCEPTS TO REMEMBER: Intuitive Notion of Limits

- The notation $f(x) \to \infty$ as $x \to \infty$ is read "$f(x)$ approaches infinity as x approaches infinity." It means that $f(x)$ grows without bound as x gets very large, and that the graph rises indefinitely at the right tail end.

Example:

For $n > 0$, $x^n \to \infty$ as $x \to \infty$

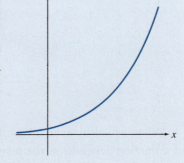

- The notation $f(x) \to -\infty$ as $x \to \infty$ is read "$f(x)$ approaches negative infinity as x approaches infinity." It means that $f(x)$ decreases without bound as x gets very large, and that the graph falls indefinitely at the right tail end.

  Example:

  For $n > 0$, $-x^n \to -\infty$ as $x \to \infty$

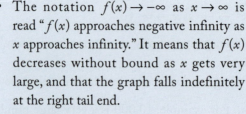

- The notation $f(x) \to c$ as $x \to \infty$ is read "$f(x)$ approaches c as x approaches infinity." It means that $f(x)$ gets arbitrarily close to c as x gets very large and that the graph approaches the line $y = c$ at the right tail end.

  Example:

  For $n > 0$, $\dfrac{1}{x^n} \to 0$ as $x \to \infty$

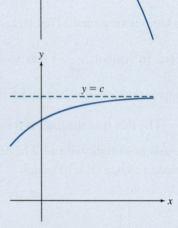

- This reasoning also applies to the case when x is negative and x gets very large in size. For this, we replace $x \to \infty$ by $x \to -\infty$, and the results describe the left tail end of the graph rather than the right.

$$f(x) \to \infty \qquad\qquad f(x) \to -\infty$$
$$\text{as } x \to -\infty \qquad\qquad \text{as } x \to -\infty$$

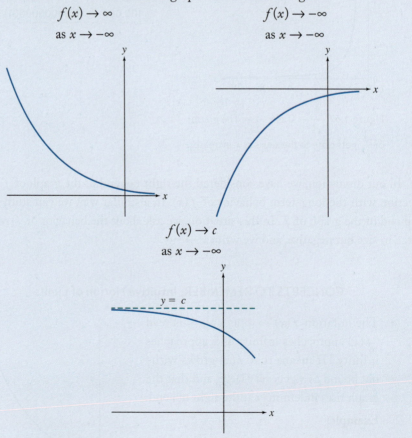

$$f(x) \to c$$
$$\text{as } x \to -\infty$$

When we calculate limits involving fractions, it is worth emphasizing that the size of a fraction depends on the relative sizes of the numerator and denominator. Just knowing, for example, that the denominator is large doesn't automatically make the fraction small, because the numerator may be much larger than the denominator. Also, not all functions have a limit at infinity.

Elementary Limit Calculations

> The intuitive notion of limits is used to perform simple limit calculations.

In the following examples, we use the results just discussed to find limits at infinity. In each example, we first explain how to find the limiting value and then support our explanation with both a table of values and a graph.

EXAMPLE 1.22 A Function That Decreases Without Bound

Find the limit as $x \to \infty$ of $y = -x^3$.

SOLUTION

We know that as x increases x^3 grows without bound. Hence, $-x^3$ grows without bound in the negative direction. We conclude that

$$-x^3 \to -\infty \text{ as } x \to \infty$$

The limit calculation is reinforced by the accompanying table. This means the graph in **Figure 1.68** falls indefinitely.

x	$-x^3$
1	−1
10	−1000
100	−1,000,000
10,000	−1,000,000,000,000

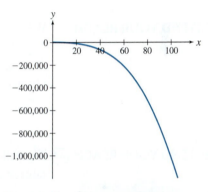

Figure 1.68 $-x^3 \to -\infty$ as $x \to \infty$

TRY IT YOURSELF 1.22 Brief answers provided at the end of the section.

Calculate the limit as $x \to -\infty$ of $y = -x^3$, and show the left-hand tail of the graph.

EXTEND YOUR REACH

Earlier in this section we gave three examples to illustrate typical behavior of limits as $x \to \infty$. Find corresponding examples to illustrate typical behavior of limits as $x \to -\infty$. The limit in Try It Yourself 1.22 shows one.

EXAMPLE 1.23 A Graph That Approaches the x-Axis

Find the limit as $x \to \infty$ of $y = \dfrac{100}{x+1}$.

SOLUTION

When x is very large, in particular much larger than 100, the denominator $x+1$ is much larger than the numerator 100. Hence, the fraction $\dfrac{100}{x+1}$ is near 0 when x is large. We conclude that

$$\frac{100}{x+1} \to 0 \text{ as } x \to \infty$$

The accompanying table of values supports our assertion that the limit is 0. As a consequence, the graph in **Figure 1.69** approaches the x-axis when x approaches its right tail end.

x	$\dfrac{100}{x+1}$
10	9.0909
100	0.9901
1000	0.0999
10,000	0.0100
100,000	0.0010

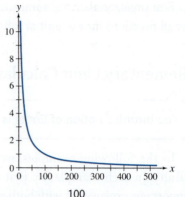

Figure 1.69 $\dfrac{100}{x+1} \to 0$ as $x \to \infty$

TRY IT YOURSELF 1.23 Brief answers provided at the end of the section.

Calculate the limit as $x \to -\infty$ of $y = \dfrac{5}{x}$, and show the left-hand tail of the graph.

EXTEND YOUR REACH   What do you think is the limit as $x \to \infty$ of the rate of change of the function $y = \dfrac{100}{x+1}$? *Suggestion*: Use a straightedge with the graph in Figure 1.69 to estimate the slopes of tangent lines as we move to the right.

EXTEND YOUR REACH The following limits are difficult to calculate exactly. Use a calculator to make a table of values and a graph. Then use those to estimate the limits.

1. Estimate the limit as $x \to \infty$ of $y = \dfrac{200}{4 + 2^{-x}}$.

2. Estimate the limit as $x \to \infty$ of $y = x(2^{1/x} - 1)$.

3. Estimate the limit as $x \to \infty$ of $y = \dfrac{x^2}{2^x}$.

Breaking Limit Calculations into Simpler Steps

> Computing limits of complicated functions is accomplished in steps.

The calculation of limits can often be broken down into simpler steps. Consider, for example, the limit as $x \to \infty$ of $y = \dfrac{1}{x^2} + 3$. We know that $\dfrac{1}{x^2} \to 0$ as $x \to \infty$. Then, because $\dfrac{1}{x^2}$ is near 0, the expression $\dfrac{1}{x^2} + 3$ is near 3. The following shorthand notation is both useful and common:

$$\text{As } x \to \infty, \quad \overset{0}{\cancel{\dfrac{1}{x^2}}} + 3 \to 3$$

Thus, $\dfrac{1}{x^2} + 3 \to 3$ as $x \to \infty$.

Both numerical and graphical evidence support our assertion that the limit is 3. Note that the right tail end of the graph shown in **Figure 1.70** approaches the line $y = 3$. The accompanying table of values also illustrates the end behavior.

x	$y = \dfrac{1}{x^2} + 3$
10	3.01
100	3.0001
1000	3.000001
100,000	3.0000000001

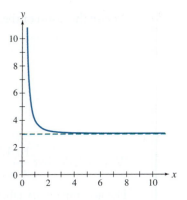

Figure 1.70 $\dfrac{1}{x^2} + 3 \to 3$ as $x \to \infty$:

The graph of $y = \dfrac{1}{x^2} + 3$ approaches the line $y = 3$ as x increases.

EXAMPLE 1.24 Steps in Limit Calculations

a. Find the limit as $x \to \infty$ of $y = 5 + \dfrac{100}{x}$.

b. Observe that $\dfrac{1+12x}{1+3x} = \dfrac{\frac{1}{x}+12}{\frac{1}{x}+3}$ when $x \neq 0$. Use this fact to find the limit as $x \to \infty$ of $y = \dfrac{1+12x}{1+3x}$.

SOLUTION

a. When x is much larger than 100, the fraction $\dfrac{100}{x}$ is near 0. Hence,

$$\text{As } x \to \infty, \; 5 + \frac{\overset{0}{\cancel{100}}}{x} \to 5$$

We conclude that

$$5 + \frac{100}{x} \to 5 \text{ as } x \to \infty$$

x	$5 + \dfrac{100}{x}$
10	15
100	6
1000	5.1
10,000	5.01
100,000	5.001

The accompanying table of values supports our calculation. This limit means that the graph in **Figure 1.71** approaches the line $y = 5$.

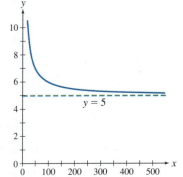

Figure 1.71 $5 + \dfrac{100}{x} \to 5$ as $x \to \infty$

b. We establish the equality by dividing both the top and the bottom of the fraction by x:

$$\frac{1+12x}{1+3x} = \frac{\left(\frac{1}{x}\right)(1+12x)}{\left(\frac{1}{x}\right)(1+3x)} \qquad \text{Divide top and bottom by } x \text{ for } x \neq 0.$$

$$= \frac{\frac{1}{x}+12}{\frac{1}{x}+3} \qquad \text{Simplify.}$$

We use this form of the fraction to find the limit:

$$\frac{\frac{1}{x}+12}{\frac{1}{x}+3} \to \frac{\overset{0}{\cancel{\frac{1}{x}}}+12}{\underset{0}{\cancel{\frac{1}{x}}}+3} \text{ as } x \to \infty \qquad \textbf{Use } \frac{1}{x} \to 0 \text{ as } x \to \infty.$$

$$= \frac{12}{3} = 4 \qquad\qquad \textbf{Simplify.}$$

We conclude that $\dfrac{1+12x}{1+3x} \to 4$ as $x \to \infty$.

The accompanying table of values supports our calculation. The limit shows that the graph in **Figure 1.72** approaches the line $y = 4$.

x	$\dfrac{1+12x}{1+3x}$
10	3.903226
100	3.990033
1000	3.999000
10,000	3.999900

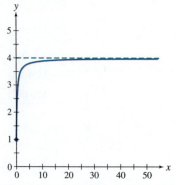

Figure 1.72 $\dfrac{1+12x}{1+3x} \to 4$ as $x \to \infty$

TRY IT YOURSELF 1.24 Brief answers provided at the end of the section.

a. Find the limit as $x \to \infty$ of $f(x) = \dfrac{2}{1+\frac{1}{x}}$.

b. Observe that $\dfrac{4x+1}{2x-1} = \dfrac{4+\frac{1}{x}}{2-\frac{1}{x}}$ when $x \neq 0$. Use this fact to find the limit as $x \to \infty$ of $y = \dfrac{4x+1}{2x-1}$.

EXTEND YOUR REACH

The approach used in Example 1.24 can be generalized to work for other limits. Use the given suggestions to calculate the following limits.

1. Find the limit as $x \to \infty$ of $y = \dfrac{2x^2+1}{x^2+3}$. *Suggestion:* Divide top and bottom by x^2.

2. Find the limit as $x \to \infty$ of $y = \dfrac{x^2+1}{x+1}$. *Suggestion:* Divide top and bottom by x.

3. Find the limit as $x \to \infty$ of $y = \dfrac{x+1}{x^2+1}$. *Suggestion:* Divide top and bottom by x^2.

4. Find the limit as $x \to \infty$ of $y = \dfrac{2x^2+ax+b}{x^2+cx+d}$. *Suggestion:* Divide top and bottom by x^2. Do the values of the constants a, b, c, and d affect the value of the limit?

Example 1.24 suggests the following strategy, which is often helpful in calculating limits of fractions. For a function of the form $y = \dfrac{ax+b}{cx+d}$, the calculations of limits at $\pm\infty$ are facilitated by dividing the top and the bottom of the fraction by x:

$$\frac{ax+b}{cx+d} = \frac{a+\frac{b}{x}}{c+\frac{d}{x}}, \quad x \neq 0$$

The concept of a limit will prove to be invaluable in describing graphs of new functions in coming chapters. For example, in Chapter 3 we will show how to calculate limits of exponential functions; in Chapter 4 we will calculate limits of logarithmic functions; and in Chapter 5 we will calculate limits of polynomials and rational functions.

MODELS AND APPLICATIONS Long-Term Behavior

In applications, limits at infinity express what happens in the long run. For example, suppose a cup of tea is left on the counter. Let $T(t)$ denote the temperature, in degrees, of the tea t minutes after it is left there. Then the limit as $t \to \infty$ of $T(t)$ is the temperature of the tea after a long time. Eventually, the temperature matches that of the room, say 70°. In terms of limits, $T(t) \to 70$ as $t \to \infty$.

EXAMPLE 1.25 The Meaning of Limits in Applied Settings

a. Water containing 20% salt flows into a tank of initially pure water, and the well-mixed solution flows out at the same rate. Let $S(t)$ denote the percentage of salt in the tank t hours after the process begins. Explain in terms of the percentage of salt the meaning of the following expression: $S(t) \to 20\%$ as $t \to \infty$.

b. A population is growing in an environment that can support 500 individuals. If $N(t)$ is the population at time t, use limit notation to express the statement "The population will eventually grow toward 500 individuals."

SOLUTION

a. The limit notation tells us that $S(t)$ tends to a value of 20% in the long run. This means that, after a long time, the solution in the tank will approach 20% salt.

b. Because the population grows to 500, the function $N(t)$ approaches 500 for large values of t. In terms of limits,

$$N(t) \to 500 \text{ as } t \to \infty$$

TRY IT YOURSELF 1.25 Brief answers provided at the end of the section.

The spacecraft *Voyager 1* is currently about 14 billion miles from the sun and is moving away at a constant velocity. In space, there is nothing to slow the craft. If $D(t)$ is the distance from *Voyager 1* to the sun at time t, what is the limit as $t \to \infty$ of $D(t)$?

EXAMPLE 1.26 Functional Response

The number P of prey taken by a predator depends on the abundance x of the prey. The relationship proposed by C. S. Holling[7] is known as the functional response:

$$P(x) = \frac{ax}{1 + bx}$$

Here a and b are constants that are determined by the specific predator and prey involved, and $b \neq 0$.

a. Find the limit as $x \to \infty$ of $P(x)$.

b. Explain the meaning of the limit you calculated in part a in terms of the number of prey taken by a predator.

SOLUTION

a. We divide the top and the bottom of the fraction by x in order to facilitate the limit calculation.

$$\frac{ax}{1 + bx} = \frac{a}{\frac{1}{x} + b} \qquad \text{Divide top and bottom by } x, x \neq 0.$$

as $x \to \infty$

$$\frac{a}{\frac{1}{x} + b} \to \frac{a}{\frac{0}{\frac{1}{x}} + b} \qquad \text{Use } \frac{1}{x} \to 0 \text{ as } x \to \infty.$$

$$= \frac{a}{b}$$

---

[7] C. S. Holling, "Some Characteristics of Simple Types of Predation and Parasitism," *The Canadian Entomologist* 91: 385-398 (1959).

We conclude that

$$P(x) \to \frac{a}{b} \text{ as } x \to \infty$$

b. To say that x is large means that the abundance of prey is large. Thus, the limit $\frac{a}{b}$ is the largest number of prey the predator will take no matter how much prey is available.

TRY IT YOURSELF 1.26 Brief answers provided at the end of the section.

If a lake is stocked with N_0 fish of the same age, then the number C of fish caught by fishing over the life span of the fish is

$$C(x) = \frac{x\,N_0}{1+x}$$

Here x represents the ratio of fish caught annually to fish that die annually of natural causes. This is one form of the catch equation.

a. Find the limit as $x \to \infty$ of $C(x)$.

b. Explain the meaning of the limit in terms of the ratio of fish caught annually to fish dying annually.

In the next chapter, we will learn how to calculate limits of exponential functions. But for now, we will rely on graphs and tables of values to make estimates.

EXAMPLE 1.27 Population Growth

The number N of a protected species in a controlled environment after t years is given by

$$N = \frac{1000}{1+19 \times 0.7^t}$$

Because the resources of the controlled environment are limited, we expect the population N to tend toward a limiting value.

a. Make a table of values showing the population in years $t = 0, 5, 10, 15, 20, 25, 30$. Round your data to the nearest whole number.

b. Use a graphing utility to plot the graph of N versus t over the first 30 years.

c. Based on the table of values from part a and the graph from part b, estimate the value of the limit as $t \to \infty$ of N.

d. Explain the meaning of the limit you calculated in part c in terms of the population.

SOLUTION

a. We used a calculator to make the accompanying table of values.

t	N
0	50
5	238
10	651
15	917
20	985
25	997
30	1000

b. The graph is in **Figure 1.73**.

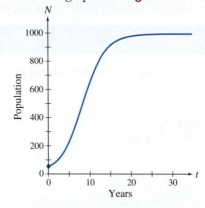

Figure 1.73 The population over a 30-year period

c. Both the graph and the table of values suggest a limiting value of around 1000.

d. In the long term, we expect the population to grow and stabilize at around 1000 individuals.

TRY IT YOURSELF 1.27 Brief answers provided at the end of the section.

Repeat the example using a growth model of $N = \dfrac{1000}{1 - 0.5 \times 0.7^t}$.

TRY IT YOURSELF ANSWERS

1.22 $-x^3 \to \infty$ as $x \to -\infty$

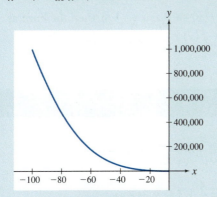

1.23 $\dfrac{5}{x} \to 0$ as $x \to -\infty$

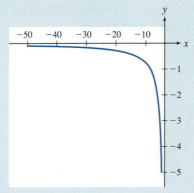

1.24 **a.** $\dfrac{2}{1 + \frac{1}{x}} \to 2$ as $x \to \infty$

b. $\dfrac{4x + 1}{2x - 1} \to 2$ as $x \to \infty$

1.25 $D(t) \to \infty$ as $t \to \infty$ (if the diameter of the universe is idealized to be infinitely large)

1.26 **a.** $C(x) \to N_0$ as $x \to \infty$

b. If the ratio of fish caught annually to fish dying annually is very large, then virtually all of the original number N_0 of fish are caught by fishing.

1.27 **a.**

t	N
0	2000
5	1092
10	1014
15	1002
20	1000
25	1000
30	1000

b.

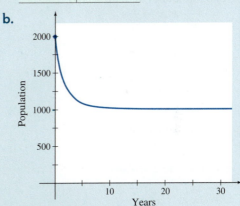

c. The limit is 1000.

d. The population declines and stabilizes at around 1000 individuals.

EXERCISE SET 1.4

CHECK YOUR UNDERSTANDING

1. The limit as $x \to \infty$ shows which of the following?

   a. The behavior of the graph at the right tail end

   b. The behavior of the graph at the left tail end

   c. The largest possible function value

   d. None of the above

2. Is it possible to have a function that is always decreasing and such that the limit as $x \to \infty$ is ∞?

3. The symbol ∞ is which of the following?

    a. The largest number there is

    b. Not as big as $\infty + 1$

    c. Not a number at all

    d. None of the above

4. The limit as $x \to -\infty$ shows which of the following?

    a. The negative of the limit as $x \to \infty$

    b. The behavior of the graph at the left tail end

    c. The same as the limit as $x \to \infty$

    d. None of the above

5. If we want to know what the graph of $f(x)$ looks like for large x, we should calculate _____.

6. **True or false:** A table of values can be useful in estimating a limit.

7. **True or false:** Every function has a limit as $x \to \infty$.

8. **True or false:** If a graph gets very close to a horizontal line, it can never move away from it.

9. **True or false:** If the denominator of a fraction is near 0, then the fraction must be very large.

10. If $f(x) \to 0$ as $x \to -\infty$, then the limit as $x \to -\infty$ of $-f(x)$ is _____.

SKILL BUILDING

In Exercises 11 through 27, find the indicated limit.

11. $y = \dfrac{1}{x^2}$ as $x \to \infty$

12. $y = \dfrac{1}{x^2}$ as $x \to -\infty$

13. $y = -\dfrac{1}{x^2}$ as $x \to -\infty$

14. $y = -\dfrac{1}{x^2}$ as $x \to \infty$

15. $y = \dfrac{1}{x}$ as $x \to -\infty$

16. $y = 3 - \dfrac{1}{x^2}$ as $x \to \infty$

17. $y = 4 + \dfrac{1}{x}$ as $x \to -\infty$

18. $y = \sqrt{x}$ as $x \to \infty$

19. $y = \dfrac{1}{\sqrt{x}} - 3$ as $x \to \infty$

20. $y = x^2 + x + 1$ as $x \to \infty$

21. $y = -3x$ as $x \to \infty$

22. $y = \dfrac{1}{x+5}$ as $x \to \infty$

23. $y = \dfrac{3}{x} - 5$ as $x \to \infty$

24. $y = \dfrac{\frac{2}{x} + 3}{\frac{1}{x} + 1}$ as $x \to \infty$

25. $y = \dfrac{2 + \frac{1}{x} + \frac{4}{x^2}}{4 - \frac{1}{x} - \frac{4}{x^2}}$ as $x \to \infty$

26. $y = \dfrac{9 - \frac{5}{x} + \frac{14}{\sqrt{x}}}{3 + \frac{1}{x}}$ as $x \to \infty$

27. $y = \dfrac{4}{x} + 3$ as $x \to -\infty$

Estimating limits based on graphs. In Exercises 28 through 31, estimate the indicated limit of $f(x)$ based on the given graph of the function f.

28. As $x \to \infty$

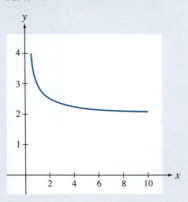

29. As $x \to \infty$

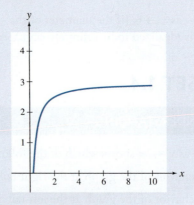

30. As $x \to \infty$ and as $x \to -\infty$

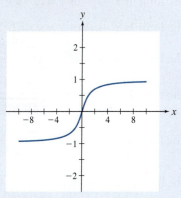

31. As $x \to \infty$

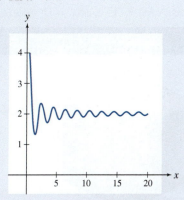

Estimating limits based on tables of values. In Exercises 32 through 35, use the given table of values for the function f to estimate the indicated limit of $f(x)$.

32. As $x \to \infty$

x	f(x)
100	1.35
200	1.79
300	1.87
400	1.96
500	1.98

34. As $x \to \infty$

x	f(x)
100	3.91
200	3.76
300	3.68
400	3.57
500	3.53

33. As $x \to \infty$

x	f(x)
100	6.33
200	6.21
300	6.15
400	6.08
500	6.04

35. As $x \to -\infty$

x	f(x)
−100	2.11
−200	2.38
−300	2.41
−400	2.47
−500	2.49

PROBLEMS

Sketching graphs. In Exercises 36 through 43, sketch a graph that satisfies the given condition. There is more than one correct answer to each of these problems. The answers provided are examples.

36. Concave down, and the limit as $x \to \infty$ is 3

37. Concave up, and the limit as $x \to -\infty$ is 2

38. The limit as $x \to \infty$ is 2, and the limit as $x \to -\infty$ is −2.

39. The limit as $x \to \infty$ is −1, and the limit as $x \to -\infty$ is ∞.

40. The limit as $x \to \infty$ is −∞, and the limit as $x \to -\infty$ is ∞.

41. The limit as $x \to \infty$ is 4, and the limit as $x \to -\infty$ is 4.

42. Increasing, and the limit as $x \to \infty$ is 5

43. Decreasing, and the limit as $x \to \infty$ is 2

Piecemeal calculations. Suppose that the limit as $x \to \infty$ of $f(x)$ is 4. In Exercises 44 through 52, use this fact to calculate the limit as $x \to \infty$ of the given expression.

44. $f(x) + 5$

45. $3f(x)$

46. $\dfrac{1}{f(x)}$

47. $\dfrac{f(x) + 1}{f(x) + 2}$

48. $\sqrt{f(x)}$

49. $(f(x))^2$

50. $1 + 3f(x)$

51. $\dfrac{25}{f(x) + 1}$

52. $(2 + f(x))^2$

Limits of functions of the form $y = \dfrac{ax + b}{cx + d}$. Exercises 53 through 59 deal with limits of certain ratios. In each case, find the limit as $x \to \infty$ of the given function.

53. $y = \dfrac{2x - 4}{x + 5}$

54. $y = \dfrac{x + 1}{x - 1}$

55. $y = \dfrac{6x+1}{2x+3}$

56. $y = \dfrac{3x+1}{x-4}$

57. $y = \dfrac{3}{x+5}$

58. $y = \dfrac{3x+4}{5}$

59. $y = \dfrac{1-x}{1+x}$

Using technology to estimate limits. For Exercises 60 through 66, use technology to produce a graph, table of values, or both to help you estimate the indicated limit.

60. Estimate the limit as $x \to \infty$ of $y = \left(1 + \dfrac{1}{x}\right)^x$.

61. Estimate the limit as $x \to \infty$ of $y = (1+x)^{1/x}$.

62. Estimate the limit as $x \to \infty$ of $y = x \times (2^{1/x} - 1)$.

63. Estimate the limit as $x \to \infty$ of $y = \sqrt{x^2 + 6x} - x$.

64. Estimate the limit as $x \to -\infty$ of $y = \dfrac{3^x - 1}{2^x - 1}$.

65. Estimate the limit as $x \to \infty$ of $y = \dfrac{12\sqrt{x}}{3 + 4\sqrt{x}}$.

66. Estimate the limit as $x \to \infty$ of $y = \dfrac{1}{x\left(\left(1+\frac{1}{x}\right)^2 - 1\right)}$.

MODELS AND APPLICATIONS

67. Carrying capacity. A breeding group of animals is introduced into a protected area. **Figure 1.74** shows the graph of the population N as a function of the time t in years since the animals were introduced. The environmental carrying capacity for a species in a given region is the largest population of the species that the given environment can sustain. Use the graph to estimate the environmental carrying capacity.

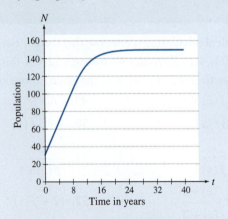

Figure 1.74 A breeding animal population

68. A box. The box shown in **Figure 1.75** has a square base of width w feet and a height of h feet. Assume that the volume is 1 cubic foot.

  a. Show that the height h of the cube is related to the width w of the base by $h(w) = \dfrac{1}{w^2}$.

  b. We know from the formula in part a that $h(w) \to 0$ as $w \to \infty$. Explain what this means in terms of the height as the width increases.

Figure 1.75 A box with a square base

69. An astronaut's view of Earth. The portion E of Earth's surface visible from a spacecraft depends on the height x above the surface of Earth. The relationship[8] is

$$E = \frac{x}{2R + 2x}$$

where R is the radius of Earth.

  a. What is the limit as $x \to \infty$ of E?

  b. Explain the meaning of the limit you calculated in part a in terms of the visible portion of Earth's surface.

70. Food consumption by sheep. The amount C of food consumed in a day by a sheep is a function of the amount V of vegetation available. The relationship is given by

$$C(V) = \frac{3V}{50 + V}$$

---

[8] *Space Mathematics*, B. Kastner, published by NASA in 1985.

Here C is measured in pounds and V in pounds per acre.

a. How much food will a sheep consume if there are 400 pounds of vegetation per acre?

b. What is the most a sheep will consume no matter what the vegetation level?

71. Ohm's law. Ohm's law says that when electric current is flowing across a resistor, the current i, measured in amperes, can be calculated from the voltage v, measured in volts, and the resistance R, measured in ohms. For a fixed voltage v, the current can be thought of as a function of the resistance R, and the relationship is

$$i(R) = \frac{v}{R}$$

What is the limit as $R \to \infty$ of $i(R)$? Explain what this limit means in terms of the current and resistance.

72. A cup of cocoa. The temperature C of a cup of cocoa t minutes after it is poured is given by

$$C = 125 \times 0.95^t + 75 \text{ degrees Fahrenheit}$$

a. Make a table of values showing the temperature of the cocoa in 20-minute intervals over the first 2 hours after it is poured.

b. Use a graphing utility to make the graph of C versus t over the first 2 hours of cooling.

c. Based on the table of values and the graph, estimate the value of the limit as $t \to \infty$ of C.

d. Explain the meaning of the limit from part c in terms of the temperature of the cocoa.

e. What is the temperature of the room in which the cocoa sits?

73. Baking a pie. A pie initially at room temperature is placed in a preheated oven to bake. Its temperature after t minutes in the oven is given by

$$P(t) = 400 - 325 \times 0.98^t \text{ degrees Fahrenheit}$$

a. Make a table of values that shows the temperature of the pie in 20-minute intervals over the first 2 hours of baking.

b. What is room temperature?

c. Use a graphing utility to make a graph of temperature versus time over the first four hours of baking.

d. Based on the graph, what is the temperature of the oven?

74. Bluebirds. Ecologists are tracking a bluebird population on a game preserve. They find that t years after observation began, the population N is given by

$$N = \frac{3450t + 1400}{10t + 7}$$

a. What was the bluebird population when tracking began?

b. According to the model, what eventual bluebird population can be expected?

75. Effective annual rate. If you borrow money, the lending institution must reveal the annual percentage rate, or APR. Interest is normally compounded, and the actual interest rate charged is called the effective annual rate, or EAR. The EAR depends on the number n of compounding periods per year. For example, $n = 1$ corresponds to yearly compounding, $n = 12$ to monthly compounding, and $n = 365$ to daily compounding. The relationship is given by

$$\text{EAR} = 100\left(\left(1 + \frac{\text{APR}}{100n}\right)^n - 1\right)$$

Here the APR and the EAR are percentages. Consider a loan with an APR of 8%.

a. Calculate the EAR in the case of yearly, monthly, and daily compounding. Report your answers correct to three decimal places.

b. Use a graphing utility to plot the graph of EAR versus APR.

c. Suppose the lending institution announced its intention to add another compounding period. Would the addition of a compounding period have more of an effect on the EAR in the case of monthly or of daily compounding?

d. What (approximately) is the maximum effective interest rate no matter how often interest is compounded? (Although not obvious, it is a fact that such a number exists.)

76. A snowball. Upon reentering the house on a winter day, your child deposits a dirty snowball on the carpet. Let $S(t)$ be the size of the snowball remaining after t minutes.

a. What is the limit as $t \to \infty$ of S?

b. Explain what has occurred when S is near its limiting value in terms of the size of the snowball.

77. Value of a dollar. Inflation I causes prices to increase. As inflation increases, the value of a dollar, in terms of what it will buy, decreases. The percentage decrease D is given by

$$D = \frac{100I}{100 + I}$$

Here I is a percentage.

a. Calculate $D(10)$, and explain in terms of the inflation rate and the value of the dollar the meaning of your calculation.

b. Calculate the limit as $I \to \infty$ of D, and explain the meaning of your calculation in terms of the inflation rate and the value of the dollar.

78. The term of a loan. If you borrow \$30,000 at an APR of 9% in order to buy a car, then your monthly payment M depends on the term t of the loan, which is the number of months required to repay the loan. The relationship is

$$M = \frac{30{,}000 \times 0.0075 \times 1.0075^t}{1.0075^t - 1} \text{ dollars}$$

a. Make a table of values for M using $t = 36, 48, 60, 200, 500,$ and 1000.

b. Based on the table of values, estimate the value of the limit as $t \to \infty$ of M.

c. Explain the meaning of the number from part b in terms of the monthly payment.

d. One month's interest on \$30,000 is \$225. Explain the relationship of this number to the limit you calculated in part b.

CHALLENGE EXERCISES FOR INDIVIDUALS OR GROUPS

Calculating limits. For Exercises 79 through 84, assume that $f(x) \to 4$ as $x \to \infty$ and that $g(x) \to \infty$ as $x \to \infty$. Assume in addition that $f(x) \neq 4$ for all x and that $g(x)$ is never 0. You may find it necessary to perform some algebraic manipulations before you can calculate the indicated limit.

79. Calculate the limit as $x \to \infty$ of $\dfrac{(f(x))^2 - 16}{f(x) - 4}$.

80. Calculate the limit as $x \to \infty$ of $\dfrac{1}{\mid f(x) - 4 \mid}$.

81. Calculate the limit as $x \to \infty$ of $\dfrac{f(x)}{g(x)}$.

82. Calculate the limit as $x \to \infty$ of $\dfrac{6g(x) + 5}{2g(x) + 9}$.

83. Calculate the limit as $x \to \infty$ of $f(x)g(x)$.

84. Calculate the limit as $x \to \infty$ of $f(g(x))$.

85. The squeeze theorem. Suppose that $f(x) \leq g(x) \leq h(x)$ for all x. Assume also that the limit of both $f(x)$ and $h(x)$ as $x \to \infty$ is 7. What is the limit as $x \to \infty$ of $g(x)$? Explain.

86. What is the limit? Figure 1.76 shows the graph of a function f. What does the graph suggest regarding the limit as $x \to \infty$ of $f(x)$?

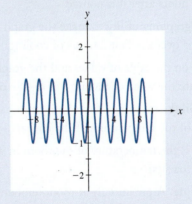

Figure 1.76 Possible limit as $x \to \infty$

REVIEW AND REFRESH: Exercises from Previous Sections

87. From Section P.1: Find the distance between $(1, 4)$ and $(2, -1)$.

Answer: $\sqrt{26}$

88. From Section P.2: Solve the inequality $3x - 1 < 7 - x$.

Answer: $x < 2$

89. From Section P.3: Solve the inequality $x^2 - 5x + 4 < 0$.

Answer: $1 < x < 4$

90. From Section 1.1: If $f(x) = \dfrac{x}{x+1}$, find $f(1-x)$.

Answer: $\dfrac{1-x}{2-x}$

91. From Section 1.2: Calculate the average rate of change of $f(x) = 3x - 2$ from $x = 10$ to $x = 20$.

Answer: 3

92. From Section 1.3: Sketch the graph of a function that is concave up on $(-\infty, 0)$ and concave down on $(0, \infty)$.

Answer:

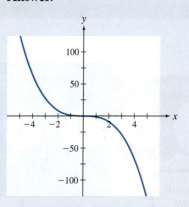

93. From Section 1.3: **True or false:** If the function f has a local maximum at $x = 1$, then $f(1)$ is the largest value the function ever has.

Answer: False. It's the largest value taken near $x = 1$, but f might take larger values farther away from $x = 1$.

CHAPTER ROADMAP AND STUDY GUIDE

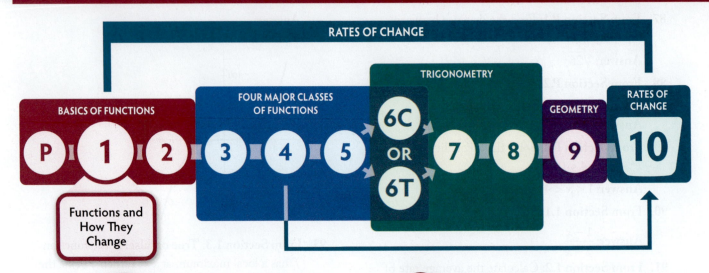

1.1   The Basics of Functions

Functions are fundamental tools for modeling real-world phenomena.

Definition of a Function

A function gives a rule for assigning some objects to others.

Evaluating Functions

Functions are evaluated by replacing the independent variable by the desired quantity.

Domain and Range

The domain comprises the allowable inputs for a function, and the range comprises the outputs.

Equations and Functions

An equation may determine a function.

Graphs and Functions

Graphs are visual representations of functions.

MODELS AND APPLICATIONS
Representing Costs of Everyday Purchases

1.2   Average Rates of Change

The average rate of change is one way to measure change over an interval.

Secant Lines

A secant line can be used to measure the rate of change of a function.

MODELS AND APPLICATIONS
Ants, Ebola, and Temperature

1.3 Graphs and Rates of Change

Rates of change are the key to a dynamic view of functions.

An Intuitive Description of Rates of Change

The rate of change is the slope of a line tangent to the graph.

Increasing and Decreasing Functions

Rates of change can show where functions are increasing and where they are decreasing.

Maximum and Minimum Values

Functions may reach maximum or minimum values where the rate of change is zero.

Concavity and Rates of Change

Concavity describes the shape of a graph and can be characterized in terms of rates of change.

MODELS AND APPLICATIONS
Sales Charts

1.4 Limits and End Behavior of Graphs

The limit at infinity describes the long-term behavior of functions.

An Intuitive Discussion of Limits at Infinity

The limit at infinity is understood in terms of the end behavior of the graph.

Examples of Limits at Infinity

A few cases illustrate common patterns for the end behavior of graphs.

Elementary Limit Calculations

The intuitive notion of limits is used to perform simple limit calculations.

Breaking Limit Calculations into Simpler Steps

Computing limits of complicated functions is accomplished in steps.

MODELS AND APPLICATIONS
Long-Term Behavior

CHAPTER QUIZ

1. If $f(x) = \dfrac{x-1}{x}$, calculate $f(2)$ and $f(x+1)$.

XR **Example 1.1**

   Answer: $f(2) = \dfrac{1}{2}$; $f(x+1) = \dfrac{x}{x+1}$

2. Find the domain and range of $f(x) = \dfrac{1}{\sqrt{x}}$.

XR **Example 1.5**

   Answer: Domain $(0, \infty)$. Range $(0, \infty)$.

3. Does the equation $\dfrac{1}{y} + x = 4$ determine y as a function of x? If so,

   what is its domain?

XR **Example 1.6**

   Answer: Yes. The domain is $(-\infty, 4) \cup (4, \infty)$.

4. Is this graph the graph of a function?

XR **Example 1.7**

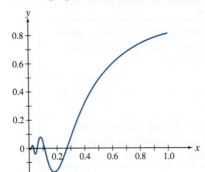

   Answer: Yes

5. Calculate the average rate of change of $f(x) = x^2 + 3x + 1$ on the interval $[x, x+h]$.

XR **Example 1.11**

   Answer: $2x + h + 3$

6. A certain bird population N after t years is given by $N(t) = 250 \times 1.1^t$. Calculate the average rate of change in N from the third through the fifth years. (Round the result to one decimal place, and be sure to include proper units.) Explain the meaning of your answer in terms of the bird population.

XR **Example 1.12**

   Answer: 34.9 birds per year. From the third through the fifth year, the bird population grew on average by 34.9 birds each year.

7. The graph of a function whose domain is $[-1, 1]$ is shown. Identify the regions of increase and decrease, locate the regions where the rate of change is positive and the regions where it is negative, and find all local maxima and minima and all absolute maxima and minima.

XR **Example 1.17**

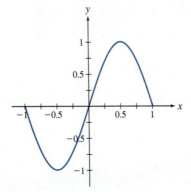

Answer: Increasing on $[-0.5, 0.5]$, decreasing on $[-1, -0.5]$ and on $[0.5, 1]$. Rate of change positive on $(-0.5, 0.5)$ and negative on $(-1, -0.5)$ and $(0.5, 1)$. Local and absolute minimum value of -1 at $x = -0.5$; local and absolute maximum value of 1 at $x = 0.5$.

8. The graph of a function is shown. Identify the intervals where the graph is concave up and where it is concave down. Also find any points of inflection.

 **Example 1.20**

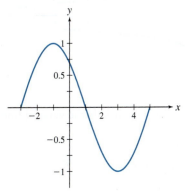

Answer: Concave down on $[-3, 1)$. Concave up on $(1, 5]$. Point of inflection at the point $(1, 0)$.

9. Find the limit as $x \to \infty$ of $y = 3 - \dfrac{10}{x^2}$. Also show a table of values and an appropriate graph.

Example 1.24

Answer: $3 - \dfrac{10}{x^2} \to 3$ as $x \to \infty$.

x	$3 - \dfrac{10}{x^2}$
10	2.9
100	2.999
1000	2.99999

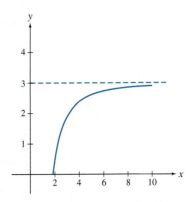

10. Find the limit as $x \to \infty$ of $y = \dfrac{2x - 3}{3x + 4}$.

Example 1.24

Answer: $\dfrac{2}{3}$

CHAPTER REVIEW EXERCISES

Section 1.1

Calculating function values. In Exercises 1 through 11, a function is given. State the domain of the function, and then calculate the indicated function value.

1. $f(x) = \dfrac{x+1}{x-1}$. Calculate $f(3)$.

   Answer: Domain $(-\infty, 1) \cup (1, \infty)$. $f(3) = 2$.

2. $f(x) = \dfrac{x}{4-x}$. Calculate $f(1-x)$.

   Answer: Domain $(-\infty, 4) \cup (4, \infty)$. $f(1-x) = \dfrac{1-x}{x+3}$.

3. $f(x) = \sqrt{x+5}$. Calculate $f(x^2 - 5)$.

   Answer: Domain $[-5, \infty)$. $f(x^2 - 5) = |x|$.

4. $f(x) = \dfrac{x+1}{\sqrt{x-3}}$. Calculate $f(7)$.

   Answer: Domain $(3, \infty)$. $f(7) = 4$.

5. $g(x) = (x+1)^2 - (x-1)^2$. Calculate $g(t+1)$.

   Answer: Domain $\mathbb{R}$. $g(t+1) = 4t+4$.

6. $g(x) = x^2$. Calculate $\dfrac{g(x+h) - g(x)}{h}$, $h \neq 0$.

   Answer: Domain $\mathbb{R}$. $\dfrac{g(x+h) - g(x)}{h} = 2x + h$.

7. $h(x) = \sqrt{x}$. Calculate $\dfrac{x-a}{h(x) - h(a)}$, $x \neq a$, $x, a > 0$.

   Answer: Domain $[0, \infty)$. $\dfrac{x-a}{h(x) - h(a)} = \sqrt{x} + \sqrt{a}$.

8. $s(t) = \dfrac{t^2 - \pi^2 + 1 + 2t}{t - \pi + 1}$. Calculate $s(p-1)$, $p \neq \pi$.

   Answer: Domain $(-\infty, \pi - 1) \cup (\pi - 1, \infty)$. $s(p-1) = p + \pi$.

9. $k(x) = \dfrac{x}{x^2 - 5x + 6}$. Calculate $k\left(\dfrac{1}{x}\right)$, $x \neq 0$.

   Answer: Domain $(-\infty, 2) \cup (2, 3) \cup (3, \infty)$. $k\left(\dfrac{1}{x}\right) = \dfrac{x}{1 - 5x + 6x^2}$.

10. $g(t) = \dfrac{1}{2 - \sqrt{t}}$. Calculate $g(9)$.

    Answer: Domain $(0, 4) \cup (4, \infty)$. $g(9) = -1$.

11. $f(z) = \dfrac{z+4}{z - \sqrt{z}}$. Calculate $f(4)$.

    Answer: Domain $(0, 1) \cup (1, \infty)$. $f(4) = 4$.

Is this the graph of a function?

12. Is the graph below the graph of a function? Explain your answer.

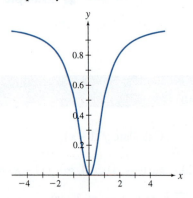

Answer: Yes, because it passes the vertical line test.

13. Is the graph below the graph of a function? Explain your answer.

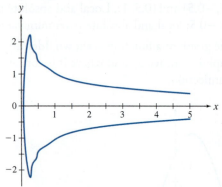

Answer: No, because it fails the vertical line test.

Finding the range.

14. Find the range of $f(x) = x^2 + 2$.

    Answer: $[2, \infty)$

15. Find the range of $u(t) = 3t - 8$.

    Answer: $\mathbb{R} = (-\infty, \infty)$

Determining y as a function of x.

16. Does the equation $x = \dfrac{y-1}{y+1}$ determine y as a function of x?

    Answer: Yes

17. Does the equation $y = \dfrac{x-1}{y-1}$ determine y as a function of x?

    Answer: No

18. Does the equation $(y+3)^2 + (2-x)^2 = 0$ determine y as a function of x?

    Answer: Yes

Section 1.2

Calculating average rates of change. In Exercises 19 through 27, calculate the average rate of change of the given function on the given interval.

19. $f(x) = x^2$. Interval $[2, 4]$.

    Answer: 6

20. $f(x) = 3x - 4$. Interval $[3, 7]$.

    Answer: 3

21. $f(x) = \dfrac{1}{x}$. Interval $[1, 3]$.

    Answer: $-\dfrac{1}{3}$

22. $g(x) = \sqrt{x}$. Interval $[9, 25]$.

Answer: $\dfrac{1}{8}$

23. $h(x) = \dfrac{x - 2}{x + 2}$. Interval $[-1, 1]$.

Answer: $\dfrac{4}{3}$

24. $f(x) = x^2 + 1$. Interval $[a, b]$.

Answer: $a + b$

25. $u(t) = t^2 + t$. Interval $[x, x + h]$.

Answer: $2x + h + 1$

26. $f(x) = \dfrac{1}{x}$. Interval $[a, b]$, $a \neq 0$, $b \neq 0$.

Answer: $-\dfrac{1}{ab}$

27. $v(s) = \dfrac{s - 1}{s + 1}$. Interval $[1, x]$.

Answer: $\dfrac{1}{x + 1}$

A flu outbreak.

28. There is a flu outbreak in a small town. By day 3, the cumulative number of flu cases reported is 30. By day 5, the cumulative number is 52. Calculate the average rate of change in the cumulative number of flu cases from day 3 to day 5, and explain what it means in terms of new cases. Then use the average rate of change to estimate the cumulative number of cases expected by day 6.

Answer: From day 5 to day 7, there were an average of 11 new cases per day. A cumulative number of 63 cases is expected by day 6.

Section 1.3

Find the rate of change.

29. The tangent line at $x = 2$ to the graph of $f(x) = x^3$ passes through the points $(2, f(2))$ and $(3, 20)$. Find the rate of change of f at $x = 2$.

Answer: 12

Features of graphs. The graphs of the functions $f(x)$ and $g(x)$ are shown in **Figure 1.77** and **Figure 1.78**. In each case, the domain of the function corresponds to the extent of the graph. Exercises 30 through 41 refer to these functions.

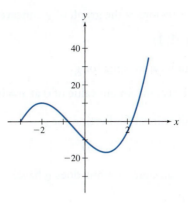

Figure 1.77 The graph of f

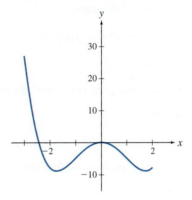

Figure 1.78 The graph of g

30. Identify all local maxima for the function f.

Answer: Local maximum value of 10 at $x = -2$

31. Give the x-value for all local minima of the function f.

Answer: $x = 1$

32. Identify the absolute maximum for f.

Answer: Absolute maximum value of 35 at $x = 3$

33. Over what regions is f increasing?

Answer: $[-3, -2]$ and $[1, 3]$

34. Over what regions is f decreasing?

Answer: $[-2, 1]$

35. Over what regions is the rate of change of f positive?

Answer: $(-3, -2)$ and $(1, 3)$

36. Over what regions is the rate of change of f negative?

Answer: $(-2, 1)$

37. Over what regions is the graph of g concave up?

Answer: $[-3, -1)$ and $(1, 2]$

38. Over what regions is the graph of g concave down?

Answer: $(-1, 1)$

39. Identify all local maxima for g.

Answer: Local maximum value of 0 at $x = 0$

40. Identify all points of inflection for g.

Answer: $(-1, -5)$ and $(1, -5)$

41. How many absolute minima does g have?

Answer: Two

Determining concavity and decrease.

42. What is the concavity of the graph of $y = x^2$?

Answer: It is concave up.

43. Over what region is the graph of $y = x^2$ decreasing?

Answer: $(-\infty, 0]$

44. Determine the concavity of the graph of $f(x) = \dfrac{1}{x}$.

Answer: Concave down on $(-\infty, 0)$, concave up on $(0, \infty)$

Sketching and analyzing graphs.

45. Sketch a graph that is increasing at a decreasing rate.

Answer:

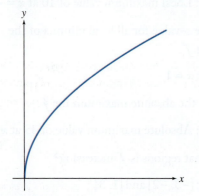

46. Sketch a graph that is concave down and has a local maximum at $x = 2$.

Answer:

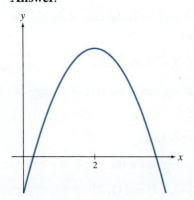

47. The following graph shows the marmot population versus time in a restricted region. Over what period was the population increasing? What was the maximum number of marmots during the period shown by the graph?

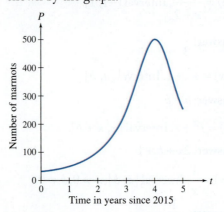

Answer: The population increased from 2015 to 2019. The maximum number of marmots was 500.

Section 1.4

Calculating limits. In Exercises 48 through 59, calculate the indicated limit.

48. $y = \dfrac{1}{x}$ as $x \to \infty$

Answer: 0

49. $y = \dfrac{7}{x}$ as $x \to -\infty$

Answer: 0

50. $y = 3 - \dfrac{1}{x^2}$ as $x \to \infty$

Answer: 3

51. $y = x + \dfrac{1}{x}$ as $x \to \infty$

Answer: ∞

52. $y = \dfrac{6 + \frac{2}{x}}{3 - \frac{5}{x}}$ as $x \to -\infty$

Answer: 2

53. $y = \dfrac{100}{x + 5}$ as $x \to \infty$

Answer: 0

54. $y = x^3 + 1$ as $x \to -\infty$

Answer: $-\infty$

55. $y = \dfrac{12 - \frac{4}{x^2}}{3 + \frac{7}{x} - \frac{5}{x^2}}$ as $x \to \infty$

Answer: 4

56. $y = \dfrac{2}{3+x^2}$ as $x \to -\infty$

Answer: 0

57. $y = \dfrac{3 + \frac{4}{x} - \frac{9}{x^2}}{1 + \frac{4}{x} + \frac{4}{x^2}}$ as $x \to \infty$

Answer: 3

58. $y = \dfrac{\frac{7}{x} + \frac{100}{x^2}}{1 + \frac{1}{x^3}}$ as $x \to -\infty$

Answer: 0

59. $y = \dfrac{30x + 2}{1 - 15x}$ as $x \to \infty$

Answer: -2

Sketching graphs. In Exercises 60 through 63, sketch a graph of a function f with the given properties. In each case, there is more than one correct answer. The answer provided is only one example.

60. The function f is decreasing, and the limit as $x \to \infty$ is 2.

Answer:

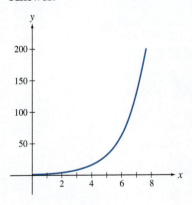

61. The limit as $x \to \infty$ is ∞.

Answer:

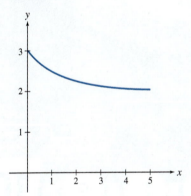

62. The graph is concave down, and the limit as $x \to \infty$ is 3.

Answer:

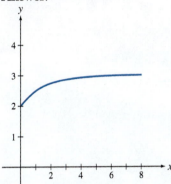

63. The function f is neither decreasing nor increasing, and the limit as $x \to \infty$ is -2.

Answer:

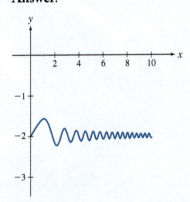

Food consumption.

64. The amount C of food consumed in a day by a sheep depends on the amount V of vegetation available. In a certain setting, this relationship is given by

$$C(V) = \frac{6V}{25 + 2V}$$

where C is measured in pounds, and V is measured in pounds per acre. We want to know the maximum amount of food a sheep will eat in a day no matter how much vegetation is available. Express your answer as a limit, and calculate its value.

Answer: The limit as $V \to \infty$ of $C = \dfrac{6V}{25 + 2V}$, which is 3 pounds.

OPERATIONS ON FUNCTIONS AND GRAPHS

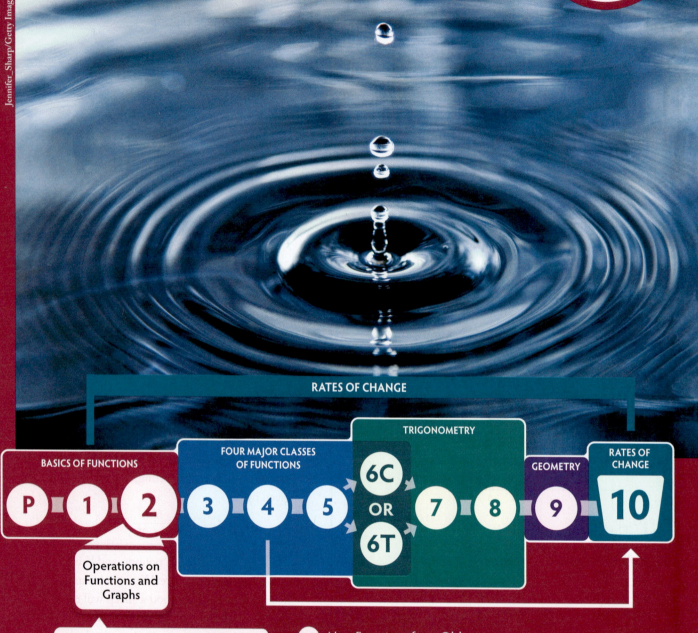

Jennifer Sharp/Getty Images

RATES OF CHANGE

BASICS OF FUNCTIONS

P 1 2

FOUR MAJOR CLASSES OF FUNCTIONS

3 4 5

TRIGONOMETRY

6C OR 6T 7 8

GEOMETRY

9

RATES OF CHANGE

10

Operations on Functions and Graphs

Functions are often built from other functions using operations such as addition and composition. In this chapter, you will learn about the effects of such operations on functions and their graphs.

2.1 New Functions from Old

2.2 Inverse Functions

2.3 Transformations of Graphs

n mathematics, we focus on some fundamental classes of functions with specific properties. These classes will be presented in subsequent chapters. But any list of functions can be combined and transformed to produce a large collection of new functions, and these operations on functions have important consequences for the graphs of the resulting functions. One of the most important operations in calculus is the calculation of the exact rate of change of a function. This calculation relies on an understanding of the operations presented in this chapter. Furthermore, in this chapter we see the effects of the transformations we study on the rate of change.

2.1 New Functions from Old

> Mathematical models are created by combining simpler functions.

2.1 New Functions from Old

2.2 Inverse Functions

2.3 Transformations of Graphs

In this section, you will learn to:

1. Perform arithmetic combinations of functions using a formula or a graph.
2. Describe the domain of an arithmetic combination of functions.
3. Calculate the composition of functions.
4. Express more complicated functions as combinations of simpler functions.
5. Solve applied problems using function composition.

Just as with numbers, real-valued functions may be combined by the operations of addition, subtraction, multiplication, and division. But another important way of combining functions is by composition, which has the effect of following one function's operation by that of another function.

Fractals are objects that show the same, or in some cases a similar, pattern at every scale. Among the most famous is the Mandelbrot set illustrated in **Figure 2.1**. Another example whose construction is easier to understand is the Menger sponge illustrated in **Figure 2.2**. To make the Menger sponge, we start with a cube and divide it into 27 smaller cubes (9 on each face). Next we remove the center cube and then the center cube from each face. The process repeated ad infinitum produces the Menger sponge. It turns out that both of these objects, as well as many other fractals, are best described using repeated function composition. Such operations are the subject of this section.

Figure 2.1 The Mandelbrot set heikeinnz/ Getty Images

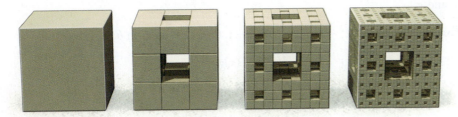

Figure 2.2 Constructing the Menger sponge

Arithmetic Combinations of Functions

> Functions can be added, subtracted, multiplied, and divided in much the same fashion as these operations are applied to real numbers.

Constant multiple: If f is a function and r is a constant, we define a new function $g = rf$ by

$$g(x) = rf(x)$$

We call the new function r times f. It has the same domain as f.

Example: If $f(x) = x^2 - 1$, then the function $g = 3f$ is defined by

$$g(x) = 3f(x)$$
$$= 3(x^2 - 1)$$
$$= 3x^2 - 3$$

Figure 2.3 provides a display. Note that to get each point on the graph of $3f$, we multiply the y-coordinate of the corresponding point on the graph of f by 3.

Addition of functions: If f and g are functions, we define $h = f + g$ by

$$h(x) = f(x) + g(x)$$

We call h the sum of f and g. Its domain is the intersection of the domains of f and g—that is, the set of numbers common to the domains of f and g.

Example: If $f(x) = x^2 - 1$ and $g(x) = x + 1$, then the function $h = f + g$ is defined by

$$h(x) = f(x) + g(x)$$
$$= (x^2 - 1) + (x + 1)$$
$$= x^2 + x$$

The graph is shown in **Figure 2.4**. Note that to get each value of $h = f + g$, we add the corresponding y-values from the graphs of f and g.

Multiplication of functions: If f and g are functions, we define $h = fg$ by

$$h(x) = f(x)g(x)$$

We call h the product of f and g. Its domain is the intersection of the domains of f and g.

Example: If $f(x) = x^2 - 1$ and $g(x) = x + 1$, then the function $h = fg$ is defined by

$$h(x) = f(x)g(x)$$
$$= (x^2 - 1)(x + 1)$$
$$= x^3 + x^2 - x - 1$$

The graphical display is in **Figure 2.5**. Note that to get each y-value of $h = fg$, we multiply the corresponding y-values from the graphs of f and g.

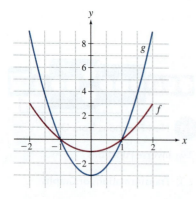

Figure 2.3 f in red and $g = 3f$ in blue

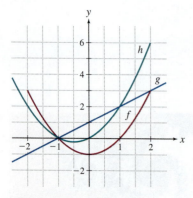

Figure 2.4 f in red, g in blue, and $h = f + g$ in green

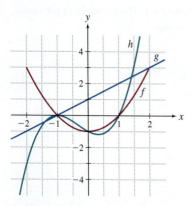

Figure 2.5 f in red, g in blue, and $h = fg$ in green

Division of functions: If f and g are functions, then we define $h = \dfrac{f}{g}$ by

$$h(x) = \frac{f(x)}{g(x)}$$

We call h the quotient of f by g. Its domain is the intersection of the domains of f and g, but with all x where $g(x) = 0$ excluded.

Example: If $f(x) = 2x$ and $g(x) = 1 - x^2$, then the function $h = \dfrac{f}{g}$ is defined by

$$h(x) = \frac{f(x)}{g(x)}$$

$$= \frac{2x}{1 - x^2}$$

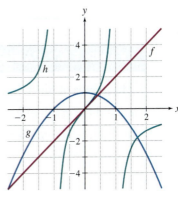

Figure 2.6 f in red, g in blue, and $h = \dfrac{f}{g}$ in green

Note that $g(x)$ is 0 when $x = \pm 1$. Thus, we must exclude these points from the domain of h. The domain is $(-\infty, -1) \cup (-1, 1) \cup (1, \infty)$.

The graphical display is in **Figure 2.6**, and the behavior of h near the excluded points is evident.

These kinds of combinations are highly intuitive and rarely lead to difficulties. When we combine functions given by formulas using arithmetic operations, all we need to do is perform the indicated arithmetic operation on the given formulas.

EXAMPLE 2.1 Arithmetic Combinations of Functions

Let $f(x) = x^2 - x$ and $g(x) = 2x + 1$.

a. Find a formula for the function $h = 2f + g$, and simplify.

b. Find a formula for the function $k = \dfrac{f}{3g}$, and simplify. What is the domain of the function k?

SOLUTION

a. We find a formula for the new function by performing the indicated operations on the formulas for f and g:

$$h(x) = 2f(x) + g(x) \qquad \blacktriangleleft \text{Apply function definition.}$$

$$= 2(x^2 - x) + (2x + 1) \qquad \blacktriangleleft \text{Apply operations to the formulas.}$$

$$= 2x^2 - 2x + 2x + 1 \qquad \blacktriangleleft \text{Apply distributive law.}$$

$$= 2x^2 + 1 \qquad \blacktriangleleft \text{Simplify.}$$

b. We proceed just as in part a:

$$k(x) = \frac{f(x)}{3g(x)} \qquad \blacktriangleleft \text{Apply function definition.}$$

$$= \frac{x^2 - x}{3(2x + 1)} \qquad \blacktriangleleft \text{Use function formulas.}$$

$$= \frac{x^2 - x}{6x + 3} \qquad \blacktriangleleft \text{Simplify.}$$

The formula makes sense except where the denominator is 0, which forces the exclusion of $-\dfrac{1}{2}$ from the domain. Therefore, the domain is

$$\left(-\infty, -\frac{1}{2}\right) \cup \left(-\frac{1}{2}, \infty\right)$$

TRY IT YOURSELF 2.1 Brief answers provided at the end of the section.

Use the functions in the example to find a formula for $m = fg + f$, and simplify.

EXAMPLE 2.2 Combining Functions Graphically

The graphs of functions f and g are shown in **Figure 2.7**.

a. Add the graph of $h = 2f$ to Figure 2.7.

b. Add the graph of $k = f + g$ to Figure 2.7.

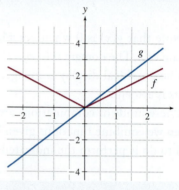

Figure 2.7 The graph of f in red and g in blue

SOLUTION

a. To get the graph of $h = 2f$, we multiply each y-value on the graph of f by 2. For example, examining the graph of f gives the values

$$f(-1) = 1, \quad f(0) = 0, \quad f(1) = 1, \quad \text{and } f(2) = 2$$

Therefore,

$$h(-1) = 2 \times 1 = 2, \quad h(0) = 2 \times 0 = 0, \quad h(1) = 2 \times 1 = 2, \quad \text{and } h(2) = 2 \times 2 = 4$$

The result is shown in **Figure 2.8**.

b. To get the graph of $k = f + g$, we add corresponding y-values from the graphs of f and g. For example, examining the graphs gives $f(2) = 2$ and $g(2) = 3$, so $k(2) = f(2) + g(2) = 2 + 3 = 5$. The result is shown in **Figure 2.9**.

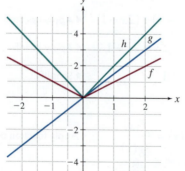

Figure 2.8 The graph of $h = 2f$ in green

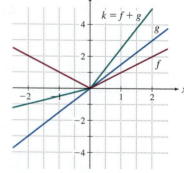

DF Figure 2.9 The graph of $k = f + g$ in green

TRY IT YOURSELF 2.2 Brief answers provided at the end of the section.

Plot the graph of $j = fg$.

Composition of Functions

> Composition of functions means that one function is applied after another.

If f and g are functions, then we can create a new function $h = f \circ g$, called the **composition** of f with g. We read $f \circ g$ as "f composed with g."

Thus, $f \circ g$ first applies the function g and then applies f to the result. The domain of $f \circ g$ is the set of those x in the domain of g for which $g(x)$ is in the domain of f.

To find $(f \circ g)(x)$, we calculate $g(x)$ first and then plug that into f. For example, if $f(x) = x^2 + x$ and $g(x) = 4x$, let's find a formula for $f \circ g$:

The **composition** of f with g is defined by
$$(f \circ g)(x) = f(g(x))$$

$$(f \circ g)(x) = f(g(x)) \quad \blacktriangleleft \text{Apply composition definition.}$$

$$= f(4x) \quad \blacktriangleleft \text{Evaluate } g.$$

$$= (4x)^2 + (4x) \quad \blacktriangleleft \text{Evaluate } f.$$

$$= 16x^2 + 4x \quad \blacktriangleleft \text{Simplify.}$$

It is important to note that $f \circ g$ and $g \circ f$ are typically not the same. Let's use these same functions and find a formula for $g \circ f$:

$$(g \circ f)(x) = g(f(x)) \quad \blacktriangleleft \text{Apply composition definition.}$$

$$= g(x^2 + x) \quad \blacktriangleleft \text{Evaluate } f.$$

$$= 4(x^2 + x) \quad \blacktriangleleft \text{Evaluate } g.$$

$$= 4x^2 + 4x \quad \blacktriangleleft \text{Simplify.}$$

Note that in this case $f \circ g$ and $g \circ f$ are different.

EXAMPLE 2.3 Composing Functions

Let $f(x) = x^2 - 2x$, and let $g(x) = x - 1$. Find a formula for the function $h = f \circ g - g \circ f$, and simplify.

SOLUTION

We begin by using the definition of composition:

$$h(x) = (f \circ g)(x) - (g \circ f)(x)$$

$$= f(g(x)) - g(f(x)) \quad \text{Apply composition definition.}$$

$$= f(x - 1) - g(x^2 - 2x) \quad \text{Apply formulas for } f \text{ and } g.$$

$$= ((x - 1)^2 - 2(x - 1)) - ((x^2 - 2x) - 1) \quad \text{Put } x - 1 \text{ in place of } x \text{ in the formula for } f \text{ and } x^2 - 2x \text{ in place of } x \text{ in the formula for } g.$$

$$= ((x^2 - 2x + 1) - 2x + 2) - (x^2 - 2x - 1) \quad \text{Expand products.}$$

$$= -2x + 4 \quad \text{Simplify.}$$

TRY IT YOURSELF 2.3 Brief answers provided at the end of the section.

If $f(x) = \dfrac{2}{x}$ and $g(x) = 3f(x)$, find a formula for $f \circ g$. Be sure to simplify.

EXTEND YOUR REACH

Consider the function $f(x) = \dfrac{1}{x}$.

a. Find formulas for $f \circ f$ and $f \circ f \circ f$.

b. What do you get if you compose f with itself an even number of times? What if you compose an odd number of times?

c. Can you find a function g with the property that composition of g with itself results in no change to the function?

EXAMPLE 2.4 Finding the Domain of Compositions

a. The domain of the function m is $[9, 27]$, and $n(t) = 3t$. What is the domain of $m \circ n$?

b. The graphs of functions f and g are shown in **Figure 2.10**. Find the domain of $f \circ g$.

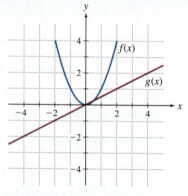

Figure 2.10 Graphs for part b

SOLUTION

a. The domain of $m \circ n$ is the set of t in the domain of n for which $n(t)$ is in the domain of m. That is all values of t so that $3t$ is in the interval $[9, 27]$. Because $9 \le 3t \le 27$ says that $3 \le t \le 9$, the domain of $m \circ n$ is $[3, 9]$.

b. The domain of f is the horizontal extent of its graph, which is $[-2, 2]$. Thus, the domain of $f \circ g$ consists of those points in the domain of g with function values between -2 and 2. The graph of g shows that $g(x)$ is between -2 and 2 when x is between -4 and 4. Thus, the domain of $f \circ g$ is $[-4, 4]$.

TRY IT YOURSELF 2.4 Brief answers provided at the end of the section.

If $f(x) = \dfrac{1}{x}$ and $g(x) = x - 1$, find the domain of $f \circ g$.

Some questions about function combinations can be answered even without formulas, as the next example shows.

EXAMPLE 2.5 Combining Functions Using Function Values

Suppose $f(2) = 3$, $f(3) = 4$, $f(5) = 7$, $g(2) = 5$, and $g(3) = 8$. In each part find the indicated function value.

a. If $h = 3f + \dfrac{f}{g}$, find $h(2)$.   **b.** If $k = f \circ g - fg$, find $k(2)$.

SOLUTION

a. By definition,

$$h = 3f + \frac{f}{g} \qquad \blacktriangleleft \text{ Apply definition of } h.$$

$$h(2) = 3f(2) + \frac{f(2)}{g(2)} \quad \blacktriangleleft \text{ Evaluate at 2.}$$

$$= 3 \times 3 + \frac{3}{5} \quad \blacktriangleleft \text{ Evaluate the functions.}$$

$$= \frac{48}{5} = 9.6 \quad \blacktriangleleft \text{ Simplify.}$$

b. For $k = f \circ g - fg$, we proceed as follows:

$$k = f \circ g - fg \qquad\qquad \blacktriangleleft \text{ Apply definition of } k.$$

$$k(2) = f(g(2)) - f(2) \times g(2) \quad \blacktriangleleft \text{ Evaluate at 2.}$$

$$= f(5) - 3 \times 5 \qquad\qquad \blacktriangleleft \text{ Evaluate the functions.}$$

$$= -8 \qquad\qquad\qquad \blacktriangleleft \text{ Simplify.}$$

TRY IT YOURSELF 2.5 Brief answers provided at the end of the section.

Use the function values in the example to find $n(2)$ if $n = f \circ f + 2f + 3g$.

Consider the function $g(x) = \sqrt{x}$.

EXTEND YOUR REACH

a. Calculate $g(2)$, $(g \circ g)(2)$, $(g \circ g \circ g)(2)$, and so on, up to 10 compositions of g with itself. Round answers to three decimal places.

b. Do the numbers you found in part a appear to be approaching a specific value? If so, identify that value.

c. Does the same thing happen if you start with 3 instead of 2?

d. Does the same thing happen if you start with 0.5 instead of 2?

e. What do you think happens if you start with any positive number?

Function Decomposition

> Function decomposition identifies how simple parts make up more complicated functions.

So far we have looked at combining functions. But there are situations in calculus where it is useful to go the other way—that is, to write a given function as a combination of simpler functions.

Suppose, for example, that $h(x) = \sqrt{x^2 + 1}$ and that we want to write h as a composition of simpler functions. (There is more than one way to do this.) The key to decomposing this function is to see that it is calculated in two steps.

Step 1: Calculate $x^2 + 1$.

Step 2: Calculate the square root of the result from step 1.

Now observe that if we let $f(x) = \sqrt{x}$ and $g(x) = x^2 + 1$, then $h(x) = f(g(x))$, so $h = f \circ g$.

We said there is more than one way to do this, and here is another way: Let $f(x) = \sqrt{x + 1}$ and $g(x) = x^2$. Then $f(g(x)) = f(x^2) = \sqrt{x^2 + 1}$, so $h = f \circ g$.

EXAMPLE 2.6 Decomposing Functions

a. If $f(x) = (3x+1)^4$, find functions g and h so that $f = g \circ h$. Find two ways to do this.

b. If $k(x) = 3\sqrt{x} + \dfrac{4}{x}$, find functions m and n and constants c_1 and c_2 so that $k = c_1 m + c_2 n$.

SOLUTION

a. We find the value of $(3x+1)^4$ by first calculating $3x+1$ and then raising the result to the fourth power. Thus, if $h(x) = 3x+1$ and $g(x) = x^4$, then

$$g(h(x)) = g(3x+1) = (3x+1)^4$$

Therefore, $f = g \circ h$.

Here is another way: Let $h(x) = 3x$ and $g(x) = (x+1)^4$. Then

$$g(h(x)) = g(3x) = (3x+1)^4$$

so $f = g \circ h$.

b. We want to find functions and constants so that

$$k(x) = 3\sqrt{x} + \frac{4}{x} = c_1 m(x) + c_2 n(x)$$

Based on this equation, we choose $c_1 = 3$, $c_2 = 4$, $m(x) = \sqrt{x}$, and $n(x) = \dfrac{1}{x}$. This choice gives $k(x) = c_1 m(x) + c_2 n(x)$ or $k = 3m + 4n$.

TRY IT YOURSELF 2.6   Brief answers provided at the end of the section.

If $f(x) = (\sqrt{x} + 1)^3$, find functions g and h so that $f = g \circ h$.

MODELS AND APPLICATIONS   Decomposing Functions in Applications

Inflating a balloon is one application of recognizing a function as a composition of other functions.

EXAMPLE 2.7 Inflating a Balloon

A child is blowing up a balloon. The volume after t seconds is $V(t) = \dfrac{4\pi}{3}(2 + 0.1t)^3$ cubic inches. Find a function f so that $V = f \circ r$, where $r(t) = 2 + 0.1t$. Interpret your answer. (Use the fact that the volume of a sphere of radius x is $\dfrac{4\pi}{3}x^3$.)

SOLUTION

We find the value of $\dfrac{4\pi}{3}(2 + 0.1t)^3$ by first calculating $2 + 0.1t$ and then raising the result to the third power and multiplying by $\dfrac{4\pi}{3}$. We recognize $2 + 0.1t$ as $r(t)$, so we choose $f(x) = \dfrac{4\pi}{3}x^3$. Then $V = f \circ r$. To interpret this, we recognize that $f(x)$ equals the volume of a sphere of radius x. Therefore, at time t the volume $V(t)$ equals the volume of a sphere of radius $r(t) = 2 + 0.1t$. We conclude that the radius of the sphere is growing according to the formula $r(t) = 2 + 0.1t$ inches.

TRY IT YOURSELF 2.7   Brief answers provided at the end of the section.

A rock tossed in a pond makes an expanding circle of water ripples. The area of that circle at time t seconds is $A(t) = \pi(1 + 2t)^2$ square inches. Find a function f so that $A = f \circ r$, where $r(t) = 1 + 2t$. Interpret your answer using the standard formula for the area of a circle.

TRY IT YOURSELF ANSWERS

2.1 $m(x) = 2x^3 - 2x$

2.2 $j = fg$ is in green.

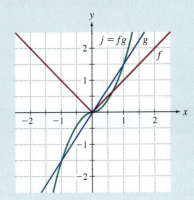

2.3 $(f \circ g)(x) = \dfrac{x}{3}$

2.4 $(-\infty, 1) \cup (1, \infty)$

2.5 $n(2) = 25$

2.6 One correct solution is $g(x) = x^3$ and $h(x) = \sqrt{x} + 1$.

2.7 $f(x) = \pi x^2$; the radius at time t is $r(t) = 1 + 2t$ inches.

EXERCISE SET 2.1

CHECK YOUR UNDERSTANDING

1. Which of the following describes the way to evaluate the composition $f \circ g$?

   a. First apply g, and then apply f to the result.

   b. First apply f, and then apply g to the result.

   c. Take the product of f and g.

   d. Take the sum of f and g.

2. **True or false:** For functions f and g, $f \circ g = g \circ f$.

3. **True or false:** For functions f and g, the product function fg is the same as gf.

4. For functions f and g, explain in words how we calculate $(f + g)(x)$.

5. **True or false:** The operation of addition for functions is associative. That is, for functions f, g, and h, $f + (g + h) = (f + g) + h$.

6. **True or false:** If the domain of a function is the same as its range, then it is possible to compose a function with itself.

7. For functions f and g, the domain of $f + g$ is _____.

8. For functions f and g, the domain of $f \circ g$ is _____.

9. For functions f and g, the domain of $\dfrac{f}{g}$ is _____.

SKILL BUILDING

Combining functions. In Exercises 10 through 15, calculate each of the following, and simplify your answer.

a. $f + g$ c. $\dfrac{f}{g}$ e. $g \circ f$

b. fg d. $f \circ g$

10. $f(x) = x + 1$, $g(x) = x - 1$

11. $f(x) = \dfrac{x}{x + 1}$, $g(x) = x^2$

12. $f(x) = \sqrt{x^2 + 1}$, $g(x) = x^2 + 1$

13. $f(x) = 2^x$, $g(x) = -\dfrac{1}{x^2}$

14. $f(x) = 3x$, $g(x) = 3x$

15. $f(x) = x^2 + 1$, $g(x) = \dfrac{1}{x} + x$

Calculating function values. The following information about functions f, g, and h is used in Exercises 16 through 29. In each case, calculate the indicated value.

$f(1) = 3 \quad f(2) = 6 \quad f(3) = 2$
$g(1) = 4 \quad g(2) = 2 \quad g(3) = 1 \quad g(4) = 7 \quad g(5) = 6$
$h(1) = 3 \qquad\qquad h(3) = 5 \quad h(4) = 2$

16. $(f + g)(1)$ 20. $2h(1)$ 25. $(g \circ h)(3)$

17. $(h - g)(4)$ 21. $(3f - 2g)(3)$ 26. $(f \circ h \circ g)(1)$

18. $gh(1)$ 22. $(f \circ g)(3)$ 27. $(f \circ f)(1)$

19. $\dfrac{f}{h}(3)$ 23. $(g \circ f)(3)$ 28. $(g \circ g \circ g)(3)$

   24. $(h \circ g)(3)$ 29. $(h \circ f - h \circ g)(1)$

Function decomposition. In Exercises 30 through 34, write the given function as a composition of functions.

30. Let $f(x) = \dfrac{x^2}{x^2 + 7}$. Write f as a composition $g \circ h$ of two functions g and h.

31. If $f(x) = \sqrt{x^3 + 1}$, find functions g and h such that $f = g \circ h$.

32. If $f(x) = \dfrac{10}{\sqrt{x} - 1}$, find functions g and h such that $f = g \circ h$.

33. If $f(x) = (x^2 - 1)^2$, find functions g and h so that $f(x) = (g \circ h)(x)$.

34. Let $f(x) = \dfrac{7}{\sqrt{x^2 - 6}}$. Write f as a composition $g \circ h$ of two functions g and h in two different ways.

PROBLEMS

Domain of combined functions.

35. The domain of f is the set of all real numbers except 7, and the domain of g is the set of all real numbers except 2. What is the domain of $f + g$?

36. The domain of f is $(3, 12]$. The domain of g is $[4, 17]$. Also, $g(4) = 0$, and $g(x) > 0$ for all other x in the domain of g. Find the domain of $\dfrac{f}{g}$.

37. The domain of f is $[6, 8]$, and $g(x) = 2x$. Find the domain of $f \circ g$.

38. The domain of f is $[1, 100]$, and $g(x) = \dfrac{1}{x}$. Find the domain of $f \circ g$.

39. The domain of f is $[1, 7]$, and the domain of g is $[3, 12]$. What is the domain of the product fg?

40. Let $f(x) = x^2$. If the domain of g is $[-7, 2]$, what is the domain of $f \circ g$?

Using graphs to find domain. In Exercises 41 through 43, use the graphs of f (in blue) and g (in red) in **Figure 2.11** to find the domain of the function indicated.

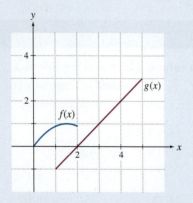

Figure 2.11 Graph for Exercises 41 through 43

41. $f + g$

42. $\dfrac{f}{g}$

43. $f \circ g$

Using graphs to find domain. In Exercises 44 through 47, use the graphs of f (in blue) and g (in red) in **Figure 2.12** to find the domain of the function indicated.

44. $\dfrac{f}{g}$

45. $f - g$

46. $f \circ g$

47. $g \circ f$

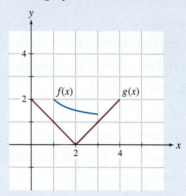

Figure 2.12 Graph for Exercises 44 through 47

Graphical combinations of functions. The graphs of f (in red) and g (in blue) are shown in **Figure 2.13**. In Exercises 48 through 53, add the graph of the indicated function to Figure 2.13.

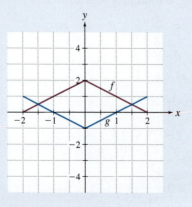

Figure 2.13 Graph for Exercises 48 through 53

48. $3g$

49. $-g$

50. $f + g$

51. $f - g$

52. fg

53. $\dfrac{f}{g}$

54. $2f$

55. $-f$

56. $f + g$

57. $f - g$

58. fg

59. $\dfrac{f}{g}$

Graphical combinations of functions. The graphs of f (in red) and g (in blue) are shown in **Figure 2.14**. In Exercises 54 through 59, add the graph of the indicated function to Figure 2.14.

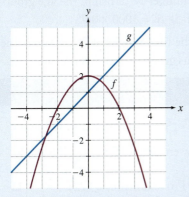

Figure 2.14 Graph for Exercises 54 through 59

Using technology to plot combinations of functions. In Exercises 60 through 63, plot f and g on the same coordinate axes. Then plot the indicated combination of f and g.

60. $f(x) = \sqrt{x},\, 0 \le x \le 4;\ g(x) = x,\, 0 \le x \le 4;\ f + g$

61. $f(x) = x^2,\, -2 \le x \le 2;\ g(x) = 1 + x,\, -2 \le x \le 2;\ f - g$

62. $f(x) = x^2,\, 0 \le x \le 4;\ g(x) = \sqrt{x} + 1,\, 0 \le x \le 4;\ f \circ g$

63. $f(x) = 1 + x,\, -2 \le x \le 2;\ g(x) = 1 + 2^x,\, -2 \le x \le 2;\ \dfrac{f}{g}$

64. Plotting function iterates. Let $f(x) = \dfrac{1}{1 + x^2}$. On the same screen, plot $y = f(x)$, $y = (f \circ f)(x)$, and $y = (f \circ f \circ f)(x)$.

MODELS AND APPLICATIONS

65. Temperature. The temperature F in degrees Fahrenheit can be calculated from the temperature C in degrees Celsius using the relation $F = \dfrac{9}{5}C + 32$. The temperature in degrees Celsius can be calculated from the temperature K in kelvins using the relation $C = K - 273.15$. Find and simplify a formula for the composition $F \circ C$, and explain what the resulting formula gives in terms of the temperature scales.

66. Growth. The weight w, in pounds, of a young boy is a function of his height h, in feet, and the formula is $w = 50h - 150$. His height can be calculated from his age t, in years, using the formula $h = \dfrac{1}{6}t + 3$. Find and simplify a formula for the composition $w \circ h$, and explain what the resulting formula gives in terms of the variables.

67. Baking bread. Bread dough is initially at a room temperature of $75°$. If the dough is put in a $400°$ oven, a typical formula for the temperature of the bread as it heats is $T(t) = 400 - 325 \times 0.955^t$ degrees, where t is the time in minutes since the dough was put in the oven. Express T as a combination of two simpler functions.

68. Fixed and variable costs. The fixed costs for a certain small business are the costs such as rent or mortgage payments that the company must pay each month, no matter how much business is conducted. The variable cost is the hourly wage paid to employees. When employees work E hours in a month, the total cost is given by

$$C = 15E + 1200 \text{ dollars}$$

Identify the fixed costs and the variable cost in the formula for the total cost function.

69. Oysters in Chesapeake Bay. One study used the following function to model stocks of market-sized oysters in Maryland's portion of Chesapeake Bay:[1]

$$N = \dfrac{5089}{1 + 0.48 \times 0.76^t}$$

Here t is the time in years since 2008, and N is the number of market-sized oysters in millions.

For a presentation, you need to recast this function in terms of the variable T, which is the time in years since 2020. Find a formula that gives N in terms of T.

[1]R. Wieland and S. Kasperski, "Estimating Net Present Value in the Northern Chesapeake Bay Oyster Fishery," prepared for NOAA Chesapeake Bay Office, Non-native Oyster Research Program (2008).

70. Present value and the discount rate. The present value P of an investment is the amount of money that must be invested now so that the investment will, in time, grow to a future value F. If the annual interest rate for the investment is r as a decimal, then the discount rate D is given by

$$D = (1+r)^{-t}$$

where t is the term, in years, of the investment. The present value is the product of the future value and the discount rate. Find a formula that gives the present value in terms of the future value, the term of the investment, and r.

71. North Sea plaice. An investigation by Beverton and Holt of plaice, a variety of flatfish in the North Sea, found the following.[2] For plaice in the same age group:

- The length L, in inches, of a plaice after t years is given by $L = 27 - 25.1 \times 0.91^t$.
- The weight W, in pounds, of a plaice is given by $W = 0.000322 L^3$.
- The total number N of fish in the age group after t years is $N = 1000 \times 0.9^t$.

The biomass is the total weight of fish. Express the biomass B for this group of fish in terms of t.

CHALLENGE EXERCISES FOR INDIVIDUALS OR GROUPS

72. A composition of functions. If f and g are increasing functions, show that the composition $f \circ g$ is also increasing.

Iteration of functions. For a positive integer n, we use the notation $f^{[n]}$ to denote f composed with itself n times. For example, $f^{[3]}(x)$ means $f(f(f(x)))$. Exercises 73 through 81 use this notation.

73. If $f(x) = x^2$, calculate $f^{[3]}(x)$.

74. If $f(x) = x^2$, calculate $f^{[n]}(x)$.

75. If $f(x) = \dfrac{1}{x}$, calculate $f^{[221]}(x)$. *Suggestion*: First calculate $f^{[n]}(x)$ for $n = 1, 2, 3, 4$ to determine a pattern.

76. If $f(x) = x^2 - 1$, calculate $f^{[21]}(-1)$. *Suggestion*: First calculate $f^{[n]}(-1)$ for $n = 1, 2, 3, 4$ to determine a pattern.

77. Let $f(x) = x^2$. As n increases, the sequence

$$f^{[1]}\left(\frac{1}{2}\right), f^{[2]}\left(\frac{1}{2}\right), f^{[3]}\left(\frac{1}{2}\right), \ldots, f^{[n]}\left(\frac{1}{2}\right)$$

gets closer and closer to a specific number. Which number is it?

The $3x + 1$ problem. For positive integers n, we define $f(n)$ as follows. If n is even, define $f(n) = \dfrac{n}{2}$. If n is odd, define $f(n) = 3n + 1$. For example, $f(22) = \dfrac{22}{2} = 11$ and $f(11) = 3 \times 11 + 1 = 34$. Exercises 78 through 81 refer to this function f. We use the notation $f^{[n]}$ to denote f composed with itself n times.

78. Find $f(1)$, $f^{[2]}(1) = f(f(1))$, and $f^{[3]}(1)$.

79. Verify that $f^{[16]}(7) = 1$.

80. Verify that for each of $n = 1, 2, 3, 4, 5, 6$ there is an integer k such that $f^{[k]}(n) = 1$.

81. As of the time of the writing of this text, no one knows how to solve the following problem: Either show that for each positive integer n there is a positive integer k such that $f^{[k]}(n) = 1$, or find a positive integer n such that $f^{[k]}(n) \neq 1$ for all positive integers k. Look on the Internet to find what progress has been made in solving this problem.

REVIEW AND REFRESH: Exercises from Previous Sections

82. From Section P.1: What is the distance from $(-3, 5)$ to $(13, 5)$?

Answer: 16

83. From Section P.2: Solve the inequality $x - 4 < 3x + 6$.

Answer: $(-5, \infty)$

84. From Section P.3: Solve the inequality $x^2 \geq 2x - 3$.

Answer: $(-\infty, \infty)$

85. From Section 1.1: If $f(x) = x^2 - 1$, find $f(\sqrt{x})$.

Answer: $x - 1$

---

[2]*On the Dynamics of Exploited Fish Populations,* Fishery Investigations, Series 2, vol. 19, R. J. H. Beverton and S. J. Holt, Ministry of Agriculture, Fisheries and Food, London, 1957.

86. From Section 1.2: If $f(x) = \dfrac{1}{x}$, find the average rate of change of f from $x = 5$ to $x = 7$.

Answer: $-\dfrac{1}{35}$

87. From Section 1.3: For a certain object, the graph of its temperature versus time is decreasing and concave up. This means that the object is (a) warming at an increasing rate, (b) warming at a decreasing rate, (c) cooling at an increasing rate, or (d) cooling at a decreasing rate.

Answer: (d)

88. From Section 1.4: Find the limit as $x \to \infty$ of

$$y = 3 - \frac{1}{\sqrt{x}}.$$

Answer: 3

2.2 Inverse Functions

> **When the effect of a function is reversed, a new function is the result.**

2.1 New Functions from Old

2.2 Inverse Functions

2.3 Transformations of Graphs

In this section, you will learn to:

1. Describe the domain and range of the inverse of a given function.
2. Explain why one function is the inverse of another.
3. Calculate the formula for the inverse of a function defined by a formula.
4. Sketch the graph of an inverse function.
5. Apply the horizontal line test to determine whether a function has an inverse.
6. Demonstrate that a function is one to one.
7. Construct an inverse function by restricting the domain of a given function.
8. Solve applied problems involving inverse functions.

Some functions have what are called inverses. We can think of the inverse as undoing the action of the function. The inverse function reverses the function process, and the roles of the domain and range are interchanged. We need to find an inverse function in a variety of applications. For example:

- If a function tells the value of a growing investment after t years, then the inverse function tells how long it takes the investment to reach a specified goal.
- If a function tells you your child's height as a function of age, then the inverse function tells how long it takes for your child to reach a given height.
- If a function gives the population of a growing town after t years, then the inverse function shows us how long it takes for the population to reach a target level.
- If a function gives the temperature of a muffin baking in the oven, then the inverse function can tell you when the muffin is done.

The list is endless, but some examples are more interesting than others. For example, it is not hard to find a model for the fraction of the surface of Earth visible at a given height above Earth, as shown in **Figure 2.15**. But if you are making a map, you need the inverse of this function. That is, you need to know what satellite orbit height will show the needed fraction of Earth's surface.

Figure 2.15 Earth as viewed from the International Space Station NASA

Defining Inverse Functions

> The inverse of a function undoes the action of that function.

To see how inverses apply to functions given by formulas, let's look at $f(x) = x^3$. In this formula, f cubes each number. Using the inverse function, we can undo that cubing by taking the cube root. Thus, $g(x) = x^{1/3}$ is the inverse of f because

$$g(f(x)) = g(x^3) = (x^3)^{1/3} = x^{3 \times 1/3} = x$$

A function g is the **inverse function** of a function f if $g(f(x)) = x$ for all x in the domain of f and $f(g(y)) = y$ for all y in the range of f.

In general, if g is the **inverse function** of f, then $g(f(x)) = x$ for each x in the domain of f. That is, f sends x to $f(x)$, and g sends $f(x)$ back to x again. The reverse is true as well: g sends y to $g(y)$, and f sends $g(y)$ back to y again.

We sometimes use the following equivalent formulation: A function g is the inverse of a function f provided that

$$g(y) = x \text{ if and only if } f(x) = y$$

One consequence of the definition is that if f is the inverse of g, then g is also the inverse of f. The inverse of f is usually denoted by f^{-1}. Note that f^{-1} is not the same as $\dfrac{1}{f}$.

Here is a simple example: If $f(x) = x + 1$ and $g(x) = x - 1$, then

$$g(f(x)) = g(x+1) = (x+1) - 1 = x$$

Also,

$$f(g(x)) = f(x-1) = (x-1) + 1 = x$$

These equations verify that g is the inverse of f, so $f^{-1}(x) = x - 1$.

If f and g are inverses of each other, then the roles of domain and range for f and g are interchanged. That is, the domain of g is the range of f, and the range of g is the domain of f.

EXAMPLE 2.8 Using the Definition of Inverse Functions

a. Let f be a function with domain $[-1, 1]$ and range $[2, 3]$. Assume that f has an inverse, and find the domain and range of f^{-1}.

b. The accompanying table defines a function $f(x)$ with inverse $g(x)$. Find $g(2)$.

x	f(x)
0	3
2	−1
5	7
9	2

SOLUTION

a. The domain of f^{-1} is the range of f. Thus, the domain of f^{-1} is $[2, 3]$. In a similar fashion, the range of f^{-1} is the domain of f. Therefore, the range of f^{-1} is $[-1, 1]$.

b. Because g "undoes" f, and $f(9) = 2$, we expect $g(2) = 9$. More formally, because g is the inverse of f, we know that $g(y) = x$ if and only if $f(x) = y$. Using $x = 9$ and $y = 2$, and noting that $f(9) = 2$, we conclude that indeed $g(2) = 9$.

TRY IT YOURSELF 2.8 Brief answers provided at the end of the section.

Use the table in part b of the example to find $g(-1)$.

EXTEND YOUR REACH

a. If the function g is the inverse of the function f, what is the inverse of the function g?

b. Can you find a function that is its own inverse?

EXAMPLE 2.9 Showing Functions are Inversese

If $f(x) = 3x + 1$, show that $f^{-1}(x) = \dfrac{x - 1}{3}$.

SOLUTION

Note that f is defined for all x, so the domain of f is $\mathbb{R}$. The range of f is also $\mathbb{R}$. Let $g(x) = \dfrac{x - 1}{3}$. According to the preceding definition, we need to establish two facts in order to show the functions are inverses.

Step 1: Show that $g(f(x)) = x$ for all x:

$$g(f(x)) = \frac{f(x) - 1}{3} \qquad \blacktriangleleft \text{ Evaluate the function } g \text{ at } f(x).$$

$$= \frac{(3x + 1) - 1}{3} \qquad \blacktriangleleft \text{ Evaluate } f.$$

$$= \frac{3x}{3} \qquad \blacktriangleleft \text{ Simplify.}$$

$$= \frac{\cancel{3}x}{\cancel{3}} \qquad \blacktriangleleft \text{ Cancel common factors.}$$

$$= x \qquad \blacktriangleleft \text{ Simplify.}$$

Step 2: Show that $f(g(x)) = x$ for all x:

$$f(g(x)) = 3g(x) + 1 \qquad \blacktriangleleft \text{ Evaluate the function } f \text{ at } g(x).$$

$$= 3 \times \frac{x - 1}{3} + 1 \qquad \blacktriangleleft \text{ Evaluate } g.$$

$$= \cancel{3} \frac{x - 1}{\cancel{3}} + 1 \qquad \blacktriangleleft \text{ Cancel common factors.}$$

$$= x \qquad \blacktriangleleft \text{ Simplify.}$$

TRY IT YOURSELF 2.9 Brief answers provided at the end of the section.

If $f(x) = x^5$ and $g(x) = x^{1/5}$, show that $g = f^{-1}$.

Finding Formulas and Graphs for Inverse Functions

> The graph of an inverse function is produced via reflection.

If a function $y = f(x)$ is given by a formula, the inverse reverses the roles of x and y and gives a formula for x in terms of y. We can calculate a formula for the inverse (if there is one) by simply solving the equation $y = f(x)$ for x. For example, consider the function $y = f(x) = 3x - 1$. Let's solve the equation $y = 3x - 1$ for x:

$$y = 3x - 1$$

$$y + 1 = 3x \qquad \blacktriangleleft \text{ Move } x\text{-terms right and all others left.}$$

$$\frac{y + 1}{3} = x \qquad \blacktriangleleft \text{ Divide by coefficient of } x.$$

This shows that the inverse of f is $f^{-1}(y) = \dfrac{y + 1}{3}$. We can use any variable name we wish in a function definition, and it is customary to rewrite the formula for f^{-1} as

$$f^{-1}(x) = \frac{x + 1}{3}.$$

EXAMPLE 2.10 Finding Inverse Functions from a Formula

Find a formula for the inverse of the function $f(x) = \dfrac{x + 1}{x - 2}$.

SOLUTION

Note that the function is not defined at $x = 2$. We solve the equation $y = \dfrac{x + 1}{x - 2}$ for x:

$$y = \frac{x + 1}{x - 2}$$

$$y(x - 2) = x + 1 \qquad \blacktriangleleft \text{ Multiply by } x - 2 \text{ to get an equation that is linear in } x.$$

$$yx - 2y = x + 1 \qquad \blacktriangleleft \text{ Apply the distributive law.}$$

$$yx - x = 2y + 1 \qquad \blacktriangleleft \text{ Move } x\text{-terms left and all others right.}$$

$$x(y - 1) = 2y + 1 \qquad \blacktriangleleft \text{ Factor out } x.$$

$$x = \frac{2y + 1}{y - 1} \qquad \blacktriangleleft \text{ Divide by coefficient of } x.$$

The solution gives the formula for the inverse function:

$$f^{-1}(y) = \frac{2y + 1}{y - 1}$$

We could also write $f^{-1}(x) = \dfrac{2x + 1}{x - 1}$.

TRY IT YOURSELF 2.10 Brief answers provided at the end of the section.

Find a formula for the inverse of $f(x) = \dfrac{x + 2}{2x + 1}$.

We reverse the roles of x and y to find the formula for the inverse of a function, and the same reversal applies to the graph. We make the graph of the inverse by interchanging the x- and y-axes. The result is to reflect the graph through the graph of the line $y = x$. Graphing the inverse function is illustrated in the next example.

EXAMPLE 2.11 Finding Inverse Functions Graphically

The graph of a function $y = f(x)$ is shown in **Figure 2.16**. Add to the figure the graph of $y = x$ and the graph of f^{-1}.

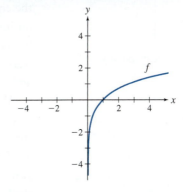

Figure 2.16 Graphing the inverse function

SOLUTION

In **Figure 2.17**, we have added the graph of $y = x$ to the graph of f. To get the graph of f^{-1} in **Figure 2.18**, we reflect the graph of f through the line $y = x$. Alternatively, we can think of getting the graph of the inverse by rotating or flipping the coordinate plane about the line $y = x$.

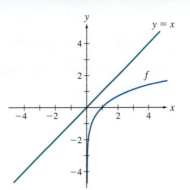

Figure 2.17 Line $y = x$ added to the graph of f

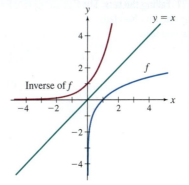

Figure 2.18 Reflecting the graph of f to get the graph of f^{-1}

TRY IT YOURSELF 2.11 Brief answers provided at the end of the section.

The graph of a function $y = f(x)$ is shown in **Figure 2.19**. Add to this figure the graphs of $y = x$ and the graph of f^{-1}.

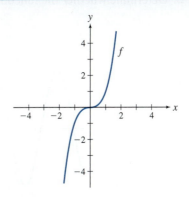

Figure 2.19 Graphing the inverse by reflecting

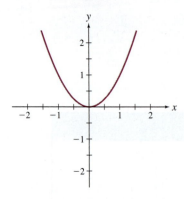

Figure 2.20 The graph of $y = x^2$

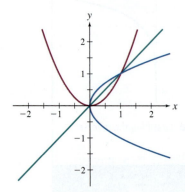

Figure 2.21 Failing the test: The reflected graph (in blue) fails the vertical line test.

The Horizontal Line Test for Inverses

> A function has an inverse if each horizontal line meets its graph no more than once.

Some functions do not have inverses. For example, let's try to find the inverse of the function $f(x) = x^2$, whose domain is the set of all real numbers. If we try to solve the equation $y = x^2$ for x, we can find a solution only if y is nonnegative. And if y is positive, we have two solutions: $x = \pm\sqrt{y}$. This fact means that f has no inverse. To be specific, note that for $y = 4$ the equation $4 = x^2$ has two solutions, 2 and -2. If g were the inverse of f, then according to the definition $g(4)$ would need to be both 2 and -2. But for a given input a function must give exactly one output—not two or more.

We can also see this fact graphically. **Figure 2.20** shows the familiar graph of $f(x) = x^2$. In **Figure 2.21** we have reflected this graph through the line $y = x$. Note that the resulting graph fails the vertical line test, so it is not the graph of a function.

The fact that the reflected graph fails the vertical line test means that some horizontal line meets the original graph more than once. This observation is the basis of the horizontal line test for inverses.

LAWS OF MATHEMATICS: The Horizontal Line Test

A function f has an inverse if and only if each horizontal line intersects the graph of f at most once.

EXAMPLE 2.12 Testing for Inverses Graphically

a. Explain why the function whose graph is shown in **Figure 2.22** has no inverse.

b. Determine whether the function whose graph is shown in **Figure 2.23** has an inverse.

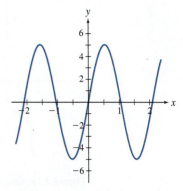

DF **Figure 2.22** Graph for part a of Example 2.12

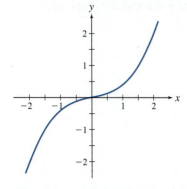

DF **Figure 2.23** Graph for part b of Example 2.12

SOLUTION

a. The function has no inverse because it fails the horizontal line test, as shown in **Figure 2.24**. Note that the horizontal line in the figure crosses the graph more than once.

b. Each horizontal line meets the graph at most once, as shown in **Figure 2.25**. Thus, the graph passes the horizontal line test, and we are assured of an inverse.

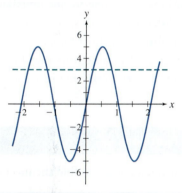

Figure 2.24 Failing the horizontal line test

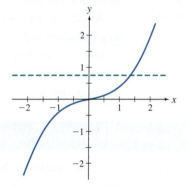

Figure 2.25 Passing the horizontal line test

TRY IT YOURSELF 2.12 Brief answers provided at the end of the section.

Determine whether the function whose graph is shown in **Figure 2.26** has an inverse.

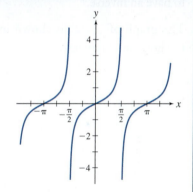

Figure 2.26 Function with possible inverse

Does the function whose graph is shown in **Figure 2.27** have an inverse even though some horizontal lines do not meet the graph at all?

EXTEND YOUR REACH

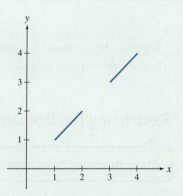

Figure 2.27 Inverse functions: Does this graph pass the horizontal line test?

One-to-One Functions

> A function has an inverse exactly when distinct elements of the domain have distinct function values.

A **one-to-one function** has a graph that satisfies the horizontal line test.

A function, such as an increasing or a decreasing function, whose graph passes the horizontal line test has a distinguishing name: **one-to-one function**.

An alternative (and equivalent) definition is this: A function is one to one if and only if distinct elements of the domain have distinct function values.

One-to-one functions are precisely the functions that have inverses. Important examples of one-to-one functions are those that are always increasing or are always decreasing.

EXAMPLE 2.13 Inverses and Increasing or Decreasing Functions

Show that the function $g(x) = 2^x$ has an inverse. Note that you are not asked to find the inverse.

SOLUTION

For the function $g(x) = 2^x$, larger values of x yield larger function values. Hence, g is an increasing function. Because g is increasing, it is guaranteed to have an inverse.[3]

The graph of $y = 2^x$ is shown in **Figure 2.28**, and it is easily seen to pass the horizontal line test.

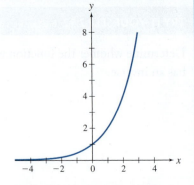

Figure 2.28 Graph of $y = 2^x$: The function is increasing, so it has an inverse.

TRY IT YOURSELF 2.13 Brief answers provided at the end of the section.

Explain why $y = -x^3$ has an inverse.

EXTEND YOUR REACH   Suppose the function f is one to one. Explain, using graphs, why the inverse f^{-1} is also a one-to-one function.

Restricting the Domain to Find Inverse Functions

> Some functions have inverses only when the domain is restricted.

Although $f(x) = x^2$ with domain $\mathbb{R}$ has no inverse, we can find an inverse if we restrict the domain. For example, if we put $h(x) = x^2$ with $x \geq 0$, we have restricted the

---

[3]The inverse of this function involves a *logarithm*, which we will study in Chapter 4.

domain of h to be $[0, \infty)$. This function *does* have an inverse: $g(x) = \sqrt{x}$ is the inverse of h because if $x \geq 0$, then

$$g(h(x)) = \sqrt{x^2} = x \quad \text{and} \quad h(g(x)) = (\sqrt{x})^2 = x$$

In summary, the function $f(x) = x^2$ has no inverse, but the function $h(x) = x^2$, $x \geq 0$, does have an inverse, namely $h^{-1}(x) = \sqrt{x}$. **Figure 2.29** shows the restricted graph and its reflection. The result is the graph of a function.

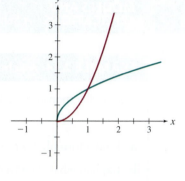

Figure 2.29 Inverse for a function with restricted domain: The reflected graph (in green) of the restricted function (in red) passes the vertical line test.

CONCEPTS TO REMEMBER: Ways to Find Inverse Functions

- If the function is given by a formula $y = f(x)$, we try to solve the equation for x.
  - If there is exactly one solution $x = g(y)$ for each y in the range of f, then interchange x and y. The result is $f^{-1}(x) = g(x)$.
  - If there is more than one solution for some y-value, then f has no inverse. (But restricting the domain may lead to an inverse.)
- A function has an inverse provided its graph passes the horizontal line test. If it passes, the graph of the inverse can be found by reflecting the graph of the original function through the line $y = x$.

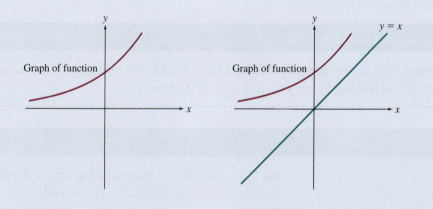

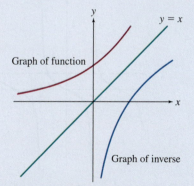

- One-to-one functions have inverses. Important examples are increasing functions and decreasing functions.

MODELS AND APPLICATIONS   A School Trip

EXAMPLE 2.14 A School Trip

A school is sponsoring a trip for students. If the school takes n students, then the cost to the school is

$$C(n) = 200 + 25n \text{ dollars}$$

a. Find a formula for the inverse function.

b. Explain what the inverse function means in terms of the number of students and the cost of the trip.

SOLUTION

a. Solving the equation $y = 200 + 25n$ for n, we find $n = \dfrac{y - 200}{25}$ or

$$C^{-1}(y) = \frac{y - 200}{25}$$

b. The function $C(n)$ gives the cost in dollars of a field trip if n students attend. Because $C(n)$ tells us the cost if we know the number of students, the inverse function $C^{-1}(y)$ tells us the number of students if we know the cost. Thus, $C^{-1}(y)$ gives the number of students the school can afford to take if its budget for the trip is y dollars.

We note that in this setting, some prefer the notation

$$n(C) = \frac{C - 200}{25}$$

because it is more suggestive of the actual relationship.

TRY IT YOURSELF 2.14   Brief answers provided at the end of the section.

Let $H(n)$ be the height of a tower a child can make with n building blocks. Explain the meaning of the inverse function H^{-1} in terms of the number of building blocks and the height of the tower.

TRY IT YOURSELF ANSWERS

2.8 $g(-1) = 2$

2.9 Calculate that $g(f(x)) = (x^5)^{1/5} = x^{5 \times (1/5)} = x$.

Also, $f(g(x)) = (x^{1/5})^5 = x^{(1/5) \times 5} = x$.

2.10 $f^{-1}(x) = \dfrac{2 - x}{2x - 1}$

2.11

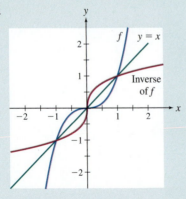

2.12 The graph fails the horizontal line test, as shown here. Hence, the function has no inverse.

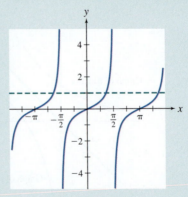

2.13 $y = -x^3$ is a decreasing function; consequently, it has an inverse.

2.14 The function $H^{-1}(h)$ tells how many blocks are needed to build a tower of height h.

EXERCISE SET 2.2

CHECK YOUR UNDERSTANDING

1. **True or false:** Every function has an inverse.

2. The horizontal line test tells us that _____.

3. Given the graph of a function with an inverse, we can make the graph of the inverse by which of the following?
   a. Reflecting through the line $y = x$
   b. Checking that the horizontal line test is passed
   c. Adding the line $y = x$
   d. Reflecting through the line $y = -x$

4. If the equation $y = f(x)$ has more than one solution x for some values of y, then _____.

5. **True or false:** If the equation $y = f(x)$ has no solution for some values of y, then f may still have an inverse.

6. If g is the inverse of f, then which is true?
   a. f is one to one.
   b. f is the inverse of g.
   c. The equation $y = f(x)$ has exactly one solution x for each y in the range of f.
   d. All of the above.

7. **True or false:** A function that does not have an inverse may have an inverse if its domain is restricted.

8. If g is the inverse of f and $f(3) = 7$, then _____.

9. If g is the inverse of f and $g(8) = 12$, then _____.

10. If g is the inverse of f and 4 is in the domain of f, then $g(f(4)) = $ _____.

11. If g is the inverse of f, how are the domains and ranges of these functions related?

SKILL BUILDING

Verifying inverses. In Exercises 12 through 24, verify that the function g is the inverse of the function f.

12. $f(x) = 3x$, $g(x) = \dfrac{x}{3}$

13. $f(x) = \dfrac{1}{x}$, $g(x) = \dfrac{1}{x}$

14. $f(x) = x + 1$, $g(x) = x - 1$

15. $f(x) = x^9$, $g(x) = x^{1/9}$

16. $f(x) = x^{1/7}$, $g(x) = x^7$

17. $f(x) = 1 + \dfrac{1}{x}$, $g(x) = \dfrac{1}{x - 1}$

18. $f(x) = \dfrac{x}{x + 2}$, $g(x) = \dfrac{2x}{1 - x}$

19. $f(x) = 3 - \dfrac{2}{x}$, $g(x) = \dfrac{2}{3 - x}$

20. $f(x) = \dfrac{2x - 1}{3x - 2}$, $g(x) = \dfrac{1 - 2x}{2 - 3x}$

21. $f(x) = 2 - \dfrac{1}{x - 1}$, $g(x) = 1 + \dfrac{1}{2 - x}$

22. $f(x) = 1 - \dfrac{x + 1}{x - 1}$, $g(x) = \dfrac{x - 2}{x}$

23. $f(x) = -x$, $g(x) = -x$

24. $f(x) = x$, $g(x) = x$

25. **Verifying inverses when domain is restricted.** If $f(x) = x^2 + 1$ with $x \geq 0$ and $g(x) = \sqrt{x - 1}$ with $x \geq 1$, verify that $g = f^{-1}$.

26. **Domain and range for inverse functions.** Let f be a function with domain $[0, 2]$ and range $[0, 7]$. If f has an inverse, identify the domain and range of f^{-1}.

27. **Domain and range for inverse functions.** Let f be a function with domain $(0, \infty)$ and range $(-\infty, \infty)$. If f has an inverse, identify the domain and range of f^{-1}.

28. **Inverse from a table.** The accompanying table defines a function $f(x)$ with inverse $g(x)$. Find $g(5)$.

x	$f(x)$
0	3
1	5
2	2
3	1

Calculating values for an inverse function. The function g is the inverse of f. Also, $f(2) = 4$ and $f(5) = 2$. In Exercises 29 through 34, use this information to calculate the given function value.

29. $g(4)$

30. $(g \circ f)(6)$, assuming 6 is in the domain of f

31. $(f \circ g)(6)$, assuming 6 is in the range of f

32. $g(2)$

33. $(g \circ g)(4)$

34. $(g \circ f \circ g)(2)$

PROBLEMS

Inverse function graphs. In Exercises 35 through 43, determine whether the given graph is the graph of a function with an inverse. If it is, then sketch the graph of the inverse function.

35.

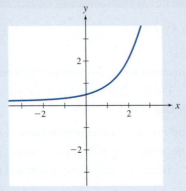

36.

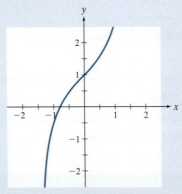

37.

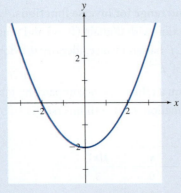

38.

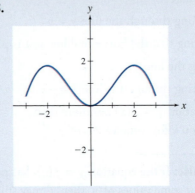

39.

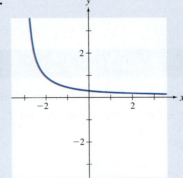

40.

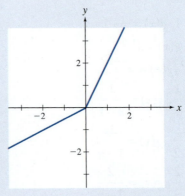

41.

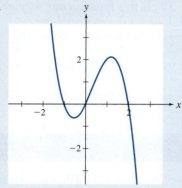

42.

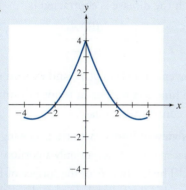

43.

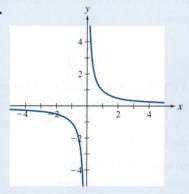

Calculating inverses. In Exercises 44 through 58, either calculate the inverse of the given function or show that no inverse exists. If you answer that no inverse exists, justify your answer.

44. $f(x) = 3x - 4$

45. $f(x) = \dfrac{1}{x-1}$

46. $f(x) = x + \dfrac{1}{x}$

47. $f(x) = 3 - x^2$

48. $f(x) = 1 + \dfrac{1}{x}$

49. $f(x) = \dfrac{2+x}{2-x}$

50. $f(x) = \sqrt{1+x^2}$

51. $f(x) = (x+1)(x+2)$

52. $f(x) = \dfrac{3x+4}{2x+5}$

53. $f(x) = 1 - \dfrac{x+1}{x-1}$

54. $f(x) = (x-1)^3$

55. $f(x) = x^5 - 1$

56. $f(x) = x^4 - 5$

57. The constant function defined by $f(x) = 2$.

58. $f(x) = 2^{x^2}$

A certain inverse. We saw in Example 2.13 that the function $g(x) = 2^x$ has an inverse function, which we call h. Exercises 59 through 63 are concerned with this function. Answer the question in each exercise. Note that you may find it helpful to sketch the graph of g first.

59. What is the domain of h?

60. What is the range of h?

61. What is $h(8)$? *Suggestion:* Solve the equation $8 = 2^x$.

62. What is $h(1)$?

63. Explain what $h(x)$ represents.

One-to-one functions. In Exercises 64 through 74, determine whether the given function is one to one. If it is not one to one, identify two specific elements of the domain that have the same function value.

64. $f(x) = x^4$

65. $f(x) = x^3$

66. $g(t) = 2^t$

67. $h(t) = t^2$

68. $k(x) = \dfrac{1}{x}$

69. $f(x) = x$

70. $h(x) = 4$

71. $f(t)$ is an increasing function.

72. $g(x)$ is a decreasing function.

73. f is the function whose graph is shown.

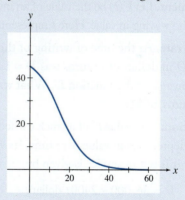

74. f is the function whose graph is shown.

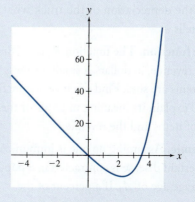

Using a graphing utility. In Exercises 75 through 80, use a graphing utility to make the graph of the given function. (Be sure to include both positive and negative values of x.) Use the graph to answer the question.

75. $f(x) = 4^x + x$. Does this function have an inverse?

76. $f(x) = 2^x - x$. Does this function have an inverse?

77. $f(x) = 3^x - 3^{-x}$. Is this function one to one?

78. $f(x) = \dfrac{2^x}{x^2 + 1}$. Is this function one to one?

79. $f(x) = x^3 - 9x + 1$. Is this function one to one?

80. $f(x) = \dfrac{10}{1 + 0.9^x}$. Is this function increasing?

MODELS AND APPLICATIONS

Inverse from a verbal description. In Exercises 81 through 84, a function is given. In each case, explain why an inverse exists, and explain what the inverse function tells us in terms of the variables.

81. **Pesos to Dollars.** Let $P(d)$ denote the function that gives the current value in Mexican pesos of d dollars.

82. **A game preserve.** Let $N(t)$ denote the population of a certain species in decline on a game preserve t years after ecologists began observing the preserve.

83. **Holling's functional response.** Holling's functional response gives the number $p(d)$ of prey a predator will consume if d is the density of prey in the region. It is a fact that p is an increasing function.

84. **An investment.** Let $V(t)$ be the value of an investment that is growing in value. Here t represents time.

85. **Exchange rate.** At the time of writing of this text, the value D in dollars of E euros was $D = 1.13E$. Find the inverse of the function D. What was the value, in euros, of \$1?

86. **Depreciation.** The value V of a truck owned by a company depreciates in value over time. Its value t years after it is put in service is given by

$$V = 36,000 - 2400t \text{ dollars}$$

Find the inverse function $t(V)$, and explain in terms of the depreciation of the truck what this function tells us.

87. **Revenue function.** The function $R(n) = 38n - 520$ gives the revenue, in dollars, a small business makes when n items are sold. Find the inverse function $n(R)$, and explain its meaning in terms of the number of items and the revenue.

88. **An epidemic.** Assuming no one is immunized at birth, a certain disease is contracted in a population at an average age of A_0. If q percent of the population is immunized at birth, then the average age at which the disease is contracted is

$$A(q) = \frac{100 A_0}{100 - q}$$

Find the inverse function $q(A)$, and explain what the inverse function tells us in terms of immunization and contracting disease.

89. **Astronauts' view of Earth.** Astronauts looking at Earth from a spacecraft can see only a portion of the surface.[4] The fraction F of the surface of Earth visible from a height h, in kilometers, above the surface is given by

$$F(h) = \frac{h}{D + 2h}$$

where D is the diameter of Earth in kilometers. Find the inverse function $h(F)$, and explain what it tells us in terms of height and how much of Earth's surface is visible.

90. **Inverse from a graph.** The graph in **Figure 2.30** shows the size of a certain population as a function of time measured in years. Determine whether the population function has an inverse. If it does have an inverse, add the graph of the inverse to the figure and explain what the inverse means in terms of the population and time. If it does not have an inverse, explain why.

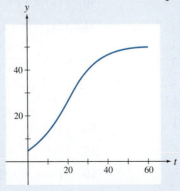

Figure 2.30 Population as a function of time

---

[4]See *Space Mathematics*, B. Kastner, published by NASA in 1985.

CHALLENGE EXERCISES FOR INDIVIDUALS OR GROUPS

91. Challenge problem. Find a function f that is one to one and has domain $(-\infty, \infty)$ but is neither an increasing nor a decreasing function. *Suggestion:* Think in terms of piecewise-defined functions.

Difficult inverses. The functions in Exercises 92 through 94 have inverses, but care is needed in finding them. State the domain of the inverse you find.

92. $f(x) = \dfrac{1}{x^2} - \dfrac{1}{x}, \quad 0 < x < 2$

93. $f(x) = x - \sqrt{x}, \quad x > 1$

94. $f(x) = x^2 - 1, \quad x < 0$

95. A composition. If f and g are one-to-one functions, show that the composition $f \circ g$ is also one to one.

96. The rate of change of the inverse. In this exercise, we find a formula for the rate of change of the inverse function.

a. Consider the line $y = 2x + 1$ and its reflection through the line $y = x$. How is the slope of the reflected line related to the slope of the original line? *Suggestion:* One way to answer this question is to graph the lines. Another way is to calculate the inverse of the function $y = 2x + 1$.

b. In general, let L be a line with nonzero slope m. Find the slope of the line obtained by reflecting L through the line $y = x$.

c. Recall that the rate of change of a function $y = f(x)$ at a value of x is the slope of the tangent line at the corresponding point on the graph. Assume that f is a function that has an inverse and that the rate of change of f at a is m, where $m \neq 0$. Use your answer to part b to show that the rate of change of f^{-1} at $f(a)$ equals $\dfrac{1}{m}$.

REVIEW AND REFRESH: Exercises from Previous Sections

97. From Section P.1: Find the distance between the points $(1, -4)$ and $(-2, 5)$.

Answer: $\sqrt{90} = 3\sqrt{10}$

98. From Section P.2: Solve the inequality $5x - 4 > 2x + 6$.

Answer: $\left(\dfrac{10}{3}, \infty\right)$

99. From Section P.3: Solve the inequality $x^4 > 4x^2$.

Answer: $(-\infty, -2) \cup (2, \infty)$

100. From Section 1.1: Find the domain of $f(x) = \dfrac{\sqrt{x-1}}{x-2}$.

Answer: $[1, 2) \cup (2, \infty)$

101. From Section 1.2: If $f(x) = x^2 + 7x - 3$, calculate the average rate of change of $f(x)$ on the interval $[x, x + h]$. Assume that $h > 0$, and simplify your answer.

Answer: $2x + 7 + h$

102. From Section 1.3: Which of the following describes the graph of the function $y = 1 - x^2$ on the interval $[0, \infty)$? (a) Increasing and concave up, (b) increasing and concave down, (c) decreasing and concave up, or (d) decreasing and concave down.

Answer: (d)

103. From Section 1.4: Find the limit of $f(x) = \dfrac{1}{x+5}$ as $x \to \infty$.

Answer: 0

104. From Section 2.1: If $f(x) = x^2 + 1$ and $g(x) = (x + 2)^2$, find $(g \circ f)(x)$.

Answer: $(x^2 + 3)^2 = x^4 + 6x^2 + 9$

105. From Section 2.1: If $h(x) = \sqrt{x^2 + 1}$, find functions $f(x)$ and $g(x)$ so that $h(x) = (f \circ g)(x)$.

Answer: One solution is $f(x) = \sqrt{x}$ and $g(x) = x^2 + 1$.

2.1 New Functions from Old

2.2 Inverse Functions

2.3 Transformations of Graphs

2.3 Transformations of Graphs

> Interpreting a graph is easier if we recognize transformations.

In this section, you will learn to:

1. Perform vertical and horizontal shifting of graphs.
2. Perform vertical stretching, compressing, and reflecting of graphs.
3. Perform horizontal stretching, compressing, and reflecting of graphs.
4. Determine whether a given function is even, odd, or neither.
5. Solve applied problems involving transformations of graphs.

In many cases, the graph of one function may be obtained by transforming the graph of another, familiar function — for example, by shifting or stretching. Recognizing this fact can simplify the process of graphing the first function. In this section we investigate such transformations. When we model real-world phenomena, examples of such transformations include changes in units of measurement and shifts in starting time.

Although it can be confusing to convert from one unit to another, units help us make sense of numbers. For example, suppose I tell you that I drove to work this morning at a speed of 880 inches per second. This statement wouldn't be a lie, but it probably wouldn't convey much information to you. In fact, 880 inches per second is 50 miles per hour, and that number seems more meaningful, at least to Americans. Many others would understand 80 kilometers per hour better.

Very large and very small numbers can be better understood when appropriate units are applied. Jupiter is 778 million kilometers from the sun. That is a long way, but it may mean more to say that it is five times as far away from the sun as Earth. As another example, in 2017 the U.S. budget deficit was \$665 billion. What does \$665 billion mean? It means that every man, woman, and child in the United States owes about \$2000. Put another way, that money could buy a \$25,000 automobile for every living soul in the states of Iowa, Kansas, Louisiana, Minnesota, Nebraska, and Oklahoma.

Visually, changing units has the effect of transforming a graph, and that fact may be an aid to understanding. The same is true for many transformations of functions and graphs.

Shifting of Graphs

> Addition and subtraction of a real number to either a function or its argument result in vertical or horizontal shifts to the graph.

The simplest transformation of a graph is the vertical shift, which occurs when a graph is moved up or down relative to the horizontal axis. For example, let's compare the graphs of $f(x) = x^2$ and $g(x) = x^2 + 1$. These two graphs are shown in **Figure 2.31**.

Note that the graph of g is obtained by shifting the graph of f up by 1 unit. This makes sense because we obtain g by adding 1 unit to f. The relationship is further illustrated by the accompanying table of values.

x	f(x) = x²	g(x) = f(x) + 1
−2	4	5
−1	1	2
0	0	1
1	1	2
2	4	5

There is nothing special about the number 1, and we can describe this process of vertical shifting for any graph.

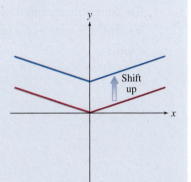

DF **Figure 2.31** Vertical shift: We sketch the graph of $f(x) + 1$ by shifting the graph of $f(x)$ up by 1 unit.

CONCEPTS TO REMEMBER: Sketching Vertically Shifted Graphs

For $c > 0$:

- To sketch the graph of $g(x) = f(x) + c$, we shift the graph of $f(x)$ *up* by c units.

- To sketch the graph of $g(x) = f(x) - c$, we shift the graph of $f(x)$ *down* by c units.

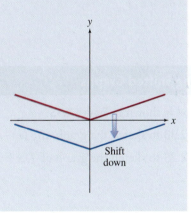

When we add (or subtract) a number inside the function parentheses, we get a shift in the horizontal direction. For example, let's compare the graphs of the functions $f(x) = x^2$ and $h(x) = f(x-1) = (x-1)^2$. Note from the accompanying table that the value of h at a point is the same as the value of f at 1 unit to the left of that point.

x	f(x) = x²	h(x) = (x − 1)²
−2	4	9
−1	1	4
0	0	1
1	1	0
2	4	1

Hence, to obtain the graph of h, we shift the graph of f by 1 unit to the right. This fact is shown in **Figure 2.32**. We can describe this process of horizontal shifting for any graph.

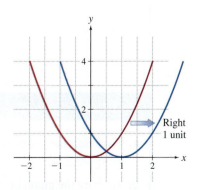

DF **Figure 2.32** Horizontal shift: To sketch the graph of $f(x-1)$, we shift the graph of $f(x)$ by 1 unit to the right.

CONCEPTS TO REMEMBER: Sketching Horizontally Shifted Graphs

For $c > 0$:

- To sketch the graph of $h(x) = f(x - c)$, we shift the graph of $f(x)$ to the *right* by c units.

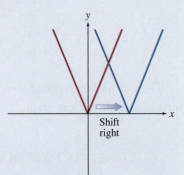

- To sketch the graph of $h(x) = f(x + c)$, we shift the graph of $f(x)$ to the *left* by c units.

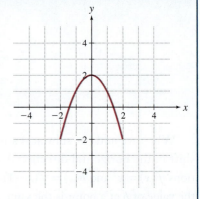

EXAMPLE 2.15 Sketching Shifted Graphs

The graph of a function $f(x)$ is shown in **Figure 2.33**.

a. Use the graph in Figure 2.33 to sketch the graph of $f(x) - 2$.

b. Use the graph in Figure 2.33 to sketch the graph of $f(x + 2)$.

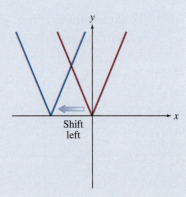

Figure 2.33 The graph of $f(x)$ for Example 2.15

SOLUTION

a. To sketch the graph of $f(x) - 2$, we shift the graph of $f(x)$ down by 2 units. The new graph is shown in **Figure 2.34**.

b. To draw the graph of $f(x + 2)$, we shift the graph of $f(x)$ to the left by 2 units. The new graph is shown in **Figure 2.35**.

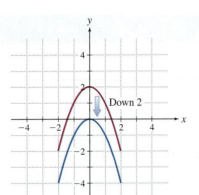

Figure 2.34 Vertical shift: To draw the graph of $f(x) - 2$, we shift the graph of $f(x)$ down by 2 units.

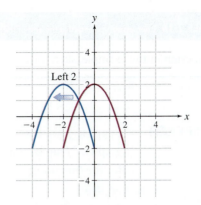

Figure 2.35 Horizontal shift: The graph of $f(x + 2)$ is obtained by shifting the graph of $f(x)$ to the left by 2 units.

TRY IT YOURSELF 2.15 Brief answers provided at the end of the section.

Sketch the graph of $f(x) + 2$.

EXAMPLE 2.16 Sketching Graphs with Multiple Shifts

Use the graph in **Figure 2.36** to sketch the graph of $f(x - 2) + 2$.

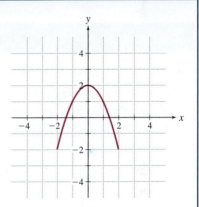

Figure 2.36 The graph of $f(x)$ for Example 2.16

SOLUTION

We find this graph in two steps. We focus first on $f(x - 2)$. Working just as we did in the previous example, we first shift the graph of $f(x)$ by 2 units to the right. The result is shown in **Figure 2.37**. Next, we deal with the vertical shift. We shift the graph of $f(x - 2)$ up by 2 units to get the graph of $f(x - 2) + 2$. The result is shown in **Figure 2.38**.

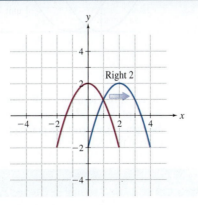

Figure 2.37 Horizontal shift: To sketch the graph of $f(x - 2)$, we shift the graph of $f(x)$ to the right by 2 units.

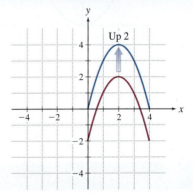

Figure 2.38 Vertical shift: To sketch the graph of $f(x - 2) + 2$, we shift the graph of $f(x - 2)$ up by 2 units.

TRY IT YOURSELF 2.16 Brief answers provided at the end of the section.

If $f(x)$ is the function in the example, sketch the graph of $f(x+2)-2$.

EXTEND YOUR REACH   Suppose $f(x)=x^2$. The graph of a function g is obtained by first shifting the graph of f up by 3 units and then shifting to the left by 4 units. Find a formula for the function g.

The next example shows how we can use the familiar graph of $y=x^2$ to make the graphs of more complicated functions.

EXAMPLE 2.17 Using Graphs of Simple Functions to Sketch Graphs of More Complicated Functions

a. Sketch the graph of $y=x^2$.

b. Use the results from part a to sketch the graph of $y=x^2-2x+2$ by observing that

$$x^2-2x+2=(x-1)^2+1$$

SOLUTION

a. The familiar graph of $y=x^2$ is shown in **Figure 2.39**.

b. We produce the graph of $y=(x-1)^2+1$ in two steps. The first step, shown in **Figure 2.40**, is to shift the graph of $y=x^2$ by 1 unit to the right to get the graph of $y=(x-1)^2$. The next step, shown in **Figure 2.41**, is to shift the graph of $y=(x-1)^2$ up by 1 unit to make the graph of $y=(x-1)^2+1$.

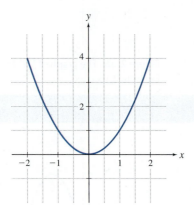

Figure 2.39 The graph of $y=x^2$

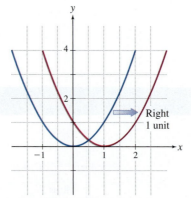

Figure 2.40 Horizontal shift: Shift the graph of $y=x^2$ by 1 unit to the right to construct the graph of $y=(x-1)^2$.

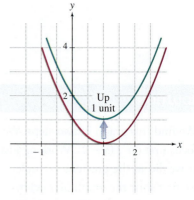

Figure 2.41 Vertical shift: Move the graph of $y=(x-1)^2$ up by 1 unit to construct the graph of $y=(x-1)^2+1$.

TRY IT YOURSELF 2.17 Brief answers provided at the end of the section.

Use the graph of $y=x^2$ to make the graph of $y=x^2-4x$. Observe that $x^2-4x=(x-2)^2-4$.

Vertical Stretching, Compressing, and Reflecting of Graphs

> Multiplication of a function by a real number results in a vertical stretch, compression, or reflection of the graph.

Some transformations lead to a vertical stretch of a graph, in which the graph is resized in the vertical direction. To illustrate vertical stretching, we compare $f(x) = x^2 - 1$ with $g(x) = 2f(x) = 2(x^2 - 1)$. The accompanying table of values shows that each function value for f is multiplied by 2. This is further illustrated by the graphs shown in **Figure 2.42**. Note that the height of each point on the graph of $2f$ is twice the height of the corresponding point on the graph of f. The result is a vertical stretch.

x	$f(x) = x^2 - 1$	$g(x) = 2f(x)$
-2	3	6
-1	0	0
0	-1	-2
1	0	0
2	3	6

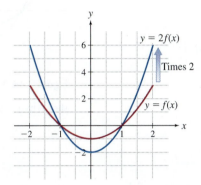

DF **Figure 2.42** Vertical stretching shown by $y = x^2 - 1$ and $y = 2(x^2 - 1)$

Figure 2.43 may clarify the relationship between the two graphs. One way to visualize this process is to think of this page as being made of rubber, except that it allows stretching only in the vertical direction. We physically pull the top of the page upward and the bottom of the page downward.

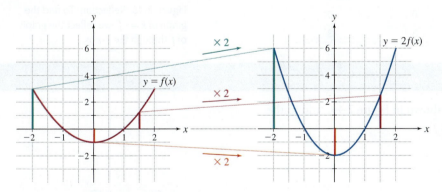

Figure 2.43 A side-by-side view of a graph stretched vertically

The effect is somewhat different if we multiply the function $f(x)$ by a number between 0 and 1 or by a negative number. We illustrate these cases in the following example.

EXAMPLE 2.18 Vertical Compression and Reflection

Use the graph of $f(x) = x^2 - 1$ to sketch graphs of the following functions.

a. $h(x) = \dfrac{1}{2}f(x) = \dfrac{1}{2}(x^2 - 1)$ **b.** $k(x) = -f(x) = -(x^2 - 1)$

SOLUTION

a. To find each value of $h(x)$, we multiply the corresponding value of $f(x)$ by $\dfrac{1}{2}$, as shown in the accompanying table. As the table indicates, we obtain the graph of $k = \dfrac{1}{2}f$ by multiplying each y-value by $\dfrac{1}{2}$. This results in a vertical compression of the graph, as shown in **Figure 2.44**.

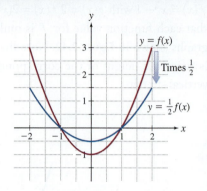

Figure 2.44 Vertical compression shown by $y = x^2 - 1$ and $y = \dfrac{1}{2}(x^2 - 1)$

x	$f(x) = x^2 - 1$	$h(x) = \dfrac{1}{2}f(x)$
−2	3	1.5
−1	0	0
0	−1	−0.5
1	0	0
2	3	1.5

b. We find each value of $k(x) = -f(x)$ by taking the negative of the corresponding value of $f(x)$, as shown in the accompanying table. The result is to reflect the graph through the x-axis, as shown in **Figure 2.45**. Intuitively, we may think of flipping the graph.

x	$f(x) = x^2 - 1$	$k(x) = -f(x)$
−2	3	−3
−1	0	0
0	−1	1
1	0	0
2	3	−3

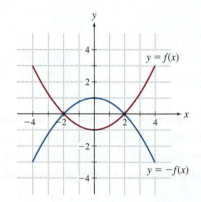

Figure 2.45 Reflection: To find the graph of $k = -f$, we reflect the graph of f through the x-axis.

TRY IT YOURSELF 2.18 Brief answers provided at the end of the section.

The graph of a function f is shown in **Figure 2.46**. Add to the illustration the graph of $h = 3f$.

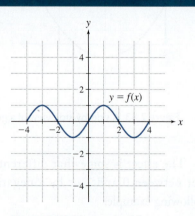

Figure 2.46 The graph of f

CONCEPTS TO REMEMBER: Sketching Graphs That Are Stretched, Compressed, or Reflected Vertically

To obtain the graph of cf, we multiply the height of each point on the graph of f by c.

- If $c > 1$, this transformation results in a vertical stretching by a factor of c.

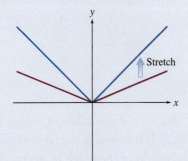

- If $0 < c < 1$, this transformation results in a vertical compression of the graph.

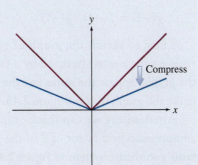

- If $c < 0$, this transformation results in a reflection of the graph about the x-axis followed by a vertical stretch or compression by a factor of $|c|$.

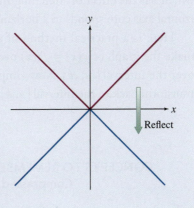

Horizontal Stretching, Compressing, and Reflecting of Graphs

Multiplication of a function argument by a real number results in a horizontal stretch, compression, or reflection of its graph.

Some transformations lead to a horizontal stretch of a graph, which means that the graph is resized in the horizontal direction. The graphs of $f(x)$ and $g(x) = f(cx)$ are related by a horizontal stretch if $c > 1$. To illustrate the relationship, we compare the functions $f(x) = x^2 - 1$ and $g(x) = f(2x) = (2x)^2 - 1$ first by looking at the two accompanying tables of values.

x	$f(x) = x^2 - 1$
-4	15
-2	3
0	-1
2	3
4	15

x	$2x$	$g(x) = (2x)^2 - 1$
-2	-4	15
-1	-2	3
0	0	-1
1	2	3
2	4	15

Note that the table of values for $g(x) = f(2x)$ on the interval $-2 \le x \le 2$, or $[-2, 2]$, is the same as the table of values for $f(x)$ on the interval $-4 \le x \le 4$, or $[-4, 4]$. The height of the graph of $g(x) = f(2x)$ for any value of x is the height of the graph of f at the value of $2x$:

$$g(-2) = f(-4), \quad g(-1) = f(-2), \quad g(0) = f(0), \quad g(1) = f(2), \; g(2) = f(4)$$

Thus, to sketch the graph of $g(x) = f(2x)$, we use the graph of $f(x)$ with the scale on the x-axis stretched by a factor of 2 (**Figure 2.47** and **Figure 2.48**). Note that the graphs appear the same, but the scale on the x-axis is stretched by a factor of 2 to make the graph of g in Figure 2.48.

Alternatively, we can think of making the graph of $g(x) = f(2x)$ via a horizontal compression by a factor of 2 of the graph of $f(x)$. The two graphs are shown on the same coordinate system in **Figure 2.49**.

To find the graph of $f(cx)$ from the graph of $f(x)$, the factor c is applied to x, which has the effect of stretching the horizontal axis by a factor of c. A stretched horizontal axis corresponds to a horizontal compression of the graph.

Here is a practical method of accomplishing horizontal stretches for $c > 0$. To make the graph of $g(x) = f(cx)$ over the interval $[a, b]$, start with the graph of $f(x)$ over the interval $[ca, cb]$. Now simply stretch or compress the x-axis as necessary to transform back to the interval $[a, b]$.

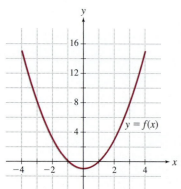

Figure 2.47 The graph of $f(x) = x^2 - 1$ over the interval $[-4, 4]$

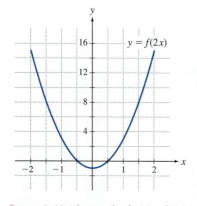

Figure 2.48 The graph of $g(x) = f(2x)$ over the interval $[-2, 2]$

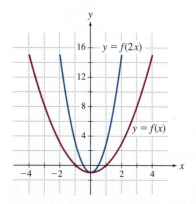

DF **Figure 2.49** The graphs of $f(x)$ and $g(x) = f(2x)$ together

CONCEPTS TO REMEMBER: Sketching Graphs That Are Stretched, Compressed, or Reflected Horizontally

To obtain the graph of $f(cx)$, we replace each function value $f(x)$ by the new function value $f(cx)$.

- If $c > 1$, the result is a horizontal compression of the graph by a factor of c. (Alternatively, one can think of this as a horizontal stretch of the x-axis.)

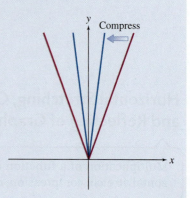

- If $0 < c < 1$, the result is a horizontal stretch of the graph or a horizontal compression of the x-axis.

- If $c < 0$, the result is a reflection of the graph about the y-axis followed by a horizontal stretch or compression by $|c|$.

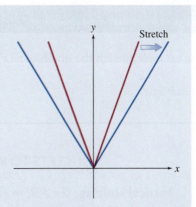

EXAMPLE 2.19 Horizontal Stretching of Graphs

The graph of a function $f(x)$ is shown in **Figure 2.50**. Use this graph to draw the graph of $g(x) = f\left(\dfrac{x}{2}\right)$ over the interval $[-2, 2]$.

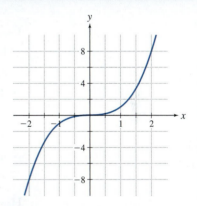

Figure 2.50 The graph of $f(x)$ for Example 2.19

SOLUTION

Because $\dfrac{x}{2} = \dfrac{1}{2}x$ and we wish to view the graph of $g(x)$ over the interval $[-2, 2]$, we start with the graph of $f(x)$ over the interval $\left[\dfrac{1}{2} \times -2, \dfrac{1}{2} \times 2\right] = [-1, 1]$. This graph is shown in **Figure 2.51**. Then we stretch that portion of the x-axis so that it becomes the interval $[-2, 2]$. The result is shown in **Figure 2.52**. Finally, we show the two graphs on the same coordinate axes in **Figure 2.53**.

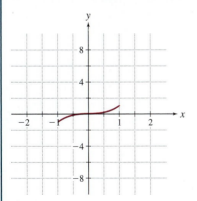

Figure 2.51 The graph of $f(x)$ over a shortened interval

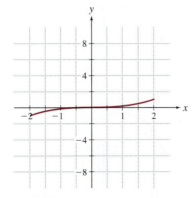

Figure 2.52 The graph over a shortened interval stretched to make the graph of $f\left(\dfrac{x}{2}\right)$

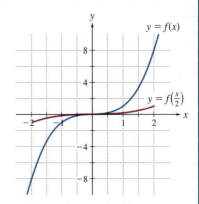

Figure 2.53 The graph of both $f(x)$ and $f\left(\dfrac{x}{2}\right)$

TRY IT YOURSELF 2.19 Brief answers provided at the end of the section.

If $f(x)$ is the function in the example, sketch the graph of $m(x) = 6f\left(\dfrac{x}{2}\right)$. *Suggestion*: Start with the graph of $g(x) = f\left(\dfrac{x}{2}\right)$ found in this example.

CONCEPTS TO REMEMBER: Transforming Graphs

Vertical shifting: If $c > 0$, to sketch the graph of $g(x) = f(x) + c$ we shift the graph of $f(x)$ up by c units, and to sketch the graph of $g(x) = f(x) - c$ we shift the graph of $f(x)$ down by c units.

Horizontal shifting: If $c > 0$, to obtain the graph of $g(x) = f(x + c)$ we shift the graph of $f(x)$ to the left by c units, and to sketch the graph of $g(x) = f(x - c)$ we shift the graph of $f(x)$ to the right by c units.

Vertical stretching, compressing, and reflecting: If $g(x) = cf(x)$, to find the height of each point on the graph of g we multiply the height of the corresponding point on the graph of f by c.

- If $c > 1$, this transformation results in a vertical stretching of the graph.

- If $0 < c < 1$, this transformation results in a vertical compression of the graph.

- If $c < 0$, this transformation results in a reflection of the graph about the x-axis followed by a vertical stretch or compression by $|c|$.

Horizontal stretching, compressing, and reflecting: Assume $c > 0$.

- To sketch the graph of $g(x) = f(cx)$ on $[a, b]$, we graph $f(x)$ on $[ac, ab]$ and then horizontally adjust the resulting graph back to the interval $[a, b]$. The adjustment is a compression if $c > 1$ and a stretch if $0 < c < 1$.

- To sketch the graph of $g(x) = f(-cx)$, we first reflect the graph about the y-axis and then horizontally stretch or compress by c.

EXAMPLE 2.20 Sketching Graphs by Combining Transformations

The graph of $f(x)$ is shown in **Figure 2.54**. Sketch the graph of $g(x) = -3f(x + 1)$.

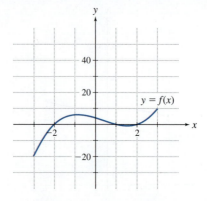

Figure 2.54 The graph of $f(x)$

SOLUTION

We break the process into three steps:

Step 1: Graph $-f(x)$. To sketch the graph of $-f(x)$, we reflect the graph of $f(x)$ about the x-axis. We can think of holding our fingertips on the ends of the x-axis and rotating the coordinate plane 180 degrees. The result is in **Figure 2.55**.

Step 2: Obtain the graph of $-3f(x)$. To do this, we start with the graph in Figure 2.55 and stretch it vertically by a factor of 3. That is, we multiply each function value by 3. The result is shown in **Figure 2.56**.

Step 3: Sketch the graph of $-3f(x+1)$ using the graph from step 2. To do this, we shift the graph in Figure 2.56 by 1 unit to the left. (Replacing x by $x+1$ shifts the graph to the left.) The result is shown in **Figure 2.57**.

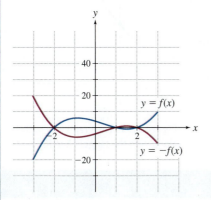

Figure 2.55 Reflection: Reflect the graph of $f(x)$ about the x-axis to sketch the graph of $-f(x)$.

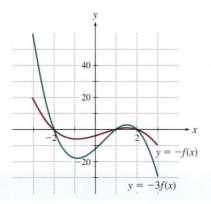

Figure 2.56 Vertical stretch: Stretch the graph of $-f$ vertically by a factor of 3 to sketch the graph of $-3f$.

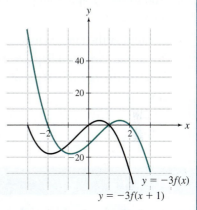

Figure 2.57 Horizontal shift: Shift the graph of $-3f(x)$ to the left by 1 unit to obtain the graph of $-3f(x+1)$.

TRY IT YOURSELF 2.20 Brief answers provided at the end of the section.

The graph of g is shown in **Figure 2.58**. Sketch the graph of $2g(x-1)$.

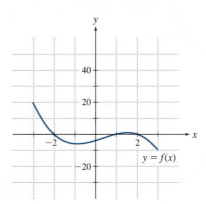

Figure 2.58 The graph of g for Try It Yourself 2.20

Let $f(x)$ be a function, and let $g(x) = -2f(3x) + 7$. Explain how you would proceed step-by-step to produce the graph of g starting with the graph of f.

EXAMPLE 2.21 Shifting and Rates of Change

The graph of a function $f(x)$ along with the tangent line for $x = 0$ is shown in **Figure 2.59**.

a. Add to the graph in Figure 2.59 the graph of $g(x) = f(x-2)$.

b. Add to your graph from part a the tangent line to the graph of $g(x)$ for $x = 2$.

c. Recall that the rate of change of a function at x is the slope of the tangent line at the corresponding point on the graph. Use your answer to part b to relate the rate of change of $f(x)$ at $x = 0$ to the rate of change of $g(x)$ at $x = 2$.

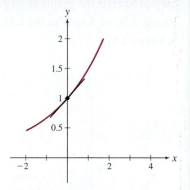

Figure 2.59 The graph of $y = f(x)$ along with the tangent line for $x = 0$

SOLUTION

a. To obtain the graph of $g(x)$, we shift the graph of $f(x)$ to the right by 2 units. The result is in **Figure 2.60**.

b. Shifting the graph of f to the right by 2 units shifts the tangent line in the same way, and $x = 0$ now corresponds to $x = 2$. The resulting graph is in **Figure 2.61**.

c. The shifted tangent line has the same slope as the original tangent line. Thus, the rate of change of $f(x)$ at $x = 0$ equals the rate of change of $g(x)$ at $x = 2$.

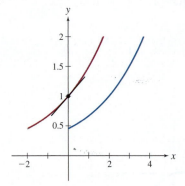

Figure 2.60 Horizontal shift: We find the graph of $y = f(x-2)$ by shifting to the right.

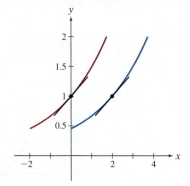

Figure 2.61 Shifting the tangent line: The tangent line shifts along with the graph of the function.

TRY IT YOURSELF 2.21 Brief answers provided at the end of the section.

If $f(x)$ is the function in the example, add to the graph in Figure 2.59 the graph of $h(x) = f(x+1)$ and its tangent line at $x = -1$.

The relationship suggested by this example holds in general: The rate of change of a function $f(x)$ at $x = a$ equals the rate of change of the function $f(x - c)$ at $x = a + c$.

Even and Odd Functions

> Reflections of graphs can display properties of functions.

An **even function** is a function f such that $f(-x) = f(x)$ for all x in the domain of f.

An **odd function** is a function f such that $f(-x) = -f(x)$ for all x in the domain of f.

Two types of symmetries in graphs come from the properties of certain functions. A function may be an **even function**, or it may be an **odd function** (or it may be neither).

EXAMPLE 2.22 Even and Odd Functions

a. Show that $f(x) = x^2$ is an even function.

b. Show that $g(x) = x^3$ is an odd function.

c. What geometric feature distinguishes the graph of an even function?

SOLUTION

a. To show that f is an even function, we need to show that $f(-x) = f(x)$:

$$f(-x) = (-x)^2 = (-1)^2 x^2 = x^2 = f(x)$$

b. To show that g is an odd function, we must show that $g(-x) = -g(x)$:

$$g(-x) = (-x)^3 = (-1)^3 x^3 = -x^3 = -g(x)$$

c. For any function f, we obtain the graph of $y = f(-x)$ by reflecting the graph of $y = f(x)$ through the y-axis. For an even function, $f(x) = f(-x)$. This means the graph of f is unchanged when we reflect through the y-axis.

TRY IT YOURSELF 2.22 Brief answers provided at the end of the section.

What geometric feature distinguishes the graph of an odd function?

Note that even functions are exactly those whose graphs are unchanged when we reflect through the y-axis. Thus, the graph of an even function is symmetric with respect to the y-axis, and vice versa. Similarly, odd functions are exactly those whose graphs are unchanged when reflected through the x-axis and then through the y-axis or, equivalently, when reflected through the origin. Thus, the graph of an odd function is symmetric with respect to the origin, and vice versa.

MODELS AND APPLICATIONS   Starting Times

When business plans change, promotional materials describing the plan must also change. Many times this involves the transformation of a graph.

EXAMPLE 2.23 A Presentation Graph

For an initiative to begin in 2022, the graph in **Figure 2.62** shows the projected daily sales $S(t)$, in thousands of dollars, in year t. But funding issues have caused the launch of the initiative to be delayed by 2 years.

a. Make a graph that incorporates this new information.

b. How is the function S adjusted to correspond to the new graph?

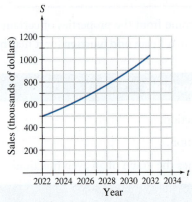

Figure 2.62 A graph of sales

SOLUTION

a. The graph needs to be shifted by 2 years to the right so that it begins at 2024 rather than 2022. The resulting graph is in **Figure 2.63**.

b. Shifting the graph by 2 units to the right corresponds to replacing t by $t - 2$. The appropriate function representation is $S(t - 2)$.

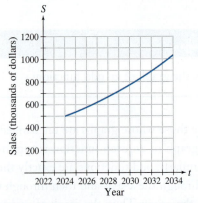

Figure 2.63 The graph starting at 2024 rather than 2022

TRY IT YOURSELF 2.23 Brief answers provided at the end of the section.

New information indicates that all sales will be about $100,000 less than originally expected. Adjust the graph in Figure 2.63 to reflect this new information.

TRY IT YOURSELF ANSWERS

2.15

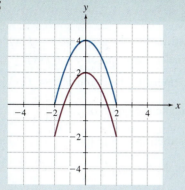

2.16

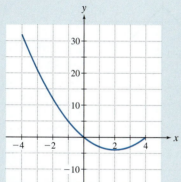

2.17

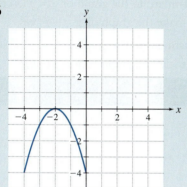

2.18

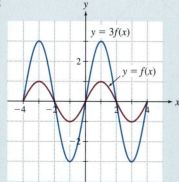

$y = 3f(x)$

$y = f(x)$

2.19

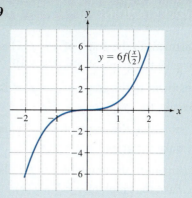

$y = 6f\left(\frac{x}{2}\right)$

2.20

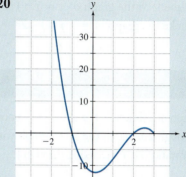

2.21

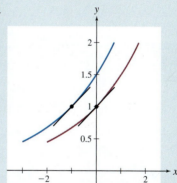

2.22 The graph of an odd function is unchanged when reflected through the *x*-axis and then the *y*-axis (equivalently, when reflected through the origin).

2.23

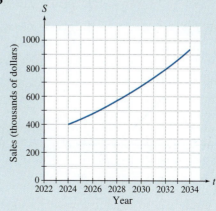

EXERCISE SET 2.3

CHECK YOUR UNDERSTANDING

1. What adjustment to the graph of $y = f(x)$ is needed to make the graph of $y = f(x + 3)$?

2. What adjustment to the graph of $y = f(x)$ is needed to make the graph of $y = f(x - 2)$?

3. What adjustment to the graph of $y = f(x)$ is needed to make the graph of $y = f(4x)$?

4. What adjustment to the graph of $y = f(x)$ is needed to make the graph of $y = 5 + f(x)$?

5. What adjustment to the graph of $y = f(x)$ is needed to make the graph of $y = f(x) - 2$?

6. What adjustment to the graph of $y = f(x)$ is needed to make the graph of $y = 6 f(x)$?

7. What adjustment to the graph of $y = f(x)$ is needed to make the graph of $y = f\left(\dfrac{x}{3}\right)$?

8. What adjustment to the graph of $y = f(x)$ is needed to make the graph of $y = \dfrac{1}{3} f(x)$?

9. What adjustment to the graph of $y = f(x)$ is needed to make the graph of $y = -f(x)$?

10. What adjustment to the graph of $y = f(x)$ is needed to make the graph of $y = f(-x)$?

SKILL BUILDING

Matching. The graph of a function f is shown in **Figure 2.64**. In Exercises 11 through 20, match the given graph with the appropriate transformation of f labeled A through J.

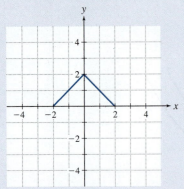

Figure 2.64 The graph of f

A. $g(x) = -f(x)$

B. $g(x) = f(x - 1)$

C. $g(x) = 2 f(x)$

D. $g(x) = \dfrac{1}{2} f(x)$

E. $g(x) = f(x) - 1$

F. $g(x) = f(x + 1)$

G. $g(x) = f\left(\dfrac{1}{2} x\right)$

H. $g(x) = f(x) + 1$

I. $g(x) = f(2x)$

J. $g(x) = f(-x)$

11.

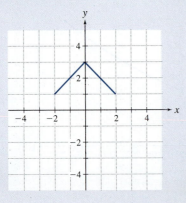

12.

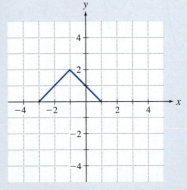

13.

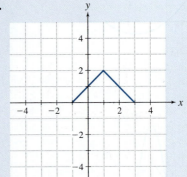

14.

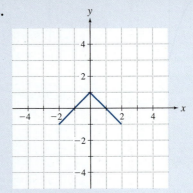

15.

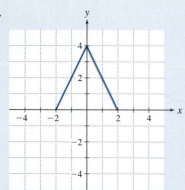

16.

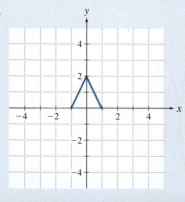

17.

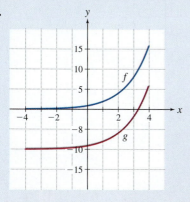

18.

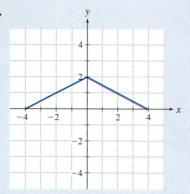

19.

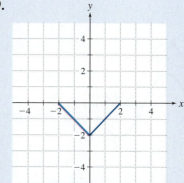

20.

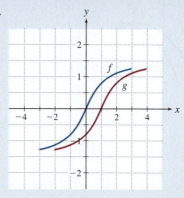

PROBLEMS

Finding transformations. In Exercises 21 through 31, you are given a graph of f and a graph of g. In each case, express g in terms of f using the transformations studied in this section. For example, $g(x) = f(x + 5)$ is the correct form of an answer.

21.

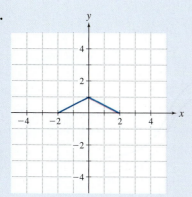

22.

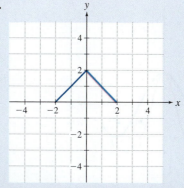

23.

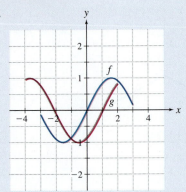

24.

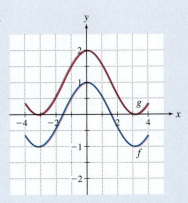

25.

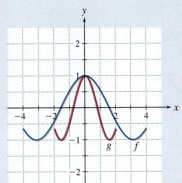

26.

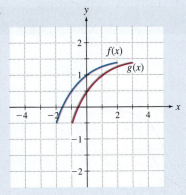

27.

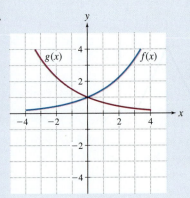

28.

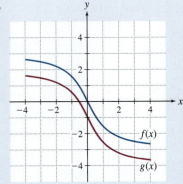

29.

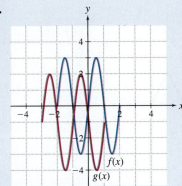

30.

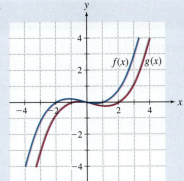

31.

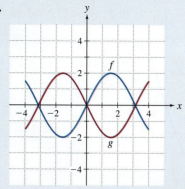

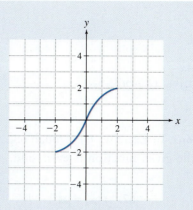

Figure 2.66 The graph of the function g

Sketching graphs involving simple transformations.

The graph of the function f is shown in **Figure 2.65**. In Exercises 32 through 41, sketch the graph of the given function.

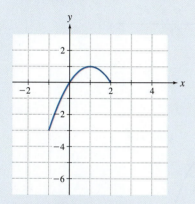

Figure 2.65 The graph of f

32. $g(x) = f(x) + 1$

33. $g(x) = f(x) - 1$

34. $g(x) = f(x - 1)$

35. $g(x) = f(x + 1)$

36. $g(x) = 2f(x)$

37. $g(x) = \dfrac{1}{2} f(x)$

38. $g(x) = f(2x)$

39. $g(x) = f\left(\dfrac{1}{2}x\right)$

40. $g(x) = -f(x)$

41. $g(x) = f(-x)$

Sketching graphs involving multiple transformations.

The graph of the function g is shown in **Figure 2.66**. In Exercises 42 through 53, sketch the indicated graph along with the graph of g on the same coordinate axes.

42. $h(x) = 2g(x) + 1$

43. $h(x) = 2g(x) - 1$

44. $h(x) = g(1 - x)$

45. $h(x) = 2g\left(\dfrac{1}{2}x\right) - 1$

46. $h(x) = \dfrac{1}{2} g(2x) - 1$

47. $h(x) = 1 - g(x)$

48. $h(x) = 1 + g(2x)$

49. $h(x) = 2g(-x)$

50. $h(x) = -g(2x)$

51. $h(x) = 2 + g(x - 1)$

52. $h(x) = g(4x + 2)$

53. $h(x) = g(2x - 4)$

Using transformations to construct graphs.

The graphs required in Exercises 54 through 63 can be constructed by transforming the familiar parabola that is the graph of $y = x^2$. Sketch the indicated graph along with the graph of $y = x^2$ on the same coordinate axes.

54. $y = x^2 - 4x + 4$

Note that $x^2 - 4x + 4 = (x - 2)^2$.

55. $y = x^2 + 4x + 4$

Note that $x^2 + 4x + 4 = (x + 2)^2$.

56. $y = 10 - x^2$

57. $y = x^2 - 2x - 9$

Note that $x^2 - 2x - 9 = (x - 1)^2 - 10$.

58. $y = 10 - 4x^2$

Note that $10 - 4x^2 = 10 - (2x)^2$.

59. $y = 2x^2 + 4x + 2$

Note that $2x^2 + 4x + 2 = 2(x + 1)^2$.

60. $y = 4x^2 + 4x + 1$

Note that $4x^2 + 4x + 1 = 4\left(x + \dfrac{1}{2}\right)^2$.

61. $y = 10 - \dfrac{x^2}{4}$

62. $y = 18 - 2x^2 + 4x$

Note that $18 - 2x^2 + 4x = 20 - 2(x-1)^2$.

63. $y = -x^2 - 4x - 4$

Even and odd functions. In Exercises 64 through 68, determine whether the given function is even, odd, or neither.

64. $f(x) = x^3$

65. $f(x) = \dfrac{1}{1 + x^2}$

66. $f(x) = x + x^2$

67. $f(x) = |x|$

68. $f(x) = -\sqrt{x^2}$

Graphs of even and odd functions. In Exercises 69 through 74, determine whether the given graph is that of an even function, an odd function, or neither.

69.

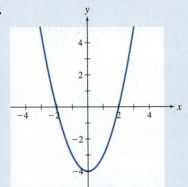

70.

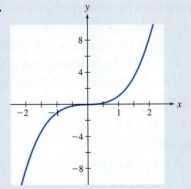

71.

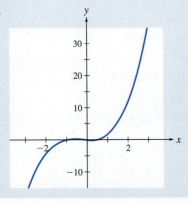

72.

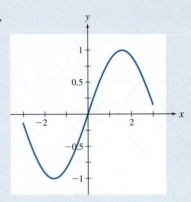

73.

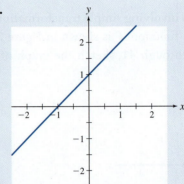

74.

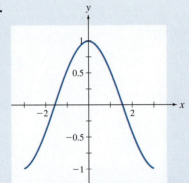

Adjustments with a graphing utility. Some adjustments to graphs are difficult to accomplish by hand. Use a graphing utility to make the graphs in Exercises 75 through 78.

75. Make the graph of $f(x) = x^2$. On the same screen, add the graph of $y = f\left(\dfrac{1}{x}\right)$, $x \neq 0$.

76. Make the graph of $f(x) = x^2$. On the same screen, add the graph of $y = \dfrac{1}{1 + f(x)}$.

77. Make the graph of $f(x) = \sqrt{x}$. On the same screen, add the graph of $y = f(1 + f(x))$.

78. On the same screen, plot the graphs of $f(x) = \sqrt{x}$ and $g(x) = x^2$, $x \geq 0$. Then add the graph of $y = g(x) - f(x)$.

MODELS AND APPLICATIONS

79. Cost overruns. The graph in **Figure 2.67** shows anticipated development costs $D(t)$ for an upcoming project. But newer estimates indicate a doubling of costs.

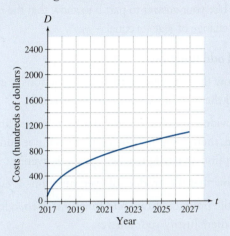

Figure 2.67 Anticipated development costs

a. Make a graph that shows the updated cost estimate.

b. Adjust the function D to get a new function whose graph matches the new estimate.

80. Torricelli's law. If a hole is punched at the base of a cylindrical container of water, Torricelli's law gives the velocity of liquid spouting from the hole. From this law we can show that the time t required to drain the cylinder is given by

$$t = c\sqrt{h} \text{ seconds}$$

where h is the initial height, in feet, of water in the cylinder, and c is a constant depending on the diameter of the cylinder and the diameter of the hole. For a certain container with a hole at its base, $t = 25\sqrt{h}$.

a. Plot the graph of t versus h for $h = 0$ to $h = 9$ feet.

b. For another container, the value of c is doubled. Add the graph of t versus h to the graph you made in part a.

81. Light-years versus parsecs. Expansion of the universe causes distances between stars to change with time t. The function $L(t)$ gives the distance in light-years from the sun to a certain star. One parsec is 3.26 light-years. Use the function L to make a new function $P(t)$ that gives the distance in parsecs.

82. A ball. The height H, in feet, of a ball t seconds after it is tossed upward from ground level is given by

$$H(t) = 64t - 16t^2$$

a. Plot the graph of H versus t from the time it is tossed until it strikes the ground.

b. Use a transformation of H to make a function Y that gives the height of the ball in yards.

c. Add the graph of Y versus t to the graph in part a.

CHALLENGE EXERCISES FOR INDIVIDUALS OR GROUPS

83. Rates of change and vertical shifting. Figure 2.68 shows the graph of $y = f(x)$ and the graph of $y = f(x) + c$ for a constant c. Tangent lines to both graphs for $x = 1$ are added.

a. How do the rates of change at $x = 1$ for the functions $y = f(x)$ and $y = f(x) + c$ compare?

b. In general, how do the rates of change of the functions $y = g(x)$ and $y = g(x) + c$ compare? Justify your answer. *Suggestion*: First think about how the graphs compare.

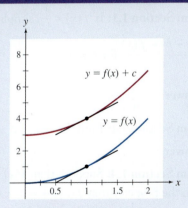

Figure 2.68 Tangent lines for $y = f(x)$ and $y = f(x) + c$

84. **Rates of change and vertical reflection.** In this exercise we examine the effect of vertical reflection on the rate of change.

   a. Graph the line $y = 2x + 1$ along with its reflection through the x-axis. How is the slope of the reflected line related to the slope of the original line?

   b. In general, how is the slope of the reflection through the x-axis of a line related to the slope of the original line?

   c. Let $f(x)$ be a function, and define $g(x) = -f(x)$. Assume that the rate of change of f at a equals m. Use your answer to part b to show that the rate of change of g at a equals $-m$.

85. **Rates of change and horizontal reflection.** In this exercise we examine the effect of horizontal reflection on the rate of change.

   a. Graph the line $y = 2x + 1$ along with its reflection through the y-axis. How is the slope of the reflected line related to the slope of the original line?

   b. In general, how is the slope of the reflection through the y-axis of a line related to the slope of the original line?

   c. Let $f(x)$ be a function, and define $g(x) = f(-x)$. Assume that the rate of change of f at a equals m. Use your answer to part b to show that the rate of change of g at $-a$ equals $-m$.

Even and odd functions.

86. Show that the sum of two even functions is even, and the sum of two odd functions is odd.

87. Provide examples to show that if f is even and g is odd, then $f + g$ may be neither even nor odd. Can you find an example where $f + g$ is even?

88. Show that the product of even functions is even.

89. Is the product of two odd functions even, odd, or neither? Justify your answer.

90. Is the product of an even function with an odd function even or odd? Justify your answer.

91. Find two functions f and g that are neither even nor odd but the function $f + g$ is even.

REVIEW AND REFRESH: Exercises from Previous Sections

92. **From Section P.2:** Solve the inequality $|2x - 1| \geq 3$.
   Answer: $(-\infty, -1] \cup [2, \infty)$

93. **From Section P.3:** Solve the inequality $\dfrac{2}{x-1} > \dfrac{1}{x}$.
   Answer: $(-1, 0) \cup (1, \infty)$

94. **From Section 1.1:** If $f(x) = \dfrac{1}{x}$, simplify
   $$\dfrac{f(x+h) - f(x)}{h} \text{ for } h \neq 0.$$
   Answer: $-\dfrac{1}{x(x+h)}$

95. **From Section 1.1:** Find the range of $y = 2 - x^4$.
   Answer: $(-\infty, 2]$

96. **From Section 1.4:** Suppose that $f(x)$ is a function with the property that $f(x) \to \infty$ as $x \to \infty$. Is there necessarily a solution x of the inequality $f(x) > 800$?
   Answer: Yes, because $f(x)$ can be made as large as we like by choosing x sufficiently large.

97. **From Section 2.1:** If $f(x) = \dfrac{2}{x}$ for $x \neq 0$, calculate $(f \circ f)(x)$.
   Answer: $(f \circ f)(x) = x$

98. **From Section 2.1:** If $f(x) = \dfrac{x}{1-x}$ and $g(x) = \dfrac{x}{x+1}$, find a formula that gives $(f \circ g)(x)$. Simplify your answer.
   Answer: $(f \circ g)(x) = x$

99. **From Section 2.2:** Find the inverse function for $f(x) = 5x - 4$.
   Answer: $f^{-1}(x) = \dfrac{x+4}{5}$

100. **From Section 2.2:** Determine whether $f(x) = x^4$ has an inverse.
   Answer: It has no inverse because for some values of y there is more than one solution for x of the equation $y = x^4$. For example, when $y = 1$, both $x = 1$ and $x = -1$ are solutions of the equation $y = x^4$.

CHAPTER ROADMAP AND STUDY GUIDE

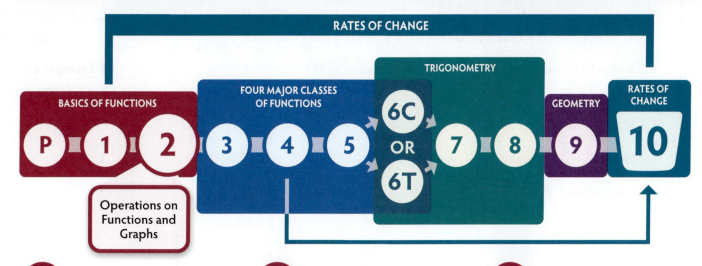

RATES OF CHANGE

TRIGONOMETRY

BASICS OF FUNCTIONS

FOUR MAJOR CLASSES OF FUNCTIONS

RATES OF CHANGE

GEOMETRY

P 1 2 3 4 5 6C OR 6T 7 8 9 10

Operations on Functions and Graphs

2.1 New Functions from Old

Mathematical models are created by combining simpler functions.

Arithmetic Combinations of Functions

Functions can be added, subtracted, multiplied, and divided in much the same fashion as these operations are applied to real numbers.

Composition of Functions

Composition of functions means that one function is applied after another.

Function Decomposition

Function decomposition identifies how simple parts make up more complicated functions.

MODELS AND APPLICATIONS
Decomposing Functions in Applications

2.2 Inverse Functions

When the effect of a function is reversed, a new function is the result.

Defining Inverse Functions

The inverse of a function undoes the action of that function.

Finding Formulas and Graphs for Inverse Functions

The graph of an inverse function is produced via reflection.

The Horizontal Line Test for Inverses

A function has an inverse if each horizontal line meets its graph no more than once.

One-to-One Functions

A function has an inverse exactly when distinct elements of the domain have distinct function values.

Restricting the Domain to Find Inverse Functions

Some functions have inverses only when the domain is restricted.

MODELS AND APPLICATIONS
A School Trip

2.3 Transformations of Graphs

Interpreting a graph is easier if we recognize transformations.

Shifting of Graphs

Addition and subtraction of a real number to either a function or its argument result in vertical or horizontal shifts to the graph.

Vertical Stretching, Compressing, and Reflecting of Graphs

Multiplication of a function by a real number results in a vertical stretch, compression, or reflection of the graph.

Horizontal Stretching, Compressing, and Reflecting of Graphs

Multiplication of a function argument by a real number results in a horizontal stretch, compression, or reflection of its graph.

Even and Odd Functions

Reflections of graphs can display properties of functions.

MODELS AND APPLICATIONS
Starting Times

CHAPTER QUIZ

1. Let $f(x) = 3 + x$ and $g(x) = 2 - x$.

   a. Find a formula for the function $h = 2f - g$. What is the domain of h?

   b. Find a formula for $h = \dfrac{f + g}{g}$. What is the domain of h?   **XR Example 2.1**

   Answer:

   a. $h(x) = 3x + 4$. The domain is $\mathbb{R} = (-\infty, \infty)$.

   b. $h(x) = \dfrac{5}{2 - x}$. The domain is $(-\infty, 2) \cup (2, \infty)$.

2. The graphs of f (red) and g (blue) are shown. Draw the graph of $f + g$.

   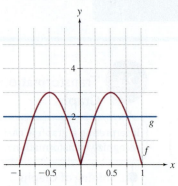

   XR Example 2.2

   Answer:

   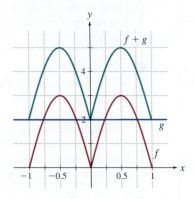

3. Let $f(x) = x - 1$ and $g(x) = \dfrac{1}{x}$. Find formulas for $f \circ g$ and $g \circ f$.

   State the domain of each.   **XR Example 2.3**

   Answer: $(f \circ g)(x) = \dfrac{1}{x} - 1$. The domain is $(-\infty, 0) \cup (0, \infty)$. $(g \circ f)(x) = \dfrac{1}{x - 1}$.

   The domain is $(-\infty, 1) \cup (1, \infty)$.

4. If $f(x) = 4 - \dfrac{1}{x - 1}$, find f^{-1}.   **XR Example 2.10**

   Answer: $f^{-1}(x) = 1 + \dfrac{1}{4 - x}$

5. The graph of a function f is given. Determine whether f has an inverse.
   If an inverse exists, add to the figure the graph of $y = x$ and the graph of f^{-1}.

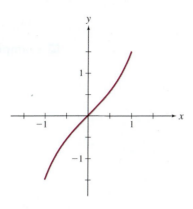

XR **Example 2.12**

Answer: The graph passes the horizontal line test, so the function has an inverse. The graph of the inverse is in blue.

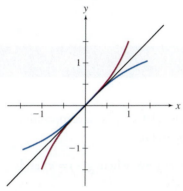

6. Explain why the function $y = 3^{-x}$ has an inverse.

XR **Example 2.13**

Answer: It is decreasing, so it is one to one.

7. The graph of $f(x)$ is shown. Add to the figure the graph of $f(x+1)-1$.

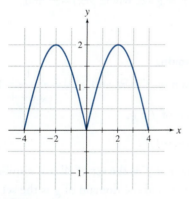

Answer:

XR **Example 2.16**

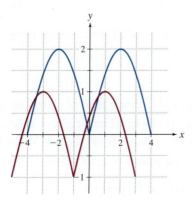

8. Refer to the graph of f from the preceding exercise. Add to the figure the graph of $\dfrac{1}{2}f(2x)$.

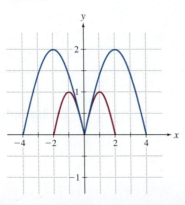

XR **Example 2.19**

Answer:

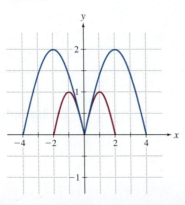

CHAPTER REVIEW EXERCISES

Section 2.1

Combining functions. In Exercises 1 through 4, functions are given. Calculate the indicated function values, and then find a formula for the indicated combination of functions.

1. $f(x) = x^2$, $g(x) = \sqrt{x}$. Calculate $(2f - 3g)(4)$. Find a formula for $(2f - 3g)(x)$.

   Answer: $(2f - 3g)(4) = 26$.
   $(2f - 3g)(x) = 2x^2 - 3\sqrt{x}$.

2. $h(t) = t^2$, $k(z) = \sqrt{z}$. Calculate $(k \circ h)(-2)$. Find a formula for $(k \circ h)(x)$.

   Answer: $(k \circ h)(-2) = 2$. $(k \circ h)(x) = |x|$

3. $f(x) = \dfrac{x}{x+1}$, $g(x) = \dfrac{1}{x}$. Calculate $\left(\dfrac{f}{g}\right)(2)$. Find a formula for $\left(\dfrac{f}{g}\right)(t)$.

   Answer: $\left(\dfrac{f}{g}\right)(2) = \dfrac{4}{3}$. $\left(\dfrac{f}{g}\right)(t) = \dfrac{t^2}{t+1}$

4. $u(t) = \dfrac{\sqrt{t}}{t-1}$, $v(t) = \dfrac{\sqrt{t}}{t+1}$. Calculate $(uv)(3)$. Find a formula for $(uv)(t)$.

   Answer: $(uv)(3) = \dfrac{3}{8}$, $(uv)(t) = \dfrac{t}{t^2 - 1}$

Calculating values.

5. If $f(2) = 5$, $g(2) = 2$, and $g(5) = 1$, calculate the value of $(g \circ f + f \circ g)(2)$.

   Answer: 6

6. If $f(2) = 5$ and $g(2) = 3$, calculate the value of $(fg - f + 2g)(2)$.

   Answer: 16

Decomposing.

7. Write $f(x) = (1 + \sqrt{x})^3$ as a composition $g \circ h$ of functions g and h.

   Answer: $f = g \circ h$, where $g(x) = x^3$ and $h(x) = 1 + \sqrt{x}$

8. Write $f(x) = \dfrac{1}{1 + x^2}$ as a composition $g \circ h$ of functions g and h.

   Answer: $f = g \circ h$, where $g(x) = \dfrac{1}{x}$ and $h(x) = 1 + x^2$

Finding the domain.

9. The domain of f is $\mathbb{R}$, and $g(x) = \dfrac{1}{\sqrt{x-4}}$. Find the domain of $f \circ g$.

   Answer: $(4, \infty)$

10. The domain of f is the set of all real numbers except $x = 3$, and the domain of g is the set of all real numbers except $x = 7$. What is the domain of $f + 2g - gf$?

    Answer: The set of all real numbers except $x = 3$ and $x = 7$.

Applications.

11. The length of a certain kind of animal of age t years is $L(t)$ inches. The weight of an animal of length x, in inches, is $W(x)$ pounds. What is the weight of an animal of age t years?

    Answer: $(W \circ L)(t)$ pounds

12. The weight of a certain kind of animal at age t years is given by the function $W(t)$, in pounds. The number of these animals in a certain area of age t years is given by the function $N(t)$. The *biomass* is the total weight of all animals of age t. Find a function that gives the total biomass of all these animals at age t in this area.

Answer: $B(t) = N(t)W(t)$ pounds

Section 2.2

Finding inverses. In Exercises 13 through 23, find the inverse of the given function, or show that none exists.

13. $f(x) = 3x - 2$

     Answer: $f^{-1}(x) = \dfrac{x + 2}{3}$

14. $f(x) = x^3$

     Answer: $f^{-1}(x) = x^{1/3}$

15. $f(x) = \dfrac{1}{x}$

     Answer: $f^{-1}(x) = \dfrac{1}{x}$

16. $f(x) = x^2$

     Answer: No inverse exists.

17. $f(x) = x^2,\ x \le 0$

     Answer: $f^{-1}(x) = -\sqrt{x},\ x \ge 0$

18. $f(x) = \dfrac{x - 1}{x + 2}$

     Answer: $f^{-1}(x) = \dfrac{2x + 1}{1 - x}$

19. $h(t) = \sqrt{t}$

     Answer: $h^{-1}(t) = t^2,\ t \ge 0$.

20. $g(x) = x + x^2$

     Answer: No inverse exists.

21. $u(t) = t^4$

     Answer: No inverse exists.

22. $f(x) = 1 - \dfrac{2}{\sqrt{x}}$

     Answer: $f^{-1}(x) = \dfrac{4}{(x - 1)^2},\ x < 1$

23. $f(x) = \dfrac{2x - 1}{3x + 2}$

     Answer: $f^{-1}(x) = \dfrac{2x + 1}{2 - 3x}$

Horizontal line test. In Exercises 24 through 27, use the horizontal line test to determine whether an inverse exists. If an inverse exists, add the graph of $y = x$ and the graph of the inverse to the given graph.

24.

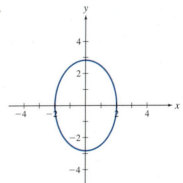

Answer: No inverse

25.

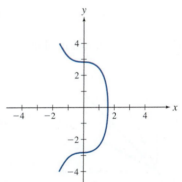

Answer: No inverse

26.

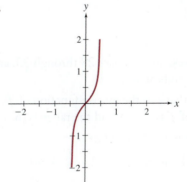

Answer: The inverse is shown in blue.

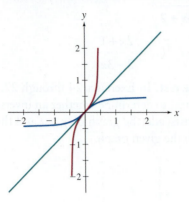

27.

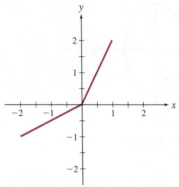

Answer: The inverse is shown in blue.

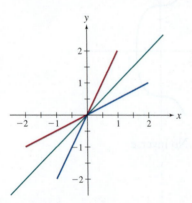

More on inverses. In Exercises 28 through 33, answer the questions as indicated.

28. Suppose that g is the inverse function of f, the domain of f is $(3, \infty)$, and the range is $(6, \infty)$. Find the domain and range of g.

Answer: Domain $(6, \infty)$. Range $(3, \infty)$.

29. It is true that the function $f(x) = \dfrac{x^3}{1 - x^3}$ has an inverse. What is the range of the inverse function?

Answer: $(-\infty, 1) \cup (1, \infty)$

30. Determine whether the function $y = x^6$ is one to one.

Answer: It is not one to one.

31. Determine whether the function $y = x^3$ is one to one.

Answer: It is one to one.

32. Suppose that g is the inverse of f and $g(2) = 3$. Find the value of $(g \circ f \circ g)(2)$.

Answer: 3

33. A population exhibits steady growth. The population t years after the year 2000 is given by the function $N(t)$. Explain why N has an inverse, and then explain the meaning of the inverse function.

Answer: The function N has an inverse because it is increasing. The inverse tells how long it takes after the year 2000 to reach a given population level.

Section 2.3

Transforming graphs. The graph of $y = f(x)$ is shown in **Figure 2.69**. In Exercises 34 through 45, add the graph of the given transformation of f.

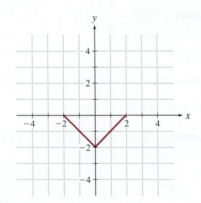

Figure 2.69 The graph of f

34. $2f(x)$

Answer:

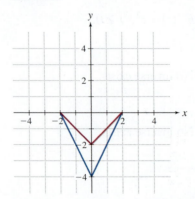

35. $\dfrac{1}{2} f(x)$

Answer:

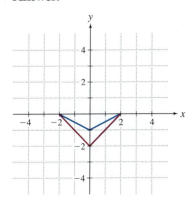

36. $-f(x)$

Answer:

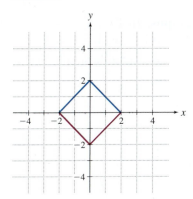

37. $f(x)+1$

Answer:

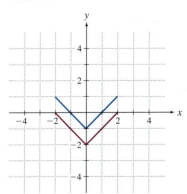

38. $f(x)-1$

Answer:

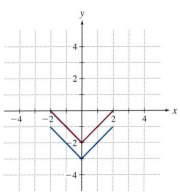

39. $f(2x)$

Answer:

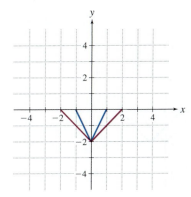

40. $f\left(\dfrac{x}{2}\right)$

Answer:

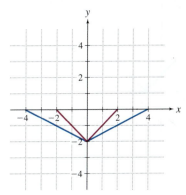

41. $f(x+1)$

Answer:

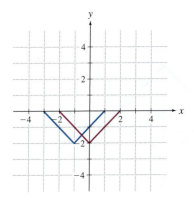

42. $f(x-1)$

Answer:

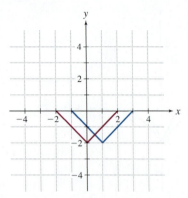

43. $f(x-1)+2$

Answer:

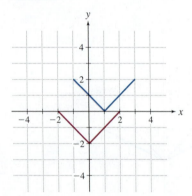

44. $-2f\left(\dfrac{x}{2}\right)$

Answer:

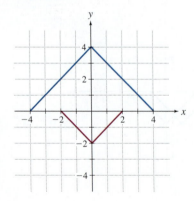

45. $f\left(\dfrac{x}{2}\right)+3$

Answer:

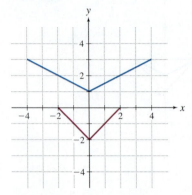

Transforming graphs. In Exercises 46 through 47, answer the questions as indicated.

46. On the same screen plot the graphs of $y = x^2$ and $y = (x-1)^2 + 2$.

Answer:

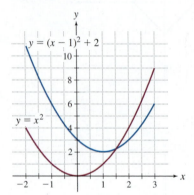

47. On the same screen plot the graphs of $y = x^3$ and $y = 2x^3 - 1$.

Answer:

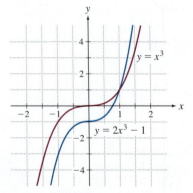

LINEAR AND EXPONENTIAL FUNCTIONS

Photo Credit: Maxiphoto/Getty Images

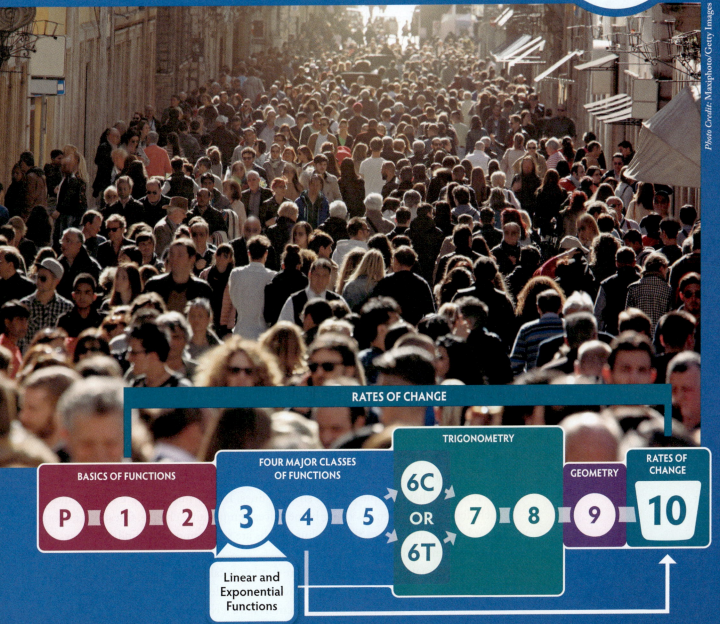

RATES OF CHANGE

FOUR MAJOR CLASSES OF FUNCTIONS

TRIGONOMETRY

BASICS OF FUNCTIONS

GEOMETRY

RATES OF CHANGE

P · 1 · 2 · 3 · 4 · 5 → 6C OR 6T → 7 · 8 · 9 · 10

Linear and Exponential Functions

This chapter explores the first of four major classes of functions—and arguably the most important functions for calculus (and all of mathematics): linear and exponential functions. These two types of functions are discussed together in this chapter because each has a simple characterization in terms of rates of change: Linear functions have a constant rate of change, and exponential functions have a constant proportional rate of change. Such characterizations will enable you to model many real-world phenomena.

3.1 Linear Functions: Constant Rate of Change

3.2 Exponential Functions: Constant Proportional Change

3.3 Modeling with Exponential Functions

I t is almost impossible to overstress the importance of linear and exponential functions in mathematics, functions that are most often recognized by their rates of change. You will encounter applications of these functions more frequently than other functions in mathematics.

As early as 1798, Thomas Robert Malthus (**Figure 3.1**) used rates of change to offer a dire warning: "Population, when unchecked, increases in a geometrical ratio. Subsistence increases only in an arithmetical ratio. A slight acquaintance with numbers will shew the immensity of the first power in comparison of the second." In modern terms, Malthus points out that population grows exponentially, but the growth of resources is linear in nature. As we shall see, this state of affairs is untenable in the long term. Malthus's claims oversimplify a complex situation, but it remains true that exponential functions are fundamental in modeling population growth, among many other phenomena.

Figure 3.1 Thomas Robert Malthus, 1766–1834 Photos.com/Getty Images

3.1 Linear Functions: Constant Rate of Change

3.2 Exponential Functions: Constant Proportional Change

3.3 Modeling with Exponential Functions

3.1 Linear Functions: Constant Rate of Change

Linear functions are widely used as models because they are the functions that show a constant rate of change.

In this section, you will learn to:

1. Calculate the average rate of change by using the definition of a linear function.
2. Explain the relationship between a linear function and a straight line.
3. Find function values for a linear function using the rate of change.
4. Calculate the equation of the secant line for a given function over a given interval.
5. Solve applied problems involving linear functions.
6. Solve applied problems involving regression lines.

In many ways, linear functions are the simplest of all functions, and these simple functions have simple graphs: straight lines. Linear functions are used to model such diverse phenomena as rising home prices, temperature change, forest litter, sports records, and much more. Straight-line graphs are so common that finding a news site without them is difficult.

In mathematics, linear functions are fundamental building blocks, and a key feature of calculus is the use of linear functions to understand more complicated functions. The crucial fact is that linear functions have a constant rate of change.

Constant Rate of Change

The average rate of change of a linear function does not depend on the interval over which it is calculated.

A tank initially contains 30 gallons of water. A tap is opened to provide a constant flow of 3 gallons per minute of water into the tank. We can use this information to find a function that gives the volume V, in gallons, of water in the tank after t minutes. We started with 30 gallons, and the volume is increasing at a constant rate of 3 gallons per minute. So the volume is 30 gallons plus 3 gallons for each elapsed minute. These words translate to a simple formula:

$$V = 3t + 30$$

The reason we are able to make the jump from a verbal description to a formula is that in this instance the volume is increasing at a constant rate. A function, like this one, with a constant rate of change is a **linear function**, and, just as in this example, the rate of change is intimately tied to the formula for linear functions.

A **linear function** is a function with a constant rate of change.

LAWS OF MATHEMATICS: Formula for a Linear Function

To write the formula for a linear function, we need the rate of change and the initial value (the value of y when $x = 0$):

$$y = \text{Rate of change} \times x + \text{Initial value}$$

If we use m for the rate of change and b for the initial value, the equation takes the familiar form

$$y = mx + b$$

The next example gives an algebraic look at the relationship between linear functions and average rates of change.

EXAMPLE 3.1 Average Rate of Change for a Linear Function

Consider the linear function

$$f(x) = mx + b$$

a. Calculate the average rate of change of f from $x = 2$ to $x = 4$.

b. One reason that m is referred to as the rate of change for this linear function is that the average rate of change over any interval has the same value m. Verify that if $p \neq q$, the average rate of change of f from $x = p$ to $x = q$ is m.

SOLUTION

a. Section 1.2 explained that the average rate of change of a function f from $x = p$ to $x = q$ is given by

$$\text{Average rate of change} = \frac{f(q) - f(p)}{q - p}$$

We use $f(x) = mx + b$, $p = 2$ and $q = 4$ in this formula:

$$\text{Average rate of change} = \frac{f(4) - f(2)}{4 - 2}$$

$$= \frac{(4m + b) - (2m + b)}{4 - 2} \quad \blacktriangleleft \text{ Apply the formula for } f.$$

$$= \frac{2m}{2} \quad \blacktriangleleft \text{ Simplify.}$$

$$= m \quad \blacktriangleleft \text{ Cancel.}$$

Thus, the average rate of change is m.

b. We perform the calculation again, this time from $x = p$ to $x = q$:

$$\text{Average rate of change} = \frac{f(q) - f(p)}{q - p}$$

$$= \frac{(mq + b) - (mp + b)}{q - p} \qquad \blacktriangleleft \textbf{Apply the formula for } f.$$

$$= \frac{m(q - p)}{q - p} \qquad \blacktriangleleft \textbf{Simplify and factor.}$$

$$= \frac{m \,\cancel{(q - p)}}{\cancel{q - p}} \qquad \blacktriangleleft \textbf{Cancel.}$$

$$= m$$

Thus, the average rate of change is m, regardless of the interval used for the calculation.

TRY IT YOURSELF 3.1　Brief answers provided at the end of the section.

What is the average rate of change on any interval of $f(x) = 5x - 4$?

EXTEND YOUR REACH

a. Assume that $y = f(x)$ is a linear function. At $x = 2$, the rate of change of f is 7. What is the rate of change of f at $x = -2$?

b. Assume that $y = g(x)$ is a function with $g(2) = 5$, $g(3) = 9$, and $g(5) = 16$. Is it possible that g is a linear function? If so, explain why. If not, find a value for $g(5)$ that would change your answer to the question.

Graphs of Linear Functions

> **The graph of a linear function is a straight line.**

The graph of the linear function $f(x) = mx + b$ is the graph of the equation $y = mx + b$. These graphs are straight lines, as described in Section P.1. A reminder is in order.

CONCEPTS TO REMEMBER: Properties of Linear Equations

The graph of the linear function $f(x) = mx + b$ is the graph of the linear equation $y = mx + b$.

- The graph is a straight line with y-intercept (initial value) b and slope (rate of change) m.

- The rate of change of the linear function is the slope m. The slope is given by

$$m = \text{Rate of change of } f = \text{Slope} = \frac{\text{Change in } y}{\text{Change in } x} = \frac{\text{Rise}}{\text{Run}} = \frac{\Delta y}{\Delta x}$$

- Each unit of run (or Δx) corresponds to m units of rise (or Δy). This fact is expressed in a rearrangement of the formula for the slope:

$$\Delta y = m \Delta x$$

The effect of varying the slope and y-intercept is illustrated in **Figure 3.2** and **Figure 3.3**.

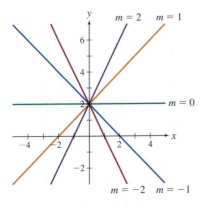

DF **Figure 3.2** Lines with various slopes

Figure 3.3 Lines with various y-intercepts

The domain of a linear function is the set of all real numbers. If $m \neq 0$, the range of the linear function $f(x) = mx + b$ is also the set of all real numbers. When $m > 0$, this function is increasing, and it increases without bound as $x \to \infty$. When $m < 0$, this function is decreasing, and it decreases without bound as $x \to \infty$.

EXAMPLE 3.2 Constructing and Graphing a Linear Function

Let f be a linear function with initial value -4 and rate of change 2.

a. Find the formula for f.

b. Plot the graph of f.

SOLUTION

a. We use the given values in the formula for a linear function:

$$f(x) = \text{Rate of change} \times x + \text{Initial value}$$

$$f(x) = 2x - 4$$

b. Section P.1 described various methods for plotting the graph of a linear equation. Perhaps the simplest is to plot two points and then produce the line joining them, as shown in **Figure 3.4** using the points $(0, -4)$ and $(2, 0)$.

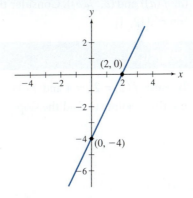

Figure 3.4 The graph of $f(x)$: a line through the points $(0, -4)$ and $(2, 0)$

TRY IT YOURSELF 3.2 Brief answers provided at the end of the section.

Find the equation for the linear function with initial value 3 and rate of change -1. Then plot the graph.

EXAMPLE 3.3 Using Rate of Change to Find Function Values

Suppose that $f(x)$ is a linear function with rate of change 2. Assume in addition that $f(4) = -1$. What is the value of $f(9)$?

SOLUTION

The change in x from $x = 4$ to $x = 9$ is $\Delta x = 9 - 4 = 5$. The first step is to use the rate of change, $m = 2$, along with $\Delta x = 5$ to calculate Δy:

$$\Delta y = m\Delta x$$

$$= 2 \times 5$$

$$= 10$$

Hence, the change in y is 10. That means y changes from $f(4) = -1$ to

$$\text{New } y\text{-value} = -1 + \Delta y$$

$$= -1 + 10$$

$$= 9$$

This gives our final answer: $f(9) = 9$.

This process is illustrated in **Figure 3.5**, which shows how the change in x and change in y lead to the desired function value.

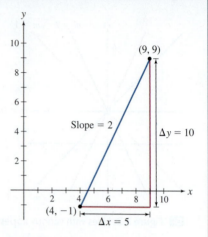

Figure 3.5 The rate of change used to find one function value from another

---

TRY IT YOURSELF 3.3 Brief answers provided at the end of the section.

Suppose that $g(x)$ is a linear function with slope -2 and that $g(5) = 9$. Find $g(8)$.

EXAMPLE 3.4 Finding Equations of Secant Lines

Section 1.2 explained that the secant line for a function f for the interval $[a, b]$ is the line through the points $(a, f(a))$ and $(b, f(b))$. Consider the function $f(x) = x^2$. Find the equation of the secant line for f for the interval $[2, 4]$.

SOLUTION

Because $f(2) = 2^2 = 4$ and $f(4) = 4^2 = 16$, the secant line passes through the points $(2, 4)$ and $(4, 16)$. We can use these points to find the slope of the secant line:

$$\text{Slope} = \frac{\Delta y}{\Delta x}$$

$$= \frac{16 - 4}{4 - 2}$$

$$= \frac{12}{2}$$

$$= 6$$

Because the slope is $m = 6$ and the line goes through the point $(2, f(2)) = (2, 4)$, we can use the point-slope form to find the equation we seek:

$$y - 4 = 6(x - 2) \quad \text{◀ Use the point-slope form.}$$

$$y - 4 = 6x - 12 \quad \text{◀ Apply the distributive law.}$$

$$y = 6x - 8 \quad \text{◀ Add 4.}$$

In **Figure 3.6** we show the graph of $y = x^2$ together with the secant line.

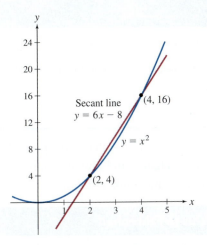

Figure 3.6 The secant line for the interval $[2, 4]$

TRY IT YOURSELF 3.4 Brief answers provided at the end of the section.

Find the equation of the secant line for $g(x) = x^3 - 1$ for the interval $[1, 3]$.

For a linear function $f(x) = mx + b$, find the equation of the secant line through two points on the graph of f. Do this in two ways: by using the point-slope form as in the example, and by picturing a secant line for the graph of a linear function.

EXTEND YOUR REACH

MODELS AND APPLICATIONS   Temperature Conversion and More

In real-world settings, we typically recognize that a function is linear by the fact that its rate of change, or slope, is constant. Understanding the meaning of the slope in such settings is a key to solving problems. The following example illustrates this point.

EXAMPLE 3.5 A Cookie Jar

My cookie jar originally contains $50. My sweet grandmother adds $10 to the jar each week.

a. Explain why the amount of money M, in dollars, in the jar is a linear function of the number w of weeks since Grandma started putting money in the jar.

b. Find a formula that gives the amount of money in the jar after w weeks.

SOLUTION

a. The amount of money in the jar is linear because it changes by the same amount, $10, each week. That is, the rate of change is constant. Thus, the amount of money in the cookie jar is a linear function of the number of weeks.

b. The rate of change of the function is $10 per week because that is the change each week. The initial value is $50. Hence, the amount of money is given by

$$M = \text{Rate of change} \times w + \text{Initial value}$$

$$M = 10w + 50$$

TRY IT YOURSELF 3.5 Brief answers provided at the end of the section.

There was originally $200 in my cookie jar. My sneaky sister steals $10 each week from the jar.

a. Explain why the amount M, in dollars, of money in the cookie jar is a linear function of the number w of weeks since my sister started stealing.

b. Express M as a linear function of w.

Example 3.5 is simple, but that is the point. In this case, it is easy to recognize the role of the rate of change and how it ensures that the formulas for linear functions apply. More interesting examples follow.

EXAMPLE 3.6 Converting from Celsius to Fahrenheit

Celsius and Fahrenheit are common temperature scales. It is a fact that a 1° increase on the Celsius scale always corresponds to a fixed increase on the Fahrenheit scale.

a. Think of the temperature F on the Fahrenheit scale as a function $F = F(C)$ of the temperature C on the Celsius scale. Explain why F is a linear function of C.

b. On the Fahrenheit scale, 32° is the freezing temperature of water at sea level. This occurs at 0° on the Celsius scale. Also, 212° Fahrenheit is the boiling temperature of water at sea level. This occurs at 100° Celsius. Find the rate of change of F as a linear function of C, and explain its meaning in terms of the temperature scales.

SOLUTION

a. Because a 1° increase on the Celsius scale always corresponds to a fixed change on the Fahrenheit scale, the rate of change of F in terms of C is constant. So F is a linear function of C.

b. The freezing and boiling points of water tell us that

$$F(0) = 32 \text{ and } F(100) = 212$$

In geometric terms, the graph of F as a function of C passes through the points $(0, 32)$ and $(100, 212)$, as shown in **Figure 3.7**.

So the slope is

$$\text{Rate of change} = \text{Slope} = \frac{\Delta F}{\Delta C}$$

$$= \frac{212 - 32}{100 - 0}$$

$$= \frac{9°}{5} \text{ Fahrenheit per degree Celsius}$$

Each 1° increase on the Celsius scale corresponds to an increase of $\dfrac{9°}{5}$ on the Fahrenheit scale. This fact is illustrated in **Figure 3.8**. Note that the slope has definite units associated with it: degrees Fahrenheit per degree Celsius. These units are often very helpful in figuring out how to interpret and use the slope.

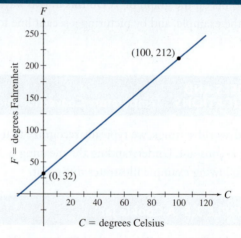

Figure 3.7 Fahrenheit versus Celsius: The graph of F versus C passes through the points $(0, 32)$ and $(100, 212)$.

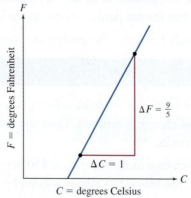

Figure 3.8 Fahrenheit versus Celsius: A 1° increase in C corresponds to a $\dfrac{9°}{5}$ increase in F.

TRY IT YOURSELF 3.6 Brief answers provided at the end of the section.

Use the slope we found in part b and the fact that 32° Fahrenheit corresponds to 0° Celsius to determine the temperature in degrees Fahrenheit when the temperature is 5° Celsius.

EXAMPLE 3.7 An Erroneous News Article

A news story released by Reuters on March 19, 2002, erroneously reported that the Antarctic peninsula had warmed by 36° Fahrenheit over the past half-century. Such temperature increases would result in catastrophic climate changes worldwide, and it is surprising that an error of this magnitude could have slipped by the editorial staff of Reuters. We can't say for certain how the error occurred, but it is likely that the British writer saw a report that the temperature had increased by 2.2° Celsius. What was the actual increase in Fahrenheit temperature? Can you suggest a plausible scenario for why the Reuters reporter got it wrong?

SOLUTION

As we noted in Example 3.6, each 1° increase on the Celsius scale means an increase of $\frac{9}{5} = 1.8°$ on the Fahrenheit scale. It follows that an increase in the Antarctic temperature of 2.2° Celsius means an increase of $2.2 \times 1.8 = 3.96°$ Fahrenheit—not the 36° increase reported by Reuters.

Because 32° Fahrenheit corresponds to 0° Celsius, a temperature of 2.2° Celsius corresponds to a temperature of $3.96 + 32 = 35.96°$ Fahrenheit. That is about 36°, the number reported by Reuters. It appears the reporter may have confused temperature with *change* in temperature.

TRY IT YOURSELF 3.7 Brief answers provided at the end of the section.

If average temperatures in a certain region decreased by 3° Celsius, what would be the corresponding Fahrenheit temperature change?

To say that one quantity is **proportional** to another means that the one is a constant multiple of the other. In a proportionality relationship, the constant multiple is called the **constant of proportionality**.

For example, a linear function $y = mx + b$ is a proportionality relationship precisely when $b = 0$, in which case the equation is $y = mx$. The slope m of a proportionality relationship is the constant of proportionality.

> The quantity A is **proportional** to the quantity B if $A = kB$ for some constant k.
>
> The **constant of proportionality** is the constant multiple k in the proportionality relationship $A = kB$.

EXAMPLE 3.8 Proportionality

Is the circumference of a circle proportional to the radius? If so, what is the constant of proportionality?

SOLUTION

We get the circumference C from the radius r using the formula $C = 2\pi r$. The circumference is indeed a constant multiple of the radius, so a proportionality relationship holds. The constant of proportionality is the multiplier (the rate of change) 2π.

A consequence of our observations is that the graph of circumference versus radius is a straight line through the origin with slope 2π. This is shown in **Figure 3.9**.

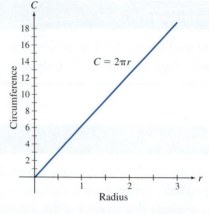

Figure 3.9 Circumference versus radius:
The circumference is proportional to the radius.

TRY IT YOURSELF 3.8 Brief answers provided at the end of the section.

Is the area of a circle proportional to the radius?

EXTEND YOUR REACH   This section has noted several cases in which linear functions apply to real-world phenomena. Discuss several other natural occurrences of linear functions.

MODELS AND APPLICATIONS   Regression Lines

Linear regression is a method of approximating data points with a line called the regression line. The method is impractical for hand calculation but is implemented on many platforms, including calculators and spreadsheet software.

EXAMPLE 3.9 Running Speed of Animals

The following table shows the running speed of various animals versus their length.[2]

Animal	L = Length (inches)	S = Speed (feet per second)
Deer mouse	3.5	8.2
Chipmunk	6.3	15.7
Desert crested lizard	9.4	24.0
Gray squirrel	9.8	24.9
Red fox	24.0	65.6
Cheetah	47.0	95.1

  a. Use a calculator or computer to find the regression line for the data.

b. Plot the data, and then add the regression line.

c. What running speed does the regression equation give for an animal that is 11 inches long?

---

[2]Data from *Size and Cycle*, J. T. Bonner, Princeton University Press, Princeton, NJ, 1965.

SOLUTION

a. A calculator provided the regression equation $S = 2.03L + 5.09$ (where we rounded to two decimal places).

b. The data from this table are graphed in **Figure 3.10**. The regression line is added in **Figure 3.11**.

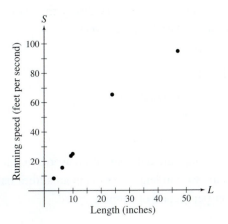

Figure 3.10 Data for running speed versus length

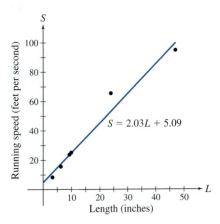

Figure 3.11 Running speed versus length: regression line added

c. We estimate the running speed of an animal that is 11 inches long using this formula:

$$S = 2.03L + 5.09$$

$$= 2.03 \times 11 + 5.09$$

$$\approx 27.42 \text{ feet per second}$$

TRY IT YOURSELF 3.9 Brief answers provided at the end of the section.

Explain the meaning of the slope of the regression line in terms of the running speed and length.

EXTEND YOUR REACH

Is it always appropriate to fit data with a regression line? What properties of data points might suggest the appropriate use of a regression line? What characteristics of the phenomenon being modeled might suggest the appropriate use of a regression line?

CONCEPTS TO REMEMBER: Nonconstant Linear Functions

Defining property: Constant rate of change.

Formula: $f(x) = mx + b$, where $m \neq 0$ is the slope and b is the initial value.

Domain: All real numbers.

Range: All real numbers.

Graph: Straight line.

In the long term:

Positive slope: $f(x) \to \infty$ as $x \to \infty$ and $f(x) \to -\infty$ as $x \to -\infty$.

Negative slope: $f(x) \to -\infty$ as $x \to \infty$ and $f(x) \to \infty$ as $x \to -\infty$.

TRY IT YOURSELF ANSWERS

3.1   The average rate of change is 5.

3.2 $f(x) = -x + 3$

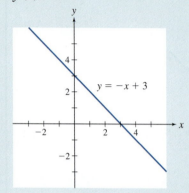

3.3 $g(8) = 3$

3.4 $y = 13x - 13$

3.5   **a.** The amount of money changes by the same amount, -10 dollars, each week.

      b. $M = -10w + 200$

3.6   41° Fahrenheit

3.7   The temperature would decrease by 5.4° Fahrenheit.

3.8   No.

3.9   The regression line tells us that on average each inch of increase in length results in an additional 2.03 feet per second of running speed.

EXERCISE SET 3.1

CHECK YOUR UNDERSTANDING

1. What are the common names for m and b for the linear function $f(x) = mx + b$?

2. How are linear functions characterized in terms of rates of change?

3. If the independent variable of a linear function is increased by 1 unit, which is true?

    a. The slope is added to the function.

    b. The function is unchanged.

    c. The function is multiplied by the slope.

    d. None of the above.

4. **True or false**: If y is proportional to x, then y is a linear function of x.

5. **True or false**: A constant rate of change always indicates a linear function.

6. **True or false**: If y is a nonconstant linear function of x, then x is a linear function of y.

7. Suppose y is proportional to x. If x is doubled, what is the effect on y?

8. The graph of a linear function is _____.

9. Which linear functions are increasing functions?

10. The graph of a linear function with slope 0 is _____.

11. A student says she wants the linear function whose graph passes through the points $(1, 3)$ and $(1, 5)$. What answer would you give?

SKILL BUILDING

Finding formulas. In Exercises 12 through 26, find the formula for the linear function $f(x)$ that satisfies the given conditions.

12. The initial value is -4, and the rate of change is 5.

13. The initial value is 9, and the rate of change is -1.

14. The initial value is 2, and the rate of change is 0.

15. The initial value is 0, and the rate of change is 0.

16. The initial value is -6, and the rate of change is 5.

17. The initial value is -6, and the rate of change is -5.

18. The initial value is 0, and the rate of change is 3.

19. The graph of f passes through the points $(2, 3)$ and $(4, 7)$.

20. The graph of f passes through the points $(1, -3)$ and $(6, 4)$.

21. The graph of f passes through the points $(1, -3)$ and $(10, -3)$.

22. The graph of f passes through the points $(-4, -5)$ and $(6, 8)$.

23. $f(2) = 7$ and $f(-1) = 4$

24. $f(-2) = -3$ and $f(0) = 9$

25. $f(8) = 7$ and $f(2) = -1$

26. $f(-2) = -7$ and $f(2) = 7$

Secant lines. In Exercises 27 through 32, find the equation of the secant line for the given function for the given interval.

27. $f(x) = x^2$ for the interval $[2, 5]$

28. $f(x) = x^3$ for the interval $[1, 4]$

29. $f(x) = \sqrt{x}$ for the interval $[9, 64]$

30. $f(x) = \dfrac{x+1}{x-1}$ for the interval $[-2, 7]$

31. $f(x) = x^2 + x - 1$ for the interval $[-2, 5]$

32. $f(x) = \dfrac{1}{x}$ for the interval $[3, 9]$

PROBLEMS

Finding function values. In Exercises 33 through 36, find the indicated function value *without* first finding a formula for the linear function.

33. f is a linear function with rate of change 2, and $f(7) = 32$. What are $f(8)$ and $f(10)$?

34. f is a linear function with rate of change 3, and $f(5) = 1$. What are $f(8)$ and $f(4)$?

35. f is a linear function with slope 0, and $f(100) = 2$. What are $f(76)$ and $f(301)$?

36. f is a linear function with slope -2, and $f(6) = 0$. What are $f(4)$ and $f(8)$?

Matching. In Exercises 37 through 42, match the given graph with the appropriate linear function from the following list:

A. $f(x) = x + 1$

B. $f(x) = -x + 1$

C. $f(x) = x$

D. $f(x) = 2x - 1$

E. $f(x) = 2$

F. $f(x) = -2x + 1$

37.

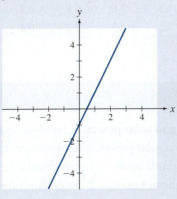

38.

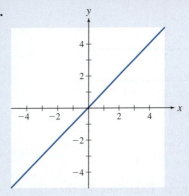

39.

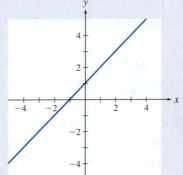

40.

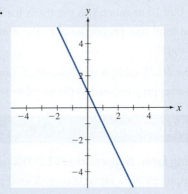

41.

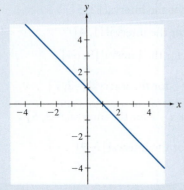

42.

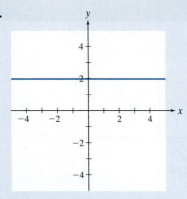

Proportionality. In Exercises 43 through 57, determine whether the relationship described is a proportionality relationship. If it is, identify the constant of proportionality.

43. If each step in a staircase rises 8 inches, is the height of the staircase above the base proportional to the number of steps?

44. If we drive for 3 hours with the cruise control set (so that the speed is always the same), is the total distance driven proportional to speed?

45. At a certain sale, you can buy CDs for $14 each. But if you buy 10 or more, you get them for $13 each. Is the total cost proportional to the number of CDs bought?

46. If sodas cost $0.75 each, is the amount of money we spend on sodas proportional to the number we buy?

47. Is the radius of a circle proportional to the circumference?

48. Is the perimeter of a square proportional to the length of a side?

49. Is the area of a square proportional to the length of a side?

50. Is the area of a square proportional to its perimeter?

51. Is the area of a triangle with a fixed base proportional to its altitude?

52. Is the volume of a balloon proportional to its diameter?

53. Is the amount (in volume) of pizza with a fixed thickness proportional to its diameter?

54. Is the area of a computer screen proportional to the length of its diagonal?

55. Is the amount of sales tax you pay proportional to the purchase price?

56. At a given time of a given day, is the value of the money in your wallet in euros proportional to its value in dollars?

57. Is length in yards proportional to length in meters?

Finding the regression line. In Exercises 58 through 60, use a calculator or computer to find the regression line for the given data. Then plot the data and the regression line on the same screen. Round to two decimal places.

58.

x	1	3	4	5	7
y	0.1	3.2	4.4	6.8	12.3

59.

x	1	3	4	5	7
y	9.3	3.6	2.1	1.4	−6.2

60.

x	1.3	2.5	5.2	6.7	8.4
y	1.2	4.3	7.7	10.4	19.3

MODELS AND APPLICATIONS

61. A struggling farm. Suppose that $125,000 is used to buy equipment for a farm. It costs $3000 per year to maintain the equipment.

   a. Explain why the total amount E, in dollars, spent on equipment is a linear function of the number t of years of operation of the farm.

   b. Find a linear formula that gives E in terms of t.

62. Converting to solar power. A family invests $2500 to convert to solar power. The conversion saves $230 per month on energy costs. Find a formula that gives the net savings N, in dollars, on energy after t months.

63. Inflating a balloon. The volume of air in a balloon is a linear function of time, with slope 10 cubic inches per minute. If the volume is 15 cubic inches at one point in time, what is the volume 3 minutes later?

64. **Circles and proportionality.** To complete this exercise, you need to know that the circumference of a circle is proportional to its radius, and that the constant of proportionality is 2π. You do not need to know either the radius of the moon's orbit or the radius of Earth.

   a. For purposes of this exercise, we assume that the moon's orbit around Earth is circular. In one trip around Earth, the moon travels approximately 2.4 million kilometers. Another satellite orbits Earth (in a circular orbit) at a distance from Earth that is one-third that of the moon. How far does this satellite travel in one trip around Earth?

   b. A rope is tied around the equator of Earth. A second rope circles Earth and is suspended 5 feet above the equator. How much longer is the second rope than the first?

65. **More on Celsius to Fahrenheit.** In Example 3.6, we found that the temperature F on the Fahrenheit scale is a linear function of the temperature C on the Celsius scale, and the slope is $\dfrac{9°}{5}$ Fahrenheit per degree Celsius.

   a. Recall that 32° Fahrenheit corresponds to 0° Celsius. Find a formula for F as a linear function of C.

   b. Solve the equation from part a for C in order to find a formula expressing the temperature on the Celsius scale as a function of the temperature on the Fahrenheit scale.

   c. Identify the slope of the function from part b, and explain its meaning in terms of the temperature scales.

66. **The Kelvin scale.** Physicists often use the Kelvin temperature scale. On this scale, 0° Fahrenheit corresponds to a temperature of 255.37 kelvins. Each 1° increase on the Fahrenheit scale corresponds to an increase of $\dfrac{5}{9}$ kelvin. Find a formula that gives the temperature K in kelvins as a function of the temperature F in degrees Fahrenheit.

67. **Aerobic power.** Aerobic power can be thought of as the maximum oxygen consumption attainable per kilogram of body mass. There are a number of ways in which physical educators estimate this. One method uses the Queens College Step Test.[3] After a prescribed exercise regimen, a 15-second pulse count P is taken. Maximum oxygen consumption M in milliliters per kilogram for men is approximated by

$$M = 111.30 - 1.68P$$

For women, the maximum oxygen consumption F in milliliters per kilogram is approximated by

$$F = 65.81 - 0.74P$$

   a. Calculate the maximum oxygen consumption for both a man and a woman showing a 15-second pulse count of 40.

   b. What 15-second pulse count will indicate the same maximum oxygen consumption for a man as for a woman?

   c. What maximum oxygen consumption is associated with your answer in part b?

68. **Rice market.** The quantity R of rice (in billions of bushels) that rice suppliers in a certain country are willing to produce in a year and offer for sale at a price P (in dollars per bushel) is determined by the relation

$$P = 2R - 1$$

The quantity T of rice (in billions of bushels) that rice consumers are willing to purchase in a year at price P is determined by the relation

$$P = 3 - T$$

The equilibrium price is the price at which the quantity supplied is the same as the quantity demanded. Find the equilibrium price for rice.

69. **Ants and temperature.** A scientist observed that the ambient temperature C in degrees Celsius can be determined from the speed S in centimeters per second at which certain ants run.[4] The proposed formula was

$$C = 5S + 13.5$$

---

[3]Data from *Nutrition, Weight Control, and Exercise*, F. I. Katch and W. D. McArdle, Lea & Febiger, Philadelphia, 1983.

[4]Data from H. Shapley, "Note on the Thermokinetics of Dolichoderine Ants," *Proceedings of the National Academy of Sciences of the United States of America* 10: 436–439 (1924).

a. Explain the meaning of the slope of the linear function C in terms of the speed of the ant and the temperature.

b. Solve for S in the preceding formula to obtain a formula expressing the running speed as a function of the temperature C.

c. Explain the meaning of the slope of the linear function you found in part b in terms of the speed of the ants and the temperature.

70. **Slowing down in a curve.** A study of average driver speed on rural highways by A. Taragin[5] found a linear relationship between average speed S (in miles per hour) and the amount D of curvature (in degrees) of the road. On a straight road ($D = 0$), the average speed was found to be 46 miles per hour. This was found to decrease by 0.75 mile per hour for each additional degree of curvature. Find a linear formula relating speed to curvature.

CHALLENGE EXERCISES FOR INDIVIDUALS OR GROUPS

71. **Satellite subscribers.** The following table shows the number D, in millions, of subscribers to a certain television satellite company at the time t years after 1995.

t = years since 1995	D = subscribers (millions)
0	1.20
4	6.68
7	11.18
9	13.00
12	16.83
14	18.08

a. Plot the data along with the regression line.

b. Give the equation of the regression line. (Round to two decimal places.)

c. Explain the meaning of the slope of the regression line in terms of the number of satellite subscribers.

d. How many satellite subscribers does the regression line predict for 2025?

72. **Crayon colors.** The following table shows the number C of crayon colors available t years after 1900.

t = years since 1900	3	49	58	72	90	98	103
C = colors	8	48	64	72	80	120	120

a. Plot the data along with the regression line.

b. Give the equation of the regression line. (Round to two decimal places.)

c. Explain the meaning of the slope of the regression line in terms of the number of crayon colors.

d. How many colors does the regression line suggest for 1980? Round your answer to the nearest whole number.

73. **National defense.** The following table shows the amount of money M, in billions of dollars, spent by the United States on national defense[6] at the time t years after 2000.

t = years since 2000	4	5	6	7	8
M = billions	455.8	495.3	521.8	551.3	616.1

a. Plot the graph of the data along with the regression line.

b. Give the equation of the regression line. (Round to two decimal places.)

c. Explain the meaning of the slope of the regression line in terms of the amount of defense spending in this time period.

d. What expenditure does the regression line predict for 2025?

74. **Important historical shift—men and women entering college.** The following table shows the percentage of high school men and women who enrolled in college within 12 months of graduation t years after 1960.

t = years since 1960	0	5	10	15
M = men	54%	57.3%	55.2%	52.6%
F = women	37.9%	45.3%	48.5%	49%

a. On the same coordinate system, plot the data for men, the data for women, and both regression lines.

b. Write the equations of the regression lines for men and women. (Round to two decimal places.)

---

[5]"Driver Performance on Horizontal Curves," *Proceedings* 33, Highway Research Board, Washington, DC, 1954, 446–466.

[6]Data from the *Statistical Abstract of the United States.*

c. The regression lines predict that the percentage of women entering college will match that of men when the regression lines cross. When does this occur? (This actually occurred for the first time in 1980.)

75. **Competing spider populations.** Consider two competing spider populations. Let s be the size of one spider population and S the size of the second. Both populations are measured in hundreds of spiders. The growth rates of the populations satisfy the following equations (note that S and s depend on each other because the populations compete).

$$\text{Growth rate for } s = 400(1000 - s - S)$$

$$\text{Growth rate for } S = 600(1000 - s/3 - 2S)$$

a. The set of points at which the growth rate for s is zero is called the isocline. Find a formula for s in terms of S that describes this isocline.

b. The set of points at which the growth rate for S is zero form another isocline. Find a formula for s in terms of S that describes this isocline.

c. A point at which the growth rates for both s and S are zero is called an equilibrium point. Use your answers to parts a and b to find the populations of each spider population at the equilibrium point.

d. Explain in terms of the two populations what happens when the populations reach an equilibrium point.

REVIEW AND REFRESH: Exercises from Previous Sections

76. **From Section P.1:** Find the midpoint of $P = (2, 7)$ and $Q = (-1, -3)$, and the distance between P and Q.

Answer: Midpoint $\left(\dfrac{1}{2}, 2\right)$, distance $\sqrt{109}$

77. **From Section P.2:** Solve the inequality $4 - x \leq 1 + x$.

Answer: $\left(\dfrac{3}{2}, \infty\right)$

78. **From Section P.3:** Solve the inequality $\dfrac{1}{x} \geq x$.

Answer: $(-\infty, -1) \cup (0, 1)$

79. **From Section 1.1:** Find the domain of

$$f(x) = \frac{x - 2}{x^2 - 5x + 4}.$$

Answer: All real numbers except 1 and 4. In interval notation: $(-\infty, 1) \cup (1, 4) \cup (4, \infty)$.

80. **From Section 1.1:** Find the range of the function $f(x) = \sqrt{x} + 4$.

Answer: $(4, \infty)$

81. **From Section 1.2:** If $f(x) = \dfrac{1}{x}$, find the average rate of change of $f(x)$ from $x = 3$ to $x = 5$.

Answer: $-\dfrac{1}{15}$

82. **From Section 1.2:** If $f(x) = x^2 + 1$, calculate the average rate of change of $f(x)$ on the interval $[x, x + h]$. Assume that $h > 0$, and simplify your answer.

Answer: $2x + h$

83. **From Section 1.3:** Suppose that $f(x)$ is a function that has a positive rate of change on the interval $[2, 7]$. What can be said about the graph of f on this interval?

Answer: It is increasing.

84. **From Section 1.4:** Calculate the limit as $x \to \infty$ of

$$f(x) = 3 + \frac{2}{x^5}.$$

Answer: 3

85. **From Section 2.1:** If $f(x) = x^2 - 1$ and $g(x) = x + 1$, find $f \circ g$.

Answer: $(f \circ g)(x) = x^2 + 2x$

86. **From Section 2.2:** Calculate the inverse of the function $f(x) = 4x - 5$.

Answer: $f^{-1}(x) = \dfrac{(x + 5)}{4}$

87. **From Section 2.3:** Explain how to get the graph of $f(x + 1)$ from the graph of $f(x)$.

Answer: The graph of $f(x + 1)$ is the graph of $f(x)$ shifted by 1 unit to the left.

3.1 Linear Functions: Constant Rate of Change

3.2 Exponential Functions: Constant Proportional Change

3.3 Modeling with Exponential Functions

3.2 Exponential Functions: Constant Proportional Change

> Exponential functions are used to model rapid increase or decrease because they show a constant proportional rate of change.

In this section, you will learn to:

1. Analyze exponential functions using constant proportional change.
2. Sketch the graphs of exponential functions using their properties.
3. Define the number e and the natural exponential function.
4. Solve applied problems involving exponential functions.
5. Compare the growth rates of exponential functions and power functions.

A case can be made that π is the most important number in mathematics. But in this section you will meet a number that challenges π for the top ranking. This is the number denoted by e. Like π, e is an irrational number whose value cannot be captured by any finite decimal. To three digits of accuracy, $e \approx 2.718$.

The designation e was coined by the Swiss mathematician Leonhard Euler (pronounced "oiler") (**Figure 3.12**). Euler (1707–1783) was among the most prolific mathematicians of all time, and his writings greatly influenced the approach of modern calculus texts.

We can use the idea of compound interest to explain e. When interest is compounded, exponential functions are involved, and the number of compounding periods plays a key role. Compounding may occur yearly, monthly, daily, hourly, or each second. As the number of compounding periods is taken to its ultimate limit, the number e makes its appearance and, surprisingly enough, complicated formulas are replaced by simpler ones that involve the number e.

Understanding how exponential functions work is crucial for success in calculus. You may wish to review the basic rules for exponents. In advanced mathematics, precise meaning is given to exponents applied to positive numbers even when the exponents are not rational. At this level, suffice it to say that a number like 2^{π} is close to $2^{3.14}$ because π is close to 3.14. We note that positive numbers raised to real exponents obey the same rules as for rational exponents.

Figure 3.12 Leonhard Euler Georgios Kollidas/Shutterstock

Constant Proportional Change

> Constant proportional change means that the rate of change is proportional to the function. Intuitively, these functions change by constant multiples.

There are initially 1000 cells in a petri dish. Reproduction is accomplished via cell division, and as a result, the number of cells in the dish doubles each hour. The accompanying table shows the number N of cells in the dish after t hours.

t = time in hours	0	1	2	3	4
N = number of cells	1000	2000	4000	8000	16,000

Note that to obtain the next hour's number we multiply this hour's number by 2. In other words, the next hour's number is proportional to last hour's number with constant of proportionality 2.

This observation allows us to find a formula for N. We start with 1000, and after t hours we multiply by 2 a total of t times:

$$N(t) = 1000 \times \overbrace{2 \times 2 \times 2 \times \cdots \times 2}^{t \text{ factors}} = 1000 \times 2^t$$

Functions like this one that grow by constant multiples show a constant proportional rate of change. That is, the rate of change is proportional to the function. Such a function is known as an **exponential function**.

The form of the preceding function N is typical of exponential functions.

An **exponential function** is a function that exhibits a constant proportional rate of change that is not equal to 0.

> ### LAWS OF MATHEMATICS: Formula for an Exponential Function
>
> An exponential function has the form $f(x) = ab^x$ where $a \neq 0, b > 0$, and $b \neq 1$.

In the formula $f(x) = ab^x$ for an exponential function, the number b is the **base**. The number a is the **initial value** because that is the value of f when $x = 0$.

It is important to be able to interpret the formula for an exponential function in terms of rates of change.

The **base** of an exponential function of the form $f(x) = ab^x$ is the number b.

The **initial value** of an exponential function of the form $f(x) = ab^x$ is the number a.

> ### CONCEPTS TO REMEMBER: How Exponential Functions Change
>
> The exponential function f with base b has the property that
>
> $$f(x+1) = bf(x)$$
>
> It may be helpful to think of $f(x)$ as the old y-value and $f(x+1)$ as the new y-value. Then
>
> $$\text{New } y\text{-value} = \text{Base} \times \text{Old } y\text{-value}$$
>
> In applications, this is often the feature that allows us to recognize exponential functions.

Note, for example, that the function $N(t) = 1000 \times 2^t$ that describes cell division is exponential with base $b = 2$ and initial value $a = 1000$.

The next example provides an algebraic derivation of the proportional change exhibited by functions of the preceding form.

EXAMPLE 3.10 How Exponential Functions Change

Let $f(x) = ab^x$ be an exponential function with base b and initial value a.

a. Show that

$$f(x+1) = bf(x)$$

b. The function g is exponential with base 3. If $g(4) = 7$, find the values of $g(5)$ and $g(6)$.

SOLUTION

a. We apply the basic rules of exponents:

$$f(x+1) = ab^{x+1} \qquad \text{◀ Apply function definition.}$$

$$f(x+1) = ab^x b^1 \qquad \text{◀ Use } A^{p+q} = A^p A^q.$$

$$f(x+1) = b(ab^x) \qquad \text{◀ Simplify.}$$

$$f(x+1) = bf(x) \qquad \text{◀ Apply function definition.}$$

b. We apply part a using the fact that the base is $b = 3$:

$$g(4+1) = 3g(4) = 3 \times 7 = 21$$

Hence, $g(5) = 21$.

The same procedure applies to find $g(6)$:

$$g(6) = g(5+1) = 3g(5) = 3 \times 21 = 63$$

TRY IT YOURSELF 3.10 Brief answers provided at the end of the section.

For the function g in part b, find $g(3)$.

EXTEND YOUR REACH   Assume that f is a function with $f(1) = 3$, $f(2) = 9$, and $f(3) = 26$. Is it possible that f is an exponential function? If so, explain why. If not, find a value for $f(3)$ that would change your answer to the question. *Suggestion:* Calculate proportional rates of change.

It is instructive to compare the way exponential functions change with the way linear functions change. Linear functions change by constant sums, but exponential functions change by constant multiples. This fact is illustrated by the following tables.

Linear with slope 2			Exponential with base 2	
x	**y = 2x**		**x**	**y = 2^x**
1	2	) +2	1	2	) ×2
2	4	) +2	2	4	) ×2
3	6	) +2	3	8	) ×2
4	8	) +2	4	16	) ×2
5	10		5	32	

EXAMPLE 3.11 Using Constant Proportional Change to Analyze Exponential Functions

a. Let $g(x)$ be an exponential function such that $g(x+1) = 4g(x)$ for all x. If $g(0) = 5$, find a formula for $g(x)$.

b. Let $f(x)$ be an exponential function with $f(1) = 8$ and $f(3) = 2$. Find a formula for $f(x)$.

SOLUTION

a. The formula $g(x+1) = 4g(x)$ tells us that g is an exponential function with base 4. Thus, $g(x) = a(4^x)$ for some initial value a. The condition $g(0) = 5$ tells us that $a = 5$. We conclude that

$$g(x) = 5(4^x)$$

b. Let $f(x) = ab^x$. The given information can be stated in terms of two equations that we need to solve in order to find a and b:

$$ab = 8 \quad \blacktriangleleft \text{Use } f(1) = 8.$$

$$ab^3 = 2 \quad \blacktriangleleft \text{Use } f(3) = 2.$$

To proceed we divide the second equation by the first:

$$\frac{ab^3}{ab} = \frac{2}{8} \quad \blacktriangleleft \text{Divide the two preceding equations.}$$

$$b^2 = \frac{1}{4} \quad \blacktriangleleft \text{Simplify.}$$

$$b = \frac{1}{2} \quad \blacktriangleleft \text{Take square roots, noting that } b > 0.$$

Putting $b = \dfrac{1}{2}$ into the equation $ab = 8$, we find $a = 16$. We conclude that

$$f(x) = 16\left(\frac{1}{2}\right)^x$$

TRY IT YOURSELF 3.11 Brief answers provided at the end of the section.

If $h(x)$ is an exponential function with $h(2) = 12$ and $h(3) = 24$, find a formula for $h(x)$.

Graphs of Exponential Functions

> Graphs of increasing exponential functions show rapid growth at the right tail end but approach the x-axis at the left tail end.

The next example gives us a first look at some of the properties of exponential functions, and it prepares us to make graphs.

EXAMPLE 3.12 Evaluating Exponential Functions

a. For $f(x) = 2^x$, make a table of values for f from $x = -20$ to $x = 20$ in steps of 5.

  i. Does your table indicate an increasing function or a decreasing function?

  ii. What limiting values are indicated by the table?

b. For $g(x) = \left(\dfrac{1}{2}\right)^x = 2^{-x}$, make a table of values for f from $x = -20$ to $x = 20$ in steps of 5.

  i. Does your table indicate an increasing function or a decreasing function?

  ii. What limiting values are indicated by the table?

SOLUTION

a. We used a calculator to produce the following table of values.

x	−20	−15	−10	−5	0	5	10	15	20
2^x	0.00000095	0.0000305	0.0009766	0.03125	1	32	1024	32,768	1,048,576

i. The table of values suggests that $f(x)$ increases as x increases. This indicates an increasing function. (We expect this because raising 2 to a larger power should give a larger result.)

ii. The table of values indicates that as x increases, 2^x increases without any bound. We expect that

$$2^x \to \infty \text{ as } x \to \infty$$

As x becomes larger but is negative, 2^x is near 0. This suggests that

$$2^x \to 0 \text{ as } x \to -\infty$$

b. Again, a calculator is required (or we can use part a):

x	−20	−15	−10	−5	0	5	10	15	20
$\left(\dfrac{1}{2}\right)^x = 2^{-x}$	1,048,576	32,768	1024	32	1	0.03125	0.0009766	0.0000305	0.00000095

i. As x gets larger, the function value gets smaller. Thus, the table indicates a decreasing function.

ii. As x increases, the table indicates that 2^{-x} approaches 0. The table suggests that $2^{-x} \to 0$ as $x \to \infty$.

When x gets larger but is negative, 2^{-x} grows ever larger. This indicates that $2^{-x} \to \infty$ as $x \to -\infty$.

TRY IT YOURSELF 3.12 Brief answers provided at the end of the section.

Make a table of values for $y = 3^x$ from −10 to 10 in steps of 5. For which values of x is 3^x larger than 2^x, and for which values of x is 3^x smaller?

The tables we made in the previous example serve as an aid to producing graphs of exponential functions. The graphs of $y = 2^x$, $y = 3^x$, and $y = 4^x$ are shown in **Figure 3.13**. The graphs display different function values, but their basic shapes are quite similar. These are typical of exponential functions with a base greater than 1.

The graphs of $y = \left(\dfrac{1}{2}\right)^x = 2^{-x}$, $y = \left(\dfrac{1}{3}\right)^x = 3^{-x}$, and $y = \left(\dfrac{1}{4}\right)^x = 4^{-x}$ are shown in **Figure 3.14**, and they are typical of the graphs of exponential functions with a base less than 1.

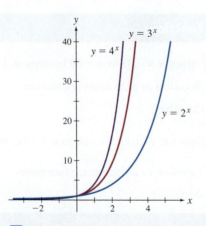

DF **Figure 3.13** The graphs of $y = 2^x$, $y = 3^x$, and $y = 4^x$: These are typical of graphs of exponential functions with a base greater than 1.

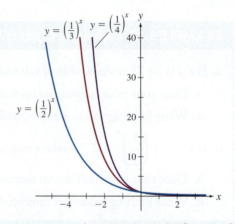

Figure 3.14 The graphs of $y = \left(\dfrac{1}{2}\right)^x$, $y = \left(\dfrac{1}{3}\right)^x$, and $y = \left(\dfrac{1}{4}\right)^x$: These are typical of graphs of exponential functions with a base less than 1.

CONCEPTS TO REMEMBER: Graphs of Exponential Functions

With base $b > 1$:	With base $0 < b < 1$:
$y = b^x$ is increasing and concave up.	$y = b^x$ is decreasing and concave up.
$b^x \to \infty$ as $x \to \infty$	$b^x \to 0$ as $x \to \infty$
$b^x \to 0$ as $x \to -\infty$	$b^x \to \infty$ as $x \to -\infty$

Note that because the base b is positive, function values for $f(x) = b^x$ are always positive. Thus, the domain of f is all real numbers, but the range is all positive numbers.

EXAMPLE 3.13 Transformations of Exponential Graphs

Plot the graphs of both $y = 2^{-x}$ and $y = 3 + 2^{-x}$ on the same coordinate axes.

SOLUTION

To obtain the graph of $y = 3 + 2^{-x}$ from the graph of $y = 2^{-x}$, we shift up by 3 units. This is shown in **Figure 3.15**.

Note that because $2^{-x} \to 0$ as $x \to \infty$, we have

$$3 + \overset{0}{2^{-x}} \to 3 \text{ as } x \to \infty$$

This limit prompts us to add the line $y = 3$ to the graph in Figure 3.15 to show the end behavior of the graph.

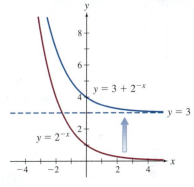

Figure 3.15 Graphs of $y = 2^{-x}$ and $y = 3 + 2^{-x}$, with the line $y = 3$ added

TRY IT YOURSELF 3.13 Brief answers provided at the end of the section.

On the same coordinate axes, plot the graphs of $y = 3^x$ and $y = 3^x - 10$.

EXTEND YOUR REACH

a. What can you say about the rate of change of $y = 2^{-x}$ as $x \to \infty$? *Suggestion:* Use a straightedge with the graph of $y = 2^{-x}$ shown in Figure 3.15 to visualize slopes of tangent lines for large values of x.

b. What can you say about the rate of change of $y = 3 + 2^{-x}$ as $x \to \infty$?

The Special Number *e*

> The special number *e* is the base for the natural exponential function.

For exponential functions there is one base that is more important than any other. This is the number designated *e*. In calculus, the number *e* is defined as a limit:

$$\left(1 + \frac{1}{x}\right)^x \to e \text{ as } x \to \infty$$

Like π, the number *e* is irrational. That is, it cannot be expressed as the quotient of two integers, and it does not have a repeating decimal expansion. We can approximate the value of *e* with a table of values that shows $\left(1 + \frac{1}{x}\right)^x$ for large values of *x*. To 20 digits of accuracy,

$$e \approx 2.7182818284590452354$$

The **natural exponential function** is the function $f(x) = e^x$.

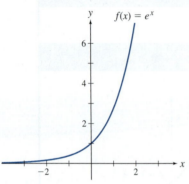

Figure 3.16 The graph of $f(x) = e^x$

x	$\left(1 + \dfrac{1}{x}\right)^x$
1	2
10	2.5937424601
100	2.7048138294215260933
1000	2.7169239322358924574
10,000	2.7181459268252248640
100,000	2.7182682371744896680
1,000,000	2.7182804693193768838

It is difficult to overstate the importance of the function $f(x) = e^x$, often called the **natural exponential function**, for both calculus and the remainder of this course.

The graphs of $f(x) = e^x$ and its companion $g(x) = e^{-x}$ are shown in **Figure 3.16** and **Figure 3.17**.

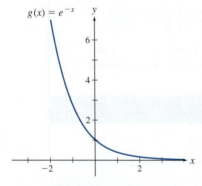

Figure 3.17 The graph of $g(x) = e^{-x}$

CONCEPTS TO REMEMBER: The Natural Exponential Function $y = e^x$

- $\left(1 + \dfrac{1}{x}\right)^x \to e$ as $x \to \infty$

- $e \approx 2.72$

- $y = e^x$ is an increasing function whose graph is concave up.
  $y = e^{-x}$ is a decreasing function whose graph is concave up.

- $e^x \to \infty$ as $x \to \infty$, and $e^x \to 0$ as $x \to -\infty$

- $e^{-x} \to 0$ as $x \to \infty$, and $e^{-x} \to \infty$ as $x \to -\infty$

EXAMPLE 3.14 The Natural Exponential Function

On the same coordinate axes, plot the graphs of $y = 2^x$, $y = e^x$, and $y = 3^x$. Comment on the relative positions of the graphs.

SOLUTION

The accompanying table of values is an aid to constructing the graph shown in **Figure 3.18**. Note that for positive *x*, the graph of $y = e^x$ is above the graph of $y = 2^x$ and below the graph of $y = 3^x$. But for negative *x*, the order is reversed: The graph of $y = 2^x$ is above the graph of $y = e^x$, which is above the graph of $y = 3^x$. These results make sense because e is between 2 and 3.

x	2^x	e^x	3^x
−3	0.13	0.05	0.04
−2	0.25	0.14	0.11
−1	0.50	0.37	0.33
0	1	1	1
1	2	2.72	3
2	4	7.39	9
3	8	20.09	27

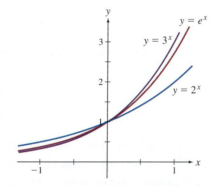

Figure 3.18 The graphs of $y = 2^x$, $y = e^x$, and $y = 3^x$

TRY IT YOURSELF 3.14 Brief answers provided at the end of the section.

On the same coordinate axes, plot the graphs of $y = 2^{-x}$, $y = e^{-x}$, and $y = 3^{-x}$.

If $1 < a < b$, discuss the relationship between the graphs of $y = a^x$ and $y = b^x$.   **EXTEND YOUR REACH**

MODELS AND APPLICATIONS Exponential Growth and Decay

When naturally occurring phenomena are modeled by increasing exponential functions, we say that exponential growth is occurring. In the case of decreasing exponential functions, we refer to exponential decay. Constant proportional change is a key property used in recognizing exponential growth and decay.

Exponential growth occurs in many settings, such as financial mathematics, as is shown in the next example.

EXAMPLE 3.15 Exponential Growth of an Investment

The value of a certain investment after t years is given by

$$B(t) = 7000e^{0.04t} \text{ dollars}$$

a. How much money was originally invested?

b. What percentage interest did the investment earn over the first year? Over the second year?

c. Plot the graph of the investment over the first 50 years.

SOLUTION

a. We use $t = 0$ to find the initial value of the investment:

$$B(0) = 7000e^{0.04 \times 0} = 7000 \times 1 = 7000 \text{ dollars}$$

b. Over the first year, the value grew from \$7000 to

$$B(1) = 7000e^{0.04 \times 1} \approx 7285.68 \text{ dollars}$$

This is an absolute increase of \$285.68 and a relative increase of

$$\frac{285.68}{7000} \approx 0.041$$

This growth represents an increase of 4.1%, which is the percentage interest earned over the first year.

The value after 2 years is

$$B(2) = 7000e^{0.04 \times 2} \approx 7583.01 \text{ dollars}$$

In the second year, the investment grew from $7285.68 to $7583.01, an absolute increase of $297.33. The relative increase is

$$\frac{297.33}{7285.68} \approx 0.041$$

So the percentage interest earned over the second year is also 4.1%. It turns out that the interest earned is 4.1% every year. The constant percentage change exhibited here turns out to be a key feature of exponential functions. We will explore this fact in more detail in the next section.

c. The following table of values is used to make the graph shown in **Figure 3.19**.

t	$B(t) = 7000e^{0.04t}$
0	$7000.00
10	$10,442.77
20	$15,578.79
30	$23,240.82
40	$34,671.23
50	$51,723.39

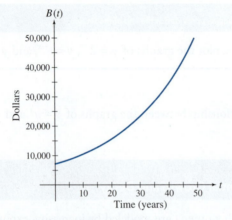

Figure 3.19 The value of an investment

TRY IT YOURSELF 3.15 Brief answers provided at the end of the section.

The value in dollars of a second investment is given by $B(t) = 4000e^{-0.15t}$, where t is the time in years. By how much did the value decline over the first 4 years?

Take note of the rapid growth of the investment in Example 3.15, from $7000 to more than $50,000. Such growth is typical of exponential functions with a base greater than 1.

One important instance of the use of decreasing exponential models concerns radioactive substances, which decay at a rate proportional to the amount present. Thus, the amount of a radioactive substance is an instance of exponential decay. The rate of decay depends on the substance and is normally given in terms of the half-life, which is the time required for half of the substance to decay.

EXAMPLE 3.16 Exponential Decay: Half-life

The radioactive element carbon-14 decays over time. If there are initially C_0 grams present, then the amount remaining after t years is given by

$$C(t) = C_0 \left(\frac{1}{2}\right)^{t/5730} \text{ grams}$$

a. As we have noted, the rate of decay of a radioactive substance is normally given in terms of the half-life. What is the half-life of carbon-14?

b. Make a graph of the amount of carbon-14 present over the first 50,000 years if there are initially 500 grams.

SOLUTION

a. To find the half-life, we need the time t when the amount remaining is half of the initial amount, $\frac{1}{2}C_0$. Thus, we seek a value of t that satisfies

$$\frac{1}{2}C_0 = C_0 \left(\frac{1}{2}\right)^{t/5730}$$

or, dividing out C_0, $\frac{1}{2} = \left(\frac{1}{2}\right)^{t/5730}$

This equation is true if the exponent $\frac{t}{5730}$ is 1. Thus, the half-life of carbon 14 is 5730 years.

b. The accompanying table is useful for sketching the graph shown in **Figure 3.20**.

t	$C(t)$
0	500
10,000	149.15
20,000	44.49
30,000	13.27
40,000	3.96
50,000	1.18

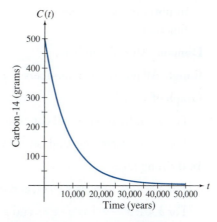

Figure 3.20 Decay of carbon-14

TRY IT YOURSELF 3.16 Brief answers provided at the end of the section.

How long does it take for three-quarters of the carbon-14 to decay? *Suggestion*: Decay of three-quarters corresponds to decay of half followed by decay of half again.

The natural exponential function and compositions of that function with others are commonly used to model many real-world phenomena. Several such models appear in the exercises.

Rapidity of Exponential Growth

> In the long term, increasing exponential functions grow more rapidly than power functions.

To emphasize the eventual rapidity of the growth of exponential functions with a base greater than 1, we compare exponential functions with power functions, which are functions of the form $y = x^k$ for some constant k. **Figure 3.21** shows the graphs of the power function $y = x^3$ and the exponential function $y = 2^x$ from $x = 0$ to $x = 9$. Note that at $x = 9$ the graph of $y = x^3$ is above the graph of $y = 2^x$, showing that the power function is larger than the exponential function. But if we extend the span to include $x = 12$, as we have done in **Figure 3.22**, we see that the exponential function has gained the dominant position.

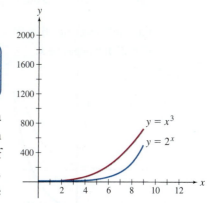

Figure 3.21 In the short term: At $x = 9$, the graph of $y = x^3$ is above the graph of $y = 2^x$.

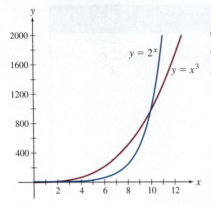

Figure 3.22 In the long term: The graph of $y = 2^x$ eventually dominates the graph of $y = x^3$.

In fact, the graph of $y = 2^x$ eventually rises above the graph of x^{1000}, or $x^{1,000,000}$, or x to any other power. In general, any increasing exponential function will eventually dominate any power function. That is, if b is greater than 1, then no matter how large k is, b^x will be greater than x^k if x is large enough. This fact is normally established in calculus, but in the exercises you will be led through an elementary proof.

CONCEPTS TO REMEMBER: Exponential Functions

Defining property: Constant (but not 0) proportional rate of change.

Formula: $f(x) = ab^x$. The constant b is the base, and we require that $b > 0$ and $b \neq 1$. The constant a is the initial value, and we require that $a \neq 0$. An important case is when $b = e$, and $a = 1$. We have the natural exponential function $y = e^x$.

Domain: All real numbers.

Range: All positive real numbers if $a > 0$. All negative real numbers if $a < 0$.

Graph of $y = b^x$:

For $b > 1$: Increasing and concave up

For $b < 1$: Decreasing and concave up

In the long term:

For $b > 1$: $b^x \to \infty$ as $x \to \infty$, and $b^x \to 0$ as $x \to -\infty$

For $b < 1$: $b^x \to 0$ as $x \to \infty$, and $b^x \to \infty$ as $x \to -\infty$

TRY IT YOURSELF ANSWERS

3.10 $g(3) = \dfrac{7}{3}$

3.11 $h(x) = 3 \times 2^x$

3.12

x	-10	-5	0	5	10
3^x	0.0000169	0.0041152	1	243	59,049

The table indicates that $3^x > 2^x$ for positive x, but $3^x < 2^x$ for negative x.

3.13

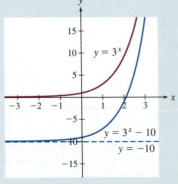

3.14

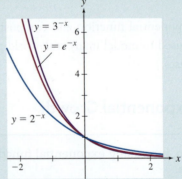

3.15 $1804.75

3.16 11,460 years

EXERCISE SET 3.2

1. Let $y = 3^x$. If x is increased by 1, how is y affected?

2. Let $y = 3^{-x}$. If x is increased by 1, how is y affected?

3. Let $y = 4^x$. If x is increased by 2, how is y affected?

4. Let $y = 2^{-x}$. If x is increased by 2, how is y affected?

5. What can you say regarding the concavity of the graph of an exponential function?

6. If $b > 0$ and $f(x) = b^x$ is an increasing function, what can you conclude about b?

7. What do you know about the base of an exponential function used to model radioactive decay?

8. **True or false**: It is possible to find a number $k > 1$ so that 1.0001^k is larger than $k^{10,000}$.

9. Which is larger, the number e or the number π?

10. Which of the following are exponential functions?

    a. $y = x^\pi$ **c.** $y = e^\pi$ **e.** $y = (e\pi)^x$

    b. $y = \pi^x$ **d.** $y = x^e$ **f.** $y = (\pi + e)^x$

11. Which of the following are exponential functions?

    a. $y = \dfrac{1}{e^x}$ **c.** $y = 4^{-x}$ **e.** $y = \dfrac{2^x}{3^x}$

    b. $y = e^{1/x}$ **d.** $y = \dfrac{1}{e^{-x}}$ **f.** $y = x^2$

12. **Linear or exponential?** Let $f(x)$ be a function with the property that adding 1 to x multiplies the value of $f(x)$ by 5. Let $g(x)$ be a function with the property that adding 1 to x adds 5 to the value of $g(x)$. Which function is linear, and which is exponential?

Finding a formula. In Exercises 13 through 17, find a formula for the exponential function $f(x)$ having the given properties.

13. $f(0) = 5$, and f is tripled when x is increased by 1.

14. $f(0) = 3$, and f is multiplied by e when x is increased by 1.

15. $f(0) = 5$, and f is divided by 4 when x is increased by 1.

16. $f(0) = -1$, and f is doubled when x is increased by 1.

17. $f(0) = -5$, and f is divided by π when x is increased by 1.

Matching functions with graphs. In Exercises 18 through 23, match the given graph with the appropriate function from the following list:

    A. $y = 2^x$ **D.** $y = \left(\dfrac{1}{3}\right)^x$

    B. $y = 2^{-x}$

    C. $y = 3^x$ **E.** $y = e^x$

                   F. $y = e^{-x}$

18.

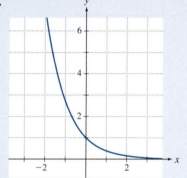

19.

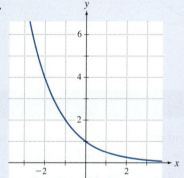

20.

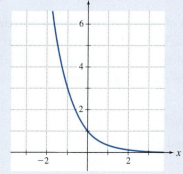

21.

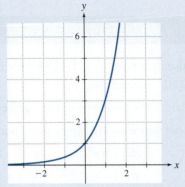

22.

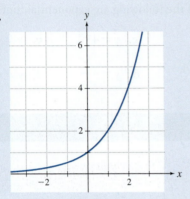

23.

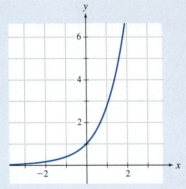

24. A graph. Sketch the graph of $y = a(b^x)$ in the case that a is negative and $b > 1$. Give a verbal description of the graph, stating where it is increasing, decreasing, concave up, and concave down.

25. A graph. Sketch the graph of $y = a(b^x)$ in the case that a is negative and $b < 1$. Give a verbal description of the graph, stating where it is increasing, decreasing, concave up, and concave down.

Finding formulas. In Exercises 26 through 32, find an exponential function having the given properties.

26. $f(0) = 6$ and $b = 7$

27. $f(0) = 3$ and $b = 0.6$

28. $f(1) = 20$ and $f(2) = 100$

29. $f(1) = 6$ and $f(3) = 54$

30. $f(1) = 9$ and $f(3) = 1$

31. $f(0) = B_0$, and increasing x by 1 multiplies f by 2

32. $f(0) = B_0$, and increasing x by 1 multiplies f by 0.8

Changing form. In Exercises 33 through 45, write the given expression in the form $a(b^x)$.

33. $3(2^x 3^x)$

34. $4^x 4^{x+1}$

35. 2^{1-x}

36. $\left(2^x\right)^3$

37. $\dfrac{1}{\pi^x}$

38. $3\left(\dfrac{2^x}{5^x}\right)$

39. $\dfrac{5^{x-1}}{7^{x+1}}$

40. 3^{2x-3}

41. $c^{-x} d^{2x+1}$

42. $\dfrac{e^{-x}}{e^{2x}}$

43. $\left(e^{3x-1}\right)^2$

44. $\dfrac{2}{e^{x-1}}$

45. $\sqrt{e^x}$

PROBLEMS

46. Domain and range. What are the domain and range of $y = a(b^x)$ if $b > 1$ and a is negative?

47. Domain and range. What are the domain and range of $y = a(b^x)$ if $b < 1$ and a is negative?

Transformations of graphs of exponential functions. In Exercises 48 through 55, use what you know about graphs of exponential functions and transformations of graphs to plot the graphs of the two given functions on the same coordinate axes.

48. $f(x) = 2^x$ and $g(x) = 2 + 2^x$

49. $f(x) = e^x$ and $g(x) = e^{-x}$

50. $f(x) = e^x$ and $g(x) = -e^x$

51. $f(x) = 2^x$ and $g(x) = 2 - 2^x$

52. $f(x) = 2^x$ and $g(x) = 2^{-x}$

53. $f(x) = 2^x$ and $g(x) = -2^{-x}$

54. $f(x) = 2^x$ and $g(x) = 2^{x-2}$

55. $f(x) = 2^x$ and $g(x) = 2^{x+2} - 4$

56. Changing forms. Show that $\dfrac{1}{e^t + e^{-t}} = \dfrac{e^t}{e^{2t} + 1}$.

Suggestion: Multiply top and bottom by e^t.

57. Changing forms. Show that $\sqrt{e^{3t} + e^{2t}} = e^t \sqrt{e^t + 1}$.

58. Logistic function. A logistic function is a function of the form $y = \dfrac{A}{1 + Be^{-rt}}$, with A, B, and r positive. Logistic functions are often used to model population growth under environmental constraints. Show that the logistic function can be rewritten as $y = \dfrac{Ae^{rt}}{e^{rt} + B}$.

59. Hyperbolic sine and cosine. The hyperbolic sine and cosine functions are denoted by sinh and cosh, respectively. They are defined as follows:

$$\sinh x = \frac{e^x - e^{-x}}{2}$$

$$\cosh x = \frac{e^x + e^{-x}}{2}$$

a. Show that $\sinh 0 = 0$ and $\cosh 0 = 1$.

b. Show the identity $\cosh^2 x - \sinh^2 x = 1$.

c. Show the identity $\sinh 2x = 2 \cosh x \sinh x$.

d. Show the identity $\cosh 2x = \cosh^2 x + \sinh^2 x$.

Solving equations graphically. Exercises 60 through 67 involve equations that are not easily solved by hand calculation. Use a graphing utility to make an appropriate graph, and solve the given equation. (Round your answers to two decimal places.)

60. $2^x + 3^x = 50 - x$

61. $2^x + 10 = 3^x$

62. $2^x = x^3$. There are two solutions.

63. $e^x = x + 2$. There are two solutions.

64. $e^{-x} + 2 = e^x$

65. $e^x + e^{-x} = 3^x$

66. $5xe^{-x} = 1$. There are two solutions.

67. $2^{x^2} - 2^x = x$. There are two solutions.

MODELS AND APPLICATIONS

68. Growth by cell division. Bacteria in a Petri dish grow by cell division. The number of bacteria cells present after t hours is given by

$$N(t) = 3000(2^t)$$

a. How many cells are initially present, and how many are present after 3 hours?

b. Explain how the bacteria population changes with time.

c. Make a graph of the bacteria population over the first 4 hours.

69. A declining population. A certain bacteria population is given by the model $N(t) = 3000\left(\dfrac{1}{2}\right)^t$, where t is time measured in hours. Explain how the population changes with time, and make a graph of the population over the first 4 hours.

70. Salary. You get raises each year, and your salary S after t years is given by an exponential function:

$$S(t) = 50,000(1.05^t) \text{ dollars}$$

a. What is your salary after 10 years?

b. What is your percentage raise each year? *Suggestion*: See part b of Example 3.15.

71. Toxic waste. A cleanup operation of a toxic substance in a dangerous waste site is in progress.

The amount A (in grams) of the toxic substance remaining after t weeks is given by

$$A = 8000e^{-0.3t}$$

a. How much of the toxic substance was present at the beginning of the cleanup operation?

b. How much did the amount decrease over the first week?

c. How much did it decrease from week 1 to week 2?

d. Plot the graph of the amount over the first 10 weeks of the cleanup operation.

e. Explain in terms of the shape of your graph in part d the difference between the answers to parts b and c.

72. U.S. population. The U.S. population from 1800 to 1860 is closely modeled by

$$N(t) = 5.34e^{0.03t} \text{ million}$$

where t is the time in years since 1800.

a. Plot the graph of the U.S. population from 1800 to 1860.

b. If the U.S. population continued to follow the same model until 2030, what would the population on that date be?

73. **Inflation.** As a result of inflation, the price of a soda is increasing according to the model $P(t) = 0.75(1.02^t)$. Here P is measured in dollars, and t is time measured in years.

   a. How is the price of a soda next year related to the price this year?

   b. By what percentage do soda prices increase each year?

   A graph of soda prices is shown in **Figure 3.23**.

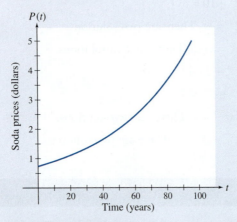

Figure 3.23 Soda prices

74. **Uranium-238.** The amount U of uranium-238 remaining after t billion years of decay is given by

$$U(t) = U_0 \left(\frac{1}{2}\right)^{t/4.5}$$

   where U_0 is the initial amount of uranium-238. What is the half-life of uranium-238? A graph showing the amount of uranium-238 remaining versus time is shown in **Figure 3.24**.

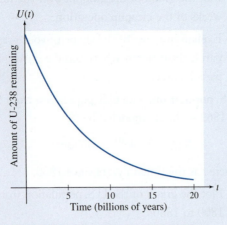

Figure 3.24 Amount of uranium-238

75. **A skydiver.** The downward velocity of a skydiver is modeled by

$$V(t) = 120(1 - e^{-0.29t})$$

Here V is measured in miles per hour, and t is the time (in seconds) into the fall.

   a. Make a graph of the velocity versus time over the first 30 seconds of the fall.

   b. Terminal velocity is the fastest speed the skydiver will attain no matter how long the fall. Use your graph to estimate terminal velocity, and then use a limit to find the value exactly. *Suggestion*: The velocity after a long time approaches the limit of $V(t)$ as $t \to \infty$.

   c. The value you found in part b determines a horizontal asymptote for the graph from part a—that is, a horizontal line that the graph approaches for large values of t. Add the asymptote to the graph.

76. **The hanging chain.** If a chain is suspended by its ends between two points, it will take the shape of a catenary curve, which is the graph of

$$y = \frac{a}{2}\left(e^{x/a} + e^{-x/a}\right)$$

for some constant $a > 0$. Plot the catenary curves for $a = 10, 15,$ and 20. Make all three plots on the horizontal span from -10 to 10.

77. **Making ice.** An ice tray filled with water at room temperature is placed in a freezer. The temperature T of the water t minutes later is given by

$$T(t) = 5 + 70e^{-0.04t} \circ \text{ Fahrenheit}$$

   a. Has the water temperature reached $32°$ Fahrenheit after 30 minutes?

   b. Make a graph of the temperature of the water in the tray over the first hour.

   c. What is room temperature? *Suggestion*: Remember that the water in the tray is initially at room temperature.

   d. What is the temperature inside the freezer? *Suggestion*: After a long time (think of the limit as $t \to \infty$), the temperature of the water will approach the temperature of the freezer.

   e. The value you found in part d determines a horizontal asymptote of the graph—that is, a horizontal line that the graph approaches for large values of t. Add the asymptote to the graph you made in part b.

78. Logistic growth. Logistic functions such as

$$N(t) = \frac{1000}{1 + 20e^{-0.3t}}$$ are often used to model

population growth. Suppose a population is modeled by this function, with time t measured in years.

a. Plot the graph of the population over the first 30 years.

b. Describe the growth of the population over the 30-year period.

c. Logistic population growth always has a limiting value, known as the environmental carrying capacity. It is the largest population that the environment can support. Find the value of the environmental carrying capacity for this population, and add the line corresponding to this value to your graph from part a.

CHALLENGE EXERCISES FOR INDIVIDUALS OR GROUPS

79. Average rate of change of e^x.

a. Calculate the average rate of change of $f(x) = e^x$ from $x = t$ to $x = t + h$ for $h \neq 0$.

b. Factor your answer from part a to write the average rate of change of $f(x) = e^x$ from $x = t$ to $x = t + h$ as a multiple of e^t.

c. Use your answer from part b to calculate the average rate of change of $f(x) = e^x$ from $x = t$ to $x = t + h$ as a multiple of e^t for $h = 10^{-2}$, $h = 10^{-3}$, and $h = 10^{-4}$. Report your answers using six digits of accuracy.

d. Use your answer from part c to describe what happens to the average rate of change of $f(x) = e^x$ from $x = t$ to $x = t + h$ as the length of the interval gets very small—that is, when h is near zero.

e. Use your answer from part d to find the instantaneous rate of change of $f(x) = e^x$ at $x = t$.

80. Exponential functions dominate power functions. Fix $c > 0$. For any positive integer x, we can expand $(1+c)^x$ using a formula known as the binomial expansion. Special cases are as follows:

$$(1+c)^2 = 1 + 2c + c^2$$

$$(1+c)^3 = 1 + 3c + 3c^2 + c^3$$

$$(1+c)^4 = 1 + 4c + 6c^2 + 4c^3 + c^4$$

For our purposes here, we need only part of the general formula:

$$(1+c)^x = 1 + cx + \text{Remainder}$$

It turns out that if x is greater than 1, the "Remainder" is positive, so

$$(1+c)^x > 1 + cx$$

This is known as Bernoulli's inequality. Bernoulli's inequality holds whenever x is any real number greater than 1.

a. In this section we stated that if $b > 1$, then we can make b^x as large as we like by choosing x to be large. According to Bernoulli's inequality, how large can we choose x in order to be sure that $b^x > 1000$?

Suggestion: Your answer should be of the form

$$x > \text{Something that depends on } b$$

First write b as $1 + c$ (so $c = b - 1 > 0$), and then use Bernoulli's inequality.

b. We discussed in this section the fact that increasing exponential functions dominate power functions. That is, if $b > 1$, then b^x is eventually larger than x^k no matter what the value of k is. In this exercise we show why this is true. To get started, note that if $b > 1$, then $b = 1 + c$ for some positive number c, namely $b - 1$. We establish the desired inequality in several steps. Throughout, we assume that $k > 0$ because otherwise the desired inequality is easy.

Step 1: Assume that $x > k + 1$. Use Bernoulli's inequality with exponent $\dfrac{x}{(k+1)}$ to show that

$$(1+c)^{x/(k+1)} > 1 + \frac{cx}{k+1}$$

and deduce that

$$(1+c)^{x/(k+1)} > \frac{cx}{k+1}$$

Step 2: Show that

$$(1+c)^x > \left(\frac{x}{k+1}\right)^{k+1} c^{k+1}$$

if $x > k+1$. *Suggestion*: Raise both sides of the inequality in step 1 to the power $k+1$.

Step 3: Show that if $x > \left(\dfrac{k+1}{c}\right)^{k+1}$ and $x > k+1$, then

$$\left(\frac{x}{k+1}\right)^{k+1} c^{k+1} > x^k$$

Suggestion: First use the fact that

$$\left(\frac{x}{k+1}\right)^{k+1} c^{k+1} = x^k x \left(\frac{c}{k+1}\right)^{k+1}.$$

Step 4: Complete the argument. That is, suppose that $b > 1$ and k is given, and show that if x is greater than both $k+1$ and $\left(\dfrac{k+1}{b-1}\right)^{k+1}$, then we have $b^x > x^k$.

c. The results of part b show us a way to make b^x dominate x^k. How large does part b tell us we can choose x to ensure that 3^x will dominate x^5?

REVIEW AND REFRESH: Exercises from Previous Sections

81. From Section P.1: Find the distance between the points $(1, 2)$ and $(3, -1)$.

Answer: $\sqrt{13}$

82. From Section P.3: Solve the inequality $x^2 - x \le 0$.

Answer: $[0, 1]$

83. From Section 1.2: If $f(x) = \dfrac{1}{x}$, find the average rate of change of f from $x = a$ to $x = b$. Assume that $a \ne b, a \ne 0$, and $b \ne 0$.

Answer: $-\dfrac{1}{ab}$

84. From Section 1.2: Calculate the average rate of change of $f(x) = mx + b$ from $x = p$ to $x = q$. Assume that $p \ne q$.

Answer: m

85. From Section 1.3: If f is increasing and concave up, explain how f changes over its domain.

Answer: The function f is increasing at an increasing rate.

86. From Section 1.4: Find the limit as $x \to \infty$ of $y = \dfrac{4}{2 + \frac{1}{x}}$.

Answer: 2

87. From Section 2.2: If $f(x) = \dfrac{1}{x}$, find a formula that gives the inverse function $f^{-1}(x)$.

Answer: $f^{-1}(x) = \dfrac{1}{x}$

88. From Section 2.3: Explain how to get the graph of $f(-x)$ from the graph of $f(x)$.

Answer: The graph of $f(-x)$ is the graph of $f(x)$ reflected about the vertical axis.

89. From Section 3.1: Find the equation of the linear function $f(x)$ so that $f(1) = 4$ and f has slope 3.

Answer: $f(x) = 3x + 1$

90. From Section 3.1: Find the equation of the secant line for $f(x) = x^2$ for the interval $[2, 4]$.

Answer: $y = 6x - 8$

91. From Section 3.1: What is the slope of the linear function $f(x)$ such that $f(1) = 5$ and $f(7) = -2$?

Answer: $-\dfrac{7}{6}$

3.1 Linear Functions: Constant Rate of Change

3.2 Exponential Functions: Constant Proportional Change

3.3 Modeling with Exponential Functions

3.3 Modeling with Exponential Functions

Constant proportional change helps us recognize when exponential models are appropriate.

In this section, you will learn to:

1. Analyze exponential functions, taking into account that they exhibit constant percentage change.
2. Apply exponential functions in everyday settings.

In this section we show applications of exponential functions in a range of settings, including population dynamics, inflation, and radioactive decay. Exponential functions are evident throughout mathematics, engineering, science, and industry—and especially prevalent in finance, where we will focus most of our examples.

Compound interest, where accrued interest is added to the principal, is virtually the only type of interest that any consumer or investor is likely to encounter, and compounding leads to exponential growth. That means virtually all substantial financial transactions you will ever make in your life involve exponential functions. This includes savings accounts, car loans, home mortgages, credit cards, student loans, cell phone contracts, and lots more. If you buy that sporty car you have your eye on, you might have to make monthly payments. The amount of your monthly payment is calculated using exponential functions.

Constant Percentage Change

> Constant proportional change can be thought of as constant percentage change, and this is the feature most often recognized in applications.

The most common descriptions of exponential functions in everyday life are given in terms of percentage increase or decrease. Suppose that you receive a 3% raise in salary each year. Then next year's salary will be 103% of this year's salary, which means we multiply the salary by 1.03:

$$\text{New salary} = \text{Old salary} \times 1.03$$

We learned in the previous section that constant proportional change leads to an exponential function. Thus, using S_0 for your initial salary, we can model the salary S after t years using

$$S(t) = S_0 \times 1.03^t$$

Now imagine a job in which the salary decreases by 3% each year. So next year's salary is 97% of this year's salary:

$$\text{New salary} = \text{Old salary} \times 0.97$$

This equation indicates the salary model

$$S(t) = S_0 \times 0.97^t$$

These ideas apply in any setting involving constant percentage change.

CONCEPTS TO REMEMBER: Constant Percentage Change

In applications, constant (but not 0) percentage change typically indicates an exponential model, $y = ab^t$.

- A constant percentage increase of r as a decimal for each unit increase in t means that the base is $b = 1 + r$, so

$$y = a(1 + r)^t$$

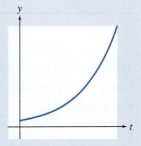

- A constant percentage decrease of r as a decimal for each unit increase in t means that the base is $b = 1 - r$, so

$$y = a(1-r)^t$$

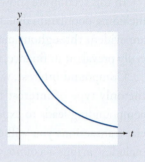

EXAMPLE 3.17 U.S. Population

a. From 1800 through 1860, the U.S. population grew at an estimated rate of 3% per year. The population in 1800 was about 5.34 million. Model the U.S. population from 1800 to 1860 using t to denote the time in years since 1800, and include a graph.

b. If population growth had continued at this rate, what would be the U.S. population in 2020?

SOLUTION

a. The yearly population growth rate of 3% is $r = 0.03$ in decimal form, so the base for our model is

$$b = 1 + 0.03 = 1.03$$

Because the initial value is 5.34 million, we have the model

$$N(t) = 5.34 \times 1.03^t \text{ million}$$

The graph is in **Figure 3.25**.

b. The year 2020 is 220 years since 1800. Thus, we use the value $t = 220$ in our model to find the indicated population:

$$N(220) = 5.34 \times 1.03^{220}$$

$$\approx 3562.30 \text{ million}$$

This value is about 3.6 billion, which is more than 10 times the current U.S. population. In fact, the percentage growth of the population slowed in the 1860s and never again reached the level of 3% per year. The current population growth rate is less than 1% per year.

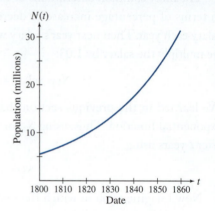

Figure 3.25 U.S. population from 1800 to 1860

TRY IT YOURSELF 3.17 Brief answers provided at the end of the section.

In 2011, Latvia had a population of 2.1 million, but the population is declining by 1% each year. Find a model for the population of Latvia.

EXTEND YOUR REACH   You see a news report about a population that grows by 2% each year. Explain briefly why an exponential model, not a linear model, would be appropriate for this population.

EXAMPLE 3.18 Inflation

Inflation is the percentage increase in the cost of a market basket of goods as defined by the U.S. Bureau of Labor Statistics. Inflation rates vary from year to year, but, for purposes of this example, let's assume inflation from 2020 to 2040 to be a constant 2% per year.

a. The cost of a suit in 2020 was $250. If the cost of the suit increases according to the inflation rate, make a model that gives the cost C of the suit t years after 2020.

b. How much would this suit cost in 2040?

SOLUTION

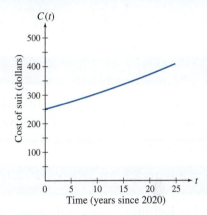

a. An increase of 2% each year expressed as a decimal is $r = 0.02$. This gives a base of

$$b = 1 + 0.02 = 1.02$$

The initial value of $250 leads to the following model for the cost C:

$$C(t) = 250 \times 1.02^t \text{ dollars}$$

b. To find the cost in 2040, we use $t = 20$ (because 2040 is 20 years after 2020):

$$C(20) = 250 \times 1.02^{20}$$

$$\approx 371.49 \text{ dollars}$$

Thus, the suit would cost $371.49 in 2040.

Figure 3.26 Cost of a suit over time

A graph of suit cost versus time is shown in **Figure 3.26**. Note that the cost rises more rapidly as time passes.

TRY IT YOURSELF 3.18 Brief answers provided at the end of the section.

Find the cost of the suit in 2040 if inflation had been 4% per year rather than 2% per year.

EXAMPLE 3.19 Uranium-238

As we saw in the preceding section, radioactive substances decay at a rate proportional to the amount present, and the half-life of such a substance is the time required for half of the substance to decay. The half-life of uranium-238 is 4.5 billion years.

a. What percentage of an initial amount of uranium-238 will remain after 4.5 billion years? After 9 billion years? After 18 billion years?

b. Make a model that gives the amount U of uranium-238 remaining after h half-lives. Use U_0 for the initial amount.

SOLUTION

a. Because 4.5 billion years is one half-life, half or 50% of the original amount will remain after 4.5 billion years. Now 9 billion years amounts to an additional half-life, so during that period half again will decay, leaving 25% of the original amount. Similar reasoning tells us that after 18 billion years or four half-lives 6.25% will remain.

b. For each half-life that passes, the amount is reduced by 50%, which is $r = 0.5$ as a decimal. This gives a base of $b = 1 - 0.5 = 0.5$. The amount remaining after h half-lives is then

$$U = U_0 \times 0.5^h$$

A graph of the amount of uranium-238 versus time measured in half-lives is shown in **Figure 3.27**. The graph assumes an initial amount of 10 grams. Note that the amount remaining decreases more slowly as time passes.

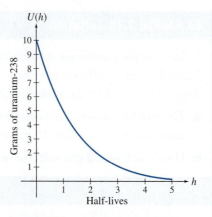

Figure 3.27 Amount of uranium-238 remaining over time

TRY IT YOURSELF 3.19 Brief answers provided at the end of the section.

Cesium-135 is present in spent nuclear reactor fuel rods. Using C_0 as the initial amount, make a model of the amount C remaining after h half-lives.

EXAMPLE 3.20 From Half-Lives to Minutes

The half-life of a drug is the time required for the amount in the bloodstream to reduce by half. Amoxicillin, a common antibiotic, has a half-life of 62 minutes. The amount of amoxicillin after h half-lives is given by $A = A_0 \times 0.5^h$. Here A_0 is the initial amount.

a. Adjust this model so that it gives the amount after t minutes.

b. If there are initially 70 milligrams of amoxicillin present, how much remains after 120 minutes?

SOLUTION

a. Because one half-life is 62 minutes, 1 minute is $\dfrac{1}{62}$ half-life. So t minutes is $\dfrac{t}{62}$ half-lives. This means that to convert half-lives to minutes, we replace h by $\dfrac{t}{62}$ in the formula $A = A_0 \times 0.5^h$. The resulting formula is

$$A = A_0 \times 0.5^{t/62}$$

b. Because the initial amount is $A_0 = 70$ milligrams, we use the model $A = 70 \times 0.5^{t/62}$ milligrams. We use $t = 120$ to find the amount remaining after 120 minutes:

$$A(120) = 70 \times 0.5^{120/62} \approx 18.30 \text{ milligrams}$$

TRY IT YOURSELF 3.20 Brief answers provided at the end of the section.

The half-life of a certain drug is 30 minutes. Find a formula that gives the amount C of the drug present after t minutes. Use C_0 for the initial amount.

EXTEND YOUR REACH   The amount of a drug in the bloodstream decays according to an exponential function $y = ab^t$, where $b < 1$. For a certain drug, it is found that 10% of the original amount has decayed after 2 hours. What percentage has decayed after 1 hour? Note that the correct answer is not 5%.

Compound Interest

> Compound interest arises in many financial transactions, and it is appropriately modeled by an exponential function.

When you invest in a savings account or borrow from a lending institution, there is a stated interest rate called the annual percentage rate (APR). But the savings institution usually compounds the interest on the balance. That is, when interest is accrued, it is added to your balance, and in the future interest is applied to the increased amount.

The accrued interest is added to the balance on a regular basis, called the compounding period, which is often monthly or daily. If there are n periods per year, then each period an interest rate of

$$\text{Period rate} = \frac{\text{APR}}{n}$$

is applied.

Suppose, for example, that you use \$5000 to open a savings account with the APR of 1.2%, and suppose interest is compounded monthly. Then $n = 12$, the interest rate is applied each month, and the monthly rate is

$$\text{Monthly rate} = \frac{\text{APR}}{12} = \frac{1.2\%}{12} = 0.1\%$$

A constant percentage increase of 0.1% per month indicates an exponential model with base $1 + 0.001 = 1.001$. The initial value is \$5000, so we can find the balance of the savings account after m months using $B(m) = 5000 \times 1.001^m$ dollars. Because t years is $12t$ months, the balance after t years is given by

$$B(t) = 5000 \times 1.001^{12t} \text{ dollars}$$

In general, the balance under compound interest is an exponential function. If the APR is written in decimal form, the percentage increase per compounding period is the period rate $\dfrac{\text{APR}}{n}$. The formula for the balance is summarized in the following.

CONCEPTS TO REMEMBER: Compound Interest Formula

If the initial balance is B_0 and interest is compounded n times each year, then the balance after t years is given by

$$B(t) = B_0 \left(1 + \frac{\text{APR}}{n}\right)^{nt}$$

Here the APR is in decimal form.

Note that this formula and others in this section are sensitive to roundoff error, so it is best to keep all the decimal places rather than to enter parts of the formula that you have rounded.

EXAMPLE 3.21 Calculating Growth with Monthly Compounding

Suppose you deposit $10,000 in a savings account that pays 2.4% APR compounded monthly.

a. Find a model that gives your account balance after t years.

b. Suppose you are saving for retirement, and you make an initial deposit of $10,000 when you are 25 years old. How much money is in your account when you reach age 65?

SOLUTION

a. The APR as a decimal is 0.024, and the initial amount is $B_0 = \$10,000$. We are compounding monthly, so $n = 12$. If B denotes the balance in dollars after t years, then

$$B = B_0 \left(1 + \frac{\text{APR}}{n}\right)^{nt}$$

$$= 10,000 \left(1 + \frac{0.024}{12}\right)^{12t}$$

$$= 10,000 \times 1.002^{12t}$$

b. When you reach age 65, you will have saved for 40 years. So the balance is

$$B(40) = 10,000 \times 1.002^{12 \times 40}$$

or about $26,091.94.

TRY IT YOURSELF 3.21 Brief answers provided at the end of the section.

You have an initial investment of $1000 that earns an APR of 1.8% compounded monthly. Find a model that gives the balance after t years.

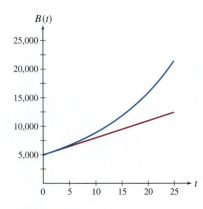

Figure 3.28 Comparing simple interest with compound interest

We can show graphically the effect of compounding. **Figure 3.28** shows the balance over a 25-year period of two investments, one earning 6% annually with no compounding (called simple interest), and the other earning an APR of 6% compounded monthly. (A rate of 6% is historically high, but we use it for illustration.) The straight line represents the first investment. As the graph indicates, the effect of compounding is greater over longer time periods.

Although many consumer transactions involve monthly compounding, other compounding periods may be used as well. For example, daily compounding is often used for savings accounts.

EXAMPLE 3.22 Calculating Growth for Other Compounding Periods

You invest $5000 in an account that pays compound interest with an APR of 2.4%. Make a table of values that shows the account balance after 10 years for each of the following compounding frequencies: semiannually, quarterly, monthly, and daily. Assume there are 365 days in every year.

SOLUTION

If there are n compounding periods in a year, then the balance after t years is

$$B = 5000\left(1 + \frac{0.024}{n}\right)^{nt} \text{ dollars}$$

To find the balance after 10 years, we use $t = 10$ in this formula. We obtain the following table.

Frequency	Number n periods per year	Formula for $B(t)$	$B(10)$
Semiannually	2	$5000\left(1 + \dfrac{0.024}{2}\right)^{2t}$	$6347.17
Quarterly	4	$5000\left(1 + \dfrac{0.024}{4}\right)^{4t}$	$6351.69
Monthly	12	$5000\left(1 + \dfrac{0.024}{12}\right)^{12t}$	$6354.72
Daily	365	$5000\left(1 + \dfrac{0.024}{365}\right)^{365t}$	$6356.20

TRY IT YOURSELF 3.22 Brief answers provided at the end of the section.

Find the balance after 10 years if interest is compounded hourly.

What happens when the number of compounding periods gets larger and larger? Certainly the amount of money you earn increases, but is there any limit to the amount of this growth? What if you compound every hour (8760 times a year) or every minute (525,600 times a year)? If we imagine letting the number of compounding periods increase indefinitely, what happens to the interest we earn? This limiting situation is called continuous compounding.

To see what happens, we write r for the APR as a decimal and make the substitution $n = rk$ in the account balance formula:

$$B = B_0\left(1 + \frac{r}{n}\right)^{nt} \qquad \blacktriangleleft \text{ Use the compound interest formula.}$$

$$= B_0\left(1 + \frac{r}{rk}\right)^{rkt} \qquad \blacktriangleleft \text{ Substitute } rk \text{ for } n.$$

$$= B_0\left(\left(1 + \frac{1}{k}\right)^{k}\right)^{rt} \qquad \blacktriangleleft \text{ Simplify using } A^{kp} = (A^k)^p.$$

We know from the previous section that $\left(1 + \dfrac{1}{k}\right)^{k} \to e$ as $k \to \infty$. Hence, the limiting value of the balance formula is

$$B = B_0 e^{rt}$$

The special number e arises naturally in the context of compound interest, and the balance formula for continuous compounding is the simplest of the balance formulas.

CONCEPTS TO REMEMBER: Formula for Continuous Compounding

If an initial balance of B_0 is subject to an APR of r as a decimal, and if interest is compounded continuously, then the balance after t years is given by

$$B(t) = B_0 e^{rt}$$

EXAMPLE 3.23 Yearly versus Continuous Compounding

You invest $5000 in an account. For the sake of illustration, take the APR to be 6%.

a. Find a formula that gives the balance M, in dollars, if interest is compounded yearly.

b. Find a formula that gives the balance C, in dollars, if interest is compounded continuously.

c. On the same coordinate axes, plot the graphs of M and C over a 50-year period.

d. What do the graphs in part c tell you about yearly versus continuous compounding?

SOLUTION

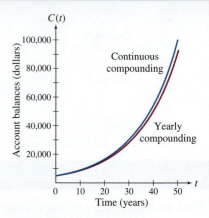

DF **Figure 3.29** Yearly versus continuous compounding

a. For yearly compounding, we use $n = 1$ in the original compound interest formula:

$$M = 5000 \times (1 + 0.06)^t$$

b. For this part, we use $r = 0.06$ in the formula involving continuous compounding:

$$C = 5000e^{0.06t}$$

c. The graphs are shown in **Figure 3.29**.

d. The graphs show that the balance grows faster with continuous compounding and that the difference is greater for longer-term investments.

TRY IT YOURSELF 3.23 Brief answers provided at the end of the section.

What is the difference between the balances of the two accounts after 50 years?

Installment Loans

> Installment loans are modeled by functions that are combinations of exponential functions.

Most consumer loans — for buying a car or a home, for example — are installment loans. With an installment loan you make fixed monthly payments, but at the same time interest is accruing on the outstanding balance of your loan. This makes the calculation of your monthly payment a bit complicated. The monthly payment M, in dollars, on an installment loan is given by the formula

$$M = \frac{B_0 r (1 + r)^p}{((1 + r)^p - 1)}$$

- $r = \dfrac{\text{APR}}{12}$ is the monthly interest rate as a decimal.

- p is the term of loan in months (the number of payments to be made).

- B_0 is the amount borrowed, in dollars.

The derivation of this formula is in Appendix 4. Note that this formula is particularly sensitive to roundoff error.

EXAMPLE 3.24 Calculating Your Monthly Payment

Suppose you secure a 30-year mortgage of $275,000 at an APR of 4.8%.

a. What is your monthly payment?

b. Compare the cost of the loan (the total interest paid) with the size of the mortgage.

SOLUTION

a. The monthly rate as a decimal is

$$\text{Monthly rate} = r = \frac{0.048}{12} = 0.004$$

The amount borrowed is $275,000, so $B_0 = 275,000$. Also, 30 years is 360 months, so we put $p = 360$ in the formula. The monthly payment is

$$\text{Monthly payment} = M = \frac{B_0 r (1+r)^p}{((1+r)^p - 1)}$$

$$= \frac{275,000 \times 0.004 \times 1.004^{360}}{(1.004^{360} - 1)}$$

To the nearest cent, this is $1442.83.

b. You make 360 payments of $1442.83 for a total of $360 \times 1442.83 = 519,418.80$ dollars paid. All but $275,000 of that is interest. This means that the loan costs $244,418.80, which is about 10% less than the amount borrowed.

TRY IT YOURSELF 3.24 Brief answers provided at the end of the section.

Many student loans are treated as installment loans. You borrow $10,000 at an APR of 6.6%, and you pay off the loan in 10 years. Find your monthly payment, and determine how much total interest you pay on this student loan.

STEP-BY-STEP STRATEGY: Monthly Payment Calculation

Step 1 Calculate r, the monthly interest rate, as $\frac{APR}{12}$. Take care to use the APR as a decimal and not to round r.

Step 2 Calculate p, the term of the loan in months. Take care to multiply by 12 if the term is given to you in years.

Step 3 Use B_0, the amount borrowed, and the results of the steps above to calculate M, the monthly payment:

$$M = \frac{B_0 r (1+r)^p}{((1+r)^p - 1)}$$

Logistic Models

Logistic models may apply when factors limit continued exponential growth.

The (quite reasonable) assumption that the percentage growth rate of a population is approximately constant leads to the use of exponential functions to model population growth. But exponential functions with a base larger than 1 eventually show rapid growth.

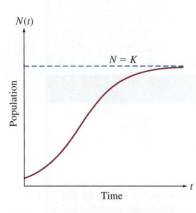

Figure 3.30 A typical logistic graph

Consequently, such a model will, in the long term, predict unrealistic population levels, so exponential models for population growth are useful only for limited time periods.

Professionals who study population dynamics use more realistic, and often more complicated, models. One important model, known as the logistic model, takes into account the environmental carrying capacity, which is the largest population the environment can support. If the environmental carrying capacity is K individuals, then the logistic model for the population as a function of time t is

$$N(t) = \frac{K}{1 + be^{-rt}}$$

Here $b = \dfrac{K}{N_0} - 1$, where N_0 is the initial population, and r is the intrinsic per capita growth rate of the population. This number is known to biologists, aptly enough, as the r-value of the population. It is also called the relative growth rate. If t is measured in years, then r is measured as per year. A typical graph is shown in **Figure 3.30**.

EXAMPLE 3.25 The George Reserve

The deer population on the George Reserve in Michigan has been studied extensively. The environmental carrying capacity of the reserve is about 177 deer, and the r-value for deer on this reserve is about 0.8 per year. Take the initial population to be 6 deer.

a. Find a logistic model for the George Reserve deer population.

b. Make a table of values, and sketch the graph of your model over the first 15 years. Include the line $y = K$ representing the environmental carrying capacity.

SOLUTION

a. The initial population is $N_0 = 6$ deer, and the carrying capacity is $K = 177$. We use these values to find the constant b:

$$b = \frac{K}{N_0 - 1}$$
$$= 28.5$$

Using the r-value of 0.8 per year, we obtain the formula

$$N(t) = \frac{K}{1 + be^{-rt}}$$
$$= \frac{177}{1 + 28.5e^{-0.8t}}$$

Here the time t is measured in years.

b. The following table is used to make the graph in **Figure 3.31**.

t	N(t)
0	6
3	49.37
6	143.37
9	173.31
12	176.66
15	176.97

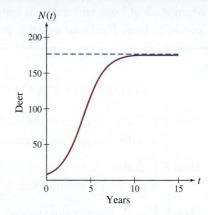

DF Figure 3.31 Deer population on the George Reserve

TRY IT YOURSELF 3.25 Brief answers provided at the end of the section.

Describe in words how the deer population changes with time.

a. Add to the graph in Figure 3.31 the graph of the deer population if there were initially 230 deer on the reserve.

b. Compare the two graphs from part a. How does the notion of environmental carrying capacity help to explain the similarities and differences between these two graphs?

EXTEND YOUR REACH

EXAMPLE 3.26 Finding Carrying Capacity

Assume that a population grows according to the logistic model

$$N = \frac{K}{1 + be^{-rt}}$$

Show that the population will eventually approach the environmental carrying capacity K. That is, show that K is the limit as $t \to \infty$ of $N(t)$. (Assume that r is positive.)

SOLUTION

We know from the previous section that $e^{-t} \to 0$ as $t \to \infty$. Then,

$$\text{as } t \to \infty, \quad e^{-rt} = \left(e^{-t}\right)^r \to \left(e^{-t^{\,0}}\right)^r = 0^r = 0$$

We use this calculation to find the limit of $N(t)$:

$$\text{as } t \to \infty, \quad \frac{K}{1 + be^{-rt}} \to \frac{K}{1 + b\,e^{-rt^{\,0}}} = \frac{K}{1 + b \times 0} = K$$

We conclude that $N(t) \to K$ as $t \to \infty$.

TRY IT YOURSELF 3.26 Brief answers provided at the end of the section.

What is the limiting size of a population that grows according to $N = \dfrac{2000}{1 + 20e^{-0.07t}}$?

Exponential functions are those having a constant proportional rate of change. Logistic functions can be characterized in terms of rates of change as well. If a population N is growing exponentially, its growth rate G is proportional to N, so we can write $G = rN$ or

$$\frac{G}{N} = r$$

for some constant r. Thus, the per capita growth rate is constant. As we noted earlier, exponential growth is an unrealistic model for the long term. Logistic growth is a more realistic model. It assumes that the per capita growth rate decreases as a linear function of the population N. This model has the form

$$\frac{G}{N} = r\left(1 - \frac{N}{K}\right)$$

It says that the growth rate is 0 at the carrying capacity K, so the population will stabilize there.

TRY IT YOURSELF ANSWERS

3.17 $N(t) = 2.1 \times 0.99^t$ million if t is time in years since 2011

3.18 $547.78

3.19 $C = C_0 \times 0.5^b$

3.20 $C = C_0 \times 0.5^{t/30}$

3.21 $B(t) = 1000 \times 1.0015^{12t}$ dollars

3.22 $6356.24

3.23 After 50 years, continuous compounding yields $8326.91 more than yearly compounding.

3.24 The monthly payment is $114.06. Total interest paid is $3687.20.

3.25 The graph shows that the population grew rapidly at first, but population growth slowed later. By year 10, the population neared the environmental carrying capacity.

3.26 The limiting size is 2000 individuals.

EXERCISE SET 3.3

CHECK YOUR UNDERSTANDING

1. Constant percentage change is indicative of what type of function?

2. Constant percentage increase of r as a decimal indicates a function of what form?

3. Constant percentage decrease of r as a decimal indicates a function of what form?

4. In one half-life, the amount is cut in half. After two half-lives, how much of the original amount remains?

5. The term used when accrued interest is periodically added to the principal is _____ interest.

6. Consider an installment loan with monthly compounding of interest. On what three factors does the monthly payment depend?

7. **True or false:** More frequent compounding increases accrued interest.

8. **True or false:** When the number of compounding periods increases without bound, the interest accrued in a year also increases without bound.

9. **True or false:** For a radioactive substance, the rate of decay is proportional to the amount present.

10. **True or false:** Logistic population growth leads to arbitrarily large populations.

SKILL BUILDING

Constant percentage change. In Exercises 11 through 16, find an exponential model that satisfies the given conditions.

11. The initial value is 400. There is a constant percentage increase of 2% for each unit increase in t.

12. The initial value is 500. There is a constant percentage decrease of 2% for each unit increase in t.

13. The initial value is 600. There is a constant percentage decrease of 8% for each unit increase in t.

14. The initial value is 400. There is a constant percentage increase of 13% for each unit increase in t.

15. The initial amount is y_0. There is a constant percentage increase of 7% for each unit increase in t.

16. The initial value is y_0. There is a constant percentage decrease of 12% for each unit increase in t.

Half-life. In Exercises 17 through 20, find an exponential model that satisfies the given conditions. Write the model first as a function of the half-life, then as a function of the given unit of time.

17. The initial value is 200. The half-life is 8 years.

18. The initial value is 18. The half-life is 6 minutes.

19. The initial value is 10. The half-life is one-third of a day.

20. The initial value is y_0. The half-life is 12 years.

Compound interest. In Exercises 21 through 28, find a function that gives the balance B of the given investment in terms of the time t in years. Use B_0 for the initial balance.

21. APR of 1.8% compounded monthly

22. APR of 0.6% compounded monthly

23. APR of 0.6% compounded semiannually

24. APR of 0.6% compounded quarterly

25. APR of 0.6% compounded continuously

26. APR of 1% compounded continuously

27. APR of r as a decimal compounded continuously

28. APR of r as a decimal compounded n times per year

Using the monthly payment formula. In Exercises 29 through 32, calculate the monthly payment for the indicated loan. Round your answers to the nearest cent.

29. You borrow $5000 at an APR of 6% and repay the loan in 24 payments.

30. You borrow $8000 at an APR of 4% and repay the loan in 36 payments. (Remember not to round r—leave it as $\dfrac{0.04}{12}$.)

31. You borrow $250,000 at an APR of 3% and repay the loan over a 30-year period.

32. You borrow $124,000 at an APR of 5% and repay the loan over a 20-year period.

Matching. In Exercise 33 through 38, match the graph with the exponential function $y = ab^t$ that has the indicated property in the following list:

A. Constant decrease of 50% per unit of time

B. Half-life of 4

C. Doubles when t increases by 3

D. Initial value of 200

E. Constant increase of 50% per unit of time

F. $y(2) = 200$

33.

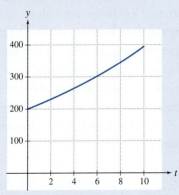

34.

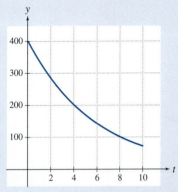

35.

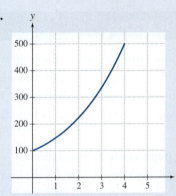

36.

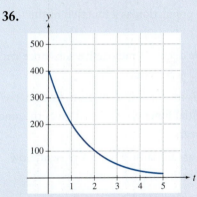

37.

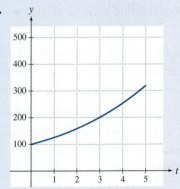

38.

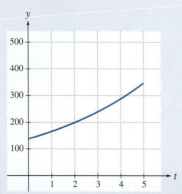

MODELS AND APPLICATIONS

39. A growing population. A population is growing at a rate of 3% per year. Using N_0 for the initial population, find an exponential function N that models the population as a function of the time t in years.

40. A declining population. A population is decreasing by 4% per year. Using N_0 for initial population, find an exponential function N that models the population as a function of the time t in years.

41. Radioactive decay. A radioactive substance decays at a rate of 2% per year.

   a. Using A_0 for the initial amount, find an exponential function $A(t)$ that gives the amount after t years.

   b. If there are initially 20 grams present, how much remains after 3 years?

42. Removing chlorine. A cleaning process reduces the amount of chlorine in contaminated water by 4% per hour.

   a. Using C_0 for the initial amount of chlorine, find an exponential function $C(t)$ that gives the amount of chlorine remaining after t hours.

   b. Make a graph of the amount of chlorine remaining if there are initially 55 grams in the water. Your graph should cover the first 30 hours of the cleaning process.

43. Salary. If your initial salary is $50,000 per year and you get a 5% raise each year, find a formula that gives your salary in dollars after t years. What is your salary after 10 years?

44. Toxic waste. There are originally 8000 grams of a toxic substance in a dangerous waste site.

The result of a cleanup operation is to decrease the amount remaining by 25% each week. Make a model of the amount T, in grams, of the toxic substance remaining after t weeks. How much is left after 3 weeks of cleanup?

45. Inflation. Suppose that inflation remains at a constant rate of 3% per year. How much will a suit that costs $500 today cost in 10 years?

46. A skydiver. The terminal velocity for an object dropped near the surface of Earth is the maximum downward velocity that will be attained by the falling object. For a man of average size, the terminal velocity is about 176 feet per second. Let $V(t)$ denote the downward velocity, in feet per second, t seconds into the fall of a skydiver who has not opened his parachute. It is found that the difference $D = 176 - V$ decreases by 25% each second. We assume that the initial velocity is 0, so the initial value of D is 176 feet per second.

   a. Find a model of the form $D = ab^t$ for D as a function of t.

   b. Find a formula for V in terms of t.

   c. What is the skydiver's downward velocity 10 seconds into the fall?

47. Newton's law of cooling. A hot pan is removed from an oven and left in the open air to cool. Let t denote the time in minutes after the pan is removed from the oven, and let $D(t)$ denote the difference between the temperature $T(t)$ of the pan and the constant temperature A of the air. (All temperatures are measured in degrees Fahrenheit.) Thus, $D = T - A$. It is found that D decreases by 12% each minute.

a. Use D_0 to denote the initial temperature difference, and find an exponential model for D as a function of t.

b. Use T_0 to denote the initial temperature, in degrees Fahrenheit, of the pan, and find a formula expressing T as a function of t.

c. Find the value of T after 10 minutes. Your answer will depend on A and T_0.

48. Spent fuel rods. Cesium-137 is produced in nuclear power plants. Suppose we start with 30 million curies of cesium-137 in a storage pool for spent fuel rods.[7]

a. Find a formula that gives the amount C, in millions of curies, of cesium-137 remaining after h half-lives.

b. How much cesium-137 remains after 3.5 half-lives?

c. The half-life of cesium-137 is 30 years. How much cesium-137 remains in the pool after 20 years?

49. A student loan. You take out a student loan of $12,000 at an APR of 5% and pay it back over 10 years as an installment loan. What is your monthly payment? How much total do you pay in interest?

50. A larger student loan. You take out a student loan of $30,000 at an APR of 6% and pay it back over 10 years as an installment loan. What is your monthly payment? How much in total do you pay in interest?

51. Comparing rates. Let's consider a $200,000 mortgage with a term of 30 years. Historical interest rates on mortgages have varied dramatically, and these can make a significant difference in monthly payment. In your calculations, remember not to round the monthly interest rate.

a. In 1981, the APR for such a mortgage reached 18%, a historic high. What would be your monthly payment if you secured this mortgage in 1981?

b. In 2000, the APR was about 8%. What would be your monthly payment if you secured this mortgage in 2000?

c. In 2012, the APR reached a historic low of about 3.5%. What would be your monthly payment if you secured this mortgage in 2012?

52. Comparing loan terms. Although a common term for a home mortgage is 30 years, some mortgages are for shorter periods. Suppose we secure a mortgage of $320,000 at an APR of 5%. (The APR may change with the term of the loan, but in this exercise we assume that it does not.) In your calculations, remember not to round $r = \dfrac{0.05}{12}$.

a. Calculate your monthly payment and the total interest paid if the term is 30 years.

b. Calculate your monthly payment and the total interest paid if the term is 20 years.

c. Calculate your monthly payment and the total interest paid if the term is 10 years.

53. Tax deductions. The interest paid on a home mortgage is deductible from your federal income tax. The actual value of your deduction depends on your marginal tax rate, which is the additional tax you pay on additional income. In this problem, we take the marginal tax rate to be 30%. In this case, a tax deduction of $1 will reduce your tax liability by 30 cents. Suppose you borrow $170,000 to buy a home at an APR of 6% and the term of the mortgage is 30 years.

a. What is your monthly payment?

b. How much interest do you pay over the life of the loan?

c. How much do you save in taxes over the life of the loan, assuming your marginal tax rate is 30%?

d. Taking into account tax savings, what is the total cost of the home? (This is the cost of the home plus the interest paid minus the tax saved.)

54. Equity. When your home is mortgaged, its value is shared between you and the lending institution. Your equity in the home is the part of the initial amount borrowed that you have paid. It is the part of the value of the home that you own. (Here we are ignoring your down payment.) If B_0 is the initial amount borrowed, the monthly rate is

---

[7]One curie is 37 billion disintegrations per second, or approximately the radioactivity of 1 gram of radium.

r as a decimal, and the term of the mortgage is p months, then after k monthly payments your equity is given by

$$E = B_0 \frac{(1+r)^k - 1}{((1+r)^p - 1)}$$

Suppose you have a 30-year mortgage of $330,000 at an APR of 6%. What is your equity after 15 years of payments? Is it half the value of the home?

55. **Rebate or dealer financing.** You need to borrow $29,000 to purchase a new car. You can get dealer financing at 4% APR, or you can take a rebate of $3000 and arrange your own financing. If you take the rebate you will need to borrow only $26,000 from your bank, but the bank's rate is 6% APR. With either option, you pay off the loan in 4 years. In your calculations, remember not to round the monthly interest rate.

   a. What is your monthly payment if you take dealer financing?

   b. What is your monthly payment if you take the rebate?

   c. Which choice is better?

   d. Which choice would be better if the bank's rate were 10%?

56. **How much can I borrow?**

   a. Your budget will allow you to make a monthly payment of M dollars for a term of t months. If the monthly interest rate is m as a decimal, find a formula that gives the amount B_0 in dollars that you can afford to borrow. *Suggestion*: Solve the monthly payment formula for B_0.

   b. Use the results from part a to determine how much you can afford to borrow if you can make payments of $275 per month for 3 years. Assume that the APR is 6%.

Logistic functions.

57. **Northern Yellowstone elk.** The carrying capacity for elk in the northern range of Yellowstone National Park was found[8] to be about 13,000. The r-value of this population was estimated to be about 0.45 per year. Take the initial population to be 3250.

   a. Make a logistic model that gives the number E of elk after t years.

   b. Make a graph of the elk population over the first 20 years. Include the horizontal line representing the carrying capacity.

58. **Varying initial populations.** A certain population grows according to the model

$$N = \frac{200}{1 + be^{-0.4t}}$$

   a. Find a formula for the model if the initial population is 50.

   b. Find a formula for the model if the initial population is 200.

   c. Find a formula for the model if the initial population is 400.

   d. Produce a picture showing the graphs of the three preceding models. Explain in words how the population changes for each of the three initial populations.

59. **An epidemic.** In a city of 2 million, there are initially 400 cases of flu. The cumulative number of infected individuals can be modeled using a logistic function. The r–value for this model is 0.3 per week. It is estimated that preventive measures can limit the cumulative number of flu cases to one-quarter of the total population.

   a. Make a logistic model that shows the cumulative number of flu cases t weeks after the outbreak of the epidemic.

   b. Plot the graph of the cumulative number of flu cases versus the time in weeks. Include the first 40 weeks of the epidemic. Add a target level of $y = 250,000$ to your graph.

   c. Estimate from the graph in part b how long it takes for the cumulative number of flu cases to reach 250,000.

60. **Tuberculosis.** The following model gives the approximate number of deaths due to tuberculosis as a fraction of all deaths in the United States t years after 1875:

$$T = \frac{0.13}{1 + 0.07e^{0.05t}}$$

Note that the r-value here is negative, which leads to a positive exponent of e. The graph of such a model is known as an inverted logistic curve.

---

[8]This exercise uses data from the study by Douglas B. Houston in *The Northern Yellowstone Elk*, Macmillan, New York, 1982.

a. Plot the graph of T over 150 years.

b. Judging from your plot, what is the limit of T as $t \to \infty$?

c. Explain what the number you found in part b means in terms of the fraction of deaths caused by tuberculosis in the United States.

Savings plans. At the end of each month, you deposit K dollars into an account that pays a monthly rate of r as a decimal. It can be shown that the balance B in dollars of the account after n months is

$$B(n) = K \left(\frac{(1+r)^n - 1}{r} \right) \text{ dollars}$$

For Exercises 61 through 64, use this formula.

61. Saving for a computer. To save for a computer, you deposit $150 at the end of each month into an account that pays 2.4% APR compounded monthly. How much do you have at the end of the year?

62. A retirement account. At age 25, you begin depositing $175 at the end of each month into an account that pays 2.4% APR compounded monthly. How much money will you have by age 65?

63. Employer contributions. At age 25, you begin depositing $175 at the end of each month into an account that pays 2.4% APR compounded monthly. Your employer matches your

contributions dollar for dollar. How much money will you have by age 65?

64. Just a bit more. Suppose that in the situation described in the preceding exercise, you deposit an extra $25 per month, but the extra amount is not matched by your employer. How much money will you have by age 65? How much will you have by age 70?

65. An amortization table. Suppose you borrow B_0 dollars at a monthly interest rate of r as a decimal, and the loan is scheduled to be paid off in p monthly payments. The balance you still owe after k payments is given by

$$B(k) = B_0 \frac{(1+r)^p - (1+r)^k}{((1+r)^p - 1)} \text{ dollars}$$

An amortization table shows the remaining balance of a loan after each payment. Suppose you borrow $12,000 at an APR of 5% and repay the loan over a period of 24 months. Use a spreadsheet to make an amortization table that shows the balance after each of the first 10 payments.

66. Finding a suitable rate. You borrow $5000 and repay it in monthly payments over a 3-year period. You can afford a monthly payment of $155.67. What is the maximum APR will you need to find to make an affordable monthly payment?

CHALLENGE EXERCISES FOR INDIVIDUALS OR GROUPS

67. Exponential population growth. It is common to use an exponential model for population growth, at least over the short term. This practice is based on the following assumptions:

- Each year approximately the same percentage of the population will die.

- Each year there is approximately the same percentage of births.

a. Explain why these assumptions indicate that it is appropriate to model the population using a multiple of an exponential function. (Ignore both immigration and emigration.)

b. Let d be the percentage (as a decimal) of deaths and b the percentage (as a decimal) of births each year. Use N_0 to denote the initial

population, and find an exponential model that describes the population.

c. Assume that the population is actually growing (rather than declining), and explain why it is unreasonable to expect the population to grow exponentially over a long period of time.

68. Logistic growth rate. At the end of the discussion of logistic functions, we stated the logistic model in terms of rates of change. If the population size is N and its growth rate is G, the model says

$$\frac{G}{N} = r \left(1 - \frac{N}{K} \right)$$

Here K is the carrying capacity. A straight-line graph showing the per capita growth rate G/N in terms of the population N is in **Figure 3.32**. Note the two points $(0, r)$ and $(K, 0)$ representing the intercepts of the line with the axes.

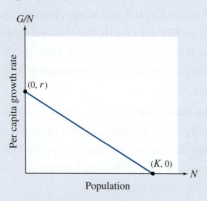

Figure 3.32 A graph of per capita growth rate as a function of population

a. Find the slope of the line in Figure 3.32 in terms of r and K.

b. Scientists measure the per capita growth rates as a certain population varies and calculate the regression line giving $\dfrac{G}{N}$ in terms of N. The slope of this line is -0.012, and the vertical intercept (or initial value) is 0.24. Use this information and your answer to part a to find the carrying capacity K.

REVIEW AND REFRESH: Exercises from Previous Sections

69. **From Section P.2:** Solve the inequality $|x - 2| \geq 4$.

Answer: $(-\infty, -2) \cup [6, \infty]$

70. **From Section 1.1:** If $f(x) = 7$ for all x, find the value of $f(4)$.

Answer: 7

71. **From Section 1.3: True or false:** If the function f has a local maximum at $x = 1$, then $f(1)$ is the largest value the function ever has.

Answer: False. It's the largest value taken near $x = 1$, but f might take larger values farther away from $x = 1$.

72. **From Section 1.3:** For a certain object, the graph of its temperature versus time is decreasing and concave up. This means that the object is (a) warming at an increasing rate, (b) warming at a decreasing rate, (c) cooling at an increasing rate, or (d) cooling at a decreasing rate.

Answer: (d)

73. **From Section 1.4:** Find the limit as $x \to \infty$ of $y = 3 - \dfrac{1}{x^2}$.

Answer: 3

74. **From Section 1.4:** Find the limit as x approaches ∞ of $f(x) = \dfrac{2x}{1 + 3x}$.

Answer: $\dfrac{2}{3}$

75. **From Section 1.4:** Suppose that $f(x)$ is a function with the property that $f(x) \to \infty$ as $x \to \infty$. Is there necessarily a solution x of the inequality $f(x) > 800$?

Answer: Yes, because $f(x)$ can be made as large as we like by choosing x sufficiently large.

76. **From Section 2.1:** If $f(x) = \dfrac{2}{x}$ for $x \neq 0$, calculate $(f \circ f)(x)$.

Answer: $(f \circ f)(x) = x$

77. **From Section 2.1:** If $h(x) = \sqrt{x^2 + 1}$, find functions $f(x)$ and $g(x)$ so that $h(x) = (f \circ g)(x)$.

Answer: One solution is $f(x) = \sqrt{x}$ and $g(x) = x^2 + 1$.

78. **From Section 2.1:** If $f(x) = \dfrac{2}{3 + x}$, find functions g and h so that $f = g \circ h$.

Answer: One solution is $g(x) = \dfrac{2}{x}$ and $h(x) = 3 + x$.

79. **From Section 2.2:** If $f(x) = \dfrac{x}{x + 1}$, find a formula for $f^{-1}(x)$.

Answer: $f^{-1}(x) = \dfrac{x}{1 - x}$

80. **From Section 2.3:** The graph of $1 + f(x + 1)$ is the same as the graph of f shifted by (a) 1 unit up and 1 unit to the right, (b) 1 unit up and 1 unit to the left, (c) 1 unit down and 1 unit to the right, or (d) 1 unit down and 1 unit to the left.

Answer: (b)

81. **From Section 3.1:** Find a formula for the linear function $f(x)$ so that $f(4) = 0$ and $f(1) = 2$.

    Answer: $y = -\dfrac{2}{3}x + \dfrac{8}{3}$

82. **From Section 3.2:** Find a formula for the exponential function $f(t)$ with base 7 and initial value 2.

    Answer: $f(t) = 2(7^t)$

83. **From Section 3.2:** Find a formula for the exponential function $f(x)$ so that $f(0) = 3$ and f is multiplied by 5 when x is increased by 1.

    Answer: $f(x) = 3(5^x)$

84. **From Section 3.2:** Calculate the limit as $x \to \infty$ of $y = e^{-x}$.

    Answer: 0

85. **From Section 3.2:** Find a formula for the exponential function $f(x)$ so that $f(1) = 6$ and $f(2) = 18$.

    Answer: $f(x) = 2(3^x)$

CHAPTER ROADMAP AND STUDY GUIDE

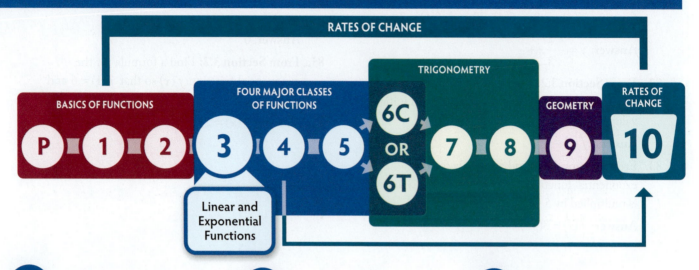

3.1 Linear Functions: Constant Rate of Change

Linear functions are widely used as models because they are the functions that show a constant rate of change.

Constant Rate of Change

The average rate of change of a linear function does not depend on the interval over which it is calculated.

Graphs of Linear Functions

The graph of a linear function is a straight line.

MODELS AND APPLICATIONS
Temperature Conversion and More

MODELS AND APPLICATIONS
Regression Lines

3.2 Exponential Functions: Constant Proportional Change

Exponential functions are used to model rapid increase or decrease because they show a constant proportional rate of change.

Constant Proportional Change

Constant proportional change means that the rate of change is proportional to the function. Intuitively, these functions change by constant multiples.

Graphs of Exponential Functions

Graphs of increasing exponential functions show rapid growth at the right tail end but approach the x-axis at the left tail end.

The Special Number e

The special number e is the base for the natural exponential function.

MODELS AND APPLICATIONS
Exponential Growth and Decay

Rapidity of Exponential Growth

In the long term, increasing exponential functions grow more rapidly than power functions.

3.3 Modeling with Exponential Functions

Constant proportional change helps us recognize when exponential models are appropriate.

Constant Percentage Change

Constant proportional change can be thought of as constant percentage change, and this is the feature most often recognized in applications.

Compound Interest

Compound interest arises in many financial transactions, and it is appropriately modeled by an exponential function.

Installment Loans

Installment loans are modeled by functions that are combinations of exponential functions.

Logistic Models

Logistic models may apply when factors limit continued exponential growth.

CHAPTER QUIZ

1. Let $f(x)$ be a linear function with rate of change -2. If $f(4) = 10$, find $f(2)$.

   XR **Example 3.3**

   Answer: $f(2) = 14$

2. Find the equation of the secant line for $y = x^3$ for the interval $[1, 3]$.

   XR **Example 3.4**

   Answer: $y = 13x - 12$

3. Scientists have determined that the running speed S, in centimeters per second, of ants depends on the temperature C, in degrees Celsius. It is found that each $1°$ increase in temperature results in an increase of 0.2 centimeter per second in running speed.

   a. Explain why the running speed of ants is a linear function of the temperature.

   b. At a temperature of $20°$ Celsius, ants run at a speed of 1.3 centimeters per second. Find a formula that gives the running speed as a linear function of the temperature.

   XR **Example 3.6**

   Answer:

   a. The rate of change is constant.

   b. $S = 0.2C - 2.7$

4. a. Let $f(x)$ be an exponential function such that $f(x+1) = 5f(x)$ for all x. If $f(0) = 2$, find a formula for $f(x)$.

   b. Let $g(x)$ be an exponential function with $g(1) = 2$ and $g(4) = 16$. Find a formula for $g(x)$.

   XR **Example 3.11**

   Answer:

   a. $f(x) = 2(5^x)$

   b. $g(x) = 2^x$

5. On the same coordinate axes, plot the graphs of $y = e^x$ and $y = e^x - 5$.

   XR **Example 3.13**

   Answer:

   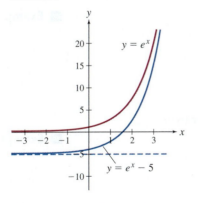

6. The balance in dollars of a certain investment after t months is given by

   $$B(t) = 300e^{0.01t}$$

   a. What is the value of the investment at the end of a year?

   b. Plot the graph of the balance over the first 36 months.

   XR **Example 3.15**

Answer:

a. $338.25

b.

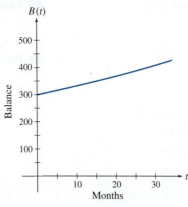

7. A certain radioactive substance decays according to the formula $R = R_0\left(\dfrac{1}{2}\right)^{t/10}$, where t is measured in days.

a. What is the half-life?

b. How much remains after 40 days?

XR Example 3.16

Answer:

a. 10 days

b. $\dfrac{1}{16}R_0$

8. The initial population in a certain country is 2.4 million. The rate of population growth is 4% per year. Find a population model of the form $N = a(b^t)$.

XR Example 3.17

Answer: $N = 2.4 \times 1.04^t$ million if t is the time in years

9. A radioactive substance has a half-life of 12 years. Using R_0 for the initial value, find a model for the amount R remaining in terms of the number of half-lives h.

XR Example 3.19

Answer: $R(h) = R_0 \times 0.5^h$.

10. Suppose you invest $5000 in an account that pays 1.8% APR compounded monthly.

a. Find a model of the form $B(t) = a(b^t)$ that gives your account balance after t years.

b. What is the account balance after 6 years?

XR Example 3.21

Answer:

a. $B(t) = 5000 \times 1.0015^{12t}$ dollars

b. $5569.79

11. Repeat the preceding exercise assuming continuous compounding rather than monthly compounding.

XR Example 3.23

Answer:

a. $B = 5000e^{0.018t}$ dollars

b. $5570.24

12. You borrow $24,000 at an APR of 6% to buy a car. Find the monthly payment if the loan is to be paid off in 4 years.

XR Example 3.24

Answer: $563.64

13. A certain population grows according to a logistic model. The intrinsic per capita growth rate is $r = 0.4$ per year, and the initial population is 700. The carrying capacity is 4900.

    a. Find a logistic model that gives the population after t years.

    b. What is the population after 4 years? (Round to the nearest whole number.)   **XR** **Example 3.25**

Answer:

    a. $N = \dfrac{4900}{1 + 6e^{-0.4t}}$

    b. 2216

CHAPTER REVIEW EXERCISES

Section 3.1

Finding formulas. In Exercises 1 through 6, find a linear function f that satisfies the given conditions.

1. The initial value is 3, and the rate of change is 2.

    Answer: $f(x) = 2x + 3$

2. The initial value is 4, and the rate of change is -3.

    Answer: $f(x) = -3x + 4$

3. The rate of change is 2, and $f(1) = 5$.

    Answer: $f(x) = 2x + 3$

4. The rate of change is -3, and $f(2) = 1$.

    Answer: $f(x) = -3x + 7$

5. $f(2) = 3$ and $f(3) = 5$.

    Answer: $f(x) = 2x - 1$

6. $f(3) = 0$ and $f(6) = 3$.

    Answer: $f(x) = x - 3$

Finding function values. In Exercises 7 through 10, information is given regarding a linear function. Use this information to calculate the indicated function value.

7. $f(2) = 5$, and the rate of change is 3. Find $f(4)$.

    Answer: $f(4) = 11$

8. $f(3) = 6$, and the rate of change is 2. Find $f(1)$.

    Answer: $f(1) = 2$

9. $f(2) = 9$, and the rate of change is -3. Find $f(4)$.

    Answer: $f(4) = 3$

10. $f(2) = 7$, and the rate of change is -2. Find $f(0)$.

    Answer: $f(0) = 11$

Proportionality.

11. Assume that y is proportional to x with a proportionality constant of 3. Find a formula that gives y in terms of x.

    Answer: $y = 3x$

12. Express the formula for the circumference of a circle as a proportionality relation.

    Answer: The circumference is proportional to the radius (or diameter).

Linear equations.

13. The temperature F in degrees Fahrenheit can be calculated from the temperature C in degrees Celsius using the formula

$$F = \frac{9}{5}C + 32$$

    Find a formula that gives C in terms of F.

    Answer: $C = \dfrac{5}{9}(F - 32)$

14. The temperature T in degrees Fahrenheit is a linear function of the number c of cricket chirps per minute. Twenty-four chirps per minute corresponds to a temperature of $43°$ Fahrenheit. Each additional chirp per minute corresponds to an increase of $0.25°$ Fahrenheit. Find a formula that gives T in terms of c.

    Answer: $T = 0.25c + 37$

15. What class of functions has a constant rate of change?

    Answer: Linear functions

Section 3.2

Finding exponential functions. In Exercises 16 through 20, find an exponential function $f(x)$ with the given properties.

16. $f(1) = 6$ and $f(3) = 54$.

Answer: $f(x) = 2 \times 3^x$

17. $f(2) = 1$ and $f(3) = \dfrac{1}{2}$.

Answer: $f(x) = 4 \times \left(\dfrac{1}{2}\right)^x$

18. $f(2) = 50$, and increasing x by 1 multiplies f by 5.

Answer: $f(x) = 2 \times 5^x$

19. $f(1) = 5$, and increasing x by 1 divides f by 2.

Answer: $f(x) = 10 \times \left(\dfrac{1}{2}\right)^x$

20. $f(2) = 18$, and decreasing x by 1 divides f by 3.

Answer: $f(x) = 2 \times 3^x$

Changing forms. In Exercises 21 through 25, write the given expression in the form $a(b^x)$.

21. $2^{x+1}3^x$

Answer: 2×6^x

22. $\dfrac{3^{x-1}}{4^{x+1}}$

Answer: $\dfrac{1}{12} \times \left(\dfrac{3}{4}\right)^x$

23. $\dfrac{1}{e^{2x}}$

Answer: $\left(e^{-2}\right)^x$

24. $\left(2^{x+1}\right)^2$

Answer: 4×4^x

25. $\sqrt{\pi^x}$

Answer: $\left(\sqrt{\pi}\right)^x$

Plotting graphs.

26. Plot the graph of $y = e^x$.

Answer:

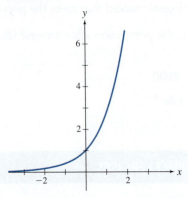

27. Plot the graph of $y = e^{-x}$.

Answer:

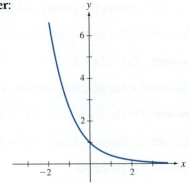

Limit.

28. What is the limit as $x \to \infty$ of $y = e^{-x}$?

Answer: 0

Domain and range.

29. What are the domain and range of $y = e^x$?

Answer: Domain $\mathbb{R} = (-\infty, \infty)$. Range $(0, \infty)$.

30. What are the domain and range of $y = e^{-x}$?

Answer: Domain $\mathbb{R} = (-\infty, \infty)$. Range $(0, \infty)$.

Evaluating an exponential function.

31. Accounting for raises, my salary after t years is given by

$$S = 75{,}000 \times 1.06^t \text{ dollars}$$

Find my salary after 5 years of work.

Answer: \$100,366.92

Section 3.3

Constant percentage change. In Exercises 32 through 35, find an exponential function $y(t)$ with the given properties.

32. The initial value is 2, and there is a constant percentage increase of 5% for each unit increase in t.

    Answer: $y = 2 \times 1.05^t$

33. The initial value is 500, and there is a constant percentage decrease of 5% for each unit increase in t.

    Answer: $y = 500 \times 0.95^t$

34. There is a constant percentage increase of 10% for each unit increase in t, and $y(1) = 3.3$.

    Answer: $y = 3 \times 1.1^t$

35. There is a constant percentage decrease of 10% for each unit increase in t, and $y(2) = 1.62$.

    Answer: $y(t) = 2 \times 0.9^t$

Evaluating an exponential function.

36. Your initial salary is $50,000, and you have a 5% raise each year. What is your salary after 10 years?

    Answer: $81,444.73

37. A certain radioactive substance has a half-life of 500 years. Use A_0 for the initial amount.

    a. Find the amount A after h half-lives.

    b. Find the amount A after t years.

Answer:

   a. $A(h) = A_0 \times 0.5^h$

   b. $A(t) = A_0 \times 0.5^{t/500}$

Exponential formulas.

38. Find the balance B after t years of an account with an initial balance of B_0 dollars if the APR is 1.2% and interest is compounded monthly.

    Answer: $B = B_0 \times 1.001^{12t}$ dollars

39. Find the monthly payment if you borrow $25,000 at an APR of 6% and repay the loan in 36 payments.

    Answer: $760.55

40. You borrow $30,000 at an APR of 6% and repay it over 5 years. What is the monthly payment, and how much interest do you pay altogether?

    Answer: $579.98; $4798.80

Logistic function.

41. A certain population is given by the logistic function

$$N = \frac{2450}{1 + 4e^{-0.07t}}$$

    where t is measured in years. What is the population after 40 years?

    Answer: 1971

LOGARITHMS

RATES OF CHANGE

BASICS OF FUNCTIONS

FOUR MAJOR CLASSES OF FUNCTIONS

TRIGONOMETRY

GEOMETRY

RATES OF CHANGE

P 1 2 3 **4** 5 6C OR 6T 7 8 9 10

Logarithms

In this chapter we introduce the logarithm. The move from exponential functions in Chapter 3 to logarithms in this chapter is natural because solving exponential equations typically requires a logarithm. But logarithms are useful in their own right to prepare you for calculus.

4.1 Logarithms: Definition and Fundamental Properties

4.2 Laws of Logarithms

4.3 Solving Exponential and Logarithmic Equations

hen the intensity of a light source doubles, our brains do not perceive a doubling of brightness. Instead, the brain perceives an increase in brightness adjusted by a logarithm. That's right—your brain has been doing logarithms all your life! The same phenomenon is true for hearing and other senses. Gustav Theodor Fechner (1801–1887; **Figure 4.1**) proposed Fechner's law (now known as the Weber-Fechner law), which explains the relationship between the magnitude of a physical stimulus and the intensity that people perceive. Subjective sensation is proportional to the logarithm of stimulus intensity.

Now that we understand that logarithms are hard-wired into our brains, it is time we understand how they work.

Figure 4.1 Gustav Fechner Science & Society Picture Library/Getty Images

4.1 Logarithms: Definition and Fundamental Properties

4.2 Laws of Logarithms

4.3 Solving Exponential and Logarithmic Equations

4.1 Logarithms: Definition and Fundamental Properties

> Whereas increasing exponential functions have an increasing rate of change, increasing logarithmic functions have a decreasing rate of change.

In this section, you will learn to:

1. Define the logarithm with base b as the inverse of the exponential function with base b.
2. Calculate the function values by applying the definition and properties of the logarithm.
3. Sketch the graphs of logarithms with a base greater than 1.
4. Analyze the long-term behavior of logarithms with a base greater than 1.
5. Calculate the function values using the change-of-base formula.
6. Solve applied problems involving logarithms.

A logarithm is the inverse of an exponential function, so there are many logarithms, one for each exponential function. In this section, we will introduce two important logarithms: the common logarithm and the natural logarithm. The natural logarithm is the logarithm used almost exclusively in mathematics, and it plays a key role in calculus, whereas the common logarithm is often used in biology and chemistry.

The Logarithm as an Inverse Function

> Every logarithm is the inverse of an exponential function.

The logarithm is defined using the notion of inverse function, which we studied in Section 2.2. Recall that the inverse of a function f is g if $f(x) = y$ precisely when $g(y) = x$. For example, $g(x) = x^{1/3}$ is the inverse function of $f(x) = x^3$ because the cube root of the cube of a number is the original number, and vice versa. Not all functions have inverses, but recall from Section 2.2 that functions that are either increasing or decreasing are one-to-one and so do have inverses. Because the exponential function $g(y) = b^y$, with $b > 0$ and $b \neq 1$, satisfies this condition, it has an inverse. The inverse is called the **base-b logarithm**, denoted by $y = \log_b x$. Thus, the logarithm with base b is the inverse of the exponential function with base b.

Thus, $\log_b x = y$ if and only if $b^y = x$. In words, $\log_b x$ is the power to which b must be raised to equal x. Always remember that the logarithm is an exponent.

The **base-b logarithm** $y = \log_b x$ is the inverse of the exponential function $g(y) = b^y$. Here $b \neq 1$ is a positive number.

The definition of the logarithm allows easy calculation of certain special function values:

$\log_b 1 = 0$	because	b raised to the power **0** gives 1:	$b^0 = 1$
$\log_b b = 1$	because	b raised to the power **1** gives b:	$b^1 = b$
$\log_{10} 100 = 2$	because	10 raised to the power **2** gives 100:	$10^2 = 100$
$\log_2 8 = 3$	because	2 raised to the power **3** gives 8:	$2^3 = 8$
$\log_7 \dfrac{1}{7} = -1$	because	7 raised to the power **−1** gives $\dfrac{1}{7}$:	$7^{-1} = \dfrac{1}{7}$

The reasoning we used to calculate these logarithms applies in a general setting. The expression $\log_b x$ asks for the power to which b must be raised to give x, so $\log_b b^t$ seeks the power to which b must be raised to equal b^t. The answer is self-evident: t is the power of b that gives b^t. That is, $\log_b b^t = t$.

In a similar fashion, because $y = \log_b t$ is the power to which b must be raised to equal t, we have $b^y = t$. Replacing y by $\log_b t$, we have $b^{\log_b t} = t$.

CONCEPTS TO REMEMBER: Properties of the Logarithm That Are Immediate Consequences of the Definition

$\log_b 1 = 0$	The power to which b is raised to equal 1 is 0.
$\log_b b = 1$	The power to which b is raised to equal b is 1.
$\log_b b^t = t$	The power to which b is raised to equal b^t is t.
$b^{\log_b t} = t$	The power to which b is raised to equal t is $\log_b t$.

In many cases, the logarithm is not easy to calculate. For example, $y = \log_{10} 5$ asks for the power to which 10 must be raised to equal 5. That is, we need y so that $10^y = 5$. We use a calculator to find the answer correct to two decimal places: $\log_{10} 5 \approx 0.70$.

Although calculators are often used to find logarithms, the following example is meant to be solved without the aid of technology.

EXAMPLE 4.1 Using the Definition of the Logarithm to Find Exact Values

a. What is the exact value of $\log_{10} 1{,}000{,}000$?

b. What is the exact value of $\log_5 1/25$?

c. What is the exact value of $\log_4 2$?

SOLUTION

a. To find the value of $\log_{10} 1{,}000{,}000$, we need to express 1 million as a power of 10. Because 1 million is 10^6, we know that 6 is the power to which 10 is raised to equal 1 million. We conclude that $\log_{10} 1{,}000{,}000 = 6$.

b. Recall that $\log_5 1/25$ asks for the power to which 5 is raised to equal $\dfrac{1}{25}$. Note that

$$\frac{1}{25} = \frac{1}{5^2} = 5^{-2}$$

Because −2 is the power to which 5 is raised to equal $\dfrac{1}{25}$, we conclude that $\log_5 1/25 = -2$.

c. To find the value of $\log_4 2$, we need to express 2 as a power of 4. We find

$$2 = \sqrt{4} = 4^{1/2}$$

We conclude that $\log_4 2 = \dfrac{1}{2}$.

TRY IT YOURSELF 4.1 Brief answers provided at the end of the section.

Find the exact value of $\log_{10} \sqrt[3]{100}$.

EXTEND YOUR REACH

a. The base-b logarithm is not defined for negative numbers. For example, $\log_2(-4)$ is not defined. Explain why we do not define this logarithm and similar logarithms. *Suggestion*: Look at the way the logarithm was defined in terms of an exponential function.

b. The logarithm is undefined for zero or negative bases. For example, $\log_{-2}$ is not defined. Explain why we do not define this logarithm. *Suggestion*: Think about what its inverse would be.

EXAMPLE 4.2 The Logarithm in the Role of Exponent

a. Calculate the exact value of $6^{\log_6 8}$.

b. Calculate the exact value of $3^{\log_3 5 + \log_3 7}$.

SOLUTION

a. We use the property noted earlier that $b^{\log_b t} = t$ by putting in $b = 6$ and $t = 8$. This gives

$$6^{\log_6 8} = 8$$

b. The law of exponents $x^{A+B} = x^A x^B$ is useful here:

$$3^{\log_3 5 + \log_3 7} = 3^{\log_3 5}\, 3^{\log_3 7} \quad \blacktriangleleft \text{Use } 3^{A+B} = 3^A 3^B.$$

$$= (5)(7) \quad \blacktriangleleft \text{Use } 3^{\log_3 t} = t.$$

$$= 35 \quad \blacktriangleleft \text{Simplify.}$$

TRY IT YOURSELF 4.2 Brief answers provided at the end of the section.

Calculate the value of $7^{\log_7 3 - \log_7 2}$.

Graphs of Logarithms

For a positive base greater than 1, the graph of the logarithm is increasing and concave down.

We defined logarithms using any positive base other than 1, but we focus on bases greater than 1. Bases less than 1 are rarely used in mathematics.

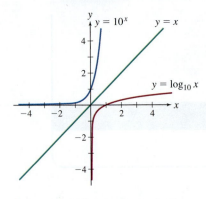

Figure 4.2 Reflecting the graph of $y = 10^x$ to make the graph of $y = \log_{10} x$

Because $f(x) = \log_b x$ is the inverse function of $g(x) = b^x$, the roles of the domain and range for these two functions are interchanged.

Function	Domain	Range
$y = b^x$	All real numbers	All positive real numbers
$y = \log_b x$	All positive real numbers	All real numbers

It is worth emphasizing that logarithms are defined only for positive real numbers. Expressions like $\log_4(-1)$ or $\log_{10} 0$ are meaningless.

Recall that to get the graph of an inverse function, we reflect the graph of the function through the line $y = x$. Thus, we reflect the graph of $y = 10^x$ through the line $y = x$ to get the graph of $y = \log_{10} x$ shown in **Figure 4.2**.

EXAMPLE 4.3 Graphing Logarithms

a. Sketch the graphs of $y = \log_2 x$ and $y = \log_{10} x$ on the same coordinate axes.

b. Comment on the features of the graphs you made. Include regions of increase or decrease and concavity, and interpret your answer in terms of rates of change.

SOLUTION

a. The accompanying table of values is used to produce the graph shown in **Figure 4.3**.

x	$\log_2 x$	$\log_{10} x$
0.25	−2	−0.60
0.5	−1	−0.30
1	0	0
2	1	0.30
4	2	0.60
8	3	0.90

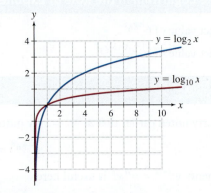

DF **Figure 4.3** Graphs of logarithms

b. The graphs show that both functions are increasing and concave down. Thus, the functions increase at a decreasing rate. This is true for the logarithm using any base greater than 1.

TRY IT YOURSELF 4.3   Brief answers provided at the end of the section.

Add the graph of $y = \log_4 x$ to Figure 4.3.

Common and Natural Logarithms

> The common logarithm has base 10, and the natural logarithm has base e.

The **common logarithm**, denoted by $\log x$, is the base-10 logarithm: $\log x = \log_{10} x$.

The **natural logarithm**, denoted by $\ln x$, is the base-e logarithm: $\ln x = \log_e x$.

The **common logarithm** $y = \log x$ and the **natural logarithm** $y = \ln x$ are not new functions, just particular cases of the logarithms already discussed.

In words, $\log x$ is the power to which 10 must be raised in order to equal x, and $\ln x$ is the power to which e must be raised in order to equal x.

CONCEPTS TO REMEMBER: Properties of the Common Logarithm and the Natural Logarithm

The common logarithm $\log x$ is the base-10 logarithm:	The natural logarithm $\ln x$ is the base-e logarithm:
$$\log x = \log_{10} x$$	$$\ln x = \log_e x$$
In words, $\log x$ is the power to which 10 must be raised in order to equal x.	In words, $\ln x$ is the power to which e must be raised in order to equal x.
• $\log x = y$ if and only if $x = 10^y$	• $\ln x = y$ if and only if $x = e^y$
• $\log 1 = 0$	• $\ln 1 = 0$
• $\log 10 = 1$	• $\ln e = 1$
• $\log 10^t = t$	• $\ln e^t = t$
• $10^{\log t} = t$	• $e^{\ln t} = t$

Figure 4.4 and **Figure 4.5** show how the graphs of the common and natural logarithms compare.

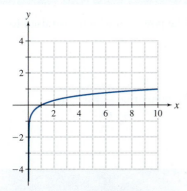

Figure 4.4 The graph of the common logarithm $y = \log x = \log_{10} x$

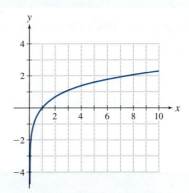

Figure 4.5 The graph of the natural logarithm $y = \ln x = \log_e x$

EXAMPLE 4.4 Calculating Values of the Common Logarithm

Find the exact values of the following.

a. $\log \dfrac{1}{10}$

b. $\log \sqrt{10}$

SOLUTION

a. Because the common logarithm is the base-10 logarithm, $\log \dfrac{1}{10}$ asks for the power to which 10 is raised to equal $\dfrac{1}{10}$. We know that $10^{-1} = \dfrac{1}{10}$, so we conclude that

$$\log \frac{1}{10} = -1$$

b. Once more, $\log \sqrt{10}$ is the power of 10 that gives $\sqrt{10}$. Because $\sqrt{10} = 10^{1/2}$, we conclude that

$$\log \sqrt{10} = \frac{1}{2}$$

The formula $\log 10^t = t$ with $t = \dfrac{1}{2}$ provides an alternative way of making this calculation:

$$\log \sqrt{10} = \log 10^{1/2} = \frac{1}{2}$$

TRY IT YOURSELF 4.4 Brief answers provided at the end of the section.

Calculate the value of $\log \dfrac{1}{100}$.

EXAMPLE 4.5 Calculations with the Natural Logarithm

Find the exact values of the following.

a. $\ln e^2$

b. $\ln \dfrac{1}{e}$

SOLUTION

a. The natural logarithm of x is the power to which e is raised to equal x. The power to which e is raised to equal e^2 is evidently 2. We conclude that

$$\ln e^2 = 2$$

b. The expression $\ln \dfrac{1}{e}$ asks for the power to which e is raised to equal $\dfrac{1}{e}$. We know that $\dfrac{1}{e} = e^{-1}$, so

$$\ln \frac{1}{e} = -1$$

TRY IT YOURSELF 4.5 Brief answers provided at the end of the section.

Find the exact value of $\ln \sqrt{e}$.

EXTEND YOUR REACH   The graphs in Figure 4.4 and Figure 4.5 show that for $x > 1$, the graph of the natural logarithm $y = \ln x$ is above the graph of the common logarithm $y = \log x$. Explain how comparison of the bases, e and 10, can account for this behavior.

EXAMPLE 4.6 The Common Logarithm as Exponent

Find the exact values of the following.

a. $10^{\log 7}$

b. $10^{\log 3} + 10^{\log 2}$

SOLUTION

a. The logarithm property $10^{\log t} = t$ with $t = 7$ applies to give

$$10^{\log 7} = 7$$

b. The expression $10^{\log 3} + 10^{\log 2}$ may appear intimidating, but it is (perhaps surprisingly) easy to evaluate. We use the logarithm property $10^{\log t} = t$. Applying this formula with $t = 3$ and again with $t = 2$, we have

$$10^{\log 3} + 10^{\log 2} = 3 + 2 = 5$$

TRY IT YOURSELF 4.6 Brief answers provided at the end of the section.

Calculate the value of $10^{\log 5}$.

EXAMPLE 4.7 Natural Logarithms and Exponents

a. Find the exact value of $e^{\ln 5}$.

b. What value of x makes $y = \ln x$ reach the target value of 10? Express your answer as a power of e, and then give a decimal approximation.

SOLUTION

a. We use the rule $e^{\ln t} = t$ with $t = 5$. It tells us that

$$e^{\ln 5} = 5$$

b. We use the fact that $\ln e^t = t$. Applying this fact with $t = 10$ tells us that $\ln e^{10} = 10$. We find that $y = \ln x$ reaches a target value of 10 if $x = e^{10} \approx 22{,}026.47$.

TRY IT YOURSELF 4.7 Brief answers provided at the end of the section.

What value of x gives $\ln x = -\dfrac{1}{2}$?

EXAMPLE 4.8 Transformations of Logarithmic Graphs

Plot the graphs of $y = \ln x$ and $y = -\ln x$ on the same coordinate axes.

SOLUTION

We find the graph of $y = -\ln x$ from the graph of $y = \ln x$ by reflecting through the x-axis. The result is shown in **Figure 4.6**.

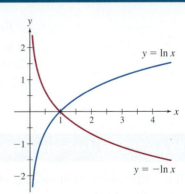

DF **Figure 4.6** Graphs of $y = \ln x$ and $y = -\ln x$

TRY IT YOURSELF 4.8 Brief answers provided at the end of the section.

On the same coordinate axes, plot the graphs of $y = \ln x$ and $y = 3 \ln x$.

Long-Term Behavior of the Logarithm

> For bases larger than 1, the logarithm increases without bound.

The shape of the graph of $y = \log x$ shown in Figure 4.3 tells us that the common logarithm function grows rapidly at first, but later the rate of increase slows. The slow rate of increase might lead one to believe that the graph eventually levels out, but this is incorrect. To see why, let's set a target height using the horizontal line $y = k$ (**Figure 4.7**). We want to show that the graph of the logarithm eventually reaches this target height no matter how large k may be. If we choose $x = 10^k$, then

$$\log x = \log 10^k = k$$

That is, we reach a target height of $y = k$ by choosing $x = 10^k$ (**Figure 4.8**).

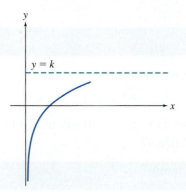

Figure 4.7 Setting a target height of $y = k$

DF **Figure 4.8** Reaching the target height: The function $g(x) = \log x$ grows slowly, but its graph eventually reaches the target height no matter how large k may be.

This shows that $y = \log x$ can be made to reach any preassigned target height by choosing x to be large enough. The logarithm function is increasing, so once it reaches a target value the function never falls back below that value. In terms of limits, this means

$$\log x \to \infty \text{ as } x \to \infty$$

To emphasize just how slowly the common logarithm grows, we note that to make the graph reach the modest target value of $y = 10$ we must use $x = 10^{10}$, which is 10 billion, on the horizontal axis. Our analysis is specific to the common logarithm, but a similar argument works for the logarithm with any base larger than 1 (see Exercise 100):

$$\text{If } b > 1, \text{ then } \log_b x \to \infty \text{ as } x \to \infty$$

EXAMPLE 4.9 Estimating Values for the Common Logarithm

If $x = 1025$, use your knowledge of powers of 10 to give a rough estimate (to the nearest whole number) of the value of $\log 1025$.

SOLUTION

First we note that 1025 is between the powers $10^3 = 1000$ and $10^4 = 10,000$. Because $y = \log x$ is an increasing function, $\log 1025$ is between $\log 10^3 = 3$ and $\log 10^4 = 4$. We conclude that $\log 1025$ is between 3 and 4. Because 1025 is relatively close to 1000, we estimate that $\log 1025$ is a little larger than 3.

Using a calculator, we find that $\log 1025 \approx 3.01$. So, we know that our estimate of 3 was pretty good.

TRY IT YOURSELF 4.9 Brief answers provided at the end of the section.

Use your knowledge of powers of 10 to estimate to the nearest whole number the value of log 10,023.

EXAMPLE 4.10 Growth of the Common Logarithm

How large must x be to make $y = \log x$ reach the target value of 7?

SOLUTION

To make $\log x = 7$, we can choose x to be $10^7 = 10,000,000$ because

$$\log x = \log 10^7 = 7$$

The conclusion is that we reach the target value of 7 by choosing x to be 10 million.

TRY IT YOURSELF 4.10 Brief answers provided at the end of the section.

If you know that $\log x$ is at least 4, what can you say about x?

EXTEND YOUR REACH

a. Use a calculator to complete the following table showing the average rate of change of the natural logarithm over the given interval. Report your answers correct to seven decimal places.

Interval	Average rate of change of ln x
$x = 10$ to $x = 11$	
$x = 100$ to $x = 101$	
$x = 1000$ to $x = 1001$	
$x = 10,000$ to $x = 10,001$	

b. Based on your answers from part a, what do you think is the limit as $x \to \infty$ of the rate of change of the natural logarithm $\ln x$? Remember that the average rate of change of $\ln x$ gives an estimate of the rate of change of $\ln x$.

c. Move a straightedge along the graph of $y = \ln x$ to visualize the rate of change for increasing values of x. Does this visualization lend support to your answer in part b? If so, explain how.

Change-of-Base Formula

> Any logarithm can be expressed in terms of the natural logarithm.

There is a simple relationship between the natural logarithm and the common logarithm. To find the relationship, we use a basic property of logarithms: $e^{\ln t} = t$ and $10^{\log t} = t$. We calculate

$$e^{(\ln 10)\log x} = \left(e^{\ln 10}\right)^{\log x} \quad \blacktriangleleft \text{Use } A^{pq} = (A^p)^q.$$

$$= 10^{\log x} \quad \blacktriangleleft \text{Use } e^{\ln t} = t.$$

$$= x \quad \blacktriangleleft \text{Use } 10^{\log t} = t.$$

This shows that $(\ln 10) \log x$ is the power to which e is raised to equal x, and we conclude that

$$(\ln 10) \log x = \ln x$$

Dividing both sides by $\ln 10$, we obtain the change-of-base formula

$$\log x = \frac{\ln x}{\ln 10}$$

This formula is used to calculate the common logarithm from the natural logarithm. It is a particular case of the general change-of-base formula.

LAWS OF MATHEMATICS: Change-of-Base Formula

If a and b are positive and not equal to 1, then for all $x > 0$ we have

$$\log_b x = \frac{\log_a x}{\log_a b}$$

A particular case of this formula is

$$\log_b x = \frac{\ln x}{\ln b}$$

The preceding particular case of the change-of-base formula shows that any logarithm can be calculated in terms of the natural logarithm. As a consequence, only one logarithm is really needed, and mathematicians prefer the natural logarithm.

EXAMPLE 4.11 Change-of-Base Formula

Use the change-of-base formula to express $\log_8 32$ in terms of the base-2 logarithm, and then calculate its value.

SOLUTION

We have

$$\log_8 32 = \frac{\log_2 32}{\log_2 8} \qquad \blacktriangleleft \text{Use the change-of-base formula.}$$

$$\log_8 32 = \frac{5}{3} \qquad \blacktriangleleft \log_2 32 = 5 \text{ because } 2^5 = 32. \text{ Similarly, } \log_2 8 = 3.$$

TRY IT YOURSELF 4.11 Brief answers provided at the end of the section.

Use the change-of-base formula to express $\log_{27} 9$ in terms of the base-3 logarithm, and then find its exact value.

Most calculators have keys only for the common and natural logarithms. The natural logarithm key together with the change-of-base formula allows the evaluation of logarithms with any base. For example, to find $\log_5 27$ we use

$$\log_5 27 = \frac{\ln 27}{\ln 5} = \frac{3.2958\ldots}{1.6094\ldots} \approx 2.05$$

EXAMPLE 4.12 Change-of-Base Formula and the Calculator

Use the change-of-base formula to express $\log_3 11 + \log_4 11$ in terms of the natural logarithm, and then approximate its value.

SOLUTION

We use the case of the change-of-base formula that involves the natural logarithm:

$$\log_3 11 + \log_4 11 = \frac{\ln 11}{\ln 3} + \frac{\ln 11}{\ln 4}$$

$$\approx 3.91$$

TRY IT YOURSELF 4.12 Brief answers provided at the end of the section.

Use the change-of-base formula and a calculator to approximate $\log_5 17$ to two decimal places.

MODELS AND APPLICATIONS Spell Checkers

If you want to check the spelling of "spelll," perhaps the most inefficient way is to check the word against every single word in the dictionary. Modern spell checkers on your computer make clever use of a tree structure to shorten this process considerably. The resulting formula for the maximum number of checks needed involves a logarithm.

EXAMPLE 4.13 Spell Checkers

If your computer has a dictionary of W words, then the maximum number of checks needed to determine whether a typed word appears in the dictionary is given by

$$C = \log_2(W + 1)$$

a. Determine the maximum number of checks required if your dictionary has only 255 words.

b. Use the change-of-base formula to express C in terms of the natural logarithm.

c. A typical word processor includes a dictionary of about 100,000 words. Find the maximum number of checks required to determine whether a typed word is in this 100,000-word dictionary. Round your answer from the formula to the next largest whole number.

SOLUTION

a. The formula tells us that at most $C = \log_2(255 + 1) = \log_2 256$ checks are required. Because $2^8 = 256$, we find that at most $C = 8$ checks are required.

b. We apply the following case of the change-of-base formula for $\log_b$:

$$\log_b x = \frac{\ln x}{\ln b}$$

For C we find

$$C = \frac{\ln(W + 1)}{\ln 2}$$

c. Substituting $W = 100,000$ into the formula from part b gives

$$C = \frac{\ln 100,001}{\ln 2} \approx 16.61$$

After rounding to the next highest whole number, we determine that at most 17 checks are required.

TRY IT YOURSELF 4.13 Brief answers provided at the end of the section.

How many checks would be required to accommodate a million-word dictionary?

> ### CONCEPTS TO REMEMBER: Logarithmic Functions
>
> **Defining property**: If $b \neq 1$ is positive, the base-b logarithm $g(x) = \log_b x$ is the inverse of the exponential function $f(x) = b^x$. That is, $\log_b x$ is the power of b that gives x.
>
> Important particular cases are as follows:
>
> | The common logarithm | The natural logarithm |
> |---|---|
> | $\log x$ is the base-10 logarithm. It is the power of 10 that gives x. | $\ln x$ is the base-e logarithm. It is the power of e that gives x. |
>
> **Domain**: All positive real numbers.
>
> **Range**: All real numbers.
>
> **Graph**: For $b > 1$: Increasing and concave down.
>
> **In the long term**: For $b > 1$: The logarithm increases very slowly, but
>
> $$\log_b x \to \infty \text{ as } x \to \infty$$

TRY IT YOURSELF ANSWERS

4.1 $\log_{10} \sqrt[3]{100} = \dfrac{2}{3}$

4.2 $\dfrac{3}{2}$

4.3
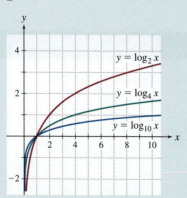

4.4 $\log \dfrac{1}{100} = -2$

4.5 $\ln \sqrt{e} = \dfrac{1}{2}$

4.6 5

4.7 $x = e^{-1/2} = \dfrac{1}{\sqrt{e}}$

4.8
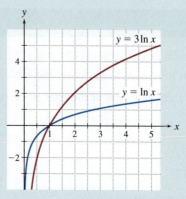

4.9 $\log 10,023$ is a little larger than 4.

4.10 x is at least $10^4 = 10,000$

4.11 $\dfrac{\log_3 9}{\log_3 27} = \dfrac{2}{3}$

4.12 $\log_5 17 = \dfrac{\ln 17}{\ln 5} \approx 1.76$

4.13 At most 20 checks

EXERCISE SET 4.1

CHECK YOUR UNDERSTANDING

1. In words, $\log_b x$ is _____.

2. In words, the common logarithm $\log x$ is _____.

3. In words, the natural logarithm $\ln x$ is _____.

4. Select all that are true. If $b > 1$, then the graph of $y = \log_b x$

   a. is concave up

   b. is concave down

   c. is increasing

   d. is decreasing

   e. has a local maximum at $x = 1$

5. The inverse of the function $f(x) = \log_7 x$ is _____.

6. **True or false:** $\ln 0 = 1$.

7. **True or false:** The change-of-base formula allows us to calculate any logarithm in terms of the natural logarithm.

8. What is the inverse of the natural logarithm function?

9. The expression $10^{\log t}$ simplifies to _____.

10. The common logarithm $\log x$ is the logarithm to which base?

11. A friend says that her calculator won't give an answer when she tries to compute $\log 0$ and $\log(-2)$. Explain why.

SKILL BUILDING

Calculating common logarithms. In Exercises 12 through 24, calculate the exact value of the logarithm without using a calculator.

12. $\log \dfrac{1}{10}$

13. $\log 0.001$

14. $\log 10^8$

15. $\log \sqrt{10}$

16. $\log 1$

17. $\log 10$

18. $\log 10,000$

19. $\log \dfrac{\sqrt{10}}{\sqrt[5]{10}}$

20. $\log \dfrac{1}{\sqrt{10}}$

21. $\log \sqrt{1000}$

22. $\log \sqrt{10}\sqrt[3]{100}$

23. $\log 100\sqrt{1000}$

24. $\log \dfrac{1}{100}$

Calculating natural logarithms. In Exercises 25 through 28, calculate the exact value of the logarithm without using a calculator.

25. $\ln 1$

26. $\ln e$

27. $\ln \dfrac{1}{e^4}$

28. $\ln \dfrac{e^2}{\sqrt{e}}$

Calculating logarithms. In Exercises 29 through 40, calculate the exact value of the logarithm without using a calculator.

29. $\log_2 16$

30. $\log_2 \sqrt{2}$

31. $\log_3 \dfrac{1}{9}$

32. $\log_{100} 10$

33. $\log_8 2$

34. $\log_5 125$

35. $\log_\pi 1$

36. $\log_{12} 12$

37. $\ln(\log 10^e)$

38. $\log(\ln e^{100})$

39. $\log(\log 10)$

40. $\ln(\ln e^e)$

Changing forms. In Exercises 41 through 53, rewrite the given expression so that no exponential or logarithmic functions are involved. The formula $b^{\log_b t} = t$ and the basic rules of exponents will be helpful. Assume that x and y are positive.

41. $e^{\ln 4}$

42. $5^{\log_5 3}$

43. $10^{\log 2x}$

44. $10^{-\log x}$

  Reminder: $10^{-A} = (10^A)^{-1}$

45. $e^{4 \ln x}$

  Reminder: $e^{4A} = (e^A)^4$

46. $3^{\log_3 x + \log_3 y}$

  Reminder: $3^{A+B} = 3^A 3^B$

47. $e^{\ln(x+1) - \ln x}$

  Reminder: $e^{A-B} = \dfrac{e^A}{e^B}$

48. $3^{2 \ln \sqrt{x}}$

49. $e^{-\ln x}$

50. $2^{-4 \log_2 x}$

51. $10^{\log x - \log y}$

52. $e^{2 \ln x}$

53. $\dfrac{10^{\log 4}}{e^{\ln 2}}$

Matching functions with graphs. In Exercises 54 through 57, match the given graph with the appropriate function from the list that follows.

A. $y = \ln x$ **C.** $y = \log_2 x$

B. $y = \log x$ **D.** $y = \log_4 x$

54.

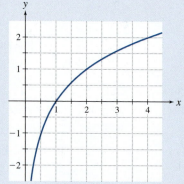

55.

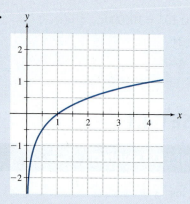

56.

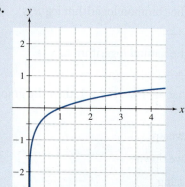

57.

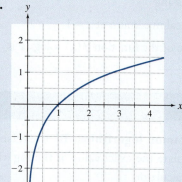

Using the change-of-base formula. In Exercises 58 through 66, use the change-of-base formula to convert to natural logarithms, then use a calculator to find the requested value. Round to two decimal places.

58. $\log_7 9$ **63.** $\log_4 e$

59. $\log_2 12$ **64.** $\log_9 77$

60. $\log_5 7$ **65.** $\log_9 4$

61. $\log_8 3$ **66.** $\log_3 19$

62. $\log_{12} 37$

PROBLEMS

67. Find the relation. Suppose $\log A = 1.4$ and $\log B = 2.8$. How are A and B related?

68. An interesting equation. Can the equation $\ln x = x$ be solved? *Suggestion:* Make the graphs of $y = x$ and $y = \ln x$.

Finding the domain. In Exercises 69 through 78, find the domain of the given function.

69. $f(x) = \log(x - 5)$

70. $f(x) = \ln(x - 3) + \log(x - 4)$

71. $f(x) = \ln\sqrt{x}$

72. $f(x) = \sqrt{\ln x}$

73. $f(x) = \dfrac{1}{\ln x}$

74. $f(x) = \ln(x - 3) + \ln(x - 4)$

75. $f(x) = \ln(x - 3) + \ln(4 - x)$

76. $\log \dfrac{x}{x - 1}$ *Suggestion:* First make a sign diagram for the rational function $y = \dfrac{x}{x - 1}$.

77. $\dfrac{\ln x - 1}{\log(2 - x)}$

78. $f(x) = \ln|x|$

79. Logarithms with base 1. Explain why we do not define logarithms with base 1.

80. Logarithms with negative bases. Explain why we do not define logarithms with negative bases.

Transforming the graph of the common logarithm. In Exercises 81 through 88, plot the graph of the common logarithm along with the graph of the given function.

81. $y = 1 + \log x$

82. $y = -\log x$

83. $y = \log(-x)$

84. $y = \log|x|$

85. $y = \log(2x)$

86. $y = 2\log x$

87. $y = \log(x - 1)$

88. $y = \log(x + 1) - 1$

Transforming the graph of the natural logarithm. In Exercises 89 through 92, plot the graph of $y = \ln x$ together with the graph of the given function.

89. $y = 1 + \ln x$

90. $y = \ln(x - 1)$

91. $y = \ln(-x)$

92. $y = |\ln x|$

Average rate of change. In Exercises 93 through 95, calculate the average rate of change of the given function on the given interval.

93. $f(x) = \log x$ on $[1, 2]$

94. $f(x) = \log x$ on $[10, 12]$

95. $f(x) = \log x$ on $[a, b]$

96. Logarithms and concavity.

   a. Calculate the average rate of change of the function $f(x) = \ln x$ on the intervals $[1, 2]$ and $[10, 11]$.

   b. Use a calculator to compare your answers in part a. Explain how the result is consistent with the concavity of the graph of the natural logarithm.

97. Secant lines. Find the equation of the secant line for $f(x) = \log x$ for the interval $[1, 3]$.

98. Secant lines. Find the equation of the secant line for $f(x) = \ln x$ for the interval $[1, 5]$.

99. Doubling the logarithm. What power of x will cause the natural logarithm of x to double? That is, for what value of k is the equation $2\ln x = \ln x^k$ true?

100. Long-term behavior of the logarithm. Show that if $b > 1$, then $\log_b x \to \infty$ as $x \to \infty$. Note that, because $y = \log_b x$ is increasing, you need only show that $y = \log_b$ reaches a target height of $y = k$ no matter how large the value of k.

101. How large? How large does x need to be to make $y = \log x$ reach the target value of $y = 100$?

102. How large? How large does x need to be to ensure that $y = \ln x$ reaches the target value of $y = 100$?

Change-of-base formula. Exercises 103 through 113 can be solved using the change-of-base formula.

103. Find the exact value of $\log_{16} 8$. *Suggestion:* Change the base to 2.

104. Find the exact value of $\log_9 27$. *Suggestion:* Change the base to 3.

105. Express $\log_7 x$ in terms of the natural logarithm.

106. Calculate $\log_{32} 128$. *Suggestion:* Change the base to 2.

107. Calculate $\log_{27} 81$. *Suggestion:* Change the base to 3.

108. Calculate $\log_{\sqrt[3]{7}} 49$. *Suggestion:* Change the base to 7.

109. Calculate $\log_{1/3} 9\sqrt{3}$. *Suggestion:* Change the base to 3.

110. Calculate $\log_{0.1} \dfrac{\sqrt{10}}{100}$. *Suggestion*: Use the common logarithm.

111. Express the natural logarithm in terms of the common logarithm.

112. Simplify the expression $2^{\log_4 x}$. Your answer should involve no logarithms. *Suggestion*: Change the base to 2.

113. Simplify the expression $9^{\log_3 e}$. Your answer should involve no logarithms. *Suggestion*: Change the base to 9.

Bases smaller than 1. Exercises 114 through 118 concern logarithms having a base between 0 and 1.

114. Find the value of $\log_{1/2} 2$.

115. Find the value of $\log_{1/3} 1$.

116. Find the value of $\log_{1/2} 8$.

117. Plot the graph of $y = \log_{1/2} x$. Comment on the concavity as well as regions of increase or decrease.

118. What is the limit as $x \to \infty$ of $y = \log_{1/3} x$?

MODELS AND APPLICATIONS

119. Decibels. The decibel is a measure of loudness. If a sound has intensity I, in watts per square meter, then its decibel level is given by

$$D = 10 \log\left(\frac{I}{I_0}\right)$$

where $I_0 = 10^{-12}$ watts per square meter. What is the decibel reading of a sound with an intensity of 5 watts per square meter? Report your answer as a whole number.

120. Apparent magnitude. If I is the intensity of light arriving at Earth from a star, then its apparent magnitude is given by

$$m = 2.5 \log\left(\frac{I_v}{I}\right)$$

where I_v is the intensity of light arriving from the star Vega. The apparent magnitude is a measure of the brightness of a star as viewed from Earth.

a. What is the apparent magnitude of Vega?

b. Does a higher apparent magnitude indicate a brighter or a dimmer star?

c. Light arriving from Sirius is 3.63 times as bright as light arriving from Vega. What is the apparent magnitude of Sirius? Note that apparent magnitudes are normally reported with a single decimal place of accuracy.

121. Hick's law. Hick's law asserts that the reaction time R required to select from among n choices is given by the formula

$$R = c \log_2(n + 1)$$

where c is a constant that depends on the individual responding and the types of choices.[1] In a certain setting the value of c is 0.5 if R is measured in seconds. A graph of R versus n is shown in **Figure 4.9**.

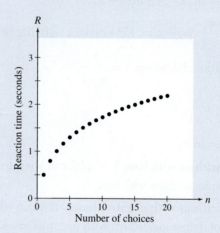

Figure 4.9 Reaction time versus number of choices

a. What is the reaction time when there are 15 choices? Note that you should be able to find the exact answer without using a calculator.

b. Express R in terms of the natural logarithm.

c. Use your answer from part b to find the reaction time when there are 9 choices.

d. Use the shape of the graph in Figure 4.9 to answer the following question. Which has a greater effect on the reaction time: adding another choice when there are already 2 choices, or adding another choice when there are already 12 choices?

[1] W. E. Hick, "On the Rate of Gain of Information," *Quarterly Journal of Experimental Psychology* 4: 11–26 (1952).

CHALLENGE EXERCISES FOR INDIVIDUALS OR GROUPS

122. General change-of-base formula. Show that, if $a, b > 0$ and $a, b \neq 1$, then $\log_a x = \dfrac{\log_b x}{\log_b a}$.

Suggestion: Look at $b^{(\log_b a)\log_a x}$. Use the fact that $A^{pq} = (A^p)^q$.

123. Rate of change of the natural logarithm. Use the graph of the natural logarithm to answer the following questions.

a. Is the rate of change of the logarithm always positive, always negative, or sometimes positive and sometimes negative? Explain.

b. Consider the rate of change of the logarithm as a function. Is the rate of change an increasing function or a decreasing function? Explain.

124. Comparing rates of change for the natural logarithm and the natural exponential function. Figure 4.10 shows the graphs of the natural logarithm and the natural exponential function. The numbers $x = 0.1$ and $x = 2$ are marked on the horizontal axis.

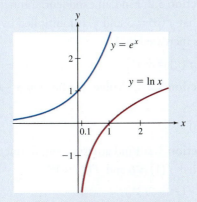

Figure 4.10 The natural logarithm and the natural exponential function

a. Which is larger, the rate of change of the natural exponential function at $x = 0.1$ or the rate of change of the natural logarithm at $x = 0.1$?

b. Which is larger, the rate of change of the natural exponential function at $x = 2$ or the rate of change of the natural logarithm at $x = 2$?

c. Do you think there is a point where the rate of change of the natural logarithm and the rate of change of the natural exponential function are the same? Explain your reasoning.

The natural logarithm and area. In calculus, it is shown that for $x > 1$ the area under the curve $y = \dfrac{1}{t}$ from $t = 1$ to $t = x$ is $\ln x$. That area is shown in Figure 4.11. This area calculation is one of the reasons we use the name "natural" for this logarithm. Exercises 125 through 128 make use of this fact.

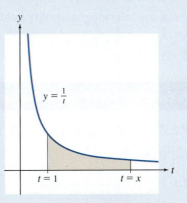

Figure 4.11 Area determined by the graph of $y = \frac{1}{t}$: The natural logarithm gives the area under the curve.

125. Find the area under the curve $y = \dfrac{1}{t}$ from $t = 1$ to $t = 3$.

126. Find the area under the curve $y = \dfrac{1}{t}$ from $t = 3$ to $t = 7$. *Suggestion:* Find the area from 1 to 7, and then subtract the area from 1 to 3.

127. Figure 4.12 illustrates an important inequality, as we will see in this exercise and the next. The figure shows clearly that

Area of the 3 rectangles > Area under the curve

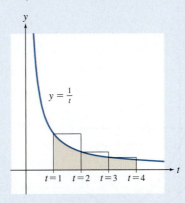

Figure 4.12 Areas of rectangles and area under the curve

a. Calculate the total area of the three rectangles. Note that the first rectangle has length 1 and height 1, and the second rectangle has length 1 and height $\dfrac{1}{2}$.

b. Show that $1 + \dfrac{1}{2} + \dfrac{1}{3} > \ln 4$.

128. The figure used in Exercise 127 can be adapted to show that in general the following inequality holds:

$$1 + \frac{1}{2} + \frac{1}{3} + \cdots + \frac{1}{n} > \ln(n+1)$$

a. Explain why the preceding inequality tells us that we can make the sum $1 + \dfrac{1}{2} + \dfrac{1}{3} + \cdots + \dfrac{1}{n}$ as large as we like by choosing n large. Note that we are increasing the sum by less and less as n gets larger and larger.

b. Use the preceding inequality to estimate how many terms of the sum we must add up to make the sum larger than 100.

REVIEW AND REFRESH: Exercises from Previous Sections

129. **From Section P.2:** Solve the inequality $|x - 1| \geq 3$.

Answer: $(-\infty, -2] \cup [4, \infty)$

130. **From Section 2.1:** If $f(x) = \dfrac{3}{(2+x)^2}$, find functions $g(x)$ and $h(x)$ so that $f = g \circ h$.

Answer: One solution is $g(x) = \dfrac{3}{x}$ and $h(x) = 2 + x$.

131. **From Section 2.3:** The graph of the function $f(x)$ is shown. Add to the picture the graph of $f(x+1)$.

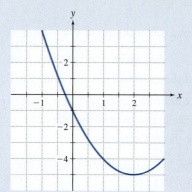

Answer:

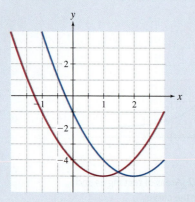

132. **From Section 3.1:** Find the formula for the linear function $f(x)$ so that $f(1) = 0$ and $f(2) = 3$.

Answer: $y = 3x - 3$

133. **From Section 3.1:** Suppose that $f(x)$ is a linear function with slope 2. If $f(3) = 5$, what is the value of $f(5)$?

Answer: 9

134. **From Section 3.2:** Find an exponential function $f(x)$ so that $f(0) = 3$ and f is multiplied by 5 when x is increased by 1.

Answer: $f(x) = 3(5^x)$

135. **From Section 3.2:** Calculate the limit as $x \to \infty$ of $y = e^{-x}$.

Answer: 0

136. **From Section 3.2:** Find an exponential function $f(x)$ so that $f(1) = 6$ and $f(2) = 18$.

Answer: $f(x) = 2(3^x)$

137. **From Section 3.3:** A savings account has an initial balance of \$300. It grows at the constant percentage rate of 1.2% per year. Find the balance after 15 years.

Answer: $300 \times 1.012^{15} \approx \$358.78$

4.2 Laws of Logarithms

Logarithms use the product, power, and quotient laws to simplify complicated expressions.

4.1 Logarithms: Definition and Fundamental Properties

4.2 Laws of Logarithms

4.3 Solving Exponential and Logarithmic Equations

In this section, you will learn to:

1. Derive the laws of logarithms using the rules for exponents.
2. Calculate values and rewrite expressions using the laws of logarithms.
3. Solve applied problems using the laws of logarithms.

Logarithms are used to measure the size of earthquakes (among many other phenomena). Earthquakes impart an acceleration to the ground, causing the movements that are so dangerous. The relative intensity of ground acceleration seems a natural measurement of earthquakes. But the Richter scale modifies relative intensity using a logarithm to determine the magnitude of an earthquake. Why use the logarithm? Doesn't that just complicate things? **Figure 4.13** shows a typical year of earthquakes measured by ground acceleration. The graph gives the impression that there were very few earthquakes; in fact, there were thousands of them. The difficulty is that the scale is very broad, stretching from 0 to 140 million, so almost all of the earthquakes are obscured by the line representing the horizontal axis.

Measuring the size of the earthquakes using the magnitude produces the graph in **Figure 4.14**, a more accurate depiction of earthquake activity. The logarithm adjusts the scale in Figure 4.13 and reveals earthquake activity that is actually obscured by direct measurement. We emphasize that the graphs in Figure 4.13 and Figure 4.14 show the same data viewed through different lenses. Charles Richter and Beno Gutenberg, inventors of the Richter magnitude scale, did not use logarithms to obfuscate geological activity; rather, logarithms were used to make it understandable.

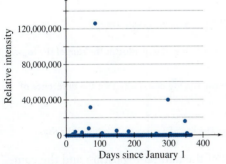

Figure 4.13 Earthquakes recorded by relative intensity

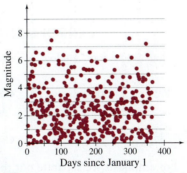

DF **Figure 4.14** Earthquakes recorded by magnitude, which is the logarithm of relative intensity

Some of the most useful properties of the logarithm are really just the familiar rules for exponents recast in terms of logarithms. In this section, we will show how this recasting is done and how to apply it.

Deriving the Laws of Logarithms

The laws of logarithms are derived using the rules followed by their inverses, exponential functions.

Consider, for example, the following rule for exponents:

$$e^{A+B} = e^A e^B$$

If we put $A = \ln x$ and $B = \ln y$ into this formula, we find

$$e^{\ln x + \ln y} = \left(e^{\ln x}\right)\left(e^{\ln y}\right) \quad \blacktriangleleft \text{ Use } e^{A+B} = e^A e^B.$$

$$e^{\ln x + \ln y} = xy \quad \blacktriangleleft \text{ Use } e^{\ln t} = t.$$

That is, the power of e that gives xy is $\ln x + \ln y$. That means this power, $\ln x + \ln y$, is the natural logarithm of xy:

$$\ln(xy) = \ln x + \ln y$$

This equation is the product law for the natural logarithm.[2]

In a similar fashion, we make use of the rule for exponents $e^{pA} = (e^A)^p$. Putting $A = \ln x$, we find

$$e^{p \ln x} = \left(e^{\ln x}\right)^p \quad \blacktriangleleft \text{ Use } e^{pA} = \left(e^A\right)^p.$$

$$e^{p \ln x} = x^p \quad \blacktriangleleft \text{ Use } e^{\ln t} = t.$$

We see that $p \ln x$ is the power of e that gives x^p. It follows that this power, $p \ln x$, is the natural logarithm of x^p:

$$\ln x^p = p \ln x$$

The rule for exponents $e^{A-B} = \dfrac{e^A}{e^B}$ leads to

$$\ln\left(\frac{x}{y}\right) = \ln x - \ln y$$

These are the laws of logarithms. The same ideas can be used to establish similar laws for the common logarithm and indeed for any logarithm.

LAWS OF MATHEMATICS: Laws of Logarithms

$$\log_b(xy) = \log_b x + \log_b y \quad \blacktriangleleft \text{Product law: The log of a product is the sum of the logs.}$$

$$\log_b\left(\frac{x}{y}\right) = \log_b x - \log_b y \quad \blacktriangleleft \text{Quotient law: The log of a quotient is the difference of the logs.}$$

$$\log_b x^p = p\log_b x \quad \blacktriangleleft \text{Power law: The log of a power } p \text{ is } p \text{ times the log.}$$

For emphasis we separate out the cases of the natural logarithm and the common logarithm.

Natural logarithm		Common logarithm
$\ln(xy) = \ln x + \ln y$	Product law	$\log(xy) = \log x + \log y$
$\ln\left(\dfrac{x}{y}\right) = \ln x - \ln y$	Quotient law	$\log\left(\dfrac{x}{y}\right) = \log x - \log y$
$\ln x^p = p\ln x$	Power law	$\log x^p = p\log x$

---

[2] The left side of this equation is defined as long as $xy > 0$, but the right side is defined only when $x > 0$ and $y > 0$. In general, when we apply the laws of logarithms we will assume that the variables have been restricted so that all quantities are defined.

Using Laws of Logarithms

> The laws of logarithms are used to analyze expressions that would otherwise be unwieldy.

EXAMPLE 4.14 Using the Laws to Find Values of Logarithms

Use the laws of logarithms to find the exact value of the following.

a. $\log_6 4 + \log_6 9$

b. $\log_2 48 - \log_2 3$

SOLUTION

a. Note that neither term of the sum is easily calculated. The laws of logarithms can help:

$$\log_6 4 + \log_6 9 = \log_6 (4 \times 9) \qquad \blacktriangleleft \text{Apply the product law.}$$

$$= \log_6 36 \qquad \blacktriangleleft \text{Simplify.}$$

$$= 2 \qquad \blacktriangleleft \text{Use } 6^2 = 36.$$

b. As in part a, neither term is easily calculated:

$$\log_2 48 - \log_2 3 = \log_2 \left(\frac{48}{3} \right) \qquad \blacktriangleleft \text{Apply the quotient law.}$$

$$= \log_2 16 \qquad \blacktriangleleft \text{Simplify.}$$

$$= 4 \qquad \blacktriangleleft \text{Use } 2^4 = 16.$$

TRY IT YOURSELF 4.14 Brief answers provided at the end of the section.

Use the laws of logarithms to find the exact value of $\log_8 4 + \log_8 2$.

EXTEND YOUR REACH

Explain in your own words the relationship between the laws of logarithms and the rules for exponents. Your explanation should be expressed in terms of inverse functions.

EXAMPLE 4.15 Calculations Using the Laws of Logarithms: Natural Logarithms

Suppose $\ln A = 1.3$, $\ln B = -2.2$, and $\ln C = 4.7$.

a. Find the value of $\ln (AB^2)$.

b. Find the value of $\ln \sqrt{A/C}$.

SOLUTION

a. We start with the law for the product:

$$\ln(AB^2) = \ln A + \ln B^2 \quad \blacktriangleleft \textbf{Apply the product law.}$$

$$= \ln A + 2\ln B \quad \blacktriangleleft \textbf{Apply the power law.}$$

$$= 1.3 + 2(-2.2) \quad \blacktriangleleft \textbf{Substitute values for logarithms.}$$

$$= -3.1 \quad \blacktriangleleft \textbf{Simplify.}$$

b. First we convert the radical to an exponent:

$$\ln\sqrt{A/C} = \ln(A/C)^{1/2} \quad \blacktriangleleft \textbf{Use the definition of rational exponent.}$$

$$= \frac{1}{2}\ln(A/C) \quad \blacktriangleleft \textbf{Apply the power law.}$$

$$= \frac{1}{2}(\ln A - \ln C) \quad \blacktriangleleft \textbf{Apply the quotient law.}$$

$$= \frac{1}{2}(1.3 - 4.7) \quad \blacktriangleleft \textbf{Substitute values for logarithms.}$$

$$= -1.7 \quad \blacktriangleleft \textbf{Simplify.}$$

TRY IT YOURSELF 4.15   Brief answers provided at the end of the section.

Use the values given in the example to find the value of $\ln\left(\dfrac{A^2 B}{C}\right)$.

EXAMPLE 4.16   Calculations Using the Laws of Logarithms: Common Logarithms

Use the laws of the common logarithm to find the exact value of $\log\left(\dfrac{\sqrt[3]{100}}{\sqrt{1000}}\right)$.

SOLUTION

We start with the law for the quotient:

$$\log\left(\frac{\sqrt[3]{100}}{\sqrt{1000}}\right) = \log\sqrt[3]{100} - \log\sqrt{1000} \quad \blacktriangleleft \textbf{Apply the quotient law.}$$

$$= \log 100^{1/3} - \log 1000^{1/2} \quad \blacktriangleleft \textbf{Use the definition of rational exponent.}$$

$$= \frac{1}{3}\log 100 - \frac{1}{2}\log 1000 \quad \blacktriangleleft \textbf{Apply the power law.}$$

$$= \frac{1}{3} \times 2 - \frac{1}{2} \times 3 \quad \blacktriangleleft \textbf{Use } \log 10^t = t.$$

$$= -\frac{5}{6} \quad \blacktriangleleft \textbf{Simplify.}$$

TRY IT YOURSELF 4.16 Brief answers provided at the end of the section.

If $\log x = 4.1$, find the value of $\log\left(\dfrac{x}{10}\right)$.

One of the common mistakes made in connection with the laws of logarithms is to apply the incorrect "equation" $\log(A + B) = \log A + \log B$. Find specific values of A and B demonstrating that this "equation" is invalid.

EXTEND YOUR REACH

The next examples show some types of calculations that are routine in calculus.

EXAMPLE 4.17 Simplifying Logarithmic Expressions

Express $\dfrac{\log x - \log(x^2 + 1)}{2}$ as the logarithm of a single quantity.

SOLUTION

First we convert the fraction to a product:

$$\frac{\log x - \log(x^2 + 1)}{2} = \frac{1}{2}(\log x - \log(x^2 + 1)) \quad \text{◀ Make the fraction a product.}$$

$$= \frac{1}{2}\log\frac{x}{x^2 + 1} \quad \text{◀ Apply the quotient law.}$$

$$= \log\sqrt{\frac{x}{x^2 + 1}} \quad \text{◀ Apply the power law.}$$

TRY IT YOURSELF 4.17 Brief answers provided at the end of the section.

Simplify the expression $\ln x + \ln\left(\dfrac{1}{x}\right)$.

EXAMPLE 4.18 Logarithms in the Exponent

Rewrite $e^{-3\ln x}$ so that no logarithms or exponential functions occur.

SOLUTION

We begin by applying the power law:

$$e^{-3\ln x} = e^{\ln(x^{-3})} \quad \text{◀ Apply the power law.}$$

$$= x^{-3} \quad \text{◀ Use } e^{\ln t} = t.$$

$$= \frac{1}{x^3} \quad \text{◀ Simplify.}$$

TRY IT YOURSELF 4.18 Brief answers provided at the end of the section.

Simplify the expression $10^{-2\log x}$.

A word of caution relating to this example: When students encounter an expression like $e^{-\ln x}$, some are tempted to do something creative with the minus sign. The proper way to deal with the minus sign is to use the power law as follows:

$$e^{-\ln x} = e^{(-1)\ln x} = e^{\ln(x^{-1})} = x^{-1} = \frac{1}{x}$$

EXAMPLE 4.19 Using Laws of Logarithms to Sketch Graphs

Use the laws of logarithms to simplify $y = \ln 2^x$, and then sketch its graph.

SOLUTION

The power rule tells us that

$$y = \ln 2^x = x \ln 2$$

It follows that this apparently complicated function is a simple linear function with slope $\ln 2 \approx 0.69$. The graph, which is a straight line, is shown in **Figure 4.15**.

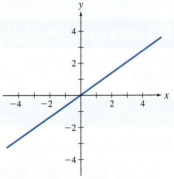

Figure 4.15 The graph of $y = \ln 2^x$: Because $\ln 2^x = x \ln 2$, the graph is a line.

TRY IT YOURSELF 4.19 Brief answers provided at the end of the section.

Simplify the function $y = \ln x^2$ for $x > 0$, and then sketch its graph.

Any exponential function can be written in terms of the natural exponential function, and this is an important skill for calculus.

EXAMPLE 4.20 Rewriting Exponential Functions

Express 7^x in the form e^{rx}. That is, find a value of r so that $7^x = e^{rx}$.

SOLUTION

We rewrite 7 as $7 = e^{\ln 7}$:

$$7^x = \left(e^{\ln 7}\right)^x \quad \blacktriangleleft \text{ Use } e^{\ln t} = t.$$

$$= e^{x \ln 7} \quad \blacktriangleleft \text{ Use } (e^A)^B = e^{AB}.$$

Hence, $7^x = e^{rx}$ if $r = \ln 7$.

TRY IT YOURSELF 4.20 Brief answers provided at the end of the section.

Write $\frac{1}{3^x}$ in the form e^{rx}.

EXTEND YOUR REACH

Explain how the logarithm can be used to express any exponential function in terms of the natural exponential function.

The **magnitude of an earthquake** is
Magnitude = log (Relative intensity).

MODELS AND APPLICATIONS   Earthquakes

The Richter scale, used for measuring earthquakes, works as follows: A baseline ground acceleration I_0 is chosen, and if an earthquake causes a ground acceleration of I, then the relative intensity of the quake is

$$\text{Relative intensity} = \frac{I}{I_0}$$

The **magnitude of an earthquake** is defined as the common logarithm of its relative intensity. Multiplying the relative intensity by 10 increases the magnitude by 1. In exponential form, the definition says

$$10^{\text{Magnitude}} = \text{Relative intensity}$$

Because the relative intensity is $10^{\text{Magnitude}}$, an increase of 1 in magnitude indicates a 10-fold multiple of relative intensity. Hence, a magnitude 6.0 earthquake is significantly more powerful, 10 times as intense, than a magnitude 5.0 quake. (Magnitudes are normally reported using one decimal place.)

Earthquakes of magnitude 2.5 or less are generally not felt. Magnitudes 6.0 through 7.0 represent major quakes that can cause serious damage. The most powerful earthquake ever recorded was a magnitude 9.5 quake in Chile on May 22, 1960.

EXAMPLE 4.21   Earthquakes: Comparing Relative Intensity

On October 17, 1989, the Loma Prieta earthquake hit Northern California, killing 67 people, injuring 3000 more, and costing more than $5 billion in damages.

a. The September 29, 2009, Samoa earthquake had a magnitude 1 higher on the Richter scale than the magnitude of the Loma Prieta earthquake. Compare the relative intensities of these two earthquakes.

b. On March 11, 2011, an earthquake struck the Pacific Ocean off the coast of Japan, causing a catastrophic tsunami. The magnitude of the Tohoku earthquake, as the geological event came to be called, was 1.9 points higher on the Richter scale than the magnitude of the Loma Prieta earthquake. Compare the relative intensity of the Tohoku earthquake with that of the Loma Prieta earthquake.

SOLUTION

a. Increasing the magnitude by 1 multiplies the relative intensity by 10. Thus, the relative intensity of the Samoa earthquake was 10 times that of the Loma Prieta earthquake.

b. In this case the increase in magnitude is not a whole number, so we must work harder. Let's use M_l and I_l to denote the magnitude and relative intensity of the Loma Prieta earthquake and M_t and I_t for the corresponding values for the Tohoku earthquake. In terms of M_t and M_l, the given information is that $M_t = M_l + 1.9$. To convert this equation into a relation involving intensity, we use the formula

$$\text{Relative intensity} = 10^{\text{Magnitude}}$$

which tells us that $I_l = 10^{M_l}$ and $I_t = 10^{M_t}$. Because $M_t = M_l + 1.9$, we have

$$10^{M_t} = 10^{M_l + 1.9} \qquad \text{◀ Exponentiate.}$$

$$10^{M_t} = 10^{M_l} 10^{1.9} \qquad \text{◀ Use } 10^{A+B} = 10^A 10^B.$$

$$I_t = I_l 10^{1.9} \qquad \text{◀ Use the intensity-magnitude relation.}$$

Because $10^{1.9} \approx 79.43$, we conclude that the Tohoku earthquake was almost 80 times as intense as the Loma Prieta earthquake.

TRY IT YOURSELF 4.21   Brief answers provided at the end of the section.

How does the relative intensity of an earthquake compare with that of the Loma Prieta earthquake if its magnitude is 2 points lower on the Richter scale than the magnitude of the Loma Prieta earthquake?

EXTEND YOUR REACH   Earthquakes of magnitude 8.0 or larger, such as the magnitude 9.5 quake that struck Chile in 1960, are truly catastrophic. Geologists have good reason to believe that, in spite of recent television movies, magnitude 10.0 earthquakes are theoretically possible but extraordinarily unlikely. Do some Internet research to find out why such quakes are deemed so unlikely.

EXAMPLE 4.22   Earthquakes: Comparing Magnitudes

On January 24, 2016, a 7.1-magnitude earthquake was reported in the Kenai Peninsula of Alaska. On December 26, 2004, an earthquake struck the Indian Ocean near Indonesia. This earthquake was 100 times as intense as the 2016 Alaska quake. What was the Richter scale reading for the Indonesian earthquake?

SOLUTION

Multiplying the relative intensity by 10 increases the magnitude by 1. Multiplying by 100 is the same as multiplying by 10 twice, thus increasing the magnitude by 2. The magnitude of the Indonesia quake was 9.1.

TRY IT YOURSELF 4.22   Brief answers provided at the end of the section.

If an earthquake is half as intense as the 2016 Alaska quake, what is its magnitude? *Suggestion*: If I is the relative intensity of the Alaska quake, then $7.1 = \log I$.

The loudness of sound is measured by using a logarithm in much the same way as is done with the Richter scale. The intensity of a sound is measured in watts per square meter. Then this intensity is divided by a "base intensity," which physicists have set at 10^{-12} watts per square meter. (This base intensity is about the softest sound anyone can hear.) This quotient is called the relative intensity. The loudness of a sound, the **decibel** level, is calculated as 10 times the common logarithm of the relative intensity.

The **decibel** level of a sound is calculated using the formula Decibels = 10 log (Relative intensity).

In exponential form, this formula can be written as

$$\text{Relative intensity} = 10^{\text{Decibels}/10}$$

Decibel readings are normally given as whole numbers. A whisper is about 20 decibels, and normal conversation is about 60 decibels. The front row at a rock concert may be 110 decibels, and continued exposure to even louder sounds can damage the ear.

EXAMPLE 4.23   Loudness: Comparing Relative Intensity

If one sound has a relative intensity of 100 times that of another, how do their decibel readings compare?

SOLUTION

Let the two sounds have relative intensities I_1 and I_2 and corresponding decibel readings of D_1 and D_2. The given information is that $I_2 = 100I_1$. We have

$$D_2 = 10 \log I_2 \qquad \text{◀ Use the decibel-intensity relation.}$$
$$= 10 \log(100I_1) \qquad \text{◀ Use } I_2 = 100I_1.$$
$$= 10(\log 100 + \log I_1) \qquad \text{◀ Apply the product law.}$$
$$= 10 \times 2 + 10 \log I_1 \qquad \text{◀ Use } \log 100 = 2.$$
$$= 20 + D_1 \qquad \text{◀ Use the decibel-intensity relation.}$$

We conclude that multiplying the relative intensity by 100 increases the decibel reading by 20.

TRY IT YOURSELF 4.23   Brief answers provided at the end of the section.

If one sound is five times as intense as another, how do their decibel readings compare?

EXAMPLE 4.24   Loudness: Comparing Decibels

Suppose we have a speaker playing music at 60 decibels. If we add a second speaker that is also playing music at 60 decibels, what is the loudness of the combined speakers?

SOLUTION

Let I denote the relative intensity of a single speaker. Because it is playing at 60 decibels, we have $10 \log I = 60$. Placing the speakers side by side doubles the relative intensity. So, the decibel reading of the pair of speakers is

$$\text{Decibels} = 10 \log(2I) \qquad \text{◀ Use the decibel-intensity relation.}$$
$$= 10(\log 2 + \log I) \qquad \text{◀ Apply the product law.}$$
$$= 10 \log 2 + 10 \log I \qquad \text{◀ Simplify.}$$
$$= 10 \log 2 + 60 \qquad \text{◀ Use } 10 \log I = 60$$

Using a calculator gives $10 \log 2 \approx 3$, so we have

$$\text{Decibels} \approx 3 + 60 = 63$$

Therefore, the loudness is about 63 decibels. Note that doubling the intensity causes only a small increase in the decibel level.

TRY IT YOURSELF 4.24   Brief answers provided at the end of the section.

What is the decibel reading if three speakers are used?

It is important to note that the decibel scale is not logarithmic in nature because physicists like to apply complicated scales to sound. It is logarithmic in nature because that is how we actually hear. As sound increases in intensity, perceived loudness increases logarithmically.

TRY IT YOURSELF ANSWERS

4.14   1

4.15   −4.3

4.16   3.1

4.17   0

4.18 $x^{-2} = \dfrac{1}{x^2}$

4.19 $y = \ln x^2 = 2\ln x$

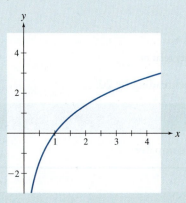

4.20 $e^{x\ln(1/3)} = e^{-x\ln 3}$

4.21   The relative intensity is 1 one-hundredth that of the Loma Prieta earthquake.

4.22   6.8

4.23   The level of the louder sound is about 7 decibels higher than that of the softer sound.

4.24   65 decibels

EXERCISE SET 4.2

CHECK YOUR UNDERSTANDING

1. In words, the logarithm of a product is _____.

2. In words, the logarithm of a quotient is _____.

3. The magnitude M of an earthquake can be calculated from the relative intensity I using the formula _____.

4. The loudness D, in decibels, of a sound can be calculated from the relative intensity I using the formula _____.

5. The expression $e^{-\ln x}$ simplifies to

   a. $-x$

   b. $\dfrac{1}{x}$

   c. $-\dfrac{1}{x}$

   d. x

   e. $\ln(-x)$

6. **True or false:** The logarithm of a sum is the product of the logarithms.

7. Criticize the following calculation:

   $\log 20 = \log(10 + 10) = \log 10 + \log 10 = 1 + 1 = 2$

8. The product law of logarithms is a consequence of which law of exponents?

9. The natural logarithm is the logarithm with base:

   a. 10

   b. e

   c. 1

   d. $\log e$

SKILL BUILDING

Calculating logarithms. In Exercises 10 through 24, use the laws of logarithms to find the exact value of the given expression without using a calculator.

10. $\ln(3e^2) - \ln 3$

11. $\log 400 - 2\log 2$

12. $2\log_6 3 + \log_6 4$

13. $\ln\left(\dfrac{e}{e+e^2}\right) + \ln(1+e)$

14. $2\ln(3e) - \ln 9$

15. $\log_2 320 - \log_2 10$

16. $\log_3 5 - \log_3 15$

17. $\ln\left(\dfrac{e^3}{\sqrt{e}}\right)$

18. $\log_6 72 - \log_6 2$

19. $\log_{15} 3 + \log_{15} 5$

20. $\log 25 + \log 4$

21. $\ln\left(\dfrac{\sqrt{e}}{\sqrt[3]{e}}\right)$

22. $\log_6 8 + 3\log_6 3$

23. $\log 3 - \log 60 - \log 5$

24. $\ln(4e) + \ln(5e) - \ln 20$

Changing forms. In Exercises 25 through 35, rewrite the given expression so that no exponential or logarithmic functions are involved.

25. $e^{-\ln x}$

26. $e^{4\ln x}$

27. $e^{\ln(x+1)-\ln x}$

28. $e^{2\ln x + \ln\sqrt{x}}$

29. $e^{-5\ln x}$

30. $e^{3\ln 2x}$

31. $e^{3(\ln 10)(\log x)}$

32. $2^{-\log_2 x}$

33. $2^{3\log_2 y - \log_2 x}$

34. $10^{2\log y + 5\log x}$

35. $10^{-2\log x}$

Combining logarithms. In Exercises 36 through 49, rewrite the given expression as the logarithm of a single quantity, and simplify.

36. $\ln(x^2 - 1) - \ln(x-1)$

37. $\ln x + 2\ln y - \ln z$

38. $\ln\left(\dfrac{e}{x}\right) + \ln x$

39. $\log(x-1) + \log(x+1)$

40. $\log(10x^2) + 2\log x$

41. $\ln x + \ln y - 2\ln z$

42. $2\ln x - \ln(1+x) + \ln(1-x)$

43. $2\ln x + 4\ln y$

44. $x\ln 2 + 2\ln x$

45. $\ln(x^2 - 3x + 2) - 2\ln(x-1)$

46. $(\ln 10)(\log x)$

47. $\dfrac{\ln e^x}{x\log 10}$

48. $\ln(\log_2 4^x) - \ln(\log_4 4^x)$

49. $x\ln 6 - x\ln 2$

Expanding logarithmic expressions. In Exercises 50 through 60, expand and simplify the given logarithmic expression.

50. $\ln(ex)$

51. $\log\left(\dfrac{100}{x}\right)$

52. $\ln\left(\dfrac{xy}{e^2}\right)$

53. $\ln x^2 + 2\ln\left(\dfrac{1}{x}\right)$

54. $\log_2(8x)$

55. $\log\left(\dfrac{xy}{z}\right)$

56. $x - \ln e^x$

57. $\ln(xe^x)$

58. $\ln\left(\dfrac{2x}{e^x + e^{2x}}\right)$

59. $\ln\left(\dfrac{\sqrt{1+x^2}}{e}\right)$

60. $\ln(\sqrt{x}\sqrt[3]{1+x})$

Calculating values. In Exercises 61 through 68, use the following information to calculate the given value:

$$\ln A = 1.4 \quad \ln B = 2.1 \quad \ln C = 6 \quad \ln b = 2$$

61. $\ln(AB)$

62. $\ln\left(\dfrac{C}{A}\right)$

63. $\ln\sqrt{B}$

64. $\ln\left(\dfrac{AB}{C}\right)$

65. C

66. $\log_b C$. *Suggestion:* Use the change-of-base formula.

67. $\log_b\sqrt{AB}$. *Suggestion:* Use the change-of-base formula.

68. $\log_b(bC)$. *Suggestion:* Use the change-of-base formula.

Converting exponential functions. In Exercises 69 through 72, write the given function in terms of the function $y = e^{rx}$.

69. $y = 9^x$

70. $y = 8^x$

71. $y = 3^{-x}$

72. $y = 2^{3x}$

PROBLEMS

73. Average rate of change. Show that the average rate of change of $f(x) = \log x$ from x to $x+h$ is $\dfrac{\log(1+h/x)}{h}$ if $x > 0$, $x+h > 0$, and $h \neq 0$.

74. Changing form. Show that $x^x = e^{x\ln x}$ for $x > 0$.

75. Average rate of change. Show that the average rate of change of $f(x) = \ln x$ from x to $x+h$ is $\dfrac{\ln(1+h/x)}{h}$ if $x > 0$, $x+h > 0$, and $h \neq 0$.

76. **Secant lines.** Find the equation of the secant line for the graph of $f(x) = \ln x$ from $x = 1$ to $x = 3$.

77. **Secant lines.** Find the equation of the secant line for the graph of $f(x) = \ln x$ from $x = e$ to $x = e^2$.

78. **Exponential functions.** Show that every exponential function $f(x) = b^x$ can be written as $f(x) = e^{rx}$ for some constant r.

Graphing logarithms. In Exercises 79 through 83, use the laws of logarithms to aid in graphing the given function.

79. $y = \ln 3^x$

80. $y = \ln\left(\dfrac{1}{x}\right)$. *Suggestion:* You may wish to write the function in a different form before sketching the graph.

81. Plot $y = \ln x$ and $y = \ln(3x)$ on the same coordinate axes. *Suggestion:* First apply the product law.

82. Plot $y = \ln x$ and $y = \ln x^3$ on the same coordinate axes. *Suggestion:* First apply the power law.

83. Plot $y = \ln\left(\dfrac{e}{x}\right)$ and $y = \ln x$ on the same coordinate axes. *Suggestion:* First apply the quotient law.

Deriving logarithm laws.

84. **Quotient law for the natural logarithm.** Use the rule $e^{A-B} = \dfrac{e^A}{e^B}$ to show that $\ln\left(\dfrac{x}{y}\right) = \ln x - \ln y$.

85. **Product law for the common logarithm.** Use the rule $10^{A+B} = 10^A 10^B$ to show that $\log(xy) = \log x + \log y$.

86. **Quotient law for the common logarithm.** Use the rule $10^{A-B} = \dfrac{10^A}{10^B}$ to show that $\log\left(\dfrac{x}{y}\right) = \log x - \log y$.

87. **Power law for the common logarithm.** Use the rule $10^{pA} = (10^A)^p$ to show that $\log x^p = p \log x$.

88. **General case of the product law.** Use the rule $b^{A+B} = b^A b^B$ to show that $\log_b(xy) = \log_b x + \log_b y$.

89. **General case of the quotient law.** Use the rule $b^{A-B} = \dfrac{b^A}{b^B}$ to show that $\log_b\left(\dfrac{x}{y}\right) = \log_b x - \log_b y$.

90. **General case of the power law.** Use the rule $b^{pA} = (b^A)^p$ to show that $\log_b x^p = p \log_b x$.

Exponential data. When data are plotted, it may be difficult to determine whether it is appropriate to model them by an exponential function or by some other function. It can be shown that data points (x_n, y_n) come from an exponential function if and only if the points $(x_n, \ln y_n)$ come from a linear function. That is, the data points (x_n, y_n) are appropriately modeled by an exponential function exactly when the points $(x_n, \ln y_n)$ lie on (or near) a straight line. In Exercises 91 through 94, plot the given data points (x_n, y_n) and the data points $(x_n, \ln y_n)$. Then determine whether an exponential model may be appropriate.

91.

x	1	3	4	6	7
y	6	24	48	192	384

92.

x	1	3	4	6	7
y	1	9	16	36	49

93.

x	1	9	16	25	49
y	1.1	2.7	4.2	5.1	7.2

94.

x	1.1	3.2	4.1	4.3	5.1
y	3.3	32.2	91.7	111.5	269.2

MODELS AND APPLICATIONS

95. **The Richter scale.** One earthquake reads 4.2 on the Richter scale, and another reads 7.2. How do the two quakes compare in terms of relative intensity?

96. **The Richter scale again.** One earthquake has a Richter scale reading of 6.5. A second is 100 times as intense. What is the Richter scale reading of the second earthquake?

97. **The decibel scale.** If one sound has a relative intensity of 1000 times that of another, how do their decibel levels compare?

98. **More decibels.** One sound has a decibel reading of 55. Another has a decibel reading of 75. How do the relative intensities of the sounds compare?

99. **Jet engines.** A jet engine close up produces sound at 155 decibels. What is the decibel reading of a pair of nearby jet engines?

100. **Earthquakes in Alaska, Chile, and Missouri.** In 1964, an earthquake measuring 9.2 on the Richter scale occurred in Alaska.

a. As we noted in this section, in 1960 an earthquake with a magnitude of 9.5 occurred in Chile. How did the relative intensity of this earthquake compare with that of the 1964 Alaska earthquake?

b. On December 16, 1811, an earthquake occurred near New Madrid, Missouri.[3] The 1964, Alaska quake was 2.5 times as intense as the one in Missouri. What was the Richter scale reading for the quake in Missouri?

101. **Deadly quakes.** In 1976, one of the world's deadliest earthquakes struck Tangshan, China. It measured 8.0 on the Richter scale and killed 255,000 people.[4]

a. As we noted in Example 4.21, on March 11, 2011, a quake occurred off the coast of Japan. The magnitude was 9.0. How many times as intense as the Tangshan quake was the Japan quake?

b. In 1908, a quake one-third as intense as the Tangshan quake struck Messina, Italy, killing 70 to 100 thousand people. What was the magnitude of the Messina quake?

102. **Speakers.** At a community event, six speakers each play at 70 decibels. What is the decibel reading of the six speakers together?

103. **Brightness of stars.** Astronomers use the magnitude scale to measure apparent brightness. The relative intensity I is calculated by dividing the intensity of light from the star Vega by that of the star of interest. The apparent magnitude m is then given by

$$m = 2.5 \log I$$

a. The light striking Earth from Vega is 2.9 times as intense as that from the star Fomalhaut. What is the apparent magnitude of Fomalhaut?

b. The star Antares has an apparent magnitude of 0.92. How does the intensity of light reaching Earth from Antares compare with that from Vega?

c. If the intensity of light striking Earth from one star is twice that of another, how do their apparent magnitudes compare?

d. Can a star have negative apparent magnitude? If so, would it be a very bright or a very dim star?

104. **Spectroscopic parallax.** The apparent magnitude m of a star is the brightness of its light reaching Earth. The absolute magnitude M of a star is the intrinsic brightness (it does not depend on the distance from Earth). The difference $S = m - M$ is the spectroscopic parallax. The spectroscopic parallax is related to the distance D from Earth in parsecs[5] by

$$S = 5 \log D - 5$$

A graph of S versus D is shown in **Figure 4.16**.

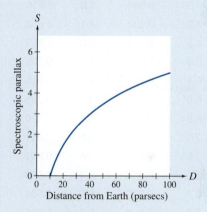

Figure 4.16 Spectroscopic parallax versus distance

a. The distance to Kaus Astralis is 38.04 parsecs. What is the spectroscopic parallax of this star?

b. The spectroscopic parallax of Rasalhague is 1.27. How far away is this star?

c. How is the spectroscopic parallax affected if the distance is multiplied by 10?

d. The star Shaula is 3.78 times as far away as the star Atria. How does the spectroscopic parallax of Shaula compare to that of Atria?

---

[3]This quake temporarily reversed the course of the Mississippi River. The area was sparsely populated at the time, and there were thought to be few fatalities.

[4]This is the official estimate. The actual death toll is thought to be much higher.

[5]One parsec is about 3.26 light-years. It is the distance from Earth that corresponds to a 1-second parallax angle (the angle subtended by the two extremes of Earth's orbit).

CHALLENGE EXERCISES FOR INDIVIDUALS OR GROUPS

105. Relationship between linear and exponential functions.

a. If $y = ab^x$, show that $z = \ln y$ is a linear function of x.

b. If $y = mx + b$, show that the substitution $z = e^y$ can be written in the form $z = cd^x$.

The common logarithm as a calculator: Part 1. We will show here how the common logarithm can be used as a calculating device, as was done before the invention of modern calculating machines. For these exercises the use of a calculator is not appropriate. At the heart of the process is the accompanying basic table of logarithms. This is what mathematicians and scientists would use in place of an electronic calculator. (This is a "cheap" calculator. A better one would use more decimal places.)

x	log x	x	log x	x	log x	x	log x	x	log x
1.0	0	1.1	0.0414	1.2	0.0792	1.3	0.114	1.4	0.146
1.5	0.176	1.6	0.204	1.7	0.230	1.8	0.255	1.9	0.279
2.0	0.301	2.1	0.322	2.2	0.342	2.3	0.362	2.4	0.380
2.5	0.398	2.6	0.415	2.7	0.431	2.8	0.447	2.9	0.462
3.0	0.477	3.1	0.491	3.2	0.505	3.3	0.519	3.4	0.531
3.5	0.544	3.6	0.556	3.7	0.568	3.8	0.580	3.9	0.591
4.0	0.602	4.1	0.613	4.2	0.623	4.3	0.633	4.4	0.643
4.5	0.653	4.6	0.663	4.7	0.672	4.8	0.681	4.9	0.690
5.0	0.699	5.1	0.708	5.2	0.716	5.3	0.724	5.4	0.732
5.5	0.740	5.6	0.748	5.7	0.756	5.8	0.763	5.9	0.771
6.0	0.778	6.1	0.785	6.2	0.792	6.3	0.799	6.4	0.806
6.5	0.813	6.6	0.820	6.7	0.826	6.8	0.833	6.9	0.839
7.0	0.845	7.1	0.851	7.2	0.857	7.3	0.863	7.4	0.869
7.5	0.875	7.6	0.881	7.7	0.886	7.8	0.892	7.9	0.898
8.0	0.903	8.1	0.908	8.2	0.914	8.3	0.919	8.4	0.924
8.5	0.929	8.6	0.934	8.7	0.940	8.8	0.944	8.9	0.949
9.0	0.954	9.1	0.959	9.2	0.964	9.3	0.968	9.4	0.973
9.5	0.978	9.6	0.982	9.7	0.987	9.8	0.991	9.9	0.996

Use this table in Exercises 106 through 113.

106. Find $\log 23$. *Suggestion*: The number 23 is not in the x-column of the table, but $23 = 2.3 \times 10$, so

$$\log 23 = \log(10 \times 2.3) = \log 10 + \log 2.3$$
$$= 1 + \log 2.3$$

107. Find $\log 0.42$.

108. Find x if $\log x = 2.14$. That is, find $10^{2.14}$. *Suggestion*: The number 2.14 is not in the $\log x$-column of the table, but if $\log x = 2.14$ then

$$\log\left(\frac{x}{100}\right) = \log x - \log 100 = \log x - 2 = 0.14$$

The number 0.14 does not appear in that column of the table, so we use the nearest value that does appear, and ultimately we find an approximation for x.

109. Find x if $\log x = -3.45$. That is, find $10^{-3.45}$.

The common logarithm as a calculator: Part 2. Suppose we wish to calculate $x = 7^{1/3}$. Then

$$\log x = \log 7^{1/3} = \frac{1}{3}\log 7 \approx \frac{1}{3} \times 0.845 \approx 0.282$$

where the first approximation comes from our basic table of logarithms. When we consult the table again, we find that the entry closest to having a logarithm of 0.282 is $x = 1.9$. Thus, $7^{1/3} \approx 1.9$. This example shows how a table of logarithms can be used as a rudimentary calculator. Use the basic table of logarithms in Exercises 110 through 113.

110. Approximate the value of $\sqrt{7}$.

111. Approximate the value of $\dfrac{\sqrt{88}}{3.7}$.

112. Approximate the value of $77^{2/5}$.

113. Approximate the value of $\dfrac{18}{\sqrt{33}}$.

---

114. **From Section P.2:** Solve the inequality $|x - 1| \geq 2$.

Answer: $(-\infty, -1] \cup [3, \infty)$

115. **From Section 2.1:** If $f(x) = \dfrac{3}{2 + x}$, find functions g and h so that $f = g \circ h$.

Answer: One solution is $g(x) = \dfrac{3}{x}$ and $h(x) = 2 + x$.

116. **From Section 2.2:** Find the inverse function of $f(x) = \dfrac{x}{x + 2}$.

Answer: $f^{-1}(x) = \dfrac{2x}{1 - x}$

117. **From Section 3.1:** Find the formula for the linear function $f(x)$ so that $f(1) = 0$ and $f(4) = 9$.

Answer: $f(x) = 3x - 3$

118. **From Section 3.2:** Calculate the limit as $x \to \infty$ of $y = 2^{-x}$.

Answer: 0

119. **From Section 3.2:** Find an exponential function whose graph goes through the points $(1, 6)$ and $(2, 18)$.

Answer: $f(x) = 2(3^x)$

120. **From Section 4.1:** Simplify the expression $10^{\log 5}$.

Answer: 5

121. **From Section 4.1:** Without using your calculator, find the exact value of $\log \sqrt{1000}$.

Answer: $\dfrac{3}{2}$

4.3 # Solving Exponential and Logarithmic Equations

Many applications require solving equations involving exponential functions and logarithms.

4.1 Logarithms: Definition and Fundamental Properties

4.2 Laws of Logarithms

4.3 Solving Exponential and Logarithmic Equations

In this section, you will learn to:

1. Solve basic exponential equations.
2. Solve logarithmic equations.
3. Solve equations involving both exponential and logarithmic functions.
4. Apply the methods of solving exponential and logarithmic equations in real-world settings.

We will use logarithms to solve equations that involve exponential functions. Finding such solutions depends on a fact we emphasized earlier: the logarithm is the inverse of an exponential function. Speaking informally, we can say that applying the logarithm undoes exponentiation, and vice versa. In a similar fashion, we can use exponential functions to solve logarithmic equations.

One important example of solving exponential equations arises in radiocarbon dating, which is used to determine the age of relatively recent organic substances. The idea is based on the assumption that the amount of carbon-14 in the air has remained stable over the past few tens of thousands of years. (There are, in fact, fairly well-understood variations.) Suppose, for example, that the remnants of a blanket are found at an archeological site. The thread used to make the blanket was made from a once-living cotton plant. Organisms such as cotton plants that exchange air through photosynthesis have a fixed percentage of carbon-14 while they are alive. Carbon-14 decays according to a well-understood exponential function into the stable form nitrogen-14, and when the plant dies the carbon-14 is not replenished. We measure the percentage of carbon-14 in the thread. (That turns out to be a technically difficult exercise.) Then determining the age of the thread is a matter of solving an exponential equation.

Other dating techniques that depend on radioactive decay are also tied to solving exponential equations. Notable examples of radiocarbon dating include the determination of glaciation dates in North America, origin dates for the Dead Sea Scrolls, and, recently, the date for the construction of Stonehenge.

Solving Exponential Equations

> The logarithm applied to many exponential equations results in a simple linear equation.

To see how to use the logarithm to solve equations involving exponential functions, let's consider $14 = 6^x$. The plan is to undo the exponential function $y = 6^x$ by applying the natural logarithm:

$$14 = 6^x$$

$$\ln 14 = \ln 6^x \qquad \blacktriangleleft \textbf{ Take the natural logarithm of both sides.}$$

$$\ln 14 = x \ln 6 \qquad \blacktriangleleft \textbf{ Apply the power law.}$$

$$\frac{\ln 14}{\ln 6} = x \qquad \blacktriangleleft \textbf{ Divide by the coefficient of x.}$$

Note that by applying the logarithm the equation became linear in x. Hence, we are able to use linear operations to complete the solution. With the aid of a calculator we find $x \approx 1.47$.

We could have used the common logarithm instead of the natural logarithm to find the equivalent solution $x = \dfrac{\log 14}{\log 6}$. We could also use the definition of the base-6 logarithm to arrive at the solution $x = \log_6 14$. The change-of-base formula shows that these expressions are the same as our earlier solution. But the solution using the natural logarithm is preferred for many reasons, including its prominent role in calculus (and throughout higher mathematics).

EXAMPLE 4.25   Solving Exponential Equations

Solve the equation $55 = 5 \times 3^x$. Give the exact answer as well as a two-digit approximation.

SOLUTION

As is shown in **Figure 4.17**, the solution of the equation is found where the graphs of $y = 55$ and $y = 5 \times 3^x$ cross.

The first step toward solving the equation is to simplify by dividing by 5:

$$55 = 5 \times 3^x$$

$$11 = 3^x \qquad \blacktriangleleft \textbf{ Divide by 5.}$$

$$\ln 11 = \ln(3^x) \qquad \blacktriangleleft \textbf{ Apply the natural logarithm.}$$

$$\ln 11 = x \ln 3 \qquad \blacktriangleleft \textbf{ Apply the power law.}$$

$$\frac{\ln 11}{\ln 3} = x \qquad \blacktriangleleft \textbf{ Divide by the coefficient of x.}$$

A calculator gives the approximate value $x \approx 2.18$.

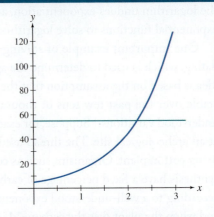

Figure 4.17 The graphs of $y = 5 \times 3^x$ and $y = 55$: The solution corresponds to the crossing point.

TRY IT YOURSELF 4.25 Brief answers provided at the end of the section.

Solve the equation $2 \times 5^x = 24$.

EXTEND YOUR REACH

We noted earlier that applying the logarithm to an exponential equation produces a linear equation. The corresponding claim in this example is that $\ln 11 = x \ln 3$ is a linear equation. Yet this equation involves the logarithm, which is a nonlinear function. Can you resolve the apparent discrepancy?

EXAMPLE 4.26 Solving Equations Involving the Natural Exponential Function

Solve the equation $7e^x = e^{3x}$. Give the exact answer as well as a two-digit approximation.

SOLUTION

The solution corresponds to the crossing point of the two graphs in **Figure 4.18**. We proceed to find it algebraically:

$$7e^x = e^{3x}$$

$$7 = \frac{e^{3x}}{e^x} \qquad \blacktriangleleft \text{ Divide to move variables right.}$$

$$7 = e^{2x} \qquad \blacktriangleleft \text{ Simplify.}$$

$$\ln 7 = \ln e^{2x} \qquad \blacktriangleleft \text{ Apply the natural logarithm.}$$

$$\ln 7 = 2x \qquad \blacktriangleleft \text{ Use } \ln e^t = t.$$

$$\frac{\ln 7}{2} = x \qquad \blacktriangleleft \text{ Divide by the coefficient of } x.$$

$$0.97 \approx x \qquad \blacktriangleleft \text{ Approximate.}$$

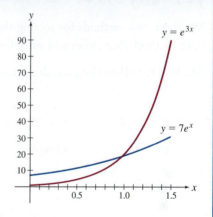

Figure 4.18 The graphs of $y = 7e^x$ and $y = e^{3x}$

TRY IT YOURSELF 4.26 Brief answers provided at the end of the section.

Solve the equation $6e^{-x} = e^{2x}$.

EXTEND YOUR REACH

a. Solve the equation $e^{2x} - 7e^x + 12 = 0$. *Suggestion*: Factor the left-hand side of the equation. To do this, think of the factorization of $X^2 - 7X + 12$. Then recall how to use this factorization to solve the equation $X^2 - 7X + 12 = 0$.

b. Solve the equation $e^{2x} - e^x - 1 = 0$. *Suggestion*: First use the quadratic formula to solve the equation $X^2 - X - 1 = 0$. Be careful: There is only one solution of the equation $e^{2x} - e^x - 1 = 0$.

Although no problem-solving strategy is effective for solving all equations involving exponential functions, the method used in the two preceding examples is often helpful.

STEP-BY-STEP STRATEGY: Solving Equations Involving Exponential Functions

Step 1: Try to use algebraic operations to write the equation in the form $a^x = b$.

Step 2: Take the logarithm of both sides: $\ln a^x = \ln b$.

Step 3: Apply the power law: $x \ln a = \ln b$.

Step 4: Divide by the coefficient of x: $x = \dfrac{\ln b}{\ln a}$.

EXAMPLE 4.27   More Exponential Equations: Alternative Methods of Solution

Solve the equation $2^{3x} = 5 \times 4^x$.

SOLUTION

Figure 4.19 shows the solution as the crossing point of the graphs of $y = 2^{3x}$ and $y = 5 \times 4^x$.

We show two methods for solving this equation. You should understand both methods, but either will work for many equations.

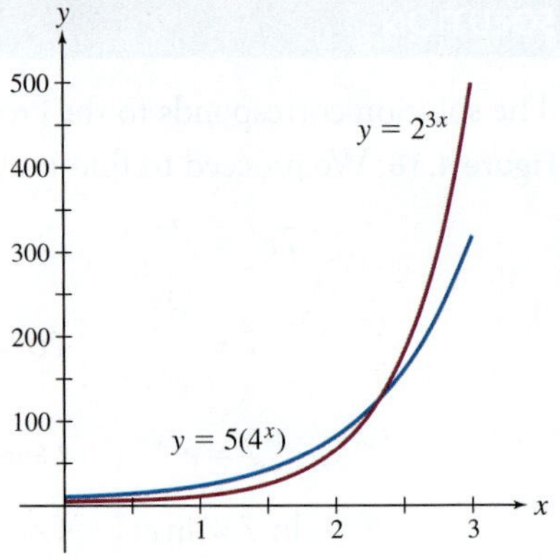

Method 1: Follow the preceding strategy.

$2^{3x} = 5 \times 4^x$

$(2^3)^x = 5 \times 4^x$ ◀ Use $2^{AB} = (2^A)^B$.

$8^x = 5 \times 4^x$ ◀ Simplify.

$\dfrac{8^x}{4^x} = 5$ ◀ Divide to move variables left.

$2^x = 5$ ◀ Use $\dfrac{A^x}{B^x} = \left(\dfrac{A}{B}\right)^x$. This is now in the form $a^x = b$.

$\ln 2^x = \ln 5$ ◀ Apply the logarithm.

$x \ln 2 = \ln 5$ ◀ Apply the power law.

$x = \dfrac{\ln 5}{\ln 2} \approx 2.32$ ◀ Divide by the coefficient of x.

Figure 4.19 The graphs of $y = 2^{3x}$ and $y = 5 \times 4^x$

Method 2: Rely more on laws of logarithms.

First we take the natural logarithm of each side, making sure that ln is applied to the entirety of each:

$$2^{3x} = 5 \times 4^x$$

$$\ln 2^{3x} = \ln(5 \times 4^x) \quad \text{◀ Apply the natural logarithm.}$$

$$\ln 2^{3x} = \ln 5 + \ln 4^x \quad \text{◀ Apply the product law.}$$

$$3x \ln 2 = \ln 5 + x \ln 4 \quad \text{◀ Apply the power law.}$$

$$3x \ln 2 - x \ln 4 = \ln 5 \quad \text{◀ Move variables left and constants right.}$$

$$x(3 \ln 2 - \ln 4) = \ln 5 \quad \text{◀ Factor.}$$

$$x = \frac{\ln 5}{3 \ln 2 - \ln 4} \quad \text{◀ Divide by the coefficient of } x.$$

A calculator gives the same approximation as for the solution found using the first method.

Solve the equation $3^{4x-1} = e^{7-x}$.

Solving Logarithmic Equations

> Applying exponential functions to many logarithmic equations results in a simple linear equation.

Just as the logarithm is used to solve equations involving exponential functions, we can exponentiate to undo the logarithm in equations involving logarithmic expressions. For example, to solve the equation $\ln(x+1) = 2$, we exponentiate both sides using the base e. That is, we raise both sides as exponents of e:

$$\ln(x+1) = 2$$

$$e^{\ln(x+1)} = e^2 \qquad \blacktriangleleft \text{ Exponentiate each side.}$$

$$x + 1 = e^2 \qquad \blacktriangleleft \text{ Use } e^{\ln t} = t.$$

$$x = e^2 - 1 \qquad \blacktriangleleft \text{ Solve for } x.$$

A calculator gives $x \approx 6.39$.

We remark that if the equation had involved the common logarithm, we would have used 10 to exponentiate both sides of the equation. To illustrate let's solve the logarithmic equation

$$\log(2x + 4) = \log(x + 20)$$

First we exponentiate both sides of the equation using 10 as the base:

$$10^{\log(2x+4)} = 10^{\log(x+20)} \qquad \blacktriangleleft \text{Exponentiate using base 10.}$$

$$2x + 4 = x + 20 \qquad \blacktriangleleft \text{Use } 10^{\log t} = t.$$

$$x = 16 \qquad \blacktriangleleft \text{Solve for } x.$$

An alternative method for solving this equation is to use the fact that $y = \log x$ is a one-to-one function to conclude from $\log(2x + 4) = \log(x + 20)$ that $2x + 4 = x + 20$ and so $x = 16$.

EXAMPLE 4.28   Solving Logarithmic Equations

Solve the equation $3 = \ln(x - 20)$.

SOLUTION

We exponentiate using base e, making sure to include the entirety of both left and right sides in the exponent:

$$3 = \ln(x - 20)$$

$$e^3 = e^{\ln(x-20)} \qquad \blacktriangleleft \text{ Exponentiate.}$$

$$e^3 = x - 20 \qquad \blacktriangleleft \text{ Use } e^{\ln t} = t.$$

$$e^3 + 20 = x \qquad \blacktriangleleft \text{ Solve for } x.$$

Using a calculator, we find $x \approx 40.09$. This solution is shown in **Figure 4.20** as the crossing point.

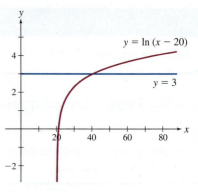

Figure 4.20 The graphs of $y = 3$ and $y = \ln(x - 20)$: The domain of $y = \ln(x - 20)$ is $(20, \infty)$.

TRY IT YOURSELF 4.28 Brief answers provided at the end of the section.

Solve the equation $\log_2(3x - 4) = 3$.

EXAMPLE 4.29   Solving Equations Using Laws of Logarithms

Solve the equation $\log x + \log 4 = 2$.

SOLUTION

This equation is easier to solve if we first simplify using the product law for logarithms:

$$\log x + \log 4 = 2$$

$$\log 4x = 2$$

The next step is to exponentiate using base 10:

$$10^{\log 4x} = 10^2 \quad \blacktriangleleft \textbf{Exponentiate.}$$

$$4x = 100 \quad \blacktriangleleft \textbf{Use } 10^{\log t} = t.$$

$$x = 25 \quad \blacktriangleleft \textbf{Divide by 4.}$$

TRY IT YOURSELF 4.29 Brief answers provided at the end of the section.

Solve the equation $\ln(10x) - \ln 2 = 4$.

Just as with exponential equations, no strategy will always work for solving logarithmic equations, but the following strategy is often helpful.

STEP-BY-STEP STRATEGY: Solving Equations Involving Logarithms

Step 1: Try to use algebraic operations and the laws of logarithms to put the equation in the form $\ln(ax + b) = c$.

Step 2: Exponentiate both sides to obtain $ax + b = e^c$.

Step 3: Solve the linear equation: $x = \dfrac{e^c - b}{a}$.

EXAMPLE 4.30 Logarithmic Equations Leading to Quadratics

Solve the equation $\ln x + \ln (x + 2) = \ln 3$.

SOLUTION

We can simplify this equation using the product law for logarithms:

$$\ln x + \ln (x + 2) = \ln 3$$

$$\ln (x(x + 2)) = \ln 3 \quad \blacktriangleleft \textbf{ Apply the product law.}$$

$$e^{\ln (x(x+2))} = e^{\ln 3} \quad \blacktriangleleft \textbf{ Exponentiate.}$$

$$x(x + 2) = 3 \quad \blacktriangleleft \textbf{ Use } e^{\ln t} = t.$$

$$x^2 + 2x - 3 = 0 \quad \blacktriangleleft \textbf{ Move all terms left.}$$

$$(x + 3)(x - 1) = 0 \quad \blacktriangleleft \textbf{ Factor.}$$

$$x = -3 \text{ or } x = 1 \quad \blacktriangleleft \textbf{ Solve the factored equation.}$$

When we solve logarithmic equations, it is important to check solutions. Because -3 is not in the domain of the logarithm, only $x = 1$ is a solution of the original equation.

TRY IT YOURSELF 4.30   Brief answers provided at the end of the section.

Solve the equation $\ln x + \ln (x - 1) = 1$. The quadratic formula will be needed to complete the solution.

Mixed Equations

> Some equations require the use of both exponential and logarithmic functions for their solution.

Sometimes solving equations requires a mixture of the strategies we have already used. Consider, for example, the equation $\ln (1 + e^x) = 3$. We first treat this as a logarithmic equation of the form $\ln (1 + u) = 3$. (In fact, some might find it easier to follow this procedure by making the formal substitution $u = e^x$.) We exponentiate using base e:

$$\ln (1 + e^x) = 3$$

$$e^{\ln (1+e^x)} = e^3 \quad \blacktriangleleft \textbf{ Exponentiate both sides.}$$

$$1 + e^x = e^3 \quad \blacktriangleleft \textbf{ Use } e^{\ln A} = A.$$

Now we have an exponential equation, which we proceed to solve using the natural logarithm:

$$1 + e^x = e^3$$

$$e^x = e^3 - 1 \quad \blacktriangleleft \textbf{ Subtract 1 from both sides.}$$

$$\ln e^x = \ln (e^3 - 1) \quad \blacktriangleleft \textbf{ Apply the logarithm to both sides.}$$

$$x = \ln (e^3 - 1) \quad \blacktriangleleft \textbf{ Use } \ln e^A = A.$$

A calculator gives $x \approx 2.95$.

EXAMPLE 4.31 Solving Mixed Equations

Solve the following equations.

a. $2^{\ln x} = 3$

b. $2 = \ln(1 + 3^x)$

SOLUTION

a. The solution of the equation $2^{\ln x} = 3$ corresponds to the crossing point in **Figure 4.21**. We will use the formal substitution $u = \ln x$. (Some may prefer to solve directly without making this substitution.) We have

$$2^{\ln x} = 3$$

$$2^u = 3 \qquad \blacktriangleleft \text{ Substitute } u = \ln x.$$

$$\ln 2^u = \ln 3 \qquad \blacktriangleleft \text{ Apply the logarithm.}$$

$$u \ln 2 = \ln 3 \qquad \blacktriangleleft \text{ Apply the power law.}$$

$$u = \frac{\ln 3}{\ln 2} \qquad \blacktriangleleft \text{ Divide by the coefficient of } u.$$

$$\ln x = \frac{\ln 3}{\ln 2} \qquad \blacktriangleleft \text{ Replace } u \text{ by } \ln x.$$

$$e^{\ln x} = e^{\ln 3/\ln 2} \qquad \blacktriangleleft \text{ Exponentiate.}$$

$$x = e^{\ln 3/\ln 2} \approx 4.88 \qquad \blacktriangleleft \text{ Use } e^{\ln t} = t.$$

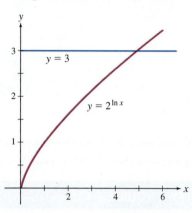

Figure 4.21 The graphs of $y = 3$ and $y = 2^{\ln x}$

b. The first step in solving $2 = \ln(1 + 3^x)$ is to exponentiate to undo the logarithm. (Some will find the work easier if the substitution $u = 3^x$ is used.) We have

$$2 = \ln(1 + 3^x)$$

$$e^2 = e^{\ln(1+3^x)} \qquad \blacktriangleleft \text{ Exponentiate.}$$

$$e^2 = 1 + 3^x \qquad \blacktriangleleft \text{ Use } e^{\ln t} = t.$$

$$e^2 - 1 = 3^x \qquad \blacktriangleleft \text{ Move constants left.}$$

$$\ln(e^2 - 1) = \ln 3^x \qquad \blacktriangleleft \text{ Apply the logarithm.}$$

$$\ln(e^2 - 1) = x \ln 3 \qquad \blacktriangleleft \text{ Apply the power law.}$$

$$\frac{\ln(e^2 - 1)}{\ln 3} = x \qquad \blacktriangleleft \text{ Divide by the coefficient of } x.$$

Using a calculator, we find $x \approx 1.69$.

TRY IT YOURSELF 4.31 Brief answers provided at the end of the section.

Solve the equation $\ln(e^x - 3) = 5$.

EXTEND YOUR REACH Try to solve the equation $e^x = \ln x$. You may find this challenging. Try plotting the graphs of $y = e^x$ and $y = \ln x$ on the same coordinate axes.

MODELS AND APPLICATIONS   Radiocarbon Dating

As was discussed in the opening of this section, radiocarbon dating is an important archeological tool that involves solving exponential equations.

EXAMPLE 4.32 Carbon Dating

Carbon-14 decays over time. If the initial amount is C_0, then the amount C remaining after t years is given by

$$C = C_0 \times 0.5^{t/5730}$$

Charcoal from an ancient campfire is found to contain one-third the percentage of carbon-14 as do living trees. How long ago was the tree cut down for use as campfire wood? Round your answer to the nearest 100 years.

SOLUTION

The amount of carbon-14 in the sample of charcoal is one-third of the original amount, $\frac{1}{3}C_0$. So, to find the age of the charcoal we need to know how long it takes for the amount of carbon-14 remaining to reach $\frac{1}{3}C_0$. That is, we need to solve for t the exponential equation

$$\frac{1}{3}C_0 = C_0 \times 0.5^{t/5730}$$

We solve the exponential equation using our earlier strategy:

$$\frac{1}{3}C_0 = C_0 \times 0.5^{t/5730}$$

$$\frac{1}{3} = 0.5^{t/5730} \qquad \blacktriangleleft \text{Divide by } C_0.$$

$$\ln\frac{1}{3} = \ln\left(0.5^{t/5730}\right) \qquad \blacktriangleleft \text{Apply the logarithm.}$$

$$\ln\frac{1}{3} = \frac{t}{5730}\ln 0.5 \qquad \blacktriangleleft \text{Apply the power law.}$$

$$\frac{5730\ln\left(\frac{1}{3}\right)}{\ln 0.5} = t \qquad \blacktriangleleft \text{Divide by the coefficient of } t.$$

This solution is the crossing point of the graph of $y = 0.5^{t/5730}$ with the line $y = \frac{1}{3}$ as shown in **Figure 4.22**.

A calculator yields the value 9081.84. We conclude that, rounded to the nearest 100 years, the charcoal is about 9100 years old.

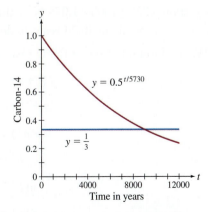

Figure 4.22 Graphs of $y = 0.5^{t/5730}$ and $y = \frac{1}{3}$

TRY IT YOURSELF 4.32 Brief answers provided at the end of the section.

A log used for roof support in an ancient building is found to have one-fifth of the percentage of carbon-14 as do living trees. How old is the log? Round your answer to the nearest 100 years.

EXAMPLE 4.33 Doubling Time

a. The value B of an investment grows according to the formula

$$B = 800 \times 1.07^t \text{ dollars}$$

Here t is the time in years since the initial investment. How long does it take for the value of the investment to double the initial value? How long does it take to double again?

b. In general, consider $f(t) = a(b^t)$ with $b > 1$. The doubling time for this exponential function is d if $f(t+d) = 2f(t)$ no matter what the value of t is. Find the doubling time for f.

SOLUTION

a. The investment doubles when $B = 1600$. Hence, we need to solve the exponential equation $1600 = 800 \times 1.07^t$ for t. We solve the exponential equation using our earlier strategy:

$$1600 = 800 \times 1.07^t$$

$$2 = 1.07^t \quad \blacktriangleleft \textbf{Divide by 800.}$$

$$\ln 2 = \ln 1.07^t \quad \blacktriangleleft \textbf{Apply the logarithm.}$$

$$\ln 2 = t \ln 1.07 \quad \blacktriangleleft \textbf{Apply the power law.}$$

$$\frac{\ln 2}{\ln 1.07} = t \quad \blacktriangleleft \textbf{Divide by the coefficient of } t.$$

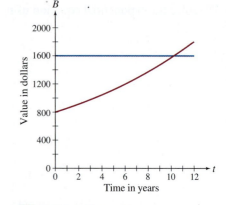

Thus, the investment doubles in $t \approx 10.24$ years. This solution is shown in **Figure 4.23** as the crossing point of the graph of $y = 800 \times 1.07^t$ with the line $y = 1600$.

Figure 4.23 Doubling time

To determine how long it takes for the value to double again, we need to know how long it takes the investment to grow from \$1600 to \$3200. That is, we need to solve the equation $3200 = 1600 \times 1.07^t$. Note that if we divide by 1600, we get the equation we just solved. So, it takes the same time, about 10.24 years, to double once more.

b. We need to find d so that for all t we have $f(t+d) = 2f(t)$:

$$f(t+d) = 2f(t)$$

$$a(b^{t+d}) = 2a(b^t) \quad \blacktriangleleft \textbf{Evaluate the function.}$$

$$a(b^t b^d) = 2ab^t \quad \blacktriangleleft \textbf{Use } A^{x+y} = A^x A^y.$$

$$b^d = 2 \quad \blacktriangleleft \textbf{Divide by } ab^t.$$

$$\ln b^d = \ln 2 \quad \blacktriangleleft \textbf{Apply the logarithm.}$$

$$d \ln b = \ln 2 \quad \blacktriangleleft \textbf{Apply the power law.}$$

$$d = \frac{\ln 2}{\ln b} \quad \blacktriangleleft \textbf{Divide by the coefficient of } d.$$

TRY IT YOURSELF 4.33 Brief answers provided at the end of the section.

How long does it take for the function in part b to triple?

TRY IT YOURSELF ANSWERS

4.25 $x = \dfrac{\ln 12}{\ln 5} \approx 1.54$

4.26 $x = \dfrac{1}{3}\ln 6 \approx 0.60$

4.27 $x = \dfrac{\ln 3 + 7}{4\ln 3 + 1} \approx 1.50$

4.28 $x = 4$

4.29 $x = \dfrac{e^4}{5} \approx 10.92$

4.30 $x = \dfrac{1 + \sqrt{1 + 4e}}{2} \approx 2.22$

4.31 $x = \ln(3 + e^5) \approx 5.02$

4.32 13,300 years

4.33 $\dfrac{\ln 3}{\ln b}$

EXERCISE SET 4.3

CHECK YOUR UNDERSTANDING

1. To solve an equation of the form $a^x = b$, the first step is _____.

2. The first step toward solving an equation of the form $\ln(ax + b) = c$ is _____.

3. If $e^x = c$, then $x =$:

   a. $\ln c$

   b. $\log c$

   c. $\log_c e$

   d. 10^c

4. If $10^x = c$, then $x =$:

   a. $\ln c$

   b. $\log c$

   c. $\log_c e$

   d. e^c

5. If $\ln x = c$, then $x =$:

   a. $\log c$

   b. 10^c

   c. e^c

   d. $\log_x c$

6. If $\log x = c$, then $x =$:

   a. $\log c$

   b. 10^c

   c. e^c

   d. $\log_x c$

7. Which of the following are correct solutions of $4^x = 7$?

   a. $x = \dfrac{\ln 7}{\ln 4}$

   b. $x = \dfrac{\log 7}{\log 4}$

   c. $x = \log_4 7$

   d. All of the above

   e. None of the above

8. Solve the equation $\ln x + \ln(x + 1) = \ln(x^2 + x)$.

9. My friend tried to solve the equation $4e^x = e^3$ as follows: First he took the natural logarithm of both sides: $\ln(4e^x) = \ln e^3$. Then he got $(\ln 4)(\ln e^x) = 3$, from which he obtained $(\ln 4)x = 3$, and finally $x = \dfrac{3}{\ln 4}$. What error did he make?

10. My friend tried to solve the equation $4 + \ln x = \ln 5$ as follows: First she exponentiated both sides: $e^{4 + \ln x} = e^{\ln 5}$. Then she got $e^4 + e^{\ln x} = e^{\ln 5}$, from which she obtained $e^4 + x = 5$, and finally $x = 5 - e^4$. What error did she make?

SKILL BUILDING

Solving exponential equations. In Exercises 11 through 29, solve the equation for x.

11. $4 \times 3^x = 17$

12. $2^x = 5$

13. $e^x = t$ if $t > 0$

14. $e^x = 2$

15. $3^x = 2$

16. $2^x = 3 \times 5^x$

17. $7^x 9^x = 2 \times 58^x$

18. $\dfrac{7^x}{6^x} = 17$

19. $3 = \dfrac{51}{1 + 4e^x}$

20. $5e^{-x} = 2^{2x}$

21. $3 = \dfrac{7e^x}{4 + e^x}$

22. $2^{3x+1} = 3^{2x+1}$

23. $e^\pi = \pi^{ex}$

24. $e^x = e^{-x}$

25. $e^{3x} = 5e^{2x}$

26. $\dfrac{e^x}{e^x + e^{-x}} = \dfrac{1}{2}$

   Suggestion: First multiply top and bottom by e^x.

27. $a = bc^x$ if $a, b, c > 0$ and $c \neq 1$

28. $ab^x = cd^x$ if $a, b, c, d > 0$ and $b \neq d$

29. $(2e^x + e^{-x})(2e^x - e^{-x}) = e^{2x}$

Solving logarithmic equations. In Exercises 30 through 47, solve the equation for x.

30. $\ln x = 6$

31. $\ln x = 2$

32. $\log x = 3$

33. $\log_2 x = 5$

34. $\ln x = t$

35. $\ln \ln x = 1$

36. $\log \log x = 2$

37. $\log x = 6$

38. $\ln (x + 1) = 4$

39. $\ln (2x - 3) = 1$

40. $\log (x + 1) = 2$

41. $\log (3x + 1) = 1$

42. $\ln (2x - 1) = \ln (7)$

43. $\ln (3x - 2) = \ln (6 - x)$

44. $\log (3x) = 2 + \log (2x - 1)$

45. $\ln (3x) - \ln (1 + x) = 1$

46. $\log (3x) - \log (1 + x) = 1$

47. $\ln (e + x) = 2$

Solving mixed equations. In Exercises 48 through 59, solve the equation for x.

48. $4^{\ln x} = 29$

49. $3^{2 + \ln x} = 16$

50. $2^{1 + \ln x} = 4^{\ln x}$

51. $2^{1 + \log x} = 4^{1 - \log x}$

52. $\log (4 + x) = 2$

53. $\log (3 + 10^x) = 1$

54. $\ln (1 + 2^x) = 5$

55. $\ln (e^x + 1) = 2 + \ln (e^x - 1)$

56. $a^{b + \ln x} = c$ if $a, c > 0$ and $a \neq 1$

57. $2^{(3^x)} = 5$

58. $a^{b + \ln x} = c^{d + \ln x}$ if $a, c > 0$ and $a \neq c$

59. $\ln (a + bc^x) = d$ if $a, b, c > 0, c \neq 1$, and $d > \ln a$

Finding inverse functions. In Exercises 60 through 65, find the inverse of the given function. Recall that we do this by solving $y = f(x)$ for x.

60. $f(x) = 3 \times 2^x$

61. $f(x) = \ln (1 + e^x)$

62. $f(x) = \log (2 - x)$

63. $f(x) = 2^{1 + \log x}$

64. $f(x) = 2 + 3 \ln x$

65. $f(x) = \ln (x + 1) - \ln (x - 1)$

Logarithms with various bases. In Exercises 66 through 71, logarithms other than natural logarithms are involved. Solve the equations.

66. $\log_2 (x + 1) = 3$

67. $\log_3 (3^x - 54) = 3$

68. $\log_2 x + \log_2 (x + 2) = 3$

69. $\log_2 (7 + \log_3 x) = 4$

70. $\log_4 (18 \times 4^x - 32) = 2x$. *Suggestion*: Exponentiate, then substitute $u = 4^x$.

71. $\log_5 (23 + \log_3 x) = 2$

Solving equations using technology. Exercises 72 through 75 involve equations that cannot easily be solved by hand. For each equation, use a graphing utility to make an appropriate graph, and report the solution correct to two decimal places.

72. $7 + \ln x = x$

73. $e^{-x} = x^2$

74. $3 + \ln x = e^x$.
    There are two solutions.

75. $\ln x = (x - 2)^2$.
    There are two solutions.

MODELS AND APPLICATIONS

76. **Time to quadruple.** An investment grows according to the formula $B = B_0 \times 1.05^t$, where t is measured in years. How long does it take for the value of the investment to quadruple? This situation is illustrated in **Figure 4.24**, in which case the initial investment is \$1000.

77. **Doubling time.** If $f(t) = a(b^t)$ and $b > 1$, then according to Example 4.33 the doubling time is $\dfrac{\ln 2}{\ln b}$. The initial size of a population is N_0. The population doubles every 7 years. Find a value of r so that the function $N = N_0 e^{rt}$ models this population.

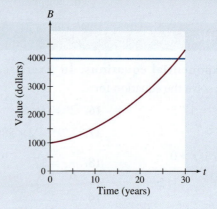

Figure 4.24 An investment that quadruples

78. Halving time. Suppose an investment is losing value. Its balance is given by $B = B_0(b^t)$, where B_0 is the initial investment, t is the time measured in years, and $0 < b < 1$. How long will it take for the value of the investment to decline to half of its original value? How much additional time is required for the investment to reach one-quarter of its original value?

79. Half-life. A population is modeled by

$$N = N_0 \times 0.96^t$$

where t is measured in years and N_0 is the initial population. What is the half-life of this population? That is, how long is required for the population to be reduced to half of its initial level?

80. Removing salt. A desalination process reduces the amount of salt in a container of water by 4% per hour. If 55 grams is the initial amount of salt, then the amount of salt remaining after t hours is modeled by

$$S = 55 \times 0.96^t \text{ grams}$$

  a. Make a graph of the amount of salt remaining after t hours. Your graph should cover the first 30 hours of the process.

  b. How long must the process continue to reduce the amount of salt to 5 grams?

81. Toxic waste. There are originally T_0 grams of a toxic substance in a dangerous waste site. The amount T of toxic waste remaining at the site after t days of a cleanup operation is modeled by

$$T = T_0 \times b^t$$

where $0 < b < 1$. How long is required for the initial amount of toxic waste at the site to be reduced by 30%?

82. Equity. When your home is mortgaged, its value is shared between you and the lending institution. Your equity in the home is the part of the initial amount borrowed that you have paid. It is the part of the value of the home that you own. (Here we are ignoring your down payment.) If B_0 is the initial amount borrowed, the monthly rate is $r = \dfrac{APR}{12}$ as a decimal, and the term of the mortgage is p months, then after k monthly payments your equity is given by

$$E = B_0 \frac{(1+r)^k - 1}{(1+r)^p - 1}$$

How many payments must you make before your equity is half the value of the home? Your answer should be expressed in terms of p and r.

83. Spell checkers. As we noted in Section 4.1, word processors commonly use a search tree structure to check for misspelled words. Using this, a spell checker can accommodate a dictionary of $2^c - 1$ words using at most c checks for each test word. A modern spell checker uses a dictionary of about 100,000 words. What is the maximum number of checks required to test whether a word appears in the computer's dictionary?

84. A skydiver. The terminal velocity for an object dropped near the surface of Earth is the maximum downward velocity that will be attained by the falling object. For a man of average size, the terminal velocity is about 176 feet per second. It is found that a skydiver's velocity v in feet per second t seconds into his fall (before he opens his parachute) is given by

$$v = 176(1 - 0.75^t)$$

How long does it take the skydiver to achieve 99% of terminal velocity? Round your answer to the nearest second.

85. Antibiotic in the bloodstream. The amount of an antibiotic in a patient's bloodstream is modeled by $A = 65 \times 0.48^t$, where t is the time in hours since injection and A is measured in milligrams. When will the amount of the antibiotic reach 7 milligrams? Round your answer to the nearest hour.

86. The George Reserve. The circumstances surrounding the George Reserve in Michigan have made it particularly easy for ecologists to monitor accurately the growth of the deer population on the reserve.[6] They have developed the logistic function

$$N = \frac{6.21}{0.035 + 0.45^t}$$

to model the population. Here N denotes the number of deer and t denotes the time in years since observation began. According to this model, when will the deer population reach 160? Round your answer to the nearest year.

---

[6]Dale R. McCullough, *The George Reserve Deer Herd*, University of Michigan Press, Ann Arbor, MI, 1979.

87. **Logistic population growth.** The logistic function

$$N = \frac{K}{1 + be^{-rt}}$$

is often used to model population growth under environmental constraints. Here N is the population, t is the time in years since observation began, and K, b, and r are positive constants determined by the environment, the species being described, and the initial population. The model of maximum sustainable yield applies to harvested resources such as trees or commercially raised catfish. It says that the population should be maintained at a level of $N = \dfrac{K}{2}$. At what time does the population reach this level? Your answer should be expressed in terms of b and r.

88. **Reaction time.** For certain decisions, the time it takes to respond is a logarithmic function of the number of choices faced.[7] One model is

$$R = 0.17 + 0.44 \log N$$

where R is the reaction time in seconds and N is the number of choices. Find a formula that gives the number of choices that can be faced in terms of the reaction time.

89. **GDP of India.** The gross domestic product (GDP) of a country is the annual market value of the goods and services produced in that country. One model for the GDP of India is $I = 2.7 \times 1.07^t$, where t is the time in years since 2018 and I is measured in trillions of U.S. dollars. According to this model, when will the GDP of India reach 5 trillion U.S. dollars? Round your answer to the nearest year.

REVIEW AND REFRESH: Exercises from Previous Sections

90. **From Section 1.2:** If $f(x) = \dfrac{1}{x}$, find the average rate of change of f from $x = 1$ to $x = 2$.

    Answer: $-\dfrac{1}{2}$

91. **From Section 1.3:** If the function f is increasing and its graph is concave up, explain how f changes over its domain.

    Answer: The function f is increasing at an increasing rate.

92. **From Section 1.4:** Find the limit as $x \to \infty$ of

    $$y = \frac{4x + 1}{2x + 3}.$$

    Answer: 2

93. **From Section 2.3:** Explain how to get the graph of $f(-x)$ from the graph of $f(x)$.

    Answer: The graph of $f(-x)$ is the graph of $f(x)$ reflected across the vertical axis.

94. **From Section 3.1:** Find the equation of the linear function $f(x)$ so that $f(1) = 5$ and f has slope 2.

    Answer: $f(x) = 2x + 3$

95. **From Section 3.2:** Find an exponential function with initial value 3 and base 2.

    Answer: $f(x) = 3(2^x)$

96. **From Section 4.1:** Without using your calculator, find the exact value of $\log \sqrt{10}$.

    Answer: $\dfrac{1}{2}$

97. **From Section 4.1:** Without using your calculator, find the exact value of $\ln\left(\dfrac{1}{e^2}\right)$.

    Answer: -2

98. **From Section 4.2:** If $x > 0$, simplify $\ln x + \ln \dfrac{1}{x}$.

    Answer: 0

99. **From Section 4.2:** If $x > 0$, simplify $e^{-\ln x}$.

    Answer: $\dfrac{1}{x}$

---

[7]The model in this exercise is based on *Space Mathematics* by B. Kastner, published by NASA, 1985.

CHAPTER ROADMAP AND STUDY GUIDE

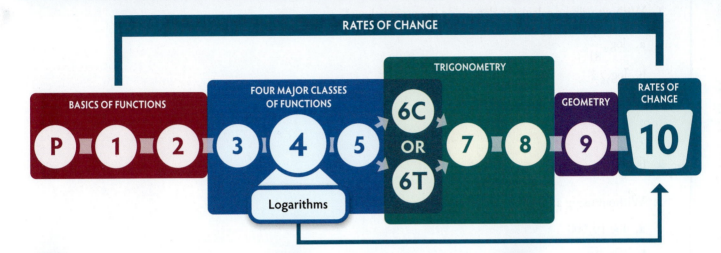

4.1 Logarithms: Definition and Fundamental Properties

Whereas increasing exponential functions have an increasing rate of change, increasing logarithmic functions have a decreasing rate of change.

The Logarithm as an Inverse Function

Every logarithm is the inverse of an exponential function.

Graphs of Logarithms

For a positive base greater than 1, the graph of the logarithm is increasing and concave down.

Common and Natural Logarithms

The common logarithm has base 10, and the natural logarithm has base e.

Long-Term Behavior of the Logarithm

For bases larger than 1, the logarithm increases without bound.

Change-of-Base Formula

Any logarithm can be expressed in terms of the natural logarithm.

MODELS AND APPLICATIONS
Spell Checkers

4.2 Laws of Logarithms

Logarithms use the product, power, and quotient laws to simplify complicated expressions.

Deriving the Laws of Logarithms

The laws of logarithms are derived using the rules followed by their inverses, exponential functions.

Using Laws of Logarithms

The laws of logarithms are used to analyze expressions that would otherwise be unwieldy.

MODELS AND APPLICATIONS
Earthquakes

4.3 Solving Exponential and Logarithmic Equations

Many applications require solving equations involving exponential functions and logarithms.

Solving Exponential Equations

The logarithm applied to many exponential equations results in a simple linear equation.

Solving Logarithmic Equations

Applying exponential functions to many logarithmic equations results in a simple linear equation.

Mixed Equations

Some equations require the use of both exponential and logarithmic functions for their solution.

MODELS AND APPLICATIONS
Radiocarbon Dating

CHAPTER QUIZ

1. Without using a calculator, determine the exact value of

   a. $\log_9 \dfrac{1}{81}$

   b. $\log_{16} 2$

   Answer

   a. -2

   b. $\dfrac{1}{4}$

   XR Example 4.1

2. Without using a calculator, determine the exact value of

   a. $\log 10{,}000$

   b. $\log \sqrt{10}$

   c. $10^{\log 7}$

   Answer:

   a. 4

   b. $\dfrac{1}{2}$

   c. 7

   XR Example 4.4

3. Without using a calculator, find the exact values of the following.

   a. $\ln \dfrac{1}{e^2}$

   b. $e^{\ln e}$

   Answer:

   a. -2

   b. e

   XR Example 4.5

4. How large must x be to make $y = \log x$ reach a target value of 4?

   Answer: 10,000

   XR Example 4.10

5. Express $\log_7 23$ in terms of the natural logarithm, and then use a calculator to obtain a two-digit approximation.

   Answer: $\dfrac{\ln 23}{\ln 7} \approx 1.61$

   XR Example 4.12

6. If $\log A = 2.8$ and $\log B = 1.6$, find the value of $\log \dfrac{A^2}{\sqrt{B}}$.

   Answer: 4.8

   XR Example 4.15

7. Simplify the following expressions.

   a. $10^{-\log x}$

   b. $e^{2 \ln x}$

Answer:

a. $x^{-1} = \dfrac{1}{x}$

b. x^2

XR **Example 4.18**

8. Write 5^x in the form e^{rx}.

Answer: $5^x = e^{(\ln 5)x}$

XR **Example 4.20**

9. One earthquake has magnitude 3.7. A second quake has relative intensity six times as large. Find the magnitude of the second quake.

Answer: 4.5

XR **Example 4.22**

10. Solve the equation $3^x = 17$.

Answer: $x = \dfrac{\ln 17}{\ln 3} \approx 2.58$

XR **Example 4.25**

11. Solve the equation $3^{2x-1} = 14$.

Answer: $x = -\dfrac{\ln 14 + \ln 3}{2\ln 3} \approx 1.70$

XR **Example 4.27**

12. Solve the equation $\ln(1 + x) = 2$.

Answer: $x = e^2 - 1 \approx 6.39$

XR **Example 4.28**

13. Solve the equation $\ln(e^x + 1) = 4$.

Answer: $x = \ln(e^4 - 1) \approx 3.98$

XR **Example 4.31**

CHAPTER REVIEW EXERCISES

Section 4.1

Calculating logarithms. In Exercises 1 through 19, find the exact value of the logarithm.

1. $\log 100$

   Answer: 2

2. $\ln e$

   Answer: 1

3. $\log_2 8$

   Answer: 3

4. $\log\left(\dfrac{1}{10}\right)$

   Answer: −1

5. $\log_4 2$

   Answer: $\dfrac{1}{2}$

6. $\ln\left(\dfrac{1}{e}\right)$

   Answer: −1

7. $\log_2\left(\dfrac{1}{16}\right)$

   Answer: −4

8. $\log \sqrt[3]{100}$

   Answer: $\dfrac{2}{3}$

9. $\ln 1$

   Answer: 0

10. $\log_3 81$

    Answer: 4

11. $\ln\left(\dfrac{1}{\sqrt{e}}\right)$

Answer: $-\dfrac{1}{2}$

12. $\log_{28} 1$

Answer: 0

13. $\log\left(\dfrac{1}{1000}\right)$

Answer: -3

14. $\log 10^{13}$

Answer: 13

15. $\log 10^e$

Answer: e

16. $\ln\left(\log 10^e\right)$

Answer: 1

17. $\ln e^{10}$

Answer: 10

18. $\log_2\left(\log_3 81\right)$

Answer: 2

19. $\log\left(\ln e^{100}\right)$

Answer: 2

Properties of logarithms.

20. Find the limit as $x \to \infty$ of $y = \ln x$.

Answer: ∞

21. Plot the graph of $y = \log x$.

Answer:

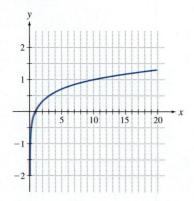

22. Plot the graph of $y = \ln x$.

Answer:

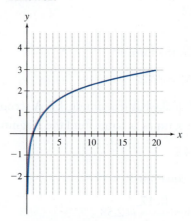

23. What are the domain and range of $y = \ln x$?

Answer: Domain $(0, \infty)$. Range $\mathbb{R} = (-\infty, \infty)$.

24. What are the domain and range of $y = \log x$?

Answer: Domain $(0, \infty)$. Range $\mathbb{R} = (-\infty, \infty)$.

25. What value of x will make $y = \ln x$ reach a target value of 200?

Answer: $x = e^{200}$

26. Find the exact value of $\log_8 32$ by changing to base 2.

Answer: $\dfrac{5}{3}$

27. Find the exact value of $\log_{27} 81$ by changing to base 3.

Answer: $\dfrac{4}{3}$

28. Use the change-of-base formula and a calculator to find $\log_7 17$ correct to two decimal places.

Answer: $\dfrac{\ln 17}{\ln 7} \approx 1.46$

29. Use the change-of-base formula and a calculator to find $\log_5 231$ correct to two decimal places.

Answer: $\dfrac{\ln 231}{\ln 5} \approx 3.38$

Section 4.2

Calculating logarithms. In Exercises 30 through 34, use the laws of logarithms to find the exact value of the given quantity.

30. $2\ln\left(4e^2\right) - \ln 16$

Answer: 4

31. $\log \sqrt{5} + \log \sqrt{2}$

   Answer: $\dfrac{1}{2}$

32. $\log_6 9 + \log_6 4$

   Answer: 2

33. $2\log_{12} 6 - \log_{12} 3$

   Answer: 1

34. $\ln(5e) + \ln(6e) - \ln 30$

   Answer: 2

Combining logarithms. In Exercises 35 through 42, combine the given expression into a single logarithm, and simplify.

35. $\ln(1 - x^2) - \ln(1 - x)$

   Answer: $\ln(1 + x)$

36. $\ln x + \ln\left(\dfrac{1}{x}\right)$

   Answer: 0

37. $\log \dfrac{10}{x^2} + 2\log x$

   Answer: 1

38. $\ln(x + 1) + \ln(x + 2)$

   Answer: $\ln(x^2 + 3x + 2)$

39. $2\ln x + 3\ln(xy) - 4\ln y$

   Answer: $\ln\left(\dfrac{x^5}{y}\right)$

40. $\ln\left(\dfrac{1}{x^2}\right) + 2\ln x$

   Answer: 0

41. $\ln(x^2 - 5x + 6) - \ln(x - 2)$

   Answer: $\ln(x - 3)$

42. $\log(5x) + \log(4x) - \log 2$

   Answer: $\log(10x^2)$

Combining exponents and logarithms. In Exercises 43 through 48, simplify the given expression so that no logarithms or exponential functions appear.

43. $e^{2\ln x}$

   Answer: x^2

44. $e^{-\ln x}$

   Answer: $x^{-1} = \dfrac{1}{x}$

45. $2^{\log_2 2}$

   Answer: 2

46. $e^{\ln x + \ln 2}$

   Answer: $2x$

47. $10^{-2\log x}$

   Answer: $x^{-2} = \dfrac{1}{x^2}$

48. $2^{3\log_2 x + \log_2 y}$

   Answer: $x^3 y$

Algebraic properties of ln.

49. If $\ln A = 2.1$ and $\ln B = 1.4$, find $\ln\left(\dfrac{A^2}{\sqrt{B}}\right)$.

   Answer: 3.5

Two applications.

50. One earthquake measures 3.2 on the Richter scale, and another reads 4.2. How do their relative intensities compare?

   Answer: The second quake is 10 times as intense as the first.

51. One earthquake measures 3.2 on the Richter scale, and another reads 3.7. How do their relative intensities compare?

   Answer: The second quake is 3.16 times as intense as the first.

Section 4.3

Solving logarithmic and exponential equations. In Exercises 52 through 71, find the exact solution, and give an approximation correct to two decimal places.

52. $2 \times 5^x = 86$

   Answer: $x = \dfrac{\ln 43}{\ln 5} \approx 2.34$

53. $\log x = 3$

   Answer: $x = 1000$

54. $\ln x = 5$

   Answer: $x = e^5 \approx 148.41$

55. $\log_2 x = 5$

   Answer: $x = 32$

56. $3 \times 4^x = 102$

Answer: $\dfrac{\ln\left(\frac{102}{3}\right)}{\ln 4} \approx 2.54$

57. $7^x = 9$

Answer: $\dfrac{\ln 9}{\ln 7} \approx 1.13$

58. $\dfrac{2^x}{3^x} = 8$

Answer: $x = \dfrac{\ln 8}{\ln\left(\frac{2}{3}\right)} \approx -5.13$

59. $5 = \dfrac{804}{1 + 2e^x}$

Answer: $x = \ln\left(\dfrac{799}{10}\right) \approx 4.38$

60. $e^x = e^{-x}$

Answer: $x = 0$

61. $\ln x = 2$

Answer: $x = e^2 \approx 7.39$

62. $\log_3 x = 3$

Answer: $x = 27$

63. $\ln(2x - 1) = 8$

Answer: $x = \dfrac{e^8 + 1}{2} \approx 1490.98$

64. $\ln(2x + 3) = 1 + \ln x$

Answer: $x = \dfrac{3}{e - 2} \approx 4.18$

65. $\ln(1 + x) = \ln(3x - 4)$

Answer: $x = \dfrac{5}{2} = 2.5$

66. $\log(95x + 25) = 2 + \log x$

Answer: $x = 5$

67. $\ln(30 - e^{-x}) = 2$

Answer: $x = -\ln(30 - e^2) \approx -3.12$

68. $\log x + \log(x - 3) = 1$

Answer: $x = 5$

69. $\ln(1 + e^x) = 2$

Answer: $x = \ln(e^2 - 1) \approx 1.85$

70. $2 + \ln x = \ln(5 + x)$

Answer: $x = \dfrac{5}{e^2 - 1} \approx 0.78$

71. $\ln x + \ln(x - 1) = 5$

Answer: $x = \dfrac{1 + \sqrt{1 + 4e^5}}{2} \approx 12.69$

The logistic equation.

72. Solve the logistic equation $N = \dfrac{K}{1 + be^{-rt}}$ for the variable t.

Answer: $t = -\dfrac{1}{r}\ln\left(\dfrac{K - N}{Nb}\right)$

POLYNOMIALS AND RATIONAL FUNCTIONS

5

Terry Oakley/Alamy Stock Photo

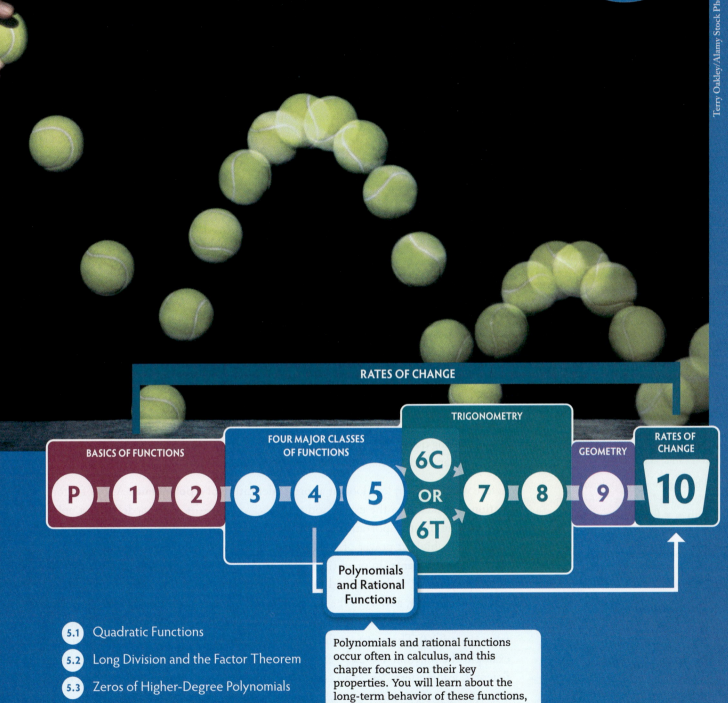

RATES OF CHANGE

BASICS OF FUNCTIONS

FOUR MAJOR CLASSES OF FUNCTIONS

TRIGONOMETRY

GEOMETRY

RATES OF CHANGE

P 1 2 3 4 5 6C OR 6T 7 8 9 10

Polynomials and Rational Functions

5.1 Quadratic Functions

5.2 Long Division and the Factor Theorem

5.3 Zeros of Higher-Degree Polynomials

5.4 Graphs of Polynomials

5.5 Rational Functions

Polynomials and rational functions occur often in calculus, and this chapter focuses on their key properties. You will learn about the long-term behavior of these functions, their local behavior (including the location of extreme values), and how to construct their graphs.

293

Modern computers complete their assigned tasks by performing many, very simple calculations. Computers are so fast today that many millions of these elementary calculations are done in the blink of an eye. But there is a limit to how quickly these calculations can be made, and this speed limit is regulated by polynomials. As a result, computers are said to function in polynomial time. This means that speed is limited by the growth rate of polynomials.

In this chapter we introduce polynomial functions. We will begin our study of polynomials by first considering the special case of quadratic functions because they are familiar and exhibit many properties common to all polynomials. Our focus in this chapter is on the properties of polynomials, including the remainder and factor theorems. We also present the fundamental theorem of algebra, which is among the most important theorems precalculus students are likely to encounter. The chapter closes with a look at rational functions, which are ratios of polynomials.

5.1 Quadratic Functions

5.2 Long Division and the Factor Theorem

5.3 Zeros of Higher-Degree Polynomials

5.4 Graphs of Polynomials

5.5 Rational Functions

A **polynomial** is a function of the form $P(x) = a_n x^n + a_{n-1} x^{n-1} + \cdots + a_1 x + a_0$, where n is a nonnegative integer, $a_0, \ldots, a_n$ are constants, and $a_n \neq 0$.

The **degree** of the polynomial $P(x) = a_n x^n + a_{n-1} x^{n-1} + \cdots + a_1 x + a_0$ with $a_n \neq 0$ is the nonnegative integer n.

The **coefficients** of the polynomial $P(x) = a_n x^n + a_{n-1} x^{n-1} + \cdots + a_1 x + a_0$ are the constants $a_0, \ldots, a_n$.

The **leading term** of the polynomial $P(x) = a_n x^n + a_{n-1} x^{n-1} + \cdots + a_1 x + a_0$ with $a_n \neq 0$ is $a_n x^n$.

The **leading coefficient** of the polynomial $P(x) = a_n x^n + a_{n-1} x^{n-1} + \cdots + a_1 x + a_0$ with $a_n \neq 0$ is a_n.

The **constant term** of the polynomial $P(x) = a_n x^n + a_{n-1} x^{n-1} + \cdots + a_1 x + a_0$ is a_0.

5.1 Quadratic Functions

Quadratic functions are the simplest polynomials whose graphs display many important features of the class, such as concavity.

In this section, you will learn to:

1. Locate the zeros of a quadratic function by factoring or using the quadratic formula.
2. Calculate the discriminant of a quadratic function to describe the zeros.
3. Apply the basics of complex numbers to find complex zeros of quadratic functions.
4. Use the factor theorem to find the factorization of quadratic polynomials.
5. Write a quadratic function in standard form.
6. Graph a parabola.
7. Use the vertex of a parabola to locate the maximum or minimum value of the corresponding quadratic function.
8. Solve applied problems involving quadratic functions.

Before we go further, let's look at a formal definition of a **polynomial**. A lot of information is packed into the expression $a_n x^n + a_{n-1} x^{n-1} + \cdots + a_1 x + a_0$ defining a polynomial. The most important number for analyzing a polynomial is the **degree** n, which is the highest power of x that appears. We will use the constants $a_0, \ldots, a_n$, called the **coefficients**, in operations such as long division. The long-term behavior of the polynomial is determined by $a_n x^n$, the **leading term**. Its coefficient a_n is the **leading coefficient**, and a_0 is the **constant term**. We will use information about the leading coefficient and the constant term to locate where the polynomial function takes the value 0. The constant function 0 is also considered to be a polynomial. For example, the function $y = 2x^3 + x - 4$ is a polynomial of degree 3 with leading coefficient 2, leading term $2x^3$, and constant term -4. On the other hand, the functions $y = 3x^2 + \sqrt{x} = 3x^2 + x^{1/2}$ and $y = 3x^2 + \dfrac{1}{x} = 3x^2 + x^{-1}$ are not polynomials because the exponents $\dfrac{1}{2}$ and -1 are not nonnegative integers. Note that a polynomial of degree 1 is a linear function and a polynomial of degree 0 is a constant function.

Quadratic functions have the form $P(x) = ax^2 + bx + c$ with $a \neq 0$, so they are polynomial functions of degree 2. We treat quadratic functions separately because of their prevalence in both mathematics and applications, and also because their behavior is a prelude to the more complicated behavior of general polynomials.

The graph of a quadratic function is a **parabola**, which is U-shaped. We see parabolas often in everyday life—for example, if air resistance is ignored, then a baseball tossed from home to second follows a parabolic path. A rifle bullet fired toward a target also follows a parabolic path.

Quadratic functions have the form $P(x) = ax^2 + bx + c$ with $a \neq 0$.

A **parabola** is a graph of a quadratic function.

Zeros of Quadratic Functions and the Discriminant

> A quadratic function may have two real zeros, one real zero, or two complex zeros.

For a polynomial $P(x)$, a **zero** is a real or complex number that is a solution of the equation $P(x) = 0$. Zeros are also called roots. Real zeros (that is, zeros that are real numbers) correspond to the x-intercepts of the graph (**Figure 5.1**).

Thus, the zeros of a quadratic polynomial $P(x) = ax^2 + bx + c$ are the solutions of the quadratic equation $ax^2 + bx + c = 0$. Factoring and using the quadratic formula are two ways to solve quadratic equations, both of which are explored in the following step-by-step strategy.

A **zero** of a polynomial $P(x)$ is a number r such that $P(r) = 0$. That is, $x = r$ is a solution of the equation $P(x) = 0$.

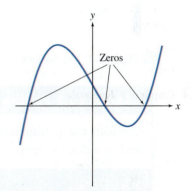

STEP-BY-STEP STRATEGY: Solving Quadratic Equations

Step 1 When the factors of $ax^2 + bx + c$ are apparent, factoring is the easiest way to solve the quadratic equation.

Step 2 When the factors are not apparent, use the quadratic formula

$$x = \frac{-b \pm \sqrt{b^2 - 4ac}}{2a}$$

to solve the equation.

Figure 5.1 Real zeros and horizontal intercepts

The expression under the square root symbol in the quadratic formula is called the **discriminant** of the quadratic function $y = ax^2 + bx + c$. When we solve a quadratic equation to find the zeros of a quadratic polynomial, we will find either two real solutions, one real solution, or no real solutions. For example:

The **discriminant** of the quadratic function $y = ax^2 + bx + c$ is $b^2 - 4ac$.

- $x^2 - 1 = 0$ or $x^2 = 1$ has two real solutions, $x = 1$ and $x = -1$.
- $(x - 2)^2 = 0$ has one real solution, $x = 2$.
- $x^2 + 1 = 0$ or $x^2 = -1$ has no real solutions.

The sign of the discriminant determines which of these three cases occurs. For a quadratic function $y = ax^2 + bx + c$ with real coefficients:

Positive discriminant	Zero discriminant	Negative discriminant
Two real zeros	**One real zero**	**No real zeros**
$b^2 - 4ac > 0$	$b^2 - 4ac = 0$	$b^2 - 4ac < 0$

The quadratic formula yields two real zeros, and the graph crosses the x-axis twice, as shown in **Figure 5.2**.

The square root in the quadratic formula vanishes, leaving a single real zero. In this case we have a perfect square, and we say that the zero is repeated. The graph touches the x-axis once, as shown in **Figure 5.3**.

The quadratic formula involves the square root of a negative number. The result is complex (not real) zeros. The graph misses the x-axis altogether, as shown in **Figure 5.4**.

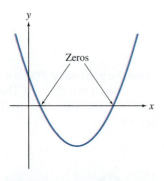

DF **Figure 5.2** Two real zeros

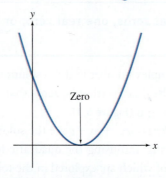

Figure 5.3 One real zero

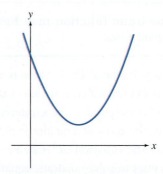

Figure 5.4 No real zeros

EXAMPLE 5.1 Positive Discriminant: Two Real Zeros

Calculate the discriminant of $y = x^2 - x - 1$. Determine how many zeros are promised by the discriminant, and then find the zeros.

SOLUTION

For $y = x^2 - x - 1$, we have $a = 1$, $b = -1$, and $c = -1$. Thus, the discriminant is

$$\text{Discriminant} = b^2 - 4ac$$

$$= (-1)^2 - 4(1)(-1)$$

$$= 5$$

Because the discriminant is positive, we expect two real zeros.

To find the zeros, we apply the quadratic formula to $x^2 - x - 1 = 0$:

$$x = \frac{-(-1) \pm \sqrt{5}}{2} = \frac{1 \pm \sqrt{5}}{2}$$

As expected, we found two zeros:

$$x = \frac{1 + \sqrt{5}}{2} \text{ and } x = \frac{1 - \sqrt{5}}{2}$$

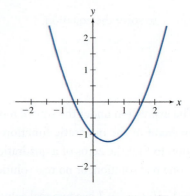

Figure 5.5 The graph of $y = x^2 - x - 1$: This graph crosses the x-axis twice, corresponding to two real zeros.

Figure 5.5 shows both zeros.

TRY IT YOURSELF 5.1 Brief answers provided at the end of the section.

Find the discriminant of $y = 2x^2 + x - 5$. How many real zeros are indicated by the discriminant?

EXAMPLE 5.2 Zero Discriminant: One Real Zero

Calculate the discriminant of $y = x^2 - \sqrt{8}x + 2$. Determine how many zeros are promised by the discriminant, and then find the zeros.

SOLUTION

For $y = x^2 - \sqrt{8}x + 2$, we have $a = 1$, $b = -\sqrt{8}$, and $c = 2$. Thus, the discriminant is

$$\text{Discriminant} = b^2 - 4ac$$
$$= (-\sqrt{8})^2 - 4(1)(2)$$
$$= 0$$

Because the discriminant is 0, we expect one real zero.

We apply the quadratic formula:

$$x^2 - \sqrt{8}x + 2 = 0$$
$$x = \frac{\sqrt{8} \pm \sqrt{0}}{2}$$
$$x = \sqrt{2}$$

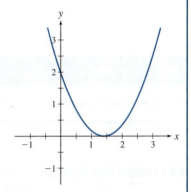

We find only one real zero, as the discriminant promised. Note that $x^2 - \sqrt{8}x + 2$ is the same as the perfect square $(x - \sqrt{2})^2$. The graph shown in **Figure 5.6** touches the x-axis, indicating the single zero.

Figure 5.6 The graph of $y = x^2 - \sqrt{8}x + 2$: The graph of this perfect square meets the x-axis only once, yielding a single zero.

TRY IT YOURSELF 5.2 Brief answers provided at the end of the section.

Find the zeros of the quadratic function $y = x^2 - \sqrt{12}x + 3$.

EXAMPLE 5.3 Negative Discriminant: No Real Zeros

Calculate the discriminant of $y = x^2 + x + 1$. Determine how many zeros are promised by the discriminant, and then find the zeros.

SOLUTION

For $y = x^2 + x + 1$, we have

$$\text{Discriminant} = b^2 - 4ac$$
$$= 1^2 - 4(1)(1)$$
$$= -3$$

Because the discriminant is negative, we expect no real zeros.

Applying the quadratic formula, we find

$$x^2 + x + 1 = 0$$
$$x = \frac{-1 \pm \sqrt{-3}}{2(1)}$$

Because we encounter the square root of a negative number, we conclude that there are no real solutions of this quadratic equation. The graph in **Figure 5.7** does not intersect the x-axis, which supports our conclusion that this quadratic function has no real zeros.

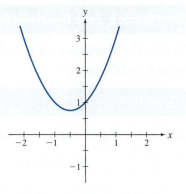

Figure 5.7 The graph of $y = x^2 + x + 1$: This graph does not meet the x-axis, so there are no real zeros.

TRY IT YOURSELF 5.3   Brief answers provided at the end of the section.

Find the discriminant of $y = 3x^2 + x + 1$. How many real zeros are indicated?

EXTEND YOUR REACH   Consider the quadratic function $y = x^2 - 4x + c$. Calculate the discriminant, find the number of zeros, and plot the graph of this function in each of the cases $c = 2$, $c = 3$, $c = 4$, and $c = 5$. Explain what happens to the graph, discriminant, and number of zeros as c begins at 2 and increases.

Complex Zeros of Quadratic Functions

> The complex zeros of a quadratic function with real coefficients occur in complex conjugate pairs.

Although quadratic functions such as $y = x^2 + 1$ have no real zeros, they do have complex zeros (zeros that are complex numbers). In fact, complex numbers were conceived specifically to provide zeros for such functions. The basics of complex numbers are provided in Appendix 2 and summarized in the following.

CONCEPTS TO REMEMBER: Complex Numbers

- The number i has the property that

$$i^2 = -1$$

- A complex number is a number of the form $a + bi$, where a and b are real numbers.

- If $z = a + bi$ (with a and b real), then the complex conjugate of z is denoted by $\bar{z}$ and is defined by

$$\bar{z} = a - bi$$

- If d is a positive real number, then there are two solutions of the equation $x^2 = -d$:

$$x = \pm\sqrt{-d} = \pm i\sqrt{d}$$

EXAMPLE 5.4 Complex Zeros of Quadratic Functions

Find the complex zeros of the quadratic function $y = x^2 - 2x + 2$.

SOLUTION

We apply the quadratic formula:

$$x^2 - 2x + 2 = 0$$

$$x = \frac{2 \pm \sqrt{(-2)^2 - 4(1)(2)}}{2(1)}$$ ◀ Use the quadratic formula with $a = 1$, $b = -2$, $c = 2$.

$$x = \frac{2 \pm \sqrt{-4}}{2}$$ ◀ Simplify.

$$x = \frac{2 \pm 2i}{2}$$ ◀ For d positive, $\sqrt{-d} = \pm i\sqrt{d}$.

$$x = 1 \pm i$$ ◀ Simplify.

Thus, there are two zeros: $x = 1 + i$ and $x = 1 - i$.

TRY IT YOURSELF 5.4   Brief answers provided at the end of the section.

Find the complex zeros of the quadratic function $y = x^2 - 4x + 5$.

Note that in this example the zeros occur in pairs as complex conjugates. The structure of the quadratic formula shows that this is always the case.

LAWS OF MATHEMATICS: Complex Zeros of Quadratic Functions

When a quadratic function (with real coefficients) has a complex zero, its complex conjugate is also a zero. Thus, complex zeros occur in conjugate pairs.

The Factor Theorem for Quadratic Polynomials

The factor theorem relates the zeros of a quadratic function to its factors.

We have seen that factoring a quadratic polynomial makes it easy to find the zeros. This process can be reversed: If we know the zeros of a quadratic polynomial, then we can factor it.

LAWS OF MATHEMATICS: Factor Theorem for Quadratic Polynomials

A number r is a zero of the quadratic polynomial $P(x)$ if and only if $x - r$ is a factor of $P(x)$. As a consequence:

- If $x = p$ and $x = q$ are distinct zeros of a quadratic polynomial $P(x)$ with leading coefficient a, then

$$P(x) = a(x - p)(x - q)$$

- If p is the only zero of a quadratic polynomial $P(x)$ with leading coefficient a, then the factorization takes the form

$$P(x) = a(x - p)^2$$

EXAMPLE 5.5 Factoring Quadratic Polynomials: Two Real Zeros

a. Find the quadratic polynomial $P(x)$ with zeros $x = 1$ and $x = 4$ and with leading coefficient 3.

b. Factor $x^2 - 4x - 4$.

SOLUTION

a. According to the factor theorem for quadratic polynomials, $x - 1$ and $x - 4$ are factors. Because the leading coefficient is 3, we must have

$$P(x) = 3(x - 1)(x - 4)$$

b. The factors are not readily apparent, but the factor theorem tells us how to find them from the zeros. Check that the quadratic formula gives the zeros $x = 2 + \sqrt{8}$ and $x = 2 - \sqrt{8}$. Thus, $x - (2 + \sqrt{8})$ and $x - (2 - \sqrt{8})$ are factors. The leading coefficient is 1, so the factorization is

$$x^2 - 4x - 4 = (x - (2 + \sqrt{8}))(x - (2 - \sqrt{8}))$$

TRY IT YOURSELF 5.5 Brief answers provided at the end of the section.

Find the factors of $3x^2 - 2x - 5$.

EXAMPLE 5.6 Factoring Quadratic Polynomials: One Real Zero

Assume that $x = 3$ is the only zero of a quadratic polynomial $P(x)$. If $P(x)$ has leading coefficient 4, find a formula for $P(x)$.

SOLUTION

Because $x = 3$ is the only zero of the quadratic polynomial, it must be a repeated zero. Thus, $x - 3$ is a factor *twice*. Because the leading coefficient is 4, we have

$$P(x) = 4(x - 3)(x - 3) = 4(x - 3)^2$$

TRY IT YOURSELF 5.6 Brief answers provided at the end of the section.

Assume that $x = 4$ is the only zero of a quadratic polynomial $Q(x)$. If $Q(x)$ has leading coefficient -1, find a formula for $Q(x)$.

EXTEND YOUR REACH   We can shift the graph of a quadratic function up or down so that it meets the x-axis exactly once. Use this understanding, along with the factor theorem, to argue that every quadratic function can be written in the form $y = a(x - h)^2 + k$.

EXAMPLE 5.7 Factoring Quadratic Polynomials: Complex Zeros

Show that $x^2 - 2x + 2$ cannot be factored over the real numbers. Then find the complex factors.

SOLUTION

The discriminant is $b^2 - 4ac = (-2)^2 - 4(1)(2) = -4$. Because it is negative, there are no real zeros. The factor theorem then assures us that we cannot factor this quadratic expression using real numbers.

But the factor theorem applies to complex zeros just as it does to real zeros. The quadratic formula yields the zeros $x = \dfrac{2 \pm \sqrt{-4}}{2} = 1 \pm i$. Hence, we can factor as follows:

$$x^2 - 2x + 2 = (x - (1 + i))(x - (1 - i))$$

TRY IT YOURSELF 5.7 Brief answers provided at the end of the section.

Find the factors of $x^2 - 4x + 5$.

Graphs of Quadratic Functions

> The graph of a quadratic function can be determined by completing the square and shifting graphs.

If the leading coefficient of a quadratic polynomial is positive, then the parabola that is its graph opens upward (concave up), as shown in **Figure 5.8**. In this case, the low point of the parabola represents an absolute minimum and is called the vertex of the parabola. If the leading coefficient is negative, then the parabola opens downward (concave down), as shown in **Figure 5.9**. In this case, the high point represents an absolute maximum and is also called the vertex. Note that in both cases the parabola is symmetric about the vertical line through the vertex.

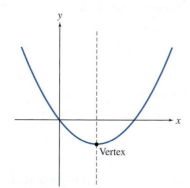

Figure 5.8 Positive leading coefficient: The parabola opens upward, and the vertex represents a minimum.

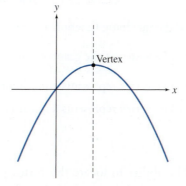

Figure 5.9 Negative leading coefficient: The parabola opens downward, and the vertex represents a maximum.

In order to make accurate graphs, it is helpful to write quadratic functions in **standard form**, which can be found by completing the square. Recall that if we add $\dfrac{p^2}{4}$ to $x^2 + px$, we obtain a perfect square. For example, let's write $y = 5x^2 - 30 + 1$ in standard form. The coefficient of x^2 is not 1, so we factor the coefficient 5 out of the first two terms:

> The **standard form** of the quadratic function $y = ax^2 + bx + c$ is $y = a(x - h)^2 + k$.

$$y = 5x^2 - 30 + 1$$
$$= 5(x^2 - 6x) + 1$$
$$= 5(x^2 - 6x + 9 - 9) + 1 \quad \blacktriangleleft \text{Add } \frac{(-6)^2}{4} = 9 \text{ and compensate.}$$
$$= 5(x - 3)^2 - 44 \quad \blacktriangleleft \text{Write as a square, and simplify.}$$

Once we have the quadratic function written in standard form, we can use what we learned about transformations of graphs in Section 2.3 to make the graph: For $K > 0$,

- To graph $f(x - K)$, we shift the graph of $f(x)$ to the right by K units.
- To graph $f(x + K)$, we shift the graph of $f(x)$ to the left by K units.
- To graph $f(x) - K$, we shift the graph of $f(x)$ down by K units.
- To graph $f(x) + K$, we shift the graph of $f(x)$ up by K units.

STEP-BY-STEP STRATEGY: Graphing Quadratic Functions

Step 1 Write the quadratic function in standard form $y = a(x - h)^2 + k$.

Step 2 Apply shifting and stretching techniques to the graph of $y = x^2$.

The vertex of $y = x^2$ occurs at the origin. If we follow the required shifts and stretches to make the graph of $y = a(x - h)^2 + k$, we find that the vertex is the point (h, k). The procedure used to find the standard form gives us the vertex for $y = ax^2 + bx + c$.

CONCEPTS TO REMEMBER: Locating the Vertex

For the quadratic function $y = ax^2 + bx + c$:

- The vertex of the graph occurs at $x = -\dfrac{b}{2a}$.
- The vertex represents a minimum of y if a is positive.
- The vertex represents a maximum of y if a is negative.

The ability to locate the vertex provides an alternative strategy for sketching parabolas that have two real zeros: Locate the vertex, and locate the two zeros on the horizontal axis. Use these three points to produce the graph.

EXAMPLE 5.8 Graphing a Quadratic Function

For the quadratic function $y = 3x^2 - 12x + 9$:

a. Find the zeros and vertex.

b. Determine whether the vertex corresponds to a maximum or a minimum.

c. Use the information you have learned to make the graph.

SOLUTION

a. We get the zeros by factoring:

$$3x^2 - 12x + 9 = 3(x^2 - 4x + 3)$$

$$= 3(x - 1)(x - 3)$$

Thus, the zeros occur at $x = 1$ and $x = 3$. We find the x-coordinate of the vertex using the preceding formula:

$$\text{Vertex occurs at } x = -\frac{b}{2a}$$

$$= -\frac{-12}{2 \times 3} \quad \blacktriangleleft \text{Use } a = 3, b = -12.$$

$$= 2 \quad \blacktriangleleft \text{Simplify.}$$

To find the y-coordinate of the vertex, we put this value for x into the formula:

$$y = 3x^2 - 12x + 9$$

$$y = 3 \times 2^2 - 12 \times 2 + 9 \quad \blacktriangleleft \text{Substitute 2 for } x.$$

$$= -3 \quad \blacktriangleleft \text{Simplify.}$$

Thus, the vertex is the point $(2, -3)$.

b. Because the leading coefficient $a = 3$ is positive, the vertex $(2, -3)$ corresponds to a minimum.

c. We offer two methods for making the graph.

Method 1: Plotting Points

In **Figure 5.10** we have located the zeros and vertex we found in parts a and b. Then in **Figure 5.11** we complete the graph by sketching the parabola through these points.

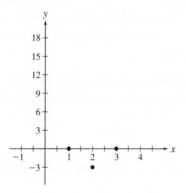

Figure 5.10 Zeros and minimum of $y = 3x^2 - 12x + 9$ plotted

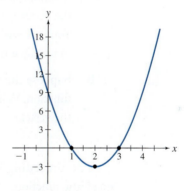

Figure 5.11 Completed graph of $y = 3x^2 - 12x + 9$

Method 2: Shifting and Stretching

The standard form of the quadratic function $y = 3x^2 - 12x + 9$ is $y = 3(x - 2)^2 - 3$. We begin with the graph of $y = x^2$ in **Figure 5.12** and use shifting and stretching in **Figure 5.13**, **Figure 5.14**, and **Figure 5.15** to make the graph we want.

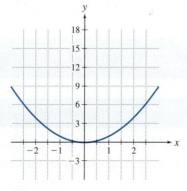

DF **Figure 5.12** Graph of $y = x^2$

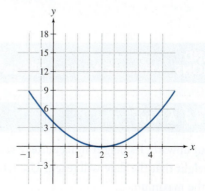

Figure 5.13 Graph of $y = x^2$ shifted right to make the graph of $y = (x - 2)^2$

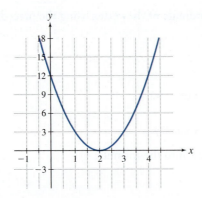

Figure 5.14 Graph of $y = (x - 2)^2$ stretched to make the graph of $y = 3(x - 2)^2$

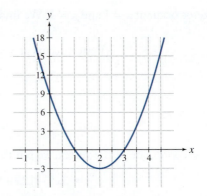

Figure 5.15 Graph of $y = 3(x - 2)^2$ shifted down to make the graph of $y = 3(x - 2)^2 - 3$

TRY IT YOURSELF 5.8   Brief answers provided at the end of the section.

Find the vertex of $y = -5x^2 + 40x + 200$. Does it correspond to a maximum or a minimum?

EXTEND YOUR REACH

a. Suppose the vertex of a parabola represents a maximum for a quadratic function. What can be said about the rate of change of the function near the vertex (just before the vertex, at the vertex, and just after the vertex)? You may wish to draw an example of such a graph and use a straightedge to visualize the rate of change.

b. Suppose the vertex of a parabola represents a minimum for a quadratic function. What can be said about the rate of change of the function near the vertex?

The preceding discussion makes evident the concavity of the graph of a quadratic function.

> **CONCEPTS TO REMEMBER: Concavity of the Graph of a Quadratic Function**
>
> Consider the quadratic function $y = ax^2 + bx + c$. If $a > 0$ the parabola that is its graph opens upward, and the graph is concave up. If $a < 0$ the parabola opens downward, and the graph is concave down.

MODELS AND APPLICATIONS   Maximum Height of Objects Propelled Upward

EXAMPLE 5.9 A Flare

A flare is shot upward. Its height in feet after t seconds is given by $h(t) = -16t^2 + 64t + 6$. What is the maximum height that the flare will reach? Use rates of change to describe the height of the flare as it approaches the maximum.

SOLUTION

Note that $h(t)$ is a quadratic function. Because the coefficient of t^2 is negative, the vertex corresponds to a maximum (as we expect physically). We locate the t-value of the vertex:

$$\text{Vertex occurs at } t = -\frac{b}{2a}$$

$$= -\frac{64}{2 \times (-16)} \quad \blacktriangleleft \text{ Use } a = -16, b = 64.$$

$$= 2 \quad \blacktriangleleft \text{ Simplify.}$$

The maximum height is the corresponding value of h:

$$h(t) = -16t^2 + 64t + 6$$

$$h(2) = -16 \times 2^2 + 64 \times 2 + 6 \quad \blacktriangleleft \text{ Substitute 2 for } t.$$

$$= 70 \quad \blacktriangleleft \text{ Simplify.}$$

Thus, the maximum height is 70 feet. The maximum is located on the graph in **Figure 5.16**. Because the graph is concave down, the height increases at a decreasing rate as the flare approaches the maximum.

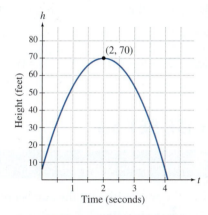

Figure 5.16 The flare at its maximum height of 70 feet after 2 seconds

---

TRY IT YOURSELF 5.9 Brief answers provided at the end of the section.

The height of a projectile shot from the surface of Mars is $p(t) = -1.9t^2 + 11.4t + 3.2$ meters after t seconds. At what time does the projectile reach its maximum height?

EXAMPLE 5.10 A Puzzle

The difference of two numbers y and x is 8 (so $y - x = 8$). Find x and y if the numbers are chosen so that their product is as small as possible.

SOLUTION

We want to make the product xy a minimum. Now because $y - x = 8$, we have $y = x + 8$. Substituting this value for y gives

$$xy = x(x + 8) = x^2 + 8x$$

The minimum value occurs at the vertex, where $x = -\dfrac{b}{2a} = -\dfrac{8}{2 \times 1} = -4$. Hence, $x = -4$ and $y = -4 + 8 = 4$.

TRY IT YOURSELF 5.10 Brief answers provided at the end of the section.

The sum of two numbers x and y is 6. Find x and y if the numbers are chosen so that their product is as large as possible.

TRY IT YOURSELF ANSWERS

5.1 The discriminant is 41. It is positive, so there are two real zeros.

5.2 There is one zero, $x = \sqrt{3}$.

5.3 The discriminant is -11. Because the discriminant is negative, there are no real zeros.

5.4 $x = 2 \pm i$

5.5 $3x^2 - 2x - 5 = 3(x+1)\left(x - \dfrac{5}{3}\right) = (x+1)(3x-5)$

5.6 $Q(x) = -(x-4)^2$

5.7 $x^2 - 4x + 5 = (x - (2+i))(x - (2-i))$

5.8 The vertex is $(4, 280)$. This vertex corresponds to a maximum.

5.9 The maximum height is reached after 3 seconds.

5.10 $x = 3, y = 3$

EXERCISE SET 5.1

CHECK YOUR UNDERSTANDING

1. List the possibilities for the number of real zeros of a quadratic function.

2. Does the parabola that is the graph of $P(x) = -5x^2 + 3x$ open upward or downward?

3. Does the parabola $y = (7 - x)(x + 3)$ open upward or downward?

4. What can be said about the zeros of a quadratic function when the discriminant is positive?

5. What can be said about the zeros of a quadratic function when the discriminant is zero?

6. What can be said about the zeros of a quadratic function when the discriminant is negative?

7. Consider a quadratic function $f(x)$ whose graph is a given parabola. If the vertex of a parabola is a maximum, then which of these are true?

   a. The parabola opens downward.

   b. The leading coefficient of f is negative.

   c. $f(x) \to -\infty$ as $x \to \infty$.

   d. All of the above.

8. If $x = a$ and $x = b$ are zeros of a quadratic function with leading coefficient c, then the quadratic function factors as _____.

9. If $x = a$ is the only real zero of a quadratic function (with real coefficients) with leading coefficient c, then the quadratic function factors as _____.

10. **True or false:** A quadratic function can have at most one maximum.

SKILL BUILDING

Using the factor theorem. In Exercises 11 through 20, write the polynomial in factored form.

11. Leading coefficient: 1; zeros: 2, 3

12. Leading coefficient: 3; zeros: 2, 3

13. Leading coefficient: 2; zeros: 2, -4

14. Leading coefficient: 7; zeros: $2 + \sqrt{8}, 2 - \sqrt{8}$

15. Leading coefficient: -1; zeros: $\dfrac{1 \pm \sqrt{5}}{2}$

16. Leading coefficient: 4; zeros: $\dfrac{-3 \pm \sqrt{17}}{9}$

17. Leading coefficient: 2; zeros: $i, -i$

18. Leading coefficient: 1; zeros: $1 + i, 1 - i$

19. Leading coefficient: -2; zeros: $\dfrac{1 \pm i\sqrt{3}}{5}$

20. Leading coefficient: 4; zeros: $\dfrac{-2 \pm i\sqrt{2}}{2}$

Getting information from graphs. Exercises 21 through 26 show graphs of quadratic functions. For each such function provide the following information.

- List the real zeros.

- Locate the vertex, and determine whether it corresponds to a maximum or a minimum.

- Provide what information you can regarding the discriminant.

21.

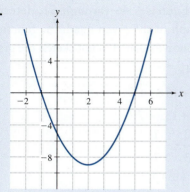

22.

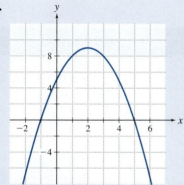

23.

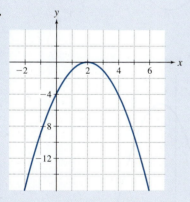

24.

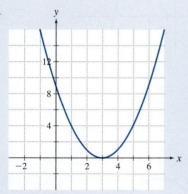

25.

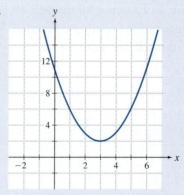

26.

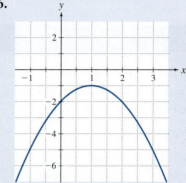

Verifying factors. In Exercises 27 through 31, verify the given factorization by direct multiplication. *Suggestion*: Rearrange the terms of the factors. For example,

$$\left(x - (2 + \sqrt{3})\right)\left(x - (2 - \sqrt{3})\right) = \left((x - 2) - \sqrt{3}\right)\left((x - 2) + \sqrt{3}\right)$$

Then use the fact that $(A - B)(A + B) = A^2 - B^2$.

27. $x^2 - 4x + 1 = \left(x - (2 + \sqrt{3})\right)\left(x - (2 - \sqrt{3})\right)$

28. $x^2 - 6x + 7 = \left(x - (3 + \sqrt{2})\right)\left(x - (3 - \sqrt{2})\right)$

29. $x^2 + 2x - 4 = \left(x - (-1 + \sqrt{5})\right)\left(x - (-1 - \sqrt{5})\right)$

30. $x^2 - 8x + 17 = (x - (4 + i))(x - (4 - i))$

31. $x^2 - 2x + 5 = (x - (1 + 2i))(x - (1 - 2i))$

Finding vertices. In Exercises 32 through 39, find the vertex of the parabola, and indicate whether it corresponds to a maximum or a minimum.

32. $y = 2x^2 + 8x - 3$

33. $y = x^2 - 6x + 3$

34. $y = -x^2 + 4x + 5$

35. $y = x^2 - 8x + 10$

36. $y = 3x^2 + 12x + 1$

37. $y = -6x^2 + 12x + 3$

38. $y = 5x^2 - 10x + 3$

39. $y = -3x^2 - 6x + 7$

The discriminant. In Exercises 40 through 45, calculate the discriminant of the quadratic function. What does the discriminant tell you about the zeros?

40. $y = x^2 + 4x - 3$

41. $y = 3x^2 + 8x + 1$

42. $y = 4x^2 + 12x + 9$

43. $y = x^2 - \dfrac{2}{5}x + \dfrac{1}{25}$

44. $y = x^2 + 3x + 3$

45. $y = 2x^2 - x + 7$

Concavity. In Exercises 46 through 51, determine whether the graph of the given quadratic function is concave up or concave down. Do not sketch the graph.

46. $y = x^2 - 4x$

47. $y = -3x^2 + 15x + 17$

48. $y = -4x^2 + 13x + 7$

49. $y = x^2 + \dfrac{2}{3}x - \dfrac{1}{7}$

50. $y = -7x^2 + x + 1$

51. $y = 2x^2 - x + 7$

PROBLEMS

Matching. Match the graph in Exercises 52 through 57 with the appropriate function from the following list. Here a, b, and c are nonzero real numbers.

A. $f(x) = (x + a)(x + b)$ with $a \neq -b$

B. $g(x) = -4(x - b)(x + c)$

C. $h(x) = x^2 - b^2$

D. $u(x) = ax^2 + bx$

E. $v(x) = x^2 + a^2$

F. $z(x) = -x^2 - a^2$

52.

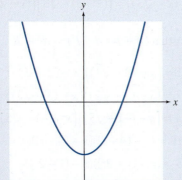

53.

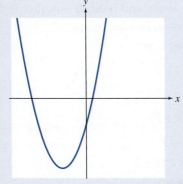

54.

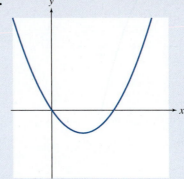

55.

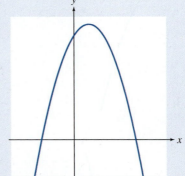

56.

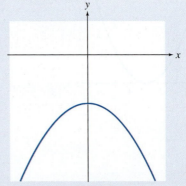

57.

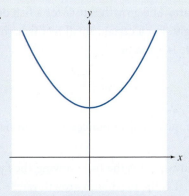

58. Standard form: general case. Consider the quadratic function $y = ax^2 + bx + c$. Follow these steps to put the function in standard form.

Step 1: Factor out a from the first two terms to obtain $y = a\left(x^2 + \dfrac{b}{a}x\right) + c$.

Step 2: Complete the square for $x^2 + \dfrac{b}{a}x$ to obtain

$$y = a\left(\left(x + \dfrac{b}{2a}\right)^2 - \dfrac{b^2}{4a^2}\right) + c.$$

Step 3: Simplify to show that

$$y = a\left(x + \dfrac{b}{2a}\right)^2 + \left(c - \dfrac{b^2}{4a}\right).$$

Standard form. In Exercises 59 through 63, write the quadratic function in standard form.

59. $y = x^2 + 2x + 5$

60. $y = x^2 + 4x + 1$

61. $y = x^2 + 6x + 13$

62. $y = x^2 + 2x$

63. $y = x^2 + 8x + 2$

Factoring. In Exercises 64 through 77, use the quadratic formula and the factor theorem to factor the given quadratic polynomial. In some cases, the factors involve complex numbers.

64. $x^2 - \dfrac{3}{2}x + \dfrac{1}{2}$

65. $8x^2 - 6x - 27$

66. $x^2 - 4x - 3$

67. $x^2 - 4x - 6$

68. $2x^2 + x - 5$

69. $x^2 + 1$

70. $x^2 + 2x + 2$

71. $x^2 - 2x + 9$

72. $x^2 - 4x + 1$

73. $3x^2 + 2$

74. $2x^2 - 5$

75. $x^2 + 3\sqrt{2}x + 4$

76. $x^2 + 4\sqrt{3}x + 9$

77. $x^2 - 2x - 1$

Direct multiplication. In Exercises 78 through 83, multiply the factors, and simplify. *Suggestion*: Use the fact that $(A - B)(A + B) = A^2 - B^2$. In some cases, it will be useful to rearrange the terms of the factors. For example,

$$\left(x - (1 + \sqrt{6})\right)\left(x - (1 - \sqrt{6})\right) = \left((x - 1) - \sqrt{6}\right)\left((x - 1) + \sqrt{6}\right)$$

78. $\left(x - \sqrt{5}\right)\left(x + \sqrt{5}\right)$

79. $(x - i)(x + i)$

80. $\left(x - (1 + \sqrt{6})\right)\left(x - (1 - \sqrt{6})\right)$

81. $\left(x + 1 - \sqrt{8}\right)\left(x + 1 + \sqrt{8}\right)$

82. $(x + 1 - i)(x + 1 + i)$

83. $2\left(x - \dfrac{-1 + i\sqrt{7}}{4}\right)\left(x - \dfrac{-1 - i\sqrt{7}}{4}\right)$

Sketching graphs. In Exercises 84 through 89, find the real zeros (if any), find the vertex, and sketch the graph. Mark the zeros and vertex on your graph.

84. $P(x) = x^2 - x - 6$

85. $P(x) = x^2 - 4x + 3$

86. $P(x) = 8 - 2x - x^2$

87. $P(x) = x^2 + 2x + 3$

88. $P(x) = -x^2 + 6x - 10$

89. $P(x) = 2x - x^2$

MODELS AND APPLICATIONS

90. A baseball. A child tosses a baseball upward and catches it as it falls back. The height above the ground of the ball t seconds after the toss is given by

$$h = -16t^2 + 24t + 3$$

where h is measured in feet.

a. How high above the ground is the ball when it is tossed?

b. How long is the ball in the air? (The ball is caught when its height above the ground is the same as in part a.)

c. How long after it is tossed does the ball reach its maximum height? What is the maximum height?

d. Use rates of change to describe the height of the ball after it reaches the maximum.

91. **Traffic engineering.** Traffic engineers need to determine safe speeds for travel. For commercial vehicles driving at night on city streets, the accident rate A can be modeled by

$$A = 7.8s^2 - 514s + 8734$$

Here s is the speed in miles per hour, and A is measured per 100,000,000 vehicle-miles. Find the speed at which the accident rate is minimized. Round your answer to the nearest whole number.

92. **Economics of school consolidation.** In both urban and rural areas, school districts have been faced with consolidation to save costs. In one region, the cost C per student of operating a high school is given by

$$C = 100,000 - 600n + n^2$$

Here n is the number of students enrolled, and C is measured in dollars per student.

   a. Find the enrollment at which the cost per student is minimized. Round your answer to the nearest whole number.

   b. Use rates of change to describe the cost per student as the enrollment increases toward the location of the minimum.

93. **Concentration of a drug.** The concentration C of a drug in the bloodstream is given by

$$C = 5 + 6t - t^2$$

Here t is the time in hours since ingestion of the drug, and C is measured in milligrams per liter.

   a. Locate the t-value of the vertex of the graph of C.

   b. For safety, the concentration should not exceed 15 milligrams per liter. Will that concentration be exceeded? Explain.

94. **Maximum product.** Two numbers x and y add to 10. Show that the product of x and y is $x(10 - x)$. What choice of x and y gives a maximum product?

95. **Finding a minimum.** The difference of two numbers is 1. Find the two numbers if their product minus their sum is a minimum.

96. **More on the minimum.** The difference of two numbers is 1. Find the two numbers if their product minus their difference is the minimum, nonnegative value. Note that there are two correct answers.

97. **Growth rate.** The growth rate G for a fish depends on its weight. For a certain small reef fish, the growth rate is given by

$$G = w - \frac{w^2}{4}$$

Here w is measured in ounces and G in ounces per year.

   a. At what weight is the fish growing the fastest, and what is the maximum growth rate?

   b. Assuming that the end of life occurs when growth stops, at what weight will death occur?

98. **A square and a circle.** A 10-inch piece of wire is cut into two pieces of lengths x and $10 - x$. The piece x inches long is bent into a circle, and the other is bent into a square (**Figure 5.17**).

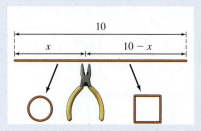

Figure 5.17 Making a circle and a square

   a. What is the radius of the circle?

   b. How long is a side of the square?

   c. What is the total area enclosed by the circle and square?

   d. How should the wire be cut so that the total area is a minimum? (That is, what value of x gives a minimum area?)

   e. How should the wire be cut so that the total area is a maximum?

99. **The Doyle log rule.** The Doyle log rule is a formula that gives the amount B, in board feet, of lumber that can be obtained from a log of diameter D inches and length L feet. The relationship is

$$B = \frac{(D - 4)^2 L}{16}$$

(Here $D \geq 4$.) Rearrange the formula so that it gives the diameter of a log of length L that will produce B board feet of lumber.

100. A projectile. A projectile fired from the origin makes a parabolic path. It follows the graph

$$y = sx - 16(1 + s^2)\left(\frac{x}{v_0}\right)^2$$ feet. Here s is the slope

of the shooting barrel, v_0 is the initial velocity in feet per second, and the variable x is the distance downrange in feet. Your answers to the following questions will involve s and v_0.

   a. How far downrange will the projectile travel?

   b. How far downrange will the projectile reach its maximum height?

   c. What initial velocity is needed to make the projectile clear a wall that is h feet high at d feet downrange? (Your answer will involve s, d, and h.)

101. A community garden plot. You are to make a rectangular garden plot using 400 yards of fencing, and you want to enclose as much area as possible. One side of the garden plot is a building, so you need fencing for only three sides of the garden plot. See **Figure 5.18**.

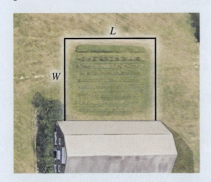

Figure 5.18 A rectangular garden plot next to a building

   a. If W is the width of the garden plot and L is the length, both in yards, use the fact that there are a total of 400 yards of fencing to find an expression giving L in terms of W.

   b. Find an expression for the area of the garden plot that involves only the variable W.

   c. Find the dimensions of the garden plot of maximum area.

102. A partitioned enclosure. A farmer wants to make a rectangular enclosure using 600 feet of fencing. She wants to partition it into three parts, as shown in **Figure 5.19**.

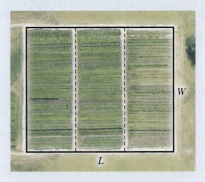

Figure 5.19 A partitioned rectangular enclosure

   a. If W is the width of the enclosure (as seen in Figure 5.19) and L is the length, find an expression giving L in terms of W.

   b. Find a formula for the area of the enclosure in terms of W.

   c. Find the dimensions of the enclosure that has the maximum area.

103. Distance from a point to a line. Consider the line $y = 7 - x$, as shown in **Figure 5.20**.

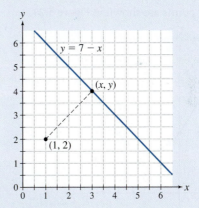

Figure 5.20 Distance from a point to a line

   a. Find the square of the distance from the point $(1, 2)$ to a point (x, y).

   b. If the point (x, y) lies on the line $y = 7 - x$, express your answer from part a in terms of x alone.

   c. Find the point on the line $y = 7 - x$ nearest to the point $(1, 2)$. Note that the same x-value that minimizes the square of the distance also minimizes the distance.

CHALLENGE EXERCISES FOR INDIVIDUALS OR GROUPS

104. Three points determine a parabola. Just as two points determine a line, three points determine a parabola (as long as the points are not collinear). Consider points (a_1, b_1), (a_2, b_2), and (a_3, b_3), where a_1, a_2, and a_3 are distinct numbers. Our goal is to find a polynomial $P(x)$ of degree at most 2 such that $P(a_1) = b_1$, $P(a_2) = b_2$, and $P(a_3) = b_3$. The first step is to make three polynomials Q_1, Q_2, and Q_3, as follows:

$Q_1(x) = (x - a_2)(x - a_3)$, $Q_2(x) = (x - a_1)(x - a_3)$, $Q_3(x) = (x - a_1)(x - a_2)$

Then we define the polynomial $P(x)$ in terms of these three:

$$P(x) = \frac{b_1}{Q_1(a_1)} Q_1(x) + \frac{b_2}{Q_2(a_2)} Q_2(x) + \frac{b_3}{Q_3(a_3)} Q_3(x)$$

a. Show that $P(a_1) = b_1$, $P(a_2) = b_2$, and $P(a_3) = b_3$.

b. Find a quadratic polynomial whose graph passes through the points $(0, 3)$, $(1, 2)$, and $(2, 3)$. These three points are shown in **Figure 5.21**.

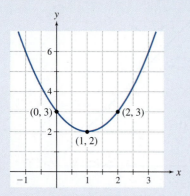

Figure 5.21 A parabola passing through three points

105. Powers of i. Calculate i^n for all nonnegative integers n. *Suggestion:* Begin by calculating powers of i for $n = 0, 1, 2, \ldots$, and then look for a pattern.

Conjugates. If $z = a + bi$ and $w = c + di$ (with a, b, c, d real), show that the statements in Exercises 106 through 111 are true. Consult Appendix 2 for a refresher on complex numbers.

106. $\overline{z + w} = \overline{z} + \overline{w}$

107. $\overline{zw} = \overline{z}\,\overline{w}$

108. $z + \overline{z} = 2a$

109. $z - \overline{z} = 2bi$

110. $z\overline{z} = a^2 + b^2$

111. Show that $(x - z)(x - \overline{z})$ is a quadratic polynomial with real coefficients if z is a given complex number.

Standard form. In Exercises 112 through 117, write the given expression in the form $a + bi$ (with a, b real). Consult Appendix 2 for a refresher on complex numbers.

112. $(1 - i)(1 + i)$

113. $(2 + i)(1 + 4i)$

114. $(3 - 2i)(3 + i)$

115. $\dfrac{1}{2 - i}$. *Suggestion:* Multiply top and bottom by $2 + i$.

116. $\dfrac{i}{i - 1}$. *Suggestion:* Multiply top and bottom by $i + 1$.

117. $\dfrac{1}{i}$. *Suggestion:* Multiply top and bottom by i.

REVIEW AND REFRESH: Exercises from Previous Sections

118. From Section P.1: Find the distance between the points $(1, 1)$ and $(3, 1)$.

Answer: 2

119. From Section P.3: Solve the inequality $x^2 - 5x < -6$.

Answer: $(2, 3)$

120. From Section 1.1: Find the domain of the function $f(x) = \sqrt{x - 4}$.

Answer: $[4, \infty)$

121. From Section 1.2: Find the average rate of change of $f(x) = \dfrac{1}{x}$ from $x = 1$ to $x = 4$.

Answer: $-\dfrac{1}{4}$

122. From Section 1.2: Find the average rate of change of $f(x) = x^2 + 2$ on the interval $[a, a + h]$ with $h > 0$. Simplify your answer.

Answer: $2a + h$

123. From Section 1.4: Find the limit as $x \to \infty$ of
$$f(x) = \frac{6x-1}{2x+5}.$$
Answer: 3

124. From Section 2.2: Find the inverse function of
$$f(x) = \frac{x-1}{x+2}.$$
Answer: $f^{-1}(x) = \dfrac{2x+1}{1-x}$

125. From Section 2.3: The graph of a function f is in **Figure 5.22**. Sketch the graph of $g(x) = f(x-1)+1$.

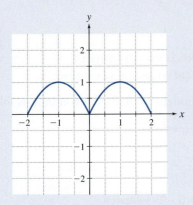

Figure 5.22

Answer:

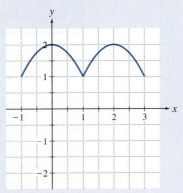

126. From Section 3.1: If $f(x) = x^2+1$, find the equation of the secant line for f for the interval $[2,4]$.

Answer: $y = 6x-7$

5.2 Long Division and the Factor Theorem

Long division is a key tool that will allow us to factor polynomials.

5.1 Quadratic Functions

5.2 Long Division and the Factor Theorem

5.3 Zeros of Higher-Degree Polynomials

5.4 Graphs of Polynomials

5.5 Rational Functions

In this section, you will learn to:

1. Implement the division algorithm for polynomials using long division.
2. Calculate the remainder on division by $x-k$ using the remainder theorem for polynomials.
3. Factor polynomials using the factor theorem.
4. Perform synthetic division.

Long division is a fundamental tool that allows us to use information about simpler polynomials to understand more complicated polynomials. In this section, we develop the idea of division and derive some of its consequences.

Long Division

> Long division of polynomials is analogous to long division of real numbers.

We can perform long division with polynomials much as we do with integers. To illustrate the procedure, we divide $3x^2 + 4x + 7$ by $x - 2$:

$$x - 2 \overline{\smash{)}3x^2 + 4x + 7}$$

Compare this example with the familiar division process, say for $3\overline{\smash{)}857}$. To begin, we think of dividing x into $3x^2$. That quotient is $3x$, and this is our starting place. We proceed just as if we were performing long division with integers. We multiply $x - 2$ by $3x$ and subtract the result from $3x^2 + 4x + 7$:

$$
\begin{array}{r}
3x \\
x - 2 \overline{\smash{)}3x^2 + 4x + 7} \\
\underline{3x^2 - 6x} \\
10x + 7
\end{array}
$$

◀ This is $3x \times (x - 2)$. Subtract it.
◀ Record the result of subtraction.

The next step is to divide x into $10x$. The quotient is 10, and we continue as follows:

$$
\begin{array}{r}
3x + 10 \\
x - 2 \overline{\smash{)}3x^2 + 4x + 7} \\
\underline{3x^2 - 6x} \\
10x + 7 \\
\underline{10x - 20} \\
27
\end{array}
$$

◀ This is the quotient.

◀ This is $10 \times (x - 2)$. Subtract it.
◀ This is the remainder, the result of subtraction.

Because we cannot divide x into 27 (which has a lower degree than x), the process stops. Thus, $3x^2 + 4x + 7$ divided by $x - 2$ gives a quotient of $3x + 10$ with a remainder of 27. As a result, we obtain

$$\frac{3x^2 + 4x + 7}{x - 2} = 3x + 10 + \frac{27}{x - 2}$$

which is the quotient form. In the exercises you are asked to verify the following equivalent form, which is known as the product form:

$$3x^2 + 4x + 7 = (x - 2)(3x + 10) + 27$$

The pieces involved in long division of polynomials have the same names as their counterparts in division of integers. In the preceding example, $3x^2 + 4x + 7$ is called the dividend, $x - 2$ is called the divisor, $3x + 10$ is called the quotient, and 27 is called the remainder. These components appear in the process as

$$
\begin{array}{r}
\text{Quotient} \\
\text{Divisor} \overline{\smash{)}\text{Dividend}} \\
\vdots \\
\overline{\text{Remainder}}
\end{array}
$$

Typically, the degree of the dividend is greater than the degree of the divisor. In that case, the degree of the quotient equals the degree of the dividend minus the degree of the divisor. Also, either the remainder is 0 or the degree of the remainder is less than the degree of the divisor. This tells us when to stop the long-division process.

EXAMPLE 5.11 Practicing Long Division

Calculate $x - 3 \overline{) x^2 + 3}$. Express your answer in product form. That is, write

$$x^2 + 3 = (x - 3) \times Quotient + Remainder$$

SOLUTION

To keep track of all of the terms, we put in $0x$ for the missing x term. This illustrates how the terms line up:

$$x - 3 \overline{) x^2 + 0x + 3}$$

We think of dividing x into x^2. That quotient is x:

$$
\begin{array}{r}
x \phantom{{}+0x+3} \\
x - 3 \overline{) x^2 + 0x + 3} \\
\underline{x^2 - 3x} \phantom{{}+3} \\
3x + 3 \phantom{{}}
\end{array}
$$

◄ This is $x \times (x - 3)$.
◄ Record the result of subtraction.

At each step, we divide the leading term of the divisor into the leading term of the difference, so now we divide x into $3x$. That quotient is 3:

$$
\begin{array}{r}
x + 3 \phantom{{}+3} \\
x - 3 \overline{) x^2 + 0x + 3} \\
\underline{x^2 - 3x} \phantom{{}+3} \\
3x + 3 \phantom{{}} \\
\underline{3x - 9} \phantom{{}} \\
12 \phantom{{}}
\end{array}
$$

◄ This is the quotient.
◄ This is $3 \times (x - 3)$.
◄ This is the remainder, the result of subtraction.

The quotient is $x + 3$, and the remainder is 12. This gives

$$x^2 + 3 = (x - 3) \times Quotient + Remainder$$

$$x^2 + 3 = (x - 3)(x + 3) + 12$$

TRY IT YOURSELF 5.11 Brief answers provided at the end of the section.

Calculate $x + 4 \overline{) x^2 + 5x - 1}$. Express your answer in product form.

CONCEPTS TO REMEMBER: The Division Algorithm

- If $P(x)$ and $S(x)$ are polynomials other than 0, when we divide $P(x)$ by $S(x)$ we obtain

$$\text{Quotient form: } \frac{P(x)}{S(x)} = Quotient + \frac{Remainder}{S(x)}$$

This can also be written as

$$\text{Product form: } P(x) = S(x) \times Quotient + Remainder$$

- Either the remainder is 0 or the degree of the remainder is less than the degree of the divisor $S(x)$. In the special case that the divisor is linear, the remainder is constant.

- When we perform long division, to keep track of all of the terms we put in 0 as the coefficient for any missing term.

The process remains the same when the divisor has a degree higher than 1.

EXAMPLE 5.12   Division of Higher-Degree Polynomials

Divide the fourth-degree polynomial $x^4 + 3x^3 + 6x^2 + 7x + 8$ by $x^2 + x + 1$. Express your answer in both quotient and product forms.

SOLUTION

As with the previous example, we find the terms of the quotient by dividing the leading terms:

$$\frac{x^4}{x^2} \quad \frac{2x^3}{x^2} \quad \frac{3x^2}{x^2}$$

◀ Indicate how to find the quotient.

$$
\begin{array}{r}
x^2 + 2x + 3 \\
x^2 + x + 1 \overline{\big)\, x^4 + 3x^3 + 6x^2 + 7x + 8} \\
x^4 + x^3 + x^2 \\
\hline
2x^3 + 5x^2 + 7x + 8 \\
2x^3 + 2x^2 + 2x \\
\hline
3x^2 + 5x + 8 \\
3x^2 + 3x + 3 \\
\hline
2x + 5
\end{array}
$$

◀ This is the quotient.

◀ This is $x^2(x^2 + x + 1)$.

◀ Record the result of subtraction.

◀ This is $2x(x^2 + x + 1)$.

◀ Record the result of subtraction.

◀ This is $3(x^2 + x + 1)$.

◀ This is the remainder.

We can write the result as

Quotient form: $\dfrac{x^4 + 3x^3 + 6x^2 + 7x + 8}{x^2 + x + 1} = x^2 + 2x + 3 + \dfrac{2x + 5}{x^2 + x + 1}$

or as

Product form: $x^4 + 3x^3 + 6x^2 + 7x + 8 = (x^2 + x + 1)(x^2 + 2x + 3) + (2x + 5)$

TRY IT YOURSELF 5.12   Brief answers provided at the end of the section.

Divide $x^4 + 3x^2 + 3x + 3$ by $x^2 + 1$. Express your answer in both quotient and product forms.

EXTEND YOUR REACH

a. Explain what happens (in terms of the quotient and the remainder) when you divide one polynomial by a polynomial of higher degree. The analogy with division by integers such as 2 divided by 7 may be helpful.

b. Long division can be carried out in some cases for functions other than polynomials. Use long division to show that $x^{3/2} - x + 3x^{1/2} + 4$ divided by $x^{1/2} - 1$ gives a quotient of $x + 3$ and a remainder of 7.

The Remainder and Factor Theorems

> The factor theorem applies to all polynomials, not just quadratic functions.

When the divisor is linear, we know that the remainder is a constant—it does not depend on x. We can say more about this remainder: Assume that the dividend $P(x)$ is nonconstant and that the divisor has the form $S(x) = x - k$. Let's write $Q(x)$ for the quotient and r for the remainder:

$$P(x) = (x - k)Q(x) + r$$

This equation is true for any value of x. In particular, it remains valid if we put in $x = k$:

$$P(x) = (x - k)Q(x) + r$$
$$P(k) = (k - k)Q(k) + r$$
$$P(k) = 0Q(k) + r$$
$$P(k) = r$$

Thus, we find the remainder r by putting $x = k$ into the polynomial $P(x)$. As a consequence, we can find the remainder without actually carrying out the long division. For example, if we divide $P(x) = x^2 + x + 5$ by $x - 2$, the remainder is

$$Remainder = P(2) = 2^2 + 2 + 5 = 11$$

You should use long division to verify that 11 is indeed the remainder.

In the case that k is a zero of $P(x)$, division by $x - k$ leaves a remainder of $P(k) = 0$. Thus, if $Q(x)$ is the quotient, then

$$P(x) = (x - k)Q(x)$$

The result is that $x - k$ is a factor of $P(x)$.

Conversely, if $x - k$ is a factor of a polynomial $P(x)$, then clearly k is a zero of $P(x)$.

LAWS OF MATHEMATICS: Remainder and Factor Theorems

Remainder theorem: If we divide the nonconstant polynomial $P(x)$ by $x - k$, the remainder is $P(k)$.

Factor theorem: The number k is a zero of a polynomial if and only if $x - k$ is a factor of that polynomial.

Note the similarity of the factor theorem with the corresponding result for quadratic functions presented in the preceding section.

EXAMPLE 5.13   Using the Remainder Theorem

Let $P(x) = x^3 - x - 6$. Use the remainder theorem to calculate the remainder when $P(x)$ is divided by $x + 3$.

SOLUTION

Note that $x + 3 = x - (-3)$. Hence, the remainder when $P(x)$ is divided by $x + 3$ is $P(-3)$:

$$P(x) = x^3 - x - 6$$
$$P(-3) = (-3)^3 - (-3) - 6$$
$$= -30$$

So, the remainder when $P(x)$ is divided by $x + 3$ is -30. Check this by actually carrying out the long division.

TRY IT YOURSELF 5.13   Brief answers provided at the end of the section.

Let $T(x) = x^3 + x - 30$. Use the remainder theorem to calculate the remainder when $T(x)$ is divided by $x - 1$.

EXAMPLE 5.14   Using the Factor Theorem

Let $P(x) = x^3 - x - 6$.

a. Use the factor theorem to show that $x - 2$ is a factor of $P(x)$.

b. Find $Q(x)$ so that $P(x) = (x - 2)Q(x)$.

SOLUTION

a. We calculate $P(2)$:

$$P(x) = x^3 - x - 6$$

$$P(2) = 2^3 - 2 - 6$$

$$= 0$$

Because $P(2) = 0$, the factor theorem tells us that $x - 2$ is a factor of $P(x)$.

b. We found in part a that $x - 2$ is a factor of $P(x)$. We are asked now to exhibit the factorization. We accomplish this by long division. Note that we put in $0x^2$ for the missing x^2 term:

$$
\require{enclose}
\begin{array}{r}
x^2 + 2x + 3 \\[-3pt]
x - 2 \enclose{longdiv}{x^3 + 0x^2 - x - 6} \\
\underline{x^3 - 2x^2} \\
2x^2 - x - 6 \\
\underline{2x^2 - 4x} \\
3x - 6 \\
\underline{3x - 6} \\
0
\end{array}
$$

Thus, $x^3 - x - 6 = (x - 2)(x^2 + 2x + 3)$.

TRY IT YOURSELF 5.14   Brief answers provided at the end of the section.

Let $T(x) = x^3 + x - 30$. Show that $x - 3$ is a factor of $T(x)$, and exhibit the factorization.

EXTEND YOUR REACH   Assume that $P(x)$ is a polynomial of degree 3 with leading coefficient 1 and that $P(1) = P(2) = P(3) = 0$. Find $P(x)$.

Synthetic Division

> Synthetic division provides an efficient method for evaluating polynomials.

Calculating quotients by long division is somewhat cumbersome, but there is a shortcut called synthetic division that makes the process quick and easy when the divisor has the form $x - k$. To illustrate this method, let's divide $3x^3 + x^2 + 1$ by $x - 2$. We illustrate the process using four steps, but in practice it is done all at once.

Step 1: The division problem:

$$Divisor\ x - 2 \enclose{longdiv}{3x^3 + x^2 + 0x + 1}$$ ◀ **This is the dividend. (Note the inclusion of 0 coefficients.)**

Step 2: List only the coefficients of the dividend, including 0 coefficients:

$$3 \quad 1 \quad 0 \quad 1$$

Step 3: Perform the actual calculation as follows. Put the number 2 (corresponding to the divisor $x - 2$) in the upper left corner. (We use positive 2 when we divide by $x - 2$.) Begin at the left-hand side and follow the arrows. The down arrows indicate addition. The diagonal arrows indicate multiplication by 2 (the number in the upper left corner).

$$
\begin{array}{c|cccc}
2 & 3 & 1 & 0 & 1 \\
& & 6 & 14 & 28 \\
\hline
& +\downarrow \times 2\nearrow & +\downarrow \times 2\nearrow & +\downarrow \times 2\nearrow & +\downarrow \\
\hline
& 3 & 7 & 14 & \underline{|29}
\end{array}
$$

Step 4: Read off the answer:

$$
\begin{array}{cccc}
3 & 1 & 0 & 1 \\
& 6 & 14 & 28 \\
\hline
3 & 7 & 14 & |29 \\
\hline
\text{\textit{Coefficients}} & & & \text{\textit{Remainder}} \\
\text{\textit{of quotient}}
\end{array}
$$

Because the dividend is a polynomial of degree 3, the quotient is a quadratic polynomial. Its coefficients appear in the last line of the result from synthetic division, beginning with the leading coefficient. Hence, the quotient is $3x^2 + 7x + 14$, and the remainder is 29. Thus,

$$3x^3 + x^2 + 1 = (x - 2)(3x^2 + 7x + 14) + 29$$

With a little practice, synthetic division goes very fast. It is so fast that we often use synthetic division and the remainder theorem to calculate function values.

EXAMPLE 5.15 Using Synthetic Division to Find Factors

a. Use the remainder theorem and synthetic division to show that $x = 1$ is a zero of $P(x) = x^3 - 6x^2 + 11x - 6$.

b. Find a polynomial $Q(x)$ so that $P(x) = (x - 1)Q(x)$.

c. Complete the factorization of $P(x)$.

SOLUTION

a. Recall that the remainder when $P(x)$ is divided by $x - 1$ is $P(1)$. Thus, synthetic division allows us to evaluate $P(1)$ at the same time as we divide. We find

$$
\begin{array}{c|cccc}
1 & 1 & -6 & 11 & -6 \\
& & 1 & -5 & 6 \\
\hline
& 1 & -5 & 6 & |\underline{0}
\end{array}
$$

Because the remainder is 0, we know that $P(1) = 0$.

b. Because $P(x)$ has degree 3 and the divisor is linear, the quotient is quadratic. The division in part a gives the quotient $Q(x) = x^2 - 5x + 6$ as well as the remainder of 0. This gives:

$$P(x) = (x - 1) \times Quotient + Remainder$$

$$P(x) = (x - 1)(x^2 - 5x + 6)$$

c. To complete the factorization of $P(x)$, we need only factor the quadratic expression $x^2 - 5x + 6$:

$$P(x) = (x - 1)(x^2 - 5x + 6) = (x - 1)(x - 2)(x - 3)$$

TRY IT YOURSELF 5.15 Brief answers provided at the end of the section.

a. For $P(x) = 2x^3 - 7x^2 + 4x + 3$, use synthetic division to calculate $P(3)$.

b. For $P(x) = x^3 - 2x^2 - x + 2$, show that $P(2) = 0$, and then write $P(x)$ in factored form.

EXTEND YOUR REACH

a. Show that the zeros of $P(x) = x^3 - 9x^2 + 26x - 24$ are 2, 3, and 4.

b. Reverse the order of the coefficients of $P(x)$ in part a to get $Q(x) = -24x^3 + 26x^2 - 9x + 1$. Show that the zeros of $Q(x)$ are $\frac{1}{2}$, $\frac{1}{3}$, and $\frac{1}{4}$.

c. Experiment with other polynomials to see whether reversing the order of the coefficients has the effect on the zeros we saw in parts a and b. Can you explain, using synthetic division or some other means, why this property holds?

TRY IT YOURSELF ANSWERS

5.11 $x^2 + 5x - 1 = (x + 4)(x + 1) - 5$

5.12 Quotient form:

$$\frac{x^4 + 3x^2 + 3x + 3}{x^2 + 1} = x^2 + 2 + \frac{3x + 1}{x^2 + 1}$$

Product form:

$$x^4 + 3x^2 + 3x + 3 = (x^2 + 1)(x^2 + 2) + 3x + 1$$

5.13 -28

5.14 $T(3) = 0$, and $T(x) = (x - 3)(x^2 + 3x + 10)$

5.15 **a.**

$$
\begin{array}{r|rrrr}
3 & 2 & -7 & 4 & 3 \\
  &   & 6 & -3 & 3 \\
\hline
  & 2 & -1 & 1 & \underline{6} \\
\end{array}
$$

Hence, $P(3) = 6$.

b. $P(x) = (x - 2)(x + 1)(x - 1)$

EXERCISE SET 5.2

CHECK YOUR UNDERSTANDING

1. The factor theorem tells us that a is a zero of a polynomial $P(x)$ if and only if _____.

2. **True or false:** If the remainder of $P(x)$ divided by $Q(x)$ is 0, then $Q(x)$ is a factor of $P(x)$.

3. Suppose that $P(x)$ is a polynomial such that the remainder upon division of $P(x)$ by $x - 4$ is 7. What is the value of $P(4)$?

4. Suppose that $P(x)$ is a polynomial such that the remainder upon division of $P(x)$ by $x + 3$ is 5. What is the value of $P(-3)$?

5. Suppose that $P(x)$ is a polynomial and $P(4) = 0$. Find one factor of $P(x)$.

6. Suppose that $P(x)$ is a polynomial and $P(-3) = 0$. Find one factor of $P(x)$.

7. Suppose that $x - 1$ is a factor of the polynomial $P(x)$. What is the value of $P(1)$?

8. Suppose that $x + 3$ is a factor of the polynomial $P(x)$. What is the value of $P(-3)$?

9. Suppose that $P(x)$ is a polynomial of degree 7 and $S(x)$ is a polynomial of degree 3. What is the degree of the quotient when we divide $P(x)$ by $S(x)$? What can you say about the degree of the remainder if it is not 0?

10. Suppose that $P(x)$ is a polynomial of degree 8 and $S(x)$ is a polynomial of degree 2. What is the degree of the quotient when we divide $P(x)$ by $S(x)$? What can you say about the degree of the remainder if it is not 0?

SKILL BUILDING

Zeros. In Exercises 11 through 16, use direct substitution to answer the questions.

11. Is 3 a zero of $x^3 - x - 24$?

12. Is 2 a zero of $x^3 + x$?

13. Is -1 a zero of $x^4 - 1$?

14. Is -2 a zero of $x^3 + 8$?

15. Is i a zero of $x^4 - 1$?

16. Is i a zero of $x^3 + x$?

Long division. Use long division to calculate the quotient and remainder in Exercises 17 through 30. Express your final answer in both quotient form and product form.

17. $x - 1 \overline{)\, x^3 - x + 1}$

18. $x + 1 \overline{)\, x^2 + 2x}$

19. $x + 1 \overline{)\, x^3 + 1}$

20. $x + 2 \overline{)\, x^3 - 2x + 5}$

21. $x - 2 \overline{)\, 2x^4 - 3x^2 + 1}$

22. $x^2 + 1 \overline{)\, x^3 + x + 1}$

23. $x^2 + 1 \overline{)\, x^5 + 2}$

24. $x^2 - x + 1 \overline{)\, x^4 + 2x^3 + x^2 - 3}$

25. $x - 3 \overline{)\, x^3 - 1}$

26. $x^2 - 3x + 4 \overline{)\, x^5 - 2x^4 + x^3 - 3x^2 + x - 3}$

27. $x^4 - 1 \overline{)\, x^6 - 1}$

28. $x - 2 \overline{)\, x^3 + x + 1}$

29. $x^2 + 1 \overline{)\, x^4 + 1}$

30. $x^2 + x \overline{)\, x^4 - x^3 + 1}$

Finding quotient and remainder using synthetic division. In Exercises 31 through 42, find the quotient and the remainder using synthetic division. Express your answer in both quotient form and product form.

31. $x - 1 \overline{)\, x^3 + x^2 + x + 1}$

32. $x + 1 \overline{)\, x^4 + 2}$

33. $x + 2 \overline{)\, x^4 - x^2 + 3}$

34. $x + 1 \overline{)\, 2x^5 - x^4 + 1}$

35. $x - 3 \overline{)\, x^3 - x + 1}$

36. $x + 3 \overline{)\, 2x^4 + x}$

37. $x - 2 \overline{)\, x^3 - 2x + 5}$

38. $x + 2 \overline{)\, x^4 + 3x^2 + 2}$

39. $x + 1 \overline{)\, x^6 + 1}$

40. $x + 2 \overline{)\, x - 2}$

41. $x + 3 \overline{)\, 4x^3 - 2x^2 + 1}$

42. $x - 1 \overline{)\, 3x^2 - x + 5}$

Evaluating with synthetic division. In Exercises 43 through 47, find the indicated function value using synthetic division.

43. $P(x) = x^3 + x + 1$ at $x = 3$

44. $P(x) = x^4 + 3x^2 - 5x + 4$ at $x = -4$

45. $P(x) = x^6 + 2x^3 - 5$ at $x = 1$

46. $P(x) = 2x^4 - 3x^3 - 4x^2 + 5x + 6$ at $x = 3$

47. $P(x) = 3x^5 - 5x^4 + x^3 + x + 1$ at $x = 2$

PROBLEMS

48. The product form. In our first example of long division, we showed that

$$\frac{3x^2 + 4x + 7}{x - 2} = 3x + 10 + \frac{27}{x - 2}$$

Multiply both sides of this equation by the remainder $x - 2$ to show that

$$3x^2 + 4x + 7 = (x - 2)(3x + 10) + 27$$

Changing forms. In Exercises 49 through 58, use long division to establish the given equality. Use synthetic division where appropriate.

49. $\dfrac{x^3 - x^2 + 2x + 1}{x^2 + 1} = x - 1 + \dfrac{x + 2}{x^2 + 1}$

50. $\dfrac{x^4 + 1}{x - 2} = x^3 + 2x^2 + 4x + 8 + \dfrac{17}{x - 2}$

51. $\dfrac{x^3 + x + 1}{x + 1} = x^2 - x + 2 - \dfrac{1}{x + 1}$

52. $\dfrac{x^3 + 3x^2 + 3x + 8}{x^2 + 2} = x + 3 + \dfrac{x + 2}{x^2 + 2}$

53. $\dfrac{x^4 + x^2}{x - 5} = x^3 + 5x^2 + 26x + 130 + \dfrac{650}{x - 5}$

54. $\dfrac{2x^4 + 1}{x^2 - 2} = 2x^2 + 4 + \dfrac{9}{x^2 - 2}$

55. $\dfrac{3x + 7}{x - 4} = 3 + \dfrac{19}{x - 4}$

56. $\dfrac{x^5 + 1}{x - 1} = x^4 + x^3 + x^2 + x + 1 + \dfrac{2}{x - 1}$

57. $x^4 + x^2 + 2 = (x^2 - x + 1)(x^2 + x + 1) + 1$

58. $2x^3 + 1 = 2(x^3 + 4) - 7$

Using the remainder theorem. In Exercises 59 through 64, use the remainder theorem to find the remainder without actually carrying out the long-division process.

59. Remainder of $x^2 + x + 1$ divided by $x - 3$

60. Remainder of $2x^2 + x - 4$ divided by $x + 1$

61. Remainder of $x^2 + 3x + 5$ divided by $x - 5$

62. Remainder of $x^2 - 2x + 1$ divided by $x - 1$

63. Remainder of $x^2 + 2x + 1$ divided by $x + 1$

64. Remainder of $x^2 + 2x + 1$ divided by $x - 1$

Using the factor theorem. In Exercises 65 through 71, use the factor theorem. (That is, do the exercise without factoring the polynomial.)

65. Show that $x - 2$ is a factor of $P(x) = x^3 - 2x^2 + x - 2$.

66. Show that $x + 2$ is a factor of $P(x) = x^3 + 2x^2 - x - 2$.

67. Show that $x - 1$ is a factor of $P(x) = x^3 - 2x^2 + x$.

68. Show that $x + 1$ is a factor of $P(x) = x^3 + x + 2$.

69. Show that $x - 1$ is a factor of $P(x) = x^3 - x^2 + 5x - 5$.

70. Show that $x + 1$ is a factor of $P(x) = x^3 + x^2 + 3x + 3$.

71. Show that $x - 2$ is a factor of $P(x) = x^3 - 2x^2 + 2x - 4$.

More on long division. In the case of dividing one quadratic polynomial by another, the process of long division is completed in a single step. For example:

$$x^2 + x + 1 \overline{)\begin{array}{l} 3 \\ 3x^2 + 5x + 4 \\ \underline{3x^2 + 3x + 3} \\ 2x + 1 \end{array}}$$

The result is that

$$\frac{3x^2 + 5x + 4}{x^2 + x + 1} = 3 + \frac{2x + 1}{x^2 + x + 1}$$

This method allows us to simplify the numerator of certain fractions. In Exercises 72 through 75, use this process to write the given fraction in the form

$$a + \frac{bx + c}{dx^2 + fx + g}$$

72. $\dfrac{7x^2 + 9x - 1}{x^2 + x + 1}$

73. $\dfrac{2x^2 + 7}{x^2 + 4}$

74. $\dfrac{x^2 + 2x + 1}{x^2 + x - 1}$

75. $\dfrac{x^2 + x + 1}{x^2 + x + 3}$

Factoring polynomials. In Exercises 76 through 87, use the given information to produce a complete factorization of the given polynomial.

76. $P(x) = x^3 + x^2 - 4x - 4$. Show that -1 is a zero of $P(x)$, and then provide a complete factorization of $P(x)$.

77. $P(x) = x^3 - 3x - 2$. Show that $P(2) = 0$, and then provide a complete factorization of $P(x)$.

78. $P(x) = x^3 - 3x^2 - 10x + 24$. Show that $P(2) = 0$, and then provide a complete factorization of $P(x)$.

79. $P(x) = x^3 + 3x^2 + 3x + 1$. Show that $P(-1) = 0$, and then provide a complete factorization of $P(x)$.

80. $P(x) = x^3 + 2x^2 - 9x - 18$. Show that $P(3) = 0$, and then provide a complete factorization of $P(x)$.

81. $P(x) = x^3 + 7x^2 + 14x + 8$. Show that $P(-1) = 0$, and then provide a complete factorization of $P(x)$.

82. $P(x) = x^3 + x^2 - 5x - 5$. Show that $P(-1) = 0$, and then provide a complete factorization of $P(x)$.

83. $P(x) = x^3 - 3x^2 + 2$. Show that $P(1) = 0$, and then provide a complete factorization of $P(x)$.

84. $P(x) = x^3 - 4x^2 + 8$. Show that $P(2) = 0$, and then provide a complete factorization of $P(x)$.

85. $P(x) = x^4 - 5x^2 + 4$. Use long division to verify that $x^2 - 1$ is a factor of $P(x)$. Then provide a complete factorization of $P(x)$.

86. $P(x) = x^4 + x^3 - 7x^2 - 13x - 6$. Use long division to verify that $(x + 1)^2$ is a factor of $P(x)$. Then provide a complete factorization of $P(x)$.

87. $P(x) = x^4 - 8x^2 + 15$. Use long division to show that $x^2 - 3$ is a factor of $P(x)$. Then provide a complete factorization of $P(x)$.

MODELS AND APPLICATIONS

88. A sphere. The volume of a sphere of radius r is given by $V = \dfrac{4}{3}\pi r^3$. The surface area is $S = 4\pi r^2$. Use division to express the volume in terms of the radius and the surface area.

CHALLENGE EXERCISES FOR INDIVIDUALS OR GROUPS

89. **Average rate of change of polynomials.** Consider the average rate of change from x to $x + h$ for a polynomial of degree n. In this setting, we will think of x as the variable and h as a constant.

    a. Show that the average rate of change from x to $x + h$ of the quadratic polynomial $P(x) = ax^2 + bx + c$ is a linear function of x.

    b. Show that the average rate of change from x to $x + h$ of the polynomial $Q(x) = ax^3 + bx^2 + cx + d$ is a quadratic function of x. You will need to use the fact that

    $$(x + h)^3 = x^3 + 3x^2 h + 3xh^2 + h^3$$

    Note that similar arguments show that the average rate of change, as a function of x, of any polynomial is a polynomial of one less degree. As a consequence, the (instantaneous) rate of change of any polynomial is also a polynomial of one less degree.

90. **Four points determine a third-degree polynomial.** Consider points (a_1, b_1), (a_2, b_2), (a_3, b_3), and (a_4, b_4), where a_1, a_2, a_3, and a_4 are distinct numbers. Our goal is to find a polynomial $P(x)$ of degree at most 3 such that $P(a_1) = b_1$, $P(a_2) = b_2$, $P(a_3) = b_3$, and $P(a_4) = b_4$. The first step is to make four polynomials as follows:

    $$Q_1(x) = (x - a_2)(x - a_3)(x - a_4),$$
    $$Q_2(x) = (x - a_1)(x - a_3)(x - a_4),$$
    $$Q_3(x) = (x - a_1)(x - a_2)(x - a_4),$$
    $$Q_4(x) = (x - a_1)(x - a_2)(x - a_3)$$

Then we define the polynomial $P(x)$ in terms of these four:

$$P(x) = \frac{b_1}{Q_1(a_1)} Q_1(x) + \frac{b_2}{Q_2(a_2)} Q_2(x) + \frac{b_3}{Q_3(a_3)} Q_3(x) + \frac{b_4}{Q_4(a_4)} Q_4(x)$$

a. Show that $P(a_1) = b_1$, $P(a_2) = b_2$, $P(a_3) = b_3$, and $P(a_4) = b_4$.

b. Find a polynomial of degree 3 whose graph passes through the points $(-1, 5)$, $(0, 2)$, $(1, 1)$, and $(2, 8)$. These points are shown in **Figure 5.23**.

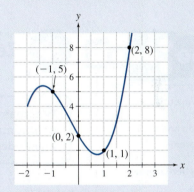

Figure 5.23 A polynomial whose graph passes through four points

91. **An alternative proof of the remainder theorem for quadratic polynomials.** Here is an alternative way to prove the remainder theorem in the case of a quadratic polynomial. Carry out the long-division process for dividing $x - k$ into $P(x) = ax^2 + bx + c$. You should get a remainder of $ak^2 + bk + c = P(k)$.

REVIEW AND REFRESH: Exercises from Previous Sections

92. **From Section P.2:** Solve the inequality
    $2x + 4 \le 5x - 8$.

    Answer: $[4, \infty)$

93. **From Section 1.3:** The graph of a function f is shown in **Figure 5.24**. Identify (approximately) the intervals where f is increasing and the intervals where f is concave down.

    Answer: f is increasing on $[-1, 0]$. f is concave down on $[-2, -1.5) \cup (-0.5, 0.5]$.

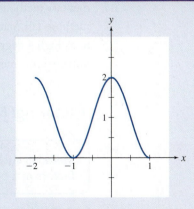

Figure 5.24 A graph for Exercise 93

94. From Section 3.1: Find the formula for the linear function $f(x)$ with slope 3 such that $f(1) = -9$.

Answer: $f(x) = 3x - 12$

95. From Section 3.1: The linear function $y = mx + b$ has slope 3. What is the effect on y if x is increased by 2?

Answer: y is increased by 6.

96. From Section 5.1: Factor the quadratic polynomial $P(x)$ having a leading coefficient of 2 and zeros 1 and 4.

Answer: $P(x) = 2(x - 1)(x - 4)$

97. From Section 5.1: Find the vertex of the parabola $y = x^2 - 8x + 4$.

Answer: $(4, -12)$

98. From Section 5.1: Use the quadratic formula to factor $x^2 - x - 1$.

Answer: $\left(x - \dfrac{1 + \sqrt{5}}{2} \right)\left(x - \dfrac{1 - \sqrt{5}}{2} \right)$

5.3 Zeros of Higher-Degree Polynomials

5.1 Quadratic Functions

5.2 Long Division and the Factor Theorem

5.3 Zeros of Higher-Degree Polynomials

5.4 Graphs of Polynomials

5.5 Rational Functions

> Whereas the quadratic formula can be used to find zeros of quadratic polynomials, several different methods are required to find zeros of higher-degree polynomials.

In this section, you will learn to:

1. Factor polynomials using methods such as substitution.
2. Solve polynomial equations by applying the factor theorem.
3. Identify candidates for zeros of polynomials by applying the rational zeros theorem.
4. Solve applied problems involving zeros of higher-degree polynomials.

In Section 5.1, we used the quadratic formula to factor quadratic polynomials by finding the zeros. There are similar, but much more complicated, formulas for zeros of cubic (third-degree) and quartic (fourth-degree) polynomials. Niels Henrik Abel proved in 1823 that there is no equivalent formula yielding the zeros of polynomials of degree five and higher. In many cases, solutions of higher-degree equations can be found by elementary methods, and we will explore some of these methods in this section. Where such methods fail, mathematicians have developed effective techniques for obtaining solutions to as high a degree of accuracy as may be wanted. These rather sophisticated methods, initially implemented on computers, may well be available today on your smartphone.

Factoring: Methods and Formulas

> Methods such as substitution and grouping terms are useful for finding the factors of a polynomial.

If a polynomial can be written in factored form, it is easy to find the zeros. For example, let's solve the polynomial equation

$$(x - 2)^2(x + 5)(2x + 1) = 0$$

If a product is 0, at least one of the factors must be 0:

$$(x-2)^2 = 0 \quad \text{or} \quad x+5 = 0 \quad \text{or} \quad 2x+1 = 0 \quad \blacktriangleleft \text{ Set each factor equal to 0.}$$

$$x = 2 \quad \text{or} \quad x = -5 \quad \text{or} \quad x = -\frac{1}{2} \quad \blacktriangleleft \text{ Solve each equation.}$$

The zeros 2, −5, and $-\dfrac{1}{2}$ are shown in **Figure 5.25** as crossing points of the graph with the x-axis. This example illustrates how the problem of finding zeros for polynomials can be reduced to the problem of finding the factors.

Although there is no simple procedure that always produces the factors of a polynomial, there are methods that often help.

Method 1: Factoring out a power of x: This method applies when the constant term is zero. In this case, we can factor out the highest power of the variable x that appears in each term of the polynomial. For example, the polynomial $x^4 + 2x^3 - 3x^2$ is a quartic, but we can factor x^2 out of each term to get $x^2(x^2 + 2x - 3)$. The expression in parentheses is a quadratic polynomial and easily factors as $(x-1)(x+3)$. This gives a complete factorization of the quartic:

$$x^4 + 2x^3 - 3x^2 = x^2(x-1)(x+3)$$

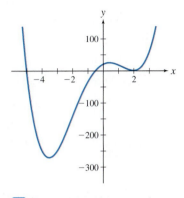

DF **Figure 5.25** The zeros of $P(x) = (x-2)^2(x+5)(2x+1)$ located at $x = 2$, $x = -5$ and $x = -\dfrac{1}{2}$

Method 2: Substitution to reduce the degree: When certain terms are missing, substitutions can turn higher-degree polynomials into quadratic polynomials. We illustrate the method by finding the factors of $x^4 - 5x^2 + 6$. If we let $y = x^2$, we obtain the quadratic polynomial $y^2 - 5y + 6$. Factoring gives $y^2 - 5y + 6 = (y-3)(y-2)$. Because $y = x^2$, this gives

$$x^4 - 5x^2 + 6 = (x^2 - 3)(x^2 - 2)$$

$$= \left(x - \sqrt{3}\right)\left(x + \sqrt{3}\right)\left(x - \sqrt{2}\right)\left(x + \sqrt{2}\right)$$

Method 3: Factoring by grouping: Sometimes we can make progress by factoring in pieces. To show the idea, let's factor $x^3 - 3x^2 - 5x + 15$. We separate this polynomial into two pieces and factor each one individually:

$$x^3 - 3x^2 - 5x + 15 = (x^3 - 3x^2) - (5x - 15)$$

$$= (x^2(x-3)) - (5(x-3))$$

Note that the two pieces share the common factor $x - 3$, which we factor out:

$$(x^2(x-3)) - (5(x-3)) = (x^2 - 5)(x-3)$$

$$= \left(x - \sqrt{5}\right)\left(x + \sqrt{5}\right)(x-3)$$

This factorization may be easier if the substitution $A = (x-3)$ is made.

Method 4: Formulas for the sum and difference of powers: The following factoring formulas are sometimes useful in solving polynomial equations. They can be verified by direct multiplication.

If n is a positive integer,

$$x^n - y^n = (x-y)(x^{n-1} + x^{n-2}y + x^{n-3}y^2 + \cdots + y^{n-1})$$

A particular case of this is

$$x^3 - y^3 = (x-y)(x^2 + xy + y^2)$$

For odd values of n,

$$x^n + y^n = (x + y)(x^{n-1} - x^{n-2}y + x^{n-3}y^2 - x^{n-4}y^3 + \cdots + y^{n-1})$$

Note that the signs alternate.

A particular case of this is

$$x^3 + y^3 = (x + y)(x^2 - xy + y^2)$$

As an example, let's factor $x^3 - 27$. Now $27 = 3^3$, so we use the factorization of $x^3 - y^3$ with $y = 3$:

$$x^3 - 27 = x^3 - 3^3$$
$$= (x - 3)(x^2 + 3x + 9)$$

Now we try to factor the remaining quadratic polynomial $x^2 + 3x + 9$. The discriminant is

$$b^2 - 4ac = 3^2 - 4 \times 1 \times 9 = -27$$

Because the discriminant is negative, $x^2 + 3x + 9$ does not have further real factors, and we conclude that $x^3 - 27 = (x - 3)(x^2 + 3x + 9)$ is the complete factorization over the real numbers.

If this list of factoring techniques seems like something of a hodgepodge, that's because it is. There is no one method that works for all polynomials.

EXAMPLE 5.16   Solving by Factoring

Find the zeros of $P(x) = x^3 - x^2 - x$.

SOLUTION

We begin by factoring x from each term:

$$x^3 - x^2 - x = x(x^2 - x - 1) \quad \blacktriangleleft \text{ Factor } x \text{ from each term.}$$

$$x = 0 \quad \text{or} \quad x^2 - x - 1 = 0 \quad \blacktriangleleft \text{ Set each factor equal to 0.}$$

We have found one zero, $x = 0$. We get the others by applying the quadratic formula to $x^2 - x - 1 = 0$:

$$x^2 - x - 1 = 0$$

$$x = \frac{1 \pm \sqrt{(-1)^2 - 4(1)(-1)}}{2(1)}$$

$$= \frac{1 \pm \sqrt{5}}{2}$$

We have found a total of three zeros: 0, $\dfrac{1 + \sqrt{5}}{2}$, and $\dfrac{1 - \sqrt{5}}{2}$.

These zeros are represented in **Figure 5.26** as points where the graph crosses the x-axis.

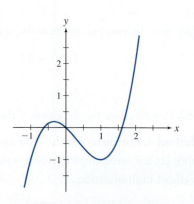

Figure 5.26 The three zeros of $P(x) = x^3 - x^2 - x$

TRY IT YOURSELF 5.16   Brief answers provided at the end of the section.

Find all zeros (both real and complex) of $P(x) = x^3 - 64$. Note that this is the difference of two cubes.

In this exercise we write ω for the complex number $-\dfrac{1}{2} + \dfrac{\sqrt{3}}{2}i$.

a. Verify by direct multiplication that ω is a cube root of unity. In other words, calculate that $\omega^3 = 1$.

b. Verify that $x = -1$, $x = -\omega$, and $x = -\omega^2$ are all zeros of $P(x) = x^3 + 1$.

c. Use part b to conclude that $x^3 + 1 = (x + 1)(x + \omega)(x + \omega^2)$.

EXAMPLE 5.17 Substitutions to Reduce the Degree

Use the substitution $y = x^2$ to find the zeros of $P(x) = x^4 - 6x^2 + 5$.

SOLUTION

We begin with the indicated substitution:

$$x^4 - 6x^2 + 5 = y^2 - 6y + 5 \qquad \blacktriangleleft \text{Substitute } y = x^2.$$

$$= (y - 5)(y - 1) \qquad \blacktriangleleft \text{Factor.}$$

$$= (x^2 - 5)(x^2 - 1) \qquad \blacktriangleleft \text{Replace } y \text{ by } x^2.$$

$$= \left(x - \sqrt{5}\right)\left(x + \sqrt{5}\right)(x - 1)(x + 1) \qquad \blacktriangleleft \text{Factor as the difference of squares.}$$

Setting each of these factors equal to 0 gives the four zeros $x = \pm\sqrt{5}$ and $x = \pm 1$. These zeros are shown in **Figure 5.27**.

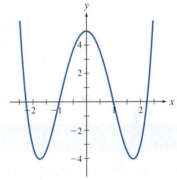

Figure 5.27 The zeros of $P(x) = x^4 - 6x^2 + 5$

TRY IT YOURSELF 5.17 Brief answers provided at the end of the section.

Find the zeros of $P(x) = x^4 - 13x^2 + 36$.

EXAMPLE 5.18 Factoring By Grouping Terms

Group terms to find the zeros of $P(x) = x^3 + 4x^2 - x - 4$. First group the terms as $(x^3 - x) + (4x^2 - 4)$.

SOLUTION

The first step is to group terms and factor:

$$x^3 + 4x^2 - x - 4 = (x^3 - x) + (4x^2 - 4) \qquad \blacktriangleleft \text{Group terms.}$$

$$= x(x^2 - 1) + 4(x^2 - 1) \qquad \blacktriangleleft \text{Factor each piece.}$$

$$= (x + 4)(x^2 - 1) \qquad \blacktriangleleft \text{Factor } x^2 - 1 \text{ from both pieces.}$$

$$= (x + 4)(x - 1)(x + 1) \qquad \blacktriangleleft \text{Factor the difference of squares.}$$

It may help to let $A = x^2 - 1$. Then the factorization from the second line to the third is $xA + 4A = (x + 4)A$.

Setting each factor equal to zero, we find the zeros $x = -4$, $x = 1$, and $x = -1$. These zeros are shown in **Figure 5.28**.

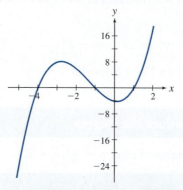

Figure 5.28 The zeros of
$P(x) = x^3 + 4x^2 - x - 4$

TRY IT YOURSELF 5.18   Brief answers provided at the end of the section.

Find the zeros of $P(x) = x^3 - x^2 - 4x + 4$ by grouping as $(x^3 - x^2) - (4x - 4)$.

Using the Factor Theorem to Solve Equations

> The factor theorem shows the relationship between a polynomial and one of lower degree.

If we happen to know one zero of a polynomial, then we can use the factor theorem to reduce the degree of the polynomial.

EXAMPLE 5.19 Using the Factor Theorem to Factor

Consider the cubic polynomial $P(x) = x^3 - 3x - 2$.

a. Show that $x = 2$ is a zero of $P(x)$, and find $Q(x)$ so that $P(x) = (x - 2)Q(x)$.

b. Find the remaining zeros of $P(x)$.

c. Write $P(x)$ as a product of linear factors.

SOLUTION

a. We can show that $x = 2$ is a zero by putting $x = 2$ into the formula for $P(x)$ or by using the remainder theorem to find $P(2)$. We choose the latter method because it will also obtain the required factor of $P(x)$. To find $P(2)$ using the remainder theorem, we divide $P(x)$ by $x - 2$. We use synthetic division, remembering to put in 0 for the missing x^2 term:

$$
\begin{array}{r|rrrr}
2 & 1 & 0 & -3 & -2 \\
  &   & 2 & 4 & 2 \\
\hline
  & 1 & 2 & 1 & \underline{0} \\
\end{array}
$$

Because the remainder is 0, we know that $P(2) = 0$. Also, the quotient is $x^2 + 2x + 1$. Hence,

$$x^3 - 3x - 2 = (x - 2) \times Quotient + Remainder$$

$$= (x - 2)(x^2 + 2x + 1)$$

b. By part a, the equation $x^3 - 3x - 2 = 0$ says

$$(x - 2)(x^2 + 2x + 1) = 0$$

Now we need to solve $x^2 + 2x + 1 = 0$, which we can do by factoring:

$$0 = x^2 + 2x + 1 = (x + 1)^2$$

We conclude that $x = 2$ and $x = -1$ are the zeros of the polynomial $P(x)$.

c. By part a,

$$P(x) = (x - 2)(x^2 + 2x + 1)$$

We found the factors of $x^2 + 2x + 1$ in part b. Hence,

$$P(x) = (x - 2)(x + 1)(x + 1) = (x - 2)(x + 1)^2$$

TRY IT YOURSELF 5.19 Brief answers provided at the end of the section.

Given that $x = 3$ is a zero of $P(x) = x^3 - 7x - 6$, find the remaining zeros, and write $P(x)$ as a product of linear factors.

STEP-BY-STEP STRATEGY: Solving Equations When One Zero Is Known

Suppose we know that $x = a$ is a zero of a polynomial $P(x)$.

Step 1 Use the factor theorem to conclude that $x - a$ is a factor of $P(x)$.

Step 2 Use long division to find a polynomial $Q(x)$ so that $P(x) = (x - a)Q(x)$.

Step 3 Find the remaining zeros of $P(x)$ by finding the zeros of the simpler polynomial $Q(x)$.

When we divide out a factor corresponding to a zero, we obtain a polynomial of lower degree. As a consequence, a polynomial of degree n has at most n zeros, real or complex. This simple observation yields an important theorem. In the next section we will generalize it to the fundamental theorem of algebra.

LAWS OF MATHEMATICS: Maximum Number of Zeros in a Polynomial

A polynomial of degree n has at most n zeros.

Note that in part c of Example 5.19, the factor $x + 1$ occurs twice. For this reason, we say that -1 is a zero of **order** 2. Some texts use multiplicity in place of order.

> The **order** of a zero k of a polynomial $P(x)$ is the highest power of $x - k$ that is a factor of $P(x)$.

Rational Zeros

> The rational zeros of a polynomial with integer coefficients can be found by analyzing the constant term and the leading coefficient.

If we know that k is a zero of a polynomial, we can divide the polynomial by $x - k$ and reduce it to a polynomial of lower degree, as we just noted. But how do we find a zero to get us started? In general this is a difficult problem, but there are helpful methods in addition to the factoring methods we discussed at the beginning of this section.

Look, for example, at $P(x) = 3x^3 - 2x^2 - 10x + 4$. Suppose $x = k$ is an integer zero (a zero that happens to be an integer). Then

$$3k^3 - 2k^2 - 10k + 4 = 0 \quad \blacktriangleleft \text{Use the fact that } k \text{ is a zero of } P(x).$$

$$3k^3 - 2k^2 - 10k = -4 \quad \blacktriangleleft \text{Move the constant term to the other side.}$$

$$k(3k^2 - 2k - 10) = -4 \quad \blacktriangleleft \text{Factor out } k.$$

$$km = -4 \quad \blacktriangleleft \text{Substitute } m = 3k^2 - 2k - 10.$$

Because m is an integer, we conclude that k is a divisor of 4. So, the only possible integer zeros of $P(x) = 3x^3 - 2x^2 - 10x + 4$ are the divisors of 4: ± 1, ± 2, and ± 4. If we check these six possibilities (by plugging them into the polynomial), we find that $x = 2$ is indeed a zero, but the other five are not.

This method works in general to give restrictions on the integer zeros of polynomials with coefficients that are integers. Additional analysis can show us the possible rational zeros of a polynomial with coefficients that are integers.

LAWS OF MATHEMATICS: Rational Zeros Theorem

If $P(x)$ is a nonconstant polynomial with integer coefficients, the only possible rational zeros have the form $\dfrac{a}{b}$, where a divides the constant term and b divides the leading coefficient.

The rational zeros theorem gives us a short list of possibilities. Some, all, or none of the candidates on the list may actually be zeros, and we must check each.

STEP-BY-STEP STRATEGY: Finding Zeros of Polynomials with Integer Coefficients

If we fail to find a factorization for a nonconstant polynomial $P(x)$ with integer coefficients, we may proceed as follows:

Step 1 Use the rational zeros theorem to make a list of candidates for rational zeros of $P(x)$. They are the numbers $\dfrac{a}{b}$, where a divides the constant term and b divides the leading coefficient.

Step 2 Proceed with checking the candidate list until a rational zero is found. (If no candidate from the list turns out to be a zero, then this strategy fails.)

Step 3 Once a zero k for a polynomial is found, we can divide the polynomial by $x - k$ and reduce it to a lower-degree polynomial.

EXAMPLE 5.20 Finding Rational Zeros

Let $P(x) = 2x^3 - 3x^2 - 3x + 2$.

a. Find all rational zeros of $P(x)$.

b. Give a complete factorization of $P(x)$.

SOLUTION

a. Note that $P(x)$ has integer coefficients. The possible rational zeros have the form $\dfrac{a}{b}$, where a divides the constant term 2 and b divides the leading coefficient, which is also 2. Thus,

$$\text{Possibilities for } a \text{ (divisors of 2):} = \pm 1, \pm 2$$

$$\text{Possibilities for } b \text{ (divisors of 2):} = \pm 1, \pm 2$$

$$\text{Possibilities for } \frac{a}{b}: = \pm 1, \pm 2, \pm\frac{1}{2}$$

All rational zeros (if there are any) must be on this list of six candidates. (There may be irrational zeros as well.) Next we check each of the six candidates.

$$P(x) = 2x^3 - 3x^2 - 3x + 2$$

$$P(1) = 2 \times 1^3 - 3 \times 1^2 - 3 \times 1 + 2 = -2$$

$$P(-1) = 2 \times (-1)^3 - 3 \times (-1)^2 - 3 \times (-1) + 2 = 0$$

$$P(2) = 2 \times 2^3 - 3 \times 2^2 - 3 \times 2 + 2 = 0$$

$$P(-2) = 2 \times (-2)^3 - 3 \times (-2)^2 - 3 \times (-2) + 2 = -20$$

$$P\left(\frac{1}{2}\right) = 2 \times \left(\frac{1}{2}\right)^3 - 3 \times \left(\frac{1}{2}\right)^2 - 3 \times \left(\frac{1}{2}\right) + 2 = 0$$

$$P\left(-\frac{1}{2}\right) = 2 \times \left(-\frac{1}{2}\right)^3 - 3 \times \left(-\frac{1}{2}\right)^2 - 3 \times \left(-\frac{1}{2}\right) + 2 = \frac{5}{2}$$

Here we evaluated the polynomial directly, but the process may be shortened using synthetic division. We find that the rational zeros are $x = -1$, $x = 2$, and $x = \dfrac{1}{2}$. Because $P(x)$ is a cubic polynomial, it has at most three zeros, so we have found all zeros.

b. Because -1, 2, and $\dfrac{1}{2}$ are zeros, we conclude that $x + 1$, $x - 2$, and $x - \dfrac{1}{2}$ are factors. Because we have found three factors for a polynomial of degree 3, we can write down the factorization. As we did for quadratic polynomials in Section 5.1, we multiply the leading coefficient (in this case, 2) and the factors to get $P(x)$:

$$P(x) = 2(x + 1)(x - 2)\left(x - \frac{1}{2}\right)$$

TRY IT YOURSELF 5.20 Brief answers provided at the end of the section.

Find all rational zeros of $P(x) = 3x^3 - 10x^2 + 9x - 2$.

Although the method used in this example can find rational zeros if they exist, the procedure may be quite tedious.

EXTEND YOUR REACH

a. Consider the polynomial $P(x) = 35x^5 - 9x + 12$. How many numbers must you check to determine whether this polynomial has any rational zeros?

b. Suppose $Q(x)$ is a polynomial with integer coefficients. Assume the coefficient of the leading term has n distinct positive integer factors, the constant term has m distinct positive integer factors, and there are no positive factors in common between these two numbers except for 1. What is the maximum possible number of candidates that need to be checked to determine whether $Q(x)$ has a rational zero?

EXAMPLE 5.21 Finding Zeros of Higher-Degree Polynomials

Find all zeros (both real and complex) of $P(x) = x^4 - 3x^3 + 3x^2 - 3x + 2$.

SOLUTION

This quartic polynomial has integer coefficients, so we can list the possible rational zeros as follows:

$$\text{Divisors of constant term 2: } a = \pm 1, \pm 2$$

$$\text{Divisors of leading coefficient 1: } b = \pm 1$$

$$\text{Rational candidates: } \frac{a}{b} = \pm 1, \pm 2$$

Checking these, we find that 1 and 2 are zeros but -1 and -2 are not. We can divide out the corresponding factors $x - 1$ and $x - 2$ one at a time, or we can divide out the product $(x-1)(x-2) = x^2 - 3x + 2$ in one step. We choose the former method because it allows us to use synthetic division, which is quicker. We first divide by $x - 1$:

$$
\begin{array}{r|rrrr}
1 & 1 & -3 & 3 & -3 & 2 \\
  &   & 1 & -2 & 1 & -2 \\
\hline
  & 1 & -2 & 1 & -2 & \underline{|0}
\end{array}
$$

The quotient is $x^3 - 2x^2 + x - 2$. Thus,

$$P(x) = (x-1)(x^3 - 2x^2 + x - 2)$$

We next divide the resulting cubic polynomial by $x - 2$:

$$
\begin{array}{r|rrrr}
2 & 1 & -2 & 1 & -2 \\
  &   & 2 & 0 & 2 \\
\hline
  & 1 & 0 & 1 & \underline{|0}
\end{array}
$$

The quotient is $x^2 + 1$, so we obtain

$$P(x) = (x-1)(x-2)(x^2 + 1)$$

We find the remaining zeros by solving $x^2 + 1 = 0$. The result is $x = \pm i$. So, the zeros we are looking for are $x = 1$, $x = 2$, $x = i$, and $x = -i$. Because $P(x)$ is a fourth-degree polynomial, these four are all of the zeros. **Figure 5.29** shows these zeros.

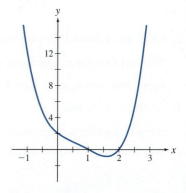

Figure 5.29 The two real zeros of $P(x) = x^4 - 3x^3 + 3x^2 - 3x + 2$

TRY IT YOURSELF 5.21   Brief answers provided at the end of the section.

Find all zeros (both real and complex) of $P(x) = x^3 + x^2 - 4x + 2$.

EXTEND YOUR REACH   Explain how the method used here can be adapted to find the rational zeros of a polynomial with rational coefficients (rather than strictly integer coefficients). To get started, is there a nonzero number you can multiply by to convert a polynomial with rational coefficients to a polynomial with integer coefficients?

Often, none of the methods we've discussed will work. The cubic polynomial $P(x) = 3x^3 + x + 2$ is an example. The possible rational zeros are ± 1, ± 2, $\pm \dfrac{1}{3}$, and $\pm \dfrac{2}{3}$, but a check shows that none of these is a zero. Even when we have no way of finding exact zeros, we can find approximate zeros using calculators and computer programs, which employ clever techniques for estimating solutions to a high degree of accuracy.

MODELS AND APPLICATIONS Poiseuille's Law

EXAMPLE 5.22 Poiseuille's Law

Poiseuille's law gives the flow rate F of a liquid through a pipe of radius r. The relationship is

$$F = cr^4$$

where r is measured in inches, F is measured in cubic inches per minute, and c is a constant that depends on the viscosity of the liquid, the pressure, and the length of the pipe. In this exercise, the viscosity, pressure, and the length of the pipe are kept constant, so the value of c does not change.

a. The pipe has a radius of 1 inch, and it is found that 16 cubic inches per minute flow through the pipe. Find the value of c.

b. The radius of the pipe is increased to accommodate a flow of 81 cubic inches per minute. Factor the resulting polynomial in r, and determine the new pipe radius.

SOLUTION

a. We know that $r = 1$ when the flow rate is $F = 16$. We put these values into Poiseuille's law and solve for c:

$$F = cr^4$$

$$16 = c \times 1^4$$

$$16 = c$$

b. We use the value of c found in part a:

$$F = 16r^4$$

We put 81 in place of F and factor the resulting polynomial in r:

$$F = 16r^4$$

$$81 = 16r^4 \qquad \blacktriangleleft \textbf{Put in 81 for } F.$$

$$0 = 16r^4 - 81 \qquad \blacktriangleleft \textbf{Move the constant right.}$$

$$0 = (4r^2 - 9)(4r^2 + 9) \qquad \blacktriangleleft \textbf{Factor the difference of squares.}$$

$$0 = (2r - 3)(2r + 3)(4r^2 + 9) \qquad \blacktriangleleft \textbf{Factor the difference of squares.}$$

We complete the solution by setting each factor equal to zero. Only one resulting equation, $2r - 3 = 0$, yields a nonnegative, real solution. We conclude that the new pipe radius is $r = \dfrac{3}{2} = 1.5$ inches. Note that in this case we could have found the new pipe radius by dividing both sides of the equation $81 = 16r^4$ by 16 and taking the fourth root.

TRY IT YOURSELF 5.22 Brief answers provided at the end of the section.

What pipe radius will accommodate a flow of 625 cubic inches per minute?

5.16 $x = 4$, $x = -2 + \sqrt{12}i = -2 + 2\sqrt{3}i$, and $x = -2 - \sqrt{12}i = -2 - 2\sqrt{3}i$

5.17 $x = 3$, $x = -3$, $x = 2$, $x = -2$

5.18 $x = 1$, $x = 2$, and $x = -2$

5.19 $x = -1$, $x = -2$, $x^3 - 7x - 6 = (x - 3)(x + 1)(x + 2)$

5.20 $x = 1$, $x = 2$, and $x = \dfrac{1}{3}$

5.21 $x = 1$, $x = -1 + \sqrt{3}$, and $x = -1 - \sqrt{3}$

5.22 $r = 2.5$ inches

EXERCISE SET 5.3

CHECK YOUR UNDERSTANDING

1. Possible rational zeros of a nonconstant polynomial with integer coefficients all have the form $\dfrac{a}{b}$, where _____.

2. **True or false:** If the leading coefficient of a polynomial with integer coefficients is 1, then all rational zeros must be integer zeros.

3. **True or false:** If one zero of a polynomial is rational, then all zeros must be rational.

4. **True or false:** If a polynomial with integer coefficients has no rational zeros, then it has no zeros at all.

5. If $x - a$ is a factor of $P(x)$, then if we divide that factor out we obtain a

  a. zero of $P(x)$

  b. polynomial of lower degree

  c. polynomial of larger degree

  d. complex zero

6. The order, or multiplicity, of a zero a of a polynomial $P(x)$ is _____.

7. **True or false:** The order, or multiplicity, of a zero is always less than the degree of the polynomial.

8. A friend says he thinks $x = \dfrac{1}{2}$ is a zero of $P(x) = x^{11} - 5x^9 - 4x^7 + 1$. What do you think?

9. A friend says her calculator gave her nine solutions of the equation $3x^8 - 5x^5 - 4x + 121 = 0$. What do you think?

SKILL BUILDING

Finding zeros of factored polynomials. In Exercises 10 through 21, find the zeros of the given polynomial. Include complex zeros if any occur.

10. $P(x) = (x - 1)(x - 2)(x - 3)$

11. $P(x) = x^2(x - 1)^2$

12. $P(x) = (x^2 - 1)(x^2 - 2)$

13. $P(x) = x(x^2 - 2x - 2)$

14. $P(x) = (x^2 - 7)(x^2 - x - 1)$

15. $P(x) = (x^2 - 1)x^2$

16. $P(x) = (x - 1)^8$

17. $P(x) = (x^2 - 2x + 1)(x^2 - 5x + 6)$

18. $P(x) = (x + 2)(x^2 - 2x + 3)$

19. $P(x) = (x^2 - 1)(x^2 + 1)$

20. $P(x) = (x^2 + 2x + 2)(x^2 + x + 1)$

21. $P(x) = (x^2 - x - 12)^3(x^2 + x - 3)^4$

Factoring. In Exercises 22 through 34, factor the given polynomial as far as possible without using complex numbers.

22. $x^3 - 1$

23. $x^6 - x^4$

24. $x^6 - x^2$

25. $x^4 - 4x^2 - 5$

26. $x^3 + 64$

27. $x^4 - x$

28. $(x + 3)(x^2 - 4) + x(x^2 - 4)$

29. $(x + 1)x^2 - (x + 1)x - 6(x + 1)$

30. $(x - 2)^2(x + 1) + (x - 2)^2(x - 1)$

31. $x^2(x + 4) - 4(x + 4)$

32. $(x - 1)(x^2 - 3) + (x - 4)(x^2 - 3)$

33. $x^4 - 1$

34. $(x^2 - 3x + 2)(x^2 - 5) + (x^2 - 3x + 2)(-9x + 25)$

Rational zeros. In Exercises 35 through 50, find all rational zeros, if any.

35. $P(x) = x^3 - 5x^2 + x - 5$

36. $P(x) = x^3 - 4x^2 + 6x - 4$

37. $P(x) = x^4 - x^3 + x^2 - 1$

38. $P(x) = x^4 + 2x^3 - 2x - 4$

39. $P(x) = x^3 + 5x^2 - 4x - 20$

40. $P(x) = x^3 + 3x^2 + 3x + 1$

41. $P(x) = x^6 - x + 1$

42. $P(x) = x^3 - x + 5$

43. $P(x) = 2x^3 - x^2 + 2x - 1$

44. $P(x) = 3x^3 + 2x^2 + 2x - 1$

45. $P(x) = 3x^3 - 2x^2 + 3x - 2$

46. $P(x) = 4x^3 - 12x^2 + 11x - 3$

47. $P(x) = 2x^4 + x^3 + x^2 + x - 1$

48. $P(x) = 9x^3 + 18x^2 - x - 2$

49. $P(x) = x^4 + x + 1$

50. $P(x) = x^3 - 7$

PROBLEMS

51. **No real zeros.** Show that the polynomial
$P(x) = 2x^6 + 7x^4 + x^2 + 1$ has no real zeros. Note that the powers of x are all even.

Solving by factoring. In Exercises 52 through 69, use the methods introduced in this section to factor the given polynomial, and then find all zeros (both real and complex).

52. $P(x) = x^3 - 1$

53. $P(x) = x^4 - 2x^2 + 1$

54. $P(x) = (x^4 - 3x^2) - (5x^2 - 15)$

55. $P(x) = x^3 + 8$

56. $P(x) = x^6 + 5x^5 + 4x^4$

57. $P(x) = x^4 - 3x^2 + 2$

58. $P(x) = (x^4 + x^2) - (4x^2 + 4)$

59. $P(x) = x^4 - 2x^2 - 3$

60. $P(x) = 8x^3 - 27$

61. $P(x) = x^4 + 3x^2 - 4$

62. $P(x) = x^6 - 9x^3 + 8$

63. $P(x) = x^8 - 5x^4 + 4$

64. $P(x) = (x^3 - 4x^2 + 4x) + (3x^2 - 12x + 12)$

65. $P(x) = x^4 - 8x$

66. $P(x) = (6x^4 - 3x^2) - (2x^2 - 1)$

67. $P(x) = x^5 - 2x^4 - 3x^3$

68. $P(x) = x^4 - 1$

69. $P(x) = x^5 - x^4 - 3x^3 + 3x^2 + 2x - 2$. *Suggestion:* First find an integer zero.

Given zeros. In Exercises 70 through 79, some zeros are given. Verify that they are indeed zeros, and find all remaining zeros. Include complex zeros if there are any.

70. $x = 5$ is a zero of $P(x) = x^3 - 5x^2 + x - 5$.

71. $x = 1$ is a zero of $P(x) = x^3 - 13x + 12$.

72. $x = 2$ is a zero of $P(x) = x^3 + 5x^2 - 4x - 20$.

73. $x = 2$ and $x = 3$ are zeros of
$P(x) = x^4 - 10x^3 + 37x^2 - 60x + 36$.

74. $x = -1$ is a zero of $P(x) = x^3 - x^2 - x + 1$.

75. $x = 2$ and $x = -2$ are zeros of
$P(x) = x^4 - 13x^2 + 36$.

76. $x = 2$ is a zero of $P(x) = x^3 - 4x^2 + 6x - 4$.

77. $x = 3$ and $x = 4$ are zeros of
$P(x) = x^4 - 5x^3 - x^2 + 17x + 12$.

78. $x = 2$ and $x = \dfrac{1}{2}$ are zeros of
$P(x) = 2x^4 - 5x^3 - 4x^2 + 15x - 6$.

79. $x = 1$ and $x = 2$ are zeros of
$P(x) = x^4 - 3x^3 + 6x^2 - 12x + 8$.

Solving by finding rational zeros. In Exercises 80 through 93, find all zeros (including complex zeros if they exist) by first finding at least one rational zero.

80. $P(x) = x^3 + x^2 - 2x - 2$

81. $P(x) = x^3 - 2x^2 - x + 2$

82. $P(x) = x^3 - 7x + 6$

83. $P(x) = 2x^3 - 3x^2 - 11x + 6$

84. $P(x) = 3x^3 + 2x^2 - 19x + 6$

85. $P(x) = 2x^3 - 3x^2 - 18x - 8$

86. $P(x) = x^3 - 2x + 1$

87. $P(x) = x^4 - 3x^3 + 3x^2 - 3x + 2$

88. $P(x) = x^3 - 2x^2 + 2x - 4$

89. $P(x) = x^5 - 2x^4 - x + 2$

90. $P(x) = 3x^3 - x^2 - 6x + 2$

91. $P(x) = 2x^3 + x^2 + x - 1$

92. $P(x) = 6x^4 - 5x^3 - 11x^2 + 10x - 2$

93. $P(x) = 6x^4 + 5x^3 + 7x^2 + 5x + 1$

Finding approximate solutions. The zeros of the polynomials in Exercises 94 through 99 may not be found by methods presented in this section. For each exercise, use a calculator or computer to make a graph and to find the real zeros. Report your answers to two decimal places. You may assume that all real zeros lie in the interval $[-3, 3]$.

94. $P(x) = x^3 + x - 1$

95. $P(x) = x^4 - x - 1$

96. $P(x) = x^3 - 3x - 1$

97. $P(x) = x^4 - 4x^2 + 2$

98. $P(x) = x^5 - 4x^3 + 2$

99. $P(x) = -x^6 + 7x^2 - 1$

MODELS AND APPLICATIONS

100. **More on Poiseuille's law.** As in Example 5.22, we assume that the constant c in Poiseuille's law remains fixed, but for this exercise we do not know the value of c. An artery is blocked by plaque so that its effective radius is cut in half. How is the flow rate through the artery affected? *Suggestion:* Let $F_0 = cr_0^4$ be the flow rate of the artery with no plaque buildup. Replace r_0 by $\dfrac{r_0}{2}$, and see what happens to F_0.

101. **Diving board.** A person weighing 75 kilograms is standing at the free end of a diving board that is 3 meters long. If we denote by x the horizontal distance, in meters, from a point on the board to the hinged end of the board, the amount $D(x)$ by which the board is deflected due to the weight of the person is given by

$$D(x) = 0.081x^2 - 0.009x^3$$

See **Figure 5.30**. Here D is measured in meters, and we assume that $0 \le x \le 3$. Round your answers in meters to two decimal places.

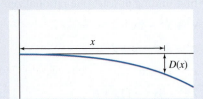

Figure 5.30 A diving board

a. What is the deflection at the free end of the board?

b. At what distance x is the deflection 0.2 meter?

102. **Builder's Old Measurement.** The Builder's Old Measurement was used in 18th-century England to estimate the total tonnage T of a wooden ship from its beam width B and length L, both measured in feet. The relationship is

$$T = \frac{(L - 0.6B)B^2}{188}$$

Consider a ship that is 60 feet long.

a. What is the tonnage if the beam width is 20 feet?

b. What beam width will yield a tonnage of 115? (Assume that the beam width is less than the length.)

103. **A balloon.** For a balloon of radius r, the volume is $V = \dfrac{4}{3}\pi r^3$, and the surface area is $S = 4\pi r^2$.

a. For what nonzero radius is the volume numerically equal to the surface area?

b. For what nonzero radius is the volume numerically equal to the radius?

104. **A puzzle.** The cube of one number plus the square of another is 17. The first number minus the second is 5. Find both numbers. Note that there are three different answers.

105. **Another puzzle.** The product of three consecutive integers is 120. Find the integers.

106. **Carry-on bag.** The bag I want to carry on my next flight has the same width as height. Its length is 10 inches more than the width. The bag holds 375 cubic inches.

a. What are the dimensions of the bag?

b. Carry-on bags must have dimensions no larger than 9 inches by 14 inches by 22 inches. Will I be able to carry on this bag, or must I check it?

107. **A wooden ball.** A ball of radius R is made of Douglas fir. If the ball floats in water, then d inches of its radius will be below the waterline, as shown

in **Figure 5.31**. Archimedes' principle can be used to show that d satisfies the cubic equation

$$d^3 - 3Rd^2 + 2R^3 = 0$$

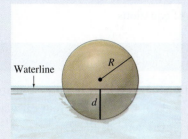

Figure 5.31 A floating ball

a. Show that $d = R$ is a solution of this equation, and explain what this solution means in terms of the floating ball.

b. There are two additional real solutions of this cubic equation. Find them.

c. Do the two solutions from part b make practical sense? Explain your answer.

108. A box. A box with a square base is to have a volume of 12 cubic feet. The height of the box is 1 foot longer than the base. (See **Figure 5.32**.) What are the dimensions of the box?

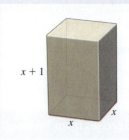

Figure 5.32 A box with a square base

109. Speed of sound in the ocean. The speed of sound S, in meters per second, in the North Atlantic depends on the water temperature T in degrees Celsius and the depth. At a depth of 100 meters, the relationship[1] is

$$S = 1449.361675 + 4.771299T - 0.054350T^2 + 0.000237T^3$$

a. What is the speed of sound at a depth of 100 meters if the temperature is 10° Celsius? Report your answer correct to two decimal places.

b. At what temperature is the speed of sound 1480 meters per second? Assume that the depth is 100 meters.

---

[1]Adapted from *Underwater Acoustic Modeling*, P. C. Etter, E & FN Spon, London, 1996, 15–16.

CHALLENGE EXERCISES FOR INDIVIDUALS OR GROUPS

110. Five points determine a quartic polynomial. Consider the points (a_1, b_1), (a_2, b_2), (a_3, b_3), (a_4, b_4), and (a_5, b_5), where $a_1, a_2, a_3, a_4,$ and a_5 are distinct numbers. Your task is to find a polynomial $P(x)$ of degree at most 4 such that $P(a_1) = b_1$, $P(a_2) = b_2$, $P(a_3) = b_3$, $P(a_4) = b_4$, and $P(a_5) = b_5$. This situation is illustrated in **Figure 5.33**. *Suggestion*: Look at the challenge exercises in Sections 5.1 and 5.2.

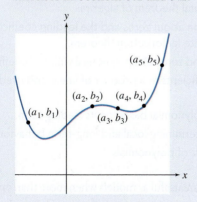

Figure 5.33 A quartic polynomial passing through five points

111. Solving cubic equations.

a. Show that if we divide both sides of the general cubic equation $ax^3 + bx^2 + cx + d = 0$ by a, we get an equation of the form $x^3 + px^2 + qx + r = 0$.

b. We now restrict our attention to the specific example of $x^3 + 3x^2 + 12x - 16 = 0$. The method can be used to solve any cubic equation.

  i. Show that the substitution $x = y - \dfrac{3}{3} = y - 1$ reduces the cubic equation to $y^3 + 9y = 26$. Note that we get $\dfrac{3}{3}$ by dividing the coefficient of x^2 by 3.

  ii. Show that if $3st = A$ and $s^3 - t^3 = B$, then $y = s - t$ is a solution of $y^3 + Ay = B$.

  iii. Show that $s = 3$, and $t = 1$ is one solution of the system of equations

$$3st = 9$$
$$s^3 - t^3 = 26$$

  iv. Use the results of part iii to find one solution of $y^3 + 9y = 26$.

v. Use the preceding results to find one solution of $x^3 + 3x^2 + 12x - 16 = 0$.

vi. Find the remaining solutions.

c. A crucial step here is the solution for s and t of the system of equations

$$3st = A$$

$$s^3 - t^3 = B$$

Show how to find at least one solution of this system of equations.

REVIEW AND REFRESH: Exercises from Previous Sections

112. **From Section P.2:** Solve the inequality $|2x - 1| < 5$.

Answer: $(-2, 3)$

113. **From Section 1.2:** Find the average rate of change of $f(x) = x^2 + 2$ on the interval $[a, b]$ with $a \neq b$. Simplify your answer.

Answer: $a + b$

114. **From Section 1.4:** Find the limit as $x \to \infty$ of $f(x) = \dfrac{1-x}{1+x}$.

Answer: -1

115. **From Section 2.1:** If $f(x) = \dfrac{2}{x^2 + 3}$, find functions $g(x)$ and $h(x)$ so that $f = g \circ h$.

Answer: One solution is $g(x) = \dfrac{2}{x}$ and $h(x) = x^2 + 3$.

116. **From Section 5.1:** Calculate the discriminant of $P(x) = 2x^2 - 3x + 4$. What does your calculation tell you about the zeros of $P(x)$?

Answer: The discriminant is -23. Because the discriminant is negative, there are no real zeros.

117. **From Section 5.2:** Perform the long division:
$$x - 2 \overline{\smash{)}\, x^2 + x + 4}.$$

Answer: The quotient is $x + 3$, and the remainder is 10.

118. **From Section 5.2:** Write $P(x) = x^3 + 1$ in the form $(x^2 - 3)Q(x) + R(x)$, where $Q(x)$ and $R(x)$ are polynomials and the degree of $R(x)$ is less than 2.

Answer: $x^3 + 1 = (x^2 - 3)x + (3x + 1)$

119. **From Section 5.2:** Use synthetic division to evaluate $P(x) = x^3 - 2x + 5$ at $x = 4$.

Answer: 61

5.4 Graphs of Polynomials

5.1 Quadratic Functions

5.2 Long Division and the Factor Theorem

5.3 Zeros of Higher-Degree Polynomials

5.4 Graphs of Polynomials

5.5 Rational Functions

Polynomial functions are the simplest functions whose graphs have a complex structure, including dramatically varying rates of change.

In this section, you will learn to:

1. Explain the importance of the fundamental theorem of algebra.
2. Construct a polynomial given information about zeros and the leading coefficient by using the zeros theorem and the complete factorization theorem.
3. Apply the conjugate zeros theorem to find zeros of a polynomial with real coefficients.
4. Write a given polynomial with real coefficients as a product of linear and irreducible quadratic factors with real coefficients.
5. Describe the long-term behavior of a polynomial based on its leading term.
6. Sketch the graph of a polynomial by determining local and long-term behavior.
7. Solve applied problems involving graphs of polynomials.

The graphs of higher-degree polynomials are useful as models when more than one local maximum or minimum is present. For example, in an uncertain economy the value of an investment may fluctuate over time. It may increase in value before its value falls, only to increase at a later time. A polynomial can be selected to show any number of ups and downs.

In this section, we explore graphs of polynomials, long-term behavior, and local behavior. We will discuss the usefulness of factorization and locating zeros for sketching graphs.

Fundamental Theorem of Algebra and Complete Factorization

> We can find a complete factorization of a polynomial given its zeros.

In Section 5.3, we observed that every polynomial of degree n has at most n zeros. But is it possible for a (nonconstant) polynomial to have no zeros at all? The legendary mathematician Carl F. Gauss is usually credited with giving in 1799 the first proof that this cannot be. This result is called the fundamental theorem of algebra.

LAWS OF MATHEMATICS: Fundamental Theorem of Algebra

Every polynomial (with real or complex coefficients) of degree 1 or greater has at least one (possibly complex) zero.

We can use this theorem to understand important facts about the zeros of a polynomial. Suppose $P_n(x)$ is a polynomial of degree $n \geq 1$. The fundamental theorem of algebra tells us that this polynomial has a zero k_1, so this polynomial has a factor that we can divide out to get a polynomial of degree $n - 1$:

$$P_n(x) = (x - k_1)P_{n-1}(x)$$

Applying the fundamental theorem of algebra to $P_{n-1}(x)$, we get another zero, another factor, and a polynomial of still lower degree:

$$P_n(x) = (x - k_1)(x - k_2)P_{n-2}(x)$$

We can continue this until we have $P_n(x)$ written as a product of n linear factors (some of which may involve complex numbers).

In Section 5.3, we defined the order of the zero of a polynomial to be the number of times it appears in a factor of the polynomial. The preceding argument then shows that every polynomial of degree n has exactly n zeros, if we count complex zeros and the order of the zeros.

LAWS OF MATHEMATICS: Factoring Polynomials

Zeros theorem: Every polynomial of degree $n \geq 1$ has exactly n zeros, if we count complex zeros and the order of the zeros. Here is another way of saying the same thing:

> The sum of the orders of all zeros is the degree of the polynomial.

Complete factorization theorem: We can write every nonconstant polynomial as a product of linear factors. In fact, if $P(x)$ has degree n and a leading coefficient a,

$$P(x) = a(x - k_1)(x - k_2) \cdots (x - k_n)$$

where $k_1, k_2, \ldots, k_n$ are the zeros of $P(x)$. Here the zeros may be complex numbers, and a zero is repeated in the list according to its order.

EXAMPLE 5.23   Constructing a Polynomial

A polynomial $P(x)$ of degree 5 has leading coefficient 4. The distinct zeros of $P(x)$ are $x = 2$, $x = 5$, and $x = 6$. The zeros $x = 5$ and $x = 6$ each have order 1.

a. What is the order of the zero $x = 2$?

b. Find a formula for $P(x)$.

SOLUTION

a. The zeros theorem tells us that the sum of the orders of the zeros of $P(x)$ is its degree, which is 5. Because $x = 5$ and $x = 6$ both have order 1, the only remaining zero, $x = 2$, must have order 3.

b. We know from part a that $(x - 2)^3$, $x - 5$, and $x - 6$ are all the factors of $P(x)$. Because the leading coefficient is 4, by the complete factorization theorem we have

$$P(x) = 4(x - 2)^3(x - 5)(x - 6)$$

The graph in **Figure 5.34** shows the zeros of $P(x)$.

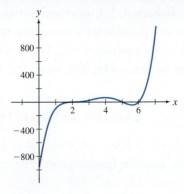

DF **Figure 5.34** A graph showing the zeros of $P(x)$

TRY IT YOURSELF 5.23 Brief answers provided at the end of the section.

Suppose that $x = -1$, $x = 3$, $x = -4$, and $x = 8$ are the distinct zeros of a polynomial $Q(x)$ of degree 10. Assume that both $x = -1$ and $x = 3$ are zeros of order 2. Assume also that the zeros $x = -4$ and $x = 8$ each have order 3 and that the leading coefficient is 2. Find a formula for $Q(x)$.

Polynomials with Real Coefficients

> Polynomials with real coefficients can be expressed as a product of linear and quadratic factors that also have real coefficients.

For the remainder of this section, we restrict our attention to polynomials with real coefficients. In Section 5.1, we noted that for quadratic polynomials with real coefficients the complex zeros occur in conjugate pairs. That is, if $a + bi$ (with a and b real) is a zero, so is its complex conjugate $a - bi$. The same property holds for all polynomials with real coefficients. (See Appendix 2 for a proof.) This property has an important consequence. Let $P(x)$ be a polynomial with real coefficients. If $a + bi$ is a complex (but not real) zero of $P(x)$, both $x - (a + bi)$ and $x - (a - bi)$ are factors of $P(x)$. Hence, their product is a factor. We calculate the product by rearranging the factors:

$$
\begin{aligned}
(x - (a + bi))(x - (a - bi)) &= ((x - a) - bi)((x - a) + bi) && \blacktriangleleft \text{ Rearrange factors.} \\
&= (x - a)^2 - (bi)^2 && \blacktriangleleft \text{ Multiply as } (A - B)(A + B). \\
&= (x^2 - 2ax + a^2) - (b^2(-1)) && \blacktriangleleft \text{ Expand squares.} \\
&= x^2 - 2ax + a^2 + b^2 && \blacktriangleleft \text{ Simplify.}
\end{aligned}
$$

We conclude that $x^2 - 2ax + a^2 + b^2$ is a quadratic polynomial with real coefficients that is also a factor of $P(x)$. Further, $x^2 - 2ax + a^2 + b^2$ is irreducible, in the sense that it has no real zeros. The result is that a nonconstant polynomial with real coefficients can be written as the product of linear and quadratic factors with real coefficients that have no real zeros, so the quadratic factors are irreducible.

LAWS OF MATHEMATICS: Polynomials with Real Coefficients

Conjugate zeros theorem: If the coefficients of a polynomial are real numbers, the complex zeros occur in conjugate pairs.

Linear and quadratic factors theorem: Every nonconstant polynomial with real coefficients is a product of linear and quadratic factors with real coefficients so that the quadratic factors are irreducible—that is, they have no real zeros.

EXAMPLE 5.24 Using the Conjugate Zeros Theorem

It is a fact that $x = 1 + i$ is a zero of $P(x) = x^4 + 2x^3 - 2x^2 + 8$. (This fact may be checked by directly evaluating $P(1 + i)$ or by using synthetic division.)

a. Find one additional zero of $P(x)$.

b. Write $P(x)$ as a product of linear and irreducible quadratic factors, each of which has real coefficients.

c. Find the remaining zeros of $P(x)$.

SOLUTION

a. Because $1 + i$ is a zero, the conjugate zeros theorem tells us that its complex conjugate $1 - i$ is also a zero.

b. Because $1 + i$ is a zero, the factor theorem tells us that $(x - (1 + i))$ is a factor. Similarly, $(x - (1 - i))$ is a factor. Because both are factors, their product is also a factor. The product is

$$(x - (1+i))(x - (1-i)) = ((x-1) - i)((x-1) + i) \quad \blacktriangleleft \text{Rearrange the factors.}$$

$$= (x-1)^2 - i^2 \quad \blacktriangleleft \text{Multiply as } (A-i)(A+i).$$

$$= (x^2 - 2x + 1) - (-1) \quad \blacktriangleleft \text{Expand the square.}$$

$$= x^2 - 2x + 2 \quad \blacktriangleleft \text{Simplify.}$$

We conclude that $x^2 - 2x + 2$ is a factor of $P(x)$. To find the remaining factors, we perform long division:

$$
\begin{array}{r}
x^2 + 4x + 4 \\
x^2 - 2x + 2 \overline{)x^4 + 2x^3 - 2x^2 + 0x + 8} \\
\underline{x^4 - 2x^3 + 2x^2} \\
4x^3 - 4x^2 + 0x + 8 \\
\underline{4x^3 - 8x^2 + 8x} \\
4x^2 - 8x + 8 \\
\underline{4x^2 - 8x + 8} \\
0
\end{array}
$$

This result is that

$$P(x) = (x^2 - 2x + 2)(x^2 + 4x + 4)$$

To complete the factorization, we address the second factor, which is a perfect square:

$$P(x) = (x^2 - 2x + 2)(x^2 + 4x + 4)$$

$$= (x^2 - 2x + 2)(x + 2)^2$$

Note that the quadratic polynomial $Q(x) = x^2 - 2x + 2$ is irreducible because its zeros $1 + i$ and $1 - i$ are not real numbers. The graph of $P(x)$ is shown in **Figure 5.35**. Note that the graph touches, but does not cross, the x-axis at $x = -2$. This behavior reflects the fact that the zero $x = -2$ is of order 2.

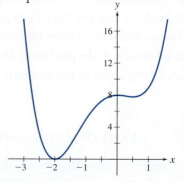

Figure 5.35 A graph showing the order-2 zero of $P(x)$

c. We know that $1 + i$ and $1 - i$ are zeros of $P(x)$. From the factorization in part b, the only other zero is $x = -2$. It is worth noting that the zero $x = -2$ has order 2. As a consequence, the degree (4) of the polynomial matches the number of zeros when order is taken into account.

TRY IT YOURSELF 5.24   Brief answers provided at the end of the section.

Given that $x = i$ is a zero of $P(x) = x^4 + 8x^3 + 17x^2 + 8x + 16$, find the remaining zeros.

CONCEPTS TO REMEMBER: Factoring When One Complex Zero Is Known

Let $P(x)$ be a polynomial with real coefficients. If a complex (but not real) zero $a + bi$ of $P(x)$ is known, then its complex conjugate $a - bi$ is also a zero. As a consequence,

$$(x - (a + bi))(x - (a - bi)) = x^2 - 2ax + a^2 + b^2$$

is an irreducible factor of $P(x)$.

Long-Term Behavior

> **All nonconstant polynomials approach plus or minus infinity at their horizontal extremes.**

Although the local picture for graphs of polynomials may be somewhat complicated, the picture for the long-term behavior is much simpler. It turns out that for values of x that are large in magnitude, a polynomial behaves like its leading term. We can illustrate this fact with the cubic polynomial

$$P(x) = x^3 - 3x^2 - x + 20$$

We factor out the leading term x^3:

$$x^3 - 3x^2 - x + 20 = x^3 \left(1 - \frac{3}{x} - \frac{1}{x^2} + \frac{20}{x^3} \right)$$

If x is a large number, then $1 - \dfrac{3}{x} - \dfrac{1}{x^2} + \dfrac{20}{x^3}$ is close to 1 because the fractions are all nearly zero. In terms of limits, this says

$$1 - \frac{3}{x} - \frac{1}{x^2} + \frac{20}{x^3} \to 1 \text{ as } x \to \infty$$

Because $1 - \dfrac{3}{x} - \dfrac{1}{x^2} + \dfrac{20}{x^3}$ is close to 1 when x is large, the product $x^3 \left(1 - \dfrac{3}{x} - \dfrac{1}{x^2} + \dfrac{20}{x^3} \right)$ is close to x^3 when x is large, in the sense that the ratio of these two functions is close to 1. A similar argument shows that $x^3 - 3x^2 - x + 20$ is close to x^3 when x is large but negative.

A graphical representation of this phenomenon may be helpful. In **Figure 5.36** we see that the graphs of $y = x^3$ and $y = x^3 - 3x^2 - x + 20$ appear quite different when displayed on the interval $[-3, 3]$. But the two graphs look much alike when viewed on the larger scale shown in **Figure 5.37**. Note that in this figure the two graphs seem to merge at the two tail ends. When we graph on a still larger scale, as we have done in **Figure 5.38**, the plots become indistinguishable.

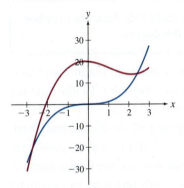

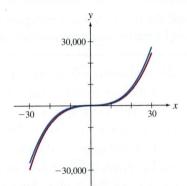

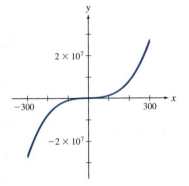

DF **Figure 5.36** The graphs of $y = x^3$ (blue) and $y = x^3 - 3x^2 - x + 20$ (red): The graphs look different on $[-3, 3]$.

Figure 5.37 The graphs of $y = x^3$ and $y = x^3 - 3x^2 - x + 20$ on a wider scale: The graphs appear nearly the same on $[-30, 30]$.

Figure 5.38 The graphs of $y = x^3$ and $y = x^3 - 3x^2 - x + 20$ on a still wider scale: The graphs are indistinguishable on $[-300, 300]$.

> **CONCEPTS TO REMEMBER: Determining the Long-Term Behavior of Polynomials**
>
> The long-term behavior of a polynomial is the same as that of its leading term.

EXAMPLE 5.25 Finding Long-Term Behavior

Use limits to describe the long-term behavior of $y = 6x^4 - 3x^2 + 2$.

SOLUTION

The long-term behavior of $y = 6x^4 - 3x^2 + 2$ is the same as that of its leading term $y = 6x^4$. Because $6x^4 \to \infty$ as $x \to \infty$, we conclude that $6x^4 - 3x^2 + 2 \to \infty$ as $x \to \infty$. Also, because $6x^4 \to \infty$ as $x \to -\infty$, we find $6x^4 - 3x^2 + 2 \to \infty$ as $x \to -\infty$. In **Figure 5.39** we show the two graphs on the interval $[-1, 1]$, where they are different. But on the interval $[-10, 10]$ shown in **Figure 5.40**, the long-term similarity of the graphs is evident.

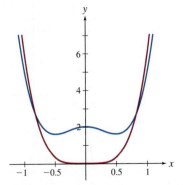

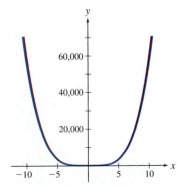

Figure 5.39 The graphs of $y = 6x^4$ (red) and $y = 6x^4 - 3x^2 + 2$ (blue) on $[-1, 1]$

Figure 5.40 The graphs of $y = 6x^4$ and $y = 6x^4 - 3x^2 + 2$ displaying similar long-term behavior

TRY IT YOURSELF 5.25 Brief answers provided at the end of the section.

Use limits to describe the long-term behavior of $y = -x^3 + x^2 + 1$.

Local Behavior

> If multiplicity is counted and complex zeros are included, then the number of zeros of a nonconstant polynomial is exactly the degree.

The local behavior of a polynomial, which includes zeros, local maxima, and local minima, is more difficult to determine than its long-term behavior. To illustrate the procedure, we compare the graph of the quartic polynomial $y = x^4 - 5x^2 + 4$ shown in **Figure 5.41** with that of $y = x^4 + x + 1$ shown in **Figure 5.42**. At the tail ends, both graphs are similar to the graph of the leading term $y = x^4$. But locally the graphs are quite different. The graph in Figure 5.41 shows two local minima, one local maximum, and four zeros, whereas the graph in Figure 5.42 shows only one local minimum, no local maxima, and no zeros.

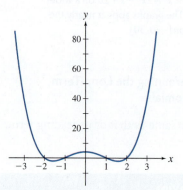

DF **Figure 5.41** A quartic polynomial with three local extreme values

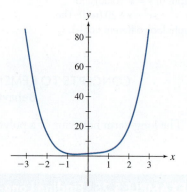

Figure 5.42 A quartic polynomial with only one local extreme value

We can learn a great deal about the local behavior of a polynomial by locating its zeros and where it changes sign. To illustrate the procedure, let's sketch the graph of $y = x^3 - 16x$. We begin by observing that the long-term behavior is the same as that of $y = x^3$, whose graph is shown in **Figure 5.43**. This allows us to begin making the graph of $y = x^3 - 16x$, as shown in **Figure 5.44**.

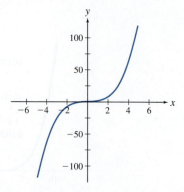

Figure 5.43 The graph of $y = x^3$

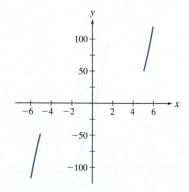

Figure 5.44 Graphing the tail ends: The tail ends of the graph of $y = x^3 - 16x$ are similar to the tail ends of $y = x^3$.

To make further progress, we factor the polynomial:

$$x^3 - 16x = x(x^2 - 16) = x(x - 4)(x + 4)$$

Hence, the zeros occur at $x = -4$, $x = 0$, and $x = 4$. We mark these points in **Figure 5.45**.

Now we want to find the regions where $y = x^3 - 16x$ is positive and where it is negative. We can determine these regions using the sign diagram, which was introduced in Section P.3 for the purpose of solving inequalities. It shows where factored functions are positive and where they are negative. The sign diagram for $y = x(x - 4)(x + 4)$ is

$$x(x - 4)(x + 4)$$

$$\begin{array}{c|c|c|c} (-)(-)(-) = - & (-)(-)(+) = + & (+)(-)(+) = - & (+)(+)(+) = + \\ \hline & -4 & 0 & 4 \end{array}$$

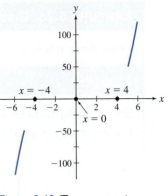

Figure 5.45 The next step in graphing $y = x^3 - 16x$: Mark the zeros.

Remember that if you are unsure about the sign of a polynomial on an interval of a sign diagram, you can simply evaluate the polynomial at any point in that interval and check the sign. We see that $y = x^3 - 16x$ is positive on $(-4, 0)$ and $(4, \infty)$ and is negative on $(-\infty, -4)$ and $(0, 4)$. This result is illustrated in **Figure 5.46**. The completed graph is shown in **Figure 5.47**.

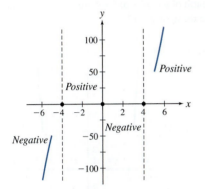

Figure 5.46 Above and below the x-axis: $y = x^3 - 16x$ is positive on $(-4, 0)$ and $(4, \infty)$ and is negative on $(-\infty, -4)$ and $(0, 4)$.

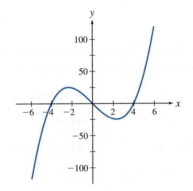

Figure 5.47 The completed graph of $y = x^3 - 16x$

The preceding discussion suggests the following strategy for graphing polynomials.

STEP-BY-STEP STRATEGY: Graphing Polynomials

Step 1 Begin the graph by using the fact that the long-term behavior of a polynomial is the same as that of its leading term.

Step 2 Find the zeros, usually by factoring, and mark them on the x-axis.

Step 3 Use a sign diagram to determine where the graph is above the x-axis and where it is below.

Step 4 Complete a sketch of the graph.

EXAMPLE 5.26 Graphing Polynomials: Factoring

Draw the graph of $y = x^4 - 5x^2 + 4$.

SOLUTION

Step 1: Long-term behavior: We know that the long-term behavior of $y = x^4 - 5x^2 + 4$ matches that of the leading term $y = x^4$, shown in **Figure 5.48**. This fact allows us to begin the graph, as shown in **Figure 5.49**.

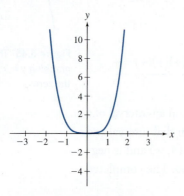

Figure 5.48 The graph of $y = x^4$

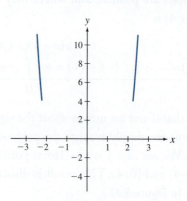

Figure 5.49 Step 1: The tail ends of the graph of $y = x^4 - 5x^2 + 4$ are similar to those of $y = x^4$.

Step 2: Find the zeros and mark them: The next step is to find the zeros. We begin by factoring the polynomial. To do this, substitute $z = x^2$:

$$x^4 - 5x^2 + 4 = z^2 - 5z + 4 \qquad \blacktriangleleft \text{Substitute } z = x^2.$$

$$= (z - 4)(z - 1) \qquad \blacktriangleleft \text{Factor.}$$

$$= (x^2 - 4)(x^2 - 1) \qquad \blacktriangleleft \text{Replace } z \text{ by } x^2.$$

$$= (x - 2)(x + 2)(x - 1)(x + 1) \qquad \blacktriangleleft \text{Factor.}$$

The zeros of $y = x^4 - 5x^2 + 4$ are at $-2, -1, 1,$ and 2. These are marked in **Figure 5.50**.

Step 3: Identify sign changes: For this part of the problem, we use the sign diagram:

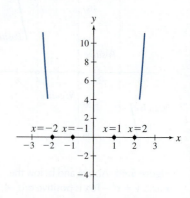

Figure 5.50 Step 2: The zeros of $y = x^4 - 5x^2 + 4$ are marked.

$$(x - 2)(x + 2)(x - 1)(x + 1)$$

$$\underbrace{(-)(-)(-)(-) = +}_{\qquad} \Big|\quad \underbrace{(-)(+)(-)(-) = -}_{\qquad} \Big|\quad \underbrace{(-)(+)(-)(+) = +}_{\qquad} \Big|\quad \underbrace{(-)(+)(+)(+) = -}_{\qquad} \Big|\quad (+)(+)(+)(+) = +$$
$$\qquad -2 \qquad\qquad -1 \qquad\qquad 1 \qquad\qquad 2$$

This diagram tells us where the graph is above the x-axis and where it is below that axis.

Step 4: Complete the graph: The completed graph is in **Figure 5.51**.

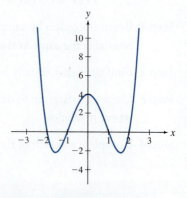

Figure 5.51 Step 4: The graph of $y = x^4 - 5x^2 + 4$ is completed.

TRY IT YOURSELF 5.26 Brief answers provided at the end of the section.

Draw the graph of $y = x^3 - 2x^2 - 8x$.

Is it necessary for the sign of the rate of change of a polynomial to change (negative to positive or positive to negative) between two zeros of a polynomial? Can the sign change more than once?

EXAMPLE 5.27 Graphing Polynomials: Rational Zeros

Draw the graph of $y = x^3 - 3x - 2$.

SOLUTION

Step 1: Long-term behavior: We know that the long-term behavior of $y = x^3 - 3x - 2$ matches that of the leading term $y = x^3$. The graph of $y = x^3$ is shown in **Figure 5.52**, and the tail ends of the graph of $y = x^3 - 3x - 2$ are shown in **Figure 5.53**.

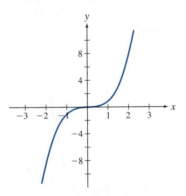

Figure 5.52 The graph of $y = x^3$

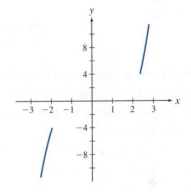

Figure 5.53 Step 1: The tail ends of the graph of $y = x^3 - 3x - 2$ are similar to the tail ends of the graph of $y = x^3$.

Step 2: Find the zeros and mark them: We begin by looking for rational zeros so we can factor $y = x^3 - 3x - 2$. The candidates are ± 1 and ± 2. Checking $x = 2$ gives

$$y(2) = 2^3 - 3 \times 2 - 2 = 0$$

(Many prefer to find the value of $y(2)$ using synthetic division because it calculates both the function value and the quotient in a single step.) Because 2 is a zero, $x - 2$ is a factor. We use synthetic division to find the quotient:

$$
\begin{array}{r|rrr}
2 & 1 \quad 0 \quad -3 \quad -2 \\
  & \quad\ \ 2 \quad\ \ 4 \quad\ \ 2 \\
\hline
  & 1 \quad 2 \quad\ \ 1 \quad \underline{|0}
\end{array}
$$

We conclude that

$$x^3 - 3x - 2 = (x - 2)(x^2 + 2x + 1) = (x - 2)(x + 1)^2$$

Hence, $x = 2$ is a zero of order 1, and $x = -1$ is a zero of order 2. We have marked these zeros in **Figure 5.54**.

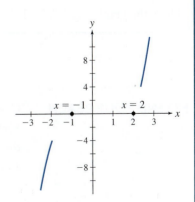

Figure 5.54 Step 2: The zeros of $y = x^3 - 3x - 2$ are marked.

Step 3: Identify sign changes: We use the sign diagram:

$$(x-2)(x+1)^2$$

$$\begin{array}{ccc} (-)(+) = - & (-)(+) = - & (+)(+) = + \end{array}$$

$$\begin{array}{c|cc} & -1 & 2 \end{array}$$

We see that $y = (x-2)(x+1)^2$ is negative on $(-\infty, -1) \cup (-1, 2)$ and positive on $(2, \infty)$. These regions are indicated in **Figure 5.55**.

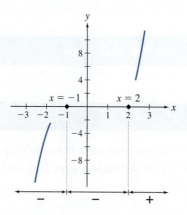

Figure 5.55 Step 3: The sign changes of $y = x^3 - 3x - 2$ are identified.

Step 4: Complete the graph: We use the information we have collected to make the graph, as shown in **Figure 5.56**. Observe that the graph touches but does not cross the x-axis at the even-order zero $x = -1$.

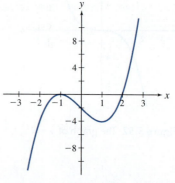

Figure 5.56 Step 4: The graph of $y = x^3 - 3x - 2$ is completed.

TRY IT YOURSELF 5.27 Brief answers provided at the end of the section.

Draw the graph of $y = x^3 - 3x^2 + 4$.

EXTEND YOUR REACH   Consider the three polynomials $P(x) = x^2(x-1)^3$, $Q(x) = -x^2(x-1)^3$, and $S(x) = x(x-1)^4$. Compare the graphs of these three functions by focusing on the local behavior, not the long-term behavior. How does the order of each zero affect the local behavior?

Local Maxima and Minima

> The number of local maxima or minima of a nonconstant polynomial is at most one less than the degree.

Another feature of the graph of a polynomial is the location of the local maxima and minima. It turns out that a polynomial of degree n can have at most $n - 1$ local maxima or minima. Determining where the local maxima and minima occur is a classic calculus problem. But the information we have already gathered can give us some idea about the location of these extreme values. For example, consider again the graph of $y = x^3 - 3x - 2$ sketched in Figure 5.56. It is clear from the graph that the function reaches a local maximum at $x = -1$. Further, the graph shows that the function reaches a local minimum at $x = 1$.

We can use what we have learned about local and long-term behavior to find the range of polynomials.

Range of odd-degree polynomials: For polynomials of odd degree, the range is $\mathbb{R} = (-\infty, \infty)$. The basic reason is that if $P(x)$ is such a polynomial, then $P(x) \to \infty$ at one tail end, and $P(x) \to -\infty$ at the other tail end.

Range of even-degree polynomials with positive leading coefficient: If a nonconstant even-degree polynomial has a positive leading coefficient, then it attains an absolute minimum value, say $y = m$. Then the range is $[m, \infty)$.

Range of even-degree polynomials with negative leading coefficient: Nonconstant even-degree polynomials with a negative leading coefficient have a range of the form $(-\infty, M]$, where the absolute maximum value is $y = M$.

EXAMPLE 5.28 Range of a Polynomial

Find the range of $P(x) = 3x^2 - 12x + 8$.

SOLUTION

This is a polynomial of even degree with positive leading coefficient, so it attains an absolute minimum value. In fact, $P(x)$ is a quadratic function, so its graph is a parabola. We now want to locate the vertex of this parabola. The x-coordinate of the vertex is

$$x = -\frac{b}{2a} = -\frac{-12}{2 \times 3} = 2$$

Thus, the vertex is $(2, P(2)) = (2, -4)$. Because the leading coefficient is positive, the parabola opens upward, as shown in **Figure 5.57**, and $(2, -4)$ is the location of the absolute minimum. Thus, $P(x)$ is a polynomial of even degree with an absolute minimum value of -4. We conclude that the range is $[-4, \infty)$.

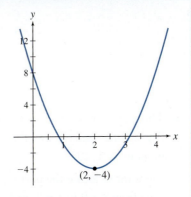

Figure 5.57 The graph of $P(x) = 3x^2 - 12x + 8$

TRY IT YOURSELF 5.28 Brief answers provided at the end of the section.

Find the range of $Q(x) = -5x^3 + 2x^2 - 4x + 9$.

EXTEND YOUR REACH   Let $P(x)$ be a polynomial of even degree. Show that there is a constant c such that the polynomial $Q(x) = P(x) + c$ has no real zeros. *Suggestion*: Think in terms of shifting the graph up or down.

MODELS AND APPLICATIONS   Changing Value of Investments

EXAMPLE 5.29 An Investment

The value V, in thousands of dollars, of an investment t years after January 1, 2020, is given by

$$V = 2t^3 - 21t^2 + 60t + 10$$

The formula is valid from 2020 through 2025.

a. Use a graphing utility to make the graph of V versus t.

b. When does the graph reach a local maximum value, and what is that local maximum value?

c. Over what time period is the value of the investment increasing at a decreasing rate?

SOLUTION

a. The graph is shown in **Figure 5.58**.

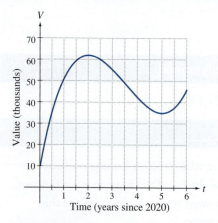

Figure 5.58 The changing value of an investment

b. The graph shows that the local maximum value occurs at about $t = 2$, which corresponds to January 1, 2022. A graphing utility shows that this is the correct value of t and that a local maximum value of $62,000 occurs on January 1, 2022.

c. We want to locate the portion of the graph that is increasing and concave down. That corresponds to the time from $t = 0$ to $t = 2$, or the years 2020 and 2021.

TRY IT YOURSELF 5.29 Brief answers provided at the end of the section.

When between 2021 and 2026 does the local minimum value occur, and what is that local minimum value?

CONCEPTS TO REMEMBER: Polynomials with Real Coefficients

Formula: $P(x) = a_n x^n + a_{n-1} x^{n-1} + \cdots + a_1 x + a_0$, where n is a nonnegative integer, $a_0, \ldots, a_n$ are real constants, and $a_n \neq 0$. The nonnegative integer n is the degree of the polynomial, a_n is the leading coefficient, $a_n x^n$ is the leading term, and a_0 is the constant term. The constant function 0 is also considered to be a polynomial.

Domain: All real numbers.

Range: For odd-degree polynomials, the range is $\mathbb{R} = (-\infty, \infty)$. Nonconstant polynomials with even degree and with a positive leading coefficient have a range of the form $[m, \infty)$, where the absolute minimum value is $y = m$. Nonconstant polynomials with even degree and with a negative leading coefficient have a range of the form $(-\infty, M]$, where $y = M$ is the absolute maximum value.

Graph: A polynomial of degree $n \geq 1$

- has n zeros (counting order); n or fewer of these zeros are real, so the graph meets the x-axis at most n times.
- has at most $n - 1$ local maxima or minima.

In the long term: The long-term behavior of a polynomial is the same as that of its leading term.

TRY IT YOURSELF ANSWERS

5.23 $Q(x) = 2(x+1)^2(x-3)^2(x+4)^3(x-8)^3$

5.24 The remaining zeros are $x = -i$ and $x = -4$, which is of order 2.

5.25 $-x^3 + x^2 + 1 \to -\infty$ as $x \to \infty$, and $-x^3 + x^2 + 1 \to \infty$ as $x \to -\infty$.

5.26

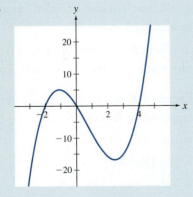

5.27 $x^3 - 3x^2 + 4 = (x+1)(x-2)^2$

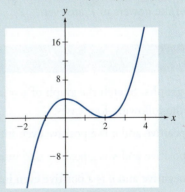

5.28 The range is $\mathbb{R} = (-\infty, \infty)$.

5.29 A local minimum value of \$35,000 occurs on January 1, 2025.

EXERCISE SET 5.4

CHECK YOUR UNDERSTANDING

In Exercises 1 through 8, assume that $P(x)$ is a polynomial of degree 6 with real coefficients.

1. The sum of the orders of the zeros of $P(x)$ is _____.

2. If $2 - i$ is a zero of $P(x)$, then _____ is also a zero.

3. Counting order, there are exactly _____ zeros of $P(x)$.

4. There are at most _____ local maxima or minima for $P(x)$. (Give the best answer you can.)

5. If the leading coefficient of $P(x)$ is positive, what is the limit as $x \to \infty$ of $P(x)$?

6. Suppose that $P(x)$ has three distinct zeros, one having order 2 and another having order 1. What is the order of the third zero?

7. The graph of $P(x)$ crosses the x-axis at most _____ times.

8. Can $P(x)$ change signs at a zero of even order?

9. Let $P(x)$ be a polynomial with real coefficients. If $P(x)$ has complex zeros, then which are true?

   a. The complex zeros come in conjugate pairs.

   b. All of the powers in P are even.

   c. There are more complex zeros than real zeros.

   d. None of the above.

10. Suppose that $P(x)$ is a polynomial of even degree. Assume that the zeros of $P(x)$ lie on the interval $[-10, 10]$ and that $P(-20)$ is positive. What can you conclude about the sign of $P(20)$?

11. Suppose that $P(x)$ is a polynomial with three distinct zeros of odd order and two distinct zeros of even order. How many sign changes can $P(x)$ exhibit?

12. A friend says she found four real zeros and three complex (not real) zeros of $P(x) = 2x^7 - 7x^5 - x + 10$. What do you think?

13. Is it possible for a polynomial to have two local maxima but no local minimum? (Calculus is required for a rigorous justification of the correct answer to this question. But you can get the right answer if you think about the graph.)

14. Is it possible for a nonconstant polynomial of even degree to have no real zeros?

SKILL BUILDING

15. **Simple graphs.** Sketch the graph of $y = ax^n$ in each of the following four cases.

   a. a is positive and n is a positive even integer.

   b. a is positive and n is a positive odd integer.

   c. a is negative and n is a positive even integer.

   d. a is negative and n is a positive odd integer.

Finding polynomials. In Exercises 16 through 19, find a polynomial with the given leading coefficient, with zeros of the given order, and with no other zeros. Leave your answer in factored form.

16. Leading coefficient 4, a zero of order 5 at $x = 1$, and a zero of order 3 at $x = 2$

17. Leading coefficient 2, a zero of order 1 at $x = 2$, and a zero of order 2 at $x = -2$

18. Leading coefficient 1, a zero of order 7 at $x = 4$, and a zero of order 3 at $x = -5$

19. Leading coefficient -5, a zero of order 5 at $x = 1$, one of order 4 at $x = 2$, and one of order 3 at $x = 3$

Matching. In Exercises 20 through 27, match the given graph with the appropriate function from the list below.

   A. $P(x) = x(x - 1)(x + 1)(x - 2)$

   B. $P(x) = (x - 1)(x^2 + 1)$

   C. $P(x) = -x(x - 1)(x + 1)(x - 2)$

   D. $P(x) = (x - 1)(x + 1)(x - 2)$

   E. $P(x) = -x^2 - 3$

   F. $P(x) = -x(x^2 + 1)$

   G. $P(x) = (x^2 + 2)(x^2 + 1)$

   H. $P(x) = (x^2 - 1)^2$

20.

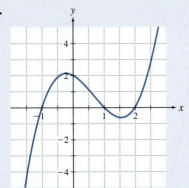

21.

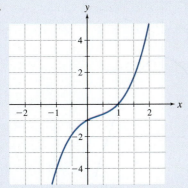

22.

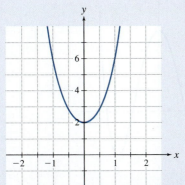

23.

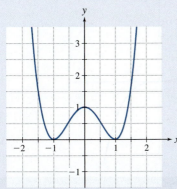

24.

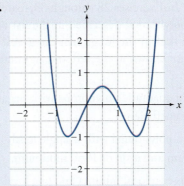

25.

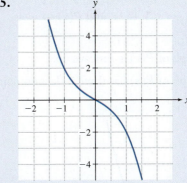

26.

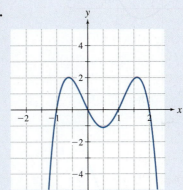

27.

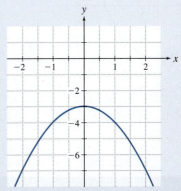

More matching. In Exercises 28 through 33, match the graph with the polynomial described in the list below. You may assume that in the long term the graph continues the behavior shown at the left and right ends of the x-axis.

A. $P(x)$ is a quadratic polynomial with negative leading coefficient.

B. $P(x)$ is a polynomial with a single local maximum and a single local minimum.

C. $P(x)$ has two local minima and one local maximum.

D. $P(x) \to -\infty$ as $x \to \infty$, and $P(x) \to \infty$ as $x \to -\infty$.

E. The graph of $P(x)$ is concave up.

F. The graph of $P(x)$ has three inflection points.

28.

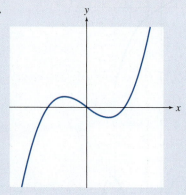

29.

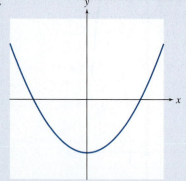

30.

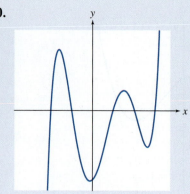

31.

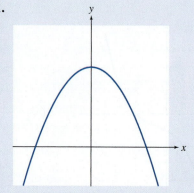

32.

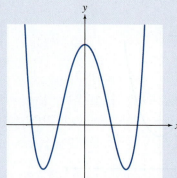

33.

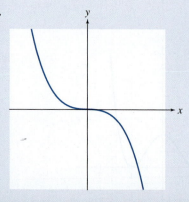

PROBLEMS

Sketching graphs. In Exercises 34 through 43, sketch the graph of the given polynomial.

34. $P(x) = (x - 2)(x - 3)(x - 4)$

35. $P(x) = (x + 2)(x + 3)(x - 1)$

36. $P(x) = x(1 - x)$

37. $P(x) = x^2(1 - x)$

38. $P(x) = x(1 - x)^2$

39. $P(x) = (x - 2)^2(x - 3)$

40. $P(x) = x^2(x - 1)^2$

41. $P(x) = x^3(x - 1)^2$

42. $P(x) = (x + 1)x(x - 1)(x - 2)$

43. $P(x) = x(x - 1)(x + 1)(x - 2)(x + 2)$

More graphs. In Exercises 44 through 47, sketch the graph of a polynomial with the given properties. In each case, there is more than one correct answer.

44. The polynomial has a local maximum at $x = 0$ and a local minimum at $x = 1$.

45. The polynomial has zeros at $x = 0$ and $x = 2$ and a local minimum at $x = 1$.

46. The polynomial has a local maximum at $x = -1$ and a local minimum at $x = 0$.

47. The polynomial has local minima at $x = 0$ and $x = 3$ and a local maximum at $x = 1$.

Complex conjugate zeros. In Exercises 48 through 55, you are given a complex zero of a polynomial $P(x)$. Write $P(x)$ as a product of linear and irreducible quadratic factors, each of which has real coefficients. Then list the zeros of $P(x)$.

48. $x = i$ is a zero of $P(x) = x^4 - 2x^3 + 2x^2 - 2x + 1$.

49. $x = i\sqrt{2}$ is a zero of $P(x) = x^4 + x^3 - x^2 + 2x - 6$.

50. $x = 2i$ is a zero of $P(x) = x^4 + 5x^2 + 4$.

51. $x = i$ is a zero of $P(x) = x^4 - 2x^3 - 4x^2 - 2x - 5$.

52. $x = 1 - i$ is a zero of $P(x) = x^4 - 2x^3 + 5x^2 - 6x + 6$.

53. $x = 2 + i$ is a zero of
$P(x) = x^4 - 2x^3 - 6x^2 + 22x - 15$.

54. $x = 2 + i$ is a zero of
$P(x) = x^4 - 4x^3 + 10x^2 - 20x + 25$.

55. $x = -i$ is a zero of $P(x) = x^4 - 2x^3 - 2x - 1$.

Making graphs. In Exercises 56 through 77, write the given polynomial as a product of linear and irreducible quadratic factors, each of which has real coefficients. Then sketch the graph.

56. $P(x) = x^3 + x^2 + x + 1$

57. $P(x) = (x^2 - x - 1) - x(x^2 - x - 1)$. *Suggestion:* Think of $A - xA$.

58. $P(x) = -x^3 - x^2 + 3x + 3$

59. $P(x) = x^3 + 4x^2 + 6x + 4$

60. $P(x) = x^3 - 1$

61. $P(x) = x^3 + 1$

62. $P(x) = x^4 - 13x^2 + 36$

63. $P(x) = x^4 - 5x^2 - 36$

64. $P(x) = x^4 + 3x^2 + 2$

65. $P(x) = (x^2 - 5)^2 - 1$. *Suggestion:* Think of $A^2 - 1$.

66. $P(x) = x^4 - 2x^3 + 2x - 1$

67. $P(x) = x^3 - 7x^2 + 12x$

68. $P(x) = x^4 - 1$

69. $P(x) = x^4 - 3x^3 + 3x^2 - 3x + 2$

70. $P(x) = x^3 - 2x^2 - x + 2$

71. $P(x) = x^4 + x^3 - 16x^2 - 4x + 48$

72. $P(x) = x^4 - 4x^3 + 6x^2 - 4x + 1$

73. $P(x) = x^4 + x^3 - 7x^2 - x + 6$

74. $P(x) = x^2(x^2 - 7) - x(x^2 - 7) - (x^2 - 7)$. *Suggestion:* Think of $x^2 A - xA - A$.

75. $P(x) = x^3 - x^2 - x + 1$

76. $P(x) = -x^3 + 2x^2 + x - 2$

77. $P(x) = x^3 - x^2 - 5x + 5$

78. A calculus problem. In calculus it is shown that $P(x) = 2x^3 - 3x^2 - 12x + 5$ is decreasing when $y = x^2 - x - 2$ is negative. Find where $P(x)$ is decreasing.

Using a graphing utility. In Exercises 79 through 85, graph the given polynomial. Report the real zeros, local maxima, and local minima correct to two decimal places. (Give both coordinates of any maxima or minima.)

79. $P(x) = x^3 - 3x + 1$

80. $P(x) = 2x^3 - 9x^2 + 12x + 4$

81. $P(x) = -x^3 + 6x^2 - 9x - 1$

82. $P(x) = 3x^4 + 4x^3 - 36x^2 + 50$

83. $P(x) = x^3 + 3x + 1$

84. $P(x) = x^3 - 3x^2 + 3x + 1$

85. $P(x) = x^4 - 4x^3 + 6x^2 - 4x - 1$

MODELS AND APPLICATIONS

86. Production function. In economics, the total product curve is a graphical representation of the total quantity of output that a firm can produce with a given amount of labor (or some other input). In this exercise, we study the total product curve that is the graph of

$$T = -n^3 + 3n^2 + n$$

where n is measured in units of labor and T is measured in units of product.

a. Use a graphing utility to make the graph of T versus n as n varies from 0 to 3.

b. By the law of diminishing returns, the total product curve has an inflection point: initially T is increasing at an increasing rate, but then that rate begins to decrease. For what values of n is T increasing at an increasing rate? Express your answer using whole numbers.

c. What is the maximum value of T? Give your answer as a whole number.

87. Another investment. The value V of an investment t years after January 1, 2022, is given by

$$V = -t^3 + 15t^2 - 63t + 200 \text{ dollars}$$

The formula is valid up to January 1, 2030.

a. Use a graphing utility to make the graph of V versus t.

b. When does the investment reach a local minimum, and what is that local minimum value?

c. When after 2024 does the value reach a local maximum, and what is that local maximum value?

d. Over what time period is the value of the investment decreasing at a decreasing rate?

88. Load on a beam. A beam that is 10 feet long is clamped at both ends and is subjected to a load that is distributed uniformly along the beam. If we denote by x the horizontal distance, in feet, from a point on the beam to the left end, the amount $D(x)$ by which the beam is deflected due to the load is

$$D = 0.002x^4 - 0.04x^3 + 0.2x^2$$

Here D is measured in feet, and we assume that $0 \le x \le 10$. Structural failure is a concern if the deflection is more than 2 feet at any point.

Is structural failure a concern for this load? *Suggestion:* Use symmetry to argue that the maximum deflection occurs at the midpoint.

89. Speed of sound in the ocean. The speed of sound S, in meters per second, in the North Atlantic depends on the water temperature t in degrees Celsius and the depth d in meters. The relationship[1] is

$$S = 1447.733 + 4.7713t - 0.05435t^2 + 0.0002374t^3$$
$$+ 0.0163d + 1.675 \times 10^{-7} d^2 - 7.139 \times 10^{-13} td^3$$

a. Find a formula for the speed of sound at a depth of 50 meters, and plot its graph for temperatures from $0°$ to $10°$ Celsius.

b. Does increasing temperature indicate an increase or a decrease in the speed of sound at a depth of 50 meters? (Assume that the temperature varies between $0°$ and $10°$ Celsius.)

90. Lumber from logs. The Scribner rule[2] for estimating the amount of lumber that can be obtained from a log is

$$V = \frac{L}{16}(0.79D^2 - 2D - 4)$$

Here V is the volume of lumber in board feet, L is the length of the log in feet, and D is the diameter of the smaller end of the log in inches. The Scribner rule is used for logs of diameter 6 to 28 inches. According to this rule, what are the possible amounts of lumber that can be milled from a 24-foot-long log? *Suggestion:* You are being asked for the range of the function V for 24-foot-long logs, so pay attention to the span over which the formula is valid.

91. A shipping container. Restrictions on how a container is constructed mean that if the base is x feet by x feet, then the height is $15 - x$ feet. This relationship is shown in **Figure 5.59**. The container is to be constructed so that the volume is a maximum.

---

[1] Adapted from *Underwater Acoustic Modeling*, P. C. Etter, E & FN Spon, London, 1996, 15–16.

[2] "Understanding Log Scales and Log Rules," https://extension.tennessee.edu/publications/documents/pb1650.pdf.

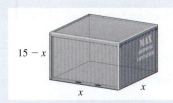

Figure 5.59 A shipping container with a square base

a. Find a formula that gives the volume V, in cubic feet, in terms of x.

 b. Use a graphing utility to plot the graph of V versus x.

c. Use the graph from part b to find the dimensions of the container of largest volume. What is the largest volume?

CHALLENGE EXERCISES FOR INDIVIDUALS OR GROUPS

92. Which is larger? Is $y = 10^{-4}x^4$ eventually larger than $y = 10^6(x^3 + x^2 + x + 1)$? Estimate how large x needs to be to make this happen. Note that when x is larger than 1, $x^3 + x^2 + x + 1 < x^3 + x^3 + x^3 + x^3 = 4x^3$.

93. Extreme values of a polynomial. In the challenge exercises of Section 5.2, we observed that the rate of change of any polynomial is a polynomial of one less degree. Also, local maxima and local minima of polynomials can occur only where the rate of change is zero. Use these facts together with what you know about zeros of polynomials to explain why a polynomial of degree n can have at most $n - 1$ local maxima and local minima.

94. The mean value theorem. Figure 5.60 shows the graph of a polynomial along with the secant line for the interval from $x = A$ to $x = G$. The mean value theorem is an important result in calculus. It tells us that there is a point between A and G where the rate of change (the slope of the tangent line) is the same as the average rate of change (the slope of the secant line) from $x = A$ to $x = G$.

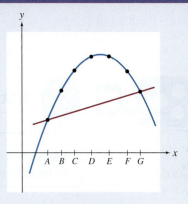

Figure 5.60 A secant line for a polynomial graph

a. At which of the points B through F is the rate of change the same as the average rate of change from $x = A$ to $x = G$?

b. You drive 100 miles in 2 hours, so your average velocity (the average rate of change in distance) is 50 miles per hour. Use the mean value theorem to explain why your speedometer must read exactly 50 miles per hour at some point in your trip. Bear in mind that velocity is the rate of change in distance.

REVIEW AND REFRESH: Exercises from Previous Sections

95. From Section P.1: Write the equation of the circle with center $(2, -1)$ and radius $\sqrt{3}$.

Answer: $(x - 2)^2 + (y + 1)^2 = 3$

96. From Section P.2: Solve $|x - 15| < 10$.

Answer: $(5, 25)$

97. From Section P.3: Make the sign diagram for $x(1 - x)(x - 2)$.

Answer:

$$x(1 - x)(x - 2)$$

$(-)(+)(-) = +$ | $(+)(+)(-) = -$ | $(+)(-)(-) = +$ | $(+)(-)(+) = -$
　　　　　　　0　　　　　　　　1　　　　　　　　2

98. From Section 5.1: Find the maximum value of the function $P(x) = 1 + 2x - x^2$.

Answer: 2

99. From Section 5.2: Given that $x^2 - 1$ is a factor of the polynomial $P(x)$, find two zeros of $P(x)$.

Answer: $x = \pm 1$

100. **From Section 5.2**: Write $P(x) = x^4 + 3$ in the form $P(x) = (x+2)Q(x) + r$, where Q is a polynomial and r is a constant.

Answer: $x^4 + 3 = (x+2)(x^3 - 2x^2 + 4x - 8) + 19$

101. **From Section 5.3**: Use factoring to find all zeros of $P(x) = x^4 - 3x^2 + 2$.

Answer: $x = \pm 1$ and $x = \pm\sqrt{2}$

102. **From Section 5.3**: Given that $x = 1$ is one zero of $P(x) = x^3 - 2x^2 - x + 2$, find all remaining zeros.

Answer: $x = -1$ and $x = 2$

103. **From Section 5.3**: List all possible rational zeros of $P(x) = x^9 + 13x + 5$.

Answer: ± 1, ± 5

104. **From Section 5.3**: If a and b are integers, list all possible rational zeros of $P(x) = 2x^3 + ax^2 + bx + 3$.

Answer: ± 1, ± 3, $\pm\dfrac{1}{2}$, $\pm\dfrac{3}{2}$

5.5 Rational Functions

- **5.1** Quadratic Functions
- **5.2** Long Division and the Factor Theorem
- **5.3** Zeros of Higher-Degree Polynomials
- **5.4** Graphs of Polynomials
- **5.5** Rational Functions

Rational functions have graphs with a complex structure that can be used to model catastrophic events as well as long-term behavior.

In this section, you will learn to:

1. Locate the domain and poles of a rational function.
2. Sketch the graph of a rational function by finding any zeros, any vertical asymptotes, any sign changes, and the long-term behavior.
3. Solve applied problems involving rational functions.

A rational function is a quotient of polynomials where the numerator and denominator have no real zeros in common. Newton's law of gravity says the gravitational attraction between two bodies is inversely proportional to the square of the distance between them. Formally,

$$F = \frac{GM_1M_2}{d^2}$$

where M_1 and M_2 are the masses of the bodies, G is the gravitational constant, d is the distance between the two bodies, and F is the force due to gravity. As a function of d, this is a classic example of a rational function, and it has an important property that is typical of rational functions. When the denominator is near 0, the function value gets large. The interpretation is that when large bodies such as planets get too close, the force of gravity tears them apart. Avoiding collision does not mean avoiding disaster. A near miss of Venus to Earth would be catastrophic. How near? The Roche limit is the distance at which one celestial body will disintegrate due to tidal forces from a second body.

Think this is all science fiction? Many astronomers today believe the moon was created by a collision of early Earth with a celestial body, Theia, which was about the size of Mars.[3] An artist's conception of this event is shown in **Figure 5.61**.

---

[3]For more information, see "Two Planets Suffer Violent Collision," *ScienceDaily* (University of California, Los Angeles), (September 24, 2008), www.sciencedaily.com/releases/2008/09/080923164646.htm.

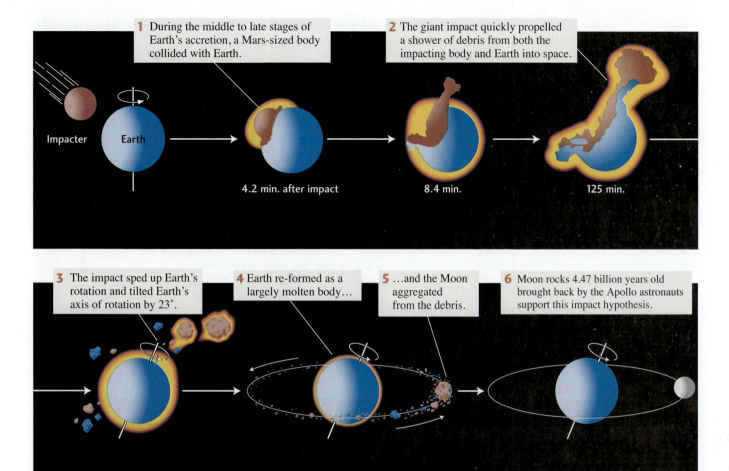

Figure 5.61 Possible formation of the moon resulting from a planetary collision. From *Understanding Earth*, 8e, by John Grotzinger and Thomas Jordan. Copyright 2020 by W. H. Freeman and Company. All rights reserved. Used by permission of the publisher Macmillan Learning. Data from Solid-Earth Sciences and Society. National Research Council, 1993.

In this section, we investigate rational functions such as gravitational attraction. Corresponding to the catastrophic event just discussed is the mathematical concept of a pole. Poles are located at zeros of the denominator and are exhibited as vertical asymptotes of the graph. Similarly, long-term behavior may be exhibited as a horizontal asymptote of the graph. Sign diagrams play a key role in the production of graphs of rational functions.

Domain and Poles

> Poles of rational functions occur at zeros of the denominator.

A **rational function** is simply a ratio of two polynomials in reduced form. The domain of a rational function is the set of all real numbers except those for which the denominator is 0, the **poles**.

Let's look, for example, at the rational function $R(x) = \dfrac{x+1}{x^3 - 2x^2 + x}$. The denominator $x^3 - 2x^2 + x$ can be factored as

$$x(x^2 - 2x + 1) = x(x-1)^2$$

This polynomial is zero when $x = 0$ and $x = 1$. Thus, the domain of the rational function consists of all real numbers except 0 and 1, which are the poles of $R(x)$. Note the

A **rational function** is a function of the form $R(x) = \dfrac{P(x)}{Q(x)}$, where $P(x)$ and $Q(x)$ are polynomials with no real zeros in common.

The **poles** of a rational function are the real numbers for which the denominator is 0.

behavior of the graph in **Figure 5.62** near these poles. Such behavior will turn out to be an important feature of graphs of rational functions.

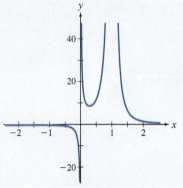

Figure 5.62 The graph of $y = \dfrac{x+1}{x^3 - 2x^2 + x}$

EXAMPLE 5.30 Finding Poles

Consider the rational function $R(x) = \dfrac{2x + 3}{x^3 - x^2 - 6x}$. Determine the domain of R, and identify the poles.

SOLUTION

The denominator factors as

$$x^3 - x^2 - 6x = x(x^2 - x - 6) = x(x - 3)(x + 2)$$

Thus, the denominator is 0 at $x = 0$, $x = -2$, and $x = 3$. We conclude that the domain consists of all real numbers except these three points, which are the poles of $R(x)$. The graph is shown in **Figure 5.63**. Note once more the appearance of the graph near the three poles.

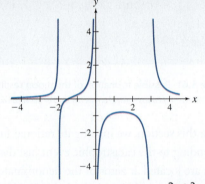

Figure 5.63 The graph of $y = \dfrac{2x + 3}{x^3 - x^2 - 6x}$

TRY IT YOURSELF 5.30 Brief answers provided at the end of the section.

Find the domain of the rational function $R(x) = \dfrac{x^2 - 1}{x^2 - 5x + 6}$, and identify the poles.

Local Behavior: Vertical Asymptotes

> Vertical asymptotes enable us to visualize the effect of poles.

Poles manifest themselves graphically in two basic types, which are exemplified by the graphs of $y = \pm \dfrac{1}{(x - 1)^2}$ (**Figure 5.64** and **Figure 5.65**) and $y = \pm \dfrac{1}{x - 1}$ (**Figure 5.66** and **Figure 5.67**) near their poles at $x = 1$.

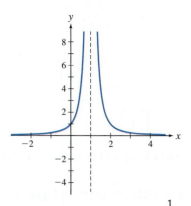

Figure 5.64 The graph of $y = \dfrac{1}{(x-1)^2}$ near its pole at $x = 1$: Both branches go up.

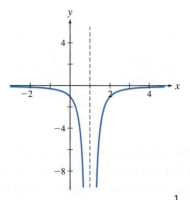

Figure 5.65 The graph of $y = -\dfrac{1}{(x-1)^2}$ near its pole at $x = 1$: Both branches go down.

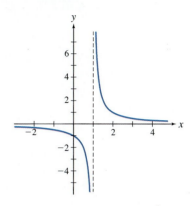

Figure 5.66 The graph of $y = \dfrac{1}{x-1}$ near its pole at $x = 1$: The left branch goes down, and the right branch goes up.

The vertical line $x = 1$ in these figures is called a **vertical asymptote**. Although the line is not actually part of the graph, it is included because it helps us visualize the behavior of the graph.

Note that the graphs in Figure 5.64 and Figure 5.65 behave in a similar way on either side of the asymptotes (the branches both go up or both go down), which is typical of zeros of even order in the denominator. But the graphs in Figure 5.66 and Figure 5.67 behave in opposite ways on either side of the asymptotes (one branch goes up and the other goes down), which is typical of zeros of odd order in the denominator.

Vertical asymptotes of a rational function occur at poles, so to find the vertical asymptotes we have to factor the denominator. Because the behavior of a rational function near a vertical asymptote may look like any of the four figures just discussed, we need to identify the sign on either side of an asymptote. Sign changes can also occur at the real zeros of the numerator.

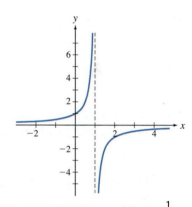

Figure 5.67 The graph of $y = -\dfrac{1}{x-1}$ near its pole at $x = 1$: The left branch goes up, and the right branch goes down.

LAWS OF MATHEMATICS: Sign Changes of Rational Functions

The rational function $R(x) = \dfrac{P(x)}{Q(x)}$ can change sign only at real zeros of the numerator and at real zeros of the denominator.

A **vertical asymptote** of the graph of a rational function is a vertical line that the graph approaches but never intersects.

For example, let's make the graph of $R(x) = \dfrac{x-2}{(x-1)(x-3)}$. First we identify the zeros. The numerator is zero at $x = 2$, so $R(x)$ has a zero there.

Next we identify the poles and the vertical asymptotes. The denominator is 0 at $x = 1$ and $x = 3$, so these are the poles. The poles correspond to vertical asymptotes, which are the lines $x = 1$ and $x = 3$. These asymptotes, along with the zero at $x = 2$, are shown in **Figure 5.68**.

Now we want to determine where $R(x)$ is positive and where it is negative. In Section P.3, we used sign diagrams to solve inequalities involving fractions, and that is the method we use here. The sign diagram for $R(x)$ is as follows. (Recall that boxes around numbers in the diagram indicate zeros of the denominator, which are poles of the rational function.)

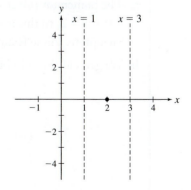

Figure 5.68 Locating zero and vertical asymptotes of $y = \dfrac{x-2}{(x-1)(x-3)}$

$$\frac{x-2}{(x-1)(x-3)}$$

$$\frac{(-)}{(-)(-)} = - \quad | \quad \frac{(-)}{(+)(-)} = + \quad | \quad \frac{(+)}{(+)(-)} = - \quad | \quad \frac{(+)}{(+)(+)} = +$$

$$\boxed{1} \qquad\qquad 2 \qquad\qquad \boxed{3}$$

Thus, $R(x)$ is positive on the intervals $(1, 2)$ and $(3, \infty)$ and is negative on the intervals $(-\infty, 1)$ and $(2, 3)$, as illustrated in **Figure 5.69**.

The information we have accumulated allows us to make the graph in **Figure 5.70**.

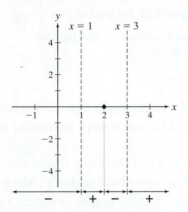

Figure 5.69 Marking the regions where $y = \dfrac{x-2}{(x-1)(x-3)}$ is positive and where it is negative

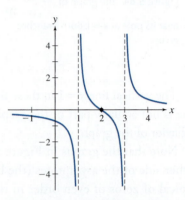

Figure 5.70 Completed graph of $y = \dfrac{x-2}{(x-1)(x-3)}$

EXAMPLE 5.31   Graphing Rational Functions Near Vertical Asymptotes

For the rational function $R(x) = \dfrac{x}{(x-2)^2}$:

a. Locate the zeros and poles, and identify all vertical asymptotes.

b. Determine where $R(x)$ is positive and where it is negative.

c. Sketch a graph of $R(x)$ showing all vertical asymptotes.

SOLUTION

a. The numerator is 0 at $x = 0$, so $R(x)$ has a zero at $x = 0$. The denominator is 0 at $x = 2$, so this is a pole of $R(x)$, and the line $x = 2$ is a vertical asymptote. These features are shown in **Figure 5.71**.

b. We get additional information from the sign diagram.

$$\frac{x}{(x-2)^2}$$

$$\frac{(-)}{(+)} = - \quad | \quad \frac{(+)}{(+)} = + \quad | \quad \frac{(+)}{(+)} = +$$

$$0 \qquad\qquad \boxed{2}$$

Thus, $R(x)$ is negative on $(-\infty, 0)$ and positive on $(0, 2) \cup (2, \infty)$.

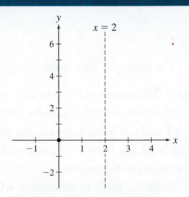

Figure 5.71 Part a: The numerator of $R(x) = \dfrac{x}{(x-2)^2}$ is zero at $x = 0$. There is a vertical asymptote at $x = 2$.

c. We have noted the information from the sign diagram in **Figure 5.72**. The completed graph is shown in **Figure 5.73**.

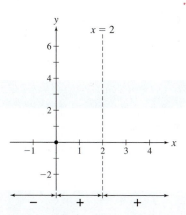

Figure 5.72 Part b: $R(x) = \dfrac{x}{(x-2)^2}$ is negative on $(-\infty, 0)$ and positive on $(0, 2)$ and $(2, \infty)$.

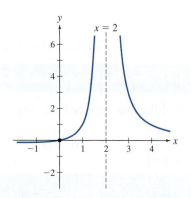

Figure 5.73 Part c: The graph of $R(x) = \dfrac{x}{(x-2)^2}$ is completed.

TRY IT YOURSELF 5.31   Brief answers provided at the end of the section.

Identify the zeros and poles of $T(x) = \dfrac{x-2}{(x-1)(x-3)^2}$, determine where T is positive and where it is negative, and then sketch a graph of T showing all vertical asymptotes.

EXTEND YOUR REACH

We have discussed the limit of functions as $x \to \infty$, but in calculus you will also encounter limits as x approaches a real number like 1 or 2. Consider the rational function $R(x)$ in Example 5.31. What would you say about the limit as $x \to 2$ of $R(x)$? That is, what does this function approach when x is near 2?

Long-Term Behavior of Rational Functions

> The long-term behavior of rational functions is determined by the ratio of the leading terms of the numerator and denominator.

To determine the long-term behavior of a rational function $R(x)$, we must calculate the limits of $R(x)$ as $x \to \infty$ and as $x \to -\infty$. In Section 5.4, we observed that the long-term behavior of a polynomial is determined by its leading term. This fact makes the calculation of limits of rational functions quite easy.

LAWS OF MATHEMATICS: Limits of Rational Functions

The limit as $x \to \infty$ or $x \to -\infty$ of a rational function $R(x) = \dfrac{P(x)}{Q(x)}$ is the same as the corresponding limit for

$$\frac{\text{Leading term of } P(x)}{\text{Leading term of } Q(x)}$$

For example, the limit as $x \to \infty$ of $R(x) = \dfrac{6x^2 - x + 4}{2x^2 - 5}$ is the same as the limit as $x \to \infty$ of

$$\frac{6x^2}{2x^2} = 3$$

Thus, $R(x) \to 3$ as $x \to \infty$.

EXAMPLE 5.32 Finding Long-Term Behavior

Find the limit of the following rational functions as $x \to \infty$ and as $x \to -\infty$.

a. $S(x) = \dfrac{x^5 - 1}{x^2 + x + 1}$ **b.** $T(x) = \dfrac{x^2 + 1}{x^3 + 1}$

SOLUTION

a. As $x \to \infty$,

$$\frac{x^5 - 1}{x^2 + x + 1} \text{ behaves like } \frac{x^5}{x^2} = x^3$$

Because $x^3 \to \infty$ as $x \to \infty$, we conclude that

$$\frac{x^5 - 1}{x^2 + x + 1} \to \infty \text{ as } x \to \infty$$

Also, because $x^3 \to -\infty$ as $x \to -\infty$, we conclude that

$$\frac{x^5 - 1}{x^2 + x + 1} \to -\infty \text{ as } x \to -\infty$$

b. Once again, we employ the ratio of the leading terms. As $x \to \pm\infty$,

$$\frac{x^2 + 1}{x^3 + 1} \text{ behaves like } \frac{x^2}{x^3} = \frac{1}{x}$$

Because $\dfrac{1}{x} \to 0$ as $x \to \pm\infty$, we conclude that

$$\frac{x^2 + 1}{x^3 + 1} \to 0 \text{ as } x \to \infty$$

$$\text{and } \frac{x^2 + 1}{x^3 + 1} \to 0 \text{ as } x \to -\infty$$

TRY IT YOURSELF 5.32 Brief answers provided at the end of the section.

Find the long-term behavior of the rational function $R(x) = \dfrac{9x + 7}{3x - 7}$.

End Behavior of Graphs: Horizontal Asymptotes

> A horizontal asymptote of the graph of a rational function reflects the long-term behavior of the function.

Let's use what we have learned about the long-term behavior of rational functions to make the graph of $R(x) = \dfrac{2x^2 + 1}{(x + 1)(x - 1)}$. The numerator has no zeros, but the denominator has two, $x = -1$ and $x = 1$. Thus, the lines $x = -1$ and $x = 1$ are vertical asymptotes. In the following sign diagram, we use the fact that $2x^2 + 1$ is always positive.

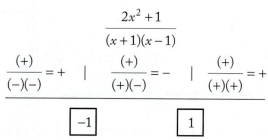

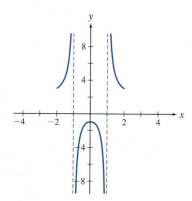

DF **Figure 5.74** The graph of
$R(x) = \dfrac{2x^2 + 1}{(x+1)(x-1)}$ near its vertical asymptotes

Using this information, we make the graph in **Figure 5.74**. But this graph does not show the long-term behavior. Limits tell us the long-term behavior of rational functions, and that in turn tells us how to draw the tail ends of graphs. As $x \to \infty$, $R(x)$ behaves like the ratio of its leading terms. Expanding the denominator gives $R(x) = \dfrac{2x^2 + 1}{x^2 - 1}$, so that ratio is $\dfrac{2x^2}{x^2} = 2$. That is,

$$R(x) \to 2 \text{ as } x \to \infty, \quad \text{and} \quad R(x) \to 2 \text{ as } x \to -\infty$$

Thus, the graph is close to the line $y = 2$ for x that is very large in size. This line is known as a **horizontal asymptote**.

The horizontal asymptote allows us to complete the graph, as shown in **Figure 5.75**. Note how the graph approaches the horizontal asymptote for very large positive x and for very large negative x. In the following step-by-step strategy, we list the steps needed to graph a rational function.

A **horizontal asymptote** of the graph of a rational function $R(x)$ is a line $y = b$ such that $R(x) \to b$ as $x \to \infty$ or $R(x) \to b$ as $x \to -\infty$.

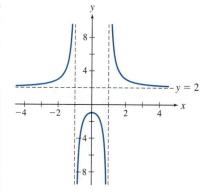

Figure 5.75 The completed graph of $R(x) = \dfrac{2x^2 + 1}{(x+1)(x-1)}$ showing the horizontal asymptote $y = 2$

STEP-BY-STEP STRATEGY: Graphing Rational Functions

Step 1 Find the zeros of the numerator. These are zeros of the function, and they show where the graph touches the x-axis.

Step 2 Find the zeros of the denominator. These are the poles, and they determine the vertical asymptotes.

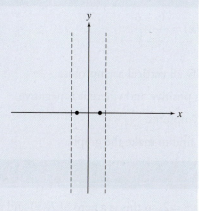

Step 3 Use a sign diagram to determine the sign changes of the function.

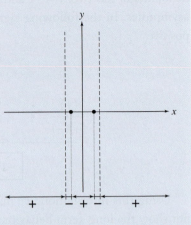

Step 4 Use limits to determine the long-term behavior of the graph. This will determine any horizontal asymptotes that may exist.

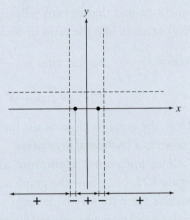

Step 5 Use the steps above to sketch the graph.

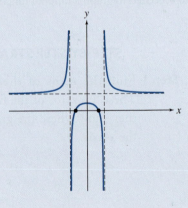

EXAMPLE 5.33 Graphing a Rational Function Using Horizontal and Vertical Asymptotes

For the rational function $R(x) = \dfrac{3x - 6}{x - 3}$:

a. Find all zeros.

b. Find all poles, and identify all vertical asymptotes.

c. Determine where $R(x)$ is positive and where it is negative.

d. Identify all horizontal asymptotes.

e. Use the information you find to make the graph.

SOLUTION

a. The numerator is zero at $x = 2$, so this is a zero of $R(x)$ and a point where the graph touches the x-axis.

b. The denominator is zero at $x = 3$, so this is a pole of $R(x)$. Because $R(x)$ has a pole at 3, the graph has the vertical asymptote $x = 3$.

c. Here is the sign diagram:

$$\frac{3x - 6}{x - 3}$$

$$\frac{(-)}{(-)} = + \quad | \quad \frac{(+)}{(-)} = - \quad | \quad \frac{(+)}{(+)} = +$$

$$\begin{array}{ccc} & 2 & \boxed{3} \end{array}$$

Thus, $R(x)$ is positive on $(-\infty, 2) \cup (3, \infty)$ and negative on $(2, 3)$. The sign diagram shows that changes in sign occur at $x = 2$ and $x = 3$.

d. The long-term behavior is the same as that of the ratio of leading terms, $\dfrac{3x}{x} = 3$. Hence, $R(x) \to 3$ as $x \to \infty$ and also as $x \to -\infty$.

Thus, the graph of $R(x)$ has the horizontal asymptote $y = 3$.

e. We know three things: where the graph crosses the x-axis, the behavior near the vertical asymptote, and the long-term behavior. That is enough to produce a pretty good sketch of the graph, which is shown in **Figure 5.76**.

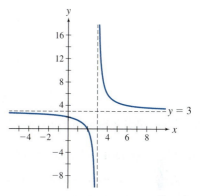

Figure 5.76 The graph of $R(x) = \dfrac{3x - 6}{x - 3}$

TRY IT YOURSELF 5.33 Brief answers provided at the end of the section.

Identify the horizontal and vertical asymptotes of $y = \dfrac{6x + 1}{3x - 12}$. Then sketch the graph.

EXTEND YOUR REACH

a. For the function $R(x)$ in this example, explain carefully why the limit as $x \to \infty$ of $R(x)$ is the same as the limit as $x \to -\infty$ of $R(x)$.

b. Let $T(x) = \dfrac{3x + 7}{5x^2 - x}$. Compare the limit as $x \to \infty$ of $T(x)$ to the limit as $x \to -\infty$ of $T(x)$.

c. Can you find an example of a rational function $S(x)$ that has a horizontal asymptote, but the limit as $x \to \infty$ is different from the limit as $x \to -\infty$? If not, are these limits always the same when the rational function has a horizontal asymptote? Remember that each limit is the same as the corresponding limit of the ratio of the leading terms of the numerator and denominator.

EXAMPLE 5.34 Using Long-Term Behavior to Graph Rational Functions

For the rational function $S(x) = \dfrac{x^2 - 2x}{x - 3}$:

a. Find all zeros.

b. Find all poles, and identify all vertical asymptotes.

c. Determine where $S(x)$ is positive and where it is negative.

d. Identify all horizontal asymptotes.

e. Use the information you find to make the graph.

SOLUTION

a. Factoring the numerator of $S(x)$, we find $S(x) = \dfrac{x(x-2)}{x-3}$. Hence, the numerator is zero at $x = 0$ and $x = 2$. These are the zeros of $S(x)$ and the places where the graph meets the x-axis.

b. There is a pole at $x = 3$, so there is a vertical asymptote $x = 3$.

c. We obtain the following sign diagram:

$$\dfrac{x(x-2)}{x-3}$$

$$\dfrac{(-)(-)}{(-)} = - \quad | \quad \dfrac{(+)(-)}{(-)} = + \quad | \quad \dfrac{(+)(+)}{(-)} = - \quad | \quad \dfrac{(+)(+)}{(+)} = +$$

$$\overline{\hspace{2cm} 0 \hspace{2cm} 2 \hspace{2cm} \boxed{3} \hspace{2cm}}$$

Thus, $S(x)$ is negative on $(-\infty, 0) \cup (2, 3)$ and positive on $(0, 2) \cup (3, \infty)$.

d. The long-term behavior of $S(x)$ is the same as that of $\dfrac{x^2}{x} = x$. Thus, $S(x) \to \infty$ as $x \to \infty$, and $S(x) \to -\infty$ as $x \to -\infty$. Hence, $S(x)$ has no horizontal asymptotes.

e. Because $S(x) \to \infty$ as $x \to \infty$, the graph increases without bound for very large x. Also, $S(x) \to -\infty$ as $x \to -\infty$, so the graph decreases without bound for x very large in size but negative (**Figure 5.77**).

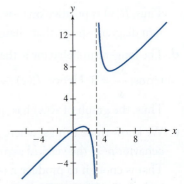

Figure 5.77 The graph of $S(x) = \dfrac{x^2 - 2x}{x - 3}$

TRY IT YOURSELF 5.34   Brief answers provided at the end of the section.

Identify the horizontal and vertical asymptotes of $y = \dfrac{x^2 + 1}{1 - x}$, and determine the long-term behavior. Then sketch the graph.

EXTEND YOUR REACH   The graph of a rational function is shown in **Figure 5.78**. Propose a rational function whose graph matches this one. *Suggestion:* You need a rational function whose zeros, horizontal asymptotes, and vertical asymptotes match those in Figure 5.78. Be careful about the sign of the function values, too.

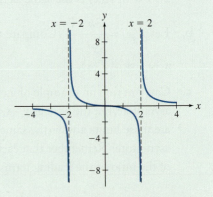

Figure 5.78 The graph of an unknown rational function

MODELS AND APPLICATIONS   Traffic Flow

EXAMPLE 5.35   Waiting on Traffic

When a feeder road meets a main road carrying a lot of cars, traffic may back up, forming a queue (a line of cars). Let g denote the average rate at which a gap in traffic on the main road appears (allowing entry for a car on the feeder road), and let a be the average rate of arrival of cars on the feeder road. (Both g and a are measured per minute.) Assume that gaps in traffic appear at random and that cars on the feeder road arrive at the intersection randomly. If $a < g$, then the average length of a queue on the feeder road is given by

$$L = \frac{a^2}{g(g - a)}$$

We think of L as a function of a.

a. Explain what the formula tells us about L when a is near 0.

b. Explain what the formula tells us about L when a is near (but less than) g.

SOLUTION

a. When a is near 0, L is also near 0. Practically, this means that if traffic on the feeder road is very light, then long lines would not be expected.

b. The function L has a vertical asymptote at $a = g$. Thus, when a is close to g (but less than g), L is large. Practically speaking, this means that when the arrival rate is close to the rate for gaps to develop, then the line on the feeder road is expected to be long.

TRY IT YOURSELF 5.35 Brief answers provided at the end of the section.

Does the formula make practical sense if $a > g$? What would you expect to happen in the case that $a > g$?

CONCEPTS TO REMEMBER: Rational Functions

Formula: $R(x) = \dfrac{P(x)}{Q(x)}$, where $P(x)$ and $Q(x)$ are polynomials with no real zeros in common.

Domain: All real numbers except where the denominator is 0.

Range: Varies greatly from function to function.

Graph: Varies greatly from function to function.

In the long term: The long-term behavior of a rational function is the same as that of the ratio of the leading terms of the numerator and denominator.

TRY IT YOURSELF ANSWERS

5.30 The domain consists of all real numbers except $x = 2$ and $x = 3$. The poles occur at $x = 2$ and $x = 3$.

5.31 There is a zero at $x = 2$. The poles are at $x = 1$ and $x = 3$. The sign diagram is

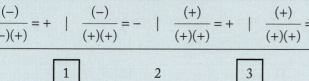

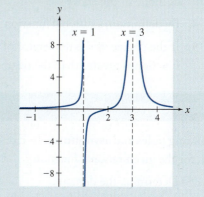

5.32 $R(x) \to 3$ as $x \to \infty$ and $R(x) \to 3$ as $x \to -\infty$.

5.33 The line $x = 4$ is a vertical asymptote, and the line $y = 2$ is a horizontal asymptote.

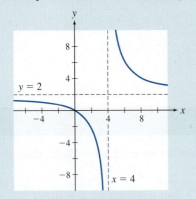

5.34 The line $x = 1$ is a vertical asymptote. There is no horizontal asymptote: $y \to -\infty$ as $x \to \infty$, and $y \to \infty$ as $x \to -\infty$.

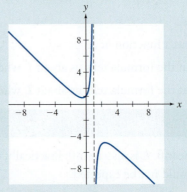

5.35 When $a > g$, the value of L is negative, so the formula makes no sense. In fact, when the arrival rate is greater than the gap rate, the line will be long and continue to grow in length.

EXERCISE SET 5.5

CHECK YOUR UNDERSTANDING

1. Where do vertical asymptotes for graphs of rational functions occur?

2. Poles of rational functions occur where
   a. the numerator is 0
   b. the denominator is 0
   c. there is a horizontal asymptote
   d. none of the above

3. For a rational function $R(x)$, if $R(x) \to \infty$ as $x \to \infty$, then the graph of $R(x)$ has
   a. a vertical asymptote
   b. a horizontal asymptote
   c. no horizontal asymptote
   d. none of the above

4. By definition, a rational function is _____.

5. If the degree of the numerator of a rational function is greater than the degree of the denominator, what can you conclude about the existence of horizontal asymptotes?

6. If the degree of the denominator of a rational function is greater than the degree of the numerator, what can you conclude about horizontal asymptotes?

7. If $x = a$ is a vertical asymptote of the graph of the rational function $\dfrac{P(x)}{Q(x)}$, what is the significance of $x = a$ for the rational function $\dfrac{Q(x)}{P(x)}$?

8. The denominator of a rational function has no real zeros. What can you conclude about vertical asymptotes of the graph?

9. The denominator of a rational function has degree n. What is the largest possible number of vertical asymptotes for the graph of this function?

10. Is there any real number that cannot be a horizontal asymptote of the graph of a rational function?

SKILL BUILDING

Finding poles and asymptotes. In Exercises 11 through 23, for the given rational function:

• Determine the domain.

• Identify any poles.

• Identify any vertical asymptotes.

• Identify any horizontal asymptote.

11. $R(x) = \dfrac{1}{x-1}$

12. $R(x) = \dfrac{(x-1)(x-2)}{x(x+2)}$

13. $R(x) = \dfrac{1}{(x-2)^2}$

14. $R(x) = \dfrac{x^2+1}{x}$

15. $R(x) = \dfrac{x^2+1}{x^2+2}$

16. $R(x) = x$

17. $R(x) = \dfrac{1}{x}$

18. $R(x) = \dfrac{2x^2}{(x+4)(x-1)}$

19. $R(x) = \dfrac{x^2+1}{x^2+2x+1}$

20. $R(x) = \dfrac{x-1}{x+1}$

21. $R(x) = \dfrac{x^2+5x+6}{x^2+8x+6}$

22. $R(x) = \dfrac{x+1}{x+5}$

23. $R(x) = \dfrac{(x-4)^2}{(x-5)^2}$

Long-term behavior. In Exercises 24 through 34, find the limit of the given rational function as $x \to \infty$ and the limit as $x \to -\infty$.

24. $R(x) = \dfrac{1}{x-1}$

25. $R(x) = \dfrac{1}{1-x}$

26. $R(x) = \dfrac{x}{x-1}$

27. $R(x) = \dfrac{x}{1-x}$

28. $R(x) = \dfrac{x+4}{x^3-1}$

29. $R(x) = \dfrac{x^2+1}{2x^2-1}$

30. $R(x) = \dfrac{3x^2+x+1}{2x-5}$

31. $R(x) = \dfrac{1+x+12x^2}{2-x-3x^2}$

32. $R(x) = \dfrac{x^4-x+5}{x^3-2x+4}$

33. $R(x) = \dfrac{3x^4-x^3+4}{4x^4+x+2}$

34. $R(x) = \dfrac{x^4+4}{3-x^3}$

Sketching graphs. In Exercises 35 through 41, sketch the graph of a rational function consistent with the given information. In each case, there is more than one correct answer.

35. Vertical asymptote $x = 2$ and horizontal asymptote $y = 3$

36. Vertical asymptotes $x = -3$ and $x = 3$ and horizontal asymptote $y = -1$

37. Vertical asymptote $x = 1$ and no horizontal asymptote

38. Vertical asymptotes $x = 2$ and $x = -2$ and no horizontal asymptote

39. Zeros at $x = -1$ and $x = 4$, vertical asymptotes $x = 2$ and $x = -3$, and horizontal asymptote $y = 2$

40. Zero at $x = 4$, vertical asymptote $x = 5$, with limit $-\infty$ as $x \to \infty$, and with limit ∞ as $x \to -\infty$

41. No zeros, no vertical asymptotes, and horizontal asymptote $y = 1$

PROBLEMS

Graphing rational functions. In Exercises 42 through 58:

- List the zeros.
- Identify any vertical asymptotes.
- Identify any horizontal asymptote.
- Sketch the graph.

42. $R(x) = \dfrac{x}{x^2-1}$

43. $R(x) = \dfrac{x+1}{(x-1)(x-3)}$

44. $R(x) = \dfrac{4x-8}{2x-1}$

45. $R(x) = \dfrac{6x-12}{x-3}$

46. $R(x) = \dfrac{x-2}{x^2-2x+1}$

47. $R(x) = \dfrac{x^2-1}{x^2-4}$

48. $R(x) = \dfrac{x-1}{x(x+1)(x-2)}$

49. $R(x) = \dfrac{(x-1)(x-3)}{x+1}$

50. $R(x) = \dfrac{x+1}{x-1}$

51. $R(x) = \dfrac{x-4}{x-5}$

52. $R(x) = \dfrac{(x-1)(x-3)}{x-2}$

53. $R(x) = \dfrac{(x+1)(x-2)}{(x-1)(x+2)}$

54. $R(x) = \dfrac{x-1}{(x-2)(x+1)^2}$

55. $R(x) = \dfrac{x^2-2x+1}{x^2+5x}$

56. $R(x) = \dfrac{(x-5)^2}{x+5}$

57. $R(x) = \dfrac{x^3-7x^2+6x}{x^2-9}$

58. $R(x) = \dfrac{2x^2-18}{x^2-25}$

Matching. In Exercises 59 through 67, select the function that best matches the given graph.

59.

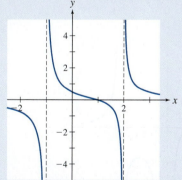

a. $\dfrac{x-1}{(x+1)(x-2)}$

b. $\dfrac{x}{(x+1)(x-2)}$

c. $\dfrac{x-1}{(x+2)(x-1)}$

d. $\dfrac{x+1}{(x-1)(x-2)}$

e. $\dfrac{x-2}{x(x-2)}$

60.

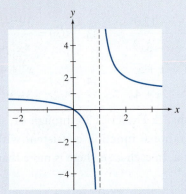

a. $\dfrac{x-1}{x}$ **d.** $\dfrac{x+1}{x-1}$

b. $\dfrac{x}{x-1}$ **e.** $\dfrac{x-1}{x+1}$

c. $\dfrac{x}{x+1}$

61.

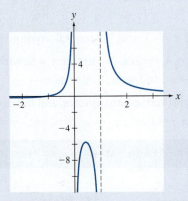

a. $\dfrac{x-1}{x(x+1)}$ **d.** $\dfrac{(x+1)(x-1)}{x}$

b. $\dfrac{x}{(x-1)(x+1)}$ **e.** $\dfrac{x}{(x+1)^2}$

c. $\dfrac{x+1}{x(x-1)}$

62.

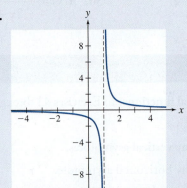

a. $\dfrac{x}{(x-1)(x-3)^2}$ **b.** $\dfrac{x}{(x-1)(x-3)}$

c. $\dfrac{x^2}{(x-1)^2(x-3)}$ **e.** $\dfrac{x-1}{x(x+3)^2}$

d. $\dfrac{x}{(x-1)^2(x-3)}$

63.

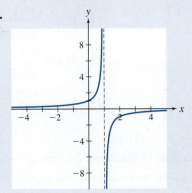

a. $\dfrac{1}{x-1}$ **d.** $\dfrac{1-x}{1+x}$

b. $\dfrac{1}{1-x}$ **e.** $\dfrac{1}{(x-1)^2}$

c. $\dfrac{1}{(1-x)^2}$

64.

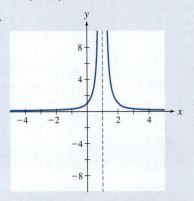

a. $\dfrac{1}{x-1}$ **d.** $\dfrac{1-x}{1+x}$

b. $\dfrac{1}{1-x}$ **e.** $\dfrac{1}{(x-1)^2}$

c. $\dfrac{1}{(1-x)^2}$

65.

a. $\dfrac{1}{x-1}$

b. $\dfrac{1}{1-x}$

c. $\dfrac{1}{(1-x)^2}$

d. $\dfrac{1-x}{1+x}$

e. $\dfrac{x}{(x-1)^2}$

66.

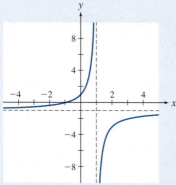

a. $\dfrac{x-1}{x+1}$

b. $\dfrac{1-x}{1+x}$

c. $\dfrac{1+x}{(1-x)^2}$

d. $\dfrac{(1-x)^2}{1+x}$

e. $\dfrac{x+1}{1-x}$

67.

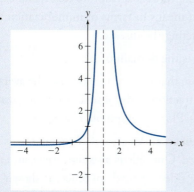

a. $\dfrac{x-1}{x+1}$

b. $\dfrac{1-x}{1+x}$

c. $\dfrac{1-x}{(1+x)^2}$

d. $\dfrac{(1-x)^2}{1+x}$

e. $\dfrac{x+1}{(1-x)^2}$

68. Inverses of rational functions. Many rational functions do not have inverses, but some do.

Let $R(x) = \dfrac{x+1}{x-5}$.

a. Find a formula for the inverse function $S(x)$.

b. List the asymptotes of both R and S, and plot the graphs of R and S, including their asymptotes.

c. Recall that we get the graph of the inverse function by reflecting the graph of the function through the line $y = x$. Use this fact, along with your solution to part b, to answer the following question: In general, how are the asymptotes of the graph of a function related to the asymptotes of the graph of the inverse function?

69. Combining fractions. If $R(x)$ and $S(x)$ are rational functions, then $R(x) + S(x)$ is, after common factors are canceled, also a rational function. Suppose the vertical asymptotes of $R(x)$ are $x = 0$ and $x = 1$ and that the only vertical asymptote $S(x)$ is $x = 2$. What are the vertical asymptotes of $R(x) + S(x)$?

70. Multiplicative inverse of rational functions: If $a \neq 0$ and the line $y = a$ is a horizontal asymptote of the graph of the rational function $\dfrac{P(x)}{Q(x)}$, show that $y = \dfrac{1}{a}$ is a horizontal asymptote of the graph of the rational function $\dfrac{Q(x)}{P(x)}$.

71. An example: Give an example to show that if $R(x)$ is a rational function, then $\sqrt{R(x)}$ need not be rational.

📈 **Using a graphing utility.** For the given function in Exercises 72 through 78:

- List any zeros (approximately if necessary).
- Identify any vertical asymptotes (approximately if necessary).
- Identify any horizontal asymptote.
- Use a graphing utility to make the graph.

72. $R(x) = \dfrac{3x^2}{x^2+1}$

73. $R(x) = \dfrac{x^2+1}{x^2-5}$

74. $R(x) = \dfrac{x^2+1}{x^3-x+1}$

75. $R(x) = \dfrac{3x^3-24x}{x^3-3x^2+3x-1}$

76. $R(x) = \dfrac{x^3+x^2+1}{x^3-x^2+1}$

77. $R(x) = \dfrac{x^3-4x^2-11x+30}{x^3+4x^2-11x-30}$

78. $R(x) = \dfrac{x}{x^4-6x^2+3}$

MODELS AND APPLICATIONS

79. Waiting time. The average number of callers served by a customer service representative is r per minute. The average number of new callers per minute is c, but the system involves randomness. Under these circumstances, the average time a caller waits on hold is given by

$$T = \frac{c}{r(r-c)} \text{ minutes}$$

This formula is valid for $0 < c < r$. We think of T as a function of c.

a. Explain, in terms of the time a caller waits on hold, what happens when c is near r.

b. Make a graph of T versus c in the case that the average number of callers served is five per minute. (Assume that c is between 0 and 5.)

80. The catch equation. If a lake is stocked with N_0 fish of the same age, then the total number C of these fish caught over the life span of the fish is given by

$$C = \frac{A}{D+A} N_0$$

Here A is the proportion of fish caught annually, and D is the proportion of fish that die annually of natural causes.

a. If $x = \dfrac{A}{D}$, show that $C = \dfrac{x}{1+x} N_0$. The graph is shown in **Figure 5.79**.

b. Find the horizontal asymptote for the function of x in part a.

c. Explain the meaning of the horizontal asymptote in terms of the number of fish caught over the life span.

81. A view from space. When Earth is viewed from a distance H from the surface, the proportion E of the surface that is visible is given by

$$E = \frac{H}{2R + 2H}$$

where R is the radius of Earth. Because R is a constant, we think of E as a function of H.

a. Identify the horizontal asymptote.

b. Explain the meaning of the horizontal asymptote in terms of how much of the surface is visible when viewed from a very high altitude.

82. Immunity. In the absence of immunization, a certain disease is contracted at an average age a. If b percent of the population is immunized at birth, then the average age at which the disease is contracted is given by

$$A = \frac{100a}{100 - b}$$

We think of a as a constant, so A is a function of b.

a. Find the vertical asymptote for A, and explain what it means in terms of immunization and contracting the disease.

b. What is the domain of A?

c. Suppose that for a certain disease, the average age of contraction in the absence of immunization is 4 years. Plot the graph of A versus b.

83. Fencing an area. An area of A square feet is to be enclosed. One side is a straight river, and the remaining sides are made of fence, as shown in **Figure 5.80**.

Figure 5.80 A fence next to a river

a. If the width is x feet, show that the length is $\dfrac{A}{x}$ feet.

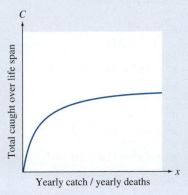

Figure 5.79 The graph of the catch equation

b. Show that the total length of fence used to enclose the area is given by

$$F = \frac{A}{x} + 2x \text{ feet}$$

c. Find the vertical asymptote of the function $F(x)$ in part b, and explain what it means in terms of the width and the amount of fence needed.

d. Calculate the limit as $x \to \infty$ of $F(x)$, and explain what it means in terms of the width and the amount of fence needed.

e. Plot the graph of F versus x in case the area to be enclosed is 200 square feet.

f. Use the graph you made in part e to find the dimensions of the enclosure that requires a minimum amount of fence. (Assume again that the area to be enclosed is 200 square feet.)

84. Power for flight. The power required for level flight of a bird or airplane is given by

$$P = \frac{v^3}{A} + \frac{B}{v}$$

where v is the velocity, and A and B are constants that depend on the flying body.

a. Find the vertical asymptote for P, and explain what it means in terms of the velocity and the amount of power needed to maintain level flight.

b. Calculate the limit as $v \to \infty$ of $P(v)$, and explain what it means in terms of the velocity and the amount of power needed to maintain level flight.

c. For a certain bird, $A = 5000$, $B = 486$, and the velocity v is measured in kilometers per hour. Plot the graph of P versus v.

d. What velocity, measured in kilometers per hour, for the bird in part c requires the minimum power?

85. Escape velocity. If a projectile is shot upward with an initial velocity v, then it will reach a height H before falling back to Earth. The following relationship between velocity and height is valid for $v < \sqrt{2gR}$, where g is the acceleration due to gravity near the surface of Earth and R is the radius of Earth:

$$H = \frac{Rv^2}{2gR - v^2}$$

We think of H as a function of v. The graph is shown in **Figure 5.81**.

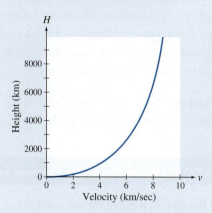

Figure 5.81 Height versus initial velocity

a. The function H has two vertical asymptotes. Only one corresponds to a positive value of v. Find this vertical asymptote, and explain its meaning in terms of the trajectory of the projectile when the initial velocity is near that value.

b. The velocity corresponding to the asymptote from part a is known as the escape velocity. Explain what happens when a projectile is fired upward at escape velocity.

c. The acceleration due to gravity g is 0.0098 kilometer per second per second, and the radius R is 6371 kilometers. Find the escape velocity in kilometers per second.

d. One kilometer is 0.62 mile. Use your answer in part c to find the escape velocity in miles per hour. (Round to a whole number.)

86. Red shift. If a stellar object is moving away from us at a high speed, then the observed wavelength λ_o of a spectrum line of light is different from the emitted wavelength λ_e. The red shift of this object is defined by the ratio

$$z = \frac{\lambda_o - \lambda_e}{\lambda_e}$$

Suppose a faraway galaxy shows an observed (positive) wavelength λ_o. We consider λ_o to be constant, so z is a function of λ_e. What is the vertical asymptote for z? What is the meaning of the vertical asymptote?

CHALLENGE EXERCISES FOR INDIVIDUALS OR GROUPS

87. More than one asymptote? Show that no rational function can have two distinct horizontal asymptotes. *Suggestion:* Recall that the limit as $x \to \pm\infty$ of a rational function is the same as the limit of the ratio of the leading terms.

Which are rational? Suppose that $R(x)$ and $S(x)$ are distinct, nonconstant rational functions. In Exercises 88 through 94, show that the given combination, after cancelation of common factors, is also rational.

88. $R(x) + S(x)$

89. $3R(x) - 2S(x)$

90. $xR(x) + \dfrac{1}{x} S(x)$

91. $R(x)S(x)$

92. $\dfrac{R(x)}{S(x)}$

93. $R \circ S$

94. $\dfrac{R(x) + S(x)}{R(x) - S(x)}$

REVIEW AND REFRESH: Exercises from Previous Sections

95. From Section P.3: Solve the inequality $x^2 - 6x + 5 \le 0$.

Answer: $[1, 5]$

96. From Section 1.4: Suppose that $f(x)$ is a function with the property that $f(x) \to -\infty$ as $x \to \infty$. What does this fact tell you about the graph of f?

Answer: Given any y-value, the graph of f is below the corresponding horizontal line for x sufficiently large.

97. From Section 5.1: Use the quadratic formula to factor $P(x) = x^2 - x - 4$.

Answer: $P(x) = \left(x - \dfrac{1 + \sqrt{17}}{2} \right)\left(x - \dfrac{1 - \sqrt{17}}{2} \right)$

98. From Section 5.2: Use synthetic division to evaluate $P(x) = x^2 - 2x + 5$ at $x = 2$.

Answer: $P(2) = 5$

99. From Section 5.3: List all possible rational zeros of $P(x) = 3x^5 - x^2 + 7$.

Answer: $\pm 1,\ \pm 7,\ \pm\dfrac{1}{3},\ \pm\dfrac{7}{3}$

100. From Section 5.3: Given that $x = 1$ is one zero of $P(x) = x^3 - 5x^2 + 8x - 4$, find all remaining zeros.

Answer: $x = 2$ with order 2

101. From Section 5.4: If $x = 1$, $x = 2$, and $x = 3$ are zeros of a cubic polynomial with leading coefficient 7, find the polynomial. Leave your answer in factored form.

Answer: $P(x) = 7(x - 1)(x - 2)(x - 3)$

102. From Section 5.4: If $4 - 3i$ is a zero of a polynomial with real coefficients, find one additional zero.

Answer: $4 + 3i$

103. From Section 5.4: What is the limit as $x \to \infty$ of $P(x) = 3x^3 - 12x^2 + 4$?

Answer: ∞

104. From Section 5.4: Find where the polynomial $P(x) = (x - 1)(x - 2)^3$ is negative.

Answer: $(1, 2)$

CHAPTER ROADMAP AND STUDY GUIDE

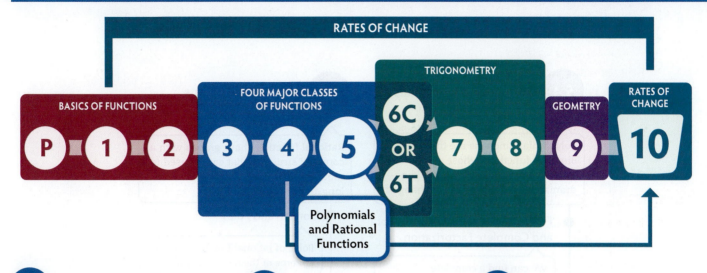

5.1 Quadratic Functions

Quadratic functions are the simplest polynomials whose graphs display many important features of the class, such as concavity.

Zeros of Quadratic Functions and the Discriminant

A quadratic function may have two real zeros, one real zero, or two complex zeros.

Complex Zeros of Quadratic Functions

The complex zeros of a quadratic function with real coefficients occur in complex conjugate pairs.

The Factor Theorem for Quadratic Polynomials

The factor theorem relates the zeros of a quadratic function to its factors.

Graphs of Quadratic Functions

The graph of a quadratic function can be determined by completing the square and shifting graphs.

MODELS AND APPLICATIONS
Maximum Height of Objects Propelled Upward

5.2 Long Division and the Factor Theorem

Long division is a key tool that will allow us to factor polynomials.

Long Division

Long division of polynomials is analogous to long division of real numbers.

The Remainder and Factor Theorems

The factor theorem applies to all polynomials, not just quadratic functions.

Synthetic Division

Synthetic division provides an efficient method for evaluating polynomials.

5.3 Zeros of Higher-Degree Polynomials

Whereas the quadratic formula can be used to find zeros of quadratic polynomials, several different methods are required to find zeros of higher-degree polynomials.

Factoring: Methods and Formulas

Methods such as substitution and grouping terms are useful for finding the factors of a polynomial.

Using the Factor Theorem to Solve Equations

The factor theorem shows the relationship between a polynomial and one of lower degree.

Rational Zeros

The rational zeros of a polynomial with integer coefficients can be found by analyzing the constant term and the leading coefficient.

MODELS AND APPLICATIONS
Poiseuille's Law

CHAPTER ROADMAP AND STUDY GUIDE (continued)

5.4 Graphs of Polynomials

Polynomial functions are the simplest functions whose graphs have a complex structure, including dramatically varying rates of change.

Fundamental Theorem of Algebra and Complete Factorization

We can find a complete factorization of a polynomial given its zeros.

Polynomials with Real Coefficients

Polynomials with real coefficients can be expressed as a product of linear and quadratic factors that also have real coefficients.

Long-Term Behavior

All nonconstant polynomials approach plus or minus infinity at their horizontal extremes.

Local Behavior

If multiplicity is counted and complex zeros are included, then the number of zeros of a nonconstant polynomial is exactly the degree.

Local Maxima and Minima

The number of local maxima or minima of a nonconstant polynomial is at most one less than the degree.

MODELS AND APPLICATIONS
Changing Value of Investments

5.5 Rational Functions

Rational functions have graphs with a complex structure that can be used to model catastrophic events as well as long-term behavior.

Domain and Poles

Poles of rational functions occur at zeros of the denominator.

Local Behavior: Vertical Asymptotes

Vertical asymptotes enable us to visualize the effect of poles.

Long-Term Behavior of Rational Functions

The long-term behavior of rational functions is determined by the ratio of the leading terms of the numerator and denominator.

End Behavior of Graphs: Horizontal Asymptotes

A horizontal asymptote of the graph of a rational function reflects the long-term behavior of the function.

MODELS AND APPLICATIONS
Traffic Flow

CHAPTER QUIZ

1. Find the zeros of $y = x^2 + 2x - 5$.

   Answer: $x = -1 \pm \sqrt{6}$

   XR **Example 5.1**

2. Find the zeros of $y = x^2 - 2x + 1$.

   Answer: 1

   XR **Example 5.2**

3. Calculate the discriminant of $y = x^2 + x + 2$. Tell how many zeros are promised by the discriminant.

   Answer: The discriminant is -7. There are no real zeros.

   XR **Example 5.3**

4. Factor the expression $x^2 - 2x - 2$.

   Answer: $(x - 1 + \sqrt{3})(x - 1 - \sqrt{3})$

   XR **Example 5.5**

5. Find the vertex of $y = x^2 - 6x + 4$. Does the vertex correspond to a maximum or a minimum?

   Answer: The vertex is $(3, -5)$. It corresponds to a minimum.

   XR **Example 5.8**

6. Calculate $x - 1 \overline{\smash{)}\,x^3 + x + 1}$. Express your answer both in quotient and product form.

   Answer: $\dfrac{x^3 + x + 1}{x - 1} = x^2 + x + 2 + \dfrac{3}{x - 1}$ and $x^3 + x + 1 = (x - 1)(x^2 + x + 2) + 3$

   XR **Example 5.11**

7. Let $P(x) = x^3 - 3x^2 + x - 3$. Use the factor theorem to show that $x - 3$ is a factor of $P(x)$. Find $Q(x)$ so that $P(x) = (x - 3)Q(x)$.

   Answer: $P(3) = 0$, so $x - 3$ is a factor. $Q(x) = x^2 + 1$.

   XR **Example 5.14**

8. Factor the following polynomials, and then find their zeros.

   a. $P(x) = x^4 - 5x^3 + 4x^2$

   b. $P(x) = x^3 - 2x^2 - 9x + 18$. Note that you may group the terms as $(x^3 - 9x) - (2x^2 - 18)$.

   Answer:

   XR **Example 5.16**

   a. $P(x) = x^4 - 5x^3 + 4x^2 = x^2(x - 1)(x - 4)$. Zeros: 0, 1, 4.

   b. $P(x) = x^3 - 2x^2 - 9x + 18 = (x - 2)(x - 3)(x + 3)$. Zeros: 2, 3, –3.

9. Find the rational zeros of $P(x) = x^3 - x^2 - 3x + 3$, and then write the polynomial in factored form.

   Answer: $x = 1$ is the only rational zero. $x^3 - x^2 - 3x + 3 = (x - 1)(x - \sqrt{3})(x + \sqrt{3})$.

   XR **Example 5.20**

10. Find all the zeros, both real and complex, of $P(x) = x^3 + x^2 + x + 1$.

    Answer: $-1, i, -i$

    XR **Example 5.21**

11. Let $P(x)$ be a polynomial with leading coefficient 3. Assume that 1 is a zero of order 2, 5 is a zero of order 2, 7 is a zero of order 1, and there are no other zeros. Find $P(x)$.

    Answer: $P(x) = 3(x - 1)^2(x - 5)^2(x - 7)$

    XR **Example 5.23**

12. Given that i is a zero of $P(x) = x^4 - 3x^2 - 4$, write $P(x)$ as a product of linear and irreducible quadratic factors, each of which has real coefficients.

    Answer: $P(x) = (x - 2)(x + 2)(x^2 + 1)$

    XR **Example 5.24**

13. For the polynomial $P(x) = x^3 - 2x^2 - 2x + 4$:

a. Find the zeros. *Suggestion*: Look first for a rational zero.

b. Make a sign diagram.

c. Sketch the graph.

Answer: XR **Example 5.26**

a. $2, \sqrt{2}, -\sqrt{2}$

$$(x + \sqrt{2})(x - \sqrt{2})(x - 2)$$

b. $(-)(-)(-) = -$ $|$ $(+)(-)(-) = +$ $|$ $(+)(+)(-) = -$ $|$ $(+)(+)(+) = +$

$$\underline{}$$
$$\qquad -\sqrt{2} \qquad\qquad\qquad \sqrt{2} \qquad\qquad\qquad\qquad 2$$

c.

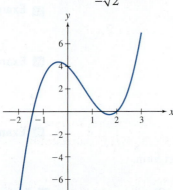

14. Find the domain of $R(x) = \dfrac{2x + 1}{x^2 + x - 2}$, and identify any poles.

Answer: The domain is all real numbers except -2 and 1. There are poles at $x = -2$ and $x = 1$.   XR **Example 5.30**

15. For $R(x) = \dfrac{x^2 - 1}{x^2 - 4}$:

a. Find the zeros and poles.

b. Identify any horizontal and vertical asymptotes.

c. Make a sign diagram.

d. Sketch the graph. Include any horizontal and vertical asymptotes.

Answer: XR **Example 5.33**

a. The zeros are at $x = \pm 1$. The poles are at $x = \pm 2$.

b. Horizontal asymptote $y = 1$. Vertical asymptotes $x = \pm 2$.

$$\frac{(x + 1)(x - 1)}{(x + 2)(x - 2)}$$

c. $\dfrac{(-)(-)}{(-)(-)} = +$ $|$ $\dfrac{(-)(-)}{(+)(-)} = -$ $|$ $\dfrac{(+)(-)}{(+)(-)} = +$ $|$ $\dfrac{(+)(+)}{(+)(-)} = -$ $|$ $\dfrac{(+)(+)}{(+)(+)} = +$

$$\underline{}$$
$$\boxed{-2} \qquad\qquad -1 \qquad\qquad 1 \qquad\qquad \boxed{2}$$

d.

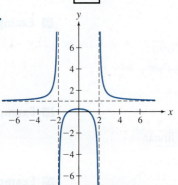

CHAPTER REVIEW EXERCISES

Section 5.1

Finding formulas. In Exercises 1 through 5, find a quadratic polynomial $P(x)$ with real coefficients that satisfies the given conditions.

1. Leading coefficient 2, with zeros 1 and 3

   Answer: $P(x) = 2x^2 - 8x + 6$

2. Leading coefficient 1, with zeros -2 and -3

   Answer: $P(x) = x^2 + 5x + 6$

3. Leading coefficient 2, with zeros $1 + i$ and $1 - i$.

   Answer: $P(x) = 2x^2 - 4x + 4$

4. Leading coefficient 3, and one zero is i

   Answer: $P(x) = 3x^2 + 3$

5. Leading coefficient 1, and the only real zero is 2

   Answer: $P(x) = x^2 - 4x + 4$

Vertices. In Exercises 6 through 9, find the vertex of the parabola, and determine whether the vertex corresponds to a maximum or a minimum.

6. $y = 6x^2 - 12x + 5$

   Answer: Vertex $(1, -1)$. Minimum.

7. $y = -x^2 + 4x + 1$

   Answer: Vertex $(2, 5)$. Maximum.

8. $y = -3x^2 + 6x + 4$

   Answer: Vertex $(1, 7)$. Maximum.

9. $y = 2x^2 + 8x + 1$

   Answer: Vertex $(-2, -7)$. Minimum.

Factoring. In Exercises 10 through 13, use the quadratic formula to factor the given quadratic polynomial.

10. $y = x^2 - 2x - 1$

    Answer: $(x - (1 + \sqrt{2}))(x - (1 - \sqrt{2}))$

11. $y = x^2 - 4x + 1$

    Answer: $(x - (2 + \sqrt{3}))(x - (2 - \sqrt{3}))$

12. $y = x^2 - 2\sqrt{2}x + 2$

    Answer: $(x - \sqrt{2})^2$

13. $y = x^2 - 2x + 2$

    Answer: $(x - (1 + i))(x - (1 - i))$

Discriminants.

14. The discriminant of a quadratic polynomial with real coefficients is 7. What does this tell you about the zeros?

    Answer: There are two real zeros.

15. The discriminant of a quadratic polynomial with real coefficients is 0. What does this tell you about the zeros?

    Answer: There is one real zero.

16. The discriminant of a quadratic polynomial with real coefficients is -5. What does this tell you about the zeros?

    Answer: There are two complex (not real) zeros, and they are complex conjugates of each other.

Maximum value.

17. One number plus twice another equals 24. Which pair of such numbers has a maximum product?

    Answer: 6 and 12

Section 5.2

Long division. In Exercises 18 through 24, use long division or synthetic division as appropriate to perform the indicated calculation. Write your answer in both quotient and product form.

18. $x + 1 \overline{)\, x^2 + 3x + 4}$

    Answer: Quotient form: $\dfrac{x^2 + 3x + 4}{x + 1} = x + 2 + \dfrac{2}{x + 1}$

    Product form: $x^2 + 3x + 4 = (x + 1)(x + 2) + 2$

19. $x - 2 \overline{)\, x^2 - 3x + 5}$

    Answer: Quotient form: $\dfrac{x^2 - 3x + 5}{x - 2} = x - 1 + \dfrac{3}{x - 2}$

    Product form: $x^2 - 3x + 5 = (x - 2)(x - 1) + 3$

20. $x - 1 \overline{)\, 2x^3 - x^2 + 4x - 3}$

    Answer: Quotient form:

    $\dfrac{2x^3 - x^2 + 4x - 3}{x - 1} = 2x^2 + x + 5 + \dfrac{2}{x - 1}$

    Product form: $2x^3 - x^2 + 4x - 3 = (x - 1)(2x^2 + x + 5) + 2$

21. $x + 2 \overline{\smash{\big)} x^3 + 3x^2 - 2x + 4}$

Answer: Quotient form:

$$\frac{x^3 + 3x^2 - 2x + 4}{x + 2} = x^2 + x - 4 + \frac{12}{x + 2}$$

Product form: $x^3 + 3x^2 - 2x + 4 = (x + 2)$
$(x^2 + x - 4) + 12$

22. $x^2 + 1 \overline{\smash{\big)} x^3 + x + 1}$

Answer: Quotient form: $\dfrac{x^3 + x + 1}{x^2 + 1} = x + \dfrac{1}{x^2 + 1}$

Product form: $x^3 + x + 1 = x(x^2 + 1) + 1$

23. $x^2 - 1 \overline{\smash{\big)} x^4 + 2x^3 - 1}$

Answer: Quotient form:

$$\frac{x^4 + 2x^3 - 1}{x^2 - 1} = x^2 + 2x + 1 + \frac{2x}{x^2 - 1}$$

Product form: $x^4 + 2x^3 - 1 = (x^2 - 1)$
$(x^2 + 2x + 1) + 2x$

24. $x^2 + x - 2 \overline{\smash{\big)} x^5 - 2x^4 + 3x^2 + x + 4}$

Answer: Quotient form:

$$\frac{x^5 - 2x^4 + 3x^2 + x + 4}{x^2 + x - 2} = x^3 - 3x^2 + 5x - 8 + \frac{19x - 12}{x^2 + x - 2}$$

Product form: $x^5 - 2x^4 + 3x^2 + x + 4 =$
$(x^2 + x - 2)(x^3 - 3x^2 + 5x - 8) + 19x - 12$

Using the remainder theorem.

25. Use the remainder theorem to find the remainder when $x^3 - 3x^2 + 7$ is divided by $x - 2$.

Answer: 3

Using the factor theorem to factor. In Exercises 26 through 31, use the given information to factor the given polynomial as far as possible without using complex numbers.

26. $x = 1$ is a zero of $P(x) = x^3 - 7x + 6$

Answer: $(x - 1)(x - 2)(x + 3)$

27. $x = 2$ is a zero of $P(x) = x^3 + x^2 - 4x - 4$

Answer: $(x - 2)(x + 2)(x + 1)$

28. $x = -1$ is a zero of $P(x) = x^3 + 6x^2 + 9x + 4$

Answer: $(x + 4)(x + 1)^2$

29. $x = -3$ is a zero of $P(x) = x^3 + 6x^2 + 11x + 6$

Answer: $(x + 3)(x + 1)(x + 2)$

30. $x = 2$ is a zero of $P(x) = x^3 - 2x^2 + x - 2$

Answer: $(x - 2)(x^2 + 1)$

31. $x = -1$ is a zero of $P(x) = x^3 + 2x^2 + 3x + 2$

Answer: $(x + 1)(x^2 + x + 2)$

The remainder and factor theorems.

32. What does the remainder theorem tell us?

Answer: The remainder when a nonconstant polynomial $P(x)$ is divided by $x - a$ is $P(a)$.

33. What does the factor theorem tell us?

Answer: $x - a$ is a factor of a polynomial $P(x)$ if and only if a is a zero of $P(x)$.

Section 5.3

Rational zero candidates. In Exercises 34 through 37, make a list of all possible rational zeros.

34. $P(x) = 6x^4 + x^3 + x - 5$

Answer: $\pm 1, \pm 5, \pm\dfrac{1}{2}, \pm\dfrac{5}{2}, \pm\dfrac{1}{3}, \pm\dfrac{5}{3}, \pm\dfrac{1}{6}, \pm\dfrac{5}{6}$

35. $P(x) = x^8 - x + 1$

Answer: ± 1

36. $P(x) = 3x^4 + x^3 - 2$

Answer: $\pm 1, \pm 2, \pm\dfrac{1}{3}, \pm\dfrac{2}{3}$

37. $P(x) = 2x^3 - 5x + 4$

Answer: $\pm 1, \pm 2, \pm 4, \pm\dfrac{1}{2}$

Factoring. The polynomials in Exercises 38 through 41 can be factored by grouping or by substitution. Factor as far as possible without using complex numbers.

38. $P(x) = x^5 - x$

Answer: $x(x - 1)(x + 1)(x^2 + 1)$

39. $P(x) = x^4 - 6x^2 + 8$

Answer: $(x - 2)(x + 2)(x - \sqrt{2})(x + \sqrt{2})$

40. $P(x) = (x^3 - 2x^2) - (9x - 18)$

Answer: $(x - 3)(x + 3)(x - 2)$

41. $P(x) = (x^3 - 4x) + (5x^2 - 20)$

Answer: $(x + 5)(x - 2)(x + 2)$

Zeros. In Exercises 42 through 48, find all zeros, both real and complex. Finding rational zeros may help, and factoring techniques may help.

42. $P(x) = x^4 - 5x^2 + 4$

   Answer: $1, -1, 2, -2$

43. $P(x) = x^3 + 3x^2 - 4$

   Answer: $1, -2$

44. $P(x) = (x^3 - 4x) - (3x^2 - 12)$

   Answer: $2, -2, 3$

45. $P(x) = x^3 + 2x^2 - x - 2$

   Answer: $1, -1, -2$

46. $P(x) = x^3 - x^2 + x - 1$

   Answer: $1, i, -i$

47. $P(x) = 2x^3 - 5x^2 + 6x - 2$

   Answer: $\dfrac{1}{2}, 1+i, 1-i$

48. $P(x) = x^4 - 6x^3 + 13x^2 - 12x + 4$

   Answer: $1, 2$

Possible rational zeros.

49. If $P(x)$ is a nonconstant polynomial with integer coefficients, then the only possible rational zeros of $P(x)$ have the form $\dfrac{p}{q}$, where _____.

   Answer: p divides the constant term and q divides the leading coefficient

Section 5.4

Zeros and their orders. In Exercises 50 through 52, find a formula for the polynomial with the given properties. You may assume there are no zeros other than those listed.

50. $x = 1$ is a zero of order 2, $x = 3$ is a zero of order 4, and the leading coefficient is 4.

   Answer: $P(x) = 4(x - 1)^2(x - 3)^4$

51. $x = 0$ is a zero of order 3, $x = -2$ is a zero of order 1, $x = 2$ is a zero of order 2, and the leading coefficient is 7.

   Answer: $P(x) = 7x^3(x + 2)(x - 2)^2$

52. $x = 1$ is a zero of order 3, $x = 3$ is a zero, the leading coefficient is 1, and the degree of the polynomial is 5.

   Answer: $P(x) = (x - 1)^3(x - 3)^2$

Sketching graphs. In Exercises 53 through 59, sketch the graph of the given polynomial.

53. $P(x) = x(x - 2)(x + 2)$

   Answer:

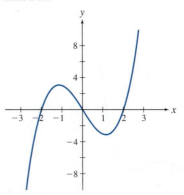

54. $P(x) = -x(x - 1)^2$

   Answer:

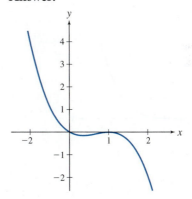

55. $P(x) = (x^2 - 1)(x - 2)$

   Answer:

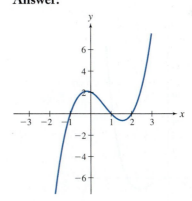

56. $P(x) = (x^2 - 4)(x^2 - 2x + 1)$

   Answer:

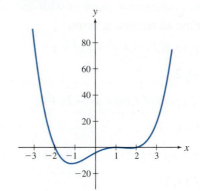

57. $P(x) = (x^2 - 1)(x^2 + 2x)$

Answer:

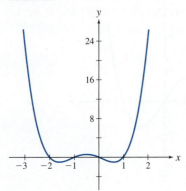

58. $P(x) = (x - 1)^3$

Answer:

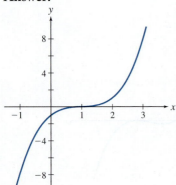

59. $P(x) = x^2(x^2 - 4)$

Answer:

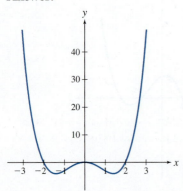

Complex conjugate zeros. In Exercises 60 through 65, you are given a polynomial together with one of its complex zeros. Find all remaining zeros.

60. $x = i$ is a zero of $P(x) = x^3 - 2x^2 + x - 2$.

Answer: $-i, 2$

61. $x = 1 + i$ is a zero of $P(x) = x^3 - 2x + 4$.

Answer: $1 - i, -2$

62. $x = 2 - i$ is a zero of $P(x) = x^3 - 5x^2 + 9x - 5$.

Answer: $2 + i, 1$

63. $x = -i$ is a zero of $P(x) = x^4 + 5x^2 + 4$.

Answer: $i, 2i, -2i$

64. $x = 1 + 2i$ is a zero of
$P(x) = x^4 - 2x^3 + x^2 + 8x - 20$.

Answer: $1 - 2i, 2, -2$

65. $x = 2 - i$ is a zero of $x^4 - 6x^3 + 12x^2 - 6x - 5$.

Answer: $2 + i, 1 + \sqrt{2}, 1 - \sqrt{2}$

Linear and quadratic factors. In Exercises 66 through 73, write the given polynomial as a product of linear and irreducible quadratic factors. Factoring techniques as well as identification of rational zeros may be helpful.

66. $P(x) = x^3 - 6x^2 + 8x$

Answer: $x(x - 4)(x - 2)$

67. $P(x) = x^4 - 16$

Answer: $(x - 2)(x + 2)(x^2 + 4)$

68. $P(x) = x^3 - 8$

Answer: $(x - 2)(x^2 + 2x + 4)$

69. $P(x) = x^4 - x^2 - 2$

Answer: $(x - \sqrt{2})(x + \sqrt{2})(x^2 + 1)$

70. $P(x) = x^3 + 2x^2 - 2x - 1$

Answer:
$$(x - 1)\left(x - \left(-\frac{3}{2} + \frac{\sqrt{5}}{2}\right)\right)\left(x - \left(-\frac{3}{2} - \frac{\sqrt{5}}{2}\right)\right)$$

71. $P(x) = x^6 - 2x^3 + 1$

Answer: $(x - 1)^2(x^2 + x + 1)^2$

72. $P(x) = x^6 - 1$

Answer: $(x + 1)(x - 1)(x^2 - x + 1)(x^2 + x + 1)$

73. $P(x) = x^2(x^2 - 2x + 1) - 4(x^2 - 2x + 1)$

Answer: $(x - 2)(x + 2)(x - 1)^2$

The fundamental theorem of algebra.

74. What does the fundamental theorem of algebra state?

Answer: Every nonconstant polynomial has at least one (possibly complex) zero.

Complex zeros of real polynomials.

75. If $P(x)$ is a polynomial with real coefficients, and if $a + bi$ is a zero with a and $b \neq 0$ real, find one other zero.

Answer: $a - bi$

A cubic application.

 76. The number of disease cases reported on day n of an epidemic is given by

$$D(n) = \frac{n^3}{3} - 13n^2 + 160n$$

a. Use a graphing utility to plot the number of cases reported per day over the first 20 days of the epidemic.

b. When did the number of cases reach a local maximum, and when did the number of cases reach a local minimum?

c. Over what time period was the number of cases increasing at a decreasing rate?

Answer:

a.

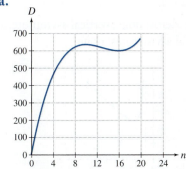

b. Local maximum on day 10 and local minimum on day 16

c. From day 0 to day 10

Section 5.5

Asymptotes. In Exercises 77 through 85, identify all zeros, horizontal asymptotes, and vertical asymptotes, and then sketch the graph.

77. $R(x) = \dfrac{x}{x-1}$

Answer: Zero: $x = 0$, vertical asymptote: $x = 1$, horizontal asymptote: $y = 1$

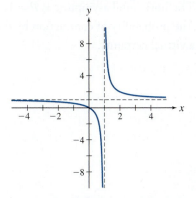

78. $R(x) = \dfrac{2x}{(x-1)^2}$

Answer: Zero: $x = 0$, vertical asymptote: $x = 1$, horizontal asymptote: $y = 0$

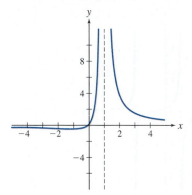

79. $R(x) = \dfrac{3(x^2 - 4)}{(x-1)^2}$

Answer: Zeros: $x = 2$ and $x = -2$, vertical asymptote: $x = 1$, horizontal asymptote: $y = 3$

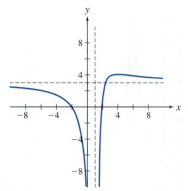

80. $R(x) = \dfrac{3x^2}{x^2 - 1}$

Answer: Zero: $x = 0$, vertical asymptotes: $x = 1$ and $x = -1$, horizontal asymptote: $y = 3$

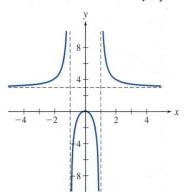

81. $R(x) = \dfrac{x-1}{(x-2)(x+1)}$

Answer: Zero: $x = 1$, vertical asymptotes: $x = -1$ and $x = 2$, horizontal asymptote: $y = 0$

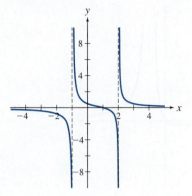

82. $R(x) = \dfrac{x-1}{(x-2)^2(x+1)}$

Answer: Zero: $x = 1$, vertical asymptotes: $x = -1$ and $x = 2$, horizontal asymptote: $y = 0$

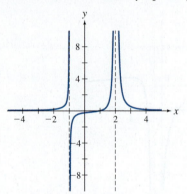

83. $R(x) = \dfrac{2x^2+1}{x^2+1}$

Answer: Zeros: none, vertical asymptotes: none, horizontal asymptote: $y = 2$

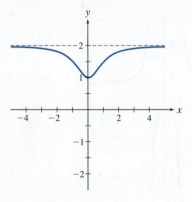

84. $R(x) = \dfrac{x^4+1}{x^2+1}$

Answer: Zeros: none, vertical asymptotes: none, horizontal asymptotes: none

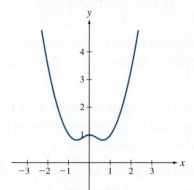

85. $R(x) = \dfrac{x^2+1}{x-1}$

Answer: Zeros: none, vertical asymptote: $x = 1$, horizontal asymptotes: none

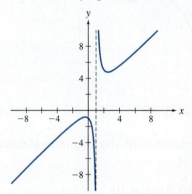

86. A disease application. The risk of contracting a certain disease depends on age. The probability of contracting the disease by age t is given by

$$P = \frac{t^2}{t^2+1}$$

Find the horizontal asymptote, and explain its meaning in terms of the probability of contraction of the disease.

Answer: The horizontal asymptote is $P = 1$. As age increases, the probability of contraction by that age becomes a virtual certainty.

INTRODUCTION TO TRIGONOMETRY

A UNIT CIRCLE APPROACH

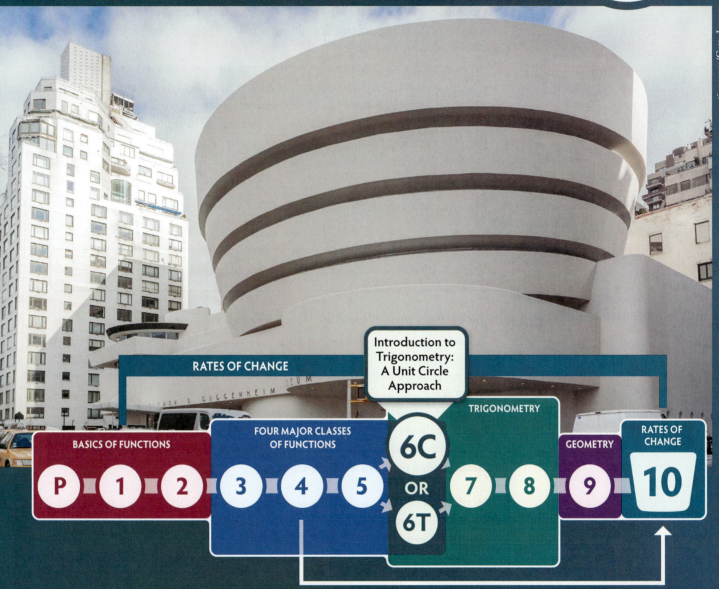

RATES OF CHANGE

Introduction to Trigonometry: A Unit Circle Approach

TRIGONOMETRY

BASICS OF FUNCTIONS

FOUR MAJOR CLASSES OF FUNCTIONS

GEOMETRY

RATES OF CHANGE

P 1 2 3 4 5 6C OR 6T 7 8 9 10

6C.1   Angles

6C.2   The Unit Circle

6C.3   The Trigonometric Functions

6C.4   Right Triangle Trigonometry

The trigonometric functions are the final class of functions we will discuss. These functions are new to many, so we devote three chapters (6, 7, and 8) to their development. This chapter introduces the trigonometric functions, which appear throughout calculus and its applications. You will gain the important skill of analyzing these functions both by using the unit circle and by referring to right triangles.

Chapters 6C and 6T offer alternative approaches to trigonometry, covering the same material in a different order. Use either one, but not both.

tinnaporn/Getty Images

After a brief refresher on angles, Chapter 6C begins with a discussion of the unit circle and then proceeds to right triangles. Chapter 6T is provided for those who prefer to introduce the trigonometric functions via right triangles and then extend their definitions via the unit circle. Students need to be able to use trigonometric functions in either context, and either chapter will provide appropriate background for students to accomplish this goal.

The word *trigonometry* comes from the Greek words for *triangle* and *measure*. As the name implies, trigonometry arose from a study of triangles—particularly right triangles (those with a 90° angle).

Early astronomy and architecture necessitated the development of trigonometry. In the third century BCE Eratosthenes used rudimentary trigonometry to give a remarkably accurate estimate of Earth's circumference.

In this chapter we lay the foundation for trigonometry by studying angles and then introduce the six trigonometric functions, which play a key role in calculus as well as applications of mathematics.

6C.1 Angles

6C.2 The Unit Circle

6C.3 The Trigonometric Functions

6C.4 Right Triangle Trigonometry

6C.1 Angles

> Trigonometry depends on the measurement of angles.

In this section, you will learn to:

1. Locate an angle with degree measure in a given range that is coterminal with a given angle.
2. Locate an angle with radian measure in a given range that is coterminal with a given angle.
3. Perform conversions between degree measure and radian measure.
4. Calculate the arc length and area determined by a sector of a circle.
5. Determine the unknown sides and angles of a pair of similar triangles.
6. Solve applied problems involving angles.

Two triangles with equal corresponding angles but possibly different side lengths are known as similar triangles. For two similar triangles, the shape is the same, but the size may be different. Understanding similar triangles is fundamental to trigonometry.

Similarity of triangles preserves angles and shapes, and it also preserves symmetry. In fact, the preservation of shape and symmetry applies in an even more general setting. Imagine the letter "A" presented using various type sizes:

ᴀ A A A A

Each presentation has the same shape and a vertical line of symmetry.

In this section we introduce angle measure and show how it applies to triangles.

Conventions for Angles

> Angles can be thought of as rotations from an initial ray to a terminal ray.

An angle consists of two rays emanating from a common vertex. More precisely, an angle is a rotation of the initial ray to the terminal ray. Sometimes these rays are called the sides of the angle. A counterclockwise rotation gives a positive angle, and a clockwise rotation gives a negative angle. These cases are illustrated in **Figure 6C.1** and **Figure 6C.2**.

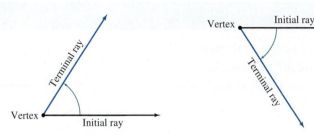

Figure 6C.1 A positive angle (counterclockwise rotation)

Figure 6C.2 A negative angle (clockwise rotation)

Usually we consider angles in a rectangular coordinate system. We say that such an angle is in standard position if its vertex is at the origin and its initial ray lies along the positive horizontal axis. Sometimes we assume that a given angle is in standard position without explicitly saying so.

There are two common ways to measure angles. The degree is the measure that is familiar to most people and is often used in applications. Another is radian measure, which mathematicians and scientists prefer. Radians, as we'll see, are a natural measure and are, by far, the more important of the two measures in preparing for calculus. Because each way of measuring angles is commonly used, we will use both throughout the text.

Degree Measure

> Degree measure of angles is probably familiar from everyday experience.

We begin by defining the **degree** as a measurement of angles. To measure an angle in degrees, we begin by assigning 360 degrees (written as 360°) to a full circle. Then we measure how many degrees are in an angle with vertex at the center of the circle by looking at what portion of a circle it cuts. For example, going counterclockwise halfway around a circle gives an angle whose measure is one half of 360° or 180°, a straight angle. Going one-quarter of the way around a circle gives one-quarter of 360° or 90°, a right angle. An acute angle is a positive angle that measures less than 90°, and an obtuse angle measures more than 90° and less than 180°. **Figure 6C.3** shows the degree measure of some common angles.

Note that assigning the number 360 to a circle is purely arbitrary. We could measure angles by assigning other numbers.

A rotation of 360° is one complete trip around the circle, so it represents a return to the initial ray. Therefore, if angles of 0° and 360° have the same initial ray, then they also have the same terminal ray. Angles that share initial and terminal rays are called **coterminal angles**.

Every angle is coterminal with an angle of degree measure between 0 and 360.

The **degree** is an angle measurement such that an angle of 1° is a positive angle with vertex at the center of a circle that cuts an arc of one-360th of a full circle.

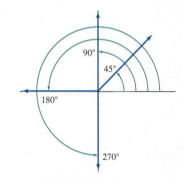

Figure 6C.3 Degree measures of some common angles

Coterminal angles are pairs of angles that have the same initial and terminal rays.

EXAMPLE 6C.1 Coterminal Angles

Find angles with degree measure between 0 and 360 that are coterminal with the following angles.

a. 750°   **b.** −400°

SOLUTION

a. An angle of 720° is two complete rotations of a circle. A rotation of 750° is two trips around the circle and an additional 30° as shown in **Figure 6C.4**. Hence, an angle of 750° is coterminal with an angle of 30°.

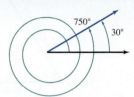

Figure 6C.4 Coterminal angles: An angle of 750° is coterminal with an angle of 30°.

b. An angle of −400° indicates a rotation in the clockwise direction of 400°. That is 360° and an additional 40°, so an angle of −400° is coterminal with an angle of 320°. This is shown in **Figure 6C.5**.

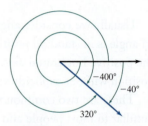

Figure 6C.5 Coterminal angles: An angle of −400° is coterminal with an angle of 320°.

TRY IT YOURSELF 6C.1   Brief answers provided at the end of the section.

Find an angle with degree measure between 0 and 360 that is coterminal with an angle of 920°.

EXTEND YOUR REACH

a. Find an angle between 0° and 360° that is coterminal with an angle of 1000°.

b. Find an angle between 0° and 360° that is coterminal with an angle of 10,000°.

c. Explain how you will find an angle between 0° and 360° that is coterminal with an angle of any given positive degree measure.

Radian Measure

> Radian measure is calculated by using the radius of a circle to measure arcs along the circle.

The **radian** is an angle measurement such that an angle of 1 radian is an angle with its vertex at the center of a circle that cuts an arc the same length as the radius.

The measure of angles used almost exclusively in calculus is the **radian**.

For a tactile understanding of radian measure, we can use a piece of string as the radius of a circle. We then lay that same piece of string on the circumference of the circle to make an arc having the same length as the radius, as shown in **Figure 6C.6**. An angle of 1 radian has its vertex at the center of the circle, and its sides pass through the endpoints of the arc made by the string.

It is remarkable that the length of the string we started with doesn't matter. The resulting angle is the same regardless of the length of the string. This fact is illustrated in **Figure 6C.7**. For this reason, radian measure is called a natural measure. Radian measure is such a natural measure that intelligent aliens would certainly understand the concept (but it is unlikely that they would use degrees).

For an angle with its vertex at the center of a circle, an angle of 1 radian is an angle that cuts an arc of length equal to the radius. An angle of 2 radians cuts an arc of length equal to twice the radius, and so on. In general, an angle of θ radians cuts an arc of length equal to θ times the radius. This observation can be expressed as follows:

$$\text{Arc length} = \text{Radius} \times \text{Radian measure}$$

or

$$\text{Radian measure} = \frac{\text{Arc length}}{\text{Radius}}$$

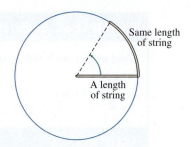

Figure 6C.6 Making an angle of 1 radian

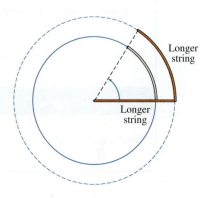

Figure 6C.7 Radian measure: Using a different length results in the same angle

Because the circumference of a circle is 2π times its radius, we get

$$\begin{aligned}
\text{Radians in a circle} &= \frac{\text{Arc length of circle}}{\text{Radius}} \\
&= \frac{\text{Circumference}}{\text{Radius}} \\
&= \frac{2\pi \times \text{Radius}}{\text{Radius}} \\
&= \frac{2\pi \times \cancel{\text{Radius}}}{\cancel{\text{Radius}}} \\
&= 2\pi
\end{aligned}$$

So, there are 2π radians in a circle. A straight angle cuts half a circle, so its radian measure is π. A right angle cuts a quarter circle, so its measure is $\dfrac{\pi}{2}$ radians. This gives us some easy comparisons with degree measure of some common angles, as shown in the following table and in **Figure 6C.8**.

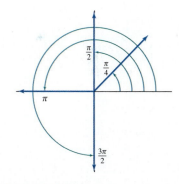

Figure 6C.8 Radian measure of some common angles

Angle name	Degrees	Radians
Full circle	360	2π (coterminal with 0)
	270	$\dfrac{3\pi}{2}$
Straight angle	180	π
Right angle	90	$\dfrac{\pi}{2}$
	45	$\dfrac{\pi}{4}$

Every angle is coterminal with an angle of radian measure between 0 and 2π.

EXAMPLE 6C.2   Coterminal Angles Using Radian Measure

Find angles with radian measure between 0 and 2π that are coterminal with the following angles.

a. $\dfrac{9\pi}{2}$ radians

b. $-\dfrac{2\pi}{3}$ radians

SOLUTION

a. There are 2π radians in a complete circle, so an angle of radian measure $\dfrac{9\pi}{2} = 4\pi + \dfrac{\pi}{2}$ is coterminal with an angle of radian measure $\dfrac{\pi}{2}$.

b. An angle of radian measure $-\dfrac{2\pi}{3}$ is coterminal with an angle of radian measure $-\dfrac{2\pi}{3} + 2\pi = \dfrac{4\pi}{3}$.

TRY IT YOURSELF 6C.2   Brief answers provided at the end of the section.

Find an angle with radian measure between 0 and 2π that is coterminal with an angle of $\dfrac{35\pi}{6}$.

Converting Angle Measures

> Radian and degree measures are related by simple formulas.

Because $180°$ is π radians, $1°$ is $\dfrac{\pi}{180}$ radian. So an angle of $d°$ is $d\dfrac{\pi}{180}$ radians. In the other direction, we find that an angle of θ radians is $\theta\dfrac{180}{\pi}$ degrees. One radian is equivalent to just under $60°$.

CONCEPTS TO REMEMBER: Converting Between Degrees and Radians

1. The radian measure of an angle is the length of the arc that the angle cuts out of a circle (centered at the vertex) divided by the radius of the circle:

$$\text{Radian measure} = \frac{\text{Arc length}}{\text{Radius}}$$

2. To convert from radians to degrees, multiply radians by $\dfrac{180}{\pi}$.

3. To convert from degrees to radians, multiply degrees by $\dfrac{\pi}{180}$.

EXAMPLE 6C.3   Converting from Degree to Radian Measure

What is the radian measure of a $60°$ angle?

SOLUTION

To convert to radians, we multiply by $\dfrac{\pi}{180}$:

$$60° = 60 \times \frac{\pi}{180} \text{ radians}$$

$$= \cancel{60} \times \frac{\pi}{\cancel{180}^{3}} \text{ radians}$$

$$= \frac{\pi}{3} \text{ radians}$$

TRY IT YOURSELF 6C.3 Brief answers provided at the end of the section.

What is the radian measure of an angle of 30°?

EXAMPLE 6C.4 Converting from Radian to Degree Measure

What is the degree measure of an angle of $\dfrac{\pi}{6}$ radian?

SOLUTION

To convert to degrees, we multiply by $\dfrac{180}{\pi}$:

$$\frac{\pi}{6} \text{ radian} = \frac{\pi}{6} \times \frac{180}{\pi} \text{ degrees}$$

$$= \frac{\cancel{\pi}}{\cancel{6}} \times \frac{\cancel{180}^{30}}{\cancel{\pi}} \text{ degrees}$$

$$= 30°$$

TRY IT YOURSELF 6C.4 Brief answers provided at the end of the section.

What is the degree measure of an angle of $\dfrac{5\pi}{6}$ radians?

---

a. Construct a circle. Using a piece of string with a length equal to the radius of the circle, construct an angle measuring 2 radians. According to the conversion formula, what is the degree measure of this angle? Check your answer with a protractor that measures angles in degrees.

b. Nairobi, Kenya, and Quito, Ecuador, are both close to the equator. Look up the longitude of each of these cities, the distance in miles between them, and the equatorial radius of Earth. (The longitude is the degree measure of the angle on the equatorial circle that the city makes with the Greenwich meridian. Longitude east of this meridian is positive, and longitude west is negative. See **Figure 6C.9**.) Are the numbers you find consistent with the conversion formula? *Suggestion*: If you are not sure how to answer this question, look at part a.

EXTEND YOUR REACH

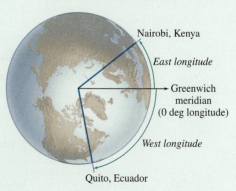

Figure 6C.9 Nairobi, Quito, and the meridian

Arc Length and Area

> Arc length and area are calculated for sectors of circles.

In discussing radian measure, we used the formula

$$\text{Arc length} = \text{Radius} \times \text{Radian measure}$$

If we use θ for the radian measure and r for the radius, this formula becomes

$$\text{Arc length} = r\theta$$

For simplicity, we assume that θ is between 0 and 2π. In a circle of radius 1, the arc length is the same as the radian measure of the angle.

We have a formula for the arc length cut by an angle. We can also work out the area determined by (or subtended by) an angle. The angle cuts a sector from the circle that looks like a slice of pie, as shown in **Figure 6C.10**. The entire circle has area πr^2. The area of the slice of pie is some fraction of that. It takes 2π radians to get all the way around the circle, so the fraction of the circle we have covered is

$$\text{Fraction covered} = \frac{\text{Radian measure of arc}}{\text{Radian measure of circle}} = \frac{\theta}{2\pi}$$

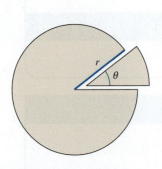

Figure 6C.10 A sector of a circle cut by an angle of radian measure θ

Hence, the area is

$$\text{Area subtended} = \text{Fraction covered} \times \text{Area of circle}$$

$$= \frac{\theta}{2\pi} \times \pi r^2$$

$$= \frac{\theta}{2\pi} \times \pi r^2$$

$$= \frac{\theta r^2}{2}$$

CONCEPTS TO REMEMBER: Area and Arc Length

Radians. Consider a circle of radius r centered at the vertex of an angle of radian measure θ, where $0 < \theta < 2\pi$. For the slice of the circle cut out by this angle,

$$\text{Arc length} = r\theta$$

and

$$\text{Area subtended} = \frac{\theta r^2}{2}$$

Degrees. Consider a circle of radius r centered at the vertex of an angle of degree measure d, where $0 < d < 360$. For the slice of the circle cut out by this angle,

$$\text{Arc length} = \frac{d\pi r}{180}$$

and

$$\text{Area subtended} = \frac{d\pi r^2}{360}$$

EXAMPLE 6C.5 Area and Arc Length

a. If a circle of radius 5 units is centered at the vertex of an angle of 2 radians, find the arc length and area subtended.

b. If a circle of radius 4 units is centered at the vertex of an angle of 25°, find the arc length and area subtended.

SOLUTION

a. We use the arc length and area formulas for radian measure with $r = 5$ and $\theta = 2$:

$$\text{Arc length} = r\theta$$

$$= 5 \times 2$$

$$= 10 \text{ units}$$

Also,

$$\text{Area subtended} = \frac{\theta r^2}{2}$$

$$= \frac{2 \times 5^2}{2}$$

$$= 25 \text{ square units}$$

b. We use the arc length and area formulas for degree measure with $r = 4$ and $d = 25$:

$$\text{Arc length} = \frac{d\pi r}{180}$$

$$= \frac{25\pi \times 4}{180}$$

$$= \frac{5\pi}{9}$$

$$\approx 1.75 \text{ units}$$

Also,

$$\text{Area subtended} = \frac{d\pi r^2}{360}$$

$$= \frac{25\pi \times 4^2}{360}$$

$$= \frac{10\pi}{9}$$

$$\approx 3.49 \text{ square units}$$

TRY IT YOURSELF 6C.5 Brief answers provided at the end of the section.

The vertex of an angle of $\dfrac{\pi}{3}$ radians is at the center of a circle. The angle cuts a sector of area 6 square units. What is the radius of the circle?

Similar Triangles

> Similarity is at the foundation of trigonometry.

Similar triangles are pairs of triangles for which the corresponding angles are the same.

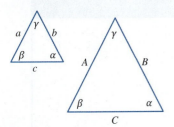

Figure 6C.11 Similar triangles: These are triangles that have equal corresponding angles, but the sides may have different lengths.

As discussed earlier, similarity of triangles is fundamental to trigonometry. **Similar triangles** often occur in applications, as in Example 6C.8, and are used throughout Section 6C.4.

When triangles are similar, as in **Figure 6C.11**, one appears to be a blown-up version of the other.

The following is the key property of similar triangles, and it is crucial to the definitions of the trigonometric functions that follow.

LAWS OF MATHEMATICS: Similar Triangles

The ratios of corresponding sides of similar triangles are the same. For the similar triangles in Figure 6C.11, we have

$$\frac{a}{A} = \frac{b}{B} = \frac{c}{C}$$

We can also write these equations as

$$\frac{A}{B} = \frac{a}{b}, \quad \frac{A}{C} = \frac{a}{c}, \quad \frac{B}{C} = \frac{b}{c}$$

EXAMPLE 6C.6 Corresponding Sides of Similar Triangles

The triangles in **Figure 6C.12** are similar. Find the lengths of all sides that are not already calculated.

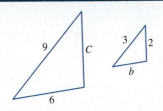

Figure 6C.12 Similar triangles for Example 6C.6

SOLUTION

We use the relationships among the sides of similar triangles:

$$\frac{9}{3} = \frac{C}{2}$$

so $C = 6$. Similarly,

$$\frac{6}{b} = \frac{9}{3}$$

so $b = 2$.

The triangles in **Figure 6C.13** are similar. Find all sides that are not already calculated.

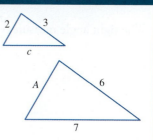

Figure 6C.13 Similar triangles for Try It Yourself 6C.6

For the two right triangles in **Figure 6C.14** it is true that

$$\frac{A}{a} = \frac{B}{b}$$

a. Use a bit of algebra to show that

$$\frac{\sqrt{A^2 + B^2}}{\sqrt{a^2 + b^2}} = \frac{A}{a}$$

b. Is the result in part a consistent with the fact that the two triangles are similar? Explain your answer.

EXTEND YOUR REACH

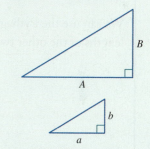

Figure 6C.14 Right triangles

An important geometric fact is that the sum of the three angles of a triangle is the same for all triangles.[1]

> **LAWS OF MATHEMATICS: Angle Sum of a Triangle**
>
> The sum of the angles of a triangle is π radians or 180°.
>
> As a consequence, when two pairs of corresponding angles from a pair of triangles are the same, then the third pair must also be the same. Hence, having two pairs of equal angles is sufficient to ensure similarity.

EXAMPLE 6C.7 Right Triangles

The triangles in **Figure 6C.15** are right triangles. The angle measures shown are in radians. Find all unmarked sides and angles.

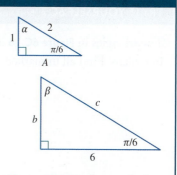

Figure 6C.15 Right triangles for Example 6C.7

---

[1]This fact is a feature of plane geometry. It does not hold for triangles on curved surfaces, such as the surface of Earth, for example. See the exercises.

SOLUTION

The right angle measures $\dfrac{\pi}{2}$ radians, and the sum of the three angles in a triangle is π radians, so

$$\alpha + \frac{\pi}{6} + \frac{\pi}{2} = \pi \qquad \text{◀ The angle sum is } \pi \text{ radians.}$$

$$\alpha = \pi - \frac{\pi}{6} - \frac{\pi}{2} \qquad \text{◀ Solve for } \alpha.$$

$$= \frac{\pi}{3} \text{ radians} \qquad \text{◀ Simplify.}$$

Similarly, $\beta = \dfrac{\pi}{3}$ radians.

We can use the Pythagorean theorem (see Section P.1) to find the length of A because we know the lengths of the other two sides:

$$A^2 + 1^2 = 2^2 \qquad \text{◀ Use the Pythagorean theorem.}$$

$$A^2 = 3 \qquad \text{◀ Solve for } A^2.$$

$$A = \sqrt{3} \qquad \text{◀ Take square roots.}$$

Because the corresponding angles are the same, the triangles are similar:

$$\frac{b}{1} = \frac{6}{A}$$

$$b = \frac{6}{\sqrt{3}}$$

In the same fashion,

$$\frac{c}{2} = \frac{6}{A}$$

$$\frac{c}{2} = \frac{6}{\sqrt{3}}$$

$$c = \frac{12}{\sqrt{3}}$$

TRY IT YOURSELF 6C.7 Brief answers provided at the end of the section.

The triangles in **Figure 6C.16** are right triangles. The angle measures shown are in radians. Find all unmarked sides and angles.

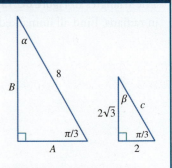

Figure 6C.16 Right triangles for Try It Yourself 6C.7

MODELS AND APPLICATIONS   Shadows

Similarity of triangles has many applications to everyday measurements.

EXAMPLE 6C.8   A Shadow

A man who is 6 feet tall casts a shadow of 2 feet. At the same time, a building casts a shadow that is 12 feet long. How tall is the building?

SOLUTION

Let's use H to denote the height of the building, as shown in **Figure 6C.17**. The angles α and β are the same because they are determined by the sun. Hence, the two right triangles have two common angles, so they are similar. Equating ratios of corresponding sides, we have

$$\frac{H}{6} = \frac{12}{2}$$

$$H = 36$$

We conclude that the building is 36 feet tall.

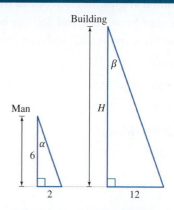

Figure 6C.17 Lengths of shadows

TRY IT YOURSELF 6C.8   Brief answers provided at the end of the section.

How long is the shadow cast by a 15-foot-tall flagpole at the same time of day as in the example?

TRY IT YOURSELF ANSWERS

6C.1 $200°$

6C.2 $\dfrac{11\pi}{6}$

6C.3 $\dfrac{\pi}{6}$

6C.4 $150°$

6C.5   The radius is $\dfrac{6}{\sqrt{\pi}}$ units.

6C.6 $A = 4$ and $c = \dfrac{7}{2} = 3.5$

6C.7 $\alpha = \beta = \dfrac{\pi}{6}$ radian, $A = 4$, $c = 4$, and $B = 4\sqrt{3}$

6C.8   5 feet

EXERCISE SET 6C.1

CHECK YOUR UNDERSTANDING

1. Coterminal angles:
   a. begin at the same place
   b. end at the same place
   c. have common initial and terminal sides
   d. None of the above.

2. **True or false:** Every angle is coterminal with an angle of degree measure between 0 and 360.

3. **True or false:** Every angle is coterminal with an angle of radian measure between 0 and π.

4. **True or false:** Degree measure depends on an arbitrary choice of numbers.

5. What is the radian measure of an angle of 0°?

6. **True or false:** An angle with vertex at the center of a circle cuts the same area whether the angle is measured in degrees or in radians.

7. Two triangles are similar if _____.

8. The radian measure of an angle of 90° is _____.

9. What is the radian measure of a straight angle?

10. The angle sum of any triangle is _____.

11. State the Pythagorean theorem.

SKILL BUILDING

Coterminal angles. In Exercises 12 through 19, find an angle either between 0 and 360° or between 0 and 2π radians that is coterminal with the given angle.

12. 600°

13. 420°

14. 750°

15. 200π radians

16. −200°

17. $\dfrac{10\pi}{3}$ radians

18. 7 radians

19. -11π radians

Converting degrees to radians. In Exercises 20 through 27, convert the given degree measure to radian measure.

20. 60°

21. 30°

22. 15°

23. 120°

24. 300°

25. −90°

26. $\pi°$

27. 70°

Converting radians to degrees. In Exercises 28 through 34, convert the given radian measure to degree measure.

28. $\dfrac{\pi}{6}$ radian

29. $\dfrac{3\pi}{4}$ radians

30. 3π radians

31. $\dfrac{4\pi}{3}$ radians

32. $\dfrac{\pi}{12}$ radian

33. 90 radians

34. $-\dfrac{5\pi}{4}$ radians

Using the Pythagorean theorem. In Exercises 35 through 38, use the Pythagorean theorem to find the missing side.

35. A right triangle has legs of length 3 and 7. Find the length of the hypotenuse.

36. A right triangle has legs of length 4 and 6. Find the length of the hypotenuse.

37. A right triangle has one leg of length 4 and a hypotenuse of length 7. Find the length of the other leg.

38. A right triangle has one leg of length 5 and a hypotenuse of length 9. Find the length of the other leg.

Angle sum. For Exercises 39 through 45, recall that, in a triangle, the angle sum is π radians or 180°.

39. One angle of a triangle is $\dfrac{\pi}{9}$ radian, and another is $\dfrac{\pi}{4}$ radian. Find the radian measure of the third angle.

40. One angle of a triangle is 20°, and another is 40°. What is the degree measure of the third angle?

41. One angle of a triangle is $\dfrac{\pi}{4}$ radian, and another is $\dfrac{\pi}{3}$ radian. What is the radian measure of the third angle?

42. One angle of a triangle is 20°. Another is 0.3 radian. Find both the radian measure and the degree measure of the third angle.

43. One angle of a triangle is 70°. Another is 10°. Find the degree measure of the third angle.

44. An equilateral triangle has all three angles equal. What is the degree measure of an angle of an equilateral triangle?

45. An equilateral triangle has all three angles equal. What is the radian measure of an angle of an equilateral triangle?

Area and arc length. In Exercises 46 through 51, you are given an angle and the radius of a circle. The angle has its vertex at the center of the circle. Find the arc length cut by the angle, and find the area subtended by the angle.

46. Angle of $\frac{\pi}{6}$ radian. Circle of radius 5 units.

47. Angle of 1 radian. Circle of radius 9 units.

48. Angle of $\frac{\pi}{3}$ radian. Circle of radius 6 units.

49. Angle of $\frac{\pi}{6}$ radian. Circle of radius 3 units.

50. Angle of 20°. Circle of radius 10 units.

51. Angle of $d°$. Circle of radius $\frac{1}{d}$ units.

PROBLEMS

52. Degrees in sports. In skateboarding and snowboarding competitions, multiple turns are measured in degrees.

   a. In 2018, Anna Gasser became the first female snowboarder to land a cab triple 1260. The number 1260 refers to a turn of 1260°. How many rotations does this indicate?

   b. Sage Kotsenburg's slopestyle gold medal run in the 2014 Winter Olympics at Sochi included a 1620. How many rotations are indicated?

53. Grads. Another type of angle measure is the grad, sometimes used on European maps as a metric equivalent of the degree. There are 400 grads in a circle.

   a. What is the grad measure of an angle of 90°?

   b. What is the grad measure of an angle of π radians?

   c. What is the grad measure of an angle of 36°?

54. Mils. The mil is a type of angle measure used in the military. This term is derived from milliradian, and 1000 mils equal 1 radian.

   a. What is the degree measure of 1 mil? Give your answer accurate to three decimal places.

   b. A target is 1 yard wide and subtends an angle of 1 mil in a soldier's field of vision. How far from the soldier is the target? *Suggestion*: Think of the target as an arc of a circle.

   c. How many mils are in a circle?[2]

Area and arc length.

55. Consider a circle of radius r centered at the vertex of an angle of degree measure d, where $0 < d < 360$. For the slice of the circle cut out by this angle, show that

$$\text{Arc length} = \frac{d\pi r}{180}$$

and

$$\text{Area subtended} = \frac{d\pi r^2}{360}$$

56. An angle has its vertex centered at the origin of a circle. It cuts an arc length of 3 units and subtends an area of 9 square units. What is the radius of the circle? What is the radian measure of the angle?

57. An angle has its vertex centered at the origin of a circle. It cuts an arc length of 3 units and subtends an area of 6 square units. What is the radius of the circle? What is the degree measure of the angle?

58. An angle has its vertex centered at the origin of a circle. It cuts an arc length of L units and subtends an area of A square units. What is the radius of the circle? What is the radian measure of the angle?

59. Find a formula that gives the area A of a sector of a circle in terms of the radius r and arc length L.

60. When area equals arc length. A circle of a certain radius has the property that if an angle has its vertex at the center of the circle, the area it subtends is numerically the same as the arc length. Find the radius of the circle.

61. Doubling. Your friend and you are cutting pieces of a pie. Your friend cuts a piece twice as big (say in area) as the one you cut for yourself. How do the central angles compare?

62. More on doubling. You have two pies, one of which has twice the diameter of the other. You cut each pie into the same number of pieces. How do the pieces of the larger pie compare (say in area) to those of the smaller pie?

Similarity. The triangles in **Figure 6C.18** are similar. Exercises 63 through 68 refer to these triangles.

---

[2]The U.S. military uses a somewhat different definition of a mil, in which there are 6400 mils in a circle. Other countries use different definitions.

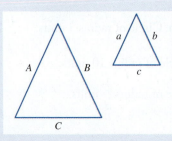

Figure 6C.18 Similar triangles

63. $A = 24$, $B = 18$, $a = 4$, and $c = 2$. Find C and b.

64. $A = 24$, $B = 16$, $C = 12$, and $c = 3$. Find a and b.

65. $A = 6$, $a = 3$, $b = 4$, and $c = 2$. Find B and C.

66. $A = 24$, $B = 16$, $b = 4$, and $c = 3$. Find a and C.

67. $A = x$, $B = 1$, and $a = 1$. Find b in terms of x.

68. $A = 1$, $C = a$, and $c = x$. Find C in terms of x.

69. **Preparing for trigonometry.** For the right triangles in **Figure 6C.19**, $s = t$.

   a. Show that the triangles are similar.

   b. Show that $\dfrac{a}{c} = \dfrac{A}{C}$.

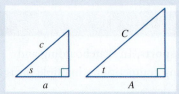

Figure 6C.19 Two right triangles, with angle s the same as angle t

MODELS AND APPLICATIONS

70. **Cartography.** Topographical maps show heights of mountains and depths of valleys. One difficulty in making such a map is that it is often impossible to travel over the entire terrain to be mapped. Instead of doing this, laser and radar measurements are taken. Such measurements indicate that the base of a sheer rock wall is 2000 feet from your observation point. Its peak is 2050 feet away. See **Figure 6C.20**. How tall is the wall?

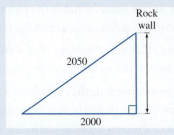

Figure 6C.20 Height of a rock wall

71. **More cartography.** A sheer rock wall is known to be 0.16 mile high. We find that its peak is 2 miles from our observation point. See **Figure 6C.21**. How far away is the base of the rock wall?

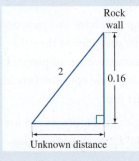

Figure 6C.21 Distance to a rock wall

72. **Latitude.** Wichita, Kansas, is due north of Fort Worth, Texas. This means that they lie on a circle whose center is that of Earth and whose radius equals the polar radius of Earth, about 3950 miles. Further, the latitude of Wichita is about 37° north, and that of Fort Worth is about 32° north. (These angles are measured along the circle described earlier, starting at the equator.) See **Figure 6C.22**. How far is it from Fort Worth to Wichita?

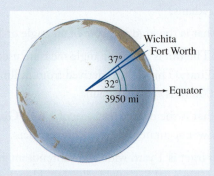

Figure 6C.22 Distance from Fort Worth to Wichita

73. **More latitude.** Refer to the preceding exercise for information about the measurement of latitude. Winnipeg, Manitoba (Canada) is due north of Fort Worth, Texas. It is about 1180 miles from Fort Worth to Winnipeg. See **Figure 6C.23**. What is the latitude of Winnipeg?

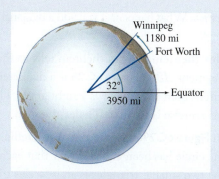

Figure 6C.23 The latitude of Winnipeg

74. **Shadows.** A 10-foot-tall vertical pole casts a 6-foot-long horizontal shadow. See **Figure 6C.24**. How tall is a tree that at the same time casts a 21-foot-long shadow? *Suggestion*: Use similar triangles.

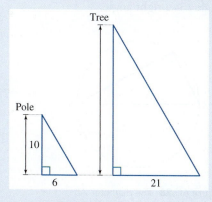

Figure 6C.24 Shadows

75. **Another shadow.** A 6-foot-tall woman casts a 2-foot-long horizontal shadow. How long a shadow is cast at the same time by an 18-foot-tall flag pole?

76. **A ladder.** A 20-foot-long ladder leans against a vertical wall, and the horizontal distance from the base of the ladder to the wall is 5 feet. One rung of the ladder is 12 feet from the base of the ladder. See **Figure 6C.25**. Use similar triangles to determine how far that rung is from the wall.

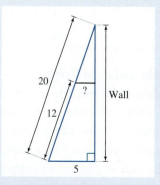

Figure 6C.25 One rung of a ladder

CHALLENGE EXERCISES FOR INDIVIDUALS OR GROUPS

77. **The inverse Pythagorean theorem.** The goal of this exercise is to establish the inverse Pythagorean theorem. In **Figure 6C.26** we have added an altitude, the segment perpendicular to the hypotenuse of the right triangle. Using a, b, and h as shown in the figure, the theorem states

$$\frac{1}{a^2} + \frac{1}{b^2} = \frac{1}{h^2}$$

a. Use

$$\text{Area} = \frac{1}{2}\,\text{Base} \times \text{Height}$$

to calculate the area of the large triangle and the areas (marked A_1 and A_2) of the two smaller triangles.

b. By equating the area of the large triangle to the sum of the areas A_1 and A_2, show that $ab = hc$. Deduce that $c = \dfrac{ab}{h}$.

c. Put the value of c calculated in part b into the Pythagorean theorem $a^2 + b^2 = c^2$. Then use a bit of algebraic manipulation to produce the inverse Pythagorean theorem.

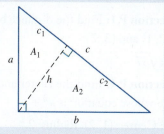

Figure 6C.26 A right triangle with an altitude added

78. **An angle and a circle.** Consider **Figure 6C.27**, in which an angle meets a circle. It can be shown that $\triangle ADB$ is similar to $\triangle ACE$. (The correspondence matches $\angle ADB$ with $\angle ACE$ and $\angle ABD$ with $\angle AEC$.) Show that $|AB| \times |AC| = |AD| \times |AE|$. (The vertical bars denote the length of a segment.)

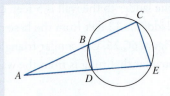

Figure 6C.27 An angle and a circle

79. **Triangles on the surface of Earth.** On the surface of Earth, "straight lines" are parts of great circles. A great circle is the intersection of the surface of Earth with a plane that goes through the center of Earth. A segment of the equator is a straight line on the surface of Earth. Similarly, a segment of a longitude line is a straight line.

    a. Let A be the North Pole, and let B and C be distinct points on the equator one-quarter of the way around Earth from each other. What is the angle sum of $\triangle ABC$?

    b. Can any triangle on the surface of Earth have angle sum 180°? *Suggestion:* To answer this question, consider that triangles on the surface of Earth are somewhat "bent" due to the curvature of Earth. What would happen to the angles if we could "iron" such a triangle out flat?

80. **Hyperbolic geometry.** In hyperbolic geometry, the angle sum of any triangle is strictly less than 180°. Show that in such a geometry rectangles (quadrilaterals with four right angles) cannot exist.

Suggestion: What happens if you add a diagonal to a rectangle?

81. **Thales' theorem.** Thales' theorem tells us that if the segment AB in **Figure 6C.28** is a diameter of the circle, then the angle t is a right angle. The goal of this exercise is to prove Thales' theorem.

    a. In **Figure 6C.29**, a segment that is a radius of the circle has been added. Show that each of the two smaller triangles is an isosceles triangle (that is, two sides have equal length).

    b. Recall from geometry that the base angles of an isosceles triangle are congruent. Use this information together with the fact that the angle sum of any triangle is 180° to prove that angle t is a right angle.

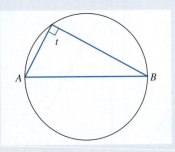

Figure 6C.28 The diameter AB

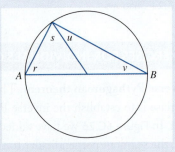

Figure 6C.29 Using a radius to divide the triangle

REVIEW AND REFRESH: Exercises from Previous Sections

82. **From Section P.1:** Find the distance between the points $(2, 1)$ and $(3, 7)$.

    Answer: $\sqrt{37}$

83. **From Section P.1:** Find the center and radius of the circle whose equation is $x^2 + y^2 + 1 = 2x + 4y$.

    Answer: Center: $(1, 2)$. Radius: 2.

84. **From Section P.3:** Solve the inequality $x^2 + 2x < 3$.

    Answer: $(-3, 1)$

85. **From Section 2.1:** If $f(x) = \dfrac{1}{x+1}$ and $g(x) = \dfrac{1}{x}$, find and simplify a formula for $(f \circ g)(x)$.

    Answer: $(f \circ g)(x) = \dfrac{x}{1+x}$

86. **From Section 4.1:** Without using your calculator, find the exact value of $\ln \sqrt{e^3}$.

    Answer: $\dfrac{3}{2}$

87. From Section 4.1: If $f(x) = \log(x - 1)$, calculate $f(1 + 10^{-x})$.

Answer: $-x$

88. From Section 4.3: Find the inverse function of $f(x) = 2 + \log x$.

Answer: $f^{-1}(x) = 10^{x-2}$

89. From Section 4.3: Solve $2^x = 3^{x-1}$.

Answer: $x = \dfrac{\ln 3}{\ln 3 - \ln 2}$

90. From Section 5.1: What can be said of the zeros of $y = ax^2 + bx + c$ if $a \neq 0$ and $b^2 - 4ac > 0$?

Answer: There are two real zeros.

91. From Section 5.5: Find the horizontal asymptote(s) of the rational function $y = \dfrac{2x + 1}{x - 1}$.

Answer: $y = 2$

6C.2 The Unit Circle

> The unit circle can be used to present the trigonometric functions.

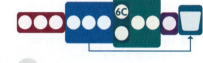

6C.1 Angles

6C.2 The Unit Circle

6C.3 The Trigonometric Functions

6C.4 Right Triangle Trigonometry

In this section, you will learn to:

1. Locate a point on the unit circle given one coordinate.
2. Find the coordinates of a trigonometric point.
3. Locate the trigonometric points for special angles.
4. Calculate the reference number for a given trigonometric point.
5. Use the reference number and quadrant to find the coordinates of a given trigonometric point.

The connection between circles and the number π represents a fundamental relationship between geometry and algebra. You will recognize the formula for the circumference of a circle:

$$\text{Circumference} = \pi \times \text{Diameter}$$

This formula is more striking if we divide by the diameter to rearrange it:

$$\frac{\text{Circumference}}{\text{Diameter}} = \pi$$

In this form, the formula tells us that no matter which circle we look at, from the equator of Earth to a coin-sized circle (**Figure 6C.30**), if we divide the circumference by the diameter, the answer is always the same: the number π.

filo/Getty Images

SimpleFoto/Deposit Photos

Figure 6C.30 Circumference divided by diameter: The ratio is the same for a dime or the equator of Earth.

The importance of π was recognized by ancient Egyptian and Babylonian mathematicians, who knew reasonably good approximations of its value. Today, supercomputers using very clever calculation techniques have produced approximations of π accurate to trillions of digits—perhaps a few more than are actually needed for scientific applications. Such calculations are used as benchmarks for testing computer speed.

Circles serve an immediate purpose. The trigonometric functions are defined in terms of coordinates of points on the circle of radius 1 centered at the origin, the unit circle.

Finding Points on the Unit Circle

> If one coordinate of a point on the unit circle is given, the other coordinate can be found using the equation of the circle.

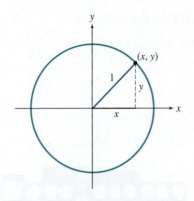

DF **Figure 6C.31** The equation for the unit circle: $x^2 + y^2 = 1$

The unit circle shown in **Figure 6C.31** is the circle in the plane that is centered at the origin and has radius 1. It consists of points 1 unit distance from the origin, and its equation is $x^2 + y^2 = 1$.

EXAMPLE 6C.9 Trigonometric Points: Points on the Unit Circle

Find the possible values of y if the point $\left(\dfrac{1}{3}, y\right)$ lies on the unit circle.

SOLUTION

Because $\left(\dfrac{1}{3}, y\right)$ lies on the unit circle, the coordinates must satisfy the equation $x^2 + y^2 = 1$, so

$$\left(\frac{1}{3}\right)^2 + y^2 = 1 \qquad \text{◄ Use the equation of unit circle.}$$

$$\frac{1}{9} + y^2 = 1 \qquad \text{◄ Simplify.}$$

$$y^2 = \frac{8}{9} \qquad \text{◄ Solve for } y^2.$$

$$y = \pm\frac{\sqrt{8}}{3} \qquad \text{◄ Take square roots.}$$

Thus, there are two solutions,

$$y = \frac{\sqrt{8}}{3} = \frac{2\sqrt{2}}{3}$$

and

$$y = -\frac{\sqrt{8}}{3} = -\frac{2\sqrt{2}}{3}$$

These points are shown on the unit circle in **Figure 6C.32**.

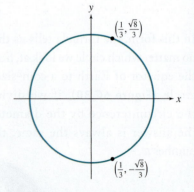

Figure 6C.32 Points on the unit circle

TRY IT YOURSELF 6C.9 Brief answers provided at the end of the section.

Find all possible values of x if $\left(x, \dfrac{2}{5}\right)$ lies on the unit circle.

Let P denote the point $\left(\dfrac{1}{3}, y\right)$ in the preceding example.

a. How will your calculation in this example be different if P lies on the circle of radius 2 centered at the origin rather than the unit circle? Do the required calculation.

b. Do the calculation in this example if P lies on the circle of radius 3 centered at the point $(1, 0)$.

c. Is it possible for P to lie on the circle of radius 1 centered at $(10, 10)$? *Suggestion*: Draw a picture.

Let t be a real number that, for the moment, we assume to be between 0 and 2π. We begin at $(1, 0)$ and move in the counterclockwise direction around the unit circle a distance t, as shown in **Figure 6C.33**. We refer to the resulting point $P(t)$ as the **trigonometric point** associated with t. We extend this idea so that it applies to any real number (**Figure 6C.34**).

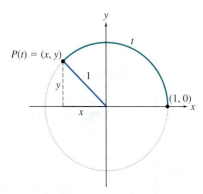

Figure 6C.33 Trigonometric point: An arc length of $t > 0$ on the unit circle makes an angle of t radians. If the angle is in standard position, the terminal side meets the circle at the point $P(t)$.

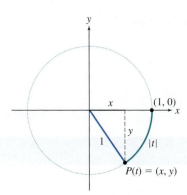

Figure 6C.34 $P(t)$ for t negative

By the definition of radian measure, an angle of $t > 0$ radians with its vertex at the center cuts an arc of length t on the unit circle. Hence, we can think of $P(t)$ equally as depending on the real number t or on an angle of radian measure t. If the angle is measured in degrees, this will be specifically indicated.

The **trigonometric point** $P(t)$ associated with the number t is defined as follows: If $t \geq 0$, we begin at the point $(1, 0)$ and move in the counterclockwise direction around the unit circle a distance t to the point $P(t)$. If t is negative, nothing changes except that we start at $(1, 0)$ and move in the clockwise direction.

Compass Points on the Unit Circle

> The coordinates of the trigonometric points corresponding to the four compass points on the unit circle are calculated.

EXAMPLE 6C.10 Trigonometric Points and Compass Points

The compass points on the unit circle are the points $(1, 0)$, $(0, 1)$, $(-1, 0)$, and $(0, -1)$.

a. Find the coordinates of the trigonometric point associated with an angle of $\dfrac{\pi}{2}$ radians.

b. Find the coordinates of the trigonometric point associated with the real number $-\dfrac{\pi}{2}$.

c. Find the coordinates of the trigonometric point associated with an angle of $270°$.

SOLUTION

a. It is important to remember that the trigonometric point associated with the angle of radian measure t is the same as the trigonometric point associated with the real number t. An arc length of $t = \dfrac{\pi}{2}$ is one-quarter of the circumference of the circle. Hence, the coordinates of the trigonometric point $P\left(\dfrac{\pi}{2}\right)$ are $(x, y) = (0, 1)$. This point is shown on the unit circle in **Figure 6C.35**.

b. Because $-\dfrac{\pi}{2}$ is negative, we begin at $(1, 0)$ and move $\dfrac{\pi}{2}$ units in a clockwise direction. Because an arc of length $\dfrac{\pi}{2}$ is one-quarter of the circumference, we land at the trigonometric point $P\left(-\dfrac{\pi}{2}\right) = (0, -1)$.

This point is shown on the unit circle in **Figure 6C.36**.

c. An angle of $270°$ covers three-quarters of the circle, so $P(270°) = (0, -1)$. Note that this is the same as the point we found in part b.

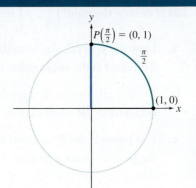

Figure 6C.35 Locating $P\left(\dfrac{\pi}{2}\right)$

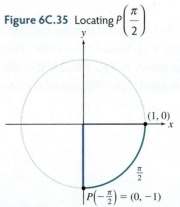

Figure 6C.36 Locating $P\left(-\dfrac{\pi}{2}\right)$

TRY IT YOURSELF 6C.10 Brief answers provided at the end of the section.

Find $P(\pi)$.

EXTEND YOUR REACH   If $P(s) = (x, y)$ and $P(t) = (z, w)$, do you think it is true that $P(s + t) = (x + z, y + w)$? You might try some specific values of s and t to answer this question. The trigonometric points calculated in this example may prove useful.

We summarize the coordinates of the trigonometric points associated with compass points in **Table 6C.1**. Many of the entries in this table come from Example 6C.10. You are asked to verify the remaining entries in the exercises.

Table 6C.1 Trigonometric Points Associated with Compass Points

Degrees	Radians	Associated figure P(t)	Trigonometric point
0	0		$P(0) = (1, 0)$
		$P(0) = (1, 0)$ 0° or 0 radians	

Table 6C.1 Trigonometric Points Associated with Compass Points (*continued*)

Degrees	Radians	Associated figure $P(t)$	Trigonometric point
90	$\dfrac{\pi}{2}$	90° or $\frac{\pi}{2}$ radians	$P\left(\dfrac{\pi}{2}\right) = (0, 1)$
180	π	180° or π radians	$P(\pi) = (-1, 0)$
270	$\dfrac{3\pi}{2}$	270° or $\frac{3\pi}{2}$ radians	$P\left(\dfrac{3\pi}{2}\right) = (0, -1)$
360	2π	360° or 2π radians	$P(2\pi) = (1, 0)$

Special Angles

> Geometric ideas are used to find coordinates of trigonometric points corresponding to certain special angles.

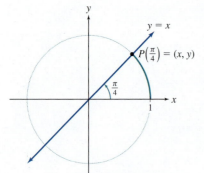

Figure 6C.37 The special angle $\dfrac{\pi}{4}$ radian: $P\left(\dfrac{\pi}{4}\right)$ lies on the line $y = x$.

Three angles between 0 and $\dfrac{\pi}{2}$ radians are particularly important: $\dfrac{\pi}{6}$ radian or $30°$, $\dfrac{\pi}{4}$ radian or $45°$, and $\dfrac{\pi}{3}$ radian or $60°$. These are called special angles because the coordinates of the corresponding trigonometric points can be calculated exactly using basic geometry. These angles appear often in mathematics and its applications.

We start by calculating $P\left(\dfrac{\pi}{4}\right)$. The line $y = x$ makes an angle of $45°$ or $\dfrac{\pi}{4}$ radian with the positive x-axis. Hence, the trigonometric point $P\left(\dfrac{\pi}{4}\right)$ lies on this line, as shown in **Figure 6C.37**.

If we put $P\left(\dfrac{\pi}{4}\right) = (x, y)$, then

$$x^2 + y^2 = 1 \qquad \blacktriangleleft \text{ Use: } (x, y) \text{ lies on the unit circle.}$$

$$x = y \qquad \blacktriangleleft \text{ Use: } (x, y) \text{ lies on the line } y = x.$$

$$x^2 + x^2 = 1 \qquad \blacktriangleleft \text{ Substitute } x \text{ for } y.$$

$$2x^2 = 1 \qquad \blacktriangleleft \text{ Simplify.}$$

$$x^2 = \frac{1}{2} \qquad \blacktriangleleft \text{ Solve for } x^2.$$

$$x = \pm\frac{1}{\sqrt{2}} \qquad \text{ Take square roots.}$$

Because we know x is positive, we conclude that

$$P\left(\frac{\pi}{4}\right) = \left(\frac{1}{\sqrt{2}}, \frac{1}{\sqrt{2}}\right)$$

To study angles of $30°$ and $60°$, we begin with the equilateral triangle shown in **Figure 6C.38** and bisect one of the $60°$ angles. The angle bisector cuts the triangle in half and produces a $30°$–$60°$–$90°$ triangle, as shown in **Figure 6C.39**.

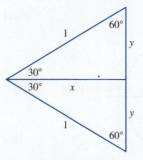

Figure 6C.38 An angle of an equilateral triangle bisected

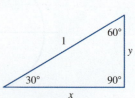

Figure 6C.39 A $30°$–$60°$–$90°$ triangle

EXAMPLE 6C.11   Finding Trigonometric Points from the 30°–60°–90° Triangle

a. Use the fact that the angle bisector in Figure 6C.38 bisects the opposite side of the triangle to find the value of y.

b. Use the Pythagorean theorem to calculate the value of x in Figure 6C.39.

c. Find the coordinates of $P\left(\dfrac{\pi}{6}\right)$.

SOLUTION

a. The bisector cuts the side of length 1 in half, so $y = \dfrac{1}{2}$.

b. We use the fact that $y = \dfrac{1}{2}$:

$$x^2 + y^2 = 1 \qquad \blacktriangleleft \text{ Use the equation of circle.}$$

$$x^2 + \left(\frac{1}{2}\right)^2 = 1 \qquad \blacktriangleleft \text{ Substitute known values.}$$

$$x^2 + \frac{1}{4} = 1 \qquad \blacktriangleleft \text{ Simplify.}$$

$$x^2 = \frac{3}{4} \qquad \blacktriangleleft \text{ Solve for } x^2.$$

$$x = \pm\frac{\sqrt{3}}{2} \qquad \blacktriangleleft \text{ Take square roots.}$$

Because x is positive, we have $x = \dfrac{\sqrt{3}}{2}$.

c. Because an angle of 30° has radian measure $\dfrac{\pi}{6}$, the trigonometric point $P\left(\dfrac{\pi}{6}\right)$ is a vertex of a 30°–60°–90° triangle, as shown in **Figure 6C.40**. From parts a and b, we know that $x = \dfrac{\sqrt{3}}{2}$ and $y = \dfrac{1}{2}$. We conclude that

$$P\left(\frac{\pi}{6}\right) = \left(\frac{\sqrt{3}}{2}, \frac{1}{2}\right)$$

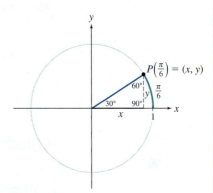

Figure 6C.40 The special angle $\dfrac{\pi}{6}$ radian: $P\left(\dfrac{\pi}{6}\right)$ is a vertex of a 30°–60°–90° triangle.

TRY IT YOURSELF 6C.11   Brief answers provided at the end of the section.

Use the information provided in **Figure 6C.41** to find the coordinates of the trigonometric point $P\left(\dfrac{\pi}{3}\right)$.

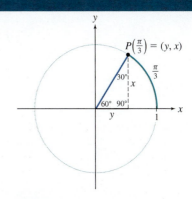

Figure 6C.41 The 30°–60°–90° triangle rotated to make the special angle $\dfrac{\pi}{3}$ radians

EXTEND YOUR REACH   It is true that

$$P\left(\frac{\pi}{8}\right) = \left(\frac{\sqrt{2+\sqrt{2}}}{2}, \frac{\sqrt{2-\sqrt{2}}}{2}\right)$$

Figure 6C.42 A triangle that may be rotated to find $P\left(\frac{3\pi}{8}\right)$

Rotate a triangle as we did in Example 6C.11 to find $P\left(\frac{3\pi}{8}\right)$. The triangle in **Figure 6C.42** may prove helpful.

We summarize what we have learned about trigonometric points of special angles in the following **Table 6C.2** and **Figure 6C.43**.

Table 6C.2 Trigonometric Points for Special Angles

Degrees	Radians	Trigonometric point
$t = 30°$	$t = \dfrac{\pi}{6}$	$P(t) = \left(\dfrac{\sqrt{3}}{2}, \dfrac{1}{2}\right)$
$t = 45°$	$t = \dfrac{\pi}{4}$	$P(t) = \left(\dfrac{1}{\sqrt{2}}, \dfrac{1}{\sqrt{2}}\right)$
$t = 60°$	$t = \dfrac{\pi}{3}$	$P(t) = \left(\dfrac{1}{2}, \dfrac{\sqrt{3}}{2}\right)$

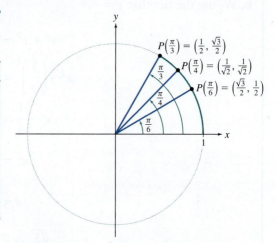

Figure 6C.43 Visual display of special angles

Quadrants and Reference Numbers

> Trigonometric points are associated with angles in the first quadrant.

The coordinate axes divide the plane into four quadrants, usually labeled I, II, III, and IV, as in **Figure 6C.44**. The signs of the coordinates of trigonometric points are determined by the quadrant in which they lie.

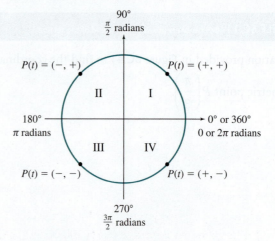

Figure 6C.44 Trigonometric points: The signs of the coordinates are different in each of the four quadrants.

We can get information about trigonometric points in Quadrants II, III, and IV from certain trigonometric points in Quadrant I, the first quadrant. We use $\frac{2\pi}{3}$ to see how to do this. A length of π is half the circumference of the unit circle, and $\frac{2\pi}{3}$ is two-thirds of that distance. Hence, $P\left(\frac{2\pi}{3}\right)$ lies in the second quadrant, as shown in **Figure 6C.45**. Observe that the x- and y-coordinates of this point are the same in absolute value as those of the trigonometric point $P\left(\frac{\pi}{3}\right)$, as shown in **Figure 6C.46**. In fact, $\frac{\pi}{3}$ is the shortest arc length from $P\left(\frac{2\pi}{3}\right)$ to the x-axis.

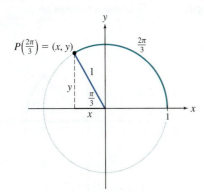

Figure 6C.45 The trigonometric point $P\left(\dfrac{2\pi}{3}\right)$

Figure 6C.46 The trigonometric point $P\left(\dfrac{\pi}{3}\right)$: The absolute values of the coordinates of $P\left(\dfrac{2\pi}{3}\right)$ are the coordinates of $P\left(\dfrac{\pi}{3}\right)$.

Because $\frac{\pi}{3}$ is a special angle, we know the coordinates of the corresponding trigonometric point:

$$P\left(\frac{\pi}{3}\right) = \left(\frac{1}{2}, \frac{\sqrt{3}}{2}\right)$$

We need only attach the proper signs for Quadrant II to find $P\left(\dfrac{2\pi}{3}\right)$:

$$P\left(\frac{2\pi}{3}\right) = \left(-\frac{1}{2}, \frac{\sqrt{3}}{2}\right)$$

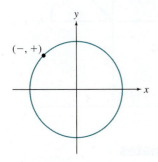

This discussion suggests a general method for finding the coordinates of the trigonometric point $P(t)$. The first step is to find the **reference number** by locating the appropriate arc in the first quadrant.

The **reference number** r for the trigonometric point $P(t)$ is the shortest arc length from $P(t)$ to the x-axis.

For example, the reference number for $P\left(\dfrac{2\pi}{3}\right)$ is $\dfrac{\pi}{3}$. A reference number of r corresponds to a reference angle of r radians. The reference angle is always between 0 and $\dfrac{\pi}{2}$ radians (0° to 90°).

The procedure for finding trigonometric points is simplified by using **Table 6C.3** in determining the reference number r. Proper signs attached to the coordinates of $P(r)$ give the coordinates of the trigonometric point $P(t)$.

Table 6C.3 A Graphical View of Reference Numbers

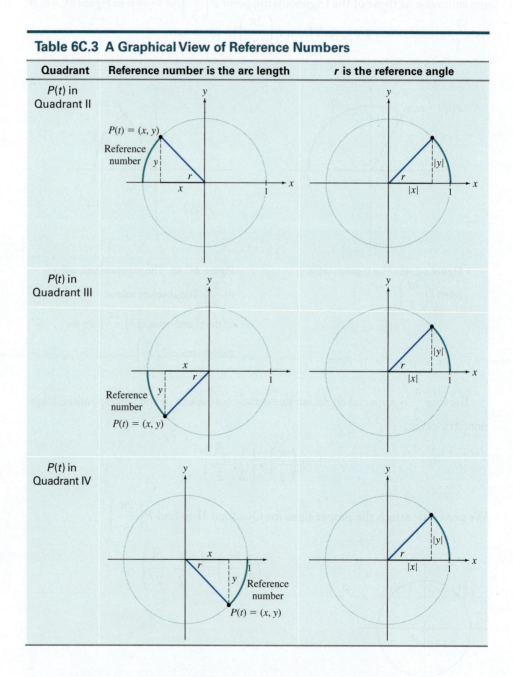

Quadrant	Reference number is the arc length	r is the reference angle
$P(t)$ in Quadrant II		
$P(t)$ in Quadrant III		
$P(t)$ in Quadrant IV		

Using Reference Numbers to Find Coordinates of Trigonometric Points

Trigonometric points in the first quadrant are used to find trigonometric points anywhere on the unit circle.

EXAMPLE 6C.12 Finding Reference Numbers

Find the reference number for $P\left(\dfrac{3\pi}{4}\right)$.

SOLUTION

An arc of length π is halfway around the circle, and $\dfrac{3\pi}{4}$ is three-quarters of that distance. The shortest arc back to the x-axis is $\pi - \dfrac{3\pi}{4} = \dfrac{\pi}{4}$. Thus, the reference number for $P\left(\dfrac{3\pi}{4}\right)$ is $\dfrac{\pi}{4}$. This is shown as the shortest arc back to the x-axis in **Figure 6C.47**.

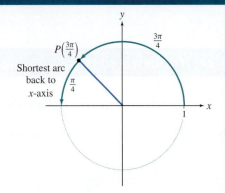

DF Figure 6C.47 Finding reference numbers: The reference number for $P\left(\dfrac{3\pi}{4}\right)$ is $\dfrac{\pi}{4}$.

TRY IT YOURSELF 6C.12 Brief answers provided at the end of the section.

Find the reference number for $P\left(\dfrac{5\pi}{6}\right)$.

EXAMPLE 6C.13 Finding and Using Reference Angles and Numbers

a. Find the degree measure of the reference angle for $P(290°)$.

b. Find the coordinates of the trigonometric point associated with $\dfrac{7\pi}{6}$.

SOLUTION

a. The trigonometric point $P(290°)$ lies in the fourth quadrant. The shortest arc leading back to the x-axis cuts an angle of $360° - 290° = 70°$, as shown in **Figure 6C.48**, so the reference angle has degree measure $70°$.

b. The trigonometric point $P\left(\dfrac{7\pi}{6}\right)$ lies in the third quadrant, as shown in **Figure 6C.49**. The shortest arc back to the x-axis is $\dfrac{\pi}{6}$, and this is the reference number. We know that $P\left(\dfrac{\pi}{6}\right) = \left(\dfrac{\sqrt{3}}{2}, \dfrac{1}{2}\right)$. Now we attach the signs.

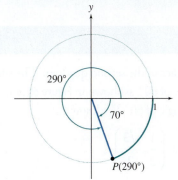

Figure 6C.48 The reference angle for $P(290°)$: $70°$

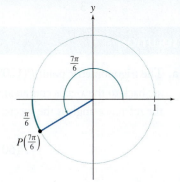

Figure 6C.49 Reference number for $P\left(\dfrac{7\pi}{6}\right)$: This point lies in the third quadrant and has reference number $\dfrac{\pi}{6}$.

In the third quadrant, both coordinates are negative, so

$$P\left(\frac{7\pi}{6}\right) = \left(-\frac{\sqrt{3}}{2}, -\frac{1}{2}\right)$$

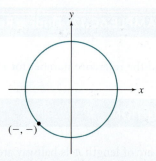

$(-, -)$

TRY IT YOURSELF 6C.13 Brief answers provided at the end of the section.

Find the reference number and coordinates of $P\left(\dfrac{3\pi}{4}\right)$.

EXTEND YOUR REACH

a. Find the reference number for $P\left(\dfrac{\pi}{6} + \pi\right)$.

b. Find the reference number for $P\left(\dfrac{\pi}{4} + \pi\right)$.

c. Find the reference number for $P\left(\dfrac{\pi}{3} + \pi\right)$.

d. If $P(t)$ lies in the first quadrant, find the reference number for $P(t + \pi)$.

EXAMPLE 6C.14 Using Reference Angles

Find the coordinates of the trigonometric points associated with the following.

a. An angle of $120°$

b. An angle of $\dfrac{11\pi}{3}$ radians

SOLUTION

a. The trigonometric point $P(120°)$ lies in the second quadrant. The shortest arc back to the x-axis cuts an angle of $60°$, as shown in **Figure 6C.50**, so the reference angle is the special angle of $60°$. We know that

$$P(60°) = \left(\frac{1}{2}, \frac{\sqrt{3}}{2}\right)$$

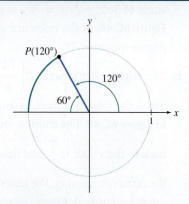

Figure 6C.50 Reference angle: The reference angle for $120°$ has degree measure $60°$.

Now we attach the signs. In the second quadrant, the first coordinate is negative and the second is positive, so

$$P(120°) = \left(-\frac{1}{2}, \frac{\sqrt{3}}{2} \right)$$

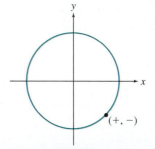

b. Now $\frac{11\pi}{3} = 2\pi + \frac{5\pi}{3}$. This is a full trip around the circle followed by an arc of length $\frac{5\pi}{3}$, so the angle lies in the fourth quadrant. The reference number is the same as in part b, namely 60° or $\frac{\pi}{3}$. We need to attach the signs.

In the fourth quadrant, the first coordinate is positive and the second is negative, so

$$P\left(\frac{11\pi}{3} \right) = \left(\frac{1}{2}, -\frac{\sqrt{3}}{2} \right)$$

TRY IT YOURSELF 6C.14 Brief answers provided at the end of the section.

Find the coordinates of $P\left(\frac{5\pi}{4} \right)$.

EXAMPLE 6C.15 Using Reference Numbers for Negative Angles

Find the coordinates of the trigonometric points associated with the following.

a. An angle of −330°

b. An angle of $-\frac{5\pi}{2}$ radians

SOLUTION

a. The trigonometric point $P(-330°)$ lies in the first quadrant, as shown in **Figure 6C.51**. The shortest arc back to the x-axis cuts an angle of 30°. We know that

$$P(30°) = \left(\frac{\sqrt{3}}{2}, \frac{1}{2} \right)$$

In the first quadrant, both coordinates are positive, so

$$P(-330°) = \left(\frac{\sqrt{3}}{2}, \frac{1}{2} \right)$$

b. Because $-\frac{5\pi}{2} = -2\pi - \frac{\pi}{2}$, the angle represented corresponds to a full trip clockwise around the circle followed by another one-quarter trip, so the trigonometric point has coordinates $(0, -1)$. See **Figure 6C.52**.

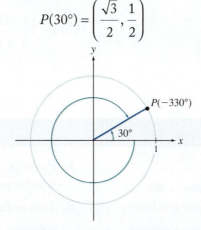

Figure 6C.51 Reference angle for $P(-330°)$: This point lies in the first quadrant and has reference angle with degree measure 30°.

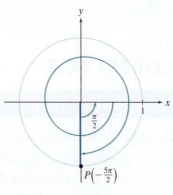

Figure 6C.52 $P\left(-\frac{5\pi}{2} \right) = (0, -1)$

Find the coordinates of $P\left(-\dfrac{11\pi}{4}\right)$.

Reference numbers are used to locate the trigonometric points shown in **Figure 6C.53**. This figure is helpful for future reference.

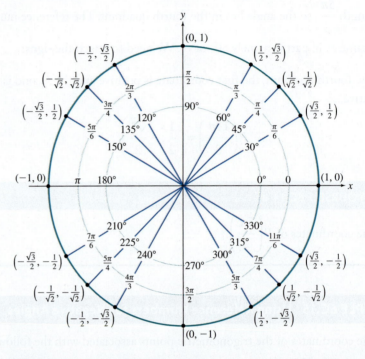

DF **Figure 6C.53** Trigonometric points corresponding to special angles

TRY IT YOURSELF ANSWERS

6C.9 $x = \pm\sqrt{21}/5$

6C.10 $P(\pi) = (-1, 0)$

6C.11 $P\left(\dfrac{\pi}{3}\right) = \left(\dfrac{1}{2}, \dfrac{\sqrt{3}}{2}\right)$

6C.12 $\dfrac{\pi}{6}$

6C.13 $\dfrac{\pi}{4}; P\left(\dfrac{3\pi}{4}\right) = \left(-\dfrac{1}{\sqrt{2}}, \dfrac{1}{\sqrt{2}}\right)$

6C.14 $P\left(\dfrac{5\pi}{4}\right) = \left(-\dfrac{1}{\sqrt{2}}, -\dfrac{1}{\sqrt{2}}\right)$

6C.15 $P\left(-\dfrac{11\pi}{4}\right) = \left(-\dfrac{1}{\sqrt{2}}, -\dfrac{1}{\sqrt{2}}\right)$

EXERCISE SET 6C.2

CHECK YOUR UNDERSTANDING

1. The radius of the unit circle is _____.

2. List the signs of each of the coordinates of the trigonometric points in each of the four quadrants.

3. In which quadrant are both coordinates negative?

4. In which quadrant is the first coordinate positive and the second negative?

5. Points (x, y) on the unit circle satisfy which equation?

6. **True or false:** For any trigonometric point, the reference angle is between 0 and $\dfrac{\pi}{2}$ radians.

7. For a real number t, the trigonometric point $P(t)$ is the same as _____.

   a. $P(t)$ for an angle of $t°$

   b. $P(t)$ for an angle of t radians

   c. a reference angle

   d. None of the above.

8. **True or false:** Reference numbers help us calculate the coordinates of trigonometric points.

9. The reference number for the trigonometric point $P(t)$ is _____.

10. **True or false:** The reference number for the trigonometric point $P(t)$ is the same as the radian measure of the reference angle.

SKILL BUILDING

Points on the unit circle. In Exercises 11 through 21, find the given point on the unit circle.

11. Find all possible values of x if $\left(x, \dfrac{1}{2}\right)$ lies on the unit circle.

12. Find all possible values of x if $\left(x, -\dfrac{2}{3}\right)$ lies on the unit circle.

13. Find all possible values of x if $\left(x, \dfrac{3}{4}\right)$ lies on the unit circle.

14. Find all possible values of x if $\left(x, \dfrac{\sqrt{3}}{2}\right)$ lies on the unit circle.

15. Find all possible values of x if $\left(x, -\dfrac{1}{\sqrt{2}}\right)$ lies on the unit circle.

16. Find all possible values of y if $\left(\dfrac{1}{4}, y\right)$ lies on the unit circle.

17. Find all possible values of y if $\left(\dfrac{\sqrt{2}}{3}, y\right)$ lies on the unit circle.

18. Find all possible values of y if $\left(-\dfrac{\pi}{4}, y\right)$ lies on the unit circle.

19. Find all possible values of y if $\left(\dfrac{5}{6}, y\right)$ lies on the unit circle.

20. Find all possible values of y if $\left(-\dfrac{1}{3}, y\right)$ lies on the unit circle.

21. Find all possible values of x if $\left(x, -\dfrac{2}{7}\right)$ lies on the unit circle and the point lies in the third quadrant.

Identifying trigonometric points. In **Figure 6C.54**, trigonometric points corresponding to compass points and special angles are marked. In Exercises 22 through 29, match the given angle with the appropriate point labeled in the figure.

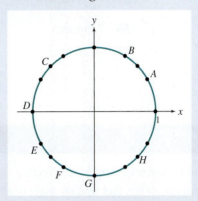

Figure 6C.54 Some trigonometric points

22. $\dfrac{3\pi}{4}$ radians

23. $30°$

24. $315°$

25. π radians

26. $\dfrac{\pi}{3}$ radians

27. $210°$

28. $240°$

29. $\dfrac{3\pi}{2}$ radians

Reference angles with degrees. In Exercises 30 through 43, find the reference angle in degrees associated with the given angle.

30. $123°$

31. $234°$

32. $284°$

33. $328°$

34. $-123°$

35. $777°$

36. $-20°$

37. $-100°$

38. $30°$

39. $-370°$

40. $377°$

41. $1236°$

42. $-277°$

43. $634°$

Reference angles with radians. In Exercises 44 through 57, find the reference angle in radians associated with the given angle.

44. $\dfrac{5\pi}{7}$ radians

45. $\dfrac{12\pi}{5}$ radians

46. $\dfrac{11\pi}{9}$ radians

47. $\dfrac{2\pi}{3}$ radians

48. $-\dfrac{2\pi}{3}$ radians

49. $-\dfrac{\pi}{6}$ radian

50. $\dfrac{5\pi}{4}$ radians

51. $-\dfrac{5\pi}{6}$ radians

52. $\dfrac{5\pi}{3}$ radians

53. $\dfrac{29\pi}{7}$ radians

54. $-\dfrac{3\pi}{7}$ radians

55. $-\dfrac{5\pi}{7}$ radians

56. 7 radians

57. −4 radians

Locating compass points. In Exercises 58 through 68, find the coordinates of the trigonometric point.

58. $P(90°)$

59. $P(0°)$

60. $P(180°)$

61. $P(270°)$

62. $P(360°)$

63. $P\left(\dfrac{\pi}{2}\right)$

64. $P(\pi)$

65. $P\left(\dfrac{3\pi}{2}\right)$

66. $P(2\pi)$

67. $P(720°)$

68. $P\left(\dfrac{-3\pi}{2}\right)$

PROBLEMS

69. **Arc length and radian measure.** In this section we have often made use of the fact that an angle of t radians whose vertex lies at the center of the unit circle cuts an arc of length t. Use the formula for arc length to verify this fact.

Points on the unit circle. In Exercises 70 through 75, find all values of x so that the given point lies on the unit circle.

70. (x, x)

71. $(x, 2x)$

72. $(x, -x)$

73. $(x, \sqrt{x})$

74. $\left(x, \dfrac{1}{x}\right)$

75. $(3x, x)$ lies on the unit circle in the third quadrant.

Calculating trigonometric points for special angles. In Exercises 76 through 94, find the coordinates of the trigonometric point associated with the given number or angle. In each case the appropriate reference number or angle is one of the special values.

76. 135°

77. $\dfrac{5\pi}{4}$ radians

78. $-\dfrac{\pi}{3}$ radians

79. $-\dfrac{\pi}{4}$ radian

80. −135°

81. 210°

82. π radians

83. 315°

84. $-\dfrac{\pi}{6}$ radian

85. 120°

86. $\dfrac{7\pi}{6}$ radians

87. 300°

88. $\dfrac{2\pi}{3}$ radians

89. 240°

90. $\dfrac{5\pi}{3}$ radians

91. $\dfrac{13\pi}{6}$ radians

92. The real number $\dfrac{11\pi}{6}$

93. The real number $\dfrac{25\pi}{4}$

94. The real number $-\dfrac{2\pi}{3}$

95. **Periodicity.** Let t be a real number such that

$$P(t) = \left(\dfrac{2}{5}, \dfrac{3}{5}\right).$$

a. Find $P(t + 2\pi)$.

b. Find $P(t + 4\pi)$.

c. Find $P(t + 2k\pi)$ for any integer k.

CHALLENGE EXERCISES FOR INDIVIDUALS OR GROUPS

96. Adding π. If $P(t) = (x, y)$, find the coordinates of $P(t + \pi)$ in terms of x and y. *Suggestion*: You may find **Figure 6C.55** helpful.

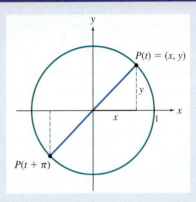

Figure 6C.55 $P(t)$ and $P(t + \pi)$

REVIEW AND REFRESH: Exercises from Previous Sections

97. From Section P.2: Solve the inequality $|3x - 4| < 10$.

Answer: $\left(-2, \dfrac{14}{3}\right)$

98. From Section 2.3: The graph of $f(x)$ is given in **Figure 6C.56**. Add to the picture the graph of $g(x) = f(x+1) - 1$.

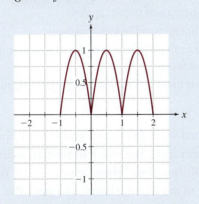

Figure 6C.56 Graph for Exercise 98

Answer:

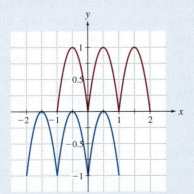

99. From Section 3.1: Find the equation of the linear function f with slope -2 if $f(3) = 4$.

Answer: $f(x) = -2x + 10$

100. From Section 3.2: Find an exponential function $f(x)$ so that $f(0) = 3$ and f is multiplied by 5 when x is increased by 1.

Answer: $f(x) = 3(5^x)$

101. From Section 4.1: If $f(x) = \ln(x + 1)$, find $f(e^x - 1)$.

Answer: x

102. From Section 4.2: Simplify the expression $e^{2\ln x}$.

Answer: x^2

103. From Section 4.3: Solve the equation $\ln(x + 4) = 7$.

Answer: $x = e^7 - 4$

104. From Section 5.1: What can you conclude about the zeros of the quadratic function $y = ax^2 + bx + c$ if the discriminant $b^2 - 4ac$ is negative?

Answer: There are no real zeros.

105. From Section 5.4: Find the zeros of $f(x) = x^3 - 3x^2 + 2x$, and then plot its graph.

Answer: Zeros: 0, 1, 2

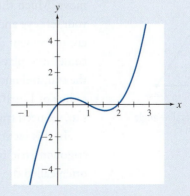

106. From Section 5.5: Find the horizontal and vertical asymptotes of $f(x) = \dfrac{2x^2}{x^2 - 1}$. Then plot the graph, and show the asymptotes.

Answer: Horizontal asymptote $y = 2$. Vertical asymptotes $x = -1$ and $x = 1$.

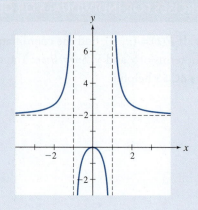

6C.1 Angles

6C.2 The Unit Circle

6C.3 The Trigonometric Functions

6C.4 Right Triangle Trigonometry

6C.3 The Trigonometric Functions

> Trigonometric points on the unit circle determine the trigonometric functions.

In this section, you will learn to:

1. Use the definitions of the six trigonometric functions to find function values.
2. Recite the values of the trigonometric functions of special angles.
3. Produce the signs of the trigonometric functions in the four quadrants.
4. Use the reference number and quadrant to find the values of the trigonometric functions in terms of the values in the first quadrant.
5. Use the Pythagorean identity and related identities to find a trigonometric function value from a known value.

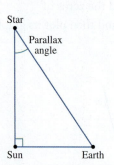

Figure 6C.57 The parallax angle

Many astronomical facts can be gleaned from persistent and careful observation, and early humans were masters of observation. But much of modern astronomy depends on the answer to a seemingly simple question: "How far away is it?" You can look online today to find accurate distances to the moon, planets, and stars. But where do those numbers come from? How does one figure out how far away the lights in the sky are? A key step in finding distances to stars uses the notion of a parallax angle. As Earth moves from one extreme of its orbit to another, relatively close stars appear to move. Much the same thing will happen if you view an object 5 feet away first with your left eye and then with your right eye. Half of the angle determined by the apparent movement of a star is the parallax angle. The right triangle shown in **Figure 6C.57** comes into play here. One side is the radius of Earth's orbit, one side is the distance to the star, and one angle is the parallax angle. We can measure the parallax angle and the radius of Earth's orbit. To find the unknown side of the triangle, the distance to the star, we need trigonometry.

In this section, we will use our understanding of trigonometric points to define the trigonometric functions. Because these functions arise from the unit circle, they are often called the circular functions.

Defining the Trigonometric Functions

> The coordinates of trigonometric points are used to define the six trigonometric functions.

The first coordinate of the trigonometric point $P(t)$ is defined to be the **cosine** function at t, and the second coordinate is defined to be the **sine** function at t. See **Figure 6C.58**.

There are four other trigonometric functions, the **tangent**, **cotangent**, **secant**, and **cosecant**, which we calculate in terms of ratios of the sine and cosine.

The **cosine** function at t, $\cos t$, is the first coordinate of the trigonometric point $P(t)$.

The **sine** function at t, $\sin t$, is the second coordinate of the trigonometric point $P(t)$.

The **tangent** function at t, $\tan t$, is defined as the ratio $\tan t = \dfrac{\sin t}{\cos t}$.

The **cotangent** function at t, $\cot t$, is defined as the ratio $\cot t = \dfrac{\cos t}{\sin t}$.

The **secant** function at t, $\sec t$, is defined as the ratio $\sec t = \dfrac{1}{\cos t}$.

The **cosecant** function at t, $\csc t$, is defined as the ratio $\csc t = \dfrac{1}{\sin t}$.

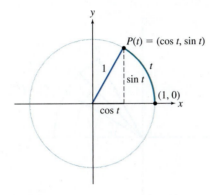

DF Figure 6C.58 $P(t) = (\cos t, \sin t)$

CONCEPTS TO REMEMBER: The Trigonometric Functions for Any Real Number or Any Angle

If the coordinates of the trigonometric point $P(t)$ are (x, y), the six trigonometric functions are calculated as follows:

sine: $\sin t = y$ cosine: $\cos t = x$

tangent: $\tan t = \dfrac{\sin t}{\cos t} = \dfrac{y}{x}$ cotangent: $\cot t = \dfrac{\cos t}{\sin t} = \dfrac{x}{y}$

secant: $\sec t = \dfrac{1}{\cos t} = \dfrac{1}{x}$ cosecant: $\csc t = \dfrac{1}{\sin t} = \dfrac{1}{y}$

The trigonometric functions evaluated at an angle of radian measure t are the same as the trigonometric functions evaluated at the real number t. If the angle is measured in degrees, this will be specifically indicated.

Trigonometric Functions of Special Angles

> Trigonometric points for special angles are used to evaluate corresponding trigonometric functions.

The coordinates of the trigonometric point $P(t)$ are $(\cos t, \sin t)$. Because we know the coordinates of the trigonometric points associated with the special angles $t = \dfrac{\pi}{6}$,

$t = \dfrac{\pi}{4}$, and $t = \dfrac{\pi}{3}$, we can evaluate the trigonometric functions at these points. See **Figure 6C.59**. For example, because $P\left(\dfrac{\pi}{6}\right) = \left(\dfrac{\sqrt{3}}{2}, \dfrac{1}{2}\right)$, we find

$$\cos\frac{\pi}{6} = \text{First coordinate of } P\left(\frac{\pi}{6}\right) = \frac{\sqrt{3}}{2}$$

$$\sin\frac{\pi}{6} = \text{Second coordinate of } P\left(\frac{\pi}{6}\right) = \frac{1}{2}$$

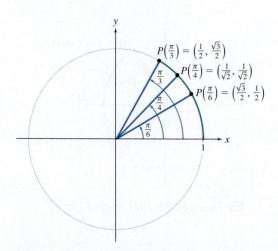

Figure 6C.59 Trigonometric points corresponding to special angles

We find the values of the other four trigonometric functions using the values of the sine and cosine:

$$\tan\left(\frac{\pi}{6}\right) = \frac{\sin\left(\dfrac{\pi}{6}\right)}{\cos\left(\dfrac{\pi}{6}\right)} = \frac{\dfrac{1}{2}}{\dfrac{\sqrt{3}}{2}} = \frac{1}{\sqrt{3}} \qquad \cot\left(\frac{\pi}{6}\right) = \frac{\cos\left(\dfrac{\pi}{6}\right)}{\sin\left(\dfrac{\pi}{6}\right)} = \frac{\dfrac{\sqrt{3}}{2}}{\dfrac{1}{2}} = \sqrt{3}$$

$$\sec\left(\frac{\pi}{6}\right) = \frac{1}{\cos\left(\dfrac{\pi}{6}\right)} = \frac{1}{\dfrac{\sqrt{3}}{2}} = \frac{2}{\sqrt{3}} \qquad \csc\left(\frac{\pi}{6}\right) = \frac{1}{\sin\left(\dfrac{\pi}{6}\right)} = \frac{1}{\dfrac{1}{2}} = 2$$

EXAMPLE 6C.16 Trigonometric Functions of Special Angles

Find the values of the six trigonometric functions of $60°$.

SOLUTION

We know the coordinates of the trigonometric point $P(60°)$ because $60°$ is one of the special angles:

$$P(60°) = \left(\frac{1}{2}, \frac{\sqrt{3}}{2}\right)$$

We use these coordinates to find the sine and cosine:

$$\cos 60° = \text{First coordinate of } P(60°) = \frac{1}{2}$$

$$\sin 60° = \text{Second coordinate of } P(60°) = \frac{\sqrt{3}}{2}$$

We use the sine and cosine to calculate the values of the remaining trigonometric functions:

$$\tan 60° = \frac{\sin 60°}{\cos 60°} = \frac{\frac{\sqrt{3}}{2}}{\frac{1}{2}} = \sqrt{3} \qquad \cot 60° = \frac{\cos 60°}{\sin 60°} = \frac{\frac{1}{2}}{\frac{\sqrt{3}}{2}} = \frac{1}{\sqrt{3}}$$

$$\sec 60° = \frac{1}{\cos 60°} = \frac{1}{\frac{1}{2}} = 2 \qquad \csc 60° = \frac{1}{\sin 60°} = \frac{1}{\frac{\sqrt{3}}{2}} = \frac{2}{\sqrt{3}}$$

TRY IT YOURSELF 6C.16 Brief answers provided at the end of the section.

Find the values of the six trigonometric functions of $\frac{\pi}{4}$.

EXTEND YOUR REACH

a. It is true that the trigonometric point $P\left(\frac{\pi}{12}\right)$ has coordinates

$$\left(\frac{\sqrt{6} + \sqrt{2}}{4}, \frac{\sqrt{6} - \sqrt{2}}{4}\right)$$

Find the values of the six trigonometric functions of $\frac{\pi}{12}$.

b. If $P(t)$ lies in the first quadrant, then its coordinates are $(x, \sqrt{1 - x^2})$ for some x between 0 and 1. Express in terms of x the values of the six trigonometric functions of t.

We summarize the sines and cosines of special angles in **Table 6C.4**. The values of the other four trigonometric functions can be easily obtained from these by using division or by taking inverses.

Table 6C.4 Special Angles

Angle in degrees	Angle in radians	sin	cos
30	$\frac{\pi}{6}$	$\frac{1}{2}$	$\frac{\sqrt{3}}{2}$
45	$\frac{\pi}{4}$	$\frac{1}{\sqrt{2}}$	$\frac{1}{\sqrt{2}}$
60	$\frac{\pi}{3}$	$\frac{\sqrt{3}}{2}$	$\frac{1}{2}$

Here is a different presentation of the same information that some will find useful as a memory aid because of the pattern of the numerators.

Angle	$\dfrac{\pi}{6}$	$\dfrac{\pi}{4}$	$\dfrac{\pi}{3}$
Sine	$\dfrac{\sqrt{1}}{2}$	$\dfrac{\sqrt{2}}{2}$	$\dfrac{\sqrt{3}}{2}$
Cosine	$\dfrac{\sqrt{3}}{2}$	$\dfrac{\sqrt{2}}{2}$	$\dfrac{\sqrt{1}}{2}$

Trigonometric Functions of Compass Points

> The trigonometric functions of the four compass points are evaluated using the coordinates that were calculated earlier.

EXAMPLE 6C.17 Trigonometric Functions of Compass Points

a. Find the values of the six trigonometric functions of $\dfrac{\pi}{2}$ radians. Note which functions are undefined.

b. Find the values of the six trigonometric functions of the real number $-\dfrac{\pi}{2}$. Note which functions are undefined.

c. Find the values of the six trigonometric functions of 270°.

SOLUTION

a. The trigonometric point $P\left(\dfrac{\pi}{2}\right)$ is $(x, y) = (0, 1)$ as shown in **Figure 6C.60**. We note first that the tangent and secant functions of $\dfrac{\pi}{2}$ are undefined because their definitions would involve division by 0. The values of the trigonometric functions are

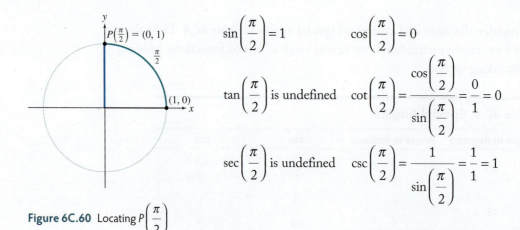

$$\sin\left(\frac{\pi}{2}\right) = 1 \qquad \cos\left(\frac{\pi}{2}\right) = 0$$

$$\tan\left(\frac{\pi}{2}\right) \text{ is undefined} \qquad \cot\left(\frac{\pi}{2}\right) = \frac{\cos\left(\dfrac{\pi}{2}\right)}{\sin\left(\dfrac{\pi}{2}\right)} = \frac{0}{1} = 0$$

$$\sec\left(\frac{\pi}{2}\right) \text{ is undefined} \qquad \csc\left(\frac{\pi}{2}\right) = \frac{1}{\sin\left(\dfrac{\pi}{2}\right)} = \frac{1}{1} = 1$$

Figure 6C.60 Locating $P\left(\dfrac{\pi}{2}\right)$

b. Now $P\left(-\dfrac{\pi}{2}\right) = (0, -1)$ as shown in **Figure 6C.61**. Once again, the tangent and secant functions are undefined because their definitions would involve division by zero. The values of the trigonometric functions are

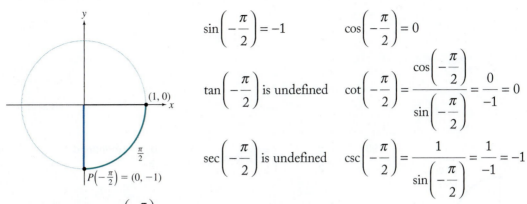

$$\sin\left(-\frac{\pi}{2}\right) = -1 \qquad \cos\left(-\frac{\pi}{2}\right) = 0$$

$$\tan\left(-\frac{\pi}{2}\right) \text{ is undefined} \quad \cot\left(-\frac{\pi}{2}\right) = \frac{\cos\left(-\dfrac{\pi}{2}\right)}{\sin\left(-\dfrac{\pi}{2}\right)} = \frac{0}{-1} = 0$$

$$\sec\left(-\frac{\pi}{2}\right) \text{ is undefined} \quad \csc\left(-\frac{\pi}{2}\right) = \frac{1}{\sin\left(-\dfrac{\pi}{2}\right)} = \frac{1}{-1} = -1$$

Figure 6C.61 Locating $P\left(-\dfrac{\pi}{2}\right)$

c. An angle of 270° covers three-quarters of the circle, so $P(270°) = (0, -1)$. Note that this is the same point that we found in part b. Hence, the trigonometric functions are the same:

$$\sin 270° = -1 \qquad \cos 270° = 0$$

$$\tan 270° \text{ is undefined} \quad \cot 270° = \frac{0}{-1} = 0$$

$$\sec 270° \text{ is undefined} \quad \csc 270° = \frac{1}{-1} = -1$$

TRY IT YOURSELF 6C.17 Brief answers provided at the end of the section.

Find the values of the six trigonometric functions of π. Note which functions are undefined.

EXTEND YOUR REACH

a. Show that the function $y = \sin t$ is increasing on the interval $\left[0, \dfrac{\pi}{2}\right]$.

Suggestion: As t moves from 0 to $\dfrac{\pi}{2}$, what happens to the second coordinate of $P(t)$?

b. Is the cosine function increasing or decreasing on the interval $\left[0, \dfrac{\pi}{2}\right]$?

c. Are there two different acute angles s and t such that $\sin s = \sin t$?

We summarize the values of the trigonometric functions at the compass points in **Table 6C.5**. Many of the entries in this table come from Example 6C.17. You are asked to verify the remaining entries in the exercises. It is not necessary to memorize the table of compass points because you should be able to calculate its contents easily.

We note that all trigonometric functions of a number t are defined except in certain cases when $P(t)$ is one of the compass points. The domains of the trigonometric functions are summarized as follows:

Table 6C.5 Values of Trigonometric Functions and Compass Points

Figure	Degrees	Radians	P(t)	Sine	Cosine	Tangent	Cotangent	Secant	Cosecant
	0	0	$P(0) = (1, 0)$	0	1	0	undefined	1	undefined
	90	$\dfrac{\pi}{2}$	$P\left(\dfrac{\pi}{2}\right) = (0, 1)$	1	0	undefined	0	undefined	1
	180	π	$P(\pi) = (-1, 0)$	0	−1	0	undefined	−1	undefined
	270	$\dfrac{3\pi}{2}$	$P\left(\dfrac{3\pi}{2}\right) = (0, -1)$	−1	0	undefined	0	undefined	−1
	360	2π	$P(2\pi) = (1, 0)$	0	1	0	undefined	1	undefined

CONCEPTS TO REMEMBER: Domains of Trigonometric Functions

Function	Domain
Sine and cosine	All real numbers
Tangent and secant	All real numbers except $\dfrac{\pi}{2}$ plus multiples of π
Cotangent and cosecant	All real numbers except multiples of π

Signs of Trigonometric Functions in the Four Quadrants

> Determining the sign of trigonometric functions is a key component of their evaluation.

The signs of the trigonometric functions are determined by the quadrant in which the trigonometric point $P(t) = (\cos t, \sin t)$ lies. Cosine, being the x-coordinate, is negative in Quadrants II and III, and sine, being the y-coordinate, is negative in Quadrants III and IV. These facts are shown in **Figure 6C.62**, which also shows a popular mnemonic device for remembering the signs. It associates the first letters of All Students Take Calculus with the first letters of the positive trigonometric functions All, Sine, Tangent, Cosine in each of the four quadrants.

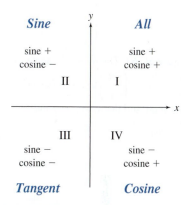

Figure 6C.62 Signs of sine and cosine in the four quadrants: *All Students Take Calculus* indicates the positive functions.

EXAMPLE 6C.18 Locating Quadrants

a. What is the sign of the cotangent in the second quadrant?

b. In what quadrant is sec t positive and cot t negative?

SOLUTION

a. In the second quadrant, the trigonometric point $P(t)$ has the form $(-, +)$. Hence, the cosine is negative and the sine is positive:

$$\cot t = \frac{\cos t}{\sin t} = \frac{-}{+} = -$$

We conclude that the cotangent is negative in the second quadrant.

b. Because $\sec t = \dfrac{1}{\cos t}$ is positive, $\cos t$ is also positive. This tells us we are in Quadrant I or IV.

Because cot t is negative, we must be in Quadrant IV.

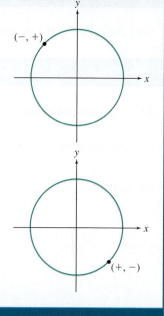

TRY IT YOURSELF 6C.18 Brief answers provided at the end of the section.

What is the sign of the cosecant in Quadrant III?

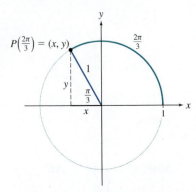

Figure 6C.63 The reference angle for $\frac{2\pi}{3}$ radians: $P\left(\frac{2\pi}{3}\right)$ lies in the second quadrant, and the reference angle is $\frac{\pi}{3}$ radian.

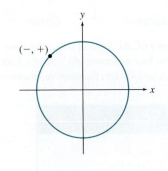

Using Reference Numbers to Evaluate Trigonometric Functions

> The values of the trigonometric functions for the first quadrant can be used to evaluate trigonometric functions for any quadrant.

Recall from the preceding section that reference numbers are helpful for finding trigonometric points. Consequently, they are important for calculating the values of the trigonometric functions. For example, let's find the values of the trigonometric functions of $\frac{2\pi}{3}$. The reference angle for $\frac{2\pi}{3}$ is $\frac{\pi}{3}$, as shown in **Figure 6C.63**. Recall that

$$\sin\left(\frac{\pi}{3}\right) = \frac{\sqrt{3}}{2} \qquad \cos\left(\frac{\pi}{3}\right) = \frac{1}{2}$$

We need only attach the proper signs ($-$, $+$) for Quadrant II to find the sine and cosine of $\frac{2\pi}{3}$:

$$\sin\left(\frac{2\pi}{3}\right) = \frac{\sqrt{3}}{2} \qquad \cos\left(\frac{2\pi}{3}\right) = -\frac{1}{2}$$

The sine and cosine are used to find the values of the other four trigonometric functions:

$$\tan\left(\frac{2\pi}{3}\right) = \frac{\sin\left(\dfrac{2\pi}{3}\right)}{\cos\left(\dfrac{2\pi}{3}\right)} = \frac{\dfrac{\sqrt{3}}{2}}{-\dfrac{1}{2}} = -\sqrt{3} \qquad \cot\left(\frac{2\pi}{3}\right) = \frac{\cos\left(\dfrac{2\pi}{3}\right)}{\sin\left(\dfrac{2\pi}{3}\right)} = \frac{-\dfrac{1}{2}}{\dfrac{\sqrt{3}}{2}} = -\frac{1}{\sqrt{3}}$$

$$\sec\left(\frac{2\pi}{3}\right) = \frac{1}{\cos(2\pi/3)} = \frac{1}{-1/2} = -2 \qquad \csc\left(\frac{2\pi}{3}\right) = \frac{1}{\sin(2\pi/3)} = \frac{1}{\sqrt{3}/2} = \frac{2}{\sqrt{3}}$$

EXAMPLE 6C.19 Using the Reference Number

Find the values of the six trigonometric functions of $\frac{7\pi}{6}$.

SOLUTION

The trigonometric point $P\left(\frac{7\pi}{6}\right)$ lies in the third quadrant and has reference number $\frac{\pi}{6}$, as shown in **Figure 6C.64**. The sine of $\frac{\pi}{6}$ is $\frac{1}{2}$, and the cosine is $\frac{\sqrt{3}}{2}$.

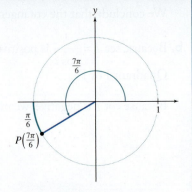

Figure 6C.64 Reference number for $\frac{7\pi}{6}$: $P\left(\frac{7\pi}{6}\right)$ lies in the third quadrant and has reference number $\frac{\pi}{6}$.

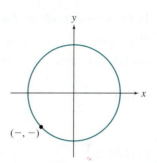

In the third quadrant, both the sine and cosine are negative.

This gives us the values of the sine and cosine, which we use to evaluate the remaining trigonometric functions:

$$\sin\left(\frac{7\pi}{6}\right) = -\frac{1}{2} \qquad \cos\left(\frac{7\pi}{6}\right) = -\frac{\sqrt{3}}{2}$$

$$\tan\left(\frac{7\pi}{6}\right) = \frac{\sin\left(\frac{7\pi}{6}\right)}{\cos\left(\frac{7\pi}{6}\right)} = \frac{-\frac{1}{2}}{-\frac{\sqrt{3}}{2}} = \frac{1}{\sqrt{3}} \qquad \cot\left(\frac{7\pi}{6}\right) = \frac{\cos\left(\frac{7\pi}{6}\right)}{\sin\left(\frac{7\pi}{6}\right)} = \frac{-\frac{\sqrt{3}}{2}}{-\frac{1}{2}} = \sqrt{3}$$

$$\sec\left(\frac{7\pi}{6}\right) = \frac{1}{\cos\left(\frac{7\pi}{6}\right)} = \frac{1}{-\frac{\sqrt{3}}{2}} = -\frac{2}{\sqrt{3}} \qquad \csc\left(\frac{7\pi}{6}\right) = \frac{1}{\sin\left(\frac{7\pi}{6}\right)} = \frac{1}{-\frac{1}{2}} = -2$$

TRY IT YOURSELF 6C.19 Brief answers provided at the end of the section.

Find the values of the six trigonometric functions of $\frac{5\pi}{4}$.

EXAMPLE 6C.20 Using Reference Angles

In each case, find the values of the six trigonometric functions of the given angle.

a. An angle of $120°$

b. An angle of $\frac{11\pi}{3}$ radians

SOLUTION

a. The trigonometric point $P(120°)$ lies in the second quadrant and has reference angle $60°$, as shown in **Figure 6C.65**. The sine of $60°$ is $\frac{\sqrt{3}}{2}$, and the cosine is $\frac{1}{2}$.

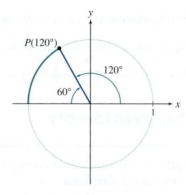

Figure 6C.65 The reference angle for $120°$: $P(120°)$ lies in the second quadrant and has reference angle $60°$.

In the second quadrant, the sine is positive and the cosine is negative.

We use the reference angle and signs to find the values of $\sin 120°$ and $\cos 120°$, and from these we find the values of all six trigonometric functions:

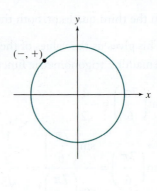

$$\sin 120° = \frac{\sqrt{3}}{2} \qquad\qquad \cos 120° = -\frac{1}{2}$$

$$\tan 120° = \frac{\sin 120°}{\cos 120°} = \frac{\frac{\sqrt{3}}{2}}{-\frac{1}{2}} = -\sqrt{3} \qquad \cot 120° = \frac{\cos 120°}{\sin 120°} = \frac{-\frac{1}{2}}{\frac{\sqrt{3}}{2}} = -\frac{1}{\sqrt{3}}$$

$$\sec 120° = \frac{1}{\cos 120°} = \frac{1}{-\frac{1}{2}} = -2 \qquad \csc 120° = \frac{1}{\sin 120°} = \frac{1}{\frac{\sqrt{3}}{2}} = \frac{2}{\sqrt{3}}$$

b. Now $\dfrac{11\pi}{3} = 2\pi + \dfrac{5\pi}{3}$. This is a full trip around the circle followed by an arc of length $\dfrac{5\pi}{3}$. We see that this angle is in Quadrant IV and the reference number is the same as in part a, namely $60°$ or $\dfrac{\pi}{3}$.

In the fourth quadrant, the sine is negative and the cosine is positive, so the values of the trigonometric functions are

$$\sin\left(\frac{11\pi}{3}\right) = -\frac{\sqrt{3}}{2} \qquad \cos\left(\frac{11\pi}{3}\right) = \frac{1}{2}$$

$$\tan\left(\frac{11\pi}{3}\right) = -\sqrt{3} \qquad \cot\left(\frac{11\pi}{3}\right) = -\frac{1}{\sqrt{3}}$$

$$\sec\left(\frac{11\pi}{3}\right) = 2 \qquad \csc\left(\frac{11\pi}{3}\right) = -\frac{2}{\sqrt{3}}$$

TRY IT YOURSELF 6C.20 Brief answers provided at the end of the section.

Find the values of the six trigonometric functions of $\dfrac{11\pi}{6}$.

EXTEND YOUR REACH   It is true that (correct to two decimal places) $\sin 20° = 0.34$ and $\cos 20° = 0.94$.

a. Using the given information, find the values of the sine and cosine of $160°$.

b. Find the values of the remaining four trigonometric functions of $160°$.

The Pythagorean Identity

> The formula for the unit circle gives a fundamental relationship between the sine and cosine functions.

Because the trigonometric point $P(t)$ lies on the unit circle, its coordinates, $(\cos t, \sin t)$, satisfy the equation $x^2 + y^2 = 1$. The result is the important Pythagorean identity connecting sine and cosine. Note that it is standard and convenient to write $\sin^2 t$ instead of $(\sin t)^2$, and similar notation is used for the other trigonometric functions.

LAWS OF MATHEMATICS: Pythagorean Identity

The Pythagorean identity is

$$\sin^2 t + \cos^2 t = 1$$

The following identities are closely related:

$$\tan^2 t + 1 = \sec^2 t$$

$$1 + \cot^2 t = \csc^2 t$$

EXAMPLE 6C.21 Finding Trigonometric Function Values from a Known Value

If $\sin t = \dfrac{2}{3}$ and $P(t)$ is in the second quadrant, find the values of the other five trigonometric functions of t.

SOLUTION

We use the Pythagorean identity to find the cosine:

$$\sin^2 t + \cos^2 t = 1 \qquad \text{◄ Use the Pythagorean identity.}$$

$$\left(\frac{2}{3}\right)^2 + \cos^2 t = 1 \qquad \text{◄ Substitute the known value.}$$

$$\frac{4}{9} + \cos^2 t = 1 \qquad \text{◄ Simplify.}$$

$$\cos^2 t = \frac{5}{9} \qquad \text{◄ Solve for } \cos^2 t.$$

$$\cos t = \pm\frac{\sqrt{5}}{3} \qquad \text{◄ Take square roots.}$$

Because we are in the second quadrant, the cosine is negative. So,

$$\cos t = -\frac{\sqrt{5}}{3}$$

We find the values of the other trigonometric functions from the sine and cosine:

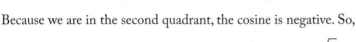

$$\tan t = \frac{\sin t}{\cos t} = \frac{\dfrac{2}{3}}{-\dfrac{\sqrt{5}}{3}} = -\frac{2}{\sqrt{5}} \qquad\qquad \cot t = \frac{\cos t}{\sin t} = \frac{-\dfrac{\sqrt{5}}{3}}{\dfrac{2}{3}} = -\frac{\sqrt{5}}{2}$$

$$\sec t = \frac{1}{\cos t} = \frac{1}{-\dfrac{\sqrt{5}}{3}} = -\frac{3}{\sqrt{5}} \qquad\qquad \csc t = \frac{1}{\sin t} = \frac{1}{\dfrac{2}{3}} = \frac{3}{2}$$

TRY IT YOURSELF 6C.21 Brief answers provided at the end of the section.

If $\cos t = -\dfrac{2}{5}$ and $P(t)$ is in the third quadrant, find the other five trigonometric functions of t.

Show that there is no number t such that $\sin t = 2$. *Suggestion*: Determine the condition on x required for the point $(x, 2)$ to lie on the unit circle.

STEP-BY-STEP STRATEGY: Finding Trigonometric Function Values from a Known Trigonometric Function Value Using Pythagorean Identities

Step 1 Use the Pythagorean identity for the known trigonometric value to obtain another trigonometric value (up to the sign).

Step 2 Use knowledge of the quadrant involved to assign the correct sign to the trigonometric value found in step 1.

Step 3 Use the definitions of the trigonometric functions and the known values to find the values of the sine and cosine.

Step 4 Use the sine and cosine to find the values of all the trigonometric functions.

EXAMPLE 6C.22 Finding Trigonometric Function Values from a Known Value Using Another Pythagorean Identity

If $\tan s = 6$ and s is an angle in standard position with terminal side in the third quadrant, find the values of the other five trigonometric functions of s. Then use a calculator to find decimal expressions correct to two decimal places.

SOLUTION

First we use the identity $\tan^2 s + 1 = \sec^2 s$ to find the secant:

$$\tan^2 s + 1 = \sec^2 s \quad \blacktriangleleft \textbf{ Use the related Pythagorean identity.}$$

$$6^2 + 1 = \sec^2 s \quad \blacktriangleleft \textbf{ Substitute the known values.}$$

$$37 = \sec^2 s \quad \blacktriangleleft \textbf{ Simplify.}$$

$$\pm\sqrt{37} = \sec s \quad \blacktriangleleft \textbf{ Take square roots.}$$

Because we are in the third quadrant, where trigonometric points have the form $(-, -)$, we find $\sec s = -\sqrt{37} \approx -6.08$.

Rearranging $\sec s = \dfrac{1}{\cos s}$, we find $\cos s = \dfrac{1}{\sec s}$. Hence, we get

$$\cos s = \frac{1}{-\sqrt{37}} = -\frac{1}{\sqrt{37}} \approx -0.16$$

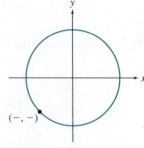

Solving the Pythagorean identity $\sin^2 s + \cos^2 s = 1$ for $\sin s$, and using the fact that the sine is negative in the third quadrant, we obtain $\sin s = -\dfrac{6}{\sqrt{37}} \approx -0.99$. Finally, the definitions of the cotangent and cosecant give

$$\cot s = \frac{\cos s}{\sin s} = \frac{-\dfrac{1}{\sqrt{37}}}{-\dfrac{6}{\sqrt{37}}} = \frac{1}{6} \approx 0.17 \qquad \csc s = \frac{1}{\sin s} = \frac{1}{-\dfrac{6}{\sqrt{37}}} = -\frac{\sqrt{37}}{6} \approx -1.01$$

TRY IT YOURSELF 6C.22 Brief answers provided at the end of the section.

If $\cot t = -2$ and $P(t)$ is in the fourth quadrant, find the other five trigonometric functions of t.

TRY IT YOURSELF ANSWERS

6C.16 $\sin\left(\dfrac{\pi}{4}\right) = \dfrac{1}{\sqrt{2}}$ $\cos\left(\dfrac{\pi}{4}\right) = \dfrac{1}{\sqrt{2}}$

$\tan\left(\dfrac{\pi}{4}\right) = 1$ $\cot\left(\dfrac{\pi}{4}\right) = 1$

$\sec\left(\dfrac{\pi}{4}\right) = \sqrt{2}$ $\csc\left(\dfrac{\pi}{4}\right) = \sqrt{2}$

6C.17 $\sin \pi = 0$ $\cos \pi = -1$

$\tan \pi = 0$ $\cot \pi$ is undefined

$\sec \pi = -1$ $\csc \pi$ is undefined

6C.18 The cosecant is negative in Quadrant III.

6C.19 $\sin\left(\dfrac{5\pi}{4}\right) = -\dfrac{1}{\sqrt{2}}$ $\cos\left(\dfrac{5\pi}{4}\right) = -\dfrac{1}{\sqrt{2}}$

$\tan\left(\dfrac{5\pi}{4}\right) = 1$ $\cot\left(\dfrac{5\pi}{4}\right) = 1$

$\sec\left(\dfrac{5\pi}{4}\right) = -\sqrt{2}$ $\csc\left(\dfrac{5\pi}{4}\right) = -\sqrt{2}$

6C.20 $\sin\left(\dfrac{11\pi}{6}\right) = -\dfrac{1}{2}$ $\cos\left(\dfrac{11\pi}{6}\right) = \dfrac{\sqrt{3}}{2}$

$\tan\left(\dfrac{11\pi}{6}\right) = -\dfrac{1}{\sqrt{3}}$ $\cot\left(\dfrac{11\pi}{6}\right) = -\sqrt{3}$

$\sec\left(\dfrac{11\pi}{6}\right) = \dfrac{2}{\sqrt{3}}$ $\csc\left(\dfrac{11\pi}{6}\right) = -2$

6C.21

$\sin t = -\dfrac{\sqrt{21}}{5}$, $\tan t = \dfrac{\sqrt{21}}{2}$, $\cot t = \dfrac{2}{\sqrt{21}}$, $\sec t = -\dfrac{5}{2}$,

$\csc t = -\dfrac{5}{\sqrt{21}}$

6C.22

$\tan t = -\dfrac{1}{2}$, $\sin t = -\dfrac{1}{\sqrt{5}}$, $\cos t = \dfrac{2}{\sqrt{5}}$, $\sec t = \dfrac{\sqrt{5}}{2}$,

$\csc t = -\sqrt{5}$

EXERCISE SET 6C.3

CHECK YOUR UNDERSTANDING

1. If $P(t) = (x, y)$, express each of the six trigonometric functions of t in terms of x and y.

2. In which quadrant is the tangent positive and the cosine negative?

3. In which quadrant is the sine negative and the cosine positive?

4. The sine of a certain angle in standard position is negative, and the cosine is positive. In which quadrant does the terminal side of the angle lie?

5. The tangent of a certain angle in standard position is negative. In which quadrants can the terminal side of the angle lie?

6. Express the tangent, cotangent, secant, and cosecant functions in terms of the sine and cosine.

7. State the Pythagorean identity.

8. State the related Pythagorean identity that connects the tangent and the secant.

9. State the related Pythagorean identity that connects the cotangent and the cosecant.

10. In which quadrants do the sine and cosine functions have the same sign?

SKILL BUILDING

Calculating trigonometric function values. Figure 6C.66 shows points on the unit circle. In each case, one coordinate is given. In Exercises 11 through 19, find the missing coordinate for the indicated point, and then list the six trigonometric function values. (Use t for the variable.)

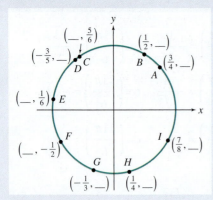

Figure 6C.66 Trigonometric points with one coordinate given

11. Trigonometric point A

12. Trigonometric point B

13. Trigonometric point C

14. Trigonometric point D

15. Trigonometric point E

16. Trigonometric point F

17. Trigonometric point G

18. Trigonometric point H

19. Trigonometric point I

Trigonometric functions of compass points. In Exercises 20 through 36, find the values of the six trigonometric functions of the given angles. In each case, one or more of the trigonometric functions may not be defined. Remember that radian measure is indicated if degrees are not explicitly stated.

20. $90°$

21. $0°$

22. $180°$

23. $270°$

24. $360°$

25. $\dfrac{\pi}{2}$

26. π

27. 5π

28. $450°$

29. $-180°$

30. $\dfrac{7\pi}{2}$

31. $-\pi$

32. -6π

33. $\dfrac{3\pi}{2}$

34. 2π

35. $900°$

36. $-\dfrac{\pi}{2}$

PROBLEMS

37. **Signs in quadrants.** Make a table that shows the signs of each of the six trigonometric functions in each of the four quadrants.

38. **Verifying a Pythagorean identity.** Use the identity $\sin^2 t + \cos^2 t = 1$ to establish the identity $\tan^2 t + 1 = \sec^2 t$. *Suggestion*: Divide both sides by $\cos^2 t$.

39. **Verifying another Pythagorean identity.** Use the identity $\sin^2 t + \cos^2 t = 1$ to establish the identity $1 + \cot^2 t = \csc^2 t$. *Suggestion*: Divide both sides by $\sin^2 t$.

Evaluating trigonometric functions at special angles. In Exercises 40 through 55, calculate the values of the six trigonometric functions of some special angles. Remember that radian measure is indicated if degree measure is not explicitly stated.

40. $135°$

41. $330°$

42. $\dfrac{5\pi}{4}$

43. $-\dfrac{\pi}{3}$

44. $-\dfrac{\pi}{4}$

45. $-135°$

46. $210°$

47. π

48. $\dfrac{13\pi}{6}$

49. $-\dfrac{\pi}{6}$

50. $120°$

51. $\dfrac{7\pi}{6}$

52. $300°$

53. $\dfrac{2\pi}{3}$

54. $240°$

55. $\dfrac{5\pi}{3}$

Getting one trigonometric function value from another. In Exercises 56 through 71, you are given the value of one trigonometric function. Find the values of the remaining five trigonometric functions.

56. $\cos t = -\dfrac{3}{4}$, and $\sin t$ is positive.

57. $\cos t = \dfrac{3}{4}$, and $\sin t$ is positive.

58. $\tan t = \sqrt{3}$, and $\sin t$ is positive.

59. $\sin t = \dfrac{5}{6}$, and $\cos t$ is negative.

60. $\sec t = -\dfrac{5}{3}$, and $\tan t$ is negative.

61. $\sin t = \dfrac{1}{6}$, and $\sec t$ is negative.

62. $\cos t = -\dfrac{\sqrt{3}}{2}$, and $\sin t$ is negative.

63. $\cos t = -\dfrac{1}{3}$, and $\sin t$ is negative.

64. $\sec t = 4$, and $\csc t$ is negative.

65. $\cot t = -\dfrac{7}{\sqrt{15}}$, and $\cos t$ is positive.

66. $\sec t = 4$, and $\sin t$ is negative.

67. $\cot t = 5$, and $P(t)$ lies in the third quadrant.

68. $\sec t = -3$, and $\cot t$ is negative.

69. $\sec t = 3$, and $\tan t$ is negative.

70. $\cot t = 1.20$, and $\sin t$ is negative. Report answers correct to two decimal places, but do not use rounded values in calculations.

71. $\cos t = 0.55$, and $P(t)$ is in the fourth quadrant. Report answers correct to two decimal places, but do not use rounded values in the calculations.

72. **Solving a trigonometric equation.** Find one solution of the equation $\sin t = \cos t$. *Suggestion:* Think about the special angles that you know.

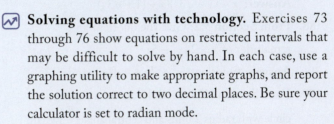 **Solving equations with technology.** Exercises 73 through 76 show equations on restricted intervals that may be difficult to solve by hand. In each case, use a graphing utility to make appropriate graphs, and report the solution correct to two decimal places. Be sure your calculator is set to radian mode.

73. $\cos t = \ln t$, $0 < t < 2$

74. $\sin t = t^2$, $0 < t < 2$

75. $\tan t = \cot t$, $0 < t < 1$

76. $1 + \sin t = \csc t$, $0 < t < 1$

CHALLENGE EXERCISES FOR INDIVIDUALS OR GROUPS

Theorem on undetermined coefficients. It is a fact that if A, B, C, and D are constants, and if the equation

$$A \cos t + B \sin t = C \cos t + D \sin t$$

is true for all values of t, then $A = C$ and $B = D$. In Exercises 77 through 81, you will be led to show this theorem is true and then apply that fact to solve problems.

77. **Showing the theorem is true.** Suppose the equation $A \cos t + B \sin t = C \cos t + D \sin t$ is true for all values of t.

   a. Put $t = 0$ into the equation. What can you conclude regarding A, B, C, and D?

   b. Put $t = \dfrac{\pi}{2}$ into the equation. What can you conclude regarding A, B, C, and D?

78. **Using the theorem.** Suppose

$$(A - 2) \cos t + 3 \sin t = 4 \cos t + (B + 7) \sin t$$

is true for all values of t. Find the values of A and B.

79. **Using the theorem.** Suppose

$$A(\cos t + \sin t) = 3 \cos t + B \sin t$$

is true for all values of t. Find the values of A and B.

80. **Using the theorem.** Suppose

$$(A + B) \cos t + (A - B) \sin t = 4 \cos t + 2 \sin t$$

is true for all values of t. Find the values of A and B.

81. **Using the theorem.** Suppose

$$2A \sin t + 4A \cos t = B \sin t + \sin t - B \cos t + 6 \cos t$$

is true for all values of t. Find the values of A and B.

REVIEW AND REFRESH: Exercises from Previous Sections

82. **From Section P.1:** Find the distance between the points $(x^2, 2x)$ and $(1, 0)$.

   Answer: $x^2 + 1$

83. **From Section 1.4:** Find the limit as $x \to \infty$ of
$$y = \dfrac{2}{1 + 1/x}.$$

   Answer: 2

84. **From Section 2.2:** Find the inverse of the function $f(x) = \dfrac{x}{x + 1}$.

   Answer: $f^{-1}(x) = \dfrac{x}{1 - x}$

85. **From Section 3.1:** Find the linear function with initial value 4 and slope 3.

   Answer: $f(x) = 3x + 4$

86. From Section 4.3: Solve the equation $e^{2x-1} = 7$.

Answer: $x = \dfrac{1 + \ln 7}{2}$

87. From Section 5.1: Factor the quadratic expression $x^2 - 2x - 2$.

Answer: $(x - (1 - \sqrt{3}))(x - (1 + \sqrt{3}))$

88. From Section 5.2: Give the quotient and remainder of $x - 2 \overline{\smash{)}x^3 - 5}$.

Answer: Quotient: $x^2 + 2x + 4$. Remainder: 3.

89. From Section 6C.1: Find the area of a sector of a circle with radius 2 units and central angle $\dfrac{\pi}{5}$ radian.

Answer: $\dfrac{2\pi}{5}$ square units

90. From Section 6C.2: Find the coordinates of the trigonometric point $P\left(\dfrac{\pi}{3}\right)$.

Answer: $\left(\dfrac{1}{2}, \dfrac{\sqrt{3}}{2}\right)$

91. From Section 6C.2: Find all possible values of x if $(x, -3x)$ lies on the unit circle in Quadrant IV.

Answer: $x = \dfrac{1}{\sqrt{10}}$

6C.1 Angles

6C.2 The Unit Circle

6C.3 The Trigonometric Functions

6C.4 Right Triangle Trigonometry

6C.4 Right Triangle Trigonometry

> Trigonometry is closely connected to the study of right triangles.

In this section, you will learn to:

1. Calculate the values of the trigonometric functions of an acute angle in a right triangle using the lengths of the sides.
2. Use a given trigonometric value of an acute angle to construct a right triangle having that angle.
3. Use a right triangle to relate the value of one trigonometric function to the other function values.
4. Solve a right triangle given one acute angle and one side.
5. Apply the ideas of trigonometry to analyze right triangles arising in practical settings.

In this section we show how the trigonometric functions apply to analyze right triangles—the very reason trigonometry was invented. Analysis of this sort lies at the heart of many applications, such as surveying and cartography.

For example, the fundamental tool of the surveyor is the transit (**Figure 6C.67**). The transit can accurately measure angles, and that measurement, together with some basic trigonometry, is the key to many practical applications. The transit is used to determine the dimensions of structures, both natural and human-made, that do not lend themselves to direct measurement.

Figure 6C.67 A transit

Trigonometric Functions of Acute Angles

> Trigonometric functions of acute angles can be defined using right triangles.

Figure 6C.68 shows a trigonometric point $P(t) = (\cos t, \sin t)$ for t between 0 and $\dfrac{\pi}{2}$.

We have added to the illustration a larger right triangle that has an acute angle of radian measure t. **Figure 6C.69** shows how we label the sides of this triangle in relation to the angle t.

If we use the angle t as a reference, the adjacent side is the leg of the triangle that is a side of the angle t, and the opposite side is the leg of the triangle across from the angle t. Because the larger and smaller right triangles have two common angles, they are similar, and similarity implies equality of ratios of corresponding sides:

$$\sin t = \frac{\sin t}{1} = \frac{\text{Opposite}}{\text{Hypotenuse}}$$

$$\cos t = \frac{\cos t}{1} = \frac{\text{Adjacent}}{\text{Hypotenuse}}$$

The result is that we can find the values of the sine and cosine of an acute angle t using the sides of any right triangle with t as an acute angle.

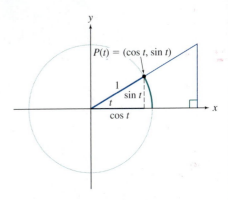

Figure 6C.68 The trigonometric point for an acute angle with a large right triangle added

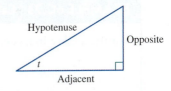

Figure 6C.69 The sides of the larger right triangle labeled

LAWS OF MATHEMATICS: Calculating Values of Trigonometric Functions Using a Right Triangle

If t is an acute angle of a right triangle, then we can calculate the values of the trigonometric functions of t as follows:

$$\sin t = \frac{\text{Opposite}}{\text{Hypotenuse}} \qquad \cos t = \frac{\text{Adjacent}}{\text{Hypotenuse}}$$

$$\tan t = \frac{\text{Opposite}}{\text{Adjacent}} \qquad \cot t = \frac{\text{Adjacent}}{\text{Opposite}}$$

$$\sec t = \frac{\text{Hypotenuse}}{\text{Adjacent}} \qquad \csc t = \frac{\text{Hypotenuse}}{\text{Opposite}}$$

A useful mnemonic for the values of sine, cosine, and tangent is SOHCAHTOA, which incorporates the first letters of "Sine Opposite over Hypotenuse, Cosine Adjacent over Hypotenuse, Tangent Opposite over Adjacent."

Using Right Triangles to Evaluate Trigonometric Functions

> Trigonometric functions of acute angles can be calculated in terms of side lengths of a right triangle.

We can calculate the values of the trigonometric functions of angles if the sides of the right triangle are known. Consider, for example, the triangle in **Figure 6C.70**, where the lengths of the sides are labeled.

For the angle s, the adjacent side has length 5 and the opposite side has length 12, so

$$\sin s = \frac{\text{Opposite}}{\text{Hypotenuse}} = \frac{12}{13} \qquad \cos s = \frac{\text{Adjacent}}{\text{Hypotenuse}} = \frac{5}{13}$$

$$\tan s = \frac{\text{Opposite}}{\text{Adjacent}} = \frac{12}{5} \qquad \cot s = \frac{\text{Adjacent}}{\text{Opposite}} = \frac{5}{12}$$

$$\sec s = \frac{\text{Hypotenuse}}{\text{Adjacent}} = \frac{13}{5} \qquad \csc s = \frac{\text{Hypotenuse}}{\text{Opposite}} = \frac{13}{12}$$

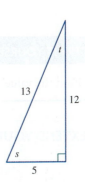

Figure 6C.70 Calculating values of trigonometric functions

For the angle t, the adjacent side has length 12 and the opposite side has length 5, so

$$\sin t = \frac{\text{Opposite}}{\text{Hypotenuse}} = \frac{5}{13} \qquad \cos t = \frac{\text{Adjacent}}{\text{Hypotenuse}} = \frac{12}{13}$$

$$\tan t = \frac{\text{Opposite}}{\text{Adjacent}} = \frac{5}{12} \qquad \cot t = \frac{\text{Adjacent}}{\text{Opposite}} = \frac{12}{5}$$

$$\sec t = \frac{\text{Hypotenuse}}{\text{Adjacent}} = \frac{13}{12} \qquad \csc t = \frac{\text{Hypotenuse}}{\text{Opposite}} = \frac{13}{5}$$

If we know two sides of a right triangle, then we can use the Pythagorean theorem to find the length of the third side. Hence, only two sides are needed to calculate the values of the trigonometric functions.

EXAMPLE 6C.23   Calculating Function Values Using Two Known Sides of a Right Triangle

For the angle t in **Figure 6C.71**, calculate the values of the six trigonometric functions.

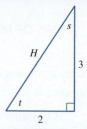

Figure 6C.71 Two sides of a right triangle given

SOLUTION

First we use the Pythagorean theorem to calculate the length of the hypotenuse H:

$$H^2 = a^2 + b^2 \qquad \blacktriangleleft \text{ Use the Pythagorean theorem.}$$

$$H^2 = 2^2 + 3^2 \qquad \blacktriangleleft \text{ Substitute known values.}$$

$$H = \sqrt{13} \qquad \blacktriangleleft \text{ Take square roots.}$$

The side adjacent to the angle t has length 2, and the opposite side has length 3. We use these lengths to calculate the values of the trigonometric functions:

$$\sin t = \frac{\text{Opposite}}{\text{Hypotenuse}} = \frac{3}{\sqrt{13}} \qquad \cos t = \frac{\text{Adjacent}}{\text{Hypotenuse}} = \frac{2}{\sqrt{13}}$$

$$\tan t = \frac{\text{Opposite}}{\text{Adjacent}} = \frac{3}{2} \qquad \cot t = \frac{\text{Adjacent}}{\text{Opposite}} = \frac{2}{3}$$

$$\sec t = \frac{\text{Hypotenuse}}{\text{Adjacent}} = \frac{\sqrt{13}}{2} \qquad \csc t = \frac{\text{Hypotenuse}}{\text{Opposite}} = \frac{\sqrt{13}}{3}$$

TRY IT YOURSELF 6C.23   Brief answers provided at the end of the section.

Find the values of the six trigonometric functions of the angle s in Figure 6C.71.

EXTEND YOUR REACH   If, as in the preceding example, s and t are the acute angles of a right triangle, how are the six trigonometric functions evaluated at s related to the six trigonometric functions evaluated at t? *Suggestion:* To get started, observe that the side opposite s is adjacent to t, and the side adjacent to s is opposite t. The hypotenuse is the same for both angles.

EXAMPLE 6C.24 Finding Values of Trigonometric Functions with a Known Reference Angle

Suppose s is an angle in standard position with reference angle t shown in **Figure 6C.72**. Suppose further that the radian measure of s is between $\dfrac{3\pi}{2}$ and 2π. Find the values of the six trigonometric functions of s.

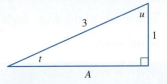

Figure 6C.72 Angle t, which is the reference angle for angle s

SOLUTION

We use the Pythagorean theorem to find the value of A:

$$A^2 + 1^2 = 3^2 \quad \blacktriangleleft \text{ Use the Pythagorean theorem.}$$

$$A^2 + 1 = 9 \quad \blacktriangleleft \text{ Simplify.}$$

$$A^2 = 8 \quad \blacktriangleleft \text{ Solve for } A^2.$$

$$A = \sqrt{8} = 2\sqrt{2} \quad \blacktriangleleft \text{ Solve for A.}$$

This information is used to find the values of the trigonometric functions of t:

$$\sin t = \frac{\text{Opposite}}{\text{Hypotenuse}} = \frac{1}{3} \qquad\qquad \cos t = \frac{\text{Adjacent}}{\text{Hypotenuse}} = \frac{\sqrt{8}}{3}$$

$$\tan t = \frac{\text{Opposite}}{\text{Adjacent}} = \frac{1}{\sqrt{8}} \qquad\qquad \cot t = \frac{\text{Adjacent}}{\text{Opposite}} = \sqrt{8}$$

$$\sec t = \frac{\text{Hypotenuse}}{\text{Adjacent}} = \frac{3}{\sqrt{8}} \qquad\qquad \csc t = \frac{\text{Hypotenuse}}{\text{Opposite}} = 3$$

Because s is between $\dfrac{3\pi}{2}$ and 2π, its terminal side lies in the fourth quadrant, so the corresponding trigonometric point has the form $(+, -)$. This fact allows us to attach the proper signs to the values of the trigonometric functions of t to find the values of the trigonometric functions of s:

$$\sin s = -\frac{1}{3} \qquad\qquad \cos s = \frac{\sqrt{8}}{3}$$

$$\tan s = -\frac{1}{\sqrt{8}} \qquad\qquad \cot s = -\sqrt{8}$$

$$\sec s = \frac{3}{\sqrt{8}} \qquad\qquad \csc s = -3$$

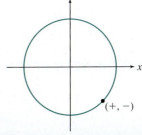

TRY IT YOURSELF 6C.24 Brief answers provided at the end of the section.

Let v be an angle in standard position with radian measure between $\dfrac{\pi}{2}$ and π. Assume that the reference angle for v is the angle u in Figure 6C.72. Find the values of the six trigonometric functions of v.

Representing Angles

> An angle with a given trigonometric function value can be represented by a right triangle.

Next we will explore an important method of representing angles by using known trigonometric values.

EXAMPLE 6C.25   Representing Angles

The cosine of the acute angle t is $\dfrac{2}{3}$. Construct a right triangle with acute angle t, and label the lengths of the sides.

SOLUTION

Because $\cos t = \dfrac{\text{Adjacent}}{\text{Hypotenuse}}$, we can find the value of $\dfrac{2}{3}$ for the cosine if the adjacent side has length 2 and the hypotenuse has length 3. This triangle is shown in **Figure 6C.73**. We can find the opposite side B using the Pythagorean theorem:

$$c^2 = a^2 + b^2 \quad \blacktriangleleft \textbf{Use the Pythagorean theorem.}$$

$$3^2 = 2^2 + B^2 \quad \blacktriangleleft \textbf{Substitute known values.}$$

$$5 = B^2 \quad \blacktriangleleft \textbf{Solve for } B^2.$$

$$\sqrt{5} = B \quad \blacktriangleleft \textbf{Take square roots.}$$

The completed triangle is shown in **Figure 6C.74**.

Note that there are many other right triangles that are correct answers. For example, we could also represent $\cos t = \dfrac{2}{3}$ by using an adjacent side of 4 and a hypotenuse of 6. The resulting opposite side would be $2\sqrt{5}$.

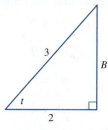

Figure 6C.73 An angle with cosine equal to $\dfrac{2}{3}$

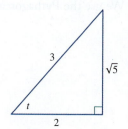

Figure 6C.74 An angle with cosine equal to $\dfrac{2}{3}$: The Pythagorean theorem is used to calculate the opposite side.

TRY IT YOURSELF 6C.25   Brief answers provided at the end of the section.

The tangent of the acute angle t is $\dfrac{4}{5}$. Construct a right triangle with acute angle t, and label the lengths of the sides.

EXTEND YOUR REACH

a. Is there more than one acute angle t whose cosine is $\dfrac{2}{3}$? *Suggestion*: Look at the Extend Your Reach accompanying Example 6C.17 in Section 6C.3.

b. We noted in the present example that there is more than one right triangle with an acute angle t such that $\cos t = \dfrac{2}{3}$. How are these triangles related? *Suggestion*: Think first about how the angles of the right triangles are related.

Using Representative Triangles to Evaluate Trigonometric Functions

Triangles representing one trigonometric function value can be used to find the values of other trigonometric functions.

EXAMPLE 6C.26   Finding One Trigonometric Function Value from Another

If t is an acute angle such that $\tan t = \dfrac{3}{4}$, find $\sin t$.

SOLUTION

Because $\tan t = \dfrac{\text{Opposite}}{\text{Adjacent}}$, we can find the desired value if we make a right triangle with the opposite side of length 3 and the adjacent side of length 4, as shown in **Figure 6C.75**. We use the Pythagorean theorem to find the length of the hypotenuse:

$$H^2 = 3^2 + 4^2$$

$$H^2 = 25$$

$$H = 5$$

The completed triangle is shown in **Figure 6C.76**. We use this figure to calculate the sine:

$$\sin t = \frac{\text{Opposite}}{\text{Hypotenuse}} = \frac{3}{5}$$

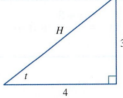

Figure 6C.75 An angle whose tangent is $\dfrac{3}{4}$

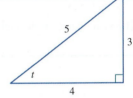

Figure 6C.76 An angle whose tangent is $\dfrac{3}{4}$: The hypotenuse is 5.

TRY IT YOURSELF 6C.26   Brief answers provided at the end of the section.

If t is an acute angle such that $\cos t = \dfrac{1}{3}$, find the sine of t.

EXAMPLE 6C.27   Finding One Trigonometric Function Value from Another Given by an Algebraic Expression

If s is an acute angle such that $\cos s = \dfrac{x}{x+2}$ for a positive number x, find $\cot s$.

SOLUTION

Because the cosine is $\dfrac{\text{Adjacent}}{\text{Hypotenuse}}$, we can get $\cos s = \dfrac{x}{x+2}$ if we use x for the side adjacent to s and $x+2$ for the hypotenuse. This configuration is shown in **Figure 6C.77**.

We find the value of the opposite side B using the Pythagorean theorem:

$$x^2 + B^2 = (x+2)^2 \qquad \blacktriangleleft \textbf{ Use the Pythagorean theorem.}$$

$$x^2 + B^2 = x^2 + 4x + 4 \qquad \blacktriangleleft \textbf{ Expand the square.}$$

$$B^2 = 4x + 4 \qquad \blacktriangleleft \textbf{ Cancel } x^2.$$

$$B = \sqrt{4x + 4} \qquad \blacktriangleleft \textbf{ Take square roots.}$$

$$B = 2\sqrt{x + 1} \qquad \blacktriangleleft \textbf{ Simplify.}$$

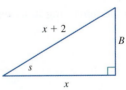

Figure 6C.77 An acute angle whose cosine is $\dfrac{x}{(x+2)}$

This new information is incorporated in **Figure 6C.78**, and we use this figure to calculate the cotangent of s:

$$\cot s = \frac{\text{Adjacent}}{\text{Opposite}}$$

$$= \frac{x}{2\sqrt{x+1}}$$

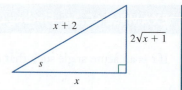

Figure 6C.78 An acute angle whose cosine is $\dfrac{x}{(x+2)}$: The opposite side is calculated.

TRY IT YOURSELF 6C.27 Brief answers provided at the end of the section.

If s is an acute angle such that $\cot s = \dfrac{x+3}{2}$, find the sine of s.

If we know the value of one trigonometric function, and if we can identify the quadrant, we can calculate the values of the other trigonometric functions.

EXAMPLE 6C.28   Calculating When One Function Value Is Known

The sine of t is $-\dfrac{2}{3}$. Find the values of the other five trigonometric functions of t if the cosine of t is positive.

SOLUTION

The reference angle for t is an acute angle r with $\sin r = \dfrac{2}{3}$. We want a triangle with acute angle r whose sine is $\dfrac{2}{3}$. We can accomplish this by choosing an opposite side of 2 and a hypotenuse of 3, as illustrated in **Figure 6C.79**.

The Pythagorean theorem shows that the adjacent side has length $\sqrt{3^2 - 2^2} = \sqrt{5}$, so:

$$\cos r = \frac{\text{Adjacent}}{\text{Hypotenuse}} = \frac{\sqrt{5}}{3}$$

Because the cosine of t is positive, we have $\cos t = \dfrac{\sqrt{5}}{3}$. Using the sine and cosine, we obtain the remaining function values:

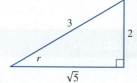

Figure 6C.79 A reference angle whose sine is $\dfrac{2}{3}$

$$\tan t = -\frac{2}{\sqrt{5}} \qquad\qquad \cot t = -\frac{\sqrt{5}}{2}$$

$$\sec t = \frac{3}{\sqrt{5}} \qquad\qquad \csc t = -\frac{3}{2}$$

TRY IT YOURSELF 6C.28 Brief answers provided at the end of the section.

If $\tan t = \dfrac{1}{2}$ and $P(t)$ lies in the third quadrant, find the values of the other five trigonometric functions of t.

STEP-BY-STEP STRATEGY: Finding Trigonometric Function Values from a Known Trigonometric Function Value

Step 1 Label two sides of a right triangle so that a reference angle in the triangle has the appropriate trigonometric function value.

Step 2 Use the Pythagorean theorem to find the third side of the triangle.

Step 3 Use the triangle to calculate the other trigonometric function values of the reference angle.

Step 4 Attach the proper signs to the function values calculated in step 3.

Solving Right Triangles

> **When one side and one acute angle of a right triangle are known, the remaining sides and angle may be calculated.**

Right triangles can be used to determine the values of the trigonometric functions, but the reverse is also true. Trigonometric functions are essential tools for analyzing triangles.

A basic problem in trigonometry is to solve a given right triangle—that is, to find the length of each side and the measure of each angle. The trigonometric functions allow us to solve a right triangle if we know one acute angle and one side. The method is illustrated in the next example.

EXAMPLE 6C.29 Solving a Triangle with a Special Angle

One acute angle of a right triangle has radian measure $\dfrac{\pi}{3}$, and the opposite side has length 3. Find the unknown angle and sides in **Figure 6C.80**.

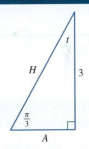

Figure 6C.80 A triangle with one special angle

SOLUTION

We can use the fact that the sum of the three angles is π radians to find the measure of the angle t:

$$t + \frac{\pi}{3} + \frac{\pi}{2} = \pi$$

$$t = \pi - \left(\frac{\pi}{3} + \frac{\pi}{2}\right)$$

$$= \frac{\pi}{6} \text{ radian}$$

We use the sine to find the value of H because it relates the unknown hypotenuse H to a known side. We know that $\sin\left(\dfrac{\pi}{3}\right) = \dfrac{\sqrt{3}}{2}$. We can also use the given triangle to calculate the sine of $\dfrac{\pi}{3}$:

$$\frac{\sqrt{3}}{2} = \sin\left(\frac{\pi}{3}\right) = \frac{\text{Opposite}}{\text{Hypotenuse}} = \frac{3}{H}$$

Solving the equation $\dfrac{3}{H} = \dfrac{\sqrt{3}}{2}$ for H, we find $H = 6/\sqrt{3} = 2\sqrt{3}$.

Next we make use of the tangent to find the value of A. (We could instead use the Pythagorean theorem.) Because $\dfrac{\pi}{3}$ is a special angle, we have

$$\tan\left(\frac{\pi}{3}\right) = \frac{\sin\left(\dfrac{\pi}{3}\right)}{\cos\left(\dfrac{\pi}{3}\right)} = \frac{\dfrac{\sqrt{3}}{2}}{\dfrac{1}{2}} = \sqrt{3}$$

Now using the triangle in Figure 6C.80 gives

$$\sqrt{3} = \tan\left(\frac{\pi}{3}\right) = \frac{\text{Opposite}}{\text{Adjacent}} = \frac{3}{A}$$

We solve the equation $\dfrac{3}{A} = \sqrt{3}$ for A to get $A = \dfrac{3}{\sqrt{3}} = \sqrt{3}$.

TRY IT YOURSELF 6C.29   Brief answers provided at the end of the section.

Find the angle t and the unknown sides in **Figure 6C.81**.

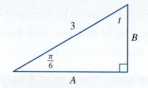

Figure 6C.81 A triangle for Try It Yourself 6C.29

If the known angles of a triangle are not special angles, then we must use technology to find approximations for the values of the trigonometric functions.

EXAMPLE 6C.30   Solving Right Triangles

Suppose the angle s in **Figure 6C.82** has a measure of 1 radian. Solve the triangle. Give answers correct to two decimal places.

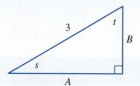

Figure 6C.82 A right triangle with an angle of measure 1 radian

SOLUTION

We first use the fact that the angle sum is π radians:

$$t + 1 + \frac{\pi}{2} = \pi$$

$$t = \pi - \left(1 + \frac{\pi}{2}\right)$$

$$\approx 0.57 \text{ radian}$$

The definition of the sine function allows us to express B in terms of $\sin s$:

$$\sin s = \frac{\text{Opposite}}{\text{Hypotenuse}}$$ ◀ Use the definition of sine.

$$\sin s = \frac{B}{3}$$ ◀ Substitute known values.

$$3\sin s = B$$ ◀ Solve for B.

Because 1 radian is not one of the special angles, we use a calculator (set to radian mode) to evaluate its sine:

$$B = 3\sin 1 \approx 2.52$$

Similarly, we can use the cosine to find A:

$$\cos s = \frac{A}{3}$$ ◀ Calculate using the triangle.

$$3\cos s = A$$ ◀ Solve for A.

$$3\cos 1 = A$$ ◀ Substitute 1 for s.

$$1.62 \approx A$$ ◀ Approximate the value using a calculator.

TRY IT YOURSELF 6C.30 Brief answers provided at the end of the section.

Solve the triangle in Figure 6C.82 if instead the angle s is $25°$. (Set your calculator to degree mode.)

The generalization of these techniques for the case of a known acute angle t and hypotenuse h is important. Our aim is to find the sides x and y in **Figure 6C.83** in terms of h and t. We use the sine of t to find the value of y and the cosine of t to find the value of x:

$$\sin t = \frac{y}{h} \quad \text{or} \quad y = h\sin t$$

$$\cos t = \frac{x}{h} \quad \text{or} \quad x = h\cos t$$

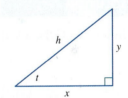

Figure 6C.83 Finding x and y in terms of t and h

CONCEPTS TO REMEMBER: Finding Legs from Angle and Hypotenuse

1. If t is an acute angle of a right triangle with hypotenuse h, we can find the legs of the triangle using

   Adjacent side $= h\cos t$
   Opposite side $= h\sin t$

   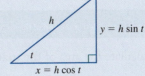

2. A hypotenuse of length 1 is a common case:

   Adjacent side $= \cos t$
   Opposite side $= \sin t$

   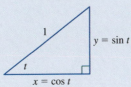

A New Formula for the Area of a Triangle

> The trigonometry of right triangles yields a useful formula for the area of a triangle.

In the next example, we use a formula for the area of a triangle that may be familiar to you from geometry: If B is the length of the base and H is the height, then

$$\text{Area} = \frac{1}{2}BH$$

A new area formula is developed, and we will find it useful as our study of trigonometry progresses.

EXAMPLE 6C.31   A New Area Formula

Show that the area of the triangle in **Figure 6C.84** is

$$\text{Area} = \frac{1}{2}AB \sin t$$

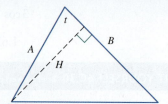

Figure 6C.84 Triangle for Example 6C.31

SOLUTION

In Figure 6C.84, B is the length of the base and H is the height, so

$$\text{Area} = \frac{1}{2}BH$$

Focus now on the smaller right triangle with sides A and H. We can use these sides to calculate the sine of t:

$$\sin t = \frac{H}{A}$$

Solving for H gives $H = A \sin t$. Putting this expression in place of H in the area formula gives the desired result:

$$\text{Area} = \frac{1}{2}AB \sin t$$

TRY IT YOURSELF 6C.31   Brief answers provided at the end of the section.

Use the new area formula to calculate the area of the triangle shown in **Figure 6C.85**.

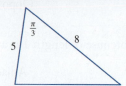

Figure 6C.85 Triangle for Try It Yourself 6C.31

MODELS AND APPLICATIONS   Transits and Shadows

These methods for solving triangles can be used in practical settings to find unknown distances. In practical applications, we rarely encounter special angles, and as a result we must use a calculator or computer. When using technology, take care that the mode for measuring angles on your device is set to degrees or radians, as appropriate. Also, many calculators have sine and cosine keys but may not have keys for the other four trigonometric functions. The basic trigonometric relationships that give the other four functions in terms of the sine and cosine are essential in that case.

EXAMPLE 6C.32   Surveying: Height of a Building

A surveyor stands 300 horizontal feet from a building and aims a transit (an instrument for measuring angles) at the top of the building, as shown in **Figure 6C.86**. The transit measures an angle of 21° from the horizontal.

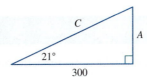

Figure 6C.86 Measuring with a transit

a. How much taller is the building than the transit?

b. How long would a straight cable that reached from the transit directly to the top of the building need to be?

SOLUTION

a. The desired height is represented by the side marked A in Figure 6C.86. We know the length of the adjacent side, so we should use the tangent function, which relates these two sides:

$$\tan 21° = \frac{\text{Opposite}}{\text{Adjacent}} = \frac{A}{300}$$

Solving for A, we find

$$A = 300 \tan 21° \approx 115.16 \text{ feet}$$

We conclude that the building is about 155.16 feet taller than the transit.

b. The distance from the transit directly to the top of the building is the hypotenuse of the triangle and is marked C in Figure 6C.86. The cosine relates this side to the adjacent side of 300 feet:

$$\cos 21° = \frac{\text{Adjacent}}{\text{Hypotenuse}} = \frac{300}{C}$$

Solving for C gives

$$C = \frac{300}{\cos 21°} \approx 321.34 \text{ feet}$$

We conclude that a straight cable reaching from the transit to the top of the building would need to be about 321.34 feet long.

TRY IT YOURSELF 6C.32   Brief answers provided at the end of the section.

How much taller is the building than the transit if instead the angle is 27°?

EXAMPLE 6C.33   Towers and Shadows

A 7-foot-tall pole casts a shadow of 4 feet at the same time a tower casts a shadow of 15 feet, as shown in **Figure 6C.87**. How tall is the tower?

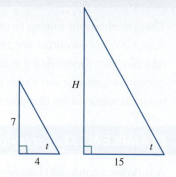

Figure 6C.87 Shadows of a pole and a tower

SOLUTION

Because the shadows are produced by the sun at the same time, the angles marked t in the two triangles are the same. Using the smaller triangle, we calculate $\tan t = \dfrac{7}{4}$. Using the larger triangle, we calculate $\tan t = \dfrac{H}{15}$.

Solving the equation $\dfrac{H}{15} = \dfrac{7}{4}$, we find $H = \dfrac{15 \times 7}{4} = 26.25$ feet. We could also solve this problem using similar triangles.

TRY IT YOURSELF 6C.33   Brief answers provided at the end of the section.

How tall is a tower that casts a shadow of 20 feet at the same time?

TRY IT YOURSELF ANSWERS

6C.23 $\sin s = \dfrac{2}{\sqrt{13}}$, $\cos s = \dfrac{3}{\sqrt{13}}$, $\tan s = \dfrac{2}{3}$,

$\cot s = \dfrac{3}{2}$, $\sec s = \dfrac{\sqrt{13}}{3}$, $\csc s = \dfrac{\sqrt{13}}{2}$

6C.24 $\sin v = \dfrac{\sqrt{8}}{3}$, $\cos v = -\dfrac{1}{3}$, $\tan v = -\sqrt{8}$,

$\cot v = -\dfrac{1}{\sqrt{8}}$, $\sec v = -3$, $\csc v = \dfrac{3}{\sqrt{8}}$

6C.25 One possibility is

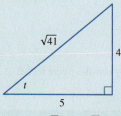

6C.26 $\sin t = \dfrac{\sqrt{8}}{3} = \dfrac{2\sqrt{2}}{3}$

6C.27 $\sin s = \dfrac{2}{\sqrt{x^2 + 6x + 13}}$

6C.28 $\sin t = -\dfrac{1}{\sqrt{5}}$, $\cos t = -\dfrac{2}{\sqrt{5}}$, $\cot t = 2$,

$\sec t = -\dfrac{\sqrt{5}}{2}$, $\csc t = -\sqrt{5}$

6C.29 $A = \dfrac{3\sqrt{3}}{2}$, $B = \dfrac{3}{2}$, $t = \dfrac{\pi}{3}$ radians

6C.30 $t = 65°$, $A \approx 2.72$, $B \approx 1.27$

6C.31 $10\sqrt{3} \approx 17.32$

6C.32 $300 \tan 27° \approx 152.86$ feet

6C.33 35 feet

EXERCISE SET 6C.4

CHECK YOUR UNDERSTANDING

1. Express sin t using the sides of the triangle in **Figure 6C.88**.

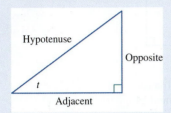

Figure 6C.88 Sides of a triangle labeled

2. Express cos t using the sides of the triangle in Figure 6C.88.

3. Express tan t using the sides of the triangle in Figure 6C.88.

4. Express cot t using the sides of the triangle in Figure 6C.88.

5. Express sec t using the sides of the triangle in Figure 6C.88.

6. Express csc t using the sides of the triangle in Figure 6C.88.

7. Express the adjacent side in Figure 6C.88 in terms of the hypotenuse and a trigonometric function of the angle t.

8. Express the opposite side in Figure 6C.88 in terms of the hypotenuse and a trigonometric function of the angle t.

9. Express the opposite side in Figure 6C.88 in terms of the adjacent side and a trigonometric function of the angle t.

10. The instruction to "solve a right triangle" means _____.

SKILL BUILDING

11. **The 3–4–5 right triangle**. Any triangle with sides 3, 4, and 5 is a right triangle, known as the 3–4–5 right triangle. It is shown in **Figure 6C.89**. Find the sine, cosine, tangent, secant, cosecant, and cotangent of the angle s.

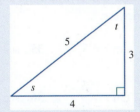

Figure 6C.89 The 3–4–5 right triangle

12. **More on the 3–4–5 right triangle.** Calculate the values of the six trigonometric functions of the angle t in Figure 6C.89.

Calculating values of trigonometric functions. In Exercises 13 through 26, calculate the values of the six trigonometric functions of the given angle.

13. The angle t in **Figure 6C.90**

14. The angle s in Figure 6C.90

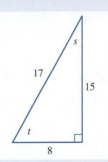

Figure 6C.90 Triangle for Exercises 13 and 14

15. The angle t in **Figure 6C.91**

16. The angle s in Figure 6C.91

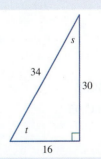

Figure 6C.91 Triangle for Exercises 15 and 16

17. The angle t in **Figure 6C.92**

18. The angle s in Figure 6C.92

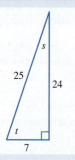

Figure 6C.92 Triangle for Exercises 17 and 18

19. The angle t in **Figure 6C.93**

20. The angle s in Figure 6C.93

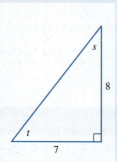

Figure 6C.93 Triangle for Exercises 19 and 20

21. The angle t in **Figure 6C.94**

22. The angle s in Figure 6C.94

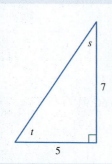

Figure 6C.94 Triangle for Exercises 21 and 22

23. The angle t in **Figure 6C.95**

24. The angle s in Figure 6C.95

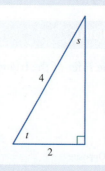

Figure 6C.95 Triangle for Exercises 23 and 24

25. The angle t in **Figure 6C.96**

26. The angle s in Figure 6C.96

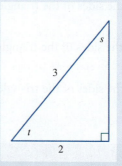

Figure 6C.96 Triangle for Exercises 25 and 26

Making triangles. In Exercises 27 through 38, construct a right triangle that exhibits the given value of a trigonometric function of an acute angle s. Be sure to label all three sides of the triangle and to mark the angle s. There are many correct answers.

27. $\sin s = \dfrac{2}{3}$ **31.** $\tan s = 5$ **35.** $\sec s = \dfrac{5}{3}$

28. $\sin s = \dfrac{3}{4}$ **32.** $\tan s = \dfrac{3}{4}$ **36.** $\sec s = 3$

29. $\cos s = \dfrac{1}{4}$ **33.** $\cot s = \dfrac{4}{5}$ **37.** $\csc s = \dfrac{5}{4}$

30. $\cos s = \dfrac{2}{5}$ **34.** $\cot s = 2$ **38.** $\csc s = 2$

Getting other trigonometric functions from a given one. In Exercises 39 through 46, you are given the value of one of the trigonometric functions of an acute angle t. Find the values of the other five trigonometric functions of t. If you have difficulty, refer to Example 6C.28.

39. $\sin t = \dfrac{2}{3}$

40. $\cos t = \dfrac{1}{5}$

41. $\tan t = \dfrac{5}{8}$

42. $\sec t = \dfrac{5}{4}$

43. $\csc t = 7$

44. $\cot t = 1$

45. $\tan t = \dfrac{2}{5}$

46. $\tan t = 4$

Getting a length. In Exercises 47 through 52, you are given the value of a trigonometric function of an acute angle and the length of one side of a right triangle with that angle. Use this information to calculate the lengths of the remaining sides of the triangle. You may find it helpful to draw a picture of the right triangle. Report answers correct to two decimal places.

47. The sine of an acute angle in a right triangle is 0.11. The length of the opposite side is 8.

48. The cosine of an acute angle in a right triangle is 0.71. The length of the adjacent side is 8.

49. The tangent of an acute angle in a right triangle is 4. The length of the adjacent side is 7.

50. The cotangent of an acute angle in a right triangle is $\dfrac{2}{5}$. The length of the opposite side is 9.

51. The secant of an acute angle in a right triangle is 5. The length of the hypotenuse is 6.

52. The cosecant of an acute angle in a right triangle is 3. The length of the opposite side is 9.

PROBLEMS

53. Basic trigonometric relationships. In this section we showed that the trigonometric functions can be defined in terms of the hypotenuse, adjacent side, and opposite side of a right triangle. Using these definitions, verify the following basic identities.

a. $\tan t = \dfrac{\sin t}{\cos t}$

b. $\cot t = \dfrac{\cos t}{\sin t}$

c. $\sec t = \dfrac{1}{\cos t}$

d. $\csc t = \dfrac{1}{\sin t}$

54. Calculating some trigonometric functions. In a certain right triangle, the side opposite one of the acute angles t is three times as long as the adjacent side. Find the sine, cosine, and tangent of t.

Getting other trigonometric function values from a given one. In Exercises 55 through 66, you are given the value (expressed in terms of x) of one of the trigonometric functions of an acute angle t. Find the values of the other five trigonometric functions of t in terms of x. If you have difficulty, you may wish to review Example 6C.28.

55. $\sin t = x$

56. $\cos t = x$

57. $\tan t = x$

58. $\sec t = x$

59. $\csc t = x$

60. $\cot t = x$

61. $\cot t = 3x$

62. $\cos t = 2x - 1$

63. $\tan t = \dfrac{x}{x+1}$

64. $\cot t = \dfrac{2x+1}{3x+1}$

65. $\sin t = \dfrac{x}{x+2}$

66. $\sin t = \dfrac{1-x}{1+x}$

Area and perimeter. In Exercises 67 through 72, some angles, and the length of a side, of a triangle are given. The altitude of the triangle is included. Find the area and perimeter of the triangle. Note that you may wish first to find some of the unlabeled sides.

67.

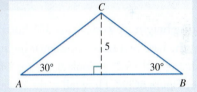

68.

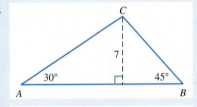

69.

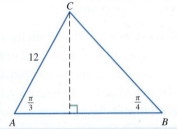

70.

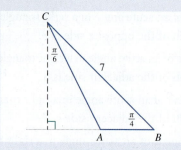

71.

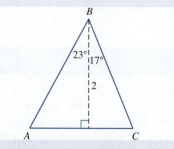

72.

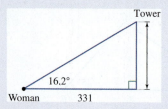

73. Complements of angles. Suppose t is an acute angle as shown in **Figure 6C.97**. The complement

of t is the angle $s = \dfrac{\pi}{2} - t$ if we use radians, or $s = 90° - t$ if we use degrees.

Show that in terms of degrees	and in terms of radians
$\tan(90° - t) = \cot t$	$\tan\left(\dfrac{\pi}{2} - t\right) = \cot t$
$\cot(90° - t) = \tan t$	
$\sec(90° - t) = \csc t$	$\cot\left(\dfrac{\pi}{2} - t\right) = \tan t$
$\csc(90° - t) = \sec t$	
	$\sec\left(\dfrac{\pi}{2} - t\right) = \csc t$
	$\csc\left(\dfrac{\pi}{2} - t\right) = \sec t$

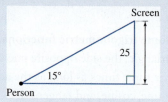

Figure 6C.97 The degree measure of s is 90° minus the degree measure of t.

MODELS AND APPLICATIONS

74. Calculating height. A woman sits 331 horizontal feet from the base of a tower. She must incline her eyes at an angle of 16.2° to look at the top of the tower. See **Figure 6C.98**. How tall is the tower?

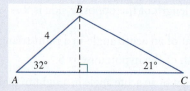

Figure 6C.98 Height of a tower

75. Home theater. The top of the screen for a home theater is 25 inches above the eye level of a person seated in a chair. For the sake of comfort, the top of the screen should be 15° above eye level. See **Figure 6C.99**. How far from the screen should the chair be placed? Round your answer to the nearest inch.

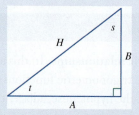

Figure 6C.99 Home theater

76. Speed of a drone. A drone is flying due north at a constant speed. An observer uses a laser to locate the drone due east at a distance of 130 feet. One minute later the observer rotates the laser by 40° to locate the drone. See **Figure 6C.100**. How fast is the drone traveling? Report your answer in feet per minute to the nearest whole number.

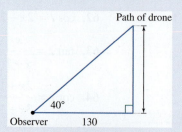

Figure 6C.100 Locating a drone

77. **Calculating distance.** A man sits 18 horizontal feet from the base of a tree. He must incline his eyes at an angle of 21° to look at the top of the tree. See **Figure 6C.101**. What is the distance from the man directly to the top of the tree?

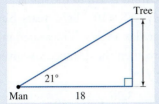

Figure 6C.101 Distance to the top of a tree

78. **The width of a river.** You are a surveyor. You stand on the north bank of a river and look due south at a tree on the opposite bank. Your friend on the opposite bank measures 35 yards due east to a second tree. You must rotate your transit through an angle of 21° to point toward the second tree. See **Figure 6C.102**. How wide is the river?

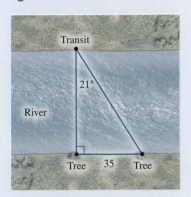

Figure 6C.102 Width of a river

79. **Grasping prey:** The diagram in **Figure 6C.103** is useful for estimating the diameter of a circular prey that would be optimal for the grasping claws of a praying mantis.[3] Find a formula for the diameter $|BC|$ of the circle in terms of the length $|AC|$ and the angle t. (The vertical bars denote the length of a segment.)

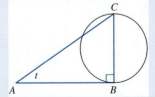

Figure 6C.103 The claws of a praying mantis

80. **Dispersal method.** When there is moisture on the ground, microorganisms are found in the thin layer of water on leaves.[4] A monolayer containing these organisms is formed on the top, and (because of the spreading pressure of this monolayer) the floating organisms spread *upward* when the surface is tilted. This is a dispersal method onto newly fallen leaves. In **Figure 6C.104**, the horizontal segment represents the ground, and the angle t measures the amount by which the surface is tilted. (We assume that t is less than 90°.) The length d is the distance the organisms move, and the length v is the change in the elevation.

   a. Find a formula giving v in terms of d and t.

   b. Assume that the distance the organisms move is 30 centimeters. Find the change in elevation if the angle t is 15° and if the angle t is 30°.

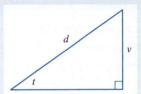

Figure 6C.104 Distance moved and elevation

81. **Parallax angle.** If we view a star now and then view it again half a year later, our position will have changed by the diameter of Earth's orbit around the sun. For nearby stars (within 100 light-years or so, such as Vega), the change in viewing location is sufficient to make the star appear to be in a slightly different location in the sky. Half of the angle from one location to the next is known as the parallax angle. In **Figure 6C.105**, the parallax angle is marked t, d is the distance from Earth to the sun, and s is the distance from the sun to the star. Show that we get the distance s using $s = d \cot t$.

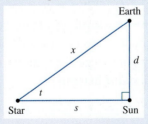

Figure 6C.105 Parallax angle

[3]Based on the work of C. S. Holling, as described by E. R. Pianka in *Evolutionary Ecology*, 6th ed. (Benjamin Cummings, San Francisco, CA, 2000).

[4]Based on the study by R. J. Bandoni and R. E. Koske, "Monolayers and Microbial Dispersal," *Science* 183: 1079–1081 (1974).

82. **More on parallax angle.** This exercise uses the description of parallax angles in the preceding exercise. Parallax angles are very small and are normally measured in seconds of arc. One second of arc is $\dfrac{1}{60^2}$ degree. At the same time, the distance s is very large and is normally measured in light-years. The angle t is so small and the distance s is so large that s is very nearly the same as the distance x marked in Figure 6C.105, and astronomers find this distance more convenient to calculate.

    a. Show that x is given by $x = \dfrac{d}{\sin t}$.

    b. Parallax angles are small enough that the approximation $\sin t \approx t$ is quite good,

    for t measured in radians. Show that if t is measured in seconds of arc, then

    $$\sin t \approx \frac{\pi t}{180 \times 60^2}$$

    c. The distance from Earth to the sun is approximately 1.58×10^{-5} light-years. Show that if the parallax angle t is measured in seconds of arc, then the distance x is given approximately by

    $$x = \frac{3.26}{t} \text{ light-years}$$

CHALLENGE EXERCISES FOR INDIVIDUALS OR GROUPS

83. **An important calculus inequality.** In calculus we calculate the instantaneous rate of change of the sine function (which turns out to be the cosine). The inequality

    $$\cos t < \frac{t}{\sin t} < \frac{1}{\cos t}$$

    for $0 < t < \dfrac{\pi}{2}$ is a key part of this calculation. The goal of this exercise is to establish this inequality. In **Figure 6C.106**, the acute angle t is measured in radians.

    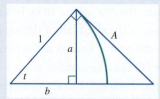

    Figure 6C.106 Two right triangles and a sector of a circle

    a. In the figure we see a small right triangle with base b and altitude a, a sector of a circle of radius 1, and a large right triangle with altitude A. Calculate a, b, and A in terms of t.

    b. In the figure

    Area of small triangle < Area of sector < Area of large triangle

    Use this together with your work from part a to show

    $$\frac{\sin t \cos t}{2} < \frac{t}{2} < \frac{\tan t}{2}$$

    c. Use your work from part b to show that if $0 < t < \dfrac{\pi}{2}$ then

    $$\cos t < \frac{t}{\sin t} < \frac{1}{\cos t}$$

84. **Approximating the sine function.** If an angle t is measured in radians, and if t is near zero, then $\sin t \approx t$. To see why this approximation is reasonable, consider **Figure 6C.107**. When t is small, the distance from A to B is very close to the arc length from A to C. Show that this comparison yields $\sin t \approx t$.

    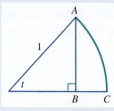

    Figure 6C.107 Approximating arc length: When t is near 0, the arc length approximates the vertical height.

85. **The rate of change of the sine function.** The goal of this exercise is to give an intuitive justification for the fact that the rate of change of the sine function is the cosine. We can think of the rate of change of $y = \sin t$ as the value that the average rate of change $\dfrac{\Delta y}{\Delta t}$ approaches when Δt is near zero. A bit of geometry can show that, when Δt is small, $\angle s$ in **Figure 6C.108** is nearly equal to $\angle t$, and the length of PQ is approximately Δt. It is a bit of a challenge to show this, but if you wish to try, consider the fact that when Δt is very small, the line $\overline{PQ}$ is nearly tangent to the unit circle. You may proceed assuming these facts to be true. **Figure 6C.109** is a blown-up version of the right triangle in Figure 6C.108. Use this triangle to show that $\cos t$ is close to $\dfrac{\Delta y}{\Delta t}$ when Δt is small.

Complete the argument.

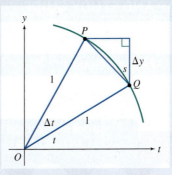

Figure 6C.108 When Δt is small: $\angle s$ is close to $\angle t$ and $|PQ|$ is close to Δt.

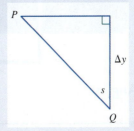

Figure 6C.109 A blown-up version of the right triangle in Figure 6C.108

REVIEW AND REFRESH: Exercises from Previous Sections

86. **From Section P.1:** Find the point midway between $(2, 7)$ and $(4, 9)$.

    Answer: $(3, 8)$

87. **From Section P.3:** Solve the inequality $\dfrac{x}{x-1} < 2$.

    Answer: $(-\infty, 1) \cup (2, \infty)$

88. **From Section 1.2:** Find the average rate of change of $f(x) = x^2$ from x to $x + h$ with $h \neq 0$. Simplify your answer.

    Answer: $2x + h$

89. **From Section 1.4:** Find the limit as $x \to \infty$ of $y = \dfrac{x+1}{x-1}$.

    Answer: 1

90. **From Section 3.1:** Find a linear function with initial value 4 and rate of change 7.

    Answer: $y = 7x + 4$

91. **From Section 4.2:** Write 2^x in the form e^{rx}.

    Answer: $2^x = e^{x \ln 2}$

92. **From Section 4.3:** Solve the equation $\ln x + \ln(x+1) = 1$.

    Answer: $x = \dfrac{-1 + \sqrt{1 + 4e}}{2}$

93. **From Section 6C.1:** Find the length of the arc of a circle of radius 4 cut by an angle of $\dfrac{\pi}{3}$ radians whose vertex lies at the center of the circle.

    Answer: $\dfrac{4\pi}{3}$

94. **From Section 6C.3:** Find the values of the remaining five trigonometric functions of t if $\sin t = \dfrac{1}{3}$ and $\cos t$ is negative.

    Answer:

    $\cos t = -\dfrac{2\sqrt{2}}{3}$, $\tan t = -\dfrac{1}{2\sqrt{2}}$, $\cot t = -2\sqrt{2}$,

    $\sec t = -\dfrac{3}{2\sqrt{2}}$, $\csc t = 3$

CHAPTER ROADMAP AND STUDY GUIDE

RATES OF CHANGE

Introduction to Trigonometry: A Unit Circle Approach

TRIGONOMETRY

BASICS OF FUNCTIONS

FOUR MAJOR CLASSES OF FUNCTIONS

GEOMETRY

RATES OF CHANGE

P — 1 — 2 — 3 — 4 — 5 — 6C OR 6T — 7 — 8 — 9 — 10

6C.1   Angles

Trigonometry depends on the measurement of angles.

Conventions for Angles

Angles can be thought of as rotations from an initial ray to a terminal ray.

Degree Measure

Degree measure of angles is probably familiar from everyday experience.

Radian Measure

Radian measure is calculated by using the radius of a circle to measure arcs along the circle.

Converting Angle Measures

Radian and degree measures are related by simple formulas.

Arc Length and Area

Arc length and area are calculated for sectors of circles.

Similar Triangles

Similarity is at the foundation of trigonometry.

MODELS AND APPLICATIONS
Shadows

6C.2   The Unit Circle

The unit circle can be used to present the trigonometric functions.

Finding Points on the Unit Circle

If one coordinate of a point on the unit circle is given, the other coordinate can be found using the equation of the circle.

Compass Points on the Unit Circle

The coordinates of the trigonometric points corresponding to the four compass points on the unit circle are calculated.

Special Angles

Geometric ideas are used to find coordinates of trigonometric points corresponding to certain special angles.

Quadrants and Reference Angles

Trigonometric points are associated with angles in the first quadrant.

Using Reference Numbers to Find Coordinates of Trigonometric Points

Trigonometric points in the first quadrant are used to find trigonometric points anywhere on the unit circle.

6C.3 The Trigonometric Functions

> Trigonometric points on the unit circle determine the trigonometric functions.

Defining the Trigonometric Functions

> The coordinates of trigonometric points are used to define the six trigonometric functions.

Trigonometric Functions of Special Angles

> Trigonometric points for special angles are used to evaluate corresponding trigonometric functions.

Trigonometric Functions of Compass Points

> The trigonometric functions of the four compass points are evaluated using the coordinates that were calculated earlier.

Signs of Trigonometric Functions in the Four Quadrants

> Determining the sign of trigonometric functions is a key component of their evaluation.

Using Reference Numbers to Evaluate Trigonometric Functions

> The values of the trigonometric functions for the first quadrant can be used to evaluate trigonometric functions for any quadrant.

The Pythagorean Identity

> The formula for the unit circle gives a fundamental relationship between the sine and cosine functions.

6C.4 Right Triangle Trigonometry

> Trigonometry is closely connected to the study of right triangles.

Trigonometric Functions of Acute Angles

> Trigonometric functions of acute angles can be defined using right triangles.

Using Right Triangles to Evaluate Trigonometric Functions

> Trigonometric functions of acute angles can be calculated in terms of side lengths of a right triangle.

Representing Angles

> An angle with a given trigonometric function value can be represented by a right triangle.

Using Representative Triangles to Evaluate Trigonometric Functions

> Triangles representing one trigonometric function value can be used to find the values of other trigonometric functions.

Solving Right Triangles

> When one side and one acute angle of a right triangle are known, the remaining sides and angle may be calculated.

A New Formula for the Area of a Triangle

> The trigonometry of right triangles yields a useful formula for the area of a triangle.

MODELS AND APPLICATIONS
Transits and Shadows

CHAPTER QUIZ

1. Find an angle with radian measure between 0 and 2π that is coterminal with an angle of $-\dfrac{16\pi}{3}$.

   Answer: $\dfrac{2\pi}{3}$

   XR Example 6C.2

2. What is the radian measure of an angle of $150°$?

   Answer: $\dfrac{5\pi}{6}$

   XR Example 6C.3

3. Find the arc length and area subtended by an angle of $\dfrac{\pi}{6}$ radian with vertex at the center of a circle of radius 3 units.

   Answer: Arc length: $\dfrac{\pi}{2}$ units. Area: $\dfrac{3\pi}{4}$ square units

   XR Example 6C.5

4. Find the unknown sides and angles of the similar triangles shown in **Figure 6C.110**.

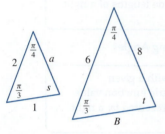

   Figure 6C.110

   Answer: $s = t = \dfrac{5\pi}{12}$, $a = \dfrac{8}{3}$, $B = 3$

   XR Example 6C.7

5. Find all possible values of x if $\left(x, \dfrac{3}{4}\right)$ lies on the unit circle.

   Answer: $x = \pm\dfrac{\sqrt{7}}{4}$

   XR Example 6C.9

6. Find the coordinates of the trigonometric point $P(3\pi)$.

   Answer: $(-1, 0)$

   XR Example 6C.10

7. Find the reference number for $\dfrac{19\pi}{6}$. Then find the coordinates of the trigonometric point $P\left(\dfrac{19\pi}{6}\right)$.

   Answer: Reference number $\dfrac{\pi}{6}$. $P\left(\dfrac{19\pi}{6}\right) = \left(-\dfrac{\sqrt{3}}{2}, -\dfrac{1}{2}\right)$

   XR Example 6C.13

8. Find the values of the six trigonometric functions of the 30° angle.

Answer: $\sin 30° = \dfrac{1}{2}$, $\cos 30° = \dfrac{\sqrt{3}}{2}$, $\tan 30° = \dfrac{1}{\sqrt{3}}$, $\cot 30° = \sqrt{3}$,

$\sec 30° = \dfrac{2}{\sqrt{3}}$, $\csc 30° = 2$

XR **Example 6C.16**

9. Find the values of the six trigonometric functions of 7π. Note which functions are undefined.

Answer: $\sin 7\pi = 0$, $\cos 7\pi = -1$, $\tan 7\pi = 0$, $\cot 7\pi$ undefined,
$\sec 7\pi = -1$, $\csc 7\pi$ undefined

XR **Example 6C.17**

10. Find the values of the six trigonometric functions of $\dfrac{5\pi}{3}$.

Answer: $\sin\left(\dfrac{5\pi}{3}\right) = -\dfrac{\sqrt{3}}{2}$, $\cos\left(\dfrac{5\pi}{3}\right) = \dfrac{1}{2}$, $\tan\left(\dfrac{5\pi}{3}\right) = -\sqrt{3}$,

$\cot\left(\dfrac{5\pi}{3}\right) = -\dfrac{1}{\sqrt{3}}$, $\sec\left(\dfrac{5\pi}{3}\right) = 2$, $\csc\left(\dfrac{5\pi}{3}\right) = -\dfrac{2}{\sqrt{3}}$

XR **Example 6C.19**

11. Assume that $\cos t = \dfrac{1}{5}$. Find the values of the remaining five trigonometric functions of t if the sine of t is negative.

Answer: $\sin t = -\dfrac{2\sqrt{6}}{5}$, $\tan t = -2\sqrt{6}$, $\cot t = -\dfrac{1}{2\sqrt{6}}$, $\sec t = 5$, $\csc t = -\dfrac{5}{2\sqrt{6}}$

XR **Example 6C.21**

12. Calculate the values of the six trigonometric functions of the angle t using **Figure 6C.111**.

Figure 6C.111

Answer: $\sin t = \dfrac{2\sqrt{2}}{3}$, $\cos t = \dfrac{1}{3}$, $\tan t = 2\sqrt{2}$, $\cot t = \dfrac{1}{2\sqrt{2}}$, $\sec t = 3$, $\csc t = \dfrac{3}{2\sqrt{2}}$

XR **Example 6C.23**

13. Let t be an acute angle with $\cot t = \dfrac{2}{3}$. Find the values of the remaining five trigonometric functions of t.

Answer: $\sin t = \dfrac{3}{\sqrt{13}}$, $\cos t = \dfrac{2}{\sqrt{13}}$, $\tan t = \dfrac{3}{2}$, $\sec t = \dfrac{\sqrt{13}}{2}$, $\csc t = \dfrac{\sqrt{13}}{3}$

XR **Example 6C.26**

14. If $\sin t = \dfrac{3}{4}$ and the trigonometric point $P(t)$ is in the second quadrant, find the values of the remaining five trigonometric functions of t.

Answer: $\cos t = -\dfrac{\sqrt{7}}{4}$, $\tan t = -\dfrac{3}{\sqrt{7}}$, $\cot t = -\dfrac{\sqrt{7}}{3}$, $\sec t = -\dfrac{4}{\sqrt{7}}$, $\csc t = \dfrac{4}{3}$

XR **Example 6C.28**

15. Find the unknown sides and angle in **Figure 6C.112**.

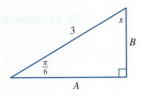

Figure 6C.112

Answer: $s = \dfrac{\pi}{3}$, $A = \dfrac{3\sqrt{3}}{2}$, $B = \dfrac{3}{2}$

 **Example 6C.29**

16. Find the area of the triangle in **Figure 6C.113**.

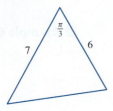

Figure 6C.113

Answer: $\dfrac{21\sqrt{3}}{2}$

XR Example 6C.31

CHAPTER REVIEW EXERCISES

Section 6C.1

Coterminal angles. In Exercises 1 through 6, find an angle between 0° and 360° or between 0 and 2π radians that is coterminal with the given angle.

1. 400°

    Answer: 40°

2. −180°

    Answer: 180°

3. 7π radians

    Answer: π radians

4. $\dfrac{9\pi}{2}$ radians

    Answer: $\dfrac{\pi}{2}$ radians

5. −30°

    Answer: 330°

6. $-\dfrac{13\pi}{6}$ radians

    Answer: $\dfrac{11\pi}{6}$ radians

Converting from radians to degrees. In Exercises 7 through 12, radian measures of angles are given. Convert these to degree measure.

7. $\dfrac{\pi}{2}$ radians

    Answer: 90°

8. $\dfrac{\pi}{5}$ radian

    Answer: 36°

9. π radians

    Answer: 180°

10. $\dfrac{\pi}{3}$ radians

    Answer: 60°

11. $\dfrac{\pi}{9}$ radian

   Answer: 20°

12. 2 radians

   Answer: $\dfrac{360}{\pi} \approx 114.59°$

Converting from degrees to radians. In Exercises 13 through 18, degree measures of angles are given. Convert these to radian measure.

13. 30°

   Answer: $\dfrac{\pi}{6}$ radian

14. 20°

   Answer: $\dfrac{\pi}{9}$ radian

15. 45°

   Answer: $\dfrac{\pi}{4}$ radian

16. 120°

   Answer: $\dfrac{2\pi}{3}$ radians

17. $\pi°$

   Answer: $\dfrac{\pi^2}{180} \approx 0.05$ radian

18. 50°

   Answer: $\dfrac{5\pi}{18}$ radian

Area and arc length. In Exercises 19 through 24, you are given the radius of a circle and an angle whose vertex is at the center of the circle. Find the area and arc length cut from the circle by the angle.

19. Radius 3 units. Angle $\dfrac{\pi}{6}$ radian.

   Answer: Arc length: $\dfrac{\pi}{2}$ units

   Area: $\dfrac{3\pi}{4}$ square units

20. Radius 3 units. Angle $\dfrac{\pi}{3}$ radians.

   Answer: Arc length: π units

   Area: $\dfrac{3\pi}{2}$ square units

21. Radius 2 units. Angle $\dfrac{\pi}{4}$ radian.

   Answer: Arc length: $\dfrac{\pi}{2}$ units

   Area: $\dfrac{\pi}{2}$ square units

22. Radius 4 units. Angle $\dfrac{\pi}{8}$ radian.

   Answer: Arc length: $\dfrac{\pi}{2}$ units

   Area: π square units

23. Radius 2 units. Angle 30°.

   Answer: Arc length: $\dfrac{\pi}{3}$ units

   Area: $\dfrac{\pi}{3}$ square units

24. Radius 1 unit. Angle 20°.

   Answer: Arc length: $\dfrac{\pi}{9}$ unit

   Area: $\dfrac{\pi}{18}$ square unit

Similar triangles. Exercises 25 through 28 show similar triangles. Find all unknown sides.

25.

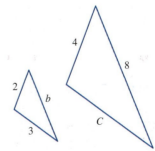

   Answer: $b = 4, C = 6$

26.

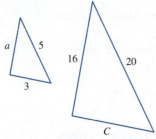

Answer: $a = 4, C = 12$

27.

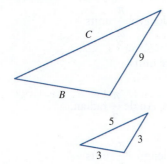

Answer: $B = 9, C = 15$

28.

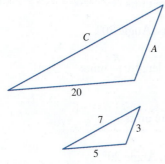

Answer: $A = 12, C = 28$

Section 6C.2

Points on the unit circle. In Exercises 29 through 34, find all values of the variable so that the indicated point lies on the unit circle and in the given quadrant.

29. $\left(x, -\dfrac{1}{2}\right)$, Quadrant IV

Answer: $x = \dfrac{\sqrt{3}}{2}$

30. $\left(\dfrac{2}{3}, y\right)$, Quadrant I

Answer: $y = \dfrac{\sqrt{5}}{3}$

31. $\left(x, \dfrac{3}{5}\right)$, Quadrant II

Answer: $x = -\dfrac{4}{5}$

32. $\left(-\dfrac{3}{4}, y\right)$, Quadrant III

Answer: $y = -\dfrac{\sqrt{7}}{4}$

33. $(x, -x)$, Quadrant IV

Answer: $x = \dfrac{1}{\sqrt{2}}$

34. $(x, 2x)$, Quadrant I

Answer: $x = \dfrac{1}{\sqrt{5}}$

Reference angles. In Exercises 35 through 40, find the reference angle for the given angle. If the angle is given in degrees, your answer should be in degrees. If the angle is given in radians, your answer should be in radians.

35. $\dfrac{5\pi}{4}$ radians

Answer: $\dfrac{\pi}{4}$ radian

36. $120°$

Answer: $60°$

37. $-\dfrac{2\pi}{3}$ radians

Answer: $\dfrac{\pi}{3}$ radians

38. $-30°$

Answer: $30°$

39. $\dfrac{7\pi}{6}$ radians

Answer: $\dfrac{\pi}{6}$ radians

40. $170°$

Answer: $10°$

Calculating trigonometric points related to special angles. In Exercises 41 through 48, find the coordinates of the trigonometric point associated with the given angle. In each case, the appropriate reference angle is one of the special angles.

41. $135°$

Answer: $\left(-\dfrac{1}{\sqrt{2}}, \dfrac{1}{\sqrt{2}}\right)$

42. $-\dfrac{\pi}{3}$

Answer: $\left(\dfrac{1}{2}, -\dfrac{\sqrt{3}}{2}\right)$

43. $210°$

Answer: $\left(-\dfrac{\sqrt{3}}{2}, -\dfrac{1}{2}\right)$

44. $\dfrac{13\pi}{6}$

Answer: $\left(\dfrac{\sqrt{3}}{2}, \dfrac{1}{2}\right)$

45. $-\dfrac{2\pi}{3}$

Answer: $\left(-\dfrac{1}{2}, -\dfrac{\sqrt{3}}{2}\right)$

46. $315°$

Answer: $\left(\dfrac{1}{\sqrt{2}}, -\dfrac{1}{\sqrt{2}}\right)$

47. $\dfrac{7\pi}{6}$

Answer: $\left(-\dfrac{\sqrt{3}}{2}, -\dfrac{1}{2}\right)$

48. $240°$

Answer: $\left(-\dfrac{1}{2}, -\dfrac{\sqrt{3}}{2}\right)$

Section 6C.3

Special angles and compass points. Use your knowledge of special angles and compass points to find the values of the six trigonometric functions of the angles or real numbers given in Exercises 49 through 58. State which, if any, are undefined. Remember that if degrees are not specifically indicated, then radian measure is assumed.

49. $0°$

Answer:
$\sin 0° = 0$, $\cos 0° = 1$, $\tan 0° = 0$, $\sec 0° = 1$. The cotangent and cosecant are undefined.

50. $210°$

Answer: $\sin 210° = -\dfrac{1}{2}$, $\cos 210° = -\dfrac{\sqrt{3}}{2}$,
$\tan 210° = \dfrac{1}{\sqrt{3}}$, $\cot 210° = \sqrt{3}$,
$\sec 210° = -\dfrac{2}{\sqrt{3}}$, $\csc 210° = -2$

51. $\dfrac{7\pi}{2}$

Answer: $\sin\left(\dfrac{7\pi}{2}\right) = -1$, $\cos\left(\dfrac{7\pi}{2}\right) = 0$,
$\cot\left(\dfrac{7\pi}{2}\right) = 0$, $\csc\left(\dfrac{7\pi}{2}\right) = -1$. The tangent and secant are undefined.

52. $180°$

Answer: $\sin 180° = 0$, $\cos 180° = -1$, $\tan 180° = 0$, $\sec 180° = -1$. The cotangent and cosecant are undefined.

53. $135°$

Answer: $\sin 135° = \dfrac{1}{\sqrt{2}}$, $\cos 135° = -\dfrac{1}{\sqrt{2}}$,
$\tan 135° = -1$, $\cot 135° = -1$,
$\sec 135° = -\sqrt{2}$, $\csc 135° = \sqrt{2}$

54. $\dfrac{5\pi}{4}$

Answer: $\sin\left(\dfrac{5\pi}{4}\right) = -\dfrac{1}{\sqrt{2}}$, $\cos\left(\dfrac{5\pi}{4}\right) = -\dfrac{1}{\sqrt{2}}$,
$\tan\left(\dfrac{5\pi}{4}\right) = 1$, $\cot\left(\dfrac{5\pi}{4}\right) = 1$,
$\sec\left(\dfrac{5\pi}{4}\right) = -\sqrt{2}$, $\csc\left(\dfrac{5\pi}{4}\right) = -\sqrt{2}$

55. $270°$

Answer: $\sin 270° = -1$, $\cos 270° = 0$,
$\cot 270° = 0$, $\csc 270° = -1$. The tangent and secant are undefined.

56. $-\dfrac{\pi}{6}$

Answer: $\sin\left(-\dfrac{\pi}{6}\right) = -\dfrac{1}{2}$, $\cos\left(-\dfrac{\pi}{6}\right) = \dfrac{\sqrt{3}}{2}$,
$\tan\left(-\dfrac{\pi}{6}\right) = -\dfrac{1}{\sqrt{3}}$, $\cot\left(-\dfrac{\pi}{6}\right) = -\sqrt{3}$,
$\sec\left(-\dfrac{\pi}{6}\right) = \dfrac{2}{\sqrt{3}}$, $\csc\left(-\dfrac{\pi}{6}\right) = -2$

57. $\dfrac{\pi}{2}$

Answer: $\sin\left(\dfrac{\pi}{2}\right) = 1$, $\cos\left(\dfrac{\pi}{2}\right) = 0$,

$\cot\left(\dfrac{\pi}{2}\right) = 0$, $\csc\left(\dfrac{\pi}{2}\right) = 1$. The tangent and secant are undefined.

58. $\dfrac{\pi}{6}$

Answer: $\sin\left(\dfrac{\pi}{6}\right) = \dfrac{1}{2}$, $\cos\left(\dfrac{\pi}{6}\right) = \dfrac{\sqrt{3}}{2}$,

$\tan\left(\dfrac{\pi}{6}\right) = \dfrac{1}{\sqrt{3}}$, $\cot\left(\dfrac{\pi}{6}\right) = \sqrt{3}$,

$\sec\left(\dfrac{\pi}{6}\right) = \dfrac{2}{\sqrt{3}}$, $\csc\left(\dfrac{\pi}{6}\right) = 2$

Finding trigonometric function values from a known value. In Exercises 59 through 66, you are given the value of one trigonometric function. Find the values of the remaining five trigonometric functions.

59. $\cos t = \dfrac{3}{4}$, and $\sin t$ is positive.

Answer: $\sin t = \dfrac{\sqrt{7}}{4}$, $\tan t = \dfrac{\sqrt{7}}{3}$, $\cot t = \dfrac{3}{\sqrt{7}}$,

$\sec t = \dfrac{4}{3}$, $\csc t = \dfrac{4}{\sqrt{7}}$

60. $\tan t = \sqrt{3}$, and $\sin t$ is positive.

Answer: $\cos t = \dfrac{1}{2}$, $\sin t = \dfrac{\sqrt{3}}{2}$, $\cot t = \dfrac{1}{\sqrt{3}}$,

$\sec t = 2$, $\csc t = \dfrac{2}{\sqrt{3}}$

61. $\sin t = \dfrac{5}{6}$, and $\cos t$ is negative.

Answer: $\cos t = -\dfrac{\sqrt{11}}{6}$, $\tan t = -\dfrac{5}{\sqrt{11}}$,

$\cot t = -\dfrac{\sqrt{11}}{5}$, $\sec t = -\dfrac{6}{\sqrt{11}}$, $\csc t = \dfrac{6}{5}$

62. $\sin t = \dfrac{1}{6}$, and $\cos t$ is negative.

Answer: $\cos t = -\dfrac{\sqrt{35}}{6}$, $\tan t = -\dfrac{1}{\sqrt{35}}$,

$\cot t = -\sqrt{35}$, $\sec t = -\dfrac{6}{\sqrt{35}}$, $\csc t = 6$

63. $\cos t = -\dfrac{1}{3}$, and $\tan t$ is negative.

Answer: $\sin t = \dfrac{\sqrt{8}}{3}$, $\tan t = -\sqrt{8}$,

$\cot t = -\dfrac{1}{\sqrt{8}}$, $\sec t = -3$, $\csc t = \dfrac{3}{\sqrt{8}}$

64. $\sec t = 4$, and $\sin t$ is negative.

Answer: $\cos t = \dfrac{1}{4}$, $\sin t = -\dfrac{\sqrt{15}}{4}$,

$\tan t = -\sqrt{15}$, $\cot t = -\dfrac{1}{\sqrt{15}}$, $\csc t = -\dfrac{4}{\sqrt{15}}$

65. $\cot t = 1.10$, and $\sin t$ is negative. Report answers correct to two decimal places.

Answer: $\sin t = -0.67$, $\cos t = -0.74$,

$\tan t = 0.91$, $\sec t = -1.35$, $\csc t = -1.49$

66. $\cos t = 0.55$, and $\sin t$ is positive. Report answers correct to two decimal places.

Answer: $\sin t = 0.84$, $\tan t = 1.52$, $\cot t = 0.66$,

$\sec t = 1.82$, $\csc t = 1.20$

Section 6C.4

Calculating trigonometric function values from triangles. In Exercises 67 through 74, calculate the values of the six trigonometric functions of the given angle.

67. The angle s in **Figure 6C.114**

Answer: $\sin s = \dfrac{4}{5}$, $\cos s = \dfrac{3}{5}$, $\tan s = \dfrac{4}{3}$,

$\cot s = \dfrac{3}{4}$, $\sec s = \dfrac{5}{3}$, $\csc s = \dfrac{5}{4}$

68. The angle t in Figure 6C.114

Answer: $\sin t = \dfrac{3}{5}$, $\cos t = \dfrac{4}{5}$, $\tan t = \dfrac{3}{4}$,

$\cot t = \dfrac{4}{3}$, $\sec t = \dfrac{5}{4}$, $\csc t = \dfrac{5}{3}$

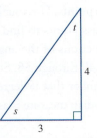

Figure 6C.114 Triangle for Exercises 67 and 68

69. The angle s in **Figure 6C.115**

Answer: $\sin s = \dfrac{\sqrt{7}}{4}$, $\cos s = \dfrac{3}{4}$, $\tan s = \dfrac{\sqrt{7}}{3}$,

$\cot s = \dfrac{3}{\sqrt{7}}$, $\sec s = \dfrac{4}{3}$, $\csc s = \dfrac{4}{\sqrt{7}}$

70. The angle t in Figure 6C.115

Answer: $\sin t = \dfrac{3}{4}$, $\cos t = \dfrac{\sqrt{7}}{4}$, $\tan t = \dfrac{3}{\sqrt{7}}$,

$\cot t = \dfrac{\sqrt{7}}{3}$, $\sec t = \dfrac{4}{\sqrt{7}}$, $\csc t = \dfrac{4}{3}$

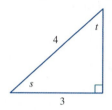

Figure 6C.115 Triangle for Exercises 69 and 70

71. The angle s in **Figure 6C.116**

Answer: $\sin s = \dfrac{2}{3}$, $\cos s = \dfrac{\sqrt{5}}{3}$, $\tan s = \dfrac{2}{\sqrt{5}}$,

$\cot s = \dfrac{\sqrt{5}}{2}$, $\sec s = \dfrac{3}{\sqrt{5}}$, $\csc s = \dfrac{3}{2}$

72. The angle t in Figure 6C.116

Answer: $\sin t = \dfrac{\sqrt{5}}{3}$, $\cos t = \dfrac{2}{3}$, $\tan t = \dfrac{\sqrt{5}}{2}$,

$\cot t = \dfrac{2}{\sqrt{5}}$, $\sec t = \dfrac{3}{2}$, $\csc t = \dfrac{3}{\sqrt{5}}$

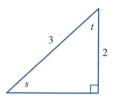

Figure 6C.116 Triangle for Exercises 71 and 72

73. The angle s in **Figure 6C.117**

Answer: $\sin s = \dfrac{1}{\sqrt{5}}$, $\cos s = \dfrac{2}{\sqrt{5}}$, $\tan s = \dfrac{1}{2}$,

$\cot s = 2$, $\sec s = \dfrac{\sqrt{5}}{2}$, $\csc s = \sqrt{5}$

74. The angle t in Figure 6C.117

Answer: $\sin t = \dfrac{2}{\sqrt{5}}$, $\cos t = \dfrac{1}{\sqrt{5}}$, $\tan t = 2$,

$\cot t = \dfrac{1}{2}$, $\sec t = \sqrt{5}$, $\csc t = \dfrac{\sqrt{5}}{2}$

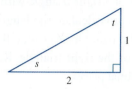

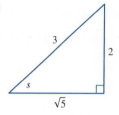

Figure 6C.117 Triangle for Exercises 73 and 74

Making triangles. In Exercises 75 through 78, construct a right triangle that exhibits the given value of a trigonometric function of an acute angle s. Be sure to label all three sides of the triangle and to mark the angle s. There are many correct answers.

75. $\sin s = \dfrac{2}{3}$

Answer:

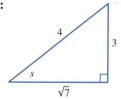

76. $\sin s = \dfrac{3}{4}$

Answer:

77. $\tan s = 5$

Answer:

78. $\csc s = \dfrac{5}{3}$

Answer:

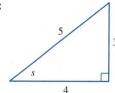

Finding lengths. In Exercises 79 through 82, you are given the value of a trigonometric function of an angle and the length of one side of a right triangle with that angle. Use this information to calculate the lengths of the remaining sides of the triangle. You may find it helpful to draw a picture of the right triangle. Report answers correct to two decimal places.

79. The sine of an acute angle in a right triangle is 0.15. The length of the opposite side is 5.

    Answer: Hypotenuse = 33.33.
    Adjacent side = 32.96.

80. The cosine of an acute angle in a right triangle is 0.50. The length of the adjacent side is 8.

    Answer: Hypotenuse = 16. Opposite side = 13.86.

81. The tangent of an acute angle in a right triangle is 5. The length of the adjacent side is 9.

    Answer: Opposite side = 45. Hypotenuse = 45.89.

82. The cotangent of an acute angle in a right triangle is $\dfrac{3}{5}$. The length of the opposite side is 25.

    Answer: Adjacent side = 15. Hypotenuse = 29.15.

83. **Calculating the distance to the top of an object.** A man sits 150 horizontal feet from the base of a wall. He must incline his eyes at an angle of 10° to look at the top of the wall. See **Figure 6C.118**. What is the distance from the man directly to the top of the wall?

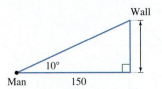

Figure 6C.118

Answer: 152.31 feet

INTRODUCTION TO TRIGONOMETRY

A RIGHT TRIANGLE APPROACH

PHILIPPE LOPEZ//AFP/Getty Images

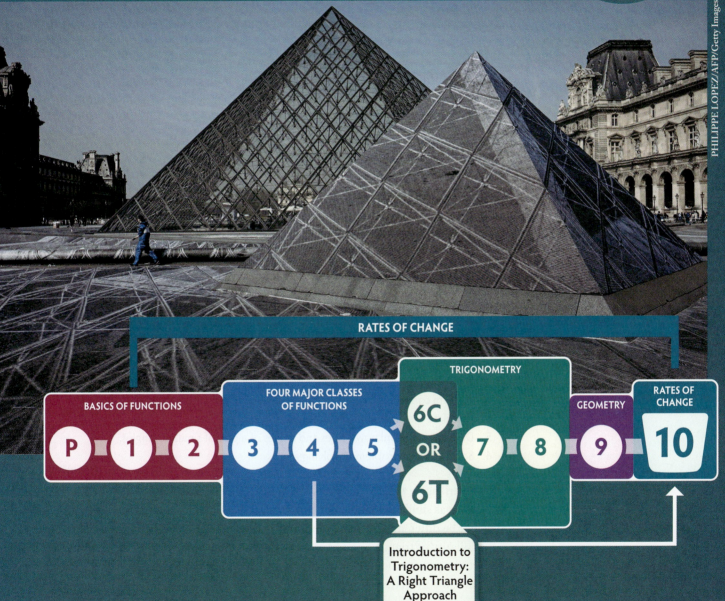

6T.1   Angles

6T.2   Definition of the Trigonometric Functions Using Right Triangles

6T.3   Analysis of Right Triangles

6T.4   Extending the Trigonometric Functions: The Unit Circle

The trigonometric functions are the final class of functions we will discuss. These functions are new to many, so we devote three chapters (6, 7, and 8) to their development. This chapter introduces the trigonometric functions, which appear throughout calculus and its applications. You will gain the important skill of analyzing these functions both by using the unit circle and by referring to right triangles.

Chapters 6C and 6T offer alternative approaches to trigonometry, covering the same material in a different order. Use either one, but not both.

After a brief refresher on angles, Chapter 6T begins with a discussion of right triangles and then proceeds to the unit circle. Chapter 6C is provided for those who prefer to introduce the trigonometric functions via the unit circle and then apply their definitions to right triangles. Students need to be able to use trigonometric functions in either context, and either chapter will provide appropriate background for students to accomplish this goal.

The word *trigonometry* comes from the Greek words for *triangle* and *measure*. As the name implies, trigonometry arose from a study of triangles—particularly right triangles (those with a 90° angle).

Early astronomy and architecture necessitated the development of trigonometry. In the third century BCE, Eratosthenes used rudimentary trigonometry to give a remarkably accurate estimate of Earth's circumference.

In this chapter we lay the foundation for trigonometry by studying angles and then introduce the six trigonometric functions, which play a key role in calculus as well as applications of mathematics.

6T.1 Angles

6T.2 Definition of the Trigonometric Functions Using Right Triangles

6T.3 Analysis of Right Triangles

6T.4 Extending the Trigonometric Functions: The Unit Circle

6T.1 Angles

> Trigonometry depends on the measurement of angles.

In this section, you will learn to:

1. Locate an angle with degree measure in a given range that is coterminal with a given angle.
2. Locate an angle with radian measure in a given range that is coterminal with a given angle.
3. Perform conversions between degree measure and radian measure.
4. Calculate the arc length and area determined by a sector of a circle.
5. Determine the unknown sides and angles of a pair of similar triangles.
6. Solve applied problems involving angles.

Two triangles with equal corresponding angles but possibly different side lengths are known as similar triangles. For two similar triangles, the shape is the same, but the size may be different. Understanding similar triangles is fundamental to trigonometry.

Similarity of triangles preserves angles and shapes, and it also preserves symmetry. In fact, the preservation of shape and symmetry applies in an even more general setting. Imagine the letter "A" presented using various type sizes:

ᴀ A A A A

Each presentation has the same shape and a vertical line of symmetry.

In this section, we introduce angle measure and show how it applies to triangles.

Conventions for Angles

> Angles can be thought of as rotations from an initial ray to a terminal ray.

An angle consists of two rays emanating from a common vertex. More precisely, an angle is a rotation of the initial ray to the terminal ray. Sometimes these rays are called the sides of the angle. A counterclockwise rotation gives a positive angle, and a clockwise rotation gives a negative angle. These cases are illustrated in **Figure 6T.1** and **Figure 6T.2**.

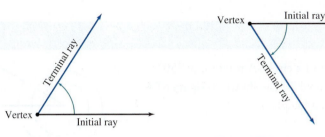

Figure 6T.1 A positive angle (counterclockwise rotation)

Figure 6T.2 A negative angle (clockwise notation)

Usually we consider angles in a rectangular coordinate system. We say that such an angle is in standard position if its vertex is at the origin and its initial ray lies along the positive horizontal axis. Sometimes we assume that a given angle is in standard position without explicitly saying so.

There are two common ways to measure angles. The degree is the measure that is familiar to most people and is often used in applications. Another is radian measure, which mathematicians and scientists prefer. Radians, as we'll see, are a natural measure and are, by far, the more important of the two measures in preparing for calculus. Because each way of measuring angles is commonly used, we will use both throughout the text.

Degree Measure

> **Degree measure of angles is probably familiar from everyday experience.**

We begin by defining the **degree** as a measurement of angles. To measure an angle in degrees, we begin by assigning 360 degrees (written as 360°) to a full circle. Then we measure how many degrees are in an angle with vertex at the center of the circle by looking at what portion of a circle it cuts. For example, going counterclockwise halfway around a circle gives an angle whose measure is one-half of 360° or 180°, a straight angle. Going one-quarter of the way around a circle gives one-quarter of 360° or 90°, a right angle. An acute angle is a positive angle that measures less than 90°, and an obtuse angle measures more than 90° and less than 180°. **Figure 6T.3** shows the degree measure of some common angles.

Note that assigning the number 360 to a circle is purely arbitrary. We could measure angles by assigning other numbers.

A rotation of 360° is one complete trip around the circle, so it represents a return to the initial ray. Therefore, if angles of 0° and 360° have the same initial ray, then they also have the same terminal ray. Angles that share initial and terminal rays are called **coterminal angles**.

Every angle is coterminal with an angle of degree measure between 0 and 360.

The **degree** is an angle measurement such that an angle of 1° is a positive angle with vertex at the center of a circle that cuts an arc of one-360th of a full circle.

Coterminal angles are pairs of angles that have the same initial and terminal rays.

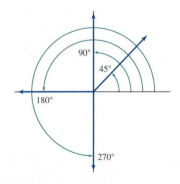

Figure 6T.3 Degree measures of some common angles

EXAMPLE 6T.1 Coterminal Angles

Find angles with degree measure between 0 and 360 that are coterminal with the following angles.

a. 750°

b. −400°

SOLUTION

a. An angle of 720° is two complete rotations of a circle. A rotation of 750° is two trips around the circle and an additional 30°, as shown in **Figure 6T.4**. Hence, an angle of 750° is coterminal with an angle of 30°.

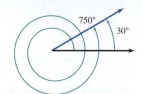

Figure 6T.4 Coterminal angles: An angle of 750° is coterminal with an angle of 30°.

b. An angle of −400° indicates a rotation in the clockwise direction of 400°. That is 360° and an additional 40°. Therefore, an angle of −400° is coterminal with an angle of 320°. This is shown in **Figure 6T.5**.

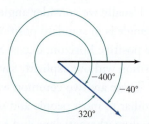

Figure 6T.5 Coterminal angles: An angle of −400° is coterminal with an angle of 320°.

TRY IT YOURSELF 6T.1 Brief answers provided at the end of the section.

Find an angle with degree measure between 0 and 360 that is coterminal with an angle of 920°.

EXTEND YOUR REACH

a. Find an angle between 0° and 360° that is coterminal with an angle of 1000°.

b. Find an angle between 0° and 360° that is coterminal with an angle of 10,000°.

c. Explain how you will find an angle between 0° and 360° that is coterminal with an angle of any given positive degree measure.

Radian Measure

> Radian measure is calculated by using the radius of a circle to measure arcs along the circle.

The **radian** is an angle measurement such that an angle of 1 radian is an angle with vertex at the center of a circle that cuts an arc the same length as the radius.

The measure of angles used almost exclusively in calculus is the **radian**. For a tactile understanding of radian measure, we can use a piece of string as the radius of a circle. We then lay that same piece of string on the circumference of the circle to make an arc having the same length as the radius, as shown in **Figure 6T.6**. An angle of 1 radian has its vertex at the center of the circle, and its sides pass through the endpoints of the arc made by the string.

It is a remarkable fact that the length of the string we started with doesn't matter. The resulting angle is the same regardless of the length of the string. This fact is illustrated in **Figure 6T.7**. For this reason, radian measure is called a natural measure. Radian measure is such a natural measure that intelligent aliens would certainly understand the concept (but it is unlikely that they would use degrees).

For an angle with vertex at the center of a circle, an angle of 1 radian is an angle that cuts an arc of length equal to the radius. An angle of 2 radians cuts an arc of length equal to twice the radius, and so on. In general, an angle of θ radians cuts an arc of length equal to θ times the radius. This observation can be expressed as follows:

$$\text{Arc length} = \text{Radius} \times \text{Radian measure}$$

or

$$\text{Radian measure} = \frac{\text{Arc length}}{\text{Radius}}$$

Because the circumference of a circle is 2π times its radius, we get

$$\text{Radians in a circle} = \frac{\text{Arc length of circle}}{\text{Radius}}$$

$$= \frac{\text{Circumference}}{\text{Radius}}$$

$$= \frac{2\pi \times \text{Radius}}{\text{Radius}}$$

$$= \frac{2\pi \times \cancel{\text{Radius}}}{\cancel{\text{Radius}}}$$

$$= 2\pi$$

So, there are 2π radians in a circle. A straight angle cuts half a circle, so its radian measure is π. A right angle cuts a quarter circle, so its measure is $\dfrac{\pi}{2}$ radians. This gives us some easy comparisons with degree measure of some common angles, as shown in the following table and in **Figure 6T.8**.

Angle name	Degrees	Radians
Full circle	360	2π (coterminal with 0)
	270	$\dfrac{3\pi}{2}$
Straight angle	180	π
Right angle	90	$\dfrac{\pi}{2}$
	45	$\dfrac{\pi}{4}$

Every angle is coterminal with an angle of radian measure between 0 and 2π.

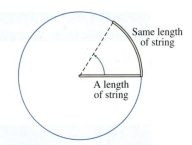

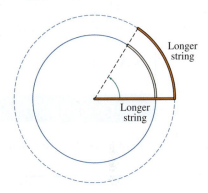

Figure 6T.6 Making an angle of 1 radian

Figure 6T.7 Radian measure: Using a different length results in the same angle.

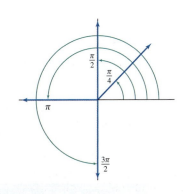

Figure 6T.8 Radian measure of some common angles

EXAMPLE 6T.2   Coterminal Angles Using Radian Measure

Find angles with radian measure between 0 and 2π that are coterminal with the following angles.

a. $\dfrac{9\pi}{2}$ radians   **b.** $-\dfrac{2\pi}{3}$ radians

SOLUTION

a. There are 2π radians in a complete circle, so an angle of radian measure $\dfrac{9\pi}{2} = 4\pi + \dfrac{\pi}{2}$ is coterminal with an angle of radian measure $\dfrac{\pi}{2}$.

b. An angle of radian measure $-\dfrac{2\pi}{3}$ is coterminal with an angle of radian measure $-\dfrac{2\pi}{3} + 2\pi = \dfrac{4\pi}{3}$.

TRY IT YOURSELF 6T.2   Brief answers provided at the end of the section.

Find an angle with radian measure between 0 and 2π that is coterminal with an angle of $\dfrac{35\pi}{6}$.

Converting Angle Measures

> **Radian and degree measures are related by simple formulas.**

Because $180°$ is π radians, $1°$ is $\dfrac{\pi}{180}$ radian. So an angle of $d°$ is $d\dfrac{\pi}{180}$ radians. In the other direction, we find that an angle of θ radians is $\theta\dfrac{180}{\pi}$ degrees. One radian is equivalent to just under $60°$.

> **CONCEPTS TO REMEMBER: Converting Between Degrees and Radians**
>
> 1. The radian measure of an angle is the length of the arc that the angle cuts out of a circle (centered at the vertex) divided by the radius of the circle:
>
> $$\text{Radian measure} = \frac{\text{Arc length}}{\text{Radius}}$$
>
> 2. To convert from radians to degrees, multiply radians by $\dfrac{180}{\pi}$.
>
> 3. To convert from degrees to radians, multiply degrees by $\dfrac{\pi}{180}$.

EXAMPLE 6T.3   Converting from Degree to Radian Measure

What is the radian measure of a $60°$ angle?

SOLUTION

To convert to radians, we multiply by $\dfrac{\pi}{180}$:

$$60° = 60 \times \dfrac{\pi}{180} \text{ radians}$$

$$= \cancel{60} \times \dfrac{\pi}{\cancel{180}^{3}} \text{ radians}$$

$$= \dfrac{\pi}{3} \text{ radians}$$

TRY IT YOURSELF 6T.3 Brief answers provided at the end of the section.

What is the radian measure of an angle of 30°?

EXAMPLE 6T.4 Converting from Radian to Degree Measure

What is the degree measure of an angle of $\dfrac{\pi}{6}$ radian?

SOLUTION

To convert to degrees, we multiply by $\dfrac{180}{\pi}$:

$$\dfrac{\pi}{6} \text{ radian} = \dfrac{\pi}{6} \times \dfrac{180}{\pi} \text{ degrees}$$

$$= \dfrac{\cancel{\pi}}{\cancel{6}} \times \dfrac{\cancel{180}^{30}}{\cancel{\pi}} \text{ degrees}$$

$$= 30°$$

TRY IT YOURSELF 6T.4 Brief answers provided at the end of the section.

What is the degree measure of an angle of $\dfrac{5\pi}{6}$ radians?

EXTEND YOUR REACH

a. Construct a circle. Using a piece of string with a length equal to the radius of the circle, construct an angle measuring 2 radians. According to the conversion formula, what is the degree measure of this angle? Check your answer with a protractor that measures angles in degrees.

b. Nairobi, Kenya, and Quito, Ecuador, are both close to the equator. Look up the longitude of each of these cities, the distance in miles between them, and the equatorial radius of Earth. (The longitude is the degree measure of the angle on the equatorial circle that the city makes with the Greenwich meridian. Longitude east of this meridian is positive, and longitude west is negative. See **Figure 6T.9**.) Are the numbers you find consistent with the conversion formula? *Suggestion:* If you are not sure how to answer this question, look at part a.

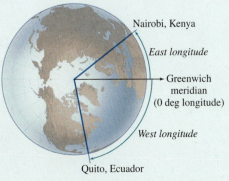

Figure 6T.9 Nairobi, Quito, and the meridian

Arc Length and Area

> Arc length and area are calculated for sectors of circles.

In discussing radian measure, we used the formula

$$\text{Arc length} = \text{Radius} \times \text{Radian measure}$$

If we use θ for the radian measure and r for the radius, this formula becomes

$$\text{Arc length} = r\theta$$

For simplicity, we assume that θ is between 0 and 2π. In a circle of radius 1, the arc length is the same as the radian measure of the angle.

We have a formula for the arc length cut by an angle. We can also work out the area determined by (or subtended by) an angle. The angle cuts a sector from the circle that looks like a slice of pie, as shown in **Figure 6T.10**. The entire circle has area πr^2. The area of the slice of pie is some fraction of that. It takes 2π radians to get all the way around the circle, so the fraction of the circle we have covered is

$$\text{Fraction covered} = \frac{\text{Radian measure of arc}}{\text{Radian measure of circle}} = \frac{\theta}{2\pi}$$

Hence, the area is

$$\text{Area subtended} = \text{Fraction covered} \times \text{Area of circle}$$

$$= \frac{\theta}{2\pi} \times \pi r^2$$

$$= \frac{\theta}{2\cancel{\pi}} \times \cancel{\pi} r^2$$

$$= \frac{\theta r^2}{2}$$

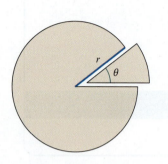

Figure 6T.10 A sector of a circle cut by an angle of radian measure θ

CONCEPTS TO REMEMBER: Area and Arc Length

Radians. Consider a circle of radius r centered at the vertex of an angle of radian measure θ, where $0 < \theta < 2\pi$. For the slice of the circle cut out by this angle,

$$\text{Arc length} = r\theta$$

and

$$\text{Area subtended} = \frac{\theta r^2}{2}$$

Degrees. Consider a circle of radius r centered at the vertex of an angle of degree measure d, where $0 < d < 360$. For the slice of the circle cut out by this angle,

$$\text{Arc length} = \frac{d\pi r}{180}$$

and

$$\text{Area subtended} = \frac{d\pi r^2}{360}$$

EXAMPLE 6T.5 Area and Arc Length

a. If a circle of radius 5 units is centered at the vertex of an angle of 2 radians, find the arc length and area subtended.

b. If a circle of radius 4 units is centered at the vertex of an angle of 25°, find the arc length and area subtended.

SOLUTION

a. We use the arc length and area formulas for radian measure with $r = 5$ and $\theta = 2$:

$$\text{Arc length} = r\,\theta$$

$$= 5 \times 2$$

$$= 10 \text{ units}$$

Also,

$$\text{Area subtended} = \frac{\theta r^2}{2}$$

$$= \frac{2 \times 5^2}{2}$$

$$= 25 \text{ square units}$$

b. We use the arc length and area formulas for degree measure with $r = 4$ and $d = 25$:

$$\text{Arc length} = \frac{d\pi r}{180}$$

$$= \frac{25\pi \times 4}{180}$$

$$= \frac{5\pi}{9}$$

$$\approx 1.75 \text{ units}$$

Also,

$$\text{Area subtended} = \frac{d\pi r^2}{360}$$

$$= \frac{25\pi \times 4^2}{360}$$

$$= \frac{10\pi}{9}$$

$$\approx 3.49 \text{ square units}$$

TRY IT YOURSELF 6T.5 Brief answers provided at the end of the section.

The vertex of an angle of $\dfrac{\pi}{3}$ radians is at the center of a circle. The angle cuts a sector of area 6 square units. What is the radius of the circle?

Similar Triangles

> Similarity is at the foundation of trigonometry.

Similar triangles are pairs of triangles for which the corresponding angles are the same.

Figure 6T.11 Similar triangles: These are triangles that have equal corresponding angles, but sides may have different lengths.

As discussed earlier, similarity of triangles is fundamental to trigonometry. **Similar triangles** often occur in applications, as in Example 6T.8, and are used throughout Section 6T.2.

When triangles are similar, as in **Figure 6T.11**, one appears to be a blown-up version of the other.

The following is the key property of similar triangles, and it is crucial to the definitions of the trigonometric functions that follow.

LAWS OF MATHEMATICS: Similar Triangles

The ratios of corresponding sides of similar triangles are the same. For the similar triangles in Figure 6T.11, we have

$$\frac{a}{A} = \frac{b}{B} = \frac{c}{C}$$

We can also write these equations as

$$\frac{A}{B} = \frac{a}{b}, \quad \frac{A}{C} = \frac{a}{c}, \quad \frac{B}{C} = \frac{b}{c}$$

EXAMPLE 6T.6   Corresponding Sides of Similar Triangles

The triangles in **Figure 6T.12** are similar. Find the lengths of all sides that are not already calculated.

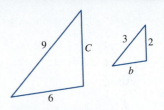

Figure 6T.12 Similar triangles for Example 6T.6

SOLUTION

We use the relationships among the sides of similar triangles:

$$\frac{9}{3} = \frac{C}{2}$$

Thus, $C = 6$. Similarly,

$$\frac{6}{b} = \frac{9}{3}$$

so $b = 2$.

TRY IT YOURSELF 6T.6 Brief answers provided at the end of the section.

The triangles in **Figure 6T.13** are similar. Find all sides that are not already calculated.

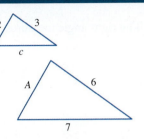

Figure 6T.13 Similar triangles for Try It Yourself 6T.6

EXTEND YOUR REACH

For the two right triangles in **Figure 6T.14** it is true that

$$\frac{A}{a} = \frac{B}{b}$$

a. Use a bit of algebra to show that

$$\frac{\sqrt{A^2 + B^2}}{\sqrt{a^2 + b^2}} = \frac{A}{a}$$

b. Is the result in part a consistent with the fact that the two triangles are similar? Explain your answer.

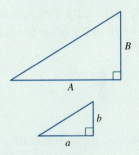

Figure 6T.14 Right triangles

An important geometric fact is that the sum of the three angles of a triangle is the same for all triangles.[1]

LAWS OF MATHEMATICS: Angle Sum of a Triangle

The sum of the angles of a triangle is π radians or 180°.

As a consequence, when two pairs of corresponding angles from a pair of triangles are the same, then the third pair must also be the same. Hence, having two pairs of equal angles is sufficient to ensure similarity.

EXAMPLE 6T.7 Right Triangles

The triangles in **Figure 6T.15** are right triangles. The angle measures shown are in radians. Find all unmarked sides and angles.

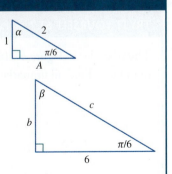

Figure 6T.15 Right triangles for Example 6T.7

---

[1]This fact is a feature of plane geometry. It does not hold for triangles on curved surfaces, such as the surface of Earth, for example. See the exercises.

SOLUTION

The right angle measures $\dfrac{\pi}{2}$ radians, and the sum of the three angles in a triangle is π radians, so

$$\alpha + \frac{\pi}{6} + \frac{\pi}{2} = \pi \qquad \blacktriangleleft \text{ The angle sum is } \pi \text{ radians.}$$

$$\alpha = \pi - \frac{\pi}{6} - \frac{\pi}{2} \qquad \blacktriangleleft \text{ Solve for } \alpha.$$

$$= \frac{\pi}{3} \text{ radians} \qquad \blacktriangleleft \text{ Simplify.}$$

Similarly, $\beta = \dfrac{\pi}{3}$ radians.

We can use the Pythagorean theorem (see Section P.1) to find the length of A because we know the lengths of the other two sides:

$$A^2 + 1^2 = 2^2 \qquad \blacktriangleleft \text{ Use the Pythagorean theorem.}$$

$$A^2 = 3 \qquad \blacktriangleleft \text{ Solve for } A^2.$$

$$A = \sqrt{3} \qquad \blacktriangleleft \text{ Take square roots.}$$

Because the corresponding angles are the same, the triangles are similar:

$$\frac{b}{1} = \frac{6}{A}$$

$$b = \frac{6}{\sqrt{3}}$$

In the same fashion,

$$\frac{c}{2} = \frac{6}{A}$$

$$\frac{c}{2} = \frac{6}{\sqrt{3}}$$

$$c = \frac{12}{\sqrt{3}}$$

TRY IT YOURSELF 6T.7   Brief answers provided at the end of the section.

The triangles in **Figure 6T.16** are right triangles. The angle measures shown are in radians. Find all unmarked sides and angles.

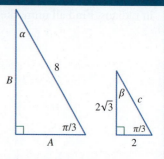

Figure 6T.16 Right triangles for Try It Yourself 6T.7

MODELS AND APPLICATIONS   Shadows

Similarity of triangles has many applications to everyday measurements.

EXAMPLE 6T.8   A Shadow

A man who is 6 feet tall casts a shadow of 2 feet. At the same time, a building casts a shadow that is 12 feet long. How tall is the building?

SOLUTION

Let's use H to denote the height of the building, as shown in **Figure 6T.17**. The angles α and β are the same because they are determined by the sun. Hence, the two right triangles have two common angles, so they are similar. Equating ratios of corresponding sides, we have

$$\frac{H}{6} = \frac{12}{2}$$

$$H = 36$$

We conclude that the building is 36 feet tall.

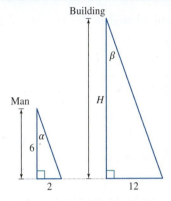

Figure 6T.17 Lengths of shadows

TRY IT YOURSELF 6T.8   Brief answers provided at the end of the section.

How long is the shadow cast by a 15-foot-tall flagpole at the same time of day as in the example?

TRY IT YOURSELF ANSWERS

6T.1 $200°$

6T.2 $\dfrac{11\pi}{6}$

6T.3 $\dfrac{\pi}{6}$

6T.4 $150°$

6T.5   The radius is $\dfrac{6}{\sqrt{\pi}}$ units.

6T.6 $A = 4$ and $c = \dfrac{7}{2} = 3.5$

6T.7 $\alpha = \beta = \dfrac{\pi}{6}$ radian, $A = 4, c = 4$, and $B = 4\sqrt{3}$

6T.8   5 feet

EXERCISE SET 6T.1

CHECK YOUR UNDERSTANDING

1. Coterminal angles:

   a. begin at the same place

   b. end at the same place

   c. have common initial and terminal sides

   d. None of the above.

2. **True or false:** Every angle is coterminal with an angle of degree measure between 0 and 360.

3. **True or false:** Every angle is coterminal with an angle of radian measure between 0 and π.

4. **True or false:** Degree measure depends on an arbitrary choice of numbers.

5. What is the radian measure of an angle of 0°?

6. **True or false:** An angle with vertex at the center of a circle cuts the same area whether the angle is measured in degrees or in radians.

7. Two triangles are similar if _____.

8. The radian measure of an angle of 90° is _____.

9. What is the radian measure of a straight angle?

10. The angle sum of any triangle is _____.

11. State the Pythagorean theorem.

SKILL BUILDING

Coterminal angles. In Exercises 12 through 19, find an angle either between 0 and 360° or between 0 and 2π radians that is coterminal with the given angle.

12. 600°

13. 420°

14. 750°

15. 200π radians

16. −200°

17. $\dfrac{10\pi}{3}$ radians

18. 7 radians

19. -11π radians

Converting degrees to radians. In Exercises 20 through 27, convert the given degree measure to radian measure.

20. 60°

21. 30°

22. 15°

23. 120°

24. 300°

25. −90°

26. $\pi°$

27. 70°

Converting radians to degrees. In Exercises 28 through 34, convert the given radian measure to degree measure.

28. $\dfrac{\pi}{6}$ radian

29. $\dfrac{3\pi}{4}$ radians

30. 3π radians

31. $\dfrac{4\pi}{3}$ radians

32. $\dfrac{\pi}{12}$ radian

33. 90 radians

34. $-\dfrac{5\pi}{4}$ radians

Using the Pythagorean theorem. In Exercises 35 through 38, use the Pythagorean theorem to find the missing side.

35. A right triangle has legs of length 3 and 7. Find the length of the hypotenuse.

36. A right triangle has legs of lrength 4 and 6. Find the length of the hypotenuse.

37. A right triangle has one leg of length 4 and a hypotenuse of length 7. Find the length of the other leg.

38. A right triangle has one leg of length 5 and a hypotenuse of length 9. Find the length of the other leg.

Angle sum. For Exercises 39 through 45, recall that, in a triangle, the angle sum is π radians or 180°.

39. One angle of a triangle is $\dfrac{\pi}{9}$ radian, and another is $\dfrac{\pi}{4}$ radian. Find the radian measure of the third angle.

40. One angle of a triangle is 20°, and another is 40°. What is the degree measure of the third angle?

41. One angle of a triangle is $\dfrac{\pi}{4}$ radian, and another is $\dfrac{\pi}{3}$ radians. What is the radian measure of the third angle?

42. One angle of a triangle is 20°. Another is 0.3 radian. Find both the radian measure and the degree measure of the third angle.

43. One angle of a triangle is 70°. Another is 10°. Find the degree measure of the third angle.

44. An equilateral triangle has all three angles equal. What is the degree measure of an angle of an equilateral triangle?

45. An equilateral triangle has all three angles equal. What is the radian measure of an angle of an equilateral triangle?

Area and arc length. In Exercises 46 through 51, you are given an angle and the radius of a circle. The angle has its vertex at the center of the circle. Find the arc length cut by the angle, and find the area subtended by the angle.

46. Angle of $\dfrac{\pi}{6}$ radian. Circle of radius 5 units.

47. Angle of 1 radian. Circle of radius 9 units.

48. Angle of $\dfrac{\pi}{3}$ radians. Circle of radius 6 units.

49. Angle of $\dfrac{\pi}{6}$ radian. Circle of radius 3 units.

50. Angle of 20°. Circle of radius 10 units.

51. Angle of $d°$. Circle of radius $\dfrac{1}{d}$ units.

PROBLEMS

52. **Degrees in sports.** In skateboarding and snowboarding competitions, multiple turns are measured in degrees.

    a. In 2018, Anna Gasser became the first female snowboarder to land a cab triple 1260. The number 1260 refers to a turn of 1260°. How many rotations does this indicate?

    b. Sage Kotsenburg's slopestyle gold medal run in the 2014 Winter Olympics at Sochi included a 1620. How many rotations are indicated?

53. **Grads.** Another type of angle measure is the grad, sometimes used on European maps as a metric equivalent of the degree. There are 400 grads in a circle.

    a. What is the grad measure of an angle of 90°?

    b. What is the grad measure of an angle of π radians?

    c. What is the grad measure of an angle of 36°?

54. **Mils.** The mil is a type of angle measure used in the military. The term mil is derived from milliradian, and 1000 mils equal 1 radian.

    a. What is the degree measure of 1 mil? Give your answer accurate to three decimal places.

    b. A target is 1 yard wide and subtends an angle of 1 mil in a soldier's field of vision. How far from the soldier is the target? *Suggestion*: Think of the target as an arc of a circle.

    c. How many mils are in a circle?[2]

Area and arc length.

55. Consider a circle of radius r centered at the vertex of an angle of degree measure d, where $0 < d < 360$. For the slice of the circle cut out by this angle, show that

$$\text{Arc length} = \frac{d\pi r}{180}$$

and

$$\text{Area subtended} = \frac{d\pi r^2}{360}$$

56. An angle has its vertex centered at the origin of a circle. It cuts an arc length of 3 units and subtends an area of 9 square units. What is the radian measure of the angle? What is the radius of the circle?

57. An angle has its vertex centered at the origin of a circle. It cuts an arc length of 3 units and subtends an area of 6 square units. What is the degree measure of the angle? What is the radius of the circle?

58. An angle has its vertex centered at the origin of a circle. It cuts an arc length of L units and subtends an area of A square units. What is the radian measure of the angle? What is the radius of the circle?

59. Find a formula that gives the area A of a sector of a circle in terms of the radius r and arc length L.

60. **When area equals arc length.** A circle of a certain radius has the property that if an angle has its vertex at the center of the circle, the area it subtends is numerically the same as the arc length. Find the radius of the circle.

61. **Doubling.** Your friend and you are cutting pieces of a pie. Your friend cuts a piece twice as big (say in area) as the one you cut for yourself. How do the central angles compare?

62. **More on doubling.** You have two pies, one of which has twice the diameter of the other. You cut each pie into the same number of pieces. How do the pieces of the larger pie compare (say in area) to those of the smaller pie?

Similarity. The triangles in **Figure 6T.18** are similar. Exercises 63 through 68 refer to these triangles.

---

[2]The U.S. military uses a somewhat different definition of a mil, in which there are 6400 mils in a circle. Other countries use different definitions.

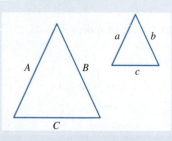

Figure 6T.18 Similar triangles

63. $A = 24$, $B = 18$, $a = 4$, and $c = 2$. Find C and b.

64. $A = 24$, $B = 16$, $C = 12$, and $c = 3$. Find a and b.

65. $A = 6$, $a = 3$, $b = 4$, and $c = 2$. Find B and C.

66. $A = 24$, $B = 16$, $b = 4$, and $c = 3$. Find a and C.

67. $A = x$, $B = 1$, and $a = 1$. Find b in terms of x.

68. $A = 1$, $C = a$, and $c = x$. Find C in terms of x.

69. Preparing for trigonometry. For the right triangles in **Figure 6T.19**, $s = t$.

  a. Show that the triangles are similar.

  b. Show that $\dfrac{a}{c} = \dfrac{A}{C}$.

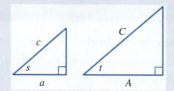

Figure 6T.19 Two right triangles, with angle s the same as angle t

MODELS AND APPLICATIONS

70. Cartography. Topographical maps show heights of mountains and depths of valleys. One difficulty in making such a map is that it is often impossible to travel over the entire terrain to be mapped. Instead, laser and radar measurements are taken. Such measurements indicate that the base of a sheer rock wall is 2000 feet from your observation point. Its peak is 2050 feet away. See **Figure 6T.20**. How tall is the wall?

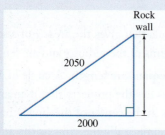

Figure 6T.20 Height of a rock wall

71. More cartography. A sheer rock wall is known to be 0.16 mile high. We find that its peak is 2 miles from our observation point. See **Figure 6T.21**. How far away is the base of the rock wall?

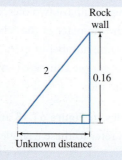

Figure 6T.21 Distance to a rock wall

72. Latitude. Wichita, Kansas, is due north of Fort Worth, Texas. This means that they lie on a circle whose center is that of Earth and whose radius equals the polar radius of Earth, about 3950 miles. Further, the latitude of Wichita is about 37° north, and that of Fort Worth is about 32° north. (These angles are measured along the circle described earlier, starting at the equator.) See **Figure 6T.22**. How far is it from Fort Worth to Wichita?

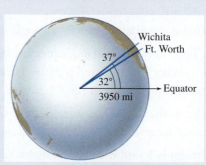

Figure 6T.22 Distance from Fort Worth to Wichita

73. More latitude. Refer to the preceding exercise for information about the measurement of latitude. Winnipeg, Manitoba (Canada), is due north of Fort Worth, Texas. It is about 1180 miles from Fort Worth to Winnipeg. See **Figure 6T.23**. What is the latitude of Winnipeg?

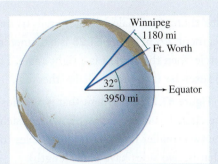

Figure 6T.23 The latitude of Winnipeg

74. **Shadows.** A 10-foot-tall vertical pole casts a 6-foot-long horizontal shadow. See **Figure 6T.24**. How tall is a tree that at the same time casts a 21-foot-long shadow? *Suggestion*: Use similar triangles.

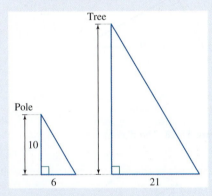

Figure 6T.24 Shadows

75. **Another shadow.** A 6-foot-tall woman casts a 2-foot-long horizontal shadow. How long a shadow is cast at the same time by an 18-foot-tall flag pole?

76. **A ladder.** A 20-foot-long ladder leans against a vertical wall, and the horizontal distance from the base of the ladder to the wall is 5 feet. One rung of the ladder is 12 feet from the base of the ladder. See **Figure 6T.25**. Use similar triangles to determine how far that rung is from the wall.

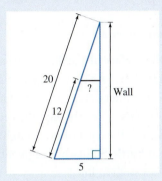

Figure 6T.25 One rung of a ladder

CHALLENGE EXERCISES FOR INDIVIDUALS OR GROUPS

77. **The inverse Pythagorean theorem.** The goal of this exercise is to establish the inverse Pythagorean theorem. In **Figure 6T.26** we have added an altitude, the segment perpendicular to the hypotenuse of the right triangle. Using a, b, and h as shown in the figure, the theorem states

$$\frac{1}{a^2} + \frac{1}{b^2} = \frac{1}{h^2}$$

   a. Use

$$\text{Area} = \frac{1}{2} \text{ Base} \times \text{Height}$$

   to calculate the area of the large triangle and the areas (marked A_1 and A_2) of the two smaller triangles.

   b. By equating the area of the large triangle to the sum of the areas A_1 and A_2, show that $ab = hc$. Deduce that $c = \dfrac{ab}{h}$.

   c. Put the value of c calculated in part b into the Pythagorean theorem $a^2 + b^2 = c^2$. Then use a bit of algebraic manipulation to produce the inverse Pythagorean theorem.

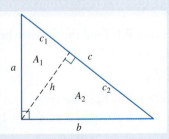

Figure 6T.26 A right triangle with an altitude added

78. **An angle and a circle.** Consider **Figure 6T.27**, in which an angle meets a circle. It can be shown that $\triangle ADB$ is similar to $\triangle ACE$. (The correspondence matches $\angle ADB$ with $\angle ACE$ and $\angle ABD$ with $\angle AEC$.) Show that $|AB| \times |AC| = |AD| \times |AE|$. (The vertical bars denote the length of a segment.)

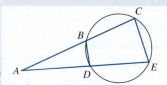

Figure 6T.27 An angle and a circle

79. **Triangles on the surface of Earth.** On the surface of Earth, "straight lines" are parts of great circles. A great circle is the intersection of the surface of Earth with a plane that goes through the center of Earth. A segment of the equator is a straight line on the surface of Earth. Similarly, a segment of a longitude line is a straight line.

    a. Let A be the North Pole, and let B and C be distinct points on the equator one-quarter of the way around Earth from each other. What is the angle sum of $\triangle ABC$?

    b. Can any triangle on the surface of Earth have angle sum $180°$? *Suggestion:* To answer this question, consider that triangles on the surface of Earth are somewhat "bent" due to the curvature of Earth. What would happen to the angles if we could "iron" such a triangle out flat?

80. **Hyperbolic geometry.** In hyperbolic geometry, the angle sum of any triangle is strictly less than $180°$. Show that in such a geometry, rectangles (quadrilaterals with four right angles) cannot exist. *Suggestion:* What happens if you add a diagonal to a rectangle?

81. **Thales' theorem.** Thales' theorem tells us that if the segment AB in **Figure 6T.28** is a diameter of the circle, then the angle t is a right angle. The goal of this exercise is to prove Thales' theorem.

    a. In **Figure 6T.29**, a segment that is a radius of the circle has been added. Show that each of the two smaller triangles is an isosceles triangle (that is, two sides have equal length).

    b. Recall from geometry that the base angles of an isosceles triangle are congruent. Use this information together with the fact that the angle sum of any triangle is $180°$ to prove that angle t is a right angle.

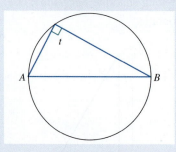

Figure 6T.28 The diameter AB

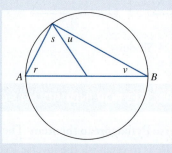

Figure 6T.29 Using a radius to divide the triangle

REVIEW AND REFRESH: Exercises from Previous Sections

82. **From Section P.1:** Find the distance between the points $(2, 1)$ and $(3, 7)$.

    Answer: $\sqrt{37}$

83. **From Section P.1:** Find the center and radius of the circle whose equation is $x^2 + y^2 + 1 = 2x + 4y$.

    Answer: Center: $(1, 2)$. Radius: 2.

84. **From Section P.3:** Solve the inequality $x^2 + 2x < 3$.

    Answer: $(-3, 1)$

85. **From Section 2.1:** If $f(x) = \dfrac{1}{x+1}$ and $g(x) = \dfrac{1}{x}$, find and simplify a formula for $(f \circ g)(x)$.

    Answer: $(f \circ g)(x) = \dfrac{x}{1+x}$

86. **From Section 4.1:** Without using your calculator, find the exact value of $\ln \sqrt{e^3}$.

    Answer: $\dfrac{3}{2}$

87. From Section 4.1: If $f(x) = \log(x-1)$, calculate $f(1+10^{-x})$.

Answer: $-x$

88. From Section 4.3: Find the inverse function of $f(x) = 2 + \log x$.

Answer: $f^{-1}(x) = 10^{x-2}$

89. From Section 4.3: Solve $2^x = 3^{x-1}$.

Answer: $x = \dfrac{\ln 3}{\ln 3 - \ln 2}$

90. From Section 5.1: What can be said of the zeros of $y = ax^2 + bx + c$ if $a \neq 0$ and $b^2 - 4ac > 0$?

Answer: There are two real zeros.

91. From Section 5.5: Find the horizontal asymptote(s) of the rational function $y = \dfrac{2x+1}{x-1}$.

Answer: $y = 2$

6T.2 Definition of the Trigonometric Functions Using Right Triangles

Trigonometry is closely connected to the study of right triangles.

6T.1 Angles

6T.2 Definition of the Trigonometric Functions Using Right Triangles

6T.3 Analysis of Right Triangles

6T.4 Extending the Trigonometric Functions: The Unit Circle

In this section, you will learn to:

1. Use the definitions of the six trigonometric functions of an acute angle of a right triangle to find function values.
2. Recite the values of the trigonometric functions of special angles.
3. Use a given trigonometric value of an acute angle to construct a right triangle having that angle.
4. Use a right triangle to relate the value of one trigonometric function to the other function values.

Many astronomical facts can be gleaned from persistent and careful observation, and early humans were masters of observation. But much of modern astronomy depends on the answer to a seemingly simple question: "How far away is it?" You can look online today to find accurate distances to the moon, planets, and stars. But where do those numbers come from? A key step in finding distances to stars uses the notion of a parallax angle. As Earth moves from one extreme of its orbit to another, "relatively close" stars appear to move. It is much the same thing that will happen if you view an object five feet away first with only your left eye and then with only your right eye. Half of the angle determined by the apparent movement of a star is the parallax angle. The right triangle shown in **Figure 6T.30** comes into play here. One side is the radius of Earth's orbit, one side is the distance to the star, and one angle is the parallax angle. We can measure the parallax angle and the radius of Earth's orbit. To find the unknown side of the triangle, the distance to the star, we need trigonometry.

In this section, we will use right triangles to define the trigonometric functions of acute angles.

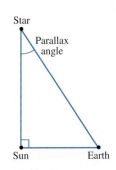

Figure 6T.30 The parallax angle

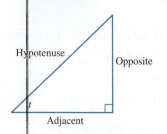

Figure 6T.31 The sides of a right triangle

The **sine** function at t, $\sin t$, is defined as $\sin t = \dfrac{\text{Opposite}}{\text{Hypotenuse}}$.

The **cosine** function at t, $\cos t$, is defined as $\cos t = \dfrac{\text{Adjacent}}{\text{Hypotenuse}}$.

The **tangent** function at t, $\tan t$, is defined as $\tan t = \dfrac{\text{Opposite}}{\text{Adjacent}}$.

The **cotangent** function at t, $\cot t$, is defined as $\cot t = \dfrac{\text{Adjacent}}{\text{Opposite}}$.

The **secant** function at t, $\sec t$, is defined as $\sec t = \dfrac{\text{Hypotenuse}}{\text{Adjacent}}$.

The **cosecant** function at t, $\csc t$, is defined as $\csc t = \dfrac{\text{Hypotenuse}}{\text{Opposite}}$.

Definitions of Trigonometric Functions of Acute Angles

> Trigonometric functions of acute angles are defined in terms of right triangles.

In a right triangle, the hypotenuse is the side opposite the right angle. The other two sides are legs. If we use one of the acute angles as a reference point, there are standard labels associated with the two legs of a right triangle. In **Figure 6T.31**, we have labeled one of the acute angles t. With this figure as a reference, the adjacent side is the leg of the triangle that is a side of the angle t, and the opposite side is the leg of the triangle across from the angle t.

We define the six trigonometric functions at t—namely **sine, cosine, tangent, cotangent, secant**, and **cosecant**—using these sides of a right triangle.

A useful mnemonic for the values of sine, cosine, and tangent is SOHCAHTOA, which incorporates the first letters of "Sine Opposite over Hypotenuse, Cosine Adjacent over Hypotenuse, Tangent Opposite over Adjacent." Observe from the definitions of the trigonometric functions that we can use the sine and cosine to find the other four trigonometric functions. **Table 6T.1** shows the relationships.

Table 6T.1 Basic Trigonometric Identities: Other Trigonometric Functions in Terms of Sine and Cosine

$\tan t = \dfrac{\sin t}{\cos t}$	$\cot t = \dfrac{\cos t}{\sin t}$
$\sec t = \dfrac{1}{\cos t}$	$\csc t = \dfrac{1}{\sin t}$

EXAMPLE 6T.9 Getting Other Function Values from the Sine and Cosine

If $\sin t = \dfrac{1}{3}$ and $\cos t = \dfrac{\sqrt{8}}{3}$, find $\tan t$, $\cot t$, $\sec t$, and $\csc t$.

SOLUTION

We have

$$\tan t = \frac{\sin t}{\cos t} = \frac{\dfrac{1}{3}}{\dfrac{\sqrt{8}}{3}} = \frac{1}{\sqrt{8}}$$

$$\cot t = \frac{\cos t}{\sin t} = \frac{\dfrac{\sqrt{8}}{3}}{\dfrac{1}{3}} = \sqrt{8}$$

$$\sec t = \frac{1}{\cos t} = \frac{1}{\dfrac{\sqrt{8}}{3}} = \frac{3}{\sqrt{8}}$$

$$\csc t = \frac{1}{\sin t} = \frac{1}{\dfrac{1}{3}} = 3$$

TRY IT YOURSELF 6T.9 Brief answers provided at the end of the section.

If $\sin t = \dfrac{3}{4}$ and $\cos t = \dfrac{\sqrt{7}}{4}$, find the values of the other four trigonometric functions of t.

Evaluating Trigonometric Functions Using Right Triangles

Side lengths of right triangles can be used to evaluate trigonometric functions of acute angles.

We can calculate the values of the trigonometric functions of angles if the sides of the right triangle are known. Consider, for example, the triangle in **Figure 6T.32**, where the lengths of the sides are labeled.

For the angle s, the adjacent side has length 5 and the opposite side has length 12, so

$$\sin s = \frac{\text{Opposite}}{\text{Hypotenuse}} = \frac{12}{13} \qquad \cos s = \frac{\text{Adjacent}}{\text{Hypotenuse}} = \frac{5}{13}$$

$$\tan s = \frac{\text{Opposite}}{\text{Adjacent}} = \frac{12}{5} \qquad \cot s = \frac{\text{Adjacent}}{\text{Opposite}} = \frac{5}{12}$$

$$\sec s = \frac{\text{Hypotenuse}}{\text{Adjacent}} = \frac{13}{5} \qquad \csc s = \frac{\text{Hypotenuse}}{\text{Opposite}} = \frac{13}{12}$$

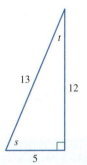

Figure 6T.32 Calculating the values of trigonometric functions

For the angle t, the adjacent side has length 12 and the opposite side has length 5, so

$$\sin t = \frac{\text{Opposite}}{\text{Hypotenuse}} = \frac{5}{13} \qquad \cos t = \frac{\text{Adjacent}}{\text{Hypotenuse}} = \frac{12}{13}$$

$$\tan t = \frac{\text{Opposite}}{\text{Adjacent}} = \frac{5}{12} \qquad \cot t = \frac{\text{Adjacent}}{\text{Opposite}} = \frac{12}{5}$$

$$\sec t = \frac{\text{Hypotenuse}}{\text{Adjacent}} = \frac{13}{12} \qquad \csc t = \frac{\text{Hypotenuse}}{\text{Opposite}} = \frac{13}{5}$$

If we know two sides of a right triangle, then we can use the Pythagorean theorem to find the length of the third side. Hence, only two sides are needed to calculate the values of the trigonometric functions.

EXAMPLE 6T.10 Calculating Function Values Using Two Known Sides of a Right Triangle

For the angle t in **Figure 6T.33**, calculate the values of the six trigonometric functions.

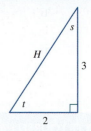

Figure 6T.33 Two given sides of a right triangle

SOLUTION

First we use the Pythagorean theorem to calculate the length of the hypotenuse H:

$$H^2 = a^2 + b^2 \quad \blacktriangleleft \text{ Use the Pythagorean theorem.}$$

$$H^2 = 2^2 + 3^2 \quad \blacktriangleleft \text{ Substitute known values.}$$

$$H = \sqrt{13} \quad \blacktriangleleft \text{ Take square roots.}$$

The side adjacent to the angle t has length 2, and the opposite side has length 3, so

$$\sin t = \frac{\text{Opposite}}{\text{Hypotenuse}} = \frac{3}{\sqrt{13}} \qquad \cos t = \frac{\text{Adjacent}}{\text{Hypotenuse}} = \frac{2}{\sqrt{13}}$$

$$\tan t = \frac{\text{Opposite}}{\text{Adjacent}} = \frac{3}{2} \qquad \cot t = \frac{\text{Adjacent}}{\text{Opposite}} = \frac{2}{3}$$

$$\sec t = \frac{\text{Hypotenuse}}{\text{Adjacent}} = \frac{\sqrt{13}}{2} \qquad \csc t = \frac{\text{Hypotenuse}}{\text{Opposite}} = \frac{\sqrt{13}}{3}$$

TRY IT YOURSELF 6T.10 Brief answers provided at the end of the section.

Find the values of the six trigonometric functions of the angle s in Figure 6T.33.

EXTEND YOUR REACH   If, as in the preceding example, s and t are the acute angles of a right triangle, how are the six trigonometric functions evaluated at s related to the six trigonometric functions evaluated at t? *Suggestion*: To get started, observe that the side opposite s is adjacent to t, and the side adjacent to s is opposite t. The hypotenuse is the same for both angles.

Consistency of the Definitions

> Any right triangle with a fixed acute angle can be used to calculate trigonometric function values.

There is more than one right triangle with a given acute angle, and it may appear that the values of the trigonometric functions depend on the triangle used to make the calculations. But properties of similar triangles ensure that if we choose two different right triangles having the same angle t as an acute angle, the ratios in the definitions of the trigonometric functions of t will be the same. To see how this works, consider the right triangles in **Figure 6T.34**. The triangles are not the same, but their acute angles t are the same. The right angles give a second shared angle, so the two triangles are similar.

If we calculate $\sin t$ using the smaller triangle we get $\dfrac{b}{c}$, and if we calculate $\sin t$ using the larger triangle we get $\dfrac{B}{C}$. But because the triangles are similar, we have

$$\sin t \text{ calculated from small triangle} = \frac{b}{c} = \frac{B}{C} = \sin t \text{ calculated from large triangle.}$$

A similar argument applies to the other trigonometric functions. The result is that, even though we use right triangles to calculate trigonometric functions, the values of the functions depend only on the angle—not on the particular triangle used for the calculation.

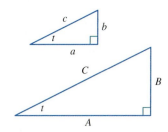

Figure 6T.34 Similar right triangles: Right triangles with a common acute angle are similar.

EXAMPLE 6T.11   Verifying Consistency of Definitions

a. Explain why the large right triangle in **Figure 6T.35** is similar to the small right triangle in the figure.

b. Use similarity to find the lengths A and H.

c. Calculate the sine and cosine of the angle s using the small right triangle, and compare your results with the calculation of these functions using the large right triangle.

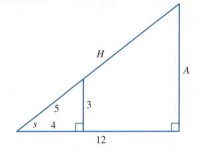

Figure 6T.35 Similar right triangles

SOLUTION

a. The two triangles share the angle s, and both have a right angle. Two common angles are sufficient to ensure similarity.

b. Because the triangles are similar, the ratios of sides are the same:

$$\frac{12}{4} = \frac{A}{3} = \frac{H}{5}$$

Solving these equations gives $A = 9$ and $H = 15$.

c. We put the results in a table for ease of comparison.

Using the small triangle	Using the large triangle
$\sin s = \dfrac{3}{5}$	$\sin s = \dfrac{A}{H} = \dfrac{9}{15} = \dfrac{3}{5}$
$\cos s = \dfrac{4}{5}$	$\cos s = \dfrac{12}{H} = \dfrac{12}{15} = \dfrac{4}{5}$

The results are the same for each triangle.

TRY IT YOURSELF 6T.11   Brief answers provided at the end of the section.

Calculate the tangent of s using the small triangle in Figure 6T.35, and compare the result with the calculation of the tangent using the large triangle.

Special Angles

> Geometry is used to evaluate trigonometric functions of certain special angles.

Generally speaking, we must use a calculator to find (approximate) values of trigonometric functions. But there are certain special angles for which we can make exact calculations. These angles are $\dfrac{\pi}{6}$ radian or 30°, $\dfrac{\pi}{4}$ radian or 45°, and $\dfrac{\pi}{3}$ radians or 60°. The special angles appear often in mathematics and its applications.

We start with the angle 45°. The angle sum of a triangle is 180°, so if one acute angle of a right triangle is 45°, then the other acute angle must also be 45°. Two equal angles means we have an isosceles triangle (two sides have equal length). This fact is illustrated in **Figure 6T.36**, where the equal sides have length 1 but the hypotenuse has unknown length H.

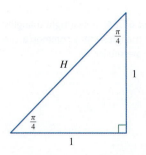

Figure 6T.36 An isosceles right triangle

We can find the value of H using the Pythagorean theorem, and this additional information is incorporated in **Figure 6T.37**:

$$a^2 + b^2 = c^2 \qquad \blacktriangleleft \text{Use the Pythagorean theorem.}$$

$$1^2 + 1^2 = H^2 \qquad \blacktriangleleft \text{Substitute known values.}$$

$$2 = H^2 \qquad \blacktriangleleft \text{Simplify.}$$

$$\sqrt{2} = H \qquad \blacktriangleleft \text{Take square roots.}$$

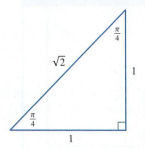

Figure 6T.37 Finding the hypotenuse

A right triangle with angles of radian measure $\dfrac{\pi}{6}$ and $\dfrac{\pi}{3}$ (or 30° and 60°) is shown in **Figure 6T.38**. In the exercises, you will be asked to verify the side lengths shown there.

We can use the triangles in Figure 6T.37 and Figure 6T.38 to find the trigonometric functions of the special angles $\dfrac{\pi}{4}$, $\dfrac{\pi}{6}$, and $\dfrac{\pi}{3}$. For example, Figure 6T.37 gives

$$\sin\left(\frac{\pi}{4}\right) = \frac{\text{Opposite}}{\text{Hypotenuse}} = \frac{1}{\sqrt{2}}$$

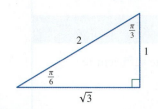

Figure 6T.38 A 30° − 60° − 90° triangle

We summarize the sines and cosines of special angles in Table 6T.2. The values of the other four trigonometric functions can be easily obtained from these by division or taking inverses.

Table 6T.2 Special Angles

Angle in degrees	Angle in radians	sin	cos
30	$\dfrac{\pi}{6}$	$\dfrac{1}{2}$	$\dfrac{\sqrt{3}}{2}$
45	$\dfrac{\pi}{4}$	$\dfrac{1}{\sqrt{2}}$	$\dfrac{1}{\sqrt{2}}$
60	$\dfrac{\pi}{3}$	$\dfrac{\sqrt{3}}{2}$	$\dfrac{1}{2}$

To the right is a different presentation of the same information that some will find useful as a memory aid because of the pattern of the numerators.

Angle	$\dfrac{\pi}{6}$	$\dfrac{\pi}{4}$	$\dfrac{\pi}{3}$
sine	$\dfrac{\sqrt{1}}{2}$	$\dfrac{\sqrt{2}}{2}$	$\dfrac{\sqrt{3}}{2}$
cosine	$\dfrac{\sqrt{3}}{2}$	$\dfrac{\sqrt{2}}{2}$	$\dfrac{\sqrt{1}}{2}$

EXAMPLE 6T.12 Other Trigonometric Functions of Special Angles

Use Table 6T.2 and the basic trigonometric relationships from Table 6T.1 to find exact values for the following.

a. $\tan\left(\dfrac{\pi}{4}\right)$ **b.** $\sec\left(\dfrac{\pi}{6}\right)$

SOLUTION

a. The basic relationships tell us that $\tan\left(\dfrac{\pi}{4}\right) = \dfrac{\sin\left(\dfrac{\pi}{4}\right)}{\cos\left(\dfrac{\pi}{4}\right)}$. The table of special angles gives the values we need:

$$\tan\left(\frac{\pi}{4}\right) = \frac{\sin\left(\dfrac{\pi}{4}\right)}{\cos\left(\dfrac{\pi}{4}\right)}$$

$$= \frac{\dfrac{1}{\sqrt{2}}}{\dfrac{1}{\sqrt{2}}}$$

$$= 1$$

b. The secant is the reciprocal of the cosine:

$$\sec\left(\frac{\pi}{6}\right) = \frac{1}{\cos\left(\dfrac{\pi}{6}\right)}$$

$$= \frac{1}{\dfrac{\sqrt{3}}{2}}$$

$$= \frac{2}{\sqrt{3}}$$

TRY IT YOURSELF 6T.12 Brief answers provided at the end of the section.

Use Table 6T.2 and the basic relationships from Table 6T.1 to find the exact value of $\cot\left(\dfrac{\pi}{3}\right)$.

EXTEND YOUR REACH

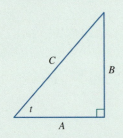

Figure 6T.39 A right triangle with an acute angle of $\frac{\pi}{12}$ radian

In **Figure 6T.39** the angle t is $\frac{\pi}{12}$ radian, the side A has length $\frac{\sqrt{6} + \sqrt{2}}{4}$, and the hypotenuse C has length 1. Find the exact values of the six trigonometric functions of $\frac{\pi}{12}$.

Representing Angles

> An angle with a given trigonometric function value can be represented by a right triangle.

Representing angles by using known trigonometric values is an important tool, illustrated in the following example.

EXAMPLE 6T.13   Representing Angles

The cosine of the acute angle t is $\frac{2}{3}$. Construct a right triangle with acute angle t, and label the lengths of the sides.

SOLUTION

Because $\cos t = \dfrac{\text{Adjacent}}{\text{Hypotenuse}}$, we can obtain the value of $\frac{2}{3}$ for the cosine if the adjacent side has length 2 and the hypotenuse has length 3. This triangle is shown in **Figure 6T.40**. We can find the opposite side B using the Pythagorean theorem:

$$3^2 = 2^2 + B^2 \quad \blacktriangleleft \text{ Use the Pythagorean theorem.}$$

$$5 = B^2 \quad \blacktriangleleft \text{ Simplify.}$$

$$\sqrt{5} = B \quad \blacktriangleleft \text{ Take square roots.}$$

The completed triangle is shown in **Figure 6T.41**.

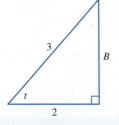

Figure 6T.40 An angle whose cosine is $\frac{2}{3}$

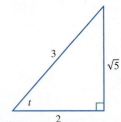

Figure 6T.41 An angle whose cosine is $\frac{2}{3}$: The Pythagorean theorem is used to calculate the opposite side.

Note that there are many other right triangles that are correct answers. For example, we could also represent $\cos t = \dfrac{2}{3}$ by using an adjacent side of 4 and an hypotenuse of 6. The resulting opposite side would be $2\sqrt{5}$.

TRY IT YOURSELF 6T.13 Brief answers provided at the end of the section.

The tangent of the acute angle t is $\dfrac{4}{5}$. Construct a right triangle with acute angle t, and label the lengths of the sides.

Using Representative Triangles to Evaluate Trigonometric Functions

Triangles representing one trigonometric function value can be used to find the values of other trigonometric functions.

EXAMPLE 6T.14 Finding One Trigonometric Function Value from Another

If t is an acute angle such that $\tan t = \dfrac{3}{4}$, find $\sin t$.

SOLUTION

Because $\tan t = \dfrac{\text{Opposite}}{\text{Adjacent}}$, we can find the desired value if we make a right triangle with the opposite side of length 3 and the adjacent side of length 4, as shown in **Figure 6T.42**. We use the Pythagorean theorem to find the length of the hypotenuse:

$$H^2 = 3^2 + 4^2$$
$$H^2 = 25$$
$$H = 5$$

The complete triangle is shown in **Figure 6T.43**. We use this figure to calculate the sine:

$$\sin t = \frac{\text{Opposite}}{\text{Hypotenuse}} = \frac{3}{5}$$

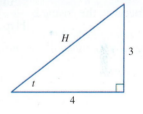

Figure 6T.42 An angle whose tangent is $\dfrac{3}{4}$

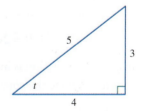

Figure 6T.43 An angle whose tangent is $\dfrac{3}{4}$: The hypotenuse is 5.

TRY IT YOURSELF 6T.14 Brief answers provided at the end of the section.

If t is an acute angle such that $\cos t = \dfrac{1}{3}$, find the sine of t.

EXTEND YOUR REACH

a. The cosine of an acute angle t is x. Find the values of the other five trigonometric functions of t in terms of x.

b. Use part a to explain how the values of all six trigonometric functions of an acute angle can be calculated from the cosine of that angle.

STEP-BY-STEP STRATEGY: Finding Trigonometric Function Values from a Known Trigonometric Function Value

Step 1 Label two sides of a right triangle so that an acute angle in the triangle has the appropriate trigonometric function value.

Step 2 Use the Pythagorean theorem to find the third side of the triangle.

Step 3 Use the triangle to calculate the other trigonometric function values of the angle.

EXAMPLE 6T.15   Finding One Trigonometric Function Value from Another Given by an Algebraic Expression

If s is an acute angle such that $\cos s = \dfrac{x}{x+2}$ for a positive number x, find $\cot s$.

SOLUTION

Because the cosine is $\dfrac{\text{Adjacent}}{\text{Hypotenuse}}$, we can get $\cos s = \dfrac{x}{x+2}$ if we use x for the side adjacent to s and $x+2$ for the hypotenuse. This configuration is shown in **Figure 6T.44**.

We find the value of the opposite side B using the Pythagorean theorem:

$$x^2 + B^2 = (x+2)^2 \qquad \blacktriangleleft \textbf{ Use the Pythagorean theorem.}$$

$$x^2 + B^2 = x^2 + 4x + 4 \qquad \blacktriangleleft \textbf{ Expand the square.}$$

$$B^2 = 4x + 4 \qquad \blacktriangleleft \textbf{ Cancel } x^2.$$

$$B = \sqrt{4x+4} \qquad \blacktriangleleft \textbf{ Take square roots.}$$

$$B = 2\sqrt{x+1} \qquad \blacktriangleleft \textbf{ Simplify.}$$

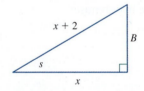

Figure 6T.44 An angle whose cosine is $\dfrac{x}{x+2}$

This new information is incorporated in **Figure 6T.45**, and we use this figure to calculate the cotangent of s:

$$\cot s = \frac{\text{Adjacent}}{\text{Opposite}}$$

$$= \frac{x}{2\sqrt{x+1}}$$

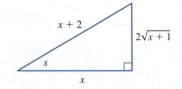

Figure 6T.45 An angle whose cosine is $\dfrac{x}{x+2}$: The opposite side is calculated.

TRY IT YOURSELF 6T.15 Brief answers provided at the end of the section.

If s is an acute angle such that $\cot s = \dfrac{x+3}{2}$, find the sine of s.

TRY IT YOURSELF ANSWERS

6T.9 $\tan t = \dfrac{3}{\sqrt{7}}, \cot t = \dfrac{\sqrt{7}}{3}, \sec t = \dfrac{4}{\sqrt{7}}, \csc t = \dfrac{4}{3}$

6T.10 $\sin s = \dfrac{2}{\sqrt{13}}, \cos s = \dfrac{3}{\sqrt{13}}, \tan s = \dfrac{2}{3}, \cot s = \dfrac{3}{2},$

$\sec s = \dfrac{\sqrt{13}}{3}, \csc s = \dfrac{\sqrt{13}}{2}$

6T.11 Using the small triangle, we find $\tan s = \dfrac{3}{4}$.

Using the large triangle, we find $\tan s = \dfrac{9}{12} = \dfrac{3}{4}$.

The two values are the same.

6T.12 $\cot\left(\dfrac{\pi}{3}\right) = \dfrac{1}{\sqrt{3}}$

6T.13 One possibility is

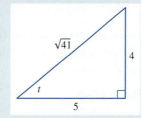

6T.14 $\sin t = \dfrac{\sqrt{8}}{3} = \dfrac{2\sqrt{2}}{3}$

6T.15 $\sin s = \dfrac{2}{\sqrt{x^2 + 6x + 13}}$

EXERCISE SET 6T.2

CHECK YOUR UNDERSTANDING

1. Express $\sin t$ using the sides of the triangle in **Figure 6T.46**.

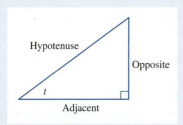

Figure 6T.46 Sides of a triangle labeled

2. Express $\cos t$ using the sides of the triangle in Figure 6T.46.

3. Express $\tan t$ using the sides of the triangle in Figure 6T.46.

4. Express $\cot t$ using the sides of the triangle in Figure 6T.46.

5. Express $\sec t$ using the sides of the triangle in Figure 6T.46.

6. Express $\csc t$ using the sides of the triangle in Figure 6T.46.

7. What property of triangles is used to show the consistency of the definitions of the trigonometric functions?

8. If two sides of a right triangle are known, what theorem allows us to find the third side?

9. List the special angles between 0 and $\dfrac{\pi}{2}$ radians.

10. What names are commonly used for the sides of a right triangle relative to a given acute angle of the triangle?

SKILL BUILDING

11. The 3-4-5 right triangle. Any triangle with sides 3, 4, and 5 is a right triangle, known as the 3-4-5 right triangle. It is shown in **Figure 6T.47**. Find the sine, cosine, tangent, secant, cosecant, and cotangent of the angle s.

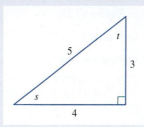

Figure 6T.47 The 3-4-5 right triangle

12. **More on the 3-4-5 right triangle.** Calculate the values of the six trigonometric functions of the angle t in Figure 6T.47.

Calculating values of trigonometric functions In Exercises 13 through 26, calculate the values of the six trigonometric functions of the given angle.

13. The angle t in **Figure 6T.48**

14. The angle s in Figure 6T.48

15. The angle t in **Figure 6T.49**

16. The angle s in Figure 6T.49

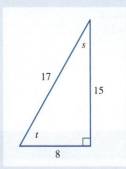

Figure 6T.48 Triangle for Exercises 13 and 14

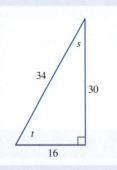

Figure 6T.49 Triangle for Exercises 15 and 16

17. The angle t in **Figure 6T.50**

18. The angle s in Figure 6T.50

19. The angle t in **Figure 6T.51**

20. The angle s in Figure 6T.51

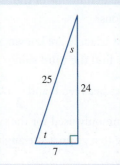

Figure 6T.50 Triangle for Exercises 17 and 18

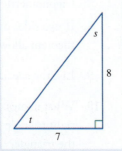

Figure 6T.51 Triangle for Exercises 19 and 20

21. The angle t in **Figure 6T.52**

22. The angle s in Figure 6T.52

23. The angle t in **Figure 6T.53**

24. The angle s in Figure 6T.53

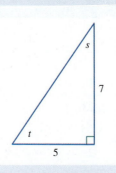

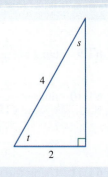

Figure 6T.52 Triangle for Exercises 21 and 22

Figure 6T.53 Triangle for Exercises 23 and 24

25. The angle t in **Figure 6T.54**

26. The angle s in Figure 6T.54

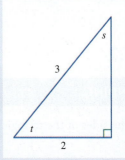

Figure 6T.54 Triangle for Exercises 25 and 26

Making triangles. In Exercises 27 through 38, construct a right triangle that exhibits the given value of a trigonometric function of an acute angle s. Be sure to label all three sides of the triangle and to mark the angle s. There are many correct answers.

27. $\sin\ s = \dfrac{2}{3}$

28. $\sin\ s = \dfrac{3}{4}$

29. $\cos\ s = \dfrac{1}{4}$

30. $\cos\ s = \dfrac{2}{5}$

31. $\tan\ s = 5$

32. $\tan\ s = \dfrac{3}{4}$

33. $\cot\ s = \dfrac{4}{5}$

34. $\cot\ s = 2$

35. $\sec\ s = \dfrac{5}{3}$

36. $\sec\ s = 3$

37. $\csc\ s = \dfrac{5}{4}$

38. $\csc\ s = 2$

Getting other trigonometric function values from a given one. In Exercises 39 through 46, you are given the value of one of the trigonometric functions of an acute angle t. Find the values of the other five trigonometric functions of t.

39. $\sin t = \dfrac{2}{3}$

40. $\cos t = \dfrac{1}{5}$

41. $\tan t = \dfrac{5}{8}$

42. $\sec t = \dfrac{5}{4}$

43. $\csc t = 7$

44. $\cot t = 1$

45. $\tan t = \dfrac{2}{5}$

46. $\tan t = 4$

PROBLEMS

47. The definition of the cosine. Show that the value of cos t depends only on the acute angle t and not on the particular right triangle used to calculate it. *Suggestion*: If you have difficulty, refer to the discussion before Example 6T.11 of this same property for the sine.

48. The definition of the tangent. Show that the value of tan t depends only on the acute angle t and not on the particular right triangle used to calculate it. *Suggestion*: If you have difficulty, refer to the discussion before Example 6T.11 of this same property for the sine.

49. The definition of the cotangent. Show that the value of cot t depends only on the acute angle t and not on the particular right triangle used to calculate it. *Suggestion*: If you have difficulty, refer to the discussion before Example 6T.11 of this same property for the sine.

50. The definition of the cosecant. Show that the value of csc t depends only on the acute angle t and not on the particular right triangle used to calculate it. *Suggestion*: If you have difficulty, refer to the discussion before Example 6T.11 of this same property for the sine.

51. The definition of the secant. Show that the value of sec t depends only on the acute angle t and not on the particular right triangle used to calculate it. *Suggestion*: If you have difficulty, refer to the discussion before Example 6T.11 of this same property for the sine.

52. Calculating some trigonometric functions. In a certain right triangle, the side opposite one of the acute angles t is three times as long as the adjacent side. Find the sine, cosine, and tangent of t.

53. A 30°-60°-90° triangle. In this exercise, we show how to get the side lengths for the 30°-60°-90° triangle. Starting with the equilateral triangle shown in **Figure 6T.55**, we bisect the top angle as in **Figure 6T.56**.

a. Find the length of A in Figure 6T.56. *Suggestion*: The angle bisector is the perpendicular bisector of the base.

b. Use the Pythagorean theorem to find the length of H.

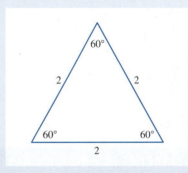

Figure 6T.55 An equilateral triangle

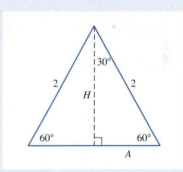

Figure 6T.56 Top angle bisected

54. Special angles. Use the basic trigonometric relationships from Table 6T.1 to complete the following table of special angles.

Table of Special Angles							
Angle in degrees	Angle in radians	sin	cos	tan	cot	sec	csc
30	$\dfrac{\pi}{6}$	$\dfrac{1}{2}$	$\dfrac{\sqrt{3}}{2}$				
45	$\dfrac{\pi}{4}$	$\dfrac{1}{\sqrt{2}}$	$\dfrac{1}{\sqrt{2}}$				
60	$\dfrac{\pi}{3}$	$\dfrac{\sqrt{3}}{2}$	$\dfrac{1}{2}$				

Obtaining other trigonometric function values from a given one. In Exercises 55 through 66, you are given the value (expressed in terms of x) of one of the trigonometric functions of an acute angle t. Find the values of the other five trigonometric functions of t in terms of x. If you have difficulty, you may wish to review Example 6T.15.

55. $\sin t = x$

56. $\cos t = x$

57. $\tan t = x$

58. $\sec t = x$

59. $\csc t = x$

60. $\cot t = x$

61. $\cot t = 3x$

62. $\cos t = 2x - 1$

63. $\tan t = \dfrac{x}{x+1}$

64. $\cot t = \dfrac{2x+1}{3x+1}$

65. $\sin t = \dfrac{x}{x+2}$

66. $\sin t = \dfrac{1-x}{1+x}$

67. Solving a trigonometric equation. Find one solution of the equation $\sin t = \cos t$. *Suggestion:* Think about the special angles that you know.

Solving equations with technology. Exercises 68 through 71 show equations on restricted intervals that may be difficult to solve by hand. In each case, use a graphing utility to make appropriate graphs, and report the solution correct to two decimal places. Be sure your calculator is set to radian mode.

68. $\cos t = \ln t, \quad 0 < t < 2$

69. $\sin t = t^2, \quad 0 < t < 2$

70. $\tan t = \cot t, \quad 0 < t < 1$

71. $1 + \sin t = \csc t, \quad 0 < t < 1$

CHALLENGE EXERCISES FOR INDIVIDUALS OR GROUPS

72. Complements of angles. Suppose t is an acute angle, as shown in **Figure 6T.57**. The complement of t is the angle $s = 90° - t$ if we use degrees or $s = \dfrac{\pi}{2} - t$ if we use radians.

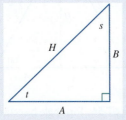

Figure 6T.57 Complementary angles: The degree measure of s is $90°$ minus the degree measure of t.

Show that in terms of degrees	and in terms of radians
$\tan(90° - t) = \cot t$	$\tan\left(\dfrac{\pi}{2} - t\right) = \cot t$
$\cot(90° - t) = \tan t$	$\cot\left(\dfrac{\pi}{2} - t\right) = \tan t$
$\sec(90° - t) = \csc t$	$\sec\left(\dfrac{\pi}{2} - t\right) = \csc t$
$\csc(90° - t) = \sec t$	$\csc\left(\dfrac{\pi}{2} - t\right) = \sec t$

Showing basic identities. In Exercises 73 through 78, use the triangle in **Figure 6T.58** to establish the given identity.

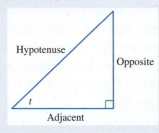

Figure 6T.58 Sides of a triangle labeled

73. Show that $\tan t = \dfrac{\sin t}{\cos t}$.

74. Show that $\cot t = \dfrac{\cos t}{\sin t}$.

75. Show that $\sec t = \dfrac{1}{\cos t}$.

76. Show that $\csc t = \dfrac{1}{\sin t}$.

77. Show that $\sin t = \dfrac{1}{\csc t}$.

78. Show that $\cos t = \dfrac{1}{\sec t}$.

REVIEW AND REFRESH: Exercises from Previous Sections

79. From Section P.2: Solve the inequality $|3x - 4| < 10$.

Answer: $\left(-2, \dfrac{14}{3}\right)$

80. From Section 2.3: The graph of $f(x)$ is given in **Figure 6T.59**. Add to the picture the graph of $g(x) = f(x + 1) - 1$.

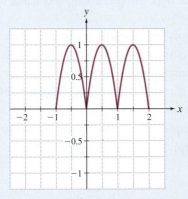

Figure 6T.59 Graph for Exercise 80

Answer:

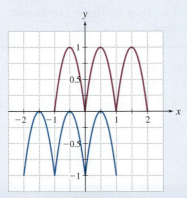

81. From Section 3.1: Find the equation of the linear function f with slope -2 if $f(3) = 4$.

Answer: $f(x) = -2x + 10$

82. From Section 3.2: Find an exponential function $f(x)$ so that $f(0) = 3$ and f is multiplied by 5 when x is increased by 1.

Answer: $f(x) = 3(5^x)$

83. From Section 4.1: If $f(x) = \ln(x + 1)$, find $f(e^x - 1)$.

Answer: x

84. From Section 4.2: Simplify the expression $e^{2 \ln x}$.

Answer: x^2

85. From Section 4.3: Solve the equation $\ln(x + 4) = 7$.

Answer: $x = e^7 - 4$

86. From Section 5.1: What can you conclude about the zeros of the quadratic function $y = ax^2 + bx + c$ if the discriminant $b^2 - 4ac$ is negative?

Answer: There are no real zeros.

87. From Section 5.4: Find the zeros of $f(x) = x^3 - 3x^2 + 2x$, and then plot its graph.

Answer: Zeros: $0, 1, 2$

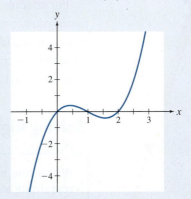

88. From Section 5.5: Find the horizontal and vertical asymptotes of $f(x) = \dfrac{2x^2}{x^2 - 1}$. Then plot the graph, and show the asymptotes.

Answer: Horizontal asymptote $y = 2$. Vertical asymptotes $x = -1$ and $x = 1$.

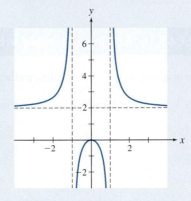

6T.1 Angles

6T.2 Definition of the Trigonometric Functions Using Right Triangles

6T.3 Analysis of Right Triangles

6T.4 Extending the Trigonometric Functions: The Unit Circle

Figure 6T.60 A transit

6T.3 Analysis of Right Triangles

Right triangles provide information about trigonometric functions—and trigonometric functions provide information about right triangles.

In this section, you will learn to:

1. Solve a right triangle given one acute angle and one side.
2. Apply the ideas of trigonometry to analyze right triangles arising in practical settings.

Trigonometric functions are essential tools for analyzing right triangles. In this section, we develop the idea of using trigonometry to study right triangles. It lies at the heart of many applications, such as surveying and cartography.

For example, the fundamental tool of the surveyor is the transit (**Figure 6T.60**). The transit can accurately measure angles, and that measurement, together with some basic trigonometry, is the key to many practical applications. The transit is used to determine the dimensions of structures, both natural and human made, that do not lend themselves to direct measurement.

Solving Right Triangles

When one side and one acute angle of a right triangle are known, the remaining sides and angle may be calculated.

A basic problem in trigonometry is to solve a given right triangle—that is, to find the length of each side and the measure of each angle. The trigonometric functions allow us to solve a right triangle if we know one acute angle and one side, as illustrated in the next example.

EXAMPLE 6T.16 Solving a Triangle with a Special Angle

One acute angle of a right triangle has radian measure $\dfrac{\pi}{3}$, and the opposite side has length 3. Find the unknown angle and sides in **Figure 6T.61**.

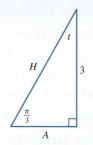

Figure 6T.61 A triangle with one special angle

SOLUTION

We can use the fact that the sum of the three angles is π radians to find the measure of the angle t:

$$t + \frac{\pi}{3} + \frac{\pi}{2} = \pi$$

$$t = \pi - \left(\frac{\pi}{3} + \frac{\pi}{2}\right)$$

$$= \frac{\pi}{6} \text{ radian}$$

We use the sine to find the value of H because it relates the unknown hypotenuse H to a known side. We know that $\sin\left(\dfrac{\pi}{3}\right) = \dfrac{\sqrt{3}}{2}$. We can also use the given triangle to calculate the sine of $\dfrac{\pi}{3}$:

$$\frac{\sqrt{3}}{2} = \sin\left(\frac{\pi}{3}\right) = \frac{\text{Opposite}}{\text{Hypotenuse}} = \frac{3}{H}$$

Solving the equation $\dfrac{3}{H} = \dfrac{\sqrt{3}}{2}$ for H, we find $H = 6/\sqrt{3} = 2\sqrt{3}$.

Next we make use of the tangent to find the value of A. (We could instead use the Pythagorean theorem.) Because $\dfrac{\pi}{3}$ is a special angle, we have

$$\tan\left(\frac{\pi}{3}\right) = \frac{\sin\left(\dfrac{\pi}{3}\right)}{\cos\left(\dfrac{\pi}{3}\right)} = \frac{\dfrac{\sqrt{3}}{2}}{\dfrac{1}{2}} = \sqrt{3}$$

Now using the triangle in Figure 6T.61 gives

$$\sqrt{3} = \tan\left(\frac{\pi}{3}\right) = \frac{\text{Opposite}}{\text{Adjacent}} = \frac{3}{A}$$

We solve the equation $\dfrac{3}{A} = \sqrt{3}$ for A to get $A = \dfrac{3}{\sqrt{3}} = \sqrt{3}$.

TRY IT YOURSELF 6T.16 Brief answers provided at the end of the section.

Find the angle t and the unknown sides in **Figure 6T.62**.

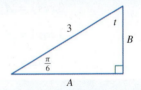

Figure 6T.62 A triangle for Try It Yourself 6T.16

If the known angles of a triangle are not special angles, then we must use technology to find approximations for the values of the trigonometric functions.

EXAMPLE 6T.17   Solving Right Triangles

Suppose the angle s in **Figure 6T.63** has a measure of 1 radian. Solve the triangle. Give answers correct to two decimal places.

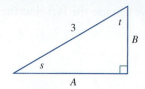

Figure 6T.63 A right triangle with an angle of measure 1 radian

SOLUTION

We first use the fact that the angle sum is π radians:

$$t + 1 + \frac{\pi}{2} = \pi$$

$$t = \pi - \left(1 + \frac{\pi}{2}\right)$$

$$\approx 0.57 \text{ radian}$$

The definition of the sine function allows us to express B in terms of $\sin s$:

$$\sin s = \frac{\text{Opposite}}{\text{Hypotenuse}} \qquad \text{◀ Use the definition of sine.}$$

$$\sin s = \frac{B}{3} \qquad \text{◀ Substitute known values.}$$

$$3 \sin s = B \qquad \text{◀ Solve for } B.$$

Because 1 radian is not one of the special angles, we use a calculator (set to radian mode) to evaluate its sine:

$$B = 3 \sin s = 3 \sin 1 \approx 2.52$$

Similarly, we can use the cosine to find A:

$$\cos s = \frac{A}{3} \qquad \text{◀ Calculate the cosine from the triangle.}$$

$$3 \cos s = A \qquad \text{◀ Solve for } A.$$

$$3 \cos 1 = A \qquad \text{◀ Substitute 1 for } s.$$

$$1.62 \approx A \qquad \text{◀ Approximate the value using a calculator.}$$

TRY IT YOURSELF 6T.17   Brief answers provided at the end of the section.

Solve the triangle in Figure 6T.63 if instead the angle s is 25°. (Set your calculator to degree mode.)

a. In **Figure 6T.64**, find the length of the side marked B in terms of a trigonometric function of the angle t.

b. Show that for acute angles t, the function $y = \sin t$ is increasing. *Suggestion:* Think about what happens to the side marked B in Figure 6T.64 as the angle t increases.

c. Is the function $y = \cos t$ increasing or decreasing for acute angles t?

EXTEND YOUR REACH

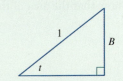

Figure 6T.64 A right triangle with a hypotenuse of length 1

We generalize these techniques for the case of a known acute angle t and hypotenuse h. Our aim is to find the sides x and y in **Figure 6T.65** in terms of h and t. We use the sine of t to find the value of y and the cosine of t to find the value of x:

$$\sin t = \frac{y}{h} \text{ or } y = h \, \sin t$$

$$\cos t = \frac{x}{h} \text{ or } x = h \, \cos t$$

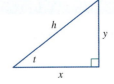

Figure 6T.65 Finding x and y in terms of t and h

CONCEPTS TO REMEMBER: Finding Legs from Angle and Hypotenuse

1. If t is an acute angle of a right triangle with hypotenuse h, we can find the legs of the triangle using

    Adjacent side $= h \, \cos t$
    Opposite side $= h \, \sin t$

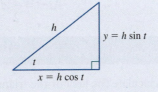

2. A hypotenuse of length 1 is a common case:

    Adjacent side $= \cos t$
    Opposite side $= \sin t$

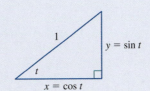

A New Formula for the Area of a Triangle

> The trigonometry of right triangles yields a useful formula for the area of a triangle.

In the next example, we use a formula for the area of a triangle that may be familiar to you from geometry: If B is the length of the base and H is the height, then

$$\text{Area} = \frac{1}{2} BH$$

A new area formula is developed, and we will find it useful as our study of trigonometry progresses.

EXAMPLE 6T.18   A New Area Formula

Show that the area of the triangle in **Figure 6T.66** is

$$\text{Area} = \frac{1}{2}\,AB\,\sin t$$

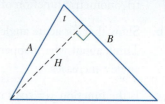

Figure 6T.66 Triangle for Example 6T.18

SOLUTION

In Figure 6T.66, B is the length of the base and H is the height, so

$$\text{Area} = \frac{1}{2}\,BH$$

Focus now on the smaller right triangle with sides A and H. We can use these sides to calculate the sine of t:

$$\sin t = \frac{H}{A}$$

Solving for H gives $H = A \sin t$. Putting this expression in place of H in the area formula gives the desired result:

$$\text{Area} = \frac{1}{2}\,AB\,\sin t$$

TRY IT YOURSELF 6T.18   Brief answers provided at the end of the section.

Use the new area formula to calculate the area of the triangle shown in **Figure 6T.67**.

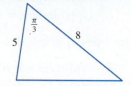

Figure 6T.67 Triangle for Try It Yourself 6T.18

EXTEND YOUR REACH

In **Figure 6T.68**, the angle s is greater than the angle t. Show that the area of the lower triangle is greater than the area of the upper triangle. *Suggestion:* Look at the Extend Your Reach accompanying Example 6T.17.

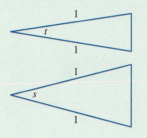

Figure 6T.68 Comparing areas

MODELS AND APPLICATIONS Transits and Shadows

These methods for solving triangles can be used in practical settings to find unknown distances. In practical applications we rarely encounter special angles, and as a result we must use a calculator or computer. When using technology, take care that the mode for measuring angles on your device is set to degrees or radians, as appropriate. Also, many calculators have sine and cosine keys but may not have keys for the other four trigonometric functions. The basic trigonometric relationships that give the other four functions in terms of the sine and cosine are essential in that case.

EXAMPLE 6T.19 Surveying: Height of a Building

A surveyor stands 300 horizontal feet from a building and aims a transit (an instrument for measuring angles) at the top of the building, as shown in **Figure 6T.69**. The transit measures an angle of 21° from the horizontal.

a. How much taller is the building than the transit?

b. How long would a straight cable that reached from the transit directly to the top of the building need to be?

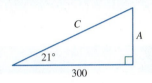

Figure 6T.69 Measuring with a transit

SOLUTION

a. The desired height is represented by the side marked A in Figure 6T.69. We know the length of the adjacent side, so we should use the tangent function, which relates these two sides:

$$\tan 21° = \frac{\text{Opposite}}{\text{Adjacent}} = \frac{A}{300}$$

Solving for A, we find

$$A = 300 \tan 21° \approx 115.16 \text{ feet}$$

We conclude that the building is about 155.16 feet taller than the transit.

b. The distance from the transit directly to the top of the building is the hypotenuse of the triangle and is marked C in Figure 6T.69. The cosine relates this side to the adjacent side of 300 feet:

$$\cos 21° = \frac{\text{Adjacent}}{\text{Hypotenuse}} = \frac{300}{C}$$

Solving for C gives

$$C = \frac{300}{\cos 21°} \approx 321.34 \text{ feet}$$

We conclude that a straight cable reaching from the transit to the top of the building would need to be about 321.34 feet long.

TRY IT YOURSELF 6T.19 Brief answers provided at the end of the section.

How much taller is the building than the transit if instead the angle is 27°?

EXAMPLE 6T.20 Towers and Shadows

A 7-foot-tall pole casts a shadow of 4 feet at the same time a tower casts a shadow of 15 feet, as shown in **Figure 6T.70**. How tall is the tower?

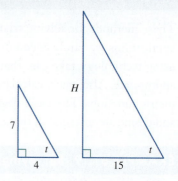

Figure 6T.70 Shadows of a pole and a tower

SOLUTION

Because the shadows are produced by the sun at the same time, the angles marked t in the two triangles are the same. Using the smaller triangle, we calculate $\tan t = \dfrac{7}{4}$. Using the larger triangle, we calculate $\tan t = \dfrac{H}{15}$.

Solving the equation $\dfrac{H}{15} = \dfrac{7}{4}$, we find $H = \dfrac{15 \times 7}{4} = 26.25$ feet.

We could also solve this problem using similar triangles.

TRY IT YOURSELF 6T.20 Brief answers provided at the end of the section.

How tall is a tower that casts a shadow of 20 feet at the same time?

TRY IT YOURSELF ANSWERS

6T.16 $A = \dfrac{3\sqrt{3}}{2}, B = \dfrac{3}{2}, t = \dfrac{\pi}{3}$ radians

6T.17 $t = 65°, A \approx 2.72, B \approx 1.27$

6T.18 $10\sqrt{3} \approx 17.32$

6T.19 $300 \tan 27° \approx 152.86$ feet

6T.20 35 feet

EXERCISE SET 6T.3

CHECK YOUR UNDERSTANDING

1. The instruction to "solve a right triangle" means _____.

2. **True or false:** If we know an acute angle of a right triangle, we can solve the triangle.

3. Express the adjacent side in **Figure 6T.71** in terms of the hypotenuse and a trigonometric function of the angle t.

4. Express the opposite side in Figure 6T.71 in terms of the hypotenuse and a trigonometric function of the angle t.

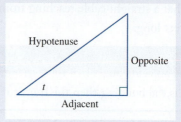

Figure 6T.71 Sides of a triangle labeled

5. Express the opposite side in Figure 6T.71 in terms of the adjacent side and a trigonometric function of the angle t.

6. Express the adjacent side in Figure 6T.71 in terms of the opposite side and a trigonometric function of the angle t.

7. Express the hypotenuse in Figure 6T.71 in terms of the adjacent side and a trigonometric function of the angle t.

8. Express the hypotenuse in Figure 6T.71 in terms of the opposite side and a trigonometric function of the angle t.

9. A transit is used in practical settings to measure _____.

10. If t is the angle between sides A and B of a triangle, then the area of the triangle is given by _____.

SKILL BUILDING

Making triangles. In Exercises 11 through 22, draw and label the sides of a right triangle with the given acute angle and with hypotenuse of length 1. Exercises 11 through 16 involve special angles, so no calculator is needed. In Exercises 17 through 22, report side lengths correct to two decimal places. Remember that radian measure is used unless degrees are specifically indicated.

11. $\dfrac{\pi}{4}$

12. $\dfrac{\pi}{3}$

13. $\dfrac{\pi}{6}$

14. 45°

15. 60°

16. 30°

17. 20°

18. 50°

19. 35°

20. 28°

21. 1 radian

22. 0.7 radian

Getting a length. In Exercises 23 through 40, you are given the value of a trigonometric function of an acute angle and the length of one side of a right triangle with that angle. Use this information to calculate the lengths of the remaining sides of the triangle. You may find it helpful to draw a picture of the right triangle. Report answers correct to two decimal places.

23. The sine of an acute angle in a right triangle is 0.11. The length of the opposite side is 8.

24. The sine of an acute angle in a right triangle is $\dfrac{1}{3}$. The length of the opposite side is 4.

25. The sine of an acute angle in a right triangle is $\dfrac{2}{5}$. The length of the hypotenuse is 10.

26. The cosine of an acute angle in a right triangle is 0.71. The length of the adjacent side is 8.

27. The cosine of an acute angle in a right triangle is $\dfrac{1}{4}$. The length of the adjacent side is 5.

28. The cosine of an acute angle in a right triangle is $\dfrac{1}{2}$. The length of the hypotenuse is 6.

29. The tangent of an acute angle in a right triangle is 4. The length of the adjacent side is 7.

30. The tangent of an acute angle in a right triangle is 3. The length of the adjacent side is 4.

31. The tangent of an acute angle in a right triangle is $\dfrac{1}{2}$. The length of the opposite side is 4.

32. The cotangent of an acute angle in a right triangle is $\dfrac{2}{5}$. The length of the opposite side is 9.

33. The cotangent of an acute angle in a right triangle is $\dfrac{4}{3}$. The length of the opposite side is 6.

34. The cotangent of an acute angle in a right triangle is 3. The length of the adjacent side is 9.

35. The secant of an acute angle in a right triangle is 5. The length of the hypotenuse is 6.

36. The secant of an acute angle in a right triangle is 24. The length of the hypotenuse is 6.

37. The secant of an acute angle in a right triangle is 2. The length of the adjacent side is 2.

38. The cosecant of an acute angle in a right triangle is 5. The length of the opposite side is 5.

39. The cosecant of an acute angle in a right triangle is $\dfrac{3}{2}$. The length of the hypotenuse is 6.

40. The cosecant of an acute angle in a right triangle is 3. The length of the opposite side is 9.

Finding area. In Exercises 41 through 47, values for A, B, and the angle t for **Figure 6T.72** are given. Use this information to find the area of the triangle.

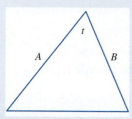

Figure 6T.72 Calculating area

41. $A = 2$, $B = 3$, and $t = \dfrac{\pi}{3}$

42. $A = 4$, $B = 5$, and $t = \dfrac{\pi}{6}$

43. $A = 6$, $B = 3$, and $t = 45°$

44. $A = 8$, $B = 3$, and $t = 30°$

45. $A = 2$, $B = 3$, and $t = 12°$. Report your answer correct to two decimal places.

46. $A = 5$, $B = 7$, and $t = 35°$. Report your answer correct to two decimal places.

47. $A = 6$, $B = 1$, and $t = 0.4$ radian. Report your answer correct to two decimal places.

48. Find a length. In Figure 6T.72, $A = 2$, the area of the triangle is 5, and $t = 60°$. Find the length of B.

49. Find a length. In Figure 6T.72, $A = 4$, the area of the triangle is 12, and $t = \dfrac{\pi}{4}$ radian. Find the length of B.

PROBLEMS

Right triangles. In Exercises 50 through 58, draw a right triangle consistent with the given information, solve for the lengths of the two unknown sides, and label all the sides.

50. An acute angle of $\dfrac{\pi}{4}$ radian and an adjacent side of length 2

51. An acute angle of $\dfrac{\pi}{3}$ radians and an opposite side of length 3

52. An acute angle of $\dfrac{\pi}{6}$ radian and a hypotenuse of length 4

53. An acute angle of 20° and an opposite side of length 2. Report side lengths correct to two decimal places.

54. An acute angle of 50° and a hypotenuse of length 3. Report side lengths correct to two decimal places.

55. An acute angle of 1 radian and an opposite side of length 5. Report side lengths correct to two decimal places.

56. An acute angle of 0.4 radian and a hypotenuse of length 7. Report side lengths correct to two decimal places.

57. An acute angle of $\dfrac{\pi}{6}$ radian and a hypotenuse of length x

58. An acute angle of 60° and a opposite side of length x

Area and perimeter. In Exercises 59 through 64, some angles, and the length of a side, of a triangle are given. The altitude of the triangle is included. Find the area and perimeter of the triangle. Note that you may wish first to find some of the unlabeled sides.

59.

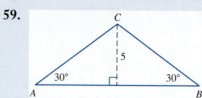

60.

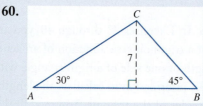

61.

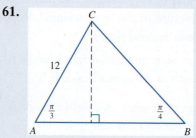

62.

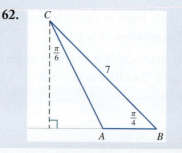

63.

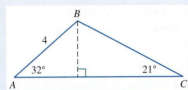

64.

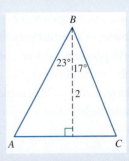

MODELS AND APPLICATIONS

65. Calculating height: A woman sits 331 horizontal feet from the base of a tower. She must incline her eyes at an angle of 16.2° to look at the top of the tower. See **Figure 6T.73**. How tall is the tower?

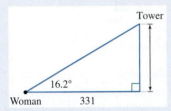

Figure 6T.73 Height of a tower

66. Home theater. The top of the screen for a home theater is 25 inches above the eye level of a person seated in a chair. For the sake of comfort, the top of the screen should be 15° above eye level. See **Figure 6T.74**. How far from the screen should the chair be placed? Round your answer to the nearest inch.

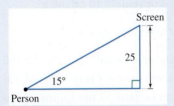

Figure 6T.74 Home theater

67. Speed of a drone. A drone is flying due north at a constant speed. An observer uses a laser to locate the drone due east at a distance of 130 feet. One minute later the observer rotates the laser by 40° to locate the drone. See **Figure 6T.75**. How fast is

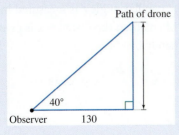

Figure 6T.75 Locating a drone

the drone traveling? Report your answer in feet per minute to the nearest whole number.

68. Calculating distance. A man sits 18 horizontal feet from the base of a tree. He must incline his eyes at an angle of 21° to look at the top of the tree. See **Figure 6T.76**. What is the distance from the man directly to the top of the tree?

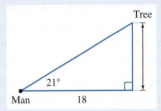

Figure 6T.76 Distance to a tree

69. The width of a river. You are a surveyor. You stand on the north bank of a river and look due south at a tree on the opposite bank. Your helper on the opposite bank measures 35 yards due east to a second tree. You must rotate your transit through an angle of 21° to point toward the second tree. See **Figure 6T.77**. How wide is the river?

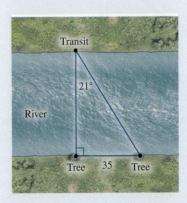

Figure 6T.77 Measuring the width of a river

70. **Grasping prey.** The diagram in **Figure 6T.78** is useful for estimating the diameter of a circular prey that would be optimal for the grasping claws of a praying mantis.[3] Find a formula for the diameter $|BC|$ of the circle in terms of the length $|AC|$ and the angle t. (The vertical bars denote the length of a segment.)

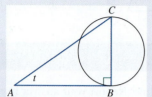

Figure 6T.78 Claws of a praying mantis

71. **Dispersal method.** When there is moisture on the ground, microorganisms are found in the thin layer of water on leaves.[4] A monolayer containing these organisms is formed on the top, and (because of the spreading pressure of this monolayer) the floating organisms spread upward when the surface is tilted. This is a dispersal method onto newly fallen leaves. In **Figure 6T.79**, the horizontal segment represents the ground, and the angle t measures the amount by which the surface is tilted. (We assume that t is less than 90°.) The length d is the distance the organisms move, and the length v is the change in the elevation.

a. Find a formula giving v in terms of d and t.

b. Assume that the distance the organisms move is 30 centimeters. Find the change in elevation if the angle t is 15° and if the angle t is 30°.

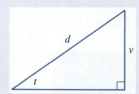

Figure 6T.79 Distance traveled and elevation

72. **Parallax angle.** If we view a star now and then view it again half a year later, our position will have changed by the diameter of Earth's orbit around

the sun. For nearby stars (within 100 light-years or so, such as Vega), the change in viewing location is sufficient to make the star appear to be in a slightly different location in the sky. Half of the angle from one location to the next is known as the parallax angle. In **Figure 6T.80**, the parallax angle is marked t, d is the distance from Earth to the sun, and s is the distance from the sun to the star. Show that we get the distance s using $s = d \cot t$.

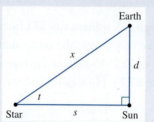

Figure 6T.80 The parallax angle

73. **More on parallax angle.** This exercise uses the description of parallax angles in the preceding exercise. Parallax angles are very small and are normally measured in seconds of arc. One second of arc is $\dfrac{1}{60^2}$ degree. At the same time, the distance s is very large and is normally measured in light-years. The angle t is so small and the distance s is so large that s is nearly the same as the distance x marked in Figure 6T.80, and astronomers find this distance more convenient to calculate.

a. Show that x is given by $x = \dfrac{d}{\sin t}$.

b. Parallax angles are small enough that the approximation $\sin t \approx t$ is quite good, for t measured in radians. (This approximation is derived in Exercise 75.) Show that if t is measured in seconds of arc, then

$$\sin t \approx \frac{\pi t}{180 \times 60^2}.$$

c. The distance from Earth to the sun is approximately 1.58×10^{-5} light-year. Show that if the parallax angle t is measured in seconds of arc, then the distance x is given approximately by

$$x = \frac{3.26}{t} \text{ light-years}.$$

---

[3]Based on the work of C. S. Holling, as described by E. R. Pianka in *Evolutionary Ecology*, 6th ed. Benjamin Cummings, San Francisco, CA, 2000.

[4]Based on the study by R. J. Bandoni and R. E. Koske, "Monolayers and Microbial Dispersal," *Science* 183: 1079–1081 (1974).

CHALLENGE EXERCISES FOR INDIVIDUALS OR GROUPS

74. An important calculus inequality. In calculus we calculate the instantaneous rate of change of the sine function (which turns out to be the cosine). The inequality

$$\cos t < \frac{t}{\sin t} < \frac{1}{\cos t}$$

for $0 < t < \dfrac{\pi}{2}$ is a key part of this calculation. The goal of this exercise is to establish this inequality. In **Figure 6T.81**, the acute angle t is measured in radians.

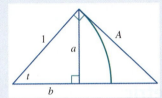

Figure 6T.81 Two right triangles and a sector of a circle

a. In the figure we see a small right triangle with base b and altitude a, a sector of a circle of radius 1, and a large right triangle with altitude A. Calculate a, b, and A in terms of t.

b. In the figure

Area of small triangle < Area of sector < Area of large triangle

Use this together with your work from part a to show

$$\frac{\sin t \, \cos t}{2} < \frac{t}{2} < \frac{\tan t}{2}$$

c. Use your work from part b to show that if $0 < t < \dfrac{\pi}{2}$, then

$$\cos t < \frac{t}{\sin t} < \frac{1}{\cos t}$$

75. Approximating the sine function. If an angle t is measured in radians, and if t is near zero, then $\sin t \approx t$. To see why this approximation is reasonable, consider **Figure 6T.82**. When t is small, the distance from A to B is very close to the arc length from A to C. Show that this comparison yields $\sin t \approx t$.

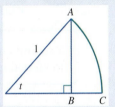

Figure 6T.82 When t is near 0: The arc length approximates the vertical height.

REVIEW AND REFRESH: Exercises from Previous Sections

76. From Section P.1: Find the distance between the points $(x^2, 2x)$ and $(1, 0)$.

Answer: $x^2 + 1$

77. From Section 1.4: Find the limit as $x \to \infty$ of

$$y = \frac{x+1}{x-1}.$$

Answer: 1

78. From Section 3.1: Find a linear function with initial value 4 and rate of change 7.

Answer: $y = 7x + 4$

79. From Section 4.3: Solve the equation $e^{2x-1} = 7$.

Answer: $x = \dfrac{1 + \ln 7}{2}$

80. From Section 5.1: Factor the quadratic expression $x^2 - 2x - 2$.

Answer: $(x - (1 - \sqrt{3}))(x - (1 + \sqrt{3}))$

81. From Section 5.2: Give the quotient and remainder of $x - 2 \overline{)x^3 - 5}$.

Answer: Quotient: $x^2 + 2x + 4$. Remainder: 3.

82. From Section 6T.1: Find the area of a sector of a circle with radius 2 units and central angle $\dfrac{\pi}{5}$ radian.

Answer: $\dfrac{2\pi}{5}$ square units

83. From Section 6T.2: List the values of the six trigonometric functions of $\dfrac{\pi}{3}$.

Answer:

$$\sin\left(\frac{\pi}{3}\right) = \frac{\sqrt{3}}{2}, \ \cos\left(\frac{\pi}{3}\right) = \frac{1}{2}, \ \tan\left(\frac{\pi}{3}\right) = \sqrt{3},$$

$$\cot\left(\frac{\pi}{3}\right) = \frac{1}{\sqrt{3}}, \ \sec\left(\frac{\pi}{3}\right) = 2, \ \csc\left(\frac{\pi}{3}\right) = \frac{2}{\sqrt{3}}$$

84. From Section 6T.2: If t is an acute angle and $\tan t = \dfrac{x}{x+1}$, find $\sin t$.

Answer: $\sin t = \dfrac{x}{\sqrt{2x^2 + 2x + 1}}$

85. From Section 6T.2: If t is an acute angle and $\sin t = \dfrac{2}{5}$, find the values of the other five trigonometric functions of t.

Answer:

$$\cos t = \frac{\sqrt{21}}{5}, \ \tan t = \frac{2}{\sqrt{21}}, \ \cot t = \frac{\sqrt{21}}{2},$$

$$\sec t = \frac{5}{\sqrt{21}}, \ \csc t = \frac{5}{2}$$

6T.1 Angles

6T.2 Definition of the Trigonometric Functions Using Right Triangles

6T.3 Analysis of Right Triangles

6T.4 Extending the Trigonometric Functions: The Unit Circle

6T.4 Extending the Trigonometric Functions: The Unit Circle

> The unit circle is used to extend the domain of the trigonometric functions to all real numbers.

In this section, you will learn to:

1. Locate a point on the unit circle given one coordinate.
2. Find the coordinates of a trigonometric point.
3. Use the definitions of the six trigonometric functions for all numbers and angles to find function values.
4. Produce the signs of the trigonometric functions in the four quadrants.
5. Calculate the reference number for a given trigonometric point.
6. Use the reference number and quadrant to find the values of the trigonometric functions in terms of the values in the first quadrant.
7. Use the Pythagorean identity and related identities to find one trigonometric function value from a known value.

The connection between circles and the number π represents a fundamental relationship between geometry and algebra. You will recognize the formula for the circumference of a circle:

$$\text{Circumference} = \pi \times \text{Diameter}$$

This formula is more striking if we divide by the diameter to rearrange it:

$$\frac{\text{Circumference}}{\text{Diameter}} = \pi$$

In this form, the formula tells us that no matter which circle we look at, from the equator of Earth to a coin-sized circle (**Figure 6T.83**), if we divide the circumference by the diameter, the answer is always the same: the number π.

filo/Getty Images

SimpleFoto/Deposit Photos

Figure 6T.83 Circumference divided by diameter: The ratio is the same for a dime or the equator of Earth.

Ancient Egyptian and Babylonian mathematicians recognized the importance of π, and they knew reasonably good approximations of its value. Today, supercomputers using clever calculation techniques have produced approximations of π accurate to trillions of digits—perhaps a few more than are actually needed for scientific applications. Such calculations are used as benchmarks for testing computer speed.

Circles serve an immediate purpose. We use the circle of radius 1 centered at the origin, the unit circle, to extend the trigonometric functions.

The Unit Circle

> The formula for the unit circle can be used to locate points on the unit circle.

The unit circle shown in **Figure 6T.84** is the circle in the plane that is centered at the origin and has radius 1. It consists of points 1 unit distance from the origin, and its equation is $x^2 + y^2 = 1$.

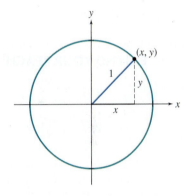

Figure 6T.84 The equation of the unit circle: $x^2 + y^2 = 1$

EXAMPLE 6T.21 Finding Points on the Unit Circle

Find the possible values of y if the point $\left(\dfrac{1}{3}, y\right)$ lies on the unit circle.

SOLUTION

Because $\left(\dfrac{1}{3}, y\right)$ lies on the unit circle, the coordinates must satisfy the equation $x^2 + y^2 = 1$, so

$$\left(\frac{1}{3}\right)^2 + y^2 = 1 \quad \blacktriangleleft \textbf{ Use the equation of unit circle.}$$

$$\frac{1}{9} + y^2 = 1 \quad \blacktriangleleft \textbf{ Simplify.}$$

$$y^2 = \frac{8}{9} \quad \blacktriangleleft \text{ Solve for } y^2.$$

$$y = \pm\frac{\sqrt{8}}{3} \quad \blacktriangleleft \text{ Take square roots.}$$

Thus, there are two solutions,

$$y = \frac{\sqrt{8}}{3} = \frac{2\sqrt{2}}{3}$$

and

$$y = -\frac{\sqrt{8}}{3} = -\frac{2\sqrt{2}}{3}$$

These points are shown on the unit circle in **Figure 6T.85**.

Figure 6T.85 Points on the unit circle

TRY IT YOURSELF 6T.21   Brief answers provided at the end of the section.

Find all possible values of x if $\left(x, \dfrac{2}{5}\right)$ lies on the unit circle.

EXTEND YOUR REACH   Let P denote the point $\left(\dfrac{1}{3}, y\right)$ in the preceding example.

a. How will your calculation in this example be different if P lies on the circle of radius 2 centered at the origin rather than the unit circle? Do the required calculation.

b. Do the calculation in this example if P lies on the circle of radius 3 centered at the point $(1, 0)$.

c. Is it possible for P to lie on the circle of radius 1 centered at $(10, 10)$? *Suggestion:* Draw a picture.

Extending the Trigonometric Functions: Trigonometric Points

The coordinates of trigonometric points are used to give alternative definitions of the trigonometric functions.

The **trigonometric point** $P(t)$ associated with the number t is defined as follows: If $t \geq 0$, we begin at the point $(1, 0)$ and move in the counterclockwise direction around the unit circle a distance t to the point $P(t)$. If t is negative, nothing changes except that we start at $(1, 0)$ and move in the clockwise direction.

Let t be a real number that, for the moment, we assume to be between 0 and $\dfrac{\pi}{2}$.

We begin at $(1, 0)$ and move in the counterclockwise direction around the unit circle a distance t, as shown in **Figure 6T.86**. We refer to the resulting point $P(t)$ as the **trigonometric point** associated with t. We extend this idea so that it applies to any real number.

We are interested in the coordinates of the trigonometric point $P(t)$ for t between 0 and $\dfrac{\pi}{2}$. The definition of radian measure tells us that an angle of t radians with vertex at the center cuts an arc of length t on the unit circle. Hence, we can think of $P(t)$ equally as depending on the real number t or on an angle of radian measure t.

We apply the definitions of the sine and cosine functions to the right triangle shown in Figure 6T.86:

$$\sin t = \frac{\text{Opposite}}{\text{Hypotenuse}} = \frac{y}{1} = y$$

and

$$\cos t = \frac{\text{Adjacent}}{\text{Hypotenuse}} = \frac{x}{1} = x$$

Thus, the coordinates of the point $P(t)$ on the unit circle turn out to be $(x, y) = (\cos t, \sin t)$, as is illustrated in **Figure 6T.87**.

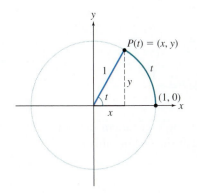

DF **Figure 6T.86** The trigonometric point $P(t)$: An arc length of t on the unit circle makes an angle of t radians. If the angle is in standard position, the terminal side meets the circle at the point $P(t)$.

DF **Figure 6T.87** Sine and cosine: The coordinates turn out to be $(\cos t, \sin t)$.

Because the coordinates of the trigonometric point $P(t)$ are $(\cos t, \sin t)$ and we know the trigonometric functions of the special angles $t = \dfrac{\pi}{6}, t = \dfrac{\pi}{4}$, and $t = \dfrac{\pi}{3}$, we can locate the trigonometric points associated with $\dfrac{\pi}{6}, \dfrac{\pi}{4}$, and $\dfrac{\pi}{3}$. These are shown in **Figure 6T.88**, which is an important figure for future reference.

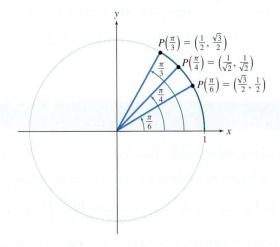

Figure 6T.88 Trigonometric points corresponding to special angles

This approach using trigonometric points suggests a way of defining the sine and cosine functions for any real number t. The first coordinate of the trigonometric point $P(t)$ is defined to be the cosine function at t, and the second coordinate is defined to be the sine function at t. The four other trigonometric functions are calculated in terms of the sine and cosine.

CONCEPTS TO REMEMBER: The Trigonometric Functions for Any Real Number or Any Angle

If the coordinates of the trigonometric point $P(t)$ are (x, y), then the six trigonometric functions are calculated as follows:

$$\sin t = y \qquad\qquad \cos t = x$$

$$\tan t = \frac{\sin t}{\cos t} = \frac{y}{x} \qquad\qquad \cot t = \frac{\cos t}{\sin t} = \frac{x}{y}$$

$$\sec t = \frac{1}{\cos t} = \frac{1}{x} \qquad\qquad \csc t = \frac{1}{\sin t} = \frac{1}{y}$$

The trigonometric functions evaluated at an angle of radian measure t are the same as the trigonometric functions evaluated at the real number t.

Trigonometric Functions of Compass Points

> The values of the trigonometric functions at the angles corresponding to the four compass points on the unit circle are calculated.

EXAMPLE 6T.22   Trigonometric Functions of Compass Points

The compass points on the unit circle are the points $(1, 0)$, $(0, 1)$, $(-1, 0)$, and $(0, -1)$.

a. Find the values of the six trigonometric functions of $\dfrac{\pi}{2}$ radians. Note which functions are undefined.

b. Find the values of the six trigonometric functions of the real number $-\dfrac{\pi}{2}$. Note which functions are undefined.

c. Find the six trigonometric functions of $270°$.

SOLUTION

a. It is important to remember that the values of the trigonometric functions of the angle of radian measure t are the same as the values of the trigonometric functions of the real number t. An arc length of $t = \dfrac{\pi}{2}$ is one-quarter of the circumference of the circle. Hence, the trigonometric point $P\left(\dfrac{\pi}{2}\right)$ is $(x, y) = (0, 1)$.

See **Figure 6T.89**. We note first that the tangent and secant functions of $\dfrac{\pi}{2}$ are undefined because their definitions would involve division by 0. The values of the trigonometric functions are

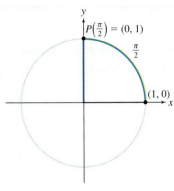

$$\sin\left(\frac{\pi}{2}\right) = 1 \qquad \cos\left(\frac{\pi}{2}\right) = 0$$

$$\tan\left(\frac{\pi}{2}\right) \text{ is undefined} \qquad \cot\left(\frac{\pi}{2}\right) = \frac{\cos\left(\frac{\pi}{2}\right)}{\sin\left(\frac{\pi}{2}\right)} = \frac{0}{1} = 0$$

$$\sec\left(\frac{\pi}{2}\right) \text{ is undefined} \qquad \csc\left(\frac{\pi}{2}\right) = \frac{1}{\sin\left(\frac{\pi}{2}\right)} = \frac{1}{1} = 1$$

Figure 6T.89 Locating $P\left(\frac{\pi}{2}\right)$

b. Because $-\frac{\pi}{2}$ is negative, we begin at $(1, 0)$ and move $\frac{\pi}{2}$ units in a clockwise direction. Because an arc of length $\frac{\pi}{2}$ is one-quarter of the circumference, we land at the trigonometric point $P\left(-\frac{\pi}{2}\right) = (0, -1)$. See **Figure 6T.90**. Once again, the tangent and secant functions are undefined because their definitions would involve division by 0. The values of the trigonometric functions are

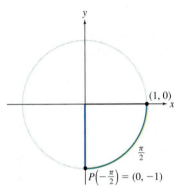

$$\sin\left(-\frac{\pi}{2}\right) = -1 \qquad \cos\left(-\frac{\pi}{2}\right) = 0$$

$$\tan\left(-\frac{\pi}{2}\right) \text{ is undefined} \quad \cot\left(-\frac{\pi}{2}\right) = \frac{\cos\left(-\frac{\pi}{2}\right)}{\sin\left(-\frac{\pi}{2}\right)} = \frac{0}{-1} = 0$$

$$\sec\left(-\frac{\pi}{2}\right) \text{ is undefined} \quad \csc\left(-\frac{\pi}{2}\right) = \frac{1}{\sin\left(-\frac{\pi}{2}\right)} = \frac{1}{-1} = -1$$

Figure 6T.90 Locating $P\left(-\frac{\pi}{2}\right)$

c. An angle of $270°$ covers three-quarters of the circle, so $P(270°) = (0, -1)$. Note that this is the same point as we found in part b. Hence, the values of the trigonometric functions are the same as in part b:

$$\sin(270°) = \sin\left(-\frac{\pi}{2}\right) = -1, \text{ and so on.}$$

TRY IT YOURSELF 6T.22 Brief answers provided at the end of the section.

Find the values of the six trigonometric functions of π. Note which functions are undefined.

If $P(s) = (x, y)$ and $P(t) = (z, w)$, do you think it is true that $P(s + t) = (x + z, y + w)$? You might try some specific values of s and t to answer this question. The trigonometric points calculated in this example may prove useful.

EXTEND YOUR REACH

We summarize the values of the trigonometric functions at the compass points in **Table 6T.3**. Many of the entries in this table come from Example 6T.22. You are asked to verify the remaining entries in the exercises. It is not necessary to memorize the table of compass points because you should be able to calculate its contents easily.

Table 6T.3 Values of Trigonometric Functions and Compass Points

Figure	Degrees	Radians	P(t)	Sine	Cosine	Tangent	Cotangent	Secant	Cosecant
	0	0	$P(0) = (1, 0)$	0	1	0	undefined	1	undefined
	90	$\dfrac{\pi}{2}$	$P\left(\dfrac{\pi}{2}\right) = (0, 1)$	1	0	undefined	0	undefined	1
	180	π	$P(\pi) = (-1, 0)$	0	−1	0	undefined	−1	undefined
	270	$\dfrac{3\pi}{2}$	$P\left(\dfrac{3\pi}{2}\right) = (0, -1)$	−1	0	undefined	0	undefined	−1
	360	2π	$P(2\pi) = (1, 0)$	0	1	0	undefined	1	undefined

We note that all trigonometric functions of a number t are defined except in certain cases when $P(t)$ is one of the compass points. The domains of the trigonometric functions are summarized as follows:

CONCEPTS TO REMEMBER: Domains of Trigonometric Functions

Function	Domain
Sine and cosine	All real numbers
Tangent and secant	All real numbers except $\frac{\pi}{2}$ plus multiples of π
Cotangent and cosecant	All real numbers except multiples of π

Quadrants

> Determining the sign of trigonometric functions using the quadrant is a key component of their evaluation.

The coordinate axes divide the plane into four quadrants, commonly labeled I, II, III, and IV, as in **Figure 6T.91**. The signs of the trigonometric functions are determined by the quadrant in which the trigonometric point $P(t) = (\cos t, \sin t)$ lies. This is shown in **Figure 6T.92**, which also shows a popular mnemonic device for remembering the signs. It associates the first letters of All Students Take Calculus with the first letters of the positive trigonometric functions All, Sine, Tangent, Cosine in each of the four quadrants.

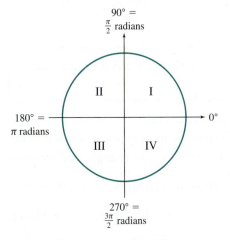

Figure 6T.91 The plane divided into four quadrants by the coordinate axes

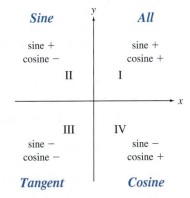

Figure 6T.92 Signs of sine and cosine in the four quadrants: All Students Take Calculus indicates the positive functions.

EXAMPLE 6T.23 Locating Quadrants

a. In what quadrant does the trigonometric point $P(3)$ lie? Is sin 3 positive or negative?

b. In what quadrant is sec t positive and cot t negative?

SOLUTION

a. Because $\pi \approx 3.14$, 3 is just a bit smaller than π. This means that the arc of length 3 from (1, 0) counterclockwise around the unit circle ends just shy of the point (−1, 0), as shown in **Figure 6T.93**. Consequently, $P(3)$ lies in the second quadrant. In the second quadrant the sine is positive, so sin 3 is positive.

b. Because $\sec t = \dfrac{1}{\cos t}$ is positive, cos t is also positive.

This tells us we are in quadrant I or IV. Because cot t is negative, we must be in quadrant IV.

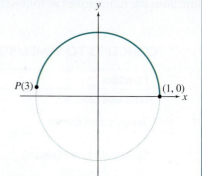

Figure 6T.93 Locating $P(3)$

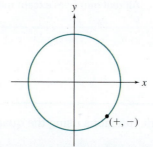

TRY IT YOURSELF 6T.23   Brief answers provided at the end of the section.

a. In what quadrant does $P(-3)$ lie? What is the sign of $\cos(-3)$?

b. What is the sign of the cosecant in quadrant III?

Reference Numbers

> Reference numbers relate values of trigonometric functions to trigonometric points in the first quadrant.

For trigonometric points in the first quadrant, we can calculate the sine and cosine using a right triangle. The same turns out to be true for trigonometric points in other quadrants, but a bit of care is needed in choosing the correct triangle, and we must pay attention to signs. We use $\dfrac{2\pi}{3}$ to see how this method works. A length of π is half the circumference of the unit circle, and $\dfrac{2\pi}{3}$ is two-thirds of that distance. Hence, $P\left(\dfrac{2\pi}{3}\right)$ lies in the second quadrant, as shown in **Figure 6T.94**. Observe that the x- and y-coordinates of this point are the same in absolute value as the sides of the right triangle with acute angle $\dfrac{\pi}{3}$ shown in **Figure 6T.95**. In fact, $\dfrac{\pi}{3}$ is the shortest arc length from $P\left(\dfrac{2\pi}{3}\right)$ to the x-axis.

Figure 6T.94 The trigonometric point $P\left(\dfrac{2\pi}{3}\right)$

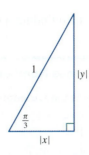

Figure 6T.95 An associated triangle: The absolute values of the coordinates of $P\left(\dfrac{2\pi}{3}\right)$ are the sides of a right triangle with a special angle.

Figure 6T.88 provides a reminder of the values of the trigonometric functions of the special angle $\dfrac{\pi}{3}$:

$$\sin\left(\frac{\pi}{3}\right) = \frac{\sqrt{3}}{2} \qquad \cos\left(\frac{\pi}{3}\right) = \frac{1}{2}$$

We need only attach the proper signs $(-, +)$ for quadrant II to find the sine and cosine of $\dfrac{2\pi}{3}$:

$$\sin\left(\frac{2\pi}{3}\right) = \frac{\sqrt{3}}{2} \qquad \cos\left(\frac{2\pi}{3}\right) = -\frac{1}{2}$$

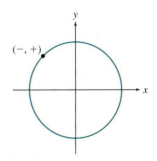

The sine and cosine are used to find the values of the other four trigonometric functions:

$$\tan\left(\frac{2\pi}{3}\right) = \frac{\sin\left(\dfrac{2\pi}{3}\right)}{\cos\left(\dfrac{2\pi}{3}\right)} = \frac{\dfrac{\sqrt{3}}{2}}{-\dfrac{1}{2}} = -\sqrt{3} \qquad \cot\left(\frac{2\pi}{3}\right) = \frac{\cos\left(\dfrac{2\pi}{3}\right)}{\sin\left(\dfrac{2\pi}{3}\right)} = \frac{-\dfrac{1}{2}}{\dfrac{\sqrt{3}}{2}} = -\frac{1}{\sqrt{3}}$$

$$\sec\left(\frac{2\pi}{3}\right) = \frac{1}{\cos\left(\dfrac{2\pi}{3}\right)} = \frac{1}{-\dfrac{1}{2}} = -2 \qquad \csc\left(\frac{2\pi}{3}\right) = \frac{1}{\sin\left(\dfrac{2\pi}{3}\right)} = \frac{1}{\dfrac{\sqrt{3}}{2}} = \frac{2}{\sqrt{3}}$$

In general, to calculate the coordinates of the trigonometric point $P(t)$ we need to find the **reference number** by locating the appropriate arc in the first quadrant.

Key facts about reference numbers are summarized in the following.

The **reference number** r for the trigonometric point $P(t)$ is the shortest arc length from $P(t)$ to the x-axis.

CONCEPTS TO REMEMBER: Properties of Reference Numbers

1. The reference number r for the trigonometric point $P(t)$ is the shortest arc length from $P(t)$ to the x-axis.

2. A reference number of r corresponds to a reference angle of r radians.

3. The reference angle is always between 0 and $\dfrac{\pi}{2}$ radians ($0°$ to $90°$).

4. Proper signs attached to the values of the trigonometric functions of the reference number r give the values of the trigonometric functions of t.

Table 6T.4 gives a visual guide to finding reference numbers.

Table 6T.4 A Graphical View of Reference Numbers

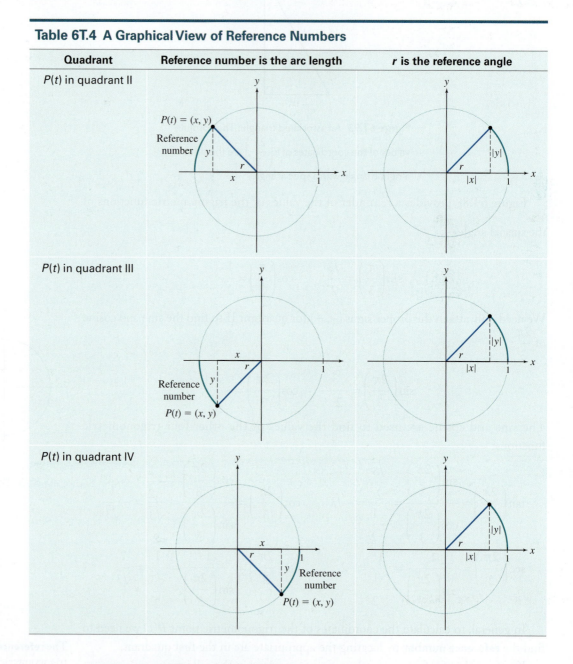

Quadrant	Reference number is the arc length	r is the reference angle
$P(t)$ in quadrant II		
$P(t)$ in quadrant III		
$P(t)$ in quadrant IV		

Using Reference Numbers to Evaluate Trigonometric Functions

The values of the trigonometric functions for the first quadrant can be used to evaluate trigonometric functions for any quadrant.

EXAMPLE 6T.24   Finding Reference Numbers

a. Find the reference number for $P\left(\dfrac{3\pi}{4}\right)$.

b. Find the reference number for $P(4)$.

SOLUTION

a. An arc of length π is halfway around the circle, and $\dfrac{3\pi}{4}$ is three-quarters of that distance. The shortest arc back to the x-axis is $\pi - \dfrac{3\pi}{4} = \dfrac{\pi}{4}$, as shown in **Figure 6T.96**. The reference number for $P\left(\dfrac{3\pi}{4}\right)$ is $\dfrac{\pi}{4}$.

b. Because 4 is greater than π and less than $\dfrac{3\pi}{2}$, the trigonometric point $P(4)$ lies in the third quadrant, as shown in **Figure 6T.97**. The reference number is the length of the arc from $P(4)$ back to the x-axis. This value is $4 - \pi$.

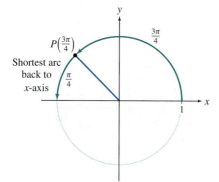

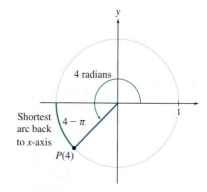

DF **Figure 6T.96** $P\left(\dfrac{3\pi}{4}\right)$ This trigonometric point lies in the second quadrant, and the reference number is $\dfrac{\pi}{4}$.

Figure 6T.97 $P(4)$: This trigonometric point lies in the third quadrant, and the reference number is $4 - \pi$.

TRY IT YOURSELF 6T.24 Brief answers provided at the end of the section.

Find the reference number for $P\left(\dfrac{7\pi}{6}\right)$.

EXTEND YOUR REACH

a. Find the reference number for $P\left(\dfrac{\pi}{6} + \pi\right)$.

b. Find the reference number for $P\left(\dfrac{\pi}{4} + \pi\right)$.

c. Find the reference number for $P\left(\dfrac{\pi}{3} + \pi\right)$.

d. If $P(t)$ lies in the first quadrant, find the reference number for $P(t + \pi)$.

EXAMPLE 6T.25 Finding and Using Reference Angles and Numbers

a. Find the degree measure of the reference angle for $P(290°)$.

b. Find the values of sine and cosine of the number $\dfrac{7\pi}{6}$.

SOLUTION

a. The trigonometric point $P(290°)$ lies in the fourth quadrant. The shortest arc leading back to the x-axis cuts an angle of $360° - 290° = 70°$, as shown in **Figure 6T.98**. Thus, the reference angle has degree measure $70°$.

b. The trigonometric point $P\left(\dfrac{7\pi}{6}\right)$ lies in the third quadrant, as shown in **Figure 6T.99**. The shortest arc back to the x-axis is $\dfrac{\pi}{6}$, and this is the reference number. The sine of $\dfrac{\pi}{6}$ is $\dfrac{1}{2}$, and the cosine is $\dfrac{\sqrt{3}}{2}$. Now we need to attach the signs.

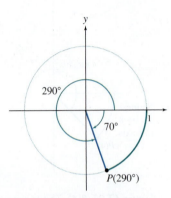

Figure 6T.98 $P(290°)$: This trigonometric point lies in the fourth quadrant, and the reference angle is $70°$.

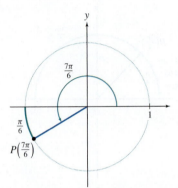

Figure 6T.99 $P\left(\dfrac{7\pi}{6}\right)$ This trigonometric point lies in the third quadrant and has reference number $\dfrac{\pi}{6}$.

In the third quadrant, both the sine and cosine are negative, so

$$\sin\left(\frac{7\pi}{6}\right) = -\frac{1}{2} \quad \text{and} \quad \cos\left(\frac{7\pi}{6}\right) = -\frac{\sqrt{3}}{2}$$

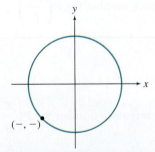

TRY IT YOURSELF 6T.25   Brief answers provided at the end of the section.

Find the values of sine and cosine of the number $\dfrac{5\pi}{4}$.

EXAMPLE 6T.26 Using Reference Angles

In each case find the values of the six trigonometric functions of the given angle.

a. An angle of $120°$ **b.** An angle of $\dfrac{11\pi}{3}$ radians

SOLUTION

a. The trigonometric point $P(120°)$ lies in the second quadrant. The shortest arc back to the x-axis cuts an angle of $60°$, as shown in **Figure 6T.100**.

Thus, the reference angle is the special angle of $60°$. The sine of $60°$ is $\dfrac{\sqrt{3}}{2}$, and the cosine is $\dfrac{1}{2}$.

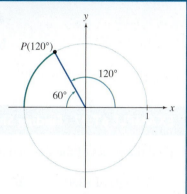

Figure 6T.100 $P(120°)$: This trigonometric point lies in the second quadrant and has reference angle $60°$.

In the second quadrant, the sine is positive and the cosine is negative, so

$$\sin 120° = \frac{\sqrt{3}}{2} \qquad\qquad \cos 120° = -\frac{1}{2}$$

$$\tan 120° = \frac{\sin 120°}{\cos 120°} = \frac{\frac{\sqrt{3}}{2}}{-\frac{1}{2}} = -\sqrt{3} \qquad \cot 120° = \frac{\cos 120°}{\sin 120°} = \frac{-\frac{1}{2}}{\frac{\sqrt{3}}{2}} = -\frac{1}{\sqrt{3}}$$

$$\sec 120° = \frac{1}{\cos 120°} = \frac{1}{-\frac{1}{2}} = -2 \qquad \csc 120° = \frac{1}{\sin 120°} = \frac{1}{\frac{\sqrt{3}}{2}} = \frac{2}{\sqrt{3}}$$

b. Now $\dfrac{11\pi}{3} = 2\pi + \dfrac{5\pi}{3}$. This is a full trip around the circle followed by an arc of length $\dfrac{5\pi}{3}$. We see that this angle is in quadrant IV and the reference number is the same as in part a, namely $60°$ or $\dfrac{\pi}{3}$. We need to attach the signs. In the fourth quadrant, the sine is negative and the cosine is positive, so

$$\sin\left(\frac{11\pi}{3}\right) = -\frac{\sqrt{3}}{2} \qquad \cos\left(\frac{11\pi}{3}\right) = \frac{1}{2}$$

$$\tan\left(\frac{11\pi}{3}\right) = -\sqrt{3} \qquad \cot\left(\frac{11\pi}{3}\right) = -\frac{1}{\sqrt{3}}$$

$$\sec\left(\frac{11\pi}{3}\right) = 2 \qquad \csc\left(\frac{11\pi}{3}\right) = -\frac{2}{\sqrt{3}}$$

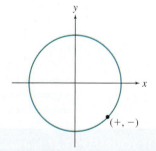

TRY IT YOURSELF 6T.26 Brief answers provided at the end of the section.

Find the values of the six trigonometric functions of the angle $300°$.

EXTEND YOUR REACH   It is true that $\sin 20° = 0.34$ and $\cos 20° = 0.94$ (correct to two decimal places).

   a. Using the given information, find the values of the sine and cosine of $160°$.

   b. Find the values of the remaining four trigonometric functions of $160°$.

EXAMPLE 6T.27 Finding Sine and Cosine of Negative Angles

Find the values of the sine and cosine of the following.

a. An angle of $-330°$ **b.** An angle of $-\dfrac{5\pi}{2}$ radians

SOLUTION

a. The trigonometric point $P(-330°)$ lies in the first quadrant, as shown in **Figure 6T.101**. The shortest arc back to the x-axis cuts an angle of $30°$. The sine of $30°$ is $\dfrac{1}{2}$, and the cosine is $\dfrac{\sqrt{3}}{2}$. In the first quadrant, both the sine and the cosine are positive, so

$$\sin(-330°) = \frac{1}{2} \qquad \cos(-330°) = \frac{\sqrt{3}}{2}$$

b. Because $-\dfrac{5\pi}{2} = -2\pi - \dfrac{\pi}{2}$, the angle represented corresponds to a full trip clockwise around the circle followed by another one-quarter trip, so the trigonometric point has coordinates $(0, -1)$. See **Figure 6T.102**. Hence, $\sin\left(-\dfrac{5\pi}{2}\right) = -1$ and $\cos\left(-\dfrac{5\pi}{2}\right) = 0$.

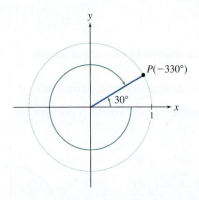

Figure 6T.101 $P(-330°)$: This trigonometric point lies in the first quadrant and has reference angle $30°$.

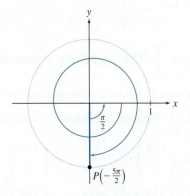

Figure 6T.102 $P\left(-\dfrac{5\pi}{2}\right) = (0, -1)$

TRY IT YOURSELF 6T.27 Brief answers provided at the end of the section.

Find the values of the sine and cosine of $P\left(-\dfrac{11\pi}{4}\right)$.

Reference numbers are used to locate the trigonometric points shown in **Figure 6T.103**. This figure is helpful for future reference.

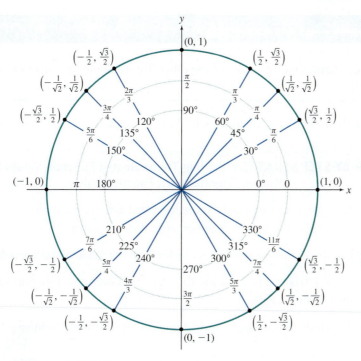

DF Figure 6T.103 Trigonometric points corresponding to special angles

The Pythagorean Identity

> The formula for the unit circle gives a fundamental relationship between the sine and cosine functions.

If we know the value of one trigonometric function, and if we can identify the quadrant, we can calculate the values of the other trigonometric functions.

EXAMPLE 6T.28 Calculating When One Function Value Is Known

The sine of t is $-\dfrac{2}{3}$. Find the values of the other five trigonometric functions of t if the cosine of t is positive.

Calculate exact values, and then give an approximation correct to two decimal places.

SOLUTION

The reference angle for t is an acute angle r with $\sin r = \dfrac{2}{3}$. We want a triangle with acute angle r whose sine is $\dfrac{2}{3}$.

We can accomplish this by choosing an opposite side of 2 and a hypotenuse of 3, as shown in **Figure 6T.104**. The

Pythagorean theorem shows that the adjacent side has length $\sqrt{3^2 - 2^2} = \sqrt{5}$. We use this to calculate the cosine:

$$\cos r = \frac{\text{Adjacent}}{\text{Hypotenuse}} = \frac{\sqrt{5}}{3}$$

Because the cosine of t is positive, we have $\cos t = \dfrac{\sqrt{5}}{3} \approx 0.75$. Using the sine and

cosine, we obtain the values of the remaining functions:

$$\tan t = -\frac{2}{\sqrt{5}} \approx -0.89 \qquad \cot t = -\frac{\sqrt{5}}{2} \approx -1.12$$

$$\sec t = \frac{3}{\sqrt{5}} \approx 1.34 \qquad \csc t = -\frac{3}{2} = -1.5$$

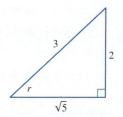

Figure 6T.104 A reference angle whose sine is $\dfrac{2}{3}$

TRY IT YOURSELF 6T.28 Brief answers provided at the end of the section.

If $\tan t = \dfrac{1}{2}$ and $P(t)$ lies in the third quadrant, find the values of the other five trigonometric functions of t.

The method we used in Example 6T.28 can be codified into a useful strategy.

STEP-BY-STEP STRATEGY: Finding Trigonometric Function Values from a Known Trigonometric Function Value

Step 1 Label two sides of a right triangle so that a reference angle in the triangle has the appropriate trigonometric function value.

Step 2 Use the Pythagorean theorem to find the third side of the triangle.

Step 3 Use the triangle to calculate the other trigonometric function values of the reference angle.

Step 4 Attach the proper signs to the function values calculated in step 3.

Because the trigonometric point $P(t)$ lies on the unit circle, its coordinates $(\cos t, \sin t)$ satisfy the equation $x^2 + y^2 = 1$. The result is the important Pythagorean identity connecting sine and cosine. Note that it is standard and convenient to write $\sin^2 t$ instead of $(\sin t)^2$, and similar notation is used for the other trigonometric functions.

LAWS OF MATHEMATICS: Pythagorean Identity

The Pythagorean identity is

$$\sin^2 t + \cos^2 t = 1$$

The following identities are closely related:

$$\tan^2 t + 1 = \sec^2 t$$

$$1 + \cot^2 t = \csc^2 t$$

TRY IT YOURSELF ANSWERS

6T.21 $x = \pm \dfrac{\sqrt{21}}{5}$

6T.22 $\sin \pi = 0 \qquad \cos \pi = -1$
$\tan \pi = 0 \qquad \cot \pi$ is undefined
$\sec \pi = -1 \qquad \csc \pi$ is undefined

6T.23 **a.** $P(-3)$ lies in the third quadrant, and $\cos(-3)$ is negative.
   b. The cosecant is negative in quadrant III.

6T.24 $\dfrac{\pi}{6}$

6T.25 $\sin\left(\dfrac{5\pi}{4}\right) = -\dfrac{1}{\sqrt{2}}, \quad \cos\left(\dfrac{5\pi}{4}\right) = -\dfrac{1}{\sqrt{2}}$

6T.26 $\sin 300° = -\dfrac{\sqrt{3}}{2} \qquad \cos 300° = \dfrac{1}{2}$
$\tan 300° = -\sqrt{3} \qquad \cot 300° = -\dfrac{1}{\sqrt{3}}$
$\sec 300° = 2 \qquad \csc 300° = -\dfrac{2}{\sqrt{3}}$

6T.27 $\sin\left(-\dfrac{11\pi}{4}\right) = -\dfrac{1}{\sqrt{2}}, \quad \cos\left(-\dfrac{11\pi}{4}\right) = -\dfrac{1}{\sqrt{2}}$

6T.28 $\sin t = -\dfrac{1}{\sqrt{5}}, \cos t = -\dfrac{2}{\sqrt{5}}, \cot t = 2,$
$\sec t = -\dfrac{\sqrt{5}}{2}, \csc t = -\sqrt{5}$

EXERCISE SET 6T.4

CHECK YOUR UNDERSTANDING

1. Points (x, y) on the unit circle satisfy which equation?

2. Find the signs of each of the coordinates of the trigonometric points in each of the four quadrants.

3. In which quadrant are both the sine and cosine negative?

4. If $P(t) = (x, y)$, express each of the six trigonometric functions of t in terms of x and y.

5. **True or false:** For any trigonometric point, the reference angle is between 0 and $\dfrac{\pi}{2}$ radians.

6. For a real number t, the trigonometric point $P(t)$ is the same as:

   a. $P(t)$ for an angle of $t°$

   b. $P(t)$ for an angle of t radians

   c. a reference angle

   d. None of the above.

7. **True or false:** Reference numbers help us calculate the values of trigonometric functions.

8. The reference number for the trigonometric point $P(t)$ is _____.

9. **True or false:** The reference number for the trigonometric point $P(t)$ is the same as the radian measure of the reference angle.

SKILL BUILDING

Points on the unit circle. In Exercises 10 through 15, find the given point on the unit circle.

10. Find all possible values of x if $\left(x, \dfrac{1}{2} \right)$ lies on the unit circle.

11. Find all possible values of x if $\left(x, -\dfrac{2}{3} \right)$ lies on the unit circle.

12. Find all possible values of x if $\left(x, \dfrac{3}{4} \right)$ lies on the unit circle.

13. Find all possible values of y if $\left(\dfrac{5}{6}, y \right)$ lies on the unit circle.

14. Find all possible values of y if $\left(-\dfrac{1}{3}, y \right)$ lies on the unit circle.

15. Find all possible values of x if $\left(x, -\dfrac{2}{7} \right)$ lies on the unit circle and the point lies in the third quadrant.

Identifying trigonometric points. In **Figure 6T.105**, trigonometric points corresponding to compass points and special angles are marked. In Exercises 16 through 23, match the given angle with the appropriate point labeled in the figure.

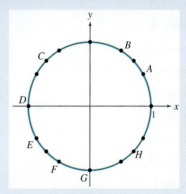

Figure 6T.105 Various trigonometric points

16. $\dfrac{3\pi}{4}$ radians

17. $30°$

18. $315°$

19. π radians

20. $\dfrac{\pi}{3}$ radians

21. $210°$

22. $240°$

23. $\dfrac{3\pi}{2}$ radians

Calculating trigonometric function values. Figure 6T.106 shows points on the unit circle. In each case, one coordinate is given. In Exercises 24 through 32, find the missing coordinate for the indicated point, and then list the six trigonometric function values. (Use t for the variable.)

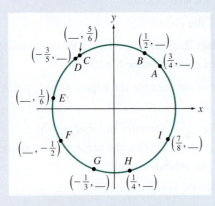

Figure 6T.106 Trigonometric points with one coordinate given

24. Trigonometric point A
25. Trigonometric point B
26. Trigonometric point C
27. Trigonometric point D
28. Trigonometric point E
29. Trigonometric point F
30. Trigonometric point G
31. Trigonometric point H
32. Trigonometric point I

Reference angles with degrees. In Exercises 33 through 42, find the reference angle in degrees associated with the given angle.

33. 123°
34. 777°
35. 377°
36. −20°
37. −100°
38. 30°
39. −370°
40. 1236°
41. −277°
42. 634°

Reference angles with radians. In Exercises 43 through 52, find the reference angle in radians associated with the given angle.

43. $\dfrac{5\pi}{7}$ radians
44. $\dfrac{12\pi}{5}$ radians
45. $\dfrac{2\pi}{3}$ radians
46. $-\dfrac{2\pi}{3}$ radians
47. $-\dfrac{\pi}{6}$ radian
48. $\dfrac{5\pi}{4}$ radians
49. $-\dfrac{5\pi}{6}$ radians
50. $\dfrac{5\pi}{3}$ radians
51. $-\dfrac{5\pi}{7}$ radians
52. −4 radians

Trigonometric functions of compass points. In Exercises 53 through 65, find the values of the six trigonometric functions of the given angles. In each case, one or more of the trigonometric functions may not be defined. Remember that radian measure is indicated if degrees are not explicitly stated.

53. 90°
54. 180°
55. 5π
56. 450°
57. −180°
58. $\dfrac{7\pi}{2}$
59. −π
60. 270°
61. $\dfrac{\pi}{2}$
62. π
63. $\dfrac{3\pi}{2}$
64. 2π
65. $-\dfrac{\pi}{2}$

PROBLEMS

Getting one trigonometric function value from another. In Exercises 66 through 81, you are given the value of one trigonometric function. Find the values of the remaining five trigonometric functions.

66. $\cos t = -\dfrac{3}{4}$, and $\sin t$ is positive.

67. $\cos t = \dfrac{3}{4}$, and $\sin t$ is positive.

68. $\tan t = \sqrt{3}$, and $\sin t$ is positive.

69. $\sin t = \dfrac{5}{6}$, and $\cos t$ is negative.

70. $\sec t = -\dfrac{5}{3}$, and $\tan t$ is negative.

71. $\sin t = \dfrac{1}{6}$, and $\sec t$ is negative.

72. $\cos t = -\dfrac{\sqrt{3}}{2}$, and $\sin t$ is negative.

73. $\cos t = -\dfrac{1}{3}$, and $\sin t$ is negative.

74. $\sec t = 4$, and $\csc t$ is negative.

75. $\cot t = -\dfrac{7}{\sqrt{15}}$, and $\cos t$ is positive.

76. $\sec t = 4$, and $\sin t$ is negative.

77. $\cot t = 5$, and $P(t)$ lies in the third quadrant.

78. $\sec t = -3$, and $\cot t$ is negative.

79. $\sec t = 3$, and $\tan t$ is negative.

80. $\cot t = 1.20$, and $\sin t$ is negative. Report answers correct to two decimal places, but do not use rounded values in calculations.

81. $\cos t = 0.55$, and $P(t)$ is in the fourth quadrant. Report answers correct to two decimal places, but do not use rounded values in calculations.

82. **Verifying a Pythagorean identity.** Use the identity $\sin^2 t + \cos^2 t = 1$ to establish the identity $\tan^2 t + 1 = \sec^2 t$. *Suggestion*: Divide both sides by $\cos^2 t$.

83. **Verifying another Pythagorean identity.** Use the identity $\sin^2 t + \cos^2 t = 1$ to establish the identity

$1 + \cot^2 t = \csc^2 t$. *Suggestion*: Divide both sides by $\sin^2 t$.

Evaluating trigonometric functions at special angles. In Exercises 84 through 99, calculate the values of the trigonometric functions of some special angles. Remember that radian measure is indicated if degree measure is not explicitly stated.

84. $135°$

85. $330°$

86. $\dfrac{5\pi}{4}$

87. $-\dfrac{\pi}{6}$

88. $120°$

89. $-\dfrac{\pi}{3}$

90. $-\dfrac{\pi}{4}$

91. $-135°$

92. $210°$

93. π

94. $\dfrac{13\pi}{6}$

95. $\dfrac{7\pi}{6}$

96. $300°$

97. $\dfrac{2\pi}{3}$

98. $240°$

99. $\dfrac{5\pi}{3}$

CHALLENGE EXERCISES FOR INDIVIDUALS OR GROUPS

Theorem on undetermined coefficients. It is a fact that if A, B, C, and D are constants, and if the equation

$$A \cos t + B \sin t = C \cos t + D \sin t$$

is true for all values of t, then $A = C$ and $B = D$. In Exercises 100 through 104, you will be led to show this theorem is true and then apply that fact to solve problems.

100. **Showing the theorem is true.** Suppose the equation $A \cos t + B \sin t = C \cos t + D \sin t$ is true for all values of t.

   a. Put $t = 0$ into the equation. What can you conclude regarding A, B, C, and D?

   b. Put $t = \dfrac{\pi}{2}$ into the equation. What can you conclude regarding A, B, C, and D?

101. **Using the theorem.** Suppose

$$(A - 2)\cos t + 3 \sin t = 4 \cos t + (B + 7) \sin t$$

is true for all values of t. Find the values of A and B.

102. **Using the theorem.** Suppose

$$A(\cos t + \sin t) = 3 \cos t + B \sin t$$

is true for all values of t. Find the values of A and B.

103. **Using the theorem.** Suppose

$$(A + B)\cos t + (A - B)\sin t = 4 \cos t + 2 \sin t$$

is true for all values of t. Find the values of A and B.

104. **Using the theorem.** Suppose

$$2A \sin t + 4A \cos t = B \sin t + \sin t - B \cos t + 6 \cos t$$

is true for all values of t. Find the values of A and B.

105. **The rate of change of the sine function.** The goal of this exercise is to give an intuitive justification for the fact that the rate of change of the sine function is the cosine. We can think of the rate of change of $y = \sin t$ as the value that the average rate of change $\dfrac{\Delta y}{\Delta t}$ approaches when Δt is near zero.

A bit of geometry can show that, when Δt is small, $\angle s$ in **Figure 6T.107** is nearly equal to $\angle t$, and the length of PQ is approximately Δt. It is a

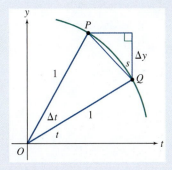

Figure 6T.107 When Δt is small: The $\angle s$ is close to $\angle t$ and PQ is close to Δt.

bit of a challenge to show this, but if you wish to try, consider the fact that when Δt is very small, the line $\overline{PQ}$ is nearly tangent to the unit circle. You may proceed assuming these facts to be true.

Figure 6T.108 is a blown-up version of the right triangle in Figure 6T.107. Use this triangle to show that $\cos t$ is close to $\dfrac{\Delta y}{\Delta t}$ when Δt is small. Complete the argument.

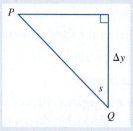

Figure 6T.108: A blown-up version of the right triangle in Figure 6T.107

REVIEW AND REFRESH: Exercises from Previous Sections

106. From Section P.1: Find the point midway between $(2, 7)$ and $(4, 9)$.

Answer: $(3, 8)$

107. From Section P.3: Solve the inequality $\dfrac{x}{x-1} < 2$.

Answer: $(-\infty, 1) \cup (2, \infty)$

108. From Section 1.2: Find the average rate of change of $f(x) = x^2$ from x to $x + h$ with $h \neq 0$. Simplify your answer.

Answer: $2x + h$

109. From Section 1.4: Find the limit as $x \to \infty$ of $\dfrac{2}{1 + \dfrac{1}{x}}$.

Answer: 2

110. From Section 2.2: Find the inverse of the function $f(x) = \dfrac{x}{x+1}$.

Answer: $f^{-1}(x) = \dfrac{x}{1-x}$

111. From Section 4.2: Write 2^x in the form e^{rx}.

Answer: $2^x = e^{x \ln 2}$

112. From Section 4.3: Solve the equation $\ln x + \ln (x + 1) = 1$.

Answer: $x = \dfrac{-1 + \sqrt{1 + 4e}}{2}$

113. From Section 6T.1: Find the length of the arc of a circle of radius 4 cut by an angle of $\dfrac{\pi}{3}$ radians whose vertex lies at the center of the circle.

Answer: $\dfrac{4\pi}{3}$

114. From Section 6T.2: Find the values of the six trigonometric functions of t if t is an acute angle of a right triangle whose side adjacent to t is 2 and whose hypotenuse is 8.

Answer: $\sin t = \dfrac{\sqrt{15}}{4}$, $\cos t = \dfrac{1}{4}$, $\tan t = \sqrt{15}$, $\cot t = \dfrac{1}{\sqrt{15}}$, $\sec t = 4$, $\csc t = \dfrac{4}{\sqrt{15}}$

115. From Section 6T.3: A right triangle has an acute angle of $\dfrac{\pi}{6}$. The hypotenuse has length 6. Find the lengths of the remaining sides of the triangle.

Answer: Adjacent side: $3\sqrt{3}$. Opposite side: 3.

CHAPTER ROADMAP AND STUDY GUIDE

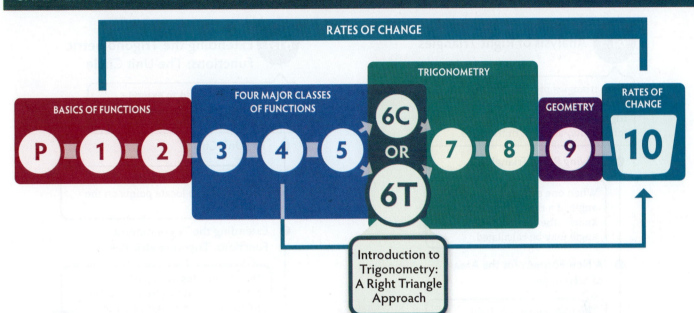

Introduction to Trigonometry: A Right Triangle Approach

6T.1 Angles

Trigonometry depends on the measurement of angles.

Conventions for Angles

Angles can be thought of as rotations from an initial ray to a terminal ray.

Degree Measure

Degree measure of angles is probably familiar from everyday experience.

Radian Measure

Radian measure is calculated by using the radius of a circle to measure arcs along the circle.

Converting Angle Measures

Radian and degree measures are related by simple formulas.

Arc Length and Area

Arc length and area are calculated for sectors of circles.

Similar Triangles

Similarity is at the foundation of trigonometry.

MODELS AND APPLICATIONS
Shadows

6T.2 Definition of the Trigonometric Functions Using Right Triangles

Trigonometry is closely connected to the study of right triangles.

Definitions of Trigonometric Functions of Acute Angles

Trigonometric functions of acute angles are defined in terms of right triangles.

Evaluating Trigonometric Functions Using Right Triangles

Side lengths of right triangles can be used to evaluate trigonometric functions of acute angles.

Consistency of the Definitions

Any right triangle with a fixed acute angle can be used to calculate trigonometric function values.

Special Angles

Geometry is used to evaluate trigonometric functions of certain special angles.

Representing Angles

An angle with a given trigonometric function value can be represented by a right triangle.

Using Representative Triangles to Evaluate Trigonometric Functions

Triangles representing one trigonometric function value can be used to find the values of other trigonometric functions.

CHAPTER ROADMAP AND STUDY GUIDE (continued)

6T.3 Analysis of Right Triangles

Right triangles provide information about trigonometric functions—and trigonometric functions provide information about right triangles.

Solving Right Triangles

When one side and one acute angle of a right triangle are known, the remaining sides and angle may be calculated.

A New Formula for the Area of a Triangle

The trigonometry of right triangles yields a useful formula for the area of a triangle.

MODELS AND APPLICATIONS
Transits and Shadows

6T.4 Extending the Trigonometric Functions: The Unit Circle

The unit circle is used to extend the domain of the trigonometric functions to all real numbers.

The Unit Circle

The formula for the unit circle can be used to locate points on the unit circle.

Extending the Trigonometric Functions: Trigonometric Points

The coordinates of trigonometric points are used to give alternative definitions of the trigonometric functions.

Trigonometric Functions of Compass Points

The values of the trigonometric functions at the angles corresponding to the four compass points on the unit circle are calculated.

Quadrants

Determining the sign of trigonometric functions using the quadrant is a key component of their evaluation.

Reference Numbers

Reference numbers relate values of trigonometric functions to trigonometric points in the first quadrant.

Using Reference Numbers to Evaluate Trigonometric Functions

The values of the trigonometric functions for the first quadrant can be used to evaluate trigonometric functions for any quadrant.

The Pythagorean Identity

The formula for the unit circle gives a fundamental relationship between the sine and cosine functions.

CHAPTER QUIZ

1. Find an angle with radian measure between 0 and 2π that is coterminal with an angle of $-\dfrac{16\pi}{3}$.

   Answer: $\dfrac{2\pi}{3}$

   XR **Example 6T.2**

2. What is the radian measure of an angle of $150°$?**3**

   Answer: $\dfrac{5\pi}{6}$

   XR **Example 6T.3**

3. Find the arc length and area subtended by an angle of $\dfrac{\pi}{6}$ radian with vertex at the center of a circle of radius 3 units.

   Answer: Arc length: $\dfrac{\pi}{2}$ units. Area: $\dfrac{3\pi}{4}$ square units.

   XR **Example 6T.5**

4. Find the unknown sides and angles of the similar triangles shown in **Figure 6T.109**.

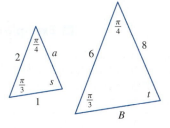

   Figure 6T.109

   Answer: $s = t = \dfrac{5\pi}{12}$, $a = \dfrac{8}{3}$, $B = 3$

   XR **Example 6T.7**

5. Calculate the values of the six trigonometric functions of the angle t in **Figure 6T.110**.

   Figure 6T.110

   Answer: $\sin t = \dfrac{2\sqrt{2}}{3}$, $\cos t = \dfrac{1}{3}$, $\tan t = 2\sqrt{2}$, $\cot t = \dfrac{1}{2\sqrt{2}}$,

   XR **Example 6T.10**

   $\sec t = 3$, $\csc t = \dfrac{3}{2\sqrt{2}}$

6. Use the known values of the sine and cosine for $\dfrac{\pi}{6}$ to get the values of the other four trigonometric functions of $\dfrac{\pi}{6}$.

   Answer: $\tan\left(\dfrac{\pi}{6}\right) = \dfrac{1}{\sqrt{3}}$, $\cot\left(\dfrac{\pi}{6}\right) = \sqrt{3}$, $\sec\left(\dfrac{\pi}{6}\right) = \dfrac{2}{\sqrt{3}}$, $\csc\left(\dfrac{\pi}{6}\right) = 2$ **XR Example 6T.12**

7. If t is an acute angle such that $\cot t = \dfrac{4}{5}$, find the values of the remaining five trigonometric functions of t.

   Answer: $\sin t = \dfrac{5}{\sqrt{41}}$, $\cos t = \dfrac{4}{\sqrt{41}}$, $\tan t = \dfrac{5}{4}$, $\sec t = \dfrac{\sqrt{41}}{4}$, $\csc t = \dfrac{\sqrt{41}}{5}$ **XR Example 6T.14**

8. Find the unknown sides and angle of the triangle in **Figure 6T.111**.

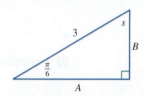

Figure 6T.111

   Answer: $s = \dfrac{\pi}{3}$, $A = \dfrac{3\sqrt{3}}{2}$, $B = \dfrac{3}{2}$ **XR Example 6T.16**

9. Find the area of the triangle in **Figure 6T.112**.

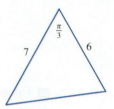

Figure 6T.112

   Answer: $\dfrac{21\sqrt{3}}{2}$ **XR Example 6T.18**

10. A transit is 200 horizontal feet from the base of a tree. An angle of $18°$ aims the transit directly at the top of the tree (**Figure 6T.113**). How much taller than the transit is the tree?

   Answer: $200 \tan 18° \approx 64.98$ feet   **XR Example 6T.19**

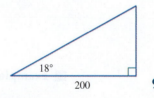

Figure 6T.113

11. Find all possible values of x if $\left(x, \dfrac{3}{4}\right)$ lies on the unit circle.

   Answer: $x = \pm\dfrac{\sqrt{7}}{4}$ **XR Example 6T.21**

12. Find the values of the six trigonometric functions of $\dfrac{3\pi}{2}$. Note which functions are undefined.

   Answer: $\sin\left(\dfrac{3\pi}{2}\right) = -1$, $\cos\left(\dfrac{3\pi}{2}\right) = 0$, $\tan\left(\dfrac{3\pi}{2}\right)$ undefined,

   $\cot\left(\dfrac{3\pi}{2}\right) = 0$, $\sec\left(\dfrac{3\pi}{2}\right)$ undefined, $\csc\left(\dfrac{3\pi}{2}\right) = -1$

   XR Example 6T.22

13. Find the reference number for $t = \dfrac{19\pi}{6}$. Then find the six trigonometric functions of $\dfrac{19\pi}{6}$.

   Answer: Reference number: $\dfrac{\pi}{6}$; $\sin t = -\dfrac{1}{2}$, $\cos t = -\dfrac{\sqrt{3}}{2}$, $\tan t = \dfrac{1}{\sqrt{3}}$,

   $\cot t = \sqrt{3}$, $\sec t = -\dfrac{2}{\sqrt{3}}$, $\csc t = -2$

   XR Example 6T.25

CHAPTER REVIEW EXERCISES

Section 6T.1

Coterminal angles. In Exercises 1 through 6, find an angle between $0°$ and $360°$ or between 0 and 2π radians that is coterminal with the given angle.

1. $400°$

   Answer: $40°$

2. $-180°$

   Answer: $180°$

3. 7π radians

   Answer: π radians

4. $\dfrac{9\pi}{2}$ radians

   Answer: $\dfrac{\pi}{2}$ radians

5. $-30°$

   Answer: $330°$

6. $-\dfrac{13\pi}{6}$ radians

   Answer: $\dfrac{11\pi}{6}$ radians

Converting from radians to degrees. In Exercises 7 through 12, you are given radian measures of angles. Convert these to degree measure.

7. $\dfrac{\pi}{2}$ radians

   Answer: $90°$

8. $\dfrac{\pi}{5}$ radian

   Answer: $36°$

9. π radians

   Answer: $180°$

10. $\dfrac{\pi}{3}$ radians

   Answer: $60°$

11. $\dfrac{\pi}{9}$ radian

   Answer: $20°$

12. 2 radians

   Answer: $\dfrac{360}{\pi} \approx 114.59°$

Converting from degrees to radians. In Exercises 13 through 18, you are given degree measures of angles. Convert these to radian measure.

13. $30°$

   Answer: $\dfrac{\pi}{6}$ radian

14. 20°

Answer: $\dfrac{\pi}{9}$ radian

15. 45°

Answer: $\dfrac{\pi}{4}$ radian

16. 120°

Answer: $\dfrac{2\pi}{3}$ radians

17. $\pi°$

Answer: $\dfrac{\pi^2}{180} \approx 0.05$ radian

18. 50°

Answer: $\dfrac{5\pi}{18}$ radian

Area and arc length. In Exercises 19 through 24, you are given the radius of a circle and an angle whose vertex is at the center of the circle. Find the area and arc length cut from the circle by the angle.

19. Radius 3 units. Angle $\dfrac{\pi}{6}$ radian.

Answer: Arc length: $\dfrac{\pi}{2}$ units

Area: $\dfrac{3\pi}{4}$ square units

20. Radius 3 units. Angle $\dfrac{\pi}{3}$ radians.

Answer: Arc length: π units

Area: $\dfrac{3\pi}{2}$ square units

21. Radius 2 units. Angle $\dfrac{\pi}{4}$ radian.

Answer: Arc length: $\dfrac{\pi}{2}$ units

Area: $\dfrac{\pi}{2}$ square units

22. Radius 4 units. Angle $\dfrac{\pi}{8}$ radian.

Answer: Arc length: $\dfrac{\pi}{2}$ units

Area: π square units

23. Radius 2 units. Angle 30°.

Answer: Arc length: $\dfrac{\pi}{3}$ units

Area: $\dfrac{\pi}{3}$ square units

24. Radius 1 unit. Angle 20°.

Answer: Arc length: $\dfrac{\pi}{9}$ unit

Area: $\dfrac{\pi}{18}$ square unit

Similar triangles. Exercises 25 through 28 show similar triangles. Find all unknown sides.

25.

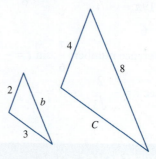

Answer: $b = 4, C = 6$

26.

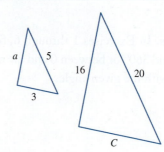

Answer: $a = 4, C = 12$

27.

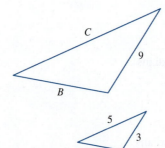

Answer: $B = 9, C = 15$

28.

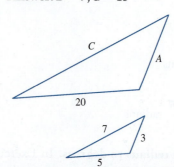

Answer: $A = 12, C = 28$

Section 6T.2

Calculating trigonometric function values from triangles. In Exercises 29 through 36, calculate the values of the six trigonometric functions of the given angle.

29. The angle s in **Figure 6T.114**

    Answer: $\sin s = \dfrac{4}{5}$, $\cos s = \dfrac{3}{5}$, $\tan s = \dfrac{4}{3}$,

    $\cot s = \dfrac{3}{4}$, $\sec s = \dfrac{5}{3}$, $\csc s = \dfrac{5}{4}$

30. The angle t in Figure 6T.114

    Answer: $\sin t = \dfrac{3}{5}$, $\cos t = \dfrac{4}{5}$, $\tan t = \dfrac{3}{4}$,

    $\cot t = \dfrac{4}{3}$, $\sec t = \dfrac{5}{4}$ $\csc t = \dfrac{5}{3}$

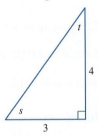

Figure 6T.114 Triangle for Exercises 29 and 30

31. The angle s in **Figure 6T.115**

    Answer: $\sin s = \dfrac{\sqrt{7}}{4}$, $\cos s = \dfrac{3}{4}$, $\tan s = \dfrac{\sqrt{7}}{3}$,

    $\cot s = \dfrac{3}{\sqrt{7}}$, $\sec s = \dfrac{4}{3}$, $\csc s = \dfrac{4}{\sqrt{7}}$

32. The angle t in Figure 6T.115

    Answer: $\sin t = \dfrac{3}{4}$, $\cos t = \dfrac{\sqrt{7}}{4}$, $\tan t = \dfrac{3}{\sqrt{7}}$,

    $\cot t = \dfrac{\sqrt{7}}{3}$, $\sec t = \dfrac{4}{\sqrt{7}}$, $\csc t = \dfrac{4}{3}$

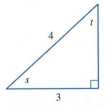

Figure 6T.115 Triangle for Exercises 31 and 32

33. The angle s in **Figure 6T.116**

    Answer: $\sin s = \dfrac{2}{3}$, $\cos s = \dfrac{\sqrt{5}}{3}$, $\tan s = \dfrac{2}{\sqrt{5}}$,

    $\cot s = \dfrac{\sqrt{5}}{2}$, $\sec s = \dfrac{3}{\sqrt{5}}$, $\csc s = \dfrac{3}{2}$

34. The angle t in Figure 6T.116

    Answer: $\sin t = \dfrac{\sqrt{5}}{3}$, $\cos t = \dfrac{2}{3}$, $\tan t = \dfrac{\sqrt{5}}{2}$,

    $\cot t = \dfrac{2}{\sqrt{5}}$, $\sec t = \dfrac{3}{2}$, $\csc t = \dfrac{3}{\sqrt{5}}$

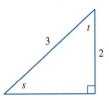

Figure 6T.116 Triangle for Exercises 33 and 34

35. The angle s in **Figure 6T.117**

    Answer: $\sin s = \dfrac{1}{\sqrt{5}}$, $\cos s = \dfrac{2}{\sqrt{5}}$, $\tan s = \dfrac{1}{2}$,

    $\cot s = 2$, $\sec s = \dfrac{\sqrt{5}}{2}$, $\csc s = \sqrt{5}$

36. The angle t in Figure 6T.117

    Answer: $\sin t = \dfrac{2}{\sqrt{5}}$, $\cos t = \dfrac{1}{\sqrt{5}}$, $\tan t = 2$,

    $\cot t = \dfrac{1}{2}$, $\sec t = \sqrt{5}$, $\csc t = \dfrac{\sqrt{5}}{2}$

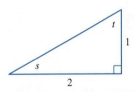

Figure 6T.117 Triangle for Exercises 35 and 36

Making triangles. In Exercises 37 through 40, construct a right triangle that exhibits the given value of a trigonometric function of an acute angle s. Be sure to label all three sides of the triangle and to mark the angle s. There are many correct answers.

37. $\sin s = \dfrac{2}{3}$

    Answer:

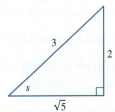

38. $\sin s = \dfrac{3}{4}$

Answer:

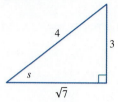

39. $\tan s = 5$

Answer:

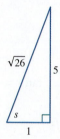

40. $\csc s = \dfrac{5}{3}$

Answer:

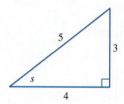

Section 6T.3

Finding lengths. In Exercises 41 through 50, you are given the value of a trigonometric function of an angle and the length of one side of a right triangle with that angle. Use this information to calculate the lengths of the remaining sides of the triangle. You may find it helpful to draw a picture of the right triangle. Report answers correct to two decimal places.

41. The sine of an acute angle in a right triangle is 0.15. The length of the opposite side is 5.

Answer: Hypotenuse = 33.33.
Adjacent side = 32.96.

42. The cosine of an acute angle in a right triangle is 0.50. The length of the adjacent side is 8.

Answer: Hypotenuse = 16. Opposite side = 13.86.

43. The tangent of an acute angle in a right triangle is 5. The length of the adjacent side is 9.

Answer: Opposite side = 45. Hypotenuse = 45.89.

44. The cosine of an acute angle in a right triangle is 0.44. The length of the adjacent side is 6.

Answer: Hypotenuse = 13.64.
Opposite side = 12.25.

45. The cosine of an acute angle in a right triangle is $\dfrac{1}{4}$. The length of the adjacent side is 5.

Answer: Hypotenuse = 20. Opposite side = 19.36.

46. The tangent of an acute angle in a right triangle is 3. The length of the adjacent side is 10.

Answer: Opposite side = 30. Hypotenuse = 31.62.

47. The tangent of an acute angle in a right triangle is $\dfrac{1}{2}$. The length of the opposite side is 4.

Answer: Adjacent side = 8. Hypotenuse = 8.94.

48. The cotangent of an acute angle in a right triangle is $\dfrac{4}{3}$. The length of the opposite side is 12.

Answer: Adjacent side = 16. Hypotenuse = 20.

49. The secant of an acute angle in a right triangle is 5. The length of the hypotenuse is 6.

Answer: Adjacent side = 1.2. Opposite side = 5.88.

50. The cotangent of an acute angle in a right triangle is $\dfrac{3}{5}$. The length of the opposite side is 25.

Answer: Adjacent side = 15. Hypotenuse = 29.15.

Right triangles. In Exercises 51 through 54, draw a right triangle consistent with the given information, solve for the lengths of the two unknown sides, and label all the sides.

51. An acute angle of $\dfrac{\pi}{4}$ radian and an adjacent side of length 2.

Answer:

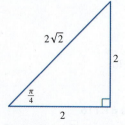

52. An acute angle of $\dfrac{\pi}{3}$ radians and an opposite side of length 3.

Answer:

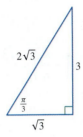

53. An acute angle of $\dfrac{\pi}{6}$ radian and a hypotenuse of length 4.

Answer:

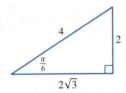

54. An acute angle of 20° and an opposite side of length 2. Report side lengths correct to two decimal places.

Answer:

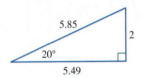

55. Calculating the distance to the top of an object. A man sits 150 horizontal feet from the base of a wall. He must incline his eyes at an angle of 10° to look at the top of the wall. See **Figure 6T.118.** What is the distance from the man directly to the top of the wall?

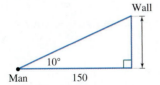

Figure 6T.118 Triangle for Exercise 55

Answer: 152.31 feet

Section 6T.4

Points on the unit circle. In Exercises 56 through 61, find all values of the variable so that the indicated point lies on the unit circle and in the given quadrant.

56. $\left(x, -\dfrac{1}{2}\right)$, quadrant IV

Answer: $x = \dfrac{\sqrt{3}}{2}$

57. $\left(\dfrac{2}{3}, y\right)$, quadrant I

Answer: $y = \dfrac{\sqrt{5}}{3}$

58. $\left(x, \dfrac{3}{5}\right)$, quadrant II

Answer: $x = -\dfrac{4}{5}$

59. $\left(-\dfrac{3}{4}, y\right)$, quadrant III

Answer: $y = -\dfrac{\sqrt{7}}{4}$

60. $(x, -x)$, quadrant IV

Answer: $x = \dfrac{1}{\sqrt{2}}$

61. $(x, 2x)$, quadrant I

Answer: $x = \dfrac{1}{\sqrt{5}}$

Reference angles. In Exercises 62 through 67, find the reference angle for the given angle. If the angle is given in degrees, your answer should be in degrees. If the angle is given in radians, your answer should be in radians.

62. $\dfrac{5\pi}{4}$ radians

Answer: $\dfrac{\pi}{4}$ radian

63. 120°

Answer: 60°

64. $-\dfrac{2\pi}{3}$ radians

Answer: $\dfrac{\pi}{3}$ radians

65. −30°

Answer: 30°

66. $\dfrac{7\pi}{6}$ radians

Answer: $\dfrac{\pi}{6}$ radian

67. 170°

   Answer: 10°

Special angles and compass points. Use your knowledge of special angles and compass points to find the values of the six trigonometric functions of the angles or real numbers given in Exercises 68 through 77. State which, if any, are undefined. Remember that if degrees are not specifically indicated, then radian measure is assumed.

68. 0°

   Answer: $\sin 0° = 0$, $\cos 0° = 1$, $\tan 0° = 0$, $\sec 0° = 1$. The cotangent and cosecant are undefined.

69. 210°

   Answer: $\sin 210° = -\dfrac{1}{2}$, $\cos 210° = -\dfrac{\sqrt{3}}{2}$,

   $\tan 210° = \dfrac{1}{\sqrt{3}}$, $\cot 210° = \sqrt{3}$,

   $\sec 210° = -\dfrac{2}{\sqrt{3}}$, $\csc 210° = -2$

70. $\dfrac{7\pi}{2}$

   Answer: $\sin\left(\dfrac{7\pi}{2}\right) = -1$, $\cos\left(\dfrac{7\pi}{2}\right) = 0$,

   $\cot\left(\dfrac{7\pi}{2}\right) = 0$, $\csc\left(\dfrac{7\pi}{2}\right) = -1$

   The tangent and secant are undefined.

71. 180°

   Answer: $\sin 180° = 0$, $\cos 180° = -1$,

   $\tan 180° = 0$, $\sec 180° = -1$

   The cotangent and cosecant are undefined.

72. 135°

   Answer: $\sin 135° = \dfrac{1}{\sqrt{2}}$, $\cos 135° = -\dfrac{1}{\sqrt{2}}$,

   $\tan 135° = -1$, $\cot 135° = -1$,

   $\sec 135° = -\sqrt{2}$, $\csc 135° = \sqrt{2}$

73. $\dfrac{5\pi}{4}$

   Answer: $\sin\left(\dfrac{5\pi}{4}\right) = -\dfrac{1}{\sqrt{2}}$, $\cos\left(\dfrac{5\pi}{4}\right) = -\dfrac{1}{\sqrt{2}}$,

   $\tan\left(\dfrac{5\pi}{4}\right) = 1$, $\cot\left(\dfrac{5\pi}{4}\right) = 1$, $\sec\left(\dfrac{5\pi}{4}\right) = -\sqrt{2}$,

   $\csc\left(\dfrac{5\pi}{4}\right) = -\sqrt{2}$

74. 270°

   Answer: $\sin 270° = -1$, $\cos 270° = 0$,

   $\cot 270° = 0$, $\csc 270° = -1$

   The tangent and secant are undefined.

75. $-\dfrac{\pi}{6}$

   Answer: $\sin\left(-\dfrac{\pi}{6}\right) = -\dfrac{1}{2}$, $\cos\left(-\dfrac{\pi}{6}\right) = \dfrac{\sqrt{3}}{2}$,

   $\tan\left(-\dfrac{\pi}{6}\right) = -\dfrac{1}{\sqrt{3}}$, $\cot\left(-\dfrac{\pi}{6}\right) = -\sqrt{3}$,

   $\sec\left(-\dfrac{\pi}{6}\right) = \dfrac{2}{\sqrt{3}}$, $\csc\left(-\dfrac{\pi}{6}\right) = -2$

76. $\dfrac{\pi}{2}$

   Answer: $\sin\left(\dfrac{\pi}{2}\right) = 1$, $\cos\left(\dfrac{\pi}{2}\right) = 0$, $\cot\left(\dfrac{\pi}{2}\right) = 0$,

   $\csc\left(\dfrac{\pi}{2}\right) = 1$

   The tangent and secant are undefined.

77. $\dfrac{\pi}{6}$

   Answer: $\sin\left(\dfrac{\pi}{6}\right) = \dfrac{1}{2}$, $\cos\left(\dfrac{\pi}{6}\right) = \dfrac{\sqrt{3}}{2}$,

   $\tan\left(\dfrac{\pi}{6}\right) = \dfrac{1}{\sqrt{3}}$, $\cot\left(\dfrac{\pi}{6}\right) = \sqrt{3}$, $\sec\left(\dfrac{\pi}{6}\right) = \dfrac{2}{\sqrt{3}}$,

   $\csc\left(\dfrac{\pi}{6}\right) = 2$

Finding trigonometric function values from a known value. In Exercises 78 through 85, you are given the value of one trigonometric function. Find the values of the remaining five trigonometric functions.

78. $\cos t = \dfrac{3}{4}$, and $\sin t$ is positive.

   Answer: $\sin t = \dfrac{\sqrt{7}}{4}$, $\tan t = \dfrac{\sqrt{7}}{3}$, $\cot t = \dfrac{3}{\sqrt{7}}$,

   $\sec t = \dfrac{4}{3}$, $\csc t = \dfrac{4}{\sqrt{7}}$

79. $\tan t = \sqrt{3}$, and $\sin t$ is positive.

   Answer: $\cos t = \dfrac{1}{2}$, $\sin t = \dfrac{\sqrt{3}}{2}$, $\cot t = \dfrac{1}{\sqrt{3}}$,

   $\sec t = 2$, $\csc t = \dfrac{2}{\sqrt{3}}$

80. $\sin t = \dfrac{5}{6}$, and $\cos t$ is negative.

Answer: $\cos t = -\dfrac{\sqrt{11}}{6}$, $\tan t = -\dfrac{5}{\sqrt{11}}$,

$\cot t = -\dfrac{\sqrt{11}}{5}$, $\sec t = -\dfrac{6}{\sqrt{11}}$, $\csc t = \dfrac{6}{5}$

81. $\sin t = \dfrac{1}{6}$, and $\cos t$ is negative.

Answer: $\cos t = -\dfrac{\sqrt{35}}{6}$, $\tan t = -\dfrac{1}{\sqrt{35}}$,

$\cot t = -\sqrt{35}$, $\sec t = -\dfrac{6}{\sqrt{35}}$, $\csc t = 6$

82. $\cos t = -\dfrac{1}{3}$, and $\tan t$ is negative.

Answer: $\sin t = \dfrac{\sqrt{8}}{3}$, $\tan t = -\sqrt{8}$,

$\cot t = -\dfrac{1}{\sqrt{8}}$, $\sec t = -3$, $\csc t = \dfrac{3}{\sqrt{8}}$

83. $\sec t = 4$, and $\sin t$ is negative.

Answer: $\cos t = \dfrac{1}{4}$, $\sin t = -\dfrac{\sqrt{15}}{4}$,

$\tan t = -\sqrt{15}$, $\cot t = -\dfrac{1}{\sqrt{15}}$, $\csc t = -\dfrac{4}{\sqrt{15}}$

84. $\cot t = 1.10$, and $\sin t$ is negative. Report answers correct to two decimal places.

Answer: $\sin t = -0.67$, $\cos t = -0.74$, $\tan t = 0.91$, $\sec t = -1.35$, $\csc t = -1.49$

85. $\cos t = 0.55$, and $\sin t$ is positive. Report answers correct to two decimal places.

Answer: $\sin t = 0.84$, $\tan t = 1.52$, $\cot t = 0.66$, $\sec t = 1.82$, $\csc t = 1.20$

GRAPHS AND PERIODICITY OF TRIGONOMETRIC FUNCTIONS

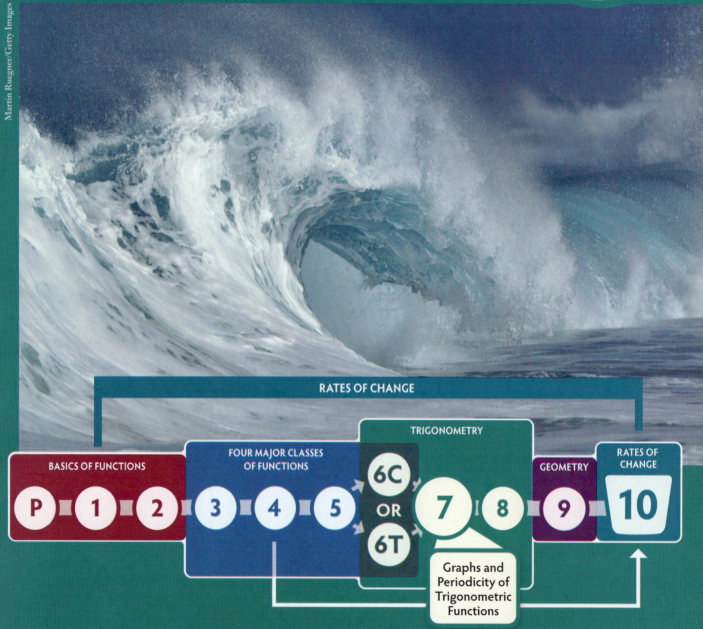

Martin Ruegner/Getty Images

- **7.1** Graphs of the Sine and Cosine
- **7.2** Graphs of Other Trigonometric Functions

Trigonometric functions repeat in a predictable fashion, and this fact facilitates the production of their graphs. The dynamic properties of trigonometric functions and their graphs are brought out especially by the emphasis on modeling.

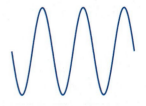

Trigonometric functions are periodic—that is, they repeat in a predictable fashion. This behavior is reflected in their graphs and is useful in many applications such as modeling natural phenomena. The graphs of the sine and cosine resemble waves, as shown in **Figure 7.1,** and similar graphs are often called sine waves. From ocean waves to alternating current, X-rays, musical notes, temperature variations, and populations of competing species, the list of phenomena trigonometric functions are used to describe is almost limitless. Note that modeling applications of trigonometric functions are largely treated in online Section ON7.3.

Figure 7.1 A sine wave

7.1 Graphs of the Sine and Cosine

> The graphs of the sine and cosine functions can be shifted, stretched, and compressed to produce familiar sine waves.

7.1 Graphs of the Sine and Cosine

7.2 Graphs of Other Trigonometric Functions

In this section, you will learn to:

1. Explain what it means for the sine and cosine to have period 2π.
2. Construct the graphs of the sine and the cosine.
3. Identify the amplitude of a given sine wave.
4. Explain the effect of changing the amplitude of a sine wave.
5. Identify the period of a given sine wave.
6. Explain the effect of changing the period of a sine wave.
7. Identify the phase shift of a given sine wave.
8. Explain the effect of changing the phase shift of a sine wave.

Many of us see sine waves as integral to engineering, science, and mathematics. But sine waves also play a role in art. An interesting example is the thesis project of Daniel Sierra called *Oscillate*, which is available online. A still of one of his animations is shown in **Figure 7.2**. Sierra's artwork illustrates the close ties among mathematics, art, and music. Trigonometry can be beautiful.

Figure 7.2 A still from one of Sierra's video animations Daniel Sierra

Periodicity of the Sine and Cosine

> The sine and cosine repeat every 2π units.

A **periodic function** is a function that repeats each p units, and the smallest p that makes this happen is the **period**. For example, if $f(t)$ is the location of the minute

A **periodic function** is a function f for which there exists a number $p > 0$ such that $f(x) = f(x + p)$ for every x in the domain of f.

The **period** of a periodic function f is the smallest number p such that $f(x) = f(x + p)$ for every x in the domain of f.

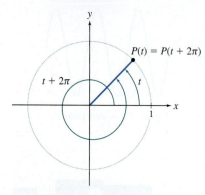

Figure 7.3 Periodicity of trigonometric points: The trigonometric point $P(t) = (\cos t, \sin t)$ repeats every 2π units, just as would the second hand of a clock moving backward.

hand of a clock t minutes after noon, then $f(t)$ is periodic with period 60 because the minute hand lands on the same position each 60 minutes. We can use this same idea to see the periodicity of the sine and cosine.

As we move around the unit circle in the counterclockwise direction, the trigonometric point $P(t)$ lands on the same spot on each trip around the circle. Because one trip around the circle corresponds to an increase of 2π units in t, we conclude that $P(t) = P(t + 2\pi)$ for all values of t. This fact is illustrated in **Figure 7.3**. Then, because $P(t) = (\cos t, \sin t)$, it follows that the functions $x = \cos t$ and $y = \sin t$ also repeat every 2π units. These functions have period 2π, or $360°$.

CONCEPTS TO REMEMBER: The Periods of the Sine and Cosine Functions

For any t:

In terms of radians	In terms of degrees
Sine and cosine are periodic with period 2π.	Sine and cosine are periodic with period $360°$.
$\sin(t + 2\pi) = \sin t$	$\sin(t° + 360°) = \sin t°$
$\cos(t + 2\pi) = \cos t$	$\cos(t° + 360°) = \cos t°$

EXAMPLE 7.1   Using Periodicity to Evaluate Sine and Cosine

Use the period and what you know about special angles to find the value of $\sin\left(\dfrac{\pi}{3} + 14\pi\right)$.

SOLUTION

We can rewrite this expression as $\sin\left(\dfrac{\pi}{3} + 7(2\pi)\right)$. This is the sine of $\dfrac{\pi}{3}$ plus 7 periods of the sine function. Because the sine function repeats each period, we conclude that

$$\sin\left(\frac{\pi}{3} + 7(2\pi)\right) = \sin\frac{\pi}{3} = \frac{\sqrt{3}}{2}$$

TRY IT YOURSELF 7.1   Brief answers provided at the end of the section.

Find the value of $\cos\left(\dfrac{\pi}{3} + 20\pi\right)$.

EXTEND YOUR REACH

a. Find the sine of $\dfrac{200\pi}{3}$.

b. Explain how you would calculate the sine of $\dfrac{k\pi}{3}$ for any positive integer k.

The Graph of the Sine Function

> The graph of the sine function can be made by following the y-coordinate as we move around the unit circle in a counterclockwise direction.

The first step in producing the graph of the sine function is to make the graph over a single period, from $t = 0$ to $t = 2\pi$. To make the graph, we begin at $(1, 0)$, which corresponds to $t = 0$, and we move around the circle in a counterclockwise direction, keeping track of the value of the sine. Because the y-coordinate of the trigonometric point $P(t)$ is $\sin t$, we can think of $\sin t$ as the signed distance from the trigonometric point $P(t)$ to the x-axis, as is illustrated in **Figure 7.4**. This observation yields the information in **Table 7.1**, which we use to construct the graph in **Figure 7.5**.

Because the sine function is periodic, its graph repeats each 2π units. So the entire graph of the sine function is made up of copies of the graph in Figure 7.5. The result is the graph in **Figure 7.6**, which suggests that the sine is 0 at integer multiples of π.

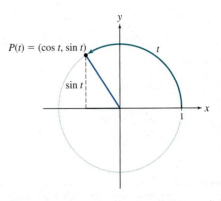

DF Figure 7.4 $\sin t$: We can think of this function as the signed distance from the trigonometric point $P(t)$ to the x-axis.

Table 7.1 Values of Sine Between 0 and 2π

t	0	$\dfrac{\pi}{6}$	$\dfrac{\pi}{4}$	$\dfrac{\pi}{3}$	$\dfrac{\pi}{2}$	$\dfrac{2\pi}{3}$	$\dfrac{3\pi}{4}$	$\dfrac{5\pi}{6}$	π	$\dfrac{7\pi}{6}$	$\dfrac{5\pi}{4}$	$\dfrac{4\pi}{3}$	$\dfrac{3\pi}{2}$	$\dfrac{5\pi}{3}$	$\dfrac{7\pi}{4}$	$\dfrac{11\pi}{6}$	2π
$\sin t$	0	$\dfrac{1}{2}$	$\dfrac{1}{\sqrt{2}}$	$\dfrac{\sqrt{3}}{2}$	1	$\dfrac{\sqrt{3}}{2}$	$\dfrac{1}{\sqrt{2}}$	$\dfrac{1}{2}$	0	$-\dfrac{1}{2}$	$-\dfrac{1}{\sqrt{2}}$	$-\dfrac{\sqrt{3}}{2}$	-1	$-\dfrac{\sqrt{3}}{2}$	$-\dfrac{1}{\sqrt{2}}$	$-\dfrac{1}{2}$	0
Approximate decimal value	0	0.5	0.71	0.87	1	0.87	0.71	0.5	0	−0.5	−0.71	−0.87	−1	−0.87	−0.71	−0.5	0

$\sin t$ increases from 0 to $\dfrac{\pi}{2}$ $\sin t$ decreases from $\dfrac{\pi}{2}$ to $\dfrac{3\pi}{2}$ $\sin t$ increases from $\dfrac{3\pi}{2}$ to 2π

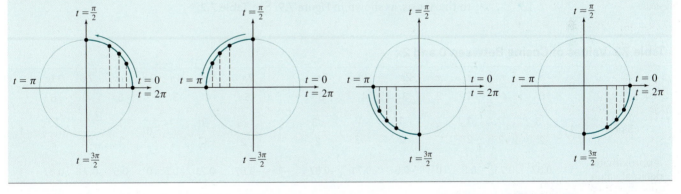

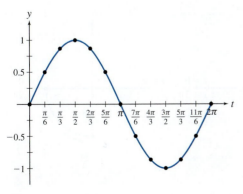

Figure 7.5 The graph of the sine function on the interval from 0 to 2π: This is one period of the sine function.

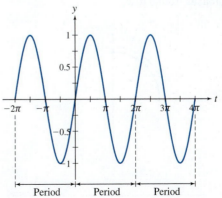

DF Figure 7.6 The graph of the sine function showing three periods

Figure 7.7 shows the intervals of increase and decrease of the sine function, and **Figure 7.8** shows the concavity along with points of inflection.

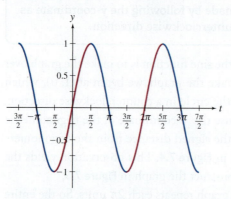

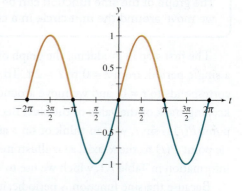

Figure 7.7 The graph of the sine function: intervals of increase in red, intervals of decrease in blue

Figure 7.8 The graph of the sine function: upward concavity in green, downward concavity in brown: Points of inflection occur at integer multiples of π.

The Graph of the Cosine Function

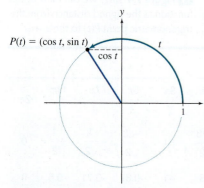

DF Figure 7.9 cos t: We can think of this function as the signed distance from the trigonometric point $P(t)$ to the y-axis.

> The graph of the cosine can be constructed by following the x-coordinate as we move around the unit circle in a counterclockwise direction.

We can make the graph of the cosine function using methods similar to those used for the sine function. Observe first that cos t is the x-coordinate of the trigonometric point $P(t) = (\cos t, \sin t)$. Thus, we can think of cos t as the signed distance from $P(t)$ to the y-axis, as shown in **Figure 7.9**. See **Table 7.2**.

Table 7.2 Values of Cosine Between 0 and 2π

t	0	$\dfrac{\pi}{6}$	$\dfrac{\pi}{4}$	$\dfrac{\pi}{3}$	$\dfrac{\pi}{2}$	$\dfrac{2\pi}{3}$	$\dfrac{3\pi}{4}$	$\dfrac{5\pi}{6}$	π	$\dfrac{7\pi}{6}$	$\dfrac{5\pi}{4}$	$\dfrac{4\pi}{3}$	$\dfrac{3\pi}{2}$	$\dfrac{5\pi}{3}$	$\dfrac{7\pi}{4}$	$\dfrac{11\pi}{6}$	2π
$\cos t$	1	$\dfrac{\sqrt{3}}{2}$	$\dfrac{1}{\sqrt{2}}$	$\dfrac{1}{2}$	0	$-\dfrac{1}{2}$	$-\dfrac{1}{\sqrt{2}}$	$-\dfrac{\sqrt{3}}{2}$	-1	$-\dfrac{\sqrt{3}}{2}$	$-\dfrac{1}{\sqrt{2}}$	$-\dfrac{1}{2}$	0	$\dfrac{1}{2}$	$\dfrac{1}{\sqrt{2}}$	$\dfrac{\sqrt{3}}{2}$	1
Approximate decimal value	1	0.87	0.71	0.5	0	−0.5	−0.71	−0.87	−1	−0.87	−0.71	−0.5	0	0.5	0.71	0.87	1

cos t decreases from 0 to π cos t increases from π to 2π

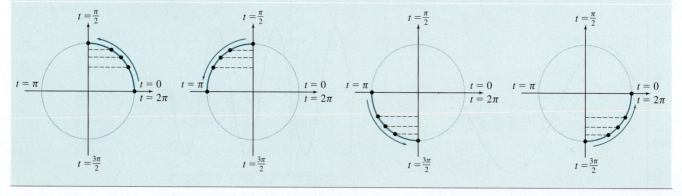

The graph over a single period is shown in **Figure 7.10**. As with the sine function, we can use the graph over a single period to make the complete graph shown in **Figure 7.11**. As Figure 7.11 suggests, the cosine is 0 at $\frac{\pi}{2}$ plus integer multiples of π.

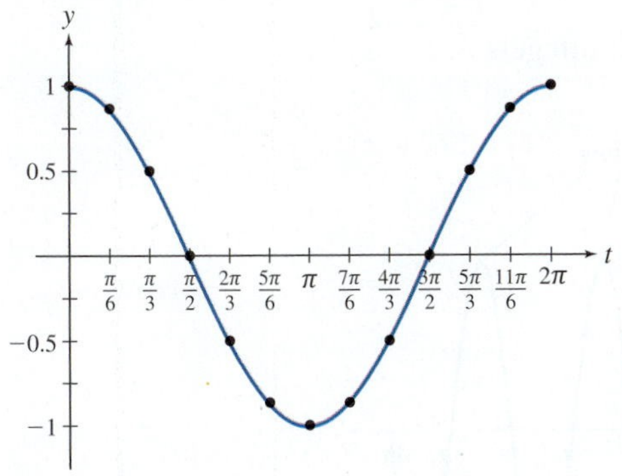

Figure 7.10 The graph of the cosine function on the interval from 0 to 2π, a single period of the cosine

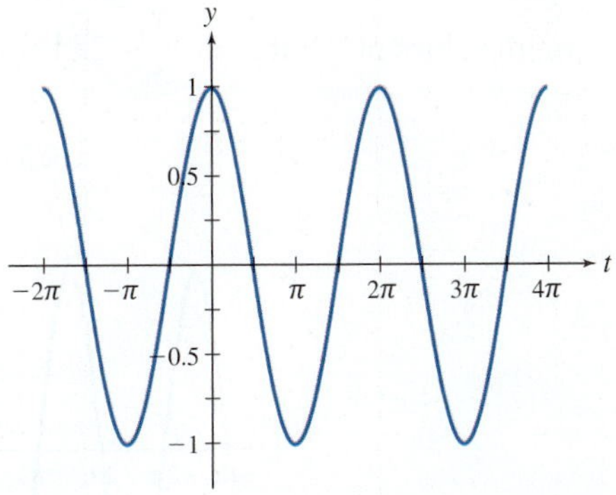

Figure 7.11 Using periodicity: The graph of the cosine function is made by repeating the graph over a single period.

Figure 7.12 shows the intervals of increase and decrease of the cosine function, and **Figure 7.13** shows the concavity along with points of inflection.

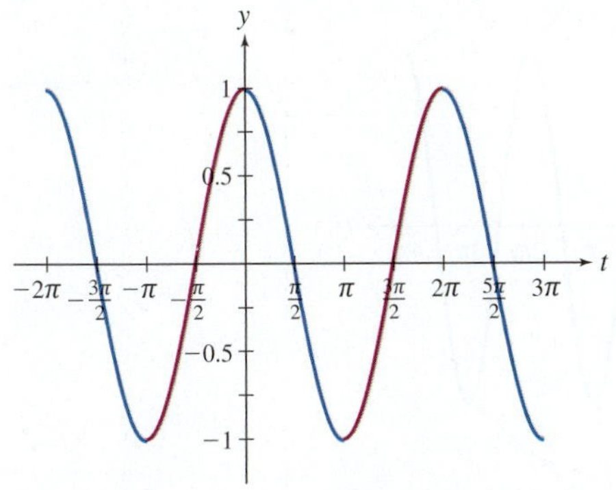

Figure 7.12 The graph of the cosine function: intervals of increase in red, intervals of decrease in blue

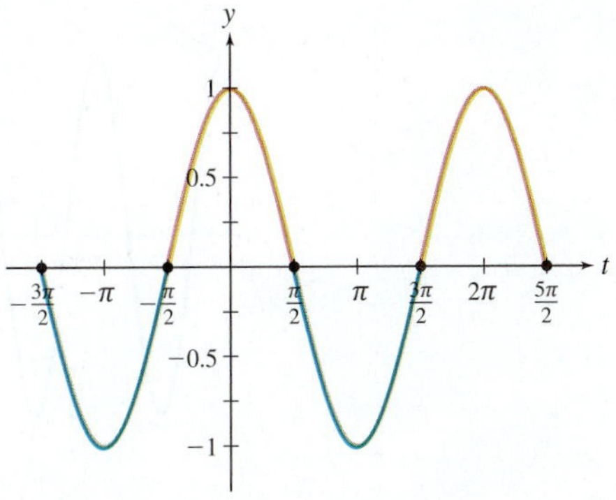

Figure 7.13 The graph of the cosine function: upward concavity shown in green, downward concavity in brown: Points of inflection occur at $\frac{\pi}{2}$ plus integer multiples of π.

In the process of making the graphs of the sine and cosine, we can observe an important feature of these functions. For any point $P(t)$ on the unit circle, both the x- and y-coordinates are between -1 and 1. This fact about the coordinates of $P(t)$ is reflected in the graphs of the sine and cosine functions.

CONCEPTS TO REMEMBER: Bounds for the Sine and Cosine

The values of the sine and cosine functions are always between -1 and 1. The range of these functions is the interval $[-1, 1]$:

The sine function has local maximum values of 1 at $t = \dfrac{\pi}{2} + 2k\pi$ and local minimum values of -1 at $t = -\dfrac{\pi}{2} + 2k\pi$ for all integers k.

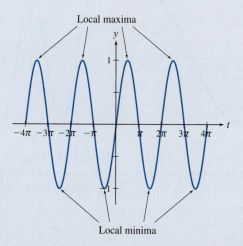

The cosine function has local maximum values of 1 at $t = 2k\pi$ and local minimum values of -1 at $t = \pi + 2k\pi$ for all integers k.

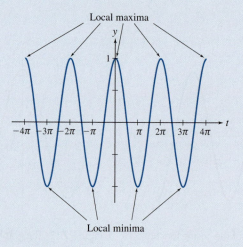

Amplitude

> **The amplitude of a sine wave is the height of the wave.**

A **sine wave** is the graph of a function of the form $y = A \sin B(x - C)$ or $y = A \cos B(x - C)$.

The **amplitude** of the sine wave $y = A \sin x$ or $y = A \cos x$ is the constant $|A|$.

Variants of the sine and cosine of the form $y = A \sin B(x - C)$ or $y = A \cos B(x - C)$ occur often, and we will examine the roles of each of the constants A, B, and C. The graph of such a function is often referred to as a **sine wave** because of its similarity to the graph of the sine function. The height of a sine wave is called the **amplitude**.

The effect of changing the amplitude is illustrated in **Figure 7.14**, where we have graphed $y = \sin x$, $y = 2 \sin x$, and $y = 3 \sin x$. Observe that the heights of the graphs change with A, but the periods do not.

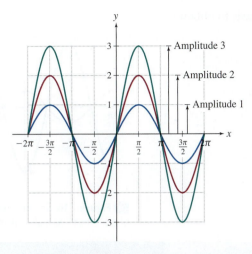

Figure 7.14 Graphs of $y = \sin x$, $y = 2 \sin x$, and $y = 3 \sin x$ showing amplitude

We learned in Section 2.3 how the graph of $y = Af(x)$ is related to the graph of $y = f(x)$. We summarize the relationship for the case of a sine wave.

CONCEPTS TO REMEMBER: Amplitude

The height of a sine wave is called the amplitude.

If $A > 0$, to make the graph of $y = A \sin x$ or $y = A \cos x$ we vertically stretch or compress the graph of $y = \sin x$ or $y = \cos x$, respectively, by a factor of A. The graph oscillates between $-A$ and A.

If $A < 0$, we vertically stretch or compress by a factor of $|A|$ and then reflect about the x-axis.

Changing the amplitude does not affect the period.

EXAMPLE 7.2 Graphing Using Amplitude

Identify the amplitude of $y = -4 \cos x$. On the same coordinate axes, plot the graphs of $y = \cos x$ and $y = -4 \cos x$.

SOLUTION

The amplitude of $y = -4 \cos x$ is $|-4| = 4$. We construct the graph in two steps. First we make the graph of $y = 4 \cos x$ by vertically stretching the graph of $y = \cos x$ by a factor of 4, as shown in **Figure 7.15**.

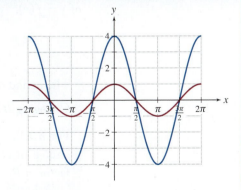

Figure 7.15 First step: Stretch the graph of $y = \cos x$ to make the graph of $y = 4 \cos x$.

Next we reflect this graph about the horizontal axis to obtain the graph of $y = -4 \cos x$ shown in **Figure 7.16**.

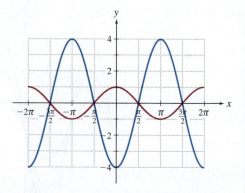

DF **Figure 7.16** Second step: Reflect the graph of $y = 4 \cos x$ to make the graph of $y = -4 \cos x$.

TRY IT YOURSELF 7.2 Brief answers provided at the end of the section.

Show the graphs of $y = \sin x$ and $y = 5 \sin x$ on the same coordinate axes.

EXTEND YOUR REACH

a. Where is the rate of change of the sine function positive, and where is it negative? Where is it zero?

b. On the interval $\left(0, \dfrac{\pi}{2}\right)$, how does the rate of change of $y = \sin x$ compare with the rate of change of $y = 5 \sin x$? *Suggestion*: Graph both functions on the same coordinate axes, and then use a straightedge to estimate the slopes of tangent lines.

The Period

> The period of a sine wave is the time required for the cycle to repeat.

The positive constant B in $\sin Bx$ or $\cos Bx$ changes the period from 2π to $\dfrac{2\pi}{B}$, as is indicated by the following calculation:

$$\sin B\left(x + \frac{2\pi}{B}\right) = \sin\left(Bx + 2\pi\right) = \sin Bx$$

We know from Section 2.3 how the graph of $y = f(Bx)$ is related to the graph of $y = f(x)$.

CONCEPTS TO REMEMBER: Period and Amplitude

The period of $y = A \sin Bx$ or $y = A \cos Bx$ is $\dfrac{2\pi}{|B|}$. The amplitude of each is $|A|$. Changing B does not change the amplitude.

For $B > 0$, to make the graph of $y = A \sin Bx$ or $y = A \cos Bx$ we horizontally stretch or compress the graph of $y = A \sin x$ or $y = A \cos x$, respectively, by a factor of B.

For $B < 0$, we horizontally stretch or compress by a factor of $|B|$ and then reflect about the y-axis.

EXAMPLE 7.3 Period and Amplitude

a. Find the period and amplitude of $y = 3 \cos 2x$.

b. On the same axes, plot $y = \cos x$ and $y = 3 \cos 2x$.

SOLUTION

a. The amplitude is $|A| = 3$. The period is $\dfrac{2\pi}{|B|} = \dfrac{2\pi}{2} = \pi$.

b. We make the graph in two steps:

First we make the graph of $y = \cos 2x$. This function has period π, so it completes its period in half the distance of the cosine. Its graph can be found from the graph of $y = \cos x$ by using a horizontal compression by a factor of 2. The resulting graph is shown in **Figure 7.17**.

Next we deal with the amplitude. To obtain the graph of $y = 3 \cos 2x$, we stretch the graph of $y = \cos 2x$ vertically by a factor of 3. The completed graph is shown in **Figure 7.18**.

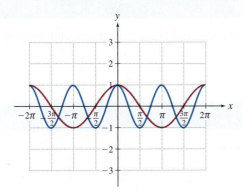

Figure 7.17 First step: Compress the graph of $y = \cos x$ horizontally to find the graph of $y = \cos 2x$ (blue).

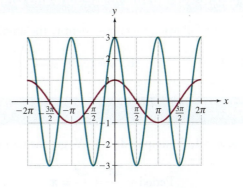

Figure 7.18 Second step: Stretch the graph of $y = \cos 2x$ vertically to find the graph of $y = 3 \cos 2x$ (green).

TRY IT YOURSELF 7.3   Brief answers provided at the end of the section.

Identify the period and amplitude of $y = 2 \sin \dfrac{x}{2}$. Plot the graph of this function along with the graph of $y = \sin x$.

EXTEND YOUR REACH

a. Explain what happens to the graph of $y = \sin(Bx)$ as the positive number B increases but the horizontal span over which the plot is made remains the same.

b. Describe what the graph of $y = \sin(1{,}000{,}000x)$ would look like if it were plotted on the interval $[0, 2\pi]$.

Phase Shift

> **The phase determines the horizontal shift of the sine wave.**

To obtain the graphs of $y = A \sin B(x - C)$ and $y = A \cos B(x - C)$, we shift horizontally the graph of $y = A \sin Bx$ or $y = A \cos Bx$, respectively. The amount by which we shift is the **phase shift**.

The **phase shift** in the functions $y = A \sin B(x - C)$ and $y = A \cos B(x - C)$ is the number C.

CONCEPTS TO REMEMBER: Phase Shift

For the functions $y = A \sin B(x - C)$ and $y = A \cos B(x - C)$, the constant C is the phase shift.

To find the graphs of these functions, we shift the graph of $y = A \sin Bx$ or $y = A \cos Bx$, respectively, by C units to the right if $C > 0$. If $C < 0$, then the shift is by $|C|$ units to the left.

For the functions $y = A \sin B(x - C)$ and $y = A \cos B(x - C)$, the amplitude is $|A|$, and the period is $\dfrac{2\pi}{|B|}$. Changing the phase shift does not affect the amplitude or the period.

EXAMPLE 7.4   Identifying Phase Shift and Graphing

For each of the following functions, identify the amplitude, period, and phase shift. Then plot the graph.

a. $y = \cos 2x$
b. $y = \cos 2\left(x - \dfrac{\pi}{2}\right)$
c. $y = 3 \cos 2\left(x - \dfrac{\pi}{2}\right)$

SOLUTION

a. Because $A = 1$, the amplitude is 1. We get the period from $B = 2$:

$$\text{Period} = \frac{2\pi}{|B|} = \frac{2\pi}{2} = \pi$$

The phase shift is $C = 0$. As we noted in the solution of Example 7.3, the graph of $y = \cos 2x$ can be found from the graph of $y = \cos x$ by using a horizontal compression by a factor of 2. The resulting graph is shown in **Figure 7.19**.

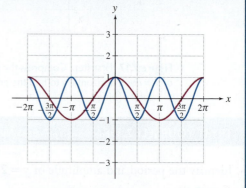

Figure 7.19 Part a: Compress the graph of $y = \cos x$ horizontally to find the graph of $y = \cos 2x$ (blue).

b. For $y = \cos 2\left(x - \dfrac{\pi}{2}\right)$, the amplitude is still 1, and the period is still π. The phase shift is $C = \dfrac{\pi}{2}$. To make the graph of $y = \cos 2\left(x - \dfrac{\pi}{2}\right)$, we shift the graph of $y = \cos 2x$ by $\dfrac{\pi}{2}$ units to the right. The graph is in **Figure 7.20**.

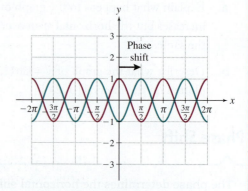

Figure 7.20 Part b: Shift the graph of $y = \cos 2x$ to the right to make the graph of $y = \cos 2\left(x - \dfrac{\pi}{2}\right)$ (green).

c. Only the amplitude has changed from part b. The new amplitude is $|A| = 3$. So we find the graph by stretching the graph from part b vertically by a factor of 3. The resulting graph is shown in **Figure 7.21**.

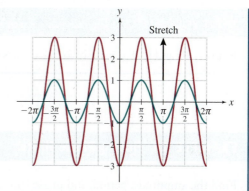

Figure 7.21 Part c: Stretch the graph from part b vertically to make the graph of $y = 3 \cos 2\left(x - \dfrac{\pi}{2}\right)$ (red).

TRY IT YOURSELF 7.4 Brief answers provided at the end of the section.

Identify the amplitude, period, and phase shift of $y = -2 \cos \dfrac{1}{3}\left(x + \dfrac{\pi}{4}\right)$. Then plot the graph.

It was easier to plot the graph in part c of Example 7.4 because we already had the graphs in parts a and b. This observation suggests the following strategy.

STEP-BY-STEP STRATEGY: Graphing $y = A \cos B(x - C)$ or $y = A \sin B(x - C)$

To graph $y = A \cos B(x - C)$, we complete the following steps:

Step 1 Graph $y = \cos Bx$ by using a horizontal stretch or compression of $y = \cos x$ by B. (If $B < 0$, horizontally stretch or compress by a factor of $|B|$, and then reflect the graph about the y-axis.)

Step 2 Graph $y = \cos B(x - C)$ by shifting the graph of $y = \cos Bx$ horizontally by C.

Step 3 Graph $y = A \cos B(x - C)$ by using a vertical stretch or compression of $y = \cos B(x - C)$ by A. (If $A < 0$, vertically stretch or compress by a factor of $|A|$, and then reflect the graph about the x-axis.)

The steps for graphing $y = A \sin B(x - C)$ are similar.

EXAMPLE 7.5 Making Sine Models

Find A, B, and C if $y = A \sin B(x - C)$ has amplitude 5, period 7, and phase shift -2. Assume that $A > 0$.

SOLUTION

Because the amplitude is 5, we take $A = 5$. The period is $\dfrac{2\pi}{B}$, so we have $\dfrac{2\pi}{B} = 7$. Solving for B gives $B = \dfrac{2\pi}{7}$. Because the phase shift is -2, we have $C = -2$.

TRY IT YOURSELF 7.5 Brief answers provided at the end of the section.

Find A, B, and C if $y = A \sin B(x - C)$ has amplitude 2, period 3π, and phase shift 1. Assume that $A > 0$.

EXAMPLE 7.6 Finding Phase Shifts

Find the amplitude, period, and phase shift for $y = 3 \sin\left(2x - \dfrac{4}{3}\pi \right)$.

SOLUTION

Because $2x - \dfrac{4}{3}\pi = 2\left(x - \dfrac{2}{3}\pi \right)$, we can write the function as $y = 3 \sin 2\left(x - \dfrac{2}{3}\pi \right)$. From this we see that the

amplitude is $A = 3$, the period is $\dfrac{2\pi}{B} = \dfrac{2\pi}{2} = \pi$, and the phase shift is $C = \dfrac{2}{3}\pi$.

TRY IT YOURSELF 7.6 Brief answers provided at the end of the section.

Find the amplitude, period, and phase shift for $y = -4 \cos\left(3x - \dfrac{\pi}{2} \right)$.

EXAMPLE 7.7 Identifying Sine Models from Graphs

In **Figure 7.22** you are given the graph of a function of the form $y = A \sin B(x - C)$ with $A > 0$. Identify the amplitude, period, and phase shift.

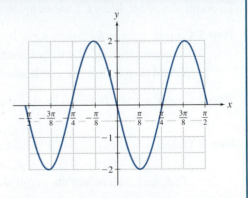

Figure 7.22 Graph for Example 7.7

SOLUTION

Because the graph oscillates from -2 to 2, the amplitude is 2. The graph repeats every $\dfrac{\pi}{2}$ units (but not over any shorter interval), so the period is $\dfrac{\pi}{2}$. The graph involves a shift of the sine function to the right by $\dfrac{\pi}{4}$, so we take the phase shift to be $\dfrac{\pi}{4}$. (The answer for the phase shift is not unique; for example, it could also be given as $\dfrac{9\pi}{4}$.)

TRY IT YOURSELF 7.7 **TRY IT YOURSELF 7.7** Brief answers provided at the end of the section.

In **Figure 7.23** you are given the graph of a function of the form $y = A \sin B(x - C)$ with $A > 0$. Identify the amplitude, period, and phase shift.

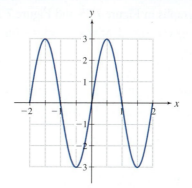

Figure 7.23 Graph for Try It Yourself 7.7

Even and Odd Functions

The cosine is an even function, and the sine is an odd function.

We can use what we know about graphing sine waves to illustrate important new trigonometric identities, such as $\sin(-x) = -\sin x$.

EXAMPLE 7.8   Using Graphs to Illustrate an Identity

a. Make the graph of $y = \sin(-x)$.

b. Make the graph of $y = -\sin x$.

c. What trigonometric identity is suggested by the graphs in parts a and b?

SOLUTION

a. For reference, the graph of $y = \sin x$ is displayed in **Figure 7.24**. To find the graph of $y = \sin(-x)$, we reflect the graph of $y = \sin x$ about the y-axis. The result is shown in **Figure 7.25**.

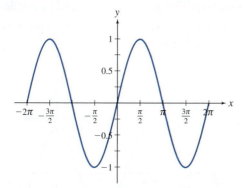

Figure 7.24 The graph of the sine

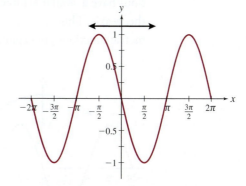

Figure 7.25 Reflecting the graph of the sine about the y-axis to find the graph of $y = \sin(-x)$

b. To obtain the graph of $y = -\sin x$, we reflect the graph of $y = \sin x$ about the x-axis. The result is shown in **Figure 7.26**.

c. The graphs in Figure 7.25 and Figure 7.26 are identical. This fact suggests the identity $\sin(-x) = -\sin x$. This is indeed an identity.

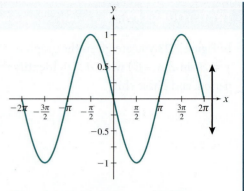

Figure 7.26 Reflecting the graph of the sine about the x-axis to find the graph of $y = -\sin x$

TRY IT YOURSELF 7.8   Brief answers provided at the end of the section.

Compare the graph of $y = \cos x$ with the graph of $y = \cos(-x)$. What trigonometric identity does the result suggest?

EXTEND YOUR REACH

a. There is exactly one function $y = f(x)$ with domain all real numbers that satisfies both identities $f(-x) = -f(x)$ and $f(-x) = f(x)$. Can you find that function?

b. Find a function that satisfies neither the identity $f(-x) = -f(x)$ nor the identity $f(-x) = f(x)$. *Suggestion*: Use a combination of sine and cosine.

Recall from Section 2.3 that a function f is called odd if $f(-x) = -f(x)$ for all x in the domain of f. Because of the identity $\sin(-x) = -\sin x$, the sine is an odd function. The graphs of odd functions have a particular symmetry property: The graph is unchanged when reflected about the x-axis and then about the y-axis or, equivalently, when rotated by $180°$ about the origin. This fact can be seen in **Figure 7.27**. Similarly, because of the identity $\cos(-x) = \cos x$, the cosine is even. The graphs of even functions have a similar symmetry property: The graph is unchanged by a reflection about the y-axis. This fact can be seen in **Figure 7.28**. Even and odd properties of trigonometric functions are explored further in Chapter 8.

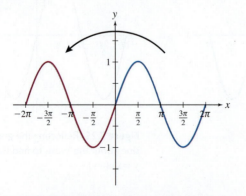

Figure 7.27 The odd function $y = \sin x$: Its graph is unchanged by rotation of $180°$ about the origin.

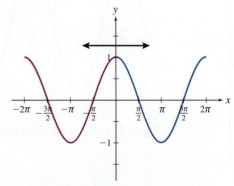

Figure 7.28 The even function $y = \cos x$: Its graph is unchanged by reflection about the y-axis.

CONCEPTS TO REMEMBER: The Sine Function

Defining property: The sine function $f(t) = \sin t$ is the signed distance from the trigonometric point $P(t)$ on the unit circle to the x-axis.

Domain: All real numbers.

Range: The interval $[-1, 1]$.

Period: The sine function has period 2π: For all x, $\sin(x + 2\pi) = \sin x$.

Symmetry: The sine function is odd—that is, $\sin(-x) = -\sin x$ for all x.

CONCEPTS TO REMEMBER: The Cosine Function

Defining property: The cosine function $f(t) = \cos t$ is the signed distance from the trigonometric point $P(t)$ on the unit circle to the y-axis.

Domain: All real numbers.

Range: The interval $[-1, 1]$.

Period: The cosine function has period 2π: For all x, $\cos(x + 2\pi) = \cos x$.

Symmetry: The cosine function is even—that is, $\cos(-x) = \cos x$ for all x.

The functions $y = \sin x$ and $y = \cos x$ do not have a limit as $x \to \infty$.

TRY IT YOURSELF ANSWERS

7.1 $\dfrac{1}{2}$

7.2

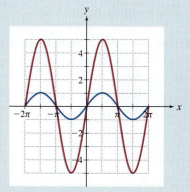

7.3 The period is 4π, and the amplitude is 2.

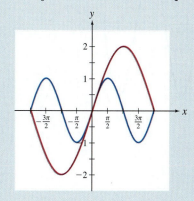

7.4 The amplitude is 2, the period is 6π, and the phase shift is $-\dfrac{\pi}{4}$.

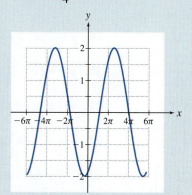

7.5 $A = 2, B = \dfrac{2}{3}, C = 1$

7.6 The amplitude is 4, the period is $\dfrac{2\pi}{3}$, and the phase shift is $\dfrac{\pi}{6}$.

7.7 The amplitude is 3, the period is 2, and one possible answer for the phase shift is 0.

7.8 The graph of $y = \cos(-x)$ is identical to the graph of $y = \cos x$, which suggests the (correct) identity $\cos(-x) = \cos x$.

EXERCISE SET 7.1

CHECK YOUR UNDERSTANDING

1. The period of the sine function is _____.

2. The period of the cosine function is _____.

3. For the trigonometric point $P(t)$, $\sin t$ can be thought of as:

   a. the directed distance to the x-axis.

   b. the directed distance to the y-axis.

   c. Either of the above.

   d. None of the above.

4. For the trigonometric point $P(t)$, $\cos t$ can be thought of as:

   a. the directed distance to the x-axis.

   b. the directed distance to the y-axis.

   c. Either of the above.

   d. None of the above.

5. For which values of x does the sine function reach a maximum, and what is the maximum value?

6. Consider the graph of the sine function on $[0, 2\pi]$. On what interval is the function decreasing?

7. For which values of x does the cosine function reach a maximum, and what is the maximum value?

8. The amplitude of a sine wave refers to _____.

9. What is the period of the function $y = \cos Bx$, $B > 0$?

10. The number C in $y = A \sin B(x - C)$ is called the:

    a. amplitude.   c. phase shift.

    b. period.   d. None of the above.

11. To say that the sine function is odd means:

    a. $\sin(-x) = \sin x$ for all x.

    b. $\sin(-x) = -\sin x$ for all x.

    c. $\sin x$ is zero for odd numbers x.

    d. None of the above.

SKILL BUILDING

Amplitude, period, and phase shift. In Exercises 12 through 24, identify the amplitude, period, and phase shift of the given function.

12. $y = 2 \cos 3x$

13. $y = 3 \sin 2x$

14. $y = -\sin \pi x$

15. $y = -2 \cos \pi x$

16. $y = \pi \sin\left(\dfrac{x}{3}\right)$

17. $y = -3 \cos\left(\dfrac{x}{\pi}\right)$

18. $y = 5 \sin\left(\dfrac{\pi x}{2}\right)$

19. $y = 3 \sin 2(x - \pi)$

20. $y = -2 \cos 3\left(x - \dfrac{\pi}{2}\right)$

21. $y = \pi \sin 2(x - 1)$

22. $y = 4 \cos \pi\left(x + \dfrac{\pi}{3}\right)$

23. $y = 3 \sin(2x - 1)$

24. $y = -2 \cos(\pi x + 2)$

Finding functions with given amplitude, phase shift, and period. In Exercises 25 through 28, find a function $y = A \sin B(x - C)$ with the given period, phase shift, and amplitude. In each case, there is more than one correct answer.

25. Period 2, phase shift 0, amplitude 4

26. Period π, phase shift 2, amplitude 1

27. Period 1, phase shift π, amplitude 3

28. Period $\dfrac{1}{2}$, phase shift 0, amplitude 2

Amplitude, period, and phase shift from graphs. In Exercises 29 through 34, you are given the graph of a function of the form $y = A \sin B(x - C)$ with $A > 0$. For each graph, identify the amplitude, period, and phase shift. (The answer for the phase shift is not unique.)

29.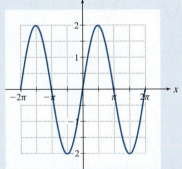

30.

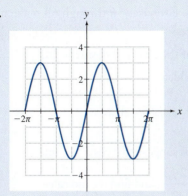

31.

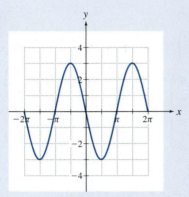

32.

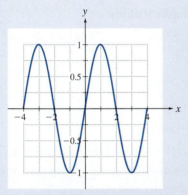

33.

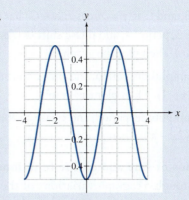

34.

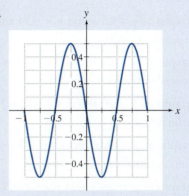

PROBLEMS

35. Graph of the sine using degrees. Sketch a graph of $y = \sin t$ if t is measured using degrees.

36. Graph of the cosine using degrees. Sketch a graph of $y = \cos t$ if t is measured using degrees.

Graphs of sine waves. In Exercises 37 through 45, provide the indicated information, and plot the given sine waves. Be sure to include at least one complete period.

37. Find the amplitude, period, and phase shift of $y = 2 \sin 2x$. Then plot the graph.

38. Find the amplitude, period, and phase shift of $y = 2 \cos 2x$. Then plot the graph.

39. Find the amplitude, period, and phase shift of $y = 3 \sin \pi(x - 1)$. Then plot the graph.

40. Find the amplitude, period, and phase shift of $y = 4 \cos(2\pi(x - 1))$. Then plot the graph.

41. Find A, B, and C if $y = A \sin B(x - C)$ has amplitude 2, period π, and phase shift $\dfrac{\pi}{2}$. (Assume that $A > 0$.) Then plot the graph.

42. Find A, B, and C if $y = A \cos B(x - C)$ has amplitude $\dfrac{1}{2}$, period 2, and phase shift 0. (Assume that $A > 0$.) Then plot the graph.

43. Find A, B, and C if $y = A \cos B(x - C)$ has amplitude 3, period 4, and phase shift 0. (Assume that $A > 0$.) Then plot the graph.

44. Find the amplitude, period, and phase shift of $y = \cos(2x - 1)$. Then plot the graph.

45. Find the amplitude, period, and phase shift of $y = 4 \sin(2x - \pi)$. Then plot the graph.

Transformations of sine and cosine functions. In Exercises 46 through 51, graph variations of the sine and cosine functions.

46. Plot the graphs of $y = \sin x$ and $y = -2 + \sin x$ on the same coordinate axes.

47. Plot the graphs of $y = \cos x$ and $y = 3 + \cos x$ on the same coordinate axes.

48. Plot the graphs of $y = \sin x$ and $y = \sin(-x)$ on the same coordinate axes.

49. Plot the graphs of $y = \sin x$ and $y = 1 + 3 \sin\left(\dfrac{x}{2}\right)$ on the same coordinate axes.

50. Plot the graphs of $y = \cos x$ and $y = -\cos(1 - x)$ on the same coordinate axes.

51. Plot the graphs of $y = \cos x$ and $y = 2 + \cos\left(\dfrac{x}{2} + 1\right)$ on the same coordinate axes.

Graphical evidence for identities. In Exercises 52 through 55, plot the indicated graphs and compare them. Then explain what trigonometric identity is suggested by your plots.

52. $y = \cos\left(x + \dfrac{\pi}{2}\right)$ and $y = -\sin x$

53. $y = \sin(x + \pi)$ and $y = -\sin x$

54. $y = \cos(x + \pi)$ and $y = -\cos x$

55. $y = \cos(x - \pi)$ and $y = -\cos x$

Using identities. Exercises 56 through 63 make use of the following facts:

- The period of the sine and cosine is 2π.
- The sine is odd, and the cosine is even.
- $\sin\left(x + \dfrac{\pi}{2}\right) = \cos x$ for all x.

For these exercises, suppose that $\cos x = \dfrac{1}{3}$ and $\cos\left(x + \dfrac{\pi}{2}\right) = -\dfrac{2\sqrt{2}}{3}$, and find the indicated function value.

56. $\sin\left(x + \dfrac{\pi}{2}\right)$

57. $\sin\left(-x + \dfrac{\pi}{2}\right)$

58. $\sin\left(-x - \dfrac{\pi}{2}\right)$

59. $\sin(x + \pi)$

60. $\sin(-x - \pi)$

61. $\sin\left(x - \dfrac{3\pi}{2}\right)$

62. $\sin\left(x + \dfrac{5\pi}{2}\right)$

63. $\cos(-x)$

📈 **Graphs.** In Exercises 64 through 69, plot the indicated graphs using a graphing utility.

64. $y = \sin\left(\dfrac{1}{x}\right)$

65. $y = \cos^2 x$

66. $y = \sin x + \cos x$

67. $y = \sin x - \cos x$

68. $y = \cos x$ and $y = x$ over the interval $[0, \pi]$. Then solve the equation $x = \cos x$, giving your answer correct to two decimal places.

69. $y = \sin x$ and $y = \ln x$ over the interval $[0, \pi]$. Then solve the equation $\ln x = \sin x$, giving your answer correct to two decimal places.

MODELS AND APPLICATIONS

70. **Phases of the moon.** Let W be the fraction of the width of the moon illuminated (as we observe it) as a function of time t in days. For example, when the moon is full we have $W = 1$. One model is

$$W = \frac{1}{2} + \frac{1}{2} \cos\left(\frac{2\pi t}{29.5}\right).$$ This model assumes that $t = 0$ corresponds to a full moon.

a. This model for W represents a sine wave that has been shifted vertically. State the period of the sine wave, and explain in terms of phases of the moon what this period means.

b. Use the model for W to determine how much of the width of the moon is illuminated 5 days after a full moon.

c. The following observation is taken from the website of the Astronomical Applications Department of the U.S. Naval Observatory:

> Although Full Moon occurs each month at a specific date and time, the Moon's disk may appear to be full for several nights in a row if it is clear. This is because the percentage of the Moon's disk that appears illuminated changes very slowly around the time of Full Moon (also around New Moon, but the Moon is not visible at all then).[1]

Explain this observation in terms of a graph of the function W and rates of change. For what phase(s) is it easiest for an untrained observer to assign a date for that phase precisely?

CHALLENGE EXERCISES FOR INDIVIDUALS OR GROUPS

71. Getting the graph of the cosine from the graph of the sine.

a. A new identity Figure 7.29 shows the relationship between the trigonometric points $P(t)$ and $P\left(t + \dfrac{\pi}{2}\right)$. Note that we simply rotate the triangle in the first quadrant counterclockwise to get the triangle in the second quadrant. This observation tells us that the first coordinate of $P(t)$ is the same as the second coordinate of $P\left(t + \dfrac{\pi}{2}\right)$. Use this fact to argue that

$$\cos t = \sin\left(t + \frac{\pi}{2}\right)$$

b. Graphing the cosine Explain how to use the identity established in part a to make the graph of the cosine from the graph of the sine.

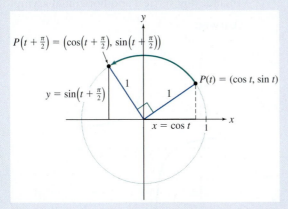

Figure 7.29 The trigonometric points $P(t)$ and $P\left(t + \dfrac{\pi}{2}\right)$: Moving from $P(t)$ to $P\left(t + \dfrac{\pi}{2}\right)$ rotates a triangle.

72. Even and odd functions. Figure 7.30 illustrates the fact that if the coordinates of the trigonometric point $P(t)$ are (x, y), then the coordinates of the trigonometric point $P(-t)$ are $(x, -y)$. Use this fact, along with the fact that $P(t) = (\cos t, \sin t)$, to show that $\sin(-t) = -\sin t$ and $\cos(-t) = \cos t$.

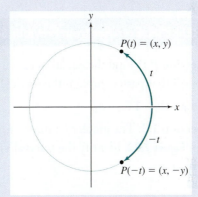

Figure 7.30 Comparing $P(t)$ and $P(-t)$

73. An interesting limit.

a. Plot the graph of $y = \dfrac{\sin x}{x}$ on the interval $[0, 50]$.

b. What does the graph indicate is the limit as $x \to \infty$ of $\dfrac{\sin x}{x}$?

74. No limiting value. Explain why the limit of $y = \sin x$ does not exist as $x \to \infty$. *Suggestion*: Does the graph of $y = \sin x$ approach any single value when x is large?

75. Rate of change of the sine function. In calculus you will show that the rate of change of the sine function is the cosine function.

---

[1] http://aa.usno.navy.mil/faq/docs/moon_phases.php

a. Verify that the cosine function is positive on intervals of the form (a, b) where the sine function increases. Also, verify that the cosine function is negative on intervals of the form (a, b) where the sine function decreases.

b. Suggest a candidate for the rate of change of the cosine function. (Think about what alteration you need to make to the sine function so that it is positive where the cosine increases and negative where the cosine decreases.)

76. An area function. In **Figure 7.31**, the sides A and B are fixed (constant), whereas the angle t is allowed to change. Let $Area(t)$ denote the area of the triangle when $0° < t < 180°$.

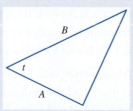

Figure 7.31 Calculating area using two sides and the included angle

a. Explain why $Area(t)$ is increasing for $0° < t < 90°$ and decreasing for $90° < t < 180°$. *Suggestion:* Apply the new area formula for triangles developed in Chapter 6 $\left(Area = \dfrac{1}{2} \, AB \sin t \right)$.

b. For what values of t is the rate of change of $Area(t)$ positive? Where is the rate of change negative, and where is it zero?

REVIEW AND REFRESH: Exercises from Previous Sections

77. From Section P.1: Find the equation of the line with slope 3 that passes through the point $(1, 2)$.

Answer: $y = 3x - 1$

78. From Section 1.3: The graph of a function $f(x)$ is shown in **Figure 7.32**. Identify the interval(s) where f is increasing.

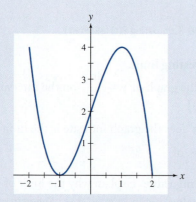

Figure 7.32 Graph for Exercise 78

Answer: $[-1, 1]$

79. From Section 3.1: Find the equation of the secant line for $f(x) = x^3$ for the interval $[0, 2]$.

Answer: $y = 4x$

80. From Section 4.1: How large must x be in order to make $\ln x$ at least 7?

Answer: $x = e^7$ or larger

81. From Section 4.1: Simplify the expression $\ln \dfrac{1}{\sqrt{e}}$.

Answer: $-\dfrac{1}{2}$

82. From Section 5.3: List all possible rational zeros of $P(x) = 3x^5 + x^2 - 4$.

Answer: $\pm 1, \ \pm 2, \ \pm 4, \pm \dfrac{1}{3}, \ \pm \dfrac{2}{3}, \ \pm \dfrac{4}{3}$

83. From Section 5.5: Plot the graph of $y = \dfrac{x-1}{x-2}$. Include any vertical and horizontal asymptotes.

Answer:

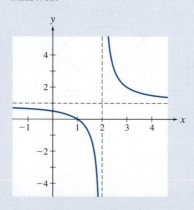

84. From Chapter 6: Find all values of x such that $(2x, x)$ lies on the unit circle.

Answer: $x = \pm \dfrac{1}{\sqrt{5}}$

85. From Chapter 6: Given that $\cos t = -\dfrac{1}{3}$ and $\tan t$ is negative, find the values of the remaining five trigonometric functions of t.

Answer: $\sin t = \dfrac{\sqrt{8}}{3}$, $\tan t = -\sqrt{8}$, $\cot t = -\dfrac{1}{\sqrt{8}}$, $\sec t = -3$, $\csc t = \dfrac{3}{\sqrt{8}}$

7.2 Graphs of Other Trigonometric Functions

> The graphs of the tangent, cotangent, secant, and cosecant are very different from the graphs of the sine and cosine.

7.1 Graphs of the Sine and Cosine

7.2 **Graphs of Other Trigonometric Functions**

In this section, you will learn to:

1. Use the period of the tangent, cotangent, secant, and cosecant to evaluate those functions.
2. Construct the graphs of the tangent, cotangent, secant, and cosecant.
3. Apply the standard transformations to the graphs of the tangent, cotangent, secant, and cosecant.
4. Identify the period of a transformed tangent, cotangent, secant, or cosecant.

Unlike the graphs of the sine and cosine, the graphs of the other four trigonometric functions, the tangent, cotangent, secant, and cosecant, all have vertical asymptotes. The asymptotes of the secant function have an important relationship to cartography.

The Mercator projection is a map made by placing a cylinder around the Earth, as shown in **Figure 7.33**, and projecting the Earth's surface onto the cylinder in a way that preserves the angles.[2] The cylinder is then rolled out flat to make the map. The course headings marked on the map align with the actual course a ship sets out to sail.

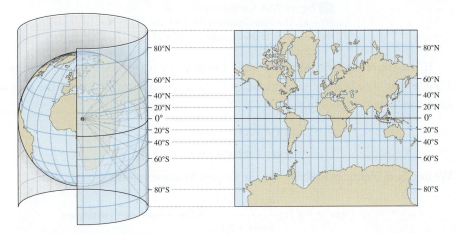

Figure 7.33 The Mercator projection

No flat map accurately represents the spherical Earth, so there are flaws in the Mercator projection. Features near the equator are accurately represented, but as we move to the north or south, features are distorted—magnified. And it turns out that the

---

[2]In modern mathematical terms, the projection is conformal.

amount of the magnification is determined by the secant function. The result is that Greenland appears on a flat map to be larger than Australia even though it is in fact less than one-third the size. The magnification is so great beyond 70° north or south that the usual Mercator projection is unusable at those latitudes. Finally, the poles are stretched infinitely, corresponding to vertical asymptotes of the secant function.

Periods of the Other Trigonometric Functions

> The periods of the secant and cosecant are the same as the period of the sine, but the periods of the tangent and cotangent are smaller.

As in the case of the sine and cosine, the periods of the other four trigonometric functions are a key feature of these functions as well as an essential tool for producing their graphs.

Because the secant, cosecant, tangent, and cotangent are constructed from the sine and cosine, it is natural to think that they inherit the period of 2π from the sine and cosine. This is correct for the secant and cosecant.

But, perhaps surprisingly, the periods of the tangent and cotangent turn out to be π. To see this, we need an identity, one for which **Figure 7.34** is useful.

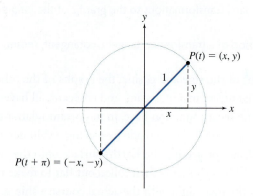

DF Figure 7.34 Comparing $P(t)$ with $P(t + \pi)$: The coordinates of $P(t + \pi)$ are the negatives of the coordinates of $P(t)$.

If the trigonometric point $P(t)$ has coordinates (x, y), then its opposite point $P(t + \pi)$ has coordinates $(-x, -y)$. Because these coordinates are in fact the sine and cosine, we have the following:

$$\sin(t + \pi) = -\sin t$$

$$\cos(t + \pi) = -\cos t$$

We use these identities to show that the tangent function repeats each π units:

$$\tan(t + \pi) = \frac{\sin(t + \pi)}{\cos(t + \pi)} \qquad \blacktriangleleft \text{ Use the definition of tangent.}$$

$$= \frac{-\sin t}{-\cos t} \qquad \blacktriangleleft \text{ Apply the preceding identity.}$$

$$= \frac{\sin t}{\cos t} \qquad \blacktriangleleft \text{ Simplify.}$$

$$= \tan t \qquad \blacktriangleleft \text{ Use the definition of tangent.}$$

The same result holds for the cotangent. The period of the tangent and the cotangent is π.

CONCEPTS TO REMEMBER: The Periods of Secant, Cosecant, Tangent, and Cotangent

For any x:

In terms of radians

Secant and cosecant are periodic with period 2π.

$$\sec(x + 2\pi) = \sec x$$

$$\csc(x + 2\pi) = \csc x$$

Tangent and cotangent are periodic with period π.

$$\tan(x + \pi) = \tan x$$

$$\cot(x + \pi) = \cot x$$

In terms of degrees

Secant and cosecant are periodic with period $360°$.

$$\sec(x° + 360°) = \sec x°$$

$$\csc(x° + 360°) = \csc x°$$

Tangent and cotangent are periodic with period $180°$.

$$\tan(x° + 180°) = \tan x°$$

$$\cot(x° + 180°) = \cot x°$$

EXAMPLE 7.9   Using the Period to Evaluate Trigonometric Functions

a. Use the period and what you know about special angles to evaluate $\tan\left(\dfrac{87\pi}{4}\right)$.

b. Use the period and what you know about special angles to find the value of $\cot 3720°$.

SOLUTION

a. Because $\dfrac{87\pi}{4} = 21\pi + \dfrac{3\pi}{4}$ and the tangent has period π, we have

$$\tan\left(\frac{87\pi}{4}\right) = \tan\left(21\pi + \frac{3\pi}{4}\right) = \tan\left(\frac{3\pi}{4}\right)$$

Note that the trigonometric point $P\left(\dfrac{3\pi}{4}\right)$ lies in the second quadrant and has reference angle $\dfrac{\pi}{4}$. Because $\tan\left(\dfrac{\pi}{4}\right) = 1$ and the tangent is negative in the second quadrant, we conclude that

$$\tan\left(\frac{87\pi}{4}\right) = \tan\left(\frac{3\pi}{4}\right) = -1$$

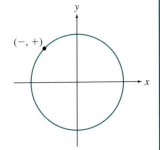

b. Because $3720 = 20 \times 180 + 120$ and the cotangent has period $180°$, we have

$$\cot 3720° = \cot(20 \times 180° + 120°) = \cot 120°$$

Note that $P(120°)$ lies in the second quadrant and has reference angle $60°$. For this reference angle, we find

$$\cot 60° = \frac{1}{\sqrt{3}}$$

We attach a minus sign to get the cotangent of $120°$:

$$\cot 3720° = \cot 120° = -\frac{1}{\sqrt{3}}$$

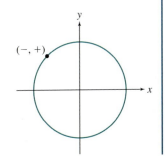

Find the exact value of $\tan\left(-\dfrac{55\pi}{3}\right)$.

Graphs of the Tangent and Cotangent

> The shorter periods of the tangent and cotangent aid in the production of their graphs.

To begin, we focus on a single period of the tangent function, from $x = -\dfrac{\pi}{2}$ to $x = \dfrac{\pi}{2}$. Because $\sin 0 = 0$ and $\cos x$ is zero at both $x = -\dfrac{\pi}{2}$ and $x = \dfrac{\pi}{2}$, the tangent function $\tan x = \dfrac{\sin x}{\cos x}$ has features similar to the rational function

$$y = \frac{-x}{\left(x - \dfrac{\pi}{2}\right)\left(x + \dfrac{\pi}{2}\right)}$$

In particular, the graph of the tangent function has vertical asymptotes at both $x = -\dfrac{\pi}{2}$ and $x = \dfrac{\pi}{2}$. Note how these asymptotes are indicated in **Table 7.3**, which shows approximate values.

Table 7.3 Approximate Values of Tangent Between $-\dfrac{\pi}{2}$ **and** $\dfrac{\pi}{2}$

x	$-\dfrac{\pi}{2}$	−1.57	−1.5	−1.2	−0.9	−0.6	−0.3	0	0.3	0.6	0.9	1.2	1.5	1.57	$\dfrac{\pi}{2}$
$\tan x$	undefined	−1255.77	−14.10	−2.57	−1.26	−0.68	−0.31	0	0.31	0.68	1.26	2.57	14.10	1255.77	undefined

This table is used to make the graph over a single period of the tangent function shown in **Figure 7.35**. Because the tangent has period π, we can use copies of the graph in Figure 7.35 to make the entire graph of the tangent. The result is shown in **Figure 7.36**.

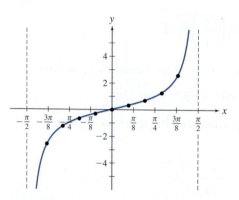

Figure 7.35 The graph of a single period of the tangent

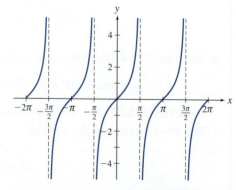

Figure 7.36 Using periodicity: Copies of a single period of the graph are used to make the graph of the tangent function.

EXAMPLE 7.10 The Graph of the Cotangent

Using Table 7.3 as an example, generate a table of values for the cotangent on the interval $(0, \pi)$. Use the table you generated to make the graph of the cotangent on the interval $(0, \pi)$. Then use copies of this graph and periodicity to make the graph of the cotangent.

SOLUTION

Because $\cos \dfrac{\pi}{2} = 0$ and the sine function is 0 at both $x = 0$ and $x = \pi$, the cotangent function $\cot x = \dfrac{\cos x}{\sin x}$ has features similar to the rational function

$$y = \frac{x - \dfrac{\pi}{2}}{x(x - \pi)}$$

on the interval $(0, \pi)$. In particular, the graph of the cotangent has vertical asymptotes at $x = 0$ and $x = \pi$.

Table 7.4 gives approximate values.

Table 7.4 Approximate Values of Cotangent Between 0 and π

x	0	0.1	0.4	0.7	1.0	1.3	$\dfrac{\pi}{2}$	1.6	1.9	2.2	2.5	2.8	3.1	π
$\cot x$	undefined	9.97	2.37	1.19	0.64	0.28	0	−0.03	−0.34	−0.73	−1.34	−2.81	−24.03	undefined

Plotting these points yields the graph in **Figure 7.37**. Because the cotangent has period π, we can use copies of this graph to make the entire graph. See **Figure 7.38**.

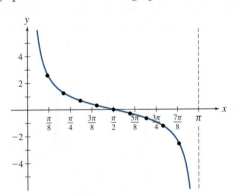

Figure 7.37 The graph of the cotangent on a single period $(0, \pi)$

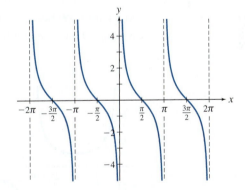

Figure 7.38 Using periodicity: Copies of a single period of the graph are used to make the graph of the cotangent function.

TRY IT YOURSELF 7.10 Brief answers provided at the end of the section.

For which values of x is the cotangent equal to 0? For which values of x does the graph of the cotangent have asymptotes?

The graphs of the tangent and cotangent have no local maxima and no local minima.

Graphs of the Secant and Cosecant

> Zeros of the sine and cosine correspond to asymptotes of the secant and cosecant.

We proceed as with the tangent and cotangent by focusing on a single period of the cosecant, namely $(0, \pi) \cup (\pi, 2\pi)$. The sine function is 0 at $x = 0$, $x = \pi$, and $x = 2\pi$, so the cosecant, the reciprocal of the sine, is undefined at these values. The graph of the cosecant has some of the same features as the graph of the rational function

$$y = \frac{1}{x(x - \pi)(x - 2\pi)}$$

In particular, it has vertical asymptotes at $x = 0$, $x = \pi$, and $x = 2\pi$.

Table 7.5 gives values for the cosecant using special angles between 0 and 2π.

Table 7.5 Values of Cosecant Between 0 and 2π

x	0	$\dfrac{\pi}{6}$	$\dfrac{\pi}{4}$	$\dfrac{\pi}{3}$	$\dfrac{\pi}{2}$	$\dfrac{2\pi}{3}$	$\dfrac{3\pi}{4}$	$\dfrac{5\pi}{6}$	π	$\dfrac{7\pi}{6}$	$\dfrac{5\pi}{4}$	$\dfrac{4\pi}{3}$	$\dfrac{3\pi}{2}$	$\dfrac{5\pi}{3}$	$\dfrac{7\pi}{4}$	$\dfrac{11\pi}{6}$	2π
csc x	undefined	2	$\sqrt{2}$	$\dfrac{2}{\sqrt{3}}$	1	$\dfrac{2}{\sqrt{3}}$	$\sqrt{2}$	2	undefined	-2	$-\sqrt{2}$	$-\dfrac{2}{\sqrt{3}}$	-1	$-\dfrac{2}{\sqrt{3}}$	$-\sqrt{2}$	-2	undefined
Approximate decimal value	undefined	2	1.41	1.15	1	1.15	1.41	2	undefined	-2	-1.41	-1.15	-1	-1.15	-1.41	-2	undefined

This table is used to make the graph of the cosecant over a single period shown in **Figure 7.39**. Because the cosecant has period 2π, we can use copies of this graph to produce the entire graph. See **Figure 7.40**.

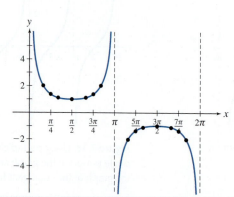

Figure 7.39 The graph of the cosecant over a single period

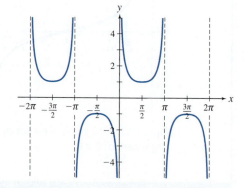

Figure 7.40 Using periodicity: Copies of a single period of the graph are used to make the graph of the cosecant function.

EXAMPLE 7.11 The Graph of the Secant Function

a. Use special angles to make a table of values for the secant over the single period $\left(\dfrac{\pi}{2}, \dfrac{3\pi}{2}\right) \cup \left(\dfrac{3\pi}{2}, \dfrac{5\pi}{2}\right)$.

b. Use the table of values to make the graph of the secant over a single period.

c. Use the graph over a single period and periodicity to make the graph of the secant function.

SOLUTION

a. Using special angles between $\dfrac{\pi}{2}$ and $\dfrac{5\pi}{2}$, we construct **Table 7.6**.

As the table suggests, over this span the secant function has vertical asymptotes at $\dfrac{\pi}{2}$, $\dfrac{3\pi}{2}$, and $\dfrac{5\pi}{2}$.

Table 7.6 Values of Secant Between $\dfrac{\pi}{2}$ and $\dfrac{5\pi}{2}$

x	$\dfrac{\pi}{2}$	$\dfrac{2\pi}{3}$	$\dfrac{3\pi}{4}$	$\dfrac{5\pi}{6}$	π	$\dfrac{7\pi}{6}$	$\dfrac{5\pi}{4}$	$\dfrac{4\pi}{3}$	$\dfrac{3\pi}{2}$	$\dfrac{5\pi}{3}$	$\dfrac{7\pi}{4}$	$\dfrac{11\pi}{6}$	2π	$\dfrac{13\pi}{6}$	$\dfrac{9\pi}{4}$	$\dfrac{7\pi}{3}$	$\dfrac{5\pi}{2}$
sec x	undefined	-2	$-\sqrt{2}$	$-\dfrac{2}{\sqrt{3}}$	-1	$-\dfrac{2}{\sqrt{3}}$	$-\sqrt{2}$	-2	undefined	2	$\sqrt{2}$	$\dfrac{2}{\sqrt{3}}$	1	$\dfrac{2}{\sqrt{3}}$	$\sqrt{2}$	2	undefined
Approximate decimal value	undefined	-2	-1.41	-1.15	-1	-1.15	-1.41	-2	undefined	2	1.41	1.15	1	1.15	1.41	2	undefined

b. We use these values to make the graph of the secant over a single period shown in **Figure 7.41**.

c. The completed graph is in **Figure 7.42**.

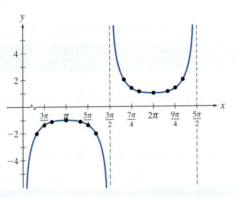

Figure 7.41 Part b: the graph of the secant over a single period

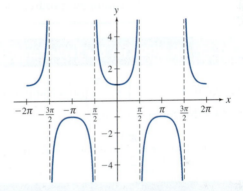

Figure 7.42 Part c: the completed graph of the secant function

TRY IT YOURSELF 7.11 Brief answers provided at the end of the section.

For which values of x does the graph of the secant have vertical asymptotes?

Unlike the tangent and cotangent, the secant and cosecant have many local maxima and minima. The secant has a local maximum value of -1 at

$$x = \pi + 2k\pi \text{ for all integers } k = \ldots, -3\pi, -\pi, \pi, 3\pi, \ldots$$

and a local minimum value of 1 at

$$x = 2k\pi \text{ for all integers } k = \ldots, -4\pi, -2\pi, 0, 2\pi, 4\pi, \ldots$$

The cosecant has a local maximum value of -1 at

$$x = -\dfrac{\pi}{2} + 2k\pi \text{ for all integers } k = \ldots, -\dfrac{5\pi}{2}, -\dfrac{\pi}{2}, \dfrac{3\pi}{2}, \dfrac{7\pi}{2}, \ldots$$

and a local minimum value of 1 at

$$x = \dfrac{\pi}{2} + 2k\pi \text{ for all integers } k = \ldots, -\dfrac{7\pi}{2}, -\dfrac{3\pi}{2}, \dfrac{\pi}{2}, \dfrac{5\pi}{2}, \ldots$$

See **Figure 7.43** and **Figure 7.44**.

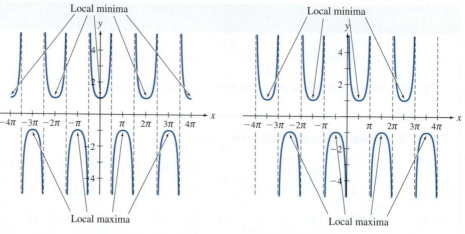

Figure 7.43 Local maximum and minimum values of the secant

Figure 7.44 Local maximum and minimum values of the cosecant

Transformations of Other Trigonometric Functions

Shifting and stretching for graphs of the secant, cosecant, tangent, and cotangent produce new graphs.

EXAMPLE 7.12 Identities for the Secant and Cosecant

a. Make the graph of $y = \csc\left(x + \dfrac{\pi}{2}\right)$, and compare it with the graph of $y = \sec x$. What trigonometric identity does this comparison suggest?

b. Use the fact that $\sin\left(x + \dfrac{\pi}{2}\right) = \cos x$ to give an algebraic derivation of the identity from part a.

SOLUTION

a. To get the graph of $y = \csc\left(x + \dfrac{\pi}{2}\right)$, we shift the graph of $y = \csc x$ by $\dfrac{\pi}{2}$ units to the left, as shown in **Figure 7.45**. The result is a graph that is identical to that of the secant shown in **Figure 7.46**. This suggests the identity $\csc\left(x + \dfrac{\pi}{2}\right) = \sec x$.

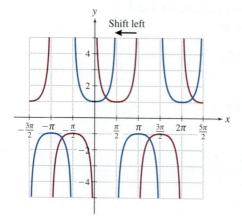

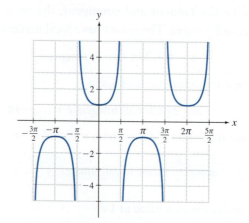

Figure 7.45 The graph of $y = \csc x$ (red) shifted to the left by $\dfrac{\pi}{2}$ units to obtain the graph of $y = \csc\left(x + \dfrac{\pi}{2}\right)$

Figure 7.46 The graph of the secant

b. We use the reciprocal relationships among the trigonometric functions:

$$\csc\left(x+\frac{\pi}{2}\right)=\frac{1}{\sin\left(x+\dfrac{\pi}{2}\right)} \qquad \blacktriangleleft \text{ The cosecant is the reciprocal of the sine.}$$

$$=\frac{1}{\cos x} \qquad \blacktriangleleft \text{ Use the given identity.}$$

$$=\sec x \qquad \blacktriangleleft \text{ The secant is the reciprocal of the cosine.}$$

TRY IT YOURSELF 7.12 Brief answers provided at the end of the section.

Make the graph of $y = \sec\left(x-\dfrac{\pi}{2}\right)$, and compare it with the graph of $y = \csc x$. What trigonometric identity does this suggest? Verify this identity using the fact that $\cos\left(x-\dfrac{\pi}{2}\right) = \sin x$.

We saw in Section 7.1 that multiplying the variable x by $B \neq 0$ divides the period of $\sin x$ and $\cos x$ by $|B|$. The other trigonometric functions have the same property.

CONCEPTS TO REMEMBER: Periods for the Transformed Secant, Cosecant, Tangent, and Cotangent Functions

- The period of $y = \sec Bx$ and $y = \csc Bx$ is $\dfrac{2\pi}{|B|}$.

- The period of $y = \tan Bx$ and $y = \cot Bx$ is $\dfrac{\pi}{|B|}$.

We handle stretching and compressing for graphs of the secant, cosecant, tangent, and cotangent in much the same way as we did for sine waves.

EXAMPLE 7.13 Horizontal and Vertical Stretching and Compressing

a. Identify the period of $y = \sec 2x$. Plot the graphs of $y = \sec x$ and $y = \sec 2x$ on the same coordinate axes.

b. On the same coordinate axes, make the graphs of $y = \cot x$ and $y = 2\cot x$.

SOLUTION

a. The period of $y = \sec 2x$ is $\dfrac{2\pi}{2} = \pi$. We find the graph of $y = \sec 2x$ via a horizontal compression of the graph of $y = \sec x$ by a factor of 2. The result is shown in **Figure 7.47**.

b. A vertical stretch by a factor of 2 of the graph of $y = \cot x$ produces the graph of $y = 2\cot x$. The graphs are shown in **Figure 7.48**.

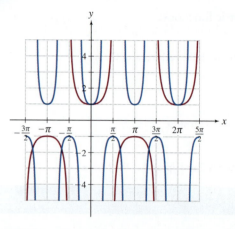

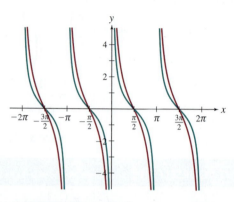

Figure 7.47 Part a: The graph of $y = \sec x$ (red) is compressed horizontally to make the graph of $y = \sec 2x$ (blue).

Figure 7.48 Part b: The graph of $y = \cot x$ (green) is stretched vertically to make the graph of $y = 2 \cot x$ (red).

TRY IT YOURSELF 7.13 Brief answers provided at the end of the section.

On the same coordinate axes, plot the graphs of $y = \tan x$ and $y = -\tan x$.

EXTEND YOUR REACH   Both the sine and the cosine have an amplitude that can be adjusted. Can the idea of amplitude be applied to any of the other four trigonometric functions? If so, how?

As we explained in Section 2.3, adding a number to, or subtracting a number from, a function has the effect of shifting the graph up or down. In the next example, we apply this fact to the transformed trigonometric functions.

EXAMPLE 7.14 Horizontal and Vertical Shifts

Transform the graph of $y = \tan x$ to make the graph of $y = 5 + \tan\left(x - \dfrac{\pi}{2}\right)$.

SOLUTION

We make the required graph in two steps. First we find the graph of $y = \tan\left(x - \dfrac{\pi}{2}\right)$ by shifting the graph of $y = \tan x$ to the right by $\dfrac{\pi}{2}$ units. This is shown in **Figure 7.49**.

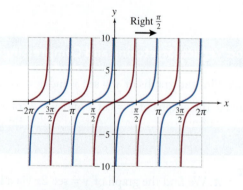

Figure 7.49 First step: Shift the graph of $y = \tan x$ (red) to the right to make the graph of $y = \tan\left(x - \dfrac{\pi}{2}\right)$ (blue).

The next step is to shift the graph of $y = \tan\left(x - \dfrac{\pi}{2}\right)$ up 5 by units to make the graph of $y = 5 + \tan\left(x - \dfrac{\pi}{2}\right)$. This is shown in **Figure 7.50**.

Finally, we show the two graphs, $y = \tan x$ and $y = 5 + \tan\left(x - \dfrac{\pi}{2}\right)$, together in **Figure 7.51**.

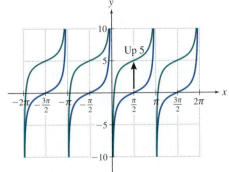

Figure 7.50 Second step: Shift the graph of $y = \tan\left(x - \dfrac{\pi}{2}\right)$ up to make the graph of $y = 5 + \tan\left(x - \dfrac{\pi}{2}\right)$ (green).

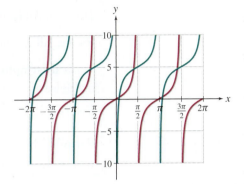

Figure 7.51 The graph of $y = \tan x$ (red) along with the graph of $y = 5 + \tan\left(x - \dfrac{\pi}{2}\right)$ (green)

TRY IT YOURSELF 7.14 Brief answers provided at the end of the section.

On the same axes, plot the graphs of $y = \tan x$ and $y = \tan\left(x + \dfrac{\pi}{2}\right) - 5$.

EXTEND YOUR REACH

a. Shift the graph of the tangent function to make the graph of $y = \tan\left(x - \dfrac{\pi}{2}\right)$.

b. Use the graph in part a to make the graph of $y = -\tan\left(x - \dfrac{\pi}{2}\right)$.

c. Compare the graph in part b with the graph of the cotangent. Does the comparison suggest a trigonometric identity?

CONCEPTS TO REMEMBER: The Tangent Function

Defining property: The tangent function is the sine divided by the cosine:

$$\tan x = \frac{\sin x}{\cos x}$$

Domain: All real numbers except odd integer multiples of $\dfrac{\pi}{2}$.

Range: All real numbers.

Vertical asymptotes: The tangent function has vertical asymptotes at odd integer multiples of $\dfrac{\pi}{2}$, or

$$\ldots, -\frac{3\pi}{2}, -\frac{\pi}{2}, \frac{\pi}{2}, \frac{3\pi}{2}, \ldots$$

Period: The tangent function has period π: For all x in the domain, $\tan(x + \pi) = \tan x$.

CONCEPTS TO REMEMBER: The Cotangent Function

Defining property: The cotangent function is the cosine divided by the sine:

$$\cot x = \frac{\cos x}{\sin x}$$

Domain: All real numbers except integer multiples of π.

Range: All real numbers.

Vertical asymptotes: The cotangent function has vertical asymptotes at integer multiples of π, or

$$\ldots, -2\pi, \ -\pi, 0, \ \pi, \ 2\pi, \ldots$$

Period: The cotangent function has period π: For all x in the domain, $\cot(x + \pi) = \cot x$.

CONCEPTS TO REMEMBER: The Secant Function

Defining property: The secant function is the reciprocal of the cosine function:

$$\sec x = \frac{1}{\cos x}$$

Domain: All real numbers except odd integer multiples of $\dfrac{\pi}{2}$.

Range: $(-\infty, -1] \cup [1, \infty)$.

Vertical asymptotes: The secant function has vertical asymptotes at odd integer multiples of $\dfrac{\pi}{2}$, or

$$\ldots, -\frac{3\pi}{2}, \ -\frac{\pi}{2}, \ \frac{\pi}{2}, \ \frac{3\pi}{2}, \ldots$$

Period: The secant function has period 2π: For all x in the domain, $\sec(x + 2\pi) = \sec x$.

CONCEPTS TO REMEMBER: The Cosecant Function

Defining property: The cosecant function is the reciprocal of the sine function:

$$\csc x = \frac{1}{\sin x}$$

Domain: All real numbers except integer multiples of π.

Range: $(-\infty, -1] \cup [1, \infty)$.

Vertical asymptotes: The cosecant function has vertical asymptotes at integer multiples of π, or

$$\ldots, -2\pi, -\pi, 0, \pi, 2\pi, \ldots$$

Period: The cosecant function has period 2π: For all x in the domain, $\csc(x + 2\pi) = \csc x$.

TRY IT YOURSELF ANSWERS

7.9 $-\sqrt{3}$

7.10 The graph is 0 at $x = \dfrac{\pi}{2} + k\pi$, where k is any integer. There are asymptotes at $x = k\pi$, where k is any integer:

$$x = \dots, -2\pi, \ -\pi, \ 0, \ \pi, \ 2\pi, \dots$$

7.11 Asymptotes occur at $x = \dfrac{\pi}{2} + k\pi$ for all integers k:

$$x = \dots, -\frac{3\pi}{2}, -\frac{\pi}{2}, \frac{\pi}{2}, \frac{3\pi}{2}, \dots$$

7.12 We shift the graph of $y = \sec x$ by $\dfrac{\pi}{2}$ units to the right to get the graph of $y = \sec\left(x - \dfrac{\pi}{2}\right)$.

The resulting graph is identical to the graph of the cosecant, which suggests the identity

$$\sec\left(x - \frac{\pi}{2}\right) = \csc x. \text{ We verify this as follows:}$$

$$\sec\left(x - \frac{\pi}{2}\right) = \frac{1}{\cos\left(x - \dfrac{\pi}{2}\right)} = \frac{1}{\sin x} = \csc x$$

7.13 In the figure, $y = \tan x$ is shown in red, and $y = -\tan x$ is in blue.

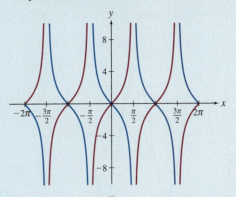

7.14

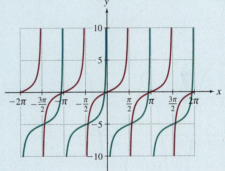

EXERCISE SET 7.2

CHECK YOUR UNDERSTANDING

1. What is the period of the tangent function?

2. What is the period of the cotangent function?

3. What is the period of the secant function?

4. What is the period of the cosecant function?

5. Where are the vertical asymptotes of the tangent function?

6. Where are the vertical asymptotes of the cotangent function?

7. Where are the vertical asymptotes of the secant function?

8. Where are the vertical asymptotes of the cosecant function?

9. Which of the six trigonometric functions have no local maxima and no local minima?

10. The tangent and secant have vertical asymptotes where:

   a. the cosine function is 0.

   b. the cosecant function reaches a local maximum or minimum.

   c. the cotangent function is 0.

   d. All of the above.

   e. None of the above.

SKILL BUILDING

Identifying graphs. The graphs shown in Exercises 11 through 16 are graphs of the six trigonometric functions. Identify the function depicted in each graph.

11.

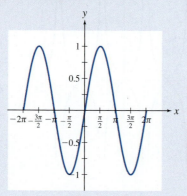

12.

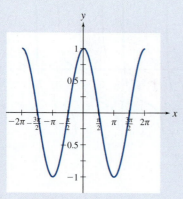

13.

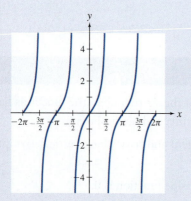

14.

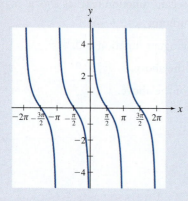

15.

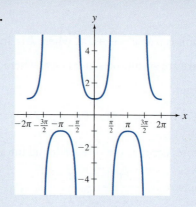

16.

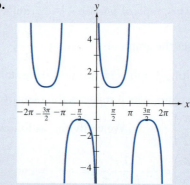

Finding periods. In Exercises 17 through 20, find the period of the given function.

17. $y = \tan 2x$

18. $y = \sec \pi x$

19. $\cot\left(\dfrac{x}{2}\right)$

20. $\csc\left(\dfrac{\pi x}{2}\right)$

Identifying transformations of trigonometric graphs. The graphs in Exercises 21 through 30 are shifted, stretched, compressed, or reflected versions of the graphs of the secant, cosecant, tangent, or cotangent. Identify a function that made the graph. Note that there is more than one correct answer.

21.

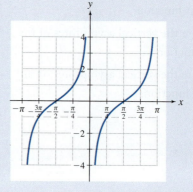

22.

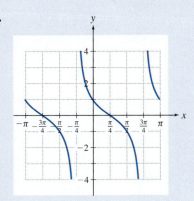

23.

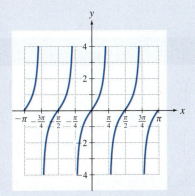

24.

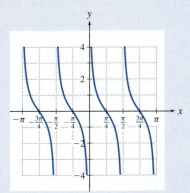

25.

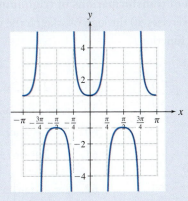

26.

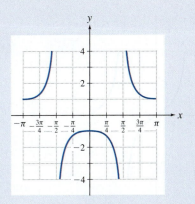

27.

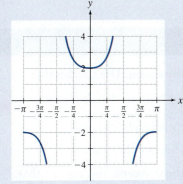

28.

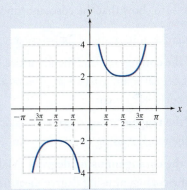

29.

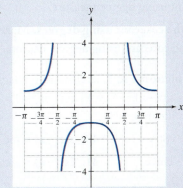

30.

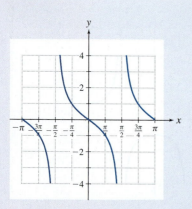

PROBLEMS

31. Period of the secant. Use the fact that
$\sec x = \dfrac{1}{\cos x}$ to show that the secant
function repeats each 2π units:
$$\sec (x + 2\pi) = \sec x.$$

32. Period of the cosecant. Use the fact that
$\csc x = \dfrac{1}{\sin x}$ to show that the cosecant function
repeats each 2π units: $\csc (x + 2\pi) = \csc x$.

33. Period of the cotangent. Use the identities
$\sin (x + \pi) = -\sin x$ and $\cos (x + \pi) = -\cos x$ to
show that $\cot (x + \pi) = \cot x$.

34. Graph of the tangent using degrees. Make a
graph of the tangent using degree measure.

35. Graph of the cotangent using degrees. Make a
graph of the cotangent using degree measure.

36. Graph of the secant using degrees. Make a graph
of the secant using degree measure.

37. Graph of the cosecant using degrees. Make a
graph of the cosecant using degree measure.

Even and odd functions. In Exercises 38 through 41,
use the identities $\sin(-x) = -\sin x$ and $\cos(-x) = \cos x$
to show the given identity.

38. $\cot(-x) = -\cot x$ **40.** $\sec(-x) = \sec x$

39. $\tan(-x) = -\tan x$ **41.** $\csc(-x) = -\csc x$

Using the period to calculate trigonometric functions.
In Exercises 42 through 45, find the indicated values.
Use special angles, together with the facts that π is the
period of the tangent and cotangent and that 2π is the
period of the secant and cosecant.

42. $\cot\left(\dfrac{11\pi}{3}\right)$ **44.** $\sec\left(-\dfrac{11\pi}{4}\right)$

43. $\tan\left(\dfrac{17\pi}{3}\right)$ **45.** $\csc\left(\dfrac{37\pi}{6}\right)$

Verifying periods. In Exercises 46 through 49, use your
knowledge of the periods of the trigonometric functions
to establish the indicated period. Assume that $B > 0$.

46. Show that the period of $y = \tan Bx$ is $\dfrac{\pi}{B}$.

47. Show that the period of $y = \cot Bx$ is $\dfrac{\pi}{B}$.

48. Show that the period of $y = \sec Bx$ is $\dfrac{2\pi}{B}$.

49. Show that the period of $y = \csc Bx$ is $\dfrac{2\pi}{B}$.

**Making graphs of transformations of trigonometric
functions.** In Exercises 50 through 64, sketch the graph
of the indicated function.

50. $y = \tan 3x$ **57.** $y = \sec 2x$

51. $y = \cot\left(x + \dfrac{\pi}{2}\right)$ **58.** $y = 2 \csc 2x$

59. $y = -\cot(2x)$

52. $y = 3 \csc x$ **60.** $y = 1 + 2 \sec x$

53. $y = \sec\left(\dfrac{1}{2}x\right)$ **61.** $y = 2 + \tan(x - 1)$

54. $y = \tan\left(x + \dfrac{\pi}{2}\right)$ **62.** $y = -3 \tan\left(x - \dfrac{\pi}{4}\right)$

55. $y = \cot 2x$ **63.** $y = \cot 3\pi(x - 1)$

56. $y = \csc(x + \pi)$ **64.** $y = 2 \csc(\pi x)$

Graphs and identities. Exercises 65 through 67 involve trigonometric identities.

65. a. Plot the graphs of $y = \sec\left(x + \dfrac{\pi}{2}\right)$ and

   $y = -\csc x$. Which trigonometric identity do the graphs suggest?

   b. Use the fact that $\cos\left(x + \dfrac{\pi}{2}\right) = -\sin x$ to verify the identity proposed in part a.

66. a. Plot the graphs of $y = \tan\left(x + \dfrac{\pi}{2}\right)$ and

   $y = -\cot x$. Which trigonometric identity do the graphs suggest?

   b. Use the facts that $\sin\left(x + \dfrac{\pi}{2}\right) = \cos x$ and

   $\cos\left(x + \dfrac{\pi}{2}\right) = -\sin x$ to verify the identity proposed in part a.

67. a. Plot the graphs of $y = \sec(-x)$ and $y = \sec x$. Which trigonometric identity do the graphs suggest?

   b. Use the fact that $\cos(-x) = \cos x$ to verify the identity proposed in part a.

Graphs and equations. In Exercises 68 through 70, use a graph to solve an equation or find a limit. Show all appropriate graphs.

68. Plot the graphs of $y = \sec x$ and $y = 3$ on the

   same coordinate axes over the interval $\left[0, \dfrac{\pi}{2}\right]$. Use

   your graph to solve the equation $\sec x = 3$ over that interval. Report your answer correct to two decimal places.

69. Plot the graphs of $y = \sec x$ and $y = \cot x$ on the

   same coordinate axes on the interval $\left[0, \dfrac{\pi}{2}\right]$. Use

   your graph to solve the equation $\sec x = \cot x$ over that interval. Report your answer correct to two decimal places.

70. a. Plot the graph of $y = \tan\left(\dfrac{1}{x}\right)$ on the interval $[1, 50]$.

   b. What does the graph indicate is the limit as

   $x \to \infty$ of $y = \tan\left(\dfrac{1}{x}\right)$?

MODELS AND APPLICATIONS

71. The Mercator projection. Under the Mercator projection, objects that are at a latitude of L degrees are stretched by a factor of $k = \sec L$.

   a. By what factor are objects at a latitude of $30°$ stretched? Round your answer to two decimal places.

   b. Look up the latitude of your home town. What is the stretching factor for your home town?

   c. At what latitude are objects stretched to twice their actual size by the Mercator projection? *Suggestion:* The answer is a special angle.

CHALLENGE EXERCISES FOR INDIVIDUALS OR GROUPS

72. More on the Mercator projection. A point on the Earth's surface is determined by latitude and longitude, both measured in degrees:

$$(\text{Longitude, Latitude}) = (\lambda, \phi)$$

On the Mercator map, we choose the horizontal line corresponding to the equator as the x-axis and the vertical line corresponding to the prime meridian ($\lambda = 0$) as the y-axis. Then the Mercator projection maps points (λ, ϕ) on the surface of the Earth to points (x, y) on the flat map using the correspondence

$$x = \lambda$$
$$y = \ln(\sec\phi + \tan\phi)$$

Our goal is to demonstrate the vertical stretching that occurs with the Mercator projection.

a. Plot the graph of $y = \ln(\sec x + \tan x)$ for $0° \le x < 90°$. (Be sure your calculator is set to degree mode.)

b. Explain in your own words how vertical stretching relates to latitude.

73. Graph of the cotangent by shifting and reflecting. In **Figure 7.52** we recall the figure we used to show that $\sin\left(t + \dfrac{\pi}{2}\right) = \cos t$. (See the exercises in Section 7.1.) This time we focus on different sides of the triangle.

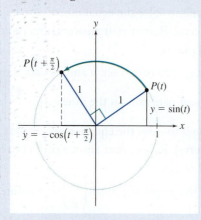

Figure 7.52 Comparing $P(t)$ with $P\left(t + \dfrac{\pi}{2}\right)$

a. Use Figure 7.52 to show that
$$\cos\left(t + \frac{\pi}{2}\right) = -\sin t.$$

b. Use part a along with the identity
$$\sin\left(t + \frac{\pi}{2}\right) = \cos t \text{ to show that}$$
$$-\tan\left(t + \frac{\pi}{2}\right) = \cot t.$$

c. Use part b to show that the graph of the cotangent can be obtained from the graph of the tangent by first shifting to the left by $\dfrac{\pi}{2}$ units and then reflecting about the horizontal axis.

74. Rate of change of the secant function. There are infinitely many values of x where the rate of change of $y = \sec x$ is zero. Identify these values.

REVIEW AND REFRESH: Exercises from Previous Sections

75. From Section P.1: Find the equation of the circle with center $(1, -1)$ and radius 3.

Answer: $(x-1)^2 + (y+1)^2 = 9$

76. From Section P.3: Solve the inequality $x^3 + 4x < 5x^2$.

Answer: $(-\infty, 0) \cup (1, 4)$

77. From Section 1.4: Find the limit as $x \to \infty$ of
$$y = \frac{6x-1}{2x+5}.$$

Answer: 3

78. From Section 2.2: Find the inverse of the function
$$f(x) = 1 + \frac{x}{x-3}.$$

Answer: $f^{-1}(x) = \dfrac{3x-3}{x-2}$

79. From Section 4.3: Solve the equation $5^x = 7$.

Answer: $x = \dfrac{\ln 7}{\ln 5}$

80. From Chapter 6: Suppose $\sin t = \dfrac{3}{5}$ and the tangent of t is negative. Find the remaining five trigonometric functions of t.

Answer: $\cos t = -\dfrac{4}{5}$, $\tan t = -\dfrac{3}{4}$, $\cot t = -\dfrac{4}{3}$, $\sec t = -\dfrac{5}{4}$, $\csc t = \dfrac{5}{3}$

81. From Section 7.1: Find a function $y = A \sin B(x - C)$ that has amplitude 3, period $\dfrac{1}{2}$, and phase shift 2. Then plot its graph.

Answer: $y = 3 \sin(4\pi(x - 2))$

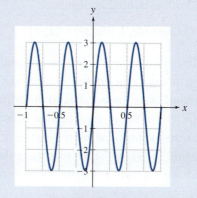

CHAPTER ROADMAP AND STUDY GUIDE

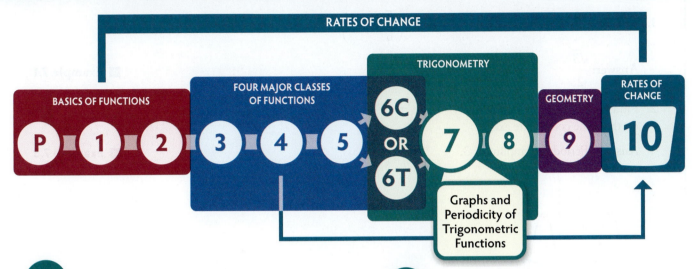

7.1 Graphs of the Sine and Cosine

The graphs of the sine and cosine functions can be shifted, stretched, and compressed to produce familiar sine waves.

Periodicity of the Sine and Cosine

The sine and cosine repeat every 2π units.

The Graph of the Sine Function

The graph of the sine function can be made by following the y-coordinate as we move around the unit circle in a counterclockwise direction.

The Graph of the Cosine Function

The graph of the cosine can be constructed by following the x-coordinate as we move around the unit circle in a counterclockwise direction.

Amplitude

The amplitude of a sine wave is the height of the wave.

The Period

The period of a sine wave is the time required for the cycle to repeat.

Phase Shift

The phase determines the horizontal shift of the sine wave.

Even and Odd Functions

The cosine is an even function, and the sine is an odd function.

7.2 Graphs of Other Trigonometric Functions

The graphs of the tangent, cotangent, secant, and cosecant are very different from the graphs of the sine and cosine.

Periods of the Other Trigonometric Functions

The periods of the secant and cosecant are the same as the period of the sine, but the periods of the tangent and cotangent are smaller.

Graphs of the Tangent and Cotangent

The shorter periods of the tangent and cotangent aid in the production of their graphs.

Graphs of the Secant and Cosecant

Zeros of the sine and cosine correspond to asymptotes of the secant and cosecant.

Transformations of Other Trigonometric Functions

Shifting and stretching for graphs of the secant, cosecant, tangent, and cotangent produce new graphs.

CHAPTER QUIZ

1. Use the period to find the exact value of $\cos\left(\dfrac{\pi}{6} + 12\pi\right)$.

   Answer: $\dfrac{\sqrt{3}}{2}$

   XR Example 7.1

2. Find the period and amplitude of $y = 3\sin\pi x$. Then plot $y = \sin x$ and $y = 3\sin\pi x$ on the same coordinate axes.

   Answer: Period: 2. Amplitude: 3.

   XR Example 7.3

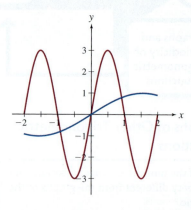

3. Identify the phase shift of $y = \sin(x - \pi)$. Then plot the graphs of $y = \sin x$ and $y = \sin(x - \pi)$ on the same coordinate axes.

   Answer: Phase shift: π

   XR Example 7.4

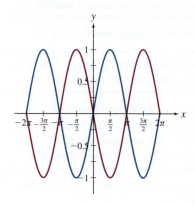

4. Use the period to calculate the exact value of $\cot\left(\dfrac{15\pi}{4}\right)$.

   Answer: -1

   XR Example 7.9

CHAPTER REVIEW EXERCISES

Section 7.1

Graphs of the sine and cosine.

1. Which of **Figure 7.53** or **Figure 7.54** is the graph of the sine function?

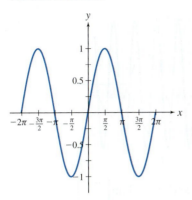

Figure 7.53 Sine or cosine?

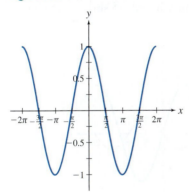

Figure 7.54 Sine or cosine?

Answer: Figure 7.53

2. Which of Figure 7.53 or Figure 7.54 is the graph of the cosine function?

Answer: Figure 7.54

3. True or false: We can think of $\sin t$ as the directed distance from the trigonometric point $P(t)$ to the x-axis.

Answer: True

4. True or false: We can think of $\cos t$ as the directed distance from the trigonometric point $P(t)$ to the y-axis.

Answer: True

5. On the interval $[0, 2\pi]$, state the intervals where the sine function is increasing.

Answer: $\left[0, \dfrac{\pi}{2}\right]$ and $\left[\dfrac{3\pi}{2}, 2\pi\right]$

6. On the interval $[0, 2\pi]$, state the intervals where the cosine function is decreasing.

Answer: $[0, \pi]$

Period, amplitude, and phase shift. In Exercises 7 through 15, identify the period, amplitude, and phase shift of the given function.

7. $y = 2 \sin 3x$

Answer: Period $\dfrac{2\pi}{3}$.
Amplitude 2.
Phase shift 0.

8. $y = 4 \sin 3x$

Answer: Period $\dfrac{2\pi}{3}$.
Amplitude 4.
Phase shift 0.

9. $y = -\sin 2\pi x$

Answer: Period 1.
Amplitude 1.
Phase shift 0.

10. $y = -2 \sin \dfrac{\pi}{2} x$

Answer: Period 4.
Amplitude 2.
Phase shift 0.

11. $y = 5 \sin(x - \pi)$

Answer: Period 2π.
Amplitude 5.
Phase shift π.

12. $y = -3 \cos\left(\dfrac{1}{\pi}\left(x - \dfrac{\pi}{2}\right)\right)$

Answer: Period $2\pi^2$.
Amplitude 3.
Phase shift $\dfrac{\pi}{2}$.

13. $y = 5 \sin\left(\dfrac{\pi}{2}(x - 2)\right)$

Answer: Period 4.
Amplitude 5.
Phase shift 2.

14. $y = 3 \sin 2(x - \pi)$

Answer: Period π.
Amplitude 3.
Phase shift π.

15. $y = -2 \cos 3\left(x - \dfrac{\pi}{2}\right)$

Answer: Period $\dfrac{2\pi}{3}$.

Amplitude 2.

Phase shift $\dfrac{\pi}{2}$.

Identifying amplitude, period, and phase shift.

16. **Figure 7.55** shows the graph of a function of the form $y = A \sin B(x - C)$ with $A > 0$. Identify the amplitude, period, and phase shift. (The answer for the phase shift is not unique.)

Answer: Amplitude 3.
Period 2.
Phase shift 1.

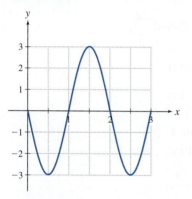

Figure 7.55 Identifying amplitude, period, and phase shift

17. **Figure 7.56** shows the graph of a function of the form $y = A \sin B(x - C)$ with $A > 0$. Identify the amplitude, period, and phase shift. (The answer for the phase shift is not unique.)

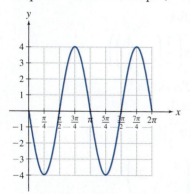

Figure 7.56 Identifying amplitude, period, and phase shift

Answer: Amplitude 4.
Period π.

Phase shift $\dfrac{\pi}{2}$.

Plotting graphs. In Exercises 18 through 21, plot the graph of the two given functions on the same coordinate axes. In each case, use the span $[-2\pi, 2\pi]$.

18. $y = \sin x$ and $y = 2 \sin(\pi x)$

Answer: $\sin x$ in red, $2 \sin(\pi x)$ in blue

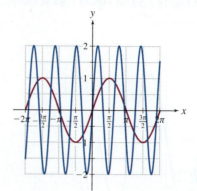

19. $y = \sin x$ and $y = -3 \sin\left(\dfrac{\pi x}{4}\right)$

Answer: $\sin x$ in red, $-3 \sin\left(\dfrac{\pi x}{4}\right)$ in blue

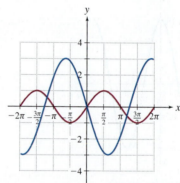

20. $y = \cos x$ and $y = \cos(x + \pi)$

Answer: $\cos x$ in red, $\cos(x + \pi)$ in blue

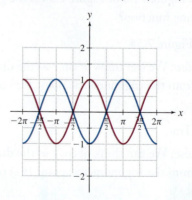

21. $y = \cos x$ and $y = 1 + \cos x$

Answer: $\cos x$ in red, $1 + \cos x$ in blue

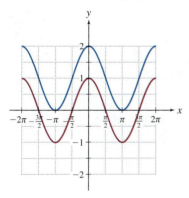

Section 7.2

Identifying the graph.

22. Identify the graph of the tangent from among **Figure 7.57** through **Figure 7.60**.

Answer: Figure 7.60

23. Identify the graph of the cotangent from among Figure 7.57 through Figure 7.60.

Answer: Figure 7.59

24. Identify the graph of the secant from among Figure 7.57 through Figure 7.60.

Answer: Figure 7.58

25. Identify the graph of the cosecant from among Figure 7.57 through Figure 7.60.

Answer: Figure 7.57

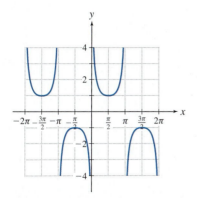

Figure 7.57 The graph of a trigonometric function

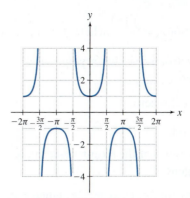

Figure 7.58 The graph of a trigonometric function

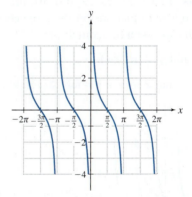

Figure 7.59 The graph of a trigonometric function

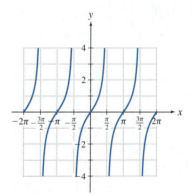

Figure 7.60 The graph of a trigonometric function

Locating zeros and asymptotes.

26. For a function $y = \dfrac{f(x)}{g(x)}$ where f and g have no common zeros, the zeros of y are found where _____.

Answer: $f(x)$ is zero

27. For a function $y = \dfrac{f(x)}{g(x)}$ where f and g have no common zeros, the vertical asymptotes of y are found where _____.

Answer: $g(x)$ is zero

Finding the period.

28. Find the periods of the secant, cosecant, tangent, and cotangent.

Answer: secant 2π, cosecant 2π, tangent π, cotangent π

Plotting graphs. In Exercises 29 and 30, plot the graphs of the two given functions on the same coordinate axes. In each case, use a horizontal span of $[-\pi, \pi]$.

29. $y = \tan x$ and $y = \tan(\pi x)$

Answer: $y = \tan x$ in red, $y = \tan(\pi x)$ in blue

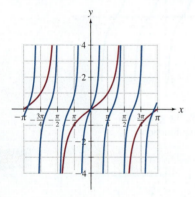

30. $y = \sec x$ and $y = 1 + \sec x$

Answer: $y = \sec x$ in red, $y = 1 + \sec x$ in blue

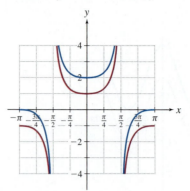

Using identities.

31. Given that $\tan x = \dfrac{3}{4}$, find $\tan(-x)$.

Answer: $-\dfrac{3}{4}$

32. Given that $\sec x = \dfrac{3}{2}$, find $\sec(-x)$.

Answer: $\dfrac{3}{2}$

THE ALGEBRA OF TRIGONOMETRIC FUNCTIONS

8

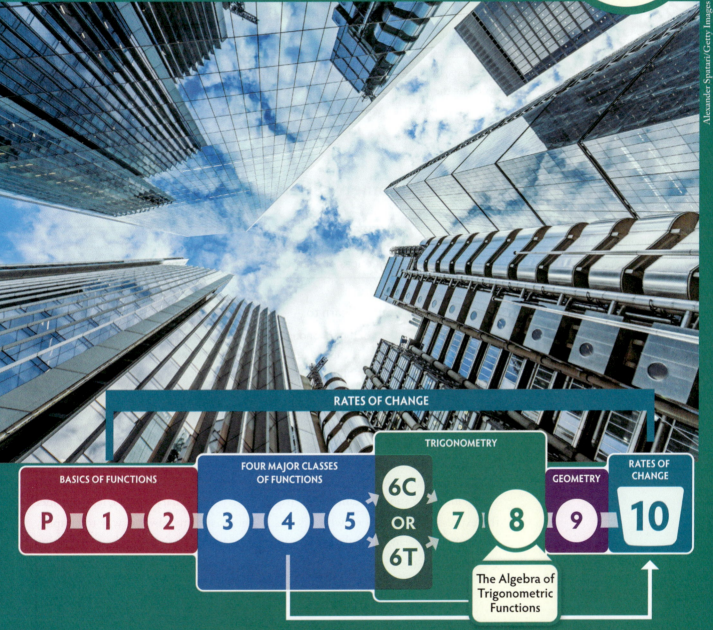

Alexander Spatari/Getty Images

RATES OF CHANGE

TRIGONOMETRY

BASICS OF FUNCTIONS

FOUR MAJOR CLASSES OF FUNCTIONS

GEOMETRY

RATES OF CHANGE

P 1 2 — 3 4 5 — 6C OR 6T — 7 8 9 — 10

The Algebra of Trigonometric Functions

8.1 Trigonometric Identities

8.2 Sum and Difference Formulas

8.3 Double-Angle and Half-Angle Formulas

8.4 Inverse Trigonometric Functions

8.5 Solving Trigonometric Equations

The trigonometric identities you will study in this chapter are used to simplify integration in calculus, and the inverse trigonometric functions studied here appear throughout mathematics and its applications.

s you prepare for calculus, it is important to feel comfortable working with trigonometric functions, just as with polynomials, exponential functions, and logarithms. In this chapter, we will become more familiar with the rules that govern trigonometric functions. Along the way, we will examine several relationships among the functions. These include basic identities as well as more complicated relationships such as sum formulas. We also introduce inverse trigonometric functions and use them to solve trigonometric equations.

Although trigonometric functions were born in geometry, they show up in traditional algebra as well. For example, various methods are available for solving cubic equations, but one of the most useful involves substituting certain trigonometric functions for the variable. Thus, the solution of the cubic equation is reduced to the problem of solving a trigonometric equation, a skill that you will develop in this chapter.

8.1　Trigonometric Identities

8.2　Sum and Difference Formulas

8.3　Double-Angle and Half-Angle Formulas

8.4　Inverse Trigonometric Functions

8.5　Solving Trigonometric Equations

8.1　Trigonometric Identities

> Relationships among the trigonometric functions extend well beyond the immediate consequences of their definition.

In this section, you will learn to:

1. Use the definitions of the trigonometric functions to establish identities.
2. Use the Pythagorean identity to establish other identities.
3. Use the cofunction identities to evaluate trigonometric functions.
4. Assess whether a given function is even, odd, or neither.
5. Use algebra to verify trigonometric identities.

There are some basic relationships known as trigonometric identities that hold among the trigonometric functions. We will recall some identities that we have already encountered and show how to derive new ones.

To illustrate an important role of trigonometry in ordinary algebra, let's consider the expression $y = \sqrt{1 - x^2}$ with $0 \le x \le 1$. There isn't any obvious way to simplify this expression—in particular, to eliminate the radical. Let's make the substitution $x = \cos t$ (with $0 \le t \le \frac{\pi}{2}$):

$$y = \sqrt{1 - x^2}$$
$$= \sqrt{1 - \cos^2 t} \qquad \blacktriangleleft \text{ Substitute } x = \cos t.$$
$$= \sqrt{\sin^2 t} \qquad \blacktriangleleft \text{ Use the Pythagorean identity.}$$
$$= \sin t \qquad \blacktriangleleft \text{ Simplify, using } \sin t \ge 0.$$

It is almost magic. The square root is gone, replaced by the sine function. Some calculus operations are difficult to perform on square roots but relatively easy for trigonometric functions. Thus, this kind of alteration using trigonometric identities is an important tool for calculus.

Basic Identities

> The definitions of trigonometric functions provide fundamental identities.

The first identities we look at are simply the definitions of tangent, secant, cotangent, and cosecant in terms of the sine and cosine functions:

$$\tan t = \frac{\sin t}{\cos t} \qquad \cot t = \frac{\cos t}{\sin t}$$

$$\sec t = \frac{1}{\cos t} \qquad \csc t = \frac{1}{\sin t}$$

From these immediately follow some related identities:

$$\tan t = \frac{1}{\cot t} \qquad \sin t = \frac{1}{\csc t} \qquad \cos t = \frac{1}{\sec t}$$

Note that in calculus you will often need to use trigonometric functions with variables other than t or x, so we will, at times, use other variables, such as α or θ.

In Chapter 6 we used the fact that the point $(\cos t, \sin t)$ lies on the unit circle (**Figure 8.1**), which has equation $x^2 + y^2 = 1$, to obtain the basic Pythagorean identity $\sin^2 t + \cos^2 t = 1$. We recall this identity along with two other identities that can be derived from it:

$$\sin^2 t + \cos^2 t = 1 \qquad \tan^2 t + 1 = \sec^2 t \qquad 1 + \cot^2 t = \csc^2 t$$

We refer to any one of these three as a Pythagorean identity.

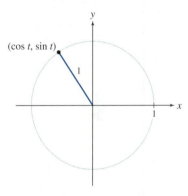

Figure 8.1 The trigonometric point $P(t) = (\cos t, \sin t)$ on the unit circle

---

EXAMPLE 8.1 Establishing Basic Identities Using Definitions of Trigonometric Functions

Use the definitions in terms of sine and cosine to establish the identity $\tan t = \dfrac{1}{\cot t}$.

SOLUTION

One way to establish identities is to begin with one side and work toward the other. In this case, we start with $\dfrac{1}{\cot t}$. When no other strategy presents itself, it is often a good idea to rewrite expressions in terms of sines and cosines.[1] The result is

$$\frac{1}{\cot t} = \frac{1}{\cos t / \sin t} \qquad \blacktriangleleft \text{ Use the definition of cotangent.}$$

$$= \frac{\sin t}{\cos t} \qquad \blacktriangleleft \text{ The reciprocal of } \frac{a}{b} \text{ is } \frac{b}{a}.$$

$$= \tan t \qquad \blacktriangleleft \text{ Use the definition of tangent.}$$

---

[1]Neither the following derivation nor the identity itself is valid in case $\sin t = 0$. We adopt the convention that trigonometric identities are valid only for angles for which they are defined.

TRY IT YOURSELF 8.1 Brief answers provided at the end of the section.

Show that $\dfrac{1}{\sec t} = \cos t$. *Suggestion*: Use the definition of $\sec t$ in terms of the cosine.

EXAMPLE 8.2 Establishing New Identities Using Known Identities

Use the Pythagorean identity $\sin^2 t + \cos^2 t = 1$ to establish the identity $\tan^2 t + 1 = \sec^2 t$.

SOLUTION

Another common method for establishing identities is to start with a known identity and work toward the desired identity. In this case, we start with the Pythagorean identity $\sin^2 t + \cos^2 t = 1$. Because we want a secant in the equation and $\sec t = \dfrac{1}{\cos t}$, we divide the Pythagorean identity by the square of the cosine:

$$\sin^2 t + \cos^2 t = 1 \qquad \blacktriangleleft \text{ Use the Pythagorean identity.}$$

$$\frac{\sin^2 t + \cos^2 t}{\cos^2 t} = \frac{1}{\cos^2 t} \qquad \blacktriangleleft \text{ Divide both sides by } \cos^2 t.$$

$$\frac{\sin^2 t}{\cos^2 t} + \frac{\cos^2 t}{\cos^2 t} = \frac{1}{\cos^2 t} \qquad \blacktriangleleft \text{ Use } \frac{A+B}{C} = \frac{A}{C} + \frac{B}{C}.$$

$$\tan^2 t + 1 = \sec^2 t \qquad \blacktriangleleft \text{ Use the definitions of tangent and secant.}$$

TRY IT YOURSELF 8.2 Brief answers provided at the end of the section.

Use the basic Pythagorean identity $\sin^2 t + \cos^2 t = 1$ to verify the identity $1 + \cot^2 t = \csc^2 t$.

EXTEND YOUR REACH   The Pythagorean identity we used to begin Example 8.2, $\sin^2 t + \cos^2 t = 1$, is true for all values of t. But the identity that we derived, $\tan^2 t + 1 = \sec^2 t$, fails to hold at infinitely many points. For example, neither the tangent nor the secant is defined for $t = \dfrac{\pi}{2}$. Examine the derivation carefully to determine the step (or steps) where certain values of t must be ruled out.

EXAMPLE 8.3 Expressing One Trigonometric Function in Terms of Others

Express the cotangent in terms of the cosecant for an angle t whose terminal side lies in the second quadrant when the angle is in standard position.

SOLUTION

One of the Pythagorean identities relates the cotangent and the cosecant:

$$1 + \cot^2 t = \csc^2 t \qquad \blacktriangleleft \text{ Use the Pythagorean identity.}$$

$$\cot^2 t = \csc^2 t - 1 \qquad \blacktriangleleft \text{ Solve for } \cot^2 t.$$

$$\cot t = \pm\sqrt{\csc^2 t - 1} \qquad \blacktriangleleft \text{ Take square roots.}$$

In the second quadrant, the cotangent is negative. Hence, $\cot t = -\sqrt{\csc^2 t - 1}$.

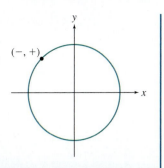

TRY IT YOURSELF 8.3 Brief answers provided at the end of the section.

Express the secant in terms of the tangent for an angle whose terminal side lies in the third quadrant when the angle is in standard position.

Cofunction Identities

> The interplay between the two acute angles of a right triangle results in cofunction identities.

Sides of a right triangle are identified as opposite or adjacent in relation to one of the acute angles. If we change angles, the identifications of the sides change, and this simple fact yields some important trigonometric identities.

For the angles s and t in **Figure 8.2**,

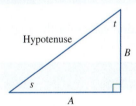

Figure 8.2 Changing angles and corresponding change in the identification of sides

$$\text{Opposite side for } s = B = \text{Adjacent side for } t$$

$$\text{Adjacent side for } s = A = \text{Opposite side for } t$$

Using these relationships, we find

$$\sin s = \frac{\text{Opposite } s}{\text{Hypotenuse}} = \frac{B}{\text{Hypotenuse}} = \frac{\text{Adjacent to } t}{\text{Hypotenuse}} = \cos t$$

But because the angle sum of the triangle is π radians, $s = \dfrac{\pi}{2} - t$. Hence, we have the identity

$$\sin\left(\frac{\pi}{2} - t\right) = \sin s = \cos t$$

Recall from geometry that two angles are complementary if their measures add up to 90° or $\dfrac{\pi}{2}$ radians. Thus, s and t are complementary angles, and because of the preceding identity, sine and cosine are called cofunctions.

This identity holds for all angles, not just acute angles, and similar arguments apply to all three pairs of trigonometric cofunctions. Here we summarize the cofunction identities.

CONCEPTS TO REMEMBER: Cofunction Identities

The sine and cosine functions take the same values at complementary angles. The same is true of the tangent and cotangent as well as the secant and cosecant. Here are the corresponding identities:

$$\sin\left(\frac{\pi}{2} - t\right) = \cos t \qquad \cos\left(\frac{\pi}{2} - t\right) = \sin t$$

$$\tan\left(\frac{\pi}{2} - t\right) = \cot t \qquad \cot\left(\frac{\pi}{2} - t\right) = \tan t$$

$$\sec\left(\frac{\pi}{2} - t\right) = \csc t \qquad \csc\left(\frac{\pi}{2} - t\right) = \sec t$$

If t is measured in degrees, use $90°$ instead of $\frac{\pi}{2}$ in these identities.

EXAMPLE 8.4   Using Cofunction Identities to Evaluate Trigonometric Functions

If $\sin t = \frac{3}{5}$ and $0 \le t \le \frac{\pi}{2}$, find the six trigonometric function values for the angle $s = \frac{\pi}{2} - t$.

SOLUTION

Using a cofunction identity, we have

$$\cos s = \cos\left(\frac{\pi}{2} - t\right) = \sin t = \frac{3}{5}$$

Now that we have the cosine of s, we can construct the right triangle shown in **Figure 8.3**. Because $0 \le t \le \frac{\pi}{2}$, we have $0 \le s \le \frac{\pi}{2}$. Using this fact, we can read the values of the remaining trigonometric function values from the figure:

$$\sin s = \frac{4}{5}, \quad \tan s = \frac{4}{3}, \quad \cot s = \frac{3}{4}, \quad \sec s = \frac{5}{3}, \quad \csc s = \frac{5}{4}$$

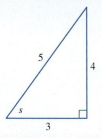

Figure 8.3 An angle whose cosine is $\frac{3}{5}$

TRY IT YOURSELF 8.4   Brief answers provided at the end of the section.

If $\sin t = \frac{3}{5}$ but now $\frac{\pi}{2} \le t \le \pi$, find the six trigonometric function values for the angle $s = \frac{\pi}{2} - t$. Note first that $-\frac{\pi}{2} \le s \le 0$.

Even and Odd Identities

> **Reflective symmetries of graphs of trigonometric functions translate into important identities.**

Recall from Section 2.3 that a function f is called odd if $f(-x) = -f(x)$ for all x. Similarly, f is said to be even if $f(-x) = f(x)$ for all x. In Section 7.1 we noted that the graphs of sine and cosine suggest that sine is odd and cosine is even. This observation corresponds to the identities $\sin(-x) = -\sin x$ and $\cos(-x) = \cos x$. We verify these identities here. As shown in **Figure 8.4**, we can find the trigonometric point $P(-t) = (\cos(-t), \sin(-t))$ from the point $P(t) = (\cos t, \sin t)$ by reflecting about the x-axis. Such a reflection changes the sign of the second coordinate, $\sin t$, but leaves the first coordinate, $\cos t$, the same. We conclude that $\sin(-t) = -\sin t$ and $\cos(-t) = \cos t$.

In the exercises, you are asked to show that additional trigonometric functions are either even or odd. Here is a summary.

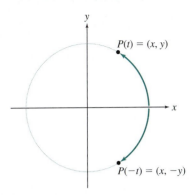

Figure 8.4 Comparing the trigonometric points $P(t)$ and $P(-t)$: The point $P(-t)$ is the reflection through the x-axis of $P(t)$.

CONCEPTS TO REMEMBER: Even and Odd Trigonometric Functions

The cosine and secant are even functions. The sine, cosecant, tangent, and cotangent are odd functions. Here are the corresponding identities:

Even functions	Odd functions	
$\cos(-t) = \cos t$	$\sin(-t) = -\sin t$	$\csc(-t) = -\csc t$
$\sec(-t) = \sec t$	$\tan(-t) = -\tan t$	$\cot(-t) = -\cot t$

EXAMPLE 8.5 Even and Odd Functions

Determine whether the function $g(x) = \sec x \tan x$ is even, odd, or neither.

SOLUTION

We look at $g(-x)$:

$$g(-x) = \sec(-x) \tan(-x)$$

$$= (\sec x)(-\tan x) \quad \blacktriangleleft \text{ The secant is even, and the tangent is odd.}$$

$$= -\sec x \tan x$$

$$= -g(x)$$

Because $g(-x) = -g(x)$, the function g is odd.

TRY IT YOURSELF 8.5 Brief answers provided at the end of the section.

Determine whether $h(x) = \csc^2 x$ is even, odd, or neither.

EXTEND YOUR REACH   If m and n are integers, define $f(x) = \sin^m x \cos^n x$. For what values of m and n is the function f even? For what values of m and n is the function f odd?

Check your answers by looking at the functions in Example 8.5 and in Try It Yourself 8.5.

Algebraic Verification of Trigonometric Identities

> Algebraic operations on the trigonometric identities we know can generate many other identities.

There is no single recipe that can guide you through the verification of every trigonometric identity. But there are a few guiding principles that can help. The next few examples are designed to illustrate these principles. Experience, as always, is the best teacher.

EXAMPLE 8.6   Reducing a Complex Expression to a Simpler One

a. Show that $\sec t \cot t = \csc t$.

b. Show that $\sin t(\csc t - \sin t) = \cos^2 t$.

SOLUTION

a. We begin with the more complicated side of the equation, $\sec t \cot t$, and reduce it to the other side. Expressing trigonometric functions in terms of sine and cosine is useful here, as it so often is:

$$\sec t \cot t = \frac{1}{\cos t} \frac{\cos t}{\sin t} \qquad \blacktriangleleft \text{ Use the definition of secant and cotangent.}$$

$$= \frac{1}{\cos t} \frac{\cos t}{\sin t} \qquad \blacktriangleleft \text{ Cancel the cosines.}$$

$$= \frac{1}{\sin t}$$

$$= \csc t \qquad \blacktriangleleft \text{ Use the definition of cosecant.}$$

b. Once again, we begin with the more complicated side of the equation:

$$\sin t(\csc t - \sin t) = \sin t \csc t - \sin^2 t \qquad \blacktriangleleft \text{ Use } A(B - A) = AB - A^2.$$

$$= \sin t \frac{1}{\sin t} - \sin^2 t \qquad \blacktriangleleft \text{ Use the definition of cosecant.}$$

$$= 1 - \sin^2 t \qquad \blacktriangleleft \text{ Cancel the sines.}$$

$$= \cos^2 t \qquad \blacktriangleleft \text{ Use the Pythagorean identity.}$$

TRY IT YOURSELF 8.6   Brief answers provided at the end of the section.

Show that $\dfrac{\cos t}{\sec t} = 1 - \sin^2 t$.

We can put together known identities to obtain new identities, as shown in Example 8.7.

EXAMPLE 8.7 Combining Known Identities to Make New Ones

Combine cofunction and even-odd identities to show the following.

a. $\cos\left(t - \dfrac{\pi}{2}\right) = \sin t$

b. $\cos\left(t + \dfrac{\pi}{2}\right) = -\sin t$

SOLUTION

a. We observe first that $t - \dfrac{\pi}{2} = -\left(\dfrac{\pi}{2} - t\right)$. Therefore,

$$\cos\left(t - \frac{\pi}{2}\right) = \cos\left(-\left(\frac{\pi}{2} - t\right)\right) \qquad ◀ \text{ Use } A - B = -(B - A).$$

$$= \cos\left(\frac{\pi}{2} - t\right) \qquad ◀ \text{ Cosine is an even function.}$$

$$= \sin t \qquad ◀ \text{ Use a cofunction identity.}$$

b. We begin with a cofunction identity:

$$\cos\left(t + \frac{\pi}{2}\right) = \sin\left(\frac{\pi}{2} - \left(t + \frac{\pi}{2}\right)\right) \qquad ◀ \text{ Use a cofunction identity.}$$

$$= \sin(-t) \qquad ◀ \text{ Simplify.}$$

$$= -\sin t \qquad ◀ \text{ The sine is odd.}$$

TRY IT YOURSELF 8.7 Brief answers provided at the end of the section.

Show that $\sin\left(t + \dfrac{\pi}{2}\right) = \cos t$.

There are many more methods used for establishing identities. We show some of the most commonly used methods.

EXAMPLE 8.8 Algebraic Combination of Terms

Show that $\dfrac{1}{1 - \sin t} + \dfrac{1}{1 + \sin t} = 2\sec^2 t$.

SOLUTION

We begin by adding the two fractions on the left side of the equation:

$$\frac{1}{1 - \sin t} + \frac{1}{1 + \sin t} = \frac{(1 + \sin t) + (1 - \sin t)}{(1 - \sin t)(1 + \sin t)} \qquad ◀ \text{ Use } \frac{1}{A} + \frac{1}{B} = \frac{B + A}{AB}.$$

$$= \frac{2}{1 - \sin^2 t} \qquad ◀ \text{ Use } (1 - A)(1 + A) = 1 - A^2.$$

$$= \frac{2}{\cos^2 t}$$ ◀ Use the Pythagorean identity.

$$= 2 \sec^2 t$$ ◀ Use the definition of secant.

TRY IT YOURSELF 8.8 Brief answers provided at the end of the section.

Show that $\dfrac{\cos t}{\sin t} + \dfrac{\sin t}{\cos t} = \csc t \sec t$.

EXTEND YOUR REACH

Here is a "derivation" of an "identity" that is not an identity at all. Your job is to identify the error.

$$\text{Suppose } \sin^2 t = \cos^2 t$$

$$\cos^2 t = \sin^2 t$$ ◀ Turn the equation around.

$$\sin^2 t + \cos^2 t = \cos^2 t + \sin^2 t$$ ◀ Add the two equations.

$$1 = 1$$ ◀ Use the Pythagorean identity.

Because this last equation is true, we conclude that the equation we began with, $\sin^2 t = \cos^2 t$, must also be true.

What is the error in this argument?

EXAMPLE 8.9   Factoring and Dividing

Show that $\dfrac{\sin^2 t}{1 - \cos t} = 1 + \cos t$.

SOLUTION

This identity uses a common but somewhat subtle trick. The numerator of the fraction doesn't factor in any helpful way as it stands, but if we first apply the basic Pythagorean identity, it does:

$$\frac{\sin^2 t}{1 - \cos t} = \frac{1 - \cos^2 t}{1 - \cos t}$$ ◀ Use the Pythagorean identity.

$$= \frac{(1 - \cos t)(1 + \cos t)}{1 - \cos t}$$ ◀ Use $1 - A^2 = (1 - A)(1 + A)$.

$$= \frac{(1 - \cos t)(1 + \cos t)}{1 - \cos t}$$ ◀ Cancel common factors.

$$= 1 + \cos t$$

TRY IT YOURSELF 8.9 Brief answers provided at the end of the section.

Show that $\dfrac{\cos^2 t}{1 + \sin t} = 1 - \sin t$.

EXAMPLE 8.10 Introducing New Terms to Simplify Trigonometric Expressions

Show that $\dfrac{\sin t}{1 + \cos t} = \dfrac{1 - \cos t}{\sin t}$.

SOLUTION

For the identity $\dfrac{\sin t}{1 + \cos t} = \dfrac{1 - \cos t}{\sin t}$, we want to transform the left side into the right side. There are two ways of looking at the strategy for accomplishing this. One way is to note that we want to introduce $1 - \cos t$ into the numerator because it appears in the right-hand fraction. Thus, we should multiply the top and bottom of the left-hand fraction by this expression.

Another way to look at the method is illustrated by the preceding example. Namely, when squares of the sine or cosine are present, we can use the Pythagorean identity. Multiplying the denominator on the left by $1 - \cos t$ produces such squares.

Either point of view suggests that we multiply the top and bottom of the left-hand fraction by $1 - \cos t$:

$$\frac{\sin t}{1 + \cos t} = \frac{\sin t(1 - \cos t)}{(1 + \cos t)(1 - \cos t)} \qquad \blacktriangleleft \text{Multiply top and bottom by } 1 - \cos t.$$

$$= \frac{\sin t(1 - \cos t)}{1 - \cos^2 t} \qquad \blacktriangleleft \text{Use } (1 + A)(1 - A) = 1 - A^2.$$

$$= \frac{\sin t(1 - \cos t)}{\sin^2 t} \qquad \blacktriangleleft \text{Use the Pythagorean identity.}$$

$$= \frac{\cancel{\sin t}(1 - \cos t)}{\sin^{\cancel{2}1} t} \qquad \blacktriangleleft \text{Cancel the sines.}$$

$$= \frac{1 - \cos t}{\sin t}$$

TRY IT YOURSELF 8.10 Brief answers provided at the end of the section.

Establish the identity in the example by multiplying the top and bottom of the left-hand fraction by $\sin t$ rather than by $1 - \cos t$.

As noted earlier, there is no recipe that will prove all trigonometric identities. We summarize the techniques we have already found useful.

> ### CONCEPTS TO REMEMBER: Techniques for Proving Trigonometric Identities
>
> - Express one trigonometric function in terms of others. (See Example 8.1 and Example 8.3.)
> - Use known identities to derive new identities. (See Example 8.2 and Example 8.7.)
> - Reduce a complex trigonometric expression to a simpler one. (See Example 8.6.)
> - Use common algebraic operations to simplify trigonometric expressions. (See Example 8.8 and Example 8.9.)
> - Introduce new terms to simplify trigonometric expressions. (See Example 8.10.)

TRY IT YOURSELF ANSWERS

8.1 $\dfrac{1}{\sec t} = \dfrac{1}{\dfrac{1}{\cos t}} = \cos t$

8.2 Divide both sides of the basic Pythagorean identity by $\sin^2 t$ and simplify.

8.3 $\sec t = -\sqrt{\tan^2 t + 1}$

8.4 $\sin s = -\dfrac{4}{5}$, $\cos s = \dfrac{3}{5}$, $\tan s = -\dfrac{4}{3}$, $\cot s = -\dfrac{3}{4}$,

$\sec s = \dfrac{5}{3}$, $\csc s = -\dfrac{5}{4}$

8.5 $h(x) = \csc^2 x$ is an even function.

8.6 Briefly, $\dfrac{\cos t}{\sec t} = \cos^2 t = 1 - \sin^2 t$

8.7 Briefly,

$$\sin\left(t + \frac{\pi}{2}\right) = \cos\left(\frac{\pi}{2} - \left(t + \frac{\pi}{2}\right)\right) = \cos(-t) = \cos t$$

8.8 Briefly,

$$\frac{\cos t}{\sin t} + \frac{\sin t}{\cos t} = \frac{\cos^2 t + \sin^2 t}{\sin t \cos t} = \frac{1}{\sin t \cos t} = \csc t \sec t$$

8.9 Briefly,

$$\frac{\cos^2 t}{1 + \sin t} = \frac{1 - \sin^2 t}{1 + \sin t} = \frac{(1 - \sin t)(1 + \sin t)}{1 + \sin t} = 1 - \sin t$$

8.10 Briefly,

$$\frac{\sin t}{1 + \cos t} = \frac{\sin^2 t}{(\sin t)(1 + \cos t)} = \frac{1 - \cos^2 t}{(\sin t)(1 + \cos t)}$$

$$= \frac{(1 - \cos t)(1 + \cos t)}{(\sin t)(1 + \cos t)} = \frac{1 - \cos t}{\sin t}$$

EXERCISE SET 8.1

CHECK YOUR UNDERSTANDING

1. State the three Pythagorean identities.

2. State the six cofunction identities.

3. One of the two functions sine and cosine is an odd function. Which one is it?

4. We find the trigonometric point $P(-t)$ by:

   a. reflecting $P(t)$ about the y-axis

   b. reflecting $P(t)$ about the x-axis

   c. reflecting $P(t)$ through the origin

   d. taking the negative of $P(t)$

5. **True or false:** If s and t are complementary angles, then $\sin s = \cos t$.

6. **True or false:** We can establish a trigonometric identity by starting with one side and reducing it to the other side.

7. **True or false:** Combining known identities can produce new identities.

8. **True or false:** The cofunction identities apply only to angles in the first quadrant.

9. **True or false:** In establishing trigonometric identities, it is often helpful to write expressions in terms of the sine and cosine.

10. **True or false:** Some trigonometric expressions can be simplified by introducing new terms.

SKILL BUILDING

Calculating using trigonometric identities. In Exercises 11 through 22, identities presented earlier in this section will be helpful.

11. If $\sin x = \dfrac{2}{3}$, find $\cos\left(\dfrac{\pi}{2} - x\right)$.

12. If $\tan x = 4$, find $\cot\left(\dfrac{\pi}{2} - x\right)$.

13. If $\cot x = 4$, find $\cot\left(\dfrac{\pi}{2} - x\right)$.

14. If $\sin x = \dfrac{3}{4}$, find $\sin(-x)$.

15. If $\tan x = -\dfrac{4}{3}$, find $\tan(-x)$.

16. Suppose $\sin \alpha = -\dfrac{5}{6}$ and the trigonometric point $P(\alpha)$ is in the third quadrant. Find $\cos(-\alpha)$.

17. Suppose $\cos \alpha = -\dfrac{3}{4}$ and the terminal side of the angle α is in the third quadrant when the angle is in standard position. Find $\sin\left(\dfrac{\pi}{2} - \alpha\right)$ and $\cos\left(\dfrac{\pi}{2} - \alpha\right)$.

18. If $\sec x = \dfrac{7}{4}$, find $\sec(-x)$.

19. If $\cos x = -\dfrac{2}{5}$, find $\cos(-x)$.

20. If $\sin x = \dfrac{2}{7}$, find $\cos\left(x - \dfrac{\pi}{2}\right)$.

21. If $\cot x = -3$, find $\tan\left(x - \dfrac{\pi}{2}\right)$. *Suggestion:* You may need to do this one in two steps.

22. If $\sec x = \dfrac{7}{5}$, find $\csc\left(x - \dfrac{\pi}{2}\right)$. *Suggestion:* You may need to do this one in two steps.

Simplifying trigonometric expressions. In Exercises 23 through 36, simplify the given trigonometric expression. It will often be helpful to express functions in terms of the sine and cosine.

23. $\sin t \dfrac{\cos t}{\sin t}$

24. $\sin t \csc t$

25. $\tan t \cot t$

26. $\dfrac{\sin t}{\csc t}$

27. $\cos t \tan t$

28. $\cot t \sin t$

29. $1 - \sin^2 x$

30. $\sec^2 y - 1$

31. $\sec x \cot x$

32. $\csc x \tan x$

33. $\sin^2 2t + \cos^2 2t$

34. $1 + \tan^2 3t$

35. $(1 - \sin x)(1 + \sin x)$

36. $\left(\dfrac{\cos x}{1 - \sin x}\right)\left(\dfrac{\cos x}{1 - \sin x}\right)$

PROBLEMS

Basic identities. In Exercises 37 through 40, use the definitions of the trigonometric functions to establish the given basic identity.

37. $\cot t = \dfrac{1}{\tan t}$

38. $\tan t = \dfrac{1}{\cot t}$

39. $\sin t = \dfrac{1}{\csc t}$

40. $\cos t = \dfrac{1}{\sec t}$

Cofunction identities. In Exercises 41 through 45, verify the cofunction identities.

41. $\cos\left(\dfrac{\pi}{2} - t\right) = \sin t$

42. $\cot\left(\dfrac{\pi}{2} - t\right) = \tan t$

43. $\tan\left(\dfrac{\pi}{2} - t\right) = \cot t$

44. $\sec\left(\dfrac{\pi}{2} - t\right) = \csc t$

45. $\csc\left(\dfrac{\pi}{2} - t\right) = \sec t$

Even and odd functions. Exercises 46 through 53 concern even and odd functions. You will need to use the fact that the sine function is odd and the cosine is even.

46. Show that the tangent is an odd function.

47. Show that the cotangent is an odd function.

48. Show that the secant is an even function.

49. Show that the cosecant is an odd function.

50. Determine whether the function $f(t) = \sin t \cos t$ is even or odd. Justify your answer.

51. Determine whether the function $f(t) = \sin^2 t$ is even or odd. Justify your answer.

52. Determine whether the function $f(t) = \tan t \sin t$ is even or odd. Justify your answer.

53. Determine whether the function $f(t) = \csc^3 t$ is even or odd. Justify your answer.

Expressing one trigonometric function in terms of others.

54. Express the cosine in terms of the cotangent and the sine.

55. Express the secant of t in terms of the tangent of t. Assume $0 < t < \dfrac{\pi}{2}$.

56. Express $\cos t$ in terms of $\sin t$. Assume $\pi < t < \dfrac{3\pi}{2}$.

57. Express $\sin t$ in terms of $\cos t$ if $\sin t$ is negative.

58. Express $\sec t$ in terms of $\sin t$ if $\cos t$ is negative.

59. Express $\tan t$ in terms of $\cos t$ if $\dfrac{\pi}{2} < t < \pi$.

60. Express $\tan t$ in terms of $\sin t$ if $\dfrac{3\pi}{2} < t < 2\pi$.

61. Express $\cot t$ in terms of $\tan t$.

62. Express $\sec t$ in terms of $\csc t$ if $0 < t < \dfrac{\pi}{2}$.

63. Express $\tan t$ in terms of $\sec t$ and $\csc t$.

64. Express $\sin t$ in terms of $\tan t$. Assume $\dfrac{\pi}{2} < t < \pi$.

65. Express $\cot t$ in terms of $\sin t$. Assume $\dfrac{\pi}{2} < t < \pi$.

Various identities. In Exercises 66 through 92, use identities from this section to show that the given identity is true.

66. $\sec x \cot x = \csc x$

   Suggestion: Express $\sec x$ and $\cot x$ in terms of sine and cosine.

67. $\tan x = \dfrac{\sec x}{\csc x}$

68. $\dfrac{\sin t}{\tan t} = \cos t$

69. $\dfrac{\sin t}{\csc t} = \sin^2 t$

   Suggestion: Express $\csc t$ in terms of the sine.

70. $\sec x \cot x = \csc x$

71. $\dfrac{\csc \theta}{\sec \theta} = \cot \theta$

72. $1 - \sin \alpha = \dfrac{\cos^2 \alpha}{1 + \sin \alpha}$

   Suggestion: Multiply the left side by $\dfrac{1 + \sin \alpha}{1 + \sin \alpha}$.

73. $\csc t - \sin t = \cot t \cos t$

74. $\sec x - \cos x = \sin x \tan x$

75. $(\sin y + \cos y)^2 = 1 + 2 \sin y \cos y$

   Suggestion: Expand the left side.

76. $\dfrac{\tan^2 t}{\sec t - 1} = \sec t + 1$

77. $\dfrac{\tan x}{\sec x} = \sin x$

78. $\sec x \tan x = \dfrac{\sin x}{1 - \sin^2 x}$

   Suggestion: Express $\sec x$ and $\tan x$ in terms of the sine and cosine.

79. $\cot^2 x = \csc^2 x - 1$

80. $\dfrac{1 + \sin x}{\cos x} = \dfrac{\cos x}{1 - \sin x}$

81. $\dfrac{\sec x + \tan x}{\sec x - \tan x} = \dfrac{1 + \sin x}{1 - \sin x}$

   Suggestion: Express functions on the left side in terms of sine and cosine. Then multiply top and bottom by the cosine.

82. $\cot x + \tan x = \sec x \csc x$

83. $\dfrac{\cos t + 1}{\tan^2 t} = \dfrac{\cos t}{\sec t - 1}$

84. $\sec t + \tan t = \dfrac{1}{\sec t - \tan t}$

   Suggestion: Multiply top and bottom of the right side by $\sec t + \tan t$.

85. $\dfrac{1}{\csc t + \cot t} = \csc t - \cot t$

86. $(\sec x - \tan x)^2 = \dfrac{1 - \sin x}{1 + \sin x}$

   Suggestion: Start by expanding the left-hand side.

87. $(\csc x + \cot x)^2 = \dfrac{1 + \cos x}{1 - \cos x}$

   Suggestion: Start by expanding the left-hand side.

88. $\dfrac{\cot x + 1}{\cot x - 1} = \dfrac{1 + \tan x}{1 - \tan x}$

89. $\dfrac{\sec x - 1}{1 - \cos x} = \sec x$

90. $\dfrac{1}{\sec x \tan x} = \csc x - \sin x$

   Suggestion: Write $\sec x$ and $\tan x$ in terms of the sine and cosine. Then simplify the fraction and use the Pythagorean identity.

91. $\dfrac{\cos x \cot x}{1 - \sin x} = 1 + \csc x$

92. $\dfrac{\cot x}{\csc x - 1} = \dfrac{\csc x + 1}{\cot x}$

   Suggestion: Multiply top and bottom of the left side by $\csc x + 1$.

Other identities. In Exercises 93 through 100, verify the given identities. For some of these exercises, you may find the following identities helpful.

$$\sin\left(x+\frac{\pi}{2}\right) = \cos x$$

$$\cos\left(x+\frac{\pi}{2}\right) = -\sin x$$

93. $\dfrac{\sin t}{\sin\left(\dfrac{\pi}{2}-t\right)} = \tan t$

94. $\sin^2 t + \sin^2\left(t-\dfrac{\pi}{2}\right) = 1$

95. $\sin\left(x-\dfrac{\pi}{2}\right) + \sin\left(x+\dfrac{\pi}{2}\right) = 0$

96. $\sin\left(x+\dfrac{3\pi}{2}\right) = -\cos x$

Suggestion: Note first that

$$\sin\left(x+\frac{3\pi}{2}\right) = \sin\left((x+2\pi)-\frac{\pi}{2}\right)$$

97. $\tan\left(x+\dfrac{\pi}{2}\right) = -\cot x$

98. $\sec\left(x-\dfrac{\pi}{2}\right) = \csc x$

99. $\cot\left(x-\dfrac{\pi}{2}\right) = -\tan x$

100. $\dfrac{\cos x}{1+\cos\left(x+\dfrac{\pi}{2}\right)} = \dfrac{1+\sin x}{\sin\left(\dfrac{\pi}{2}-x\right)}$

CHALLENGE EXERCISES FOR INDIVIDUALS OR GROUPS

Challenging identities. In Exercises 101 through 104, verify the given identities.

101. $\dfrac{\sin t + \cos t + \tan t - \sec t}{\cos t - \sin t + 1} = \tan t$

102. $\dfrac{2\sin x \cos x}{\sin x + \cos x - 1} = \sin x + \cos x + 1$

103. $(\sin x + \cos x)(\tan x + \cot x) = \sec x + \csc x$

104. $\dfrac{\tan x + \tan y}{\cot x + \cot y} = \tan x \tan y$

REVIEW AND REFRESH: Exercises from Previous Sections

105. From Section 1.2: Find the average rate of change of $y = x^2$ on the interval $[a, a+h]$ if $h > 0$.

Answer: $2a + h$

106. From Section 4.1: Simplify the expression $e^{1+\ln e^2}$.

Answer: e^3

107. From Section 4.3: Solve the equation $e^{2x-1} = 3$.

Answer: $x = \dfrac{1+\ln 3}{2}$

108. From Section 5.1: The discriminant of the quadratic function $y = ax^2 + bx + c$ with real coefficients is 0. What can you conclude about its zeros?

Answer: There is exactly one zero, and it is real.

109. From Section 5.3: Find all zeros of $P(x) = x^2(x-1) - 4(x-1)$.

Answer: 1, 2, and -2

110. From Chapter 6: Find the area of the triangle shown in **Figure 8.5**.

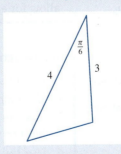

Figure 8.5 Triangle for Exercise 110

Answer: 3

111. **From Chapter 6:** If $\cos t = \dfrac{5}{6}$ and $\sin t$ is negative, find the values of the remaining five trigonometric functions of t.

Answer: $\sin t = -\dfrac{\sqrt{11}}{6}$, $\tan t = -\dfrac{\sqrt{11}}{5}$,

$\cot t = -\dfrac{5}{\sqrt{11}}$, $\sec t = \dfrac{6}{5}$, $\csc t = -\dfrac{6}{\sqrt{11}}$

112. **From Chapter 6:** Find the reference number for $\dfrac{7\pi}{6}$.

Answer: $\dfrac{\pi}{6}$

113. **From Section 7.1:** For $y = 3\sin 2t$, identify the period and amplitude, and then plot the graph.

Answer: Period: π. Amplitude: 3.

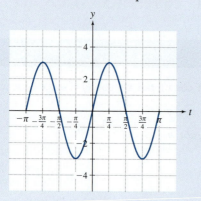

114. **From Section 7.2:** Plot the graph of

$$y = 1 + \tan\left(t - \dfrac{\pi}{2}\right).$$

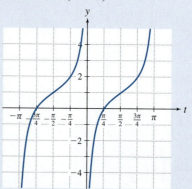

8.1 Trigonometric Identities

8.2 Sum and Difference Formulas

8.3 Double-Angle and Half-Angle Formulas

8.4 Inverse Trigonometric Functions

8.5 Solving Trigonometric Equations

8.2 Sum and Difference Formulas

> Alterations of trigonometric expressions using sum and difference formulas allow for easier application of certain calculus operations.

In this section, you will learn to:

1. Use the sum and difference formulas for the sine and cosine to find function values.
2. Use a variety of identities to derive the sum formulas.
3. Use the sum and difference formulas for the tangent and cotangent to find function values.
4. Use the product-to-sum formulas and the sum-to-product formulas to find function values.

In 1807, Jean-Baptiste Joseph Fourier showed that many periodic functions can be decomposed into sums of simpler periodic functions (in what is now called the

Fourier transform). For example, the periodic function shown in **Figure 8.6** can be expressed as the sum of the three simpler functions shown in **Figure 8.7**, **Figure 8.8**, and **Figure 8.9**. This type of decomposition is fundamental to many areas of mathematics and its applications, from music to audiology to astronomy.

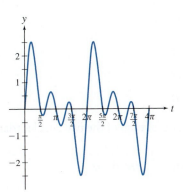

Figure 8.6 A complicated periodic function

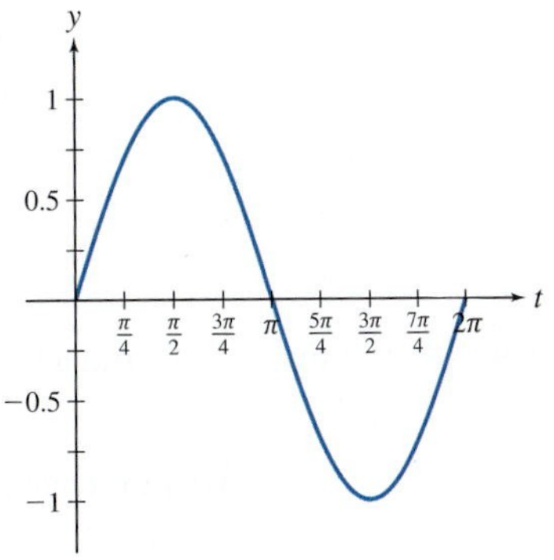

Figure 8.7 A complicated periodic function: the first term of the sum

For some graphs, such as a typical electrocardiogram, the decomposition into simpler functions may require some advanced mathematics, but in many cases all that is required is a formula for the sine or cosine of the sum of two angles.

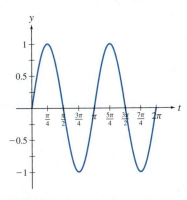

Figure 8.8 A complicated periodic function: the second term of the sum

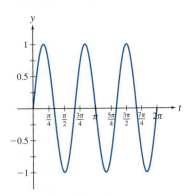

Figure 8.9 A complicated periodic function: the third term of the sum

In this section, we develop formulas for trigonometric functions of sums and differences. We begin with the sine and use that result to derive other formulas. We also derive product-to-sum and sum-to-product formulas.

Sum and Difference Formulas for Sine and Cosine

> The sine formula for the area of a triangle is used to establish the sum formula for the sine function.

The formula for the area of a triangle that we developed in Chapter 6 will help us find the sum formula for the sine. Recall that the sine area formula states that the area of the triangle in **Figure 8.10** is

$$\text{Area} = \frac{1}{2} ab \sin \theta$$

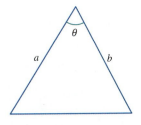

Figure 8.10 Triangle for area formula

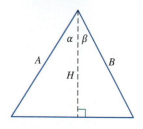

Figure 8.11 Calculating the area in two ways to find a sum formula for the sine

Let's see how to use the sine area formula to produce a sum formula for the sine function—that is, a formula for the sine of the sum of two angles. We work with the triangle in **Figure 8.11**, using the labels provided there. Note first that we can use the right triangle on the left to find the cosine of α and the right triangle on the right to find the cosine of β:

$$\cos \alpha = \frac{H}{A}$$

$$\cos \beta = \frac{H}{B}$$

Now we use the sine area formula to get the area of each of the two smaller triangles:

$$\text{Area of triangle on left} = \frac{1}{2} AH \sin \alpha$$

$$\text{Area of triangle on right} = \frac{1}{2} BH \sin \beta$$

The largest triangle has sides of length A and B and an included angle of $\alpha + \beta$. Hence, the sine area formula gives

$$\text{Area of largest triangle} = \frac{1}{2} AB \sin(\alpha + \beta)$$

Combining these two area calculations, we find

Area of largest triangle = Left area + Right area

$$\frac{1}{2} AB \sin(\alpha + \beta) = \frac{1}{2} AH \sin \alpha + \frac{1}{2} BH \sin \beta \qquad \blacktriangleleft \textbf{Perform area calculations.}$$

$$\sin(\alpha + \beta) = \frac{AH}{AB} \sin \alpha + \frac{BH}{AB} \sin \beta \qquad \blacktriangleleft \textbf{Divide by } \frac{1}{2} AB.$$

$$\sin(\alpha + \beta) = \frac{\cancel{A}H}{\cancel{A}B} \sin \alpha + \frac{\cancel{B}H}{A\cancel{B}} \sin \beta \qquad \blacktriangleleft \textbf{Cancel common factors.}$$

$$\sin(\alpha + \beta) = \frac{H}{B} \sin \alpha + \frac{H}{A} \sin \beta \qquad \blacktriangleleft \textbf{Simplify.}$$

$$\sin(\alpha + \beta) = \sin \alpha \cos \beta + \cos \alpha \sin \beta \qquad \blacktriangleleft \textbf{Use } \frac{H}{B} = \cos \beta, \frac{H}{A} = \cos \alpha.$$

This derivation of the sum formula for the sine function has the advantage of showing geometrically why the sum formula is true. The disadvantage is that the demonstration applies only when $\alpha + \beta$ is an acute angle, but the formula is true no matter the values of α and β.

One sum formula gives rise to others. Example 8.14 and the exercises will lead you through the verification of additional formulas.

CONCEPTS TO REMEMBER: Sum and Difference Formulas for the Sine and Cosine

The sine or cosine of a sum (or difference) of two angles can be written in terms of the sine and cosine of each angle:

$$\sin(\alpha + \beta) = \sin \alpha \cos \beta + \cos \alpha \sin \beta \qquad \cos(\alpha + \beta) = \cos \alpha \cos \beta - \sin \alpha \sin \beta$$

$$\sin(\alpha - \beta) = \sin \alpha \cos \beta - \cos \alpha \sin \beta \qquad \cos(\alpha - \beta) = \cos \alpha \cos \beta + \sin \alpha \sin \beta$$

EXAMPLE 8.11 Using the Sum Formula for the Sine

Find the exact value of $\sin 105°$.

SOLUTION

We note that $105°$ is the sum of the special angles $45°$ and $60°$. Applying the sum formula for the sine, we find

$$\sin 105° = \sin(45° + 60°)$$

$$= \sin 45° \cos 60° + \cos 45° \sin 60° \qquad \blacktriangleleft \text{ Apply the sum formula for the sine.}$$

$$= \left(\frac{1}{\sqrt{2}}\right)\left(\frac{1}{2}\right) + \left(\frac{1}{\sqrt{2}}\right)\left(\frac{\sqrt{3}}{2}\right) \qquad \blacktriangleleft \text{ Substitute values for sine and cosine.}$$

$$= \frac{1 + \sqrt{3}}{2\sqrt{2}} \text{ or } \frac{\left(\sqrt{2} + \sqrt{6}\right)}{4} \qquad \blacktriangleleft \text{ Simplify.}$$

TRY IT YOURSELF 8.11 Brief answers provided at the end of the section.

Use the difference formula for the sine along with your knowledge of special angles to find the exact value of $\sin\left(\dfrac{\pi}{12}\right)$. Note that $\dfrac{\pi}{12} = \dfrac{\pi}{3} - \dfrac{\pi}{4}$.

EXTEND YOUR REACH

Before seeing the sum formula for the sine function, one might suppose that $\sin(x + y) = \sin x + \sin y$. Find specific values of x and y showing that this "equation" is not true. *Suggestion*: One way to do this is to find x and y so that $\sin x + \sin y$ is greater than 1. (Recall from Section 7.1 that the value of the sine function is never greater than 1.)

EXAMPLE 8.12 Using the Difference Formula for the Cosine

Find the exact value of $\cos\left(\dfrac{\pi}{12}\right)$. Note that $\dfrac{\pi}{12} = \dfrac{\pi}{3} - \dfrac{\pi}{4}$.

SOLUTION

Applying the difference formula for the cosine gives

$$\cos\left(\frac{\pi}{12}\right) = \cos\left(\frac{\pi}{3} - \frac{\pi}{4}\right)$$

$$= \cos\left(\frac{\pi}{3}\right)\cos\left(\frac{\pi}{4}\right) + \sin\left(\frac{\pi}{3}\right)\sin\left(\frac{\pi}{4}\right)$$ ◀ Apply the difference formula for the cosine.

$$= \left(\frac{1}{2}\right)\left(\frac{1}{\sqrt{2}}\right) + \left(\frac{\sqrt{3}}{2}\right)\left(\frac{1}{\sqrt{2}}\right)$$ ◀ Substitute values for sine and cosine.

$$= \frac{1+\sqrt{3}}{2\sqrt{2}}$$ ◀ Simplify.

TRY IT YOURSELF 8.12 Brief answers provided at the end of the section.

Use the sum formula for the cosine along with your knowledge of special angles to find the exact value of $\cos\left(\frac{5\pi}{12}\right)$. Note that $\frac{5\pi}{12} = \frac{\pi}{6} + \frac{\pi}{4}$.

EXAMPLE 8.13 Sum and Difference Formulas with Given Values

Suppose $\sin s = \frac{2}{3}$ and $\sin t = -\frac{3}{4}$, where $\frac{\pi}{2} < s < \pi$ and $\pi < t < \frac{3\pi}{2}$. Calculate the value of $\sin(s-t)$.

SOLUTION

The difference formula for the sine requires both the sine and the cosine of both s and t. Applying the Pythagorean identity $\sin^2 x + \cos^2 x = 1$ gives

$$\cos s = \pm\frac{\sqrt{5}}{3} \quad \text{and} \quad \cos t = \pm\frac{\sqrt{7}}{4}$$

Because the trigonometric point $P(s)$ is in the second quadrant and $P(t)$ is in the third quadrant, both $\cos s$ and $\cos t$ are negative. Therefore, we must choose the minus sign in both of these equations:

$$\cos s = -\frac{\sqrt{5}}{3} \quad \text{and} \quad \cos t = -\frac{\sqrt{7}}{4}$$

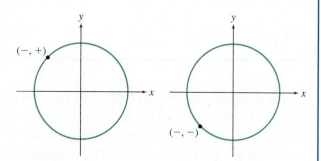

Now we have the components we need for the difference formula for the sine:

$$\sin(s-t) = \sin s \cos t - \cos s \sin t$$ ◀ Use the difference formula for the sine.

$$= \left(\frac{2}{3}\right)\left(-\frac{\sqrt{7}}{4}\right) - \left(-\frac{\sqrt{5}}{3}\right)\left(-\frac{3}{4}\right)$$ ◀ Substitute known values.

$$= -\frac{2\sqrt{7}+3\sqrt{5}}{12}$$ ◀ Simplify.

TRY IT YOURSELF 8.13 Brief answers provided at the end of the section.

Suppose that $\sin s = -\frac{2}{5}$ and $\cos t = \frac{\sqrt{2}}{3}$ and that both $P(s)$ and $P(t)$ are in the fourth quadrant. Find $\cos(s+t)$.

EXAMPLE 8.14 Deriving Sum and Difference Formulas

Use the sum formula for the sine together with the identities

$$\sin\left(x + \frac{\pi}{2}\right) = \cos x$$

$$\cos\left(x + \frac{\pi}{2}\right) = -\sin x$$

to derive the sum formula for the cosine.

SOLUTION

We begin by using the identity $\cos x = \sin\left(x + \frac{\pi}{2}\right)$ to write

$$\cos(s + t) = \sin\left((s + t) + \frac{\pi}{2}\right) = \sin\left(\left(s + \frac{\pi}{2}\right) + t\right)$$

Now we use the sum formula for the sine

$$\sin(A + B) = \sin A \cos B + \cos A \sin B$$

with $A = s + \dfrac{\pi}{2}$ and $B = t$. The result is

$$\cos(s + t) = \sin\left(\left(s + \frac{\pi}{2}\right) + t\right) \qquad \blacktriangleleft \text{ Use } \cos A = \sin\left(A + \frac{\pi}{2}\right).$$

$$= \sin\left(s + \frac{\pi}{2}\right)\cos t + \cos\left(s + \frac{\pi}{2}\right)\sin t \qquad \blacktriangleleft \text{ Use the sum formula for the sine.}$$

$$= \cos s \cos t - \sin s \sin t \qquad \blacktriangleleft \text{ Use } \sin\left(s + \frac{\pi}{2}\right) = \cos s, \cos\left(s + \frac{\pi}{2}\right) = -\sin s.$$

TRY IT YOURSELF 8.14 Brief answers provided at the end of the section.

Use the sum formula for the sine together with the facts that the sine is an odd function and the cosine is an even function to derive the difference formula for the sine. *Suggestion*: Begin with $\sin(s - t) = \sin(s + (-t))$.

Sum and Difference Formulas for the Tangent and Cotangent

> Identities that we know give rise to additional sum and difference formulas.

We can combine the sum formulas for the sine and cosine to get a sum formula for the tangent:

$$\tan(\alpha + \beta) = \frac{\sin(\alpha + \beta)}{\cos(\alpha + \beta)} \qquad \blacktriangleleft \text{ Use the definition of the tangent.}$$

$$= \frac{\sin \alpha \cos \beta + \cos \alpha \sin \beta}{\cos \alpha \cos \beta - \sin \alpha \sin \beta} \qquad \blacktriangleleft \text{ Use sum formulas for sine and cosine.}$$

Now we divide each term of the top and bottom by $\cos\alpha\cos\beta$. (This results in an equality because we are dividing the top and bottom of the fraction by the same thing.) The result is

$$\tan(\alpha+\beta)=\dfrac{\dfrac{\sin\alpha\cos\beta}{\cos\alpha\cos\beta}+\dfrac{\cos\alpha\sin\beta}{\cos\alpha\cos\beta}}{\dfrac{\cos\alpha\cos\beta}{\cos\alpha\cos\beta}-\dfrac{\sin\alpha\sin\beta}{\cos\alpha\cos\beta}}$$ ◀ Divide top and bottom by $\cos\alpha\cos\beta$.

$$=\dfrac{\dfrac{\sin\alpha\,\cos\beta}{\cos\alpha\,\cos\beta}+\dfrac{\cos\alpha\,\sin\beta}{\cos\alpha\,\cos\beta}}{\dfrac{\cos\alpha\,\cos\beta}{\cos\alpha\,\cos\beta}-\dfrac{\sin\alpha\sin\beta}{\cos\alpha\cos\beta}}$$ ◀ Cancel common terms.

$$=\dfrac{\dfrac{\sin\alpha}{\cos\alpha}+\dfrac{\sin\beta}{\cos\beta}}{1-\left(\dfrac{\sin\alpha}{\cos\alpha}\right)\left(\dfrac{\sin\beta}{\cos\beta}\right)}$$ ◀ Simplify.

$$=\dfrac{\tan\alpha+\tan\beta}{1-\tan\alpha\tan\beta}$$ ◀ Use the definition of tangent.

In the exercises, you are guided through the derivation of additional formulas.

CONCEPTS TO REMEMBER: Sum and Difference Formulas for the Tangent and Cotangent

The tangent or cotangent of a sum (or difference) of two angles can be written in terms of the tangent or cotangent of each angle:

$$\tan(\alpha+\beta)=\dfrac{\tan\alpha+\tan\beta}{1-\tan\alpha\tan\beta} \qquad \cot(\alpha+\beta)=\dfrac{\cot\alpha\cot\beta-1}{\cot\alpha+\cot\beta}$$

$$\tan(\alpha-\beta)=\dfrac{\tan\alpha-\tan\beta}{1+\tan\alpha\tan\beta} \qquad \cot(\alpha-\beta)=\dfrac{\cot\alpha\cot\beta+1}{\cot\beta-\cot\alpha}$$

EXAMPLE 8.15 Using the Sum Formula for the Tangent

Find the exact value of $\tan\left(\dfrac{5\pi}{12}\right)$. Note that $\dfrac{5\pi}{12}=\dfrac{\pi}{6}+\dfrac{\pi}{4}$.

SOLUTION

We use the sum formula for the tangent:

$$\tan\left(\dfrac{5\pi}{12}\right)=\tan\left(\dfrac{\pi}{6}+\dfrac{\pi}{4}\right)$$

$$=\dfrac{\tan\left(\dfrac{\pi}{6}\right)+\tan\left(\dfrac{\pi}{4}\right)}{1-\tan\left(\dfrac{\pi}{6}\right)\tan\left(\dfrac{\pi}{4}\right)}$$ ◀ Use the sum formula for the tangent.

$$= \frac{1/\sqrt{3} + 1}{1 - (1/\sqrt{3})(1)}$$ ◀ Use $\tan\dfrac{\pi}{6} = \dfrac{1}{\sqrt{3}}$ and $\tan\dfrac{\pi}{4} = 1$.

$$= \frac{1 + \sqrt{3}}{\sqrt{3} - 1}$$ ◀ Multiply top and bottom by $\sqrt{3}$.

This can also be written as $2 + \sqrt{3}$.

TRY IT YOURSELF 8.15 Brief answers provided at the end of the section.

Use the sum formula for the tangent to find the exact value of $\tan 165°$. Note that $165 = 120 + 45$.

EXTEND YOUR REACH

a. In Chapter 7 we noted that π is the period of the tangent function. Use the sum formula for the tangent to verify that $\tan(x + \pi) = \tan x$.

b. It is true that $\tan\left(x - \dfrac{\pi}{2}\right) = -\cot x$. Explain why the sum formula for the tangent cannot be used to establish this identity.

c. Establish the identity in part b using the difference formula for the cotangent.

EXAMPLE 8.16 Using the Difference Formula for the Cotangent

If $\cot s = 2$ and $\cot t = -3$, calculate the value of $\cot(s - t)$.

SOLUTION

Applying the difference formula for the cotangent, we find

$$\cot(s - t) = \frac{\cot s \cot t + 1}{\cot t - \cot s}$$ ◀ Use the difference formula for the cotangent.

$$= \frac{(2)(-3) + 1}{-3 - 2}$$ ◀ Use $\cot s = 2$ and $\cot t = -3$.

$$= 1$$ ◀ Simplify.

TRY IT YOURSELF 8.16 Brief answers provided at the end of the section.

Use the sum formula for the cotangent to find the exact value of $\cot 150°$. Note that $150 = 120 + 30$.

Product-to-Sum and Sum-to-Product Formulas

> Sum and difference formulas for the sine and cosine can be combined to produce formulas for products involving sine and cosine.

In certain settings in calculus, it is important to rewrite products of the sine and cosine functions in terms of sums. We use the sum and difference formulas for the sine to accomplish this:

$$\sin(s + t) = \sin s \cos t + \cos s \sin t$$
$$\sin(s - t) = \sin s \cos t - \cos s \sin t$$

If we add these two equations and divide by 2, we get

$$\frac{1}{2}\sin(s+t) + \frac{1}{2}\sin(s-t) = \sin s \cos t$$

Similar calculations using the sum and difference formulas for the cosine give additional sum-to-product formulas. We can also go the other way, rewriting sums in terms of products.

CONCEPTS TO REMEMBER: Product-to-Sum Formulas and Sum-to-Product Formulas

Products of sines and cosines can be written in terms of sums, and sums of sines and cosines can be written in terms of products:

$$\sin s \cos t = \frac{1}{2}\sin(s+t) + \frac{1}{2}\sin(s-t) \qquad \sin s + \sin t = 2\sin\left(\frac{s+t}{2}\right)\cos\left(\frac{s-t}{2}\right)$$

$$\cos s \cos t = \frac{1}{2}\cos(s+t) + \frac{1}{2}\cos(s-t) \qquad \sin s - \sin t = 2\cos\left(\frac{s+t}{2}\right)\sin\left(\frac{s-t}{2}\right)$$

$$\sin s \sin t = \frac{1}{2}\cos(s-t) - \frac{1}{2}\cos(s+t) \qquad \cos s + \cos t = 2\cos\left(\frac{s+t}{2}\right)\cos\left(\frac{s-t}{2}\right)$$

$$\cos s - \cos t = -2\sin\left(\frac{s+t}{2}\right)\sin\left(\frac{s-t}{2}\right)$$

EXAMPLE 8.17 Using Product-to-Sum Formulas

Use the product-to-sum formulas and your knowledge of special angles to find the value of $\sin 52.5° \cos 7.5°$.

SOLUTION

We use the product-to-sum formula

$$\sin s \cos t = \frac{1}{2}\sin(s+t) + \frac{1}{2}\sin(s-t)$$

with $s = 52.5°$ and $t = 7.5°$:

$$\sin 52.5° \cos 7.5° = \frac{1}{2}\sin(52.5° + 7.5°) + \frac{1}{2}\sin(52.5° - 7.5°) \qquad \blacktriangleleft \text{ Use the product-to-sum formula.}$$

$$= \frac{1}{2}\sin 60° + \frac{1}{2}\sin 45° \qquad \blacktriangleleft \text{ Simplify.}$$

$$= \frac{1}{2}\frac{\sqrt{3}}{2} + \frac{1}{2}\frac{1}{\sqrt{2}} \qquad \blacktriangleleft \text{ Use } \sin 60° = \frac{\sqrt{3}}{2} \text{ and } \sin 45° = \frac{1}{\sqrt{2}}.$$

$$= \frac{\sqrt{3}}{4} + \frac{1}{2\sqrt{2}} \qquad \blacktriangleleft \text{ Simplify.}$$

$$= \frac{\sqrt{3} + \sqrt{2}}{4} \qquad \blacktriangleleft \text{ Combine fractions.}$$

TRY IT YOURSELF 8.17 Brief answers provided at the end of the section.

Use the product-to-sum formulas and your knowledge of special angles to find the value of $\sin 7.5° \cos 52.5°$.

EXTEND YOUR REACH

Show that if $s + t = \pi$ then $\sin s = \sin t$. *Suggestion*: Use the sum-to-product formula for $\sin s - \sin t$.

EXAMPLE 8.18 Using Sum-to-Product Formulas

Use a sum-to-product formula and your knowledge of special angles to find $\sin 75° + \sin 15°$.

SOLUTION

We use the first of the sum-to-product formulas listed:

$$\sin s + \sin t = 2 \sin\left(\frac{s+t}{2}\right)\cos\left(\frac{s-t}{2}\right) \quad \blacktriangleleft \text{ Use the sum-to-product formula.}$$

$$\sin 75° + \sin 15° = 2 \sin\left(\frac{75° + 15°}{2}\right)\cos\left(\frac{75° - 15°}{2}\right) \quad \blacktriangleleft \text{ Substitute known values.}$$

$$= 2 \sin 45° \cos 30° \quad \blacktriangleleft \text{ Simplify.}$$

$$= 2 \frac{1}{\sqrt{2}} \frac{\sqrt{3}}{2} \quad \blacktriangleleft \text{ Use } \sin 45° = \frac{1}{\sqrt{2}} \text{ and } \cos 30° = \frac{\sqrt{3}}{2}.$$

$$= \frac{\sqrt{3}}{\sqrt{2}} \quad \blacktriangleleft \text{ Simplify.}$$

TRY IT YOURSELF 8.18 Brief answers provided at the end of the section.

Use a sum-to-product formula and your knowledge of special angles to find the value of $\cos 15° - \cos 75°$.

TRY IT YOURSELF ANSWERS

8.11 $\dfrac{1}{2\sqrt{2}}(\sqrt{3} - 1)$ or $\dfrac{\sqrt{2}}{4}(\sqrt{3} - 1)$

8.12 $\dfrac{1}{2\sqrt{2}}(\sqrt{3} - 1)$ or $\dfrac{\sqrt{2}}{4}(\sqrt{3} - 1)$

8.13 $\dfrac{\sqrt{42} - 2\sqrt{7}}{15}$

8.14 Briefly, $\sin(s - t) = \sin(s + (-t)) = \sin s \cos(-t) + \cos s \sin(-t) = \sin s \cos t - \cos s \sin t$.

8.15 $\dfrac{-\sqrt{3} + 1}{1 + \sqrt{3}}$ or $\sqrt{3} - 2$

8.16 $-\sqrt{3}$

8.17 $\dfrac{\sqrt{3} - \sqrt{2}}{4}$

8.18 $\dfrac{1}{\sqrt{2}}$

EXERCISE SET 8.2

CHECK YOUR UNDERSTANDING

1. State the sum and difference formulas for the sine.

2. State the sum and difference formulas for the cosine.

3. State the sum and difference formulas for the tangent.

4. State the sum and difference formulas for the cotangent.

5. **True or false:** $\sin(x + y) = \sin x + \sin y$.

6. **True or false:** A sum of cosine functions can be written as a product of cosine functions.

7. **True or false:** A product of sine functions can be written as a difference of cosine functions.

8. **True or false:** A sum of sine functions can be written as a product of sine and cosine functions.

9. **True or false:** A product of cosine functions can be written as a sum of cosine functions.

SKILL BUILDING

Using the area formula. In Exercises 10 through 15, you are given sides A and B (in feet) and the angle t of the triangle shown in **Figure 8.12**. Find the area of the triangle. In each case, the angle involved is either a special angle or has a special angle as its reference angle.

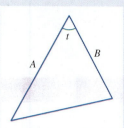

Figure 8.12 A triangle

10. $A = 2$, $B = 3$, $t = \dfrac{\pi}{4}$ radian

11. $A = 3$, $B = 5$, $t = \dfrac{2\pi}{3}$ radians

12. $A = 7$, $B = 2$, $t = 30°$

13. $A = 3$, $B = 9$, $t = 150°$

14. $A = 12$, $B = 9$, $t = \dfrac{5\pi}{6}$ radians

15. $A = 2$, $B = 4$, $t = \dfrac{3\pi}{4}$ radians

Sum formulas with special angles. Exercises 16 through 29 can be solved using sum or difference formulas involving special angles or angles whose reference angles are special angles.

16. Find the sine and cosine of $75°$. Note that $75 = 45 + 30$.

17. Find the tangent and cotangent of $75°$. Note that $75 = 45 + 30$.

18. Find the sine and cosine of $15°$. Note that $45 - 30 = 15$.

19. Find the tangent and cotangent of $15°$. Note that $45 - 30 = 15$.

20. Find the sine and cosine of $\dfrac{13\pi}{12}$. Note that $\dfrac{13\pi}{12} = \dfrac{5\pi}{6} + \dfrac{\pi}{4}$.

21. Find the tangent and cotangent of $\dfrac{\pi}{12}$. Note that $\dfrac{\pi}{12} = \dfrac{\pi}{4} - \dfrac{\pi}{6}$.

22. Find the sine and cosine of $\dfrac{19\pi}{12}$. Note that $\dfrac{19\pi}{12} = \dfrac{4\pi}{3} + \dfrac{\pi}{4}$.

23. Find the tangent and cotangent of $\dfrac{11\pi}{12}$. Note that $\dfrac{11\pi}{12} = \dfrac{2\pi}{3} + \dfrac{\pi}{4}$.

24. Find the sine and cosine of $\dfrac{7\pi}{12}$. Note that $\dfrac{7\pi}{12} = \dfrac{\pi}{3} + \dfrac{\pi}{4}$.

25. Find the tangent and cotangent of $105°$. Note that $105 = 60 + 45$.

26. Find the sine and cosine of $\dfrac{11\pi}{12}$. Note that $\dfrac{11\pi}{12} = \dfrac{5\pi}{4} - \dfrac{\pi}{3}$.

27. Find the tangent and cotangent of $165°$. Note that $165 = 225 - 60$.

28. Find the tangent and cotangent of $195°$. Note that $195 = 150 + 45$.

29. Find the tangent and cotangent of $\dfrac{19\pi}{12}$. Note that $\dfrac{19\pi}{12} = \dfrac{4\pi}{3} + \dfrac{\pi}{4}$.

Sum formulas with given values. In Exercises 30 through 38, assume s is an angle with $\sin s = \dfrac{1}{4}$, t is an angle with $\cos t = \dfrac{2}{3}$, and both angles are between 0 and $\dfrac{\pi}{2}$. Use this information, together with appropriate identities, to calculate the value of the function.

30. $\sin(s + t)$ **35.** $\tan(s + t)$

31. $\sin(s - t)$ **36.** $\tan(s - t)$

32. $\cos(s + t)$ **37.** $\cot(s + t)$

33. $\cos(s - t)$ **38.** $\cot(t - s)$

34. $\sin(t - s)$

Sum formulas with angles from other quadrants. For Exercises 39 through 46, assume that $\sin x = \dfrac{3}{5}$ and $\sin y = -\dfrac{2}{5}$, where $\dfrac{\pi}{2} < x < \pi$ and $\dfrac{3\pi}{2} < y < 2\pi$. Calculate the indicated function value.

39. $\sin(x + y)$ **43.** $\tan(x + y)$

40. $\sin(x - y)$ **44.** $\cot(x + y)$

41. $\cos(x + y)$ **45.** $\tan(x - y)$

42. $\cos(x - y)$ **46.** $\cot(x - y)$

Product-to-sum and sum-to-product formulas. In Exercises 47 through 53, use a sum-to-product or product-to-sum formula together with your knowledge of special angles to calculate the given function value.

47. $\sin\dfrac{5\pi}{24}\cos\dfrac{\pi}{24}$ **51.** $\cos\dfrac{5\pi}{24}\cos\dfrac{\pi}{24}$

48. $\sin\dfrac{5\pi}{12} - \sin\dfrac{\pi}{12}$ **52.** $\sin 105° - \sin 15°$

49. $\cos\dfrac{5\pi}{12} + \cos\dfrac{\pi}{12}$ **53.** $\sin\dfrac{5\pi}{24}\sin\dfrac{\pi}{24}$

50. $\cos\dfrac{5\pi}{24}\sin\dfrac{\pi}{24}$

PROBLEMS

Using sum formulas to establish identities. In Exercises 54 through 66, use appropriate sum formulas to show the given identity. Some of these have been established by other methods in previous sections.

54. $\sin\left(x + \dfrac{\pi}{2}\right) = \cos x$

55. $\cos\left(x + \dfrac{\pi}{2}\right) = -\sin x$

56. $\sin\left(x - \dfrac{\pi}{2}\right) = -\cos x$

57. $\cos\left(x - \dfrac{\pi}{2}\right) = \sin x$

58. $\sin(x + \pi) = -\sin x$

59. $\cos(x + \pi) = -\cos x$

60. $\cos(\pi - x) = -\cos x$

61. $\sin(\pi - x) = \sin x$

62. $\sin\left(x + \dfrac{\pi}{4}\right) = \dfrac{1}{\sqrt{2}}(\sin x + \cos x)$

63. $\cos\left(x + \dfrac{\pi}{4}\right) = \dfrac{1}{\sqrt{2}}(\cos x - \sin x)$

64. $\cos\left(x - \dfrac{\pi}{4}\right) = \sin\left(x + \dfrac{\pi}{4}\right)$

65. $\tan(x + \pi) = \tan x$

66. $\cot(x + \pi) = \cot x$

67. The tangent. Use the results of Exercises 54 and 55 to show that $\tan\left(x + \dfrac{\pi}{2}\right) = -\cot x$. Explain why this identity cannot be verified by using the sum formula for the tangent.

Finding other sum formulas. In Exercises 68 through 73, you will be led through derivations of some of the sum and difference formulas.

68. Difference formula for the cosine. Use the sum formula for the cosine and the facts that the sine function is odd and the cosine function is even to derive the difference formula for the cosine. *Suggestion:* Note that $\cos(\alpha - \beta) = \cos(\alpha + (-\beta))$, and then apply the sum formula for the cosine.

69. **Difference formula for the tangent.** Use the sum formula for the tangent and the fact that the tangent is an odd function to derive the difference formula for the tangent. *Suggestion:* Note that $\tan(\alpha - \beta) = \tan(\alpha + (-\beta))$, and then apply the sum formula for the tangent.

70. **Sum formula for the cotangent.** Use the sum formulas for the sine and cosine to derive the sum formula for the cotangent.

71. **Difference formula for the cotangent.** Use the sum formula for the cotangent and the fact the cotangent is an odd function to derive the difference formula for the cotangent.

72. **Sum formula for the secant.** Show that
$$\sec(\alpha + \beta) = \frac{\sec \alpha \sec \beta}{1 - \tan \alpha \tan \beta}$$

Suggestion: First write $\sec(\alpha + \beta) = \dfrac{1}{\cos(\alpha + \beta)}$ and use the sum formula for the cosine. Next divide both top and bottom of the fraction by $\cos \alpha \cos \beta$.

73. **Sum formula for the cosecant.** Show that
$$\csc(\alpha + \beta) = \frac{\csc \alpha \csc \beta}{\cot \beta + \cot \alpha}$$

Suggestion: First write $\csc(\alpha + \beta) = \dfrac{1}{\sin(\alpha + \beta)}$, and apply the sum formula for the sine. Next divide both top and bottom of the fraction by $\sin \alpha \sin \beta$.

Product-to-sum and sum-to-product formulas.
In Exercises 74 through 76, you are asked to derive a product-to-sum or sum-to-product formula.

74. Use the sum and difference formulas for the cosine to show that $\cos s \cos t = \dfrac{1}{2}\cos(s + t) + \dfrac{1}{2}\cos(s - t)$.

Suggestion: Add equations corresponding to $\cos(s + t)$ and $\cos(s - t)$, and then divide by 2.

75. Use the product-to-sum formulas to show that
$$\cos x + \cos y = 2\cos\left(\frac{x + y}{2}\right)\cos\left(\frac{x - y}{2}\right)$$

76. Use the sum and difference formulas for the cosine to show that $\sin s \sin t = \dfrac{1}{2}\cos(s - t) - \dfrac{1}{2}\cos(s + t)$.

Suggestion: Subtract equations corresponding to $\cos(s - t)$ and $\cos(s + t)$, and divide by 2.

77. **Adding three angles.** Show that
$$\sin(\alpha + \beta + \gamma) = \sin \alpha \cos \beta \cos \gamma + \cos \alpha$$
$$\sin \beta \cos \gamma + \cos \alpha \cos \beta \sin \gamma - \sin \alpha \sin \beta \sin \gamma$$

Suggestion: Note that $\sin(\alpha + \beta + \gamma) = \sin(A + \gamma)$, where $A = \alpha + \beta$, and then use sum formulas to expand this expression.

78. **Harmonic form.** Electrical engineers sometimes need to write a sum of trigonometric functions in a different form, known as the harmonic form:
$$a \cos \omega t + b \sin \omega t = A \sin(\omega t + \phi)$$

where $A = \sqrt{a^2 + b^2}$ and ϕ is the solution of $\tan \phi = \dfrac{a}{b}$ on the quadrant determined by (sign of b, sign of a). The goal of this problem is to derive this important form. For simplicity, we assume that $b \neq 0$.

a. Use the sum formula for the sine to expand $A \sin(\omega t + \phi)$.

b. Replace the right-hand side in the harmonic form with the expansion from part a, and then equate the coefficients of $\sin \omega t$ on both sides and the coefficients of $\cos \omega t$ on both sides to show that
$$a = A \sin \phi$$
$$b = A \cos \phi$$

c. Use the result of part b to show that $\dfrac{a}{b} = \tan \phi$.

d. Use the results of part b to show that $A^2 = a^2 + b^2$, so if we take $A \geq 0$, then $A = \sqrt{a^2 + b^2}$.

79. **Using the harmonic form.** The goal of this exercise is to use the harmonic form derived in the preceding exercise to express $-2 \cos t + 2\sqrt{3} \sin t$ in the form $A \sin(t + \phi)$. (Note that here $\omega = 1$.)

a. Find the value of A.

b. Determine the quadrant to be used in the solution of $\tan \phi = \dfrac{a}{b}$.

c. Find ϕ if we take $-\pi < \phi \leq \pi$. (Note that the reference angle for ϕ is one of the special angles.)

d. Complete the solution.

MODELS AND APPLICATIONS

80. Area of an isosceles triangle. Consider the isosceles triangle with base a shown in **Figure 8.13**. In **Figure 8.14** we have constructed an altitude, which bisects the angle α because the triangle is isosceles.

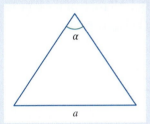

Figure 8.13 An isosceles triangle

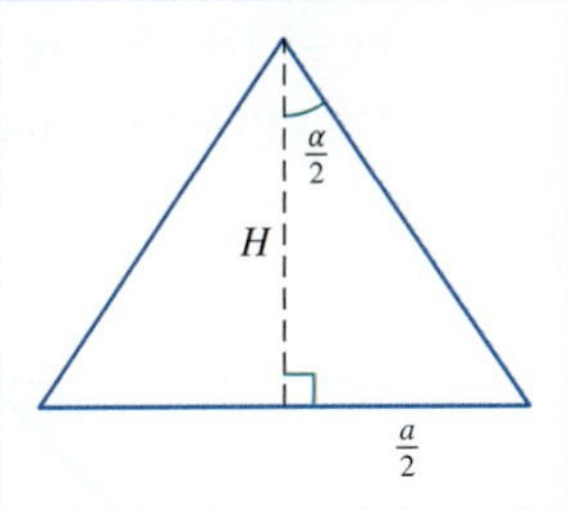

Figure 8.14 An isosceles triangle: adding an altitude

a. Show that the length of the altitude H is
$$\frac{a}{2}\cot\frac{\alpha}{2}.$$

b. Show that the area of the triangle in Figure 8.13 is $\frac{1}{4}a^2\cot\frac{\alpha}{2}$.

81. Area of a regular polygon. This exercise uses the results of the preceding exercise.

Show that the area of a regular polygon, each side of which is of length a, is
$$\text{Area} = \frac{1}{4}na^2\cot\frac{180°}{n}$$
where n is the number of sides of the polygon. *Suggestion:* Beginning at the center point of the polygon, we can slice it (like a pie) into n isosceles triangles, each with angle of $\frac{360}{n}$ degrees opposite a base of length a. This is shown for a pentagon in **Figure 8.15**.

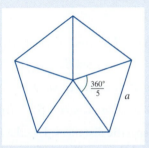

Figure 8.15 A pentagon cut into five isosceles triangles

82. Area of a hexagon. This exercise uses the formula derived in the preceding exercise. Find the area of a regular hexagon if each side has length 3.

CHALLENGE EXERCISES FOR INDIVIDUALS OR GROUPS

83. Sum formula for the cosine. Our earlier derivation of the sum formula for the sine function works only when $\alpha + \beta$ is an acute angle. In this exercise, you are led through a derivation of the sum formula for the cosine that is not limited to acute angles.

a. A distance calculation. Show that if (x_1, y_1) and (x_2, y_2) are points on the unit circle as shown in **Figure 8.16**, then the square of the distance between them is given by
$$\text{Square of distance} = 2 - 2(x_1x_2 + y_1y_2)$$

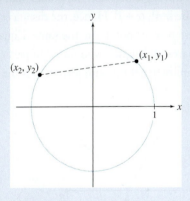

Figure 8.16 Two points on the unit circle

b. Coordinates of points. Figure 8.17 shows angles α, $\alpha + \beta$, and $-\beta$ on the unit circle.

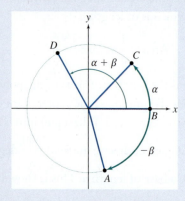

Figure 8.17 Angles α, $\alpha + \beta$, and $-\beta$

Show that

$$\text{Coordinates of } A = (\cos(-\beta), \sin(-\beta))$$
$$= (\cos \beta, -\sin \beta)$$
$$\text{Coordinates of } C = (\cos \alpha, \sin \alpha)$$
$$\text{Coordinates of } D = (\cos(\alpha + \beta), \sin(\alpha + \beta))$$

c. Calculating specific distances. Use the formula from part a and the coordinates from part b to show the following.

  a. Square of distance from A to
$$C = 2 - 2(\cos \alpha \cos \beta - \sin \alpha \sin \beta)$$

  b. Square of distance from B to
$$D = 2 - 2\cos(\alpha + \beta)$$

d. Equating distances. We see in **Figure 8.18** that the angle between the points marked A and C and the angle between the points marked B and D are both $\alpha + \beta$. Hence, the distances between these pairs of points are the same. Equate the squares of distances you found in part c and simplify to get the sum formula for the cosine.

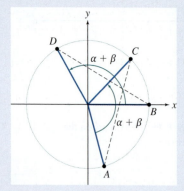

Figure 8.18 Comparing distances: The distance from A to C equals the distance from B to D.

84. Another geometric demonstration of the sum formula for the sine. This exercise shows an alternative geometric demonstration of the sum formula for the sine of acute angles. The demonstration uses the theorem from geometry that if parallel lines are cut by a transversal, alternate interior angles are congruent (denoted by $\cong$). (See **Figure 8.19**.) All parts of the exercise refer to the rectangle in **Figure 8.20**.

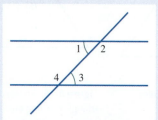

Figure 8.19 Parallel lines cut by a transversal: $\angle 1 \cong \angle 3$ and $\angle 2 \cong \angle 4$

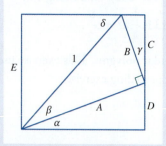

Figure 8.20 A rectangle to demonstrate the sum formula for the sine

a. Identifying angles. Show that $\gamma = \alpha$ and $\delta = \alpha + \beta$.

b. Calculating first side lengths. Show that $B = \sin \beta$ and $A = \cos \beta$.

c. Calculating next side lengths. Show the following three equalities:

$$D = A \sin \alpha = \sin \alpha \cos \beta$$
$$C = B \cos \gamma = \sin \beta \cos \alpha$$
$$E = \sin \delta = \sin(\alpha + \beta)$$

d. Final calculation. Equate the left and right sides of the rectangle to get the sum formula for the sine.

85. A geometric demonstration of the sum formula for the cosine. This exercise shows a geometric demonstration of the sum formula for the cosine of acute angles. The demonstration uses a familiar theorem from geometry that if parallel lines are cut by a transversal, alternate interior angles are congruent. (See Figure 8.19.) All parts of the exercise refer to the rectangle in **Figure 8.21**.

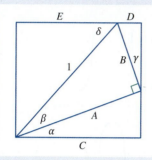

Figure 8.21 A rectangle to demonstrate the sum formula for the cosine

a. **Identifying angles.** Show that $\gamma = \alpha$ and $\delta = \alpha + \beta$.

b. **Calculating first side lengths.** Show that $B = \sin \beta$ and $A = \cos \beta$.

c. **Calculating next side lengths.** Show the following three equalities:

$$D = B \sin \gamma = \sin \alpha \sin \beta$$
$$C = A \cos \alpha = \cos \beta \cos \alpha$$
$$E = \cos \delta = \cos(\alpha + \beta)$$

d. **Final calculation.** Equate the top and bottom sides of the rectangle, and simplify to get the sum formula for the cosine.

86. A sum of tangents. Suppose $x + y + z = \pi$. Show that $\tan x + \tan y + \tan z = \tan x \tan y \tan z$. *Suggestion*: Start with $\tan(x + y + z) = \tan \pi$, and then apply the sum formula for the tangent.

REVIEW AND REFRESH: Exercises from Previous Sections

---

87. From Section 2.1: If $f(x) = x^2 - 1$ and $g(x) = x + 1$, find a formula that gives the composition $(f \circ g)(x)$. Simplify your answer.

Answer: $(f \circ g)(x) = x^2 + 2x$

88. From Section 2.2: The graph of a function f is shown in **Figure 8.22**. Add the line $y = x$ to the figure, and then add the graph of the inverse function.

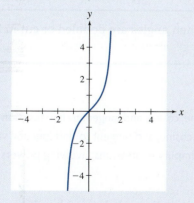

Figure 8.22 Graph for Exercise 88

Answer:

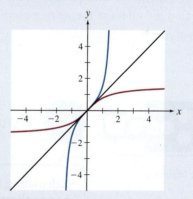

89. From Section 4.1: What is the limit as $x \to \infty$ of $f(x) = \ln x$?

Answer: $\ln x \to \infty$ as $x \to \infty$

90. From Section 4.1: Express $\log_7 3$ in terms of the natural logarithm, and then use a calculator to get its value (accurate to two decimal places).

Answer: $\log_7 3 = \dfrac{\ln 3}{\ln 7} \approx 0.56$

91. From Section 4.3: Find the exact solution of $3^x = 4 \times 2^x$, and then use a calculator to find the answer correct to two decimal places.

Answer: $x = \dfrac{\ln 4}{\ln \frac{3}{2}} \approx 3.42$

92. From Section 5.5: Plot the graph of $R(x) = \dfrac{x^2 - 4}{x^2 - 4x + 3}$. Show any horizontal or vertical asymptotes that occur.

Answer:

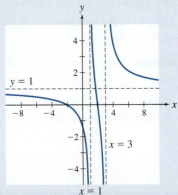

93. From Chapter 6: The point $P(t) = \left(x, \dfrac{3}{4} \right)$ lies on the unit circle in the first quadrant. Find x, and then find the values of the six trigonometric functions of t.

Answer:

$$x = \frac{\sqrt{7}}{4}, \ \cos t = \frac{\sqrt{7}}{4}, \ \sin t = \frac{3}{4},$$

$$\tan t = \frac{3}{\sqrt{7}}, \ \cot t = \frac{\sqrt{7}}{3}, \ \sec t = \frac{4}{\sqrt{7}}, \ \csc t = \frac{4}{3}$$

94. From Section 7.2: Identify the trigonometric function whose graph is shown in **Figure 8.23**.

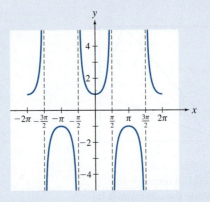

Figure 8.23

Answer: $y = \sec x$

95. From Section 8.1: Express $\tan t$ in terms of $\sin t$ if $\dfrac{\pi}{2} < t < \pi$.

Answer: $\tan t = -\dfrac{\sin t}{\sqrt{1 - \sin^2 t}}$

8.1 Trigonometric Identities

8.2 Sum and Difference Formulas

8.3 Double-Angle and Half-Angle Formulas

8.4 Inverse Trigonometric Functions

8.5 Solving Trigonometric Equations

8.3 Double-Angle and Half-Angle Formulas

Calculating trigonometric functions when angles are doubled or cut in half has applications in engineering.

In this section, you will learn to:

1. Use the double-angle formulas for the sine, cosine, and tangent to find function values.
2. Use the half-angle formulas for the sine, cosine, and tangent to find function values.
3. Use the power reduction formulas to simplify expressions involving powers of sine and cosine.

If a roof line makes an angle of $t°$ with the horizontal, then its slope is given by

$$\text{Slope} = \frac{\text{Rise}}{\text{Run}} = \tan t$$

as illustrated in **Figure 8.24**. Suppose a carpenter looks at plans for a roof inclined at an angle of t but wants to build a roof inclined at an angle $\dfrac{t}{2}$. The proper cut on the rafters is determined by the slope. How is the carpenter to determine the new slope? Because the old slope is $\tan t$, the new slope is $\tan\left(\dfrac{t}{2}\right)$. Thus, a half-angle formula for the tangent is needed. That is, we need a formula that allows the calculation of $\tan\left(\dfrac{t}{2}\right)$ in terms of trigonometric functions of t. This sort of calculation is common in carpentry, architecture, and engineering.

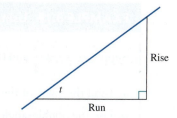

Figure 8.24 The slope of a roof line

In this section, we use the formulas from the preceding sections to find formulas that let us calculate trigonometric functions of multiples of angles. Most important are the double-angle and half-angle formulas. These formulas have important applications, and they are essential tools for calculus.

Double-Angle Formulas

> **The sum and difference formulas for the sine and cosine produce double-angle formulas.**

Double-angle formulas can be thought of as special cases of sum formulas. For example, to derive the double-angle formula for the sine, we put $x = y = t$ in the sum formula for the sine, $\sin(x + y) = \sin x \cos y + \cos x \sin y$:

$$\sin(2t) = \sin(t + t)$$
$$\sin(2t) = \sin t \cos t + \cos t \sin t$$
$$\sin(2t) = 2\sin t \cos t$$

Similar methods can be used to find double-angle formulas for the cosine and tangent.

CONCEPTS TO REMEMBER: Double-Angle Formulas

The sine or cosine of twice an angle can be expressed in terms of the sine and cosine of the angle. The tangent of twice an angle can be expressed in terms of the tangent of the angle. Here are the identities:

- $\sin 2t = 2\sin t \cos t$

- $\cos 2t = \begin{cases} \text{Formula 1:} & \cos^2 t - \sin^2 t \\ \text{Formula 2:} & 1 - 2\sin^2 t \\ \text{Formula 3:} & 2\cos^2 t - 1 \end{cases}$

- $\tan 2t = \dfrac{2\tan t}{1 - \tan^2 t}$

EXAMPLE 8.19   Using Double-Angle Formulas for the Sine and Cosine

Suppose $\sin t = \dfrac{3}{4}$ and $0 < t < \dfrac{\pi}{2}$.

a. Find the value of the cosine of t.

b. Use the double-angle formula for the sine to find the value of $\sin 2t$.

c. Use the double-angle formula for the cosine to find the value of $\cos 2t$.

SOLUTION

a. We are given the value of the sine, so we can use the Pythagorean identity to find the value of the cosine:

$$\sin^2 t + \cos^2 t = 1 \qquad \blacktriangleleft \text{ Use the Pythagorean identity.}$$

$$\left(\frac{3}{4}\right)^2 + \cos^2 t = 1 \qquad \blacktriangleleft \text{ Substitute known values.}$$

$$\frac{9}{16} + \cos^2 t = 1 \qquad \blacktriangleleft \text{ Simplify.}$$

$$\cos^2 t = \frac{7}{16} \qquad \blacktriangleleft \text{ Solve for } \cos^2 t.$$

$$\cos t = \pm\frac{\sqrt{7}}{4} \qquad \blacktriangleleft \text{ Take square roots.}$$

In the first quadrant, all trigonometric functions are positive. We conclude that $\cos t = \dfrac{\sqrt{7}}{4}$.

b. We use the double-angle formula for the sine:

$$\sin 2t = 2\sin t \cos t \qquad \blacktriangleleft \text{ Use the double-angle formula.}$$

$$= 2\left(\frac{3}{4}\right)\left(\frac{\sqrt{7}}{4}\right) \qquad \blacktriangleleft \text{ Substitute known values.}$$

$$= \frac{3\sqrt{7}}{8} \qquad \blacktriangleleft \text{ Simplify.}$$

c. For this part, we use the double-angle formula for the cosine that involves only the sine:

$$\cos 2t = 1 - 2\sin^2 t \qquad \blacktriangleleft \text{ Use the double-angle formula.}$$

$$= 1 - 2\left(\frac{3}{4}\right)^2 \qquad \blacktriangleleft \text{ Substitute known values.}$$

$$= 1 - \frac{9}{8} \qquad \blacktriangleleft \text{ Simplify.}$$

$$= -\frac{1}{8} \qquad \blacktriangleleft \text{ Simplify.}$$

TRY IT YOURSELF 8.19   Brief answers provided at the end of the section.

Find $\cos 2t$ if $\cos t = -\dfrac{1}{3}$.

a. Use the sum formula for the cosine to show that

$$\cos(nx) = \cos((n-1)x)\cos x - \sin((n-1)x)\sin x$$

Suggestion: Use the fact that $nx = (n-1)x + x$.

b. Show that

$$\cos((n-2)x) = \cos((n-1)x)\cos x + \sin((n-1)x)\sin x$$

Suggestion: Use the fact that $(n-2)x = (n-1)x - x$.

c. Add the equations from parts a and b to show that

$$\cos(nx) = 2\cos x \cos((n-1)x) - \cos((n-2)x)$$

d. Use the formula from part c with $n = 3$ to find a triple-angle formula for the cosine. (This formula is derived in a different way in Example 8.25.)

e. Can you find a quadruple-angle formula for the cosine? Explain how the formula from part c could be used to find $\cos(nx)$ in terms of $\cos x$ for any positive integer n.

EXAMPLE 8.20 Using the Double-Angle Formula for the Tangent

If $\tan t = 3$, find $\tan 2t$.

SOLUTION

We use the double-angle formula for the tangent:

$$\tan 2t = \frac{2\tan t}{1 - \tan^2 t} \qquad \blacktriangleleft \text{Use the double-angle formula.}$$

$$= \frac{2 \times 3}{1 - 3^2} \qquad \blacktriangleleft \text{Substitute known values.}$$

$$= -\frac{3}{4} \qquad \blacktriangleleft \text{Simplify.}$$

TRY IT YOURSELF 8.20 Brief answers provided at the end of the section.

Find the value of $\tan 2t$ if $\tan t = -2$.

Half-Angle Formulas

> The double-angle formula for the cosine can be rearranged to show the effect on the sine or cosine when an angle is halved.

We have given three versions of the double-angle formula for the cosine. To find a half-angle formula for the sine function, we make use of the version $\cos 2x = 1 - 2\sin^2 x$. The first step is to solve this equation for $\sin x$:

$$1 - 2\sin^2 x = \cos 2x \qquad \blacktriangleleft \text{ Use the double-angle formula.}$$

$$\sin^2 x = \frac{1 - \cos 2x}{2} \qquad \blacktriangleleft \text{ Solve for } \sin^2 x.$$

$$\sin x = \pm\sqrt{\frac{1 - \cos 2x}{2}} \qquad \blacktriangleleft \text{ Take square roots.}$$

The next step is to make the substitution $x = \dfrac{t}{2}$ or $t = 2x$. This yields

$$\sin\frac{t}{2} = \pm\sqrt{\frac{1 - \cos t}{2}}$$

As is usual in such formulas, the sign is determined from the quadrant in which the trigonometric point $P\left(\dfrac{t}{2}\right)$ is located. There are half-angle formulas for cosine and tangent as well.

CONCEPTS TO REMEMBER: Half-Angle Formulas

The sine, cosine, or tangent of half an angle can be expressed in terms of the cosine of the angle. Where $\pm$ occurs in the following identities, the sign is determined by the quadrant in which the trigonometric point $P\left(\dfrac{t}{2}\right)$ lies.

$$\sin\frac{t}{2} = \pm\sqrt{\frac{1 - \cos t}{2}}$$

$$\cos\frac{t}{2} = \pm\sqrt{\frac{1 + \cos t}{2}}$$

$$\tan\frac{t}{2} = \begin{cases} \text{Formula 1:} \quad \pm\sqrt{\dfrac{1 - \cos t}{1 + \cos t}} \\[2mm] \text{Formula 2:} \quad \dfrac{1 - \cos t}{\sin t} \\[2mm] \text{Formula 3:} \quad \dfrac{\sin t}{1 + \cos t} \\[2mm] \text{Formula 4:} \quad \csc t - \cot t \end{cases}$$

EXAMPLE 8.21 Using the Half-Angle Formula for the Cosine

Find the exact value of $\cos 15°$.

SOLUTION

Observing that $15° = \dfrac{30°}{2}$, we see that the half-angle formula for the cosine applies:

$$\cos\frac{t}{2} = \pm\sqrt{\frac{1 + \cos t}{2}} \qquad \blacktriangleleft \text{ Use the half-angle formula.}$$

$$\cos 15° = \cos\frac{30°}{2} = \pm\sqrt{\frac{1 + \cos 30°}{2}} \qquad \blacktriangleleft \text{ Use } t = 30°.$$

$$= \pm\sqrt{\frac{1 + \sqrt{3}/2}{2}} \quad \blacktriangleleft \text{ Use } \cos 30° = \frac{\sqrt{3}}{2}.$$

$$= \pm\sqrt{\frac{2 + \sqrt{3}}{4}} \quad \blacktriangleleft \text{ Simplify.}$$

An angle of 15° puts us in the first quadrant, where all trigonometric functions are positive. Therefore,

$$\cos 15° = \sqrt{\frac{2 + \sqrt{3}}{4}} \text{ or } \sqrt{\frac{2 + \sqrt{3}}{2}}$$

TRY IT YOURSELF 8.21 Brief answers provided at the end of the section.

Find the sine of $\dfrac{\pi}{12}$.

EXTEND YOUR REACH

a. Show that the half-angle formula for the cosine can be rewritten as

$$\cos\frac{t}{2} = \frac{1}{2}\sqrt{2 + 2\cos t}$$

if $0 \le t \le \pi$.

b. Use this form of the half-angle formula to establish the following values of the cosine:

- $\cos\dfrac{\pi}{4} = \dfrac{1}{2}\sqrt{2}$

- $\cos\dfrac{\pi}{8} = \dfrac{1}{2}\sqrt{2 + \sqrt{2}}$

- $\cos\dfrac{\pi}{16} = \dfrac{1}{2}\sqrt{2 + \sqrt{2 + \sqrt{2}}}$

c. Propose a formula for $\cos\dfrac{\pi}{2^n}$ if n is a positive integer.

EXAMPLE 8.22 Using a Half-Angle Formula for the Tangent

If $\pi < t < \dfrac{3\pi}{2}$ and $\tan t = 2$, find the value of $\tan\dfrac{t}{2}$.

SOLUTION

There are four listed formulas that apply to $\tan\dfrac{t}{2}$. We choose $\tan\dfrac{t}{2} = \csc t - \cot t$ because it most closely matches the given information. We obtain the value of the cotangent from the tangent:

$$\cot t = \frac{1}{\tan t} = \frac{1}{2}$$

To find the cosecant we use the version of the Pythagorean identity that relates the cosecant to the cotangent:

$$1 + \cot^2 t = \csc^2 t \quad \blacktriangleleft \text{ Use the Pythagorean identity.}$$

$$1 + \left(\frac{1}{2}\right)^2 = \csc^2 t \quad \blacktriangleleft \text{ Substitute } \cot t = \frac{1}{2}.$$

$$\frac{5}{4} = \csc^2 t \quad \blacktriangleleft \text{ Simplify.}$$

$$\pm \frac{\sqrt{5}}{2} = \csc t \quad \blacktriangleleft \text{ Take square roots.}$$

Because $\pi < t < \dfrac{3\pi}{2}$, we are in the third quadrant, where the cosecant is negative.

Hence,

$$\csc t = -\frac{\sqrt{5}}{2}$$

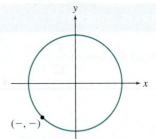

We now have the components we need to find the tangent of $\dfrac{t}{2}$:

$$\tan \frac{t}{2} = \csc t - \cot t = -\frac{\sqrt{5}+1}{2}$$

TRY IT YOURSELF 8.22 Brief answers provided at the end of the section.

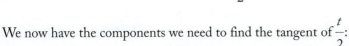

Find $\tan \dfrac{t}{2}$ if $\cos t = \dfrac{2}{3}$ and $0 \le t \le \dfrac{\pi}{2}$.

EXAMPLE 8.23 Using a Half-Angle Formula for the Secant

If $\dfrac{3\pi}{2} < s < 2\pi$ and $\sec s = 4$, find the value of $\sec \dfrac{s}{2}$.

SOLUTION

We have stated no half-angle formula for the secant, but the secant is the reciprocal of the cosine:

$$\cos s = \frac{1}{\sec s} = \frac{1}{4}$$

We apply the half-angle formula for the cosine:

$$\cos \frac{s}{2} = \pm \sqrt{\frac{1 + \cos s}{2}}$$

$$= \pm \sqrt{\frac{1 + \dfrac{1}{4}}{2}}$$

$$= \pm \sqrt{\frac{5}{8}}$$

Because $\dfrac{3\pi}{2} < s < 2\pi$, we conclude that $\dfrac{3\pi}{4} < \dfrac{s}{2} < \pi$. Hence, $P\left(\dfrac{s}{2}\right)$ lies in the

second quadrant, where the cosine is negative. So,

$$\cos \dfrac{s}{2} = -\sqrt{\dfrac{5}{8}}$$

$$\sec \dfrac{s}{2} = \dfrac{1}{\cos\left(\dfrac{s}{2}\right)} = -\sqrt{\dfrac{8}{5}}$$

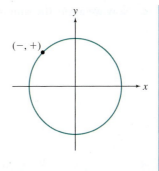

TRY IT YOURSELF 8.23 Brief answers provided at the end of the section.

Find the exact value of $\csc \dfrac{\pi}{12}$.

Combining Sum and Multiple-Angle Formulas

> The sum and double-angle formulas can be combined to produce triple-angle formulas and other identities.

EXAMPLE 8.24 Sum and Double-Angle Formulas

Let α and β be acute angles with

$$\sin \alpha = \dfrac{2}{5} \text{ and } \cos \beta = \dfrac{3}{5}$$

a. Find $\cos \alpha$ and $\sin \beta$.

b. Find $\sin 2\alpha$ and $\cos 2\alpha$.

c. Find the exact value of $\sin(2\alpha + \beta)$, and then give an approximation correct to two decimal places.

SOLUTION

a. We use the Pythagorean identity to get the cosine of α and the sine of β. Because all trigonometric functions are positive in the first quadrant, we use plus signs:

$$\cos \alpha = \sqrt{1 - \sin^2 \alpha} = \sqrt{1 - (2/5)^2} = \dfrac{\sqrt{21}}{5}$$

$$\sin \beta = \sqrt{1 - \cos^2 \beta} = \sqrt{1 - (3/5)^2} = \dfrac{4}{5}$$

b. We use double-angle formulas to find the values of $\sin 2\alpha$ and $\cos 2\alpha$:

$$\sin 2\alpha = 2 \sin \alpha \cos \alpha = 2\left(\dfrac{2}{5}\right)\left(\dfrac{\sqrt{21}}{5}\right) = \dfrac{4\sqrt{21}}{25}$$

$$\cos 2\alpha = \cos^2 \alpha - \sin^2 \alpha = \left(\dfrac{\sqrt{21}}{5}\right)^2 - \left(\dfrac{2}{5}\right)^2 = \dfrac{17}{25}$$

c. Now we apply the sum formula for the sine:

$$\sin(x + y) = \sin x \cos y + \cos x \sin y$$

$$\sin(2\alpha + \beta) = \sin 2\alpha \cos \beta + \cos 2\alpha \sin \beta$$

$$= \left(\frac{4\sqrt{21}}{25}\right)\left(\frac{3}{5}\right) + \left(\frac{17}{25}\right)\left(\frac{4}{5}\right)$$

$$= \frac{12\sqrt{21} + 68}{125} \approx 0.98$$

TRY IT YOURSELF 8.24 Brief answers provided at the end of the section.

In the setting of the example—that is, for the same angles α and β—calculate the value of $\cos\left(\dfrac{\alpha}{2} + \beta\right)$.

EXAMPLE 8.25 A Triple-Angle Formula

Show that $\cos 3x = 4\cos^3 x - 3\cos x$.

SOLUTION

We use the fact that $3x = 2x + x$ and apply the sum formula for the cosine:

$$\cos 3x = \cos(2x + x)$$

$$= \cos 2x \cos x - \sin 2x \sin x \qquad \blacktriangleleft \text{Use the sum formula for the cosine.}$$

$$= (2\cos^2 x - 1)\cos x - (2\sin x \cos x)\sin x \qquad \blacktriangleleft \text{Use the double-angle formulas for the sine and cosine.}$$

$$= (2\cos^2 x - 1)\cos x - 2\sin^2 x \cos x \qquad \blacktriangleleft \text{Simplify.}$$

$$= (2\cos^2 x - 1)\cos x - 2(1 - \cos^2 x)\cos x \qquad \blacktriangleleft \text{Use the Pythagorean identity.}$$

$$= (2\cos^3 x - \cos x) - 2(\cos x - \cos^3 x) \qquad \blacktriangleleft \text{Simplify.}$$

$$= 4\cos^3 x - 3\cos x$$

TRY IT YOURSELF 8.25 Brief answers provided at the end of the section.

Show that $\sin 4x = 4\cos x(\sin x - 2\sin^3 x)$. *Suggestion:* Use $\sin 4x = \sin(2(2x))$.

Power Reduction Formulas

> The double-angle formula for the cosine is rearranged to write powers of sine or cosine functions in terms of lower powers.

Two versions of the double-angle formula for the cosine are useful in writing integral powers of the sine or cosine in simpler terms. This reduction of powers is necessary in certain calculus operations.

Let's start with $\cos 2x = 2\cos^2 x - 1$ and solve for $\cos^2 x$:

$$2\cos^2 x - 1 = \cos 2x \qquad \text{◀ Use the double-angle formula for the cosine.}$$

$$2\cos^2 x = 1 + \cos 2x \qquad \text{◀ Move 1 to the right side.}$$

$$\cos^2 x = \frac{1}{2}(1 + \cos 2x) \qquad \text{◀ Divide by 2.}$$

There is a similar power reduction formula for the square of the sine.

CONCEPTS TO REMEMBER: Power Reduction Formulas

The square of the sine or cosine can be written in terms of the cosine of twice the angle, as in the following identities:

$$\cos^2 x = \frac{1}{2}(1 + \cos 2x)$$

$$\sin^2 x = \frac{1}{2}(1 - \cos 2x)$$

EXAMPLE 8.26 Using a Power Reduction Formula

Use power reduction formulas to express $\cos^4 x$ without using powers.

SOLUTION

We start with the reduction formula for $\cos^2 x$ and square it to obtain a formula for the fourth power:

$$\cos^2 x = \frac{1}{2}(1 + \cos 2x) \qquad \text{◀ Use the power reduction formula.}$$

$$\cos^4 x = \left(\frac{1}{2}(1 + \cos 2x)\right)^2 \qquad \text{◀ Square both sides.}$$

$$\cos^4 x = \frac{1}{4}(1 + 2\cos 2x + \cos^2 2x) \qquad \text{◀ Expand the square.}$$

The power reduction formula applied to $\cos^2 2x$ yields

$$\cos^2 2x = \frac{1}{2}(1 + \cos(2(2x))) = \frac{1}{2}(1 + \cos 4x)$$

We substitute this value into the preceding expression:

$$\cos^4 x = \frac{1}{4}\left(1 + 2\cos 2x + \frac{1}{2}(1 + \cos 4x)\right)$$

This can also be written as

$$\frac{1}{4}\left(\frac{3}{2} + 2\cos 2x + \frac{1}{2}\cos 4x\right) = \frac{3}{8} + \frac{1}{2}\cos 2x + \frac{1}{8}\cos 4x$$

TRY IT YOURSELF 8.26 Brief answers provided at the end of the section.

Use power reduction formulas to express $\sin^4 x$ without using powers.

EXTEND YOUR REACH   Here is a power reduction formula for the tangent:

$$\tan^2 t = 1 - \frac{2\tan t}{\tan(2t)}$$

Use the double-angle formula for the tangent to derive this formula.

TRY IT YOURSELF ANSWERS

8.19 $-\dfrac{7}{9}$

8.20 $\dfrac{4}{3}$

8.21 $\dfrac{\sqrt{2-\sqrt{3}}}{2}$

8.22 $\dfrac{1}{\sqrt{5}}$

8.23 $\dfrac{2}{\sqrt{2-\sqrt{3}}}$

8.24 $\dfrac{3}{5}\sqrt{\dfrac{1+\sqrt{21}/5}{2}} - \dfrac{4}{5}\sqrt{\dfrac{1-\sqrt{21}/5}{2}}$

8.25 Briefly, $\sin 4x = \sin(2(2x)) = 2\sin 2x \cos 2x = 4\sin x \cos x (1 - 2\sin^2 x) = 4\cos x(\sin x - 2\sin^3 x)$.

8.26 $\dfrac{1}{4}\left(1 - 2\cos 2x + \dfrac{1}{2}(1 + \cos 4x)\right)$ or

$\dfrac{3}{8} - \dfrac{1}{2}\cos 2x + \dfrac{1}{8}\cos 4x$

EXERCISE SET 8.3

CHECK YOUR UNDERSTANDING

1. State the double-angle formula for the sine.

2. Complete the identity: $\cos 2t = $ _____.
   a. $\cos^2 t - \sin^2 t$
   b. $1 - 2\sin^2 t$
   c. $2\cos^2 t - 1$
   d. All of the above.
   e. None of the above.

3. State the double-angle formula for the tangent.

4. State the half-angle formula for the sine.

5. State the half-angle formula for the cosine.

6. State a half-angle formula for the tangent that involves only the cosine.

7. **True or false:** Because we did not derive a half-angle formula for the cotangent, we can't find the cotangent of half angles.

8. State the power reduction formula for the square of the cosine.

9. State the power reduction formula for the square of the sine.

SKILL BUILDING

First quadrant. In Exercises 10 through 29, take both s and t to be acute angles so that $\sin s = \dfrac{1}{3}$ and $\sin t = \dfrac{4}{5}$. Find the indicated value exactly. In Exercises 22 through 29, also give an approximation correct to three decimal places.

10. $\sin 2s$

11. $\sin 2t$

12. $\cos 2s$

13. $\cos 2t$

14. $\tan 2t$

15. $\cot 2s$

16. $\csc 2t$

17. $\sin\left(\dfrac{s}{2}\right)$

18. $\cos\left(\dfrac{t}{2}\right)$

19. $\tan\left(\dfrac{t}{2}\right)$

20. $\cot\left(\dfrac{s}{2}\right)$

21. $\sec\left(\dfrac{t}{2}\right)$

22. $\sin(2t - s)$

23. $\sin(2s + 2t)$

24. $\cos(s + 2t)$

25. $\sin 4t$.
  Note that $4t = 2(2t)$.

26. $\cos 4t$.
  Note that $4t = 2(2t)$.

27. $\tan 4t$.
  Note that $4t = 2(2t)$.

28. $\sin\left(t + \dfrac{s}{2}\right)$

29. $\sin\left(\dfrac{3s}{2}\right)$.
  Note that $\dfrac{3s}{2} = s + \dfrac{s}{2}$.

Other quadrants. In Exercises 30 through 37, use double-angle or half-angle formulas to make the required calculation.

30. $\pi < t < \dfrac{3\pi}{2}$. $\sin t = -\dfrac{2}{3}$.
  Find $\sin 2t$.

31. $\pi < t < \dfrac{3\pi}{2}$. $\sin t = -\dfrac{2}{3}$.
  Find $\cos 2t$.

32. $\pi < t < \dfrac{3\pi}{2}$. $\tan t = 2$.
  Find $\tan 2t$.

33. $\pi < t < \dfrac{3\pi}{2}$. $\tan t = 2$.
  Find $\tan\left(\dfrac{t}{2}\right)$.

34. $\dfrac{3\pi}{2} < t < 2\pi$. $\cos t = \dfrac{1}{4}$.
  Find $\cos\left(\dfrac{t}{2}\right)$.

35. $\dfrac{3\pi}{2} < t < 2\pi$. $\cos t = \dfrac{1}{4}$.
  Find $\sin\left(\dfrac{t}{2}\right)$.

36. $\dfrac{3\pi}{2} < t < 2\pi$. $\cos t = \dfrac{1}{4}$.
  Find $\sec\left(\dfrac{t}{2}\right)$.

37. $\dfrac{3\pi}{2} < t < 2\pi$. $\sin t = -\dfrac{3}{4}$.
  Find $\cot\left(\dfrac{t}{2}\right)$.

Special angles. Exercises 38 through 44 can be solved using special angles. Find the indicated value exactly, and give an approximation correct to three decimal places.

38. Find the sine and cosine of $22.5°$. Note that
  $22.5 = \dfrac{45}{2}$.

39. Find the tangent and cotangent of $22.5°$. Note that
  $22.5 = \dfrac{45}{2}$.

40. Find the secant and cosecant of $22.5°$. Note that
  $22.5 = \dfrac{45}{2}$.

41. Find the sine, cosine, and tangent of $\dfrac{7\pi}{12}$. Note that

$$\dfrac{7\pi}{12} = \dfrac{\pi}{3} + \dfrac{\pi}{4}.$$

42. Half angle. Find the sine, cosine, and tangent of $\dfrac{\pi}{8}$.

43. Half angle twice. Use the half-angle formulas to find the cosine of $\dfrac{\pi}{24}$.

44. Sum and half-angle. Use the sum and half-angle formulas to get the sine of $\dfrac{105}{2}$ degrees.

PROBLEMS

Identities. In Exercises 45 through 53, use double-angle and half-angle formulas to establish the given identity.

45. $\dfrac{\cos 2x}{\cos x + \sin x} = \cos x - \sin x$

46. $\tan 2x = \dfrac{2\cot x}{\cot^2 x - 1}$

47. $\dfrac{1}{4}\sin^2 2x = \sin^2 x - \sin^4 x$

48. $\cos x = 1 - 2\sin^2 \dfrac{x}{2}$

49. $2\csc 2x = \sec x \csc x$

50. $1 + \cos 2x = \dfrac{2}{\sec^2 x}$

51. $\dfrac{1}{2}\sin 2x \cos 2x = \sin x \cos^3 x - \sin^3 x \cos x$

52. $\cos x + \sin x = \dfrac{\cos 2x}{\cos x - \sin x}$

53. $\sin 2t \tan \dfrac{t}{2} = 2(\sin^2 t + \cos t - 1)$

Double-angle formulas. In this section, we used the sum formula for the sine to get the double-angle formula for the sine. In Exercises 54 through 58, you will be led to find other double-angle formulas.

54. Double-angle formula for the cosine.

　a. Use the sum formula for the cosine to show that $\cos(2t) = \cos^2 t - \sin^2 t$. *Suggestion:* Note that $\cos(2t) = \cos(t + t)$, and apply the sum formula for the cosine.

　b. Use part a and the Pythagorean identity to show that $\cos(2t) = 1 - 2\sin^2 t$.

　c. Use part a and the Pythagorean identity to show that $\cos(2t) = 2\cos^2 t - 1$.

55. Double-angle formula for the tangent. Show that

$$\tan(2\alpha) = \dfrac{2\tan \alpha}{1 - \tan^2 \alpha}$$

Suggestion: Note that $\tan 2\alpha = \tan(\alpha + \alpha)$, and use the sum formula for the tangent.

56. Double-angle formula for the cotangent. Show that

$$\cot(2\alpha) = \dfrac{\cot^2 \alpha - 1}{2\cot \alpha}$$

Suggestion: Note that $\cot(2\alpha) = \dfrac{1}{\tan(2\alpha)}$, and apply the double-angle formula for the tangent from the preceding exercise. Next divide both top and bottom of the resulting fraction by $\tan^2 \alpha$.

57. Double-angle formula for the cosecant. Show that

$$\csc(2\alpha) = \dfrac{\csc^2 \alpha}{2\cot \alpha}$$

Suggestion: Start with the fact that the cosecant is the reciprocal of the sine.

58. Double-angle formula for the secant. Show that

$$\sec(2\alpha) = \dfrac{\sec^2 \alpha}{2 - \sec^2 \alpha}$$

Suggestion: Start with the fact that the secant is the reciprocal of the cosine.

Half-angle formulas.

59. Half-angle formula for the cosine. Mimic our derivation of the half-angle formula for the sine to get the half-angle formula for the cosine.

60. Half-angle formula for the tangent. Use the half-angle formulas for the sine and cosine to show that

$$\tan \dfrac{t}{2} = \pm\sqrt{\dfrac{\sec t - 1}{\sec t + 1}}$$

61. Another half-angle formula for the tangent.
Follow the indicated procedure to establish the
identity $\tan \dfrac{t}{2} = \dfrac{1 - \cos t}{\sin t}$.

a. Use the double-angle formulas for sine and
cosine to show that $\dfrac{1 - \cos 2x}{\sin 2x}$ can be simplified
to $\tan x$.

b. Put $x = \dfrac{t}{2}$ in your work from part a.

62. One more half-angle formula for the tangent.
Use the formula $\tan \dfrac{t}{2} = \dfrac{1 - \cos t}{\sin t}$ to show that
$\tan \dfrac{t}{2} = \csc t - \cot t$.

63. Half-angle formula for the cotangent. Show that
$\cot \dfrac{x}{2} = \csc x + \cot x$. *Suggestion*: Start with the
identity $\tan \dfrac{x}{2} = \dfrac{1 - \cos x}{\sin x}$, and then multiply top
and bottom of the fraction by $1 + \cos x$.

64. Half-angle formula for the secant. Show that
$\sec \dfrac{x}{2} = \pm \sqrt{\dfrac{2 \sec x}{\sec x + 1}}$. *Suggestion*: Start with the
half-angle formula for the cosine.

65. Half-angle formula for the cosecant. Show that
$\csc \dfrac{x}{2} = \pm \sqrt{\dfrac{2 \sec x}{\sec x - 1}}$. *Suggestion*: Start with the
half-angle formula for the sine.

66. Deriving a power reduction formula. Use the
double-angle formula $\cos 2x = 1 - 2 \sin^2 x$ to show
that $\sin^2 x = \dfrac{1}{2}(1 - \cos 2x)$.

Various multiple-angle formulas. In Exercises 67
through 71, use formulas developed in this section to
establish the given identity.

67. Show that $\sin(4x) = 4 \sin x (2 \cos^3 x - \cos x)$.
Suggestion: Note that $\sin(4x) = \sin(2(2x))$.

68. Show that $\sin(3x) = 3 \sin x - 4 \sin^3 x$. *Suggestion*:
Note that $\sin(3x) = \sin(x + 2x)$, and then apply
sum and double-angle formulas.

69. Show that $\cos(4x) = 8 \cos^4 x - 8 \cos^2 x + 1$.
Suggestion: Note that $\cos(4x) = \cos(2(2x))$, and
then apply double-angle formulas.

70. Use the fact that $\cot \dfrac{t}{2} = \csc t + \cot t$ to show that
$\cot \dfrac{x}{4} = \csc \dfrac{x}{2} + \csc x + \cot x$.

71. Show that $\tan 3x = \dfrac{3 \tan x - \tan^3 x}{1 - 3 \tan^2 x}$. *Suggestion*:
Note that $\tan 3x = \tan(x + 2x)$, and then apply
sum and double-angle formulas.

72. Power reduction. Use power reduction formulas
for both the sine and the cosine to obtain a
formula for $\sin^4 x$ that does not involve powers.

**73. Geometric derivation of the double-angle
formula for the sine.** This problem shows a
geometric derivation of the double-angle formula
for the sine in the case that 2α is an acute angle.
Figure 8.25 shows a right triangle with an acute
angle α. In **Figure 8.26** we have flipped a copy of
the triangle over and matched the two along a side.

a. Calculate the area of the triangle in Figure 8.26
using the formula $\dfrac{1}{2} AB \sin \theta$.

b. Calculate the area of the triangle in Figure 8.26
by doubling the area of the triangle in
Figure 8.25.

c. Equate the results from parts a and b to
conclude that $\sin(2\alpha) = 2 \sin \alpha \cos \alpha$.

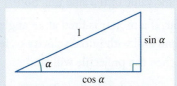

Figure 8.25 A right triangle with an
acute angle α

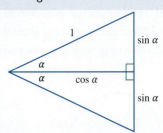

Figure 8.26 Two copies of the right
triangle joined along a side

Solving equations. In Exercises 74 through 77, plot
an appropriate graph and then find the solution of the
equation on the specified interval. Report your answer
correct to two decimal places.

74. Solve the equation $\sin 2x = \cos \dfrac{x}{2}$ if $0 \le x \le \dfrac{\pi}{4}$.

75. Solve the equation $\cot \dfrac{x}{3} = \sec x$ if $0 < x < \dfrac{\pi}{2}$.

76. Solve the equation $\tan 2x = \cos 3x$ if $0 < x < \dfrac{\pi}{4}$.

77. Solve the equation $x = \cos \dfrac{x}{2}$ if $0 < x < \pi$.

MODELS AND APPLICATIONS

78. A rectangle in a semicircle. A rectangle is inscribed in a semicircle of radius 1, as shown in **Figure 8.27**.

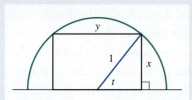

Figure 8.27 A rectangle inscribed in a semicircle of radius 1

a. Find the lengths of the sides of the rectangle in terms of the angle t.

b. Show that the area of the rectangle is given by $A = \sin 2t$.

c. What is the largest area that can be achieved for a rectangle inscribed in this semicircle? *Suggestion*: What angle produces the largest possible value of the sine function?

79. A mortar. If a mortar is fired at an angle of θ from the ground and with initial velocity v_0 feet per second, then the projectile will land

$$d = \frac{2v_0^2 \sin \theta \cos \theta}{g}$$

horizontal feet downrange, as shown in **Figure 8.28**. Here $g = 32$ feet per second per second is the acceleration due to gravity near the surface of Earth.

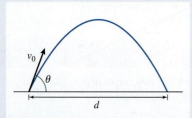

Figure 8.28 Angle and firing velocity for a mortar

a. How far downrange will the projectile strike if the angle θ is 70° and the initial velocity is 250 feet per second? Give your answer correct to two decimal places.

b. Use the double-angle formula for the sine to give an expression for d that involves only the sine function.

c. What angle will produce the maximum distance downrange? *Suggestion*: First determine what angle gives the maximum value for the sine.

80. A television screen. The diagonal measurement of the rectangular television screen shown in **Figure 8.29** is d inches.

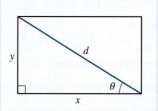

Figure 8.29 A television screen

a. Express the dimensions x and y in terms of d and the angle θ.

b. Show that the area of the television screen is given by $d^2 \cos \theta \sin \theta$.

c. Use the double-angle formula for the sine to express the area using only the sine function.

d. What angle θ gives the largest viewing area for a given value of d?

e. What geometric shape describes the television screen of maximum viewing area?

CHALLENGE EXERCISES FOR INDIVIDUALS OR GROUPS

Challenging identities. Establish the identities in Exercises 81 and 82.

81. $\dfrac{\cos 2t}{1 + \sin 2t} = \dfrac{\cot t - 1}{\cot t + 1}$

82. $(\cot t)\left(\cot \dfrac{t}{2}\right) = \dfrac{1}{\sec t - 1}$

83. Alternative derivation of the half-angle formula for cosine. Here is a way to obtain the half-angle formula for the cosine from the half-angle formula for the sine. In **Figure 8.30** we have a right triangle with an acute angle of $\dfrac{\alpha}{2}$. We know that $\sin \dfrac{\alpha}{2} = \sqrt{\dfrac{1 - \cos \alpha}{2}}$. So in **Figure 8.31** we have labeled the vertical side with this length and labeled the horizontal side x.

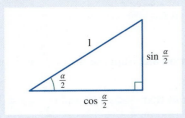

Figure 8.30 Standard right triangle with an angle of $\dfrac{\alpha}{2}$

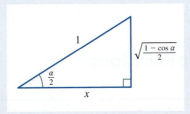

Figure 8.31 Standard right triangle with an angle of $\dfrac{\alpha}{2}$: sides relabeled

a. Use the Pythagorean theorem to find x.

b. Explain how your answer from part a can be used to give a formula for $\cos \dfrac{\alpha}{2}$.

84. Alternative derivation of half-angle formula for sine. Starting from the formula $\cos \dfrac{\alpha}{2} = \sqrt{\dfrac{1 + \cos \alpha}{2}}$, use the method outlined in the preceding exercise to give a formula for $\sin \dfrac{\alpha}{2}$.

85. Alternative derivation of the double-angle cosine formula. Refer to the unit circle in **Figure 8.32**.

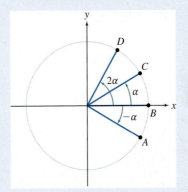

Figure 8.32 The double-angle formula for the cosine

a. Show that if (x_1, y_1) and (x_2, y_2) lie on the unit circle, then the square of the distance between these points is given by the formula

$$2 - 2(x_1 x_2 + y_1 y_2)$$

b. Find the coordinates of the points marked A, B, C, and D in Figure 8.32.

c. Explain why the distance from A to C is the same as the distance from B to D.

d. Calculate the square of the distance from A to C and the square of the distance from B to D.

e. Equate the quantities found in part d to get the double-angle formula for the cosine.

REVIEW AND REFRESH: Exercises from Previous Sections

86. From Section 1.1: Does the equation $x^2 + y^2 = xy$ determine y as a function of x?

Answer: It does not.

87. From Section 1.2: Calculate and simplify the average rate of change of $f(x) = \dfrac{1}{x}$ from $x = a$ to $x = a + h$ if $a > 0$ and $h > 0$.

Answer: $-\dfrac{1}{a(a + h)}$

88. From Section 2.3: The graph of $y = f(x)$ is shown in **Figure 8.33**. Add to the figure the graph of $y = f(x+1)$.

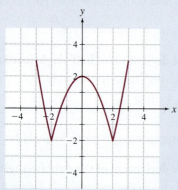

Figure 8.33 Graph for Exercise 88

Answer:

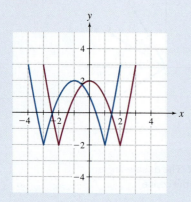

89. From Section 4.2: Simplify the expression $e^{-\ln x}$.

Answer: $\dfrac{1}{x}$

90. From Section 5.4: Given that $x = i$ is one zero of the polynomial $P(x) = x^3 - x^2 + x - 1$, find all remaining zeros (both real and complex).

Answer: $-i$ and 1

91. From Chapter 6: An arc of a circle of radius 4 subtends an angle of $\dfrac{\pi}{4}$ radian. Find the length of the arc and the area of the sector of the circle determined by the arc.

Answer: Arc length: π units. Area: 2π square units.

92. From Chapter 6: What is the reference angle for an angle of $\dfrac{9\pi}{7}$ radians?

Answer: $\dfrac{2\pi}{7}$

93. From Section 8.1: Express $\tan t$ in terms of the secant and cosecant.

Answer: $\tan t = \dfrac{\sec t}{\csc t}$

94. From Section 8.2: Suppose $\sin s = \dfrac{2}{3}$ and $\cos t = \dfrac{3}{4}$. Suppose also that $\dfrac{\pi}{2} < s < \pi$ and $\dfrac{3\pi}{2} < t < 2\pi$. Find the value of $\sin(s+t)$.

Answer: $\dfrac{6 + \sqrt{35}}{12}$

8.4 Inverse Trigonometric Functions

8.1 Trigonometric Identities

8.2 Sum and Difference Formulas

8.3 Double-Angle and Half-Angle Formulas

8.4 Inverse Trigonometric Functions

8.5 Solving Trigonometric Equations

> Domains of trigonometric functions must be restricted to produce inverses.

In this section, you will learn to:

1. Calculate function values of the arcsine, arccosine, and arctangent.
2. Calculate values of trigonometric functions composed with inverse trigonometric functions.
3. Construct the graphs of the inverse trigonometric functions.
4. Use the arcsine, arccosine, and arctangent functions in practical applications.
5. Calculate function values of the arccosecant, arcsecant, and arccotangent.

The trigonometric functions, with suitable restrictions on their domains, all have inverses. We will investigate them in this section. These inverse functions have important applications to fields from astronomy to basketball. They are useful almost any time an angle is required, and they play a key role in calculus.

We can't express the exact value of π as a rational number. On the other hand, modern computers can quickly produce values of π accurate to many decimal places. For example, with 75-place accuracy

$$\pi \approx 3.14159265358979323846264338327950288419$$
$$716939937510582097494459230781640629$$

How is that approximation done? The most obvious method of approximating π is to measure the circumference and diameter of a circle and divide. You will be doing very well indeed if you can make those measurements accurately enough to produce two correct digits beyond the decimal place.

Better methods involve very long sums. For example, a remarkable formula due to Leonhard Euler states that

$$\frac{\pi}{4} = 1 - \frac{1}{3} + \frac{1}{5} - \frac{1}{7} + \frac{1}{9} - \frac{1}{11} + \cdots$$

This formula is derived from the inverse of the tangent function, and variants of this formula are used today to calculate many digits of π—and this calculation is, in turn, used as a benchmark for modern computers.

Arcsine and Arccosine

> The inverse sine reverses the effect of the sine by producing an angle whose sine is a given number. A similar statement applies to the inverse cosine.

The sine function $y = \sin \theta$ accepts an input angle θ and gives an output number y. The inverse of this function, which we will identify as the arcsine, should accept a number y as an input and output an angle θ. This process is illustrated in **Figure 8.34**. That is, the arcsine of y is the angle θ whose sine is y.

But, for example, if we ask for the angle whose sine is $\dfrac{1}{\sqrt{2}}$, there is more than one correct answer. You could correctly answer $\dfrac{\pi}{4}, \dfrac{3\pi}{4}, -\dfrac{5\pi}{4}$, or any one of infinitely many other angles. This situation is illustrated in **Figure 8.35**, where we have graphed the sine function along with the line $y = \dfrac{1}{\sqrt{2}}$. We see that the sine function fails the horizontal line test for inverses (see Section 2.2). Each crossing of the line with the graph of the sine represents an angle θ whose sine is $\dfrac{1}{\sqrt{2}}$.

We can remedy this failure if we insist that the angle θ come from the interval $\left[-\dfrac{\pi}{2}, \dfrac{\pi}{2}\right]$. With this restriction, $\theta = \dfrac{\pi}{4}$ is the only angle whose sine is $\dfrac{1}{\sqrt{2}}$ and hence the only correct answer. This is illustrated in **Figure 8.36**. In fact, restricting sine to the domain $\left[-\dfrac{\pi}{2}, \dfrac{\pi}{2}\right]$ gives a function that satisfies the horizontal line test, so it has an inverse.

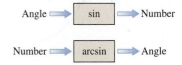

Figure 8.34 Visualizing the arcsine function: The sine function inputs an angle and outputs a number, but the arcsine function inputs a number and outputs an angle.

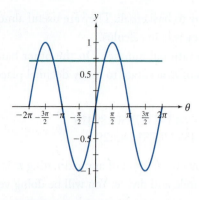

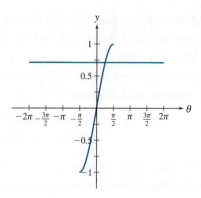

Figure 8.35 Difficulty for constructing the inverse of the sine function: There are infinitely many angles θ whose sine is $\dfrac{1}{\sqrt{2}}$.

Figure 8.36 Solution for the difficulty: There is only one angle θ in the interval $\left[-\dfrac{\pi}{2}, \dfrac{\pi}{2}\right]$ whose sine is $\dfrac{1}{\sqrt{2}}$.

The **arcsine** is the inverse of the sine function restricted to the domain $\left[-\dfrac{\pi}{2}, \dfrac{\pi}{2}\right]$.

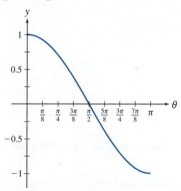

Figure 8.37 A domain restriction that allows for an inverse function: The cosine function is one-to-one on the interval $[0, \pi]$.

The **arccosine** is the inverse of the cosine function restricted to the domain $[0, \pi]$.

More formally, Figure 8.35 illustrates the fact that the sine function is not one-to-one. But as Figure 8.36 shows, the sine function is increasing on the interval $\left[-\dfrac{\pi}{2}, \dfrac{\pi}{2}\right]$, so it is one-to-one on that interval. Because the sine function with this restricted domain is a one-to-one function, it has an inverse, which is known as the **arcsine** function.

The notation is $y = \arcsin x$. Thus, $\arcsin x = \theta$ if and only if $\sin \theta = x$ and $-\dfrac{\pi}{2} \le \theta \le \dfrac{\pi}{2}$. The domain of the arcsine is $[-1, 1]$, and the range is $\left[-\dfrac{\pi}{2}, \dfrac{\pi}{2}\right]$. Intuitively, $\arcsin x$ is the angle between $-\dfrac{\pi}{2}$ and $\dfrac{\pi}{2}$ whose sine is x.

The inverse sine function arcsin is sometimes denoted by $\sin^{-1}$ in the same way that the inverse of a function f is, in general, denoted by f^{-1}. To avoid confusing $\sin^{-1} x$ with $\dfrac{1}{\sin x}$, we will always use arc to indicate the inverse of a trigonometric function.

In a similar fashion, the cosine function is not one-to-one, but it is decreasing on the interval $[0, \pi]$ (see **Figure 8.37**), so the cosine with its domain restricted to this interval also has an inverse. This inverse function is known as the **arccosine** function.

The notation is $y = \arccos x$. Thus, $\arccos x = \theta$ if and only if $\cos \theta = x$ and $0 \le \theta \le \pi$. The domain of the arccosine is $[-1, 1]$, and the range is $[0, \pi]$. Intuitively, $\arccos x$ is the angle between 0 and π whose cosine is x.

EXAMPLE 8.27   Finding Function Values of Arcsine and Arccosine

a. Find the exact value of $\arcsin \dfrac{1}{2}$.

b. Find the exact value of $\arccos 0$.

c. Use a calculator to find the approximate value of $\arcsin 0.4$.

SOLUTION

a. The arcsine of $\dfrac{1}{2}$ is the angle between $-\dfrac{\pi}{2}$ and $\dfrac{\pi}{2}$ whose sine is $\dfrac{1}{2}$. The special angle $\theta = \dfrac{\pi}{6}$ satisfies these conditions. That is,

$$\sin\frac{\pi}{6} = \frac{1}{2} \quad\text{and}\quad \frac{\pi}{6} \text{ is in the range } \left[-\frac{\pi}{2}, \frac{\pi}{2}\right] \text{ of the arcsine function}$$

We conclude that $\arcsin\dfrac{1}{2} = \dfrac{\pi}{6}$.

b. For arccos 0, we require an angle θ in the interval $[0, \pi]$ whose cosine is 0. Because $\theta = \dfrac{\pi}{2}$ satisfies these conditions, $\arccos 0 = \dfrac{\pi}{2}$.

c. The arcsine key on your calculator may look like SIN^{-1}. Be sure your calculator is set to radian mode. The result is $\arcsin 0.4 \approx 0.41$.

TRY IT YOURSELF 8.27 Brief answers provided at the end of the section.

Find the exact value of arcsin 1.

Because the arcsine is the inverse of the sine function on a restricted interval, it is true that $\sin(\arcsin x) = x$ for x in the range of the sine. This reflects the self-evident fact that the sine of the angle whose sine is x is indeed x.

But one must be careful going the other way: $\arcsin(\sin x) = x$ is true only when x is between $-\dfrac{\pi}{2}$ and $\dfrac{\pi}{2}$. For example, $\sin\dfrac{5\pi}{2} = 1$ but

$$\arcsin\left(\sin\frac{5\pi}{2}\right) = \arcsin 1 = \frac{\pi}{2}$$

Thus, $\arcsin\left(\sin\dfrac{5\pi}{2}\right)$ equals $\dfrac{\pi}{2}$, not $\dfrac{5\pi}{2}$. Special care must be taken when composing the arcsine with the sine, and a similar caution applies to the inverse cosine.

CONCEPTS TO REMEMBER: Basic Properties of Arcsine and Arccosine

- $\sin(\arcsin x) = x$ for all x in the domain of the arcsine.

  $\arcsin(\sin x) = x$ only when x is in the range of the arcsin, so $-\dfrac{\pi}{2} \le x \le \dfrac{\pi}{2}$.

- $\cos(\arccos x) = x$ for all x in the domain of the arccosine.

  $\arccos(\cos x) = x$ only when x is in the range of the arccos, so $0 \le x \le \pi$.

EXAMPLE 8.28 Composing a Trigonometric Function with Its Inverse

In each part, determine the exact value if it exists.

a. $\sin\left(\arcsin\dfrac{1}{2}\right)$

b. $\cos(\arccos(-4))$. *Suggestion: Focus first on the meaning of arccos(-4).*

SOLUTION

a. We use the relationship $\sin(\arcsin x) = x$, which is true for any x in the domain of the arcsine function. Thus,

$$\sin\left(\arcsin\dfrac{1}{2}\right) = \dfrac{1}{2}$$

b. Although it is tempting to think that the cosine and arccosine "cancel," that is true only for values in the domain of the arccosine, which is $[-1, 1]$. In point of fact, arccos(-4) is undefined: a value for arccos(-4) would be an angle whose cosine is -4. That is not possible because the range of cosine is $[-1, 1]$. So, $\cos(\arccos(-4))$ is undefined.

TRY IT YOURSELF 8.28 Brief answers provided at the end of the section.

Find the value of $\cos\left(\arccos\dfrac{2}{3}\right)$.

EXAMPLE 8.29 The Importance of the Range in Evaluating Inverse Trigonometric Functions

In each part, determine the exact value.

a. $\arccos\left(\cos\dfrac{\pi}{3}\right)$

b. $\arccos\left(\cos\dfrac{4\pi}{3}\right)$

SOLUTION

a. We show two methods for determining the value.

For the first, note that $\cos\dfrac{\pi}{3} = \dfrac{1}{2}$. Then

$$\arccos\left(\cos\dfrac{\pi}{3}\right) = \arccos\dfrac{1}{2} = \dfrac{\pi}{3}$$

At the end we used the fact that $\dfrac{\pi}{3}$ is the angle between 0 and π whose cosine is $\dfrac{1}{2}$. The second method uses the equation $\arccos(\cos x) = x$ when x is in the range $[0, \pi]$ of the arccosine function. In this case, $\arccos\left(\cos\dfrac{\pi}{3}\right) = \dfrac{\pi}{3}$, which holds because $\dfrac{\pi}{3}$ is in the range $[0, \pi]$ of the arccosine function.

b. Once again, it is tempting to think that $\arccos\left(\cos\dfrac{4\pi}{3}\right) = \dfrac{4\pi}{3}$ by the reasoning in part a, but the equation $\arccos(\cos x) = x$ applies only when x is in the range $[0, \pi]$ of the arccosine function. Because $\dfrac{4\pi}{3}$ is not in this range, the formula does not apply. Instead, we calculate directly:

$$\cos\frac{4\pi}{3} = -\frac{1}{2} \qquad \text{◀ Calculate using the special angle.}$$

$$\arccos\left(\cos\frac{4\pi}{3}\right) = \arccos\left(-\frac{1}{2}\right) \qquad \text{◀ Apply the arccosine.}$$

The solution, then, is the angle between 0 and π whose cosine is $-\dfrac{1}{2}$. That angle is $\dfrac{2\pi}{3}$, so $\arccos\left(\cos\dfrac{4\pi}{3}\right) = \dfrac{2\pi}{3}$.

We emphasize that the answer is not $\dfrac{4\pi}{3}$, as one might conclude if a formula is misapplied.

TRY IT YOURSELF 8.29 Brief answers provided at the end of the section.

Find the value of $\arcsin(\sin\pi)$.

Explain how you go about finding $\arcsin(\sin t)$ for any value of t.

EXTEND YOUR REACH

EXAMPLE 8.30 An Example for Discovery of Identities

a. Compare the exact values of $\arcsin\dfrac{1}{2}$ and $\arcsin\left(-\dfrac{1}{2}\right)$.

b. Compare the exact values of $\arccos\dfrac{1}{2}$ and $\arccos\left(-\dfrac{1}{2}\right)$.

SOLUTION

a. Because $\sin\dfrac{\pi}{6} = \dfrac{1}{2}$ and $\dfrac{\pi}{6}$ is in the range $\left[-\dfrac{\pi}{2}, \dfrac{\pi}{2}\right]$ of the arcsine function, we have

$$\arcsin\frac{1}{2} = \frac{\pi}{6}$$

To find the arcsine of $-\dfrac{1}{2}$, we use the fact that sine is an odd function:

$$\sin\left(-\frac{\pi}{6}\right) = -\sin\frac{\pi}{6} = -\frac{1}{2}$$

Then, because $-\dfrac{\pi}{6}$ is in the range $\left[-\dfrac{\pi}{2}, \dfrac{\pi}{2}\right]$ of the arcsine function , we have

$$\arcsin\left(-\dfrac{1}{2}\right) = -\dfrac{\pi}{6}$$

Observe that a result of our calculation is $\arcsin\left(-\dfrac{1}{2}\right) = -\arcsin\dfrac{1}{2}$.

b. Because $\cos\dfrac{\pi}{3} = \dfrac{1}{2}$ and $\dfrac{\pi}{3}$ is in the range $[0, \pi]$ of the arccosine function , we have $\arccos\dfrac{1}{2} = \dfrac{\pi}{3}$.

For $\arccos\left(-\dfrac{1}{2}\right)$, we need an angle θ in the range $[0, \pi]$ of the arccosine function so that $\cos\theta = -\dfrac{1}{2}$. The special angle $\theta = \dfrac{2\pi}{3}$ fills the bill, so

$$\arccos\left(-\dfrac{1}{2}\right) = \dfrac{2\pi}{3}$$

Observe that, because $\dfrac{2\pi}{3} = \pi - \dfrac{\pi}{3}$, we have $\arccos\left(-\dfrac{1}{2}\right) = \pi - \arccos\dfrac{1}{2}$.

TRY IT YOURSELF 8.30   Brief answers provided at the end of the section.

Compare the values of $\arcsin\dfrac{\sqrt{3}}{2}$ and $\arcsin\left(-\dfrac{\sqrt{3}}{2}\right)$.

EXTEND YOUR REACH

It is true that $\sin\left(\dfrac{\pi}{2} - x\right) = \cos x$. Use this identity to show that if $0 \le x \le \pi$,

then $\arcsin(\cos x) = \dfrac{\pi}{2} - x$.

The concepts presented in this example can be extended.

CONCEPTS TO REMEMBER: Identities for Arcsine and Arccosine

The arcsine is an odd function, but the arccosine is not. The identities are

$$\arcsin(-x) = -\arcsin x$$
$$\arccos(-x) = \pi - \arccos x$$

Graphs of the Arcsine and Arccosine

We find the graphs of the inverse functions by reflecting specific parts of the sine and cosine graphs about the line $y = x$.

Recall that we construct the graph of the inverse of a function f by reflecting the graph of f through the line $y = x$. The graphs in **Figure 8.38** and **Figure 8.39** show clearly how the roles of domain and range are reversed for the inverse, arcsin x, of the restricted sine function. A similar observation can be made for the cosine and arccosine, as shown in **Figure 8.40** and **Figure 8.41**.

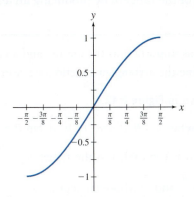

DF **Figure 8.38** The graph of the sine on $\left[-\dfrac{\pi}{2}, \dfrac{\pi}{2}\right]$

Figure 8.39 Reflecting the graph in Figure 8.38 through the line $y = x$ to make the graph of the arcsine

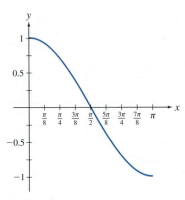

DF **Figure 8.40** The graph of the cosine on $[0, \pi]$

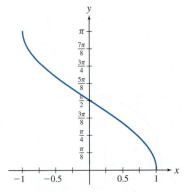

Figure 8.41 Reflecting the graph in Figure 8.40 through the line $y = x$ to make the graph of the arccosine

---

EXAMPLE 8.31 Properties of Arcsine and Arccosine Graphs

a. On what intervals is the arcsine function increasing?

b. On what intervals is the graph of the arcsine concave up, and on what intervals is it concave down?

SOLUTION

a. Referring to Figure 8.39, we see that the arcsine function increases on its domain, which is the interval $[-1, 1]$.

b. The graph in Figure 8.39 is concave up on $(0, 1]$ and concave down on $[-1, 0)$.

TRY IT YOURSELF 8.31 Brief answers provided at the end of the section.

Determine where the arccosine increases or decreases, and where its graph is concave up or concave down.

Where is the rate of change of the arccosine positive, and where is it negative? Is the rate of change ever zero?   **EXTEND YOUR REACH**

The Arctangent Function

> The inverse tangent reverses the effect of the tangent by producing an angle whose tangent is a given number.

The **arctangent** is the inverse of the tangent function restricted to the domain $\left(-\dfrac{\pi}{2}, \dfrac{\pi}{2}\right)$.

The inverse tangent function is at least as important as the arcsine and arccosine, and it is used extensively in calculus. To define the **arctangent** function, we restrict the domain of the tangent to $\left(-\dfrac{\pi}{2}, \dfrac{\pi}{2}\right)$, as shown in **Figure 8.42**.

The notation is $y = \arctan x$. Thus, $\arctan x = \theta$ if and only if $\tan \theta = x$ and $-\dfrac{\pi}{2} < \theta < \dfrac{\pi}{2}$. The domain of the arctangent is $(-\infty, \infty)$, and the range is $\left(-\dfrac{\pi}{2}, \dfrac{\pi}{2}\right)$. Intuitively, $\arctan x$ is the angle θ between $-\dfrac{\pi}{2}$ and $\dfrac{\pi}{2}$ whose tangent is x.

Just as with the arcsine and arccosine, we obtain the graph of the arctangent in **Figure 8.43** by reflecting the graph of the tangent in Figure 8.42 through the line $y = x$.

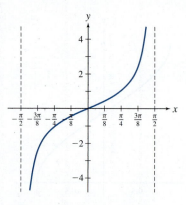

DF **Figure 8.42** Restricting the domain of the tangent function: The tangent restricted to $\left(-\dfrac{\pi}{2}, \dfrac{\pi}{2}\right)$ is a one-to-one function, so it has an inverse.

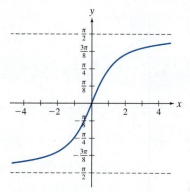

Figure 8.43 Reflecting the graph in Figure 8.42 through the line $y = x$ to make the graph of the arctangent: Note how the vertical asymptotes for the tangent become horizontal asymptotes for the arctangent.

The arctangent has several properties similar to those of the arcsine. The horizontal asymptotes in Figure 8.43 can be described in terms of limits.

CONCEPTS TO REMEMBER: Properties of the Arctangent

- $\tan(\arctan x) = x$ for all x

- $\arctan(\tan x) = x$ only when x is in the range of the arctangent, so
$$-\frac{\pi}{2} < x < \frac{\pi}{2}$$

- $\arctan(-x) = -\arctan x$

- $\arctan x \to \dfrac{\pi}{2}$ as $x \to \infty$

- $\arctan x \to -\dfrac{\pi}{2}$ as $x \to -\infty$

EXAMPLE 8.32 Basic Calculations with the Arctangent

a. Find the exact value of arctan 1.

b. Find the exact value of $\arctan(-\sqrt{3})$.

SOLUTION

a. Because $\tan\dfrac{\pi}{4} = 1$ and $\dfrac{\pi}{4}$ lies between $-\dfrac{\pi}{2}$ and $\dfrac{\pi}{2}$, we conclude that

$$\arctan 1 = \frac{\pi}{4}$$

b. We note that $\tan\left(-\dfrac{\pi}{3}\right) = -\sqrt{3}$ and that $-\dfrac{\pi}{3}$ lies in the range $\left[-\dfrac{\pi}{2}, \dfrac{\pi}{2}\right]$ of the arctangent function. Therefore,

$$\arctan(-\sqrt{3}) = -\frac{\pi}{3}$$

We could also find this value using the identity $\arctan(-x) = -\arctan x$ and the value of $\arctan\sqrt{3}$.

TRY IT YOURSELF 8.32 Brief answers provided at the end of the section.

Find the exact value of $\arctan\left(-\dfrac{1}{\sqrt{3}}\right)$.

EXTEND YOUR REACH

We have noted that $\arctan x \to \dfrac{\pi}{2}$ as $x \to \infty$. Judging from the graph of the arctangent, what do you think is the limit as $x \to \infty$ of the rate of change of $\arctan x$? *Suggestion:* A straightedge laid along the graph of the arctangent may help you to visualize the rate of change.

EXAMPLE 8.33 Solving Equations

Find one solution for each of the following equations.

a. $\sin\theta = \dfrac{\sqrt{3}}{2}$. Give an exact solution.

b. $\sin\theta = 3\cos\theta$. Give a solution in terms of inverse trigonometric functions, and provide an approximation correct to two decimal places.

c. $\arctan x = 2$. Give a solution in terms of the tangent, and provide an approximation correct to two decimal places.

SOLUTION

As we shall see in the next section, there are many solutions of the equations in parts a and b. In this instance, we need only find one.

a. The equation $\sin \theta = \dfrac{\sqrt{3}}{2}$ requires an angle θ whose sine is $\dfrac{\sqrt{3}}{2}$. One such angle is $\theta = \arcsin \dfrac{\sqrt{3}}{2}$. Using our knowledge of special angles, we find

$$\theta = \frac{\pi}{3}$$

b. We begin by dividing both sides by the cosine so that we are dealing with a single trigonometric function, the tangent, rather than both the sine and the cosine:

$$\sin \theta = 3 \cos \theta$$

$$\frac{\sin \theta}{\cos \theta} = \frac{3 \cos \theta}{\cos \theta} \qquad \blacktriangleleft \textbf{Divide by } \cos \theta.$$

$$\tan \theta = 3 \qquad \blacktriangleleft \textbf{Use } \frac{\sin \theta}{\cos \theta} = \tan \theta.$$

This equation asks for an angle θ whose tangent is 3. One such angle is $\theta = \arctan 3 \approx 1.25$.

c. We solve this equation by taking the tangent of both sides:

$$\arctan x = 2$$

$$\tan(\arctan x) = \tan 2 \qquad \blacktriangleleft \textbf{Apply the arctangent function to both sides.}$$

$$x = \tan 2 \approx -2.19 \qquad \blacktriangleleft \textbf{Use } \tan(\arctan x) = x.$$

TRY IT YOURSELF 8.33 Brief answers provided at the end of the section.

Find one solution of the equation $2 \arccos x = 5$. Give an answer in terms of the cosine, and provide an approximation correct to two decimal places.

The next example previews an important role played by inverse trigonometric functions in calculus.

EXAMPLE 8.34 Composing Trigonometric Functions with Inverse Trigonometric Functions

Simplify $\cos\left(\arctan \dfrac{4}{5}\right)$. Your final answer should not involve any trigonometric or inverse trigonometric functions.

SOLUTION

As we solve this problem, we will develop a strategy for solving similar problems. We proceed in three steps.

First we build a right triangle with an acute angle θ whose tangent is $\dfrac{4}{5}$. To get this value for the tangent, we make the opposite side 4 and adjacent side 5. This is shown in **Figure 8.44**.

Next we calculate the length of the third side. In this case, the Pythagorean theorem gives $\sqrt{41}$ for the length of the hypotenuse. This is shown in **Figure 8.45**.

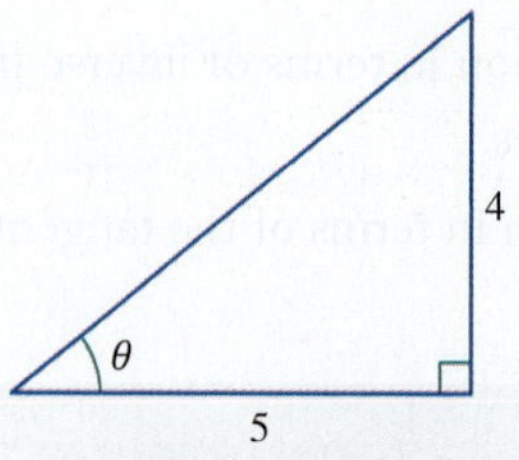

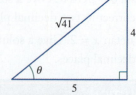

Figure 8.44 Making a triangle with $\tan \theta = \dfrac{4}{5}$

Figure 8.45 Finding the third side using the Pythagorean theorem

Finally, we use the triangle we have constructed to find the cosine of θ. The cosine is the adjacent over the hypotenuse, and this yields our final answer:

$$\cos\left(\arctan\frac{4}{5}\right) = \cos\theta = \frac{5}{\sqrt{41}}$$

TRY IT YOURSELF 8.34 Brief answers provided at the end of the section.

Simplify $\tan\left(\arccos\dfrac{4}{5}\right)$.

STEP-BY-STEP STRATEGY: Calculating Trigonometric Functions Composed with Inverse Trigonometric Functions

Step 1 Make a right triangle, labeling two sides to display the appropriate trigonometric value.

Step 2 Use the Pythagorean theorem to find the third side of the triangle.

Step 3 Use the sides of the triangle to find the value of the required trigonometric function.

EXAMPLE 8.35 Applying a Double-Angle Formula

Simplify $\sin(2\arccos t)$, writing it as an expression in t without trigonometric or inverse trigonometric functions. Assume that $0 < t < 1$.

SOLUTION

First, the expression $\arccos t$ signals us that we need a triangle showing an acute angle θ whose cosine is t. This triangle is shown in **Figure 8.46**.

Next we use the Pythagorean theorem to find the length $\sqrt{1-t^2}$ of the third side. This is shown in **Figure 8.47**.

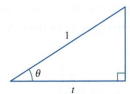

Figure 8.46 Making an angle whose cosine is t

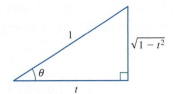

Figure 8.47 Finding the third side using the Pythagorean theorem

Now we use the triangle in Figure 8.47 to facilitate our calculations. We begin with the double-angle formula:

$$\sin(2\arccos t) = 2\sin(\arccos t)\cos(\arccos t) \qquad \blacktriangleleft \text{ Use the double-angle formula.}$$
$$= 2t\,\sin(\arccos t) \qquad \blacktriangleleft \text{ Use } \cos(\arccos t) = t.$$
$$= 2t\,\sin\theta \qquad \blacktriangleleft \theta \text{ is the angle whose cosine is } t.$$
$$= 2t\sqrt{1-t^2} \qquad \blacktriangleleft \text{ Use Figure 8.47.}$$

TRY IT YOURSELF 8.35 Brief answers provided at the end of the section.

Simplify the expression $\cos(2\arccos t)$, where $0 \le t \le 1$.

The method used in the preceding examples applies to a wide variety of combinations of trigonometric and inverse trigonometric functions, and it allows us to make quick and accurate calculations. **Figure 8.48** and **Figure 8.49** show triangles useful in this connection.

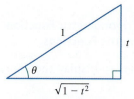

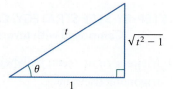

Figure 8.48 A triangle with $\sin\theta = t$: This triangle is used to show that $\cos(\arcsin t) = \cos\theta = \sqrt{1-t^2}$.

Figure 8.49 A triangle with $\cos\theta = \dfrac{1}{t}$: This triangle is used to show that $\sec\left(\arccos\dfrac{1}{t}\right) = \sec\theta = t$.

MODELS AND APPLICATIONS Flying and Basketball

Inverse trigonometric functions arise naturally in applications where there are unknown angles. In calculus, inverse trigonometric functions will always use radians, but many practical applications use degrees instead. Happily, inverse trigonometric functions can be defined equally well using degrees or radians.

EXAMPLE 8.36 Flying to Washington, D.C.

As shown in **Figure 8.50**, St. Louis is approximately 675 miles due north of New Orleans, and Washington, D.C., is approximately 818 miles due east of St. Louis. Assuming north corresponds to 0° and east corresponds to 90°, what heading (direction of flight) should an airplane take in order to fly from New Orleans directly to Washington, D.C.?

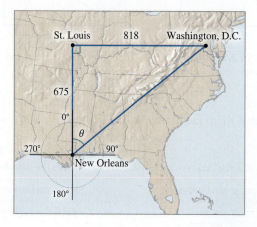

Figure 8.50 From New Orleans to Washington, D.C.

SOLUTION

The heading we want is the degree measure of the angle θ in Figure 8.50. Using this triangle, we calculate $\tan \theta = \dfrac{818}{675}$. Thus, $\theta = \arctan \dfrac{818}{675}$. Making certain that our calculator is set to degree mode, we find

$$\text{Heading} = \theta = \arctan \frac{818}{675} \approx 50.47°$$

TRY IT YOURSELF 8.36 Brief answers provided at the end of the section.

Kansas City, Missouri, is approximately 250 miles due west of St. Louis. What heading should an airplane take to fly from New Orleans directly to Kansas City?

EXAMPLE 8.37 Shooting a Basket

The rim of a standard basketball goal is 10 feet high. A point guard's eyes are 6 feet above the floor. She stands 25 feet from a point directly below the front of the rim (see **Figure 8.51**). At what angle t (measured in degrees) must she incline her eyes to look directly at the front of the rim?

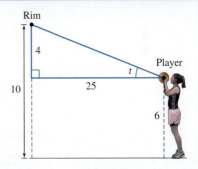

Figure 8.51 A basketball hoop

SOLUTION

The rim is $10 - 6 = 4$ feet higher than the player's eyes. The tangent of the angle t in Figure 8.51 is $\dfrac{4}{25}$. Because the angle t is acute, $t = \arctan \dfrac{4}{25}$. The angle t is not one of our special angles, so we use a calculator to get the inverse tangent: $t = \arctan \dfrac{4}{25} \approx 9.09°$.

TRY IT YOURSELF 8.37 Brief answers provided at the end of the section.

The same point guard must shoot a desperation shot from near midcourt, at a distance of 43 feet from the front of the rim. At what angle should she look to directly view the rim?

The Other Inverse Trigonometric Functions: Arccosecant, Arcsecant, and Arccotangent

> Finding the inverse of cosecant, secant, and cotangent requires restricting their domains.

Just as with the sine, cosine, and tangent functions, we can find inverse functions for the cosecant, secant, and cotangent functions, provided we restrict the domain of the functions so that the restricted functions satisfy the horizontal line test.

These inverses are used far less commonly than those for the sine, cosine, and tangent. One reason is that they can be found using the inverse functions we have already studied. For example, $\operatorname{arccsc} x = \arcsin \dfrac{1}{x}$. The domain restrictions needed for the inverse functions of cosecant, secant, and cotangent are a little more complicated than those for sine, cosine, and tangent. Consider, for example, the graph of cosecant, as shown in **Figure 8.52**.

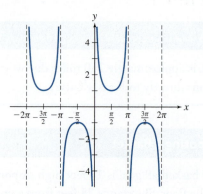

Figure 8.52 The graph of cosecant

There is more than one way to restrict the domain to satisfy the horizontal line test. The cosecant domain restriction we use, and the one you will most likely see in calculus, is $\left(0, \dfrac{\pi}{2} \right] \cup \left(\pi, \dfrac{3\pi}{2} \right]$, as shown in **Figure 8.53**.

The graph of arccosecant is obtained by reflection of the graph in Figure 8.53 through the line $y = x$ and is shown in **Figure 8.54**.

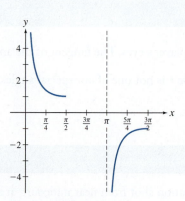

Figure 8.53 The cosecant with domain restricted to $\left(0, \dfrac{\pi}{2} \right] \cup \left(\pi, \dfrac{3\pi}{2} \right]$

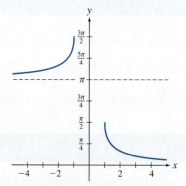

Figure 8.54 Reflecting the graph in Figure 8.53 through the line $y = x$ to make the graph of the arccosecant: The range is $\left(0, \dfrac{\pi}{2} \right] \cup \left(\pi, \dfrac{3\pi}{2} \right]$.

In a similar fashion, we can restrict the domains of the secant and cotangent to make the restricted function be one-to-one.

- For the secant function, we restrict the domain to $\left[0, \dfrac{\pi}{2}\right) \cup \left[\pi, \dfrac{3\pi}{2}\right)$, as shown in Figure 8.55. The inverse function is the arcsecant. Its graph is shown in Figure 8.56.

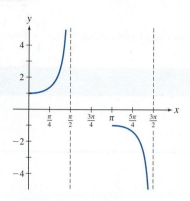

Figure 8.55 The secant with domain restricted to $\left[0, \dfrac{\pi}{2}\right) \cup \left[\pi, \dfrac{3\pi}{2}\right)$

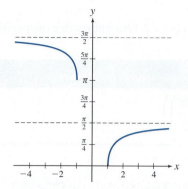

Figure 8.56 Reflecting the graph in Figure 8.55 through the line $y = x$ to make the graph of the arcsecant: The range is $\left[0, \dfrac{\pi}{2}\right) \cup \left[\pi, \dfrac{3\pi}{2}\right)$.

- For the cotangent, we restrict the domain to $(0, \pi)$, as shown in Figure 8.57. The inverse function is the arccotangent. Its graph is shown in Figure 8.58.

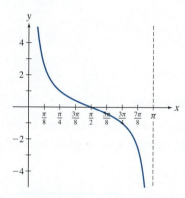

Figure 8.57 The cotangent with domain restricted to $(0, \pi)$

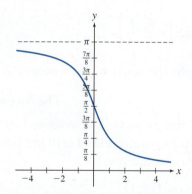

Figure 8.58 Reflecting the graph in Figure 8.57 through the line $y = x$ to make the graph of the arccotangent: The range is $(0, \pi)$.

Observe that in each case vertical asymptotes of the trigonometric function correspond to horizontal asymptotes of the inverse trigonometric function.

EXAMPLE 8.38 Evaluating the Arcsecant

Find the value of $\operatorname{arcsec}\left(\sec \dfrac{3\pi}{4}\right)$.

SOLUTION

We know that $\sec\dfrac{3\pi}{4} = -\sqrt{2}$. Thus, $\text{arcsec}\left(\sec\dfrac{3\pi}{4}\right) = \text{arcsec}(-\sqrt{2})$. This expression calls for a number in

$\left[0, \dfrac{\pi}{2}\right) \cup \left[\pi, \dfrac{3\pi}{2}\right)$ whose secant is $-\sqrt{2}$. Our knowledge of special angles then yields $\text{arcsec}\left(\sec\dfrac{3\pi}{4}\right) = \dfrac{5\pi}{4}$.

TRY IT YOURSELF 8.38 Brief answers provided at the end of the section.

Find the value of $\text{arccot}\left(\cot\left(\dfrac{3\pi}{2}\right)\right)$.

CONCEPTS TO REMEMBER: The Arcsine, Arccosine, and Arctangent

The Arcsine Function

Defining property: The arcsine function $g(x) = \arcsin x$ is the inverse function of $f(x) = \sin x$ restricted to $\left[-\dfrac{\pi}{2}, \dfrac{\pi}{2}\right]$. Intuitively, $\arcsin x$ is the angle between $-\dfrac{\pi}{2}$ and $\dfrac{\pi}{2}$ whose sine is x.

Domain: $[-1, 1]$

Range: $\left[-\dfrac{\pi}{2}, \dfrac{\pi}{2}\right]$

Graph: The function $g(x) = \arcsin x$ is increasing.

The Arccosine Function

Defining property: The arccosine function $g(x) = \arccos x$ is the inverse function of $f(x) = \cos x$ restricted to $[0, \pi]$. Intuitively, $\arccos x$ is the angle between 0 and π whose cosine is x.

Domain: $[-1, 1]$

Range: $[0, \pi]$

Graph: The function $g(x) = \arccos x$ is decreasing.

The Arctangent Function

Defining property: The arctangent function $g(x) = \arctan x$ is the inverse function of $f(x) = \tan x$ restricted to $\left(-\dfrac{\pi}{2}, \dfrac{\pi}{2}\right)$. Intuitively, $\arctan x$ is the angle between $-\dfrac{\pi}{2}$ and $\dfrac{\pi}{2}$ whose tangent is x.

Domain: All real numbers

Range: $\left(-\dfrac{\pi}{2}, \dfrac{\pi}{2}\right)$

Graph: The function $g(x) = \arctan x$ is increasing.

In the long term: $\arctan x \to \dfrac{\pi}{2}$ as $x \to \infty$, and $\arctan x \to -\dfrac{\pi}{2}$ as $x \to -\infty$

TRY IT YOURSELF ANSWERS

8.27 $\arcsin 1 = \dfrac{\pi}{2}$

8.28 $\dfrac{2}{3}$

8.29 0

8.30 $\arcsin\left(-\dfrac{\sqrt{3}}{2}\right) = -\dfrac{\pi}{3} = -\arcsin\dfrac{\sqrt{3}}{2}$

8.31 The arccosine decreases on $[-1, 1]$. The graph is concave up on $[-1, 0)$ and concave down on $(0, 1]$.

8.32 $-\dfrac{\pi}{6}$

8.33 $x = \cos\dfrac{5}{2} \approx -0.80$

8.34 $\dfrac{3}{4}$

8.35 $2t^2 - 1$

8.36 $339.68°$

8.37 $5.31°$

8.38 $\dfrac{\pi}{2}$

EXERCISE SET 8.4

CHECK YOUR UNDERSTANDING

1. State the domain and range of the arcsine function.

2. State the domain and range of the arccosine function.

3. State the domain and range of the arctangent function.

4. **True or false:** $\sin(\arcsin x) = x$ for all x in the range of the sine.

5. **True or false:** $\arccos(\cos x) = x$ for all x in the domain of the cosine.

6. Describe in words the meaning of arctan x.

7. Describe in words the meaning of arcsin x.

8. If $-1 < c < 1$, one solution for x of the equation $\sin x = c$ is _____.

9. One solution for x of the equation $\tan x = c$ is _____.

10. Describe how to simplify the expression $\sin(\arccos x)$.

SKILL BUILDING

Inverse trigonometric functions involving special angles. In Exercises 11 through 26, calculate the exact value of the given inverse trigonometric function. In each case, special angles are involved, so a calculator is not necessary.

11. $\arcsin\left(\dfrac{\sqrt{3}}{2}\right)$

12. $\arcsin\left(-\dfrac{\sqrt{3}}{2}\right)$

13. $\arccos\left(\dfrac{\sqrt{3}}{2}\right)$

14. $\arccos\left(-\dfrac{\sqrt{3}}{2}\right)$

15. $\arccos 0$

16. $\arccos\left(-\dfrac{1}{2}\right)$

17. $\arcsin\left(-\dfrac{1}{2}\right)$

18. $\arcsin\left(\dfrac{1}{\sqrt{2}}\right)$

19. $\arcsin\left(-\dfrac{1}{\sqrt{2}}\right)$

20. $\arccos\left(-\dfrac{1}{\sqrt{2}}\right)$

21. $\arccos\left(\dfrac{1}{\sqrt{2}}\right)$

22. $\arctan 0$

23. $\arctan \sqrt{3}$

24. $\arctan 1$

25. $\arctan(-1)$

26. $\arctan(-\sqrt{3})$

Graphs of inverse trigonometric functions. In Exercises 27 through 32, match the given graph with the appropriate function from the list that follows.

A: $y = \arcsin x$ B: $y = \arccos x$ C: $y = \arctan x$

D: $y = \text{arcsec } x$ E: $y = \text{arccot } x$ F: $y = \text{arccsc } x$

27.

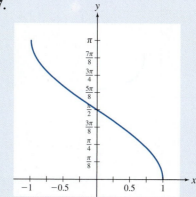

28.

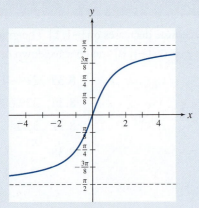

29.

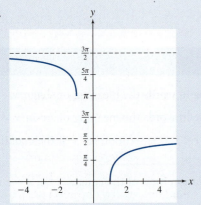

30.

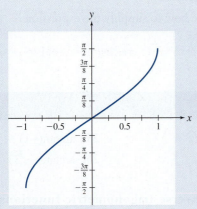

31.

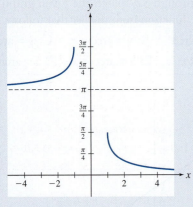

32.

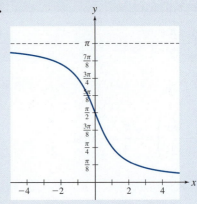

One solution of trigonometric equations. In Exercises 33 through 38, use inverse trigonometric functions to find *one* solution of the given equation. Report your answer to two decimal places.

33. $\sin x = 0.78$ **36.** $\cos x = -0.54$

34. $\sin x = -0.63$ **37.** $\tan x = 5$

35. $\cos x = 0.82$ **38.** $\tan x = -7$

Solving inverse trigonometric equations. In Exercise 39 through 44, solve the given equation. Report the exact answer, and (where needed) give a decimal answer rounded to two decimal places.

39. $\arctan x = 1$ **42.** $\arcsin x = -1$

40. $\arcsin x = -\dfrac{1}{\sqrt{2}}$ **43.** $\arccos x = -1$

41. $\arccos x = 0$ **44.** $\arctan x = 2$

Composing trigonometric functions with inverse trigonometric functions. In Exercises 45 through 58, find the exact value indicated. In exercises where x appears, you may assume that $0 < x < 1$.

45. $\tan\left(\arcsin \dfrac{1}{3}\right)$ **51.** $\tan\left(\arccos\left(-\dfrac{3}{4}\right)\right)$

46. $\cot\left(\arccos \dfrac{2}{3}\right)$ **52.** $\cot(\arctan(-3))$

47. $\sin(\arctan 5)$ **53.** $\cot(\arcsin x)$

48. $\csc\left(\arccos \dfrac{5}{8}\right)$ **54.** $\tan(\arccos x)$

55. $\sin(\arccos x)$

49. $\sin\left(\arccos \dfrac{5}{8}\right)$ **56.** $\cos(\arctan x)$

57. $\tan\left(\arcsin \dfrac{x}{x+1}\right)$

50. $\cos\left(\arctan \dfrac{6}{5}\right)$ **58.** $\cot\left(\arcsin \dfrac{1-x}{1+x}\right)$

Using multiple-angle formulas. In Exercises 59 through 68, simplify the expression by first using an appropriate sum or multiple-angle formula. In exercises where x appears, you may assume that $0 < x < 1$ unless another restriction is stated. Your eventual answer should not involve any trigonometric functions.

59. $\sin\left(2\arctan\dfrac{2}{3}\right)$

60. $\cos\left(2\arccos\dfrac{1}{3}\right)$

61. $\sin(2\arcsin x)$

62. $\sin(2\arctan x)$

63. $\cos(2\arccos x)$

64. $\sin(\arccos x + \arcsin x)$

65. $\cos(\arctan x + \arcsin x)$

66. $\tan(2\arcsin x)$

67. $\sin(4\arctan x)$

68. $\sin\left(2\arccos\dfrac{1}{x}\right)$, assuming $x > 1$

PROBLEMS

Identities involving inverse trigonometric functions. In Exercises 69 through 75, establish the given identity.

69. Show that if $0 < x < 1$, then $\arcsin x = \arccos\sqrt{1-x^2}$. *Suggestion*: Draw a right triangle with an acute angle whose sine is x.

70. Show that if $\arcsin x = x$, then $x = \sin x$.

71. Show the identity $\arcsin(-x) = -\arcsin x$. *Suggestion*: Use the identity $\sin(-\theta) = -\sin\theta$.

72. a. Use the difference formula for the cosine to show the identity $\cos(\pi - \theta) = -\cos\theta$.

   b. Use the result from part a to show the identity

$$\arccos(-x) = \pi - \arccos(x).$$

73. Use the facts that the sine is an odd function and that the cosine is an even function to show the identity $\tan(-\theta) = -\tan\theta$. Use this fact to show the identity $\arctan(-x) = -\arctan x$.

74. Show that if $0 < x < 1$, then $\arccos x + \arcsin x = \dfrac{\pi}{2}$. *Suggestion*: Draw a right triangle with an acute angle whose cosine is x.

75. Show that if $0 < x < 1$, then $\sin(\arccos x) = \cos(\arcsin x)$. *Suggestion*: Draw triangles with appropriate cosine and sine.

Solving equations involving special angles. In Exercises 76 through 80, find one exact solution of the given equation.

76. $\sin x = \cos x$. *Suggestion*: First divide each side of the equation by $\cos x$.

77. $\cot x = \sqrt{3}$. *Suggestion*: Take the reciprocal of each side.

78. $2\tan x = \sec x$. *Suggestion*: First multiply each side of the equation by $\cos x$.

79. $\tan x = 3\cot x$. *Suggestion*: First multiply each side of the equation by $\tan x$.

80. $2\sin x = \csc x$. *Suggestion*: First multiply each side of the equation by $\sin x$.

Solving equations. In Exercises 81 through 88, find one solution of the given equation. Report an exact answer using inverse trigonometric functions, and provide an approximation correct to two decimal places.

81. $\sin x = 3\cos x$. *Suggestion*: First divide both sides by the cosine.

82. $\tan x = 4\cot x$. *Suggestion*: First multiply both sides by the tangent.

83. $4\sin 3x = 1$

84. $8\sin x \cos x = 3$. *Suggestion*: Use the double-angle formula for the sine.

85. $7\csc x = \sec x$

86. $\sin 2t = \sin(t+1)$. Assume that both $2t$ and $t+1$ lie between 0 and $\dfrac{\pi}{2}$.

87. $\sin\dfrac{1}{t} = \dfrac{1}{\sqrt{2}}$. Assume $0 < \dfrac{1}{t} < \dfrac{\pi}{2}$. *Suggestion*: First solve $\sin y = \dfrac{1}{\sqrt{2}}$, and then put $y = \dfrac{1}{t}$.

88. $5\cos^2 x = 1 + 5\sin^2 x$. *Suggestion*: Think about the double-angle formula for the cosine.

Solving inverse trigonometric equations. In Exercises 89 through 91, find all solutions of the given equation for x in the given interval. Report your answers as exact solutions.

89. $\arcsin x = \arccos 3x$. Assume $0 < x < \dfrac{1}{3}$.

90. $\arctan 2x = \arcsin x$. Assume $0 \le x < 1$. Note that there are two solutions.

91. $\arcsin\dfrac{1}{2x} = \arctan\dfrac{1}{x}$. Assume $x > \dfrac{1}{2}$.

Transformations of graphs of inverse trigonometric functions.

92. Plot the graphs of $y = \arcsin x$ and $y = 1 + \arcsin x$ on the same coordinate axes.

93. Plot the graphs of $y = \arccos x$ and $y = \arccos(x + 1)$ on the same coordinate axes.

94. Plot the graphs of $y = \arcsin x$ and $y = 1 + \arcsin(x - 1)$ on the same coordinate axes.

95. Plot the graphs of $y = \arctan x$ and $y = \arctan 2x$ on the same coordinate axes.

Inverse trigonometric functions in terms of degrees. Applications of inverse trigonometric functions often involve degrees rather than radians. In terms of degrees, the ranges are

- arcsin: $[-90°, 90°]$
- arccos: $[0°, 180°]$
- arctan: $[-90°, 90°]$

For Exercises 96 through 98, set your calculator to degree mode, and report all answers correct to two decimal places.

96. Calculate arcsin 0.4 in degrees.

97. Calculate arccos 0.32 in degrees.

98. Calculate arctan 7 in degrees.

Solving a right triangle in degrees. In Exercises 99 through 101, use **Figure 8.59**.

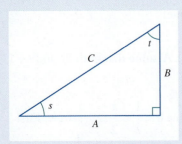

Figure 8.59 The right triangle for Exercises 99 through 101

99. If $B = 3$ and $C = 5$, find A, s, and t. Report s and t in degrees correct to two decimal places.

100. If $B = 4$ and $C = 7$, find A, s, and t. Report s and t in degrees correct to two decimal places.

101. If $A = 3$ and $B = 2$, find C, s, and t. Report s and t in degrees correct to two decimal places.

102. Relating inverse trigonometric functions.

a. If $x > 1$, show that $\operatorname{arcsec} x = \arccos \dfrac{1}{x}$.

   Suggestion: Draw a right triangle that represents an angle whose secant is x.

b. If $x > 1$, show that $\operatorname{arccsc} x = \arcsin \dfrac{1}{x}$.

   Suggestion: Draw a right triangle that represents an angle whose cosecant is x.

c. If $x > 0$, show that $\operatorname{arccot} x = \arctan \dfrac{1}{x}$.

   Suggestion: Draw a triangle that represents an angle whose cotangent is x.

103. Plot the graphs of $y = \arctan x$ and $y = x - 1$ on the same coordinate axes. Then solve the equation $\arctan x = x - 1$. Report your answer correct to two decimal places.

104. Plot the graphs of $y = \arcsin x$ and $y = e^{-x}$ on the same coordinate axes. Then solve the equation $\arcsin x = e^{-x}$. Report your answer correct to two decimal places.

105. Plot the graph of $\arctan \dfrac{1}{x}$. Estimate, based on your graph, the limit as $x \to \infty$ of $\arctan \dfrac{1}{x}$. Does your graph show a verical asymptote at $x = 0$?

MODELS AND APPLICATIONS

106. A cliff. You stand 100 horizontal feet from a vertical cliff. The top of the cliff is 30 feet above eye level (see **Figure 8.60**). At what angle above the horizontal must you incline your eyes to look directly at the top of the cliff? Report your answer in degrees correct to two decimal places.

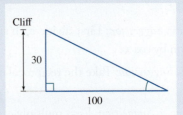

Figure 8.60 A cliff

107. A directional antenna. A radio transmitter is located 40 miles west and 13 miles north of a directional antenna (see **Figure 8.61**). To receive an optimal signal, the antenna must be aimed directly at the transmitter. How many degrees north of west should the antenna be aimed? Report your answer in degrees correct to two decimal places.

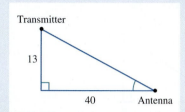

Figure 8.61 A transmitter and an antenna

108. A searchlight. A searchlight initially points due west. The light rotates clockwise at a constant speed so that it completes one revolution every 2 minutes. A building is located 200 yards west and 50 yards north of the searchlight (see **Figure 8.62**). How many minutes into the first revolution of the searchlight will the building be illuminated by the searchlight?

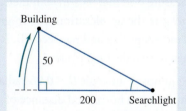

Figure 8.62 A searchlight and a building

109. A projectile. A projectile fired from ground level will strike the ground at a distance of $d = \dfrac{v^2 \sin(2\theta)}{g}$ feet downrange. Here, θ is the angle the shooting barrel makes with the ground, $g = 32$ feet per second per second is the acceleration due to gravity, and v is the muzzle velocity in feet per second.

  a. If the projectile strikes the ground d feet downrange, find the angle θ in terms of d, v, and g.

  b. If the muzzle velocity is 320 feet per second, what angle would you use to make the projectile land 1200 feet downrange? Report your answer in degrees to two decimal places.

110. Dallas to Kansas City. Dallas is 451 miles due south of Salina, Kansas, and Kansas City, Missouri, is 175 miles due east of Salina. An airplane flies on a direct trip from Dallas to Kansas City.

  a. What is the tangent of the angle that the flight path makes with Interstate 35, which runs due north from Dallas? Round your answer to two decimal places.

  b. Use your answer to part a to determine the angle that the flight path makes with Interstate 35. Report your answer in degrees to two decimal places.

  c. How far does the airplane fly? Round your answer to the nearest mile.

111. Parallax angle. Suppose we view a star now and then view it again half a year later. Our viewing location will have changed by the diameter of Earth's orbit around the sun. For nearby stars (within 100 light-years or so), the star appears to be in a slightly different location in the sky because of the change in our position. Half of the angle from one location to the next is known as the parallax angle. In **Figure 8.63**, p is the parallax angle, d is half of the diameter of Earth's orbit about the sun, and s is the distance from the sun to the star. Show how to find the parallax angle p if d and s are known.

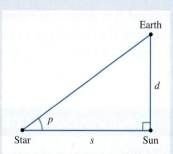

Figure 8.63 p is the parallax angle

112. Intensity of sunlight. When incident rays of sunlight form an angle θ with a flat object, such as a leaf, then the intensity of sunlight is reduced by a factor of $\sin \theta$, assuming that θ is between 0° and 90° (see **Figure 8.64**). Round your answers to two decimal places.

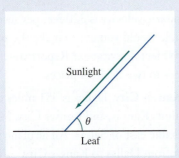

Figure 8.64 Sunlight striking a leaf

a. By what factor is the intensity reduced if the angle formed is 45°?

b. What is the angle (in degrees) θ if the intensity is reduced by a factor of 0.4?

113. Pinnate muscle tissue. In a pinnate muscle, the muscle fibers are short and are inserted along a common axis such as a tendon.[2] In this exercise, we determine the distance d by which the tendon moves when the muscle contracts. Let b be the distance between the bones to which the muscle is attached, let α be the angle between the muscle fibers and the tendon before contraction, and let α' be the angle after contraction. Then it turns out that $d = \dfrac{b}{2}(\cot \alpha - \cot \alpha')$. In practice, both α and α' are acute angles.

a. Use the inverse tangent to express the angle α' in terms of α, d, and b.

b. Assume b is 1 unit and that the angle before contraction is 10°. Find the angle α' if the distance that the tendon moves is 1 unit. Round your answer in degrees to two decimal places.

114. Size of prey seized. The diagram in **Figure 8.65** is useful for estimating the diameter of a circular prey that would be optimal for the grasping claws of a praying mantis.[3]

a. Find a formula for the angle t in terms of the diameter $|BC|$ of the circle and the length $|AC|$. (The vertical bars denote the length of a segment.)

b. If $|AC|$ is 4 centimeters, what angle gives a diameter of 3 centimeters? Report your answer in degrees to two decimal places.

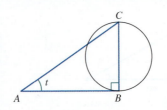

Figure 8.65 The claws of a praying mantis

115. Jumping frogs. If an animal jumps at an angle θ radians to the horizontal plane with initial velocity m, then the horizontal distance d that it will travel is given by

$$d = \frac{m^2 \sin (2\theta)}{g}$$

Here we measure d in meters and m in meters per second, and g is the acceleration due to gravity (about 9.8 meters per second per second). Round your answers to two decimal places.

a. If a frog jumps at an angle $\theta = 0.96$ radian and the jump covers a horizontal distance of 0.8 meter, what is the initial velocity?

b. Find the angle θ if the distance that an animal jumps is 1 meter and the initial velocity is 3.2 meters per second. (Assume that θ is between 0 and $\dfrac{\pi}{4}$ radians.)

CHALLENGE EXERCISES FOR INDIVIDUALS OR GROUPS

116. A cistern. A lid that is a portion of a sphere is to be built for a cylindrical cistern that is 8 feet in diameter. See **Figure 8.66**. The height of the lid at the center of the cistern is to be 3 feet. To make the lid, we must first cut rebar the length of the circular arc over the cistern. How long is the rebar? You are challenged to solve this problem without further help. If you have difficulty, the following steps may prove useful.

---

[2]Pinnate muscles are found in lobster claws, for example.

[3]Based on the work of C.S. Holling, as described by E.R. Pianka in *Evolutionary Ecology*, 6th ed., Benjamin Cummings, 2000, San Francisco.

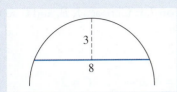

Figure 8.66 The spherical lid of a cistern

a. Use the Pythagorean theorem with **Figure 8.67** to calculate exactly the radius r of the circle.

b. Calculate the sine of the angle t, and use that information to find the angle t in radians rounded to two decimal places.

c. Use the arc length formula and your answer to part b to find the length of the rebar. Round your answer in feet to two decimal places.

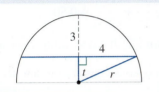

Figure 8.67 The spherical lid of a cistern: a solution aid

117. Machin's formula. The formula

$$\frac{\pi}{4} = 4\arctan\frac{1}{5} - \arctan\frac{1}{239}$$

is known as Machin's formula.[4] This formula is a bit challenging to establish, but it turns out to be an important tool for computing decimal approximations of π. Taking the tangent of both sides, we can rewrite this formula as

$$1 = \tan\left(4\arctan\frac{1}{5} - \arctan\frac{1}{239}\right)$$

We focus our attention on this version of the identity.

a. Apply the double-angle formula for the tangent twice to show that

$$\tan 4x = \frac{4\tan x(1 - \tan^2 x)}{(1 - \tan^2 x)^2 - 4\tan^2 x}$$

b. Use part a to show that $\tan\left(4\arctan\frac{1}{5}\right) = \frac{120}{119}$.

c. Use the sum formula for the tangent together with part b to show that

$$1 = \tan\left(4\arctan\frac{1}{5} - \arctan\frac{1}{239}\right)$$

REVIEW AND REFRESH: Exercises from Previous Sections

118. **From Section 1.3:** The graph of a function f is shown in **Figure 8.68**. State the interval(s) where f is increasing and the interval(s) where f is concave down.

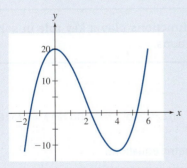

Figure 8.68 A graph for Exercise 118

Answer: Increasing on $[-2, 0]$ and $[4, 6]$. Concave down on $[-2, 2)$.

119. **From Section 4.3:** Solve the equation $\ln(e^{3x} + 1) = 2$.

Answer: $x = \frac{1}{3}\ln(e^2 - 1)$

120. **From Section 5.3:** List all possible rational zeros of $P(x) = 2x^4 + ax^3 + 6$ if a is an integer.

Answer: $\pm 1, \ \pm 2, \ \pm 3, \ \pm 6, \pm\frac{1}{2}, \ \pm\frac{3}{2}$

121. **From Section 5.3:** Is it possible that $f(x) = 7x^4 - 13x^3 + 6x^2 - x + 256$ has five distinct zeros?

Answer: No

---

[4] John Machin (1686 to June 9, 1651) was a professor of astronomy at Gresham College, London.

122. From Chapter 6: Make a table that shows the values of the sine and cosine of $0, \dfrac{\pi}{6}, \dfrac{\pi}{4}, \dfrac{\pi}{3}$, and $\dfrac{\pi}{2}$ radians.

Answer:

t	0	$\dfrac{\pi}{6}$	$\dfrac{\pi}{4}$	$\dfrac{\pi}{3}$	$\dfrac{\pi}{2}$
$\sin t$	0	$\dfrac{1}{2}$	$\dfrac{1}{\sqrt{2}}$	$\dfrac{\sqrt{3}}{2}$	1
$\cos t$	1	$\dfrac{\sqrt{3}}{2}$	$\dfrac{1}{\sqrt{2}}$	$\dfrac{1}{2}$	0

123. From Chapter 6: Find the lengths of the unknown sides of the right triangle shown in **Figure 8.69**.

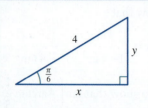

Figure 8.69 A triangle for Exercise 123

Answer: $x = 2\sqrt{3}, y = 2$

124. From Section 7.1: Choose A, B, and C so that $y = A \sin B(x - C)$ has amplitude 4, period 3, and phase shift 2.

Answer: $A = 4$, $B = \dfrac{2\pi}{3}$, $C = 2$

125. From Section 8.1: Express $\cot t$ in terms of $\sin t$. Assume that $\dfrac{\pi}{2} < t < \pi$.

Answer: $\cot t = -\dfrac{\sqrt{1 - \sin^2 t}}{\sin t}$

126. From Section 8.2: Suppose $\sin x = \dfrac{2}{3}$ and $\sin y = \dfrac{1}{5}$. Suppose further that both x and y are between 0 and $\dfrac{\pi}{2}$. Calculate $\sin(x + y)$.

Answer: $\dfrac{2}{3} \dfrac{2\sqrt{6}}{5} + \dfrac{\sqrt{5}}{3} \dfrac{1}{5} \approx 0.80$

127. From Section 8.3: If t is an acute angle and $\sin t = \dfrac{3}{7}$, find $\sin 2t$, $\cos 2t$, and $\tan 2t$.

Answer:

$\sin 2t = \dfrac{12\sqrt{10}}{49} \approx 0.77$, $\cos 2t = \dfrac{31}{49} \approx 0.63$,

$\tan 2t = \dfrac{12\sqrt{10}}{31} \approx 1.22$

8.1   Trigonometric Identities

8.2   Sum and Difference Formulas

8.3   Double-Angle and Half-Angle Formulas

8.4   Inverse Trigonometric Functions

8.5   Solving Trigonometric Equations

8.5 Solving Trigonometric Equations

Trigonometric identities and inverse functions play a key role in solving equations involving trigonometric functions.

In this section, you will learn to:

1. Solve the basic trigonometric equations $\sin t = c$, $\cos t = c$, and $\tan t = c$ for t on one full period.
2. Locate all solutions of the basic trigonometric equations.
3. Use identities to solve trigonometric equations by reducing them to basic equations.
4. Use substitutions and polynomial factorization to solve trigonometric equations.

Typical models of population growth involve exponential functions or logistic functions. But models for predator and prey populations may be more complex. For example, on the African savanna, the cheetah preys on the gazelle. When the gazelle population is large, the sufficiency of food contributes to a rise in the cheetah population. But an increasing cheetah population leads to increased predation, which reduces the gazelle population. Decreasing food supply causes a decline in the cheetah population, and thus an increase in the gazelle population.

The result is that both predator and prey populations rise and fall periodically, which indicates a sine wave model. A typical model for a predator population model is of the form

$$N = A \sin(\omega t + C) + D$$

If a wildlife manager needs to know when the population is at a certain level, say N_0, then the trigonometric equation

$$N_0 = A \sin(\omega t + C) + D$$

must be solved for t. As is illustrated by the crossing points in **Figure 8.70**, there are infinitely many solutions of equations of this type. The wildlife manager needs to know all of them.

We will cover such trigonometric equations, which often have an infinite number of solutions, in this section. The inverse trigonometric functions are the basic tools needed to solve such equations.

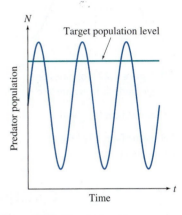

Figure 8.70 The times when a predator population is at a given level

Basic Trigonometric Equations

> Many complicated trigonometric equations are equivalent to simpler trigonometric equations that, in turn, are amenable to a straightforward solution.

We can solve many trigonometric equations by relating them to one of the following three basic equations:

$$\sin t = c \qquad \cos t = c \qquad \tan t = c$$

Let's see how to solve one of these basic equations, $\cos t = \dfrac{1}{2}$. **Figure 8.71** illustrates the fact that this equation has infinitely many solutions—one for each crossing of the line $y = \dfrac{1}{2}$ with the graph of the cosine.

We will find them all, but we begin with a single solution. The equation asks for an angle t whose cosine is $\dfrac{1}{2}$. Consequently, the inverse cosine gives one solution:

$$\text{First solution:} \quad t = \arccos \frac{1}{2} = \frac{\pi}{3}$$

But because the cosine is an even function, $\cos\left(-\dfrac{\pi}{3}\right) = \cos\dfrac{\pi}{3}$. Thus, we have a second solution:

$$\text{Second solution:} \quad t = -\arccos \frac{1}{2} = -\frac{\pi}{3}$$

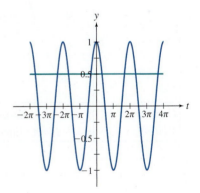

Figure 8.71 The equation $\cos t = \dfrac{1}{2}$: The solutions correspond to the (infinitely many) crossing points of $y = \dfrac{1}{2}$ and $y = \cos t$.

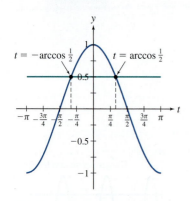

DF **Figure 8.72** Solutions on a single period: There are two solutions of $\cos t = \dfrac{1}{2}$ in the full period $[-\pi, \pi]$ of the cosine function.

As is illustrated in **Figure 8.72**, these two are the only solutions on one full period $[-\pi, \pi]$ of the cosine function. Because the cosine repeats every 2π units, we find the remaining solutions by adding integer multiples of 2π to these two solutions:

$$t = \frac{\pi}{3} + 2k\pi \text{ and } t = -\frac{\pi}{3} + 2k\pi, \text{ where } k \text{ is any integer,}$$

or

$$\dots, \frac{\pi}{3} - 4\pi, \ \frac{\pi}{3} - 2\pi, \ \frac{\pi}{3}, \ \frac{\pi}{3} + 2\pi, \ \frac{\pi}{3} + 4\pi, \dots$$

$$\text{and} \dots, -\frac{\pi}{3} - 4\pi, \ -\frac{\pi}{3} - 2\pi, \ -\frac{\pi}{3}, \ -\frac{\pi}{3} + 2\pi, \ -\frac{\pi}{3} + 4\pi, \dots$$

We use a similar method to find the solutions of the remaining two basic equations, $\sin t = c$ and $\tan t = c$. That is, we find the solutions in one full period and use those to generate all other solutions.

CONCEPTS TO REMEMBER: Solutions of Basic Trigonometric Equations

(k is any integer)

Basic equation	$\sin t = c, \ -1 \le c \le 1$	$\cos t = c, \ -1 \le c \le 1$	$\tan t = c$
Solutions on one full period	The full period $\left[-\dfrac{\pi}{2}, \dfrac{3\pi}{2}\right]$ $t = \arcsin c$ $t = \pi - \arcsin c$	The full period $[-\pi, \pi]$ $t = \arccos c$ $t = -\arccos c$	The full period $\left[-\dfrac{\pi}{2}, \dfrac{\pi}{2}\right]$ $t = \arctan c$
Complete solution	$t = \arcsin c + 2k\pi$ $t = \pi - \arcsin c + 2k\pi$	$t = \arccos c + 2k\pi$ $t = -\arccos c + 2k\pi$	$t = \arctan c + k\pi$

Note that for sine and cosine there are typically two solutions on one full period, but there is only one for the tangent. Also, we use $2k\pi$ for the sine and cosine but $k\pi$ for the tangent. This difference reflects the fact that the period for the sine and cosine is 2π, but the period of the tangent is π.

It turns out that our solutions of these basic equations provide considerable help in solving many other trigonometric equations. Many trigonometric equations can be solved by first using algebraic operations and trigonometric identities to relate them to one of the three basic trigonometric equations.

EXAMPLE 8.39 Solving Basic Trigonometric Equations

Solve the equation $2\cos t = -\sqrt{3}$.

SOLUTION

We can change the equation $2\cos t = -\sqrt{3}$ into the form of the basic equation $\cos t = c$ by dividing both sides of the equation by 2:

$$\cos t = \frac{-\sqrt{3}}{2}$$

One solution on a single period $[-\pi, \pi]$ of the cosine is

$$t_1 = \arccos\left(\frac{-\sqrt{3}}{2}\right) = \frac{5\pi}{6}$$

and the other is

$$t_2 = -\arccos\left(\frac{-\sqrt{3}}{2}\right) = -\frac{5\pi}{6}$$

These two solutions are shown in **Figure 8.73** as crossing points of the line $y = -\dfrac{\sqrt{3}}{2}$ with the graph of the cosine. We can use these two solutions to find all of the solutions of the equation:

$$t = \frac{5\pi}{6} + 2k\pi \text{ and } t = -\frac{5\pi}{6} + 2k\pi, \text{ where } k \text{ is any integer,}$$

or

$$\ldots, \frac{5\pi}{6} - 4\pi, \frac{5\pi}{6} - 2\pi, \frac{5\pi}{6}, \frac{5\pi}{6} + 2\pi, \frac{5\pi}{6} + 4\pi, \ldots$$

and $\ldots, -\dfrac{5\pi}{6} - 4\pi, -\dfrac{5\pi}{6} - 2\pi, -\dfrac{5\pi}{6}, -\dfrac{5\pi}{6} + 2\pi, -\dfrac{5\pi}{6} + 4\pi, \ldots$

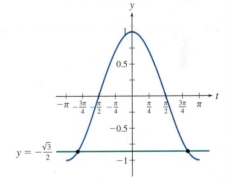

Figure 8.73 Solutions of $\cos t = -\dfrac{\sqrt{3}}{2}$ on one complete period of the cosine

TRY IT YOURSELF 8.39 Brief answers provided at the end of the section.

Solve the equation $-\tan t = \sqrt{3}$.

EXTEND YOUR REACH

Suppose we are trying to solve an equation of the form

$$\text{Trigonometric function} = 0$$

Suppose further that we know the period p of the trigonometric function on the left side of the equation. Explain how we use the solutions on one full period to find all solutions of the equation.

EXAMPLE 8.40 Using a Calculator to Solve Trigonometric Equations

Solve the trigonometric equation $\sin t = 0.4$. Write the exact solution, and then use a calculator to give approximate solutions.

SOLUTION

This is the basic equation for the sine:

$$\sin t = 0.4$$

$$t = \arcsin 0.4 \text{ and } t = \pi - \arcsin 0.4 \qquad \blacktriangleleft \text{ Solve on one full period } \left[-\frac{\pi}{2}, \frac{3\pi}{2}\right] \text{ of the sine.}$$

$$t = \arcsin 0.4 + 2k\pi \text{ and } t = \pi - \arcsin 0.4 + 2k\pi \qquad \blacktriangleleft \text{ Find the complete solution.}$$

$$t \approx 0.41 + 2k\pi \text{ and } x \approx 2.73 + 2k\pi \qquad \blacktriangleleft \text{ Take } k \text{ to be any integer.}$$

The approximate solutions can also be written as

$$\dots, \ 0.41 - 4\pi, \ 0.41 - 2\pi, \ 0.41, \ 0.41 + 2\pi, \ 0.41 + 4\pi, \dots$$

and

$$\dots, \ 2.73 - 4\pi, \ 2.73 - 2\pi, \ 2.73, \ 2.73 + 2\pi, \ 2.73 + 4\pi, \dots$$

The solutions on one full period of the sine are shown in **Figure 8.74**.

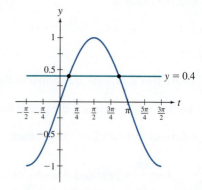

Figure 8.74 Solutions of $\sin t = 0.4$ on one full period $\left[-\frac{\pi}{2}, \frac{3\pi}{2}\right]$ of the sine

---

TRY IT YOURSELF 8.40   Brief answers provided at the end of the section.

Solve the equation $-2\tan t = \sqrt{3}$. Write the exact solution, and then use a calculator to give approximate solutions.

EXAMPLE 8.41 Using Simple Identities and Substitution to Solve Trigonometric Equations

Solve the trigonometric equation $\csc 2x = 2$.

SOLUTION

It is convenient first to make the substitution replacing $2x$ by t to get the equation $\csc t = 2$. Now, to transform $\csc t = 2$ into one of the basic equations, we use the fact that the cosecant is the reciprocal of the sine:

$$\csc t = 2$$

$$\frac{1}{\csc t} = \frac{1}{2} \qquad \blacktriangleleft \text{ Take reciprocals.}$$

$$\sin t = \frac{1}{2} \qquad \blacktriangleleft \text{ Use } \sin t = \frac{1}{\csc t}.$$

$$t = \arcsin \frac{1}{2} = \frac{\pi}{6} \text{ and}$$

$$t = \pi - \arcsin \frac{1}{2} = \pi - \frac{\pi}{6} = \frac{5\pi}{6} \qquad \blacktriangleleft \text{ Solve on one full period } \left[-\frac{\pi}{2}, \frac{3\pi}{2} \right] \text{ of the sine.}$$

$$t = \frac{\pi}{6} + 2k\pi \text{ and } t = \frac{5\pi}{6} + 2k\pi \qquad \blacktriangleleft \text{ Find the complete solution, where } k \text{ is any integer.}$$

$$2x = \frac{\pi}{6} + 2k\pi \text{ and } 2x = \frac{5\pi}{6} + 2k\pi \qquad \blacktriangleleft \text{ Replace } t \text{ by } 2x.$$

$$x = \frac{\pi}{12} + k\pi \text{ and } x = \frac{5\pi}{12} + k\pi \qquad \blacktriangleleft \text{ Solve for } x.$$

The final solution is $x = \dfrac{\pi}{12} + k\pi$ and $x = \dfrac{5\pi}{12} + k\pi$, where k is any integer, or

$$\ldots, \frac{\pi}{12} - 2\pi, \frac{\pi}{12} - \pi, \frac{\pi}{12}, \frac{\pi}{12} + \pi, \frac{\pi}{12} + 2\pi, \ldots$$

$$\text{and } \ldots, \frac{5\pi}{12} - 2\pi, \frac{5\pi}{12} - \pi, \frac{5\pi}{12}, \frac{5\pi}{12} + \pi, \frac{5\pi}{12} + 2\pi, \ldots$$

These solutions are illustrated in **Figure 8.75**.

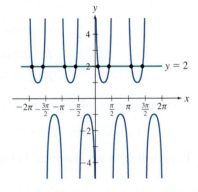

Figure 8.75 Some of the solutions of $\csc 2x = 2$

TRY IT YOURSELF 8.41 Brief answers provided at the end of the section.

Solve the equation $\cot(2x) = 1$.

EXTEND YOUR REACH

Here is a slightly different kind of trigonometric equation. **Figure 8.76** shows the graphs of the sine and cosine on the interval $[-\pi, \pi]$. There are two points in this plot where the rate of change of the sine is the same as the rate of change of the cosine. Use the graphs together with a straightedge to estimate these points.

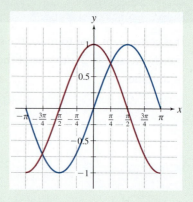

Figure 8.76 Graphs of the sine and cosine

EXAMPLE 8.42 Using a Calculator and Identities to Solve Trigonometric Equations

Solve the trigonometric equation $\sin t = 5\cos t$. Give the exact solution, and then use a calculator to give approximate solutions.

SOLUTION

Because the tangent is the sine over the cosine, we can transform this equation into the form of the basic equation $\tan t = c$ by dividing by the cosine:

$$\sin t = 5\cos t$$

$$\frac{\sin t}{\cos t} = 5 \qquad \blacktriangleleft \text{ Divide by the cosine.}$$

$$\tan t = 5 \qquad \blacktriangleleft \text{ Simplify.}$$

$$t = \arctan 5 \qquad \blacktriangleleft \text{ Solve on one full period } \left[-\frac{\pi}{2}, \frac{\pi}{2}\right] \text{ of the tangent.}$$

$$t = \arctan 5 + k\pi \qquad \blacktriangleleft \text{ Find the complete solution.}$$

$$t \approx 1.37 + k\pi \qquad \blacktriangleleft \text{ Take } k \text{ to be any integer.}$$

The approximate solutions can also be written as

$$\dots, 1.37 - 2\pi, \ 1.37 - \pi, \ 1.37, \ 1.37 + \pi, 1.37 + 2\pi, \dots$$

These solutions are illustrated in **Figure 8.77**.

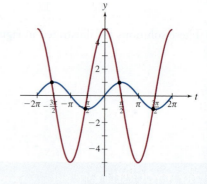

Figure 8.77 The equation $\sin t = 5\cos t$: The solutions occur where the graph of $y = \sin t$ crosses the graph of $y = 5\cos t$.

TRY IT YOURSELF 8.42 Brief answers provided at the end of the section.

Solve the equation $\sec t = 12\csc t$. Give the exact solution and then a decimal approximation.

Sometimes double-angle formulas provide the identity needed to solve a trigonometric equation.

EXAMPLE 8.43 Using Double-Angle Formulas to Solve Trigonometric Equations

Solve the equation $4 \sin x \cos x = \sqrt{2}$.

SOLUTION

The key to solving $4 \sin x \cos x = \sqrt{2}$ is to recognize the product $\sin x \cos x$ as part of the double-angle formula for the sine:

$$4 \sin x \cos x = \sqrt{2}$$

$$2(2 \sin x \cos x) = \sqrt{2} \qquad \blacktriangleleft \text{ Split 4 into a product.}$$

$$2 \sin 2x = \sqrt{2} \qquad \blacktriangleleft \text{ Use the double-angle formula for the sine.}$$

At this point, it is convenient to make the substitution replacing $2x$ by t, just as we did in Example 8.41:

$$2 \sin t = \sqrt{2} \qquad \blacktriangleleft \text{ Replace } 2x \text{ by } t.$$

$$\sin t = \frac{\sqrt{2}}{2} \qquad \blacktriangleleft \text{ Divide by 2.}$$

$$\sin t = \frac{1}{\sqrt{2}} \qquad \blacktriangleleft \text{ Simplify.}$$

$$t = \frac{\pi}{4} \text{ and } t = \pi - \frac{\pi}{4} = \frac{3\pi}{4} \qquad \blacktriangleleft \text{ Solve on one full period } \left[-\frac{\pi}{2}, \frac{3\pi}{2} \right] \text{ of the sine.}$$

$$t = \frac{\pi}{4} + 2k\pi \text{ and } t = \frac{3\pi}{4} + 2k\pi \qquad \blacktriangleleft \text{ Find the complete solution, where } k \text{ is integer.}$$

$$2x = \frac{\pi}{4} + 2k\pi \text{ and } 2x = \frac{3\pi}{4} + 2k\pi \qquad \blacktriangleleft \text{ Replace } t \text{ by } 2x.$$

$$x = \frac{\pi}{8} + k\pi \text{ and } x = \frac{3\pi}{8} + k\pi \qquad \blacktriangleleft \text{ Solve for } x.$$

The final solution is

$$x = \frac{\pi}{8} + k\pi \text{ and } x = \frac{3\pi}{8} + k\pi \text{ for any integer } k$$

or

$$\ldots, \frac{\pi}{8} - 2\pi, \frac{\pi}{8} - \pi, \frac{\pi}{8}, \frac{\pi}{8} + \pi, \frac{\pi}{8} + 2\pi, \ldots$$

$$\ldots, \frac{3\pi}{8} - 2\pi, \frac{3\pi}{8} - \pi, \frac{3\pi}{8}, \frac{3\pi}{8} + \pi, \frac{3\pi}{8} + 2\pi, \ldots$$

These solutions are illustrated in **Figure 8.78**.

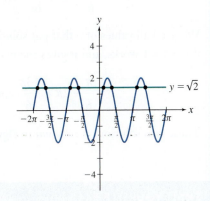

Figure 8.78 The equation $4 \sin x \cos x = \sqrt{2}$: The solutions occur where the graph of $y = 4 \sin x \cos x$ crosses the line $y = \sqrt{2}$.

TRY IT YOURSELF 8.43 Brief answers provided at the end of the section.

Find the exact solutions of $3\cos^2 t = 3\sin^2 t + 1$, and then use a calculator to find the approximate value of the solutions. *Suggestion*: Recall that $\cos 2t = \cos^2 t - \sin^2 t$.

EXAMPLE 8.44 Solving Equations on a Specified Interval

Find the solutions of $\sec t = 2 \tan t$ with $2\pi \le t \le 3\pi$.

SOLUTION

To relate this equation to one of the basic equations, we express both the secant and the tangent in terms of the sine and cosine:

$$\sec t = 2 \tan t$$

$$\frac{1}{\cos t} = 2\frac{\sin t}{\cos t} \qquad \blacktriangleleft \textbf{Convert to sines and cosines.}$$

$$\frac{1}{\cos t}\cos t = 2\frac{\sin t}{\cos t}\cos t \qquad \blacktriangleleft \textbf{Multiply by } \cos t.$$

$$1 = 2\sin t \qquad \blacktriangleleft \textbf{Simplify.}$$

$$\frac{1}{2} = \sin t \qquad \blacktriangleleft \textbf{Divide by 2.}$$

$$t = \arcsin\frac{1}{2} = \frac{\pi}{6} \text{ and } t = \pi - \arcsin\frac{1}{2} = \pi - \frac{\pi}{6} = \frac{5\pi}{6} \qquad \blacktriangleleft \textbf{Solve on one full period } \left[-\frac{\pi}{2}, \frac{3\pi}{2}\right] \textbf{of the sine.}$$

$$t = \frac{\pi}{6} + 2k\pi \text{ and } t = \frac{5\pi}{6} + 2\pi k \qquad \blacktriangleleft \textbf{Find the complete solution, where } k \textbf{ is any integer.}$$

The solutions can also be written as

$$\ldots, \frac{\pi}{6} - 2\pi, \ \frac{\pi}{6} - \pi, \ \frac{\pi}{6}, \ \frac{\pi}{6} + \pi, \ \frac{\pi}{6} + 2\pi, \ldots$$

$$\text{and} \ldots, \frac{5\pi}{6} - 2\pi, \ \frac{5\pi}{6} - \pi, \ \frac{5\pi}{6}, \ \frac{5\pi}{6} + \pi, \ \frac{5\pi}{6} + 2\pi, \ldots$$

We need all values of k that put solutions in the interval $[2\pi, 3\pi]$. Only the value $k = 1$ works, and it gives the two solutions:

$$t = \frac{\pi}{6} + 2\pi = \frac{13\pi}{6} \text{ and } t = \frac{5\pi}{6} + 2\pi = \frac{17\pi}{6}$$

These two solutions are shown in **Figure 8.79**.

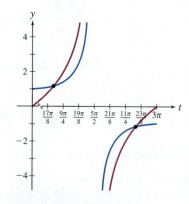

Figure 8.79 Solutions on a specified interval: The solutions are the crossing points of $y = \sec t$ (blue) with $y = 2\tan t$ (red), graphed over $[2\pi, 3\pi]$.

TRY IT YOURSELF 8.44 Brief answers provided at the end of the section.

Use a calculator to find the approximate value of the solution of $\cot x = 5$ with $3\pi \le x \le 4\pi$.

Solving by Factoring

> Some trigonometric equations can be solved by factoring and setting each factor equal to zero.

Often we can treat trigonometric equations like algebraic equations. For example, we may be able to use factoring, where the key observation is that if $AB = 0$ then $A = 0$ or $B = 0$.

EXAMPLE 8.45 Solving Factored Trigonometric Equations

Solve the trigonometric equation $(\sin x)(2\cos x - 1) = 0$.

SOLUTION

Because the product $(\sin x)(2\cos x - 1)$ is zero, one of the factors must be zero. That is, $\sin x = 0$ or $2\cos x - 1 = 0$. Just as with quadratic equations, this gives two separate equations:

$$\sin x = 0 \text{ and } 2\cos x - 1 = 0 \qquad \blacktriangleleft \text{ Set each factor equal to 0.}$$

$$\cos x = \frac{1}{2} \qquad \blacktriangleleft \text{ Solve for } \cos x.$$

$$x = k\pi \text{ and } x = \frac{\pi}{3} + 2k\pi \text{ and } x = -\frac{\pi}{3} + 2k\pi \qquad \blacktriangleleft \text{ Take } k \text{ to be any integer.}$$

The set of solutions of $\sin x(2\cos x - 1) = 0$ consists of all of these solutions, which can also be written as

$$\ldots, -2\pi, \ -\pi, \ 0, \ \pi, \ 2\pi, \ldots$$

$$\text{and } \ldots, \frac{\pi}{3} - 4\pi, \ \frac{\pi}{3} - 2\pi, \ \frac{\pi}{3}, \ \frac{\pi}{3} + 2\pi, \ \frac{\pi}{3} + 4\pi, \ldots$$

$$\text{and } \ldots, -\frac{\pi}{3} - 4\pi, \ -\frac{\pi}{3} - 2\pi, \ -\frac{\pi}{3}, \ -\frac{\pi}{3} + 2\pi, \ -\frac{\pi}{3} + 4\pi, \ldots$$

These solutions are shown in **Figure 8.80**.

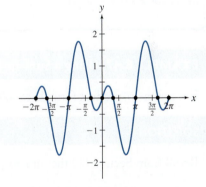

Figure 8.80 The equation $(\sin x)(2\cos x - 1) = 0$: The solutions occur where the graph crosses the x-axis.

TRY IT YOURSELF 8.45 Brief answers provided at the end of the section.

Solve $(\sec x - 2)(\tan x - 1) = 0$.

EXAMPLE 8.46 Solving by Factoring

Solve the trigonometric equation $\sin x \tan x = 6 \tan x$.

SOLUTION

We can solve $\sin x \tan x = 6 \tan x$ by factoring and then following the pattern of Example 8.45:

$$\sin x \tan x = 6 \tan x$$

$$\sin x \tan x - 6 \tan x = 0 \quad \blacktriangleleft \text{Move all terms left.}$$

$$\tan x(\sin x - 6) = 0 \quad \blacktriangleleft \text{Factor as } AB - AC = A(B - C).$$

We set each factor equal to zero and solve the resulting equations separately:

$$\tan x = 0 \text{ and } \sin x - 6 = 0 \quad \blacktriangleleft \text{Set each factor equal to 0.}$$

$$\sin x = 6 \quad \blacktriangleleft \text{Solve for } \sin x.$$

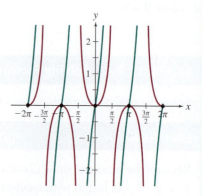

The solution of $\tan x = 0$ is $x = k\pi$ for any integer k. But the equation $\sin x = 6$ has no solution because the sine is always between -1 and 1. The final set of solutions is $x = k\pi$ for any integer k, or

$$\ldots, -2\pi, -\pi, 0, \pi, 2\pi, \ldots$$

Figure 8.81 illustrates the solution.

Figure 8.81 The equation $\sin x \tan x = 6 \tan x$: The solutions correspond to the intersections of $y = \sin x \tan x$ (red) with $y = 6 \tan x$ (green).

TRY IT YOURSELF 8.46 Brief answers provided at the end of the section.

Solve $\cos x \sin x = \sqrt{2} \cos x$.

EXAMPLE 8.47 Using a Sum-to-Product Formula

Use a sum-to-product formula to solve $\sin 5t + \sin 3t = 0$.

SOLUTION

Recall from Section 8.2 the sum-to-product formula:

$$\sin A + \sin B = 2 \sin\left(\frac{A + B}{2}\right) \cos\left(\frac{A - B}{2}\right)$$

We use this formula to express $\sin 5t + \sin 3t$ as a product:

$$\sin 5t + \sin 3t = 0$$

$$2 \sin\left(\frac{5t + 3t}{2}\right) \cos\left(\frac{5t - 3t}{2}\right) = 0 \quad \blacktriangleleft \text{Use the sum-to-product formula.}$$

$$2 \sin 4t \cos t = 0 \quad \blacktriangleleft \text{Simplify.}$$

$$\sin 4t \cos t = 0 \quad \blacktriangleleft \text{Divide by 2.}$$

$$\sin 4t = 0 \text{ and } \cos t = 0 \quad \blacktriangleleft \text{Set each factor equal to 0.}$$

$$4t = k\pi \text{ and } t = \frac{\pi}{2} + 2k\pi \text{ and } t = -\frac{\pi}{2} + 2k\pi \quad \blacktriangleleft \text{Find the complete solution, where } k \text{ is any integer.}$$

Divide the first equation by 4 to arrive at the final solution:

$$t = \frac{k\pi}{4} \text{ and } t = \frac{\pi}{2} + 2k\pi \text{ and } t = -\frac{\pi}{2} + 2k\pi$$

where k is any integer. Note that this can be simplified to $t = \frac{k\pi}{4}$, where k is any integer, or

$$\ldots, -\frac{\pi}{2}, -\frac{\pi}{4}, 0, \frac{\pi}{4}, \frac{\pi}{2}, \ldots$$

This solution is illustrated in **Figure 8.82**.

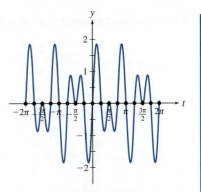

Figure 8.82 The equation $\sin 5t + \sin 3t = 0$: The solutions occur where the graph meets the t-axis.

TRY IT YOURSELF 8.47 Brief answers provided at the end of the section.

Solve $\cos 5t - \cos 3t = 0$.

Analogy with Quadratic Equations

Some trigonometric equations are solved in a fashion that reminds us of quadratic equations.

EXAMPLE 8.48 Solving Trigonometric Equations by Factoring Quadratics

Solve the equation $2\sin^2 x + \sin x = 1$.

SOLUTION

We treat the equation $2\sin^2 x + \sin x = 1$ just as we would the quadratic equation $2A^2 + A = 1$:

$$2\sin^2 x + \sin x = 1$$

$$2\sin^2 x + \sin x - 1 = 0 \quad \text{◀ Move all terms left.}$$

$$(2\sin x - 1)(\sin x + 1) = 0 \quad \text{◀ Factor as } 2A^2 + A - 1 = (2A - 1)(A + 1).$$

$$2\sin x - 1 = 0 \text{ and } \sin x + 1 = 0 \quad \text{◀ Set each factor equal to 0.}$$

$$\sin x = \frac{1}{2} \text{ and } \sin x = -1 \quad \text{◀ Solve each equation for } \sin x.$$

$$x = \frac{\pi}{6} \text{ and } x = \frac{5\pi}{6} \text{ and } x = -\frac{\pi}{2} \text{ and } x = \frac{3\pi}{2} \quad \text{◀ Find the solutions on one full period } \left[-\frac{\pi}{2}, \frac{3\pi}{2}\right] \text{ of the sine.}$$

$$x = \frac{\pi}{6} + 2k\pi \text{ and } x = \frac{5\pi}{6} + 2k\pi \text{ and } x = -\frac{\pi}{2} + 2k\pi \quad \text{◀ Find the complete solution, where } k \text{ is any integer.}$$

The solutions can also be written as

$$\dots, \frac{\pi}{6} - 4\pi, \ \frac{\pi}{6} - 2\pi, \ \frac{\pi}{6}, \ \frac{\pi}{6} + 2\pi, \ \frac{\pi}{6} + 4\pi, \dots$$

$$\text{and} \ \dots, \frac{5\pi}{6} - 4\pi, \ \frac{5\pi}{6} - 2\pi, \ \frac{5\pi}{6}, \ \frac{5\pi}{6} + 2\pi, \ \frac{5\pi}{6} + 4\pi, \dots$$

$$\text{and} \ \dots, -\frac{\pi}{2} - 4\pi, \ -\frac{\pi}{2} - 2\pi, \ -\frac{\pi}{2}, -\frac{\pi}{2} + 2\pi, \ -\frac{\pi}{2} + 4\pi, \dots$$

These solutions are shown in **Figure 8.83**.

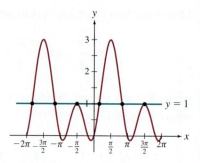

Figure 8.83 The equation $2\sin^2 x + \sin x = 1$: The solutions occur where the two graphs $y = 2\sin^2 x + \sin x$ and $y = 1$ meet.

TRY IT YOURSELF 8.48 Brief answers provided at the end of the section.

Solve the equation $\cos^2 x = 2\cos x - 1$.

TRY IT YOURSELF ANSWERS

8.39 $-\dfrac{\pi}{3} + k\pi$, where k is any integer

8.40 $\arctan\left(-\dfrac{\sqrt{3}}{2}\right) + k\pi \approx -0.71 + k\pi$, where k is any integer

8.41 $x = \dfrac{\pi}{8} + k\dfrac{\pi}{2}$, where k is any integer

8.42 $t = \arctan 12 + k\pi \approx 1.49 + k\pi$, where k is any integer

8.43 $t = \dfrac{1}{2}\arccos\left(\dfrac{1}{3}\right) + k\pi \approx 0.62 + k\pi$ and

$t = -\dfrac{1}{2}\arccos\left(\dfrac{1}{3}\right) + k\pi \approx -0.62 + k\pi$, where k is any integer

8.44 $x = \arctan\left(\dfrac{1}{5}\right) + 3\pi \approx 9.62$

8.45 $x = -\dfrac{\pi}{3} + 2k\pi$, $x = \dfrac{\pi}{3} + 2k\pi$, and $x = \dfrac{\pi}{4} + k\pi$, where k is any integer

8.46 $x = \dfrac{\pi}{2} + 2k\pi$ and $x = -\dfrac{\pi}{2} + 2k\pi$, where k is any integer. This simplifies to $x = \dfrac{\pi}{2} + k\pi$, where k is any integer.

8.47 $t = 2k\pi$, $t = \pi + 2k\pi$, $t = \dfrac{k\pi}{2}$, and $t = \dfrac{\pi}{4} + \dfrac{k\pi}{2}$, where k is any integer. This simplifies to $t = \dfrac{k\pi}{4}$, where k is any integer.

8.48 $x = 2k\pi$, where k is any integer

EXERCISE SET 8.5

CHECK YOUR UNDERSTANDING

1. State the complete solution for t of the basic equation $\sin t = c$.

2. State the complete solution for t of the basic equation $\cos t = c$.

3. State the complete solution for t of the basic equation $\tan t = c$.

4. **True or false:** In general, we expect a trigonometric equation without additional restrictions to have infinitely many solutions if the equation has at least one solution.

5. **True or false:** Inverse trigonometric functions serve as a basic tool for solving trigonometric equations.

6. If t_1 with $0 < t_1 < \dfrac{\pi}{2}$ is one solution of the equation $\sin t = c$, then the complete solution is _____.

7. If t_1 with $0 < t_1 < \dfrac{\pi}{2}$ is one solution of the equation $\cos t = c$, then the complete solution is _____.

8. If t_1 with $0 < t_1 < \dfrac{\pi}{2}$ is one solution of the equation $\tan t = c$, then the complete solution is _____.

SKILL BUILDING

Solving trigonometric equations involving special angles. In Exercises 9 through 27, find all solutions of the trigonometric equations.

9. $2\sin t = 1$

10. $2\cos t = 1$

11. $2\sin t = -1$

12. $2\cos t = -1$

13. $\sqrt{3}\tan t = 1$

14. $\sqrt{3}\tan t = -1$

15. $\tan t = \sqrt{3}$

16. $\tan t = -\sqrt{3}$

17. $\sqrt{2}\sin t = -1$

18. $2\cos t = \sqrt{2}$

19. $\sec t = 2$

20. $\csc t = -2$

21. $\cot t = -1$

22. $2 = \sqrt{3}\sec t$

23. $\sqrt{2}\sin 2x = 1$

24. $\sin\left(\dfrac{t}{3}\right) = -1$

25. $\tan(2t - 1) = 1$

26. $\sqrt{3}\cot 4t = -1$

27. $\sec 2t = -\sqrt{2}$

Solving factored equations. In Exercises 28 through 38, solve the factored equation.

28. $\sin t(2\cos t - 1) = 0$

29. $\cos t(2\sin t - 1) = 0$

30. $\tan t(\sqrt{3}\sec t - 2) = 0$

31. $(\tan t - 1)(\sec t - 2) = 0$

32. $2\sin^2 x = 1$

33. $4\cos^2 x = 3$

34. $\cot^2 x = 3$

35. $(\sin t - 1)(2\sin t + 1) = 0$

36. $(\sec t - 2)(\cot t + 1) = 0$

37. $(\cot t - 1)(\sqrt{3} + 2\sin t) = 0$

38. $(2\cos 2t - 1)(\csc 2t - 2) = 0$

Solving trigonometric equations. In Exercises 39 through 44, use inverse trigonometric functions to give the exact solutions of the given equation, and then use a calculator to get a decimal approximation of the solutions. (Round to two places.)

39. $\sin x = 0.78$

40. $\sin x = -0.63$

41. $\cos x = \dfrac{3}{4}$

42. $\cos x = -\dfrac{2}{3}$

43. $\tan x = 5$

44. $\tan x = -7$

Finding solutions on a given interval. In Exercises 45 through 51, find all solutions of the given equation on the given interval.

45. $\sin x = -\dfrac{1}{\sqrt{2}}$ on $[\pi, 2\pi]$

46. $2\cos x = \sqrt{3}$ on $[-2\pi, -\pi]$

47. $\tan x = 1$ on $[3\pi, 4\pi]$

48. $\cos x = -\dfrac{1}{2}$ on $[\pi, 3\pi]$

49. $\sin 2x = \dfrac{\sqrt{3}}{2}$ on $[2\pi, 3\pi]$

50. $\tan x = 4$ on $[\pi, 2\pi]$. Report the exact solution using inverse trigonometric functions, and then give a decimal approximation. (Round to two places.)

51. $\sin x = 0.2$ on $[2\pi, 3\pi]$. Report the exact solution using inverse trigonometric functions, and then give a decimal approximation. (Round to two places.)

Solutions using degrees. In Exercises 52 through 57, solve the given equation, reporting your answer in degrees.

52. $2\cos x = -1$

53. $2\sin x = \sqrt{2}$

54. $2\cos x = 1$

55. $\csc 2x = 2$

56. $\sqrt{3}\tan 3x = -1$

57. $\sqrt{3}\sec 4x = 2$

PROBLEMS

58. The basic equation $\sin t = c$. Assume that $-1 \le c \le 1$.

a. Show that $t = \arcsin c$ is one solution of $\sin t = c$.

b. Use the sum formula for the sine to show that $\sin(\pi - x) = \sin x$.

c. Use part b to show that $t = \pi - \arcsin c$ is a solution of $\sin t = c$.

d. Use the fact that the period of the sine function is 2π to complete the solution of $\sin t = c$.

59. The basic equation $\tan t = c$.

a. Show that $t = \arctan c$ is one solution of $\tan t = c$.

b. Use the fact that π is the period of the tangent to complete the solution of $\tan t = c$.

60. The basic equation $\cos t = c$. Show that if $t = t_1$ is a solution of $\cos t = c$, then $t = -t_1$ is also a solution.

Using identities to solve equations. In Exercises 61 through 67, use trigonometric identities to find the solutions.

61. $4 \sin x \cos x = 1$

62. $2\cos^2 x = 2\sin^2 x + \sqrt{2}$

63. $\sec x \csc x = 2$. *Suggestion*: First rewrite the equation in terms of the sine and cosine.

64. $\sin x \cos 2x = 1 - \cos x \sin 2x$. *Suggestion*: Use the sum formula for the sine.

65. $\sec t = \csc t$. *Suggestion*: First rewrite the equation in terms of the sine and cosine.

66. $1 + \cot x = \csc x$. *Suggestion*: Remember that $\tan\left(\dfrac{t}{2}\right) = \csc t - \cot t$.

67. $2\cos x = \csc x$. *Suggestion*: First multiply by $\sin x$.

Solutions by factoring. For Exercises 68 through 84, solve the equation through a combination of factoring and the use of trigonometric identities.

68. $2\sin x \tan x = \tan x$

69. $2\cos x \cot x = \sqrt{2}\cot x$

70. $\sin x = \sin 2x$

71. $2\cos x \sin x + \cos x = 0$

72. $2\sin\left(\dfrac{x}{2}\right)\sin x = \sqrt{2}\sin x$

73. $\csc^2 \theta = 2$

74. $4\cos^2 x = 1$

75. $\sec 2x = 2\cos 2x$

76. $2\sin \alpha = \tan \alpha$

77. $2\cos^2 x + 7\cos x - 4 = 0$

78. $\tan x = 3\cot x$

79. $\tan t \tan 2t - \tan t - \tan 2t + 1 = 0$
Suggestion: Group the left side as $(\tan t \tan 2t - \tan 2t) - (\tan t - 1)$.

80. $\csc t \sec t - 2\csc t - 2\sec t + 4 = 0$

81. $\cos t + 2\sin t \cos t = 1 + 2\sin t$

82. $\sin t \tan t + 4\tan t - \sin t - 4 = 0$

83. $3 + 3\sin t = 2\cos^2 t$. *Suggestion*: Use the basic Pythagorean identity to change the form of $\cos^2 t$.

84. $\sec^2 x + (1 + \sqrt{3})\tan x + \sqrt{3} = 1$. *Suggestion*: Use the identity $\tan^2 x + 1 = \sec^2 x$ to change the form of $\sec^2 x$.

Using sum-to-product formulas. In Exercises 85 through 88, use a sum-to-product formula to solve the given equation.

85. $\sin 5x + \sin 3x = 0$

86. $\sin 10x - \sin 4x = 0$

87. $\cos 9x + \cos 3x = 0$

88. $\cos 7x - \cos 3x = 0$

89. Squaring equations. Some trigonometric equations can be solved by squaring both sides of the equation. But you are cautioned that squaring can introduce extraneous roots. In parts a and b, solve the given equation. Be sure to plug potential solutions into the original equation to check whether they are indeed solutions.

a. $\sin x - \cos x = 1$

b. $\sin x + \cos x = \sqrt{\dfrac{3}{2}}$

90. Finding angles. Find all possible values of θ if $\sec \theta = 2$ and the sine of θ is negative.

Solving by graphing. The equations in Exercises 91 through 94 are not easily solved using the techniques presented in this section. Your answer should include appropriate graphs and answers correct to two decimal places.

91. $\tan x = \sin x + 2$

92. $2 \sin x + 3 \cos x = 1$

93. $\cot x = x$. Solve on the interval $[-\pi, \pi]$.

94. $\tan x + \csc x = 3$

MODELS AND APPLICATIONS

95. **Hours of daylight in Boston.** The number H of hours of daylight in Boston t days after the vernal equinox in March is given by

$$H = 12 + 3\sin\frac{2\pi t}{365}$$

a. How many hours of daylight are there in Boston on the longest day of the year, and when does the longest day occur? Report your answer to the nearest whole day.

b. Over a period of 1 year, when are there 10 or more hours of daylight in Boston? Round your answer to the nearest whole day.

96. **Weevils and wasps.** The braconid wasp preys on the azuki bean weevil. This predator-prey

relationship causes the wasp population to fluctuate. The wasp population after t generations is given by

$$w = 3500 + 2000\sin\frac{\pi t}{3}$$

a. After how many generations is the wasp population at a minimum for the first time? Round your answer to the nearest whole number. What is that minimum number?

b. When is the first time the wasp population reaches 5000? Round your answer to the nearest whole number.

CHALLENGE EXERCISES FOR INDIVIDUALS OR GROUPS

Challenging trigonometric equations. Exercises 97 through 100 involve challenging trigonometric equations. In each case, find all solutions of the given equation.

97. $\cos^4 x = \sin^4 x$

98. $1 - \cos t = \sin t$. *Suggestion:* Use a half-angle formula for the tangent if $\sin t \neq 0$.

99. $2\cos t - 1 = \sin t$. *Suggestion:* Square both sides, but watch for extraneous roots.

100. $2\cos^2 x - \cos x = 2 - \sec x$

Using the quadratic formula. In Exercises 101 through 106, use the quadratic formula to find the exact solution.

Then use a calculator to get the approximate value of the solution. (Round to two decimal places.)

101. $\sin^2 t + \sin t - 1 = 0$

102. $\sin^2 t - \sin t - 1 = 0$

103. $\tan^2 t + \tan t - 3 = 0$

104. $\cot^2 t - 2\cot t - 1 = 0$

105. $\csc^2 x - \csc x - 1 = 0$

106. $3\cos x = \sin^2 x$. *Suggestion:* First write $\sin^2 x$ as $1 - \cos^2 x$.

REVIEW AND REFRESH: Exercises from Previous Sections

107. **From Section P.3:** Solve the inequality $x^3 - x < 0$.

Answer: $(-\infty, -1) \cup (0, 1)$

108. **From Section 4.3:** Solve the equation $12 \times 5^x = 7^x$. Give the exact answer as well as an approximation accurate to two decimal places.

Answer: $x = \dfrac{\ln 12}{\ln\left(\dfrac{7}{5}\right)} \approx 7.39$

109. **From Chapter 6:** What is the degree measure of an angle of $\dfrac{\pi}{6}$ radian?

Answer: $30°$

110. **From Chapter 6:** For a circle of a certain radius r, the area of a sector with central angle θ is always the same numerically as the arc length cut by that angle. Find r.

Answer: $r = 2$

111. **From Section 7.2:** Identify the trigonometric function whose graph is shown in **Figure 8.84**.

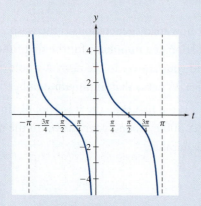

Figure 8.84 Graph for Exercise 111

Answer: $y = \cot t$

112. **From Section 8.1:** Express the tangent in terms of the secant and cosecant.

Answer: $\tan t = \dfrac{\sec t}{\csc t}$

113. **From Section 8.3:** If $\sin x = \dfrac{3}{7}$ and $\cos x$ is negative, calculate the exact value of $\sin 2x$. Then give a decimal approximation correct to two places.

Answer: $-\dfrac{12\sqrt{10}}{49} \approx -0.77$

114. **From Section 8.3:** Use a half-angle formula to calculate the exact value of $\sin \dfrac{\pi}{12}$. Then give a decimal approximation correct to two places.

Answer: $\dfrac{\sqrt{2-\sqrt{3}}}{2} \approx 0.26$

115. **From Section 8.4:** Find the exact value of $\arctan 1$.

Answer: $\dfrac{\pi}{4}$

116. **From Section 8.4:** Simplify the expression $\sin\arctan\left(\dfrac{1}{x}\right)$, assuming $x > 0$.

Answer: $\dfrac{1}{\sqrt{1+x^2}}$

CHAPTER ROADMAP AND STUDY GUIDE

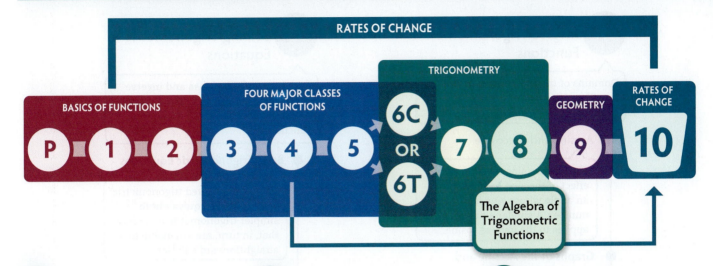

8.1 Trigonometric Identities

Relationships among the trigonometric functions extend well beyond the immediate consequences of their definition.

Basic Identities

The definitions of trigonometric functions provide fundamental identities.

Cofunction Identities

The interplay between the two acute angles of a right triangle results in cofunction identities.

Even and Odd Identities

Reflective symmetries of graphs of trigonometric functions translate into important identities.

Algebraic Verification of Trigonometric Identities

Algebraic operations on the trigonometric identities we know can generate many other identities.

8.2 Sum and Difference Formulas

Alterations of trigonometric expressions using sum and difference formulas allow for easier application of certain calculus operations.

Sum and Difference Formulas for the Sine and Cosine

The sine formula for the area of a triangle is used to establish the sum formula for the sine function.

Sum and Difference Formulas for the Tangent and Cotangent

Identities that we know give rise to additional sum and difference formulas.

Product-to-Sum and Sum-to-Product Formulas

Sum and difference formulas for the sine and cosine can be combined to produce formulas for products involving sine and cosine.

8.3 Double-Angle and Half-Angle Formulas

Calculating trigonometric functions when angles are doubled or cut in half has applications in engineering.

Double-Angle Formulas

The sum and difference formulas for the sine and cosine produce double-angle formulas.

Half-Angle Formulas

The double-angle formula for the cosine can be rearranged to show the effect on the sine or cosine when an angle is halved.

Combining Sum and Multiple-Angle Formulas

The sum and double-angle formulas can be combined to produce triple-angle formulas and other identities.

Power Reduction Formulas

The double-angle formula for the cosine is rearranged to write powers of sine or cosine functions in terms of lower powers.

CHAPTER ROADMAP AND STUDY GUIDE (continued)

8.4 Inverse Trigonometric Functions

Domains of trigonometric functions must be restricted to produce inverses.

Arcsine and Arccosine

The inverse sine reverses the effect of the sine by producing an angle whose sine is a given number. A similar statement applies to the inverse cosine.

Graphs of the Arcsine and Arccosine

We find the graphs of the inverse functions by reflecting specific parts of the sine and cosine graphs about the line $y = x$.

The Arctangent Function

The inverse tangent reverses the effect of the tangent by producing an angle whose tangent is a given number.

MODELS AND APPLICATIONS
Flying and Basketball

The Other Trigonometric Functions: Arccosecant, Arcsecant, and Arccotangent

Finding the inverse of cosecant, secant, and cotangent requires restricting their domains.

8.5 Solving Trigonometric Equations

Trigonometric identities and inverse functions play a key role in solving equations involving trigonometric functions.

Basic Trigonometric Equations

Many complicated trigonometric equations are equivalent to simpler trigonometric equations that, in turn, are amenable to a straightforward solution.

Solving by Factoring

Some trigonometric equations can be solved by factoring and setting each factor equal to zero.

Analogy with Quadratic Equations

Some trigonometric equations are solved in a fashion that reminds us of quadratic equations.

CHAPTER QUIZ

1. Show the identity $\cot t = \dfrac{1}{\tan t}$.

   Answer: $\dfrac{1}{\tan t} = \dfrac{1}{\dfrac{\sin t}{\cos t}} = \dfrac{\cos t}{\sin t} = \cot t$

   XR Example 8.1

2. Express $\tan t$ in terms of $\sin t$ if $\dfrac{\pi}{2} < t < \pi$.

   Answer: $\tan t = -\dfrac{\sin t}{\sqrt{1 - \sin^2 t}}$

   XR Example 8.3

3. Determine whether $f(x) = \sin^2 x \cos x$ is even, odd, or neither.

   Answer: It is even.

   XR Example 8.5

4. Show the identity $\csc t \tan t = \sec t$.

   Answer: $\csc t \tan t = \dfrac{1}{\sin t} \dfrac{\sin t}{\cos t} = \dfrac{1}{\cos t} = \sec t$

   XR Example 8.6

5. Find the exact value of $\cos 75°$. Note that $75 = 30 + 45$.

   Answer: $\dfrac{\sqrt{3} - 1}{2\sqrt{2}} \approx 0.26$

   XR Example 8.11

6. If $\tan s = 3$ and $\tan t = 4$, find the exact value of $\tan(s + t)$.

   Answer: $-\dfrac{7}{11}$

   XR Example 8.16

7. Use a product-to-sum formula to find the exact value of $\sin 37.5° \sin 7.5°$.

   Answer: $\dfrac{\sqrt{3}}{4} - \dfrac{1}{2\sqrt{2}} \approx 0.08$

   XR Example 8.17

8. If $\cos t = -\dfrac{2}{3}$, find the exact value of $\cos 2t$.

   Answer: $-\dfrac{1}{9}$

   XR Example 8.19

9. Use a half-angle formula to find the exact value of $\sin 15°$.

   Answer: $\sqrt{\dfrac{1}{2} - \dfrac{\sqrt{3}}{4}}$

   XR Example 8.21

10. Suppose s and t are acute angles with $\sin s = \dfrac{3}{5}$ and $\sin t = \dfrac{2}{5}$. Find the exact value of $\cos(s + 2t)$, and then give an approximation correct to two decimal places.

    Answer: $\dfrac{68 - 12\sqrt{21}}{125} \approx 0.10$

    XR Example 8.24

11. Find the exact values of arcsin $\dfrac{\sqrt{3}}{2}$ and arccos(-1).

Answer: arcsin $\dfrac{\sqrt{3}}{2} = \dfrac{\pi}{3}$; arccos$(-1) = \pi$

XR Example 8.27

12. Find the exact value of arctan 1.

Answer: $\dfrac{\pi}{4}$

XR Example 8.32

13. Find the exact value of $\tan\left(\arcsin\dfrac{2}{3}\right)$.

Answer: $\dfrac{2}{\sqrt{5}}$

XR Example 8.34

14. City B is 200 miles north of city A. City C is 50 miles east of city B. Assuming north corresponds to 0° and east corresponds to 90°, what heading should an airplane fly to go directly from city A to city C? Give your answer in degrees rounded to two decimal places.

Answer: 14.04°

XR Example 8.36

15. Solve the equation $2\sin(2t) = \sqrt{2}$.

Answer: $\dfrac{\pi}{8} + k\pi$ and $\dfrac{3\pi}{8} + k\pi$, where k is any integer

XR Example 8.41

16. Find the solutions of $\csc t = 2\cot t$ with $\pi \le t \le 3\pi$.

Answer: $\dfrac{5\pi}{3}$ and $\dfrac{7\pi}{3}$

XR Example 8.44

17. Solve the equation $\sin t + \sin^2 t - \cos t - \sin t \cos t = 0$. *Suggestion:* Write the left side as $(\sin t)(1 + \sin t) - (\cos t)(1 + \sin t)$.

Answer: $\dfrac{\pi}{4} + k\pi$ and $-\dfrac{\pi}{2} + 2k\pi$, where k is an integer

XR Example 8.45

CHAPTER REVIEW EXERCISES

Section 8.1

Using trigonometric identities. In Exercises 1 through 6, use identities that you know to find the given function value. Cofunction and even-odd identities will be helpful.

1. If $\sin x = \dfrac{1}{3}$, find $\sin(-x)$.

Answer: $-\dfrac{1}{3}$

2. If $\tan x = -4$, find $\cot(-x)$.

Answer: $\dfrac{1}{4}$

3. If $\sin\left(\dfrac{\pi}{2} - x\right) = \dfrac{3}{4}$, find $\cos(-x)$.

Answer: $\dfrac{3}{4}$

4. If $\cos\left(x - \dfrac{\pi}{2}\right) = \dfrac{7}{8}$, find $\sin x$.

Answer: $\dfrac{7}{8}$

5. If $\sec(-x) = -3$, find $\sec x$.

Answer: -3

6. If $\cot x = 7$, find $\tan\left(\dfrac{\pi}{2} - x\right)$.

Answer: 7

Simplifying trigonometric expressions. In Exercises 7 through 12, simplify the given expression.

7. $\tan t \cot t$

Answer: 1

8. $\sin t \cot t$

Answer: $\cos t$

9. $1 - \cos^2 t$

Answer: $\sin^2 t$

10. $1 + \cot^2 t$

Answer: $\csc^2 t$

11. $\sec t \cot t$

Answer: $\csc t$

12. $\csc^2 t - 1$

Answer: $\cot^2 t$

Expressing one trigonometric function in terms of others. In Exercises 13 through 17, express a trigonometric function in terms of given functions.

13. Express $\sin t$ in terms of $\cos t$. Assume $0 \le t \le \dfrac{\pi}{2}$.

Answer: $\sin t = \sqrt{1 - \cos^2 t}$

14. Express $\cos t$ in terms of $\sec t$.

Answer: $\cos t = \dfrac{1}{\sec t}$

15. Express $\cot t$ in terms of $\sec t$ and $\csc t$.

Answer: $\cot t = \dfrac{\csc t}{\sec t}$

16. Express $\tan t$ in terms of $\cot t$.

Answer: $\tan t = \dfrac{1}{\cot t}$

17. Express $\sec t$ in terms of $\sin t$. Assume $\dfrac{\pi}{2} < t < \pi$.

Answer: $\sec t = -\dfrac{1}{\sqrt{1 - \sin^2 t}}$

Establishing identities. In Exercises 18 through 23, establish the given identity.

18. $\csc t \tan t = \sec t$

Answer: Answers will vary.

19. $\dfrac{\cos t}{\sec t} = \cos^2 t$

Answer: Answers will vary.

20. $\sec t - \cos t = \sin t \tan t$

Answer: Answers will vary.

21. $(\sin t + \cos t)^2 = 2\sin t \cos t + 1$

Answer: Answers will vary.

22. $\dfrac{1}{\sec t + \tan t} = \sec t - \tan t$

Answer: Answers will vary.

23. $\dfrac{\cos t}{\sin t} + \dfrac{\sin t}{\cos t} = \sec t \csc t$

Answer: Answers will vary.

Section 8.2

Finding the area.

24. Find the area of the triangle in **Figure 8.85**.

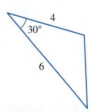

Figure 8.85 Triangle for Exercise 24

Answer: 6

25. Find the area of the triangle in **Figure 8.86**.

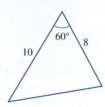

Figure 8.86 Triangle for Exercise 25

Answer: $20\sqrt{3}$

Finding a trigonometric value.

26. Use your knowledge of special angles to find $\sin(75°)$.

Answer: $\dfrac{1+\sqrt{3}}{2\sqrt{2}}$ or $\dfrac{\sqrt{2+\sqrt{3}}}{2}$

27. Use your knowledge of special angles to find $\cos(15°)$.

Answer: $\dfrac{1+\sqrt{3}}{2\sqrt{2}}$ or $\dfrac{\sqrt{2+\sqrt{3}}}{2}$

Using sum formulas. In Exercises 28 through 39, use a sum or difference formula to calculate the indicated function value.

28. If $\sin x = \dfrac{1}{2}$ and $\sin y = \dfrac{1}{3}$, calculate $\sin(x+y)$.

Assume $0 \le x \le \dfrac{\pi}{2}$ and $0 \le y \le \dfrac{\pi}{2}$.

Answer: $\dfrac{2\sqrt{2}+\sqrt{3}}{6}$

29. If $\sin x = \dfrac{1}{2}$ and $\sin y = \dfrac{1}{3}$, calculate $\sin(x-y)$.

Assume $0 \le x \le \dfrac{\pi}{2}$ and $0 \le y \le \dfrac{\pi}{2}$.

Answer: $\dfrac{2\sqrt{2}-\sqrt{3}}{6}$

30. If $\sin x = \dfrac{1}{2}$ and $\sin y = \dfrac{1}{3}$, calculate $\cos(x+y)$.

Assume $0 \le x \le \dfrac{\pi}{2}$ and $0 \le y \le \dfrac{\pi}{2}$.

Answer: $\dfrac{2\sqrt{6}-1}{6}$

31. If $\sin x = \dfrac{1}{2}$ and $\sin y = \dfrac{1}{3}$, calculate $\cos(x-y)$.

Assume $0 \le x \le \dfrac{\pi}{2}$ and $0 \le y \le \dfrac{\pi}{2}$.

Answer: $\dfrac{2\sqrt{6}+1}{6}$

32. If $\sin x = \dfrac{2}{3}$ and $\cos y = \dfrac{1}{3}$, calculate $\sin(x+y)$.

Assume $\dfrac{\pi}{2} \le x \le \pi$ and $\dfrac{3\pi}{2} \le y \le 2\pi$.

Answer: $\dfrac{2+2\sqrt{10}}{9}$

33. If $\sin x = \dfrac{2}{3}$ and $\cos y = \dfrac{1}{3}$, calculate $\sin(x-y)$.

Assume $\dfrac{\pi}{2} \le x \le \pi$ and $\dfrac{3\pi}{2} \le y \le 2\pi$.

Answer: $\dfrac{2-2\sqrt{10}}{9}$

34. If $\sin x = \dfrac{2}{3}$ and $\cos y = \dfrac{1}{3}$, calculate $\cos(x+y)$.

Assume $\dfrac{\pi}{2} \le x \le \pi$ and $\dfrac{3\pi}{2} \le y \le 2\pi$.

Answer: $\dfrac{4\sqrt{2}-\sqrt{5}}{9}$

35. If $\sin x = \dfrac{2}{3}$ and $\cos y = \dfrac{1}{3}$, calculate $\cos(x-y)$.

Assume $\dfrac{\pi}{2} \le x \le \pi$ and $\dfrac{3\pi}{2} \le y \le 2\pi$.

Answer: $\dfrac{-4\sqrt{2}-\sqrt{5}}{9}$

36. If $\tan x = 4$ and $\tan y = 3$, calculate $\tan(x+y)$.

Answer: $-\dfrac{7}{11}$

37. If $\tan x = 4$ and $\tan y = 3$, calculate $\tan(x-y)$.

Answer: $\dfrac{1}{13}$

38. If $\tan x = 4$ and $\tan y = 3$, calculate $\cot(x+y)$.

Answer: $-\dfrac{11}{7}$

39. If $\tan x = 4$ and $\tan y = 3$, calculate $\cot(x - y)$.

Answer: 13

Sum-to-product and product-to-sum formulas. In Exercises 40 through 43, use a sum-to-product or product-to-sum formula to calculate the indicated function value.

40. If $\sin x = \dfrac{1}{2}$ and $\sin y = \dfrac{2}{3}$, calculate

$$\sin\left(\frac{x + y}{2}\right)\cos\left(\frac{x - y}{2}\right).$$

Answer: $\dfrac{7}{12}$

41. If $\cos x = \dfrac{1}{4}$ and $\cos y = \dfrac{1}{5}$, calculate

$$\cos(x + y) + \cos(x - y).$$

Answer: $\dfrac{1}{10}$

42. If $\cos x = \dfrac{3}{5}$ and $\cos y = \dfrac{4}{5}$, calculate

$$\sin\left(\frac{x + y}{2}\right)\sin\left(\frac{x - y}{2}\right).$$

Answer: $\dfrac{1}{10}$

43. If $\cos x = \dfrac{3}{5}$ and $\cos y = \dfrac{4}{5}$, calculate

$$\cos(x + y) + \cos(x - y).$$

Answer: $\dfrac{24}{25}$

Section 8.3

Using double-angle and half-angle formulas in the first quadrant. In Exercises 44 through 51, assume $\cos x = \dfrac{3}{5}$ and $0 \le x \le \dfrac{\pi}{2}$.

44. $\sin 2x$

Answer: $\dfrac{24}{25}$

45. $\cos 2x$

Answer: $-\dfrac{7}{25}$

46. $\sin \dfrac{x}{2}$

Answer: $\dfrac{1}{\sqrt{5}}$

47. $\cos \dfrac{x}{2}$

Answer: $\sqrt{\dfrac{8}{10}} = \sqrt{\dfrac{4}{5}}$

48. $\tan \dfrac{x}{2}$

Answer: $\dfrac{1}{2}$

49. $\tan 2x$

Answer: $-\dfrac{24}{7}$

50. $\sin 4x$

Answer: $-\dfrac{336}{625}$

51. $\cos 4x$

Answer: $-\dfrac{527}{625}$

Using double-angle and half-angle formulas in other quadrants. For Exercises 52 through 59, assume $\cos x = -\dfrac{2}{3}$ and $\dfrac{\pi}{2} \le x \le \pi$.

52. $\sin 2x$

Answer: $-\dfrac{4\sqrt{5}}{9}$

53. $\cos 2x$

Answer: $-\dfrac{1}{9}$

54. $\sin \dfrac{x}{2}$

Answer: $\sqrt{\dfrac{5}{6}}$

55. $\cos \dfrac{x}{2}$

Answer: $\dfrac{1}{\sqrt{6}}$

56. $\tan 2x$

Answer: $4\sqrt{5}$

57. $\tan \dfrac{x}{2}$

Answer: $\sqrt{5}$

58. $\cot \dfrac{x}{2}$

Answer: $\dfrac{1}{\sqrt{5}}$

59. $\sec 2x$

Answer: -9

Verifying an identity.

60. Show that $\sin 4t = 4(\sin t \cos^3 t - \sin^3 t \cos t)$.

Answer: Answers will vary.

61. Show that $\dfrac{\cos 2x}{\cos x - \sin x} = \cos x + \sin x$.

Answer: Answers will vary.

Section 8.4

Graphs of inverse trigonometric functions. In Exercises 62 through 67, match the given function with the appropriate graph from **Figure 8.87** through **Figure 8.92**.

62. arcsine

Answer: D

63. arccosine

Answer: A

64. arctangent

Answer: B

65. arcsecant

Answer: C

66. arccotangent

Answer: F

67. arccosecant

Answer: E

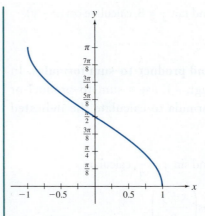

Figure 8.87 A

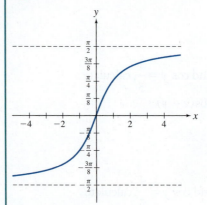

Figure 8.88 B

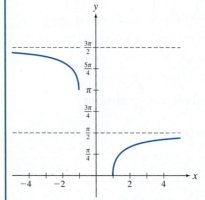

Figure 8.89 C

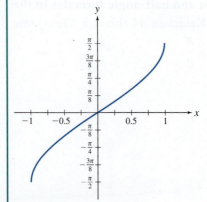

Figure 8.90 D

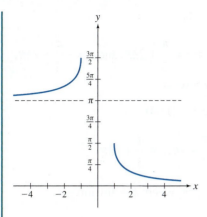

Figure 8.91 E

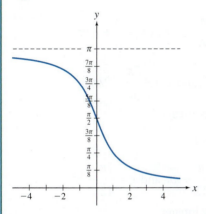

Figure 8.92 F

Inverse trigonometric functions of special angles. In Exercises 68 through 75, calculate the exact value of the given inverse trigonometric function.

68. $\arcsin\left(\dfrac{\sqrt{3}}{2}\right)$

Answer: $\dfrac{\pi}{3}$

69. $\arcsin\left(-\dfrac{\sqrt{3}}{2}\right)$

Answer: $-\dfrac{\pi}{3}$

70. $\arccos\left(\dfrac{\sqrt{3}}{2}\right)$

Answer: $\dfrac{\pi}{6}$

71. $\arccos 0$

Answer: $\dfrac{\pi}{2}$

72. $\arcsin\left(-\dfrac{1}{2}\right)$

Answer: $-\dfrac{\pi}{6}$

73. $\arctan 1$

Answer: $\dfrac{\pi}{4}$

74. $\arctan \sqrt{3}$

Answer: $\dfrac{\pi}{3}$

75. $\arctan(-\sqrt{3})$

Answer: $-\dfrac{\pi}{3}$

Composing trigonometric functions with inverse trigonometric functions. In Exercises 76 through 85, find the exact value indicated. In exercises where x appears, you may assume that $0 < x < 1$.

76. $\tan\left(\arcsin\dfrac{2}{3}\right)$

Answer: $\dfrac{2}{\sqrt{5}}$

77. $\sec\left(\arccos\dfrac{2}{3}\right)$

Answer: $\dfrac{3}{2}$

78. $\sin(\arctan 5)$

Answer: $\dfrac{5}{\sqrt{26}}$

79. $\sin\left(\arccos\dfrac{5}{8}\right)$

Answer: $\dfrac{\sqrt{39}}{8}$

80. $\tan(\arcsin x)$

Answer: $\dfrac{x}{\sqrt{1-x^2}}$

81. $\sin(\arccos x)$

Answer: $\sqrt{1-x^2}$

82. $\cot\left(\arcsin\dfrac{x}{x+1}\right)$

Answer: $\dfrac{\sqrt{2x+1}}{x}$

83. $\sin\left(2\arctan\dfrac{2}{3}\right)$

Answer: $\dfrac{12}{13}$

84. $\sin\left(2\arcsin x\right)$

Answer: $2x\sqrt{1-x^2}$

85. $\csc\left(2\arctan x\right)$

Answer: $\dfrac{1+x^2}{2x}$

Section 8.5

Solving trigonometric equations.

86. $2\sin t = 1$

Answer: $t = \dfrac{\pi}{6} + 2k\pi$ and $t = \dfrac{5\pi}{6} + 2k\pi$,

where k is any integer

87. $2\cos t = -1$

Answer: $t = \dfrac{2\pi}{3} + 2k\pi$ and $t = -\dfrac{2\pi}{3} + 2k\pi$,

where k is any integer

88. $\sqrt{3}\tan t = 1$

Answer: $t = \dfrac{\pi}{6} + k\pi$, where k is any integer

89. $\sqrt{2}\sin t = -1$

Answer: $t = -\dfrac{\pi}{4} + 2k\pi$ and $t = \dfrac{5\pi}{4} + 2k\pi$,

where k is any integer

90. $\sqrt{2}\sin 2x = 1$

Answer: $x = \dfrac{\pi}{8} + k\pi$ and $x = \dfrac{3\pi}{8} + k\pi$,

where k is any integer

91. $\sqrt{3}\cot 4t = -1$

Answer: $t = -\dfrac{\pi}{12} + \dfrac{k\pi}{4}$, where k is any integer

92. $(\sin t)(2\cos t - 1) = 0$

Answer: $t = k\pi$, $t = \dfrac{\pi}{3} + 2k\pi$, and $t = -\dfrac{\pi}{3} + 2k\pi$,

where k is any integer

93. $(\tan t - 1)(\sec t - 2) = 0$

Answer: $t = \dfrac{\pi}{4} + k\pi$, $t = \dfrac{\pi}{3} + 2k\pi$,

and $t = -\dfrac{\pi}{3} + 2k\pi$, where k is any integer

94. $4\cos^2 x = 3$

Answer: $x = \dfrac{\pi}{6} + 2k\pi$, $x = -\dfrac{\pi}{6} + 2k\pi$,

$x = \dfrac{5\pi}{6} + 2k\pi$, and $x = -\dfrac{5\pi}{6} + 2k\pi$,

where k is any integer

95. $4\sin x \cos x = 1$

Answer: $x = \dfrac{\pi}{12} + k\pi$ and $x = \dfrac{5\pi}{12} + k\pi$,

where k is any integer

96. $2\cos^2 x = 2\sin^2 x + \sqrt{2}$

Answer: $x = \dfrac{\pi}{8} + k\pi$ and $x = -\dfrac{\pi}{8} + k\pi$,

where k is any integer

97. $\sin x \cos 2x = 1 - \cos x \sin 2x$

Suggestion: Use the sum formula for the sine.

Answer: $x = \dfrac{\pi}{6} + \dfrac{2k\pi}{3}$, where k is any integer

98. $\sec t = \csc t$

Answer: $t = \dfrac{\pi}{4} + k\pi$, where k is any integer

99. $2\cos x = \csc x$

Answer: $x = \dfrac{\pi}{4} + k\pi$, where k is any integer

100. $2\sin x \tan x = \tan x$

Answer: $x = k\pi$, $x = \dfrac{\pi}{6} + 2k\pi$, and $x = \dfrac{5\pi}{6} + 2k\pi$,

where k is any integer

101. $2 \cos x \cot x = \sqrt{2} \cot x$

Answer: $x = \dfrac{\pi}{2} + k\pi$, $x = \dfrac{\pi}{4} + 2k\pi$, and

$x = -\dfrac{\pi}{4} + 2k\pi$, where k is any integer

More trigonometric equations. In Exercises 102 through 105, use inverse trigonometric functions to give the exact solutions of the given equation, and then use a calculator to get a decimal approximation of the solutions. (Round to two places.)

102. $\sin x = 0.53$

Answer: $x = \arcsin 0.53 + 2k\pi \approx 0.56 + 2k\pi$ and $x = \pi - \arcsin 0.53 + 2k\pi \approx 2.58 + 2k\pi$, where k is any integer

103. $\sin x = -0.63$

Answer: $x = \arcsin(-0.63) + 2k\pi \approx -0.68 + 2k\pi$ and $\pi - \arcsin(-0.63) + 2k\pi \approx 3.82 + 2k\pi$, where k is any integer

104. $\cos x = \dfrac{2}{3}$

Answer: $\arccos \dfrac{2}{3} + 2k\pi \approx 0.84 + 2k\pi$ and $-\arccos \dfrac{2}{3} + 2k\pi \approx -0.84 + 2k\pi$, where k is any integer

105. $\cos x = -\dfrac{5}{6}$

Answer: $\arccos\left(-\dfrac{5}{6}\right) + 2k\pi \approx 2.56 + 2k\pi$ and $x = -\arccos\left(-\dfrac{5}{6}\right) + 2k\pi \approx -2.56 + 2k\pi$, where k is any integer

TOPICS IN GEOMETRY

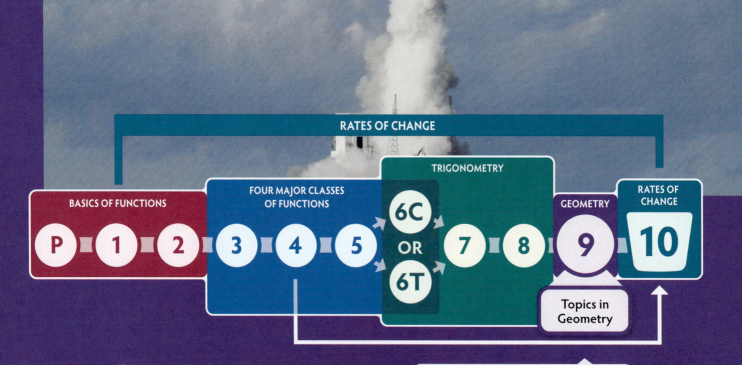

9.1 Law of Cosines

9.2 Law of Sines

9.3 Vectors

9.4 Vectors in the Plane and in Three Dimensions

This chapter applies what we have learned thus far to a variety of topics connected to trigonometry and geometry.

The topics in this chapter, which may seem unrelated, are all aspects of geometry that are tied together by the basic ideas from trigonometry. At first glance, it may appear that trigonometry is an artifact of right triangles and circles. But trigonometry is intrinsic to all triangles, and indeed to virtually every geometric construct. Any time a distance and direction arise, trigonometry plays a role.

For example, think of the great circle route that airplanes fly from New York to London. If you open a flat map and draw a "straight line" from New York to London, you may be surprised to learn that this is not the shortest route—not by a long shot. We must take into account the nature of Earth as a sphere. If we cut a spherical body by a plane that meets the center, then the intersection of the plane with the surface is the great circle shown in **Figure 9.1**. On the surface of Earth, great circles give the shortest distance between two points.

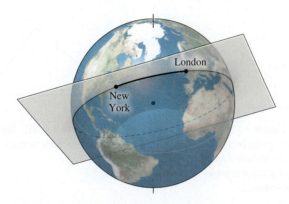

Figure 9.1 A great circle: The intersection of such a plane with the sphere gives the path airplanes fly.

Geometry provides many ways of describing the universe we live in. Some of these are shown in this chapter.

9.1 Law of Cosines

> The law of cosines uses trigonometry to analyze all triangles, not just right triangles.

9.1 Law of Cosines

9.2 Law of Sines

9.3 Vectors

9.4 Vectors in the Plane and in Three Dimensions

In this section, you will learn to:

1. Solve triangles using the law of cosines and side-angle-side.
2. Solve triangles using the law of cosines and side-side-side.
3. Calculate the area of a triangle using Heron's formula.
4. Solve applied problems using the law of cosines.

The Pythagorean theorem is an invaluable tool for studying right triangles, and it played a crucial role in the early development of mathematics. You may be surprised to learn that there is a version of the Pythagorean theorem, the law of cosines, that applies to all triangles—not just right triangles. The development of this idea, and its applications, is the focus of this section. Note that angles are commonly measured in degrees.

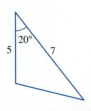

Figure 9.2 An example of SAS: There is only one triangle with one side 5, a second side 7, and the included angle 20°.

Recalling Congruence Criteria

> **Side-angle-side and side-side-side are criteria for triangle congruence.**

Consider the triangle shown in **Figure 9.2**, where one side has length 5, another has length 7, and the angle between the two sides is 20°. There is exactly one triangle that satisfies these conditions. Stated formally, this is the fundamental geometry theorem side-angle-side (SAS), which states that two sides and the included angle determine a triangle.

The SAS theorem is illustrated in **Figure 9.3**. Its companion, the side-side-side (SSS) theorem, is illustrated in **Figure 9.4**.

Two additional conditions for triangle congruence will be addressed in the next section.

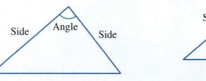

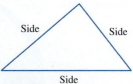

DF **Figure 9.3** SAS: If we are given two sides and the included angle, then a unique triangle is determined.

DF **Figure 9.4** SSS: Three sides determine at most one triangle.

Derivation of the Law of Cosines

> **The law of cosines is established using the Pythagorean theorem.**

The law of cosines expresses the length of one side of a triangle in terms of the lengths of the other two sides and the included angle.

LAWS OF MATHEMATICS: The Law of Cosines

For the triangle in **Figure 9.5**, where the angle C is between the sides of lengths a and b,

$$c^2 = a^2 + b^2 - 2ab \cos C$$

This formula can be rearranged to give the alternative form

$$C = \arccos\left(\frac{a^2 + b^2 - c^2}{2ab}\right)$$

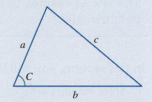

Figure 9.5 Finding c using lengths a and b and the angle C

The term $-2ab \cos C$ is the "fudge factor" that accounts for the fact that C may not be a right angle.

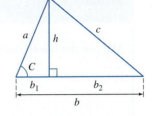

A bit of algebra is required to show that the law of cosines is true. We give a proof in the case that the angle C is acute. The obtuse case is handled in an exercise later in the chapter. We have added an altitude h to the triangle from Figure 9.5 to make the triangle in **Figure 9.6**. The altitude h divides the big triangle into two right triangles and splits the side of length b into two pieces, b_1 and b_2. Using these two right triangles, we observe that

$$\cos C = \frac{\text{Adjacent}}{\text{Hypotenuse}} = \frac{b_1}{a}$$

Figure 9.6 Preparing to prove the law of cosines

and

$$\sin C = \frac{\text{Opposite}}{\text{Hypotenuse}} = \frac{h}{a}.$$

Rearranging these gives

$$b_1 = a \cos C$$

$$h = a \sin C$$

Now because $b_2 = b - b_1$, we have $b_2 = b - a \cos C$. Using the Pythagorean theorem on the right-hand triangle, we find

$$c^2 = h^2 + b_2^2 \qquad \qquad \text{◀ Apply the Pythagorean theorem.}$$

$$= a^2 \sin^2 C + (b - a \cos C)^2 \qquad \text{◀ Substitute } h = a \sin C \text{ and } b_2 = b - a \cos C.$$

$$= a^2 \sin^2 C + b^2 - 2ab \cos C + a^2 \cos^2 C \qquad \text{◀ Expand squares.}$$

$$= a^2(\sin^2 C + \cos^2 C) + b^2 - 2ab \cos C \qquad \text{◀ Simplify.}$$

$$c^2 = a^2 + b^2 - 2ab \cos C \qquad \qquad \text{◀ Use the Pythagorean identity.}$$

Using the Law of Cosines to Solve Triangles

> The law of cosines allows us to find unknown sides and angles of triangles.

The following examples show how the law of cosines can be used to solve triangles—that is, find the unknown sides and angles.

EXAMPLE 9.1 Law of Cosines with Radian Measure: SAS Case

Solve the triangle in **Figure 9.7**, where the angles are measured in radians.

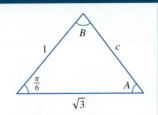

Figure 9.7 Triangle for Example 9.1

SOLUTION

We use the law of cosines with $a = 1$, $b = \sqrt{3}$, and $C = \dfrac{\pi}{6}$ to find c:

$$c^2 = a^2 + b^2 - 2ab \cos C \qquad \blacktriangleleft \text{Apply the law of cosines.}$$

$$c^2 = 1^2 + \sqrt{3}^2 - 2 \times 1 \times \sqrt{3} \times \cos \frac{\pi}{6} \qquad \blacktriangleleft \text{Substitute known values.}$$

$$c^2 = 4 - 2\sqrt{3}\,\frac{\sqrt{3}}{2} \qquad \blacktriangleleft \text{Use } \cos \frac{\pi}{6} = \frac{\sqrt{3}}{2}.$$

$$c^2 = 1 \qquad \blacktriangleleft \text{Simplify.}$$

Finally, we take square roots to find $c = 1$.

We can use the alternative form of the law of cosines with $c = 1$ to find the angle A:

$$A = \arccos\left(\frac{b^2 + c^2 - a^2}{2bc} \right)$$

$$A = \arccos\left(\frac{\sqrt{3}^2 + 1^2 - 1^2}{2 \times \sqrt{3} \times 1} \right)$$

$$A = \arccos \frac{\sqrt{3}}{2}$$

$$A = \frac{\pi}{6}$$

We use the fact that the angle sum of the triangle is π radians to find the angle B:

$$B = \pi - \frac{\pi}{6} - \frac{\pi}{6} = \frac{2\pi}{3}$$

TRY IT YOURSELF 9.1 Brief answers provided at the end of the section.

Use the law of cosines to solve the triangle in **Figure 9.8**.

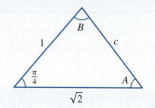

Figure 9.8 Triangle for Try It Yourself 9.1

EXTEND YOUR REACH

a. Interpret the law of cosines in the case that the angle C is a right angle.

b. Can you make sense of the law of cosines when the cosine of the angle C is 1?

c. Can you make sense of the law of cosines when the cosine of the angle C is −1?

As shown in the next two examples, it is more often the case that triangles are solved using degree measure rather than radian measure. Two-digit approximations with a calculator are typical, and such approximations sometimes lead to small errors that usually are inconsequential.

EXAMPLE 9.2 Solving Triangles Using the Law of Cosines: SAS Case

Solve the triangle in **Figure 9.9**.

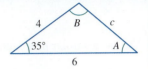

Figure 9.9 Triangle for Example 9.2

SOLUTION

First we use the law of cosines to find c:

$$c^2 = a^2 + b^2 - 2ab\cos C \qquad \blacktriangleleft \textbf{Apply the law of cosines.}$$

$$c^2 = 4^2 + 6^2 - 2 \times 4 \times 6\cos 35° \qquad \blacktriangleleft \textbf{Substitute known values.}$$

$$c = \sqrt{52 - 48\cos 35°} \approx 3.56 \qquad \blacktriangleleft \textbf{Simplify.}$$

We use the alternative form of the law of cosines to find the angle A:

$$A = \arccos\left(\frac{6^2 + c^2 - 4^2}{2 \times 6 \times c}\right) \qquad \blacktriangleleft \textbf{Apply the law of cosines.}$$

$$A \approx \arccos\left(\frac{6^2 + 3.56^2 - 4^2}{2 \times 6 \times 3.56}\right) \qquad \blacktriangleleft \textbf{Substitute 3.56 for } c.$$

$$A \approx 40.11° \qquad \blacktriangleleft \textbf{Simplify.}$$

Finally, we use the fact that the angle sum of the triangle is 180° to find B:

$$B = 180° - A - 35° \approx 180° - 40.11° - 35° = 104.89°$$

TRY IT YOURSELF 9.2 Brief answers provided at the end of the section.

Solve the triangle in **Figure 9.10**.

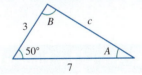

Figure 9.10 Triangle for Try It Yourself 9.2

EXAMPLE 9.3 Solving Triangles with the Law of Cosines: SSS Case

Solve the triangle in **Figure 9.11**.

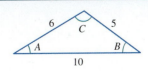

Figure 9.11 Triangle for Example 9.3

SOLUTION

We can use the alternative form of the law of cosines to find any of the angles we choose. We begin with the angle C:

$$C = \arccos\left(\frac{5^2 + 6^2 - 10^2}{2 \times 5 \times 6}\right) \quad \blacktriangleleft \textbf{Apply the law of cosines.}$$

$$C \approx 130.54° \quad \blacktriangleleft \textbf{Simplify.}$$

We use a similar procedure to find the angle A:

$$A = \arccos\left(\frac{6^2 + 10^2 - 5^2}{2 \times 6 \times 10}\right) \quad \blacktriangleleft \textbf{Apply the law of cosines.}$$

$$A \approx 22.33° \quad \blacktriangleleft \textbf{Simplify.}$$

Using the fact that the angle sum of the triangle is 180°, we find B:

$$B = 180° - A - C \approx 180° - 22.33° - 130.54° = 27.13°$$

TRY IT YOURSELF 9.3   Brief answers provided at the end of the section.

Solve the triangle in **Figure 9.12**.

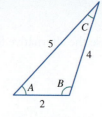

Figure 9.12
Triangle for Try It
Yourself 9.3

Note that not every choice of three numbers determines a triangle with those sides. For example, a triangle with sides 1, 1, and 100 is obviously impossible. The two short sides just can't be put together so that they reach from one end to the other of the long side. You will be led in an exercise to show the triangle inequality, which states that the sum of the lengths of any two sides of a triangle must be greater than the length of the third side.

EXAMPLE 9.4 Finding Irregular Areas

Find the area of the region in **Figure 9.13**.

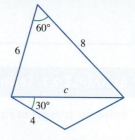

Figure 9.13 Region for
Example 9.4

SOLUTION

The region in Figure 9.13 is composed of two triangles, a top triangle and a bottom triangle. We can use the formula

$$\text{Area} = \frac{1}{2} ab \sin \theta$$

with $a = 6$, $b = 8$, and $\theta = 60°$ to find the area of the top triangle:

$$\text{Area of top triangle} = \frac{1}{2} \times 6 \times 8 \times \sin 60°$$

$$= \frac{1}{2} \times 6 \times 8 \times \frac{\sqrt{3}}{2}$$

$$= 12\sqrt{3}$$

$$\approx 20.78$$

To find the area of the bottom triangle, we first use the law of cosines for the top triangle to find the side marked c:

$$c^2 = a^2 + b^2 - 2ab \cos C \qquad \blacktriangleleft \textbf{Apply the law of cosines.}$$

$$c^2 = 6^2 + 8^2 - 2 \times 6 \times 8 \times \cos 60° \qquad \blacktriangleleft \textbf{Substitute known values.}$$

$$c^2 = 52 \qquad \blacktriangleleft \textbf{Simplify.}$$

$$c = \sqrt{52} \approx 7.21 \qquad \blacktriangleleft \textbf{Take square roots.}$$

With this information, we can use the sine area formula to find the area of the bottom triangle:

$$\text{Area of bottom triangle} = \frac{1}{2} \times 4 \times c \times \sin 30° \qquad \blacktriangleleft \textbf{Use the sine area formula.}$$

$$\approx \frac{1}{2} \times 4 \times 7.21 \times \frac{1}{2} \qquad \blacktriangleleft \textbf{Substitute known values.}$$

$$= 7.21 \qquad \blacktriangleleft \textbf{Simplify.}$$

Adding the two areas gives a final answer of $20.78 + 7.21 = 27.99$.

TRY IT YOURSELF 9.4 Brief answers provided at the end of the section.

Find the area of the region in **Figure 9.14**.

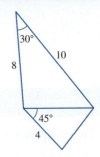

Figure 9.14
Region for Try It
Yourself 9.4

Heron's Formula for the Area of a Triangle

> Heron's formula allows us to calculate the area of a triangle using only the sides.

Because SSS is a criterion for congruence of triangles, we should be able to get the area of a triangle directly from the lengths of the sides. This is indeed the case, and the resulting formula is known as Heron's formula.[1]

LAWS OF MATHEMATICS: Heron's Formula

For a triangle with sides of lengths a, b, and c, the semiperimeter is

$$S = \frac{a + b + c}{2}$$

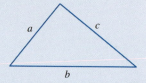

The area of the triangle is given by

$$\text{Area} = \sqrt{S(S - a)(S - b)(S - c)}$$

Heron's formula is a consequence of the law of cosines, but its derivation is a fairly serious algebraic challenge. One exercise will lead the stouthearted through the proof, and alternative versions are shown in succeeding exercises.

Although the proof of Heron's formula is difficult, heuristic arguments can be offered in its support. Note that Heron's formula multiplies four lengths together, which gives units in terms of a fourth power. When the square root is applied, fourth power units become square units, which are the proper units for area.

EXAMPLE 9.5 Using Heron's Formula

Find the area of the triangle in **Figure 9.15**.

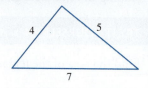

Figure 9.15 Triangle for Example 9.5

SOLUTION

The semiperimeter is

$$S = \frac{4 + 5 + 7}{2} = 8$$

---

[1]This formula is attributed to Heron of Alexandria, who lived in the first century A.D.

Then Heron's formula gives

$$\text{Area} = \sqrt{S(S-a)(S-b)(S-c)} \qquad \blacktriangleleft \textbf{Use Heron's formula.}$$

$$= \sqrt{8(8-4)(8-5)(8-7)} \qquad \blacktriangleleft \textbf{Substitute known values.}$$

$$= \sqrt{96} \qquad\qquad\qquad\quad\; \blacktriangleleft \textbf{Simplify.}$$

$$\approx 9.80 \qquad\qquad\qquad\quad\; \blacktriangleleft \textbf{Use approximation.}$$

TRY IT YOURSELF 9.5   Brief answers provided at the end of the section.

Find the area of the triangle whose sides are $a = 6$, $b = 9$, and $c = 11$.

EXTEND YOUR REACH

Show that if a, b, and c are the sides of a triangle where $a \geq b \geq c$, then the area is given by

$$\text{Area} = \frac{1}{2}\sqrt{a^2c^2 - \left(\frac{a^2 + c^2 - b^2}{2}\right)^2}$$

Suggestion: One approach is to first apply the sine area formula using the angle between a and c and then apply the law of cosines. Another approach is to use Heron's formula.

MODELS AND APPLICATIONS   A Soccer Player

EXAMPLE 9.6 A Soccer Player

The distance from the left post of a soccer goal to the right is 24 feet. The soccer player depicted in **Figure 9.16** stands 40 feet from the left goalpost and 20 feet from the right goalpost. The player kicks the ball, hoping to score. How wide a variation in the angle of the kick is allowed? That is, what is the angle between the ray from the ball to the right post and the ray from the ball to the left post?

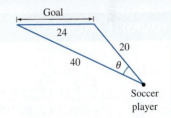

Figure 9.16 Scoring a soccer goal

SOLUTION

We need to find the angle θ shown in Figure 9.16, and we can do that using the alternative form of the law of cosines:

$$\theta = \arccos\left(\frac{a^2 + b^2 - c^2}{2ab}\right) \qquad \blacktriangleleft \textbf{Apply the law of cosines.}$$

$$\theta = \arccos\left(\frac{40^2 + 20^2 - 24^2}{2 \times 20 \times 40}\right) \qquad \blacktriangleleft \textbf{Substitute known values.}$$

$$\theta = \arccos\left(\frac{1424}{1600}\right) \qquad\qquad\quad \blacktriangleleft \textbf{Simplify.}$$

$$\theta \approx 27.13° \qquad\qquad\qquad\qquad \blacktriangleleft \textbf{Use approximation.}$$

We conclude that there is a variation of 27.13°.

> **TRY IT YOURSELF 9.6** Brief answers provided at the end of the section.
>
> Solve this problem if the soccer player is 25 feet from the left post and 45 feet from the right post.

EXAMPLE 9.7 Measuring a Hill

We want to measure the length of a hillside. Our laser locates the base of the hill 300 yards away. Elevating the laser 20°, we find that the top is 500 yards away. What is the length of the straight sloped side of the hill from the base to the top? (See **Figure 9.17**.)

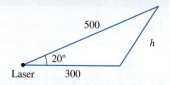

Figure 9.17 A hill and a laser

SOLUTION

We use the law of cosines to find the length marked h in Figure 9.17:

$$h^2 = 300^2 + 500^2 - 2 \times 300 \times 500 \cos 20°$$

$$h \approx 241.02$$

We conclude that the length of the hillside is about 241.02 yards.

> **TRY IT YOURSELF 9.7** Brief answers provided at the end of the section.
>
> Solve this problem if the laser is elevated 15° rather than 20°.

TRY IT YOURSELF ANSWERS

9.1 $c = 1$, $A = \dfrac{\pi}{4}$, $B = \dfrac{\pi}{2}$

9.2 $c = \sqrt{3^2 + 7^2 - 2 \times 3 \times 7 \cos 50°} \approx 5.57$,

$A = \arccos\left(\dfrac{7^2 + c^2 - 3^2}{2 \times 7 \times c}\right)$

$\approx \arccos\left(\dfrac{7^2 + 5.57^2 - 3^2}{2 \times 7 \times 5.57}\right) \approx 24.38°$,

$B \approx 180° - 24.38° - 50° = 105.62°$

9.3 $A \approx 49.46°$, $B \approx 108.21°$, $C \approx 22.33°$

9.4 27.13

9.5 $\sqrt{728} \approx 26.98$

9.6 22.81°

9.7 224.10 yards

EXERCISE SET 9.1

CHECK YOUR UNDERSTANDING

1. State the law of cosines.

2. State Heron's formula.

3. Refer to **Figure 9.18**. If $a = p$, $b = q$, and $C = R$, what can you conclude about r and c?

4. Refer to Figure 9.18. If $a = p$, $b = q$, and $c = r$, what can you conclude about R and C?

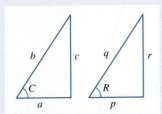

Figure 9.18 Two triangles

5. True or false: It is possible to have two noncongruent triangles whose corresponding sides all have the same length.

6. The area of a triangle is determined by:

  a. its angles

  b. the lengths of its sides

  c. the law of cosines

  d. None of the above.

SKILL BUILDING

Many of the exercises in this section require decimal approximations, so different methods of finding a solution may lead to answers slightly different from those given here (which is perfectly acceptable).

Solving triangles using the law of cosines. In Exercises 7 through 20, solve the given triangle. Use degree measure.

7.

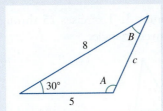

8.

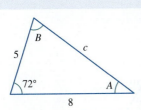

9.

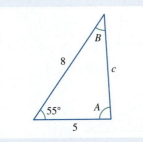

10.

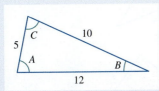

11.

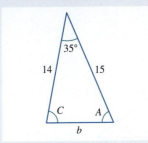

12.

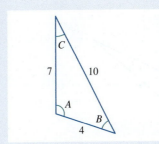

13.

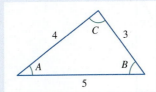

14.

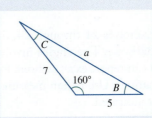

15.

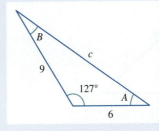

16.

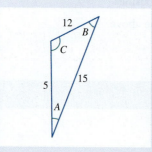

17.

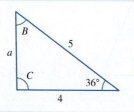

18.

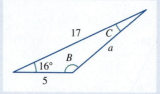

19.

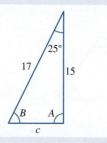

20.

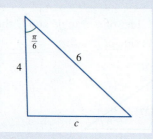

Exact answers. In Exercises 21 through 24, find the length of the side marked c. The angles involved are special angles, so you can report exact answers followed by decimal approximations. Use radian measure where degree measure is not indicated.

21.

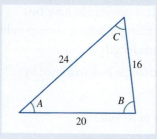

22.

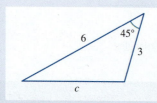

23.

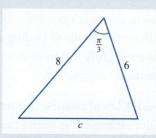

24.

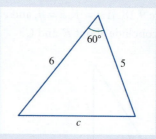

Using Heron's formula. In Exercises 25 through 30, find the area of the given triangle.

25.

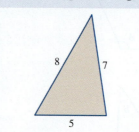

26.

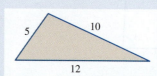

27.

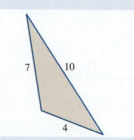

28.

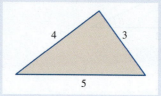

29.

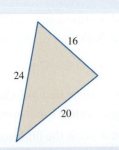

30.

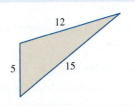

PROBLEMS

Areas of irregular regions. In Exercises 31 through 35, find the area of the given region. In each case, the area can be expressed as a sum of areas of rectangles, triangles, and circles.

31.

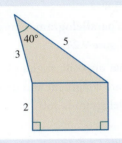

32. Note that the arc in the figure is a semicircle.

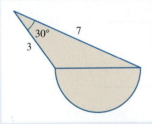

33. Find the area inside the square and outside the triangle.

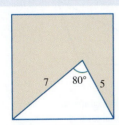

34.

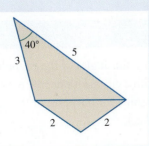

35.

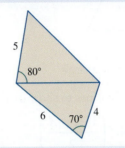

36. Alternative form for the law of cosines.

  a. Rearrange the law of cosines to show that

$$\cos C = \frac{a^2 + b^2 - c^2}{2ab}$$

  b. Explain why C is between $0°$ and $180°$.

  c. With part b in mind, explain why

$$C = \arccos\left(\frac{a^2 + b^2 - c^2}{2ab}\right)$$

  is the only solution for C of the equation in part a.

37. An inequality. Show that if a, b, and c are the lengths of the sides of a triangle, then

$$\frac{a^2 + b^2 - c^2}{2ab} \le 1$$

Suggestion: Let C be the angle opposite the side of length c. In the alternative form for the law of cosines, what is the largest value that $\cos C$ can have?

38. The triangle inequality. Let a, b, and c be the lengths of the sides of a triangle and C the angle opposite to the side of length c.

  a. Use the fact that $\cos C > -1$ to show that $-2ab \cos C < 2ab$.

  b. Use part a to show that $(a + b)^2 > a^2 + b^2 - 2ab \cos C$.

c. Use part b and the law of cosines to show that $c < a + b$. This is known as the triangle inequality.

39. The base of an isosceles triangle. This exercise refers to the isosceles triangle pictured in **Figure 9.19**.

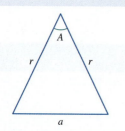

Figure 9.19 An isosceles triangle

a. Use the law of cosines to show that

$$a^2 = 4r^2\left(\frac{1 - \cos A}{2}\right)$$

b. Apply the half-angle formula for sines to show that

$$a = 2r \sin \frac{A}{2}$$

40. An inscribed polygon. *This is a continuation of the preceding exercise.* Consider a regular n-sided polygon inscribed in a circle of radius r, as in **Figure 9.20**.

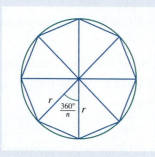

Figure 9.20 A regular polygon

a. Use part b of the preceding exercise to show that the perimeter of the polygon is $2nr \sin \dfrac{180°}{n}$. *Suggestion*: Cut the circle into n pieces of pie, as illustrated in Figure 9.20. This cuts the polygon into n isosceles triangles.

b. What do you think is the limiting value (as the number of sides increases) of the perimeters of inscribed polygons?

41. Area of an inscribed polygon. Show that the area of the n-sided polygon inscribed in a circle of radius r is given by $\dfrac{1}{2}nr^2 \sin \dfrac{360°}{n}$. *Suggestion*: Refer to Figure 9.20. First find the area of each of the n isosceles triangles.

42. The diagonal of a parallelogram. Consider the parallelogram in **Figure 9.21**.

a. Use the fact that opposite angles of a parallelogram are congruent and that the angle sum of a parallelogram is 360° to show that $\beta = 180° - \alpha$.

b. Consider the diagonal c marked in **Figure 9.22**. Show that $c^2 = a^2 + b^2 + 2ab \cos \alpha$.

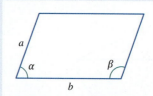

Figure 9.21 Parallelogram for Exercise 42

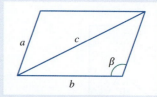

Figure 9.22 Parallelogram for Exercise 42: diagonal added

MODELS AND APPLICATIONS

43. Flying angle. The flying distance from Paris to London is 213 miles. From London to New York it is 3470 miles, and from New York to Paris it is 3635 miles. Find the angles of the triangle made by joining these three cities. Use degree measure. (See **Figure 9.23**. Assume for purposes of this exercise that Earth is flat.)

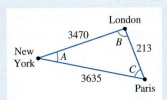

Figure 9.23 New York, London, and Paris

44. Throwing a baseball. In every Major League Baseball field, the bases form a square with sides of length 90 feet. In Fenway Park, the distance from home plate to the fence at straight-away center is about 390 feet. The center fielder catches a ball with his back against the fence at this point. He wants to make a play at third base. (See **Figure 9.24.**) How far does he have to throw the ball? Round your answer to the nearest foot.

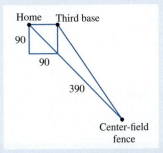

Figure 9.24 Distances at Fenway Park

45. Two airplanes. Two airplanes are flying side by side. One airplane continues on the same course at 250 miles per hour, and the other veers at an angle of 70° and flies at a speed of 300 miles per hour. (See **Figure 9.25.**) How far apart are the airplanes after 2 hours of flying?

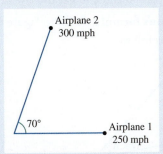

Figure 9.25 Distance between two airplanes

46. A model airplane. A model airplane is traveling at a constant speed. A laser locates the plane at a distance of 100 feet and 30° above the western horizon. One minute later the same laser finds

that the plane is 120 feet away at an angle of 70° with the eastern horizon. (See **Figure 9.26.**) How fast is the airplane traveling? Report your answer in feet per minute.

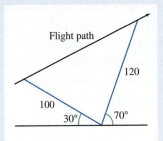

Figure 9.26 Flight path of a model airplane

47. Venus, Jupiter, and Mars. In October 2015, observers of the night sky were treated to a rare sight. Venus, Jupiter, and Mars appeared very close together, as shown in **Figure 9.27**. But appearances can be deceiving. **Figure 9.28** shows the distances (in astronomical units) from Earth to Mars and Venus in that month.

Figure 9.27 A view of the night sky in October 2015

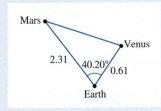

Figure 9.28 Distances from Earth to Mars and Venus in October 2015

a. Find the distance from Mars to Venus in October 2015. Report your answer in astronomical units.

b. By definition, 1 astronomical unit is the mean distance from Earth to the sun, about 92.96 million miles. Use your answer to part a to determine the distance in miles from Mars to Venus in October 2015.

48. **Jupiter and Venus.** In the night sky of October 2015, Jupiter and Venus appeared to be very close neighbors. **Figure 9.29** shows the distances (in astronomical units) from Earth to Venus and Jupiter then. How far (in miles) was Venus from Jupiter at that time? (By definition, 1 astronomical unit is the mean distance from Earth to the sun, about 92.96 million miles.)

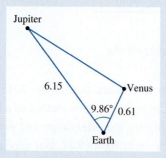

Figure 9.29 Distance from Jupiter to Venus

49. **An adjacent building.** The building adjacent to your professor's office is 80 feet tall. A wire stretched from her office window to the base of the

building is 240 feet long. (See **Figure 9.30**.) A wire from the same point to the top of the building is 230 feet long. Find the angle made by the two wires. Use degree measure.

Figure 9.30 View from your professor's office window

50. **A shark.** While diving off the coast of California, a diver spots a large shark. Using a laser device, she finds that the distance to the tail of the shark is 30 feet and the distance to the head is 32 feet. The angle between the two measurements is 20°. (See **Figure 9.31**.) How long is the shark?

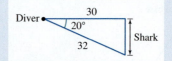

Figure 9.31 The length of a shark

CHALLENGE EXERCISES FOR INDIVIDUALS OR GROUPS

51. **Demonstration of the law of cosines when the angle C is obtuse. Figure 9.32** shows a triangle with an obtuse angle C. We drop a perpendicular to get **Figure 9.33**.

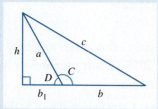

Figure 9.32 A triangle with obtuse angle C

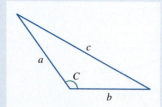

Figure 9.33 A triangle with obtuse angle C: perpendicular added

a. Show that $b_1 = a \cos D$ and $h = a \sin D$.

b. Apply the Pythagorean theorem to the large right triangle in Figure 9.33 and simplify to show that $a^2 + b^2 + 2ab \cos D = c^2$.

c. Use the sum formula for the cosine to show that $\cos D = \cos(180° - C) = -\cos C$.

d. Combine parts b and c to get the law of cosines.

52. **Proof of Heron's formula.** Refer to **Figure 9.34**, and proceed as follows.

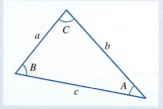

Figure 9.34 Figure for Exercises 52 through 55

a. Use the sine formula for the area of a triangle to show that if C is the angle between sides a and b, then the square of the area of the triangle is given by

$$\text{Area}^2 = \frac{a^2 b^2 \sin^2 C}{4} = \frac{a^2 b^2 (1 - \cos^2 C)}{4}$$

b. Use the law of cosines to express the cosine of the angle C in terms of a, b, and c. Deduce the formula

$$\text{Area}^2 = \frac{4a^2 b^2 - (a^2 + b^2 - c^2)^2}{16}$$

c. Use the algebraic identity $X^2 - Y^2 = (X + Y)(X - Y)$ to get

$$\text{Area}^2 = \frac{(2ab + (a^2 + b^2 - c^2))(2ab - (a^2 + b^2 - c^2))}{16}$$

d. Simplify to get

$$\text{Area}^2 = \frac{((a + b)^2 - c^2)(c^2 - (a - b)^2)}{16}$$

e. Use the algebraic identity $X^2 - Y^2 = (X - Y)(X + Y)$ to get

$$\text{Area}^2 = \frac{(a + b - c)(a + b + c)(c - a + b)(c + a - b)}{16}$$

f. Use $S = \dfrac{a + b + c}{2}$ to show that this last expression reduces to Heron's formula.

53. An identity. Refer to Figure 9.34. Let S denote the semiperimeter $S = \dfrac{1}{2}(a + b + c)$.

a. Show that $S - c = \dfrac{1}{2}(a + b - c)$.

b. Show that $S(S - c) = \dfrac{1}{4}(a^2 + b^2 + 2ab - c^2)$.

Suggestion: It may make your calculations easier if you note that

$$((a + b) - c)((a + b) + c) = (a + b)^2 - c^2$$

c. Apply the law of cosines to part b to show that

$$S(S - c) = \frac{1}{2} ab(1 + \cos C)$$

d. Apply the half-angle formula for the cosine to part c to show that

$$\sqrt{\frac{S(S - c)}{ab}} = \cos\frac{C}{2}$$

54. Another identity. Refer to Figure 9.34. Let $S = \dfrac{1}{2}(a + b + c)$.

a. Show that

$$(S - a)(S - b) = \frac{1}{4}(c + (b - a))(c - (b - a))$$

b. Show that

$$(S - a)(S - b) = \frac{1}{4}(c^2 - b^2 - a^2 + 2ab)$$

c. Apply the law of cosines to show that

$$(S - a)(S - b) = \frac{1}{2} ab(1 - \cos C)$$

d. Apply the half-angle formula for the sine to show that

$$\sqrt{\frac{(S - a)(S - b)}{ab}} = \sin\frac{C}{2}$$

55. Alternative proof of Heron's area formula. Here is an alternative demonstration of Heron's formula, which says: If a, b, and c are the lengths of the sides of a triangle and $S = \dfrac{1}{2}(a + b + c)$ then

$$\text{Area of triangle} = \sqrt{S(S - a)(S - b)(S - c)}$$

This demonstration depends on the two preceding exercises.

a. Combine part c of each of the two preceding exercises to show that

$$S(S - a)(S - b)(S - c) = \frac{1}{4} a^2 b^2 \sin^2 C$$

Here C is the angle opposite c, as in Figure 9.34.

b. Use part a to prove Heron's area formula.

56. From Section P.1: Calculate the midpoint of $P = (1, 4)$ and $Q = (-1, 6)$.

Answer: $(0, 5)$

57. From Section 4.2: Simplify the expression $e^{3 \ln 2}$.

Answer: 8

58. From Section 4.3: Solve the equation $12 = 3^{2x}$.

Answer: $x = \dfrac{\ln 12}{2 \ln 3} \approx 1.13$

59. From Chapter 6: Find the missing parts of the triangle in **Figure 9.35**.

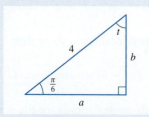

Figure 9.35 Triangle for Exercise 59

Answer: $t = \dfrac{\pi}{3}, \quad a = 2\sqrt{3}, \quad b = 2$

60. From Section 7.1: Figure 9.36 shows the graph of a function of the form $y = A \sin B(x - C)$ with $A > 0$. Find the amplitude, period, and phase shift.

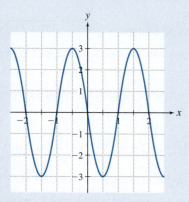

Figure 9.36 A graph for Exercise 60

Answer: Amplitude 3, period 2, phase shift 1

61. From Section 8.2: Use the sum formula for the cosine to find the exact value of $\cos 105°$.

Answer: $\dfrac{1 - \sqrt{3}}{2\sqrt{2}} = \dfrac{\sqrt{2} - \sqrt{6}}{4}$

62. From Section 8.3: Use the half-angle formula for the cosine to find the exact value of $\cos \dfrac{\pi}{12}$.

Answer: $\sqrt{\dfrac{1 + \sqrt{3}/2}{2}} = \sqrt{\dfrac{2 + \sqrt{3}}{4}} = \dfrac{\sqrt{2 + \sqrt{3}}}{2}$

63. From Section 8.4: Find the missing parts of the triangle in **Figure 9.37**. Give exact answers followed by decimal approximations in degrees.

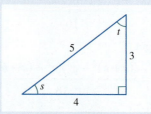

Figure 9.37 Triangle for Exercise 63

Answer: $s = \arcsin \dfrac{3}{5} \approx 36.87°,$

$t = \arcsin \dfrac{4}{5} \approx 53.13°$

64. From Section 8.5: Solve the equation $4 \sin x \cos x = \sqrt{3}$.

Answer: $x = \dfrac{\pi}{6} + k\pi$ and $x = \dfrac{\pi}{3} + k\pi$, where k is any integer

65. From Section 8.5: Give the exact solution, and then provide a decimal approximation: $\sin t = 4 \cos t$.

Answer: $t = \arctan(4) + k\pi \approx 1.33 + k\pi$, where k is any integer

9.2 Law of Sines

9.1 Law of Cosines

9.2 Law of Sines

9.3 Vectors

9.4 Vectors in the Plane and in Three Dimensions

The law of sines allows us to solve triangles that we cannot solve using the law of cosines.

In this section, you will learn to:

1. Solve triangles using the law of sines and side-angle-angle.
2. Solve triangles using the law of sines and angle-side-angle.
3. Solve triangles using the law of sines in the ambiguous case of side-side-angle.
4. Identify whether the law of sines or the law of cosines (or both) is appropriate for solving a triangle.
5. Solve applied problems after choosing whether to employ the law of sines or the law of cosines (or both).

There is a sister theorem to the law of cosines known as the law of sines. As with the law of cosines, the law of sines applies to all triangles.

The law of sines relates the ratio of sines of angles in a triangle to the ratio of sides. A ratio of sines of angles occurs in many areas of mathematics and physics and helps us understand natural phenomena. You may notice, for example, that a stick or a straw dipped into clear water appears to be broken. This is due to refraction of light, which simply means that light changes direction when it crosses from air into water. **Figure 9.38** shows a ray of light crossing from air into water.

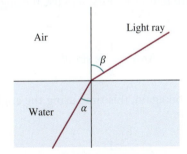

Figure 9.38 Light rays changing direction when moving from air into water

Snell's law tells us that if C_A is the speed of light in air, and C_W is the speed of light in water, then

$$\frac{\sin \alpha}{\sin \beta} = \frac{C_W}{C_A}$$

This law allows us to determine precisely how light bends and hence to understand the properties of refraction of light.

Derivation of the Law of Sines

> The law of sines is found by equating two area calculations for the same triangle.

The law of cosines allows us to solve triangles when we are given side-angle-side (SAS) or side-side-side (SSS). However, **Figure 9.39** and **Figure 9.40** show two additional congruence criteria that the law of cosines cannot handle: side-angle-angle (SAA) and angle-side-angle (ASA). For these, we need the law of sines.

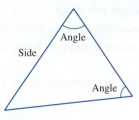

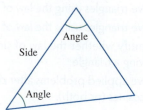

DF **Figure 9.39** SAA: If we are given one side, an opposite angle, and an adjacent angle, then at most one triangle is determined.

DF **Figure 9.40** ASA: If we are given one side and two adjacent angles, then at most one triangle is determined.

The law of sines relates the angles of a triangle to the lengths of their opposite sides. To begin the derivation, recall that we can find the area of the triangle as half the product of two side lengths with the sine of the angle between the two sides. Let's use this formula, which gives the area in terms of the sine, to calculate the area of the triangle in **Figure 9.41** in two different ways. First let's use the angle A, which is between sides b and c. The area is

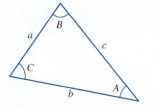

Figure 9.41 A triangle with angles and sides labeled

$$\text{Area} = \frac{1}{2}bc \sin A$$

Now let's do the calculation again, this time using the angle B, which is between sides a and c:

$$\text{Area} = \frac{1}{2}ac \sin B$$

Because both calculations give the same area, we have

$$\frac{1}{2}bc \sin A = \frac{1}{2}ac \sin B \qquad \blacktriangleleft \text{Equate area calculations.}$$

$$b \sin A = a \sin B \qquad \blacktriangleleft \text{Divide out } \frac{1}{2}c.$$

$$\frac{b \sin A}{\sin A \sin B} = \frac{a \sin B}{\sin A \sin B} \qquad \blacktriangleleft \text{Divide by } \sin A \sin B.$$

$$\frac{b \, \cancel{\sin A}}{\cancel{\sin A} \sin B} = \frac{a \, \cancel{\sin B}}{\sin A \, \cancel{\sin B}} \qquad \blacktriangleleft \text{Cancel common terms.}$$

$$\frac{b}{\sin B} = \frac{a}{\sin A} \qquad \blacktriangleleft \text{Simplify.}$$

The last equation is the law of sines, and it applies to any pair of opposite angles and sides.

LAWS OF MATHEMATICS: The Law of Sines

For a triangle labeled as in the accompanying figure,

$$\frac{a}{\sin A} = \frac{b}{\sin B} = \frac{c}{\sin C}$$

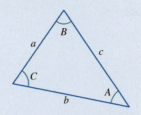

An equivalent statement is that

$$\frac{\sin A}{a} = \frac{\sin B}{b} = \frac{\sin C}{c}$$

In words, the ratio of a side to the sine of its opposite angle is the same for all three sides.

Solving Triangles Using the Law of Sines

> The law of sines is used to find unknown sides and angles of triangles.

EXAMPLE 9.8 Solving Triangles Using the Law of Sines: SAA

Solve the triangle in **Figure 9.42**.

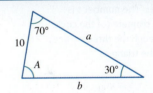

Figure 9.42 A case of SAA

SOLUTION

Our first step is to find the angle A using the fact that the angle sum of the triangle in Figure 9.42 is 180°:

$$A = 180° - (70° + 30°) = 80°$$

Next we use the law of sines to find b:

$$\frac{b}{\sin 70°} = \frac{10}{\sin 30°} \qquad ◀ \text{ Apply the law of sines.}$$

$$b = \sin 70° \times \frac{10}{\sin 30°} \qquad ◀ \text{ Solve for } b.$$

$$b \approx 18.79$$

We find a in a similar way:

$$\frac{a}{\sin 80°} = \frac{10}{\sin 30°} \qquad ◀ \text{ Apply the law of sines.}$$

$$a = \sin 80° \times \frac{10}{\sin 30°} \qquad ◀ \text{ Solve for } a.$$

$$a \approx 19.70$$

TRY IT YOURSELF 9.8 Brief answers provided at the end of the section.

Solve the triangle in **Figure 9.43**.

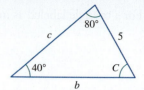

Figure 9.43 Triangle for Try It Yourself 9.8

EXTEND YOUR REACH

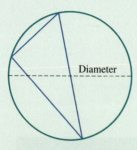

Figure 9.44 The law of sines: The number t is the diameter of the circle through the three vertices of the triangle.

The law of sines tells us that the number

$$t = \frac{\text{Side length}}{\sin(\text{Opposite angle})}$$

is the same no matter which of the three sides of a triangle we choose. Thus, the number t is a characteristic of the triangle. Does this number have a geometric meaning for the triangle? Yes. It is the diameter of the circle through the three vertices of the triangle, as shown in **Figure 9.44**.

a. To test this assertion, construct a circle, and then pick three points on the circle. Calculate the ratio t (using a protractor to measure the angle), and compare it with the diameter of the circle.

b. Read Appendix 5 to see why this assertion is true. The proof uses the fact that every angle inscribed in a semicircle is a right angle.

EXAMPLE 9.9 Solving Triangles Using the Law of Sines: ASA

Solve the triangle in **Figure 9.45**.

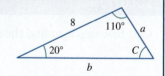

Figure 9.45 A case of ASA

SOLUTION

Just as in the preceding example, to find the angle C we use the fact that the angles in Figure 9.45 must sum to 180°:

$$C = 180° - (110° + 20°) = 50°$$

Next we can use the law of sines to find a and b:

$$\frac{a}{\sin 20°} = \frac{8}{\sin 50°} \qquad \blacktriangleleft \textbf{Apply the law of sines.}$$

$$a = \frac{8\sin 20°}{\sin 50°} \qquad \blacktriangleleft \textbf{Solve for } a.$$

$$a \approx 3.57$$

and

$$\frac{b}{\sin 110°} = \frac{8}{\sin 50°} \quad \blacktriangleleft \text{Apply the law of sines.}$$

$$b = \frac{8 \sin 110°}{\sin 50°} \quad \blacktriangleleft \text{Solve for } b.$$

$$b \approx 9.81$$

TRY IT YOURSELF 9.9 Brief answers provided at the end of the section.

Solve the triangle in **Figure 9.46**.

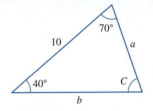

Figure 9.46 Triangle for Try It Yourself 9.9

EXTEND YOUR REACH

Using the same notation as with the law of sines, the law of tangents tells us that

$$\frac{a-b}{a+b} = \frac{\tan\left(\frac{1}{2}(A-B)\right)}{\tan\left(\frac{1}{2}(A+B)\right)}$$

Verify that the law of tangents is true for the triangle in this example. In addition, derive the law of tangents above by starting with the law of sines and using trigonometric identities. (This derivation is challenging.)

EXAMPLE 9.10 Solving Triangles with Radian Measure

Solve the triangle in **Figure 9.47**.

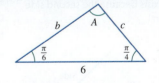

Figure 9.47 A case of ASA

SOLUTION

This is a case of ASA. The first step is to find the angle A using the fact that the angle sum is π radians:

$$A = \pi - \left(\frac{\pi}{6} + \frac{\pi}{4}\right) = \frac{7\pi}{12}$$

Next we use the law of sines to find first b and then c:

$$\frac{b}{\sin(\pi/4)} = \frac{6}{\sin(7\pi/12)}$$ ◀ Apply the law of sines.

$$b = \frac{6\sin(\pi/4)}{\sin(7\pi/12)}$$ ◀ Solve for b.

$$b \approx 4.39$$

and

$$\frac{c}{\sin(\pi/6)} = \frac{6}{\sin(7\pi/12)}$$ ◀ Apply the law of sines.

$$c = \frac{6\sin(\pi/6)}{\sin(7\pi/12)}$$ ◀ Solve for c.

$$c \approx 3.11$$

TRY IT YOURSELF 9.10 Brief answers provided at the end of the section.

Solve the triangle in **Figure 9.48**.

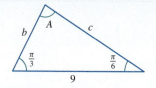

Figure 9.48 Triangle for Try It Yourself 9.10

The Ambiguous Case: Side-Side-Angle

> When presented with side-side-angle, one triangle, two triangles, or no triangle at all may be described.

We have noted four criteria for triangle congruence: SSS, SAS, SAA, and ASA. But a fifth candidate, side-side-angle (SSA), is not a valid criterion for triangle congruence. As a consequence, if we are given two sides and an angle not between those two sides, then it may be difficult to solve the triangle (and there may be no such triangle).

To understand what can happen in the SSA case, we think of fixing an angle and a side, then trying to complete the triangle by adding another side. Three possibilities arise.

Case 1: No solution: This case is illustrated in **Figure 9.49**, where we see that the second side is too short to complete a triangle.

Case 2: Exactly one solution: This case may occur when the second side is long in relation to the first side. As shown in **Figure 9.50**, there is exactly one way to place the second side and complete a triangle.

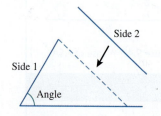

DF **Figure 9.49** SSA case 1: No solution exists because the second side is too short.

Figure 9.50 SSA case 2: There is exactly one way to place side 2 to complete the triangle, and hence exactly one solution.

Case 3: Exactly two solutions: In this case there are two ways to use the second side to complete a triangle. This case is shown in **Figure 9.51**. This case never occurs when the given angle is obtuse.

The following examples show how we address these three cases.

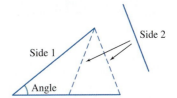

Figure 9.51 SSA case 3: There are two possible placements of side 2, yielding exactly two solutions.

EXAMPLE 9.11 SSA: The Case of No Solution

Try to solve the triangle in **Figure 9.52**.

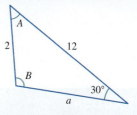

Figure 9.52 Triangle for Example 9.11

SOLUTION

We attempt to find the angle B using the law of sines:

$$\frac{12}{\sin B} = \frac{2}{\sin 30°} \qquad \blacktriangleleft \textbf{Apply the law of sines.}$$

$$\frac{12}{\sin B} = 4 \qquad \blacktriangleleft \textbf{Simplify.}$$

$$\sin B = 3 \qquad \blacktriangleleft \textbf{Solve for sin } B.$$

Because the sine function is always between -1 and 1, the equation $\sin B = 3$ has no solution. Consequently, there is no triangle with sides and angle marked as in Figure 9.52. Thus, Figure 9.52 is deceptive. An accurate picture is shown in **Figure 9.53**. This is an example of the case of no solution shown in Figure 9.49.

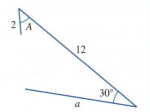

Figure 9.53 An accurate representation of the information for Example 9.11

TRY IT YOURSELF 9.11 Brief answers provided at the end of the section.

Solve the triangle in **Figure 9.54**.

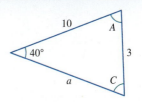

Figure 9.54 Triangle for Try It Yourself 9.11

The calculations in Example 9.11 are typical. When proposed values for SSA do not correspond to any triangle, an equation of the form $\sin x = y$ where $y > 1$ will always be encountered. This is the signal that there is no solution.

In the next example we make use of the following important facts, which can save you a lot of headaches: A triangle can have at most one obtuse angle. If there is an obtuse angle, it is across from the longest side of the triangle.

EXAMPLE 9.12 SSA: Unique Solution

Solve the triangle in **Figure 9.55**.

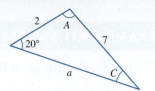

Figure 9.55 Triangle for Example 9.12

SOLUTION

To begin, we note that only the angle opposite the longest side can be obtuse. Because the side of length 2 is not the longest, the angle C must be acute (**Figure 9.56**). This observation will help us with the solution.

Apply the law of sines to find the angle C:

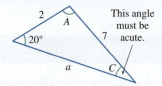

Figure 9.56 Triangle for Example 9.12: The angle across from the side of length 2 must be acute because there is a longer side.

$$\frac{2}{\sin C} = \frac{7}{\sin 20°} \qquad \blacktriangleleft \text{Apply the law of sines.}$$

$$\frac{2\sin 20°}{7} = \sin C \qquad \blacktriangleleft \text{Solve for sin C.}$$

$$0.098 \approx \sin C$$

(In these problems we round intermediate calculations and use the rounded result in subsequent calculations.)

Recall from Section 8.5 that the equation $\sin C = 0.098$ has two solutions on the interval $[0°, 180°]$, and hence the equation might represent two possible angles of a triangle, one acute and one obtuse. But because we know that the angle C is acute, there is only one solution, and we can use the inverse sine to find it:

$$C \approx \arcsin 0.098 \approx 5.62°$$

We use the fact that the angle sum of the triangle is $180°$ to find the angle A:

$$A \approx 180° - 20° - 5.62° = 154.38°$$

Finally, we use the law of sines to find a:

$$\frac{a}{\sin A} = \frac{7}{\sin 20°} \qquad \blacktriangleleft \text{Apply the law of sines.}$$

$$\frac{a}{\sin 154.38°} \approx \frac{7}{\sin 20°} \qquad \blacktriangleleft \text{Use } A \approx 154.38°.$$

$$a \approx \frac{7 \sin 154.38°}{\sin 20°} \qquad \blacktriangleleft \text{Multiply by sin 154.38°.}$$

$$a \approx 8.85$$

We have found a unique solution, which is the case shown in Figure 9.50.

TRY IT YOURSELF 9.12   Brief answers provided at the end of the section.

Solve the triangle in **Figure 9.57**.

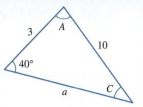

Figure 9.57 Triangle for Try It Yourself 9.12

If we hadn't noted that the angle C in Example 9.12 is acute, we would have needed to consider a second solution of $\sin c = 0.098$, which is $C = 180° -$ arcsin $0.098 \approx 174.38°$. But then c plus 20° would exceed 180° (the angle sum of the triangle), which is not possible. Again, we conclude that $C = 5.62°$ is the only possibility. We arrive at the same result, but observing which side of the triangle is longest can save time.

EXAMPLE 9.13   SSA: The Case of Two Solutions

Solve the triangle in **Figure 9.58**.

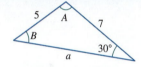

Figure 9.58 Triangle for Example 9.13

SOLUTION

To begin, we find the angle B using the law of sines:

$$\frac{7}{\sin B} = \frac{5}{\sin 30°} \qquad \text{◀ Apply the law of sines.}$$

$$\frac{7 \sin 30°}{5} = \sin B \qquad \text{◀ Solve for sin B.}$$

$$0.70 = \sin B$$

Because we don't know which side of the triangle is the longest, we cannot be certain, as in the previous example, that B is an acute angle. So we must allow for all solutions of $\sin B = 0.70$ on $[0°, 180°]$. These solutions are

$$B = \arcsin 0.70 \approx 44.43° \quad \text{and} \quad B = 180° - \arcsin 0.70 \approx 135.57°$$

We find two triangles—one for each of these solutions. This is the case shown in Figure 9.51.

The first triangle: This triangle has angles of 30°, $B \approx 44.43°$, and $A \approx 180° - 30° - 44.43° = 105.57°$. Using the law of sines, we find

$$a \approx \frac{5 \sin A}{\sin 30°} \approx \frac{5 \sin 105.57°}{\sin 30°} \approx 9.63$$

The second triangle: The second triangle has angles 30°, $B \approx 135.57°$, and $A \approx 180° - 30° - 135.57° = 14.43°$. The law of sines gives

$$a = \frac{5 \sin A}{\sin 30°} \approx \frac{5 \sin 14.43°}{\sin 30°} \approx 2.49$$

Our two solutions yield the two triangles shown in **Figure 9.59** and **Figure 9.60**.

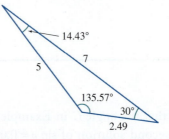

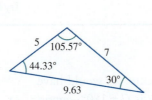

Figure 9.59 The first triangle for Example 9.13

Figure 9.60 The second triangle for Example 9.13

TRY IT YOURSELF 9.13 Brief answers provided at the end of the section.

Solve the triangle in **Figure 9.61**.

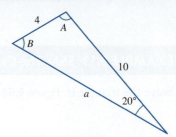

Figure 9.61 Triangle for Try It Yourself 9.13

EXTEND YOUR REACH

We have stated that, in the ambiguous case, there may be exactly one solution, exactly two solutions, or no solutions at all. Explain why it is never the case that there are three or more solutions.

Choosing a Law to Solve a Triangle

A strategy for choosing laws to solve a triangle allows us to use the appropriate solution tool.

You now have two triangle-solving tools available, the law of sines and the law of cosines. If you are presented with a triangle, how do you choose which tool to use? The answer is provided in the following strategy for solving triangles.

STEP-BY-STEP STRATEGY: Solving Triangles

Step 1 Identify which congruence criterion applies: SAS, SSS, SAA, ASA, or SSA.

Step 2 If SAS or SSS applies, solve using the law of cosines.

Step 3 If SAA or ASA applies, solve using the law of sines.

Step 4 If the ambiguous case SSA applies, use the law of sines. There may be two, one, or no solutions. Begin, if possible, by finding an angle that is not across from the longest side. If such an angle can be identified, it must be an acute angle, so the arcsine function yields the only possible angle.

EXAMPLE 9.14 Choosing a Method

For each of the triangles in **Figure 9.62** through **Figure 9.65**, determine the proper method to be used for solving the triangle. Note that you are not asked to complete the solution.

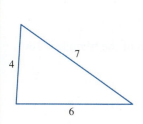

Figure 9.62 Triangle I

Figure 9.63
Triangle II

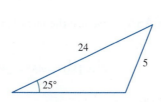

Figure 9.64 Triangle III

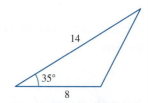

Figure 9.65 Triangle IV

SOLUTION

a. Figure 9.62 is a case of SSS. We use the law of cosines to solve the triangle.

b. Figure 9.63 is a case of ASA. The law of sines is the appropriate choice.

c. Figure 9.64 is a case of SSA. This is the ambiguous case, and we attempt a solution using the law of sines. There may be two, one, or no solutions.

d. Figure 9.65 shows a case of SAS. We use the law of cosines to solve the triangle.

TRY IT YOURSELF 9.14 Brief answers provided at the end of the section.

What method should be used to solve the triangle in **Figure 9.66**?

Figure 9.66 Triangle for Try It Yourself 9.14

MODELS AND APPLICATIONS   A Lost Airplane

EXAMPLE 9.15 A Lost Airplane

Will Rogers World Airport (OKC) in Oklahoma City is 200 miles directly north of Dallas–Fort Worth International Airport (DFW). The two airports cooperate to locate a lost aircraft. OKC locates the direction of the airplane by radio beacon to be 44° south of west. At the same time, DFW finds the airplane to be 62° north of west. (See **Figure 9.67**.) How far from DFW is the airplane? How far is it from OKC?

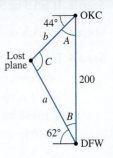

Figure 9.67 A lost airplane

SOLUTION

The distance from the airplane to DFW is marked a in Figure 9.67, and b is the distance from the airplane to OKC. We begin by finding the angles A and B:

$$A = 90° - 44° = 46°$$

$$B = 90° - 62° = 28°$$

We are faced now with a case of ASA, so we use the law of sines. Because the angle sum of the triangle is 180°, we have

$$C = 180° - 46° - 28° = 106°$$

We apply the law of sines to find a:

$$\frac{a}{\sin 46°} = \frac{200}{\sin 106°}$$

$$a = \frac{200 \sin 46°}{\sin 106°}$$

$$a \approx 149.67$$

Hence, the distance from the plane to DFW is 149.67 miles. In a similar fashion, we find b:

$$\frac{b}{\sin 28°} = \frac{200}{\sin 106°}$$

$$b = \frac{200 \sin 28°}{\sin 106°}$$

$$b \approx 97.68$$

The airplane is 97.68 miles from OKC.

TRY IT YOURSELF 9.15 Brief answers provided at the end of the section.

A ship is located in relation to two lighthouses, as shown in **Figure 9.68**. The two lighthouses are 10 miles apart. How far is the ship from each of the lighthouses?

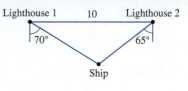

Figure 9.68 A ship and two lighthouses

TRY IT YOURSELF ANSWERS

9.8 $C = 60°,\ b = \dfrac{5 \sin 80°}{\sin 40°} \approx 7.66,$

$c = \dfrac{5 \sin 60°}{\sin 40°} \approx 6.74$

9.9 $C = 70°,\ b = 10,\ a \approx 6.84$

9.10 $A = \dfrac{\pi}{2},\ b = \dfrac{9}{2} = 4.5,\ c = \dfrac{9\sqrt{3}}{2} \approx 7.79$

9.11 There is no solution.

9.12 $C \approx 11.12°,\ A \approx 128.88°,\ a \approx 12.11$

9.13 There are two solutions. First triangle: $B \approx 58.77°,\ A \approx 101.23°,\ a \approx 11.47$. Second triangle: $B \approx 121.23°,\ A \approx 38.77°,\ a \approx 7.32$.

9.14 This is a case of SAA. The law of sines should be used.

9.15 5.98 miles from lighthouse 1 and 4.84 miles from lighthouse 2

EXERCISE SET 9.2

CHECK YOUR UNDERSTANDING

1. If the angles of a triangle are A, B, C, and the corresponding opposite sides are a, b, c, then the law of sines states _____.

2. If you attempt to solve a triangle given one side and two angles of a triangle, then:

  a. there is no solution

  b. there is a unique solution

  c. there are exactly two distinct solutions

  d. None of the above.

3. **True or false:** In the ambiguous case, you are given two sides and an adjacent angle.

4. Should you use the law of sines or the law of cosines to solve a triangle in the SSS case?

5. Should you use the law of sines or the law of cosines to solve a triangle in the SSA case?

6. Should you use the law of sines or the law of cosines to solve a triangle in the SAS case?

7. Should you use the law of sines or the law of cosines to solve a triangle in the ASA case?

8. Should you use the law of sines or the law of cosines to solve a triangle in the SAA case?

SKILL BUILDING

Many of the exercises in this section require decimal approximations, so different methods for finding a solution may lead to answers slightly different from those given here (which is perfectly acceptable).

Solving triangles: ASA and SAA. For Exercises 9 through 20, solve the triangles. Each is a case of ASA or SAA.

9.

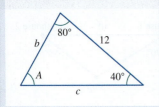

10.

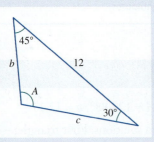

11.

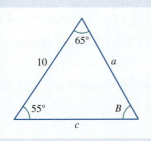

12.

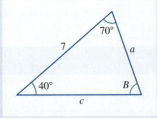

13.

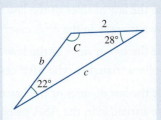

14.

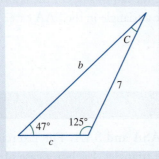

15.

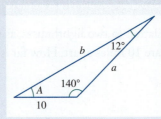

16.

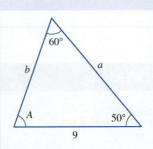

17.

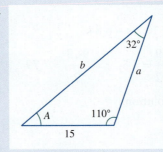

18.

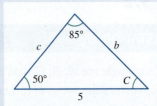

19. In this exercise, report exact answers rather than decimal approximations.

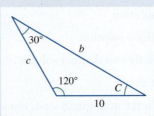

20. In this exercise, report exact answers rather than decimal approximations.

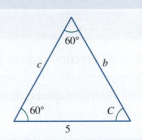

PROBLEMS

Solving triangles: the ambiguous case, SSA. For Exercises 21 through 32, solve the triangles. Each exercise is a case of SSA. Some have one solution, some have two, and some have none.

21.

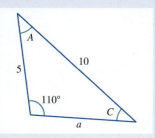

22.

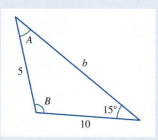

23.

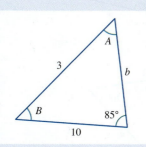

24.

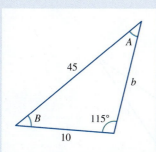

25.

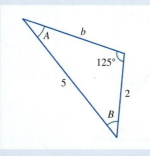

26.

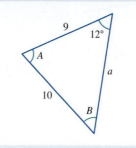

27.

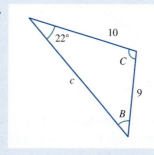

28.

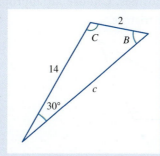

29.

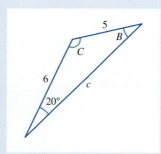

30.

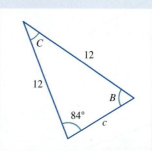

31.

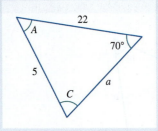

32.

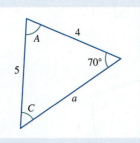

Mixed methods. Solve the triangles in Exercises 33 through 44. Some require the law of sines, and some require the law of cosines. Choose an appropriate method, and solve the triangle.

33.

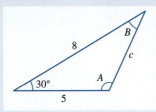

34.

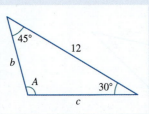

35.

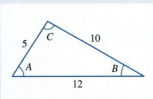

36.

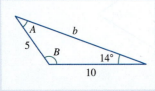

37.

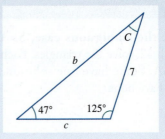

38.

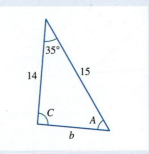

39.

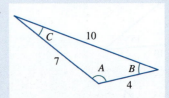

40.

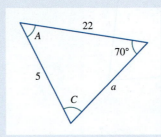

41.

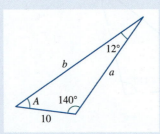

42.

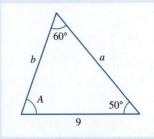

43.

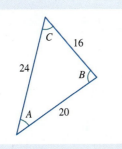

44.

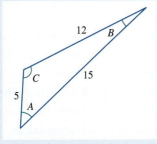

45. A case of SSA. Consider the triangle in **Figure 9.69**.

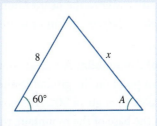

Figure 9.69 Triangle for Exercise 45

a. Show that $\sin A = \dfrac{4\sqrt{3}}{x}$.

b. What is the minimum length of x that will guarantee such a triangle exists? *Suggestion:* What condition on x will ensure that there is a solution for the equation in part a?

MODELS AND APPLICATIONS

46. Finding distance. City 1 is 200 miles from city 2. We draw straight lines from city 1 to city 2, from city 1 to city 3, and from city 2 to city 3. The angle at city 1 is 80°, and the angle at city 2 is 70° (see **Figure 9.70**). Find the distances from city 1 to city 3 and from city 2 to city 3.

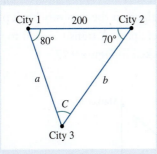

Figure 9.70 Distance between cities

47. Surveying. Two trees T_1 and T_2 lie on one side of a river, and a third tree T_3 is on the other side. A surveyor measures the distance between T_1 and T_2 and finds it to be 80 feet. Using a transit, he finds that the angle at T_1 is 40° and that the angle at T_2 is 55°. Find the distance from T_1 to T_3, c in **Figure 9.71**, and the distance from T_2 to T_3, b in Figure 9.71.

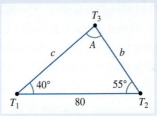

Figure 9.71 Distance between trees

48. An irregular area. Ecologists surveying a location often need to calculate the area of irregular regions. (For example, they may need to find the area of the home range of a marked animal with known multiple locations.) This calculation is done by recording certain lengths and angles associated with the region. An example of such a region is shown in **Figure 9.72**. Lengths are measured in meters.

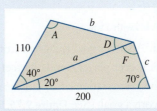

Figure 9.72 Irregular region for Exercise 48

a. Find the unknown sides and angles in Figure 9.72.

b. Find the area of the region in Figure 9.72. Round your answer to the nearest hundred square meters. *Suggestion*: Find the areas of the two triangles.

49. **The height of a mountain.** A transit shows that the line from a point on the ground to the top of a mountain makes an angle of 30°. At a point 1000 feet closer to the base of the mountain, the angle is 40°. (See **Figure 9.73.**) How tall is the mountain? Round your answer to the nearest foot.

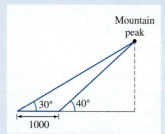

Figure 9.73 Angles from the ground to the mountain peak

50. **A tower.** A hill rises in a straight line and makes an angle of 30° with the horizontal. The base of a vertical tower is 50 feet up the hill from the base. A cable stretching from the base of the hill to the top of the tower makes an angle of 20° with the hill.

a. Use the fact that the angle sum of a triangle is 180° to find the degree measure of the angle A in **Figure 9.74.**

b. Find the degree measure of the angle B in Figure 9.74.

c. Find the height of the tower.

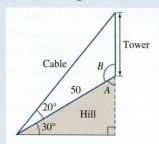

Figure 9.74 A hill with a tower

51. **Scoring a goal.** A soccer goal is 24 feet from left post to right. A soccer player stands 45 feet from the left goal post. A variation of 26° in the angle of the kick will score a goal. (See **Figure 9.75.**) How far is the player from the right goal post? Note that there are two possible solutions. Round your answers to the nearest foot.

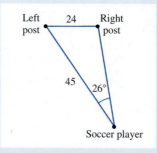

Figure 9.75 A soccer player

52. **A pop fly.** A pop fly ball is hit directly down the third base line. Third base is 90 feet from home plate. The ball reaches the peak of its flight between home and third. At the peak, a line from the ball to home makes an angle of 40° with the ground, and a line from the ball to third base makes an angle of 60° with the ground. (See **Figure 9.76.**) How high above the ground is the ball at the peak of its flight? Round your answer to the nearest foot.

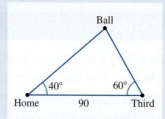

Figure 9.76 Fielding a fly ball

53. **Width of a river.** A surveyor stands on one bank of a river. A helper on the opposite bank places markers 30 feet apart. A transit is used to measure the angles marked in **Figure 9.77.**

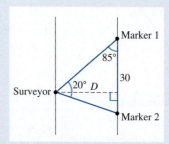

Figure 9.77 A surveyor's measurements

a. Find the distance from the surveyor to Marker 1.

b. Find the distance D, which represents the width of the river. *Suggestion*: The sine of 85° is related to the distance D and the distance from the surveyor to Marker 1. Use your answer to part a.

54. **A service ramp.** A service ramp is 20 feet long and makes an angle of 10° with the horizontal. (See **Figure 9.78**.) It is found to be out of compliance with Americans with Disabilities Act (ADA) regulations and must be replaced by a new ramp that makes an angle of 5° with the horizontal. How long is the new ramp?

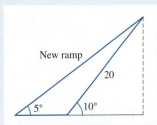

Figure 9.78 Redesigning a service ramp

CHALLENGE EXERCISES FOR INDIVIDUALS OR GROUPS

55. **The obtuse angle case of SSA.** Suppose that in **Figure 9.79** the sides a and c and angle C are known. Our goal is to show that if angle C is obtuse, then there is at most one triangle satisfying these conditions.

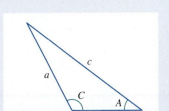

Figure 9.79 The obtuse angle case of SSA

a. Show that $\sin A = a\dfrac{\sin C}{c}$.

b. If C is obtuse as the figure indicates, then the angle A must be acute. With this in mind, how many solutions for A of the equation in part a occur in the interval $(0°, 90°)$?

c. Use the results of part b to show that there is at most one triangle with given sides a and c and given obtuse angle C.

56. **A discovery.**

a. Plot the graph of the circle $x^2 + y^2 = 1$.

b. Add to the graph from part a the triangle with vertices at $\left(-\dfrac{1}{2}, \dfrac{\sqrt{3}}{2}\right)$, $\left(-\dfrac{1}{2}, -\dfrac{\sqrt{3}}{2}\right)$, and $(1, 0)$.

c. State in words the relationship between the triangle and the circle.

d. Show that the triangle is equilateral.

e. Calculate the length of a side of the triangle divided by the sine of the opposite angle.

f. How does the number from part e compare with the diameter of the circle? For more information, see Appendix 5.

REVIEW AND REFRESH: Exercises from Previous Sections

57. **From Section P.2:** Solve the inequality $|x + 2| < 3$. Report your answer using interval notation.

Answer: $(-5, 1)$

58. **From Section 3.2:** Find an exponential function $f(x)$ such that $f(1) = 6$ and $f(2) = 12$.

Answer: $f(x) = 3 \times 2^x$

59. **From Section 4.1:** Find the exact value of $\log_2 16$.

Answer: 4

60. **From Section 4.3:** Solve the equation $\ln(1 + e^x) = 2$.

Answer: $x = \ln(e^2 - 1)$

61. **From Section 5.3:** Find the zeros of $P(x) = x^3 - 3x^2 - 4x$.

Answer: $x = 0$, $x = 4$, $x = -1$

62. From Chapter 6: If $\cos t = \dfrac{1}{3}$ and $0 \le t \le \dfrac{\pi}{2}$, find the value of $\sin t$.

Answer: $\dfrac{2\sqrt{2}}{3}$

63. From Chapter 6: Find the value of $\tan\left(\dfrac{\pi}{6}\right)$.

Answer: $\dfrac{1}{\sqrt{3}}$

64. From Section 7.1: Plot the graphs of $y = \sin x$ and $y = \cos x$ on the same coordinate axes.

Answer: The red graph is for sine, and the blue graph is for cosine.

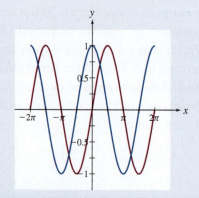

65. From Section 8.1: If $0 < t < \dfrac{\pi}{2}$, express $\tan t$ in terms of $\sin t$.

Answer: $\tan t = \dfrac{\sin t}{\sqrt{1 - \sin^2 t}}$

9.3 Vectors

9.1 Law of Cosines

9.2 Law of Sines

9.3 Vectors

9.4 Vectors in the Plane and in Three Dimensions

> Situations that involve more than one feature are often described using vectors.

In this section, you will learn to:

1. Define *vector*.
2. Use scalar multiplication to adjust the magnitude of a vector.
3. Calculate the sum of two vectors.
4. Calculate the components of a given vector in the directions of two given perpendicular vectors.
5. Solve applied problems using vectors.

Sometimes we can appropriately describe physical phenomena with a single quantity—a rock weighs 10 pounds, an airplane can fly 350 miles per hour, or the wind is blowing at 35 miles per hour. But in other situations, we need both a quantity and a direction: A rock leans against a wall, exerting a force of 10 pounds in the horizontal direction, the airplane is flying at 350 miles per hour in a westerly direction, or the wind is blowing at 35 miles per hour from the north. In such situations, we can use the concept of a **vector**, which incorporates both magnitude and direction into a single entity.

A **vector** consists of a magnitude and a direction.

Vectors may combine any number of separate features into a single entity. For example, there are two basic types of images that appear on your computer screen. Picture files such as those in JPEG format are made up of many individual pixels and fall into the general category of raster graphics. But many figures, such as those produced by popular drawing packages, rely on vector graphics. Vector graphics are made up of individual paths that are vector quantities. A path may consist of a line segment with starting and ending points, a thickness, and a color. Or it may be a circle with center, radius, and various fill properties. All of the individual features of a path are combined into a single vector. These vectors can be combined to construct complex figures. A key feature of vector graphics is its scalability. If you blow up a raster image, it may appear grainy. The only fix is to put more pixels into the original image. But vector graphics do not lose quality when scaled to larger sizes.

Overview of Vectors

> **Vectors are represented geometrically.**

We commonly represent a vector using an arrow, as shown in **Figure 9.80**. The length of the arrow represents the magnitude, and the arrowhead indicates the direction. A longer arrow indicates a larger magnitude, and we will often use the terms *magnitude* and *length* interchangeably. We can envision this geometric representation of a vector as being located anywhere, so all the arrows in **Figure 9.81** represent the same vector.

Figure 9.80 Typical vectors: Longer vectors indicate greater magnitude.

Figure 9.81 Arrows that have the same length and direction: All such arrows represent the same vector.

When we are talking about vectors, we often use the term **scalar** to to differentiate between numbers and vectors.

A **scalar** is a real number.

There is nothing new to learn about scalars—they are just numbers as usual. It is common to indicate vectors using boldface type, such as $\mathbf{v}$, to distinguish them from scalars. We use $\|\mathbf{v}\|$ to indicate the magnitude of $\mathbf{v}$.

Scalar Multiplication

> **Multiplication of a vector by a scalar changes the length and, in the case of negative scalars, reverses the direction.**

To indicate a wind of 35 miles per hour blowing toward the north, we would use a vector $\mathbf{v}$ of length 35 pointing north. If the wind speed increases to 70 miles per hour but continues to blow toward the north, we would use an arrow in the same direction but with twice the magnitude (length). It is natural to think of this doubled vector as $2\mathbf{v}$. It is also natural to indicate a reversal in wind direction using $-\mathbf{v}$. **Scalar multiplication** is the multiplication of a vector by a real number.

The magnitude of $\mathbf{v}$ is multiplied by the absolute value of the scalar a, so

$$\|a\mathbf{v}\| = |a| \|\mathbf{v}\|$$

Scalar multiplication of a vector $\mathbf{v}$ by a scalar a is the vector $a\mathbf{v}$ whose magnitude is $|a|$ times the magnitude of $\mathbf{v}$ and whose direction is the same as $\mathbf{v}$ if a is positive and opposite if a is negative.

Several scalar multiples are shown in **Figure 9.82**.

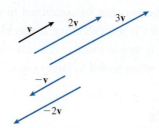

Figure 9.82 Some scalar multiples

A careful look at the definition of scalar multiplication shows that we have left out the case that $a = 0$. For this case, we have a special vector, the zero vector, indicated by **0**. The zero vector has a magnitude of 0, and it turns out to be convenient to say that the zero vector points in all directions. With this in mind,

$$0\mathbf{v} = \mathbf{0}$$

EXAMPLE 9.16 Unit Vectors

A unit vector is a vector of length 1. Suppose **v** is a vector of magnitude 7.

a. Find a unit vector with the same direction as **v**.

b. Find a unit vector in the direction opposite to **v**.

c. Find a vector of length 10 with the same direction as **v**.

SOLUTION

a. A unit vector in the same direction as **v** is a scalar multiple of **v** that has length 1. That is, its length is one-seventh the length of **v**. The vector $\mathbf{w} = \dfrac{1}{7}\mathbf{v}$ is in the same direction as **v**, and we can verify that it is indeed a unit vector:

$$\| \mathbf{w} \| = \left\| \frac{1}{7}\mathbf{v} \right\| = \frac{1}{7}\| \mathbf{v} \| = \frac{1}{7} \times 7 = 1$$

b. The vector we need has the same length, 1, as the unit vector **w** from part a, but it points in the opposite direction. A negative sign reverses the direction of a vector, so the required vector is

$$-\mathbf{w} = -\frac{1}{7}\mathbf{v}$$

c. To make a vector in the same direction as **v** but with length 10, we multiply the unit vector **w** from part a by 10. The result is

$$10\mathbf{w} = \frac{10}{7}\mathbf{v}$$

The vectors $\mathbf{v}, \dfrac{1}{7}\mathbf{v}$, and $-\dfrac{1}{7}\mathbf{v}$ are shown in **Figure 9.83**.

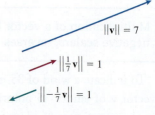

Figure 9.83 $\mathbf{v}, \dfrac{1}{7}\mathbf{v}$, and $-\dfrac{1}{7}\mathbf{v}$

TRY IT YOURSELF 9.16 Brief answers provided at the end of the section.

For the vector **v** in the example, find a vector of length 5 in the direction opposite to **v**.

Vector Addition

> **Vectors are added using parallelograms.**

Figure 9.84 illustrates the method for adding vectors v and w. We use the two given vectors to make a parallelogram, and the sum is the diagonal between them. In applications, this is how forces represented by vectors add.

There is an alternative way of thinking about the sum of vectors. If we put the base (the initial point) of one vector at the tip (the terminal point) of the other, then we get the sum by going from the base of the first to the tip of the second. This way is shown in **Figure 9.85**. We can use this idea repeatedly to see how to get the sum of several vectors, as shown in **Figure 9.86**. If we string vectors together base to tip, then the sum is the arrow going from the base of the first to the tip of the last.

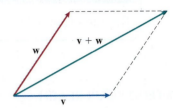

Figure 9.84 Adding vectors: The sum of two vectors is the diagonal of the parallelogram.

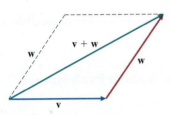

Figure 9.85 An alternative way of thinking of vector addition

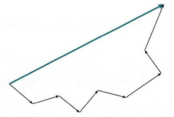

Figure 9.86 The sum of several vectors

Let's look again at Figure 9.85. If we relabel the vectors as $\mathbf{x} = \mathbf{v} + \mathbf{w}$ and $\mathbf{y} = \mathbf{v}$, then $\mathbf{x} - \mathbf{y} = \mathbf{w}$. This picture with new labels is shown in **Figure 9.87**. We see that if two vectors share a common base, then we get their difference by connecting the tips of the arrows. Note that the resulting arrow points away from the tip of the vector that is being subtracted.

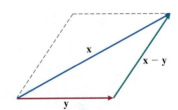

Figure 9.87 Relabeling vectors in Figure 9.85 to get the difference of vectors

---

EXAMPLE 9.17 Combinations of Vectors

Figure 9.88 shows various vectors. Refer to the figure to match the vectors in the left column with the appropriate vector in the right column.

I. $\mathbf{A} + \mathbf{B}$	**i.** **D**
II. $\mathbf{B} - \mathbf{A}$	**ii.** **F**
III. $2\mathbf{A} + \mathbf{B}$	**iii.** **C**

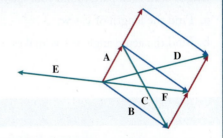

Figure 9.88 Various vectors

SOLUTION

The vector $\mathbf{A} + \mathbf{B}$ is the diagonal of the parallelogram determined by $\mathbf{A}$ and $\mathbf{B}$. That is the vector $\mathbf{F}$. Thus, I. matches with ii.

The difference vector $\mathbf{B} - \mathbf{A}$ begins at the tip of $\mathbf{A}$ and ends at the tip of $\mathbf{B}$. This is the vector $\mathbf{C}$. We conclude that II. matches with iii.

To find $2\mathbf{A} + \mathbf{B}$, we look for the diagonal of the parallelogram determined by the vector $\mathbf{B}$ and a vector twice as long as $\mathbf{A}$. This is the vector $\mathbf{D}$, so III. matches with i.

TRY IT YOURSELF 9.17 Brief answers provided at the end of the section.

Refer to Figure 9.88 to express the vector $\mathbf{E}$ in terms of the vectors $\mathbf{A}$ and $\mathbf{B}$.

EXTEND YOUR REACH

a. Suppose $\mathbf{v}$ and $\mathbf{w}$ are vectors that point in the same direction. Describe the vector $\mathbf{v} + \mathbf{w}$. That is, give its length and direction. *Suggestion*: You might first examine the case where the two vectors point in nearly the same direction.

b. Solve the same problem if the two vectors point in opposite directions.

To calculate sums of vectors, we need some basic properties of parallelograms.

CONCEPTS TO REMEMBER: Facts About Parallelograms

- Opposite sides are congruent.
- Diagonally opposite angles are congruent.
- Two angles sharing a common side add to $180°$.

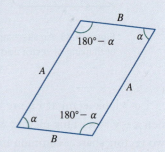

EXAMPLE 9.18 Adding Vectors

The vector $\mathbf{v}$ in **Figure 9.89** has length 4, and the vector $\mathbf{w}$ in Figure 9.89 has length 6. In each part, round your answer to two decimal places.

a. Find the length of $\mathbf{v} + \mathbf{w}$.

b. Find the acute angle $\mathbf{v} + \mathbf{w}$ makes with the vector $\mathbf{w}$.

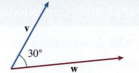

Figure 9.89 The vectors $\mathbf{v}$ and $\mathbf{w}$

SOLUTION

a. To find $\mathbf{v} + \mathbf{w}$, we must complete Figure 9.89 to make the parallelogram shown in **Figure 9.90**. The facts about parallelograms noted earlier tell us that

$$\beta = 180° - 30° = 150°$$

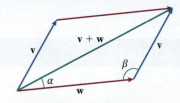

Figure 9.90 Completing to a parallelogram

Recall the law of cosines: If C is the angle between sides of lengths a and b, and side c is opposite C, then $c^2 = a^2 + b^2 - 2ab \cos C$. We apply the law of cosines to the lower triangle in Figure 9.90 to find the length $\|\mathbf{v} + \mathbf{w}\|$ of $\mathbf{v} + \mathbf{w}$:

$$\|\mathbf{v} + \mathbf{w}\|^2 = \|\mathbf{v}\|^2 + \|\mathbf{w}\|^2 - 2\|\mathbf{v}\|\,\|\mathbf{w}\| \cos 150° \quad \blacktriangleleft \textbf{Apply the law of cosines.}$$

$$\|\mathbf{v} + \mathbf{w}\|^2 = 4^2 + 6^2 - 2 \times 4 \times 6 \times \left(-\frac{\sqrt{3}}{2}\right) \quad \blacktriangleleft \textbf{Substitute known values.}$$

$$\|\mathbf{v} + \mathbf{w}\| = \sqrt{4^2 + 6^2 - 2 \times 4 \times 6 \times \left(-\frac{\sqrt{3}}{2}\right)} \quad \blacktriangleleft \textbf{Take square roots.}$$

$$\|\mathbf{v} + \mathbf{w}\| \approx 9.67 \quad \blacktriangleleft \textbf{Simplify.}$$

Thus, the length of $\mathbf{v} + \mathbf{w}$ is approximately 9.67.

b. We need to find the angle α in Figure 9.90. Recall the law of sines: If angle A is opposite side a and angle B is opposite side b, then $\dfrac{a}{\sin A} = \dfrac{b}{\sin B}$. Applying this law gives

$$\frac{\sin \alpha}{\|\mathbf{v}\|} = \frac{\sin \beta}{\|\mathbf{v} + \mathbf{w}\|} \quad \blacktriangleleft \textbf{Apply the law of sines.}$$

$$\frac{\sin \alpha}{4} \approx \frac{\sin 150°}{9.67} \quad \blacktriangleleft \textbf{Substitute known values.}$$

$$\sin \alpha \approx 4\frac{\sin 150°}{9.67} \quad \blacktriangleleft \textbf{Solve for } \sin \alpha.$$

$$\sin \alpha \approx 0.207 \quad \blacktriangleleft \textbf{Simplify.}$$

Because the angle α is less than 30°, it is acute. Thus, we can use the inverse sine function to solve for α: $\alpha \approx \arcsin 0.207 \approx 11.95°$. Note that another approach to calculating α is to use the alternative form of the law of cosines.

TRY IT YOURSELF 9.18 Brief answers provided at the end of the section.

The vector $\mathbf{v}$ in **Figure 9.91** has length 5, and the vector $\mathbf{w}$ has length 9. Find the magnitude of $\mathbf{v} + \mathbf{w}$ and the acute angle it makes with the vector $\mathbf{v}$.

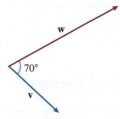

Figure 9.91 Two vectors and the angle between

EXAMPLE 9.19 Subtracting Vectors

The vector **v** in **Figure 9.92** has length 4, and the vector **w** has length 5. Find the magnitude of the difference $\|\,\mathbf{w} - \mathbf{v}\,\|$ and the acute angle **w** − **v** makes with the vector **w**.

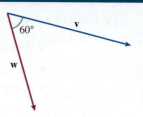

Figure 9.92 The vectors **v** and **w**

SOLUTION

The difference vector **w** − **v** starts at the tip of **v** and ends at the tip of **w**, as shown in **Figure 9.93**. The acute angle that **w** − **v** makes with **w** is marked α in Figure 9.93.

As in the previous example, we find the required magnitude and direction by solving the triangle in Figure 9.93. The result using the law of cosines and then the law of sines is

$$\|\,\mathbf{w} - \mathbf{v}\,\| \approx 4.58$$

$$\alpha \approx 49.11°$$

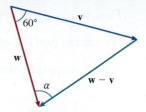

Figure 9.93 The difference vector **w** − **v** added

TRY IT YOURSELF 9.19 Brief answers provided at the end of the section.

Find the length and direction of **w** − **v** in Figure 9.92 if the length of **w** is changed from 5 to 6.

CONCEPTS TO REMEMBER: Properties of Scalar Multiplication and Vector Addition

Addition and scalar multiplication of vectors satisfy some basic and quite unsurprising properties:

- $(a \pm b)\mathbf{v} = a\mathbf{v} \pm b\mathbf{v}$
- $a(\mathbf{v} \pm \mathbf{w}) = a\mathbf{v} \pm a\mathbf{w}$
- $a(b\mathbf{v}) = (ab)\mathbf{v}$
- $1\mathbf{v} = \mathbf{v}$ and $(-1)\mathbf{v} = -\mathbf{v}$
- $0\mathbf{v} = 0$

Components of Vectors

> A vector can be written as a sum of vectors in specified directions.

Sometimes we wish to reverse the process of adding vectors. That is, we want to find particular vectors that add up to a given vector. There are lots of ways to do this, but if we choose directions that are perpendicular to each other, then there is only one way, and the solution is particularly elegant. For example, suppose we wish to write the vector **v** in **Figure 9.94** as the sum of a vector in the direction of **x**, the component in the **x**-direction, and a vector in the direction of **y**, the component in the **y**-direction. (These component vectors are in blue and red in **Figure 9.95**.) The directions of the component vectors are predetermined, so we need only find the magnitudes. Applying some basic trigonometry to the lower right triangle in Figure 9.95, we see that the length of the component in the x-direction is $F \cos \theta$. Similarly, the length of the component in the y-direction is $F \sin \theta$.

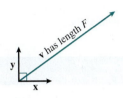

Figure 9.94 Writing the vector **v** as a sum of vectors in the directions of **x** and **y**, which are perpendicular vectors

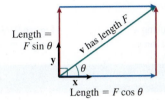

Figure 9.95 Using the sine and cosine to find magnitudes

CONCEPTS TO REMEMBER: Component Vectors

If **x** and **y** are nonzero vectors with **x** perpendicular to **y**, and if **v** has magnitude F and makes an angle of θ with **x**, then

- **v** is the sum of the component of **v** in the **x**-direction and the component of **v** in the **y**-direction.

- The magnitudes of the components are

  Magnitude of component in the **x**-direction $= F \cos \theta$

  Magnitude of component in the **y**-direction $= F \sin \theta$

EXAMPLE 9.20 A Ramp

A 200-pound weight is to be dragged by a cable up a 20° ramp, as shown in **Figure 9.96**. We want to know how hard to pull on the cable to move the weight up the ramp (if friction is ignored). That is, we need to know the magnitude of the component of the weight vector in the direction opposite to the cable.

SOLUTION

The force in the direction opposite to the cable is the component of the weight **w** in the direction of the vector **x** marked in **Figure 9.97**. Because the angle sum of the right-hand right triangle in Figure 9.97 is 180°, the angle α measures $180° - 90° - 20° = 70°$. We find the magnitude of the component of **w** in the direction of **x** as follows:

$$\text{Magnitude} = \| \mathbf{w} \| \cos \alpha \qquad \blacktriangleleft \text{Use the formula for component magnitude.}$$

$$= 200 \cos 70° \qquad \blacktriangleleft \text{Substitute known values.}$$

$$\approx 68.40 \qquad \blacktriangleleft \text{Approximate using a calculator.}$$

Thus, we need to pull with a force of 68.40 pounds on the cable.

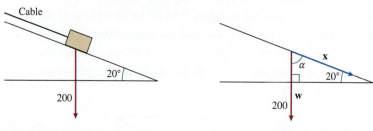

Figure 9.96 A cable pulling a 200-pound weight up a ramp

Figure 9.97 Components of the weight vector

TRY IT YOURSELF 9.20 Brief answers provided at the end of the section.

How hard do we need to pull if the ramp angle is elevated from 20° to 30°?

MODELS AND APPLICATIONS An Airplane

EXAMPLE 9.21 An Airplane

An airplane is flying with an airspeed of 350 miles per hour on a heading of 33° north of east. This information describes a velocity vector. The wind begins blowing at 75 miles per hour in an easterly direction, giving a wind vector. These vectors are shown in **Figure 9.98**. The sum of these two vectors gives a vector whose magnitude and direction determine the resultant velocity of the airplane. Find the resultant speed and direction (that is, the velocity) of the airplane.

SOLUTION

The velocity and wind vectors are shown in Figure 9.98. The resultant speed and direction of the airplane are given by the sum of these two vectors, as shown in **Figure 9.99**. One of the preceding facts about parallelograms tells us that

$$\alpha = 180° - 33° = 147°$$

We get the information we need by solving the lower triangle in Figure 9.99. Using the law of cosines, we find the resultant speed S:

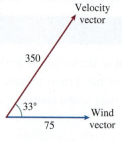

Figure 9.98 The airplane and the wind

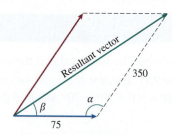

Figure 9.99 The airplane and the wind: The resultant speed and direction are the sum of the wind and velocity vectors.

$$S^2 = 75^2 + 350^2 - 2 \times 75 \times 350 \cos 147°$$ ◀ Apply the law of cosines.

$$S = \sqrt{75^2 + 350^2 - 2 \times 75 \times 350 \cos 147°}$$ ◀ Take square roots.

$$S \approx 414.92$$ ◀ Approximate using a calculator.

Thus, the resultant speed of the airplane is about 414.92 miles per hour.

Next we use the law of sines to get the angle β in Figure 9.99:

$$\frac{\sin \beta}{350} = \frac{\sin \alpha}{S}$$ ◀ Apply the law of sines.

$$\sin \beta = 350 \frac{\sin \alpha}{S}$$ ◀ Solve for $\sin \beta$.

$$\sin \beta \approx 350 \frac{\sin 147°}{414.92}$$ ◀ Substitute known values.

$$\sin \beta \approx 0.459$$ ◀ Approximate using a calculator.

Because β is acute, we can use the inverse sine to find its value:

$$\beta \approx \arcsin 0.459 \approx 27.32°$$

The result is that the airplane flies at a speed of 414.92 miles per hour at a heading of 27.32° north of east.

TRY IT YOURSELF 9.21 Brief answers provided at the end of the section.

The plane changes direction to 40° north of east, and the airspeed decreases to 330 miles per hour. Find the speed and direction of the airplane under these new conditions.

Forces have a magnitude and direction, so they are described by vectors. They combine via vector addition.

EXAMPLE 9.22 Dragging a Rock

The two chains shown in **Figure 9.100** are attached to a heavy rock. The chains make a 45° angle. A force of magnitude 100 pounds is applied to chain number 1, and a force of magnitude 150 pounds is applied to chain number 2. These forces are represented by the vectors in Figure 9.100. The resultant force is represented by the sum of these two vectors. Find the magnitude and direction (relative to chain 1) of the resultant force.

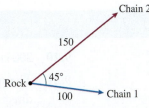

Figure 9.100 Two chains pulling on a rock

SOLUTION

To find α, we use the fact that angles that share a common side of a parallelogram add to 180°: $\alpha = 180° - 45° = 135°$. Now we apply the law of cosines to the lower triangle in **Figure 9.101** to find the magnitude F of the force:

$$F = \sqrt{100^2 + 150^2 - 2 \times 100 \times 150 \cos 135°}$$

$$\approx 231.76 \text{ pounds}$$

We use the law of sines to find the angle β:

$$\sin \beta \approx 150 \times \frac{\sin 135°}{231.76}$$

$$\sin \beta \approx 0.458$$

$$\beta \approx 27.26°$$

Thus, the resultant force has a magnitude of 231.76 pounds and makes an angle of 27.26° with chain number 1.

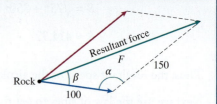

Figure 9.101 Two chains pulling on a rock: The resultant force is the vector sum.

TRY IT YOURSELF 9.22 Brief answers provided at the end of the section.

Find the resultant force on the rock if we pull twice as hard on chain number 1.

EXTEND YOUR REACH

Four forces important to the flight of an airplane are typically referred to as thrust, drag, lift, and weight.

a. Match these forces with the arrows shown in **Figure 9.102**.

b. If the airplane is in level flight at a constant speed, how is lift related to weight, and how is thrust related to drag?

c. One of these forces is typically named "weight." Is weight actually a force? *Suggestion*: Think of the components that make up a force vector.

Forces of Flight

Figure 9.102 Four forces important to flight

TRY IT YOURSELF ANSWERS

9.16 $-\dfrac{5}{7}\mathbf{v}$

9.17 $\mathbf{E} = -\mathbf{A} - \mathbf{B}$

9.18 $\| \mathbf{v} + \mathbf{w} \| \approx 11.70$. The angle is 46.30°.

9.19 $\| \mathbf{w} - \mathbf{v} \| \approx 5.29$, $\alpha \approx 40.92°$

9.20 $200 \cos 60° = 100$ pounds

9.21 The speed is about 390.44 miles per hour, and the direction is about 32.89° north of east.

9.22 The magnitude of the resultant force is 323.92 pounds. The force makes an angle of about 19.09° with chain 1.

EXERCISE SET 9.3

1. In the context of vectors, real numbers are called _____.

2. In applications, vectors may represent:

   a. force

   b. speed and direction

   c. downward weight

   d. All of the above.

3. The sum of two vectors is the _____ of the parallelogram determined by the vectors.

4. A vector of length 1 is called a _____ vector.

5. If a vector is multiplied by a scalar a, what is the effect on the magnitude of the vector?

6. The difference $\mathbf{v} - \mathbf{w}$ is the vector:

   a. emanating from the tip of $\mathbf{v}$ and ending at the tip of $\mathbf{w}$

   b. emanating from the tip of $\mathbf{w}$ and ending at the tip of $\mathbf{v}$

   c. emanating from the base of $\mathbf{v}$ and ending at the tip of $\mathbf{w}$

   d. that is the diagonal of the parallelogram determined by $\mathbf{v}$ and $\mathbf{w}$

7. **True or false:** A scalar multiple of a vector $\mathbf{v}$ always points in the same direction as $\mathbf{v}$.

8. The sum of a vector and a scalar is:

   a. a vector

   b. a scalar

   c. not defined

   d. None of the above.

9. The product of a vector and a scalar is:

   a. a scalar

   b. a vector

   c. not defined

   d. None of the above.

10. When we write a vector as the sum of components in two directions, it is best if the directions are _____.

Magnitude. A vector $\mathbf{v}$ of magnitude 4 is used in Exercises 11 through 17.

11. Find a unit vector in the same direction as $\mathbf{v}$.

12. Find a unit vector in the direction opposite to $\mathbf{v}$.

13. Find a vector of length 7 in the same direction as $\mathbf{v}$.

14. Find a vector of length 3 in the direction opposite to $\mathbf{v}$.

15. Find the length of $2\mathbf{v}$.

16. Find the length of $-6\mathbf{v}$.

17. Is it possible to find a vector of magnitude -1 in the same direction as $\mathbf{v}$?

Matching vectors. Various vectors are shown in **Figure 9.103**. In Exercises 18 through 23, match the given vector with an appropriate vector from the figure.

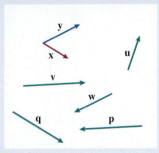

Figure 9.103 Various vectors

18. $x + y$

19. $-y$

20. $y - x$

21. $-x - y$

22. $2x$

23. $\dfrac{1}{2}(2x + 2y)$

Representing vectors. In Exercises 24 through 30, add a sketch of the given vector to **Figure 9.104**.

Figure 9.104 Figure for Exercises 24 through 30

24. $-2\mathbf{v}$

25. $\mathbf{w} - \mathbf{v}$

26. $\dfrac{1}{2}(\mathbf{v} - \mathbf{w})$

27. $2\mathbf{v} + 3\mathbf{w}$

28. $2\mathbf{v} - \mathbf{w}$

29. $-\mathbf{v} - \mathbf{w}$

30. $-\dfrac{1}{2}\mathbf{v} - \dfrac{1}{2}\mathbf{w}$

An octagon. Figure 9.105 shows an octagon. Some of its sides and diagonals are represented by vectors. In Exercises 31 through 34, express the given vector in terms of the vectors $\mathbf{u}$, $\mathbf{v}$, $\mathbf{w}$, and $\mathbf{z}$.

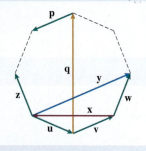

Figure 9.105 Vectors associated with an octagon

31. $\mathbf{p}$

32. $\mathbf{x}$

33. $\mathbf{y}$

34. $\mathbf{q}$

Calculating vector sums. Exercises 35 through 42 refer to the vectors in **Figure 9.106**. In each exercise, give the length of the required vector, and indicate the direction by giving the acute angle β with $\mathbf{w}$. When calculating angles between vectors, be sure they share a common base. Report all answers correct to two decimal places. These exercises require decimal approximations, so different methods of finding a solution may lead to slightly different answers (which is perfectly acceptable).

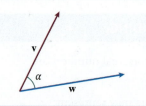

Figure 9.106 Two vectors and the angle between them

35. Find the length and direction of $\mathbf{v} + \mathbf{w}$ if $\|\mathbf{v}\| = 2$, $\|\mathbf{w}\| = 3$, and $\alpha = 20°$.

36. Find the length and direction of $\mathbf{v} + \mathbf{w}$ if $\|\mathbf{v}\| = 1$, $\|\mathbf{w}\| = 2$, and $\alpha = 30°$.

37. Find the length and direction of $\mathbf{v} + \mathbf{w}$ if $\|\mathbf{v}\| = 2$, $\|\mathbf{w}\| = 3$, and $\alpha = \dfrac{\pi}{6}$ radian.

38. Find the length and direction of $\mathbf{v} + \mathbf{w}$ if $\|\mathbf{v}\| = 1$, $\|\mathbf{w}\| = 2$, and $\alpha = \dfrac{\pi}{3}$ radians.

39. Find the length and direction of $\mathbf{w} - \mathbf{v}$ if $\|\mathbf{v}\| = 3$, $\|\mathbf{w}\| = 4$, and $\alpha = 50°$.

40. Find the length and direction of $\mathbf{w} - \mathbf{v}$ if $\|\mathbf{v}\| = 3$, $\|\mathbf{w}\| = 4$, and $\alpha = \dfrac{\pi}{4}$ radian.

41. Find the length and direction of $2\mathbf{v} + 3\mathbf{w}$ if $\|\mathbf{v}\| = 2$, $\|\mathbf{w}\| = 4$, and $\alpha = \dfrac{\pi}{6}$ radian.

42. Find the length and direction of $\mathbf{w} - 2\mathbf{v}$ if $\|\mathbf{v}\| = 3$, $\|\mathbf{w}\| = 2$, and $\alpha = 80°$.

Components. Exercises 43 through 46 refer to the vectors in **Figure 9.107**. In each exercise, find the magnitude of the component of $\mathbf{v}$ in the $\mathbf{x}$- and the $\mathbf{y}$-direction. Express your answers in terms of the sine and cosine, and then give an approximation to two decimal places.

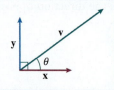

Figure 9.107 Components of the vector $\mathbf{v}$

43. $\|\mathbf{v}\| = 2$ and $\theta = 35°$

44. $\|\mathbf{v}\| = 3$ and $\theta = 55°$

45. $\|\mathbf{v}\| = 4$ and $\theta = \dfrac{\pi}{4}$ radian

46. $\|\mathbf{v}\| = 5$ and $\theta = \dfrac{\pi}{6}$ radian

PROBLEMS

Properties of scalar multiplication. In Exercises 47 through 49, provide justification for the properties of scalar multiplication in specific cases.

47. It is true that $(a + b)\mathbf{v} = a\mathbf{v} + b\mathbf{v}$. Make a vector diagram to show that $(2 + 3)\mathbf{v} = 2\mathbf{v} + 3\mathbf{v}$.

48. It is true that $a(\mathbf{v} + \mathbf{w}) = a\mathbf{v} + a\mathbf{w}$. Make a vector diagram to show that $2(\mathbf{v} + \mathbf{w}) = 2\mathbf{v} + 2\mathbf{w}$.

49. It is true that $a(b\mathbf{v}) = (ab)\mathbf{v}$. Make a vector diagram to show that $2(3\mathbf{v}) = (2 \times 3)\mathbf{v}$.

50. **Sum of unit vectors.** Is it true that if $\mathbf{v}$ and $\mathbf{w}$ are unit vectors, then $\mathbf{v} + \mathbf{w}$ is a unit vector? Either show this is always true or provide an example to show that it is not always true.

51. **More on the sum of unit vectors.** In the preceding exercise, we leave open the possibility that there may be specific examples where the sum of unit vectors is a unit vector. In fact, it is true only for one specific value of the angle θ shown in **Figure 9.108**. Our goal is to find that angle. Suppose then that $\mathbf{v}$, $\mathbf{w}$, and $\mathbf{v} + \mathbf{w}$ are all unit vectors. Figure 9.108 shows $\mathbf{v}$ and $\mathbf{w}$ and the included angle θ. In **Figure 9.109** we have completed the figure to a parallelogram and shown the lengths of the various sides. Use the fact that Figure 9.109 shows two equilateral triangles to show that $\theta = 120°$.

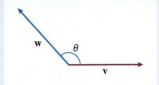

Figure 9.108 Figure for Exercise 51

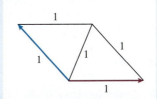

Figure 9.109 The completed parallelogram divided into two equilateral triangles

52. **Opposite directions.** The vectors $\mathbf{v}$ and $\mathbf{w}$ are of equal length but point in opposite directions.

    a. What is the sum $\mathbf{v} + \mathbf{w}$?

    b. If $\mathbf{v}$ and $\mathbf{w}$ represent forces, explain your answer to part a in terms of the net force represented by $\mathbf{v} + \mathbf{w}$.

MODELS AND APPLICATIONS

53. **North and east.** A tug pulls a barge due north. A second tug pulls the barge with the same magnitude due east. In what direction does the barge move?

54. **A drone.** A drone is initially flying on a heading 40° south of west at a speed of 275 knots. (A knot is about 1.15 miles per hour.) A wind begins to blow due west at a speed of 50 knots. Find the resultant speed and direction of the drone. Round your answer for the direction to the nearest degree.

55. **Moving a boulder.** A chain is attached to a boulder and pulled due north by a monster truck with a force of magnitude 2000 newtons. (A newton is about a quarter of a pound.) A second monster truck pulls on a chain pointed 20° west of north with a force of magnitude 3000 newtons. Find the resultant force on the boulder. Round your answer for the angle to the nearest degree.

56. **A ship.** A ship is originally sailing at 12 knots but encounters a 3-knot current heading 15° away from the original heading. (A knot is about 1.15 miles per hour.) By how much do the heading and speed of the ship change when it encounters the current? Round your answer for the angle to the nearest degree.

57. **A falling rock.** The rock shown in **Figure 9.110** is falling in the presence of a horizontal wind. The magnitude of the downward force on the rock is mg, where m is the mass of the rock and g is the acceleration due to gravity. The wind exerts a horizontal force of magnitude F on the rock. Let R denote the magnitude of the resultant force, and let α denote the angle between the resultant force and the vertical.

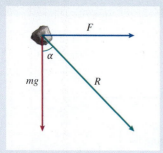

Figure 9.110 A falling rock affected by the wind

a. Show that $R^2 = F^2 + m^2 g^2$.

b. Show that $\alpha = \arctan \dfrac{F}{mg}$.

58. Jumping locust. Zoologists have studied the mechanics of a locust jumping. When a locust jumps, as depicted in **Figure 9.111**, the resultant force **R** exerted by the hind legs is the sum of two forces. The first force **W** acts directly downward to support the locust's weight, which is about 0.02 newton. The second force **F** is for takeoff; its magnitude is about 0.29 newton, and the direction makes an angle of 55° with the horizontal. Find the magnitude of the resultant force **R = F + W**.

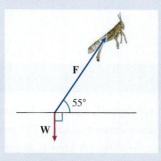

Figure 9.111 A jumping locust

59. Guppies. Guppies use both gravitation and light sensation to control their equilibrium orientation in the water. If **W** is the force due to gravity and **L** is the force caused by light sensation, then the resultant force **R** determines the equilibrium orientation of the guppy. (See **Figure 9.112**.) Show that

$$\frac{\sin \beta}{\sin \alpha} = \frac{\|\mathbf{W}\|}{\|\mathbf{L}\|}$$

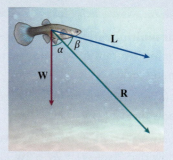

Figure 9.112 The equilibrium orientation of a guppy

60. A drone. A drone is flying at 330 knots on a heading of 10° west of north. (A knot is about 1.15 miles per hour.) Its speed and direction are the result of an original northerly heading and a westerly wind. Use the idea of components of force to find the original northerly speed of the drone on its original heading and the speed of the westerly wind.

61. A gliding bird. A gliding bird must bank to turn—that is, the bird must tilt toward the inside of the turn. This is because the horizontal component of the lift provides the centripetal force that pushes the bird into the path. The diagram in **Figure 9.113** shows a bird that is banking at an angle ϕ to the vertical. The lift force **L** (mainly due to the wings) points perpendicular to the wings. The horizontal component of **L** provides the horizontal centripetal force **F**. The vertical component **W** of **L** has magnitude equal to the weight mg of the bird. (Here m is the mass of the bird, and g is the acceleration due to gravity.)

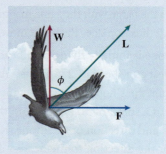

Figure 9.113 A bird in flight

a. Show that $\|\mathbf{L}\| \sin \phi = \|\mathbf{F}\|$ and $\|\mathbf{L}\| \cos \phi = mg$.

b. Find a formula for $\|\mathbf{L}\|$ in terms of m, g, and $\|\mathbf{F}\|$.

c. If a bird's wings can sustain a lift having a magnitude of at most 1 unit and the bird weighs 0.6 unit, use part a to determine how large the banking angle should be. Use degree measure.

62. **More on gliding birds.** This is a continuation of the preceding exercise. From physics we know that the magnitude $||\mathbf{F}||$ of the centripetal force can be expressed in terms of the mass m, the radius r of the turn, and the speed u:

$$||\mathbf{F}|| = \frac{mu^2}{r}$$

   a. Use this expression for $||\mathbf{F}||$, along with the equations in part a of the preceding exercise, to find a formula for $\tan\phi$ in terms of r, u, and g.

   b. Use your formula from part a to answer the following: As the angle of banking increases, what happens to the radius of the turn, assuming the speed is fixed?

63. **Pulling a weight up a ramp.** A 300-pound weight is to be dragged by a cable up a 30° ramp, as shown in **Figure 9.114.** What is the magnitude (ignoring friction) of the force that must be applied to the cable to move the weight up the ramp?

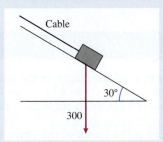

Figure 9.114 A cable pulling a weight

64. **A drifting barge.** A barge is adrift. Two tug boats attach cables to the barge to tow it to safety. (See **Figure 9.115.**) Tug 1 pulls due east at a speed of 2 miles per hour. Tug 2 pulls 25° north of east at a speed of 3 miles per hour. What are the resultant speed and direction of the barge? Round your answer for the direction to the nearest degree.

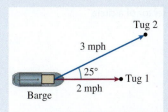

Figure 9.115 A barge pulled by two boats

65. **A drifting barge with a current.** This is a continuation of the preceding exercise. Suppose now there is a current of 1 mile per hour in a southerly direction. Find the speed and direction of the barge under these circumstances. Round your answer for the direction to the nearest degree.

66. **A car driving up a hill.** A 4000-pound car travels up a 5° grade, as shown in **Figure 9.116.** What is the magnitude of the force produced by the engine to move the car up the grade?

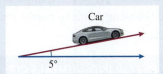

Figure 9.116 Driving up a hill

67. **A swimmer.** A river that is one-quarter of a mile wide is flowing at a speed of 1 mile per hour. A swimmer enters the river and swims directly toward the opposite bank at a speed of 3 miles per hour. The current alters the swimmer's direction and causes her to reach the opposite bank some distance downstream. (See **Figure 9.117.**)

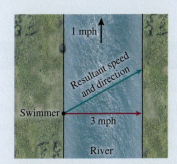

Figure 9.117 Swimming in a river

   a. Find the angle the swimmer's resultant path makes with the line directly across the river.

   b. How far downstream does the swimmer make landfall?

68. **A missile.** A missile is attempting to fly due east, but a wind of 50 knots 30° north of east has caused the resultant speed and direction of the missile to be 340 knots in a direction 4.22° north of east. (For comparison, 1 knot is about 1.15 miles per hour.) What will be the missile's speed if the wind stops? (See **Figure 9.118.**) Round your answer in knots to the nearest whole number.

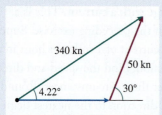

Figure 9.118 The effect of wind on a missile

69. **A storm.** A jet is flying due east across the Atlantic at a speed of 400 miles per hour. A storm brings a southerly wind (blowing south to north) that blows the jet off course by 5°.

a. What is the wind speed?

b. As the storm intensifies, the wind speed doubles. How far off course is the jet now? Round your answer to the nearest degree.

70. **A baseball.** A baseball is hit directly over second base with a velocity of 85 miles per hour. A wind of 20 miles per hour is blowing from third base to home. Describe the resultant speed and direction of the baseball. (Recall that the bases form a square.) Round your answer for the direction to the nearest degree.

CHALLENGE EXERCISES FOR INDIVIDUALS OR GROUPS

71. **Projection of vectors.** If we think of the sun as shining in the direction perpendicular to the vector **x** in **Figure 9.119**, then the projection of the vector **v** onto the vector **x** is the shadow cast by **v** on the vector **x**. Find a formula that gives the magnitude of the projection in terms of the magnitude of **v** and the angle θ. *Suggestion:* A hint is given by the vector added in **Figure 9.120**.

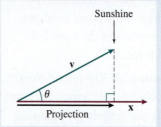

Figure 9.119 Vector projection: The projection of **v** onto **x** is a shadow.

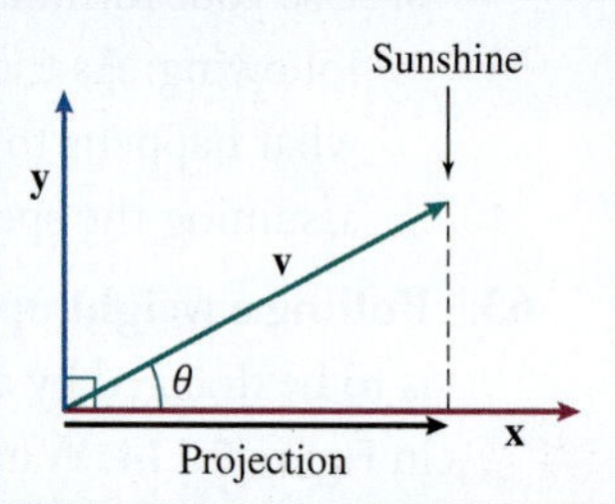

Figure 9.120 Vector projection: Adding a vector gives a hint.

REVIEW AND REFRESH: Exercises from Previous Sections

72. **From Section 3.3:** A savings account has an initial balance of $300. It grows at the constant percentage rate of 5% per year. Find the balance after 10 years.

    Answer: $300 \times 1.05^{10} \approx \$488.67$

73. **From Section 4.1:** What value of x will give a value of 200 for $\ln x$?

    Answer: $x = e^{200}$

74. **From Section 4.2:** Simplify the expression $e^{2 \ln \sqrt{x}}$.

    Answer: x

75. **From Section 5.4:** A polynomial $P(x)$ of degree 7 has the following zeros:

    • 3 is a zero of order 2.

    • 4 is a zero of order 3.

    • 5 is a zero of order 2.

    The leading coefficient of $P(x)$ is 9. Find an expression for $P(x)$.

    Answer: $P(x) = 9(x-3)^2(x-4)^3(x-5)^2$

76. **From Chapter 6:** Find the length of the arc of the circle of radius 4 cut by an angle of $\dfrac{\pi}{3}$ radians whose vertex lies at the center of the circle. (See **Figure 9.121.**)

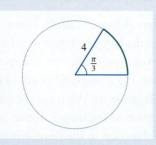

Figure 9.121 A sector of a circle

Answer: $\dfrac{4\pi}{3}$

77. From Chapter 6: Find the lengths x and y shown in **Figure 9.122** in terms of the length r and the angle θ.

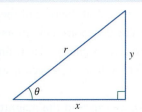

Figure 9.122 A triangle for Exercise 77

Answer: $x = r \cos \theta, \quad y = r \sin \theta$

78. From Section 8.1: Express the cotangent in terms of the secant and cosecant.

Answer: $\cot t = \dfrac{\csc t}{\sec t}$

79. From Section 9.1: Find the length y in terms of the lengths x and r and the angle θ shown in **Figure 9.123**.

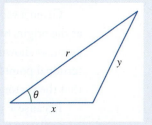

Figure 9.123 A triangle for Exercise 79

Answer: $y = \sqrt{x^2 + r^2 - 2xr \cos \theta}$

9.4   # Vectors in the Plane and in Three Dimensions

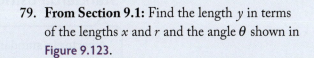

> Locating vectors in two dimensions or in three dimensions allows for the use of coordinates.

9.1   Law of Cosines

9.2   Law of Sines

9.3   Vectors

9.4   Vectors in the Plane and in Three Dimensions

In this section, you will learn to:

1. Calculate magnitudes, scalar multiples, and sums of vectors by using the coordinates of vectors in standard position in the plane.
2. Calculate the dot product of two vectors.
3. Use the dot product to calculate magnitudes and angles between vectors.
4. Calculate magnitudes, scalar multiples, sums, and dot products of three-dimensional vectors.
5. Solve applied problems using coordinates of vectors in the plane.

Vectors emanating from the origin of the plane can be identified with the coordinates of their terminal points. Once this is done, addition of vectors and scalar multiplication of vectors are easy processes. The laws of sines and cosines are no longer necessary, at least for some vector calculations.

Vectors in the plane are represented by two coordinates, so they are thought of as two-dimensional objects. It turns out that we can add as many coordinates as we like while maintaining the fundamental properties of vectors. The result is three-, four-, or higher-dimensional vectors. These are just as easy to use as are two-dimensional vectors, but their visual representation becomes problematic when the dimension is greater than three.

Standard Position in the Plane

> **A vector in standard position emanates from the origin.**

Given a vector in the plane, it is convenient to assume that the base of the vector is at the origin. Such a vector is completely determined by the coordinates of its terminal point, as shown in **Figure 9.124**. Because such a vector is completely determined by its terminal point (a, b), we use the notation $\mathbf{v} = \langle a, b \rangle$ to indicate this vector, and we say that the vector is in standard position.

Because $\mathbf{v} = \langle a, b \rangle$ has its tip at the point (a, b), we can find the magnitude of $\mathbf{v}$ immediately from the distance formula, as shown in **Figure 9.125**:

$$\| \mathbf{v} \| = \sqrt{a^2 + b^2}$$

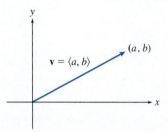

Figure 9.124 Standard position: The vector $\mathbf{v} = \langle a, b \rangle$ in standard position in the plane emanates from the origin and has its tip at the point (a, b).

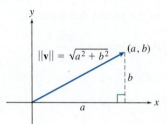

Figure 9.125 The magnitude of $\mathbf{v}$ found from the distance formula

If $\mathbf{v} = \langle a, b \rangle$, let's look at the length of $\langle 3a, 3b \rangle$:

$$\| \langle 3a, 3b \rangle \| = \sqrt{(3a)^2 + (3b)^2} = \sqrt{9(a^2 + b^2)} = 3\sqrt{a^2 + b^2} = 3 \| \langle a, b \rangle \|$$

Also, the direction from the origin to the point (a, b) is the same as the direction from the origin to the point $(3a, 3b)$. Thus, the vector $\langle 3a, 3b \rangle$ has the same direction as the vector $\langle a, b \rangle$ and has three times the length. By the definition of scalar multiples in the preceding section, we conclude that $3\mathbf{v} = \langle 3a, 3b \rangle$. This same fact holds for every scalar a:

$$\text{If } \mathbf{v} = \langle x, y \rangle, \text{ then } a\mathbf{v} = \langle ax, ay \rangle$$

Next we want to see how to add vectors in the plane in terms of the coordinates of their terminal points. To this end, we show vectors $\mathbf{v} = \langle c, d \rangle$ and $\mathbf{w} = \langle a, b \rangle$ in **Figure 9.126**. We have adjoined a right triangle to the vector $\mathbf{v}$ that will help us keep track of coordinates. In completing the parallelogram, we move the base of the vector $\mathbf{v}$ to the tip of $\mathbf{w}$. In **Figure 9.127** we have moved the right triangle along with the vector $\mathbf{v}$. This figure shows that to get from the tip of $\mathbf{w}$ to the tip of $\mathbf{v} + \mathbf{w}$, we add c to the x-coordinate and d to the y-coordinate. The result is that the tip of $\mathbf{v} + \mathbf{w}$ is located at the point $(a + c, b + d)$. That is, $\langle a, b \rangle + \langle c, d \rangle = \langle a + c, b + d \rangle$.

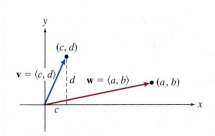

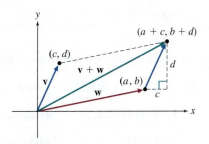

Figure 9.126 Vectors **v** and **w**

Figure 9.127 Tracking the coordinates of the tip of **v** + **w**

CONCEPTS TO REMEMBER: Vectors in the Plane

1. A vector in standard position in the plane has its base at the origin. Such a vector is determined by the coordinates of its terminal point, and we write $\mathbf{v} = \langle a, b \rangle$.

2. If $\mathbf{v} = \langle x, y \rangle$, then

$$\|\mathbf{v}\| = \sqrt{x^2 + y^2}$$

The length of a vector in standard position is the distance from the origin to its tip.

3. If $\mathbf{v} = \langle x, y \rangle$, then

$$a\mathbf{v} = \langle ax, ay \rangle$$

To find the scalar multiple of a vector, we multiply each coordinate of the vector by the scalar.

4. If $\mathbf{v} = \langle a, b \rangle$ and $\mathbf{w} = \langle c, d \rangle$, then

$$\mathbf{v} + \mathbf{w} = \langle a + c, b + d \rangle$$

To add vectors, we add their coordinates.

EXAMPLE 9.23 Calculating with Vectors in the Plane

Let $\mathbf{v} = \langle 3, -5 \rangle$ and $\mathbf{w} = \langle 2, 4 \rangle$.

a. Calculate $\|\mathbf{v}\|$.

b. Find a unit vector in the direction opposite to **v**. (Recall that a unit vector is a vector of length 1.)

c. Calculate $3\mathbf{v} - 2\mathbf{w}$.

SOLUTION

a. To find the magnitude of **v**, we use the formula $\|\langle a, b \rangle\| = \sqrt{a^2 + b^2}$:

$$\|\mathbf{v}\| = \|\langle 3, -5 \rangle\| = \sqrt{3^2 + (-5)^2} = \sqrt{34}$$

b. To find a unit vector in the same direction as **v**, we divide by its magnitude:

$$\frac{1}{\sqrt{34}}\mathbf{v} = \frac{1}{\sqrt{34}}\langle 3, -5 \rangle = \left\langle \frac{3}{\sqrt{34}}, -\frac{5}{\sqrt{34}} \right\rangle$$

The negative of this vector, $\left\langle -\dfrac{3}{\sqrt{34}}, \dfrac{5}{\sqrt{34}} \right\rangle$, is a unit vector in the direction opposite to **v**.

c. Calculations of this sort are particularly easy. We just work coordinate-wise:

$$3\mathbf{v} - 2\mathbf{w} = 3\langle 3, -5 \rangle - 2\langle 2, 4 \rangle \qquad \text{◀ Use vector definitions.}$$

$$= \langle 9, -15 \rangle - \langle 4, 8 \rangle \qquad \text{◀ Multiply coordinates by scalars.}$$

$$= \langle 5, -23 \rangle \qquad \text{◀ Subtract coordinates.}$$

TRY IT YOURSELF 9.23 Brief answers provided at the end of the section.

Calculate $||\mathbf{v} + 2\mathbf{w}||$.

Compare the ease of performing algebraic operations on vectors in standard position in the plane with similar operations in the previous section.

Dot Product

> The dot product focuses on the angle between two vectors.

The **dot product** of $\mathbf{v} = \langle a, b \rangle$ and $\mathbf{w} = \langle c, d \rangle$ is denoted by $\mathbf{v} \cdot \mathbf{w}$ and is defined by $\langle a, b \rangle \cdot \langle c, d \rangle = ac + bd$.

There are no surprises when it comes to addition and scalar multiplication of vectors in the plane, but thus far we have not defined a product for vectors. We do so now, but the **dot product** is not defined as may be expected.

To find the dot product, multiply corresponding coordinates and add. For example,

$$\langle 3, 2 \rangle \cdot \langle 5, -2 \rangle = 3 \times 5 + 2 \times -2 = 11$$

It is very important to note that the dot product of two vectors is a real number, or scalar—not a vector. The importance of the dot product is not apparent from its definition but will emerge as we establish some of its properties.

EXAMPLE 9.24 Calculating Dot Products

a. Calculate the dot product $\langle 2, 5 \rangle \cdot \langle 3, -1 \rangle$.
b. If $\mathbf{v} = \langle a, b \rangle$, show that $\mathbf{v} \cdot \mathbf{v} = ||\mathbf{v}||^2$.

SOLUTION

a. The dot product is the product of the first coordinates plus the product of the second coordinates:

$$\langle 2, 5 \rangle \cdot \langle 3, -1 \rangle = 2 \times 3 + 5 \times (-1) = 1$$

b. This formula comes from direct calculation of the dot product:

$$\mathbf{v} \cdot \mathbf{v} = \langle a, b \rangle \cdot \langle a, b \rangle \qquad \text{◀ Use vector definitions.}$$

$$= a \times a + b \times b \qquad \text{◀ Calculate the dot product.}$$

$$= a^2 + b^2 \qquad \text{◀ Simplify.}$$

$$= ||\mathbf{v}||^2 \qquad \text{◀ Use the definition of magnitude.}$$

Calculate the dot product $\langle 3, 5 \rangle \cdot \langle 2, 0 \rangle$.

To understand more about the dot product, we need a different formula for its calculation. In **Figure 9.128** we show two vectors, $\mathbf{v} = \langle a, b \rangle$ and $\mathbf{w} = \langle c, d \rangle$, and the angle α between them. We calculate the square of the distance marked C in two ways. First we apply the distance formula to the terminal points, (a, b) and (c, d):

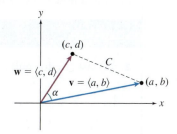

$$C^2 = (a - c)^2 + (b - d)^2 \qquad \blacktriangleleft \text{Apply the distance formula to terminal points.}$$

$$= a^2 - 2ac + c^2 + b^2 - 2bd + d^2 \qquad \blacktriangleleft \text{Expand the squares.}$$

$$= (a^2 + b^2) + (c^2 + d^2) - 2(ac + bd) \qquad \blacktriangleleft \text{Group the terms.}$$

$$= \|\mathbf{v}\|^2 + \|\mathbf{w}\|^2 - 2\mathbf{v} \cdot \mathbf{w} \qquad \blacktriangleleft \text{Express in terms of the dot product.}$$

Figure 9.128 Two vectors and the angle between them

For our second calculation, we note that the triangle in Figure 9.128 has sides of length $\|\mathbf{v}\|$, $\|\mathbf{w}\|$, and C. Then the law of cosines gives

$$C^2 = \|\mathbf{v}\|^2 + \|\mathbf{w}\|^2 - 2\|\mathbf{v}\| \, \|\mathbf{w}\| \cos \alpha$$

Equating these two calculations of C^2, we find an alternative formula for the dot product:

$$\|\mathbf{v}\|^2 + \|\mathbf{w}\|^2 - 2\mathbf{v} \cdot \mathbf{w} = \|\mathbf{v}\|^2 + \|\mathbf{w}\|^2 - 2\|\mathbf{v}\| \, \|\mathbf{w}\| \cos \alpha$$

$$\mathbf{v} \cdot \mathbf{w} = \|\mathbf{v}\| \, \|\mathbf{w}\| \cos \alpha$$

We note that the cosine of an angle α is zero exactly when $\alpha = 90°$ or $\alpha = 270°$ (if we assume $0° \le \alpha < 360°$). Consequently, two nonzero vectors are perpendicular exactly when the cosine of the angle between them is 0. By the alternative formula for the dot product, that is precisely when their dot product is 0.

This is the single most important fact about the dot product. The dot product of two vectors is 0 if and only if the two vectors are perpendicular. (We define the zero vector to be perpendicular to all other vectors.) We note that in the case of vectors it is common to use the term *orthogonal* in place of *perpendicular*.

CONCEPTS TO REMEMBER: The Dot Product

1. If $\mathbf{v} = \langle a, b \rangle$ and $\mathbf{w} = \langle c, d \rangle$, then $\mathbf{v} \cdot \mathbf{w} = ac + bd$.
2. $\mathbf{v} \cdot \mathbf{v} = \|\mathbf{v}\|^2$
3. $\mathbf{v} \cdot \mathbf{w} = \|\mathbf{v}\| \, \|\mathbf{w}\| \cos \alpha$, where α is the angle between the nonzero vectors $\mathbf{v}$ and $\mathbf{w}$.
4. Two vectors are orthogonal (perpendicular) precisely when their dot product is 0.
5. The dot product satisfies some basic algebraic properties:

   - $\mathbf{v} \cdot (\mathbf{z} + \mathbf{w}) = \mathbf{v} \cdot \mathbf{z} + \mathbf{v} \cdot \mathbf{w}$

   - $a(\mathbf{v} \cdot \mathbf{w}) = (a\mathbf{v}) \cdot \mathbf{w} = \mathbf{v} \cdot (a\mathbf{w})$

   - $\mathbf{v} \cdot \mathbf{w} = \mathbf{w} \cdot \mathbf{v}$

EXAMPLE 9.25 Using the Dot Product

Let $\mathbf{v} = \langle 1, 2 \rangle$.

a. Use the dot product to calculate $\| \mathbf{v} \|$.

b. Find a nonzero vector orthogonal to $\mathbf{v}$.

SOLUTION

a. We use the formula $\| \mathbf{v} \|^2 = \mathbf{v} \cdot \mathbf{v}$ to do this calculation:

$$\| \mathbf{v} \|^2 = \langle 1, 2 \rangle \cdot \langle 1, 2 \rangle$$

$$\| \mathbf{v} \|^2 = 1 \times 1 + 2 \times 2 = 5$$

$$\| \mathbf{v} \| = \sqrt{5}$$

b. The vector we seek is a nonzero vector $\mathbf{w} = \langle a, b \rangle$ that is orthogonal to the vector $\langle 1, 2 \rangle$. That means the dot product with $\langle 1, 2 \rangle$ is 0:

$$\mathbf{w} \cdot \langle 1, 2 \rangle = 0$$

$$a + 2b = 0$$

There are lots of correct answers for a and b, but an easy one is to reverse the coordinates and take the negative of one of them—for example, $a = 2$ and $b = -1$. Check that

$$\langle 2, -1 \rangle \cdot \langle 1, 2 \rangle = 0$$

These two vectors are shown in **Figure 9.129**.

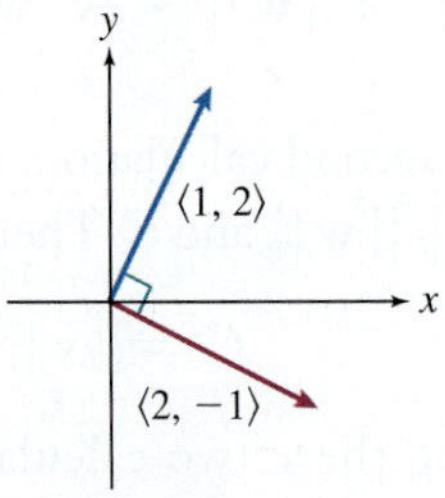

Figure 9.129
Orthogonal vectors:
The vector $\langle 2, -1 \rangle$
is orthogonal to the
vector $\langle 1, 2 \rangle$.

TRY IT YOURSELF 9.25 Brief answers provided at the end of the section.

Find a nonzero vector orthogonal to $\langle 5, -6 \rangle$.

EXTEND YOUR REACH   For any numbers a and b, show how to find a nonzero vector orthogonal to $\langle a, b \rangle$.

EXAMPLE 9.26 The Angle Between Two Vectors

Find the acute angle α shown in **Figure 9.130** between the vectors $\mathbf{v} = \langle 1, 2 \rangle$ and $\mathbf{w} = \langle 2, 3 \rangle$.

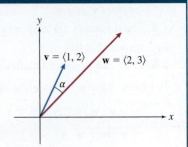

Figure 9.130 The angle between two vectors

SOLUTION

Rearranging the equation $\mathbf{v} \cdot \mathbf{w} = \| \mathbf{v} \| \, \| \mathbf{w} \| \cos \alpha$ gives the cosine of the angle α in terms of the dot product:

$$\cos \alpha = \frac{\mathbf{v} \cdot \mathbf{w}}{\| \mathbf{v} \| \, \| \mathbf{w} \|}$$

We use this equation to find α:

$$\cos \alpha = \frac{1 \times 2 + 2 \times 3}{\sqrt{1^2 + 2^2} \, \sqrt{2^2 + 3^2}}$$

$$\cos \alpha = \frac{8}{\sqrt{5} \, \sqrt{13}}$$

Because α is acute, we can use the inverse cosine to solve this equation:

$$\alpha = \arccos \left(\frac{8}{\sqrt{5} \, \sqrt{13}} \right) \approx 7.13° \text{ or } 0.12 \text{ radian}$$

TRY IT YOURSELF 9.26 Brief answers provided at the end of the section.

Find the acute angle between the vectors $\langle 2, -1 \rangle$ and $\langle 3, 1 \rangle$.

EXTEND YOUR REACH

Let $\mathbf{v}$ and $\mathbf{w}$ be the vectors in Example 9.26.

a. Calculate the length of the vector $\mathbf{v} - \mathbf{w}$.

b. Use the law of cosines together with the lengths of $\mathbf{v}$ and $\mathbf{w}$ to calculate the length of $\mathbf{v} - \mathbf{w}$ as we did in the preceding section.

c. Do your answers to parts a and b agree?

CONCEPTS TO REMEMBER: The Angle Between Two Vectors

For nonzero vectors $\mathbf{v}$ and $\mathbf{w}$ in standard position, the cosine of the angle α between the two vectors is given by

$$\cos \alpha = \frac{\mathbf{v} \cdot \mathbf{w}}{\| \mathbf{v} \| \, \| \mathbf{w} \|}$$

Three-Dimensional Vectors

Vectors in three dimensions share many properties with vectors in the plane.

Because vectors in the plane have two coordinates, they are represented as two-dimensional objects. We say that these vectors are in $\mathbb{R}^2$. We can add a coordinate to make a three-dimensional vector such as $\mathbf{v} = \langle 1, 2, 3 \rangle$. Such a vector is illustrated in Figure 9.131, and we say these vectors are in $\mathbb{R}^3$. In analogy with vectors in the plane, a vector $\mathbf{v} = \langle a, b, c \rangle$ in $\mathbb{R}^3$ is in standard position if it emanates from the origin and has its tip at the point (a, b, c).

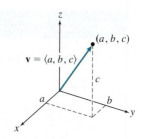

Figure 9.131 A vector with three coordinates shown in three-dimensional space

Addition, scalar multiplication, and the dot product for three-dimensional vectors behave analogously to the corresponding operations for vectors in the plane.

CONCEPTS TO REMEMBER: Three-Dimensional Vectors

Let $\mathbf{v} = \langle a, b, c \rangle$ and $\mathbf{w} = \langle d, e, f \rangle$.

Vector addition:

$$\mathbf{v} + \mathbf{w} = \langle a, b, c \rangle + \langle d, e, f \rangle = \langle a+d, b+e, c+f \rangle$$

For vector addition, we add coordinates.

Scalar multiplication: For a real number r,

$$r\mathbf{v} = r\langle a, b, c \rangle = \langle ra, rb, rc \rangle$$

For scalar multiplication, we multiply each coordinate by the scalar.

Dot product:

$$\mathbf{v} \cdot \mathbf{w} = \langle a, b, c \rangle \cdot \langle d, e, f \rangle = ad + be + cf$$

To find the dot product, multiply corresponding coordinates and add.

Magnitude:

$$\| \mathbf{v} \| = \sqrt{a^2 + b^2 + c^2}$$

$$\| \mathbf{v} \|^2 = \mathbf{v} \cdot \mathbf{v}$$

The magnitude is the square root of the sum of the squares of the coordinates.

Angle:

$$\mathbf{v} \cdot \mathbf{w} = \| \mathbf{v} \| \, \| \mathbf{w} \| \cos \alpha$$

where α is the angle between the nonzero vectors $\mathbf{v}$ and $\mathbf{w}$.

Orthogonal vectors: Two vectors are orthogonal precisely when their dot product is 0.

EXAMPLE 9.27 Operations on Vectors in $\mathbb{R}^3$

Let $\mathbf{v} = \langle 1, 2, -1 \rangle$ and $\mathbf{w} = \langle 3, 4, 1 \rangle$. Calculate the following.

a. $2\mathbf{v} + 3\mathbf{w}$

b. $\mathbf{v} \cdot \mathbf{w}$

c. $\| \mathbf{v} \|$

SOLUTION

a. We start by using the definitions:

$$2\mathbf{v} + 3\mathbf{w} = 2\langle 1, 2, -1 \rangle + 3\langle 3, 4, 1 \rangle \qquad \blacktriangleleft \text{Use the vector definitions.}$$

$$= \langle 2, 4, -2 \rangle + \langle 9, 12, 3 \rangle \qquad \blacktriangleleft \text{Multiply the coordinates by scalars.}$$

$$= \langle 11, 16, 1 \rangle \qquad \blacktriangleleft \text{Add the coordinates.}$$

b. We calculate the dot product here in the same way we would calculate the dot product of vectors in the plane:

$$\mathbf{v} \cdot \mathbf{w} = \langle 1, 2, -1 \rangle \cdot \langle 3, 4, 1 \rangle = 1 \times 3 + 2 \times 4 + (-1) \times 1 = 10$$

c. To find the magnitude, we add the squares of the coordinates and then take the square root:

$$\| \mathbf{v} \| = \| \langle 1, 2, -1 \rangle \| = \sqrt{1^2 + 2^2 + (-1)^2} = \sqrt{6}$$

Alternatively, we could have calculated $\mathbf{v} \cdot \mathbf{v}$ and taken the square root.

TRY IT YOURSELF 9.27 Brief answers provided at the end of the section.

Calculate $(\mathbf{v} + \mathbf{w}) \cdot (\mathbf{v} - \mathbf{w})$.

For vectors $\mathbf{u}$, $\mathbf{v}$, and $\mathbf{w}$, is it true that $\mathbf{u} \cdot (\mathbf{v} \cdot \mathbf{w}) = (\mathbf{u} \cdot \mathbf{v}) \cdot \mathbf{w}$? Be careful. Does the equation really make sense?   **EXTEND YOUR REACH**

EXAMPLE 9.28 More Vectors in $\mathbb{R}^3$

Let $\mathbf{v} = \langle 2, 3, 1 \rangle$.

a. Find a unit vector in the direction of $\mathbf{v}$.

b. Find a nonzero vector that is orthogonal to $\mathbf{v}$.

SOLUTION

a. Just as with vectors in the plane, to make a unit vector we divide the vector by its magnitude. First we find the magnitude:

$$\| \mathbf{v} \| = \sqrt{2^2 + 3^2 + 1^2} = \sqrt{14}$$

Multiplying by the reciprocal of the magnitude gives the desired unit vector:

$$\frac{1}{\sqrt{14}} \mathbf{v} = \frac{1}{\sqrt{14}} \langle 2, 3, 1 \rangle = \left\langle \frac{2}{\sqrt{14}}, \frac{3}{\sqrt{14}}, \frac{1}{\sqrt{14}} \right\rangle$$

You should check that this vector does indeed have magnitude 1.

b. There are lots of correct answers. Any nonzero vector whose dot product with $\mathbf{v}$ is zero will suffice. One easy way to get an answer is to choose the first coordinate to be zero, and then treat the last two coordinates as if they were plane vectors. Swap these coordinates and take the negative of one of them. This gives the vector

$$\langle 0, -1, 3 \rangle$$

You should check that the dot product of this vector with $\mathbf{v}$ is zero.

TRY IT YOURSELF 9.28 Brief answers provided at the end of the section.

Find a vector of length 2 in the direction opposite to $\mathbf{v}$.

EXTEND YOUR REACH

1. For any numbers a, b, and c, show how to find a nonzero vector orthogonal to the vector $\mathbf{v} = \langle a, b, c \rangle$.

2. For a nonzero vector $\mathbf{v}$ in $\mathbb{R}^3$, give a geometric description of the collection of all vectors in $\mathbb{R}^3$ orthogonal to $\mathbf{v}$.

We can add as many coordinates as we like and make n-dimensional vectors, vectors in $\mathbb{R}^n$. Definitions of vector addition, scalar multiplication, and the dot product are analogous to the corresponding operations for vectors in $\mathbb{R}^3$. One bonus is that the dot product can be used to define angles in spaces of higher dimension.

MODELS AND APPLICATIONS   Rotating an Antenna

The use of vectors in applications is simplified by using the coordinates introduced in this section. The dot product is a key tool.

EXAMPLE 9.29 A Directional Antenna

Dallas is 205 miles north of Houston and 80 miles west. Atlanta is 275 miles north of Houston and 650 miles east. Let's think of Houston as sitting at the origin, with east corresponding to the positive x-axis and north to the positive y-axis. (See **Figure 9.132**.)

a. Find a vector whose tip is at Dallas.

b. Find a vector whose tip is at Atlanta.

c. A directional antenna in Houston is pointed toward a source in Dallas. How must the antenna be rotated to receive a signal from Atlanta? Round your answer to the nearest degree.

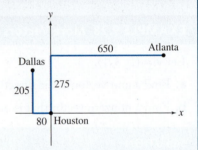

Figure 9.132 Rotating a directional antenna

SOLUTION

a. The coordinates can be read directly from the information provided in Figure 9.132. The only thing to be careful of is that the distance 80 miles corresponds to a negative x-coordinate. The required vector is $\mathbf{v} = \langle -80, 205 \rangle$.

b. As in part a, we can read the coordinates directly from Figure 9.132. We find $\mathbf{w} = \langle 650, 275 \rangle$.

c. The antenna must be rotated clockwise through the angle θ between the vectors $\mathbf{v}$ and $\mathbf{w}$. We find this angle using the dot product:

$$\cos \theta = \frac{\mathbf{v} \cdot \mathbf{w}}{\|\mathbf{v}\| \, \|\mathbf{w}\|}$$

$$= \frac{\langle -80, 205 \rangle \cdot \langle 650, 275 \rangle}{\|\langle -80, 205 \rangle\| \, \|\langle 650, 275 \rangle\|}$$

$$= 0.028\ldots$$

We use the inverse cosine to find the angle θ:

$$\theta = \arccos 0.028\ldots \approx 88°$$

We conclude that the antenna must be rotated by 88° in the clockwise direction.

TRY IT YOURSELF ANSWERS

9.23 $\sqrt{58}$

9.24 6

9.25 One correct answer is $\langle 6, 5 \rangle$.

9.26 $45°$ or $\dfrac{\pi}{4}$ radian

9.27 -20

9.28 $\left\langle -\dfrac{4}{\sqrt{14}}, -\dfrac{6}{\sqrt{14}}, -\dfrac{2}{\sqrt{14}} \right\rangle$

9.29 The antenna must be rotated counterclockwise by $74°$.

EXERCISE SET 9.4

CHECK YOUR UNDERSTANDING

1. **True or false:** A vector in standard position emanates from the origin.

2. Two vectors are orthogonal provided their dot product is _____.

3. **True or false:** Vectors that are perpendicular are also called orthogonal vectors.

4. **True or false:** The angle between two nonzero vectors can be found using the dot product.

5. The dot product of two vectors is:
   a. a vector
   b. a scalar
   c. a complex number
   d. None of the above.

6. Explain how the sum of vectors in standard position is calculated.

7. Explain how the scalar multiple of a vector in standard position is calculated.

8. **True or false:** Vectors in $\mathbb{R}^3$ are represented as three-dimensional objects.

9. Explain how to obtain a unit vector in the same direction as a given nonzero vector.

10. How is the dot product of a vector with itself related to the magnitude of the vector?

SKILL BUILDING

Vectors in the plane. In Exercises 11 through 27, let $\mathbf{u} = \langle 2, -1 \rangle$, $\mathbf{v} = \langle 3, 2 \rangle$, and $\mathbf{w} = \langle -4, 5 \rangle$. Calculate the following.

11. $\|\mathbf{u}\|$

12. $2\mathbf{u} + 3\mathbf{w}$

13. $\|\mathbf{u} + \mathbf{w}\|$

14. $2\mathbf{u} - 3\mathbf{v} + \mathbf{w}$

15. $-2\mathbf{u} + 4\mathbf{v} - 5\mathbf{w}$

16. $\mathbf{u} \cdot \mathbf{v}$

17. $\dfrac{1}{\mathbf{u} \cdot \mathbf{u}} \mathbf{u}$

18. $-2\mathbf{u} \cdot 4\mathbf{v}$

19. $\mathbf{u} \cdot (4\mathbf{v} - 2\mathbf{w})$

20. $(\mathbf{u} \cdot \mathbf{v})\mathbf{w}$

21. $(2\mathbf{u} + 3\mathbf{v}) \cdot (\mathbf{u} - 2\mathbf{w})$

22. $(\mathbf{u} \cdot \mathbf{u})\mathbf{u} - (\mathbf{u} \cdot \mathbf{v})\mathbf{v}$

23. $(\mathbf{u} \cdot \mathbf{v})(\mathbf{u} \cdot \mathbf{w})$

24. $\| (\mathbf{u} \cdot \mathbf{v})\mathbf{w} \|$

25. The cosine of the angle between $\mathbf{u}$ and $\mathbf{v}$

26. The cosine of the angle between $\mathbf{u}$ and $\mathbf{w}$

27. The cosine of the angle between $\mathbf{v} + \mathbf{w}$ and $\mathbf{v} - \mathbf{w}$

Vectors in the plane. In Exercises 28 through 32, let $\mathbf{u} = \langle 2, -1 \rangle$, $\mathbf{v} = \langle 3, 2 \rangle$, and $\mathbf{w} = \langle -4, 5 \rangle$.

28. Find a unit vector in the same direction as $\mathbf{v}$.

29. Find a unit vector in the direction opposite to $\mathbf{w}$.

30. Find a unit vector in the same direction as $-\mathbf{u} + \mathbf{v} - \mathbf{w}$.

31. Find a vector of length 5 in the same direction as $\mathbf{u}$.

32. Find a vector of length 4 in the direction opposite to $\mathbf{w}$.

Vectors in $\mathbb{R}^3$. In Exercises 33 through 42, let $\mathbf{u} = \langle 1, 2, 3 \rangle$, $\mathbf{v} = \langle 2, 0, -1 \rangle$, and $\mathbf{w} = \langle 1, 1, 2 \rangle$. Calculate the following.

33. $3\mathbf{v} + 2\mathbf{w}$

34. $4\mathbf{u} - 3\mathbf{w}$

35. $\| \mathbf{v} \|$

36. $\| \mathbf{u} \|^2$

37. $\mathbf{u} \cdot \mathbf{v}$

38. $\mathbf{u} \cdot (\mathbf{v} - \mathbf{w})$

39. $\mathbf{u} + \mathbf{v} + \mathbf{w}$

40. $\| \mathbf{u} + \mathbf{v} \|$

41. $\dfrac{1}{\mathbf{v} \cdot \mathbf{v}} \mathbf{v}$

42. $\dfrac{\mathbf{u} \cdot \mathbf{v}}{\mathbf{u} \cdot \mathbf{u}} \mathbf{u}$

Vectors in $\mathbb{R}^3$. In Exercises 43 and 44, let $\mathbf{u} = \langle 1, 2, 3 \rangle$, $\mathbf{v} = \langle 2, 0, -1 \rangle$, and $\mathbf{w} = \langle 1, 1, 2 \rangle$.

43. Find a unit vector in the direction of $\mathbf{v}$.

44. Find a vector of length 2 in the direction opposite to $\mathbf{u}$.

PROBLEMS

45. **Dot products and angles.** We learned that if $\mathbf{v}$ and $\mathbf{w}$ are nonzero vectors, and if θ is the angle between them, then $\mathbf{v} \cdot \mathbf{w} = \| \mathbf{v} \| \ \| \mathbf{w} \| \cos \theta$. In our earlier discussion, we gave a precise meaning to the angle between the vectors. But a naive identification of the angle might be any one of the four angles $\alpha, \beta, \gamma,$ or δ marked in **Figure 9.133**. Explain why all four of these angles yield the same value for the dot product.

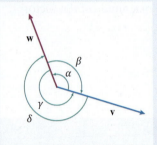

Figure 9.133 Various angles between two vectors

46. **Unit vectors.**

 a. Show that for any angle θ, the vector $\langle \cos \theta, \sin \theta \rangle$ is a unit vector.

 b. Show that if $\mathbf{v}$ is a unit vector, then $\mathbf{v} = \langle \cos \theta, \sin \theta \rangle$ for some angle θ. *Suggestion*: Because $\mathbf{v}$ is a unit vector, its tip lies on the unit circle.

47. **Vectors of length r.** If $r > 0$, show that the magnitude of the vector $\langle r \cos \theta, r \sin \theta \rangle$ is r.

48. **An interesting equation.** Comment on the equation $\mathbf{u}(\mathbf{v}\mathbf{w}) = (\mathbf{u}\mathbf{v}) \cdot \mathbf{w}$. Look carefully. Does the equation make sense?

49. **Components of vectors.** Let $\mathbf{v}$ and $\mathbf{w}$ be vectors as shown in **Figure 9.134**. We want to get a formula for the component of $\mathbf{v}$ in the direction of $\mathbf{w}$. First of all, recall that the component of $\mathbf{v}$ in the direction of $\mathbf{w}$ is a multiple $a\mathbf{w}$ of $\mathbf{w}$. We need to find the value of a. We know that

$$\mathbf{v} = a\mathbf{w} + \mathbf{z}$$

for some vector $\mathbf{z}$ that is orthogonal to $\mathbf{w}$. Thus, $\mathbf{v} - a\mathbf{w}$ is orthogonal to $\mathbf{w}$. This is shown in **Figure 9.135**. In this exercise, we show how to find the component. To find the component of $\mathbf{v}$ in the direction of $\mathbf{w}$, we need to find the scalar a that makes $\mathbf{v} - a\mathbf{w}$ orthogonal to $\mathbf{w}$. We assume that $\mathbf{w} \neq \mathbf{0}$.

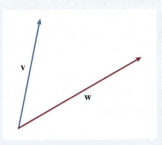

Figure 9.134 Two vectors

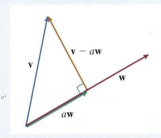

Figure 9.135 The component of **v** in the direction of **w**

a. Show that $\mathbf{w} \cdot (\mathbf{v} - a\mathbf{w}) = 0$.

b. Show that $a = \dfrac{\mathbf{v} \cdot \mathbf{w}}{\|\mathbf{w}\|^2}$.

c. Conclude that the component of **v** in the direction of **w** is given by

$$\frac{\mathbf{v} \cdot \mathbf{w}}{\mathbf{w} \cdot \mathbf{w}}\,\mathbf{w}$$

Calculating components. Exercises 50 through 53 use the formula derived in the preceding exercise (see part c).

50. Calculate the component of $\langle 1, 3 \rangle$ in the direction of $\langle 3, 4 \rangle$.

51. Calculate the component of $\langle 2, -1 \rangle$ in the direction of $\langle -2, 3 \rangle$.

52. Calculate the component of $\langle 4, 3 \rangle$ in the direction of $\langle -3, 4 \rangle$.

53. Calculate the component of $\langle 4, -5 \rangle$ in the direction of $\langle -2, -1 \rangle$.

Distance from a point to a line. To find the perpendicular distance d from a point $P = (r, s)$ to a line with equation $ax + by = c$, we use the vectors

$$\mathbf{n} = \langle a, b \rangle$$

$$\mathbf{p} = \langle r, s \rangle$$

(See **Figure 9.136**.) The distance d is given by the formula

$$d = \frac{|\mathbf{p} \cdot \mathbf{n} - c|}{\|\mathbf{n}\|}$$

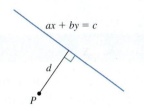

Figure 9.136 Distance from a point to a line

This formula is used in Exercises 54 through 56.

54. Find the distance from the point $(3, 1)$ to the line $2x + 5y = 12$.

55. Find the distance from the point $(2, -3)$ to the line $x + 2y = 9$.

56. Find the distance from the point $(1, -1)$ to the line $x + y = 0$.

Orthonormal vectors. Two vectors **v** and **w** are orthonormal if they are unit vectors and **v** is orthogonal to **w**. Exercises 57 through 60 concern orthonormal vectors.

57. Suppose **v** and **w** are orthonormal. Show that $\mathbf{v} \cdot \mathbf{v} = 1$, $\mathbf{w} \cdot \mathbf{w} = 1$, and $\mathbf{v} \cdot \mathbf{w} = 0$.

58. Suppose **v** and **w** are orthonormal. Show that if $a\mathbf{v} + b\mathbf{w} = 0$, then $a = b = 0$. *Suggestion:* Take the dot product of both sides of the equation with **v**, and use the results of the preceding exercise. Repeat this procedure using **w**.

59. Suppose **v** and **w** are orthonormal, and let **u** be a vector such that $\mathbf{u} = a\mathbf{v} + b\mathbf{w}$. Show that $a = \mathbf{u} \cdot \mathbf{v}$ and $b = \mathbf{u} \cdot \mathbf{w}$. *Suggestion:* Take the dot product of both sides of the equation, first with **v** and then with **w**.

60. a. Show that the vectors $\mathbf{v} = \left\langle \dfrac{1}{\sqrt{2}}, \dfrac{1}{\sqrt{2}} \right\rangle$ and $\mathbf{w} = \left\langle \dfrac{1}{\sqrt{2}}, -\dfrac{1}{\sqrt{2}} \right\rangle$ are orthonormal.

b. Use the results of the preceding exercise to solve the equation $\langle 4, 2 \rangle = a\mathbf{v} + b\mathbf{w}$ for a and b.

c. Verify that the answer you got in part b is actually correct—that is, calculate $a\mathbf{v} + b\mathbf{w}$ for the values of a and b you found in part b and verify that the result is $\langle 4, 2 \rangle$.

61. Orthogonal to fixed vector in plane. What geometric object is made up of all vectors in $\mathbb{R}^2$ orthogonal to a fixed nonzero vector $\mathbf{n}$?

62. Orthogonal to fixed vector in $\mathbb{R}^3$. What geometric object is made up of all vectors in $\mathbb{R}^3$ orthogonal to a fixed nonzero vector $\mathbf{n}$?

MODELS AND APPLICATIONS

In Exercises 63 through 66, east corresponds to the positive x-axis, and north corresponds to the positive y-axis.

63. Skirting an island. Let's put the home port of a ship at the origin. The ship's destination is a point 200 miles east of port along the x-axis. To skirt the island shown in **Figure 9.137**, the ship initially sails along the vector $\mathbf{v} = \langle 50, 20 \rangle$. From the tip of $\mathbf{v}$, the ship sails directly to its destination. Its heading and distance to the destination are given by a vector $\mathbf{w}$.

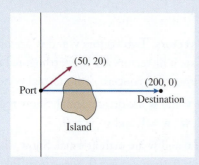

Figure 9.137 Changing course to avoid an island

a. Use coordinates to define $\mathbf{w}$. *Suggestion*: The vector $\mathbf{w}$ can be obtained by calculating a difference.

b. Use a magnitude and direction to describe $\mathbf{w}$.

c. Explain what the vector $\mathbf{w}$ means to the ship captain in terms of the heading and distance the ship must sail.

64. Avoiding a storm. An airplane leaves San Francisco en route to Seattle, which is 680 miles to the north. We locate the origin at San Francisco. To avoid a storm, the airplane follows the vector $\mathbf{v} = \langle -50, 100 \rangle$ from the origin to its tip, as shown in **Figure 9.138**. From the tip of $\mathbf{v}$ it heads straight to Seattle, following a vector $\mathbf{w}$ to its tip.

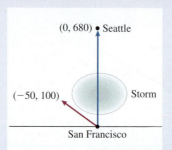

Figure 9.138 An airplane avoiding a storm

a. Find $\mathbf{w}$ in terms of coordinates.

b. What are the length and direction of $\mathbf{w}$?

c. Explain what the vector $\mathbf{w}$ means to the airplane pilot in terms of the distance and heading the airplane should fly.

65. A searchlight. A searchlight is focused on a distressed swimmer 300 yards east and 20 yards north of the light source. (See **Figure 9.139**.) Once this swimmer is rescued, the searchlight needs to focus on a second swimmer who is 250 yards east and 40 yards south of the light source. How should the light be rotated in order to focus on the second swimmer?

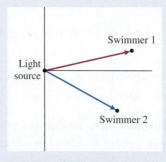

Figure 9.139 Rotating a searchlight

66. A Coast Guard rescue. A fishing boat has lost engine power and is adrift. The vessel is drifting along a vector that emanates from the Coast Guard station and ends at a dangerous shoal. The shoal is 3 miles east and 1 mile north of the Coast Guard station. A cutter is sent to rescue the fishing boat. It is determined that it is best to tow the vessel in a direction perpendicular to its line of drift. To the south is the Gulf of Mexico.

a. Use coordinates to describe a vector **v** that gives the direction the boat is to be towed. *Suggestion:* Recall that vectors are orthogonal when their dot product is 0. Also, be careful not to tow the vessel toward the Gulf.

b. Give the direction of the vector **v** from part a.

67. Muzzle velocity. Modern rifles have a very high muzzle velocity, which is the speed of the bullet as it exits the barrel. Let's think of a rifle fired from the origin into the first quadrant. The path taken by the bullet depends on both the muzzle velocity and the angle of elevation of the rifle. That is, it depends on an initial velocity vector **v**. Suppose the rifle barrel is tilted upward at an angle of θ and the muzzle velocity is m feet per second. Our goal is to find the initial velocity vector **v**.

a. The first step is to find the unit vector that makes an angle of θ with the positive x-axis. **Figure 9.140** shows the needed unit vector. Express x and y in terms of θ. *Suggestion:*

The ratio $\dfrac{x}{1}$ has an important relationship with the angle θ.

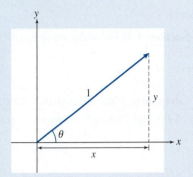

Figure 9.140 Angle of elevation and muzzle velocity

b. Use your work from part a to find the unit vector **u** that makes an angle of θ with the positive x-axis.

c. Use the vector **u** from part b to construct the vector **v**.

d. Use your work to find the initial velocity vector for a rifle with muzzle velocity 1200 feet per second if the rifle barrel is elevated at an angle of 30°. Round the coordinates of the vector to two decimal places.

CHALLENGE EXERCISES FOR INDIVIDUALS OR GROUPS

Verifying properties of vectors in the plane. In Exercises 68 through 74, you are asked to verify properties of vector operations in the plane. Suppose for example that we are asked to verify that if $\mathbf{v} = \langle a, b \rangle$, then $(r + s)\mathbf{v} = r\mathbf{v} + s\mathbf{v}$. We proceed as follows. First we calculate the left side of the equation:

$$(r + s)\mathbf{v} = (r + s)\langle a, b \rangle$$
$$= \langle (r + s)a, (r + s)b \rangle$$
$$= \langle ra + sa, rb + sb \rangle$$

Next we calculate the right side of the equation:

$$r\mathbf{v} + s\mathbf{v} = r\langle a, b \rangle + s\langle a, b \rangle$$
$$= \langle ra, rb \rangle + \langle sa, sb \rangle$$
$$= \langle ra + sa, rb + sb \rangle$$

Because our calculations show the left and right sides of the equation to be identical, the equation is verified.

68. If $\mathbf{v} = \langle a, b \rangle$ and $\mathbf{w} = \langle c, d \rangle$, verify that $r(\mathbf{v} + \mathbf{w}) = r\mathbf{v} + r\mathbf{w}$.

69. If $\mathbf{v} = \langle a, b \rangle$, show that $|| \mathbf{v} ||^2 = \mathbf{v} \cdot \mathbf{v}$.

70. If $\mathbf{v} = \langle a, b \rangle$, show that $|| r\mathbf{v} || = | r | \, || \mathbf{v} ||$.

71. If $\mathbf{u} = \langle a_1, b_1 \rangle$, $\mathbf{v} = \langle a_2, b_2 \rangle$, and $\mathbf{w} = \langle a_3, b_3 \rangle$, show that $\mathbf{u} \cdot (\mathbf{v} + \mathbf{w}) = \mathbf{u} \cdot \mathbf{v} + \mathbf{u} \cdot \mathbf{w}$.

72. If $\mathbf{u} = \langle a, b \rangle$ and $\mathbf{v} = \langle c, d \rangle$, show that $\mathbf{u} \cdot \mathbf{v} = \mathbf{v} \cdot \mathbf{u}$.

73. If **v** and **w** are vectors, use the results of the preceding exercises to show that

$$(\mathbf{v} + \mathbf{w}) \cdot (\mathbf{v} + \mathbf{w}) = \mathbf{v} \cdot \mathbf{v} + \mathbf{w} \cdot \mathbf{w} + 2\mathbf{v} \cdot \mathbf{w}$$

74. If **v** and **w** are vectors, use the results of the preceding exercises to show that

$$|| \mathbf{v} + \mathbf{w} ||^2 = || \mathbf{v} ||^2 + || \mathbf{w} ||^2 + 2\mathbf{v} \cdot \mathbf{w}$$

75. From Section 1.2: Calculate the average rate of change of $y = x^2 + x + 1$ from $x = a$ to $x = b$, $a \neq b$.

Answer: $b + a + 1$

76. From Section 4.3: Solve the equation $5^x = e$.

Answer: $x = \dfrac{1}{\ln 5}$

77. From Chapter 6: Find the missing sides and angles of the right triangle shown in **Figure 9.141**.

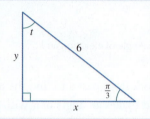

Figure 9.141 A triangle for Exercise 77

Answer: $t = \dfrac{\pi}{6}$, $x = 3$, $y = 3\sqrt{3}$

78. From Section 7.1: Find the period, amplitude, and phase shift of $y = 3\sin(2t)$.

Answer: The period is π, the amplitude is 3, and the phase shift is 0.

79. From Section 8.1: Which of the six trigonometric functions are even functions, and which are odd?

Answer: The cosine and secant are even. The sine, tangent, cotangent, and cosecant are odd.

80. From Section 8.3: Express $\cos^2 t$ in terms of $\cos(2t)$.

Answer: $\cos^2 t = \dfrac{1}{2}(1 + \cos(2t))$

81. From Section 8.5: Solve the equation $2\sin t = \sec t$.

Answer: $t = \dfrac{\pi}{4} + k\pi$, where k is any integer

82. From Section 9.1: Find the area of the triangle shown in **Figure 9.142**.

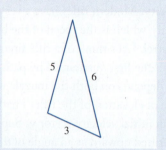

Figure 9.142 A triangle for Exercise 82

Answer: $2\sqrt{14}$

CHAPTER ROADMAP AND STUDY GUIDE

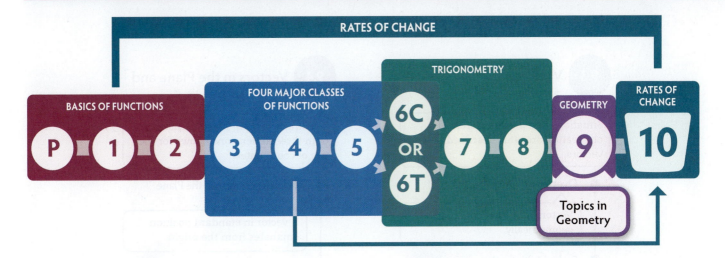

9.1 Law of Cosines

The law of cosines uses trigonometry to analyze all triangles, not just right triangles.

Recalling Congruence Criteria

Side-angle-side and side-side-side are criteria for triangle congruence.

Derivation of the Law of Cosines

The law of cosines is established using the Pythagorean theorem.

Using the Law of Cosines to Solve Triangles

The law of cosines allows us to find unknown sides and angles of triangles.

Heron's Formula for the Area of a Triangle

Heron's formula allows us to calculate the area of a triangle using only the sides.

MODELS AND APPLICATIONS

A Soccer Player

9.2 Law of Sines

The law of sines allows us to solve triangles that we cannot solve using the law of cosines.

Derivation of the Law of Sines

The law of sines is found by equating two area calculations for the same triangle.

Solving Triangles Using the Law of Sines

The law of sines is used to find unknown sides and angles of triangles.

The Ambiguous Case: Side-Side-Angle

When presented with side-side-angle, one triangle, two triangles, or no triangle at all may be described.

Choosing a Law to Solve a Triangle

A strategy for choosing laws to solve a triangle allows us to use the appropriate solution tool.

MODELS AND APPLICATIONS

A Lost Airplane

CHAPTER ROADMAP AND STUDY GUIDE (continued)

9.3 Vectors

Situations that involve more than one feature are often described using vectors.

Overview of Vectors

Vectors are represented geometrically.

Scalar Multiplication

Multiplication of a vector by a scalar changes the length and, in the case of negative scalars, reverses the direction.

Vector Addition

Vectors are added using parallelograms.

Components of Vectors

A vector can be written as a sum of vectors in specified directions.

MODELS AND APPLICATIONS
An Airplane

9.4 Vectors in the Plane and in Three Dimensions

Locating vectors in two dimensions or in three dimensions allows for the use of coordinates.

Standard Position in the Plane

A vector in standard position emanates from the origin.

Dot Product

The dot product focuses on the angle between two vectors.

Three-Dimensional Vectors

Vectors in three dimensions share many properties with vectors in the plane.

MODELS AND APPLICATIONS
Rotating an Antenna

CHAPTER QUIZ

1. Solve the triangle in **Figure 9.143**.

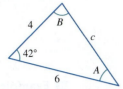

Figure 9.143
Answer: $A \approx 41.48°$, $B \approx 96.52°$, $c \approx 4.04$

XR Example 9.2

2. Solve the triangle in **Figure 9.144**.

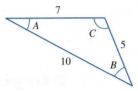

Figure 9.144
Answer: $A \approx 27.66°$, $B \approx 40.54°$, $C \approx 111.80°$

XR Example 9.3

3. Find the area of the triangle in **Figure 9.145**.

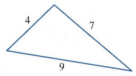

Figure 9.145
Answer: $6\sqrt{5} \approx 13.42$

XR Example 9.5

4. A laser locates the base of a hill 300 yards away. Elevating the laser 20°, we find that the top is 400 yards away. What is the length of the straight sloped side of the hill (marked h in the figure) from the base to the top? (See **Figure 9.146**.)

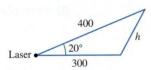

Figure 9.146 Measuring a hill
Answer: 156.44 yards

XR Example 9.7

5. Solve the triangle in **Figure 9.147**.

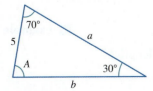

Figure 9.147
Answer: $a \approx 9.85$, $b \approx 9.40$, $A = 80°$

XR Example 9.8

6. Solve the triangle in **Figure 9.148**.

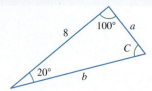

Figure 9.148

Answer: $C = 60°$, $a \approx 3.16$, $b \approx 9.10$

XR Example 9.9

7. Solve the triangle in **Figure 9.149**.

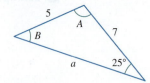

Figure 9.149

Answer: Solution 1: $B \approx 36.28°$, $A \approx 118.72°$, $a \approx 10.38$

Solution 2: $B \approx 143.72°$, $A \approx 11.28°$, $a \approx 2.31$

XR Example 9.13

8. A hiker is located in relation to towers as shown in **Figure 9.150**. The two towers are 8 miles apart. How far is the hiker from each of the towers?

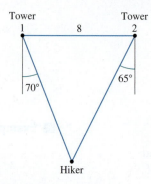

Figure 9.150

Answer: 4.78 miles from tower 1 and 3.87 miles from tower 2

XR Example 9.15

9. In **Figure 9.151**, identify the vectors $\mathbf{x} + \mathbf{z}$ and $\mathbf{u} - \mathbf{x}$.

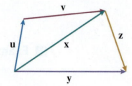

Figure 9.151

Answer: $\mathbf{x} + \mathbf{z} = \mathbf{y}$; $\mathbf{u} - \mathbf{x} = -\mathbf{v}$

XR Example 9.17

10. The vector $\mathbf{v}$ in **Figure 9.152** has length 3, and the vector $\mathbf{w}$ in Figure 9.152 has length 5.

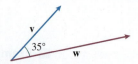

Figure 9.152

a. Find the length of $\mathbf{v} + \mathbf{w}$.

b. Find the acute angle $\mathbf{v} + \mathbf{w}$ makes with the vector $\mathbf{w}$. Round your answer to the nearest degree.

Answer:

a. 7.65

b. 13°

XR **Example 9.18**

11. A vector $\mathbf{v}$ of length 10 makes an angle of 60° with a vector $\mathbf{u}$. Find the magnitude of the component of $\mathbf{v}$ in the direction of $\mathbf{u}$.

Answer: 5

XR **Example 9.20**

12. A ship is sailing due east at a speed of 12 miles per hour. It encounters a current of 3 miles per hour heading 50° north of east. What are the resulting speed and heading (to the nearest degree) of the ship?

Answer: Speed: 14.12 miles per hour, heading: about 9° north of east

XR **Example 9.21**

13. Let $\mathbf{v} = \langle 1, 3 \rangle$ and $\mathbf{w} = \langle 2, 1 \rangle$. Find the following.

a. $\|\mathbf{v}\|$

b. $2\mathbf{v} + 3\mathbf{w}$

Answer:

a. $\sqrt{10} \approx 3.16$

b. $\langle 8, 9 \rangle$

XR **Example 9.23**

14. Calculate the dot product $\langle 2, 4 \rangle \cdot \langle 1, 3 \rangle$.

Answer: 14

XR **Example 9.24**

15. Find the acute angle between $\langle 2, 3 \rangle$ and $\langle 4, 1 \rangle$.

Answer: 42.27°

XR **Example 9.26**

16. Calculate $\langle 1, 3, -1 \rangle \cdot \langle 0, 2, 5 \rangle$.

Answer: 1

XR **Example 9.27**

CHAPTER REVIEW EXERCISES

Section 9.1

Solving triangles in the SAS and SSS cases. Solve the triangles in Exercises 1 through 8.

1.

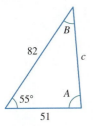

Answer: $c \approx 4.06$, $A \approx 120.51°$, $B \approx 29.49°$

2.

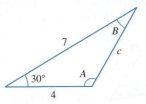

Answer: $c \approx 67.29$, $A \approx 86.62°$, $B \approx 38.38°$

3.

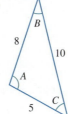

Answer: $A \approx 65.93°$, $B \approx 30.94°$, $C \approx 83.13°$

4.

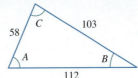

Answer: $A \approx 97.90°$, $B \approx 29.69°$, $C \approx 52.41°$

5.

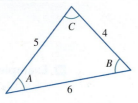

Answer: $A \approx 41.41°$, $B \approx 55.77°$, $C \approx 82.82°$

6.

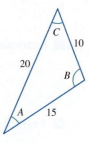

Answer: $A \approx 28.96°$, $B \approx 104.48°$, $C \approx 46.57°$

7.

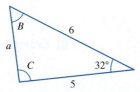

Answer: $a \approx 3.18$, $B \approx 56.41°$, $C \approx 91.59°$

8.

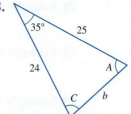

Answer: $b \approx 14.77$, $A \approx 68.80°$, $C \approx 76.20°$

Using Heron's formula. In Exercises 9 through 12, find the area of the given triangle.

9.

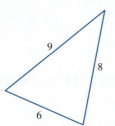

Answer: 23.53

10.

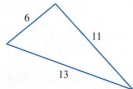

Answer: 32.86

11.

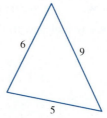

Answer: 14.14

12.

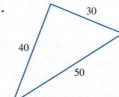

Answer: 600

13. Flying missiles. Two missiles are flying side by side. One missile continues on the same course at 300 miles per hour, and the other veers at an angle of 60° and flies at a speed of 325 miles per hour. (See **Figure 9.153**.) How far apart are the missiles after 3 hours of flying?

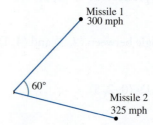

Figure 9.153

Answer: 939.75 miles

Section 9.2

Solving triangles in the ASA and SAA cases. In Exercises 14 through 21, solve the given triangle.

14.

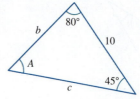

Answer: $A = 55°$, $b \approx 8.63$, $c \approx 12.02$

15.

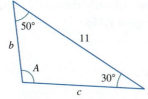

Answer: $A = 100°$, $b ≈ 5.58$, $c ≈ 8.56$

16.

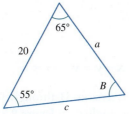

Answer: $B = 60°$, $a ≈ 18.92$, $c ≈ 20.93$

17.

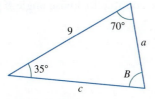

Answer: $B = 75°$, $a ≈ 5.34$, $c ≈ 8.76$

18.

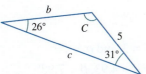

Answer: $C = 123°$, $b ≈ 5.87$, $c ≈ 9.57$

19.

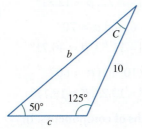

Answer: $C = 5°$, $b ≈ 10.69$, $c ≈ 1.14$

20.

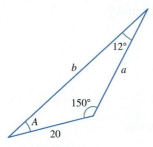

Answer: $A = 18°$, $a ≈ 29.73$, $b ≈ 48.10$

21.

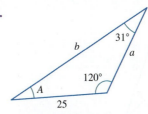

Answer: $A = 29°$, $a ≈ 23.53$, $b ≈ 42.04$

Solving triangles in the ambiguous case. In Exercises 22 through 27, solve the given triangle. For each exercise, there may be one solution, two solutions, or no solution.

22.

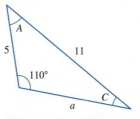

Answer: $a ≈ 8.24$, $A ≈ 44.71°$, $C ≈ 25.29°$

23.

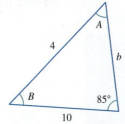

Answer: No solution

24.

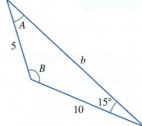

Answer:
Solution 1: $A ≈ 31.17°$, $B ≈ 133.83°$, $b ≈ 13.94$
Solution 2: $A ≈ 148.83°$, $B ≈ 16.17°$, $b ≈ 5.38$

25.

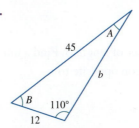

Answer: $A ≈ 14.51°$, $B ≈ 55.49°$, $b ≈ 39.46$

26.

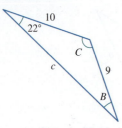

Answer: Solution 1:
$B \approx 24.60°, C \approx 133.40°, c \approx 17.46$
Solution 2: $B \approx 155.40°, C \approx 2.60°, c \approx 1.09$

27.

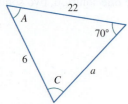

Answer: No solution

28. A surveyor. A surveyor stands on one bank of a river. A helper on the opposite bank places markers 40 feet apart. A transit is used to measure the angles marked in **Figure 9.154**.

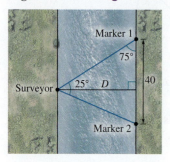

Figure 9.154

a. Find the distance from the surveyor to Marker 1.

b. Find the distance D, which represents the width of the river.

Answer:

a. 93.21 feet

b. 90.03 feet

Section 9.3

Finding a unit vector.

29. Suppose **v** is a vector of length 5. Find a unit vector in the direction opposite to **v**.

Answer: $-\dfrac{1}{5}\mathbf{v}$

30. Identifying vectors. Identify the vectors **x** + **y**, **x** − **y**, and −**x** in **Figure 9.155**.

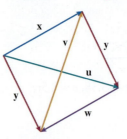

Figure 9.155

Answer: $\mathbf{x} + \mathbf{y} = \mathbf{u}, \quad \mathbf{x} - \mathbf{y} = \mathbf{v}, \quad -\mathbf{x} = \mathbf{w}$

Calculating vector sums. Exercises 31 through 34 give lengths of **v** and **w** and the angle α in **Figure 9.156**. In each exercise, give the length of **v** + **w**, and indicate the direction by giving the counterclockwise angle β from **w** to **v** + **w**.

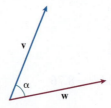

Figure 9.156

31. $\| \mathbf{v} \| = 2, \| \mathbf{w} \| = 3$, and $\alpha = 20°$
Answer: $\| \mathbf{v} + \mathbf{w} \| \approx 4.93, \beta \approx 7.98°$

32. $\| \mathbf{v} \| = 3, \| \mathbf{w} \| = 4$, and $\alpha = 30°$
Answer: $\| \mathbf{v} + \mathbf{w} \| \approx 6.77, \beta \approx 12.81°$

33. $\| \mathbf{v} \| = 8, \| \mathbf{w} \| = 6$, and $\alpha = 70°$
Answer: $\| \mathbf{v} + \mathbf{w} \| \approx 11.53, \beta \approx 40.71°$

34. $\| \mathbf{v} \| = 5, \| \mathbf{w} \| = 10$, and $\alpha = 60°$
Answer: $\| \mathbf{v} + \mathbf{w} \| \approx 13.23, \beta \approx 19.11°$

35. Finding the lengths of components. In **Figure 9.157**, the vector **x** is orthogonal to **y**. Find the lengths of the components of **v** in the directions of **x** and **y**.

Figure 9.157 Vectors for Exercise 35

Answer: In the direction of **x**, 38.30. In the direction of **y**, 32.14.

36. **Finding the speed and direction.** A drone is initially flying on a heading 40° south of west at a speed of 250 miles per hour. A wind begins to blow due west at a speed of 30 miles per hour. Find the resultant speed and direction of the drone.

    Answer:
    Speed $\approx$ 273.66 miles per hour. Direction $\approx$ 35.96° south of west.

Section 9.4

Calculations with vectors. Let $\mathbf{v} = \langle 1, 2 \rangle$ and $\mathbf{w} = \langle 2, 5 \rangle$. In Exercises 37 through 49, make the indicated calculation.

37. $\| \mathbf{v} \|$

    Answer: $\sqrt{5}$

38. $\mathbf{v} + 2\mathbf{w}$

    Answer: $\langle 5, 12 \rangle$

39. $3\mathbf{v} - \mathbf{w}$

    Answer: $\langle 1, 1 \rangle$

40. $-2\mathbf{v} + 3\mathbf{w}$

    Answer: $\langle 4, 11 \rangle$

41. $\mathbf{v} \cdot \mathbf{w}$

    Answer: 12

42. $(\mathbf{v} + \mathbf{w}) \cdot (\mathbf{v} - \mathbf{w})$

    Answer: -24

43. $\mathbf{v} \cdot \mathbf{v} + \mathbf{w} \cdot \mathbf{w}$

    Answer: 34

44. $\dfrac{1}{\mathbf{v} \cdot \mathbf{v}} (\mathbf{v} + \mathbf{w})$

    Answer: $\left\langle \dfrac{3}{5}, \dfrac{7}{5} \right\rangle$

45. Find the cosine of the angle between $\mathbf{v}$ and $\mathbf{w}$.

    Answer: $\dfrac{12}{\sqrt{145}}$

46. Find a nonzero vector orthogonal to $\mathbf{v}$.

    Answer: One correct answer is $\langle -2, 1 \rangle$.

47. Find a unit vector in the direction of $\mathbf{v}$.

    Answer: $\left\langle \dfrac{1}{\sqrt{5}}, \dfrac{2}{\sqrt{5}} \right\rangle$

48. Find a unit vector in the direction opposite to $\mathbf{v}$.

    Answer: $\left\langle -\dfrac{1}{\sqrt{5}}, -\dfrac{2}{\sqrt{5}} \right\rangle$

49. Find the magnitude of the vector $\dfrac{(\mathbf{v} \cdot \mathbf{w})}{(\mathbf{w} \cdot \mathbf{w})} \mathbf{w}$.

    Answer: $\dfrac{12}{\sqrt{29}}$

Vectors in $\mathbb{R}^3$. In Exercises 50 through 53, use $\mathbf{u} = \langle 1, 1, -1 \rangle$, $\mathbf{v} = \langle 2, 0, 3 \rangle$, and $\mathbf{w} = \langle 2, 1, 2 \rangle$.

50. Calculate $\mathbf{u} + \mathbf{v} + \mathbf{w}$.

    Answer: $\langle 5, 2, 4 \rangle$

51. Calculate $\| \mathbf{u} \|$.

    Answer: $\sqrt{3}$

52. Calculate $\mathbf{v} \cdot \mathbf{w}$.

    Answer: 10

53. Calculate $(2\mathbf{u}) \cdot (3\mathbf{w})$.

    Answer: 6

A QUALITATIVE EXPLORATION OF RATES OF CHANGE

10

RATES OF CHANGE

TRIGONOMETRY

FOUR MAJOR CLASSES
OF FUNCTIONS

BASICS OF FUNCTIONS

GEOMETRY

P · 1 · 2 · 3 · 4 · 5 · 6C OR 6T · 7 · 8 · 9

10
A Qualitative
Exploration of
Rates of
Change

10.1 An Introduction to the
Rate of Change as a Function

10.2 Change Equations: Linear
and Exponential Functions

10.3 Graphical Solutions of
Change Equations

This capstone chapter is the launching-off
point to calculus. In earlier chapters, we
studied functions using rates of change.
This chapter shows that this dynamic view
of functions provides key insights for their
deeper analysis and for the development
of mathematical models of the real world.
The methods are qualitative, emphasizing
essential concepts rather than detailed
calculations.

Throughout this book we have stressed that rates of change are fundamental to calculus. But rates of change actually serve as the gateway to all advanced mathematics and its applications. Rates of change provide us with a dynamic, holistic view of functions and their graphs, allowing for a much deeper understanding than we can get with only the tools of algebra.

Rates of change are vital to our daily lives. For example, the rate of change with respect to time of the position of an object is its velocity. Think of the challenge of describing what a jet airplane does without using words that mean rate of change. You couldn't use words like *speed*, *velocity*, or *fast*. Rates of change are part of the language we use in everyday communication.

This chapter is a springboard into calculus. It begins with a summary of what we have already learned about rates of change and emphasizes their utility. We discuss how to interpret the graph of a function in terms of its rate of change. Then we explore methods of solving equations that involve rates of change. The approach in this chapter is largely qualitative, helping you to master some of the central concepts of calculus. This conceptual approach will prepare you for understanding the quantitative aspects of calculus as well. For example, in calculus you will be asked to locate the maximum and minimum values of a function by carrying out certain computations. The concepts covered in this chapter will prepare you for interpreting the results of those calculations in terms of rates of change and so make sense of them.

10.1 An Introduction to the Rate of Change as a Function

> Rates of change are used to determine the properties of functions and their graphs.

10.1 An Introduction to the Rate of Change as a Function

10.2 Change Equations: Linear and Exponential Functions

10.3 Graphical Solutions of Change Equations

In this section, you will learn to:

1. Explain the rate of change of a function using units.
2. Sketch the graph of the rate of change of a function given the graph of the function.
3. Determine how the rate of change varies using the concavity of the graph.
4. Use rates of change to locate maximum and minimum values.

Although rates of change are important in an abstract, mathematical sense, they are easier to understand and appreciate in the context of familiar applications. As we discussed in the chapter introduction, rates of change can be found in many real-world situations. Examples include any sort of motion, profit or loss in a business, growth or decline of populations or the value of investments, and temperature change, to name only a few. Different disciplines use different terminology for the rate of change. For example, in physics the rate of change of distance as a function of time is called velocity. **Table 10.1** notes some applications in which the rate of change has a specific meaning. We'll explore many of these applications in the coming sections.

Table 10.1 Applications Involving Rates of Change

Application	Common name of rate of change	Meaning of rate of change
Distance as a function of time	Velocity	The speed (and direction) you are traveling
Velocity as a function of time	Acceleration	The rate at which you are speeding up or slowing down
Profit as a function of items produced	Marginal profit	The additional profit from production of each additional item
Tax as a function of your taxable income	Marginal tax rate	The additional tax you owe on each additional dollar of taxable income
Population as a function of time	Growth rate	The expected increase or decrease in population over a unit of time

Units of the Rate of Change

> The rate of change of a function is expressed using units.

Recall from Section 1.2 that for a function $y = f(x)$, the average rate of change is denoted by $\dfrac{\Delta y}{\Delta x}$. A variation on this notation uses $\dfrac{dy}{dx}$ to represent the instantaneous rate of change we introduced in Section 1.3. In this book, as in calculus, we normally omit the word *instantaneous* and just say that $\dfrac{dy}{dx}$, or $\dfrac{df}{dx}$, represents the rate of change of the function $y = f(x)$. This is not a new concept—just new notation. When we use this notation, we will include the words *rate of change* as a reminder.

We defined the rate of change of the function $y = f(x)$ as the slope of the tangent line to the graph of f at the point $(x, f(x))$. Because the rate of change $\dfrac{dy}{dx}$ is a slope, in applications there are always units associated with it: the units of the function divided by the units of the independent variable. This observation helps us to interpret the rate of change.

EXAMPLE 10.1   Units of Marginal Profit

The weekly profit P, in dollars, of a toy manufacturer depends on the number x of toys produced in a week. The rate of change of the function $y = P(x)$ is called the marginal profit. State the units associated with this rate of change $\dfrac{dy}{dx}$, and explain what the marginal profit means in terms of the weekly profit.

SOLUTION

The units of the rate of change $\dfrac{dy}{dx}$ are the units of the function divided by the units of the independent variable, so in this setting the units are dollars per toy produced. Thus, the marginal profit is the additional weekly profit we expect from producing one additional toy each week. For example, a marginal profit of $5 per toy means that for each additional toy produced in a week we expect an additional profit of $5.

TRY IT YOURSELF 10.1   Brief answers provided at the end of the section.

The height H, in feet, of a projectile fired upward depends on the time t, in seconds, since it was fired. State the units associated with the rate of change $\dfrac{dH}{dt}$, and explain what this rate of change means in terms of the height of the projectile.

State the units of the rates of change from the following list, and use the units to interpret the rate of change.

a. The distance traveled, in miles, after t hours is $D(t)$, and the rate of change is $\dfrac{dD}{dt}$.

b. The velocity in miles per hour t hours after movement started is $V(t)$, and the rate of change is $\dfrac{dV}{dt}$.

c. The tax you owe, in dollars, on an income of I dollars is $T(I)$, and the rate of change is $\dfrac{dT}{dI}$.

d. The number of animals present on a game reserve after t years is $N(t)$, and the rate of change is $\dfrac{dN}{dt}$.

e. Find your own application involving rates of change, give the units, and use the units to interpret the rate of change.

Sketching the Graph of the Rate of Change

> Graphs of rates of change are produced by sketching tangent lines.

In Chapter 1, we learned some important features of the rate of change of a function. The rate of change of a function $y = f(x)$ is the slope of the tangent line to the graph of f at the point $(x, f(x))$. A positive rate of change, $\dfrac{dy}{dx} > 0$, indicates an increasing function, and a negative rate of change, $\dfrac{dy}{dx} < 0$, indicates a decreasing function. A zero rate of change, $\dfrac{dy}{dx} = 0$, indicates that the function is not changing. This may occur at maxima and minima of a function.

We illustrate the utility of these ideas in the next examples.

EXAMPLE 10.2 Stopping for Coffee on a Walk to Class

You leave home a bit early so that you can stop for a cup of coffee along the way. **Figure 10.1** shows your trip to class using a graph of distance D west of home as a function of time t.

a. Add segments of several tangent lines to the graph in Figure 10.1.

b. Your westward velocity on your trip to class is the rate of change in distance west, $\dfrac{dD}{dt}$. Sketch a graph showing your westward velocity along the way.

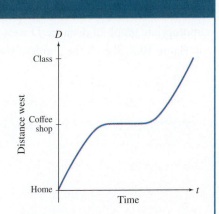

Figure 10.1 Stopping for coffee: graph of distance west

SOLUTION

a. In **Figure 10.2** we have added segments of several tangent lines to the graph.

b. In preparation for sketching the graph, we make the following observations regarding the tangent lines.

Region	Slopes	Meaning for velocity $\dfrac{dD}{dt}$
A: On the way to the coffee shop	Slopes are positive and nearly constant.	Velocity is positive and nearly constant.
B: Near the coffee shop	Slopes decrease quickly to zero, where they remain while you're at the coffee shop.	Velocity decreases to zero and remains there while you're at the coffee shop.
C: Coffee shop to class	Slopes return to the nearly constant, positive value from the first part.	Velocity resumes at about the earlier positive rate.

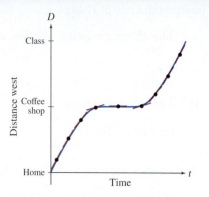

DF Figure 10.2 Stopping for coffee: tangent lines added

This description makes sense because you walk at a roughly constant speed to the coffee shop, stop for coffee, and then resume your walk to class.

The information in the preceding table is illustrated in **Figure 10.3**, where we use the labels *A*, *B*, and *C* as in the table. This information is used to make the graph in **Figure 10.4**.

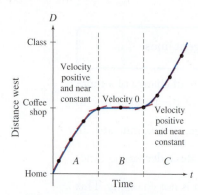

Figure 10.3 Stopping for coffee: regions marked for velocity

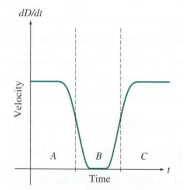

DF Figure 10.4 Stopping for coffee: the completed graph of velocity

TRY IT YOURSELF 10.2 Brief answers provided at the end of the section.

At the coffee shop, you decide to cut class and walk back home. The appropriate graph of distance *D* west as a function of time *t* is shown in **Figure 10.5**. Sketch the graph of your westward velocity.

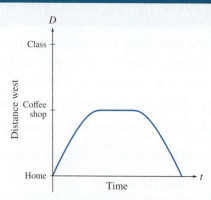

Figure 10.5 Stopping for coffee: cutting class

a. Sketch a graph showing a point where the rate of change is zero, but where the graph does not show a local maximum or minimum. *Suggestion*: Look carefully at the graph of $y = x^3$.

b. Sketch the graph of a function $y = f(x)$ that is always decreasing, but for which the rate of change $\dfrac{df}{dx}$ is not always negative. *Suggestion*: An adjustment of the graph from part a might prove useful.

EXTEND YOUR REACH

STEP-BY-STEP STRATEGY: Using the Graph of $y = f(x)$ to Sketch the Graph of the Rate of Change $\dfrac{dy}{dx}$

Step 1 Mark segments of several tangent lines on the graph of $y = f(x)$.

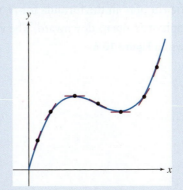

Step 2 Using the fact that the slopes of the tangent lines are the rates of change, mark sections of the graph where $\dfrac{dy}{dx}$ is positive, where it is negative, and where it is zero.

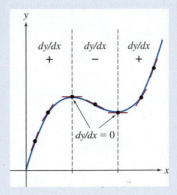

Step 3 Use the information from step 2 to sketch the graph of $\dfrac{dy}{dx}$.

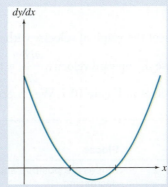

EXAMPLE 10.3 A Rock Tossed Upward

The height H, in feet, of a rock t seconds after being tossed upward from ground level is given by

$$H = 64t - 16t^2$$

a. Draw the graph of the height of the rock from the time it is tossed until it strikes the ground.

b. Add to your graph a sketch of segments of several tangent lines.

c. The rate of change in the height with respect to time, $\dfrac{dH}{dt}$, is the upward velocity. Sketch the graph of the upward velocity of the rock during the period of its flight.

SOLUTION

a. A straightforward calculation shows that the zeros of H are at $t = 0$ and $t = 4$. Thus, the flight of the rock extends from $t = 0$ to $t = 4$ seconds. Because the coefficient of t^2 in the formula for H is negative, the parabola that is the graph of H opens downward. The vertex occurs at $t = 2$. The graph is shown in **Figure 10.6**.

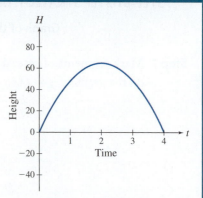

Figure 10.6 A rock tossed upward: graph of height

b. In **Figure 10.7** we have added segments of tangent lines at several points.

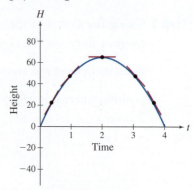

Figure 10.7 A rock tossed upward: tangent lines added

c. We cannot accurately represent the graph of velocity without using calculus, but we can sketch important features of the graph. Because the upward velocity, $\dfrac{dH}{dt}$, is the rate of change of the height, it is represented by the slopes of the tangent lines in Figure 10.7. We observe the following.

Region	Slopes	Meaning for velocity $\dfrac{dH}{dt}$
From $t = 0$ to $t = 2$	Slopes are positive and decreasing.	Velocity is positive and decreasing.
At $t = 2$	Tangent line is horizontal, so the slope is zero.	Velocity is zero.
From $t = 2$ to $t = 4$	Slopes are negative and increasing in magnitude.	Velocity is negative and decreasing.

These observations make sense because, as the rock rises, gravity slows its ascent. Thus, the upward velocity decreases to zero at the peak of the rock's flight. When the rock reverses and begins falling back toward Earth, the upward velocity is negative. Acceleration due to gravity increases the speed. The result is that the upward velocity takes on negative values that are larger in magnitude.

This information is noted in **Figure 10.8**, and it helps us to produce the graph shown in **Figure 10.9**.

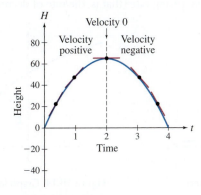 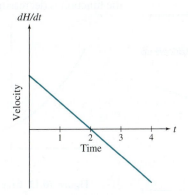

Figure 10.8 A rock tossed upward: Positive slopes mean positive velocity, and negative slopes mean negative velocity.

Figure 10.9 A rock tossed upward: graph of upward velocity, $\dfrac{dH}{dt}$

It is interesting to note that the rock reaches its highest point when the rate of change in upward motion, namely the upward velocity $\dfrac{dH}{dt}$, is zero.

TRY IT YOURSELF 10.3   Brief answers provided at the end of the section.

A spacecraft is on a mission to the moon. The mission is aborted just as the lander touches down, and so the lander immediately takes off. The height H, in feet, of the lander above the moon as a function of time t, in seconds, is shown in **Figure 10.10**. Thinking of velocity as the rate of change in height, $\dfrac{dH}{dt}$, sketch the graph of velocity.

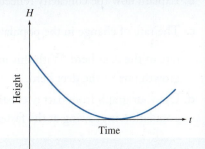

Figure 10.10 The height of a moon lander above the moon as a function of time

---

Concavity Revisited

> **Concavity is directly related to the rate of change.**

Recall from Section 1.3 that concavity tells us how a graph is bent—a graph that is concave up is bent upward, and a graph that is concave down is bent downward. At a point of inflection, the direction of bending changes.

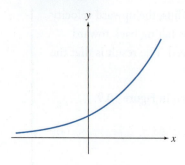

Figure 10.11 Graph rising and concave up

We observed in Section 1.3 that concavity provides important additional information about the rate of change of increasing or decreasing functions. If a graph is rising and concave up (**Figure 10.11**), the function is increasing at an increasing rate. If a graph is rising and concave down (**Figure 10.12**), the function is increasing at a decreasing rate. If a graph is falling and concave up (**Figure 10.13**), the function is decreasing at a decreasing rate (that is, the rate of decrease is decreasing). If a graph is falling and concave down (**Figure 10.14**), the function is decreasing at an increasing rate (that is, the rate of decrease is increasing).

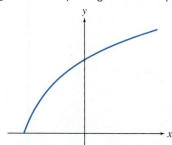

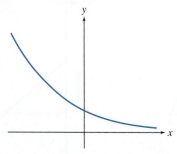

Figure 10.13 Graph falling and concave up

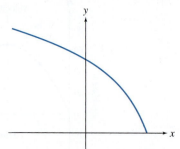

Figure 10.14 Graph falling and concave down

Figure 10.12 Graph rising and concave down

We illustrate the practical meaning of concavity in the next example.

EXAMPLE 10.4 The George Reserve

The deer population of the George Reserve in Michigan has been studied extensively by ecologists.[1] **Figure 10.15** shows the deer population N as a function of time t in years.

a. Estimate the time periods when the graph is concave up and when it is concave down.

b. Explain how the concavity reflects the growth rate of the deer herd.

c. The rate of change in the population function, $\dfrac{dN}{dt}$, is the growth rate of the deer herd. With this in mind, sketch the graph of the growth rate of the deer herd.

d. Use your graph from part c to estimate the time when the deer population is growing at the fastest rate.

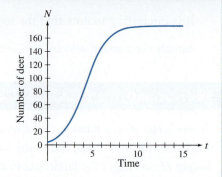

Figure 10.15 The George Reserve deer population

SOLUTION

a. Using the graph, we can only estimate the location of the point of inflection. But it appears from the graph that this point occurs at about year 4. Thus, we estimate that the graph is concave up from $t = 0$ to $t = 4$ and concave down from $t = 4$ to $t = 15$. The regions of concavity are marked in **Figure 10.16**.

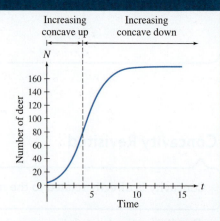

Figure 10.16 The George Reserve deer population: the regions of concavity

[1]*The George Reserve Deer Herd*, D. R. McCullough, University of Michigan Press, Ann Arbor, MI, 1979.

b. The graph is concave up at first but later is concave down. We make the following observations.

- Where the graph is increasing and concave up, from $t = 0$ to $t = 4$ years, the population is increasing at an increasing rate. That is, the growth rate $\dfrac{dN}{dt}$ is positive and increasing.

- Where the graph is increasing and concave down, from $t = 4$ to $t = 15$ years, the population is increasing at a decreasing rate. This means that the growth rate $\dfrac{dN}{dt}$ is still positive but is decreasing. The connection between the concavity and the growth rate is illustrated in **Figure 10.17.**

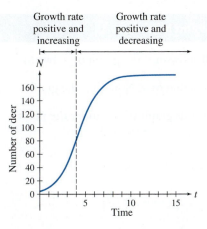

Figure 10.17 The George Reserve deer population: how concavity reflects the growth rate

c. We have noted that the growth rate is positive and increasing from 0 to 4 years and positive but decreasing from 4 to 15 years. This fact allows us to sketch the graph in **Figure 10.18**.

d. The maximum growth rate occurs where the graph in Figure 10.18 shows a maximum. That is at about year 4. It is worth noting that this time corresponds to the inflection point from part a.

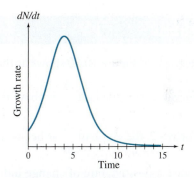

Figure 10.18 The George Reserve deer population: the growth rate

TRY IT YOURSELF 10.4   Brief answers provided at the end of the section.

Figure 10.19 shows a population in decline. Identify the times when the graph is concave up and when the graph is concave down. Then use the graph to explain how the concavity reflects population change.

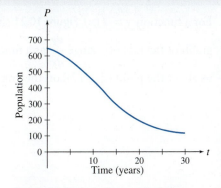

Figure 10.19 A declining population

Using Rates of Change to Analyze Functions

> Local maxima and minima of functions may occur where the rate of change is zero.

Rates of change can be used to tell us a great deal about functions. For example, a positive rate of change indicates an increasing function, and a negative rate of change indicates a decreasing function. These facts can help us to locate maximum and minimum values.

EXAMPLE 10.5   Finding Maxima and Minima

For a function $y = f(x)$, **Figure 10.20** shows not the graph of f but the graph of the rate of change $\dfrac{dy}{dx}$ as a function of x. Note that the graph crosses the x-axis at $x = 2$. What does this graph tell us about the point $(2, f(2))$ on the graph of $y = f(x)$?

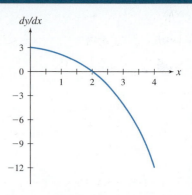

Figure 10.20 The graph of the rate of change $\dfrac{dy}{dx}$ as a function of x

SOLUTION

The graph in Figure 10.20 shows us the following.

- For $x < 2$, the graph of $\dfrac{dy}{dx}$ is above the x-axis, so the rate of change is positive.

- For $x > 2$, the graph of $\dfrac{dy}{dx}$ is below the x-axis, so the rate of change is negative.

Because a positive rate of change indicates an increasing function and a negative rate of change indicates a decreasing function, we conclude that the function f increases up to $x = 2$ and decreases afterward. Thus, f has a local maximum at the point $(2, f(2))$.

TRY IT YOURSELF 10.5   Brief answers provided at the end of the section.

For a function $y = f(x)$, **Figure 10.21** shows not the graph of f but the graph of the rate of change $\dfrac{dy}{dx}$ as a function of x. What does this graph tell us about the point $(2, f(2))$ on the graph of $y = f(x)$?

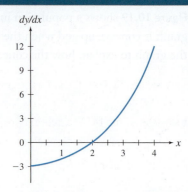

Figure 10.21 The rate of change for a new function

TRY IT YOURSELF ANSWERS

10.1 The units of the rate of change $\dfrac{dH}{dt}$ are feet per second. This rate of change is the upward velocity of the projectile, the change in height we expect over each additional second of flight.

10.2

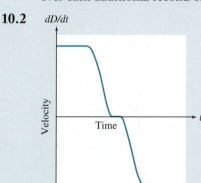

10.3

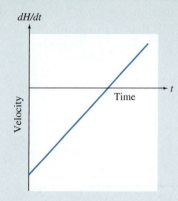

10.4 The graph is concave down from about year 0 to year 10. It is concave up afterward. From year 0 to year 10, the population is decreasing at an increasing rate. Afterward, the population decreases at a decreasing rate.

10.5 f has a local minimum at the point $(2, f(2))$.

EXERCISE SET 10.1

CHECK YOUR UNDERSTANDING

1. Given a function $y = f(x)$, what are the units of the rate of change $\dfrac{dy}{dx}$?

2. What does a positive rate of change indicate?

3. If the graph of a function is increasing and concave up, which of the following can we conclude?

   a. The function is increasing at an increasing rate.

   b. The function is increasing at a decreasing rate.

   c. The function is decreasing at an increasing rate.

   d. The function is decreasing at a decreasing rate.

4. **True or false:** For a function $y = f(x)$, $\dfrac{dy}{dx}$ indicates the slope of the tangent line to the graph of f at the point (x, y).

5. A point of inflection is _____.

6. **True or false:** If a function value is negative, then the rate of change must also be negative.

SKILL BUILDING

Units of the rate of change. In Exercises 7 through 11, state the units associated with the rate of change of the given function, and explain what the rate of change means in terms of the function.

7. $I(t)$ is the price, in dollars, of a stock as a function of time t, in days.

8. $N(m)$ is the number of birds, in thousands, in an isolated area as a function of time m, in months.

9. $T(I)$ is the tax, in dollars, you owe if your taxable income is I dollars.

10. $V(t)$ is the speed, in miles per hour, of an airplane t minutes after takeoff.

11. $F(d)$ is the cumulative number of flu cases reported as of day d.

Matching. In Exercises 12 through 17, you are given the graph of a function $y = f(x)$. Select one of the graphs A through F that is the graph of the rate of change $\dfrac{dy}{dx}$.

A. *dy/dx*

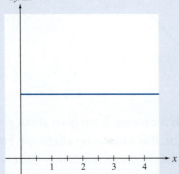

B. *dy/dx*

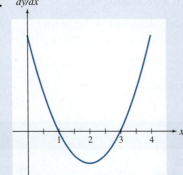

C. *dy/dx*

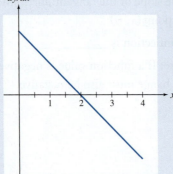

D. *dy/dx*

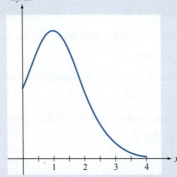

E. *dy/dx*

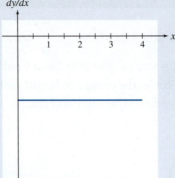

F. *dy/dx*

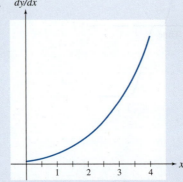

12. *y*

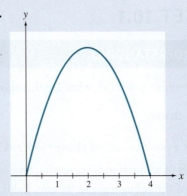

13. *y*

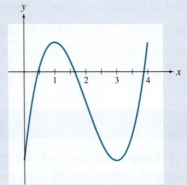

14.

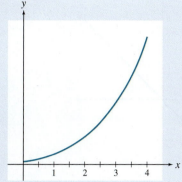

15.

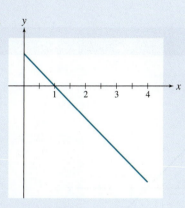

16.

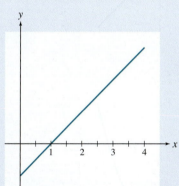

17.

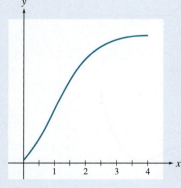

Sketching tangent lines. In Exercises 18 through 23, you are given the graph of a function $y = f(x)$. Trace the graph onto your own paper, and add segments of several tangent lines to the graph.

18.

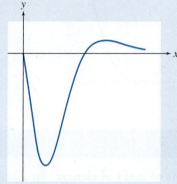

19.

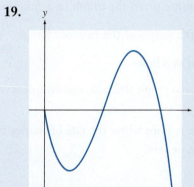

20.

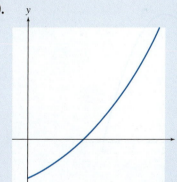

21.

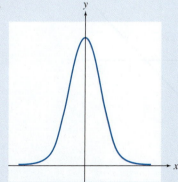

22.

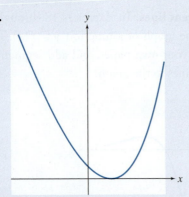

23.

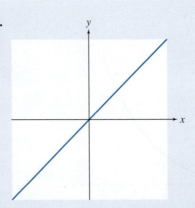

PROBLEMS

Sketching graphs of rates of change. In Exercises 24 through 31, you are given the graph of a function $y = f(x)$. Sketch the graph of the rate of change $\dfrac{dy}{dx}$. You should proceed as follows.

- First, trace the given graph and add segments of several tangent lines.
- Next, identify the regions where the rate of change is positive, negative, or zero.

24.

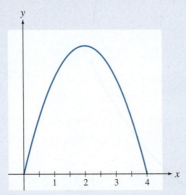

25.

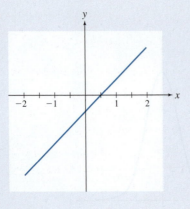

26.

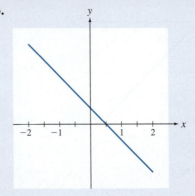

27.

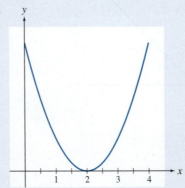

28.

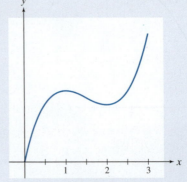

29.

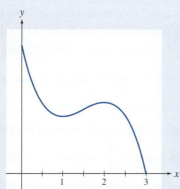

30.

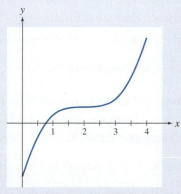

31.

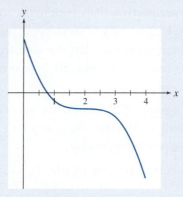

MODELS AND APPLICATIONS

32. A wolverine population. A wolverine population on a wildlife reserve in Alaska is modeled by the function $N(t)$ whose graph is in **Figure 10.22**.

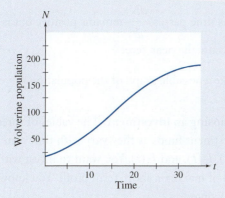

Figure 10.22 A wolverine population

a. Estimate the time periods over which the graph is concave up and over which it is concave down.

b. Explain how the concavity reflects the growth rate of the wolverine population.

c. The rate of change in the population function, $\dfrac{dN}{dt}$, is the growth rate. With this in mind, sketch the graph of the growth rate of the wolverine population.

d. Use your graph from part c to estimate the time when the wolverine population is growing at the fastest rate.

33. A model rocket. A model rocket is shot upward from a height of 10 feet. Its height H, in feet, above the ground after t seconds is given by

$$H = 10 + 159t - 16t^2$$

a. Determine the time at which the rocket strikes the ground.

b. Plot the graph of height versus time over the period of the rocket's flight.

c. Sketch the graph of upward velocity $\dfrac{dH}{dt}$ of the rocket.

34. **Changing perspective.** Refer back to the scenario in Example 10.2, where you are stopping for coffee on your way to class. But this time, think of watching your progress from the perspective of your classroom.

    a. Sketch a graph of E, the distance *east* of class, as you walk to the coffee shop, stop for a cup, and then proceed on to class.

    b. Sketch a graph of eastward velocity, $\dfrac{dE}{dt}$, versus time t.

35. **A forgotten book.** You leave home walking west toward class. You stop at the coffee shop for a cup of coffee. Upon finishing your coffee, you return home to retrieve a forgotten book. Then you resume your walk to class.

    a. Sketch a graph of your distance west D from home versus time.

    b. Sketch a graph of westward velocity, $\dfrac{dD}{dt}$, versus time.

36. **Stopping to see two friends.** You drive west from home and stop for a while to see a friend. After a short visit, you drive farther west to see another friend. Then you drive back home.

    a. Sketch a graph of your distance D west of home versus time t.

    b. Sketch a graph of your westward velocity, $\dfrac{dD}{dt}$, versus time t.

37. **A leaky water tank.** A water tank is initially full of water but springs a small leak. After a time, you notice the leak, fix it, and add water to refill the tank. Let V denote the volume of water in the tank at time t. The graph of V is shown in **Figure 10.23**.

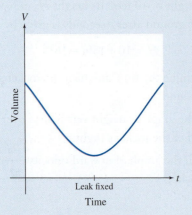

Figure 10.23 The volume of water in a tank versus time

a. Assume that the volume is measured in gallons and time is measured in minutes. State the units associated with the rate of change in volume, $\dfrac{dV}{dt}$, and explain what this rate of change means in terms of the volume of water.

b. What do you think is the value of the rate of change in volume, $\dfrac{dV}{dt}$, at the time when the leak is fixed? *Suggestion:* Use a straightedge to show a tangent line to the graph of V versus t at the time when the leak is fixed.

c. Sketch the graph of the rate of change in volume, $\dfrac{dV}{dt}$, as a function of time t.

38. **Bighorn sheep.** A breeding group of bighorn sheep is introduced into a protected area. If N denotes the sheep population at time t, then the rate of change in N with respect to t, $\dfrac{dN}{dt}$, is the growth rate of the population. Ecologists observe the following:

- Initially, $\dfrac{dN}{dt}$ is positive but small.
- After a few years, $\dfrac{dN}{dt}$ is not only positive but also larger than it was initially.
- As time passes, $\dfrac{dN}{dt}$ remains positive but is eventually near zero.

Make a possible graph of the population N versus time t.

39. **Choosing an investment.** The values of three investment funds as they vary with time t are $I_1(t)$, $I_2(t)$, and $I_3(t)$. You want to invest with the goal of making a short-term profit. You know the following:

- $\dfrac{dI_1}{dt}$ is a large positive number.
- $\dfrac{dI_2}{dt}$ is near zero.
- $\dfrac{dI_3}{dt}$ is a large negative number.

Which investment should you choose? Be sure to explain your answer.

40. **Growth in height.** The height H, in inches, of a female of age t years is shown in **Figure 10.24**. This person's growth rate (in height) is $\dfrac{dH}{dt}$.

   a. Sketch a graph of this person's growth rate versus time.

   b. What are the units of $\dfrac{dH}{dt}$?

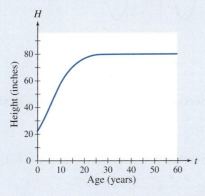

Figure 10.24 A graph of height versus age

41. **Marginal profit.** The CEO of a company that produces and sells thumb drives has kept extensive records of production and sales. One part of her records is the graph in **Figure 10.25**, which shows the marginal profit versus the number of thumb drives produced per week. Informally, in this context, the marginal profit is the additional profit that can be expected from the production of one additional drive per week. More formally, if T is the profit from producing n thumb drives per week, then the marginal profit is the rate of change $\dfrac{dT}{dn}$.

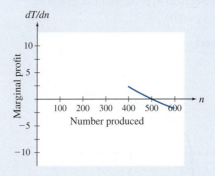

Figure 10.25 A graph of marginal profit

   a. Sketch a graph of the profit T versus the number of thumb drives produced per week. *Suggestion*: Remember that a positive marginal profit (i.e., a positive rate of change for T) means that T is increasing, and a negative marginal profit (i.e., a negative rate of change for T) means that T is decreasing.

   b. The company is currently producing 450 thumb drives per week. Should production be increased or decreased if the goal is to increase profits?

CHALLENGE EXERCISES FOR INDIVIDUALS OR GROUPS

42. **An unusual driving event.** The caped crusader is driving north when he spots the trap laid for him by the Joker. He immediately reverses course and heads back to the Batcave. **Figure 10.26** shows the graph of the distance D north of the Batcave as a function of time t.

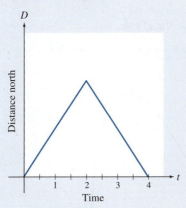

Figure 10.26 Distance D north of the Batcave

   a. Sketch the graph of the northward velocity $\dfrac{dD}{dt}$ versus time t.

   b. Considering your answer to part a, do you think it is possible, even for a superhero, to drive exactly as shown in Figure 10.26?

43. **A question for discussion.** Consider the graph of distance D versus time t shown in Figure 10.26 of the preceding exercise. What do you think is the rate of change $\dfrac{dD}{dt}$ at time $t = 2$? *Suggestion*: Think carefully about this to see that the answer is not obvious. Note that, in fact, a definitive answer to this question can be given only by using the precise definition of the rate of change that you will see in calculus. But it is still instructive to think about and discuss the question even before you have calculus tools.

REVIEW AND REFRESH: Exercises from Previous Sections

44. From Section 3.2: What is the limit as $x \to \infty$ of
$y = e^{-x}$?

Answer: 0

45. From Section 4.1: Sketch the graph of
$y = \ln(x - 10)$.

Answer:

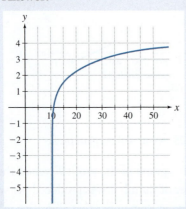

46. From Section 4.3: Solve the equation
$\ln(1 + 2^x) = 1$.

Answer: $x = \dfrac{\ln(e - 1)}{\ln 2}$

47. From Section 5.5: Plot the graph of the rational
function $y = \dfrac{x}{x^2 - 4}$. Show any vertical asymptotes
that occur.

Answer:

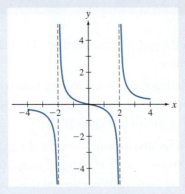

48. From Section 7.1: Make the graph of
$y = 2 + \cos\left(x - \dfrac{\pi}{4}\right)$.

Answer:

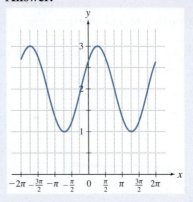

49. From Section 8.4: What is the value of arctan 1?

Answer: $\dfrac{\pi}{4}$

50. From Section 9.4: If $\mathbf{u} = \langle 1, 2 \rangle$ and $\mathbf{v} = \langle -1, 1 \rangle$,
calculate $\dfrac{\mathbf{u} \cdot \mathbf{v}}{\mathbf{v} \cdot \mathbf{v}} \mathbf{v}$.

Answer: $\left\langle -\dfrac{1}{2}, \dfrac{1}{2} \right\rangle$

51. From Section ON9.7: Parametrize the unit circle
starting at $(0, 1)$ and traversing the circle once in a
counterclockwise direction.

Answer:
$$x = \cos t$$
$$y = \sin t$$
$$\frac{\pi}{2} \le t \le \frac{5\pi}{2}$$

52. From Section ON9.7: Eliminate the parameter
from the parametric equations
$$x = \sin t$$
$$y = \sin(2t)$$
$$0 \le t \le \frac{\pi}{2}$$

Then make the graph.

Answer: $y = 2x\sqrt{1 - x^2}$, $0 \le x \le 1$

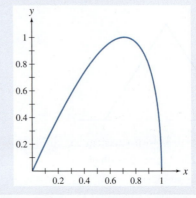

10.2 Change Equations: Linear and Exponential Functions

> Rates of change determine linear and exponential functions.

10.1 An Introduction to the Rate of Change as a Function

10.2 Change Equations: Linear and Exponential Functions

10.3 Graphical Solutions of Change Equations

In this section, you will learn to:

1. Solve change equations in the linear case.
2. Solve change equations in the exponential case.
3. Model real-world phenomena using linear and exponential change equations.

As we have established, it is often necessary to study rates of change as a way to understand physical phenomena. You will learn in calculus that typically a function can be recovered from its rate of change together with one value, such as the initial value. But even before you study calculus, you will discover that our work in Chapter 3 has already explained how to perform such calculations in the case of linear and exponential functions.

Constant Rate of Change

> A constant rate of change is characteristic of a linear function.

A change equation is nothing more than an equation that involves rates of change.[2] For example, $\frac{dy}{dx} = 3$ is a change equation. This equation means that y is a function with a constant rate of change of 3. We know from Section 3.1 that such a function is linear with slope 3. That is, it has the form $y = 3x + b$ for an (unknown) initial value b. All the change equations we consider in this chapter can be expressed simply in words, and we will always accompany the symbolic equation with its meaning in words.

LAWS OF MATHEMATICS: Change Equations—The Linear Case

The change equation

$$\frac{dy}{dx} = m$$

asks for a function whose rate of change is the constant m. That is a linear function with slope m:

$$y = mx + b$$

for an initial value b.

Understanding linear functions will serve you well when dealing with change equations of the form $\frac{dy}{dx} = m$. In fact, only the notation is new. We are simply working with the familiar fact that a linear function has a constant rate of change.

---

[2]In calculus such equations are known as differential equations.

EXAMPLE 10.6 How Objects Fall

A key development in the history of science was Galileo's discovery that the acceleration due to gravity near the surface of Earth is constant (namely, 9.8 meters per second per second). For a falling object, this acceleration is the rate of change, $\dfrac{dV}{dt}$, in downward velocity V with respect to time t. Here we measure V in meters per second and t in seconds. Thus, Galileo's discovery can be stated in terms of rates of change as

$$\text{Rate of change in velocity} = 9.8 \text{ meters per second per second}$$

which can be represented in symbols as

$$\frac{dV}{dt} = 9.8 \text{ meters per second per second}$$

a. Find a formula that gives the downward velocity V of an object t seconds after it begins to fall because it is thrown downward.

b. Your equation from part a should involve an unknown constant. Explain the meaning of this constant in terms of the velocity.

SOLUTION

a. Because the rate of change in velocity is constant, velocity must be a linear function of time. Furthermore, the slope of the linear function is the rate of change, 9.8 meters per second per second. Thus,

$$V = 9.8t + b \text{ meters per second}$$

b. The constant b from part a is the velocity of the object at time $t = 0$. That is the velocity at which the object is thrown downward. This quantity is commonly termed the initial velocity.

TRY IT YOURSELF 10.6 Brief answers provided at the end of the section.

Find the downward velocity after t seconds if the object is thrown downward with an initial velocity of 10 meters per second.

EXTEND YOUR REACH   Before Galileo's discoveries, it was widely believed that (neglecting air resistance) heavier objects fall faster than lighter objects. Is this belief consistent with the fact that the rate of change in velocity is constant for all falling objects?

EXAMPLE 10.7 Tossing a Baseball

A baseball is tossed upward from ground level with an initial velocity of 19.6 meters per second. We are thinking of up as the positive direction, so the acceleration due to gravity is $-g = -9.8$ meters per second per second.

a. Write a change equation for the upward velocity V, in meters per second, of the baseball t seconds after it is tossed. Express the equation both in symbolic form and in words.

b. Find a formula for V.

c. How long does it take for the baseball to reach the peak of its flight?

d. How long after it is tossed does the baseball strike the ground?

SOLUTION

a. In words, the rate of change in upward velocity, namely the acceleration, is $-g$. Because the acceleration is $\dfrac{dV}{dt}$, we express this fact in symbols as

$$\frac{dV}{dt} = -g$$

b. Because V has the constant rate of change of $-g$, it is linear with slope $-g$:

$$V = -gt + b$$

Here b is the initial velocity of 19.6 meters per second, so

$$V = -gt + 19.6 = -9.8t + 19.6$$

c. At the peak of its flight, the velocity of the baseball is 0. Thus, in order to find the time to the peak of the flight, we solve the equation

$$-9.8t + 19.6 = 0$$

We find that the baseball reaches its peak 2 seconds after it is tossed.

d. The baseball takes the same amount of time, 2 seconds, to fall back to the ground as it does to rise to its peak. Thus, the ball strikes the ground 4 seconds after it is tossed.

TRY IT YOURSELF 10.7   Brief answers provided at the end of the section.

Repeat Example 10.7 using 29.4 meters per second for the initial velocity of the baseball.

EXTEND YOUR REACH

In this example, we state that it takes the baseball the same amount of time to rise to its peak as it does to fall back to the ground. Explain why this is true.

The Exponential Case

A constant proportional rate of change is characteristic of an exponential function.

The change equation $\dfrac{dy}{dt} = ry$ asks for a function y whose rate of change is proportional to y. That is, y shows a constant proportional rate of change. We know from Section 3.2 that y must be an exponential function, which has the form $y = ab^t$ for some initial value a. But it is not clear what the base b should be. To answer this question, consider the case where y denotes the balance at time t of an investment that pays an annual percentage rate (APR) of r as a decimal, compounded continuously. In Section 3.3 we found that $y = ae^{rt}$ for the initial value a. But to say that interest is compounded continuously at an APR of r is just another way of saying that the proportional rate of change in y is the constant r. That is, $\dfrac{dy}{dt} = ry$. Thus, the solution of this change equation is $y = ae^{rt}$.

LAWS OF MATHEMATICS: Change Equations—The Exponential Case

The change equation

$$\frac{dy}{dt} = ry$$

can be rewritten as

$$\frac{dy/dt}{y} = r$$

This equation asks for a function whose proportional rate of change is the constant r. That is the exponential function

$$y = ae^{rt}$$

for the initial value a.

As with the linear case of change equations, only the notation is new here. Your knowledge of exponential functions (see Section 3.2) will guide you through the remainder of this section.

EXAMPLE 10.8   Change Equations and Exponential Functions

a. Find a function y that satisfies the change equation $\dfrac{dy}{dt} = 2y$. Your answer will involve an unknown constant.

b. Suppose we know that the function y from part a satisfies the initial condition $y(0) = 5$. Use this information to evaluate the unknown constant from part a.

SOLUTION

a. We have the exponential case of a change equation with $r = 2$. Thus,

$$y = ae^{2t}$$

for the initial value a.

b. The condition $y(0) = 5$ tells us that the initial value a is in fact 5. Thus,

$$y = 5e^{2t}$$

TRY IT YOURSELF 10.8   Brief answers provided at the end of the section.

Write a change equation that is satisfied by the function $y = 4e^{-3t}$.

EXAMPLE 10.9   A Leaky Balloon

A balloon initially holds 3 liters of air, but at time $t = 0$ it springs a leak. Let V denote the volume, in liters, of air in the balloon at time t measured in seconds. As the balloon leaks air, its volume changes at a rate proportional to the volume of air currently in the balloon.

a. Using r as the constant of proportionality, write a change equation that determines V.

b. Find a formula for V.

c. If the balloon loses a third of its initial volume in 5 seconds, find the exact value of r, and then use a calculator to approximate r to two decimal places.

d. Use the approximate value of r from part c to determine how long it takes for the balloon to lose two-thirds of the initial volume. Round your answer to the nearest second.

SOLUTION

a. To say that the volume changes at a rate proportional to the current volume means that the rate of change, $\dfrac{dV}{dt}$, is rV. In symbols, this is the change equation

$$\frac{dV}{dt} = rV$$

b. This is the exponential case of a change equation, and we know how to solve it: $V = ae^{rt}$. Because the initial volume is 3 liters, we know that $a = 3$. Thus,

$$V = 3e^{rt}$$

c. To say that the balloon loses a third of its initial volume of 3 liters after 5 seconds means that 2 liters are left—that is $V = 2$ when $t = 5$. We can then find r by solving the equation $3e^{5r} = 2$:

$$3e^{5r} = 2$$

$$e^{5r} = \frac{2}{3} \qquad \blacktriangleleft \textbf{Divide by 3.}$$

$$\ln e^{5r} = \ln(2/3) \qquad \blacktriangleleft \textbf{Apply the natural logarithm.}$$

$$5r = \ln(2/3) \qquad \blacktriangleleft \textbf{Use } \ln e^A = A.$$

$$r = \frac{\ln(2/3)}{5} \qquad \blacktriangleleft \textbf{Divide by 5.}$$

A calculator yields $r \approx -0.08$ per second.

d. Using the approximation for r, we get the formula

$$V = 3e^{-0.08t}$$

We want to know when the volume is 1 liter. That is, we need to solve the equation $3e^{-0.08t} = 1$. The volume is 1 liter when the graph of V crosses the line $y = 1$. This situation is shown in **Figure 10.27**.

We solve the equation using the natural logarithm:

$$3e^{-0.08t} = 1$$

$$e^{-0.08t} = \frac{1}{3} \qquad \blacktriangleleft \textbf{Divide by 3.}$$

$$\ln e^{-0.08t} = \ln(1/3) \qquad \blacktriangleleft \textbf{Apply the natural logarithm.}$$

$$-0.08t = \ln(1/3) \qquad \blacktriangleleft \textbf{Use } \ln e^A = A.$$

$$t = \frac{\ln(1/3)}{-0.08} \qquad \blacktriangleleft \textbf{Divide by } -0.08.$$

$$t \approx 13.73 \qquad \blacktriangleleft \textbf{Use a calculator.}$$

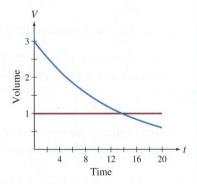

Figure 10.27 The crossing point representing the solution of the equation $3e^{-0.08t} = 1$

Thus, the balloon loses two-thirds of its initial volume after about 14 seconds, as you might have guessed by looking at Figure 10.27.

TRY IT YOURSELF 10.9   Brief answers provided at the end of the section.

Using the same balloon scenario as in Example 10.9, find the appropriate change equation if the constant of proportionality is −0.03 per second. Then find a formula for V if the initial value is 6 liters.

EXAMPLE 10.10   Newton's Law of Cooling

Suppose a hot object is placed in the open air to cool. We let $D(t)$ denote the difference between the temperature of the object and the temperature of the air t minutes after it is removed from the oven. (In this example, all temperatures are measured in degrees Fahrenheit.) Newton's law of cooling tells us that the rate of change in D is proportional to D. Let r denote the constant of proportionality. In terms of a change equation, Newton's law of cooling holds that, in words,

$$\text{Rate of change in } D \text{ is proportional to } D$$

which is the same as the following, in symbols:

$$\frac{dD}{dt} = rD$$

A cake with a temperature of 210° is taken from the oven and left to cool on the kitchen table. The temperature of the air in the kitchen is 70°. After 5 minutes, the temperature of the cake has cooled down to 140°.

a. Find a formula that gives the temperature difference D as a function of the time t in minutes since the cake was taken from the oven. (Round r to two decimal places.)

b. Find a formula that gives the temperature T of the cake t minutes after it was removed from the oven.

SOLUTION

a. Because the rate of change in D is proportional to D with constant of proportionality r, we have

$$D = ae^{rt}$$

where a is initial temperature difference. We can use the given information to evaluate the constants a and r.

The temperature of the cake when it was removed from the oven was 210°. Thus, the initial temperature difference is $a = 210 - 70 = 140$. We conclude that

$$D = 140e^{rt}$$

After 5 minutes, the temperature difference is $140 - 70 = 70$. That is, $D = 70$ when $t = 5$. So, we can find the value of r by solving the equation $140e^{5r} = 70$:

$$140e^{5r} = 70 \qquad \blacktriangleleft \text{ Use } D = 70 \text{ when } t = 5.$$

$$e^{5r} = \frac{1}{2} \qquad \blacktriangleleft \text{ Divide by 140.}$$

$$\ln e^{5r} = \ln(1/2) \qquad \blacktriangleleft \text{ Apply the natural logarithm.}$$

$$5r = \ln(1/2) \qquad \blacktriangleleft \text{ Use } \ln e^A = A.$$

$$r = \frac{\ln(1/2)}{5} \qquad \blacktriangleleft \text{ Divide by 5.}$$

A calculator yields $r \approx -0.14$ per minute. This result gives the formula

$$D = 140e^{-0.14t}$$

b. Recall that D is the difference between the temperature T of the cake and air temperature, 70:

$$T - 70 = D$$

$$T = D + 70$$

Substituting the formula we have for D gives

$$T = 140e^{-0.14t} + 70$$

The graph of the temperature of the cake is shown in **Figure 10.28**. Note that the temperature decreases more slowly as time passes.

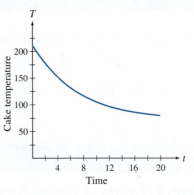

Figure 10.28 The temperature of a cooling cake

TRY IT YOURSELF 10.10 Brief answers provided at the end of the section.

How long will it take for the cake to cool to a temperature of 80°? Round your answer to the nearest minute.

Newton's law of cooling applies in the same way to heating. For example, if an object is placed in an oven whose temperature remains constant, then, just as in the cooling case, we have the change equation $\dfrac{dD}{dt} = rD$. In the preceding example of the cooling cake, the constant r turned out to be negative, which is typical of cooling objects. Explain why r is expected to be positive in the heating case.

EXTEND YOUR REACH

TRY IT YOURSELF ANSWERS

10.6 $V = 9.8t + 10$

10.7 **a.** $\dfrac{dV}{dt} = -9.8$

  b. $V = -9.8t + 29.4$

  c. The ball reaches its peak after 3 seconds.

  d. The ball strikes the ground 6 seconds after it is tossed.

10.8 $\dfrac{dy}{dt} = -3y$

10.9 The change equation is $\dfrac{dV}{dt} = -0.03V$. Then $V = 6e^{-0.03t}$.

10.10 The temperature of the cake is 80° at about 19 minutes after it is removed from the oven.

EXERCISE SET 10.2

CHECK YOUR UNDERSTANDING

1. The solution of the change equation $\dfrac{dy}{dx} = m$ is _____ .

2. The solution of the change equation $\dfrac{dy}{dt} = ry$ is _____ .

3. State in words the meaning of the change equation $\dfrac{dy}{dt} = ry$.

4. Write a change equation that means y is a linear function with slope m.

SKILL BUILDING

Matching. In Exercises 5 through 8, you are given a change equation. Which one of the graphs A through D is the solution of that equation?

A.

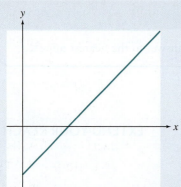

B.

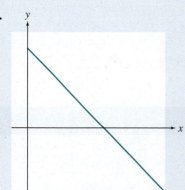

C.

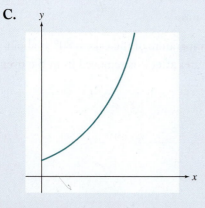

D.

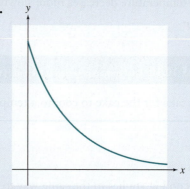

5. $\dfrac{dy}{dx} = -y$

6. $\dfrac{dy}{dx} = -1$

7. $\dfrac{dy}{dx} = 1$

8. $\dfrac{dy}{dx} = y$

Solving change equations. In Exercises 9 through 14, you are given a change equation together with an initial value. Give the solution.

9. $\dfrac{dy}{dx} = 2$, $y(0) = 3$

10. $\dfrac{dy}{dx} = -4$, $y(0) = 1$

11. $\dfrac{dy}{dx} = 2y$, $y(0) = 3$

12. $\dfrac{dy}{dx} = -3y$, $y(0) = 2$

13. $\dfrac{dy}{dx} = 0$, $y(0) = 3$

14. $\dfrac{dy}{dt} = y$, $y(0) = 1$

PROBLEMS

More change equations. In Exercises 15 through 21, you are given a change equation together with a function value that is not the initial value. Solve the equation.

15. $\dfrac{dy}{dx} = 3$, $y(2) = 11$

16. $\dfrac{dy}{dx} = 2$, $y(2) = 1$

17. $\dfrac{dy}{dx} = 2y$, $y(1) = 3e^2$

18. $\dfrac{dy}{dx} = (\ln 2)y$, $y(1) = 6$

19. $\dfrac{dy}{dx} = m$, $y(2) = 3m$

20. $\dfrac{dy}{dx} = -y$, $y(\ln 2) = 2$

21. $\dfrac{dy}{dx} = y$, $y(-\ln 3) = \dfrac{1}{3}$

MODELS AND APPLICATIONS

22. Retirement options. When you start work you are offered two retirement options. When you retire you will be given the balance of a retirement account.

Retirement option A: The initial balance is zero. The growth rate of your retirement fund A, in dollars, is given by

$$\frac{dA}{dt} = 20{,}000$$

Retirement option B: The initial balance is $10,000. The growth rate of your retirement fund B, in dollars, is given by

$$\frac{dB}{dt} = 0.1B$$

In each case, time t is measured in years.

a. Find a formula that gives the balances A and B for each retirement fund after t years.

b. Plot the graphs of both balances versus time on the same coordinate axes. Show the graphs over a 50-year period.

c. Which option should you choose if you plan to work for 40 years?

d. Which option should you choose if you plan to work for 50 years?

23. The Beer-Lambert-Bouguer law. According to the Beer-Lambert-Bouguer law, the intensity of light in the ocean decreases with depth. Let I denote the intensity of light at depth D in meters.

Then the rate of change in intensity with respect to depth $\dfrac{dI}{dD}$ satisfies the change equation

$$\frac{dI}{dD} = -rI$$

where r is a positive constant.

a. Using I_0 for light intensity at the surface, find a formula that gives the intensity I at depth D.

b. The constant r depends on the clarity of the water as well as the wavelength of the light. But in clear ocean water, the value of r is about 0.1 per meter. Restate the formula from part a using this information.

c. At what depth of clear ocean water is the light intensity half of the surface intensity?

24. Decibels. The decibel is a measure of the loudness of sound. The decibel reading D is related to the relative intensity R of sound by the change equation

$$\frac{dR}{dD} = \frac{\ln 10}{10} R$$

a. Find a formula that gives R as a function of D. Use R_0 for the relative intensity at a decibel reading of 0.

b. Restate the formula from part a using the fact that $R = 1$ when $D = 0$.

c. Find a formula that gives decibels D in terms of relative intensity R.

25. **Throwing rocks on Titan.** The acceleration due to gravity on Titan, Saturn's largest moon, is 1.4 meters per second per second. Imagine that a spacecraft with astronauts has landed on Titan.

    a. Write a change equation that describes the downward velocity V, in meters per second, of a rock t seconds after it is thrown downward on Titan.

    b. Solve the equation in part a.

    c. An astronaut on Titan throws a rock downward with an initial velocity of 3 meters per second. How fast is the rock traveling after 2 seconds?

    d. If a rock is tossed upward from the surface of Titan with a speed of 28 meters per second, how long does it take for the rock to fall back to the surface?

26. **Cesium-137.** Cesium-137 is an isotope of cesium that is a common product of fission in a nuclear reactor. The rate of change in the amount of any radioactive substance is proportional to the amount present. Let C denote the amount of cesium-137 remaining in spent nuclear fuel rods after t years.

    a. Using r for the proportionality constant, write an equation of change for C.

    b. Solve the equation from part a. Use C_0 for the initial value.

    c. The half-life of cesium-137 is 30.14 years. Use this information to find the value of the proportionality constant r from parts a and b. Give the exact value, and then round it to three decimal places. (See Section 3.2 for information about the half-life.)

    d. How long does it take for the amount of cesium-137 to decay to one-tenth of the initial amount? Round your answer to the nearest year.

27. **An investment.** You invest B_0 dollars in an account that pays an APR of r as a decimal. Interest on the account is compounded continuously. Consequently, if B is the balance of the account after t years, then the rate of change in B is proportional to B with constant of proportionality r.

    a. Write a change equation that describes B.

    b. Solve the equation in part a to find an explicit formula for B.

    c. How long will it take for the account balance to double?

    d. If your initial investment is \$5000 and the APR is 4%, find the balance of the account after 10 years.

28. **Competing newspapers.** The local *Times* newspaper has an initial circulation of 2000. The circulation T of this newspaper satisfies the change equation $\dfrac{dT}{dt} = 50$, where time t is measured in months. A competing paper, the *Post*, has an initial circulation of 500. The circulation P of the *Post* satisfies the change equation $\dfrac{dP}{dt} = 0.04P$.

    a. Find a formula that gives the circulation of the *Times* after t months.

    b. Find a formula that gives the circulation of the *Post* after t months.

    c. How long will it take for the circulation of the *Post* to match that of the *Times*? Round your answer in months to two decimal places.

29. **Exponential growth of populations.** Exponential models are often used to model population growth, at least over limited time spans. This exercise is designed to show why such models are used. The rate of change of a population is the number of births minus the number of deaths, both measured per unit of time. Exponential growth models are based on the following assumptions:

    - The number of births is proportional to the population size, with constant of proportionality b.

    - The number of deaths is proportional to the population size, with constant of proportionality d.

    Use N for the population size and t for time.

    a. Find a change equation that describes N.

    b. Solve the equation in part a to find a formula for N. Use N_0 for the initial population.

CHALLENGE EXERCISES FOR INDIVIDUALS OR GROUPS

30. Rate of change of the natural exponential function.

   a. Find the solution of the change equation

$$\frac{dy}{dx} = y \text{ with the initial value } y(0) = 1.$$

   b. In view of part a, what is the rate of change of the natural exponential function $y = e^x$?

   c. Find all nonconstant functions for which the rate of change equals the function itself.

REVIEW AND REFRESH: Exercises from Previous Sections

31. From Section P.1: Find the distance between the points $P = (1, 3)$ and $Q = (-2, 7)$.

   Answer: 5

32. From Section 2.2: The graph of a function f is shown in **Figure 10.29**. Add to the graph the line $y = x$ and the graph of f^{-1}.

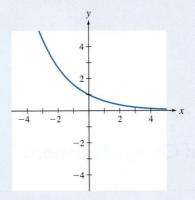

Figure 10.29 Graph for Exercise 32

Answer:

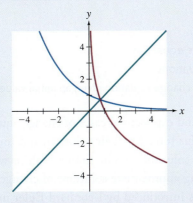

33. From Chapter 6: The trigonometric point $P(t) = (x, y)$ lies on the unit circle in the third quadrant. If $x = -\dfrac{2}{3}$, find the six trigonometric functions of t.

Answer: $\sin t = -\dfrac{\sqrt{5}}{3}, \quad \cos t = -\dfrac{2}{3}, \quad \tan t = \dfrac{\sqrt{5}}{2},$

$\cot t = \dfrac{2}{\sqrt{5}}, \quad \sec t = -\dfrac{3}{2}, \quad \csc t = -\dfrac{3}{\sqrt{5}}$

34. From Chapter 6: Referring to **Figure 10.30**, find x and y in terms of r and θ.

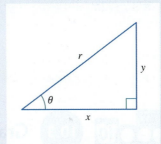

Figure 10.30 Triangle for Exercise 34

Answer: $x = r \cos \theta$ and $y = r \sin \theta$

35. From Section 9.2: Solve the triangle shown in **Figure 10.31**.

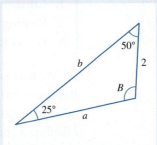

Figure 10.31 Triangle for Exercise 35

Answer: $a \approx 3.63, \quad B = 105°, \quad b \approx 4.57$

36. **From Section 9.3:** Vectors **v** and **w** have the same base, and the counterclockwise angle from **v** to **w** is 40°. Also, $\|\mathbf{v}\| = 5$ and $\|\mathbf{w}\| = 3$. Let $\mathbf{z} = \mathbf{v} + \mathbf{w}$. Find the length of **z** and the counterclockwise angle α that it makes with **v**.

   Answer: $\|\mathbf{z}\| \approx 7.55,\ \alpha \approx 14.80°$

37. **From Section ON9.6:** Find the polar coordinates of the point whose rectangular coordinates are $(-3, 3\sqrt{3})$.

   Answer: $\left[6, \dfrac{2\pi}{3} \right]$

38. **From Section ON9.7:** Eliminate the parameter from the following parametric equations, and then plot the graph:

$$x = \cos t$$
$$y = \cos(2t)$$
$$\pi \le t \le 2\pi$$

Answer: $y = 2x^2 - 1,\ -1 \le x \le 1$

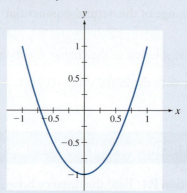

39. **From Section 10.1:** Describe in terms of rates of change a function whose graph is increasing and concave down.

   Answer: The function is increasing at a decreasing rate.

10.1 An Introduction to the Rate of Change as a Function

10.2 Change Equations: Linear and Exponential Functions

10.3 Graphical Solutions of Change Equations

10.3 Graphical Solutions of Change Equations

> The analysis of certain types of change equations is simplified by using graphs.

In this section, you will learn to:

1. Calculate equilibrium solutions of change equations.
2. Sketch graphs of solutions of change equations.
3. Predict the behavior of the solution of a change equation given an initial value.

We now know how to solve the linear and exponential cases of change equations. But many change equations require calculus tools for their solution, and for many other change equations only approximate solutions can be found, with or without calculus. A fundamental method for finding approximate solutions of change equations is known as the Euler method. It was introduced by Leonhard Euler in 1768. The method relies on the approximation over short intervals of a function by its tangent line. Generalizations of this method are used today in popular mathematics and engineering software to provide approximate solutions of change equations. In this section we show how to approximate solutions of certain change equations not by producing a formula but by sketching a graph.

Change Equations: Equilibrium Solutions

> Equilibrium solutions occur where the rate of change is zero.

In this section we focus on change equations of the form $\frac{dy}{dx} = g(y)$. For these equations,[3] the rate of change is expressed using only the dependent variable y and not the independent variable x. One example of such an equation is the exponential change equation $\frac{dy}{dx} = ry$. Other change equations of this type can yield important information even when their explicit solutions are not available.

Consider, for example, a forest where limited logging is allowed. Let F denote the number (in thousands) of trees in the forest at time t measured in years. The forest replenishes itself at a rate proportional to F, with constant of proportionality r. A constant harvest of L (in thousands) trees per year is allowed. These facts yield the following change equation:

$$\frac{dF}{dt} = rF - L$$

The solution of this equation requires calculus tools, but there is important information to be had in the absence of an explicit solution. For example, let's consider the solutions for which the rate of change $\frac{dF}{dt}$ is zero. Plugging this value into the change equation gives the simple linear equation $0 = rF - L$. When we solve for F, we find

$$F = \frac{L}{r}$$

When the rate of change is zero, the number of trees in the forest is not changing. Thus, if at any time the number of trees in the forest is at the level $F = L/r$, then the number of trees in the forest will remain at that same level indefinitely. This is all the information the forest manager needs in order to preserve the forest in the presence of logging.

This example from forestry is typical of change equations of the form $\frac{dy}{dx} = g(y)$. To find a solution for which $\frac{dy}{dx}$ is always zero, we solve the equation $g(y) = 0$. This is an **equilibrium solution** of the change equation. Because the rate of change of such solutions is always zero, these are the solutions of the change equation that remain constant. Graphs of equilibrium solutions are always horizontal lines.

An **equilibrium solution** of the change equation $\frac{dy}{dx} = g(y)$ is a solution of the equation $g(y) = 0$.

EXAMPLE 10.11 A Catfish Farm

Let C denote the number of catfish in a commercial pond at time t measured in months. The catfish reproduce at a rate proportional to the number present, with constant of proportionality $r = 0.2$ per month. Each month 500 catfish are harvested. These facts are codified in the following change equation:

$$\frac{dC}{dt} = 0.2C - 500$$

---

[3]Change equations in which the rate of change is expressed using only the dependent variable are known as autonomous equations.

a. Find the equilibrium solution of this equation.

b. Explain the meaning of the solution you found in part a in terms of the number of catfish and rate of change of the number of catfish.

c. Plot the graph of the equilibrium solution you found in part a.

SOLUTION

a. The equilibrium solution occurs where $0.2C - 500 = 0$. Solving, we find $C = 2500$ catfish.

b. At this population level, the rate of change in the number of catfish is zero. Thus, at the level of 2500 catfish, the population remains the same indefinitely. This is the ideal population level for this commercial fish farm.

c. The graph of the equilibrium solution is the horizontal line shown in **Figure 10.32**.

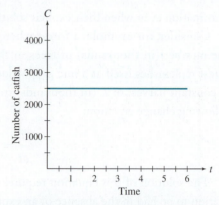

Figure 10.32 The equilibrium solution $C = 2500$ catfish

TRY IT YOURSELF 10.11   Brief answers provided at the end of the section.

Find the equilibrium solution of the change equation in Example 10.11 if the constant of proportionality r is 0.1 per month instead of 0.2.

EXTEND YOUR REACH

Can you give an example of a change equation of the form $\dfrac{dy}{dx} = g(y)$ that has no equilibrium solution?

EXAMPLE 10.12   Setting Hunting Limits

A population D of ducks in a protected area reproduces at a rate proportional to the current population. The constant of proportionality is 0.08 per year. Limited hunting is allowed. If H ducks per year are allowed to be taken by hunters, then the duck population is governed by the change equation

$$\frac{dD}{dt} = 0.08D - H$$

Here t is time measured in years. How many ducks per year should be taken by hunters if the duck population is to remain at the level of 800?

SOLUTION

We want the population to remain stable at 800 ducks. That is, $\dfrac{dD}{dt}$ is 0 when D is 800, so $D = 800$ is an equilibrium solution:

$$\frac{dD}{dt} = 0.08D - H$$

$$0 = 0.08 \times 800 - H \qquad \blacktriangleleft \text{ Substitute } \frac{dD}{dt} = 0 \text{ and } D = 800.$$

$$64 = H \qquad \blacktriangleleft \text{ Solve for } H.$$

Thus, we should restrict hunters to 64 ducks per year.

TRY IT YOURSELF 10.12 Brief answers provided at the end of the section.

How many ducks should hunters be allowed to take if the constant of proportionality is 0.06 per year and the population is to remain stable at 1000 ducks?

EXTEND YOUR REACH

Refer to the preceding example.

a. How should hunting limits be set if the goal is to increase the duck population?

b. What is expected to happen to the duck population if poachers take more than 64 ducks each year?

Sketching Graphs of Solutions

> The key to sketching graphs is to identify equilibrium solutions and then determine where the rate of change is positive and where it is negative.

We know that a positive rate of change indicates an increasing function, and a negative rate of change indicates a decreasing function. This fundamental fact allows us to make a rough sketch of the solutions of change equations of the form $\dfrac{dy}{dx} = g(y)$. We will explore this procedure in the next example.

EXAMPLE 10.13 The Catfish Farm: Graphing Solutions

Let's return to the catfish farm from Example 10.11. Recall from that example that the number C of catfish in the pond obeys the change equation

$$\frac{dC}{dt} = 0.2C - 500$$

a. Plot the graph of $\dfrac{dC}{dt}$ versus C.

b. Discuss the features of the graph of C versus t in each of the following regions: $C < 2500, C = 2500$, and $C > 2500$.

c. Add to the graph of the equilibrium solution you made in Example 10.11 a sketch of the graph of the solution C versus t in the case that the initial value is $C(0) = 3000$ catfish.

SOLUTION

a. The graph of $\dfrac{dC}{dt}$ versus C is the straight line determined by $y = 0.2C - 500$. The graph is shown in **Figure 10.33**.

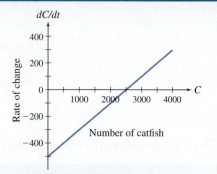

Figure 10.33 The catfish farm: the graph of $\dfrac{dC}{dt}$ versus C

b. The graph in Figure 10.33 tells us the following.

Region	$\dfrac{dC}{dt}$	Graph of C versus t
$C < 2500$	Negative	C is decreasing, so the graph falls.
$C = 2500$	Zero	Equilibrium solution: The graph is a horizontal line.
$C > 2500$	Positive	C is increasing, so the graph rises.

c. If the initial value is 3000 catfish, we are in the region where $\dfrac{dC}{dt}$ is positive, so the graph of C versus t rises. A sketch of the graph is shown in **Figure 10.34**.

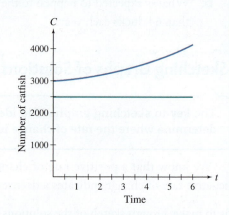

Figure 10.34 The catfish farm: number of catfish in case the initial value is 3000

TRY IT YOURSELF 10.13 Brief answers provided at the end of the section.

Add to the graph you made in Example 10.13 the graph of the solution C with initial value $C(0) = 2000$ catfish.

Note that Figure 10.33 does not show the graph of C versus t. Rather, it shows the graph of $\dfrac{dC}{dt}$ versus C, which is the growth rate that can be expected for a given population level.

The method used in this example applies to all change equations of the form we are studying.

STEP-BY-STEP STRATEGY: Sketching Solutions of
Change Equations of the Form

$$\frac{dy}{dx} = g(y)$$

Step 1 Find and plot the equilibrium solutions.

Step 2 Plot the graph of $\dfrac{dy}{dx}$ versus y.

Step 3 Determine the regions where $\dfrac{dy}{dx}$ is positive and where it is negative. This
information shows the regions of increase and regions of decrease for y as
a function of x.

Step 4 Given an initial value, use the information from step 3 to sketch the graph
of y versus x.

EXAMPLE 10.14 More on Graphing Solutions

For the change equation $\dfrac{dN}{dt} = (N - 2)(N - 4)$:

a. Find the equilibrium solutions, and plot them on coordinate axes showing N versus t.

b. Plot the graph of $\dfrac{dN}{dt}$ versus N.

c. Add the graph of N versus t to the graph you made in part a if the initial value is $N(0) = 3$.

SOLUTION

a. The equilibrium solutions are the solutions of the equation

$$(N - 2)(N - 4) = 0$$

This is a quadratic equation, and it is already factored. Thus, $N = 2$ and
$N = 4$ are the equilibrium solutions. These are plotted in **Figure 10.35**.

b. The graph of $\dfrac{dN}{dt}$ versus N is the graph of $y = (N - 2)(N - 4)$ shown
in **Figure 10.36**.

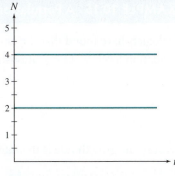

DF Figure 10.35 Graphing
solutions: the equilibrium solutions
$N = 2$ and $N = 4$

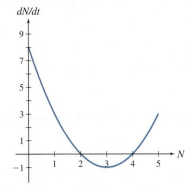

Figure 10.36 Graphing solutions:
the graph of $\dfrac{dN}{dt}$ versus N

c. The graph in Figure 10.36 shows that $\dfrac{dN}{dt}$ is negative between $N = 2$ and $N = 4$, so the initial value $N(0) = 3$ is in the region where $\dfrac{dN}{dt}$ is negative. Thus, the graph of N versus t begins at $N(0) = 3$ and decreases toward the equilibrium solution $N = 2$. The graph is shown in **Figure 10.37**.

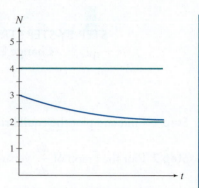

Figure 10.37 Graphing solutions: the graph of the solution of N versus t with initial value $N(0) = 3$

TRY IT YOURSELF 10.14   Brief answers provided at the end of the section.

Add to the graph you made in part c of Example 10.14 the graph of N versus t in case the initial value is $N(0) = 1$.

EXTEND YOUR REACH

a. Construct a change equation of the form $\dfrac{dy}{dx} = g(y)$ that has three equilibrium solutions.

b. Is there an upper limit to the number of equilibrium solutions that a change equation can have?

EXAMPLE 10.15   A Population of Elk

Ecologists have found that the annual growth rate of an elk population in a certain area depends on the population level N. The relationship is

$$\text{Growth rate} = 0.25N\left(1 - \frac{N}{600}\right)$$

Because the growth rate is the rate of change in the population $\dfrac{dN}{dt}$, this is the same as the change equation[4]

$$\frac{dN}{dt} = 0.25N\left(1 - \frac{N}{600}\right)$$

Here t is time in years.

a. Find the equilibrium solutions, and explain the meaning of the equilibrium solutions in terms of the elk population at those population levels.

b. Plot the graph of the equilibrium solutions.

---

[4] In this setting, 0.25 is known as the intrinsic annual exponential growth rate of the elk population. Populations following a change equation of the type in this example are said to experience logistic growth.

SOLUTION

a. To find the population levels where the growth rate is zero, we need to solve for N the equation

$$0 = 0.25N\left(1 - \frac{N}{600}\right)$$

As in the preceding example, this is a quadratic equation that is already factored. We find the solutions $N = 0$ and $N = 600$. At either of these population levels, the population is not changing. Thus, when the elk population is either 0 or 600, the population will not change but will remain stable.

b. The graph of the equilibrium solutions is shown in **Figure 10.38**.

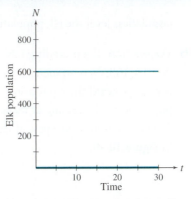

Figure 10.38 Equilibrium solutions for the elk population

TRY IT YOURSELF 10.15 Brief answers provided at the end of the section.

A certain caribou population grows according to the change equation $\dfrac{dN}{dt} = 0.29N(1 - N/700)$. Find the equilibrium solutions, and plot them on a graph.

EXAMPLE 10.16 More on the Elk Population: Graphical Solutions

In this example, we look further at the elk population from Example 10.15.

a. Plot the graph of the growth rate $\dfrac{dN}{dt}$ versus the population N, and explain what the graph reveals about the elk population N as a function of time t. In particular, determine the population level at which the population is growing at the fastest rate.

b. Assume now that the initial population is $N(0) = 100$ elk. Use the information from part a to add to the graph of equilibrium solutions (Figure 10.38) a sketch of the graph of the population N versus time t.

SOLUTION

a. The graph of $0.25N\left(1 - \dfrac{N}{600}\right)$ as a function of N is a parabola that crosses the horizontal axis at $N = 0$ and $N = 600$. The parabola opens downward, as shown in **Figure 10.39**.

This graph yields the following information.

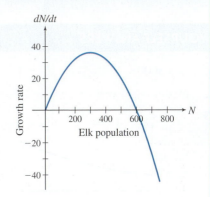

Figure 10.39 The elk population: the graph of $\dfrac{dN}{dt}$ versus N

Region	$\dfrac{dN}{dt}$	Graph of N versus t
$N = 0$	Zero	Equilibrium solution: The graph is a horizontal line.
$0 < N < 600$	Positive	N is increasing, so the graph rises.
$N = 600$	Zero	Equilibrium solution: The graph is a horizontal line.
$N > 600$	Negative	N is decreasing, so the graph falls.

We note additionally that $\dfrac{dN}{dt}$ is a maximum at $N = 300$. Thus, at this population level the elk population is increasing at its fastest rate.

b. We see from the preceding table that when the initial population is $N(0) = 100$, N is an increasing function of t, so the graph of N versus t is rising toward the equilibrium solution $N = 600$. Because the population is increasing at its fastest rate at $N = 300$, the graph of N versus t is at its steepest at this population level. The graph is shown in **Figure 10.40**.

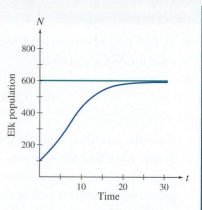

Figure 10.40 The elk population: a sketch of the graph of population versus time with an initial population of 100 elk

TRY IT YOURSELF 10.16   Brief answers provided at the end of the section.

Add to the graph you made a sketch of the graph of population versus time if the initial population is $N(0) = 800$ elk.

EXTEND YOUR REACH   Make a change equation with solutions matching the graph in **Figure 10.41**. Note that the graph indicates an equilibrium solution of $y = E$.

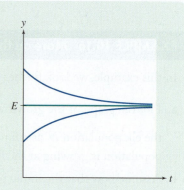

Figure 10.41 Equilibrium solution and two additional initial values for a change equation

TRY IT YOURSELF ANSWERS

10.11 $C = 5000$

10.12   60 ducks per year

10.13

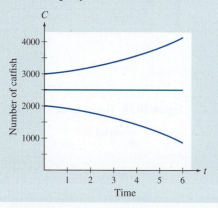

10.14

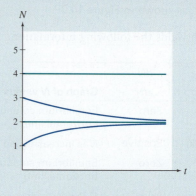

10.15 The equilibrium solutions are $N = 0$ and $N = 700$.

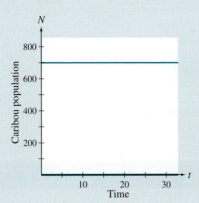

10.16

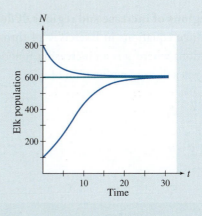

EXERCISE SET 10.3

CHECK YOUR UNDERSTANDING

1. The equilibrium solutions of the change equation $\frac{dy}{dx} = g(y)$ are the solutions of which equation?

2. Equilibrium solutions of the change equation $\frac{dy}{dx} = g(y)$:

   a. are constant functions

   b. are exponential functions

   c. show where the function y is increasing

   d. None of the above.

3. Graphs of equilibrium solutions are:

   a. graphs of increasing functions

   b. graphs of decreasing functions

   c. horizontal lines

   d. None of the above.

4. When $\frac{dy}{dx}$ is positive, the graph of y versus x is:

   a. rising

   b. falling

   c. a horizontal line

   d. None of the above.

5. When $\frac{dy}{dx}$ is negative, the graph of y versus x is:

   a. rising

   b. falling

   c. a horizontal line

   d. None of the above.

SKILL BUILDING

Finding equilibrium solutions. In Exercises 6 through 13, you are given a change equation of the form $\frac{dy}{dx} = g(y)$.

   a. Plot the graph of $\frac{dy}{dx}$ versus y.

   b. Find the equilibrium solutions, if any exist.

   c. Plot the graphs of the equilibrium solutions you find.

6. $\frac{dy}{dx} = 2 - y$

7. $\frac{dy}{dx} = 3y - 9$

8. $\frac{dy}{dx} = (y - 2)(3 - y)$

9. $\frac{dy}{dx} = y^2 - 2y + 1$

10. $\frac{dy}{dx} = y^2 - 4y + 3$

11. $\frac{dy}{dx} = y(y - 2)(y - 3)$

12. $\frac{dy}{dx} = \ln y$

13. $\frac{dy}{dx} = e^y$

Finding regions of increase and regions of decrease.
For the change equations in Exercises 14 through 19,
find the regions where y is an increasing function of x
and where y is a decreasing function of x.

14. $\dfrac{dy}{dx} = 7 - y$

15. $\dfrac{dy}{dx} = y - 5$

16. $\dfrac{dy}{dx} = (y - 2)(y - 5)$

17. $\dfrac{dy}{dx} = y^2 + 1$

18. $\dfrac{dy}{dx} = -e^{-y}$

19. $\dfrac{dy}{dx} = (y - 1)(3 - y)$

PROBLEMS

Graphing solutions. In Exercises 20 through 27, plot
the graphs of the equilibrium solutions along with the
graphs of y versus x for the given initial values (or other
function value).

20. $\dfrac{dy}{dx} = y - 3$; $y(0) = 1$ and $y(0) = 4$

21. $\dfrac{dy}{dx} = 3 - y$; $y(0) = 1$ and $y(0) = 4$

22. $\dfrac{dy}{dx} = (y - 2)(y - 4)$; $y(0) = 1$, $y(0) = 3$,
and $y(0) = 5$

23. $\dfrac{dy}{dx} = (y - 2)(4 - y)$; $y(0) = 1$, $y(0) = 3$,
and $y(0) = 5$

24. $\dfrac{dy}{dx} = y(y - 2)(y - 4)$; $y(0) = -1$, $y(0) = 1$, $y(0) = 3$,
and $y(0) = 5$

25. $\dfrac{dy}{dx} = -y(y - 2)(y - 4)$; $y(0) = -1$, $y(0) = 1$, $y(0) = 3$,
and $y(0) = 5$

26. $\dfrac{dy}{dx} = -y \ln y$; $y(0) = 2$

27. $\dfrac{dy}{dx} = -y^2$; $y(1) = 1$

28. Using technology to analyze a change equation.
Consider the following change equation:

$$\frac{dy}{dx} = e^y - y^3$$

   a. Use a graphing utility to plot the graph of $\dfrac{dy}{dx}$
versus y. Use a horizontal span of $y = 0$ to
$y = 5$.

   b. Find the equilibrium solutions correct to two
decimal places.

   c. On the span $y = 0$ to $y = 2$, what value of
y gives the largest rate of increase of y as a
function of x? Report your answer to two
decimal places.

MODELS AND APPLICATIONS

29. Merino sheep. The amount C of food consumed
by a merino sheep in a day depends on the
amount V of vegetation available. We measure C
in pounds and V in pounds per acre. Then C and
V are related by the change equation

$$\frac{dC}{dV} = 0.01(3 - C)$$

   a. Plot the graph of $\dfrac{dC}{dV}$ versus C.

   b. Find the equilibrium solution.

   c. Explain the meaning of the initial condition
$C(0) = 0$ in terms of the amount of food
consumed by a merino sheep.

   d. Plot the graph of the equilibrium solution along
with the graph of C versus V if $C(0) = 0$.

   e. What is the most a merino sheep can be
expected to consume in a day no matter how
much vegetation is available?

30. Forest litter. Plant litter such as leaves accumulates on the forest floor, where it decays. Let L be the amount of litter, in grams per square meter, after t years. We assume that litter falls at a constant rate F and decays at a rate proportional to the amount present. Then the amount of litter is described by the change equation

$$\frac{dL}{dt} = F - rL$$

Here r is the (positive) constant of proportionality.

a. Find the equilibrium solution, and explain its meaning in terms of the amount of forest floor litter.

b. Sketch the graph of the equilibrium solution. Add to this plot the graph of L versus t in the case that the initial value is a number L_0, where

$$0 < L_0 < \frac{F}{r}.$$

c. Assume that the initial amount is the same as in part b. As the amount of litter increases, the rate of decay increases. The effect is to slow the accumulation of litter so that a balance is eventually reached. What is the eventual amount of litter in the forest?

d. A study[5] of sand dunes with vegetation near Lake Michigan found the decay rate to be $r = 0.003$ per year. If a long-term study finds that litter eventually stabilizes at 2000 grams per square meter, find the value of the constant F.

31. Terminal velocity. As an object falls, gravity causes it to accelerate. But air resistance retards the acceleration, so we anticipate an eventual balance between gravity and the force due to air resistance. The velocity at this balance is known as terminal velocity. The downward velocity V in feet per second after t seconds satisfies the change equation

$$\frac{dV}{dt} = g - rV$$

Here g is the acceleration due to gravity near the surface of Earth, and r is a positive constant that depends on the object. The constant r measures the drag on the object.

a. Find the equilibrium solution. Your answer will involve the constants g and r.

b. The object falls from a great height with initial velocity 0. Sketch the graph of the equilibrium solution along with the graph of V versus t with initial condition $V(0) = 0$.

c. In view of the graph you made in part b, explain why the equilibrium solution is known as the terminal velocity.

d. For an average-sized man who jumps from an airplane, terminal velocity is about 176 feet per second. (That is about 120 miles per hour. We are assuming that the man reaches terminal velocity before he opens his parachute.) Find the value of r for a man of average size. Recall that g equals 32 feet per second per second.

32. More on terminal velocity: a meteor. A meteor may enter Earth's atmosphere at a velocity of up to 60,000 miles per hour. Earth's atmosphere slows the meteor as it falls. The downward velocity V of objects moving at such high speeds in Earth's atmosphere satisfies the change equation

$$\frac{dV}{dt} = g - 2 \times 10^{-14} V^4$$

The acceleration due to gravity g equals 78,973 miles per hour per hour. The terminal velocity of the meteor is the minimum speed that the meteor will ever attain before it strikes the ground. Assuming the meteor attains terminal velocity, how fast is it traveling when it strikes the ground? Round your answer in miles per hour to the nearest whole number.

33. A mortgage. You secure a mortgage at an APR of 4% compounded continuously. You repay the loan in the amount of $12,000 per year. Your balance B in dollars after t years satisfies the following change equation:

$$\frac{dB}{dt} = 0.04B - 12,000$$

a. Find the equilibrium solution.

b. Explain the meaning of the equilibrium solution in terms of the amount you owe at that balance level.

c. What are the long-term implications if the balance B is less than the equilibrium solution? What if it is greater than the equilibrium solution?

[5]J. S. Olson, "Rates of Succession and Soil Changes on Southern Lake Michigan Sand Dunes," *Botanical Gazette* 119:125–170 (1958).

34. **Logistic growth with a threshold.** Each species has a threshold level T, and populations with fewer individuals cannot survive. Let K denote the carrying capacity of the environment, which is the maximum population of this species that the local environment can support. Many populations grow according to a change equation of the form

$$\frac{dN}{dt} = -rN\left(1 - \frac{N}{T}\right)\left(1 - \frac{N}{K}\right)$$

Here t is time in years, and r is a positive constant, the intrinsic annual exponential growth rate of the species.

a. Find the equilibrium solutions of the equation.

b. Sketch a graph of $\frac{dN}{dt}$ versus N. (Assume that $0 < T < K$.) *Suggestion:* Note that $\frac{dN}{dt}$ is a polynomial function of N, and use the method for graphing polynomials presented in Section 5.4.

c. For the remainder of this exercise, take $r = 0.4$ per year, $T = 500$, and $K = 3000$. Use a graphing utility to make a graph of $\frac{dN}{dt}$ versus N.

d. Use your graph in part c to determine the population level at which the population is growing at the fastest rate. Round your answer to the nearest 100.

e. Make a graph of the equilibrium solutions, and add to it sketches of graphs of N versus t for the initial conditions $N(0) = 400$, $N(0) = 800$, and $N(0) = 4000$.

f. Explain in words how the population progresses for each of the three initial conditions from part e.

35. **North Sea cod.** Let w be the weight in pounds of a North Sea cod at age t years. The weight is described by the change equation

$$\frac{dw}{dt} = 2.1w^{2/3} - 0.6w$$

a. Use a graphing utility to make a graph of $\frac{dw}{dt}$ versus w.

b. Use your graph to determine the weight of a North Sea cod when it is growing at its fastest rate. Round your answer to the nearest pound.

c. Determine the weight to which the cod grows. Round your answer to the nearest pound.

d. What is happening to a North sea cod that weighs 45 pounds?

CHALLENGE EXERCISES FOR INDIVIDUALS OR GROUPS

36. **Concavity from a change equation.** A careful analysis can show us how to determine the concavity, at least in the short term, of solutions to change equations of the form $\frac{dy}{dx} = g(y)$. This exercise covers two cases, and the next exercise covers the other two cases. Suppose the graph of $\frac{dy}{dx}$ versus y is negative and increasing, as shown in **Figure 10.42**. We consider an initial value in the domain of this graph. Because $\frac{dy}{dx}$ is negative, y is a decreasing function of x. Thus, when x increases, y decreases, as shown in **Figure 10.43**. It follows that $\frac{dy}{dx}$ is decreasing as a function of x. This

means that y as a function of x is decreasing and concave down.

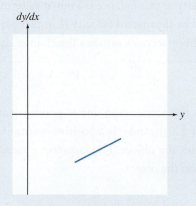

Figure 10.42 Graph of $\frac{dy}{dx}$ versus y

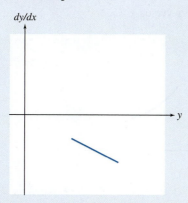

Figure 10.43 Graph of $\dfrac{dy}{dx}$ versus y:
Because $y(x)$ is decreasing, increasing
values of x mean decreasing values of y.

Now it is your turn to make such an analysis:

Show that if the graph of $\dfrac{dy}{dx}$ is as in **Figure 10.44**,
then the graph of y versus x is decreasing and
concave up.

Figure 10.44 Graph of $\dfrac{dy}{dx}$ versus y

37. **Completing a concavity list.** In the preceding
exercise, the last two items in the accompanying
bulleted list were established. Establish the first two
items in the bulleted list by connecting the given
properties of $\dfrac{dy}{dx}$ to properties of the graph of y.

Determining concavity from the rate of change

- When $\dfrac{dy}{dx}$ is positive and increasing as a function
  of y: The graph of y versus x is increasing and
  concave up.

$\dfrac{dy}{dx}$ versus y

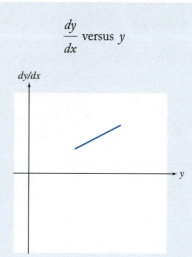

y versus x

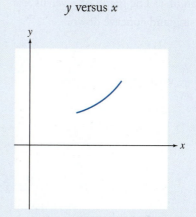

- When $\dfrac{dy}{dx}$ is positive and decreasing as a function
  of y: The graph of y versus x is increasing and
  concave down.

$\dfrac{dy}{dx}$ versus y

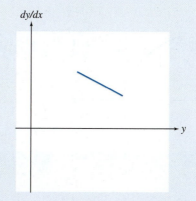

y versus *x*

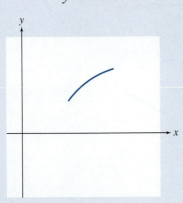

- When $\dfrac{dy}{dx}$ is negative and increasing as a function of *y*: The graph of *y* versus *x* is decreasing and concave down.

$\dfrac{dy}{dx}$ versus *y*

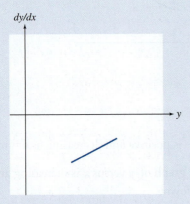

y versus *x*

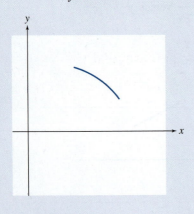

- When $\dfrac{dy}{dx}$ is negative and decreasing as a function of *y*: The graph of *y* versus *x* is decreasing and concave up.

$\dfrac{dy}{dx}$ versus *y*

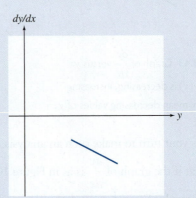

y versus *x*

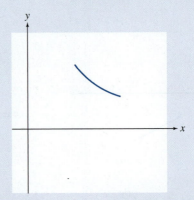

Applying concavity. Exercises 38 through 41 use the bulleted list in the preceding exercise. In each exercise, sketch the graph of *y* versus *x* for the indicated initial value, and state the concavity of the graph.

38. $\dfrac{dy}{dx} = e^y$. Use the initial condition $y(0) = 2$.

39. $\dfrac{dy}{dx} = -e^y$. Use the initial condition $y(0) = 2$.

40. $\dfrac{dy}{dx} = e^{-y}$. Use the initial condition $y(0) = 2$.

41. $\dfrac{dy}{dx} = -e^{-y}$. Use the initial condition $y(0) = 2$.

REVIEW AND REFRESH: Exercises from Previous Sections

42. From Section 4.3: Solve the equation $\ln(1 + e^x) = 2$.

Answer: $x = \ln(e^2 - 1)$

43. From Chapter 6: Using the triangle in **Figure 10.45,** express the values of the six trigonometric functions of the angle t.

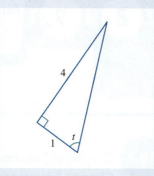

Figure 10.45 Triangle for Exercise 43

Answer: $\sin t = \dfrac{4}{\sqrt{17}}$, $\cos t = \dfrac{1}{\sqrt{17}}$, $\tan t = 4$,

$\cot t = \dfrac{1}{4}$, $\sec t = \sqrt{17}$, $\csc t = \dfrac{\sqrt{17}}{4}$

44. From Section 7.2: Sketch the graph of the tangent function.

Answer:

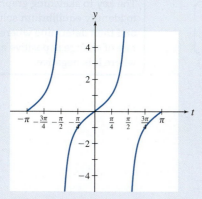

45. From Section 8.1: Express the tangent in terms of the secant and cosecant for $t \neq k\pi$.

Answer: $\tan t = \dfrac{\sec t}{\csc t}$

46. From Section 8.4: If $0 < x < 1$, simplify the expression $\sin(2\arcsin x)$.

Answer: $2x\sqrt{1 - x^2}$

47. From Section 9.1: Solve the triangle in **Figure 10.46**.

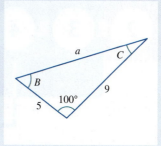

Figure 10.46 Triangle for Exercise 47

Answer: $a = 11.03$, $B = 53.48°$, $C = 26.52°$

48. From Section 9.4: If $\mathbf{v} = \langle 1, 2 \rangle$ and $\mathbf{w} = \langle 1, 4 \rangle$, calculate $\mathbf{v} \cdot (3\mathbf{v} - \mathbf{w})$.

Answer: 6

49. From Section ON9.6: Express the polar point $\left[6, \dfrac{\pi}{3} \right]$ in rectangular coordinates.

Answer: $(3, 3\sqrt{3})$

50. From Section 10.1: The graph of the rate of change of a function $f(x)$ is below the x-axis. What can we conclude about the function $f(x)$?

Answer: The function $f(x)$ is decreasing.

51. From Section 10.2: Find the solution of the change equation $\dfrac{dy}{dx} = 6$ where $y(0) = -3$.

Answer: $y = 6x - 3$

52. From Section 10.2: Find the solution of the change equation $\dfrac{dy}{dx} = 6y$ where $y(0) = -3$.

Answer: $y = -3e^{6x}$

CHAPTER ROADMAP AND STUDY GUIDE

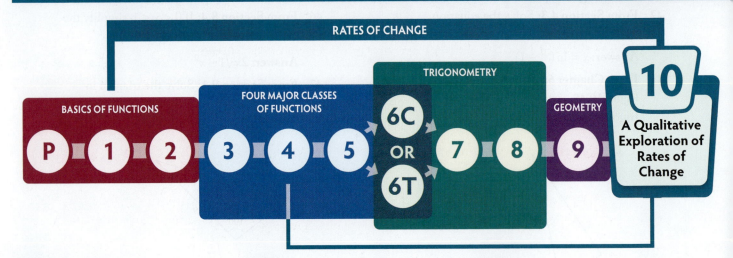

10.1 An Introduction to the Rate of Change as a Function

Rates of change are used to determine the properties of functions and their graphs.

Units of the Rate of Change

The rate of change of a function is expressed using units.

Sketching the Graph of the Rate of Change

Graphs of rates of change are produced by sketching tangent lines.

Concavity Revisited

Concavity is directly related to the rate of change.

Using Rates of Change to Analyze Functions

Local maxima and minima of functions may occur where the rate of change is zero.

10.2 Change Equations: Linear and Exponential Functions

Rates of change determine linear and exponential functions.

Constant Rate of Change

A constant rate of change is characteristic of a linear function.

The Exponential Case

A constant proportional rate of change is characteristic of an exponential function.

10.3 Graphical Solutions of Change Equations

The analysis of certain types of change equations is simplified by using graphs.

Change Equations: Equilibrium Solutions

Equilibrium solutions occur where the rate of change is zero.

Sketching Graphs of Solutions

The key to sketching graphs is to identify equilibrium solutions and then determine where the rate of change is positive and where it is negative.

CHAPTER QUIZ

1. The weight W, in pounds, of an infant is a function of her age t in days. State the units associated with the rate of change $\dfrac{dW}{dt}$, and explain what the rate of change means in terms of her weight.

   Answer: Pounds per day. The rate of growth in weight—that is, the number of pounds we expect her to gain in a day.

   ▣ **Example 10.1**

2. After class, you walk home but stop along the way to get cash at the ATM. **Figure 10.47** shows your distance D west from home on your trip as a function of time t. Your westward velocity on your trip home is the rate of change in distance west, $\dfrac{dD}{dt}$. Sketch a graph showing your westward velocity as a function of time.

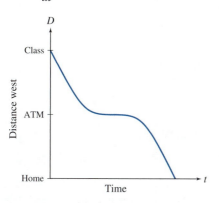

Figure 10.47 Walking home from class

Answer:

   ▣ **Example 10.2**

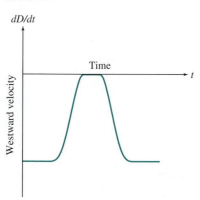

3. The height H, in feet, of a projectile t seconds after being shot upward is given by

$$H = 128t - 16t^2$$

   a. Make a graph of the height of the projectile from the time it is shot upward until it strikes the ground.

   b. The rate of change in the distance upward with respect to time, $\dfrac{dH}{dt}$, is the upward velocity. Sketch a graph of the upward velocity of the projectile during the period of its flight as a function of time.

Answer: **XR Example 10.3**

a.

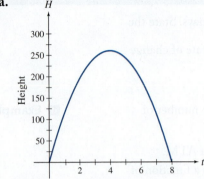

b.

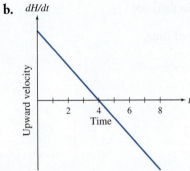

4. The population of a herd of deer is declining. The graph of the population N as
   a function of time t in years is shown in **Figure 10.48**.

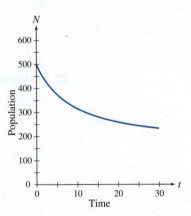

Figure 10.48 A declining population

a. Determine whether the graph is concave up or concave down, and explain
   how the concavity reflects the growth rate of the deer herd.

b. The rate of change in the population function $\dfrac{dN}{dt}$ is the growth rate of the
   deer herd. Sketch a graph of the growth rate of the deer herd.

Answer:

XR Example 10.4

a. The graph is concave up. The population is declining, but the rate of decline is slowing.

b.

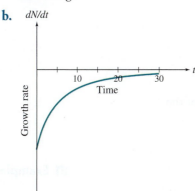

5. A stone is tossed downward with an initial velocity of 5 meters per second. Thinking of down as the positive direction, we know that the acceleration due to gravity is $g = 9.8$ meters per second per second.

   a. Write a change equation for downward velocity V, in meters per second, of the stone t seconds after it is tossed. Express the equation both in symbolic form and in words.

   b. Find a formula for V.

   Answer:

XR Example 10.7

   a. The rate of change in velocity is the acceleration, $g = 9.8$ meters per second per second:

   $$\frac{dV}{dt} = 9.8$$

   b. $V = 9.8t + 5$

6. Find a function y that satisfies the change equation $\dfrac{dy}{dt} = 3y$ if y satisfies the initial condition $y(0) = 5$.

   Answer: $y = 5e^{3t}$

XR Example 10.8

7. Lemonade drains from a jug at a rate proportional to the amount W of lemonade remaining in the jug after t minutes. The constant of proportionality is $r = -0.05$ per minute.

   a. Write a change equation that describes W as a function of time t.

   b. Using W_0 for the initial amount of lemonade in the jug, find a formula that gives W in terms of t.

   c. How long does it take for the jug to lose half its initial volume?

   Answer:

XR Example 10.9

   a. $\dfrac{dW}{dt} = -0.05W$

   b. $W = W_0 e^{-0.05t}$

   c. 13.86 minutes

8. Let F denote the number of trout in a fish hatchery at time t measured in months. Trout reproduce at a rate proportional to the number of fish present, but 35 trout are released into the wild each month. We have the following change equation:

$$\frac{dF}{dt} = 0.05F - 35$$

a. Find the equilibrium solution of the equation.

b. Explain the meaning of the solution you found in part a in terms of the change in population of the trout at that population level.

c. Plot the graph of the equilibrium solution.

Answer:

a. $F = 700$

b. If the population is 700 trout, the number of fish in the hatchery will remain the same indefinitely.

c.
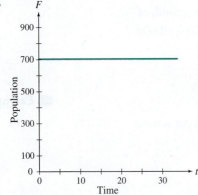

9. For the trout hatchery in the preceding exercise, make the graph of the growth rate $\frac{dF}{dt}$ versus the number of trout F. Then add to the graph you made in the preceding exercise the graph of F versus t in case the initial value is $F(0) = 500$ trout.

Answer: The graph of $\frac{dF}{dt}$ versus F is shown in **Figure 10.49.** The graph of F versus t in case $F(0) = 500$ is shown in **Figure 10.50.**

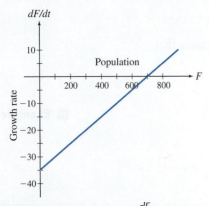

Figure 10.49 The graph of $\frac{dF}{dt}$ versus F

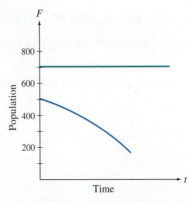

Figure 10.50 A sketch of the graph of F versus t if $F(0) = 500$

XR **Example 10.11**

XR **Example 10.13**

10. A certain population N is governed by the following change equation:

$$\frac{dN}{dt} = 0.03N\left(1 - \frac{N}{200}\right)$$

a. Plot the graph of $\frac{dN}{dt}$ versus N.

b. What population level produces the maximum growth rate?

c. Find the equilibrium solutions, and plot them.

d. Add to the graph you made in part c a graph of N versus t in case the initial condition is $N(0) = 30$, and also the graph of N versus t in case the initial condition is $N(0) = 300$.

Answer: XR **Example 10.16**

a. The graph is shown in **Figure 10.51**.

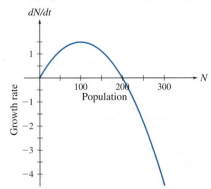

Figure 10.51 The graph of growth rate versus population

b. The growth rate is a maximum at $N = 100$.

c. The equilibrium solutions are $N = 0$ and $N = 200$. They are plotted in **Figure 10.52**.

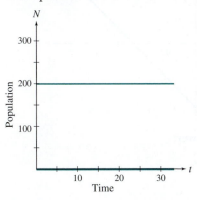

Figure 10.52 Equilibrium solutions

d.

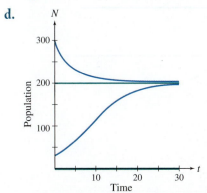

CHAPTER REVIEW EXERCISES

Section 10.1

Units of the rate of change. In Exercises 1 through 3, state the units associated with the rate of change of the given function, and explain what the rate of change means in terms of the function.

1. $H(t)$ is the height, in feet, of a flare t seconds after it is shot upward.

   Answer: Feet per second. The upward velocity—that is, the change in height we expect over a second.

2. $B(t)$ is the balance, in dollars, of a savings account after t months.

   Answer: Dollars per month. The increase in the balance we expect over a month.

3. $A(d)$ is the area in square feet of a circle of diameter d measured in feet.

   Answer: Square feet per foot. The additional area we expect for a 1-foot increase in diameter.

Sketching graphs of rates of change. In Exercises 4 through 10, you are given the graph of a function $y = f(x)$. Sketch the graph of the rate of change $\dfrac{dy}{dx}$.

4.

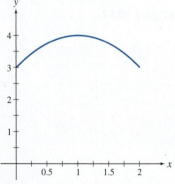

   Answer:

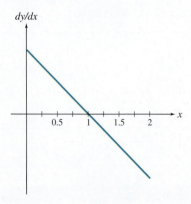

5.

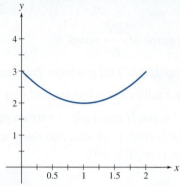

   Answer:

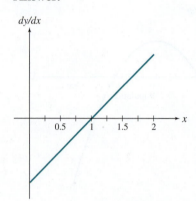

6.

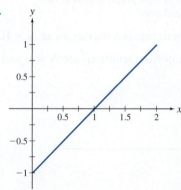

   Answer:

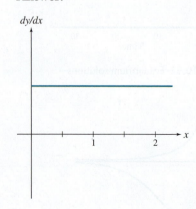

7.

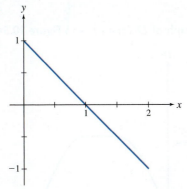

Answer:

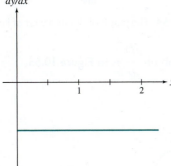

8.

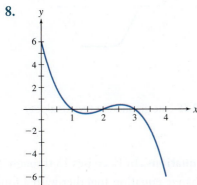

Answer:

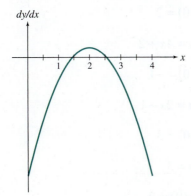

9.

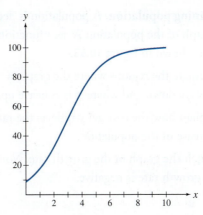

Answer:

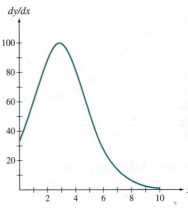

10.

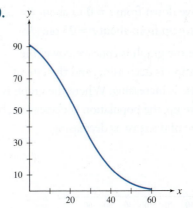

Answer:

11. **A declining population.** A population is declining. The graph of the population N as a function of time t is shown in **Figure 10.53**.

   a. Estimate the regions where the graph is concave down and where it is concave up.

   b. Explain how the concavity reflects the rate of decrease of the population.

   c. Sketch the graph of the growth rate. Note that the growth rate is negative.

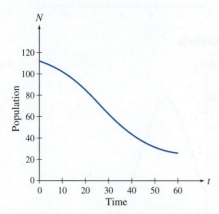

Figure 10.53 A declining population

Answer:

a. Concave down from $t = 0$ to about $t = 25$. Concave up from about $t = 25$ on.

b. Where the graph is concave down, the population is decreasing, and the rate of decrease is increasing. Where the graph is concave up, the population is decreasing, but the rate of decrease is decreasing.

c. *dN/dt*

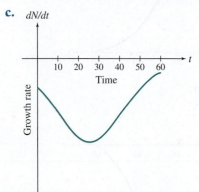

12. **A walk.** You walk west to a friend's house, where you stay for a while before walking back home.

   a. Sketch a graph of your distance west D from home versus time t.

   b. Sketch a graph of westward velocity $\dfrac{dD}{dt}$.

Answer:

a. The graph of D versus t is in **Figure 10.54**.

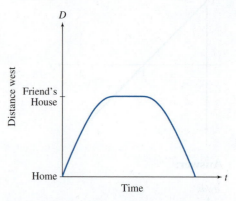

Figure 10.54 The graph of distance west of home

b. The graph of $\dfrac{dD}{dt}$ is in **Figure 10.55**.

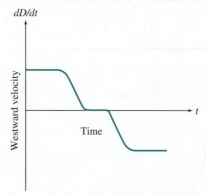

Figure 10.55 The graph of westward velocity

Section 10.2

Solving change equations. In Exercises 13 through 21, you are given a change equation together with a function value. Find the solution.

13. $\dfrac{dy}{dx} = 4$, $y(0) = 2$

   Answer: $y = 4x + 2$

14. $\dfrac{dy}{dx} = 2$, $y(0) = -1$

   Answer: $y = 2x - 1$

15. $\dfrac{dy}{dx} = 0$, $y(0) = 3$

   Answer: $y = 3$

16. $\dfrac{dy}{dx} = 4y$, $y(0) = 2$

   Answer: $y = 2e^{4x}$

17. $\dfrac{dy}{dx} = -3y,\ y(0) = 1$

    Answer: $y = e^{-3x}$

18. $\dfrac{dy}{dx} = y,\ y(0) = -1$

    Answer: $y = -e^{x}$

19. $\dfrac{dy}{dx} = 2y,\ y(1) = 5e^{2}$

    Answer: $y = 5e^{2x}$

20. $\dfrac{dy}{dx} = 3,\ y(1) = 5$

    Answer: $y = 3x + 2$

21. $\dfrac{dy}{dx} = -2y,\ y\!\left(\ln\dfrac{1}{2}\right) = 12$

    Answer: $y = 3e^{-2x}$

22. **Tossing a rock.** The (downward) acceleration due to gravity near the surface of planet P is the constant g_P, measured in meters per second per second. Let V denote the upward velocity, in meters per second, of a rock t seconds after it is tossed upward.

    a. Write a change equation that describes V.

    b. Solve the equation in part a if the initial upward velocity is $2g_P$.

    c. How long does it take the rock to reach the peak of its flight?

    Answer:

    a. $\dfrac{dV}{dt} = -g_P$

    b. $V = 2g_P - g_P t$

    c. 2 seconds

23. **An injection.** After an injection, the concentration $C(t)$ of a drug in the bloodstream changes at a rate proportional to the current concentration. Here time t is measured in hours.

    a. Using r as the constant of proportionality, write a change equation that describes C as a function of t.

    b. Using C_0 as the initial concentration, solve the equation in part a.

    c. If the concentration of the drug decreases by half over the first 6 hours, find the constant of proportionality r. Round your answer to two decimal places.

Answer:

a. $\dfrac{dC}{dt} = rC$

b. $C = C_0 e^{rt}$

c. -0.12 per hour

Section 10.3

Equilibrium solutions. In Exercises 24 through 29, find the equilibrium solutions of the given change equation.

24. $\dfrac{dy}{dx} = 6 - 2y$

    Answer: $y = 3$

25. $\dfrac{dy}{dx} = (y - 1)(2 - y)$

    Answer: $y = 1,\quad y = 2$

26. $\dfrac{dy}{dx} = y(y - 1)(y - 2)$

    Answer: $y = 0,\quad y = 1,\ y = 2$

27. $\dfrac{dy}{dx} = y^{2} - 1$

    Answer: $y = 1,\quad y = -1$

28. $\dfrac{dy}{dx} = \sqrt{y} - 3$

    Answer: $y = 9$

29. $\dfrac{dy}{dx} = 3 - \ln y$

    Answer: $y = e^{3}$

Graphing solutions. In Exercises 30 through 34, plot the graphs of all equilibrium solutions along with the graph of the solution with the given initial value.

30. $\dfrac{dy}{dx} = 6 - 2y,\ y(0) = 4$

    Answer:

    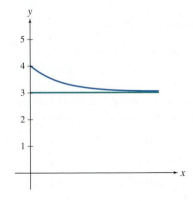

31. $\dfrac{dy}{dx} = 2y - 6$, $y(0) = 2$

Answer:

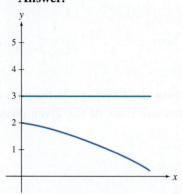

32. $\dfrac{dy}{dx} = (y - 1)(y - 3)$, $y(0) = 2$

Answer:

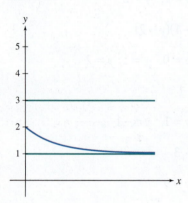

33. $\dfrac{dy}{dx} = (y - 1)(y - 3)$, $y(0) = 4$

Answer:

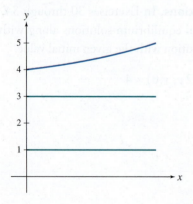

34. $\dfrac{dy}{dx} = -(y + 1)\ln y$, $y(0) = 2$

Answer:

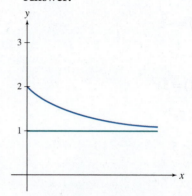

35. **The balance of a loan.** You make regular payments on a loan that accumulates interest. The balance B, in dollars, of the loan after t years satisfies the change equation

$$\frac{dB}{dt} = 0.05B - 5000$$

a. Find the equilibrium solution, and explain what it means in terms of the balance at that level.

b. Plot the graph of the equilibrium solution along with the graph of the solution with initial value $B(0) = \$125{,}000$.

c. Explain what is happening in the case of the solution plotted in part b.

Answer:

a. $B = \$100{,}000$ is the equilibrium solution. If the balance is $100,000, then payments match accrued interest. Consequently, the balance never changes.

b.

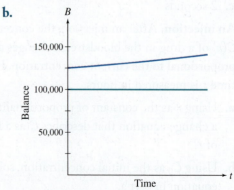

c. Payments are not keeping up with interest, so the balance grows indefinitely.

APPENDIX

USING TECHNOLOGY TO SOLVE EQUATIONS

G raphs can show us features that are difficult to see from a formula, but accurate hand-drawn graphs are difficult to produce. Graphing calculators and computers can not only make accurate graphs quickly and easily but also help us analyze them. They are particularly useful in solving equations, finding maximum and minimum values, and estimating long-term behavior.

Let's see how a graphing utility can be used to solve an equation such as $2^x = x^3$. The first step is to graph the two functions $y = 2^x$ and $y = x^3$ on the same screen. A table of values such as the one in **Figure A.1** is helpful in determining where to look for solutions and hence how to choose a graphing window. The table shows that 2^x is larger than x^3 at $x = 0$. But by the time we get to $x = 1.5$, x^3 is larger. So a solution is somewhere between $x = 0$ and $x = 1.5$. We use a horizontal span from 0 to 2 for the viewing window. This same table prompts us to choose a vertical span from 0 to 8.

These graphs are shown in **Figure A.2**, where the darker curve represents $y = 2^x$. A solution of the equation $2^x = x^3$ occurs where the two graphs cross. The coordinates of the crossing point are shown in the figure, where we see that $x \approx 1.37$ is an approximate solution of the equation. The two graphs cross for a second time, so there is a second solution of the equation. The value is $x \approx 9.94$, as shown in **Figure A.3**.

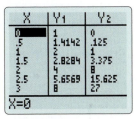

Figure A.1 A table of values for $y = 2^x$ (second column) and $y = x^3$ (third column)

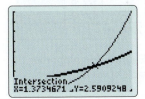

Figure A.2 The graph of $y = 2^x$ (darker curve) and $y = x^3$ on the horizontal span $[0, 2]$: The solution of the equation is the crossing point.

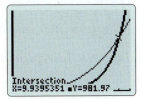

Figure A.3 A second solution: This solution occurs between $x = 9$ and $x = 10$.

This procedure is easily applied even to quite complicated equations. All that is needed is a bit of familiarity with the technology being used.

The same strategy can be used to find maximum and minimum values for functions. For example, the graph of $y = \dfrac{3x + 1}{x^2 + 1}$ is shown in **Figure A.4**. The graph shows that the function reaches a local maximum value of $y \approx 2.08$ at $x \approx 0.72$. In a similar fashion we learn from **Figure A.5** that $y = x^4 - x + 1$ reaches a local minimum value of $y \approx 0.53$ at $x \approx 0.63$.

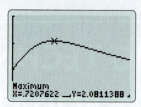

Figure A.4 The maximum value of $y = \dfrac{3x+1}{x^2+1}$ on the interval $[0,2]$

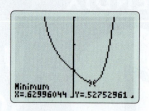

Figure A.5 The minimum value of $y = x^4 - x + 1$ on the interval $[-2,2]$

Finally, graphs can be used to estimate limiting values. For example, a certain population is given by

$$N = \frac{6.21}{0.035 + 0.45^t} \quad \text{deer}$$

where t is time measured in years. We wish to find the eventual population level. That is, we want to know the limit as $t \to \infty$ of $N(t)$ if the limit exists. The graph in **Figure A.6** shows that the population increases rapidly over the first 10 or so years but appears to level out after that. Tracing the graph, as we have done in Figure A.6, suggests that the population stabilizes at around 177 individuals. The table of values in **Figure A.7** reinforces this assertion. Calculus techniques can be used to show that in fact $N(t) \to \dfrac{6.21}{0.035} \approx 177.43$ as $t \to \infty$.

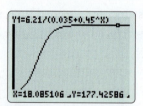

Figure A.6 Tracing a graph to estimate a limit

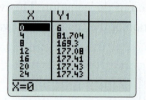

Figure A.7 A supporting table of values

Modern technology, used properly, allows analysis of functions simply not possible via hand calculation. Powerful computer programs known as computer algebra systems can perform the calculations shown here along with many others. (Examples of computer algebra systems include Maple, Mathematica, MATLAB, and Wolfram Alpha.) Computer algebra systems sometimes provide exact answers, as opposed to the approximate solutions given by most calculators. Anyone who is likely to have dealings with mathematical calculations should become familiar with the operation of these programs.

APPENDIX

2

COMPLEX NUMBERS

A complex number is a number of the form $z = a + bi$, where a and b are real numbers and $i^2 = -1$. This is called the standard form of z. Here a is called the real part of z, and b is called the imaginary part of z. Note that a complex number z is a real number precisely when its imaginary part is zero.

For addition and multiplication of complex numbers such as $a + bi$, we treat the expression as if it were $a + bx$. The analogy is emphasized in the following side-by-side calculations:

The **sum** of the complex numbers $a + bi$ and $c + di$ in standard form is defined by

$$(2 + 3i) + (5 + 7i) = (2 + 5) + (3 + 7)i \qquad (2 + 3x) + (5 + 7x) = (2 + 5) + (3 + 7)x$$

$$= 7 + 10i \qquad\qquad = 7 + 10x$$

$$(a + bi) + (c + di)$$
$$= (a + c) + (b + d)i$$

$$(2 + 3i)(5 + 7i) = 10 + 14i + 15i + 21i^2 \qquad (2 + 3x)(5 + 7x) = 10 + 14x + 15x + 21x^2$$

$$= 10 + 29i + 21i^2 \qquad\qquad = 10 + 29x + 21x^2$$

$$= 10 + 29i + 21(-1)$$

$$= -11 + 29i$$

The **product** of the complex numbers $a + bi$ and $c + di$ in standard form is defined by

$$(a + bi)(c + di)$$
$$= (ac - bd) + (ad + bc)i$$

This analogy suggests how to define the **sum** and the **product** of two complex numbers.

EXAMPLE A.1 Calculations with Complex Numbers

Let $z = 2 - i$ and $w = 1 + 4i$. Calculate $2z - w$ and zw. All answers should be expressed in standard form.

SOLUTION

We use the definitions of addition and multiplication:

$$2z - w = 2(2 - i) - (1 + 4i)$$

$$= 4 - 2i - 1 - 4i$$

$$= (4 - 1) + (-2 - 4)i$$

$$= 3 - 6i$$

Compare the calculation of $2(2 - i) - (1 + 4i)$ with that for $2(2 - x) - (1 + 4x)$.

Also,

$$zw = (2 - i)(1 + 4i)$$

$$= 2 + 8i - i - 4i^2$$

$$= 2 + 7i - 4i^2$$

$$= 2 + 7i - 4(-1)$$

$$= 6 + 7i$$

Compare the calculation of $(2 - i)(1 + 4i)$ with that for $(2 - x)(1 + 4x)$.

If $z = a + bi$ is in standard form, the **complex conjugate** of z is denoted $\bar{z}$ and is defined by

$$\bar{z} = a - bi.$$

The **complex conjugate** of a complex number has the same real part, but the imaginary part is multiplied by -1.

Complex conjugates arise when we find zeros of quadratics that have real coefficients and a negative discriminant (as defined in Section 5.1). For example, if we use the quadratic formula to find the zeros of $x^2 + 2x + 2$, we obtain

$$x = \frac{-2 \pm \sqrt{2^2 - 4(1)(2)}}{2}$$

$$= \frac{-2 \pm \sqrt{-4}}{2}$$

$$= \frac{-2 \pm 2i}{2}$$

$$= -1 \pm i$$

Hence, the zeros are $x = -1 + i$ and $x = -1 - i$. Note that each is the complex conjugate of the other. The structure of the quadratic formula tells us that this is typical. If a quadratic function with real coefficients has a negative discriminant, the zeros of the function always occur in conjugate pairs, $x = a \pm bi$.

In fact, complex zeros of all polynomials with real coefficients occur in conjugate pairs (if at all). To see this, we need further information regarding complex conjugates. For complex numbers z and w:

Property 1. $\overline{z + w} = \bar{z} + \bar{w}$

Property 2. $\overline{zw} = \bar{z}\bar{w}$

Property 3. If $z = a + bi$ is in standard form, then

$$z + \bar{z} = 2a$$

$$z\bar{z} = a^2 + b^2$$

In particular, both $z + \bar{z}$ and $z\bar{z}$ are real numbers.

We will verify property 2. You are asked to verify the rest.

Let $z = a + bi$ and $w = c + di$, both in standard form. First we calculate $\overline{zw}$:

$$\overline{zw} = \overline{(a + bi)(c + di)}$$

$$= \overline{(ac - bd) + (ad + bc)i}$$

$$= (ac - bd) - (ad + bc)i$$

Next we calculate $\bar{z}\,\bar{w}$:

$$\bar{z}\,\bar{w} = \overline{(a + bi)}\,\overline{(c + di)}$$

$$= (a - bi)(c - di)$$

$$= (ac - bd) - (ad + bc)i$$

Because the two calculations yield identical results, we conclude that $\overline{zw} = \bar{z}\,\bar{w}$.

EXAMPLE A.2 Using Complex Conjugates

Write $\dfrac{3+5i}{1+i}$ in standard form $a+bi$.

SOLUTION

Our immediate goal is to turn the denominator of this fraction into a real number. Property 3 tells us how to do this. We multiply top and bottom of the fraction by the conjugate of the denominator:

$$\frac{3+5i}{1+i} = \frac{(3+5i)(1-i)}{(1+i)(1-i)}$$

$$= \frac{8+2i}{1^2+1^2}$$

$$= \frac{8+2i}{2}$$

$$= 4+i$$

We are now prepared to show the following.

LAWS OF MATHEMATICS: Conjugate Zeros of Polynomials

If z is a zero of a polynomial $P(x)$ with real coefficients, then $\bar{z}$ is also a zero.

The immediate goal is to show that if $P(x)$ has real coefficients and w is any complex number, then $P(\overline{w}) = \overline{P(w)}$. Let $P(x) = a_n x^n + a_{n-1} x^{n-1} + \cdots + a_1 x + a_0$.

Step 1: Apply property 2 repeatedly to show that for any positive integer k, $\overline{w^k} = \overline{w}^k$.

Step 2: Apply property 1 to show that

$$\overline{P(w)} = \overline{a_n w^n} + \overline{a_{n-1} w^{n-1}} + \cdots + \overline{a_1 w} + \overline{a_0}$$

Step 3: Apply step 1 to show that

$$\overline{P(w)} = \overline{a_n}\ \overline{w}^n + \overline{a_{n-1}}\ \overline{w}^{n-1} + \cdots + \overline{a_1}\ \overline{w} + \overline{a_0}$$

Step 4: Use the fact that for a real number a we have $\bar{a} = a$ to conclude that

$$\overline{P(w)} = a_n \overline{w}^n + a_{n-1} \overline{w}^{n-1} + \cdots + a_1 \overline{w} + a_0 = P(\overline{w})$$

We use what we have established to verify the law. If $P(z) = 0$, then

$$P(\bar{z}) = \overline{P(z)} = \bar{0} = 0$$

This establishes the result.

APPENDIX

3

THE COMPLEX EXPONENTIAL FUNCTION

The **complex exponential function** of a real number (or angle in radians) θ is defined by

$$e^{i\theta} = \cos\theta + i\sin\theta$$

In Section ON9.5, we introduced the trigonometric form of complex numbers and saw rules for multiplying and dividing numbers in trigonometric form. These rules remind us of the way exponents for real numbers behave. This fact at least partially motivates the following definition of the **complex exponential function** for real numbers. This definition is known as the Euler formula.

Several values of $e^{i\theta}$ are shown in **Figure A.8**. If $s(t) = e^{it}$, then $s(0) = 1$. As t increases, $s(t)$ moves around the unit circle in the counterclockwise direction.

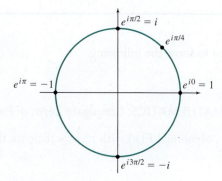

Figure A.8 Locating values of the complex exponential function

Thus, we can think of $s(t)$ as a parameterization (in the sense discussed in Section ON9.7) of the unit circle. Furthermore, because both the sine and the cosine have period 2π, $s(t)$ completes a circuit of the circle each 2π units. As t changes in the negative direction, $s(t)$ traverses the unit circle in a clockwise direction.

We emphasize that the Euler formula is valid only when the angle is expressed in radian measure. Adjustments are required to accommodate degree measure. Some elementary texts use the notation cis (θ) in place of $e^{i\theta}$. The cis notation is not used in advanced mathematics.

Applying what we know of algebraic operations on complex numbers written in trigonometric form (see Section ON9.5), we have the identity

$$e^{i\alpha}e^{i\beta} = (\cos\alpha + i\sin\alpha)(\cos\beta + i\sin\beta)$$
$$= \cos(\alpha + \beta) + i\sin(\alpha + \beta)$$
$$= e^{i(\alpha+\beta)}$$

This identity tells us that the complex exponential follows the familiar product rule for real exponents $x^a x^b = x^{a+b}$. In a similar fashion we can show additional rules:

$$\frac{e^{i\alpha}}{e^{i\beta}} = e^{i(\alpha-\beta)}$$

and, for integers n, $\left(e^{i\theta}\right)^n = e^{in\theta}$

This last equation is in fact de Moivre's theorem (for powers) couched in terms of the complex exponential function. In this form the theorem is much more intuitively appealing because it simply says that the exponents involved behave as we would expect from dealing with exponents in the case of real numbers. It is important to remember that the formula $\left(e^{i\theta}\right)^n = e^{in\theta}$ applies only when n is an integer.

EXAMPLE A.3 Algebraic Operations on the Complex Exponential

Simplify $\left(e^{i\pi/3}\right)^5 e^{i\pi/6}$. Write your final answer in standard form.

SOLUTION

We apply the preceding rules:

$$\left(e^{i\pi/3}\right)^5 e^{i\pi/6} = e^{i5\pi/3} e^{i\pi/6}$$

$$= e^{i(5\pi/3 + \pi/6)}$$

$$= e^{i11\pi/6}$$

$$= \cos\frac{11\pi}{6} + i\sin\frac{11\pi}{6}$$

$$= \frac{\sqrt{3}}{2} - i\frac{1}{2}$$

Every nonzero complex number can be written in the form $z = re^{i\theta}$, where $z = r(\cos\theta + i\sin\theta)$ is the trigonometric form for z. Consider for example $z = 6 + 6i$. We have $|z| = \sqrt{6^2 + 6^2} = 6\sqrt{2}$. Hence,

$$z = 6\sqrt{2}\left(\frac{1}{\sqrt{2}} + i\frac{1}{\sqrt{2}}\right)$$

The trigonometric form (using radians) of z is then

$$z = 6\sqrt{2}\left(\cos\frac{\pi}{4} + i\sin\frac{\pi}{4}\right)$$

This gives the exponential form

$$z = 6\sqrt{2}e^{i\pi/4}$$

EXAMPLE A.4   Writing in Complex Exponential Form

Write $z = \dfrac{3}{2} + i\,\dfrac{3\sqrt{3}}{2}$ in the form $re^{i\theta}$.

SOLUTION

To write z in trigonometric form, we need its modulus and argument:

$$|z| = \sqrt{\left(\frac{3}{2}\right)^2 + \left(3\,\frac{\sqrt{3}}{2}\right)^2} = \sqrt{\frac{9}{4} + \frac{27}{4}} = \sqrt{9} = 3$$

Thus, $z = 3\left(\dfrac{1}{2} + i\,\dfrac{\sqrt{3}}{2}\right)$. If $\theta = \arg(z)$, then $\cos\theta = \dfrac{1}{2}$ and $\sin\theta = \dfrac{\sqrt{3}}{2}$. Because z lies in the first quadrant, $\theta = \dfrac{\pi}{3}$. Therefore,

$$z = 3\left(\cos\frac{\pi}{3} + i\sin\frac{\pi}{3}\right) = 3e^{i\pi/3}$$

Many operations, including finding roots, are simpler to perform using the complex exponential form. For example, the nth roots of $e^{i\theta}$ are $e^{i(\theta + 2k\pi)/n}$, where $k = 0, 1, \ldots, n-1$.

EXAMPLE A.5   Finding Roots

Use the complex exponential to find the sixth roots of -1.

SOLUTION

We note that $-1 = e^{i\pi}$, so we take $\theta = \pi$. Applying the preceding formula, we obtain the roots

$$e^{i(\pi + 2k\pi)/6} \quad \text{where} \quad k = 0, 1, \ldots, 5$$

This gives the following six roots:

Root 1: $e^{i\pi/6} = \dfrac{\sqrt{3}}{2} + i\,\dfrac{1}{2}$

Root 2: $e^{i(\pi + 2\pi)/6} = e^{i\pi/2} = i$

Root 3: $e^{i(\pi + 4\pi)/6} = e^{i5\pi/6} = -\dfrac{\sqrt{3}}{2} + i\,\dfrac{1}{2}$

Root 4: $e^{i(\pi + 6\pi)/6} = e^{i7\pi/6} = -\dfrac{\sqrt{3}}{2} - i\,\dfrac{1}{2}$

Root 5: $e^{i(\pi + 8\pi)/6} = e^{i3\pi/2} = -i$

Root 6: $e^{i(\pi + 10\pi)/6} = e^{i11\pi/6} = \dfrac{\sqrt{3}}{2} - i\,\dfrac{1}{2}$

In order to derive the monthly payment formula, we need the geometric sum formula from Section ON13.2. We recall the pertinent information here.

For a given number a, the sum

$$1 + a + a^2 + a^3 + \cdots + a^n$$

is a geometric sum. If $a \neq 1$, the sum is given by the formula

$$1 + a + a^2 + a^3 + \cdots + a^n = \frac{1 - a^{n+1}}{1 - a}$$

For example, taking $a = 3$ gives

$$1 + 3 + 3^2 + 3^3 + \cdots + 3^n = \frac{1 - 3^{n+1}}{1 - 3} = \frac{3^{n+1} - 1}{2}$$

In particular, to evaluate $1 + 3 + 3^2 + 3^3 + 3^4$, we put $n = 4$ into the last equation to find

$$1 + 3 + 3^2 + 3^3 + 3^4 = \frac{3^{4+1} - 1}{2} = 121$$

For the derivation of the monthly payment formula, we need the following variables:

- B_0: The amount borrowed
- B_n: The balance after n months
- r: The monthly interest rate as a decimal
- M: The monthly payment
- p: The term of the loan in months. (This is the number of payments to be made.)

First, we find a formula for B_n. At the end of the month we add accrued interest and then subtract the payment:

$$\text{New balance} = \text{Old balance} + \text{Interest} - \text{Payment}$$
$$B_{n+1} = B_n + rB_n - M$$
$$= B_n(1 + r) - M$$

It is convenient to put $a = 1 + r$, so

$$B_{n+1} = B_n a - M$$

We apply this formula each month to get

$$B_1 = B_0 a - M \qquad \blacktriangleleft \text{Balance after first month}$$

$$B_2 = B_1 a - M = (B_0 a - M)a - M = B_0 a^2 - M(1 + a) \qquad \blacktriangleleft \text{Balance after second month}$$

$$B_3 = B_2 a - M = (B_0 a^2 - M(1+a))a - M = B_0 a^3 - M(1 + a + a^2) \qquad \blacktriangleleft \text{Balance after third month}$$

$$\vdots$$

$$B_n = B_0 a^n - M(1 + a + a^2 + \cdots + a^{n-1}) \qquad \blacktriangleleft \text{Balance after } n \text{ months}$$

Now we apply the geometric sum formula to obtain

$$B_n = B_0 a^n - M\left(\frac{1 - a^n}{1 - a}\right)$$

If we replace a by $1 + r$, we have

$$\text{Balance after } n \text{ months} = B_n = B_0(1 + r)^n - M\left(\frac{(1 + r)^n - 1}{r}\right)$$

Finally, we can use the formula for the balance to derive the formula for the monthly payment. We know that the balance after n months is

$$B_n = B_0(1 + r)^n - M\left(\frac{(1 + r)^n - 1}{r}\right)$$

To say that the loan is paid off in p months means that when $n = p$, the balance B_n is zero. We find

$$0 = B_0(1 + r)^p - M\left(\frac{(1 + r)^p - 1}{r}\right) \qquad \blacktriangleleft \text{Balance is zero after } p \text{ months.}$$

$$M\left(\frac{(1 + r)^p - 1}{r}\right) = B_0(1 + r)^p \qquad \blacktriangleleft \text{Move term to other side.}$$

$$M = \frac{B_0 r(1 + r)^p}{((1 + r)^p - 1)} \qquad \blacktriangleleft \text{Divide by coefficient of } M.$$

This is the desired formula for the monthly payment.

The law of sines in Section 9.2 tells us that if we take the length of any side of a triangle and divide it by the sine of its opposite angle, we get the same number no matter which of the three sides we choose. This fact suggests that this number means something special about the triangle. Indeed it does, but we need to review a bit of the geometry of circles before we can explain what the number means.

Let $\overset{\frown}{AB}$ denote the arc of the circle shown in **Figure A.9**, and let C be a point of the circle not on that arc. Then we say that $\angle ACB$ is subtended by the arc $\overset{\frown}{AB}$. In the special case that the arc is exactly half the circumference as shown in **Figure A.10**, the corresponding chord AB is a diameter, and we say the angle is inscribed in the semicircle.

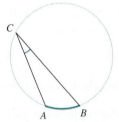

Figure A.9 An angle subtended by an arc

Figure A.10 An angle inscribed in a semicircle (AB is a diameter)

LAWS OF MATHEMATICS: Thales' Theorem

Every angle inscribed in a semicircle is a right angle.[1]

It is easy to show that Thales' theorem is true. In **Figure A.11** we have added the radius OC to the picture and labeled four of the angles. We use $\angle^\circ\alpha$ to denote the degree measure of an angle α. The angle sum of the big triangle is $180°$. Thus,

$$\angle^\circ\beta + \angle^\circ\gamma + \angle^\circ\delta + \angle^\circ\theta = 180°$$

But OA, OC, and OB are all radii of the circle, so the two triangles $\triangle AOC$ and $\triangle BOC$ are isosceles.

Because base angles of isosceles triangles are congruent (that is, have equal measure), we have

$$\angle^\circ\beta = \angle^\circ\gamma$$
$$\angle^\circ\delta = \angle^\circ\theta$$

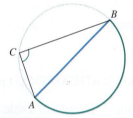

Figure A.11 Adding a radius to make isosceles triangle

---

[1]This result is associated with Thales of Miletus, c. 624–546 BCE, who was a pre-Socratic Greek philosopher.

This gives

$$180° = \angle°\beta + \angle°\gamma + \angle°\delta + \angle°\theta = 2\angle°\gamma + 2\angle°\delta$$

so

$$\angle°\gamma + \angle°\delta = 90°$$

But because $\angle°\gamma + \angle°\delta = \angle° ACB$, we conclude that $\angle° ACB = 90°$, as claimed.

A given arc of a circle subtends many angles, as illustrated in **Figure A.12**. But, perhaps surprisingly, all of these angles are congruent (that is, have the same measure).

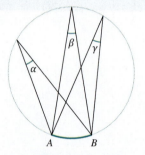

Figure A.12 Angles subtended by the same arc: The angles α, β, γ (and many others) are congruent.

> ### LAWS OF MATHEMATICS: Angles Subtended by the Same Arc
>
> Angles subtended by the same arc are congruent.

The proof of this result uses the same ideas as does the proof of Thales' theorem but is significantly more challenging. The proof can be found in any text on plane geometry.

We are now prepared to explain the geometric meaning of the ratio of a side of a triangle to the sine of its opposite angle. Three points that do not lie on a straight line determine a circle. Thus, there is exactly one circle passing through the vertices of any triangle. It is known as the circumscribed circle of the triangle.

> ### LAWS OF MATHEMATICS: Law of Sines and the Circumscribed Circle
>
> For any triangle, the ratio of a side to the sine of its opposite angle is the diameter of its circumscribed circle.

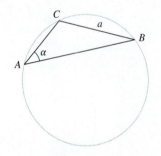

Figure A.13 The circumscribed circle

Figure A.13 shows the circumscribed circle, and in **Figure A.14** we have added a diameter of the circle. Now angles α and β are subtended by the same arc, $\overset{\frown}{BC}$, so they are congruent. Hence, $\sin\alpha = \sin\beta$. But also Thales' theorem tells us that $\triangle CBD$ is a right triangle, so we can use it to calculate $\sin\beta$:

$$\sin\beta = \frac{\text{Opposite}}{\text{Hypotenuse}} = \frac{a}{d}$$

Thus,

$$\sin\alpha = \frac{a}{d}$$

and hence,

$$d = \frac{a}{\sin\alpha}$$

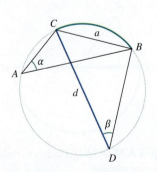

Figure A.14 Adding a diameter

This is the result we wanted to prove.

MATRIX INVERSES

For a real number $a \neq 0$, the number $a^{-1} = \dfrac{1}{a}$ is called the (multiplicative) inverse of a. The key property of an inverse is that $aa^{-1} = 1$. Intuitively, the inverse of an n by n matrix A plays an analogous role for matrix multiplication as does the number a^{-1} in multiplication of real numbers. The n by n identity matrix

$$I_n = \begin{pmatrix} 1 & 0 & 0 & \cdots & 0 \\ 0 & 1 & 0 & \cdots & 0 \\ 0 & 0 & 1 & \cdots & 0 \\ & & & \ddots & \\ 0 & 0 & 0 & \cdots & 1 \end{pmatrix}$$

plays the role in matrix multiplication of the real number 1:

$$AI_n = I_n A = A$$

We use the identity matrix to define the **matrix inverse**.

For example, if $B = \begin{pmatrix} 2 & 1 \\ 3 & 2 \end{pmatrix}$, then the inverse of B is $B^{-1} = \begin{pmatrix} 2 & -1 \\ -3 & 2 \end{pmatrix}$. You can easily verify that

$$BB^{-1} = B^{-1}B = I_2$$

Not all n by n matrices have inverses. Look, for example, at the matrix $C = \begin{pmatrix} 0 & 0 \\ 0 & 1 \end{pmatrix}$. Observe that

$$\begin{pmatrix} 0 & 0 \\ 0 & 1 \end{pmatrix}\begin{pmatrix} x & y \\ z & w \end{pmatrix} = \begin{pmatrix} 0 & 0 \\ z & w \end{pmatrix}$$

The row of zeros across the top means that the product of C with any matrix cannot yield I_2, which does not have a row of zeros at the top. Consequently, C has no inverse.

> The **matrix inverse** of an n by n matrix A is an n by n matrix A^{-1} such that
>
> $$AA^{-1} = A^{-1}A = I_n$$

For 2 by 2 matrices, the determinant can be used to find the inverse quickly (if it exists). Section ON11.4 noted that the determinant of the matrix $\begin{pmatrix} a & b \\ c & d \end{pmatrix}$ is defined as

$$\begin{vmatrix} a & b \\ c & d \end{vmatrix} = ad - bc$$

Here is the formula for the inverse, if it exists.

CONCEPTS TO REMEMBER: Inverses of 2 by 2 Matrices

For a 2 by 2 matrix $A = \begin{pmatrix} a & b \\ c & d \end{pmatrix}$:

- If $|A| = 0$, then A has no inverse.
- If $|A| \neq 0$, then $A^{-1} = \dfrac{1}{|A|}\begin{pmatrix} d & -b \\ -c & a \end{pmatrix}$.

EXAMPLE A.6 Inverses of 2 by 2 Matrices

Determine whether the following matrices have inverses. If the inverse exists, find it.

a. $A = \begin{pmatrix} 6 & 3 \\ 4 & 2 \end{pmatrix}$

b. $B = \begin{pmatrix} 2 & 2 \\ 3 & 4 \end{pmatrix}$

SOLUTION

a. First we calculate the determinant: $|A| = 6 \times 2 - 3 \times 4 = 0$. Because the determinant is 0, A has no inverse.

b. The determinant is $|B| = 2 \times 4 - 2 \times 3 = 2$. Because the determinant is not zero, we know that B has an inverse. We find it using the preceding formula:

$$B^{-1} = \frac{1}{2}\begin{pmatrix} 4 & -2 \\ -3 & 2 \end{pmatrix} = \begin{pmatrix} 2 & -1 \\ -\dfrac{3}{2} & 1 \end{pmatrix}$$

You should verify that $BB^{-1} = B^{-1}B = I_2$.

For larger matrices, the inverse can be found using determinants, but that formula is very difficult to use. Instead, it is more efficient to use a process involving row reduction to find the inverse (if it exists). The procedure is as follows.

STEP-BY-STEP STRATEGY: Procedure for Finding Inverses

For an n by n matrix A, we proceed as follows to determine whether the inverse of A exists and to find the inverse if it does exist.

Step 1 Augment the matrix A by adjoining the identity matrix I_n to make the matrix $(\ A \mid I_n \ )$.

Step 2 Perform the row operations on $(\ A \mid I_n \ )$ that put the matrix A in reduced form.

Step 3 If, at any point in the reduction process, a row of zeros appears before the bar, as in the illustration that follows, then A has no inverse.

$$\begin{pmatrix} \vdots & \vdots & \vdots & \vdots & \vdots & \vdots & \vdots & \vdots \\ 0 & 0 & \cdots & 0 & a & b & \cdots & z \\ \vdots & \vdots & \vdots & \vdots & \vdots & \vdots & \vdots & \vdots \end{pmatrix}$$

Step 4 If the reduced form of A is I_n, then the reduction process yields

$$(\ I_n \mid B \ )$$

The matrix B is the inverse of A.

EXAMPLE A.7 Finding Inverses of Larger Matrices

For the following matrices, determine whether an inverse exists. If the inverse exists, find it.

a. $A = \begin{pmatrix} 1 & 2 & 3 \\ 0 & 1 & 2 \\ 1 & 4 & 7 \end{pmatrix}$

b. $B = \begin{pmatrix} 1 & 1 & 2 \\ 1 & 2 & 3 \\ 2 & 3 & 6 \end{pmatrix}$

SOLUTION

a. The augmented matrix we need is

$$(\ A \mid I_3 \ ) = \begin{pmatrix} 1 & 2 & 3 & 1 & 0 & 0 \\ 0 & 1 & 2 & 0 & 1 & 0 \\ 1 & 4 & 7 & 0 & 0 & 1 \end{pmatrix}$$

Next we use row operations to put A in reduced form:

$$\begin{pmatrix} 1 & 2 & 3 & 1 & 0 & 0 \\ 0 & 1 & 2 & 0 & 1 & 0 \\ 0 & 2 & 4 & -1 & 0 & 1 \end{pmatrix}$$ ◀ **Row 3 – Row 1** ↦ new **Row 3**
to make zero in row 3, column 1.

$$\begin{pmatrix} 1 & 2 & 3 & 1 & 0 & 0 \\ 0 & 1 & 2 & 0 & 1 & 0 \\ 0 & 0 & 0 & -1 & -2 & 1 \end{pmatrix}$$ ◀ **Row 3 – 2Row 2** ↦ new **Row 3**
to make zero in row 3, column 2.

In the third row, we see all zeros before the bar. This signals us that A has no inverse, and no further calculations are required.

b. The augmented matrix we need is

$$(\; B \mid I_3 \;) = \begin{pmatrix} 1 & 1 & 2 & 1 & 0 & 0 \\ 1 & 2 & 3 & 0 & 1 & 0 \\ 2 & 3 & 6 & 0 & 0 & 1 \end{pmatrix}$$

We proceed with row reduction:

$$\begin{pmatrix} 1 & 1 & 2 & 1 & 0 & 0 \\ 0 & 1 & 1 & -1 & 1 & 0 \\ 0 & 1 & 2 & -2 & 0 & 1 \end{pmatrix}$$

◀ **Row 2 – Row 1** ↦ new **Row 2**
◀ **Row 3 – 2Row 1** ↦ new **Row 3**
  to make zeros in column 1, rows 2 and 3.

$$\begin{pmatrix} 1 & 0 & 1 & 2 & -1 & 0 \\ 0 & 1 & 1 & -1 & 1 & 0 \\ 0 & 0 & 1 & -1 & -1 & 1 \end{pmatrix}$$

◀ **Row 1 – Row 2** ↦ new **Row 1**
◀ **Row 3 – Row 2** ↦ new **Row 3**
  to make zeros in column 2, rows 1 and 3.

$$\begin{pmatrix} 1 & 0 & 0 & 3 & 0 & -1 \\ 0 & 1 & 0 & 0 & 2 & -1 \\ 0 & 0 & 1 & -1 & -1 & 1 \end{pmatrix}$$

◀ **Row 1 – Row 3** ↦ new **Row 1**
◀ **Row 2 – Row 3** ↦ new **Row 2**
  to make zeros in column 3, rows 1 and 2.

Now the 3 by 3 identity matrix appears to the left of the bar, so the inverse of B is the matrix to the right of the bar:

$$B^{-1} = \begin{pmatrix} 3 & 0 & -1 \\ 0 & 2 & -1 \\ -1 & -1 & 1 \end{pmatrix}$$

You should verify that $BB^{-1} = B^{-1}B = I_3$.

Mathematical induction is a method of proof that applies to theorems about positive integers. As an example, for each positive integer n, let $T(n)$ denote the proposition that

$$1 + 2 + \cdots + n = \frac{n(n+1)}{2}$$

It is a theorem that this formula holds for every positive integer n. The principle of induction says that we can verify the truth of this theorem in two steps.

Step 1: Verify that the theorem is true when $n = 1$—that is, $T(1)$ is true.

Step 2: Verify that if the theorem is true for $n = k$, then it is also true for $n = k + 1$—that is, the truth of $T(k)$ implies the truth of $T(k+1)$.

Let's see how to apply this method to show that the preceding formula is true for all positive integers. Many students will find it helpful to write down what the pertinent statements are.

Statement of $T(1)$: We put 1 in place of n in the statement of $T(n)$. We must verify this (even though verification may be trivial):

$$1 = \frac{1(1+1)}{2}$$

Statement of $T(k)$: We put k in place of n in the statement of $T(n)$. This is what we assume to be true:

$$1 + 2 + 3 + \cdots + k = \frac{k(k+1)}{2}$$

Statement of $T(k+1)$: We put $k + 1$ in place of n in the statement of $T(n)$. This is what we must verify under the assumption that $T(k)$ is true:

$$1 + 2 + 3 + \cdots + k + (k+1) = \frac{(k+1)((k+1)+1)}{2}$$

We proceed with an inductive proof that $T(n)$ is true for every positive integer n.

Step 1: For $n = 1$, the right side of the equation is $\dfrac{1(1+1)}{2} = \dfrac{2}{2} = 1$, which equals the left side of the equation. We conclude that $T(1)$ is true.

We remark that in many cases, like this one, the first step in the induction is quite easy to verify, and beginning students may overlook its importance. This step is essential, and it is important when you write it down that you actually write enough to verify that $T(1)$ is true. Every mathematics teacher has seen students write something like "$1 = 1$ so $T(1)$ is true." It might be possible to guess what this student has in mind, but we are supposed to give a proof, not a calculation. It is necessary to explicitly communicate why $T(1)$ is true—not just verify it in your own mind.

Step 2: Suppose $T(k)$ is true—that is, assume that

$$1 + 2 + 3 + \cdots + k = \frac{k(k+1)}{2}$$

Our goal is now to show that $T(k+1)$ is true—that is, we want to prove that

$$1 + 2 + 3 + \cdots + k + (k+1) = \frac{(k+1)((k+1)+1)}{2}$$

We are allowed to use the truth of $T(k)$ in order to verify $T(k+1)$. We proceed as follows:

$$1 + 2 + 3 + \cdots + k = \frac{k(k+1)}{2}$$

$$1 + 2 + 3 + \cdots + k + (k+1) = \frac{k(k+1)}{2} + (k+1) \qquad \text{◀ Add } k+1 \text{ to both sides.}$$

$$1 + 2 + 3 + \cdots + k + (k+1) = \frac{k(k+1) + 2(k+1)}{2} \qquad \text{◀ Combine into a single fraction.}$$

$$1 + 2 + 3 + \cdots + k + (k+1) = \frac{(k+1)(k+2)}{2} \qquad \text{◀ Factor out } k+1.$$

$$1 + 2 + 3 + \cdots + k + (k+1) = \frac{(k+1)((k+1)+1)}{2}$$

This argument verifies that $T(k)$ implies $T(k+1)$.

By induction, $T(n)$ is true for all values of n.

This completes the proof, but we pause to remark that the last step in the preceding calculation is specifically the conclusion that $T(n)$ is true when $n = k+1$. Note that we didn't just stop with $\dfrac{(k+1)(k+2)}{2}$. You are encouraged to do the same.

EXAMPLE A.8 The Geometric Sum Formula

Use induction to prove that if $b \neq 1$, then for all positive integers n

$$1 + b + b^2 + \cdots + b^n = \frac{1 - b^{n+1}}{1 - b}$$

SOLUTION

Let $b \neq 1$. For each positive integer n, let $T(n)$ denote the proposition that

$$1 + b + b^2 + \cdots + b^n = \frac{1 - b^{n+1}}{1 - b}$$

Step 1: Verify that $T(1)$ is true—that is, show that $1 + b = \dfrac{1 - b^2}{1 - b}$.

We begin with the right side:

$$\frac{1 - b^2}{1 - b} = \frac{(1 + b)(1 - b)}{1 - b}$$

$$= \frac{(1 + b)(1 - b)}{1 - b}$$

$$= 1 + b$$

This is the left side of the original equation for step 1, and we conclude that $T(1)$ is true.

Step 2: Show that $T(k)$ implies $T(k + 1)$—that is, assume that

$$1 + b + b^2 + \cdots + b^k = \frac{1 - b^{k+1}}{1 - b}$$

and use this to prove

$$1 + b + b^2 + \cdots + b^k + b^{k+1} = \frac{1 - b^{(k+1)+1}}{1 - b}$$

We proceed as follows:

$$1 + b + b^2 + \cdots + b^k = \frac{1 - b^{k+1}}{1 - b}$$

$$1 + b + b^2 + \cdots + b^k + b^{k+1} = \frac{1 - b^{k+1}}{1 - b} + b^{k+1} \qquad \blacktriangleleft \text{Add } b^{k+1} \text{ to both sides.}$$

$$1 + b + b^2 + \cdots + b^k + b^{k+1} = \frac{1 - b^{k+1} + (1 - b)b^{k+1}}{1 - b} \qquad \blacktriangleleft \text{Combine into a single fraction.}$$

$$1 + b + b^2 + \cdots + b^k + b^{k+1} = \frac{1 - b^{k+1} + b^{k+1} - b^{k+2}}{1 - b}$$

$$1 + b + b^2 + \cdots + b^k + b^{k+1} = \frac{1 - b^{k+2}}{1 - b}$$

$$1 + b + b^2 + \cdots + b^k + b^{k+1} = \frac{1 - b^{(k+1)+1}}{1 - b}$$

This last equation shows that $T(k + 1)$ is true. By induction, $T(n)$ is true for each positive integer n. We conclude that the theorem is true.

Why induction works: Many students find induction to be a curious method of proof. They may be uncomfortable assuming $T(k)$ and then claiming to have proved it, and some ask for a "proof that it works." Bear in mind that in step 2 you are actually proving an implication. You are proving that $T(k)$ implies $T(k + 1)$ rather than just assuming that what you wish to prove is true.

In advanced mathematics, induction is actually an axiom that is used to define what we mean by a positive integer. Consequently, a "proof that it works" is not possible. There are, however, a number of common illustrations that are intuitively quite appealing.

The fact that $T(1)$ is true gives us a starting point. But we also know it's true that $T(k)$ implies $T(k+1)$. Putting in $k=1$, we get that it's true that $T(1)$ implies $T(2)$. Because we know already that $T(1)$ is true, we must conclude that $T(2)$ is true. Next we put in $k=2$ to get that it's true that $T(2)$ implies that $T(3)$. Because we have just verified the truth of $T(2)$, we get that $T(3)$ is true. Continuing, we get the string of implications:

$T(1)$ is true.

If $T(1)$ is true, then $T(2)$ is true.

If $T(2)$ is true, then $T(3)$ is true.

If $T(3)$ is true, then $T(4)$ is true.

This can be continued indefinitely to show that $T(n)$ is true for every positive integer n.

The situation is very much like a line of people waiting to get into a sports event. Suppose we know that the first person in line has a ticket. Suppose we also know that everyone immediately behind a ticket holder also has a ticket. Then, clearly, everyone in line has a ticket.

Certain sums, such as the preceding ones, are obvious candidates for inductive proof. The next example illustrates the fact that many theorems about positive integers lend themselves to an inductive proof. We say that a nonzero integer b is a divisor of an integer a (or that b divides a) if a/b is an integer.

EXAMPLE A.9 Divisors

Prove that 3 is a divisor of $7^n - 4^n$ for every positive integer n.

SOLUTION

For each positive integer n, let $T(n)$ denote the proposition that 3 is a divisor of $7^n - 4^n$.

Step 1: Verify that $T(1)$ is true—that is, show that 3 divides $7^n - 4^n$ when $n=1$.

When $n=1$, $7^n - 4^n = 7^1 - 4^1 = 3$. Because 3 divides itself, we conclude that $T(1)$ is true.

Step 2: Suppose $T(k)$ is true, and prove $T(k+1)$ is true—that is, assume 3 is a divisor of $7^k - 4^k$, and prove that 3 is a divisor of $7^{k+1} - 4^{k+1}$.

Because 3 divides $7^k - 4^k$, there is an integer m such that $3m = 7^k - 4^k$. In order to draw a conclusion regarding $7^{k+1} - 4^{k+1}$, we need to relate it to $7^k - 4^k$. Subtracting and adding $7(4^k)$ does the trick:

$$7^{k+1} - 4^{k+1} = 7^{k+1} - 7(4^k) + 7(4^k) - 4^{k+1} \qquad \blacktriangleleft \text{ Subtract and add } 7(4^k).$$

$$7^{k+1} - 4^{k+1} = (7(7^k) - 7(4^k)) + (7(4^k) - 4(4^k)) \qquad \blacktriangleleft \text{ Group terms.}$$

$$7^{k+1} - 4^{k+1} = 7(7^k - 4^k) + 3(4^k) \qquad \blacktriangleleft \text{ Factor the first term, simplify the second.}$$

$$7^{k+1} - 4^{k+1} = 7(3m) + 3(4^k) \qquad \blacktriangleleft \text{ Use } 7^k - 4^k = 3m.$$

$$7^{k+1} - 4^{k+1} = 3(7m + 4^k) \qquad \blacktriangleleft \text{ Factor.}$$

Now because $7m + 4^k$ is an integer, we conclude that 3 is a divisor of $7^{k+1} - 4^{k+1}$. This proves that if $T(k)$ is true then $T(k+1)$ is also true. So, by induction, 3 divides $7^n - 4^n$ for all positive integers n.

We sometimes define sequences of integers recursively. In other words, rather than giving a formula for x_n, we say what x_1 is and then tell how to get x_{n+1} from x_n. For example, suppose we define $x_1 = 0$ and $x_{n+1} = \sqrt{x_n + 1}$ for $n \geq 1$. Let's write down a few of the terms of this recursive sequence:

$$x_1 = 0$$

$$x_2 = \sqrt{x_1 + 1} = \sqrt{1} = 1$$

$$x_3 = \sqrt{x_2 + 1} = \sqrt{2}$$

$$x_4 = \sqrt{x_3 + 1} = \sqrt{\sqrt{2} + 1}$$

$$x_5 = \sqrt{x_4 + 1} = \sqrt{\sqrt{\sqrt{2} + 1} + 1}$$

EXAMPLE A.10 Proving an Inequality

Prove by induction that for the sequence defined previously, $x_n < 2$ for every positive integer n.

SOLUTION

For each positive integer n let $T(n)$ denote the proposition that $x_n < 2$.

Step 1: Verify that $T(1)$ is true—that is, show that $x_1 < 2$.

Because $x_1 = 0$, which is less than 2, we conclude that $T(1)$ is true.

Step 2: Suppose $T(k)$ is true, and prove $T(k+1)$ is true—that is, we assume that $x_k < 2$, and we must prove that $x_{k+1} < 2$.

We use the definition:

$$x_{k+1} = \sqrt{x_k + 1}$$
$$< \sqrt{2 + 1} \qquad \blacktriangleleft \text{ Use } x_k < 2.$$
$$< \sqrt{3} < 2$$

This shows that $T(k+1)$ is true whenever $T(k)$ is true. By induction, $T(n)$ is true for all positive integers n, so the theorem is true.

Another inductive argument can be used to show, for the sequence x_n shown previously, $x_n < x_{n+1}$ for all positive integers n. Then it is possible to use some advanced mathematics to obtain a result that may be surprising: $x_n \to \dfrac{1 + \sqrt{5}}{2}$ as $n \to \infty$.

APPENDIX

8

PROBABILITY

Probability is a mathematical concept designed to measure the likelihood of the occurrence of an event. When we perform an experiment such as tossing a coin, there are two equally likely outcomes, heads and tails. Among these equally likely outcomes we label one favorable and the others unfavorable. If we are interested in the probability of getting heads, we say that heads is favorable and tails is unfavorable. The probability of an event is defined by

$$\text{Probability} = \frac{\text{Number of favorable outcomes}}{\text{Total number of equally likely outcomes}}$$

For the probability of heads on one coin toss there are two equally likely outcomes, heads and tails, and only one, heads, is favorable. Using $P(\text{Heads})$ to denote the probability of heads, we have

$$P(\text{Heads}) = \frac{\text{Number of favorable outcomes}}{\text{Total number of equally likely outcomes}} = \frac{1}{2}$$

When we toss a coin we are certain to get either heads or tails, so the probability of getting either heads or tails is 1. Similarly, the probability of getting neither heads nor tails is 0. In general,

$$P(\text{A certainty}) = 1$$

$$P(\text{An impossibility}) = 0$$

Suppose we toss a coin twice and ask for the probability of getting heads at least once. There are four equally likely outcomes: HH, HT, TH, and TT. (Here we use H for heads and T for tails.) Three of these outcomes, namely HH, HT, and TH, are favorable, so

$$P(\text{Heads at least once}) = \frac{3}{4}$$

This particular probability provides the opportunity to clarify what we mean by equally likely outcomes through the examination of the following erroneous calculation: If we toss a coin twice, the possible outcomes are heads twice, heads once, or no heads. That is a total of three possible outcomes, two of which are favorable. This gives the incorrect probability of $\frac{2}{3}$. The error here is that the three listed outcomes are not equally likely. Indeed, when we toss two coins there is only one way to get heads twice, but there are two ways (HT and TH) to get heads exactly once. The second outcome (heads exactly once) is twice as likely as the first. Because these outcomes are not equally likely, they cannot be used to calculate the probability. In many probability calculations it is easy to discern equally likely events but, as this example shows, some care is necessary to get it right.

There is an alternative way of finding the probability of getting heads at least once, and this alternative approach is often useful. Because an event either does or does not occur, we have

$$P(\text{Event occurring}) + P(\text{Event not occurring}) = P(\text{A certainty}) = 1$$

Rearranging, we get

$$P(\text{Event occuring}) = 1 - P(\text{Event not occurring})$$

Applying this formula to the case of tossing a coin twice, we have

$$P(\text{Heads at least once}) = 1 - P(\text{No heads})$$

To calculate the probability of getting no heads, note that there are four equally likely outcomes (HH, HT, TH, TT), only one of which (TT) has no heads. Thus, the probability of no heads is $\frac{1}{4}$, and

$$P(\text{Heads at least once}) = 1 - \frac{1}{4} = \frac{3}{4}$$

CONCEPTS TO REMEMBER: Basics of Probability

We list all possibilities for the outcome of an experiment and verify that the outcomes are equally likely. This list is divided into two categories, favorable outcomes and unfavorable outcomes.

1. $P(\text{Event occurring}) = \dfrac{\text{Number of favorable outcomes}}{\text{Total number of equally likely outcomes}}.$

2. The probability of a certainty is 1, and the probability of an impossibility is 0.

3. $P(\text{Event occurring}) = 1 - P(\text{Event not occurring}).$

EXAMPLE A.11   Tossing Dice

We toss a pair of dice.

a. What is the probability that the dots facing up total to 4?

b. What is the probability that the dots facing up total to something other than 4?

SOLUTION

a. When we toss a pair of dice, it is a good strategy to think of one as being blue (say) and the other red, even if the two dice are in fact identical. This device helps us list equally likely outcomes.

There are six possible outcomes for the red die and six for the blue. That gives a total of $6 \times 6 = 36$ possible outcomes.

The outcomes that give a 4 are as follows.

Red	Blue
1	3
2	2
3	1

That is three favorable outcomes out of 36 total, so we have

$$P(\text{Adding to 4}) = \frac{3}{36} = \frac{1}{12}$$

b. In this case it is easier to find the probability that the given event does not occur:

$$P(\text{Not adding to 4}) = 1 - P(\text{Adding to 4}) = 1 - \frac{1}{12} = \frac{11}{12}$$

It is worth considering the difficulty of calculating this probability by actually listing all of the outcomes for which the dots do not add to 4.

Making a list of all outcomes in order to calculate probabilities is often impractical. In such a situation, the counting techniques we learned in Section ON13.3 may be useful. Suppose, for example, that we toss a fair coin 10 times. What is the probability of getting exactly 5 heads?

We can think of an outcome as a list of 10 letters consisting of H's and T's. One such is HHTHTTTHTH. So the number of outcomes is the number of 10-letter words we can make using the letters H and T. There are two choices for the first letter, two for the second, and so on, giving a total of 2^{10} equally likely outcomes. We get 5 heads and 5 tails if we make the 10-letter word using exactly 5 H's and 5 T's. Selecting 5 out of 10 slots for the H's, we get a total of $\binom{10}{5}$ favorable outcomes. So the probability of exactly 5 heads in 10 coin tosses is

$$P(\text{5 heads}) = \frac{\binom{10}{5}}{2^{10}}$$

This probability comes out to about 0.246, or 24.6%, which shows that this outcome is somewhat unlikely. It may be surprising to some that when we flip a coin many times the probability that exactly half of the tosses come out heads is quite small.

EXAMPLE A.12 Cards

You draw 5 cards at random from a standard 52-card deck.

a. What is the probability you get all hearts?

b. What is the probability you get four aces?

c. What is the probability you get a pair of kings and three queens?

SOLUTION

a. The total number of possible 5-card hands is the number of ways of choosing 5 cards from 52 cards. That is $\binom{52}{5}$. There are 13 hearts, so the number of 5-card hands consisting of all hearts is $\binom{13}{5}$. This gives

$$P(\text{All hearts}) = \frac{\binom{13}{5}}{\binom{52}{5}}$$

This comes out to about 0.000495.

b. There are only four aces in the deck, so there is only one way to choose all four of them. The last card of the five can be any of the 48 remaining cards. So there are 48 ways to get a hand with four aces. The total number of outcomes is the same as in part a:

$$P(\text{Four aces}) = \frac{48}{\binom{52}{5}}$$

This comes out to about 0.0000185.

c. There are four kings, so there are $\binom{4}{2}$ ways of choosing two of them. Similarly, there are $\binom{4}{3}$ ways of choosing three queens from the four queens in the deck. So there are a total of $\binom{4}{2}\binom{4}{3}$ ways to get two kings and three queens. This gives a probability of

$$P(\text{Two kings and three queens}) = \frac{\binom{4}{2}\binom{4}{3}}{\binom{52}{5}}$$

This comes out to about 0.00000923.

Although the calculation of probabilities discussed in this appendix is fairly straightforward, many probability calculations are famously confusing. Some are quite difficult. But even in complicated settings, the ideas presented here form the basis for most probability calculations.[2]

---

[2]A significantly more sophisticated approach is required for probabilities that involve infinitely many possible outcomes.

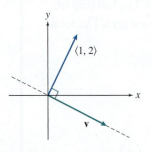

Figure A.15 Vectors **v** orthogonal to $\langle 1, 2\rangle$: Such vectors comprise a line through the origin.

A **normal vector** to a line is a vector perpendicular to the line.

The dot product can be helpful in finding equations of lines. For example, let's give a geometric description of all vectors **v** that satisfy the equation

$$\langle 1, 2\rangle \cdot \mathbf{v} = 0$$

Because the dot product is zero, any vector v that is orthogonal to $\langle 1, 2\rangle$ will satisfy the equation. A picture of such vectors is shown in **Figure A.15**. A vector in standard position is orthogonal to $\langle 1, 2\rangle$ if (and only if) its tip lies on the line shown there. Thus, the equation $\langle 1, 2\rangle \cdot \mathbf{v} = 0$ describes the line perpendicular to the vector $\langle 1, 2\rangle$. The vector $\langle 1, 2\rangle$ is often called a **normal vector** to the line.

If we put $\mathbf{v} = \langle x, y\rangle$, then we can rearrange this equation into a more familiar form:

$$\langle 1, 2\rangle \cdot \mathbf{v} = 0$$

$$\langle 1, 2\rangle \cdot \langle x, y\rangle = 0$$

$$x + 2y = 0$$

If the line does not go through the origin, then the equation is only a bit more complicated. Let's use vectors to get the equation of the line through the tip of the vector $\mathbf{w} = \langle 3, 2\rangle$ and normal to the vector $\mathbf{n} = \langle 1, 2\rangle$. This line is shown in **Figure A.16**. If **v** is a variable vector whose tip is on the line then, as shown in **Figure A.17**, $\mathbf{v} - \langle 3, 2\rangle$ is orthogonal to the normal vector $\langle 1, 2\rangle$. Thus, the dot product is zero. This gives the equation

$$\langle 1, 2\rangle \cdot (\mathbf{v} - \langle 3, 2\rangle) = 0$$

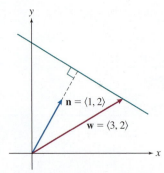

Figure A.16 The line through the tip of $\mathbf{w} = \langle 3, 2\rangle$ that is normal to $\mathbf{n} = \langle 1, 2\rangle$

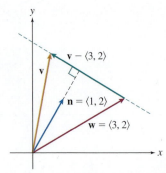

Figure A.17 Vectors **v** whose tips lie on the line: The difference $\mathbf{v} - \langle 3, 2\rangle$ is orthogonal to the vector $\mathbf{n} = \langle 1, 2\rangle$.

This reasoning works in the general case.

CONCEPTS TO REMEMBER: Equations of Lines

If we use **v** as the variable vector, the equation of the line through the tip of **w** and normal (orthogonal) to the nonzero vector **n** is

$$\mathbf{n} \cdot (\mathbf{v} - \mathbf{w}) = 0$$

We refer to the equation of the line given here as the **dot product equation** of the line. The line can also be written in the usual way, in terms of x and y, and this is the **algebraic form** of the line.

The **dot product equation** of the line through the tip of **w** and normal to the nonzero vector **n** is

$$\mathbf{n} \cdot (\mathbf{v} - \mathbf{w}) = 0$$

The **algebraic form** of the line through the tip of a vector and normal to a nonzero vector is $ax + by = c$.

EXAMPLE A.13 Finding Equations of Lines Given a Point and Normal Vector

Find both the dot product and algebraic forms of the equation of the line through the tip of $\mathbf{w} = \langle 1, -1 \rangle$ (the point $(1, -1)$) and normal to the vector $\mathbf{n} = \langle 3, 4 \rangle$. (See **Figure A.18**.)

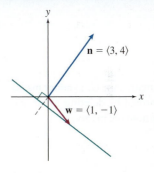

Figure A.18 The line through the tip of $\mathbf{w} = \langle 1, -1 \rangle$ and normal to $\mathbf{n} = \langle 3, 4 \rangle$

SOLUTION

Using **v** as the variable vector, we have the equation

$$\mathbf{n} \cdot (\mathbf{v} - \mathbf{w}) = 0$$
$$\langle 3, 4 \rangle \cdot (\mathbf{v} - \langle 1, -1 \rangle) = 0$$

This is the dot product form. If we put $\mathbf{v} = \langle x, y \rangle$, then we can write the equation in the more familiar algebraic form:

$$\langle 3, 4 \rangle \cdot (\mathbf{v} - \langle 1, -1 \rangle) = 0 \qquad \blacktriangleleft \text{Use the vector equation of the line.}$$
$$\langle 3, 4 \rangle \cdot (\langle x, y \rangle - \langle 1, -1 \rangle) = 0 \qquad \blacktriangleleft \text{Substitute } \langle x, y \rangle \text{ for } \mathbf{v}.$$
$$\langle 3, 4 \rangle \cdot \langle x, y \rangle - \langle 3, 4 \rangle \cdot \langle 1, -1 \rangle = 0 \qquad \blacktriangleleft \text{Distribute the dot product.}$$
$$3x + 4y - (-1) = 0 \qquad \blacktriangleleft \text{Calculate the dot products.}$$
$$3x + 4y = -1 \qquad \blacktriangleleft \text{Simplify.}$$

We remark that, although the equation $3x + 4y = -1$ is the more familiar form, the dot product equation $\langle 3, 4 \rangle \cdot (\mathbf{v} - \langle 1, -1 \rangle) = 0$ is preferred by many. The reason is that the dot product equation describes the line at a glance. It is the line through the point $(1, -1)$ that is normal to the vector $\langle 3, 4 \rangle$. This information is a bit more difficult to glean from the equation $3x + 4y = -1$.

n Section ON11.4, we defined determinants for 2 by 2 and 3 by 3 matrices. Our goal here is to provide a definition of the determinant that applies to any square matrix—that is, any n by n matrix.

Cofactor Expansion

For a square matrix A, the determinant of A, denoted $|A|$, has a complicated definition. We define the determinant of a 3 by 3 matrix in terms of 2 by 2 determinants. Then we define the determinant of a 4 by 4 matrix in terms of 3 by 3 determinants, and so on. We begin with the definition of the **minor matrix**.

If A is a matrix, the (i, j) **minor matrix** $A_{i,j}$ is formed by deleting row i and column j from the matrix A.

Suppose, for example, that we are given

$$A = \begin{pmatrix} 1 & 2 & 3 & 4 \\ 5 & 6 & 7 & 8 \\ 0 & 8 & 7 & 6 \\ 0 & 0 & 4 & 2 \end{pmatrix}$$

We get the minor matrix $A_{2,3}$ from A by deleting the second row and the third column:

$$A_{2,3} = \begin{pmatrix} 1 & 2 & \cancel{3} & 4 \\ \cancel{5} & \cancel{6} & \cancel{7} & \cancel{8} \\ 0 & 8 & \cancel{7} & 6 \\ 0 & 0 & \cancel{4} & 2 \end{pmatrix} = \begin{pmatrix} 1 & 2 & 4 \\ 0 & 8 & 6 \\ 0 & 0 & 2 \end{pmatrix}$$

Note that the minor matrix is a smaller matrix.

We use the minor matrix to define the **cofactor**, which is plus or minus the determinant of the minor matrix.

The factor $(-1)^{i+j} = (-1)^{\text{row} + \text{column}}$ determines the sign attached to the cofactor. If A is the preceding matrix, then

The (i, j) **cofactor** of a square matrix A is

$$\text{Cof}_{i,j} = (-1)^{i+j} |A_{i,j}|$$

$$\text{Cof}_{2,3} = (-1)^{2+3} |A_{2,3}| = - \begin{vmatrix} 1 & 2 & 4 \\ 0 & 8 & 6 \\ 0 & 0 & 2 \end{vmatrix}$$

EXAMPLE A.14 Finding a Cofactor

If $A = \begin{pmatrix} 1 & 2 & 3 \\ 4 & 5 & 6 \\ 7 & 8 & 9 \end{pmatrix}$, find $\text{Cof}_{3,2}$, the cofactor corresponding to the third row and the second column.

SOLUTION

We get the minor matrix $A_{3,2}$ by deleting the third row and the second column:

$$A_{3,2} = \begin{pmatrix} 1 & 2 & 3 \\ 4 & 5 & 6 \\ 7 & 8 & 9 \end{pmatrix} = \begin{pmatrix} 1 & 3 \\ 4 & 6 \end{pmatrix}$$

The sign attached to the cofactor is $-$ because

$$(-1)^{\text{row + column}} = (-1)^{3+2} = -1$$

Therefore,

$$\begin{aligned} \text{Cof}_{3,2} &= (-1)^{3+2} |A_{3,2}| \\ &= - \begin{vmatrix} 1 & 3 \\ 4 & 6 \end{vmatrix} \\ &= -(1 \times 6 - 3 \times 4) \\ &= 6 \end{aligned}$$

Now we can define the **determinant** of an n by n matrix. We calculate the determinant of an n by n matrix A with entries $a_{i,j}$ by cofactor expansion. We expand by selecting any row, say, row i, and using the following formula:

*The **determinant** is calculated using an n by n matrix A with entries $a_{i,j}$ by cofactor expansion.*

$$|A| = a_{i,1} \text{Cof}_{i,1} + a_{i,2} \text{Cof}_{i,2} + \cdots + a_{i,n} \text{Cof}_{i,n}$$

Alternatively, we can expand down any column, say column j:

$$|A| = a_{1,j} \text{Cof}_{1,j} + a_{2,j} \text{Cof}_{2,j} + \cdots + a_{n,j} \text{Cof}_{n,j}$$

It is a fact that the value of this cofactor expansion does not depend on the row or column that is chosen.

EXAMPLE A.15 Using the Cofactor Expansion to Calculate Determinants

Use the cofactor expansion to calculate the determinants of the following matrices.

a. $A = \begin{pmatrix} 3 & 2 & 3 \\ 4 & 2 & 0 \\ 2 & 1 & 2 \end{pmatrix}$

b. $B = \begin{pmatrix} 5 & 3 & 2 & 3 \\ 3 & 4 & 2 & 0 \\ 2 & 0 & 0 & 0 \\ 1 & 2 & 1 & 2 \end{pmatrix}$

SOLUTION

a. We can choose to go across any row (or down any column) we like. In this case we choose the second row. The appropriate minor matrices are as follows:

$$A_{2,1} = \begin{pmatrix} 3 & 2 & 3 \\ 4 & 2 & 0 \\ 2 & 1 & 2 \end{pmatrix} = \begin{pmatrix} 2 & 3 \\ 1 & 2 \end{pmatrix}$$

$$A_{2,2} = \begin{pmatrix} 3 & 2 & 3 \\ 4 & 2 & 0 \\ 2 & 1 & 2 \end{pmatrix} = \begin{pmatrix} 3 & 3 \\ 2 & 2 \end{pmatrix}$$

$$A_{2,3} = \begin{pmatrix} 3 & 2 & 3 \\ 4 & 2 & 0 \\ 2 & 1 & 2 \end{pmatrix} = \begin{pmatrix} 3 & 2 \\ 2 & 1 \end{pmatrix}$$

The sign attached to the first cofactor in the second row is − because

$$(-1)^{row+column} = (-1)^{2+1} = -1$$

As we go across the second row, the signs attached to the cofactor will alternate:

$$- + -$$

Thus,

$$\mathrm{Cof}_{2,1} = -\,|\,A_{2,1}\,| = -\begin{vmatrix} 2 & 3 \\ 1 & 2 \end{vmatrix} = -(2 \times 2 - 3 \times 1) = -1$$

$$\mathrm{Cof}_{2,2} = |\,A_{2,2}\,| = \begin{vmatrix} 3 & 3 \\ 2 & 2 \end{vmatrix} = 3 \times 2 - 3 \times 2 = 0$$

$$\mathrm{Cof}_{2,3} = -\,|\,A_{2,3}\,| = \begin{vmatrix} 3 & 2 \\ 2 & 1 \end{vmatrix} = -(3 \times 1 - 2 \times 2) = 1$$

Finally,

$$|\,A\,| = 4\mathrm{Cof}_{2,1} + 2\mathrm{Cof}_{2,2} + 0\mathrm{Cof}_{2,3}$$

$$= -4 + 0 + 0$$

$$= -4$$

b. The third row of the matrix B has lots of zeros. As we shall see, the presence of zeros can make our calculation easier, so we choose to expand across the third row. The sign attached to the cofactor $\mathrm{Cof}_{3,1}$ is + because

$$(-1)^{row+column} = (-1)^{3+1} = 1$$

Successive signs alternate: $+ \; - \; + \; -$. Therefore,

$$|\,B\,| = a_{3,1}\mathrm{Cof}_{3,1} + a_{3,2}\mathrm{Cof}_{3,2} + a_{3,3}\mathrm{Cof}_{3,3} + a_{3,4}\mathrm{Cof}_{3,4}$$

$$= 2\,|\,B_{3,1}\,| - 0\,|\,B_{3,2}\,| + 0\,|\,B_{3,3}\,| - 0\,|\,B_{3,4}\,|$$

This last equation clearly justifies our choice of the third row. When coefficients are zero, we don't have to calculate the determinant of the minor matrix because we are going to multiply it by zero. We have

$$|\,B\,| = 2\,|\,B_{3,1}\,|$$

We get the minor matrix $B_{3,1}$ by deleting the third row and the first column:

$$B_{3,1} = \begin{pmatrix} \cancel{5} & 3 & 2 & 3 \\ \cancel{3} & 4 & 2 & 0 \\ \cancel{2} & \cancel{0} & \cancel{0} & \cancel{0} \\ \cancel{1} & 2 & 1 & 2 \end{pmatrix} = \begin{pmatrix} 3 & 2 & 3 \\ 4 & 2 & 0 \\ 2 & 1 & 2 \end{pmatrix}$$

We calculated the determinant of this matrix in part a, so we find

$$|\,B\,| = 2\,|\,B_{3,1}\,| = 2(-4) = -8$$

Determinants and Row Operations

In Section ON11.4, we showed that, for 2 by 2 and 3 by 3 matrices, the determinant of an upper or a lower triangular matrix is the product of the entries on the main diagonal. This result applies to any determinant.

> **LAWS OF MATHEMATICS: Determinants of Triangular Matrices**
>
> The determinant of an upper or a lower triangular matrix is the product of the entries on the main diagonal.

EXAMPLE A.16 Calculating Determinants of Triangular Matrices

Calculate the determinant of the matrix

$$A = \begin{pmatrix} 2 & 3 & 5 & 6 \\ 0 & 7 & 8 & 9 \\ 0 & 0 & 10 & 11 \\ 0 & 0 & 0 & 12 \end{pmatrix}$$

SOLUTION

The matrix A is upper triangular, so its determinant is the product of the entries on the main diagonal:

$$|A| = 2 \times 7 \times 10 \times 12 = 1680$$

We can transform any square matrix into an upper triangular matrix by using appropriate row operations. Hence, we can calculate determinants for all square matrices if we can determine the effect of the row operations on the determinant.

One row plus a multiple of another: If B is obtained from A by the row operation **Row i** $+ a$ **Row j** $\mapsto$ new **Row i**, with $i \neq j$, then the determinant is unchanged. That is, $|B| = |A|$:

$$\begin{vmatrix} \text{Row } 1 \\ \vdots \\ \text{Row } i \\ \vdots \\ \text{Row } n \end{vmatrix} = \begin{vmatrix} \text{Row } 1 \\ \vdots \\ \text{Row } i + a\,\text{Row } j \\ \vdots \\ \text{Row } n \end{vmatrix}$$

Row interchange: If B is obtained from A by interchanging distinct rows (so **Row i** $\leftrightarrow$ **Row j**), then the sign of the determinant changes. That is, $|B| = -|A|$:

$$\begin{vmatrix} \text{Row } 1 \\ \vdots \\ \text{Row } i \\ \vdots \\ \text{Row } j \\ \vdots \\ \text{Row } n \end{vmatrix} = \begin{vmatrix} \text{Row } 1 \\ \vdots \\ \text{Row } j \\ \vdots \\ \text{Row } i \\ \vdots \\ \text{Row } n \end{vmatrix}$$

Multiplication of a row by a nonzero number: If B is obtained from A by multiplying a row by the nonzero number a (so a **Row i** $\mapsto$ new **Row i**), then the determinant is multiplied by a. That is, $|B| = a|A|$:

$$\begin{vmatrix} \text{Row } 1 \\ \vdots \\ a\text{Row } i \\ \vdots \\ \text{Row } n \end{vmatrix} = a \begin{vmatrix} \text{Row } 1 \\ \vdots \\ \text{Row } i \\ \vdots \\ \text{Row } n \end{vmatrix}$$

We can think of this as "factoring out" a number from a row.

EXAMPLE A.17 Calculating Determinants Using Row Operations

Use row operations to calculate $|B| = \begin{vmatrix} 3 & 6 & 9 & 3 \\ 1 & 1 & 2 & 1 \\ 0 & 2 & 1 & 0 \\ 3 & 8 & 0 & 2 \end{vmatrix}$.

SOLUTION

The goal is to obtain an upper triangular matrix by using row operations. We need to keep track of the effect on the determinant of each operation.

Note that the first row is a multiple of 3. Hence, we begin by "factoring" the 3 from the first row:

$$|B| = \begin{vmatrix} 3 & 6 & 9 & 3 \\ 1 & 1 & 2 & 1 \\ 0 & 2 & 1 & 0 \\ 3 & 8 & 0 & 2 \end{vmatrix} = 3 \begin{vmatrix} 1 & 2 & 3 & 1 \\ 1 & 1 & 2 & 1 \\ 0 & 2 & 1 & 0 \\ 3 & 8 & 0 & 2 \end{vmatrix}$$

Next we use the leading entry in the first row to eliminate the nonzero entries below it. We use row operations of the form **Row** $i + a$ **Row** $j \mapsto$ new **Row** i, which do not change the value of the determinant:

$$|B| = 3 \begin{vmatrix} 1 & 2 & 3 & 1 \\ 1 & 1 & 2 & 1 \\ 0 & 2 & 1 & 0 \\ 3 & 8 & 0 & 2 \end{vmatrix}$$

$$|B| = 3 \begin{vmatrix} 1 & 2 & 3 & 1 \\ 0 & -1 & -1 & 0 \\ 0 & 2 & 1 & 0 \\ 0 & 2 & -9 & -1 \end{vmatrix}$$
◀ Row 2 – Row 1 ↦ new Row 2
◀ Row 4 – 3Row 1 ↦ new Row 4
to make zeros in column 1, rows 2 and 4.

$$|B| = 3 \begin{vmatrix} 1 & 2 & 3 & 1 \\ 0 & -1 & -1 & 0 \\ 0 & 0 & -1 & 0 \\ 0 & 0 & -11 & -1 \end{vmatrix}$$
◀ Row 3 + 2Row 2 ↦ new Row 3
◀ Row 4 + 2Row 2 ↦ new Row 4
to make zeros in column 2, rows 3 and 4.

$$|B| = 3 \begin{vmatrix} 1 & 2 & 3 & 1 \\ 0 & -1 & -1 & 0 \\ 0 & 0 & -1 & 0 \\ 0 & 0 & 0 & -1 \end{vmatrix}$$
◀ Row 4 – 11 Row 3 ↦ new Row 4
to make zero in column 3, row 4.

This matrix is upper triangular, so the determinant is the product of the diagonal entries:

$$|B| = 3(1)(-1)(-1)(-1) = -3$$

The next example shows a mixed strategy for calculating determinants.

EXAMPLE A.18 Using Cofactors and Row Operations

Calculate the determinant of

$$A = \begin{pmatrix} 1 & 2 & 7 & 0 \\ 1 & 3 & 8 & -1 \\ 0 & 0 & 9 & 0 \\ 1 & 4 & 6 & 2 \end{pmatrix}$$

SOLUTION

Because of the three zeros in the third row, we choose to begin with cofactor expansion across that row. Only the cofactor $\text{Cof}_{3,3}$ will have a nonzero coefficient:

$$|A| = (-1)^{3+3} \, 9 \begin{vmatrix} 1 & 2 & 0 \\ 1 & 3 & -1 \\ 1 & 4 & 2 \end{vmatrix}$$

From this point, we could complete our calculations using the method for calculating 3 by 3 determinants given in Section ON11.4. But to emphasize the use of row operations, we show that method:

$$|A| = 9 \begin{vmatrix} 1 & 2 & 0 \\ 1 & 3 & -1 \\ 1 & 4 & 2 \end{vmatrix}$$

$$|A| = 9 \begin{vmatrix} 1 & 2 & 0 \\ 0 & 1 & -1 \\ 0 & 2 & 2 \end{vmatrix} \quad \begin{array}{l} \blacktriangleleft \textbf{Row 2 – Row 1} \mapsto \textbf{new Row 2} \\ \blacktriangleleft \textbf{Row 3 – Row 1} \mapsto \textbf{new Row 3} \\ \textit{to make zeros in column 1, rows 2 and 3.} \end{array}$$

$$|A| = 9 \begin{vmatrix} 1 & 2 & 0 \\ 0 & 1 & -1 \\ 0 & 0 & 4 \end{vmatrix} \quad \begin{array}{l} \blacktriangleleft \textbf{Row 3 – 2Row 2} \mapsto \textbf{new Row 3} \\ \textit{to make zero in column 2, row 3.} \end{array}$$

$$|A| = 9 \times 4 = 36$$

CHAPTER P

Section P.1

1. A circle centered at (a, b) with radius r
3. It is the single point $(1, 2)$.
5. The distance from $(5, 8)$ to $(2, 3)$
7. It tells how much the line rises or falls for each unit of run.
9. It may consist of one point, two points, or no points.
11. F
13. A
15. D
17. Midpoint $= (0, 3)$. Distance $= 2$.
19. Midpoint $= (-2, -3)$. Distance $= 2\sqrt{2}$.
21. Midpoint $= \left(t - \dfrac{1}{2}, \dfrac{t+1}{2} \right)$. Distance $= \sqrt{t^2 - 2t + 2}$.
23. Midpoint $= \left(x + \dfrac{y}{2}, \dfrac{x}{2} \right)$. Distance $= \sqrt{x^2 - 4xy + 5y^2}$.
25. Midpoint $= \left(\dfrac{x+y}{2}, \dfrac{x+y}{2} \right)$. Distance $= \sqrt{2}\,|x - y|$.
27. -2
29. $\dfrac{9}{4}$
31. 0
33. $\dfrac{y+4}{x-1}$
35. $y = 2x - 5$
37. $y = 7x - 11$
39. $y = 5x$
41. $y = 3tx + 2t - 3t^2$
43. $y = 2x - 3$
45. $y = -x + p + q$
47. **a.** $y = 2$
    b. $y = -x + 2$
    c. $y = x + 3$
    d.
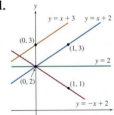
    Answers will vary.
49. $(x - 3)^2 + y^2 = 1$
51. $(x - p)^2 + (y - p)^2 = p^2$
53. $(-2, 0)$, $r = \sqrt{5}$
55. $(0, 0)$, $r = \sqrt{3}\,|t|$
57. $\left(0, \dfrac{1}{2} \right)$, $r = \dfrac{1}{2}$
59. $(1, 2)$, $r = 3$
61. $(p, 0)$, $r = |p|$

63. Center $(0, 1)$, radius 2.
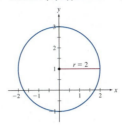

65. Center $(1, 2)$, radius 1
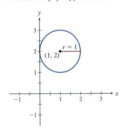

67. $(x - 1)^2 + y^2 = 8$
69. $(1, 2)$
71. $\left(-\dfrac{1}{5}, \dfrac{3}{5} \right)$ and $(-1, -1)$
73. $\left(\dfrac{12 - 3\sqrt{86}}{10}, \dfrac{4 + 9\sqrt{86}}{10} \right)$ and $\left(\dfrac{12 + 3\sqrt{86}}{10}, \dfrac{4 - 9\sqrt{86}}{10} \right)$
75. $\left(\dfrac{-1 + \sqrt{61}}{5}, \dfrac{13 + 2\sqrt{61}}{5} \right)$ and $\left(\dfrac{-1 - \sqrt{61}}{5}, \dfrac{13 - 2\sqrt{61}}{5} \right)$
77. $y = -3x + 5$
79. $y = \dfrac{1}{4}x + \dfrac{1}{4}$
81. $y = -\dfrac{1}{3}x + 6$
83. $y = -\dfrac{1}{9}x + 7$
85. $y = \dfrac{2}{3}x - \dfrac{1}{3}$
87. 0.03 foot per foot
89. **a.** -0.63 foot per mile
    b. 981.08 feet
91. Answers will vary.
93. Answers may vary.

Section P.2

1. The inequality is preserved if the multiplier is positive. The inequality reverses if the multiplier is negative.
3. The inequality is preserved if the divisor is positive. The inequality reverses if the divisor is negative.
5. No
7. **c.** $-a < x < a$
9. $|x - 1| < 4$
11. $x \geq 1$ or $x \leq 0$

13. $\dfrac{1}{x} > \dfrac{1}{y}$

15. $[11, 13)$

17. $[-2, 1]$

19. $[15, \infty)$

21. $(7, \infty)$

23. $(-\infty, -5]$

25. $(-\infty, -1] \cup [4, \infty)$

27.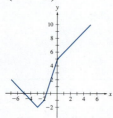
(number line: open circles at 0 and 4)

29.
(number line: closed circles at 1 and 5)

31.
(number line: open circle at 4, closed at 5)

33.
(number line: closed circle at -3, ray to right)

35.
(number line: open circle at 2, ray to left)

37. $(-\infty, 1)$

39. $\left(-\infty, \dfrac{7}{4}\right]$

41. No solution

43. $[0, \infty)$

45. $\left[-2, -\dfrac{1}{3}\right]$

47. $[4, \infty)$

49. False if $x = -2$ and $y = 1$, for example

51. True

53. The distance from x to 4 is less than 7.

55. The distance from x to y is less than the distance from x to z.

57. If the distance from x to 2 is less than 3, then the distance from y to 4 is less than 5.

59. $(1, 9)$

61. $[-13, 5]$

63. $(-\infty, -2] \cup [0, \infty)$

65. $\left[-\dfrac{14}{3}, \dfrac{4}{3}\right]$

67. $(-\infty, -1] \cup [4, \infty)$

69. $\left(-5, -\dfrac{5}{3}\right)$

71. $(-1, 1)$

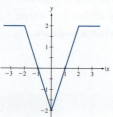

73. During the first 3 hours and 20 minutes

75. 51 or more

77. Answers will vary.

79. Answers will vary.

81. $x > 3$ or $(3, \infty)$

83. No solution

Section P.3

1. Division by x is not allowed because we don't know its sign.

3. a. Move all terms to one side and factor.

5. False

7. True

9. Answers will vary.

11.

Region	$x < -1$	$-1 < x < 4$	$4 < x$
Sign of $x + 1$	$-$	$+$	$+$
Sign of $x - 4$	$-$	$-$	$+$
Sign of $(x - 4)(x + 1)$	$+$	$-$	$+$

13.

Region	$x < -1$	$-1 < x < 4$	$4 < x$
Sign of $x + 1$	$-$	$+$	$+$
Sign of $x - 4$	$-$	$-$	$+$
Sign of $\dfrac{x - 4}{x + 1}$	$+$	$-$	$+$

15.

Region	$x < 1$	$1 < x < 2$	$2 < x < 3$	$x > 3$
Sign of $(x - 1)^2$	$+$	$+$	$+$	$+$
Sign of $x - 2$	$-$	$-$	$+$	$+$
Sign of $x - 3$	$-$	$-$	$-$	$+$
Sign of $\dfrac{x - 2}{(x - 1)^2(x - 3)}$	$+$	$+$	$-$	$+$

17. $(-1, 4)$

19. $(-\infty, -1) \cup (4, \infty)$

21. $(-1, 4)$

23. $(-\infty, -1) \cup (4, \infty)$

25. $(-1, 4)$

27. $(-\infty, -1) \cup (4, \infty)$

29. $(1, 2) \cup (3, \infty)$

31. $(-\infty, 1) \cup (2, 3)$

33. $(-\infty, 1) \cup (1, 2] \cup (3, \infty)$

35. $(2, 3)$

37. $[0, 2) \cup (3, \infty)$

39. $(-\infty, 0] \cup (2, 3)$

41.
$$(x - 2)(x - 5)$$
$$\frac{(-)(-) = + \;\;|\;\; (+)(-) = - \;\;|\;\; (+)(+) = +}{25}$$

43.
$$\frac{x - 2}{x - 5}$$
$$\frac{\dfrac{(-)}{(-)} = + \;\;|\;\; \dfrac{(+)}{(-)} = - \;\;|\;\; \dfrac{(+)}{(+)} = +}{2\boxed{5}}$$

45.

$$\dfrac{(x-1)^2(x-2)^3}{x}$$

$$\dfrac{(+)(-)}{(-)}=+ \;\Big|\; \dfrac{(+)(-)}{(+)}=- \;\Big|\; \dfrac{(+)(-)}{(+)}=- \;\Big|\; \dfrac{(+)(+)}{(+)}=+$$

$$\underline{\quad\boxed{0}\qquad\qquad 1\qquad\qquad 2\quad}$$

47. $(-\infty, 2]\cup[5, \infty)$

49. $[2, 5]$

51. $[2, 5]$

53. $(-\infty, 2]\cup[5, \infty)$

55. $(-\infty, 2]\cup(5, \infty)$

57. $[2, 5)$

59. $(-\infty, 5]$

61. $[5, \infty)$

63. $(0, 1]\cup[2, \infty)$

65. $(-\infty, 0)\cup[1, 2]$

67. $(-\infty, 0)\cup[2, \infty)$

69. $(0, 2]$

71. $[2, 3]$

73. $\mathbb{R}$ or $(-\infty, \infty)$

75. $\left(-\infty, -\dfrac{1}{4}\right)\cup\left(\dfrac{1}{3}, \infty\right)$

77. $\left(-\infty, -\sqrt{\dfrac{2}{3}}\,\right]\cup\left(\sqrt{\dfrac{2}{3}}, \infty\right)$

79. $\left(\dfrac{1}{5}, \dfrac{1}{4}\right)$

81. $(0, 1)\cup(2, \infty)$

83. $\{0\}\cup[1, \infty)$

85. $\{0\}\cup[2, 3]$

87. All real numbers except 0, 2, and 3

89. $(0, 1)\cup(2, \infty)$

91. $(-\infty, 0)\cup(1, 5)$

93. We can't have an interval with the larger number on the left.

95. It refers to numbers less than 1 *and* greater than 2. But there are no such numbers. It should say the set of all numbers that are less than 1 *or* greater than 2.

97. $0.43 < x < 8.29$

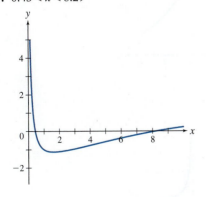

99. $-1 < x < 0.68$

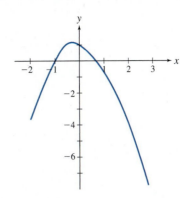

101. $w \le \sqrt{250} \approx 15.81$ miles per hour

103. From 1 thousand to 2 thousand individuals

105. a.

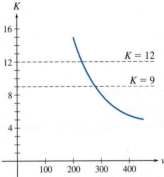

b. Between 230 and 277 words

107. a.

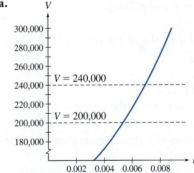

b. Between $r = 0.0053$ and $r = 0.0069$

109. $\left(\dfrac{(3-\sqrt{5})}{2}, \dfrac{(3+\sqrt{5})}{2}\right)\cup(3, \infty)$

111. $[1, 2]$

CHAPTER 1

Section 1.1

1. 1

3. The vertical line test

5. The function f is defined for t.

7. True

9. $(-\infty, 1) \cup (1, \infty)$

11. $\mathbb{R}$

13. $\left[-\dfrac{1}{2}, 0\right) \cup (0, \infty)$

15. $\mathbb{R}$

17. $(-\infty, 0) \cup (0, \infty)$

19. $\dfrac{2}{3}$

21. 2

23. $|\omega|$

25. $|z|$

27. $\dfrac{1}{x} - x$

29. $x + 1$

31. $\dfrac{x}{2x + 1}$ if $x \neq -1$ and $x \neq -\dfrac{1}{2}$

33. 2

35. $2x + h$

37. Answers will vary.

39. $-\dfrac{1}{xy}$

41. No. It fails the vertical line test.

43. No. It fails the vertical line test.

45. Yes. Each vertical line meets the graph at most once.

47. $[0, \infty)$

49. $(-\infty, 1) \cup (1, \infty)$

51. $(-\infty, 0) \cup (0, \infty)$

53. $[1, \infty)$

55. $g(7) = 49$, $g(1) = 2$, and $g(2) = 9$

57. $f(1) = 2$, $f(2) = 3$

59. $f(x + 1) = x + 2$ if $x \leq 4$ and $f(x + 1) = x + 7$ if $x > 4$.

61. 8

63. 29

65. Yes; answers will vary.

67. Yes. For each value of $x \neq 0$, the equation determines exactly one value of y.

69. No. For some values of x there is more than one value of y. For example if $x = 0$ then $y = \pm 1$.

71. Yes. For each value of $x \neq 0$, the equation determines exactly one value of y.

73. It is not a function. Some people have more than one uncle.

75. It is not a function. Some mothers have more than one child.

77. It is not a function. There is more than one number a distance of 1 or less from each number.

79. It is not a function. Most presidents served for more than 1 year.

81.

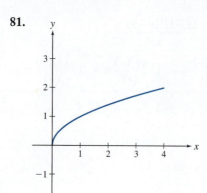

83.

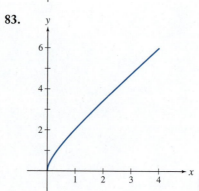

85.

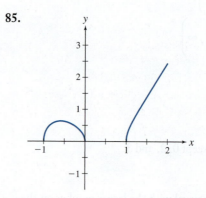

87.

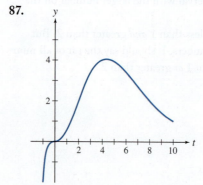

89.

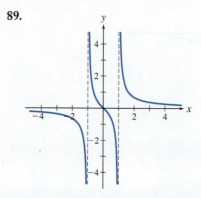

91. $[-1, 1]$

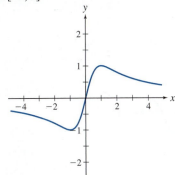

93. $[0, 2.5]$

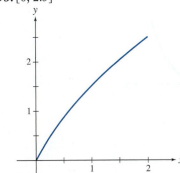

95. a. $t = 4$

   b. The domain is 0 to 4 seconds.

   c. The range is 0 to 256 feet.

97. Domain $[0, 13,000]$. Range \$0 to \$16,250.

99. This is the account balance, in dollars, after 5 years.

101. $H(T) = \begin{cases} 11 + 0.5T & \text{if } 0 \le T \le 25 \\ 23.5 - 0.5 \times (T - 25) & \text{if } 25 \le T \le 50 \end{cases}$

103. a. Larger values of D correspond to larger values of H.

   b. $D = 6$

   c. From 0 inches to 6 inches

   d. 1 inch

   e. 1 inch to 10 inches

105. It is periodic with period 60 minutes.

107. It is not periodic because the time required to traverse the circle decreases with each circuit.

109. a. Periodic with period p

   b. Periodic with period p

   c. Periodic with period p

   d. Periodic with period $\dfrac{p}{2}$

   e. Periodic with period p

111. It is periodic with period 1.

Section 1.2

1. True

3. The secant line

5. a. is positive

7. No

9. 2

11. $-\dfrac{1}{2}$

13. $-\dfrac{1}{4}$

15. $-\dfrac{1}{6}$

17. 0

19. -1

21. Dollars per month. It is the average increase per month in the balance of the account.

23. Miles per hour. The average rate of change is the average speed.

25. Miles per hour per minute. The average rate of change is the average acceleration of the airplane.

27. Dollars per item. The average rate of change is the average increase in profit per each additional item produced. (It is known as the marginal profit.)

29. Square feet per foot. It is the average increase in area per foot increase in diameter.

31. 0

33. -5

35. $2a + h$

37. $-2a - h$

39. $\dfrac{2}{\sqrt{a + h} + \sqrt{a}}$

41. $\dfrac{-1}{a(a + h)}$

43. The average rates of change are the same.

45. $2 + \dfrac{12}{19} \approx 2.63$

47. $4 + 0.4 \times 4 = 5.6$

49. 7

51. Answers will vary.

53. a.

Period	2007 to 2013	2013 to 2014	2014 to 2015
Average rate of change	0.08	0.09	0.09

   b. From 2007 to 2013

   c. 4.62 billion

55. a. 2015 to 2018

   b. $-\$0.17$ billion per year

   c. \$3.26 billion

57. 7 bats

59. 43 pounds

61. a.

Period	2013 to 2014	2014 to 2015	2015 to 2016	2016 to 2017
Average rate of change	0.10	0.15	0.11	0.38

   b. 2016 to 2017

63. Answers will vary.

Section 1.3

1. True

3. c. The graph becomes less steep.

5. d. Any of the above

7. b. Concave down

9. True

11. 0

13.

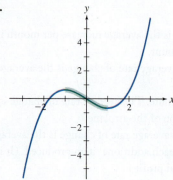

15.

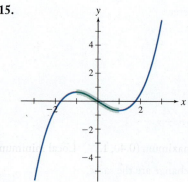

17.

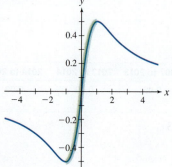

19.

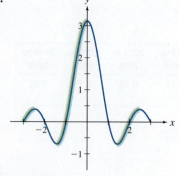

21.

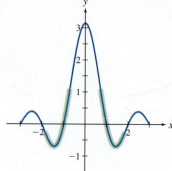

23.

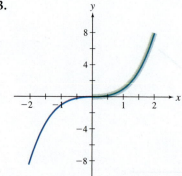

25. Increasing on $[-2, 2]$, nowhere decreasing, concave up on $[-2, 0]$, concave down on $(0, 2]$, inflection point at $(0, 0)$, no local maximum or minimum, absolute minimum at $x = -2$, absolute maximum at $x = 2$

27. Decreasing on $[-3, 3]$, nowhere increasing, concave up on $[-3, 3]$, nowhere concave down, no inflection point, no local maximum or minimum, absolute maximum at $x = -3$, absolute minimum at $x = 3$

29. Increasing on $[0, 4]$, nowhere decreasing, concave down on $[0, 4]$, nowhere concave up, no inflection point, no local maximum or minimum, absolute minimum at $x = 0$, absolute maximum at $x = 4$.

31. Decreasing on $[0, 3]$, increasing on $[3, 4]$, not concave up or down anywhere, no inflection point, local and absolute minimum at $x = 3$, no local maximum, absolute maximum at $x = 0$.

33. Rate of change positive on $(-3, -2)$ and $(2, 3)$. Rate of change negative on $(-2, 2)$. Rate of change 0 at $x = -2$ and $x = 2$.

35. Rate of change positive on $(-2, 2)$. Rate of change negative on $(-3, -2)$ and $(2, 3)$. Rate of change 0 at $x = 2$ and $x = -2$.

37. Rate of change positive on $(-4, -3)$, $(-1, 1)$, and $(3, 4)$. Rate of change negative on $(-3, -1)$ and $(1, 3)$. Rate of change 0 at $x = -3$, $x = -1$, $x = 1$, and $x = 3$.

39. 3

41. $\dfrac{1}{2}$

43.

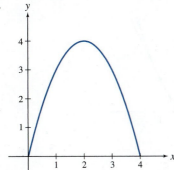

45.

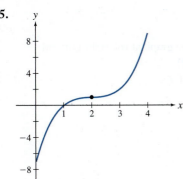

47.

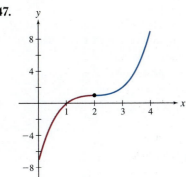

49.

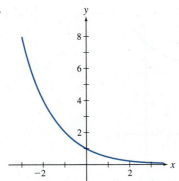

51.

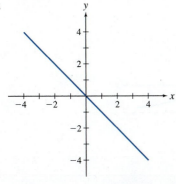

53.

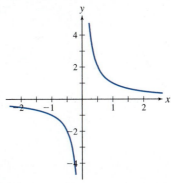

55. Rate of change positive on $(0, \infty)$. Rate of change negative on $(-\infty, 0)$.

57. No

59. The rate of change is positive for all x.

61. There is a point of inflection at $(0, 0)$.

63. The graph is concave down on $(-\infty, 0)$ and concave up on $(0, \infty)$.

65.

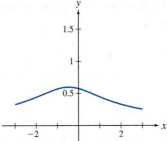

Local and absolute maximum $(0.40, 1.14)$. Local minimum $(2.48, 0.78)$. Absolute minimum $(-2, 0.05)$.

67.

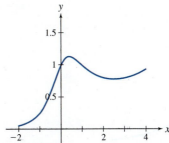

Local and absolute maximum $(-0.50, 0.60)$. No local minimum. Absolute minimum $(3, 0.26)$.

69. Local and absolute maximum at $(9, 9)$. Absolute minimum at $(0, 0)$.

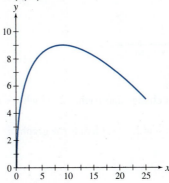

71.

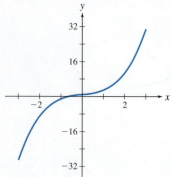

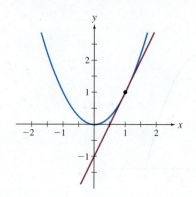

No local maxima or minima. Absolute minimum $(-3, -29)$.
Absolute maximum $(3, 31)$.

73. a. 40 degrees
   b. Dimmer light
75. a. Lower
   b. On a calm day
77. a.

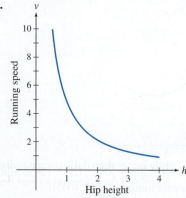

   b. The error is larger for dinosaurs with shorter hip height.
79. a. About 6000 kelvins
   b. Increasing
81. The rate of change is 0.

83. It is possible. One possible function is $f(x) = \dfrac{x^2}{1 + x^2}$.

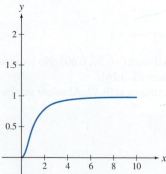

85. a. Answers will vary.
   b. The average rate of change approaches 2, which is the rate of change.
   c. The tangent line is $y = 2x - 1$. Here is the graph.

Section 1.4

1. a. The behavior of the graph at the right tail-end
3. c. Not a number at all
5. The limit as $x \to \infty$ of f
7. False
9. False
11. 0
13. 0
15. 0
17. 4
19. -3
21. $-\infty$
23. -5
25. $\dfrac{1}{2}$
27. 3
29. The limit is about 3.
31. The limit is about 2.
33. The limit is about 6.
35. The limit is about 2.5.
37.

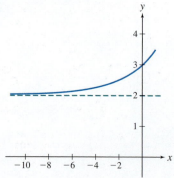

39.

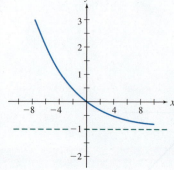

41.

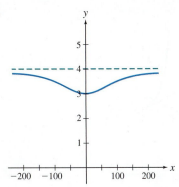

43.

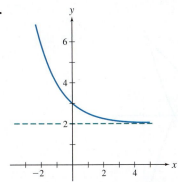

45. 12

47. $\dfrac{5}{6}$

49. 16

51. 5

53. 2

55. 3

57. 0

59. −1

61. Any answer near 1 is reasonable.

63. Any answer near 3 is reasonable.

65. Any answer near 3 is reasonable.

67. About 150 animals

69. a. $\dfrac{1}{2}$

   b. At most half of Earth's surface is visible no matter how high the spacecraft.

71. The limit is 0. When resistance is large, current is near 0.

73. a.

t	P
0	75
20	183.03
40	225.15
60	303.30
80	335.44
100	356.90
120	371.23

b. 75°

c.

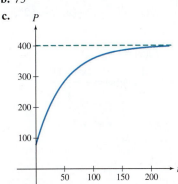

d. About 400°

75. a. Yearly 8.000%. Monthly 8.300%. Daily 8.328%.

   b. *EAR*

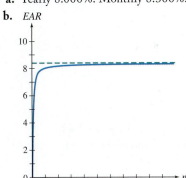

c. Monthly compounding

d. Any number slightly larger than daily compounding is acceptable. The actual number, correct to seven decimal places, is 8.3287068%.

77. a. 9.09. An inflation rate of 10% causes a 9.09% decrease in the value of a dollar.

   b. 100. When inflation is very large, the value of the dollar decreases by nearly 100%, so the dollar is virtually worthless.

79. 8

81. 0

83. ∞

85. 7; answers will vary.

CHAPTER 2

Section 2.1

1. a. First apply g and then apply f to the result.

3. True

5. True

7. Elements of the domain of f that are also in the domain of g.

9. Elements of the domain of f that are also in the domain of g, and for which g is not zero.

11. $(f + g)(x) = \dfrac{x}{x+1} + x^2,\ fg(x) = \dfrac{x^3}{x+1},\ \dfrac{f}{g}(x) = \dfrac{1}{x(x+1)},$

$(f \circ g)(x) = \dfrac{x^2}{x^2+1},\ (g \circ f)(x) = \dfrac{x^2}{(x+1)^2}$

13. $(f + g)(x) = 2^x - \dfrac{1}{x^2}$, $fg(x) = -\dfrac{2^x}{x^2}$, $\dfrac{f}{g}(x) = -x^2 2^x$,

$(f \circ g)(x) = 2^{-\frac{1}{x^2}}$, $(g \circ f)(x) = -4^{-x}$

15. $(f + g)(x) = x^2 + x + 1 + \dfrac{1}{x}$, $fg(x) = \dfrac{(x^2 + 1)^2}{x}$,

$\dfrac{f}{g}(x) = \dfrac{x^3 + x}{x^2 + 1}$, $(f \circ g)(x) = \dfrac{1}{x^2} + x^2 + 3$,

$(g \circ f)(x) = \dfrac{x^4 + 2x^2 + 2}{x^2 + 1}$

17. -5

19. $\dfrac{2}{5}$

21. 4

23. 2

25. 6

27. 2

29. 3

31. One solution is $g(x) = \sqrt{x}$ and $h(x) = x^3 + 1$.

33. One solution is $g(x) = x^2$ and $h(x) = x^2 - 1$.

35. $(-\infty, 2) \cup (2, 7) \cup (7, \infty)$

37. $[3, 4]$

39. $[3, 7]$

41. $[1, 2]$

43. $[2, 4]$

45. $[1, 3]$

47. $[1, 3]$

49.

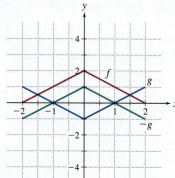

51.

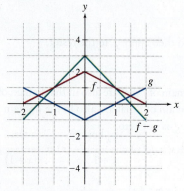

53.

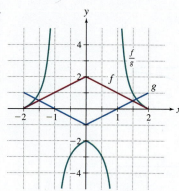

55.

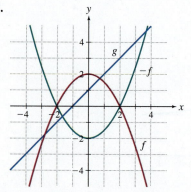

57.

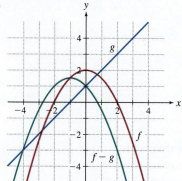

59.

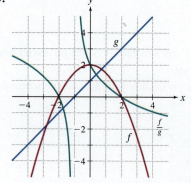

61.

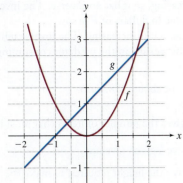

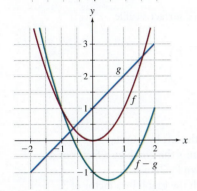

63.

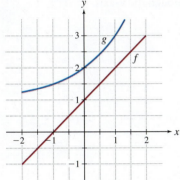

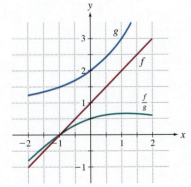

65. $F = \frac{9}{5}K - 459.67$; this gives the temperature in degrees Fahrenheit in terms of the temperature in kelvins.

67. One solution is $T(t) = R(S(t))$ where $R(x) = 400 - x$ and $S(t) = 325 \times 0.955^t$.

69. $N = \dfrac{5089}{1 + 0.48 \times 0.76^{T+12}}$

71. $B = 1000 \times 0.9^t \times 0.000322(27 - 25.1 \times 0.91^t)^3$

73. $f^{[3]}(x) = x^8$

75. $f^{[221]}(x) = \dfrac{1}{x}$

77. 0

79. $7 \to 22 \to 11 \to 34 \to 17 \to 52 \to 26 \to 13 \to 40 \to 20 \to 10 \to 5 \to 16 \to 8 \to 4 \to 2 \to 1$

81. Answers will vary.

Section 2.2

1. False

3. **a.** Reflecting through the line $y = x$

5. True

7. True

9. $f(12) = 8$

11. The domain of g is the range of f, and the range of g is the domain of f.

13. If $x \ne 0$, $f(g(x)) = x$ and $g(f(x)) = x$.

15. $f(g(x)) = x$ and $g(f(x)) = x$

17. If $x \ne 1$, $f(g(x)) = x$ and if $x \ne 0$ $g(f(x)) = x$.

19. If $x \ne 3$, $f(g(x)) = x$ and if $x \ne 0$, $g(f(x)) = x$.

21. If $x \ne 2$, $f(g(x)) = x$ and if $x \ne 1$, $g(f(x)) = x$.

23. $f(g(x)) = x$ and $g(f(x)) = x$

25. If $x \ge 1$, $f(g(x)) = x$. If $x \ge 0$, $g(f(x)) = x$.

27. The domain is $(-\infty, \infty)$ and the range is $(0, \infty)$.

29. $g(4) = 2$

31. $(f \circ g)(6) = 6$

33. $(g \circ g)(4) = 5$

35. It has an inverse.

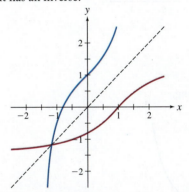

37. It has no inverse because it fails the horizontal line test.

39. It has an inverse.

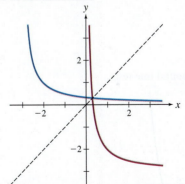

41. It has no inverse because it fails the horizontal line test.

43. It has an inverse. The graph of the inverse is the same as the graph of the function.

45. $g(x) = \dfrac{1}{x} + 1$

47. f has no inverse because for some values of y, the equation $y = 3 - x^2$ has more than one solution for x.

49. $g(x) = \dfrac{2x - 2}{x + 1}$

51. f has no inverse because for some values of y, the equation $y = (x + 1)(x + 2)$ has more than one solution for x.

53. $g(x) = \dfrac{x - 2}{x}$

55. $g(x) = (x + 1)^{1/5}$

57. f has no inverse because for $y = 2$, the equation $y = f(x)$ has more than one solution for x. (Every value of x is a solution when $y = f(x)$.)

59. $(0, \infty)$

61. $h(8) = 3$

63. It is the exponent of 2 that gives x.

65. It is one-to-one.

67. It is not one-to-one. One correct answer is $h(-1) = 1 = h(1)$.

69. It is one-to-one.

71. It is one-to-one.

73. It is one-to-one.

75. Yes; it is increasing.

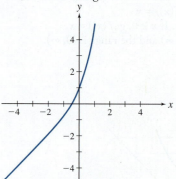

77. Yes; it is increasing.

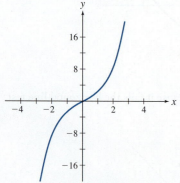

79. No; it fails the horizontal line test.

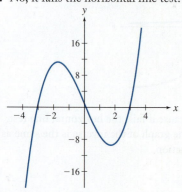

81. The function is increasing, so it has an inverse. The inverse shows the current value in dollars of P pesos.

83. p has an inverse because it is increasing. The inverse tells the density of prey that results in the predator consuming p prey.

85. $E = \dfrac{D}{1.13}$. \$1 is $\dfrac{1}{1.13}$, or about 0.88, euro.

87. $n(R) = \dfrac{R + 520}{38}$. It gives the number of items that must be sold in order to achieve a given revenue.

89. $h(D) = \dfrac{FD}{1 - 2F}$. It gives the height required to make the fraction F of Earth's surface visible.

91. Answers will vary.

93. $f^{-1}(x) = \left(\dfrac{1 + \sqrt{4x + 1}}{2} \right)^2$, $x > 0$

95. Answers will vary.

Section 2.3

1. Shift the graph to the left by 3 units.
3. Horizontally compress the graph by a factor of 4.
5. Shift the graph down by 2 units.
7. Horizontally stretch the graph by a factor of 3.
9. Reflect the graph through the x-axis.
11. H
13. B
15. C
17. I
19. A
21. $g(x) = f(x) - 10$
23. $g(x) = f(x - 1)$
25. $g(x) = f(2x)$
27. $g(x) = f(-x)$
29. $g(x) = f(x + 1) - 1$
31. $g(x) = -f(x)$
33.

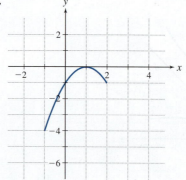

35.

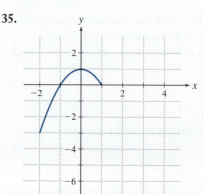

37.

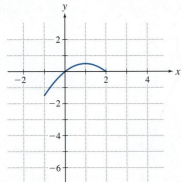

39.

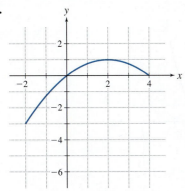

41.

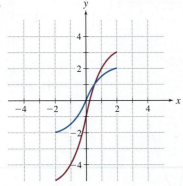

43.

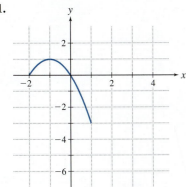

45.

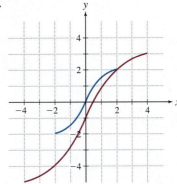

47.

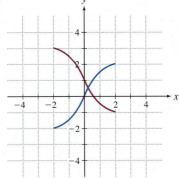

49.

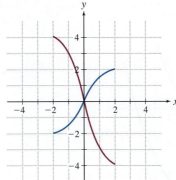

51.

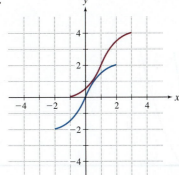

53.

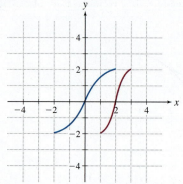

55.

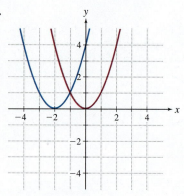

$y = x^2$ is in red.

57.

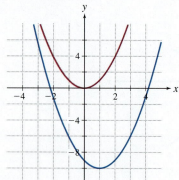

$y = x^2$ is in red.

59.

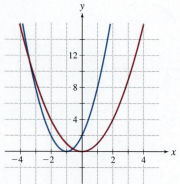

$y = x^2$ is in red.

61.

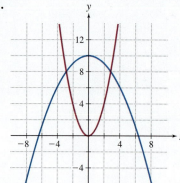

$y = x^2$ is in red.

63.

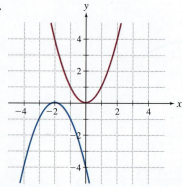

$y = x^2$ is in red.

65. Even

67. Even

69. Even

71. Neither

73. Neither

75.

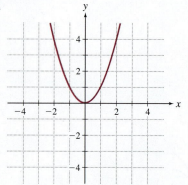

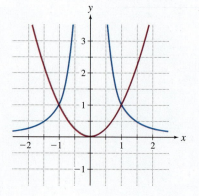

77.

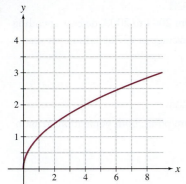

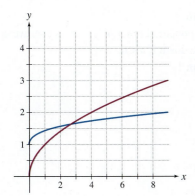

79. a.

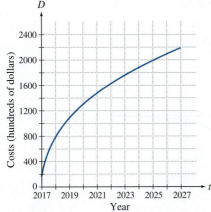

b. $y = 2D(t)$

81. $P(t) = \dfrac{1}{3.26} L(t)$

83. a. They are the same.

    b. They are the same. Answers will vary.

85. a. Graph of $y = 2x + 1$ is in red. Reflected graph is in green.

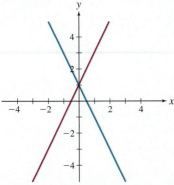

The slope of the reflected line is -2, which is the negative of the slope of the original line.

    b. The slope of the reflected line is the negative of the slope of the original line.

    c. Answers will vary.

87. Answers will vary.

89. Even. Answers will vary.

91. Answers will vary.

CHAPTER 3

Section 3.1

1. m is the slope, and b is the initial value or y-intercept.
3. **a.** The slope is added to the function.
5. True
7. y is doubled.
9. Linear functions with positive slope
11. There is no function that satisfies the condition.
13. $f(x) = -x + 9$
15. $f(x) = 0$
17. $f(x) = -5x - 6$
19. $f(x) = 2x - 1$
21. $f(x) = -3$
23. $f(x) = x + 5$
25. $f(x) = \dfrac{4}{3}x - \dfrac{11}{3}$
27. $y = 7x - 10$
29. $y = \dfrac{1}{11}x + \dfrac{24}{11}$
31. $y = 4x + 9$
33. $f(8) = 34, f(10) = 38$
35. $f(76) = f(301) = 2$
37. D
39. A
41. B
43. Yes. The constant of proportionality is 8 inches per step.
45. No
47. Yes. The constant of proportionality is $\dfrac{1}{2\pi}$.
49. No
51. Yes. The constant of proportionality is one-half the length of the base.
53. No
55. Yes. The constant of proportionality is the tax rate.
57. Yes. The constant of proportionality is the conversion factor from yards to meters.
59. $y = -2.44x + 11.78$

Horizontal span 0 to 8. Vertical span -10 to 12.

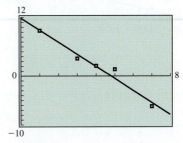

61. a. The total amount is linear because the rate of change has a constant value of $3000 per year.

b. $E = 3000t + 125,000$

63. 45 cubic inches

65. a. $F = \dfrac{9}{5}C + 32$

b. $C = \dfrac{5}{9}F - \dfrac{160}{9}$

c. The slope is $\dfrac{5}{9}$ degree Fahrenheit per degree Celsius. This means that each 1° increase on the Fahrenheit scale corresponds to an increase of $\dfrac{5}{9}$ degree on the Celsius scale.

67. a. 44.1 milliliters per kilogram for male; 36.21 milliliters per kilogram for female

b. $P = 48.39$, or about 48 beats

c. About 30 milliliters per kilogram

69. a. The slope of 5° per centimeter per second tells us that an increase of 1 centimeter per second in running speed results in an increase of 5° in temperature.

b. $S = 0.2C - 2.7$

c. The slope of 0.2 centimeter per second per degree tells us that an increase of 1° in temperature results in a 0.2 centimeter per second increase in running speed.

71. a.

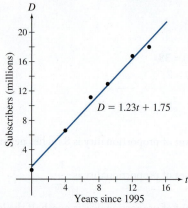

b. $D = 1.23t + 1.75$

c. Each year from 1995 through 2009, on average the number of subscribers increased by 1.23 million.

d. 38.65 million subscribers

73. a.

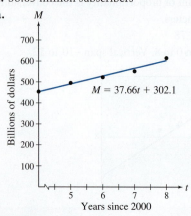

b. $M = 37.66t + 302.1$

c. In the period from 2004 to 2008, defense spending increased by an average of $37.66 billion per year.

d. $1243.6 billion

75. a. $s = 1000 - S$

b. $s = 3000 - 6S$

c. $S = 400$ and $s = 600$

d. At the equilibrium point, both populations remain stable.

Section 3.2

1. y is multiplied by 3.

3. y is multiplied by 16.

5. The graph of an exponential function $y = ab^x$ is concave up if $a > 0$ and concave down if $a < 0$.

7. The base is less than 1.

9. π

11. a, c, d, and e

13. $f(x) = 5(3^x)$

15. $f(x) = 5\left(\dfrac{1}{4}\right)^x$

17. $f(x) = -5\left(\dfrac{1}{\pi}\right)^x$

19. B

21. C

23. E

25.

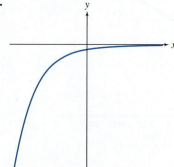

The graph is increasing and concave down.

27. $f(x) = 3 \times 0.6^x$

29. $f(x) = 2 \times 3^x$

31. $f(x) = B_0 \times 2^x$

33. $3(6^x)$

35. $2\left(\dfrac{1}{2}\right)^x$

37. $\left(\dfrac{1}{\pi}\right)^x$

39. $\dfrac{1}{35}\left(\dfrac{5}{7}\right)^x$

41. $d\left(\dfrac{d^2}{c}\right)^x$

43. $e^{-2}(e^6)^x$

45. $e^{x/2} = (e^{1/2})^x$

47. Domain: All real numbers. Range: $(-\infty, 0)$.

49.

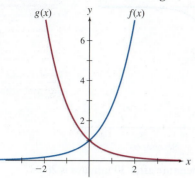

51.

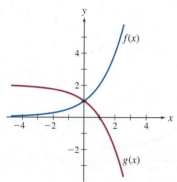

53.

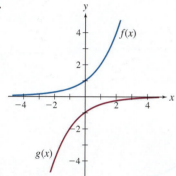

55.

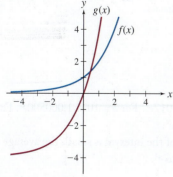

57. Answers will vary.

59. Answers will vary.

61. $x \approx 2.51$

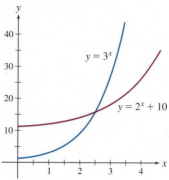

63. $x \approx -1.84$, $x \approx 1.15$

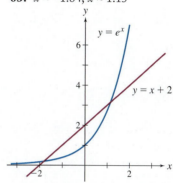

65. $x \approx 1.09$

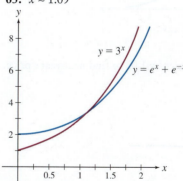

67. $x = 0$, $x \approx 1.43$

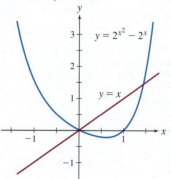

69. The population is cut in half each hour. Here is the graph.

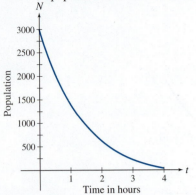

71. a. 8000 grams
  b. 2073.45 grams
  c. 1536.05 grams
  d.

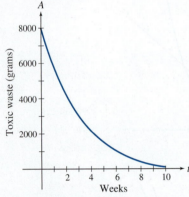

  e. Answers will vary.
73. a. Multiply this year's price by 1.02 to find next year's price.
  b. 2% per year
75. a.

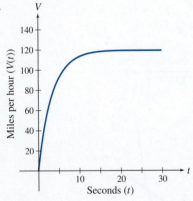

  b. 120 miles per hour

c.

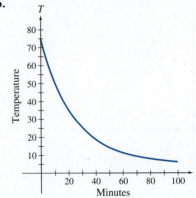

77. a. Yes, the temperature after 30 minutes is 26.08°.
  b.

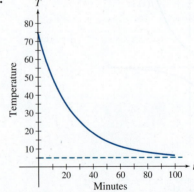

  c. 75°
  d. 5°
  e.

79. a. $\dfrac{e^{t+h} - e^t}{h}$.

  b. $e^t\left(\dfrac{e^h - 1}{h}\right)$

  c. For $h = 10^{-2}$, $1.005017e^t$. For $h = 10^{-3}$, $1.000500e^t$. For $h = 10^{-4}$, $1.000050e^t$.

  d. When the length of the interval is small, the average rate of change is close to e^t.

  e. e^t

Section 3.3

1. An exponential function
3. $y = a(1-r)^t$
5. Compound
7. True
9. True
11. $y = 400 \times 1.02^t$
13. $y = 600 \times 0.92^t$
15. $y = y_0 \times 1.07^t$
17. Half-lives: $y = 200 \times 0.5^b$. Years: $y = 200 \times 0.5^{t/8}$.
19. Half-lives: $y = 10 \times 0.5^b$. Days: $y = 10 \times 0.5^{3t}$.
21. $B = B_0 \times 1.0015^{12t}$
23. $B = B_0 \times 1.003^{2t}$
25. $B = B_0 e^{0.006t}$
27. $B = B_0 e^{rt}$
29. \$221.60
31. \$1054.01
33. D
35. E
37. C
39. $N = N_0 \times 1.03^t$
41. a. $A = A_0 \times 0.98^t$
    b. 18.82 grams
43. $S(t) = 50,000 \times 1.05^t$, \$81,444.73
45. \$671.96
47. a. $D = D_0 \times 0.88^t$
    b. $T = A + (T_0 - A)0.88^t$
    c. $T(10) = A + (T_0 - A)0.88^{10}$
49. Monthly payment: \$127.28. Interest: \$3273.60.
51. a. \$3014.17
    b. \$1467.53
    c. \$898.09
53. a. \$1019.24
    b. \$196,926.40
    c. \$59,077.92
    d. \$307,848.48
55. a. \$654.79
    b. \$610.61
    c. It is better to take the rebate.
    d. It is better to take dealer financing.
57. a. $E = \dfrac{13,000}{1 + 3e^{-0.45t}}$
    b.

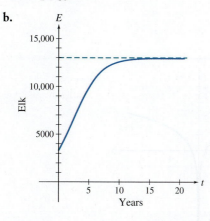
Years

59. a. $F = \dfrac{500,000}{1 + 1249e^{-0.3t}}$
    b.

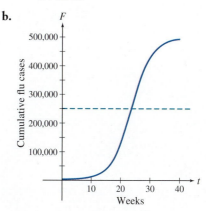
Weeks

    c. Around 24 weeks
61. \$1819.93
63. \$281,608.91
65.

Payment number	Balance
0	\$12,000.00
1	\$11,523.54
2	\$11,045.10
3	\$10,564.67
4	\$10,082.23
5	\$9597.78
6	\$9111.32
7	\$8622.82
8	\$8132.29
9	\$7639.72
10	\$7145.10

67. a. Under these assumptions, the population grows by the same percentage each year.
    b. $N = N_0(1 + b - d)^t$ if t is time in years.
    c. Exponential functions with positive base grow indefinitely large.

CHAPTER 4

Section 4.1

1. The power to which b must be raised to equal x
3. The power to which e must be raised to equal x
5. $g(x) = 7^x$
7. True
9. t
11. Neither 0 nor -2 is in the domain of the common logarithm.
13. -3
15. $\dfrac{1}{2}$
17. 1
19. $\dfrac{3}{10}$

21. $\dfrac{3}{2}$

23. $\dfrac{7}{2}$

25. 0

27. −4

29. 4

31. −2

33. $\dfrac{1}{3}$

35. 0

37. 1

39. 0

41. 4

43. $2x$

45. x^4

47. $\dfrac{x+1}{x}$

49. $\dfrac{1}{x}$

51. $\dfrac{x}{y}$

53. 2

55. D

57. A

59. 3.58

61. 0.53

63. 0.72

65. 0.63

67. $A^2 = B$

69. $(5, \infty)$

71. $(0, \infty)$

73. $(0, 1) \cup (1, \infty)$

75. $(3, 4)$

77. $(0, 1) \cup (1, 2)$

79. Answers will vary.

81.
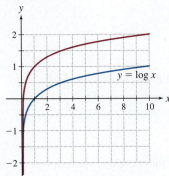

83.
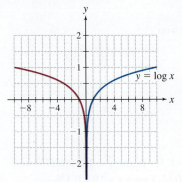

85.
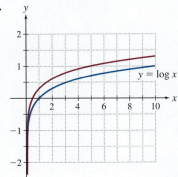

87.
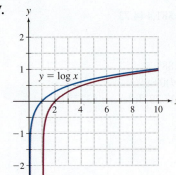

89.
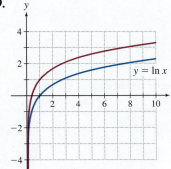

91.
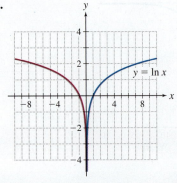

93. $\log 2$

95. $\dfrac{\log b - \log a}{b - a}$

97. $y = \dfrac{\log 3}{2}(x - 1)$

99. 2

101. $x = 10^{100}$

103. $\dfrac{3}{4}$

105. $\dfrac{\ln x}{\ln 7}$

107. $\dfrac{4}{3}$

109. $-\dfrac{5}{2}$

111. $\ln x = \dfrac{\log x}{\log e}$

113. e^2

115. 0

117.

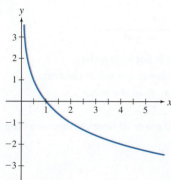

The graph is concave up and always decreasing.

119. 127 decibels

121. a. 2 seconds

  b. $R = 0.5 \dfrac{\ln(n + 1)}{\ln 2}$

  c. 1.66 seconds

  d. Adding another choice when there are already two choices

123. a. Always positive

  b. Decreasing

125. $\ln 3$

127. a. $1 + \dfrac{1}{2} + \dfrac{1}{3} = \dfrac{11}{6}$

  b. Answers will vary.

Section 4.2

1. The sum of the logarithms

3. $M = \log I$

5. b. $\dfrac{1}{x}$

7. $\log(10 + 10)$ is not the same as $\log 10 + \log 10$.

9. b. e

11. 2

13. 0

15. 5

17. $\dfrac{5}{2}$

19. 1

21. $\dfrac{1}{6}$

23. -2

25. x^{-1}

27. $\dfrac{x + 1}{x}$

29. x^{-5}

31. x^3

33. $\dfrac{y^3}{x}$

35. x^{-2}

37. $\ln \dfrac{xy^2}{z}$

39. $\log(x^2 - 1)$

41. $\ln \dfrac{xy}{z^2}$

43. $\ln(x^2 y^4)$

45. $\ln \dfrac{x - 2}{x - 1}$

47. 1

49. $\ln 3^x$

51. $2 - \log x$

53. 0

55. $\log x + \log y - \log z$

57. $x + \ln x$

59. $\dfrac{1}{2}\ln(1 + x^2) - 1$

61. 3.5

63. 1.05

65. e^6

67. 0.875

69. $y = e^{x \ln 9}$

71. $y = e^{-x \ln 3}$

73. Answers will vary.

75. Answers will vary.

77. $y = \dfrac{1}{e^2 - e}(x - e) + 1$

79.

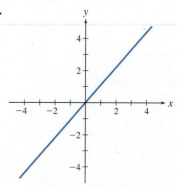

81.

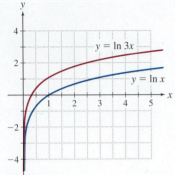

83.

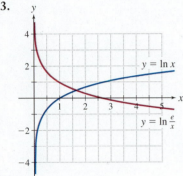

85. Answers will vary.
87. Answers will vary.
89. Answers will vary.
91.

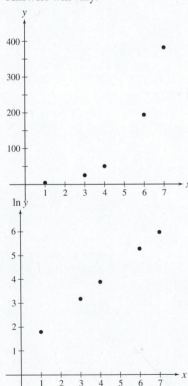

An exponential model is appropriate.

93.

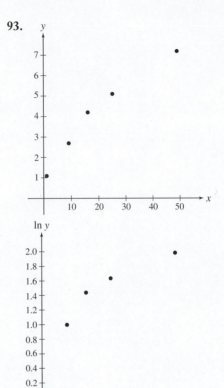

An exponential model is not appropriate.
95. The second is 1000 times as intense as the first.
97. One is 30 decibels more than the other.
99. 158 decibels
101. a. The Tangshan quake was 10 times as intense as the Japan quake.
   b. 7.5
103. a. 1.16
   b. Light from Vega is 2.33 times as intense as that from Antares.
   c. The dimmer star has a magnitude of 0.75 more than the brighter star.
   d. Very bright stars have negative magnitude.
105. Answers will vary.
107. -0.377
109. 0.00036
111. 2.5
113. 3.1

Section 4.3

1. To take the logarithm of both sides
3. a. $\ln c$
5. c. e^c
7. d. all of the above
9. $\ln(4e^x) = \ln 4 + \ln e^x$, not $\ln 4 \ln e^x$
11. $x = \dfrac{\ln\left(\frac{17}{4}\right)}{\ln 3}$
13. $x = \ln t$
15. $x = \log_3 2 = \dfrac{\ln 2}{\ln 3}$

17. $x = \dfrac{\ln 2}{\ln 63 - \ln 58}$

19. $x = \ln 4$

21. $x = \ln 3$

23. $x = \dfrac{\pi}{e \ln \pi}$

25. $x = \ln 5$

27. $x = \dfrac{\ln a - \ln b}{\ln c}$

29. $x = -\dfrac{\ln 3}{4}$

31. $x = e^2$

33. $x = 32$

35. $x = e^e$

37. $x = 10^6 = 1,000,000$

39. $x = \dfrac{e + 3}{2}$

41. $x = 3$

43. $x = 2$

45. $x = \dfrac{e}{3 - e}$

47. $x = e^2 - e$

49. $x = e^{\ln 16 / \ln 3 - 2}$

51. $x = 10^{1/3}$

53. $x = \log 7$

55. $x = \ln \dfrac{e^2 + 1}{e^2 - 1}$

57. $x = \dfrac{\ln(\ln 5 / \ln 2)}{\ln 3}$

59. $x = \dfrac{\ln((e^d - a)/b)}{\ln c}$

61. $g(x) = \ln(e^x - 1)$

63. $g(x) = 10^{\log x / \log 2 - 1}$

65. $g(x) = \dfrac{e^x + 1}{e^x - 1}$

67. $x = 4$

69. $x = 19{,}683$

71. $x = 9$

73. $x \approx 0.70$

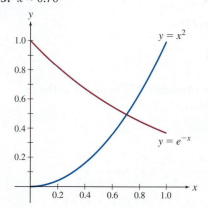

75. $x \approx 1.41$ and $x \approx 3.06$

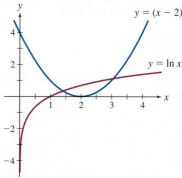

77. $r = \dfrac{\ln 2}{7}$ so that $N = N_0 e^{t \ln 2/7}$ if t is in years

79. $\dfrac{\ln 0.5}{\ln 0.96} \approx 16.98$ years

81. $\dfrac{\ln 0.7}{\ln b}$

83. $\dfrac{\ln 100{,}001}{\ln 2}$ or about 17 checks

85. After 3 hours

87. $t = \dfrac{\ln b}{r}$

89. 9 years after 2018—that is, 2027

CHAPTER 5

Section 5.1

1. Zero, one, or two real zeros

3. Downward

5. There is one real zero.

7. d. All of the above.

9. $c(x - a)^2$

11. $(x - 2)(x - 3)$

13. $2(x - 2)(x + 4)$

15. $-\left(x - \dfrac{1 + \sqrt{5}}{2}\right)\left(x - \dfrac{1 - \sqrt{5}}{2}\right)$

17. $2(x - i)(x + i)$

19. $-2\left(x - \dfrac{1 + i\sqrt{3}}{5}\right)\left(x - \dfrac{1 - i\sqrt{3}}{5}\right)$

21. Zeros at $x = -1$ and $x = 5$. Minimum at $(2, -9)$. The discriminant is positive.

23. Zero at $x = 2$. Maximum at $(2, 0)$. The discriminant is 0.

25. No real zeros. Minimum at $(3, 2)$. The discriminant is negative.

27. Answers will vary.

29. Answers will vary.

31. Answers will vary.

33. The vertex is at $(3, -6)$. Minimum.

35. The vertex is at $(4, -6)$. Minimum.

37. The vertex is at $(1, 9)$. Maximum.

39. The vertex is at $(-1, 10)$. Maximum.

41. The discriminant is 52. There are two real zeros.

43. The discriminant is 0. There is one real zero.

45. The discriminant is −55. There are no real zeros.

47. Concave down

49. Concave up

51. Concave up

53. A

55. B

57. E

59. $(x+1)^2 + 4$

61. $(x+3)^2 + 4$

63. $(x+4)^2 - 14$

65. $8\left(x - \dfrac{9}{4}\right)\left(x + \dfrac{3}{2}\right)$

67. $\left(x - (2 + \sqrt{10})\right)\left(x - (2 - \sqrt{10})\right)$

69. $(x+i)(x-i)$

71. $\left(x - (1 + 2i\sqrt{2})\right)\left(x - (1 - 2i\sqrt{2})\right)$

73. $(\sqrt{3}x + \sqrt{2}i)(\sqrt{3}x - \sqrt{2}i)$

75. $(x + \sqrt{2})(x + 2\sqrt{2})$

77. $(x - (1 + \sqrt{2}))(x - (1 - \sqrt{2}))$

79. $x^2 + 1$

81. $x^2 + 2x - 7$

83. $2x^2 + x + 1$

85. Zeros at $x = 1$ and $x = 3$. Vertex $(2, -1)$.

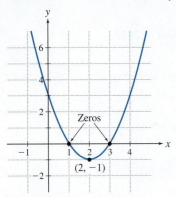

87. No real zeros. Vertex $(-1, 2)$.

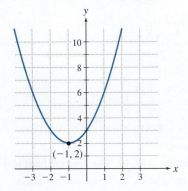

89. Zeros at $x = 0$ and $x = 2$. Vertex $(1, 1)$.

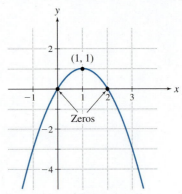

91. 33 miles per hour

93. a. $t = 3$ hours

 b. No, because by part a the maximum concentration is $C(3) = 14$ milligrams per liter.

95. $\dfrac{1}{2}$ and $\dfrac{3}{2}$

97. a. Maximum growth rate of 1 ounce per year at a weight of 2 ounces

 b. 4 ounces

99. $D = 4 + 4\sqrt{\dfrac{B}{L}}$

101. a. $L = 400 - 2W$

 b. $A = W(400 - 2W)$

 c. $L = 200$ and $W = 100$

103. a. $(x-1)^2 + (y-2)^2$

 b. $(x-1)^2 + (5-x)^2$

 c. $(3, 4)$

105. $i^0 = 1$, $i^1 = i$, $i^2 = -1$, $i^3 = -i$, and this pattern repeats so that $i^4 = 1$, $i^5 = i$, and so on.

107. Answers will vary.

109. Answers will vary.

111. Answers will vary.

113. $-2 + 9i$

115. $\dfrac{2}{5} + \dfrac{1}{5}i$

117. $-i$

Section 5.2

1. $x - a$ is a factor of $P(x)$

3. 7

5. $x - 4$

7. $P(1) = 0$

9. The degree of the quotient is 4. The degree of the remainder is 2 or less.

11. It is a zero.

13. It is a zero.

15. It is a zero.

17. Quotient form: $\dfrac{x^3 - x + 1}{x - 1} = x^2 + x + \dfrac{1}{x - 1}$

 Product form: $x^3 - x + 1 = (x - 1)(x^2 + x) + 1$

19. Quotient form: $\dfrac{x^3+1}{x+1}=x^2-x+1$

Product form: $x^3+1=(x+1)(x^2-x+1)$

21. Quotient form: $\dfrac{2x^4-3x^2+1}{x-2}=2x^3+4x^2+5x+10+\dfrac{21}{x-2}$

Product form: $2x^4-3x^2+1=(x-2)(2x^3+4x^2+5x+10)+21$

23. Quotient form: $\dfrac{x^5+2}{x^2+1}=x^3-x+\dfrac{x+2}{x^2+1}$

Product form: $x^5+2=(x^2+1)(x^3-x)+x+2$

25. Quotient form: $\dfrac{x^3-1}{x-3}=x^2+3x+9+\dfrac{26}{x-3}$

Product form: $x^3-1=(x-3)(x^2+3x+9)+26$

27. Quotient form: $\dfrac{x^6-1}{x^4-1}=x^2+\dfrac{x^2-1}{x^4-1}$

Product form: $x^6-1=x^2(x^4-1)+x^2-1$

29. Quotient form: $\dfrac{x^4+1}{x^2+1}=x^2-1+\dfrac{2}{x^2+1}$

Product form: $x^4+1=(x^2+1)(x^2-1)+2$

31. Quotient form: $\dfrac{x^3+x^2+x+1}{x-1}=x^2+2x+3+\dfrac{4}{x-1}$

Product form: $x^3+x^2+x+1=(x-1)(x^2+2x+3)+4$

33. Quotient form: $\dfrac{x^4-x^2+3}{x+2}=x^3-2x^2+3x-6+\dfrac{15}{x+2}$

Product form: $x^4-x^2+3=(x+2)(x^3-2x^2+3x-6)+15$

35. Quotient form: $\dfrac{x^3-x+1}{x-3}=x^2+3x+8+\dfrac{25}{x-3}$

Product form: $x^3-x+1=(x-3)(x^2+3x+8)+25$

37. Quotient form: $\dfrac{x^3-2x+5}{x-2}=x^2+2x+2+\dfrac{9}{x-2}$

Product form: $x^3-2x+5=(x-2)(x^2+2x+2)+9$

39. Quotient form: $\dfrac{x^6+1}{x+1}=x^5-x^4+x^3-x^2+x-1+\dfrac{2}{x+1}$

Product form: $x^6+1=(x+1)(x^5-x^4+x^3-x^2+x-1)+2$

41. Quotient form: $\dfrac{4x^3-2x^2+1}{x+3}=4x^2-14x+42-\dfrac{125}{x+3}$

Product form: $4x^3-2x^2+1=(x+3)(4x^2-14x+42)-125$

43. $P(3)=31$

45. $P(1)=-2$

47. $P(2)=27$

49. The quotient of x^3-x^2+2x+1 divided by x^2+1 is $x-1$, and the remainder is $x+2$.

51. The quotient of x^3+x+1 divided by $x+1$ is x^2-x+2, and the remainder is -1.

53. The quotient of x^4+x^2 divided by $x-5$ is $x^3+5x^2+26x+130$, and the remainder is 650.

55. The quotient of $3x+7$ divided by $x-4$ is 3, and the remainder is 19.

57. The quotient of x^4+x^2+2 divided by x^2-x+1 is x^2+x+1, and the remainder is 1.

59. 13

61. 45

63. 0

65. $P(2)=0$

67. $P(1)=0$

69. $P(1)=0$

71. $P(2)=0$

73. $2-\dfrac{1}{x^2+4}$

75. $1-\dfrac{2}{x^2+x+3}$

77. $P(x)=(x-2)(x+1)^2$

79. $P(x)=(x+1)^3$

81. $P(x)=(x+1)(x+2)(x+4)$

83. $P(x)=(x-1)(x-(1+\sqrt3))(x-(1-\sqrt3))$

85. $P(x)=(x-1)(x+1)(x-2)(x+2)$

87. $P(x)=(x-\sqrt3)(x+\sqrt3)(x-\sqrt5)(x+\sqrt5)$

89. **a.** Answers will vary.
b. Answers will vary.

91. Answers will vary.

Section 5.3

1. a divides the constant term and b divides the leading coefficient

3. False

5. b. polynomial of lower degree

7. False

9. This can't be. The polynomial can have at most eight zeros.

11. $0,1$

13. $0,1+\sqrt3,1-\sqrt3$

15. $0,1,-1$

17. $1,2,3$

19. $1,-1,i,-i$

21. $-\dfrac12+\dfrac{\sqrt{13}}{2},-\dfrac12-\dfrac{\sqrt{13}}{2},-3,4$

23. $x^4(x-1)(x+1)$

25. $(x-\sqrt5)(x+\sqrt5)(x^2+1)$

27. $x(x-1)(x^2+x+1)$

29. $(x+1)(x+2)(x-3)$

31. $(x+4)(x-2)(x+2)$

33. $(x-1)(x+1)(x^2+1)$

35. 5

37. 1

39. $2,-2,-5$

41. None

43. $\dfrac12$

45. $\dfrac23$

47. $-1,\dfrac12$

49. None

51. Answers will vary.

53. $P(x)=(x-1)^2(x+1)^2$. Zeros: $1,-1$.

55. $P(x)=(x+2)(x^2-2x+4)$. Zeros $-2,1\pm i\sqrt3$.

57. $P(x) = (x - \sqrt{2})(x + \sqrt{2})(x - 1)(x + 1)$. Zeros: $\pm\sqrt{2}, \pm 1$.

59. $P(x) = (x - \sqrt{3})(x + \sqrt{3})(x^2 + 1)$. Zeros: $\pm\sqrt{3}, \pm i$.

61. $P(x) = (x^2 + 4)(x - 1)(x + 1)$. Zeros: $\pm 2i, \pm 1$.

63. $P(x) = (x - \sqrt{2})(x + \sqrt{2})(x^2 + 2)(x - 1)(x + 1)(x^2 + 1)$.
Zeros: $\pm\sqrt{2}, \pm\sqrt{2}i, \pm 1, \pm i$.

65. $P(x) = x(x - 2)(x^2 + 2x + 4)$. Zeros: $0, 2, -1 \pm \sqrt{3}i$.

67. $P(x) = x^3(x - 3)(x + 1)$. Zeros: $0, 3, -1$.

69. $P(x) = (x - 1)^2(x + 1)(x - \sqrt{2})(x + \sqrt{2})$. Zeros: $\pm 1, \pm\sqrt{2}$.

71. Additional zeros are $3, -4$.

73. There are no additional zeros.

75. Additional zeros are $3, -3$.

77. -1 is the only additional zero.

79. Additional zeros are $2i, -2i$.

81. $1, -1, 2$

83. $\dfrac{1}{2}, -2, 3$

85. $-2, 4, -\dfrac{1}{2}$

87. $1, 2, \pm i$

89. $2, 1, -1, \pm i$

91. $\dfrac{1}{2}, \dfrac{-1 \pm i\sqrt{3}}{2}$

93. $-\dfrac{1}{2}, -\dfrac{1}{3}, \pm i$

95.

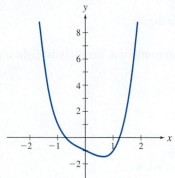

$-0.72, 1.22$

97.

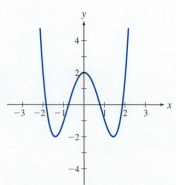

$-1.85, -0.77, 0.77, 1.85$

99.

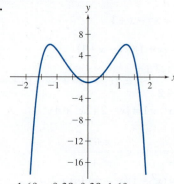

$-1.60, -0.38, 0.38, 1.60$

101. a. 0.49 meter
b. $x = 1.75$ meters

103. a. $r = 3$
b. $r = \sqrt{\dfrac{3}{4\pi}}$

105. 4, 5, and 6

107. a. Answers will vary.
b. $d = R(1 + \sqrt{3})$ and $d = R(1 - \sqrt{3})$.
c. Answers will vary.

109. a. 1491.88 meters per second
b. 6.96° Celsius

111. One solution of $y^3 + 9y = 26$ is $y = 2$. The solutions of $x^3 + 3x^2 + 12x - 16 = 0$ are $1, -2 \pm 2i\sqrt{3}$.

Section 5.4

1. 6

3. 6

5. ∞

7. 6

9. a. The complex zeros come in conjugate pairs.

11. 3

13. No

15. a.

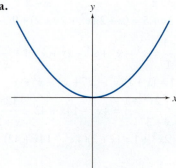

b.

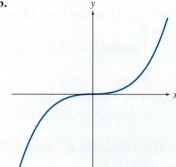

c.

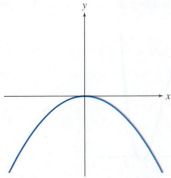

d.

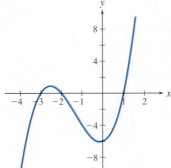

17. $2(x-2)(x+2)^2$

19. $-5(x-1)^5(x-2)^4(x-3)^3$

21. B

23. H

25. F

27. E

29. E

31. A

33. D

35.

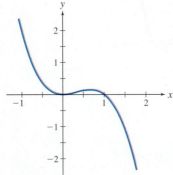

37.

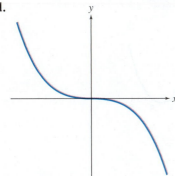

39.

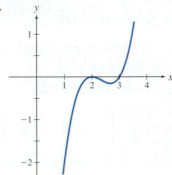

41.

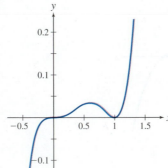

43.

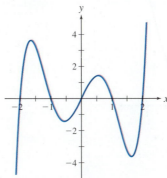

45. One correct solution is

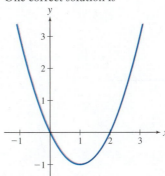

47. One correct solution is

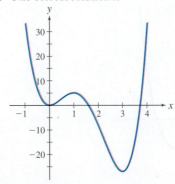

49. $P(x) = (x^2 + 2)(x^2 + x - 3)$. Zeros $i\sqrt{2}, -i\sqrt{2}, \dfrac{-1 \pm \sqrt{13}}{2}$.

51. $P(x) = (x^2 + 1)(x^2 - 2x - 5)$. Zeros $i, -i, 1 + \sqrt{6}$, and $1 - \sqrt{6}$.

53. $(x^2 - 4x + 5)(x - 1)(x + 3)$. Zeros $2 \pm i, 1, -3$.

55. $(x^2 + 1)(x - 1 + \sqrt{2})(x - 1 - \sqrt{2})$. Zeros $\pm i, 1 \pm \sqrt{2}$.

57. $(1 - x)\left(x - \dfrac{1 + \sqrt{5}}{2}\right)\left(x - \dfrac{1 - \sqrt{5}}{2}\right)$

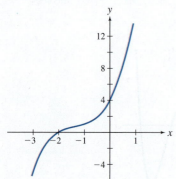

59. $(x + 2)(x^2 + 2x + 2)$

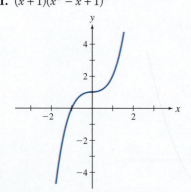

61. $(x + 1)(x^2 - x + 1)$

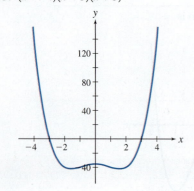

63. $(x^2 + 4)(x - 3)(x + 3)$

65. $(x - \sqrt{6})(x + \sqrt{6})(x - 2)(x + 2)$

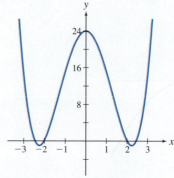

67. $x(x - 3)(x - 4)$

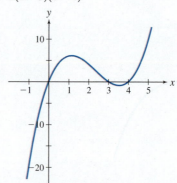

69. $(x - 2)(x - 1)(x^2 + 1)$

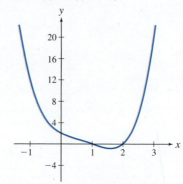

71. $(x - 3)(x - 2)(x + 2)(x + 4)$

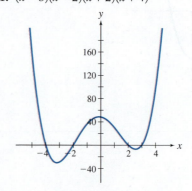

73. $(x-2)(x-1)(x+1)(x+3)$

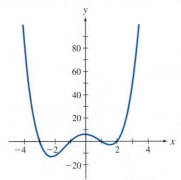

75. $(x+1)(x-1)^2$

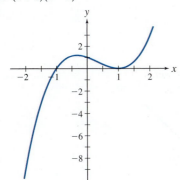

77. $(x-1)(x-\sqrt{5})(x+\sqrt{5})$

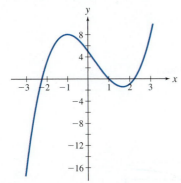

79. Zeros: $-1.88, 0.35, 1.53$
Local maximum: $(-1, 3)$
Local minimum: $(1, -1)$

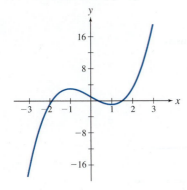

81. Zero: -0.10
Local maximum: $(3, -1)$
Local minimum: $(1, -5)$

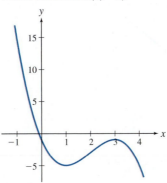

83. Zero: -0.32
Local maximum: None
Local minimum: None

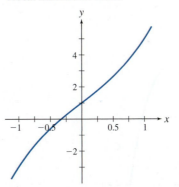

85. Zeros: $-0.19, 2.19$
Local maximum: None
Local minimum: $(1, -2)$

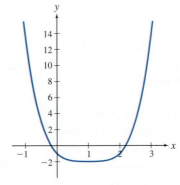

87. a.

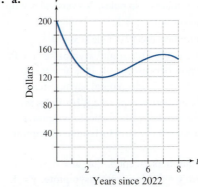

b. The minimum value of $119 occurs on January 1, 2025.
c. The maximum value of $151 occurs on January 1, 2029.
d. The years 2022 through 2024

89. a. $S = 1448.548419 + 4.771299911t - 0.05435t^2 + 0.0002374t^3$

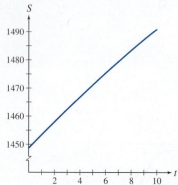

Note the unusual scale on the vertical axis.

b. Increasing temperature indicates an increase in the speed of sound.

91. a. $V = x^2(15 - x)$

b.

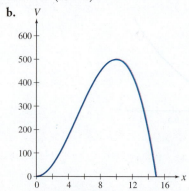

c. 10 feet by 10 feet by 5 feet. The volume is 500 cubic feet.

93. Answers will vary.

Section 5.5

1. Where the denominator of the function is 0

3. c. no horizontal asymptote

5. No horizontal asymptotes

7. It is zero at $x = a$.

9. There are at most n vertical asymptotes.

11. Domain: All real numbers except 1. Pole at $x = 1$. Vertical asymptote: $x = 1$. Horizontal asymptote: $y = 0$.

13. Domain: All real numbers except 2. Pole at $x = 2$. Vertical asymptote: $x = 2$. Horizontal asymptote: $y = 0$.

15. Domain: All real numbers. No poles. No vertical asymptote. Horizontal asymptote: $y = 1$.

17. Domain: All real numbers except 0. Pole at $x = 0$. Vertical asymptote: $x = 0$. Horizontal asymptote: $y = 0$.

19. Domain: All real numbers except -1. Pole at $x = -1$. Vertical asymptote: $x = -1$. Horizontal asymptote: $y = 1$.

21. Domain: All real numbers except $-4 \pm \sqrt{10}$. Poles at $x = -4 + \sqrt{10}$ and $x = -4 - \sqrt{10}$. Vertical asymptotes: $x = -4 + \sqrt{10}$ and $x = -4 - \sqrt{10}$. Horizontal asymptote: $y = 1$.

23. Domain: All real numbers except 5. Pole at $x = 5$. Vertical asymptote: $x = 5$. Horizontal asymptote: $y = 1$.

25. $R(x) \to 0$ as $x \to \pm\infty$.

27. $R(x) \to -1$ as $x \to \pm\infty$.

29. $R(x) \to \dfrac{1}{2}$ as $x \to \pm\infty$.

31. $R(x) \to -4$ as $x \to \pm\infty$.

33. $R(x) \to \dfrac{3}{4}$ as $x \to \pm\infty$.

35. Here is one correct answer.

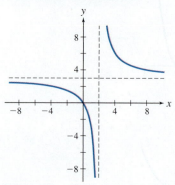

37. Here is one correct answer.

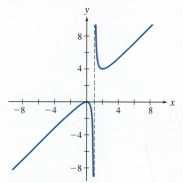

39. Here is one correct answer.

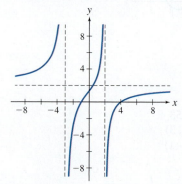

41. Here is one correct answer.

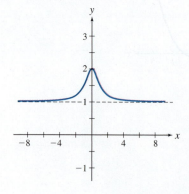

43. Zeros: −1. Vertical asymptotes: $x = 1$, $x = 3$. Horizontal asymptote: $y = 0$.

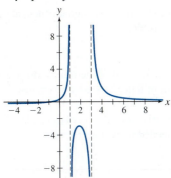

45. Zeros: 2. Vertical asymptotes: $x = 3$. Horizontal asymptote: $y = 6$.

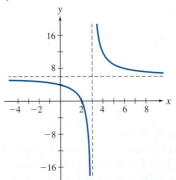

47. Zeros: 1, −1. Vertical asymptotes: $x = 2$, $x = −2$. Horizontal asymptote: $y = 1$.

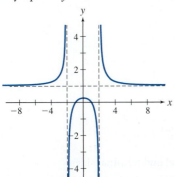

49. Zeros: 1, 3. Vertical asymptotes: $x = −1$. Horizontal asymptote: None.

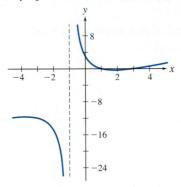

51. Zeros: 4. Vertical asymptotes: $x = 5$. Horizontal asymptote: $y = 1$.

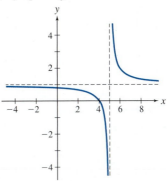

53. Zeros: −1, 2. Vertical asymptotes: $x = 1$, $x = −2$. Horizontal asymptote: $y = 1$.

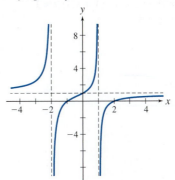

55. Zeros: 1. Vertical asymptotes: $x = −5$, $x = 0$. Horizontal asymptote: $y = 1$.

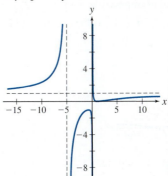

57. Zeros: 0, 1, 6. Vertical asymptotes: $x = −3$, $x = 3$. Horizontal asymptote: None.

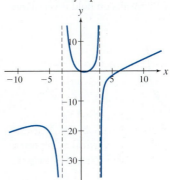

59. a
61. c
63. b
65. c
67. e

69. $x = 0$, $x = 1$, and $x = 2$

71. Answers will vary.

73. Zeros: None. Vertical asymptotes: $x = -\sqrt{5}$, $x = \sqrt{5}$. Horizontal asymptote: $y = 1$.

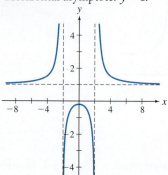

75. Zeros: $-\sqrt{8}$, 0, and $\sqrt{8}$. Vertical asymptotes: $x = 1$. Horizontal asymptote: $y = 3$.

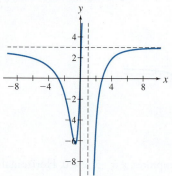

77. Zeros: -3, 2, 5. Vertical asymptotes: $x = -5$, $x = -2$, $x = 3$. Horizontal asymptote: $y = 1$.

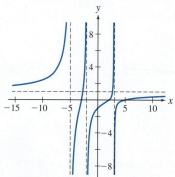

79. a. The average time a caller waits on hold is very large.

b.

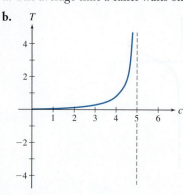

81. a. $E = \dfrac{1}{2}$

b. When Earth is viewed from a very high altitude, about half of the surface is visible.

83. a. Answers will vary.

b. Answers will vary.

c. The vertical asymptote is $x = 0$. When the width is very small, a large amount of fence is needed to enclose the area.

d. The limit is ∞. When the width is very large, a large amount of fence is needed to enclose the area.

e.

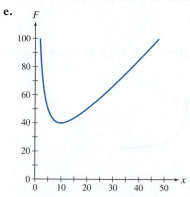

f. Width 10 feet. Length 20 feet.

85. a. $v = \sqrt{2gR}$. When the initial velocity is near $\sqrt{2gR}$, the projectile rises very high before falling back to Earth.

b. The projectile keeps rising and never falls back to Earth.

c. 11.17 kilometers per second

d. 24,931 miles per hour

87. Answers will vary.

89. Answers will vary.

91. Answers will vary.

93. Answers will vary.

CHAPTER 6C

Section 6C.1

1. c. have common initial and terminal sides

3. False

5. 0 radians

7. Corresponding angles are the same.

9. π radians

11. For a right triangle with legs of length a and b and hypotenuse c, $a^2 + b^2 = c^2$.

13. $60°$

15. 0 radians

17. $\dfrac{4\pi}{3}$ radians

19. π radians

21. $\dfrac{\pi}{6}$ radian

23. $\dfrac{2\pi}{3}$ radians

25. $-\dfrac{\pi}{2}$ radians

27. $\dfrac{7\pi}{18}$ radians

29. $135°$

31. $240°$

33. $\dfrac{16{,}200}{\pi}$ degrees

35. $\sqrt{58}$

37. $\sqrt{33}$

39. $\dfrac{23\pi}{36}$ radians

41. $\dfrac{5\pi}{12}$

43. $100°$

45. $\dfrac{\pi}{3}$ radians

47. Arc length 9 units. Area $\dfrac{81}{2}$ square units

49. Arc length $\dfrac{\pi}{2}$ units. Area $\dfrac{3\pi}{4}$ square units

51. Arc length $\dfrac{\pi}{180}$ unit. Area $\dfrac{\pi}{360d}$ square units

53. **a.** 100 grads
    b. 200 grads
    c. 40 grads

55. Answers will vary.

57. The radius is 4 units. The angle is $\dfrac{135}{\pi}$ degrees.

59. $A = \dfrac{rL}{2}$

61. The angle for your friend is twice as big as yours.

63. $C = 12,\ b = 3$

65. $B = 8,\ C = 4$

67. $b = \dfrac{1}{x}$

69. Answers will vary.

71. 1.99 miles

73. About $49°$ north

75. 6 feet

77. Answers will vary.

79. **a.** $270°$
    b. The angle sum of every triangle on the surface of Earth is greater than $180°$.

81. Answers will vary.

Section 6C.2

1. 1

3. Quadrant III

5. $x^2 + y^2 = 1$

7. b. $P(t)$ for an angle of t radians

9. The shortest arc length to the x-axis

11. $x = \pm \dfrac{\sqrt{3}}{2}$

13. $x = \pm \dfrac{\sqrt{7}}{4}$

15. $x = \pm \dfrac{1}{\sqrt{2}}$

17. $y = \pm \dfrac{\sqrt{7}}{3}$

19. $y = \pm \dfrac{\sqrt{11}}{6}$

21. $x = -\dfrac{\sqrt{45}}{7}$

23. A

25. D

27. E

29. G

31. $54°$

33. $32°$

35. $57°$

37. $80°$

39. $10°$

41. $24°$

43. $86°$

45. $\dfrac{2\pi}{5}$ radians

47. $\dfrac{\pi}{3}$ radians

49. $\dfrac{\pi}{6}$ radian

51. $\dfrac{\pi}{6}$ radian

53. $\dfrac{\pi}{7}$ radian

55. $\dfrac{2\pi}{7}$ radian

57. $4 - \pi$ radian

59. $(1, 0)$

61. $(0, -1)$

63. $(0, 1)$

65. $(0, -1)$

67. $(1, 0)$

69. Answers will vary.

71. $x = \pm \dfrac{1}{\sqrt{5}}$

73. $x = \dfrac{-1 + \sqrt{5}}{2}$

75. $x = -\dfrac{1}{\sqrt{10}}$

77. $\left(-\dfrac{1}{\sqrt{2}}, -\dfrac{1}{\sqrt{2}} \right)$

79. $\left(\dfrac{1}{\sqrt{2}}, -\dfrac{1}{\sqrt{2}} \right)$

81. $\left(-\dfrac{\sqrt{3}}{2}, -\dfrac{1}{2}\right)$

83. $\left(\dfrac{1}{\sqrt{2}}, -\dfrac{1}{\sqrt{2}}\right)$

85. $\left(-\dfrac{1}{2}, \dfrac{\sqrt{3}}{2}\right)$

87. $\left(\dfrac{1}{2}, -\dfrac{\sqrt{3}}{2}\right)$

89. $\left(-\dfrac{1}{2}, -\dfrac{\sqrt{3}}{2}\right)$

91. $\left(\dfrac{\sqrt{3}}{2}, \dfrac{1}{2}\right)$

93. $\left(\dfrac{1}{\sqrt{2}}, \dfrac{1}{\sqrt{2}}\right)$

95. $P(t + 2k\pi) = \left(\dfrac{2}{5}, \dfrac{3}{5}\right)$ for any integer k.

Section 6C.3

1. $\cos t = x$, $\sin t = y$, $\tan t = \dfrac{y}{x}$, $\cot t = \dfrac{x}{y}$, $\sec t = \dfrac{1}{x}$, $\csc t = \dfrac{1}{y}$

3. Quadrant IV

5. Quadrant II or IV

7. $\sin^2 t + \cos^2 t = 1$

9. $\cot^2 t + 1 = \csc^2 t$

11. y-coordinate $\dfrac{\sqrt{7}}{4}$

$\cos t = \dfrac{3}{4}$, $\sin t = \dfrac{\sqrt{7}}{4}$, $\tan t = \dfrac{\sqrt{7}}{3}$, $\cot t = \dfrac{3}{\sqrt{7}}$, $\sec t = \dfrac{4}{3}$, $\csc t = \dfrac{4}{\sqrt{7}}$

13. x-coordinate $-\dfrac{\sqrt{11}}{6}$

$\cos t = -\dfrac{\sqrt{11}}{6}$, $\sin t = \dfrac{5}{6}$, $\tan t = -\dfrac{5}{\sqrt{11}}$, $\cot t = -\dfrac{\sqrt{11}}{5}$, $\sec t = -\dfrac{6}{\sqrt{11}}$, $\csc t = \dfrac{6}{5}$

15. x-coordinate $-\dfrac{\sqrt{35}}{6}$

$\cos t = -\dfrac{\sqrt{35}}{6}$, $\sin t = \dfrac{1}{6}$, $\tan t = -\dfrac{1}{\sqrt{35}}$, $\cot t = -\sqrt{35}$, $\sec t = -\dfrac{6}{\sqrt{35}}$, $\csc t = 6$

17. y-coordinate $-\dfrac{\sqrt{8}}{3}$

$\cos t = -\dfrac{1}{3}$, $\sin t = -\dfrac{\sqrt{8}}{3}$, $\tan t = \sqrt{8}$, $\cot t = \dfrac{1}{\sqrt{8}}$, $\sec t = -3$, $\csc t = -\dfrac{3}{\sqrt{8}}$

19. y-coordinate $-\dfrac{\sqrt{15}}{8}$

$\cos t = \dfrac{7}{8}$, $\sin t = -\dfrac{\sqrt{15}}{8}$, $\tan t = -\dfrac{\sqrt{15}}{7}$, $\cot t = -\dfrac{7}{\sqrt{15}}$, $\sec t = \dfrac{8}{7}$, $\csc t = -\dfrac{8}{\sqrt{15}}$

21. $\sin 0° = 0$, $\cos 0° = 1$, $\sec 0° = 1$, $\tan 0° = 0$. The cosecant and cotangent are not defined.

23. $\sin 270° = -1$, $\cos 270° = 0$, $\cot 270° = 0$, $\csc 270° = -1$. The secant and tangent are not defined.

25. $\sin \dfrac{\pi}{2} = 1$, $\cos \dfrac{\pi}{2} = 0$, $\cot \dfrac{\pi}{2} = 0$, $\csc \dfrac{\pi}{2} = 1$. The secant and tangent are not defined.

27. $\sin 5\pi = 0$, $\cos 5\pi = -1$, $\tan 5\pi = 0$, $\sec 5\pi = -1$. The cosecant and cotangent are not defined.

29. $\sin (-180°) = 0$, $\cos (-180°) = -1$, $\tan (-180°) = 0$, $\sec (-180°) = -1$. The cosecant and cotangent are not defined.

31. $\sin (-\pi) = 0$, $\cos (-\pi) = -1$, $\tan (-\pi) = 0$, $\sec (-\pi) = -1$. The cosecant and cotangent are not defined.

33. $\sin \dfrac{3\pi}{2} = -1$, $\cos \dfrac{3\pi}{2} = 0$, $\cot \dfrac{3\pi}{2} = 0$, $\csc \dfrac{3\pi}{2} = -1$. The secant and tangent are not defined.

35. $\sin 900° = 0$, $\cos 900° = -1$, $\tan 900° = 0$, $\sec 900° = -1$. The cosecant and cotangent are not defined.

37.

Quadrant	I	II	III	IV
sine	+	+	−	−
cosine	+	−	−	+
tangent	+	−	+	−
cotangent	+	−	+	−
secant	+	−	−	+
cosecant	+	+	−	−

39. Answers will vary.

41. $\sin 330° = -\dfrac{1}{2}$, $\cos 330° = \dfrac{\sqrt{3}}{2}$, $\tan 330° = -\dfrac{1}{\sqrt{3}}$, $\cot 330° = -\sqrt{3}$, $\sec 330° = \dfrac{2}{\sqrt{3}}$, $\csc 330° = -2$

43. $\cos\left(-\dfrac{\pi}{3}\right) = \dfrac{1}{2}$, $\sin\left(-\dfrac{\pi}{3}\right) = -\dfrac{\sqrt{3}}{2}$, $\tan\left(-\dfrac{\pi}{3}\right) = -\sqrt{3}$, $\cot\left(-\dfrac{\pi}{3}\right) = -\dfrac{1}{\sqrt{3}}$, $\sec\left(-\dfrac{\pi}{3}\right) = 2$, $\csc\left(-\dfrac{\pi}{3}\right) = -\dfrac{2}{\sqrt{3}}$

45. $\cos (-135°) = -\dfrac{1}{\sqrt{2}}$, $\sin (-135°) = -\dfrac{1}{\sqrt{2}}$, $\tan (-135°) = 1$, $\cot (-135°) = 1$, $\sec (-135°) = -\sqrt{2}$, $\csc (-135°) = -\sqrt{2}$

47. $\cos \pi = -1$, $\sin \pi = 0$, $\tan \pi = 0$, $\cot \pi$ undefined, $\sec \pi = -1$, $\csc \pi$ undefined

49. $\sin\left(-\dfrac{\pi}{6}\right) = -\dfrac{1}{2}, \cos\left(-\dfrac{\pi}{6}\right) = \dfrac{\sqrt{3}}{2}, \tan\left(-\dfrac{\pi}{6}\right) = -\dfrac{1}{\sqrt{3}},$

$\cot\left(-\dfrac{\pi}{6}\right) = -\sqrt{3}, \sec\left(-\dfrac{\pi}{6}\right) = \dfrac{2}{\sqrt{3}}, \csc\left(-\dfrac{\pi}{6}\right) = -2$

51. $\sin\left(\dfrac{7\pi}{6}\right) = -\dfrac{1}{2}, \cos\left(\dfrac{7\pi}{6}\right) = -\dfrac{\sqrt{3}}{2}, \tan\left(\dfrac{7\pi}{6}\right) = \dfrac{1}{\sqrt{3}},$

$\cot\left(\dfrac{7\pi}{6}\right) = \sqrt{3}, \sec\left(\dfrac{7\pi}{6}\right) = -\dfrac{2}{\sqrt{3}}, \csc\left(\dfrac{7\pi}{6}\right) = -2$

53. $\sin\left(\dfrac{2\pi}{3}\right) = \dfrac{\sqrt{3}}{2}, \cos\left(\dfrac{2\pi}{3}\right) = -\dfrac{1}{2}, \tan\left(\dfrac{2\pi}{3}\right) = -\sqrt{3},$

$\cot\left(\dfrac{2\pi}{3}\right) = -\dfrac{1}{\sqrt{3}}, \sec\left(\dfrac{2\pi}{3}\right) = -2, \csc\left(\dfrac{2\pi}{3}\right) = \dfrac{2}{\sqrt{3}}$

55. $\sin\left(\dfrac{5\pi}{3}\right) = -\dfrac{\sqrt{3}}{2}, \cos\left(\dfrac{5\pi}{3}\right) = \dfrac{1}{2}, \tan\left(\dfrac{5\pi}{3}\right) = -\sqrt{3},$

$\cot\left(\dfrac{5\pi}{3}\right) = -\dfrac{1}{\sqrt{3}}, \sec\left(\dfrac{5\pi}{3}\right) = 2, \csc\left(\dfrac{5\pi}{3}\right) = -\dfrac{2}{\sqrt{3}}$

57. $\sin t = \dfrac{\sqrt{7}}{4}, \tan t = \dfrac{\sqrt{7}}{3}, \cot t = \dfrac{3}{\sqrt{7}}, \sec t = \dfrac{4}{3},$

$\csc t = \dfrac{4}{\sqrt{7}}$

59. $\cos t = -\dfrac{\sqrt{11}}{6}, \tan t = -\dfrac{5}{\sqrt{11}}, \cot t = -\dfrac{\sqrt{11}}{5},$

$\sec t = -\dfrac{6}{\sqrt{11}}, \csc t = \dfrac{6}{5}$

61. $\cos t = -\dfrac{\sqrt{35}}{6}, \tan t = -\dfrac{1}{\sqrt{35}}, \cot t = -\sqrt{35},$

$\sec t = -\dfrac{6}{\sqrt{35}}, \csc t = 6$

63. $\sin t = -\dfrac{\sqrt{8}}{3}, \tan t = \sqrt{8}, \cot t = \dfrac{1}{\sqrt{8}},$

$\sec t = -3, \csc t = -\dfrac{3}{\sqrt{8}}$

65. $\cos t = \dfrac{7}{8}, \sin t = -\dfrac{\sqrt{15}}{8}, \tan t = -\dfrac{\sqrt{15}}{7},$

$\sec t = \dfrac{8}{7}, \csc t = -\dfrac{8}{\sqrt{15}}$

67. $\sin t = -\dfrac{1}{\sqrt{26}}, \cos t = -\dfrac{5}{\sqrt{26}}, \tan t = \dfrac{1}{5},$

$\sec t = -\dfrac{\sqrt{26}}{5}, \csc t = -\sqrt{26}$

69. $\sin t = -\dfrac{2\sqrt{2}}{3}, \cos t = \dfrac{1}{3}, \tan t = -2\sqrt{2},$

$\cot t = -\dfrac{1}{2\sqrt{2}}, \csc t = -\dfrac{3}{2\sqrt{2}}$

71. $\sin t = -0.84, \tan t = -1.52, \cot t = -0.66,$
$\sec t = 1.82, \csc t = -1.20$

73. $t \approx 1.30$

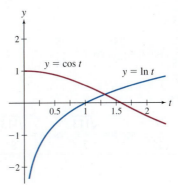

75. $t \approx 0.79$

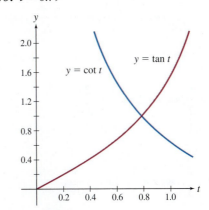

77. a. $A = C$
b. $B = D$

79. $A = B = 3$

81. $A = \dfrac{7}{6}, B = \dfrac{4}{3}$

Section 6C.4

1. $\sin t = \dfrac{\text{Opposite}}{\text{Hypotenuse}}$

3. $\tan t = \dfrac{\text{Opposite}}{\text{Adjacent}}$

5. $\sec t = \dfrac{\text{Hypotenuse}}{\text{Adjacent}}$

7. Adjacent $=$ Hypotenuse $\times \cos t$

9. Opposite $=$ Adjacent $\times \tan t$

11. $\sin s = \dfrac{3}{5}, \cos s = \dfrac{4}{5}, \tan s = \dfrac{3}{4}, \cot s = \dfrac{4}{3}, \sec s = \dfrac{5}{4},$

$\csc s = \dfrac{5}{3}$

13. $\sin t = \dfrac{15}{17}, \cos t = \dfrac{8}{17}, \sec t = \dfrac{17}{8}, \csc t = \dfrac{17}{15},$

$\tan t = \dfrac{15}{8}, \cot t = \dfrac{8}{15}$

15. $\sin t = \dfrac{15}{17}$, $\cos t = \dfrac{8}{17}$, $\sec t = \dfrac{17}{8}$, $\csc t = \dfrac{17}{15}$,

$\tan t = \dfrac{15}{8}$, $\cot t = \dfrac{8}{15}$

17. $\sin t = \dfrac{24}{25}$, $\cos t = \dfrac{7}{25}$, $\sec t = \dfrac{25}{7}$, $\csc t = \dfrac{25}{24}$,

$\tan t = \dfrac{24}{7}$, $\cot t = \dfrac{7}{24}$

19. $\sin t = \dfrac{8}{\sqrt{113}}$, $\cos t = \dfrac{7}{\sqrt{113}}$, $\sec t = \dfrac{\sqrt{113}}{7}$, $\csc t = \dfrac{\sqrt{113}}{8}$,

$\tan t = \dfrac{8}{7}$, $\cot t = \dfrac{7}{8}$

21. $\sin t = \dfrac{7}{\sqrt{74}}$, $\cos t = \dfrac{5}{\sqrt{74}}$, $\sec t = \dfrac{\sqrt{74}}{5}$, $\csc t = \dfrac{\sqrt{74}}{7}$,

$\tan t = \dfrac{7}{5}$, $\cot t = \dfrac{5}{7}$

23. $\sin t = \dfrac{\sqrt{3}}{2}$, $\cos t = \dfrac{1}{2}$, $\sec t = 2$, $\csc t = \dfrac{2}{\sqrt{3}}$,

$\tan t = \sqrt{3}$, $\cot t = \dfrac{1}{\sqrt{3}}$

25. $\sin t = \dfrac{\sqrt{5}}{3}$, $\cos t = \dfrac{2}{3}$, $\sec t = \dfrac{3}{2}$, $\csc t = \dfrac{3}{\sqrt{5}}$,

$\tan t = \dfrac{\sqrt{5}}{2}$, $\cot t = \dfrac{2}{\sqrt{5}}$

27. One correct answer is

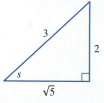

29. One correct answer is

31. One correct answer is

33. One correct answer is

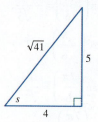

35. One correct answer is

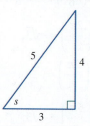

37. One correct answer is

39. $\cos t = \dfrac{\sqrt{5}}{3}$, $\tan t = \dfrac{2}{\sqrt{5}}$, $\cot t = \dfrac{\sqrt{5}}{2}$, $\sec t = \dfrac{3}{\sqrt{5}}$,

$\csc t = \dfrac{3}{2}$

41. $\sin t = \dfrac{5}{\sqrt{89}}$, $\cos t = \dfrac{8}{\sqrt{89}}$, $\cot t = \dfrac{8}{5}$, $\sec t = \dfrac{\sqrt{89}}{8}$,

$\csc t = \dfrac{\sqrt{89}}{5}$

43. $\sin t = \dfrac{1}{7}$, $\cos t = \dfrac{4\sqrt{3}}{7}$, $\sec t = \dfrac{7}{4\sqrt{3}}$, $\tan t = \dfrac{1}{4\sqrt{3}}$,

$\cot t = 4\sqrt{3}$

45. $\sin t = \dfrac{2}{\sqrt{29}}$, $\cos t = \dfrac{5}{\sqrt{29}}$, $\cot t = \dfrac{5}{2}$, $\sec t = \dfrac{\sqrt{29}}{5}$,

$\csc t = \dfrac{\sqrt{29}}{2}$

47. Hypotenuse = 72.73. Adjacent side = 72.29.
49. Opposite side = 28. Hypotenuse = 28.86.
51. Adjacent side = 1.2. Opposite side = 5.88.
53. Answers will vary.

55. $\cos t = \sqrt{1 - x^2}$, $\tan t = \dfrac{x}{\sqrt{1 - x^2}}$, $\cot t = \dfrac{\sqrt{1 - x^2}}{x}$,

$\sec t = \dfrac{1}{\sqrt{1 - x^2}}$, $\csc t = \dfrac{1}{x}$

57. $\sin t = \dfrac{x}{\sqrt{x^2 + 1}}$, $\cos t = \dfrac{1}{\sqrt{1 + x^2}}$, $\cot t = \dfrac{1}{x}$,

$\sec t = \sqrt{1 + x^2}$, $\csc t = \dfrac{\sqrt{1 + x^2}}{x}$

59. $\sin t = \dfrac{1}{x}$, $\cos t = \dfrac{\sqrt{x^2 - 1}}{x}$, $\tan t = \dfrac{1}{\sqrt{x^2 - 1}}$,

$\cot t = \sqrt{x^2 - 1}$, $\sec t = \dfrac{x}{\sqrt{x^2 - 1}}$

61. $\sin t = \dfrac{1}{\sqrt{1 + 9x^2}}$, $\cos t = \dfrac{3x}{\sqrt{1 + 9x^2}}$, $\tan t = \dfrac{1}{3x}$,

$\sec t = \dfrac{\sqrt{1 + 9x^2}}{3x}$, $\csc t = \sqrt{1 + 9x^2}$

63. $\cos t = \dfrac{x + 1}{\sqrt{2x^2 + 2x + 1}}$, $\sin t = \dfrac{x}{\sqrt{2x^2 + 2x + 1}}$, $\cot t = \dfrac{x + 1}{x}$,

$\sec t = \dfrac{\sqrt{2x^2 + 2x + 1}}{x + 1}$, $\csc t = \dfrac{\sqrt{2x^2 + 2x + 1}}{x}$

65. $\cot t = \dfrac{2\sqrt{x + 1}}{x}$, $\cos t = \dfrac{2\sqrt{x + 1}}{x + 2}$, $\tan t = \dfrac{x}{2\sqrt{x + 1}}$,

$\sec t = \dfrac{x + 2}{2\sqrt{x + 1}}$, $\csc t = \dfrac{x + 2}{x}$

67. Area $= 25\sqrt{3}$. Perimeter $= 20 + 10\sqrt{3}$.
69. Area $= 18\sqrt{3} + 54$. Perimeter $= 18 + 6\sqrt{3} + 6\sqrt{6}$.
71. Area $= 9.45$. Perimeter $= 18.83$.
73. Answers will vary.
75. 93 inches
77. 19.28 feet
79. $|BC| = |AC| \sin t$
81. Answers will vary.
83. **a.** $a = \sin t$, $b = \cos t$, $A = \tan t$
   b. Answers will vary.
   c. Answers will vary.
85. Answers will vary.

CHAPTER 6T

Section 6T.1

1. c. have common initial and terminal sides
3. False
5. 0 radians
7. Corresponding angles are the same.
9. π radians
11. For a right triangle with legs of length a and b and hypotenuse c, $a^2 + b^2 = c^2$.
13. $60°$
15. 0 radians
17. $\dfrac{4\pi}{3}$ radians
19. π radians
21. $\dfrac{\pi}{6}$ radian

23. $\dfrac{2\pi}{3}$ radians
25. $-\dfrac{\pi}{2}$ radians
27. $\dfrac{7\pi}{18}$ radians
29. $135°$
31. $240°$
33. $\dfrac{16{,}200}{\pi}$ degrees
35. $\sqrt{58}$
37. $\sqrt{33}$
39. $\dfrac{23\pi}{36}$ radians
41. $\dfrac{5\pi}{12}$ radians
43. $100°$
45. $\dfrac{\pi}{3}$ radians
47. Arc length 9 units. Area $\dfrac{81}{2}$ square units.
49. Arc length $\dfrac{\pi}{2}$ units. Area $\dfrac{3\pi}{4}$ square units.
51. Arc length $\dfrac{\pi}{180}$ unit. Area $\dfrac{\pi}{360d}$ square units.
53. **a.** 100 grads
   b. 200 grads
   c. 40 grads
55. Answers will vary.
57. The radius is 4 units. The angle is $\dfrac{135}{\pi}$ degrees.
59. $A = \dfrac{rL}{2}$
61. The angle for your friend is twice as big as yours.
63. $C = 12$, $b = 3$
65. $B = 8$, $C = 4$
67. $b = \dfrac{1}{x}$
69. Answers will vary.
71. 1.99 miles away
73. About $49°$ north
75. 6 feet
77. Answers will vary.
79. **a.** $270°$
   b. The angle sum of every triangle on the surface of Earth is greater than $180°$.
81. Answers will vary.

Section 6T.2

1. $\sin t = \dfrac{\text{Opposite}}{\text{Hypotenuse}}$

3. $\tan t = \dfrac{\text{Opposite}}{\text{Adjacent}}$

5. $\sec t = \dfrac{\text{Hypotenuse}}{\text{Adjacent}}$

7. Similarity

9. $\dfrac{\pi}{6},\ \dfrac{\pi}{4},\ \dfrac{\pi}{3}$

11. $\sin s = \dfrac{3}{5},\ \cos s = \dfrac{4}{5},\ \tan s = \dfrac{3}{4},\ \cot s = \dfrac{4}{3},$

$\sec s = \dfrac{5}{4},\ \csc s = \dfrac{5}{3}$

13. $\sin t = \dfrac{15}{17},\ \cos t = \dfrac{8}{17},\ \sec t = \dfrac{17}{8},\ \csc t = \dfrac{17}{15},$

$\tan t = \dfrac{15}{8},\ \cot t = \dfrac{8}{15}$

15. $\sin t = \dfrac{15}{17},\ \cos t = \dfrac{8}{17},\ \sec t = \dfrac{17}{8},\ \csc t = \dfrac{17}{15},$

$\tan t = \dfrac{15}{8},\ \cot t = \dfrac{8}{15}$

17. $\sin t = \dfrac{24}{25},\ \cos t = \dfrac{7}{25},\ \sec t = \dfrac{25}{7},\ \csc t = \dfrac{25}{24},$

$\tan t = \dfrac{24}{7},\ \cot t = \dfrac{7}{24}$

19. $\sin t = \dfrac{8}{\sqrt{113}},\ \cos t = \dfrac{7}{\sqrt{113}},\ \sec t = \dfrac{\sqrt{113}}{7},\ \csc t = \dfrac{\sqrt{113}}{8},$

$\tan t = \dfrac{8}{7},\ \cot t = \dfrac{7}{8}$

21. $\sin t = \dfrac{7}{\sqrt{74}},\ \cos t = \dfrac{5}{\sqrt{74}},\ \sec t = \dfrac{\sqrt{74}}{5},\ \csc t = \dfrac{\sqrt{74}}{7},$

$\tan t = \dfrac{7}{5},\ \cot t = \dfrac{5}{7}$

23. $\sin t = \dfrac{\sqrt{3}}{2},\ \cos t = \dfrac{1}{2},\ \sec t = 2,\ \csc t = \dfrac{2}{\sqrt{3}},$

$\tan t = \sqrt{3},\ \cot t = \dfrac{1}{\sqrt{3}}$

25. $\sin t = \dfrac{\sqrt{5}}{3},\ \cos t = \dfrac{2}{3},\ \sec t = \dfrac{3}{2},\ \csc t = \dfrac{3}{\sqrt{5}},$

$\tan t = \dfrac{\sqrt{5}}{2},\ \cot t = \dfrac{2}{\sqrt{5}}$

27. One correct answer is

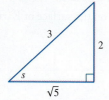

29. One correct answer is

31. One correct answer is

33. One correct answer is

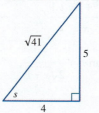

35. One correct answer is

37. One correct answer is

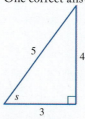

39. $\cos t = \dfrac{\sqrt{5}}{3},\ \tan t = \dfrac{2}{\sqrt{5}},\ \cot t = \dfrac{\sqrt{5}}{2},\ \sec t = \dfrac{3}{\sqrt{5}},\ \csc t = \dfrac{3}{2}$

41. $\sin t = \dfrac{5}{\sqrt{89}},\ \cos t = \dfrac{8}{\sqrt{89}},\ \cot t = \dfrac{8}{5},\ \sec t = \dfrac{\sqrt{89}}{8},$

$\csc t = \dfrac{\sqrt{89}}{5}$

43. $\sin t = \dfrac{1}{7},\ \cos t = \dfrac{4\sqrt{3}}{7},\ \sec t = \dfrac{7}{4\sqrt{3}},\ \tan t = \dfrac{1}{4\sqrt{3}},$

$\cot t = 4\sqrt{3}$

45. $\sin t = \dfrac{2}{\sqrt{29}}$, $\cos t = \dfrac{5}{\sqrt{29}}$, $\cot t = \dfrac{5}{2}$, $\sec t = \dfrac{\sqrt{29}}{5}$,

$\csc t = \dfrac{\sqrt{29}}{2}$

47. Answers will vary.

49. Answers will vary.

51. Answers will vary.

53. **a.** $A = 1$
  b. $H = \sqrt{3}$

55. $\cos t = \sqrt{1 - x^2}$, $\tan t = \dfrac{x}{\sqrt{1 - x^2}}$, $\cot t = \dfrac{\sqrt{1 - x^2}}{x}$.

$\sec t = \dfrac{1}{\sqrt{1 - x^2}}$, $\csc t = \dfrac{1}{x}$

57. $\sin t = \dfrac{x}{\sqrt{x^2 + 1}}$, $\cos t = \dfrac{1}{\sqrt{1 + x^2}}$, $\cot t = \dfrac{1}{x}$, $\sec t = \sqrt{1 + x^2}$,

$\csc t = \dfrac{\sqrt{1 + x^2}}{x}$

59. $\sin t = \dfrac{1}{x}$, $\cos t = \dfrac{\sqrt{x^2 - 1}}{x}$, $\tan t = \dfrac{1}{\sqrt{x^2 - 1}}$,

$\cot t = \sqrt{x^2 - 1}$, $\sec t = \dfrac{x}{\sqrt{x^2 - 1}}$,

61. $\sin t = \dfrac{1}{\sqrt{1 + 9x^2}}$, $\cos t = \dfrac{3x}{\sqrt{1 + 9x^2}}$, $\tan t = \dfrac{1}{3x}$,

$\sec t = \dfrac{\sqrt{1 + 9x^2}}{3x}$, $\csc t = \sqrt{1 + 9x^2}$

63. $\cos t = \dfrac{x + 1}{\sqrt{2x^2 + 2x + 1}}$, $\sin t = \dfrac{x}{\sqrt{2x^2 + 2x + 1}}$, $\cot t = \dfrac{x + 1}{x}$,

$\sec t = \dfrac{\sqrt{2x^2 + 2x + 1}}{x + 1}$, $\csc t = \dfrac{\sqrt{2x^2 + 2x + 1}}{x}$

65. $\cot t = \dfrac{2\sqrt{x + 1}}{x}$, $\cos t = \dfrac{2\sqrt{x + 1}}{x + 2}$, $\tan t = \dfrac{x}{2\sqrt{x + 1}}$,

$\sec t = \dfrac{x + 2}{2\sqrt{x + 1}}$, $\csc t = \dfrac{x + 2}{x}$

67. $t = \dfrac{\pi}{4}$

69. $t \approx 0.88$

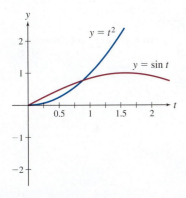

71. $t \approx 0.67$

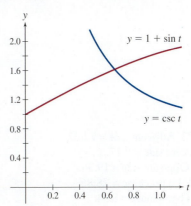

73. Answers will vary.

75. Answers will vary.

77. Answers will vary.

Section 6T.3

1. to find the unknown sides and angles.

3. Adjacent = Hypotenuse $\times \cos t$

5. Opposite = Adjacent $\times \tan t$

7. Hypotenuse = Adjacent $\times \sec t$

9. angles

11.

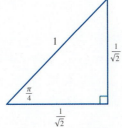

13.

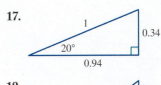

15.

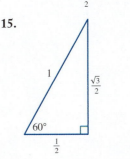

17.

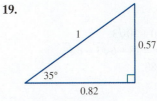

19.

21.

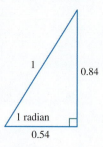

23. Hypotenuse = 72.73. Adjacent side = 72.29.
25. Opposite = 4. Adjacent side = 9.17.
27. Hypotenuse = 20. Opposite side = 19.36.
29. Opposite side = 28. Hypotenuse = 28.86.
31. Adjacent side = 8. Hypotenuse = 8.94.
33. Adjacent side = 8. Hypotenuse = 10.
35. Adjacent side = 1.2. Opposite side = 5.88.
37. Hypotenuse = 4. Opposite side = 3.46.
39. Opposite side = 4. Adjacent side = 4.47.

41. $\dfrac{3\sqrt{3}}{2}$

43. $\dfrac{9}{\sqrt{2}}$

45. 0.62
47. 1.17
49. $B = 6\sqrt{2}$

51.

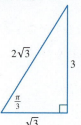

53.

55.

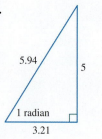

57.

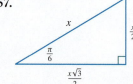

59. Area = $25\sqrt{3}$. Perimeter = $20 + 10\sqrt{3}$.
61. Area = $18\sqrt{3} + 54$. Perimeter = $18 + 6\sqrt{3} + 6\sqrt{6}$.
63. Area = 9.45. Perimeter = 18.83.
65. 96.16 feet
67. 109 feet per minute

69. 91.18 yards
71. a. $v = d \sin t$
    b. 7.76 centimeters; 15 centimeters.
73. Answers will vary.
75. Answers will vary.

Section 6T.4

1. $x^2 + y^2 = 1$
3. Quadrant III
5. True
7. True
9. True

11. $x = \pm \dfrac{\sqrt{5}}{3}$

13. $y = \pm \dfrac{\sqrt{11}}{6}$

15. $x = -\dfrac{\sqrt{45}}{7}$

17. A
19. D
21. E
23. G
25. y-coordinate $\dfrac{\sqrt{3}}{2}$

$\cos t = \dfrac{1}{2}$, $\sin t = \dfrac{\sqrt{3}}{2}$, $\tan t = \sqrt{3}$, $\cot t = \dfrac{1}{\sqrt{3}}$,

$\sec t = 2$, $\csc t = \dfrac{2}{\sqrt{3}}$

27. y-coordinate $\dfrac{4}{5}$

$\cos t = -\dfrac{3}{5}$, $\sin t = \dfrac{4}{5}$, $\tan t = -\dfrac{4}{3}$, $\cot t = -\dfrac{3}{4}$,

$\sec t = -\dfrac{5}{3}$, $\csc t = \dfrac{5}{4}$

29. x-coordinate $-\dfrac{\sqrt{3}}{2}$

$\cos t = -\dfrac{\sqrt{3}}{2}$, $\sin t = -\dfrac{1}{2}$, $\tan t = \dfrac{1}{\sqrt{3}}$, $\cot t = \sqrt{3}$,

$\sec t = -\dfrac{2}{\sqrt{3}}$, $\csc t = -2$

31. y-coordinate $-\dfrac{\sqrt{15}}{4}$

$\cos t = \dfrac{1}{4}$, $\sin t = -\dfrac{\sqrt{15}}{4}$, $\tan t = -\sqrt{15}$, $\cot t = -\dfrac{1}{\sqrt{15}}$,

$\sec t = 4$, $\csc t = -\dfrac{4}{\sqrt{15}}$

33. 57°
35. 17°
37. 80°
39. 10°
41. 83°

43. $\dfrac{2\pi}{7}$ radian

45. $\dfrac{\pi}{3}$ radians

47. $\dfrac{\pi}{6}$ radian

49. $\dfrac{\pi}{6}$ radian

51. $\dfrac{2\pi}{7}$ radian

53. $\cos 90° = 0$, $\sin 90° = 1$, $\cot 90° = 0$, $\csc 90° = 1$. The tangent and secant are not defined.

55. $\sin 5\pi = 0$, $\cos 5\pi = -1$, $\tan 5\pi = 0$, $\sec 5\pi = -1$. The cosecant and cotangent are not defined.

57. $\sin(-180°) = 0$, $\cos(-180°) = -1$, $\tan(-180°) = 0$, $\sec(-180°) = -1$. The cosecant and cotangent are not defined.

59. $\sin(-\pi) = 0$, $\cos(-\pi) = -1$, $\tan(-\pi) = 0$, $\sec(-\pi) = -1$. The cosecant and cotangent are not defined.

61. $\sin \dfrac{\pi}{2} = 1$, $\cos \dfrac{\pi}{2} = 0$, $\cot \dfrac{\pi}{2} = 0$, $\csc \dfrac{\pi}{2} = 1$. The secant and tangent are not defined.

63. $\sin \dfrac{3\pi}{2} = -1$, $\cos \dfrac{3\pi}{2} = 0$, $\cot \dfrac{3\pi}{2} = 0$, $\csc \dfrac{3\pi}{2} = -1$. The secant and tangent are not defined.

65. $\sin\left(-\dfrac{\pi}{2}\right) = -1$, $\cos\left(-\dfrac{\pi}{2}\right) = 0$, $\cot\left(-\dfrac{\pi}{2}\right) = 0$, $\csc\left(-\dfrac{\pi}{2}\right) = -1$. The secant and tangent are not defined.

67. $\sin t = \dfrac{\sqrt{7}}{4}$, $\tan t = \dfrac{\sqrt{7}}{3}$, $\cot t = \dfrac{3}{\sqrt{7}}$, $\sec t = \dfrac{4}{3}$, $\csc t = \dfrac{4}{\sqrt{7}}$

69. $\cos t = -\dfrac{\sqrt{11}}{6}$, $\tan t = -\dfrac{5}{\sqrt{11}}$, $\cot t = -\dfrac{\sqrt{11}}{5}$, $\sec t = -\dfrac{6}{\sqrt{11}}$, $\csc t = \dfrac{6}{5}$

71. $\cos t = -\dfrac{\sqrt{35}}{6}$, $\tan t = -\dfrac{1}{\sqrt{35}}$, $\cot t = -\sqrt{35}$, $\sec t = -\dfrac{6}{\sqrt{35}}$, $\csc t = 6$

73. $\sin t = -\dfrac{\sqrt{8}}{3}$, $\tan t = \sqrt{8}$, $\cot t = \dfrac{1}{\sqrt{8}}$, $\sec t = -3$, $\csc t = -\dfrac{3}{\sqrt{8}}$

75. $\cos t = \dfrac{7}{8}$, $\sin t = -\dfrac{\sqrt{15}}{8}$, $\tan t = -\dfrac{\sqrt{15}}{7}$, $\sec t = \dfrac{8}{7}$, $\csc t = -\dfrac{8}{\sqrt{15}}$

77. $\sin t = -\dfrac{1}{\sqrt{26}}$, $\cos t = -\dfrac{5}{\sqrt{26}}$, $\tan t = \dfrac{1}{5}$, $\sec t = -\dfrac{\sqrt{26}}{5}$, $\csc t = -\sqrt{26}$.

79. $\sin t = -\dfrac{2\sqrt{2}}{3}$, $\cos t = \dfrac{1}{3}$, $\tan t = -2\sqrt{2}$, $\cot t = -\dfrac{1}{2\sqrt{2}}$, $\csc t = -\dfrac{3}{2\sqrt{2}}$.

81. $\sin t = -0.84$, $\tan t = -1.52$, $\cot t = -0.66$, $\sec t = 1.82$, $\csc t = -1.20$

83. Answers will vary.

85. $\sin 330° = -\dfrac{1}{2}$, $\cos 330° = \dfrac{\sqrt{3}}{2}$, $\tan 330° = -\dfrac{1}{\sqrt{3}}$, $\cot 330° = -\sqrt{3}$, $\sec 330° = \dfrac{2}{\sqrt{3}}$, $\csc 330° = -2$

87. $\sin\left(-\dfrac{\pi}{6}\right) = -\dfrac{1}{2}$, $\cos\left(-\dfrac{\pi}{6}\right) = \dfrac{\sqrt{3}}{2}$, $\tan\left(-\dfrac{\pi}{6}\right) = -\dfrac{1}{\sqrt{3}}$, $\cot\left(-\dfrac{\pi}{6}\right) = -\sqrt{3}$, $\sec\left(-\dfrac{\pi}{6}\right) = \dfrac{2}{\sqrt{3}}$, $\csc\left(-\dfrac{\pi}{6}\right) = -2$

89. $\cos\left(-\dfrac{\pi}{3}\right) = \dfrac{1}{2}$, $\sin\left(-\dfrac{\pi}{3}\right) = -\dfrac{\sqrt{3}}{2}$, $\tan\left(-\dfrac{\pi}{3}\right) = -\sqrt{3}$, $\cot\left(-\dfrac{\pi}{3}\right) = -\dfrac{1}{\sqrt{3}}$, $\sec\left(-\dfrac{\pi}{3}\right) = 2$, $\csc\left(-\dfrac{\pi}{3}\right) = -\dfrac{2}{\sqrt{3}}$

91. $\cos(-135°) = -\dfrac{1}{\sqrt{2}}$, $\sin(-135°) = -\dfrac{1}{\sqrt{2}}$, $\tan(-135°) = 1$, $\cot(-135°) = 1$, $\sec(-135°) = -\sqrt{2}$, $\csc(-135°) = -\sqrt{2}$

93. $\cos \pi = -1$, $\sin \pi = 0$, $\tan \pi = 0$, $\cot \pi$ undefined, $\sec \pi = -1$, $\csc \pi$ undefined

95. $\sin\left(\dfrac{7\pi}{6}\right) = -\dfrac{1}{2}$, $\cos\left(\dfrac{7\pi}{6}\right) = -\dfrac{\sqrt{3}}{2}$, $\tan\left(\dfrac{7\pi}{6}\right) = \dfrac{1}{\sqrt{3}}$, $\cot\left(\dfrac{7\pi}{6}\right) = \sqrt{3}$, $\sec\left(\dfrac{7\pi}{6}\right) = -\dfrac{2}{\sqrt{3}}$, $\csc\left(\dfrac{7\pi}{6}\right) = -2$

97. $\sin\left(\dfrac{2\pi}{3}\right) = \dfrac{\sqrt{3}}{2}$, $\cos\left(\dfrac{2\pi}{3}\right) = -\dfrac{1}{2}$, $\tan\left(\dfrac{2\pi}{3}\right) = -\sqrt{3}$, $\cot\left(\dfrac{2\pi}{3}\right) = -\dfrac{1}{\sqrt{3}}$, $\sec\left(\dfrac{2\pi}{3}\right) = -2$, $\csc\left(\dfrac{2\pi}{3}\right) = \dfrac{2}{\sqrt{3}}$

99. $\sin\left(\dfrac{5\pi}{3}\right) = -\dfrac{\sqrt{3}}{2}$, $\cos\left(\dfrac{5\pi}{3}\right) = \dfrac{1}{2}$, $\tan\left(\dfrac{5\pi}{3}\right) = -\sqrt{3}$, $\cot\left(\dfrac{5\pi}{3}\right) = -\dfrac{1}{\sqrt{3}}$, $\sec\left(\dfrac{5\pi}{3}\right) = 2$, $\csc\left(\dfrac{5\pi}{3}\right) = -\dfrac{2}{\sqrt{3}}$

101. $A = 6$, $B = -4$

103. $A = 3$, $B = 1$

105. Answers will vary.

CHAPTER 7

Section 7.1

1. 2π

3. a. The directed distance to the x-axis.

5. The sine reaches a maximum value of 1 at $x = \dfrac{\pi}{2} + 2k\pi$ for any integer k.

7. The cosine reaches a maximum value of 1 at $x = 2k\pi$ for any integer k.

9. $\dfrac{2\pi}{B}$

11. **b.** $\sin(-x) = -\sin x$ for all x.

13. Period π. Amplitude 3. Phase shift 0.

15. Period 2. Amplitude 2. Phase shift 0.

17. Period $2\pi^2$. Amplitude 3. Phase shift 0.

19. Period π. Amplitude 3. Phase shift π.

21. Period π. Amplitude π. Phase shift 1.

23. Period π. Amplitude 3. Phase shift $\dfrac{1}{2}$

25. $y = 4 \sin \pi x$

27. $y = 3 \sin 2\pi(x - \pi)$

29. Amplitude 2. Period 2π. Phase shift 0.

31. Amplitude 3. Period 2π. Phase shift π.

33. Amplitude $\dfrac{1}{2}$. Period 4. Phase shift 1.

35.

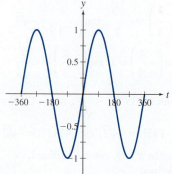

37. Amplitude 2. Period π. Phase shift 0.

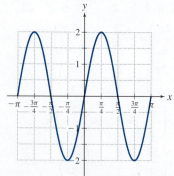

39. Amplitude 3. Period 2. Phase shift 1.

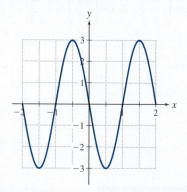

41. $A = 2, B = 2, C = \dfrac{\pi}{2}$.

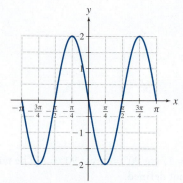

43. $A = 3, B = \dfrac{\pi}{2}, C = 0$.

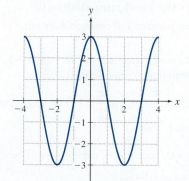

45. Amplitude 4. Period π. Phase shift $\dfrac{\pi}{2}$.

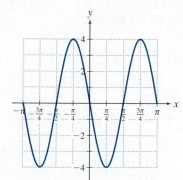

47.

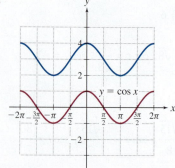

49.

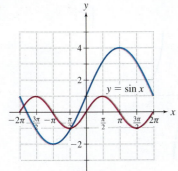

51.

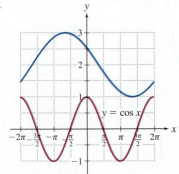

53.

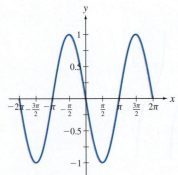

The two graphs are identical, suggesting the identity $\sin(x + \pi) = -\sin x$.

55.

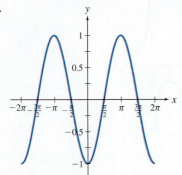

The two graphs are identical, suggesting the identity $\cos(x - \pi) = -\cos x$.

57. $\dfrac{1}{3}$

59. $-\dfrac{2\sqrt{2}}{3}$

61. $\dfrac{1}{3}$

63. $\dfrac{1}{3}$

65.

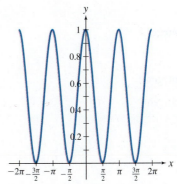

67.

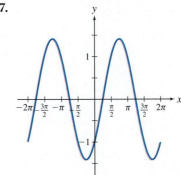

69.

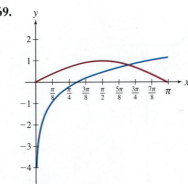

$x \approx 2.22$

71. Answers will vary.

73. a.

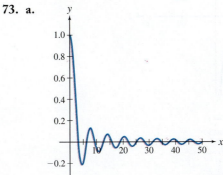

b. $\dfrac{\sin x}{x} \to 0$ as $x \to \infty$

75. Answers will vary.

Section 7.2

1. π

3. 2π

5. $x = \dfrac{\pi}{2} + k\pi$ for any integer k.

7. $x = \dfrac{\pi}{2} + k\pi$ for any integer k.

9. The tangent and cotangent

11. $\sin x$

13. $\tan x$

15. $\sec x$

17. $\dfrac{\pi}{2}$

19. 2π

21. $\tan\left(x - \dfrac{\pi}{2}\right)$

23. $y = \tan 2x$

25. $y = \sec 2x$

27. $y = 2 \sec x$

29. $y = -\sec x$

31. Answers will vary.

33. Answers will vary.

35.

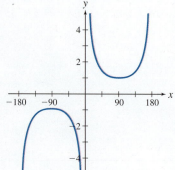

37.

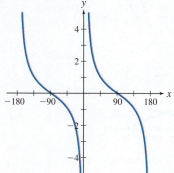

39. Answers will vary.

41. Answers will vary.

43. $-\sqrt{3}$

45. 2

47. Answers will vary.

49. Answers will vary.

51.

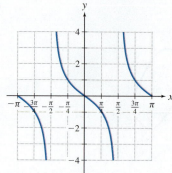

53.

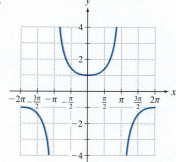

55.

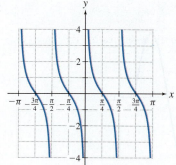

57.

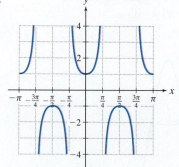

59.

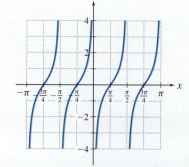

61.

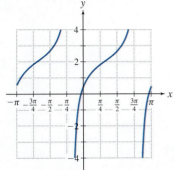

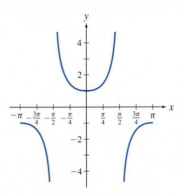

63.

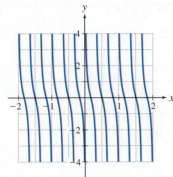

65. a. The graph of $y = \sec\left(x + \dfrac{\pi}{2}\right)$ is shown. The graph

of $y = -\csc x$ is identical, suggesting the identity

$$\sec\left(x + \dfrac{\pi}{2}\right) = -\csc x.$$

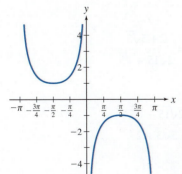

b. $\sec\left(x + \dfrac{\pi}{2}\right) = \dfrac{1}{\cos\left(x + \dfrac{\pi}{2}\right)} = \dfrac{1}{-\sin x} = -\csc x$

67. a. The graph of $y = \sec(-x)$ is shown. The graph
of $y = \sec x$ is identical, suggesting the identity
$\sec(-x) = \sec x$.

b. $\sec(-x) = \dfrac{1}{\cos(-x)} = \dfrac{1}{\cos x} = \sec x$

69.

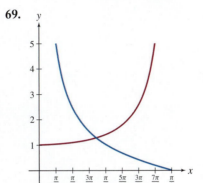

$x \approx 0.67$

71. a. 1.15
b. Answers will vary.
c. 60°
73. Answers will vary.

CHAPTER 8

Section 8.1

1. $\sin^2 t + \cos^2 t = 1,\;\; 1 + \cot^2 t = \csc^2 t,\;\; \tan^2 t + 1 = \sec^2 t$
3. sine
5. True
7. True
9. True
11. $\dfrac{2}{3}$
13. $\dfrac{1}{4}$
15. $\dfrac{4}{3}$
17. $\sin\left(\dfrac{\pi}{2} - \alpha\right) = -\dfrac{3}{4}.\; \cos\left(\dfrac{\pi}{2} - \alpha\right) = -\dfrac{\sqrt{7}}{4}.$
19. $-\dfrac{2}{5}$
21. 3
23. $\cos t$
25. 1
27. $\sin t$
29. $\cos^2 x$

31. $\csc x$

33. 1

35. $\cos^2 x$

37. Answers will vary.

39. Answers will vary.

41. Answers will vary.

43. Answers will vary.

45. Answers will vary.

47. Answers will vary.

49. Answers will vary.

51. It is even.

53. It is odd.

55. $\sec t = \sqrt{1 + \tan^2 t}$

57. $\sin t = -\sqrt{1 - \cos^2 t}$

59. $\tan t = \dfrac{\sqrt{1 - \cos^2 t}}{\cos t}$

61. $\cot t = \dfrac{1}{\tan t}$

63. $\tan t = \dfrac{\sec t}{\csc t}$

65. $\cot t = -\dfrac{\sqrt{1 - \sin^2 t}}{\sin t}$

67. Answers will vary.

69. Answers will vary.

71. Answers will vary.

73. Answers will vary.

75. Answers will vary.

77. Answers will vary.

79. Answers will vary.

81. Answers will vary.

83. Answers will vary.

85. Answers will vary.

87. Answers will vary.

89. Answers will vary.

91. Answers will vary.

93. Answers will vary.

95. Answers will vary.

97. Answers will vary.

99. Answers will vary.

101. Answers will vary.

103. Answers will vary.

Section 8.2

1. $\sin(x + y) = \sin x \cos y + \cos x \sin y$
and $\sin(x - y) = \sin x \cos y - \cos x \sin y$

3. $\tan(x + y) = \dfrac{(\tan x + \tan y)}{(1 - \tan x \tan y)}$ and

$\tan(x - y) = \dfrac{(\tan x - \tan y)}{(1 + \tan x \tan y)}$

5. False

7. True

9. True

11. $\dfrac{15\sqrt{3}}{4}$ square feet

13. $\dfrac{27}{4}$ square feet

15. $\dfrac{4}{\sqrt{2}}$ or $2\sqrt{2}$ square feet

17. $\tan 75° = \sqrt{3} + 2$, $\cot 75° = 2 - \sqrt{3}$

19. $\tan 15° = 2 - \sqrt{3}$, $\cot 15° = \sqrt{3} + 2$

21. $\tan \dfrac{\pi}{12} = 2 - \sqrt{3}$, $\cot \dfrac{\pi}{12} = \sqrt{3} + 2$

23. $\tan \dfrac{11\pi}{12} = \sqrt{3} - 2$, $\cot \dfrac{11\pi}{12} = -\sqrt{3} - 2$

25. $\tan 105° = -\sqrt{3} - 2$, $\cot 105° = \sqrt{3} - 2$

27. $\tan 165° = \sqrt{3} - 2$, $\cot 165° = -\sqrt{3} - 2$

29. $\tan \dfrac{19\pi}{12} = -\sqrt{3} - 2$,

$\cot \dfrac{19\pi}{12} = \sqrt{3} - 2$

31. $\dfrac{2 - \sqrt{75}}{12}$

33. $\dfrac{2\sqrt{15} + \sqrt{5}}{12}$

35. $\dfrac{1/\sqrt{15} + \sqrt{5}/2}{1 - \dfrac{1}{\sqrt{15}} \dfrac{\sqrt{5}}{2}} = \dfrac{2 + 5\sqrt{3}}{2\sqrt{15} - \sqrt{5}}$

37. $\dfrac{\sqrt{15}\dfrac{2}{\sqrt{5}} - 1}{\sqrt{15} + 2/\sqrt{5}} = \dfrac{2\sqrt{15} - \sqrt{5}}{5\sqrt{3} + 2}$

39. $\dfrac{3\sqrt{21} + 8}{25}$

41. $\dfrac{-4\sqrt{21} + 6}{25}$

43. $\dfrac{-3/4 - 2/\sqrt{21}}{1 - (-3/4)(-2/\sqrt{21})} = \dfrac{-3\sqrt{21} - 8}{4\sqrt{21} - 6}$

45. $\dfrac{-3/4 - (-2/\sqrt{21})}{1 + (-3/4)(-2/\sqrt{21})} = \dfrac{-3\sqrt{21} + 8}{4\sqrt{21} + 6}$

47. $\dfrac{\sqrt{2} + 1}{4}$

49. $\sqrt{3/2}$

51. $\dfrac{\sqrt{2} + \sqrt{3}}{4}$

53. $\dfrac{\sqrt{3}-\sqrt{2}}{4}$

55. Answers will vary.

57. Answers will vary.

59. Answers will vary.

61. Answers will vary.

63. Answers will vary.

65. Answers will vary.

67. Answers will vary.

69. Answers will vary.

71. Answers will vary.

73. Answers will vary.

75. Answers will vary.

77. Answers will vary.

79.　**a.** $A = 4$

　　b. Quadrant IV

　　c. $\phi = -\dfrac{\pi}{6}$

　　d. $-2\cos t + 2\sqrt{3}\sin t = 4\sin\left(t - \dfrac{\pi}{6}\right)$

81. Answers will vary.

83. Answers will vary.

85. Answers will vary.

Section 8.3

1. $\sin 2t = 2\sin t\cos t$

3. $\tan 2t = \dfrac{2\tan t}{1 - \tan^2 t}$

5. $\cos\dfrac{t}{2} = \pm\sqrt{\dfrac{1 + \cos t}{2}}$

7. False

9. $\sin^2 x = \dfrac{1}{2}(1 - \cos 2x)$

11. $\dfrac{24}{25}$

13. $-\dfrac{7}{25}$

15. $\dfrac{7\sqrt{2}}{8}$

17. $\sqrt{(3 - 2\sqrt{2})/6}$

19. $\dfrac{1}{2}$

21. $\dfrac{\sqrt{5}}{2}$

23. $2\dfrac{1}{3}\dfrac{2\sqrt{2}}{3}\left(1 - 2\left(\dfrac{4}{5}\right)^2\right) + \left(1 - 2\left(\dfrac{1}{3}\right)^2\right)2\dfrac{4}{5}\dfrac{3}{5} = \dfrac{4\sqrt{2}}{9}\left(\dfrac{-7}{25}\right)$

$\quad + \dfrac{7}{9}\left(\dfrac{24}{25}\right) \approx 0.571$

25. $2\left(2\dfrac{4}{5}\dfrac{3}{5}\right)\left(2\left(\dfrac{3}{5}\right)^2 - 1\right) = 2\left(\dfrac{24}{25}\right)\left(\dfrac{-7}{25}\right) \approx -0.538$

27. $\dfrac{2\left(\dfrac{2(4/3)}{1 - (4/3)^2}\right)}{1 - \left(\dfrac{2(4/3)}{1 - (4/3)^2}\right)^2} = \dfrac{2(-24/7)}{1 - (-24/7)^2} \approx 0.638$

29. $\dfrac{1}{3}\sqrt{\dfrac{1 + 2\sqrt{2}/3}{2}} + \dfrac{2\sqrt{2}}{3}\sqrt{\dfrac{1 - 2\sqrt{2}/3}{2}} \approx 0.488$

31. $\dfrac{1}{9}$

33. $\dfrac{-2}{\sqrt{5} - 1}$

35. $\sqrt{3/8}$

37. $-\dfrac{4 + \sqrt{7}}{3}$

39. $\tan 22.5° = \dfrac{1}{\sqrt{2} + 1} \approx 0.414,\ \ \cot 22.5° = \sqrt{2} + 1 \approx 2.414$

41. $\sin\dfrac{7\pi}{12} = \dfrac{1 + \sqrt{3}}{2\sqrt{2}} \approx 0.966,\ \ \cos\dfrac{7\pi}{12} = \dfrac{1 - \sqrt{3}}{2\sqrt{2}} \approx -0.259,$

$\quad \tan\dfrac{7\pi}{12} = \dfrac{1 + \sqrt{3}}{1 - \sqrt{3}} \approx -3.732$

43. $\cos\dfrac{\pi}{24} = \sqrt{\dfrac{1 + \sqrt{\dfrac{1 + \sqrt{3}/2}{2}}}{2}} \approx 0.991$

45. Answers will vary.

47. Answers will vary.

49. Answers will vary.

51. Answers will vary.

53. Answers will vary.

55. Answers will vary.

57. Answers will vary.

59. Answers will vary.

61. Answers will vary.

63. Answers will vary.

65. Answers will vary.

67. Answers will vary.

69. Answers will vary.

71. Answers will vary.

73.　**a.** $\dfrac{1}{2}\sin 2\alpha$

　　b. $\sin\alpha\cos\alpha$

　　c. Answers will vary.

75. $x \approx 1.15$

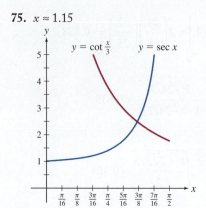

77. $x \approx 0.90$

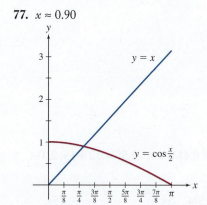

79. a. 1255.44 feet downrange

b. $d = \dfrac{v_0^2 \sin 2\theta}{g}$

c. $\theta = 45°$

81. Answers will vary.

83. a. $x = \sqrt{\dfrac{1 + \cos \alpha}{2}}$

b. Answers will vary.

85. a. Answers will vary.

b. A: $(\cos(-\alpha), \sin(-\alpha))$, B: $(1,0)$, C: $(\cos(\alpha), \sin(\alpha))$,
D: $(\cos(2\alpha), \sin(2\alpha))$

c. Answers will vary.

d. From A to C: $2 - 2(\cos(-\alpha)\cos(\alpha) + \sin(-\alpha)\sin(\alpha))$;
from B to D: $2 - 2\cos(2\alpha)$

e. Answers will vary.

Section 8.4

1. Domain $[-1, 1]$. Range $\left[-\dfrac{\pi}{2}, \dfrac{\pi}{2}\right]$

3. Domain $(-\infty, \infty)$. Range $\left(-\dfrac{\pi}{2}, \dfrac{\pi}{2}\right)$

5. False

7. It is the angle between $-\dfrac{\pi}{2}$ and $\dfrac{\pi}{2}$ whose sine is x.

9. $x = \arctan c$

11. $\dfrac{\pi}{3}$

13. $\dfrac{\pi}{6}$

15. $\dfrac{\pi}{2}$

17. $-\dfrac{\pi}{6}$

19. $-\dfrac{\pi}{4}$

21. $\dfrac{\pi}{4}$

23. $\dfrac{\pi}{3}$

25. $-\dfrac{\pi}{4}$

27. B

29. D

31. F

33. 0.89

35. 0.61

37. 1.37

39. $x = \tan 1 \approx 1.56$

41. $x = 1$

43. $x = \cos(-1) \approx 0.54$

45. $\dfrac{1}{2\sqrt{2}}$

47. $\dfrac{5}{\sqrt{26}}$

49. $\dfrac{\sqrt{39}}{8}$

51. $-\dfrac{\sqrt{7}}{3}$

53. $\dfrac{\sqrt{1 - x^2}}{x}$

55. $\sqrt{1 - x^2}$

57. $\dfrac{x}{\sqrt{2x + 1}}$

59. $\dfrac{12}{13}$

61. $2x\sqrt{1 - x^2}$

63. $2x^2 - 1$

65. $\dfrac{\sqrt{1 - x^2}}{\sqrt{1 + x^2}} - \dfrac{x^2}{\sqrt{1 + x^2}}$

67. $\dfrac{4x(1 - x^2)}{(1 + x^2)^2}$

69. Answers will vary.

71. Answers will vary.

73. Answers will vary.

75. Answers will vary.

77. $x = \dfrac{\pi}{6}$

79. $x = \dfrac{\pi}{3}$

81. $x = \arctan 3 \approx 1.25$

83. $x = \dfrac{1}{3}\arcsin\dfrac{1}{4} \approx 0.08$

85. $x = \arctan(7) \approx 1.43$

87. $t = \dfrac{4}{\pi}$

89. $x = \dfrac{1}{\sqrt{10}}$

91. $x = \dfrac{1}{\sqrt{3}}$

93.

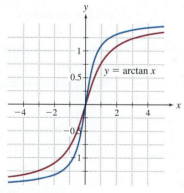

95.

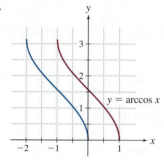

97. $71.34°$

99. $A = 4,\ s = 36.87°,\ t = 53.13°$

101. $C = \sqrt{13},\ s = 33.69°,\ t = 56.31°$

103.

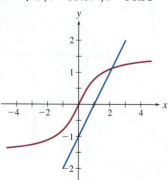

$x \approx 2.13$

105.

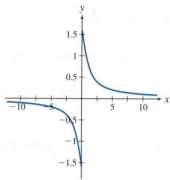

$\arctan \dfrac{1}{x} \to 0$ as $x \to \infty$. There is no vertical asymptote at $x = 0$.

107. $18.00°$

109. **a.** $\theta = \dfrac{1}{2}\arcsin\dfrac{gd}{v^2}$

  b. $11.01°$

111. $p = \arctan\dfrac{d}{s}$

113. **a.** $\alpha' = \arctan\left(\dfrac{b}{b\cot\alpha - 2d}\right)$

  b. $15.24°$

115. **a.** 2.89 meters per second

  b. 0.64 radian

117. Answers will vary.

Section 8.5

1. $t = \arcsin c + 2\pi k$ and $t = \pi - \arcsin c + 2\pi k$, where k is any integer

3. $t = \arctan c + \pi k$, where k is any integer

5. True

7. $t_1 + 2k\pi$ and $-t_1 + 2k\pi$, where k is any integer

9. $t = \dfrac{\pi}{6} + 2k\pi$ and $t = \dfrac{5\pi}{6} + 2k\pi$, where k is any integer

11. $t = -\dfrac{\pi}{6} + 2k\pi$ and $t = \dfrac{7\pi}{6} + 2k\pi$, where k is any integer

13. $t = \dfrac{\pi}{6} + k\pi$, where k is any integer

15. $t = \dfrac{\pi}{3} + k\pi$, where k is any integer

17. $t = -\dfrac{\pi}{4} + 2k\pi$ and $t = \dfrac{5\pi}{4} + 2k\pi$, where k is any integer

19. $t = \dfrac{\pi}{3} + 2k\pi$ and $t = -\dfrac{\pi}{3} + 2k\pi$, where k is any integer

21. $t = -\dfrac{\pi}{4} + k\pi$, where k is any integer

23. $x = \dfrac{\pi}{8} + k\pi$ and $x = \dfrac{3\pi}{8} + k\pi$, where k is any integer

25. $t = \dfrac{\pi + 4}{8} + \dfrac{k\pi}{2}$, where k is any integer

27. $t = \dfrac{3\pi}{8} + k\pi$ and $t = -\dfrac{3\pi}{8} + k\pi$, where k is any integer

29. $t = \dfrac{\pi}{2} + k\pi$, $t = \dfrac{\pi}{6} + 2k\pi$, and $t = \dfrac{5\pi}{6} + 2k\pi$, where k is any integer

31. $t = \dfrac{\pi}{4} + k\pi$, $t = \dfrac{\pi}{3} + 2k\pi$, and $t = -\dfrac{\pi}{3} + 2k\pi$, where k is any integer

33. $x = \dfrac{\pi}{6} + 2k\pi$, $x = -\dfrac{\pi}{6} + 2k\pi$, $x = \dfrac{5\pi}{6} + 2k\pi$, and $x = -\dfrac{5\pi}{6} + 2k\pi$, where k is any integer

35. $t = \dfrac{\pi}{2} + 2k\pi$, $t = -\dfrac{\pi}{6} + 2k\pi$, and $t = \dfrac{7\pi}{6} + 2k\pi$, where k is any integer

37. $t = \dfrac{\pi}{4} + k\pi$, $t = \dfrac{4\pi}{3} + 2k\pi$, and $t = -\dfrac{\pi}{3} + 2k\pi$, where k is any integer

39. $x = \arcsin 0.78 + 2k\pi \approx 0.89 + 2k\pi$ and $x = \pi - \arcsin 0.78 + 2k\pi \approx 2.25 + 2k\pi$, where k is any integer

41. $\arccos \dfrac{3}{4} + 2k\pi \approx 0.72 + 2k\pi$ and $-\arccos \dfrac{3}{4} + 2k\pi \approx -0.72 + 2k\pi$, where k is any integer

43. $x = \arctan 5 + k\pi \approx 1.37 + k\pi$, where k is any integer

45. $x = \dfrac{5\pi}{4}$ and $x = \dfrac{7\pi}{4}$

47. $x = 3\pi + \dfrac{\pi}{4} = \dfrac{13\pi}{4}$

49. $x = 2\pi + \dfrac{\pi}{6} = \dfrac{13\pi}{6}$ and $x = 2\pi + \dfrac{\pi}{3} = \dfrac{7\pi}{3}$

51. $x = \arcsin 0.2 + 2\pi \approx 6.48$ and $3\pi - \arcsin 0.2 \approx 9.22$

53. $x = 45° + k360°$ and $x = 135° + k360°$, where k is any integer

55. $x = 15° + k180°$ and $x = 75° + k180°$, where k is any integer

57. $x = 7.5° + k90°$ and $x = -7.5° + k90°$, where k is any integer

59. Answers will vary.

61. $x = \dfrac{\pi}{12} + k\pi$ and $x = \dfrac{5\pi}{12} + k\pi$, where k is any integer

63. $x = \dfrac{\pi}{4} + k\pi$, where k is any integer

65. $t = \dfrac{\pi}{4} + k\pi$, where k is any integer

67. $x = \dfrac{\pi}{4} + k\pi$, where k is any integer

69. $x = \dfrac{\pi}{2} + k\pi$, $x = \dfrac{\pi}{4} + 2k\pi$, and $x = -\dfrac{\pi}{4} + 2k\pi$, where k is any integer

71. $x = \dfrac{\pi}{2} + 2k\pi$, $x = -\dfrac{\pi}{2} + 2k\pi$, $x = -\dfrac{\pi}{6} + 2k\pi$, and $x = \dfrac{7\pi}{6} + 2k\pi$, where k is any integer

73. $\theta = \dfrac{\pi}{4} + 2k\pi$, $\theta = \dfrac{3\pi}{4} + 2k\pi$, $\theta = -\dfrac{\pi}{4} + 2k\pi$, and $\theta = \dfrac{5\pi}{4} + 2k\pi$, where k is any integer

75. $x = \dfrac{\pi}{8} + k\pi$, $x = -\dfrac{\pi}{8} + k\pi$, $x = \dfrac{3\pi}{8} + k\pi$, and $x = -\dfrac{3\pi}{8} + k\pi$, where k is any integer

77. $x = \dfrac{\pi}{3} + 2k\pi$ and $x = -\dfrac{\pi}{3} + 2k\pi$, where k is any integer

79. $t = \dfrac{\pi}{8} + k\dfrac{\pi}{2}$ and $t = \dfrac{\pi}{4} + k\pi$, where k is any integer

81. $t = 2k\pi$, $t = -\dfrac{\pi}{6} + 2k\pi$, and $t = \dfrac{7\pi}{6} + 2k\pi$, where k is any integer

83. $t = -\dfrac{\pi}{6} + 2k\pi$, $t = \dfrac{7\pi}{6} + 2k\pi$, $t = \dfrac{3\pi}{2} + 2k\pi$, and $t = -\dfrac{\pi}{2} + 2k\pi$, where k is any integer

85. $x = \dfrac{k\pi}{4}$, where k is any integer

87. $x = \dfrac{\pi}{12} + \dfrac{k\pi}{6}$ and $x = \dfrac{\pi}{6} + \dfrac{k\pi}{3}$, where k is any integer

89. **a.** $x = \dfrac{\pi}{2} + 2k\pi$ and $x = \pi + 2k\pi$, where k is any integer

   b. $x = \dfrac{\pi}{12} + 2k\pi$ and $x = \dfrac{5\pi}{12} + 2k\pi$, where k is any integer

91.

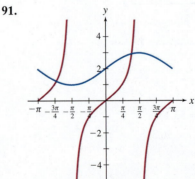

$x \approx 1.24 + 2k\pi$ and $x \approx -2.26 + 2k\pi$, where k is any integer

93.

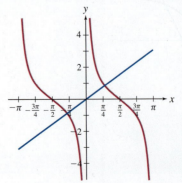

$x \approx -0.86$ and $x \approx 0.86$

95. a. The longest day of 15 hours of daylight occurs 91 days after the vernal equinox.

   b. 42 days before and 225 days after the vernal equinox

97. $x = \dfrac{\pi}{4} + k\pi$ and $x = -\dfrac{\pi}{4} + k\pi$, where k is any integer

99. $t = \arccos \dfrac{4}{5} + 2k\pi$ and $t = \dfrac{-\pi}{2} - 2k\pi$, where k is any integer

101. $t = \arcsin\left(\dfrac{-1+\sqrt{5}}{2}\right) + 2k\pi \approx 0.67 + 2k\pi$, and

$t = \pi - \arcsin\left(\dfrac{-1+\sqrt{5}}{2}\right) + 2k\pi \approx 2.48 + 2k\pi$, where k is any integer

103. $t = \arctan\left(\dfrac{-1+\sqrt{13}}{2}\right) + k\pi \approx 0.92 + k\pi$ and

$t = \arctan\left(\dfrac{-1-\sqrt{13}}{2}\right) + k\pi \approx -1.16 + k\pi$, where k is any integer

105. $x = \arcsin\left(\dfrac{-1+\sqrt{5}}{2}\right) + 2k\pi \approx 0.67 + 2k\pi$ and

$x = \pi - \arcsin\left(\dfrac{-1+\sqrt{5}}{2}\right) + 2k\pi \approx 2.48 + 2k\pi$, where k is any integer

CHAPTER 9

Section 9.1

1. If C is the angle of a triangle between sides of lengths a and b, and if c is the length of the side opposite C, then $c^2 = a^2 + b^2 - 2ab \cos C$.

3. $r = c$

5. False

7. $c \approx 4.44$, $A \approx 115.74°$, $B \approx 34.26°$

9. $c \approx 8.02$, $A \approx 36.38°$, $B \approx 71.62°$

11. $b \approx 8.77$, $A \approx 66.26°$, $C \approx 78.74°$

13. $A \approx 36.87°$, $B \approx 53.13°$, $C = 90°$

15. $c \approx 13.49$, $A \approx 32.19°$, $B \approx 20.81°$

17. $A \approx 45.04°$, $B \approx 17.15°$, $C \approx 117.82$

19. $a \approx 12.27$, $B \approx 157.55°$, $C \approx 6.45°$

21. $c = \sqrt{52 - 24\sqrt{3}} \approx 3.23$

23. $c = \sqrt{45 - 18\sqrt{2}} \approx 4.42$

25. 17.32

27. 10.93

29. 158.75

31. 11.46

33. 44.61

35. 25.96

37. Answers will vary.

39. Answers will vary.

41. Answers will vary.

43. $A \approx 2.17°$, $B \approx 139.67°$, $C \approx 38.15°$

45. 636.23 miles

47. a. 1.89 astronomical units

   b. 175.69 million miles

49. 19.45°

51. Answers will vary.

53. Answers will vary.

55. Answers will vary.

Section 9.2

1. $\dfrac{a}{\sin A} = \dfrac{b}{\sin B} = \dfrac{c}{\sin C}$

3. True

5. Law of sines

7. Law of sines

9. $A = 60°$, $b \approx 8.91$, $c \approx 13.65$

11. $B = 60°$, $a \approx 9.46$, $c \approx 10.47$

13. $C = 130°$, $b \approx 2.51$, $c \approx 4.09$

15. $A = 28°$, $a \approx 22.58$, $b \approx 30.92$

17. $A = 38°$, $a \approx 17.43$, $b \approx 26.60$

19. $C = 30°$, $c = 10$, $b = 10\sqrt{3}$

21. $a \approx 7.12$, $A \approx 41.98°$, $C \approx 28.02°$

23. No solution

25. $A \approx 19.13°$, $B \approx 35.87°$, $b \approx 3.58$

27. Solution 1: $B \approx 24.60°$, $C \approx 133.40°$, $c \approx 17.46$

   Solution 2: $B \approx 155.40°$, $C \approx 2.60°$, $c \approx 1.09$

29. Solution 1: $B \approx 24.23°$, $C \approx 135.77°$, $c \approx 10.20$

   Solution 2: $B \approx 155.77°$, $C \approx 4.23°$, $c \approx 1.08$

31. No solution

33. $c \approx 4.44$, $A \approx 115.74°$, $B \approx 34.26°$

35. $A \approx 54.90°$, $B \approx 24.15°$, $C \approx 100.95°$

37. $C = 8°$, $b \approx 7.84$, $c \approx 1.33$

39. $A \approx 128.68°$, $B \approx 33.12°$, $C \approx 18.19°$

41. $A = 28°$, $a \approx 22.58$, $b \approx 30.92$

43. $A \approx 41.41°$, $B \approx 82.82°$, $C \approx 55.77°$

45. a. Answers will vary.

   b. $x = 4\sqrt{3}$

47. $c \approx 65.78$, $b \approx 51.62$, both in feet

49. 1851 feet

51. Either 27 feet or 54 feet

53. a. 84.73 feet

   b. 84.41 feet

55. a. Answers will vary.

   b. At most one

   c. Answers will vary.

Section 9.3

1. scalars

3. diagonal

5. The magnitude is multiplied by $|a|$.

7. False

9. b. a vector

11. $\dfrac{1}{4}\mathbf{v}$

13. $\dfrac{7}{4}\mathbf{v}$

15. 8

17. No

19. **w**

21. **p**

23. **v**

25.

27.
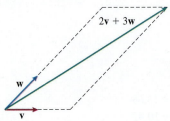

29.
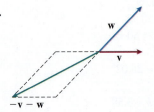

31. $\mathbf{p} = -\mathbf{v}$

33. $\mathbf{y} = \mathbf{u} + \mathbf{v} + \mathbf{w}$

35. $||\mathbf{v} + \mathbf{w}|| \approx 4.93.\ \beta \approx 7.98°.$

37. $||\mathbf{v} + \mathbf{w}|| \approx 4.84.\ \beta \approx 0.21$ radian.

39. $||\mathbf{w} - \mathbf{v}|| \approx 3.09.\ \beta \approx 47.97°.$

41. $||2\mathbf{v} + 3\mathbf{w}|| \approx 15.59.\ \beta \approx 0.13$ radian.

43. Magnitude of component in x-direction $= 2\cos 35° \approx 1.64.$
    Magnitude of component in y-direction $= 2\sin 35° \approx 1.15.$

45. Magnitude of component in x-direction $= 4\cos\dfrac{\pi}{4} \approx 2.83.$

    Magnitude of component in y-direction $= 4\sin\dfrac{\pi}{4} \approx 2.83.$

47. Answers will vary.

49. Answers will vary.

51. Answers will vary.

53. Northeast

55. Magnitude ≈ 4927.10 newtons. Direction $\approx 12°$ west of north.

57. Answers will vary.

59. Answers will vary.

61. **a.** Answers will vary.
    b. $||L|| = \sqrt{m^2 g^2 + ||F||^2}$
    c. $\phi \approx 53.13°$

63. 150 pounds

65. 4.73 miles per hour at a heading of 3° north of east.

67. **a.** 18.43°
    b. 0.08 mile downstream

69. **a.** 35 miles per hour
    b. 10°

71. Magnitude of projection $= ||\mathbf{v}|| \times \cos\theta$

Section 9.4

1. True

3. True

5. **b.** a scalar

7. The coordinates are multiplied by the scalar.

9. Multiply by the reciprocal of the magnitude.

11. $\sqrt{5}$

13. $2\sqrt{5}$

15. $\langle 28, -15 \rangle$

17. $\left\langle \dfrac{2}{5}, -\dfrac{1}{5} \right\rangle$

19. 42

21. 86

23. −52

25. $4/\left(\sqrt{5}\sqrt{13}\right)$

27. $-28/\left(\sqrt{50}\sqrt{58}\right)$

29. $\left\langle \dfrac{4}{\sqrt{41}}, \dfrac{-5}{\sqrt{41}} \right\rangle$

31. $\langle 2\sqrt{5}, -\sqrt{5} \rangle$

33. $\langle 8, 2, 1 \rangle$

35. $\sqrt{5}$

37. −1

39. $\langle 4, 3, 4 \rangle$

41. $\left\langle \dfrac{2}{5}, 0, -\dfrac{1}{5} \right\rangle$

43. $\left\langle \dfrac{2}{\sqrt{5}}, 0, -\dfrac{1}{\sqrt{5}} \right\rangle$

45. Answers will vary.

47. Answers will vary.

49. Answers will vary.

51. $\left\langle \dfrac{14}{13}, -\dfrac{21}{13} \right\rangle$

53. $\left\langle \dfrac{6}{5}, \dfrac{3}{5} \right\rangle$

55. $\dfrac{13}{\sqrt{5}}$

57. Answers will vary.

59. Answers will vary.

61. A line through the origin

63. **a.** $\mathbf{w} = \langle 150, -20 \rangle$
    b. **w** has length 151.33; it points 7.59° south of east.
    c. The ship must sail 151.33 miles on a heading 7.59° south of east.

65. The light must be rotated 12.90° in the clockwise direction.

67. **a.** $x = \cos\theta$ and $y = \sin\theta$
    b. $\mathbf{u} = \langle \cos\theta, \sin\theta \rangle$
    c. $\mathbf{v} = m\langle \cos\theta, \sin\theta \rangle$
    d. $\langle 1039.23, 600 \rangle$

69. Answers will vary.

71. Answers will vary.

73. Answers will vary.

CHAPTER 10

Section 10.1

1. The units of the function divided by the units of the independent variable.
3. **a.** The function is increasing at an increasing rate.
5. A point where the direction of bending of the graph changes.
7. Dollars per day. The change in the price we expect over a day.
9. Dollars per dollar. The additional tax you owe on each additional dollar of taxable income, called the marginal tax rate.
11. Cases per day. The number of new cases we expect in a day.
13. B
15. E
17. D
19. Answers will vary.
21. Answers will vary.
23. Answers will vary.
25.
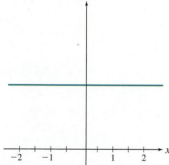
27. *dy/dx*
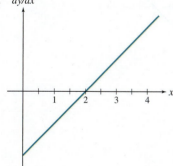
29. *dy/dx*
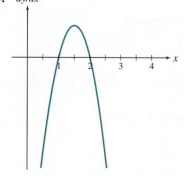

31. *dy/dx*
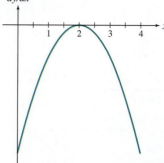

33. **a.** The rocket strikes the ground at 10 seconds.
b.
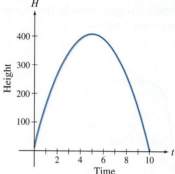

c. *dH/dt*
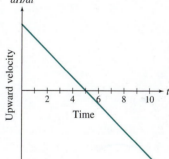

35. **a.**
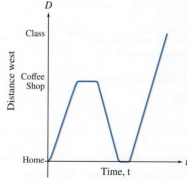

b. *dD/dt*

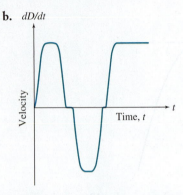

37. a. Gallons per minute. It is the rate at which the volume of water in the tank changes—that is, the change in volume we expect over 1 minute.

b. 0

c. *dV/dt*

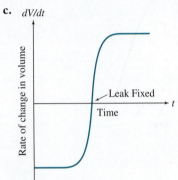

39. Answers will vary.

41. a. *T*

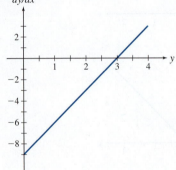

b. Production should be increased.

43. Answers will vary.

Section 10.2

1. $y = mx + b$

3. The rate of change in y is proportional to y with constant of proportionality r. Or, the proportional rate of change in y is the constant r.

5. D

7. A

9. $y = 2x + 3$

11. $y = 3e^{2x}$

13. $y = 3$

15. $y = 3x + 5$

17. $y = 3e^{2x}$

19. $y = mx + m$

21. $y = e^x$

23. a. $I = I_0 e^{-rD}$

b. $I = I_0 e^{-0.1D}$

c. 6.93 meters

25. a. $\dfrac{dV}{dt} = 1.4$

b. $V = 1.4t + b$

c. 5.8 meters per second

d. 40 seconds

27. a. $\dfrac{dB}{dt} = rB$

b. $B = B_0 e^{rt}$

c. $\dfrac{\ln 2}{r}$ years

d. \$7459.12

29. a. $\dfrac{dN}{dt} = (b - d)N$

b. $N = N_0 e^{(b-d)t}$

Section 10.3

1. $g(y) = 0$

3. c. horizontal lines

5. b. falling

7. a. *dy/dx*

b. $y = 3$

c. *y*

9. a. *dy/dx*

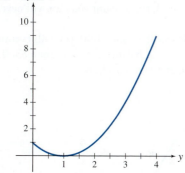

b. $y = 1$

c.

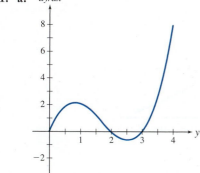

11. a. *dy/dx*

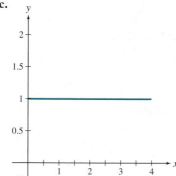

b. $y = 0$, $y = 2$, and $y = 3$

c.

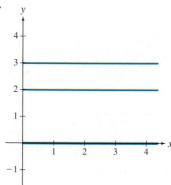

13. a. *dy/dx*

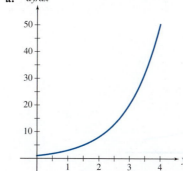

b. There are no equilibrium solutions.

c. Not applicable

15. y is increasing for $y > 5$. y is decreasing for $y < 5$.

17. y is increasing for all y.

19. y is increasing for $1 < y < 3$. y is decreasing for $y < 1$ and for $y > 3$.

21.

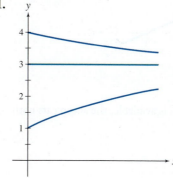

23.

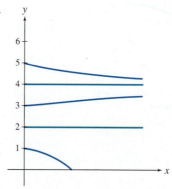

25.

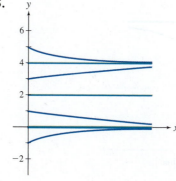

27.

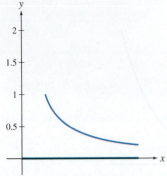

29. a.

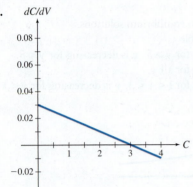

b. $C = 3$ pounds
c. When no vegetation is available, the sheep eats nothing.
d.

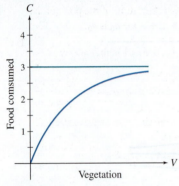

e. 3 pounds

31. a. $V = \dfrac{g}{r}$

b.

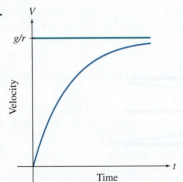

c. Answers will vary.

d. $\dfrac{32}{176} \approx 0.18$ per second

33. a. $B = \$300,000$
 b. At this balance level, the amount you owe will never change.
 c. If the balance is less than \$300,000, you will eventually pay off the loan. If the balance is greater than \$300,000, your debt will increase indefinitely.

35. a.

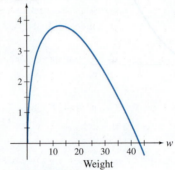

 b. About 13 pounds
 c. About 43 pounds
 d. The fish is losing weight.

37. Answers will vary.

39. Concave up

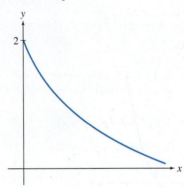

41. Concave down

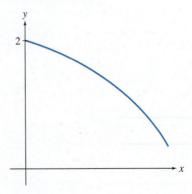

GLOSSARY

Note: Chapter references to key terms appear in parentheses.
"ON" chapters and sections are available online only.

A

absolute maximum (1) A function f reaches an absolute maximum value of $f(x_0)$ at $x = x_0$ if $f(x_0) \geq f(x)$ for every x in the domain of f.

absolute minimum (1) A function f reaches an absolute minimum value of $f(x_0)$ at $x = x_0$ if $f(x_0) \leq f(x)$ for every x in the domain of f.

absolute value (0) The "positive value" of a number.

algebraic form (App. 9) For the line through the tip of a vector and normal to a nonzero vector, it is $ax + by = c$.

amplitude (7) The constant $|A|$ of the sine wave $y = A \sin x$ or $y = A \cos x$.

arccosine (8) The inverse of the cosine function restricted to the domain $[0; \pi]$.

arcsine (8) The inverse of the sine function restricted to the domain $\left[\dfrac{-\pi}{2}, \dfrac{\pi}{2} \right]$.

arctangent (8) The inverse of the tangent function restricted to the domain $\left(\dfrac{-\pi}{2}, \dfrac{\pi}{2} \right)$.

argument (ON9.5) For arg(z) of $z \neq 0$, it is the counterclockwise angle in $[0, 2\pi)$ made by the positive real axis and the ray joining the origin to z.

arithmetic sequence (ON13) A sequence of the form

$$\{mn + b\}_{n=k}^{\infty} = \{mk + b, m(k+1) + b, m(k+2) + b, \ldots\}$$

where m and b are constants.

arithmetic sum (ON13) A sum of the form $\displaystyle\sum_{k=1}^{n}(ak + b) = (a + b) +$, $(2a + b) + \ldots + (ak + b) + \ldots + (an + b)$, where a and b are constants.

average rate of change (1) Of a function $y = f(x)$ on the interval $[a, b]$, the average rate of change is the slope of the secant line for $[a, b]$.

B

base (3) In an exponential function of the form $f(x) = ab^x$, the *base* is the number b.

base-b logarithm (4) Denoted by $y = \log_b x$, this is the inverse of the exponential function $g(y) = b^y$. Here, $b \, 6 \neq 1$ is a positive number.

bifolium (ON9.6) The graph of $r = a \sin \theta \cos^2 \theta$.

C

cardioid (ON9.6) The graph of $r = a(1 - \cos \theta)$.

center of a hyperbola (ON12) The point halfway between the foci.

center of an ellipse (ON12) The point halfway between the foci.

coefficients (5) The coefficients of the polynomial $P(x) = a_n x^n + a_{n-1}x^{n-1} + \ldots + a_1 x + a_0$ are the constants $a_0, \ldots, a_n$.

cofactor (App. 10) The (i, j) cofactor of a square matrix A is $\text{Cof}_{i,j} = (-1)^{i+j}|A_{i,j}|$.

combinations (ON13) The number of combinations of n objects taken k at a time is the number of ways to select k objects from n distinct objects without regard to order.

common difference (ON13) For the arithmetic sequence $\{mn + b\}_{n=k}^{\infty}$, it is the number m.

common logarithm (4) Denoted by $\log x$, this is the base-10 logarithm: $\log x = \log_{10} x$.

common ratio (ON13) For the geometric sequence $\{ab^n\}_{n=k}^{\infty}$, it is the number b.

complement (P) Consists of all real numbers that are *not* in the interval.

complex conjugate (App. 2) If $z = a + bi$ is in standard form, the complex conjugate of z is denoted $\bar{z}$ and is defined by $\bar{z} = a - bi$.

complex exponential function (App. 3) For a real number (or angle in radians) θ, it is defined by $e^{i\theta} = \cos \theta + i \sin \theta$.

composition (2) Results when one function is applied after another. The composition of f with g is defined by $(f \circ g)(x) = f(g(x))$.

concave down (1) A graph of a function that looks like a piece of wire whose ends are bent downward.

concave up (1) A graph of a function that looks like a piece of wire whose ends are bent upward.

conchoid of Nicomedes (ON9.6) The graph of $r = a \csc \theta - b$, where $b > a$.

constant of proportionality (3) The constant multiple k in the proportionality relationship $A = kB$.

constant term (5) The constant term of the polynomial $P(x) = a_n x^n + a_{n-1}x^{n-1} + \ldots + a_1 x + a_0$ is a_0.

cosecant (6C) At t, $\csc t$ is defined as the ratio $\csc t = \dfrac{1}{\sin t}$.

cosecant (6T) At t, $\csc t$ is defined as $\csc t = \dfrac{\text{Hypotenuse}}{\text{Opposite}}$.

cosine (6C) At t, $\cos t$ is the first coordinate of the trigonometric point $P(t)$.

cosine (6T) At t, $\cos t$ is defined as $\cos t = \dfrac{\text{Adjacent}}{\text{Hypotenuse}}$.

cotangent (6C) At t, $\cot t$ is defined as the ratio $\cot t = \dfrac{\cos t}{\sin t}$.

cotangent (6T) At t, $\cot t$ is defined as $\cot t = \dfrac{\text{Adjacent}}{\text{Opposite}}$.

coterminal angles (6C, 6T) Pairs of angles that have the same initial and terminal rays.

D

decibel (4) The level of a sound calculated by using the following formula: Decibels = 10 log(Relative intensity).

decreasing function (1) On an interval I, the decreasing function is a function f such that $f(x) > f(y)$ whenever $x < y$ and both x and y are in the interval I.

degree (5) The degree of the polynomial $P(x) = a_n x^n + a_{n-1}x^{n-1} + \ldots + a_1 x + a_0$ with $a_n \neq 0$ is the nonnegative integer n.

degree (6C, 6T) An angle measurement such that an angle of 1 degree is a positive angle with a vertex at the center of a circle that cuts an arc of one-360th of a full circle.

determinant (App. 10) We calculate the determinant of an n-by-n matrix A with entries $a_{i,j}$ by cofactor expansion.

determinant of a 2-by-2 matrix (ON11) Defined as

$$\begin{pmatrix} a & b \\ c & d \end{pmatrix}$$

$$\begin{vmatrix} a & b \\ c & d \end{vmatrix} = ad - bc$$

determinant of a 3-by-3 matrix (ON11) Defined to be

$$\begin{pmatrix} a & b & c \\ d & e & f \\ g & h & i \end{pmatrix}$$

$$\begin{vmatrix} a & b & c \\ d & e & f \\ g & h & i \end{vmatrix} = (aei + bfg + cdh) - (gec + hfa + idb)$$

directrix of a parabola (ON12) The given line from which distances are measured.

discriminant (5) The discriminant of the quadratic function $y = ax^2 + bx + c$ is $b^2 - 4ac$.

discriminant of the conic equation (ON12) This is stated as $Ax^2 + Bxy + Cy^2 + Dx + Ey + F = 0$ is $B^2 - 4AC$.

domain (1) The collection of objects for which the rule of assignment is defined.

dot product (9) For v $= (a, b)$ and w $= (c, d)$, it is denoted by v $\cdot$ w and is defined by $(a, b) \cdot (c, d) = ac + bd$.

dot product equation (App. 9) For the line through the tip of w and normal to the nonzero vector n, it is $\mathbf{n} \cdot (\mathbf{v} - \mathbf{w}) = 0$.

E

eccentricity of a conic section (ON12) This is written in the form $r = ed$ is the number e.

eccentricity of an ellipse (ON12) This is stated as $e = \dfrac{c}{a}$. Here, a is half of the length of the major axis, and c is half of the distance between the foci.

ellipse (ON12) The collection of all points such that the sum of the distances to two given points equals a fixed number.

equal as matrices (ON11) Two matrices that have the same numbers of rows and columns, and the corresponding entries are the same.

equilibrium solution (10) For the change equation

$$\frac{dy}{dx} = g(y)$$

it is a solution of the equation $g(y) = 0$.

even function (2) A function f such that $f(-x) = f(x)$ for all x in the domain of f.

exponential function (3) A function that exhibits a constant proportional rate of change that is not equal to 0.

F

factorial (ON13) For a positive integer n the factorial is denoted by $n!$ and defined by $n! = n(n-1)(n-2) \times \ldots \times 3 \times 2 \times 1$. It turns out to be convenient to define $0!$ to be 1.

foci (ON12) The two given points on an ellipse to which distances are measured.

foci of a hyperbola (ON12) The two given points to which distances are measured.

focus of a parabola (ON12) The given point from which distances are measured.

frequency (ON7.3) The reciprocal of the period of a sound wave.

function (1) A rule that assigns to each object in one collection *exactly one* object in another collection.

G

Gaussian elimination (ON11) The process of solving a linear system by elimination of variables and back substitution.

Gauss-Jordan elimination (ON11) The process of solving a system of equations by using row operations to put the augmented matrix into reduced form.

general conic equation (ON12) This equation is stated as $Ax^2 + Bxy + Cy^2 + Dx + Ey + F = 0$. Here, A, B, and C are not all zero.

geometric sequence (ON13) A sequence of the form

$$\{ab^n\}_{n=k}^{\infty} = \{ab^k, ab^k+1, ab^k+2, \ldots\}$$

where a and $b \neq 0$ are constants.

geometric sum (ON13) A sum of the form $\displaystyle\sum_{k=0}^{n} ab^k = a + ab + ab^2 + \ldots + ab^k + \ldots + ab^n$, where a and b are constants.

graph (P) In an equation, the result of plotting many points, so the graph in x and y consists of all points (x, y) in the coordinate plane that satisfy the equation.

graph of a function (1) The graph of a function f is the graph of the equation $y = f(x)$.

H

horizontal asymptote (5) A horizontal asymptote of the graph of a rational function $R(x)$ is a line $y = b$ such that $R(x) \to b$ as $x \to \infty$ or $R(x) \to b$ as $x \to -\infty$.

hyperbola (ON12) The collection of all points such that the absolute value of the difference of the distances to the two given points equals the fixed distance $2a$.

I

imaginary part (ON9.5) The part of the complex number $z = a + bi$ written in standard form is b.

increasing function (1) On an interval I, the increasing function is a function f such that $f(x) < f(y)$ whenever $x < y$ and both x and y are in the interval I.

inflection point (1) The point on the graph of a function where the direction of bending changes.

initial value (3) In an exponential function of the form $f(x) = ab^x$, the *initial value* is the number a.

integers (0) These include the counting numbers, their negatives, and zero.

inverse function (2) Undoes the action of that function. A function g is the inverse function of a function f if $g(f(x)) = x$ for all x in the domain of f and $f(g(y)) = y$ for all y in the range of f.

irrational numbers (0) Numbers that are not rational—that is, they cannot be expressed as a ratio of integers.

L

leading coefficient (5) The leading coefficient of the polynomial $P(x) = a_n x^n + a_{n-1}x^{n-1} + \ldots + a_1 x + a_0$ with $a_n \neq 0$ is a_n.

leading entry (ON11) For an augmented matrix, it is a matrix entry to the left of the vertical bar that is the first nonzero entry in its row.

leading term (5) The leading term of the polynomial $P(x) = a_n x^n + a_{n-1}x^{n-1} + \ldots + a_1 x + a_0$ with $a_n \neq 0$ is $a_n x^n$.

leading variable (ON11) For a system of equations, it is a variable corresponding to a leading entry for the augmented matrix.

limaçon of Pascal (ON9.6) The graph of $r = b + a \cos \theta$.

linear equation (P) In x and y, this is an equation of the form $y = mx + b$.

linear equation (0) In x, it is an equation that can be put in the form $ax = b$ with $a \neq 0$ using the basic operations on equations.

linear function (3) A function with a constant rate of change.

linear inequality (P) The same as a linear equation, where the equal sign is replaced by an inequality sign and can therefore can be put in the form $ax + b > 0$ or $ax + b \geq 0$. Here, a $a \neq 0$.

local maximum (1) A function f reaches a local maximum value of $f(x_0)$ at $x = x_0$ if there is an interval (a, b) containing x_0 so that $f(x_0) \geq f(x)$ for every x in the interval (a, b).

local minimum (1) A function f reaches a local minimum value of $f(x_0)$ at $x = x_0$ if there is an interval (a, b) containing x_0 so that $f(x_0) \leq f(x)$ for every x in the interval (a, b).

lower triangular (ON11) An n-by-n matrix for which all entries above the main diagonal are zero.

M

magnitude of an earthquake (4) Denoted as follows: Magnitude = log(Relative intensity).

matrix (ON11) A rectangular array of numbers.

matrix addition (ON11) For two matrices $(a_{i,j})$ and $(b_{i,j})$ having the same number of rows and columns, it is defined by $(a_{i,j}) + (b_{i,j}) = (a_{i,j}) + (b_{i,j})$.

matrix inverse (App. 6) For an n-by-n matrix A, it is an n-by-n matrix A^{-1} such that $AA^{-1} = A^{-1}A = I_n$.

matrix multiplication (ON11) For an m-by-n matrix A by an n-by-k matrix B, it is defined by first representing A by its rows and B by its columns. Then the product AB is defined to be the m-by-k matrix.

minor matrix (App. 10) If A is a matrix, the (i, j) minor matrix $A_{i,j}$ is formed by deleting row i and column j from the matrix A.

modulus (ON9.5) For $z = a + bi$ in standard form, it is $|z| = \sqrt{a^2 + b^2}$.

N

natural exponential function (3) The function $f(x) = e^x$.

natural logarithm (4) Denoted by ln x, this is the base-e logarithm: $\ln x = \log_e x$.

normal vector (App. 9) A vector perpendicular to the line.

nth root (0) Of x, it is denoted $n\sqrt{x}$ and is defined as follows: If n is an odd positive integer and x is a real number, then $n\sqrt{x}$ is defined to be the (unique) real number y such that $y^n = x$. If x is nonnegative and n is an even integer, then $n\sqrt{x}$ is the (unique) nonnegative real number y such that $y^n = x$. If n is an even positive integer, then the nth root is defined only for nonnegative real numbers.

O

odd function (2) A function f such that $f(-x) = -f(x)$ for all x in the domain of f.

one-to-one function (2) A function whose graph satisfies the horizontal line test.

order (5) The order of a zero k of a polynomial $P(x)$ is the highest power of $x - k$ that is a factor of $P(x)$.

P

parabola (5) A graph of a quadratic function.

parabola (ON12) The collection of all points equidistant from a given point not on a given line.

parametric form (ON9.7) For describing a parametric curve, it is

$$x = x(t)$$
$$y = y(t)$$

period (7) In a periodic function f, it is the smallest number p such that $f(x) = f(x + p)$ for every x in the domain of f.

periodic function (7) A function f for which there exists a number $p > 0$ such that $f(x) = f(x + p)$ for every x in the domain of f.

permutations (ON13) The number of permutations of n objects taken k at a time is the number of ways to select k objects from n distinct objects and arrange them in order.

phase shift (7) The number C in the functions $y = A \sin B(x - C)$ and $y = A \cos B(x - C)$.

poles (5) The real numbers of a rational function for which the denominator is 0.

polynomial (5) A function of the form $P(x) = a_n x^n + a_{n-1} x^{n-1} + \ldots + a^1 x + a_0$, where n is a nonnegative integer, $a_0, \ldots; a_n$ are constants; and $a_n \neq 0$.

product (App. 2) For the complex numbers $a + bi$ and $c + di$ in standard form, the product is defined by $(a + bi)(c + di) = (ac - bd) + (ad + bc)i$.

proportional (3) A quantity is *proportional* to another if one is a constant multiple of the other—that is, the quantity A is proportional to the quantity B if $A = kB$ for some constant k.

Q

quadratic (P) An inequality is quadratic if the rules for manipulating inequalities can be used to put it in the form $ax^2 + bx + c < 0$ or $ax^2 + bx + c \leq 0$ with $a \neq 0$.

quadratic equation (0) In x, it is an equation that can be put in the form $ax^2 + bx + c = 0$ with $a \neq 0$ using the basic operations on equations.

quadratic functions (5) Quadratic functions have the form $P(x) = ax^2 + bx + c$ with $a \neq 0$.

R

radian (6C, 6T) An angle measurement such that an angle of 1 radian is an angle with a vertex at the center of a circle that cuts an arc the same length as the radius.

range (1) The set of all function outputs.

rate of change (1) At x of the function f, the rate of change is the slope of the tangent line to the graph $y = f(x)$ at the point $(x; f(x))$.

rational exponents (0) In the form $x^{p/q}$, they are defined as follows: If the integer q is even, then x must be nonnegative. If q is a positive integer and p is any integer, then we define

$$x^{p/q} = q\sqrt{x^p}$$

We emphasize that, in the case that q is even, $x^{p/q}$ is defined only for nonnegative numbers x.

rational function (5) A function of the form $R(x) = \dfrac{P(x)}{Q(x)}$,

where $P(x)$ and $Q(x)$ are polynomials with no real zeros in common.

rational numbers (0) Numbers of the form $\dfrac{p}{q}$, where p and q are integers with $q \, 6 \neq 0$.

real number line (0) The geometric representation of the real numbers as points on a line.

real numbers (0) Numbers that consist of the rational numbers together with the irrational numbers.

real part (ON9.5) The part of the complex number $z = a + bi$ written in standard form is a.

recurrence relation (ON13) This tells how to obtain the next term of a sequence from earlier terms.

reduced form (ON11) For an augmented matrix, it is in row–echelon form and satisfies the additional condition that the only other entries in a column that contains a leading entry are zeros.

reference number (6C, 6T) A number such that r for the trigonometric point $P(t)$ is the shortest arc length from $P(t)$ to the x-axis.

Riemann sum (ON13) For a function f over the interval $[a, b]$ using left-hand endpoints, it is $R(a, b, n) = h \sum_{k=0}^{n-1} f(a + kh)$. Here, n is the number of pieces of equal length into which we divide $[a, b]$, and $h = \dfrac{b - a}{n}$.

row-echelon form (ON11) For an augmented matrix, it satisfies the following conditions: (1) each row of zeros (before the bar) is at the bottom, (2) each leading entry is 1, and (3) each leading entry occurs strictly to the right of all leading entries in rows above.

row operations (ON11) The three types are adding a multiple of one equation to another, multiplying one of the equations by a nonzero constant, and interchanging two equations.

S

scalar (9) A real number.

scalar multiplication (9) For a vector v by a scalar a, it is the vector av whose magnitude is $|a|$ times the magnitude of v and whose direction is the same as v if a is positive and opposite if a is negative.

scalar multiplication (ON11) Of a matrix by a scalar (real number) c, it is defined by $c(a_{i,j}) = (ca_{i,j})$.

secant (6C) At t, sec t is defined as the ratio $\sec t = \dfrac{1}{\cos t}$.

secant (6T) At t, sec t is defined as $\sec t = \dfrac{\text{Hypotenuse}}{\text{Adjacent}}$.

secant line (1) For a t function f for the interval $[a, b]$, the secant line is the line through the points $(a, f(a))$ and $(b, f(b))$.

sequence (ON13) An ordered list of numbers.

series (ON13) In $\sum_{k=0}^{\infty} a_k$, it is defined as the limit as $n \to \infty$ $\sum_{k=0}^{n} a_k$.

sigma notation (ON13) The notation $\sum_{k=p}^{q} a_k$ stands for the sum $a_p + a_{p+1} + \ldots + aq$.

similar triangles (6C, 6T) Pairs of triangles for which the corresponding angles are the same.

simple harmonic motion (ON7.3) Motion that is modeled by a sine wave.

sine (6C) At t, sin t is the second coordinate of the trigonometric point $P(t)$.

sine (6T) At t, sin t is defined as $\sin t = \dfrac{\text{Opposite}}{\text{Hypotenuse}}$.

sine wave (7) The graph of a function of the form $y = A \sin B(x - C)$ or $y = A \cos B(x - C)$.

slope (P) In the line $y = mx + b$ the slope is the constant m.

solution of a system (ON11) Consists of values for the variables that satisfy all of the equations involved.

solution region (ON11) For a system of inequalities, it consists of those points that satisfy all the inequalities.

standard form (5) The standard form of the quadratic function $y = ax^2 + bx + c$ is $y = a(x - h)^2 + k$.

sum (App. 2) For the complex numbers $a + bi$ and $c + di$ in standard form, the sum is defined by $(a + bi) + (c + di) = (a + c) = (b + d)i$.

system of equations (ON11) A collection of equations involving the same unknowns.

T

tangent (6C) At t, tan t is defined as the ratio $\tan t = \dfrac{\sin t}{\cos t}$.

tangent (6T) At t, tan t is defined as $\tan t = \dfrac{\text{Opposite}}{\text{Adjacent}}$.

terms of a sequence (ON13) The numbers as they appear in the list.

triangle inequality (0) The inequality $|x + y| \le |x| + |y|$ for real numbers x and y.

trigonometric form (ON9.5) For the complex number $z \ne 0$, it is $z = |z| (\cos \theta + i \sin \theta)$ if θ is the argument of z.

$$x = x(t)$$
$$y = y(t)$$

trigonometric point (6C, 6T) A point such that $P(t)$ associated with the number t is defined as follows: If $t \ge 0$, we begin at the point $(1, 0)$ and move in the counterclockwise direction around the unit circle a distance t to the point $P(t)$. If t is negative, nothing changes except that we start at $(1, 0)$ and move in the clockwise direction.

U

union (P) In two sets, the union consists of all elements belonging to one or both of the sets.

upper triangular (ON11) An n-by-n matrix for which all entries below the main diagonal are zero.

V

vector (9) Consists of a magnitude and a direction.

vector form (ON9.7) For describing a parametric curve, it is $s(t) = (x(t), y(t))$.

vertex (ON12) In a parabola, it is the point halfway between the focus and the directrix.

vertical asymptote (5) A vertical line that the graph of a rational function approaches but never intersects.

vertices of a hyperbola (ON12) The points on the hyperbola that are closest to the foci.

Y

y-intercept (P) In the line $y = mx + b$, the y-intercept is the constant b.

Z

zero (5) A zero of a polynomial $P(x)$ is a number r such that $P(r) = 0$—that is, $x = r$ is a solution of the equation $P(x) = 0$.

INDEX

Page numbers followed by f and t indicate figures and tables, respectively.

A

absolute maximum, 82
absolute minimum, 82
absolute value, 22–24, 24f
absolute value inequalities, 22–26, 23f–24f, 43
acceleration, 771, 772t
acute angles, 438–41, 439f, 459, 488–89, 535
addition
  of functions, 124
  vector, 731–34, 731f–32f, 752, 762
aerobic power, 195
airplanes
  flight paths, 705, 705f
  forces of flight, 738, 738f
  locating, 12–13, 12f
  and wind, 736–37, 736f
Alexander's formula, 94
algebra: fundamental theorem of, 339, 378
algebraic expressions, 443–44, 443f–44f
algebraic operations, 598–601, 679
Americans with Disabilities Act (ADA), 727
amplitude, 552–54, 553f, 585
  definition, 552
  graphing with, 553–54, 553f–54f
  period and, 554–55
angles, 388–99, 458, 470–81, 535
  acute, 438–41, 459, 488–89, 535
  conventions for, 388–89, 388f, 458, 470–71, 535
  converting measures, 392–93, 458, 474–75, 535
  coterminal, 389–90, 390f, 392, 471–72, 472f, 474
  degree measures, 389–90, 389f, 458, 471–72, 471f, 535
  finding legs from hypotenuse and, 447
  flying, 704–5, 705f
  negative, 389, 389f, 417, 417f, 471, 471f, 528, 528f
  parallax, 422, 422f, 455, 455f, 456, 487, 487f, 512, 512f, 659
  positive, 389, 389f, 471, 471f
  radian measures, 390–91, 391f, 458, 472–74, 473f, 535
  reference, 415–17, 415f–17f, 458
  representing, 441–42, 442f, 459, 494–95, 494f, 535
  right, 391, 473
  special, 410–12, 410f–12f, 412t, 423–26, 445–46, 445f, 458–59, 492–94, 492t, 502–3, 502f, 535
  straight, 391, 473
  between vectors, 748–51, 749f–50f, 756, 756f
angle-side-angle (ASA) theorem, 710, 710f, 712f

angle sum of a triangle, 397, 479
annual percentage rate (APR), 219
antennae, 754
ants, 68–71
APR (annual percentage rate), 219
arccosecant, 651–54, 652f, 680
arccosine, 639–44, 640f, 680
  basic properties, 641, 654
  definition, 640
  graphs, 644–45, 645f
  identities, 644
arccotangent, 651–54, 653f, 680
arc length, 391, 394–95, 458, 473, 476–77, 535
arcsecant, 651–54, 653f, 680
arcsine, 639–44, 639f–40f, 680
  basic properties, 641, 654
  definition, 640
  graphs, 644–45, 645f
  identities, 644
arctangent, 646–50, 646f, 680
  basic calculations, 647
  basic properties, 646, 654
  definition, 646
area
  of circle sector, 394–95, 458, 476–77, 535
  irregular, 696–97, 696f, 725, 725f
  subtended, 394, 476
  of triangle, 448, 448f, 459, 505–6, 506f, 536, 607, 607f, 761
arithmetic combinations, 124–26, 173
ASA (angle-side-angle) theorem, 710, 710f, 712f–13f
asymptotes
  horizontal, 364–69, 364f, 366–67, 367f, 378
  vertical, 360–63, 361f, 366–67, 367f, 378
autonomous equations, 801
average rate of change, 65–72, 66f

B

baseball, 705, 705f, 726, 726f, 743, 790
base-b logarithms, 241
basketball, 650–51, 651f
Beer-Lambert-Bouguer law, 797
bighorn sheep, 786
birds in flight, 742–43, 742f
Builder's Old Measurement, 336

C

calculators
  change-of-base formula, 251
  solving trigonometric equations with, 666, 666f, 668, 668f
carbon dating, 281
carrying capacity, 225
cartography, 402, 484
catch equation, 374
catfish farming, 801–4

Celsius: converting to Fahrenheit from, 188–89, 188f
center, 10, 12f
cesium-137, 798
change
  constant percentage, 215–16, 234
  constant proportional, 198–201, 214, 234
  rates of. See rates of change
change equations, 789–95, 816
  concavity from, 812–13, 812f–13f
  equilibrium solutions, 801–3, 816
  Euler method, 800
  exponential, 791–92, 816
  graphical solutions, 800–808, 805f, 816
  linear, 789–91, 816
change-of-base formula, 249–51, 287
circles, 10–12, 12f, 43
  center, 10, 12f
  circumference, 189–90, 190f, 405, 514
  completing the square to identify, 11–12
  equations of, 10–11
  full, 391, 473
  graphs of, 11
  great, 691, 691f
  radius, 10, 12f, 189–90, 190f
  sectors, 394–95, 394f, 476–77, 476f, 535
  unit, 405–18, 458, 514–15, 536
circumference
  formula for, 405, 514
  vs radius, 189–90, 190f
coefficient, 294
  definition, 294
  leading, 294
  real, 340–42, 351, 378
cofunction identities, 595–96, 679
common logarithms, 244–47, 245f, 252, 287
  calculating values of, 245–46
  calculations with laws of logarithms, 262–63
  definition, 244
  estimating values, 248–49
  as exponents, 246–47
  growth, 249
  power law, 260
  product law, 260
  properties, 245
  quotient law, 260
compass points
  example, 407–8, 408f
  trigonometric functions, 426–29, 426f–27f, 428t, 459, 518–21, 519f, 520t, 536
  trigonometric points, 408t–9t
  on unit circle, 407–9, 458
complement
  of interval, 19, 19f–20f
  solving absolute value inequalities with, 24–25

complete factorization theorem, 339
complex expressions, 598
complex numbers, 298
complex zeros, 298–99, 377
    factoring with, 300–301, 342
component vectors, 734–35, 735f, 762
composition of functions, 127–29
compound interest, 219–22, 234
    continuous compounding, 221
    continuous vs yearly compounding,
        222, 222f
    formula for, 219
    monthly compounding, 220
    other compounding periods, 220–21,
        221t
    vs simple interest, 220, 220f
    yearly compounding, 222, 222f
compression
    horizontal, 157–59, 159f, 160, 173,
        575–76, 576f
    vertical, 155–57, 156f, 160, 173,
        575–76, 576f
concavity, 777–79, 778f, 816
    from change equations, 812–13,
        812f–13f
    definition, 84
    of graphs of quadratic functions, 304
    and rates of change, 84–87, 84f–86f,
        115, 813–14
concavity lists, 813–14
congruence, triangle, 692, 692f, 710, 710f, 761
conjugate zeros theorem, 341–42
constant multiple, 124
constant of proportionality, 189
constant percentage change, 215–16,
    216f, 234
constant proportional change, 198–201,
    214, 234
constant rate of change, 182–84, 234,
    789–91, 816
constant terms, 294
continuous compounding
    formula for, 221
    vs yearly compounding, 222, 222f
conventions for angles, 388–89, 388f, 458,
    470–71, 535
cooling: Newton's law of cooling, 794–95
coordinate plane, 2–13, 43
coordinates, 2–3, 43
cosecant, 423, 488
    concepts to remember, 578
    definition, 423, 488, 578
    double-angle formula, 634
    graphs, 572–74, 572f, 574f, 585
    half-angle formula, 635
    periodicity, 569, 578
    transformed, 575
    values between 0 and 2π, 572t
    vertical asymptotes, 578
cosine, 423, 488, 517f
    bounds, 552
    concepts to remember, 561
    definition, 423, 488, 561

double-angle formula, 624, 634, 637f
    half-angle formula, 626–27, 634
    law of cosines, 691–700, 703, 761
    of negative angles, 528, 528f
    periodicity, 547–48, 551f, 561, 585
    signs, 429, 429f, 521, 521f
    sum and difference formulas, 607–11,
        679
    symmetry, 561
    trigonometric functions in terms of,
        488–89, 488t
    values between 0 and 2π, 550t
cosine graphs, 550–52, 551f, 553f, 585, 668f
cotangent, 423, 488
    concepts to remember, 578
    definition, 423, 488, 578
    double-angle formula, 634
    graphs, 570–71, 571f, 585
    half-angle formula, 635
    periodicity, 569, 578
    sum and difference formulas, 611–13,
        679
    transformed, 575
    vertical asymptotes, 578
coterminal angles, 389, 471
    definition, 389, 471
    examples, 390, 390f, 472, 472f
    with radian measure, 392, 474
cubic equations, 337–38

D
Dallas–Fort Worth International Airport
    (DFW), 720
decay. *See* exponential growth and decay
decibels, 256, 266–67, 797
decomposing functions, 129–30
decreasing functions, 79–81, 115
    definition, 79
    inverses and, 142
    without bound, 101, 101f
degree measure, 389–90, 389f, 458, 471–72,
    471f, 535
    converting from radian measure to,
        393, 475
    converting to radian measure from,
        392–93, 474–75
    definition, 389, 471
degree of an angle, 389, 471
degree of a polynomial, 294
demand curves, 95, 95f
difference: sum and difference formulas,
    606–15, 679
directional antennae, 754
discriminants, 295
    definition, 295
    negative, 296, 296f, 297–98, 298f
    positive, 296, 296f
    zero, 296, 296f, 297, 297f, 377
dispersal method, 455, 512
distance
    absolute value and, 23–24, 24f
    between cities, 725, 725f
    graph of, 773–74, 773f

between planets, 705–6, 705f–6f
    between two points, 3–5, 4f
distance formula, 4, 43
division, 600
    of functions, 125
    of higher-degree polynomials, 316
    long, 313–16, 377
    synthetic, 318–20, 377
division algorithm, 315
domain
    of compositions, 128
    of functions, 50, 52–55, 58–59, 114,
        142–43, 143f, 173
    of rational functions, 359–60, 378
    restricted, 142–43, 143f, 173
    of trigonometric functions, 429, 521
dot product, 748–51, 749f–50f, 762
    definition, 748
    three-dimensional vectors, 752
double-angle formulas, 622–25, 679
    applying, 649
    concepts to remember, 623
    for cosecant, 634
    for cosine, 634, 637f
    for cotangent, 634
    for sine and cosine, 624
    solving trigonometric equations with,
        669, 669f
    sum and double-angle formulas,
        629–30, 679
    for tangent, 634
doubling time, 282
Dow Jones Industrial Average, 49, 49f
Doyle log rule, 310
drug concentration, 310
drug half-life, 218

E
e, 198, 204, 234
earthquakes, 265–66, 270–71
    magnitude, 259, 259f, 265–66
    relative intensity, 265–66
Ebola, 68–70, 69f
economics of school consolidation, 310
effective annual rate (EAR), 111
equations
    change, 789–95
    of circles, 10–11
    cubic, 337–38
    exponential, 273–76, 287
    for functions, 55–56, 114
    graph of, 3, 3f
    linear, 6–12
    logarithmic, 273, 277–79, 287
    mixed, 279–80, 287
    with natural exponential function, 275
    solving with factor theorem, 328–29
    when one zero is known, 329
equilibrium solutions, 801–3
    definition, 801
    graphing, 805–8, 805f, 807f–8f, 816
equity, 285
Eratosthenes, 388

escape velocity, 375
Euler, Leonhard, 198, 198f, 639, 800
Euler method, 800
even-degree polynomials, 349
even functions, 163, 173, 559–61, 560f, 585
    concepts to remember, 597
    definition, 163
even identities, 597–98, 679
exponential equations
    change equations, 791–93, 816
    mixed equations, 279–80, 287
    solving, 273–77, 279–80, 287
exponential functions, 181, 198–208, 234
    applications, 205–7, 234
    base, 199
    change properties, 199–200
    concepts to remember, 208f
    definition, 199
    equations involving, 276
    evaluating, 201–2
    examples, 199–201
    formula for, 199
    graphs of, 201–3, 202f–3f, 234
    initial value, 199
    modeling with, 214–26, 234
    natural, 204–5, 204f
    rewriting, 264
exponential growth and decay, 205–7
    carrying capacity, 225
    examples, 205–7, 206f–7f, 216–18,
        216f–18f, 224, 224f
    intrinsic annual rate, 806–7
    logistic models, 223–25, 224f, 234
    with monthly compounding, 220
    with other compounding periods,
        220–21, 221t
    population growth, 216, 216f, 231–32,
        232f, 798
    rapidity of, 207–8, 207f–8f, 234
exponents, 243
    logarithms, 246–47, 263

F
factoring, 600
    complete factorization, 339
    with factor theorem, 328–29
    formulas for sum and difference and of
        powers, 325–26
    graphing polynomials by, 346, 346f
    by grouping, 325, 327–28
    with one known complex zero, 342
    power of x, 325
    solving by, 29, 34–35, 35f, 326, 328–29,
        377, 671–74, 671f–72f, 674f, 680
    with substitutions, 325, 327
    with synthetic division, 319–20
factoring higher-degree polynomials,
    324–28, 377
factoring quadratic polynomials, 673–74,
    674f
    with complex zeros, 300–301
    with one real zero, 300
    with two real zeros, 300

factor theorem, 299–301, 316–18, 377
    example use, 318
    factoring with, 328–29
Fahrenheit: converting from Celsius to,
    188–89, 188f
falling objects, 790
Fechner, Gustav Theodor, 241, 241f
flares, 304–5, 305f
Flesch-Kincaid Grade Level Formula, 42
flight, 650–51
    forces of, 738, 738f
    power required for, 375
flight paths, 705, 705f, 742–43, 742f
force, resultant, 737–38, 738f
forest litter, 811
formulas
    change-of-base, 249–51
    for compound interest, 219
    for continuous compounding, 221
    for distance, 4
    double-angle, 623
    for exponential functions, 199
    half-angle, 626
    for inverse functions, 138–39, 173
    for linear functions, 183
    for lines, 9
    for midpoint, 4
    new area formula for triangles, 448
    power reduction, 630–32
    product-to-sum, 613–15
    sum and difference formulas for sine
        and cosine, 607–11
    sum-to-product formula, 613–15
Fourier, Jean-Baptiste Joseph, 606–7
Fourier transform, 606–7
fractals, 123, 123f
fractions
    combining, 373
    inequalities involving, 35–38, 36f–37f, 43
frogs, jumping, 660
functional response, 105–6
functions
    addition of, 124
    alternative presentations, 52
    analyzing, with rates of change, 780–81,
        816
    arithmetic combinations, 124–26, 173
    basics, 49–59, 114
    combining with function values,
        128–29
    composition, 127–29, 173
    decomposition, 129–30, 173
    decreasing, 79–81, 115, 142
    decreasing without bound, 101, 101f
    definition, 50, 50f, 114
    division of, 125
    domain, 50, 52–55, 58–59, 114, 142–43
    equations for, 55–56, 114
    evaluating, 50–52, 51t, 114
    even, 163, 173, 559–61, 560f, 585
    exponential, 181, 198–208, 234
    graphs of, 56–58, 56f–57f, 114
    increasing, 79–81, 115, 142, 142f

inverse, 135–44, 173
inverse trigonometric, 638–55
iteration, 134
linear, 181–91, 234
logarithmic, 252
maximum values, 81–84, 81f, 115
minimum values, 81–84, 81f, 115
models and applications, 58–59
multiplication, 124
new from old, 123–30, 173
odd, 163, 173, 559–61, 560f, 585
one-to-one, 142–43, 173
periodic, 64, 546–48, 606–7, 607f
piecewise-defined, 52
polynomial, 338
quadratic, 294–305
range, 52–55, 58–59, 114
rates of change, 78, 771–80, 816
rational, 293, 358–69
with restricted domain, 142–43, 143f,
    173
trigonometric, 422–35, 423f, 487–97
fundamental theorem of algebra, 339, 378

G
Galileo, 791
geometry, 691
    hyperbolic, 404, 486
George Reserve (Michigan): deer
    population, 224, 224f, 285, 778–79,
    778f–79f
graphs
    with amplitude, 553–54, 553f–54f
    arcsine and arccosine, 644–45, 645f, 680
    of change equations, 800–808, 816
    of circles, 11
    concavity, 84–87, 84f–86f, 115, 304,
        777–79, 778f
    cosecant, 572–74, 572f, 574f, 585
    cosine, 550–52, 551f, 553f, 585, 668f
    cotangent, 570–71, 571f, 585
    end behavior, 97–107, 115, 364–69, 378
    of equations, 3, 3f
    of exponential functions, 201–3,
        202f–3f, 234
    features, 82–83
    of functions, 56–58, 56f–57f, 114
    of height, 776–77, 776f
    horizontal shift, 151–53, 151f, 153f, 160
    horizontal stretch, compression, and
        reflection, 157–59, 159f, 160, 173
    identifying sine models from, 558
    to illustrate identity, 559–60
    of intervals, 19, 19f–20f
    of inverse functions, 138–41, 139f–40f,
        173
    of linear functions, 184–91, 185f, 187f,
        234
    logarithmic, 243–44, 244f, 247, 247f,
        287
    logistic, 224f
    long-term behavior, 98–99, 98f–99f,
        105

graphs (*continued*)
of maximum values, 83–84, 83f
of minimum values, 83–84, 83f
models and applications, 87, 87f
multiple shifts, 153
of natural exponential function,
204, 204f
operations on, 122–73
of phase shift, 556–57, 556f–57f
of polynomials, 338–51, 343f, 346f–48f,
378
presentation, 164, 164f
of quadratic functions, 301–4, 301f,
303f–4f, 377
of rate of change, 77–88, 115, 773–77,
816
of rational functions, 363f, 365–68,
365f, 367f–68f
secant, 572–74, 573f–74f, 585
shifting, 150–54, 173
sine, 547, 547f, 549–50, 549f–50f, 553f,
559f, 585, 668f
sketching, 80–81, 81f
sketching, with laws of logarithms, 264
solving quadratic inequalities with,
30, 30f
tangent, 570–71, 571f, 585
that approach *x*-axis, 101–2, 102f
transformations, 150–65, 173
of trigonometric functions, 546–47,
547f, 549–52, 549f–51f, 553f, 559f,
567–79, 585
of velocity, 773f, 774
vertical shift, 151–53, 151f,
153f, 160
vertical stretch, compression, and
reflection, 155–57, 155f–56f,
160, 173
great circles, 691, 691f
greatest integer function, 61
Greenwich meridian, 393, 393f, 475, 475f
gross domestic product (GDP), 286
grouping: factoring by, 325, 327–28
growth. *See* exponential growth and decay
growth rate, 310, 771, 772t
guppies, 742, 742f
Gutenberg, Beno, 259

H
half-angle formulas, 622–23, 625–29, 679
for cosecant, 635
for cosine, 626–27, 634
for cotangent, 635
for secant, 628–29, 635
for tangent, 627–28, 634–35
half-life, 206–7, 207f, 285
examples, 217–18, 218f
height
growth in, 787, 787f
measuring, with transit, 449, 449f, 507,
507f
Heron's formula, 698–99, 698f, 706–7, 761
Hertzsprung-Russell diagram
(H-R diagram), 95, 95f

hexagons, 619
Hick's law, 256
higher-degree polynomials
division, 316
factoring, 324–28
zeros, 324–33, 377
horizontal asymptotes, 364–69, 364f, 378
definition, 365
graphing rational functions with,
366–67, 367f
horizontal line test, 140
for inverses, 140–41, 143, 173
horizontal shift, 151–53, 151f, 153f
concepts to remember, 160
examples, 161–62, 161f–62f
and rates of change, 162, 162f
of trigonometric functions, 576–77,
576f–77f
horizontal stretch, compression, and
reflection, 157–59, 159f, 173
concepts to remember, 160
example, 160–61, 161f
of trigonometric functions, 575–76,
576f
hunting limits, 802–3
hyperbolic geometry, 404, 486
hypotenuse, 3, 3f, 492, 492f
finding legs from angle and, 447, 505

I
immunity, 374
increasing functions, 79–81, 115
definition, 79
inverses and, 142, 142f
inequalities
absolute value, 22–26, 23f–24f, 43
fractions, 35–38, 36f–37f, 43
linear, 18–26, 43
more than two factors, 33–35,
33f–35f, 43
nonlinear, 28–38, 43
quadratic, 29–32, 30f–32f, 43
rules for manipulating, 21
triangle, 28
infinity: limits at, 97–101, 115
inflation, 217, 217f
inflection points, 84
installment loans, 222–23, 234
integer coefficients, 330
interest
compound, 219–22, 221t,
222f, 234
simple, 220, 220f
intervals, 18–19
complement of, 19, 19f–20f
graphs of, 19, 19f–20f
linear, 18–20, 43
notation for, 19
union of, 19
intrinsic annual exponential growth rate,
806–7
inverse functions, 135–44, 173
definition, 136, 173
formulas for, 138–39, 173

for functions with restricted domain,
142–43, 143f, 173
graphs for, 138–41, 139f–40f, 173
horizontal line test for, 140–41,
143, 173
for increasing or decreasing functions,
142, 142f
logarithms as, 241–43, 287
models and applications, 144
for rational functions, 373
showing, 137
trigonometric, 638–55, 680
ways to find, 143
inverse Pythagorean theorem, 403, 485
investments
changing value, 350
compound interest, 219–22, 221t, 222f,
234
exponential growth, 205–6, 206f
simple interest, 220, 220f
irregular areas, 696–97, 696f
isosceles triangles, 619, 619f, 704, 704f
right, 492, 492f
iteration of functions, 134

J
jumping frogs, 660
jumping locusts, 742, 742f

K
Kelvin scale, 195
kinetic energy, 42

L
latitude, 402, 403f, 484, 485f
law of cosines, 691–700, 692f, 761
alternative form, 703
derivation, 692–93, 761
with radian measure, 693–94, 693f
solving triangles with, 693–97,
693f–94f, 761
law of sines, 709–21, 711f–12f, 761
derivation, 710–11, 761
solving triangles with, 711, 711f,
712–13, 712f, 761
laws of logarithms, 259–68, 287
calculations with, 261–62
deriving, 259–60, 287
finding values of logarithms
with, 261
sketching graphs with, 264
solving equations with, 278
using, 261–68, 287
leading coefficients, 294
definition, 294
negative, 349
positive, 349
leading terms, 294
leaky balloons, 792–94
light
apparent magnitude, 256
Beer-Lambert-Bouguer law, 797
in ocean, 797
light rays, 709, 709f

limits, 97–107, 115
  in applied settings, 105
  calculation steps, 102–7, 115
  elementary calculations, 101–2, 115
  at infinity, 97–101, 105, 115
  intuitive notion, 99–101
  of rational functions, 363–64
line(s), 6–12, 43
  formulas for, 9
  intervals, 18–20
  parallel, 10, 16
  perpendicular, 16
  regression, 190–91, 191f
  secant, 65–71, 66f, 68f, 78, 78f, 114,
    186–87, 187f
  sketching, 7–8
  slope, 6–12, 6f, 43, 185, 185f
  straight, 6–12
  tangent, 78, 78f
  vertical line test, 56
  y-intercept, 6
linear equations, 6–12
  change equations, 789–91, 816
  common forms, 8–10, 43
  definition, 6
  examples, 9
  finding, 7–8
  point-slope form, 9
  properties, 184
  slope-intercept form, 9
  two-point form, 9
linear functions, 181–91, 234
  applications, 187–91, 234
  average rate of change, 183–84
  constructing, 185
  definition, 183
  finding, with rate of change, 185–86,
    186f
  formula for, 183
  graphs of, 184–91, 185f, 187f, 234
  nonconstant, 191
linear inequalities, 18–26, 43
  absolute value, 22–26, 23f–24f, 43
  applications, 25–26
  definition, 21
  and intervals, 18–19, 43
  solving, 21–22, 25–26, 43
linear regression, 190
loans, installment, 222–23, 234
local maxima and minima, 81,
  349–50, 378
  definitions, 81
  for sine and cosine, 552
location
  coordinate plane, 2–13, 43
  coordinates, 2–3, 43
  through triangulation, 12–13, 12f
locusts, jumping, 742, 742f
logarithmic equations
  mixed, 279–80
  solving, 273, 277–80, 287
logarithmic expressions, 263
logarithmic functions, 252
logarithmic graphs, 247, 247f

logarithms, 240
  base-b, 241
  change-of-base formula, 249–51, 287
  common, 244–47, 245f, 248–49, 252, 287
  definition, 241
  in exponent, 263
  finding values of, 261
  fundamental properties, 242, 287
  graphs of, 243–44, 244f, 287
  as inverse functions, 241–43, 287
  laws, 259–68, 287
  long-term behavior, 248–49, 287
  natural, 244–47, 245f, 252, 287
  power law, 260
  product law, 260
  quotient law, 260
logistic graphs, 224f
logistic growth, 41, 286, 806–7
  with threshold, 42, 812
logistic models, 223–25, 224f, 234
long division, 313–16, 377
longitude, 393, 475, 475f
long-term behavior, 367–68, 368f
loudness
  comparing decibels, 267
  comparing relative intensity, 267

M

Machin's formula, 661
magnitude of an earthquake, 265
magnitude of three-dimensional vectors, 752
main-sequence stars, 95
Malthus, Thomas Robert, 182, 182f
Mandelbrot set, 123, 123f
marginal profit, 771, 772t, 787, 787f
  units of, 772
marginal tax rate, 771, 772t
maximum height, 304–5, 305f
maximum values, 81–84, 81f, 115
  absolute, 82
  finding, 780, 780f
  graphs of, 83–84, 83f
  local, 81, 349–50, 378, 552
  relative, 81
mean value theorem, 357
Menger sponge, 123, 123f
Mercator projection, 567, 567f, 583
meridian, 393, 393f, 475f
Merino sheep, 810
meteors, 811
midpoint, 4
midpoint formula, 4, 43
minimum values, 81–84, 81f, 115
  absolute, 82
  finding, 780, 780f
  graphs of, 83–84, 83f
  local, 81, 349–50, 378, 552
  relative, 81
mixed equations, 279–80, 287
model airplanes, 705, 705f
modeling
  with exponential functions, 214–26, 234
  logistic models, 223–25, 224f, 234
  sine models, 557–58

model rockets, 785
monthly payments, 223
mortgages, 811
multiple-angle formulas
  double-angle formulas, 622–25, 679
  sum and multiple-angle formulas,
    629–30, 679
  triple-angle formula, 630
multiple shifts, 153
multiplication
  of functions, 124
  scalar, 729–30, 730f, 734, 752, 762
muzzle velocity, 759

N

natural exponential function, 204
  definition, 204
  examples, 204–5
  graphs, 204, 204f
  solving equations that involve, 275
natural logarithms, 244–47, 252, 287
  calculations with, 246, 261–62
  definition, 244
  graphs, 245f
  power law, 260
  product law, 260
  properties, 245
  quotient law, 260
negative angles, 389, 389f, 471, 471f
  reference numbers, 417, 417f
  sine and cosine, 528, 528f
negative discriminants, 296, 296f
  example, 297–98, 298f
negative rates of change, 80–81, 81f
new area formula for triangles,
  448, 448f
newspapers, 798
Newton's law of cooling, 70–71, 228–29,
  794–95
nonconstant linear functions, 191
nonlinear inequalities, 28–38, 43
  applications, 37–38
  involving fractions, 35–38, 36f–37f, 43
  involving several factors, 33–35,
    33f–35f, 43
  quadratic, 29–32, 43
North Sea cod, 812
notation, 18
  interval, 19
numbers
  complex, 298
  reference, 412–14

O

ocean
  light in, 797
  speed of sound in, 356
odd-degree polynomials, 349
odd functions, 163, 173, 559–61,
  560f, 585
  concepts to remember, 597
  definition, 163
odd identities, 597–98, 679
Ohm's law, 111

one-to-one functions, 142–43, 173
origin, 2
orthogonal vectors, 750, 750f
    three-dimensional, 752
orthonormal vectors, 757–58
Oscillate (Sierra), 547, 547f

P

parabolas, 295, 312
parallax, 659
    spectroscopic, 271
parallax angles, 422, 422f, 455, 455f, 456,
    487, 487f, 512, 512f
parallel lines, 10, 16
parallelograms, 704, 704f, 731, 731f, 732
percentage change, constant, 215–16, 234
period, 547–48, 554–55
periodic functions, 64, 546–48, 606–7, 607f
periodicity
    of cosine, 547–48, 551f, 561, 585
    of sine, 547–48, 549f, 561, 585
    of trigonometric functions, 547–48,
        548f–49f, 551f, 561, 585
    of trigonometric points, 548, 548f
perpendicular lines, 16
phase shift, 555–59, 585
    definition, 555
    graphing, 556–57, 556f–57f
pinnate muscle tissue, 660
π, 198, 405–6, 514–15
    approximation, 639
    formulas, 405, 514
π function, 62
plane
    coordinate, 2–13, 2f, 43
    points in, 2–3, 2f
    standard position in, 746–48,
        746f, 762
    vectors in, 745–51, 762
planetary distance, 705–6, 705f–6f
plotting points, 3f, 303, 303f
points
    compass, 407–9, 408f, 408t–9t, 426–29,
        426f–27f, 428t, 458–59, 518–21,
        536
    in plane, 2f
    plotting, 2f–3f
    trigonometric, 406–8, 406f–8f, 408t–9t,
        411–14, 411f–12f, 412t, 413f,
        516–18, 517f, 522f, 536
    on unit circle, 406, 406f, 458, 515–16,
        516f
Poiseuille's law, 333, 336
poles, 359–60, 378
polygons
    inscribed, 704, 704f
    regular, 619, 704, 704f
polynomials, 293, 338
    coefficients, 294
    constant terms, 294
    constructing, 340
    definition, 294
    degree, 294

division, 316
even-degree, 349
extreme values, 357
factoring, 324–28, 339
factor theorem for, 316–18
graphs, 338–51, 343f, 346f, 378
higher-degree, 316, 324–33
with integer coefficients, 330
leading coefficients, 294
leading term, 294
local behavior, 344–48, 344f, 378
local maxima and minima, 349–50
long division, 313–16
long-term behavior, 342–43, 343f, 346,
    346f, 347, 347f, 378
maximum number of zeros, 329
with negative leading coefficients, 349
with no zeros, 296–98, 296f, 298f
odd-degree, 349
with positive leading coefficients, 349
quadratic functions, 294–305
quartic, 337, 344f
range, 349
with real coefficients, 340–42, 351, 378
remainder theorem, 316–18, 323
sign changes, 346, 348, 348f
synthetic division, 318–20
zeros, 295–98, 295f, 296f, 346, 346f,
    347, 347f
population growth, 41, 106–7
    per capita rates, 231–32, 232f
    carrying capacity, 225
    equilibrium solutions, 806–7, 807f
    exponential, 216, 216f, 231, 798
    graphical solutions, 807–8, 807f–8f
    graphs, 778–79, 778f–79f
    intrinsic annual exponential, 806–7
    logistic, 224, 224f, 286, 806–7
positive angles, 389, 389f, 471, 471f
positive discriminants, 296, 296f
positive rates of change, 80–81, 81f
power law, 260
    calculations with, 262
    simplifying logarithmic expressions
        with, 263
power reduction formulas, 630–32, 679
presentation graphs, 164, 164f
product law, 260
    calculations with, 262
    finding values of logarithms with, 261
products of more than two factors, 33–35,
    33f–35f, 43
product-to-sum formula, 613–15, 679
profit, 37–38
    marginal, 771, 772t, 787, 787f
projectiles, 311, 659
    graphs of height, 776–77, 776f
    maximum height, 304–5, 305f
projection, vector, 743, 743f
proportional change, constant, 198–201, 214,
    234
proportionality, 189
    constant of, 189

definition, 189
    example, 189–90, 190f
puzzles, 305
Pythagorean identity, 432–35, 459, 529–30,
    529f, 536, 593
Pythagorean theorem, 3, 691
    inverse, 403, 485

Q

quadrants, 412–14, 429, 429f, 458, 521–22,
    521f, 536
quadratic equations
    analogy with trigonometric equations,
        673–74, 680
    factoring, 673–74, 674f
quadratic functions, 294–305, 377. *See also*
    polynomials
    with complex zeros, 298–301, 377
    definition, 295
    discriminants, 295–96
    factoring, 300–301
    factor theorem for, 299–301, 377
    graphs, 301–4, 301f, 303f–4f, 377
    with one real zero, 300
    remainder theorem for, 323
    standard form, 301
    step-by-step strategy for solving, 295
    with two real zeros, 300
    vertex, 302
    zeros of, 295–98, 377
quadratic inequalities, 29, 43
    definition, 29
    graphical display, 30f
    method 1: sign table, 29–30
    method 2: sign diagram, 30–32
    solving, 29–32, 30f–32f
    step-by-step strategy, 30
quartic polynomials, 337, 344f
quotient law, 260
    calculations with, 262
    finding values of logarithms with, 261
    simplifying logarithmic expressions
        with, 263

R

radian(s), 390–91, 391f, 394, 472–73,
    473f, 476
    converting between degrees and,
        392–93
    definition, 390, 472
radian measure, 390–91, 391f, 458, 472–74,
    473f, 535
    converting from degree to, 392–93,
        474–75
    converting to degree measure from,
        393, 475
    coterminal angles with, 392, 474
    law of cosines with, 693–94, 693f
    solving triangles with, 713–14,
        713f
radiocarbon dating, 281
radius, 10, 12f
    vs circumference, 189–90, 190f

ramps, 735–36
range of a function, 52–55, 58, 114
    applications, 59
    definition, 54
    in inverse trigonometric functions,
        642–43
range of a polynomial, 349
range of even-degree polynomials with
    negative leading coefficients, 349
range of even-degree polynomials with
    positive leading coefficients, 349
range of odd-degree polynomials, 349
rates of change, 771
    analyzing functions with, 780–81, 816
    applications that involve, 771, 772t
    average, 65–72, 66f, 114
    concavity and, 84–87, 84f–86f, 115,
        813–14
    concepts to remember, 80
    constant, 182–84, 234, 789–91, 816
    definition, 78
    finding function values with, 185–86,
        186f
    finding maxima and minima with,
        780, 780f
    as functions, 771–80, 816
    graphs, 77–88, 115, 773–77, 780,
        780f, 816
    intuitive description, 78, 115
    models and applications, 87
    negative, 80–81, 81f
    positive, 80–81, 81f
    properties, 78
    shifting and, 162
    units of, 772–73, 816
rational functions, 293, 358–69, 378
    definition, 359
    domain, 359–60, 369, 378
    formula, 369
    graphing, 362–63, 363f, 365–69, 365f,
        367f
    horizontal asymptotes, 364–69
    inverses, 373
    limits, 363–64
    local behavior, 360–63, 378
    long-term behavior, 363–64, 367–68,
        368f, 369, 378
    near vertical asymptotes, 362–63, 363f
    poles, 359–60, 378
    range, 369
    sign changes, 361
    vertical asymptotes, 360–63,
        366–67
rational zeros, 329–33, 377
    graphing polynomials with, 347–48,
        347f–48f
rational zeros theorem, 330
rays of light, 709, 709f
real coefficients, 340–42, 351, 378
real zeros, 295, 295f
    one, 300
    two, 296, 296f, 300
red shift, 95, 375

reference angles, 415–17, 415f–17f, 458
    finding trigonometric functions with,
        431–32, 431f, 441, 441f, 525–27,
        526f–27f, 529, 529f
reference numbers, 412–14, 522–24, 536
    definition, 413, 523
    finding, 415–16, 415f, 524–25, 525f
    finding trigonometric functions with,
        430–32, 430f, 459, 524–29, 526f,
        536
    finding trigonometric points with,
        414–18, 415f, 418f, 458
    graphical view, 414t, 524t
    for negative angles, 417, 417f
    properties, 523
    for trigonometric points, 413–15, 414t,
        415f
reflection
    horizontal, 157–59, 159f, 160–61, 161f,
        173
    vertical, 155–57, 156f, 160, 173
regression, linear, 190
regression lines, 190–91, 191f
relative growth rate or r-value, 224
relative maximum, 81
relative minimum, 81
remainder theorem, 316–18, 377
    alternative proof, 323
    example use, 317
representative triangles, 442–45, 443f–44f,
    459, 495–96, 495f, 535
retirement options, 797
Richter, Charles, 259
Richter scale, 259
right angles, 391, 473
right triangles, 397f–98f, 459, 479–80, 479f,
    494, 494f
    analysis, 502–8, 536
    example, 397–98, 397f
    finding trigonometric functions with,
        439–41, 439f–40f, 459, 487–97,
        489f–90f, 535
    isosceles, 492, 492f
    sides, 488, 488f
    similar, 490–91, 491f
    solving, 445–47, 446f, 459, 502–5, 504f,
        536
    with special angles, 445–46, 445f,
        502–3, 502f
    3-4-5 triangles, 497, 497f, 498
right triangle trigonometry, 438–50
rise and run, 6f
rise vs run, 7f
rockets, 785
roof lines, 622–23, 623f
r-value or relative growth rate, 224

S
SAA (side-angle-angle) theorem, 710,
    710f–11f
sales charts, 87, 87f
SAS (side-angle-side) theorem,
    692, 692f

scalar multiplication, 729–30, 730f, 762
    definition, 729
    properties, 734
    three-dimensional vectors, 752
secant, 423, 488
    concepts to remember, 578
    definition, 423, 488, 578
    graphs, 572–74, 573f–74f, 585
    half-angle formula, 628–29, 635
    identities, 574–75
    periodicity, 569, 578
    transformed, 575
    values, 573t
    vertical asymptotes, 578
secant lines, 65–71, 66f, 68f, 78, 78f, 114
    definition, 65
    finding equations of, 186–87, 187f
    models and applications, 68
service ramps, 727, 727f
sets: union of, 19
shadows, 399, 399f, 449–50, 450f, 481, 481f,
    507–8, 508f
sharks, 706
shifts, 150–54, 173
    example, 303–4, 303f
    horizontal, 151–53, 151f, 153f, 160–62,
        161f–62f, 576–77, 576f–77f
    multiple, 153
    phase, 555–59
    vertical, 151–53, 151f, 153f, 160,
        576–77, 576f–77f
side-angle-angle (SAA) theorem, 710,
    710f–11f
side-angle-side (SAS) theorem, 692, 692f
side-side-angle (SSA) triangles, 714–18,
    714f–17f, 761
    obtuse case, 727, 727f
side-side-side (SSS) theorem, 692, 692f
Sierra, Daniel, 547, 547f
sign, 429, 429f
sign changes, 346, 361
sign diagrams, 30–34, 34f
sign tables, 29–30, 33–35, 35f
similar triangles, 396–98, 396f, 458, 478–81,
    478f, 535
    corresponding sides, 396, 478–79
    definition, 396, 478
    right triangles, 490–91, 491f
sine, 423, 488, 517f
    bounds, 552
    concepts to remember, 561
    definition, 423, 488, 561
    double-angle formula, 624
    law of sines, 709–21, 761
    of negative angles, 528, 528f
    other trigonometric functions in terms
        of, 488–89, 488t
    periodicity, 547–48, 549f, 561, 585
    signs, 429, 429f, 521, 521f
    sum and difference formulas, 607–11,
        608f, 679
    symmetry, 561
    values between 0 and 2π, 549t

sine area formula, 607, 607f
sine graphs, 549–50, 549f–50f, 553f, 559f, 585, 668f
sine models, 557–58
sine waves, 547, 547f, 552
   amplitude, 552–54, 553f, 585
   definition, 552
skydivers, 97, 97f, 285
slope, 6–12, 6f, 43
   calculating, 6–7
   concepts to remember, 7
   definition, 6
   of roof lines, 622–23, 623f
   varying, 185, 185f
Snell's law, 709
soccer, 699–700, 699f, 726, 726f
SOHCAHTOA mnemonic, 488
sound: speed in the ocean, 337, 356
special angles, 410–12, 410f, 458, 492–94, 492t, 535
   right triangles with, 445–46, 445f, 502–3, 502f
   triangles with, 445–47, 445f
   trigonometric functions, 423–26, 425t, 459, 493
   trigonometric points, 411–12, 411f–12f, 412t, 415f, 423–24, 424f, 517, 517f, 528, 529f
spectroscopic parallax, 271
speed
   of running animals, 190–91, 191f
   of sound in the ocean, 337, 356
spell checkers, 251–52, 285
spheres, 322
squaring equations, 676
SSA (side-side-angle) triangles, 714–18, 714f–17f, 761
   obtuse case, 727, 727f
SSS (side-side-side) theorem, 692, 692f
stars, 271
straight angles, 391, 473
stretch
   example, 303–4, 303f
   horizontal, 157–59, 159f, 160, 173, 575–76, 576f
   vertical, 155–57, 155f, 160–61, 161f, 173, 575–76, 576f
substitution
   factoring with, 325, 327
   solving trigonometric equations with, 666–67, 667f
subtraction, vector, 731, 731f, 734, 734f
sum and difference formulas, 607–15, 679
   with given values, 610
   for sine and cosine, 607–11, 608f, 679
   for tangent and cotangent, 611–13, 679
sum and double-angle formulas, 629–30
sum and multiple-angle formulas, 629–30, 679
sum-to-product formula, 613–15, 679
   concepts to remember, 614
   solving trigonometric equations with, 672–73, 673f

sunlight, 659–60, 660f
supply and demand, 25–26
supply curves, 95, 95f
surveying, 449, 449f, 725, 725f, 768, 768f
symbols, 18
synthetic division, 318–20, 377

T
tangent, 423, 488
   concepts to remember, 577
   definition, 423, 488, 577
   double-angle formula, 625, 634
   graphs, 570–71, 571f, 585
   half-angle formulas, 627–28, 634–35
   periodicity, 568–69, 577
   sum and difference formulas, 611–13, 679
   transformed, 575
   values, 570t
   vertical asymptotes, 577
tangent lines, 78, 78f
   shifting, 162, 162f
tax rates, marginal, 771, 772t
temperature conversion, 133, 187–90, 188f
terminal velocity, 811
Thales' theorem, 404, 486
3-4-5 right triangle, 497, 497f, 498
three-dimensional vectors, 751–54, 751f, 762
   addition, 752
   angle between, 752
   dot product, 752
   magnitude, 752
   operations on, 752–53
   scalar multiplication, 752
threshold: logistic growth with, 42, 812
Titan, 798
Torricelli's law, 171
towers, 450, 450f, 508, 508f
traffic engineering, 310
traffic flow, 368–69
transformations
   of exponential graphs, 203, 203f
   of graphs, 150–65, 173
   of logarithmic graphs, 247, 247f
   of trigonometric functions, 574–78, 585
transits, 438, 438f, 502, 502f
   measuring height with, 449, 449f, 507, 507f
   and shadows, 449–50, 507
   surveying with, 768, 768f
triangle inequality, 28, 703–4
triangles
   angle-side-angle (ASA) theorem, 710, 710f, 712f–13f
   angle sum, 397, 479
   area, 698–99, 761
   area formula, 607, 607f
   choosing laws to solve, 718–19, 761
   congruence criteria, 692, 692f, 710, 710f, 761
   Heron's formula, 698–99, 698f
   isosceles, 492, 492f, 619, 619f, 704, 704f

new area formula, 448, 448f, 459, 505–6, 506f, 536
representative, 442–45, 443f, 459, 495–96
right, 397–98, 397f–98f, 438–50, 439f–40f, 446f, 459, 479–80, 479f, 488, 488f, 492, 492f, 494, 494f, 502–8, 502f–4f, 535–36
side-angle-angle (SAA) theorem, 710, 710f–11f
side-angle-side (SAS) theorem, 692, 692f
side-side-angle (SSA) theorem, 714–18, 714f–17f, 727, 727f, 761
side-side-side (SSS) theorem, 692, 692f
similar, 396–98, 396f, 458, 478–81, 478f, 535
solving, 719, 761
solving, choosing laws for, 718–19, 761
solving, with law of cosines, 693–97, 693f–94f, 761
solving, with law of sines, 761
solving, with radian measure, 713–14
with special angles, 445–46, 445f
3-4-5 right triangle, 497, 497f, 498
triangulation, 12–13, 12f
trigonometric equations
   analogy with quadratic equations, 673–74, 680
   basic, 663–70, 663f–65f, 680
   factored, 671, 671f
   solutions on one full period, 664, 664f, 666, 666f
   solving, 662–74, 663f–65f, 680
   solving by factoring, 671–74, 671f–72f, 674f, 680
   solving with calculator, 666, 666f, 668, 668f
   solving with double-angle formulas, 669, 669f
   solving with simple identities and substitution, 666–67, 667f
   solving with sum-to-product formula, 672–73, 673f
   on specified intervals, 670, 670f
   squaring, 676
trigonometric expressions, 601
trigonometric functions, 422–35, 459. *See also specific functions*
   of acute angles, 438–41, 459, 488–89, 535
   for any real number or any angle, 423, 518
   of compass points, 426–29, 426f–27f, 428t, 459, 518–21, 519f, 520t, 536
   consistency, 490–91, 535
   definitions, 423, 423f, 459, 535
   domains, 429, 521
   even and odd, 597
   finding, with algebraic expressions, 443–44, 443f–44f, 496, 496f
   finding, with Pythagorean identity, 433–34, 529, 529f

finding, with reference angles, 431–32, 431f, 441, 441f, 525–27, 526f–27f, 529, 529f

finding, with reference numbers, 430–32, 430f, 459, 524–29, 526f, 536

finding, with representative triangles, 442–45, 443f–44f, 459, 495–96, 495f, 535

finding, with right triangles, 439–41, 439f–40f, 459, 487–97, 489f–90f, 535

graphs, 546–47, 547f, 549–52, 549f–51f, 553f, 559f, 567–79, 585

horizontal and vertical shift, 576–77, 576f, 577f

horizontal and vertical stretch and compression, 575–76, 576f

inverse, 638–55

inverse functions, 680

periodicity, 546–47, 568–69, 585

reciprocal relationships, 575

signs, 429, 429f, 459, 521, 521f

from sine and cosine, 488–89

SOHCAHTOA mnemonic for, 488

of special angles, 423–26, 425t, 459

in terms of others, 594–95

in terms of sine and cosine, 488, 488t

transformations, 574–78, 585

trigonometric identities, 592–601, 679

algebraic verification, 598–601, 679

basic, 593–95, 679

cofunction, 595–96, 679

combining, 599

even and odd, 597–98, 679

techniques for proving, 601

trigonometric points, 406–7, 406f–7f, 412, 412f–13f, 516–18, 517f, 522f, 536, 593f

of acute angles, 438, 439f

associated with compass points, 408t–9t

comparing, 597, 597f

coordinates, 414–18

definition, 407, 516

examples, 407–8, 408f

finding, with reference numbers, 414–18, 415f, 418f, 458

periodicity, 548, 548f

reference numbers for, 413–14, 414t, 415, 415f

for special angles, 411–12, 411f–12f, 412t, 415f, 423–24, 424f, 517, 517f, 528, 529f

trigonometry, 388

right triangle, 438–50, 459

unit circle, 405–18, 458

triple-angle formula, 630

U

undetermined coefficients: theorem on, 533

union

of intervals, 19

of two sets, 19

unit

of marginal profit, 772

of rate of change, 772–73, 816

unit circle, 405–18, 458, 515–16, 536

compass points on, 407–9, 458

equation for, 406, 406f, 515, 515f

points on, 406, 406f, 458, 515–16, 516f

quadrants, 412–14, 458

reference numbers, 412–14

unit vectors, 730

uranium-238, 217–18, 218f

U.S. population, 216, 216f

V

vectors, 728–38, 731f, 762

addition, 731–34, 731f–32f, 752, 762

angles between, 748–51, 749f–50f, 752, 756, 756f

combinations, 731–32

component, 734–35, 735f, 762

definition, 728

orthogonal, 750, 750f, 752

orthonormal, 757–58

overview, 729, 762

planar, 745–51, 762

projection, 743, 743f

scalar multiplication, 729–30, 730f, 752, 762

standard position, 746–48, 746f, 762

subtraction, 731, 731f, 734, 734f

three-dimensional, 751–54, 751f, 762

unit, 730

velocity, 77, 77f, 771, 772t

of falling objects, 790

formula for, 77

graphs of, 773f, 774

muzzle, 759

of skydivers, 97–98, 97f–98f

terminal, 811

vertical asymptotes, 360–63, 361f, 378

definition, 361

graphing rational functions with, 366–67, 367f

vertical line test, 56, 57f

vertical shift, 151–53, 151f, 153f

concepts to remember, 160

of trigonometric functions, 576–77, 576f–77f

vertical stretch, compression, and reflection, 155–57, 155f–56f, 173

concepts to remember, 160

example, 160–61, 161f

sketching, 157

of trigonometric functions, 575–76, 576f

W

wasps and weeevils, 677

wave height, 41

weevils and wasps, 677

Will Rogers World Airport (OKC), 720

wind, 736–37, 736f

windchill, 94

X

x-axis, 2

graphs that approach, 101–2, 102f

Y

y-axis, 2

yearly compounding, 222, 222f

y-intercept, 6, 6f, 185, 185f

Z

zero discriminants, 296, 296f, 377

example, 297, 297f

zeros

complex, 298–99, 377

conjugate zeros theorem, 341–42

finding, 332, 346, 346f

of higher-degree polynomials, 324–33, 377

maximum number of, 329

one, 296, 296f, 297, 297f

order of, 329

of polynomials, 295, 330

of quadratic functions, 295–98, 377

rational, 329–33, 347–48, 347f–48f, 377

real, 295, 295f

two, 296, 296f

zeros theorem, 339